SECOND EDITION

Essential Mathematics for College Students

ARNOLD R. STEFFENSEN
L. MURPHY JOHNSON
Northern Arizona University

HarperCollinsPublishers

To Barbara, Barbara, Becky, Cindy, and Pam

To the Student

A *Student's Solutions Manual,* by Joseph Mutter, provides complete, worked-out solutions to all of the exercises in set A, the chapter review exercises, and the chapter tests. Your college bookstore either has this book or can order it for you.

Sponsoring Editor: Anne Kelly
Project Editor: Randee Wire
Assistant Art Director: Julie Anderson
Text and Cover Design: Lucy Lesiak Design/Lucy Lesiak
Cover Photo: Don and Pat Valenti
Photo Researcher: Nina Page
Production: Linda Murray/Helen Driller
Compositor: York Graphic Services
Printer and Binder: Courier Corporation
Cover Printer: Lehigh Press Lithographers

Essential Mathematics for College Students, Second Edition

ISBN 0-673-46453-9

91 92 93 94 9 8 7 6 5 4 3 2 1

PREFACE

With varied levels of mathematical skill, maturity, and motivation, many students today require substantial review of arithmetic, geometry, and algebra before they can be successful with more advanced algebra topics, discrete mathematics, trigonometry, calculus, or computer science courses. *Essential Mathematics for College Students, Second Edition,* provides in one text the topics that many college mathematics courses demand as prerequisites. Also, *Essential Mathematics* has been developed with special attention to topics required by certain state-administered mathematical competency exams, including the TASP™ skill requirements relative to graph reading, geometry, inductive and deductive reasoning, and other logic requirements.

Informal yet carefully worded explanations, detailed examples with accompanying practice exercises, pedagogical color, abundant exercises, and comprehensive chapter reviews are hallmarks of the book. The text has been written for maximum instructor flexibility. Both core and peripheral topics can be selected to fit individual course needs. The text is ideally suited to a two-semester course, but by flexible selection of chapters, instructors can adapt the text to a one-semester course in prealgebra and elementary algebra, a one-semester course in intermediate algebra, or a fast-paced, one-semester course covering topics in algebra and geometry. Various other combinations are also possible.

FEATURES

STUDENT GUIDEPOSTS Designed to help students locate important concepts as they study or review, student guideposts specify the major topics, rules, and procedures in each section. The guideposts are listed at the beginning of the section, then each is repeated as the corresponding material is discussed in the section.

EXAMPLES More than 800 carefully selected examples include detailed, step-by-step solutions and side annotations in color. Each example is headed by a brief descriptive title to help students focus on the concept being developed and to aid in review.

PRACTICE EXERCISES These parallel each example and keep students involved with the presentation by allowing them to check immediately their understanding of ideas. Answers immediately follow these exercises.

CAUTIONS This feature calls students' attention to common mistakes and special problems to avoid.

FIGURES AND ILLUSTRATIONS Abundant use has been made of figures and illustrations in examples and exercises, particularly applications.

COLOR Pedagogical color highlights important information throughout the book. Key definitions, rules, and procedures are set off in colored boxes for increased emphasis. Figures and graphs utilize color to clarify the concepts presented. Examples present important steps and helpful side comments in color.

EXERCISES As a key feature of the text, more than 8000 exercises, including more than 2000 review exercises, are provided. Two parallel exercise sets (A and B) and a collection of extension exercises (C) follow each section and offer a wealth of practice for students and flexibility for instructors. Exercises ranging from the routine to the more challenging, including application and calculator problems, are provided.

Exercises A This set of exercises includes space for working the problems, with answers immediately following the exercises. Some of these problems, identified by a colored circle 21, have their solutions worked out at the back of the book. Many of these solutions are to exercises that students frequently have difficulty solving.

Exercises B This set parallels the exercises in set A problem for problem, but is presented without work space or answers.

Exercises C This set is designed to give students an extra challenge. These problems extend the concepts of the section or demand more thought than exercises in sets A and B. Answers or hints are given for selected exercises in this set.

REVIEW EXERCISES To provide ample opportunities for review, the text features a variety of review exercises.

For Review exercises are located at the end of most A and B exercise sets. They not only encourage continuous review of previously covered material, but also often provide special review preparation for topics covered in the upcoming section.

Chapter Review Exercises and a practice **Chapter Test** conclude each chapter. The Chapter Review Exercises are divided into two parts: the problems in Part I are ordered and marked by section. Those in Part II are not referenced to the source section, and are thoroughly mixed to help students prepare for examinations. Answers to all review and text exercises are provided in the text.

Final Review Exercises, referenced to each chapter and with answers supplied, are located at the back of the book.

CHAPTER REVIEWS In addition to the Chapter Review Exercises and Chapter Tests, comprehensive chapter reviews also include **Key Words** and **Key Concepts**. Key Words, listed by section, have brief definitions. Key Concepts summarize the major points of each section.

CALCULATORS It is assumed that most students have a hand-held scientific calculator. Therefore, use of a calculator, while not emphasized throughout the text, is integrated with such topics as exponents, radicals, and logarithms. Step-by-step key entries for algebraic calculators are shown in those sections. Some exercises, many in Exercises C, are marked for calculator use with the symbol .

INSTRUCTIONAL FLEXIBILITY

Essential Mathematics for College Students, Second Edition, offers proven flexibility for a variety of teaching situations, such as individualized instruction, lab instruction, lecture classes, or a combination of methods.

Material in each section of the book is presented in a well-paced, easy-to-follow sequence. Students in a tutorial or lab instruction setting, aided by the student guideposts, can work through a section completely by reading the explanation, following the detailed steps in the examples, working the practice exercises, and then doing the exercises in set A.

The book can also serve as the basis for, or as a supplement to, classroom lectures. The straightforward presentation of material, numerous examples, practice exercises, and three sets of exercises offer the traditional lecture class an alternative approach within the convenient workbook format.

NEW IN THIS EDITION

In a continuing effort to make this text even better suited to the needs of instructors and students, the following are some of the enhanced features of the new edition.

- The student guideposts are more visible.
- Explanations have been polished, reworded, and streamlined where appropriate. More figures and illustrations are used in discussions.
- Exercise sets have been reviewed for grading the balance of coverage. The number and variety of practical applications have increased. Additional challenging exercises and a greater emphasis on calculators have been incorporated into Exercises C.
- Additional For Review exercises review topics in preparation for the next section.
- Chapter Review Exercises are presented in two parts, one with sectional references and one without, to help students recognize problem types and better prepare for tests.
- Key Words given in the Chapter Review are expanded to include brief definitions.
- Greater emphasis has been given to the use of figures and illustrations in examples and exercises, particularly applications.
- Instructions for using a calculator have been expanded.
- Interval notation is introduced in Chapter 2 to help prepare students for later courses.
- Where appropriate, sections from the previous edition were recombined to form new sections more easily manageable in one class period.
- Graphing in the coordinate plane remains early in the text; graphing by plotting points from a table of values is given greater emphasis.
- The Pythagorean theorem and the distance and midpoint formulas are now discussed in Chapter 7.
- Complex numbers are introduced in Chapter 7, allowing for a more thorough discussion of quadratic equations in Chapter 8.
- An expanded treatment of functions and their properties has been added in new Chapter 9.
- The use of a calculator for computing logarithms replaces the previous dual presentation of both the calculator and logarithmic tables. The use of logarithmic tables and interpolation are still presented in the Appendix.
- Chapter 11 is expanded to include a more complete treatment of the conics, especially the circle and the parabola, with an increased emphasis on standard forms and graphing.
- The topics in geometry have been moved to a separate chapter, Chapter 12, providing the opportunity for greater depth of coverage. A comparison of inductive and deductive reasoning, as well as a brief treatment of logical systems, are included in this chapter.
- A brief introduction to probability has been added in Appendix B.

SUPPLEMENTS

An expanded supplemental package is available for use with *Essential Mathematics for College Students, Second Edition*.

For the Instructor

The **INSTRUCTOR'S TESTING MANUAL** contains a series of **ready-to-duplicate tests,** including a Placement Test, six different but equivalent tests for each chapter, and two final exams, all with answers supplied. More than 1000 **additional problems,** grouped by section, are also included for use as in-class examples or on quizzes and tests. Section-by-section **teaching tips** provide suggestions for content implementation that an instructor, tutor, or teaching assistant might find helpful.

The **INSTRUCTOR'S SOLUTIONS MANUAL** contains complete, worked-out solutions to every exercise in sets B and C of the text.

HARPERCOLLINS TEST GENERATOR FOR MATHEMATICS Available in IBM and Macintosh versions, the test generator enables instructors to select questions by objective, section, or chapter, or to use a ready-made test for each chapter. Instructors may generate tests in multiple-choice or open-response formats, scramble the order of questions while printing, and produce multiple versions of each test (up to 25). The system features printed graphics and accurate mathematics symbols. It also features a preview option that allows instructors to view questions before printing, to regenerate variables, and to replace or skip questions if desired. The program MAC will also have an editor that allows instructors to add their own problems to existing data disks.

VIDEOTAPES A new videotape series, *Algebra Connection: Essentials* has been developed to accompany *Essential Mathematics for College Students, Second Edition.* Produced by an Emmy Award-winning team in consultation with a task force of academicians from both two-year and four-year colleges, the tapes cover all objectives, topics, and problem-solving techniques within the text. In addition, each lesson is preceded by motivational ''launchers'' that connect classroom activity to real-world applications.

For the Student

The **STUDENT'S SOLUTIONS MANUAL** contains complete, worked-out solutions to each practice exercise in the text, to every exercise in set A, and to all Chapter Review Exercises and Chapter Tests.

INTERACTIVE TUTORIAL SOFTWARE This innovative package is also available in Apple, IBM, and Macintosh versions. It offers interactive modular units, specifically linked to the text, for reinforcement of selected topics. The tutorial is self-paced and provides unlimited opportunities to review lessons and to practice problem solving. When students give a wrong answer, they can request to see the problem worked out. The program is menu-driven for ease of use, and on-screen help can be obtained at any time with a single keystroke. Students' scores are automatically recorded and can be printed for a permanent record.

ACKNOWLEDGMENTS

We extend our sincere gratitude to the students and instructors who used the previous edition of this book and offered many suggestions for improvement. Special thanks go to the instructors at Northern Arizona University and Yavapai Community College. In particular, the assistance given over the years by James Kirk and Michael Ratliff is most appreciated. Also, we sincerely appreciate the support and encouragement of the Northern Arizona University administration, especially President Eugene M. Hughes. It is a pleasure and privilege to serve on the faculty of a university that recognizes quality teaching as its primary role.

We are also indebted to the following reviewers for their countless beneficial suggestions at various stages of the book's revision:

Brenda Allen, *Georgia College*
Jane Edgar, *Brevard Community College*
Paul Eenigenburg, *Western Michigan University*
Ronald D. Faulstich, *Moraine Valley Community College*
Al Giambrone, *Sinclair Community College*
Linda Holden, *Indiana University, Bloomington*
Phyllis H. Jore, *Valencia Community College*
Maureen Kelley, *Northern Essex Community College*
Rebecca R. Kitto, *Antelope Valley College*
Susan C. Meyers, *Sinclair Community College*

Lloyd C. Neitling, *University of Nevada, Las Vegas*
Marilyn P. Persson, *University of Kansas*
Paul Pontius, *Pan American University*
Donna Fields Rochon, *Western Washington University*
Mark Serebransky, *Camden County College*
Mark L. Sigfrids, *Kalamazoo Valley Community College*
Anthony Soychak, *University of Southern Maine*
Barbara W. Worley, *Des Moines Area Community College*
Vicky Young, *Motlow State Community College*

We extend special appreciation to Joseph Mutter for writing the *Student's Solutions Manual* and for his countless suggestions and support since the text's first edition.

We thank our editors, Jack Pritchard, Anne Kelly, Randee Wire, and Sarah Joseph, whose support has been most appreciated.

Finally, we are indebted to our families and in particular our wives, Barbara and Barbara, whose encouragement over the years cannot be measured.

Arnold R. Steffensen
L. Murphy Johnson

CONTENTS

3 GRAPHS, RELATIONS, AND FUNCTIONS

4 SYSTEMS OF LINEAR EQUATIONS AND INEQUALITIES

5 POLYNOMIALS AND FACTORING

6 RATIONAL EXPRESSIONS

7 RADICALS AND EXPONENTS

8 QUADRATIC EQUATIONS AND INEQUALITIES

9 MORE ON FUNCTIONS

10 EXPONENTIAL AND LOGARITHMIC FUNCTIONS

11 CONIC SECTIONS

12 INTRODUCTION TO GEOMETRY

TO THE STUDENT

During the past several years we have taught mathematics at this level to more than 1500 students and have heard the following comments many times. "I've always been afraid of math and have avoided it as much as possible. Now my major requires algebra and I'm petrified." "I don't like math, but it's required to graduate." "I can't do word problems!" If you have ever made a similar statement, now is the time to think positively and start down the path toward success in mathematics. Don't worry about this course as a whole. The material in the text is presented in a way that lets you take one small step at a time. As you begin, keep in mind these guidelines that are both necessary and helpful.

GENERAL GUIDELINES

1. Mastering mathematics requires motivation and dedication. Just as an athlete does not improve without commitment to his or her goal, a mathematics student must be prepared to work hard and spend time studying. Make it a point to keep up with your assignments. If you need extra help, *ask* for it right away.
2. Mathematics is not learned simply by watching, listening, or reading; *it is learned by doing*. Use your pencil and practice. When your thoughts are organized and written in a neat and orderly way, you have taken a giant step toward success. Be complete and write out all details. The following are samples of two students' work on a word problem. Can you tell which one was more successful in the course?

STUDENT A

Let n = number of units produced during the month

C = total cost per month

$$C = 10n^2 - 100n - 2000$$
$$10{,}000 = 10n^2 - 100n - 2000$$
$$0 = 10n^2 - 100n - 12{,}000$$
$$0 = n^2 - 10n - 1200$$
$$0 = (n-40)(n+30)$$
$$n - 40 = 0 \text{ or } n + 30 = 0$$
$$n = 40 \text{ or } n = -30$$

Thus, 40 units were produced.

STUDENT F

$$10(10{,}000)^2 - 100(10{,}000) - 2000$$
$$=$$
n = units
$$n = 10n^2 - 100n - 2000$$
$$0 = 10n^2 - 101n - 2000$$

3. If you are allowed to use a calculator in your course, consult your owner's manual to become familiar with its features. When you are computing with decimals or approximating irrational numbers, a calculator can be a time-saving device. On the other hand, you should not be so dependent on a calculator that you use it for simple calculations that can be done mentally. For example, it would be ridiculous to use a calculator to solve an equation such as $2x = 8$, while it would be helpful to solve $1.12x = 7343.84$. It is important that you learn when to use and when not to use your calculator; some problems for which a calculator may be especially helpful are marked with a calculator symbol, .

SPECIFIC GUIDELINES

1. As you begin to study each section, look through the material for a preview of what is coming.
2. Return to the beginning of the section and start reading slowly. The STUDENT GUIDEPOSTS will help you find important concepts as you progress or as you review.
3. Read through each EXAMPLE and make sure you understand every step. The side comments in color will help you if something is not clear.
4. After reading each example, work the parallel PRACTICE EXERCISE. This will reinforce what you have just read and start the process of practice. (*Note:* Complete solutions to the Practice Exercises are available in the *Student's Solutions Manual.*)
5. Now and then you will see a CAUTION. These warn you of common mistakes and special problems to avoid.
6. After you finish the material in the body of the section, you need to practice, practice, practice! Begin with the exercises in set A. Answers to all of these problems are at the end of the set for easy reference. Some of the problems, identified by a blue circle on the number, 21, have step-by-step solutions at the back of the book. After trying these problems, refer to the step-by-step solutions if you have difficulty. To practice more, do the exercises in set B; to challenge yourself, try the exercises in set C. (*Note:* Solutions to all problems in set A are available in the *Student's Solutions Manual.*)
7. After you have completed all of the sections in the chapter, read the CHAPTER REVIEW that contains Key Words and Key Concepts. The Review Exercises provide more practice before you take the CHAPTER TEST. Answers to the tests are at the back of the book. Solutions are provided in the *Student's Solutions Manual.*
8. To help you study for your final examination, we have concluded the book with a comprehensive set of FINAL REVIEW EXERCISES

If you follow these steps and work closely with your instructor, you will greatly improve your chances for success in algebra.

Best of luck, and remember that you can do it!

CHAPTER 0

Prealgebra

0.1 FRACTIONS

STUDENT GUIDEPOSTS

1 Sets of Numbers	3 Reducing to Lowest Terms
2 Properties of Fractions	4 Prime Numbers and Factors

1 SETS OF NUMBERS

Two terms that are used repeatedly in algebra are *set* and *element*. A **set** is a collection of objects. These objects, the **elements** of the set, are often listed within braces { }.

Most sets we deal with are sets of *numbers*. A **number** is a human invention, an idea formed when questions about "how many" or "in what order" are asked. Symbols for numbers are called **numerals.** For example, "4" is a symbol or numeral that represents the number four. The most basic numbers are the **natural** or **counting numbers.**

$N = \{1, 2, 3, 4, . . .\}$ Natural (counting) numbers

We use three dots to mean that a sequence or pattern continues on in the same manner. When zero is included with the natural numbers, we have the set of **whole numbers.**

$W = \{0, 1, 2, 3, 4, . . .\}$ Whole numbers

2 PROPERTIES OF FRACTIONS

Numbers written as the quotient of a whole number and a natural number, such as

$$\frac{1}{2}, \frac{3}{4}, \frac{7}{3}, \frac{5}{5}, \frac{5}{1}, \frac{0}{1}, \frac{15}{1}, \frac{61}{21},$$

are called **fractions.** The **numerator** of a fraction is the number above the fraction bar, and the **denominator** is the number below the fraction bar. In the fraction $\frac{3}{4}$, 3 is the numerator and 4 is the denominator.

When the numerator is a smaller number than the denominator, the fraction is called a **proper fraction;** otherwise it is an **improper fraction.** For example, $\frac{1}{2}$ and $\frac{3}{4}$ are proper fractions, while $\frac{7}{5}$, $\frac{5}{5}$, $\frac{5}{1}$, and $\frac{61}{27}$ are improper fractions.

Every whole number can be written as a fraction by letting the whole number be the numerator and 1 be the denominator. For example,

$$5 = \frac{5}{1}, \quad 2 = \frac{2}{1}, \quad 0 = \frac{0}{1}, \quad \text{and} \quad 15 = \frac{15}{1}.$$

Five could also be written as

$$\frac{10}{2}, \quad \frac{15}{3}, \quad \frac{20}{4},$$

and so on. Similarly, the fraction $\frac{1}{4}$ could be expressed as $\frac{1}{4}$, $\frac{2}{8}$, $\frac{3}{12}$, $\frac{4}{16}$, $\frac{5}{20}$, and so on. We can always find another name for a fraction by multiplying both the numerator and the denominator by the same nonzero number. Fractions that are names for the same number or have the same value are called **equivalent fractions.**

Fundamental Principle of Fractions

If both the numerator and denominator of a fraction are multiplied or divided by the same nonzero number, the resulting fraction is **equivalent** to the original fraction.

3 REDUCING TO LOWEST TERMS

We **build up** fractions when we multiply both the numerator and denominator by the same natural number other than 1 and **reduce** fractions when we divide both by the same natural number other than 1. For example,

$$\frac{3}{4} = \frac{3 \cdot 2}{4 \cdot 2} = \frac{6}{8}, \qquad \frac{3}{4} = \frac{3 \cdot 5}{4 \cdot 5} = \frac{15}{20}, \qquad \frac{3}{4} = \frac{3 \cdot 10}{4 \cdot 10} = \frac{30}{40}.$$

Thus, $\frac{3}{4}$ has been built up to the equivalent fractions $\frac{6}{8}$, $\frac{15}{20}$, $\frac{30}{40}$. Also,

$$\frac{12}{18} = \frac{12 \div 2}{18 \div 2} = \frac{6}{9}, \qquad \frac{12}{18} = \frac{12 \div 3}{18 \div 3} = \frac{4}{6}, \qquad \frac{12}{18} = \frac{12 \div 6}{18 \div 6} = \frac{2}{3}.$$

Thus, $\frac{12}{18}$ has been reduced to the equivalent fractions $\frac{6}{9}$, $\frac{4}{6}$, and $\frac{2}{3}$.

Reducing Fractions

A fraction is **reduced to lowest terms** when 1 is the only natural number that divides both numerator and denominator.

4 PRIME NUMBERS AND FACTORS

We reduced $\frac{12}{18}$ to $\frac{6}{9}$, $\frac{4}{6}$, and $\frac{2}{3}$ above. Notice that $\frac{6}{9}$ and $\frac{4}{6}$ can each be reduced further, while $\frac{2}{3}$ cannot. Thus, $\frac{2}{3}$ is reduced to lowest terms. The concept of a *prime number* is useful for reducing fractions to lowest terms. A **prime number** is a natural number greater than 1 whose only divisors are 1 and itself. Since 1 and 5 are the only divisors of 5, 5 is prime; since 2 and 3 are divisors of 6 along with 1 and 6, 6 is not prime.

The first few primes are

$$2, \quad 3, \quad 5, \quad 7, \quad 11, \quad 13, \quad 17, \quad 19, \quad 23, \ldots.$$

Prime Numbers and Factors

Every natural number greater than 1 either is prime or can be expressed as a product of primes.

The primes in the product are called **prime factors,** and the process of expressing a number as a product of primes is called **factoring into primes.**

EXAMPLE 1 Factoring into Primes

Express as a product of primes.

(a) $15 = 3 \cdot 5$ Why not $15 = 15 \cdot 1$?

(b) $28 = 2 \cdot 2 \cdot 7$ Why not $28 = 4 \cdot 7$?

(c) $41 = 41$ 41 is a prime

Practice Exercise 1

Express as a product of primes.

(a) 21

(b) 36

(c) 53

Answers: (a) $3 \cdot 7$ (b) $2 \cdot 2 \cdot 3 \cdot 3$ (c) 53 is a prime

To Reduce a Fraction to Lowest Terms

1. Factor the numerator and denominator into products of primes.
2. Divide both the numerator and denominator by all common factors.
3. Multiply remaining factors in the numerator and multiply the remaining factors in the denominator.

Dividing both the numerator and denominator of a fraction by common factors is sometimes called **canceling factors.** This is shown by crossing out the common factors. For example,

$$\frac{9 \cdot \cancel{5}}{\cancel{5}} = \frac{9}{1} \quad \text{and} \quad \frac{\cancel{2} \cdot \cancel{7}}{\cancel{2} \cdot 3 \cdot \cancel{7}} = \frac{1}{3}.$$

Notice that when we cancel *all* of the factors out of a numerator or a denominator, it makes the numerator or denominator 1, *not* 0.

EXAMPLE 2 Reducing Fractions

Reduce the fractions to lowest terms.

(a) $\frac{6}{15} = \frac{2 \cdot \cancel{3}}{\cancel{3} \cdot 5} = \frac{2}{5}$ Cancel common factor 3 from both numerator and denominator

(b) $\frac{12}{42} = \frac{\cancel{2} \cdot 2 \cdot \cancel{3}}{\cancel{2} \cdot \cancel{3} \cdot 7} = \frac{2}{7}$ Cancel common factors 2 and 3 from both numerator and denominator

(c) $\frac{14}{15} = \frac{2 \cdot 7}{3 \cdot 5}$ There are no common factors, so $\frac{14}{15}$ is already reduced to lowest terms

Practice Exercise 2

Reduce the fractions to lowest terms.

(a) $\frac{2}{4}$

(b) $\frac{70}{25}$

(c) $\frac{15}{77}$

Answers: (a) $\frac{1}{2}$ (b) $\frac{14}{5}$ (c) already in lowest terms

CAUTION

Divide out or cancel only *factors* (numbers that are multiplied) and never cross out parts of sums. For example,

$$\frac{2\cdot 7}{2}=\frac{\not{2}\cdot 7}{\not{2}}=7,\quad \text{but}\quad \frac{2+7}{2}\ \text{is } not \text{ equal to } \frac{\not{2}+7}{\not{2}}.$$

With practice, we can shorten the process of reducing fractions, if we divide out common factors without first factoring into primes. For example, if we notice that 6 is a common factor of both 12 and 42, we can write

$$\frac{12}{42}=\frac{2\cdot\not{6}}{7\cdot\not{6}}=\frac{2}{7}.$$

Fractions describe parts of quantities in many applied situations. If a baseball player got three hits in five times at bat, we would say he had a hit $\frac{3}{5}$ (three-fifths) of the time. Thus, fractions are used to indicate what part one number is of another.

EXAMPLE 3 Fractional Parts

A group of 100 students was made up of 60 girls and 40 boys. Exactly 75 students were taking math, and exactly 35 students were taking history.

(a) What fractional part of the total number of students were girls?

There were 60 girls among 100 students for a fractional part of $\frac{60}{100}$, or $\frac{3}{5}$, reduced to lowest terms.

(b) What fraction of the students were taking math?

There were 75 math students out of the 100 total for a fraction of $\frac{75}{100}$, or $\frac{3}{4}$, reduced to lowest terms.

(c) What fraction of the students were *not* taking history?

Since 35 students were taking history, 65 were not. Thus, $\frac{65}{100}$, or $\frac{13}{20}$, of the students were not taking history.

Practice Exercise 3

In a sample of 200 pens, 120 are red and 80 are green. Exactly 65 of the pens have a fine point, and exactly 185 have retractable points.

(a) What fractional part of the total number of pens is green?

(b) What fraction of the total have fine points?

(c) What fraction do not have retractable points?

Answers: (a) $\frac{2}{5}$ (b) $\frac{13}{40}$ (c) $\frac{3}{40}$

0.1 EXERCISES A

Answer true *or* false *in Exercises 1–10. If the statement is false, tell why.*

1. A collection of objects is also called a set of objects.

2. The name or symbol for a number is a numeral.

3. {1, 2, 3, . . .} is the set of whole numbers.

4. In the fraction $\frac{5}{7}$, the number 7 is the numerator.

5. If the numerator of a fraction is a smaller number than the denominator, it is a proper fraction.

6. Two fractions that have the same value are equivalent fractions.

7. We build up a fraction when we divide both numerator and denominator by a natural number other than 1.

8. A natural number other than 1 that is divisible only by itself and 1 is a prime number.

9. The process of canceling factors from a numerator and denominator is really the process of adding common factors.

10. An idea that comes to mind when considering questions such as "how many" or "in what order" is called a number.

11. Consider the fractions $\frac{1}{5}$, $\frac{3}{8}$, $\frac{9}{9}$, $\frac{10}{2}$, $\frac{6}{6}$, $\frac{13}{15}$, and $\frac{8}{1}$.
(a) Which of these are proper fractions? **(b)** Which of these are improper fractions?

Write three fractions that are equivalent to the following.

12. $\frac{2}{5}$ **13.** $\frac{5}{2}$ **14.** 4

15. 0 **16.** $\frac{105}{2}$ **17.** $\frac{1}{100}$

18. List all prime numbers between 1 and 20.

Express as a product of primes.

19. 70 **20.** 45 **21.** 47

22. 81 **23.** 180 **24.** 1470

Reduce to lowest terms.

25. $\frac{9}{36}$ **26.** $\frac{8}{28}$ **27.** $\frac{25}{30}$

28. $\frac{12}{35}$ **29.** $\frac{48}{20}$ **30.** $\frac{72}{16}$

31. $\frac{72}{72}$ **32.** $\frac{72}{114}$ **33.** $\frac{120}{162}$

34. Write an equivalent fraction for $\frac{2}{3}$ with 15 as the denominator.

35. Write an equivalent fraction for $\frac{7}{8}$ with 32 as the denominator.

36. Write an equivalent fraction for $\frac{5}{4}$ with 15 as the numerator.

37. Write an equivalent fraction for $\frac{6}{5}$ with 36 as the numerator.

Supply the missing numerator.

38. $\frac{3}{4} = \frac{\quad}{12}$

39. $\frac{5}{7} = \frac{\quad}{28}$

40. $\frac{10}{3} = \frac{\quad}{24}$

Supply the missing denominator.

41. $\frac{3}{7} = \frac{6}{\quad}$

42. $\frac{2}{11} = \frac{10}{\quad}$

43. $\frac{21}{2} = \frac{63}{\quad}$

44. A softball pitcher struck out 12 of the 21 batters she faced. What fractional part of the total number of batters **(a)** did she strike out? **(b)** did she not strike out?

45. Marvin received \$150 from his dad. He spent \$55 for a coat and deposited the rest into his savings account. What fractional part of the \$150 went for **(a)** the coat? **(b)** savings?

ANSWERS: 1. true 2. true 3. false (natural numbers) 4. false (denominator) 5. true 6. true 7. false (reduce) 8. true 9. false (dividing) 10. true 11. (a) $\frac{1}{5}$, $\frac{3}{8}$, and $\frac{13}{15}$ (b) $\frac{9}{9}$, $\frac{10}{2}$, $\frac{6}{6}$, and $\frac{8}{1}$ *Answers to 12–17 will vary; some possibilities are given.* 12. $\frac{4}{10}$, $\frac{6}{15}$, $\frac{8}{20}$ 13. $\frac{10}{4}$, $\frac{15}{6}$, $\frac{20}{8}$ 14. $\frac{4}{1}$, $\frac{8}{2}$, $\frac{12}{3}$ 15. $\frac{0}{1}$, $\frac{0}{2}$, $\frac{0}{3}$ 16. $\frac{210}{4}$, $\frac{315}{6}$, $\frac{420}{8}$ 17. $\frac{2}{200}$, $\frac{3}{300}$, $\frac{4}{400}$ 18. 2, 3, 5, 7, 11, 13, 17, 19 19. $2 \cdot 5 \cdot 7$ 20. $3 \cdot 3 \cdot 5$ 21. 47 (prime) 22. $3 \cdot 3 \cdot 3 \cdot 3$ 23. $2 \cdot 2 \cdot 3 \cdot 3 \cdot 5$ 24. $2 \cdot 3 \cdot 5 \cdot 7 \cdot 7$ 25. $\frac{1}{4}$ 26. $\frac{2}{7}$ 27. $\frac{5}{6}$ 28. $\frac{12}{35}$ (already reduced to lowest terms) 29. $\frac{12}{5}$ 30. $\frac{9}{2}$ 31. 1 32. $\frac{12}{19}$ 33. $\frac{20}{27}$ 34. $\frac{10}{15}$ 35. $\frac{28}{32}$ 36. $\frac{15}{12}$ 37. $\frac{36}{30}$ 38. 9 39. 20 40. 80 41. 14 42. 55 43. 6 44. (a) $\frac{12}{21}$, or $\frac{4}{7}$ when reduced to lowest terms (b) $\frac{9}{21}$, or $\frac{3}{7}$ 45. (a) $\frac{55}{150}$, or $\frac{11}{30}$ when reduced (b) $\frac{95}{150}$, or $\frac{19}{30}$ when reduced

0.1 EXERCISES B

Answer true *or* false *in Exercises 1–10. If the statement is false, tell why.*

1. The objects that belong to a set are the elements of the set.

2. A numeral is the name or symbol for a set.

3. $\{0, 1, 2, 3, \ldots\}$ is the set of natural numbers.

4. The denominator of the fraction $\frac{5}{7}$ is 5.

5. If the numerator of a fraction is equal to or larger than the denominator, it is a proper fraction.

6. Equivalent fractions name the same number or have the same value.

7. We reduce a fraction when we divide both numerator and denominator by a natural number other than 1.

8. Prime numbers are divisible only by themselves and 1.

9. The process of dividing common factors from a numerator and denominator is sometimes called reducing common factors.

10. A number expressed as a quotient of a whole number and a natural number is a fraction.

11. Consider the fractions $\frac{1}{7}$, $\frac{3}{11}$, $\frac{8}{8}$, $\frac{15}{3}$, $\frac{17}{17}$, $\frac{21}{23}$, and $\frac{5}{1}$.
(a) Which of these are proper fractions? **(b)** Which of these are improper fractions?

Write three fractions that are equivalent to the following.

12. $\frac{3}{5}$

13. $\frac{7}{2}$

14. 8

15. 2

16. $\frac{95}{3}$

17. $\frac{1}{1000}$

18. List all prime numbers between 20 and 40.

Express as a product of primes.

19. 42

20. 20

21. 53

22. 625

23. 1000

24. 2541

Reduce to lowest terms.

25. $\frac{7}{42}$

26. $\frac{12}{28}$

27. $\frac{20}{32}$

28. $\frac{15}{77}$

29. $\frac{60}{35}$

30. $\frac{88}{16}$

31. $\frac{54}{54}$

32. $\frac{90}{147}$

33. $\frac{168}{228}$

34. Write an equivalent fraction for $\frac{3}{4}$ with 20 as the denominator.

35. Write an equivalent fraction for $\frac{5}{8}$ with 48 as the denominator.

36. Write an equivalent fraction for $\frac{7}{4}$ with 21 as the numerator.

37. Write an equivalent fraction for $\frac{8}{5}$ with 32 as the numerator.

Supply the missing numerator.

38. $\frac{2}{3} = \frac{?}{12}$

39. $\frac{3}{7} = \frac{?}{35}$

40. $\frac{12}{5} = \frac{?}{45}$

Supply the missing denominator.

41. $\frac{4}{7} = \frac{28}{?}$

42. $\frac{8}{11} = \frac{24}{?}$

43. $\frac{52}{3} = \frac{104}{?}$

44. A basketball player hit 21 free throws in 28 attempts during a recent tournament. What fractional part of the attempts **(a)** did he make? **(b)** did he miss?

45. Bill received \$250 from his grandmother. He spent \$120 to repair his car and the rest went for gasoline. What fractional part of the \$250 went for **(a)** car repairs? **(b)** gasoline?

0.1 EXERCISES C

Reduce to lowest terms.

1. $\frac{632}{492}$

2. $\frac{246}{1230}$

3. $\frac{780}{5005}$ [Answer: $\frac{12}{77}$]

0.2 OPERATIONS ON FRACTIONS

STUDENT GUIDEPOSTS

1. Division by Zero
2. Multiplying Fractions
3. Reciprocals
4. Dividing Fractions
5. Least Common Denominator
6. Adding Fractions
7. Mixed Numbers
8. Subtracting Fractions

1 DIVISION BY ZERO

The four basic operations on fractions are addition (+), subtraction (−), multiplication (·), and division (÷ or a fraction bar). We assume that these four operations applied to whole numbers are well understood. Division by zero, however, causes a problem. Why? The quotient of two numbers such as 8 and 2, written $8 \div 2$ or $\frac{8}{2}$, is the number which, when multiplied by 2, equals 8.

That is, $$\frac{8}{2} = \mathbf{4} \quad \text{because} \quad 8 = 2 \cdot \mathbf{4}.$$

Similarly, $$\frac{20}{4} = \mathbf{5} \quad \text{because} \quad 20 = 4 \cdot \mathbf{5}.$$

If $$\frac{20}{0} = \textbf{(a number)}, \quad \text{then} \quad 20 = 0 \cdot \textbf{(a number)}.$$

But zero times any number is zero, not 20, so there is no number equal to $\frac{20}{0}$. Since we could have used any number in our example, no number can be divided by zero (except possibly zero itself). What would $\frac{0}{0}$ equal?

$$\frac{0}{0} \text{ could be } \mathbf{5} \quad \text{since} \quad 0 = 0 \cdot \mathbf{5}$$

$$\frac{0}{0} \text{ could be } \mathbf{259} \quad \text{since} \quad 0 = 0 \cdot \mathbf{259}$$

$$\frac{0}{0} \text{ could be } \textbf{(any number)} \quad \text{since} \quad 0 \cdot \textbf{(any number)} = 0$$

Since $\frac{0}{0}$ cannot be determined, we agree never to divide 0 by 0. Thus, *any division by 0 is said to be undefined* and is excluded from mathematics. However, $0 \div 5$ or $\frac{0}{5}$ does make sense, and in fact $\frac{0}{5} = 0$ (why?).

2 MULTIPLYING FRACTIONS

We now review the four operations on fractions starting with multiplication.

To Multiply Two (or More) Fractions

1. Factor all numerators and denominators into products of primes.
2. Place all numerator factors over all denominator factors.
3. Divide out or cancel all common factors from both numerator and denominator.
4. Multiply the remaining factors in the numerator and multiply the remaining factors in the denominator.
5. The resulting fraction is the product (reduced to lowest terms) of the original fractions.

EXAMPLE 1 MULTIPLYING FRACTIONS

Multiply.

(a) $\frac{3}{8} \cdot \frac{20}{9} = \frac{3}{2 \cdot 2 \cdot 2} \cdot \frac{2 \cdot 2 \cdot 5}{3 \cdot 3}$ Factor all numerators and denominators

$= \frac{3 \cdot 2 \cdot 2 \cdot 5}{2 \cdot 2 \cdot 2 \cdot 3 \cdot 3}$ All numerator factors over all denominator factors

$= \frac{\cancel{3} \cdot \cancel{2} \cdot \cancel{2} \cdot 5}{\cancel{2} \cdot \cancel{2} \cdot 2 \cdot \cancel{3} \cdot 3}$ Divide out common factors

$= \frac{5}{6}$ Multiply the remaining factors

(b) $\frac{3}{7} \cdot 14 = \frac{3}{7} \cdot \frac{14}{1} = \frac{3}{7} \cdot \frac{2 \cdot 7}{1} = \frac{3 \cdot 2 \cdot \cancel{7}}{\cancel{7} \cdot 1} = \frac{6}{1} = 6$

PRACTICE EXERCISE 1

Multiply.

(a) $\frac{2}{15} \cdot \frac{5}{14}$

(b) $\frac{3}{8} \cdot 24$

Answers: (a) $\frac{1}{21}$ (b) 9

With practice we can shorten our work by not factoring completely to primes when larger common factors can be recognized. For example, in Example 1(a) we might multiply as follows:

$$\frac{\overset{1}{\cancel{3}}}{\underset{2}{\cancel{8}}} \cdot \frac{\overset{5}{\cancel{20}}}{\underset{3}{\cancel{9}}} = \frac{5}{6}.$$ Divide out 4 and 3

3 RECIPROCALS

To divide fractions, we need to understand the idea of a *reciprocal*. The **reciprocal** of a fraction is the fraction formed by interchanging its numerator and denominator. Some fractions and their reciprocals are given in the following table.

Fraction	*Reciprocal*
$\frac{2}{3}$	$\frac{3}{2}$
$\frac{7}{8}$	$\frac{8}{7}$
$5 \left(\text{or } \frac{5}{1}\right)$	$\frac{1}{5}$
$\frac{1}{3}$	$3 \left(\text{or } \frac{3}{1}\right)$

Any fraction multiplied by its reciprocal gives us the number 1. For example,

$$\frac{2}{3} \cdot \frac{3}{2} = 1, \quad \frac{7}{8} \cdot \frac{8}{7} = 1, \quad \frac{5}{1} \cdot \frac{1}{5} = 1, \quad \frac{1}{3} \cdot \frac{3}{1} = 1.$$

4 DIVIDING FRACTIONS

When one number is divided by a second, the first is the **dividend,** the second is the **divisor,** and the result is the **quotient.** In $10 \div 2 = 5$, 10 is the dividend, 2 the divisor, and 5 the quotient.

To Divide Two Fractions

1. Replace the divisor with its reciprocal and change the division sign to multiplication.
2. Multiply the fractions.

EXAMPLE 2 Dividing Fractions

Divide.

(a) $\frac{13}{15} \div \frac{39}{5} = \frac{13}{15} \cdot \frac{5}{39}$ Replace divisor with its reciprocal and multiply

$= \frac{13}{3 \cdot 5} \cdot \frac{5}{3 \cdot 13}$ Factor

$= \frac{\cancel{13} \cdot \cancel{5}}{3 \cdot \cancel{5} \cdot 3 \cdot \cancel{13}}$ Indicate products and divide out common factors

$= \frac{1}{9}$ A 1, not 0, remains in the numerator

(b) $\frac{4}{5} \div 44 = \frac{4}{5} \cdot \frac{1}{44} = \frac{2 \cdot 2}{5} \cdot \frac{1}{2 \cdot 2 \cdot 11} = \frac{\cancel{2} \cdot \cancel{2} \cdot 1}{5 \cdot \cancel{2} \cdot \cancel{2} \cdot 11} = \frac{1}{55}$

(c) $\frac{3}{7} \div \frac{3}{7} = \frac{3}{7} \cdot \frac{7}{3} = \frac{\cancel{3} \cdot \cancel{7}}{\cancel{7} \cdot \cancel{3}} = \frac{1}{1} = 1$ Does this answer seem reasonable?

Practice Exercise 2

Divide.

(a) $\frac{11}{6} \div \frac{44}{21}$

(b) $\frac{3}{7} \div 6$

(c) $\frac{9}{13} \div \frac{9}{13}$

Answers: (a) $\frac{7}{8}$ (b) $\frac{1}{14}$ (c) 1

To add or subtract fractions with the same denominators, simply add numerators and place the result over the common denominator. For example,

$$\frac{3}{7} + \frac{5}{7} = \frac{3+5}{7} = \frac{8}{7}.$$

CAUTION

When adding fractions, we add numerators but *not* denominators.

5 LEAST COMMON DENOMINATOR

To add or subtract fractions with different denominators, change the fractions to equivalent fractions having a common denominator. For this it is a good idea to find the **least common denominator (LCD)** of the fractions, that is, the *smallest* number that has both denominators as factors. For example,

$$\frac{1}{2} + \frac{1}{3} \quad \text{can be changed to} \quad \frac{3}{6} + \frac{2}{6}.$$

Here 6 is the least common denominator of $\frac{1}{2}$ and $\frac{1}{3}$ since there is no smaller number that has both 2 and 3 as factors. Notice we could have used 12 or 18 as a common denominator, but 6 is the *least* common denominator.

To Find the LCD of Two (or More) Fractions

1. Factor each denominator into a product of primes.
2. If there are no common factors in the denominators, the LCD is the product of *all* denominators.
3. If there are common factors in the denominators, each factor must appear in the LCD as many times as it appears in the denominator where it is found the greatest number of times.

EXAMPLE 3 FINDING THE LCD

Find the LCD of the fractions.

(a) $\frac{1}{6}$ and $\frac{4}{15}$

Factor the denominators: $6 = \overset{1}{2} \cdot \overset{1}{3}$ and $15 = \overset{1}{3} \cdot \overset{1}{5}$. The LCD must consist of one 2, one 3, and one 5. Thus, the LCD $= 2 \cdot 3 \cdot 5 = 30$.

(b) $\frac{13}{90}$ and $\frac{7}{24}$

Factor the denominators: $90 = \overset{1}{2} \cdot \overset{2}{3 \cdot 3} \cdot \overset{1}{5}$ and $24 = \overset{3}{2 \cdot 2 \cdot 2} \cdot \overset{1}{3}$. The LCD must consist of three 2's, two 3's, and one 5. Thus, the LCD $=$ $2 \cdot 2 \cdot 2 \cdot 3 \cdot 3 \cdot 5 = 360$.

(c) $\frac{3}{10}$ and $\frac{5}{21}$

Factor the denominators: $10 = \overset{1}{2} \cdot \overset{1}{5}$ and $21 = \overset{1}{3} \cdot \overset{1}{7}$. Since there are no common factors, the LCD is the product of all factors. Thus, the LCD $=$ $2 \cdot 3 \cdot 5 \cdot 7 = 210$.

(d) $\frac{1}{2}$, $\frac{5}{6}$, and $\frac{7}{25}$

Factor the denominators: $2 = \overset{1}{2}$, $6 = \overset{1}{2} \cdot \overset{1}{3}$, and $75 = \overset{1}{3} \cdot \overset{2}{5 \cdot 5}$. Thus, the LCD $= 2 \cdot 3 \cdot 5 \cdot 5 = 150$.

PRACTICE EXERCISE 3

Find the LCD of the fractions.

(a) $\frac{5}{6}$ and $\frac{1}{21}$

(b) $\frac{1}{45}$ and $\frac{7}{20}$

(c) $\frac{2}{33}$ and $\frac{5}{14}$

(d) $\frac{1}{3}$, $\frac{4}{15}$, and $\frac{2}{75}$

Answers: (a) 42 (b) 180 (c) 462 (d) 75

To see why the rule above works, let us look again at the fractions $\frac{1}{6}$ and $\frac{4}{15}$ in Example 3(a). By definition, each denominator must be a factor of the LCD. Since 6 is a factor of the LCD, the LCD must contain the factors 2 and 3. Also, with 15 a factor of the LCD, the LCD must contain the factors 3 and 5. Since we already have 3 as a factor from the denominator 6, we only need to supply the factor 5. Thus,

$$\text{LCD} = 2 \cdot 3 \cdot 5 = 30,$$

which is what we found in Example 3(a) using the rule above.

6 ADDING FRACTIONS

We now use the LCD to add fractions with different denominators.

To Add Two (or More) Fractions with Different Denominators

1. Rewrite the sum with each denominator expressed as a product of primes.
2. Find the LCD.
3. Multiply the numerators and denominators of each fraction by all those factors present in the LCD but missing in the denominator of the particular fraction.
4. Place the sum of all numerators over the LCD.
5. Simplify the resulting numerator and reduce the fraction to lowest terms.

EXAMPLE 4 Adding Fractions

Add.

(a) $\frac{5}{6}+\frac{7}{15}=\frac{5}{2\cdot 3}+\frac{7}{3\cdot 5}$ — Factor denominators; the LCD is $2\cdot 3\cdot 5$

$=\frac{5\cdot 5}{2\cdot 3\cdot 5}+\frac{7\cdot 2}{3\cdot 5\cdot 2}$ — Multiply numerators and denominators by missing factors

$=\frac{25+14}{2\cdot 3\cdot 5}$ — Simplify and add over LCD

$=\frac{39}{2\cdot 3\cdot 5}$ — Simplify

$=\frac{\cancel{3}\cdot 13}{2\cdot \cancel{3}\cdot 5}$ — Factor numerator and divide out common factors

$=\frac{13}{10}$ — The desired sum

(b) $\frac{1}{2}+\frac{5}{6}+\frac{7}{25}$

$=\frac{1}{2}+\frac{5}{2\cdot 3}+\frac{7}{5\cdot 5}$ — Factor: the LCD $= 2\cdot 3\cdot 5\cdot 5$

$=\frac{1\cdot 3\cdot 5\cdot 5}{2\cdot 3\cdot 5\cdot 5}+\frac{5\cdot 5\cdot 5}{2\cdot 3\cdot 5\cdot 5}+\frac{7\cdot 2\cdot 3}{5\cdot 5\cdot 2\cdot 3}$ — Multiply numerators and denominators by missing factors

$=\frac{75+125+42}{2\cdot 3\cdot 5\cdot 5}$ — Simplify and add over LCD

$=\frac{242}{2\cdot 3\cdot 5\cdot 5}$ — Simplify

$=\frac{\cancel{2}\cdot 11\cdot 11}{\cancel{2}\cdot 3\cdot 5\cdot 5}$ — Factor numerator and divide out common factors

$=\frac{121}{75}$ — The desired sum

PRACTICE EXERCISE 4

Add.

(a) $\frac{7}{15}+\frac{8}{21}$

(b) $\frac{2}{3}+\frac{1}{5}+\frac{5}{6}$

Answers: (a) $\frac{89}{105}$ (b) $\frac{17}{10}$

7 MIXED NUMBERS

It is sometimes helpful to write an improper fraction, such as $\frac{121}{75}$, as a *mixed number*. A **mixed number** is the sum of a whole number and a proper fraction. For example,

$$\frac{121}{75} \text{ can be written as } 1 + \frac{46}{75} \text{ or } 1\frac{46}{75}.$$

This is read "one and forty-six seventy-fifths." To change an improper fraction to a mixed number, reduce the fraction to lowest terms and divide the denominator into the numerator. The quotient is the whole number part of the mixed number, and the fractional part is the remainder over the divisor.

EXAMPLE 5 Improper Fraction to Mixed Number

Change each improper fraction to a mixed number.

(a) $\frac{13}{4}$

Divide 4 into 13.

$$\begin{array}{r} 3 \\ 4\overline{)13} \\ \underline{12} \\ 1 \end{array}$$

The mixed number is $3\frac{1}{4}$, read "three and one-fourth."

(b) $\frac{27}{6}$

Reduce $\frac{27}{6}$ to lowest terms and then divide.

$$\frac{27}{6} = \frac{9 \cdot \cancel{3}}{2 \cdot \cancel{3}} = \frac{9}{2} \qquad \begin{array}{r} 4 \\ 2\overline{)9} \\ \underline{8} \\ 1 \end{array}$$

Thus, $\frac{27}{6} = 4\frac{1}{2}$.

Practice Exercise 5

Change each improper fraction to a mixed number.

(a) $\frac{25}{8}$

(b) $\frac{40}{15}$

Answers: (a) $3\frac{1}{8}$ (b) $2\frac{2}{3}$

To change a mixed number to an improper fraction, we can think of the mixed number as a sum of two fractions. For example, the mixed number $2\frac{11}{18} = 2 + \frac{11}{18}$. Adding, we have

$$\frac{2}{1} + \frac{11}{18} = \frac{2}{1} + \frac{11}{2 \cdot 3 \cdot 3} = \frac{2 \cdot 2 \cdot 3 \cdot 3}{2 \cdot 3 \cdot 3} + \frac{11}{2 \cdot 3 \cdot 3} = \frac{36 + 11}{2 \cdot 3 \cdot 3} = \frac{47}{18}.$$

Rather than going through all of these steps, we can use a shortcut. Consider $2\frac{11}{18}$ again. Notice that if we multiply the denominator, 18, by the whole number, 2, and add the numerator, 11, we obtain

$$(18 \cdot 2) + 11 = 47.$$

When we put this result over the denominator, 18, we have the desired improper fraction $\frac{47}{18}$. We can show this as

$$\frac{\textbf{(denominator × whole number) + numerator}}{\textbf{denominator}}.$$

EXAMPLE 6 Mixed Number to Improper Fraction

Change each mixed number to an improper fraction.

(a) $4\frac{3}{5} = \frac{(5 \cdot 4) + 3}{5} = \frac{20 + 3}{5} = \frac{23}{5}$

(b) $6\frac{7}{10} = \frac{(10 \cdot 6) + 7}{10} = \frac{60 + 7}{10} = \frac{67}{10}$

Practice Exercise 6

Change each mixed number to an improper fraction.

(a) $6\frac{2}{7}$

(b) $3\frac{23}{100}$

Answers: (a) $\frac{44}{7}$ (b) $\frac{323}{100}$

To add two or more mixed numbers or mixed numbers and proper fractions, convert all mixed numbers to improper fractions and proceed as before.

EXAMPLE 7 Adding Mixed Numbers

Add.

$$3\frac{1}{4} + 6\frac{7}{10} + \frac{1}{3} = \frac{13}{4} + \frac{67}{10} + \frac{1}{3}$$ Convert to improper fractions

$$= \frac{13}{2 \cdot 2} + \frac{67}{2 \cdot 5} + \frac{1}{3}$$ The LCD = $2 \cdot 2 \cdot 3 \cdot 5$

$$= \frac{13 \cdot 3 \cdot 5}{2 \cdot 2 \cdot 3 \cdot 5} + \frac{67 \cdot 2 \cdot 3}{2 \cdot 5 \cdot 2 \cdot 3} + \frac{1 \cdot 2 \cdot 2 \cdot 5}{3 \cdot 2 \cdot 2 \cdot 5}$$

$$= \frac{195 + 402 + 20}{2 \cdot 2 \cdot 3 \cdot 5}$$

$$= \frac{617}{60} \text{ or } 10\frac{17}{60}$$

Practice Exercise 7

Add.

$$2\frac{2}{3} + 5\frac{1}{10} + \frac{1}{5}$$

Answer: $\frac{239}{30}$, or $7\frac{29}{30}$

8 SUBTRACTING FRACTIONS

The fourth operation we consider is subtraction.

To Subtract Two Fractions

1. Rewrite the difference with each denominator expressed as a product of prime factors.
2. Find the LCD.
3. Supply missing factors just as when adding.
4. Place the difference of the numerators over the LCD.
5. Simplify the resulting numerator and reduce the fraction to lowest terms.

EXAMPLE 8 Subtracting Fractions

Subtract.

(a) $\frac{7}{12} - \frac{5}{9} = \frac{7}{2 \cdot 2 \cdot 3} - \frac{5}{3 \cdot 3}$ Factor; LCD is $2 \cdot 2 \cdot 3 \cdot 3$

$= \frac{7 \cdot 3}{2 \cdot 2 \cdot 3 \cdot 3} - \frac{5 \cdot 2 \cdot 2}{3 \cdot 3 \cdot 2 \cdot 2}$ Supply missing factors

$= \frac{21 - 20}{2 \cdot 2 \cdot 3 \cdot 3}$ Subtract and simplify

$= \frac{1}{36}$ The difference in reduced form

(b) $4\frac{1}{2} - 3\frac{7}{8}$

First convert the mixed numbers to improper fractions.

$$4\frac{1}{2} - 3\frac{7}{8} = \frac{9}{2} - \frac{31}{8} = \frac{9 \cdot 4}{2 \cdot 4} - \frac{31}{8} = \frac{36 - 31}{8} = \frac{5}{8}$$

Practice Exercise 8

Subtract.

(a) $\frac{8}{35} - \frac{1}{21}$

(b) $6\frac{1}{4} - 2\frac{6}{7}$

Answers: (a) $\frac{19}{105}$ (b) $\frac{95}{28}$, or $3\frac{11}{28}$

Notice in Example 8(b) that we took a shortcut and did not factor denominators into primes. It was clear that the LCD of the two fractions is 8, so we simply supplied the factor of 4 to both numerator and denominator of $\frac{9}{2}$. With practice, we can save time with shortcuts such as this.

The rules for adding and subtracting fractions can be extended to include combinations of these operations, as shown in the next example.

EXAMPLE 9 Combination of Operations

Perform the indicated operations.

$2\frac{1}{15} + 4\frac{1}{5} - 3\frac{1}{2}$

$= \frac{31}{15} + \frac{21}{5} - \frac{7}{2}$ Convert to improper fractions

$= \frac{31}{3 \cdot 5} + \frac{21}{5} - \frac{7}{2}$ The LCD $= 2 \cdot 3 \cdot 5$

$= \frac{31 \cdot 2}{3 \cdot 5 \cdot 2} + \frac{21 \cdot 2 \cdot 3}{5 \cdot 2 \cdot 3} - \frac{7 \cdot 3 \cdot 5}{2 \cdot 3 \cdot 5}$ Supply missing factors

$= \frac{62 + 126 - 105}{2 \cdot 3 \cdot 5}$

$= \frac{83}{30}$ or $2\frac{23}{30}$

Practice Exercise 9

Perform the indicated operations.

$7\frac{1}{15} - 2\frac{2}{3} + 1\frac{1}{2}$

Answer: $\frac{59}{10}$, or $5\frac{9}{10}$

0.2 EXERCISES A

1. $\frac{0}{3} = $ ____
2. $\frac{3}{0} = $ ____
3. $\frac{0}{0} = $ ____

Multiply.

4. $\frac{1}{3} \cdot \frac{6}{5}$
5. $\frac{1}{2} \cdot \frac{3}{7}$
6. $\frac{6}{35} \cdot \frac{20}{12}$
7. $\frac{12}{5} \cdot 10$
*8. $4\frac{2}{5} \cdot \frac{3}{11}$
9. $2\frac{2}{3} \cdot 1\frac{1}{8} \cdot \frac{1}{6}$

Divide.

10. $\frac{3}{7} \div \frac{9}{28}$
11. $\frac{\frac{20}{9}}{\frac{2}{3}}$
12. $\frac{20}{27} \div \frac{35}{36}$
13. $\frac{8}{9} \div 4$
14. $3\frac{1}{5} \div \frac{2}{5}$
15. $3\frac{1}{3} \div 1\frac{3}{5}$

Find the LCD of the following fractions.

16. $\frac{2}{7}$ and $\frac{4}{33}$
17. $\frac{7}{12}$ and $\frac{1}{45}$
18. $\frac{3}{16}$ and $\frac{7}{20}$
19. 7 and $\frac{9}{2}$
20. $\frac{2}{3}$, $\frac{3}{4}$, and $\frac{1}{5}$
21. $\frac{1}{9}$, $\frac{5}{6}$, and $\frac{1}{75}$

Add.

22. $\frac{1}{2} + \frac{1}{2}$
23. $\frac{1}{6} + \frac{3}{10}$
24. $\frac{3}{28} + \frac{13}{70}$
25. $4 + \frac{3}{5}$
26. $2\frac{1}{5} + 1\frac{2}{3}$
27. $\frac{2}{3} + \frac{1}{6} + \frac{3}{4}$

Change each improper fraction to a mixed number.

28. $\frac{23}{8}$
29. $\frac{135}{4}$
30. $\frac{142}{11}$

*A colored circle on the number means there is a complete solution in the back of the text. See *To the Student* for more details.

Change each mixed number to an improper fraction.

31. $4\frac{2}{5}$

32. $66\frac{2}{3}$

33. $9\frac{7}{11}$

Subtract.

34. $\frac{3}{4} - \frac{3}{4}$

35. $\frac{7}{11} - \frac{2}{7}$

36. $\frac{7}{15} - \frac{13}{35}$

37. $\frac{19}{2} - 3$

38. $5\frac{1}{3} - 2\frac{3}{4}$

39. $11\frac{4}{5} - 5\frac{2}{3}$

Perform the indicated operations.

40. $\frac{7}{20} + \frac{3}{8} - \frac{1}{4}$

41. $\frac{14}{15} - \frac{2}{5} - \frac{1}{3}$

42. $\frac{7}{3} + \frac{1}{7} - 2$

Solve.

43 If $\frac{5}{6}$ of a class consists of girls and there are 72 students in the class, how many students are girls? How many are boys?

44. On a map, 1 inch represents $\frac{1}{2}$ of a mile. How many miles are represented by $4\frac{1}{4}$ inches?

45. A piece of rope 20 m long is to be cut into pieces, each of whose length is $\frac{2}{3}$ m. How many pieces can be cut?

46 A fuel tank holds $12\frac{1}{2}$ gallons when it is $\frac{3}{4}$ full. What is the capacity of the tank?

47. Alphonso hiked from Whiskey Creek to Bald Mountain, a distance of $2\frac{1}{3}$ mi. From there he hiked to Spook Hollow, a distance of $4\frac{1}{5}$ mi. How far did he hike?

48. To get the right shade of paint for her living room, Tracy Bell mixed $\frac{7}{8}$ of a gallon of white paint with $\frac{2}{3}$ of a gallon of light blue paint. How much paint did she have?

49. The weight of one cubic foot of water is $62\frac{1}{2}$ lb. How much do $3\frac{1}{5}$ cubic feet of water weigh?

50. Fahrenheit temperature can be found by multiplying Celsius temperature by $\frac{9}{5}$ and adding 32°. If Celsius temperature is 15°, what is Fahrenheit temperature?

FOR REVIEW

51. Write an equivalent fraction for $\frac{7}{8}$ with
(a) 21 for the numerator.
(b) 72 for the denominator.

52. Max received \$125 for doing a particular job. If he had to pay \$15 for income tax, what fractional part of the \$125 went for tax?

ANSWERS: **1.** 0 **2.** undefined **3.** undefined **4.** $\frac{2}{5}$ **5.** $\frac{3}{14}$ **6.** $\frac{2}{7}$ **7.** 24 **8.** $\frac{6}{5}$ **9.** $\frac{1}{2}$ **10.** $\frac{4}{3}$ **11.** $\frac{10}{3}$ **12.** $\frac{16}{21}$ **13.** $\frac{2}{9}$ **14.** 8 **15.** $\frac{25}{12}$ **16.** 231 **17.** 180 **18.** 80 **19.** 2 **20.** 60 **21.** 450 **22.** 1 **23.** $\frac{7}{15}$ **24.** $\frac{41}{140}$ **25.** $\frac{23}{5}$ **26.** $\frac{58}{15}$ **27.** $\frac{19}{12}$ **28.** $2\frac{7}{8}$ **29.** $33\frac{3}{4}$ **30.** $12\frac{10}{11}$ **31.** $\frac{22}{5}$ **32.** $\frac{200}{3}$ **33.** $\frac{106}{11}$ **34.** 0 **35.** $\frac{27}{77}$ **36.** $\frac{2}{21}$ **37.** $\frac{13}{2}$ **38.** $2\frac{7}{12}$ **39.** $6\frac{2}{15}$ **40.** $\frac{19}{40}$ **41.** $\frac{1}{5}$ **42.** $\frac{10}{21}$ **43.** 60 girls, 12 boys **44.** $2\frac{1}{8}$ mi **45.** 30 **46.** $16\frac{2}{3}$ gallons **47.** $6\frac{8}{15}$ mi **48.** $1\frac{13}{24}$ gallons **49.** 200 lb **50.** 59° **51.** (a) $\frac{21}{24}$ (b) $\frac{63}{72}$ **52.** $\frac{3}{25}$

0.2 EXERCISES B

1. $\frac{7}{0} = \underline{\quad ? \quad}$

2. $\frac{0}{0} = \underline{\quad ? \quad}$

3. $\frac{0}{7} = \underline{\quad ? \quad}$

Multiply.

4. $\frac{2}{3} \cdot \frac{6}{5}$

5. $\frac{3}{8} \cdot \frac{1}{2}$

6. $\frac{7}{35} \cdot \frac{20}{14}$

7. $\frac{11}{4} \cdot 8$

8. $4\frac{4}{5} \cdot \frac{1}{12}$

9. $1\frac{1}{3} \cdot 3\frac{3}{8} \cdot \frac{2}{3}$

Divide.

10. $\frac{2}{7} \div \frac{6}{28}$

11. $\frac{\frac{15}{8}}{\frac{3}{4}}$

12. $\frac{10}{27} \div \frac{15}{36}$

13. $6 \div \frac{8}{11}$

14. $2\frac{3}{5} \div \frac{26}{5}$

15. $2\frac{2}{3} \div 1\frac{1}{5}$

Find the LCD of the following fractions.

16. $\frac{1}{9}$ and $\frac{3}{22}$

17. $\frac{5}{12}$ and $\frac{2}{27}$

18. $\frac{1}{18}$ and $\frac{3}{20}$

19. $\frac{2}{9}$ and 5

20. $\frac{1}{3}, \frac{3}{4}$, and $\frac{2}{7}$

21. $\frac{1}{6}, \frac{5}{9}$, and $\frac{7}{30}$

Add.

22. $\frac{1}{4} + \frac{3}{4}$

23. $\frac{1}{8} + \frac{5}{12}$

24. $\frac{5}{24} + \frac{7}{60}$

25. $\frac{1}{8} + 2$

26. $4\frac{1}{5} + 2\frac{1}{3}$

27. $\frac{1}{3} + \frac{5}{6} + \frac{1}{5}$

Change each improper fraction to a mixed number.

28. $\frac{81}{10}$

29. $\frac{245}{9}$

30. $\frac{167}{11}$

Change each mixed number to an improper fraction.

31. $10\frac{1}{5}$

32. $42\frac{3}{10}$

33. $8\frac{4}{11}$

Subtract.

34. $\frac{5}{8} - \frac{1}{4}$

35. $\frac{8}{15} - \frac{12}{35}$

36. $5 - \frac{1}{9}$

37. $\frac{9}{2} - 1\frac{1}{4}$

38. $6\frac{2}{3} - 1\frac{3}{4}$

39. $11\frac{3}{5} - 4\frac{1}{3}$

Perform the indicated operations.

40. $\frac{9}{20} + \frac{5}{8} - \frac{3}{4}$

41. $\frac{13}{15} - \frac{1}{5} - \frac{1}{3}$

42. $\frac{8}{3} + \frac{2}{7} - 2$

Solve.

43. Raoul took out a loan for $\frac{2}{3}$ of the cost of a motorbike. If the bike cost $744, how much did he borrow?

44. Sandi can hike $3\frac{1}{2}$ mi in one hour. How many miles can she hike in $\frac{3}{4}$ of an hour?

45. When Fred Vickers pounds a nail into a board, he sinks it $1\frac{1}{4}$ inches with each blow. How many times will he have to hit a 5-in nail to drive it completely into the board?

46. After driving 285 miles, the Stotts had completed $\frac{5}{8}$ of their trip. What was the total length of the trip?

47. In July, Memphis had two rain storms, one that dropped $1\frac{1}{4}$ in of rain and another that dropped $\frac{3}{5}$ in. How much rain did Memphis receive in July?

48. It took Burford $8\frac{1}{2}$ hr to fix his car himself. If he had the money, he could have hired a mechanic to do the job in $1\frac{1}{3}$ hr. How many hours of work could Burford have saved?

49. Cindy Brown is paid $4\frac{1}{2}$ dollars per hr to type a manuscript. If she types for a period of $20\frac{2}{3}$ hr, how much is she paid?

50. Celsius temperature can be found by subtracting 32° from Fahrenheit temperature and multiplying the result by $\frac{5}{9}$. If Fahrenheit temperature is 122°, what is the corresponding Celsius temperature?

FOR REVIEW

51. Write an equivalent fraction for $\frac{8}{9}$ with **(a)** 32 as the numerator, and **(b)** 81 as the denominator.

52. A farmer collected 280 bushels of grain from his field. If he had to give 40 bushels to his landlord for rent, what fractional part of the 280 bushels went for rent?

0.2 EXERCISES C

Perform the indicated operations.

1. $22\frac{5}{8} + 17\frac{3}{10} - 8\frac{3}{4}$
[Answer: $31\frac{7}{40}$]

2. $42\frac{7}{15} - 18\frac{5}{6} - 11\frac{3}{10}$

3. $125\frac{16}{25} + 37\frac{2}{15} - 83\frac{7}{20}$

0.3 DECIMALS

STUDENT GUIDEPOSTS

1. Place-Value Number System
2. Decimal Notation
3. Operations on Decimals

1 PLACE-VALUE NUMBER SYSTEM

Remember that our number system is a **place-value system.** That is, each **digit** (0, 1, 2, 3, 4, 5, 6, 7, 8, or 9) in a numeral has a particular value determined by its location or place in the symbol. For example, the numeral

$$125.378$$

is a shorthand symbol for the **expanded numeral**

$$100 + 20 + 5 + \frac{3}{10} + \frac{7}{100} + \frac{8}{1000},$$

or $$\mathbf{1} \cdot 100 + \mathbf{2} \cdot 10 + \mathbf{5} \cdot 1 + \mathbf{3} \cdot \frac{1}{10} + \mathbf{7} \cdot \frac{1}{100} + \mathbf{8} \cdot \frac{1}{1000},$$

where each of the digits, 1, 2, 5, 3, 7, and 8 is multiplied by 100, 10, 1, $\frac{1}{10}$, $\frac{1}{100}$, or $\frac{1}{1000}$, depending upon the location of the digit in the symbol 125.378. We read 125.378 as "one hundred twenty-five *and* three hundred seventy-eight thousandths." This results from thinking of the number as a mixed number

$$125 + \frac{378}{1000}, \quad \text{or} \quad 125\frac{378}{1000}.$$

2 DECIMAL NOTATION

The numeral 125.378, representing the mixed number $125\frac{378}{1000}$, is in **decimal notation,** and numbers expressed in this manner are referred to simply as **decimals.** Decimal notation is based on the number 10, and the period used in a decimal is the **decimal point.** If there are no digits to the right of the decimal point we usually omit it; we write 53 rather than 53., for example. The diagram below gives the value of each position. Notice that the decimal point separates the ones digit from the tenths digit.

Hundreds	Tens	Ones	Decimal Point	Tenths	Hundredths	Thousandths
1	2	5	.	3	7	8

The following table expresses each decimal in expanded notation, as a mixed number, and as a proper or improper fraction.

Decimal	*Expanded notation*	*Mixed number*	*Fraction*
1.25	$1 + \frac{2}{10} + \frac{5}{100}$	$1\frac{25}{100}$	$\frac{125}{100}$
23.46	$20 + 3 + \frac{4}{10} + \frac{6}{100}$	$23\frac{46}{100}$	$\frac{2346}{100}$
43.278	$40 + 3 + \frac{2}{10} + \frac{7}{100} + \frac{8}{1000}$	$43\frac{278}{1000}$	$\frac{43{,}278}{1000}$
562.34	$500 + 60 + 2 + \frac{3}{10} + \frac{4}{100}$	$562\frac{34}{100}$	$\frac{56{,}234}{100}$
0.23	$\frac{2}{10} + \frac{3}{100}$	$0\frac{23}{100}$	$\frac{23}{100}$
47 or 47.0	$40 + 7$ or $40 + 7 + \frac{0}{10}$	47 or $47\frac{0}{10}$	$\frac{47}{1}$ or $\frac{470}{10}$

3 OPERATIONS ON DECIMALS

We have already reviewed the operations of adding, subtracting, multiplying, and dividing numbers in fractional notation. Now we review these operations when the numbers are in decimal notation.

To Add Two (or More) Decimals

1. Arrange the numbers in a column so that the decimal points line up vertically.
2. Add in columns from right to left as if there were no decimal points.
3. Place the decimal point in the sum in line with the other decimal points.

EXAMPLE 1 Adding Decimals

Add.

(a)
$$\begin{array}{r} 21.3 \\ +\ \ 4.2 \\ \hline 25.5 \end{array}$$

(b)
$$\begin{array}{r} 123.417 \\ 45. \\ +\quad 1.14 \\ \hline 169.557 \end{array}$$

PRACTICE EXERCISE 1

Add.

(a)
$$\begin{array}{r} 43.8 \\ +\ \ 6.1 \\ \hline \end{array}$$

(b)
$$\begin{array}{r} 10.001 \\ 0.52 \\ +\ 129.007 \\ \hline \end{array}$$

Answers: (a) **49.9** (b) **139.528**

To Subtract Two Decimals

1. Arrange the numbers in a column so that the decimal points line up vertically.
2. Subtract in columns from right to left as if there were no decimal points. (Additional zeros may be placed to the right of the decimal points if desired.)
3. Place the decimal point in the difference in line with the other decimal points.

EXAMPLE 2 SUBTRACTING DECIMALS

Subtract.

(a)
$$\begin{array}{r} 27.5 \\ -\ 2.4 \\ \hline 25.1 \end{array}$$

(b)
$$\begin{array}{r} 421.3000 \\ -\ 8.9999 \\ \hline 412.3001 \end{array}$$

(In (b), the three zeros to the right of 3 have been supplied.)

PRACTICE EXERCISE 2

Subtract.

(a)
$$\begin{array}{r} 51.2 \\ -\ 9.8 \\ \hline \end{array}$$

(b)
$$\begin{array}{r} 654.2 \\ -\ 26.888 \\ \hline \end{array}$$

Answers: (a) 41.4 (b) 627.312

We will agree to say that 42.7 has *one decimal place,* 12.16 has *two decimal places,* 5.006 has *three decimal places,* and so forth.

To Multiply Two Decimals

1. Ignore the decimal points and multiply as if the numbers were whole numbers.
2. Place the decimal point so the product has the same number of decimal places as the *sum* of the number of decimal places in the factors.

EXAMPLE 3 MULTIPLYING DECIMALS

Multiply.

(a)
$$\begin{array}{rl} 1.23 & \text{2 decimal places} \\ \times\ 2.1 & \text{1 decimal place} \\ \hline 123 & \\ 246\ \ & \\ \hline 2.583 & 2 + 1 = \text{3 decimal places} \end{array}$$

(b)
$$\begin{array}{rl} 42.103 & \text{3 decimal places} \\ \times\ 2.15 & \text{2 decimal places} \\ \hline 210515 & \\ 42103\ \ & \\ 84206\ \ \ \ & \\ \hline 90.52145 & 3 + 2 = \text{5 decimal places} \end{array}$$

(c) $(0.02)(0.013) = 0.00026$ 5 decimal places

PRACTICE EXERCISE 3

Multiply.

(a)
$$\begin{array}{r} 41.3 \\ \times\ 2.11 \\ \hline \end{array}$$

(b)
$$\begin{array}{r} 6.003 \\ \times\ 1.09 \\ \hline \end{array}$$

(c) $(0.004)(0.0255)$

Answers: (a) 87.143 (b) 6.54327 (c) 0.000102

To Divide One Decimal into Another

1. Move the decimal point in the divisor (the dividing number) to the right until the divisor becomes a whole number.
2. Move the decimal point in the dividend (the number divided into) to the right the same number of places adding zeros as necessary.
3. Divide as if the numbers were whole numbers.
4. Place the decimal point in the quotient directly above the decimal point in the dividend.

EXAMPLE 4 Dividing Decimals

Divide.

(a) $0.12\overline{)14.4}$

$0.12\overline{)14.40}$ (decimal points each moved 2 places) becomes

$$\begin{array}{r} 120 \\ 12\overline{)1440} \\ \underline{12} \\ 24 \\ \underline{24} \\ 0 \end{array}$$

(b) $0.003\overline{)42.3}$

$0.003\overline{)42.300}$ (decimal points each moved 3 places) becomes

$$\begin{array}{r} 14{,}100 \\ 3\overline{)42{,}300} \end{array}$$

Practice Exercise 4

Divide.

(a) $2.1\overline{)28.35}$

(b) $0.0002\overline{)1.5448}$

Answers: (a) 13.5 (b) 7724

We need to know how to operate on decimals since these techniques are used often in algebra. However, using a calculator can eliminate much of the tedious, time-consuming work with decimal operations. If you have a calculator, be sure to read the instruction manual that comes with your model. You might use a calculator to check part of your work in some of the following exercises.

0.3 EXERCISES A

Express in expanded notation.

1. 3.47 **2.** 107.28 **3.** 203.005

Express in decimal notation.

4. $400 + 20 + 7 + \frac{3}{10}$

5. $500 + 3 + \frac{2}{10} + \frac{8}{1000}$

6. $4000 + 30 + \frac{2}{100} + \frac{5}{1000}$

Add.

7. $\begin{array}{r} 31.4 \\ +\ 1.07 \\ \hline \end{array}$

8. $\begin{array}{r} 403.1 \\ +\ 21.08 \\ \hline \end{array}$

9. $\begin{array}{r} 201.005 \\ 30.2 \\ +\ 1.07 \\ \hline \end{array}$

10. 23.4 + 9.75

11. 512.3 + 2.009 + 13

12. 15.0001 + 9.3 + 0.004

Subtract.

13. $\begin{array}{r} 4.0003 \\ -\ 1.1 \\ \hline \end{array}$

14. $\begin{array}{r} 427.006 \\ -\ 135.04 \\ \hline \end{array}$

15. $\begin{array}{r} 36.5 \\ -\ 0.0007 \\ \hline \end{array}$

16. 46.01 − 13.4

17. 923.006 − 23

18. 47.2 − 0.0003

Multiply.

19. $\begin{array}{r} 3.42 \\ \times\ 2.1 \\ \hline \end{array}$

20. $\begin{array}{r} 15.207 \\ \times\ 3.12 \\ \hline \end{array}$

21. $\begin{array}{r} 0.0034 \\ \times\ 0.209 \\ \hline \end{array}$

22. (0.003)(0.003)

23. (0.02)(0.02)(0.02)

24. (0.0001)(0.00001)

Divide.

25. $2.1\overline{)70.14}$

26. $0.007\overline{)38.717}$

27. $20.4\overline{)0.687072}$

28. $8\overline{)7.0}$

29. $2\overline{)1.0}$

30. $9\overline{)1.0}$

Solve.

31. Before leaving on a trip, Harley's odometer read 23,411.8. If he drove 723.9 mi on the trip, what did it read when he returned?

32. The Northern Arizona four-mile relay team members recorded times of 4.23, 4.17, 3.99, and 4.08 minutes. What was their combined time for the race?

33. Roseanne Strauch buys a pair of hose for $3.95. If she pays with a $20 bill and the sales tax amounts to $0.20, how much change should she receive?

34. When Kimmie was sick, she ran a temperature of 103.2° Fahrenheit. If normal body temperature is 98.6° Fahrenheit, how many degrees above normal was her temperature?

35. Jose is paid $7.35 per hr. If he worked a total of 35.2 hr last week, how much was he paid?

36. How many months will it take to pay off a loan of $6562.80 if the monthly payments are $182.30?

FOR REVIEW

Perform the indicated operations.

37. $\frac{2}{9} + \frac{5}{12}$

38. $\frac{23}{36} - \frac{4}{9}$

39. $\frac{3}{4} \cdot \frac{16}{27}$

40. $1\frac{1}{12} \div 9\frac{3}{4}$

Solve.

41 Nancy Chandler plans a two-day hike from Gopher Gulch to Aspen Glen by way of North Face. It is $7\frac{3}{4}$ mi from Gopher Gulch to North Face and $9\frac{1}{8}$ mi from there to Aspen Glen. If she plans to travel half the total distance on each day, how far will she hike on the first day?

42. Julio has a rope $26\frac{2}{3}$ meters in length. He wishes to cut the rope into 3 pieces of equal length. What will be the length of each piece?

ANSWERS: 1. $3 + \frac{4}{10} + \frac{7}{100}$ 2. $100 + 7 + \frac{2}{10} + \frac{8}{100}$ 3. $200 + 3 + \frac{5}{1000}$ 4. 427.3 5. 503.208 6. 4030.025 7. 32.47 8. 424.18 9. 232.275 10. 33.15 11. 527.309 12. 24.3041 13. 2.9003 14. 291.966 15. 36.4993 16. 32.61 17. 900.006 18. 47.1997 19. 7.182 20. 47.44584 21. 0.0007106 22. 0.000009 23. 0.000008 24. 0.000000001 25. 33.4 26. 5531 27. 0.03368 28. 0.875 29. 0.5 30. 0.111 . . . 31. 24,135.7 32. 16.47 minutes 33. \$15.85 34. 4.6° 35. \$258.72 36. 36 months 37. $\frac{23}{36}$ 38. $\frac{7}{36}$ 39. $\frac{4}{9}$ 40. $\frac{1}{9}$ 41. $8\frac{7}{16}$ mi 42. $8\frac{8}{9}$ meters

0.3 EXERCISES B

Express in expanded notation.

1. 35.6

2. 257.663

3. 3215.802

Express in decimal notation.

4. $700 + 30 + 8 + \frac{6}{10}$

5. $4000 + 6 + \frac{7}{100}$

6. $300 + 40 + \frac{2}{100} + \frac{7}{1000}$

Add.

7. $\begin{array}{r} 52.3 \\ +\ 2.09 \\ \hline \end{array}$

8. $\begin{array}{r} 602.9 \\ +\ 43.07 \\ \hline \end{array}$

9. $\begin{array}{r} 421.544 \\ 2.8 \\ +\ 43.09 \\ \hline \end{array}$

10. $42.6 + 8.41$

11. $631.9 + 5.007 + 21$

12. $12.0003 + 7.2 + 0.003$

Subtract.

13. $\begin{array}{r} 6.0004 \\ -\ 2.3 \\ \hline \end{array}$

14. $\begin{array}{r} 637.008 \\ -\ 123.05 \\ \hline \end{array}$

15. $\begin{array}{r} 51.6 \\ -\ 0.0004 \\ \hline \end{array}$

16. $43.02 - 17.1$

17. $843.002 - 51$

18. $39.4 - 0.0006$

Multiply.

19. $\begin{array}{r} 2.91 \\ \times\ 3.4 \\ \hline \end{array}$

20. $\begin{array}{r} 17.603 \\ \times\ \ 4.15 \\ \hline \end{array}$

21. $\begin{array}{r} 0.0053 \\ \times\ 0.402 \\ \hline \end{array}$

22. (0.002)(0.002)

23. (0.03)(0.03)(0.03)

24. (0.0002)(0.00002)

Divide.

25. $4.2\overline{)94.5}$

26. $0.003\overline{)13.014}$

27. $30.1\overline{)6.52267}$

28. $8\overline{)5.0}$

29. $4\overline{)1.0}$

30. $9\overline{)4.0}$

Solve.

31. The Saturday and Sunday receipts at Waldo's Waffle Palace were $621.13 and $743.19, respectively. How much money did Waldo take in on that weekend?

32. What is the **perimeter** of (the distance around) a four-sided figure with sides of 6.031 m, 5.411 m, 4.003 m, and 5.772 m?

33. Sam Passamonte had a balance of $635.86 in his checking account when he wrote a check for $249.93. What was his new balance?

34. When George filled the gas tank on his Bronco, the odometer read 62,421.8. The next time he filled it, the odometer read 62,914.2. How far had he driven?

35. Betty Sue bought 13.4 pounds of peaches at $0.55 per pound. How much did she pay for the peaches?

36. Kate Williams paid $58.00 to rent a car for two days. If the rental fee amounts to $12.50 per day plus $0.15 per mile driven, how far did she drive the car?

FOR REVIEW

Perform the indicated operations.

37. $\frac{12}{15} + \frac{8}{35}$

38. $\frac{13}{24} - \frac{3}{8}$

39. $1\frac{2}{3} \cdot 4\frac{1}{5}$

40. $\frac{6}{5} \div \frac{24}{45}$

Solve.

41. A wheel on a child's car is turning at a rate of $40\frac{1}{2}$ revolutions per minute. If the car is driven for $3\frac{1}{9}$ minutes, how many revolutions does the wheel make?

42. Marguerite is proofing a computer program that is $14\frac{1}{3}$ pages in length. If she has completed $10\frac{1}{5}$ pages, how much remains to be done?

0.3 EXERCISES C

Perform the indicated operations. A calculator would be useful for these problems.

1. $\begin{array}{r} 374.26 \\ \times\ 8.603 \\ \hline \end{array}$

2. $0.0257\overline{)0.02584392}$ [Answer: 1.0056]

0.4 CONVERTING FRACTIONS, DECIMALS, AND PERCENTS

STUDENT GUIDEPOSTS

1. Converting Fractions to Decimals
2. Rational and Irrational Numbers
3. Converting Terminating Decimals to Fractions
4. Conversions Involving Percent

1 CONVERTING FRACTIONS TO DECIMALS

Every fraction can be converted to a decimal. Several problems in the previous section gave a preview of this process.

To Convert a Fraction to a Decimal

1. Divide the denominator of the fraction into the numerator.
2. Place the decimal point as in any decimal division problem.

EXAMPLE 1 Converting a Fraction to a Decimal

Convert each fraction to a decimal.

(a) $\frac{3}{20}$ We divide 20 into 3 as follows:

$$\begin{array}{r} .15 \\ 20\overline{)3.00} \\ \underline{2\ 0} \\ 1\ 00 \\ \underline{1\ 00} \\ 0 \end{array} \qquad \frac{3}{20} = 0.15$$

(b) $\frac{1}{3}$

$$\begin{array}{r} .333\ldots \\ 3\overline{)1.000} \\ \underline{9} \\ 10 \\ \underline{9} \\ 10 \\ \underline{9} \\ 1 \end{array}$$

It is clear that we will continue to obtain "3" on each division. When this happens we write $\frac{1}{3} = 0.333\ldots = 0.\overline{3}$, where the bar denotes the digit (sometimes digits) that continues to repeat. For the mixed decimal number $3\frac{1}{3}$ we write $3.\overline{3}$ (not $\overline{3}$).

(c) $\frac{2}{7}$

$$\begin{array}{r} .285714285714\ldots \\ 7\overline{)2.000000000000} \end{array}$$

$$\frac{2}{7} = 0.285714285714\ldots$$
$$= 0.\overline{285714}$$

Practice Exercise 1

Convert each fraction to a decimal.

(a) $\frac{1}{2}$

(b) $\frac{13}{40}$

(c) $\frac{2}{3}$

Answers: (a) 0.5 (b) 0.325 (c) $0.\overline{6}$

It is a property of every fraction that in decimal notation, the digits after the decimal either terminate, as in

$$\frac{1}{2} = 0.5, \qquad \frac{3}{20} = 0.15, \qquad \frac{13}{40} = 0.325,$$

or contain a block of digits that repeats without ending, as in

$$\frac{1}{3} = 0.\overline{3}, \qquad \frac{2}{7} = 0.\overline{285714}, \qquad \frac{2}{3} = 0.\overline{6}.$$

2 RATIONAL AND IRRATIONAL NUMBERS

It is also true that if a number has a decimal representation that terminates or repeats, the number can always be expressed as a fraction. This fact is often used to define the set of **rational numbers** or fractions. The rational numbers include all of the numbers we have considered thus far. Numbers that are not rational numbers, called **irrational numbers,** have decimal representations that neither terminate nor repeat. One of the most famous irrational numbers is π (the ratio of the circumference of a circle to its diameter).

3 CONVERTING TERMINATING DECIMALS TO FRACTIONS

We now turn our attention to converting a terminating decimal to a fraction. For example,

$$0.5 \quad \text{is} \quad \frac{5}{10},$$

$$0.23 \quad \text{is} \quad \frac{2}{10} + \frac{3}{100} = \frac{20}{100} + \frac{3}{100} = \frac{23}{100},$$

$$0.537 \quad \text{is} \quad \frac{5}{10} + \frac{3}{100} + \frac{7}{1000} = \frac{500}{1000} + \frac{30}{1000} + \frac{7}{1000} = \frac{537}{1000}.$$

To Convert a Terminating Decimal to a Fraction

1. Express the decimal in expanded notation.
2. Add the whole number terms and add the fractional terms to obtain a mixed number.
3. Convert the mixed number to a single fraction by addition.

EXAMPLE 2 Converting a Decimal to a Fraction

Convert each decimal to a fraction.

(a) $0.47 = \frac{4}{10} + \frac{7}{100} = \frac{40}{100} + \frac{7}{100} = \frac{47}{100}$

(b) $2.35 = 2 + \frac{3}{10} + \frac{5}{100} = 2 + \frac{30}{100} + \frac{5}{100}$

$= 2 + \frac{35}{100} = \frac{2 \cdot 100}{100} + \frac{35}{100} = \frac{235}{100}$

(c) $43.619 = 40 + 3 + \frac{6}{10} + \frac{1}{100} + \frac{9}{1000} = 43 + \frac{619}{1000} = \frac{43{,}619}{1000}$

Practice Exercise 2

Convert each decimal to a fraction.

(a) 0.11

(b) 3.86

(c) 51.015

Answers: (a) $\frac{11}{100}$ (b) $\frac{193}{50}$ (c) $\frac{10{,}203}{200}$

4 CONVERSIONS INVOLVING PERCENT

The concept *percent* is closely related to decimals and is used in a variety of applied problems. The word **percent** literally means *per hundred;* it refers to the number of parts in one hundred parts. A five percent tax means five parts per one hundred parts, or a tax of 5¢ on every 100¢. Five percent would be written

$$5\%.$$

Since the % symbol means "divide by 100" or "multiply by $\frac{1}{100} = 0.01$,"

$$5\% = 5(0.01) = 0.05 \quad \text{or} \quad 5\% = \frac{5}{100} = \frac{1}{20}.$$

To Convert a Percent to a Decimal

Remove the % symbol and multiply by 0.01. Or remove the % symbol and move the decimal point two places to the *left,* adding zeros as necessary.

To Convert a Percent to a Fraction

Remove the % symbol and divide by 100. Reduce the resulting fraction to lowest terms.

EXAMPLE 3 Converting Percent to a Decimal

Convert each percent to a decimal.

(a) $43\% = (43)(0.01) = 0.43$ 43% = 0.43

(b) $325.5\% = (325.5)(0.01) = 3.255$ 325.5% = 3.255

(c) $0.05\% = (0.05)(0.01) = 0.0005$ 0.05% = 0.0005

(d) $0.2\% = (0.2)(0.01) = 0.002$ 0.2% = 0.002

Practice Exercise 3

Convert each percent to a decimal.

(a) 79%

(b) 640.5%

(c) 0.09%

(d) 0.8%

Answers: (a) 0.79 (b) 6.405 (c) 0.0009 (d) 0.008

EXAMPLE 4 Converting Percent to a Fraction

Convert each percent to a fraction.

(a) $28\% = \frac{28}{100} = \frac{4 \cdot 7}{4 \cdot 25} = \frac{7}{25}$

(b) $37.5\% = \frac{37.5}{100} = \frac{(37.5)(10)}{(100)(10)}$ Clear decimal by multiplying by 10

$$= \frac{375}{1000}$$

$$= \frac{3 \cdot 125}{8 \cdot 125}$$

$$= \frac{3}{8}$$ Reduce fraction

Practice Exercise 4

Convert each percent to a fraction.

(a) 65%

(b) 42.5%

(c) $66\frac{2}{3}\% = \left(66\frac{2}{3}\right)\left(\frac{1}{100}\right)$ Dividing by 100 is the same as multiplying by $\frac{1}{100}$

$= \left(\frac{200}{3}\right)\left(\frac{1}{100}\right)$ $66\frac{2}{3} = \frac{(3 \cdot 66) + 2}{3} = \frac{200}{3}$

$= \frac{2 \cdot \cancel{100}}{3 \cdot \cancel{100}} = \frac{2}{3}$ Reduce fraction

(d) $100\% = \frac{100}{100} = 1$

(c) $16\frac{1}{6}\%$

(d) 500%

Answers: (a) $\frac{13}{20}$ (b) $\frac{17}{40}$ (c) $\frac{97}{600}$ (d) 5

To Convert a Decimal to a Percent

Multiply by 100 and attach the % symbol. Or move the decimal point two places to the *right*, adding zeros as necessary, and attach the % symbol.

To Convert a Fraction to a Percent

Convert the fraction to a decimal by dividing the numerator by the denominator. Then change the resulting decimal to a percent.

EXAMPLE 5 Converting a Decimal to Percent

Convert each decimal to a percent.

(a) $0.37 = (0.37)(100\%) = 37\%$ $0.37 = 37.\% = 37\%$

(b) $5.81 = (5.81)(100)\% = 581\%$ $5.81 = 581.\% = 581\%$

(c) $0.0004 = (0.0004)(100)\% = 0.04\%$ $0.0004 = 000.04\% = 0.04\%$

(d) $0.\overline{3} = (0.\overline{3})(100)\% = (0.333\ldots)(100)\%$

$= 33.33\ldots\%$

$= 33.\overline{3}\%$ or $33\frac{1}{3}\%$ $0.\overline{3} = \frac{1}{3}$

Practice Exercise 5

Convert each decimal to a percent.

(a) 0.81

(b) 9.05

(c) 0.003

(d) $0.\overline{7}$

Answers: (a) 81% (b) 905% (c) 0.3% (d) $77.\overline{7}\%$

EXAMPLE 6 Converting a Fraction to Percent

Convert each fraction to a percent.

(a) $\frac{3}{4} = 0.75$ Divide 4 into 3

$= (0.75)(100)\% = 75\%$ $0.75 = 075.\% = 75\%$

(b) $\frac{1}{8} = 0.125$ Divide 8 into 1

$= (0.125)(100)\% = 12.5\%$ $0.125 = 012.5\% = 12.5\%$

Practice Exercise 6

Convert each fraction to a percent.

(a) $\frac{4}{5}$

(b) $\frac{7}{10}$

(c) $\frac{5}{3} = 1.\overline{6}$ — Divide 3 into 5

$= (1.\overline{6})(100)\%$

$= (1.666\ldots)(100)\%$

$= 166.666\ldots\%$

$= 166.\overline{6}\%$ or $166\frac{2}{3}\%$ — $0.\overline{6} = \frac{2}{3}$

(c) $\frac{1}{9}$

Answers: (a) 80% (b) 70% (c) $11.\overline{1}\%$

Keeping a simple example in mind will help you remember whether you should multiply by 0.01 or by 100 when making percent conversions. 50% of some quantity is 0.5 or $\frac{1}{2}$ of that quantity. Thus, to convert

50% to 0.5, multiply by 0.01

and to convert 0.5 to 50%, multiply by 100.

Many applied problems involve finding a percent of some number. For example, we might ask

What is 5% of $40.00?

To solve such problems we change the percent to a decimal or fraction and then multiply it by the given number. The word *of* means "multiply" or "times" in this setting.

$$\begin{array}{ccc} 5\% & \text{of} & \$40.00 \\ \downarrow & \downarrow & \downarrow \\ (0.05) & \cdot & (40.00) \end{array}$$

Thus, 5% of $40.00 is (0.05)(40.00) = $2.00.

0.4 EXERCISES A

Convert each fraction to a decimal.

1. $\frac{1}{8}$ **2.** $\frac{7}{20}$ **3.** $\frac{5}{9}$

4. $\frac{8}{3}$ **5.** $\frac{5}{6}$ **6.** $2\frac{1}{7}$

Convert each decimal to a fraction.

7. 0.3 **8.** 0.007 **9.** 0.48

10. 27.95 **11.** 3.207 **12.** 25.001

Convert each percent to a decimal.

13. 65.5%

14. $\frac{1}{2}\%$

15. 0.6%

16. 200%

17. $8\frac{1}{4}\%$

18. $\frac{1}{3}\%$

Convert each percent to a fraction.

19. 57%

20. 100%

21. $\frac{1}{4}\%$

22. $3\frac{2}{3}\%$

23. 1000%

24. $\frac{1}{10}\%$

Convert each decimal to a percent.

25. 2.37

26. 0.06

27. 0.005

28. $0.62\overline{3}$

29. $0.1\overline{6}$

30. 1.1

Convert each fraction to a percent.

31. $\frac{1}{2}$

32. $\frac{6}{5}$

33. $\frac{10}{3}$

34. $\frac{7}{50}$

35. $\frac{5}{6}$

36. $\frac{50}{3}$

Solve.

37 A basketball player made 15 shots in 20 attempts during a recent game. What fractional part of his shots did he make? What percent of his shots did he make?

38. Bill Schulz earned $24,000 last year and gave $3000 to charity. What fractional part of his income was given to charity? What percent of his income was given to charity?

39 There were 5000 votes cast in an election. Mr. Gomez received 55% of the votes. How many votes did he receive?

40. Snow, Montana has a normal annual snowfall of 120 in. This year the city received 150% of normal. How many inches of snow fell this year?

FOR REVIEW

41. Express 32.51 in expanded notation.

42. Add: $\begin{array}{r} 403.2 \\ 1.006 \\ +\ \ 0.04 \\ \hline \end{array}$

43. Subtract: $3.21 - 0.0005$

44. Multiply: $\begin{array}{r} 0.0014 \\ \times\ \ 1.06 \\ \hline \end{array}$

45. Divide: $0.004\overline{)0.086}$

46. John Hagood rented an Escort for 3 days. If the rental fee was \$8.95 per day plus \$0.14 per mi, how much was he charged if he drove a total of 185.5 mi?

ANSWERS: 1. 0.125 2. 0.35 3. $0.\overline{5}$ 4. $2.\overline{6}$ 5. $0.8\overline{3}$ 6. $2.\overline{142857}$ 7. $\frac{3}{10}$ 8. $\frac{7}{1000}$ 9. $\frac{12}{25}$ 10. $\frac{559}{20}$ 11. $\frac{3207}{1000}$ 12. $\frac{25,001}{1000}$ 13. 0.655 14. 0.005 15. 0.006 16. 2 17. 0.0825 18. $0.00\overline{3}$ 19. $\frac{57}{100}$ 20. 1 21. $\frac{1}{400}$ 22. $\frac{11}{300}$ 23. 10 24. $\frac{1}{1000}$ 25. 237% 26. 6% 27. 0.5% 28. $62.\overline{3}\%$ or $62\frac{1}{3}\%$ 29. $16.\overline{6}\%$ or $16\frac{2}{3}\%$ 30. 110% 31. 50% 32. 120% 33. $333\frac{1}{3}\%$ 34. 14% 35. $83\frac{1}{3}\%$ 36. $1666\frac{2}{3}\%$ 37. $\frac{3}{4}$, 75% 38. $\frac{1}{8}$, 12.5% 39. 2750 votes 40. 180 in 41. $30 + 2 + \frac{5}{10} + \frac{1}{100}$ 42. 404.246 43. 3.2095 44. 0.001484 45. 21.5 46. \$52.82

0.4 EXERCISES B

Convert each fraction to a decimal.

1. $\frac{5}{8}$ **2.** $\frac{5}{16}$ **3.** $\frac{11}{15}$

4. $\frac{4}{5}$ **5.** $\frac{1}{6}$ **6.** $2\frac{3}{7}$

Convert each decimal to a fraction.

7. 0.09 **8.** 0.002 **9.** 2.6

10. 31.85 **11.** 20.02 **12.** 35.001

Convert each percent to a decimal.

13. 35.5% **14.** $3\frac{1}{2}\%$ **15.** 26.54%

16. 300% **17.** $7\frac{3}{4}\%$ **18.** $\frac{2}{3}\%$

Convert each percent to a fraction.

19. 90% **20.** 500% **21.** 275%

22. $4\frac{1}{3}\%$ **23.** 2000% **24.** $\frac{1}{100}\%$

Convert each decimal to a percent.

25. 4.29

26. 0.01

27. 0.007

28. $0.57\overline{3}$

29. 7

30. 2.2

Convert each fraction to a percent.

31. $\frac{3}{5}$

32. $\frac{11}{3}$

33. $\frac{9}{50}$

34. $\frac{3}{10}$

35. $\frac{5}{8}$

36. $\frac{25}{3}$

Solve.

37. A football quarterback completed 21 passes in 35 attempts during a game last fall. What fraction of his attempts did he complete? What percent of his attempts did he complete?

38. Amy received $12,000 from her grandfather. She spent $8000 on a new car. What fraction of the money went for the car? What percent of the money went for the car?

39. There are 0.2% impurities in a laboratory solution weighing 485 grams. What is the weight of the impurities?

40. Diane received a raise giving her a salary equal to 120% of her previous salary. If she earned $16,000 last year, how much will she earn this year?

FOR REVIEW

41. Express 29.65 in expanded notation.

42. Add: $\begin{array}{r} 209.6 \\ 4.007 \\ +\ 0.01 \\ \hline \end{array}$

43. Subtract: $4.81 - 0.0002$

44. Multiply: $\begin{array}{r} 0.0015 \\ \times\ 1.08 \\ \hline \end{array}$

45. Divide: $0.003\overline{)0.1488}$

46. Ben Whitney rented a midsize car for two days. If the rental fee was $11.95 per day plus $0.18 per mi, how much was he charged if he drove a total of 230.5 mi?

0.4 EXERCISES C

Solve. A calculator would help in solving these problems.

1. A retailer paid $92.00 for a dress and marked it up 30% to sell in her store. When the dress did not sell, she sold it at a 25% discount on the marked price. What was the percent profit or loss based on the original cost? [Answer: 2.5% loss]

2. A salesman receives a 3% commission on all sales of $50,000 or less and a 4% commission on the amount he sells over $50,000. What is his commission on sales of $85,000? What is the effective total commission rate on the $85,000 sales?

0.5 STATISTICS AND GRAPHS

STUDENT GUIDEPOSTS

1. Statistics and Numerical Data
2. Circle Graphs
3. Bar Graphs
4. Broken-Line Graphs
5. Histograms
6. Frequency Polygons

1 STATISTICS AND NUMERICAL DATA

Statistics is the branch of mathematics that deals with collecting, organizing, and summarizing numerical data in order to describe or interpret it. The world of sports relies on statistics such as batting averages, pass completions, free-throw percentages, and points scored. The number of people under age 30 in the United States, the average study time of junior college students, and the Michigan election results are all statistics.

Tables provide one way to present data in a manner that is easy to interpret. Suppose Tanya, a college student, has decided to keep a record of her monthly expenses. She might summarize the data in a table as follows.

Monthly expenses	
Item	*Amount spent*
Rent	\$200
Food	\$150
Entertainment	\$100
Clothes	\$ 35
Other	\$ 15
TOTAL	\$500

2 CIRCLE GRAPHS

Sometimes it is hard to interpret a large set of data. One way to simplify the process is to "picture" the data in some way. A **circle graph** is often used to represent parts of a whole unit. For example, the data collected by Tanya and presented in the table above could be organized and displayed in a circle graph. The circle graph in Figure 0.1 represents her entire budget, and each pie-shaped part, or sector, represents the number of dollars she spent on each item.

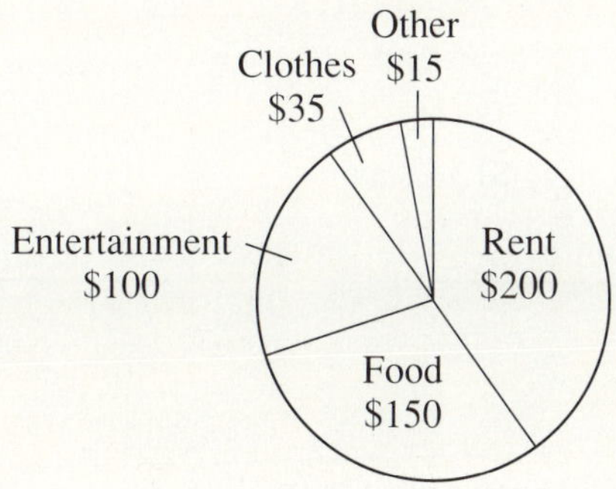

Figure 0.1 Circle Graph

By reading information from a circle graph, we can make various comparisons. Consider Figure 0.1 again. Since the total amount Tanya spent during the month was \$500, the ratio of rent to total expenses is

$$\frac{\$200}{\$500} = \frac{2}{5}.$$

Converting this fraction to a percent, $\frac{2}{5} = 0.40 = 40\%$. Tanya sees that she spent 40% of her budget on rent.

Tanya can also find the ratio of one expense to another. For example, the ratio of food expenses to entertainment expenses is

$$\frac{\$150}{\$100} = \frac{3}{2}.$$

Thus, for every \$3 spent on food, Tanya spends \$2 on entertainment.

EXAMPLE 1 Reading a Circle Graph

The circle graph in Figure 0.2 shows the distribution of Mr. Green's monthly income.

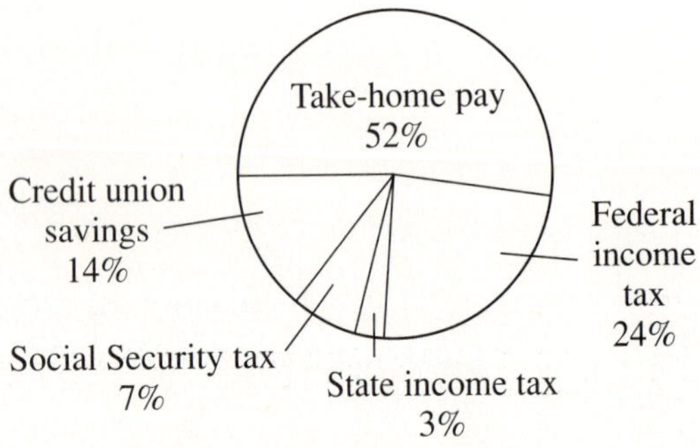

Figure 0.2 Circle Graph

(a) If Mr. Green's income is \$2800 each month, determine his take-home pay.

Since the percent of take-home pay is 52%, find

$$\begin{array}{ccc} 52\% & \text{of} & \$2800. \\ \downarrow & \downarrow & \downarrow \\ 0.52 & \cdot & 2800 = 1456 \end{array}$$

Thus, Mr. Green takes home \$1456 each month.

(b) What is the ratio of federal income tax to state income tax?

The ratio is

$$\frac{24\%}{3\%} = \frac{8}{1}.$$

Thus, for every \$8 paid in federal income tax Mr. Green pays \$1 in state income tax.

Practice Exercise 1

Use the circle graph in Figure 0.2 to answer the following.

(a) If Mr. Green's income is \$2800 each month, what amount does he pay in federal income tax?

(b) What is the ratio of credit union savings to Social Security tax?

Answers: (a) \$672 (b) $\frac{2}{1}$

3 BAR GRAPHS

Another way to present data is with a **bar graph.** Suppose 100 people have been polled regarding their preference among four brands of toothpaste, Brand A, Brand B, Brand C, and Brand D. The results of the poll are given in the bar graph in Figure 0.3.

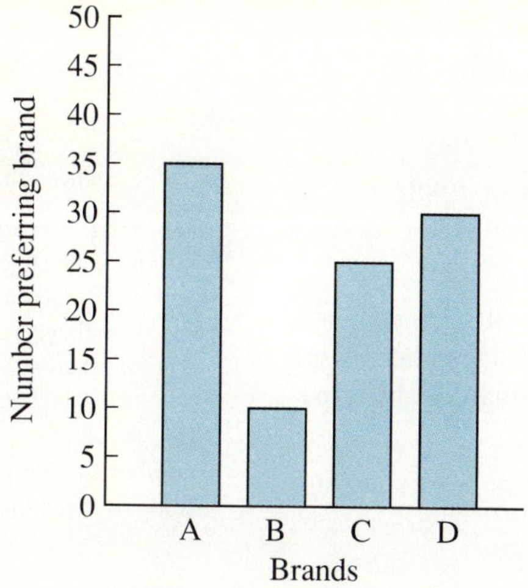

Figure 0.3 Bar Graph

The graph shows that 35 people chose Brand A, 10 chose Brand B, 25 chose Brand C, and 30 chose Brand D.

Bar graphs are quite useful for comparing data. It is easy to see at a glance which brand was most preferred (Brand A) and which was least preferred (Brand B). Also, Brand A was preferred to Brand B by a ratio of

$$\frac{35}{10} = \frac{7}{2}.$$

That is, for every 7 people who chose Brand A, 2 people chose Brand B.

EXAMPLE 2 Reading a Bar Graph

Use the bar graph in Figure 0.3 to determine the ratio of people preferring Brand C to those preferring Brand B.

Since 25 people preferred Brand C and 10 people preferred Brand B, the ratio is $\frac{25}{10} = \frac{5}{2}$.

Practice Exercise 2

Use the bar graph in Figure 0.3 to determine the ratio of those preferring Brand A over Brand C.

Answer: $\frac{7}{5}$

A double-bar graph is often used to compare two sets of data. The double-bar graph in Figure 0.4 shows the total enrollment in beginning algebra classes compared to intermediate algebra classes at a community college during a four-year period.

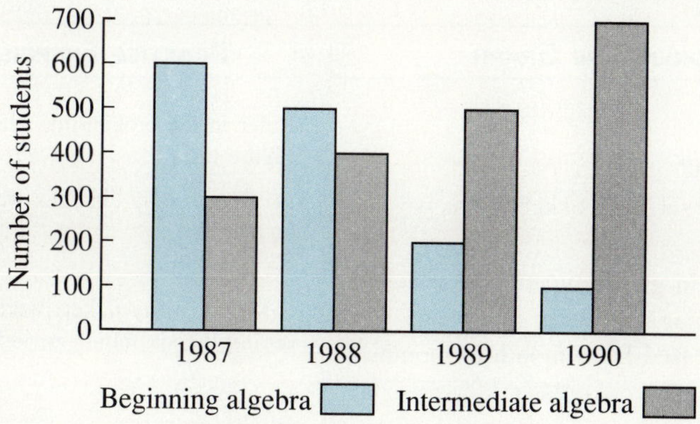

Figure 0.4 Double-Bar Graph

EXAMPLE 3 Reading a Double-Bar Graph

Use the double-bar graph in Figure 0.4 to answer the following.

(a) What was the number of beginning algebra students in 1988?

Since the bar on the left in 1988 rises to the 500 level, there were 500 students in beginning algebra in 1988.

(b) Which year had the greatest number of intermediate algebra students?

The right bar rises highest in 1990. Thus, the greatest number of students in intermediate algebra occurred in 1990. In fact, that number was 700 students.

(c) What could you say about the number of beginning algebra students over the four-year period?

Since the number went from 600 to 500 to 200 to 100, we can see that the number of beginning algebra students declined over the four-year period.

Practice Exercise 3

Use the double-bar graph in Figure 0.4 to answer the following.

(a) What was the number of intermediate algebra students in 1989?

(b) Which year had more beginning algebra students, 1988 or 1989?

(c) What could you say about the number of intermediate algebra students over the four-year period?

Answers: (a) 500 (b) 1988 (c) The number increased.

4 BROKEN-LINE GRAPHS

A **broken-line graph** is often used to show a trend in data collected over a period of time. For example, the broken-line graph in Figure 0.5 shows the number of traffic citations issued by a police department over a six-month period. Each dot shows the number of tickets given during that month.

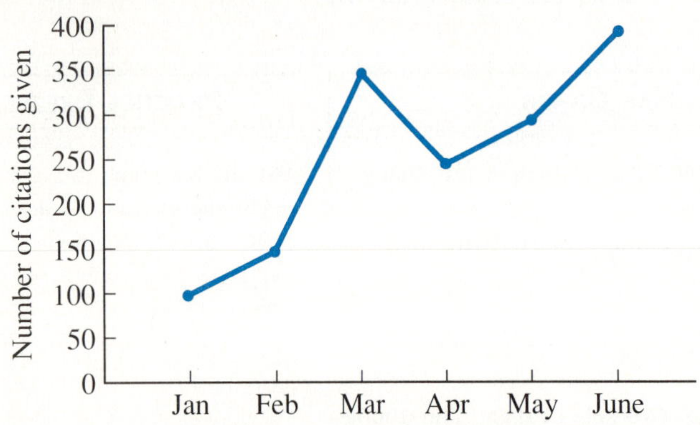

Figure 0.5 Broken-Line Graph

EXAMPLE 4 Reading a Broken-Line Graph

Refer to the broken-line graph in Figure 0.5.

(a) In which month were 350 tickets given out?

Since the dot above March is at the 350 level, 350 tickets were issued in March.

(b) How would you briefly describe the trend in giving out tickets over the six-month period?

Basically the number of tickets given increased from month-to-month with the exception of March to April.

Practice Exercise 4

Refer to the broken-line graph in Figure 0.5.

(a) How many citations were issued in April?

(b) How many tickets were given out during the six-month period?

Answers: (a) 250 (b) 1550

Two broken-line graphs in the same figure are often used to show a comparison between two sets of data. For example, the broken-line graphs in Figure 0.6 show the quarterly profits realized by a company during the years 1990 and 1991. By showing both graphs together, it is easy to see at a glance how one year compared with the other.

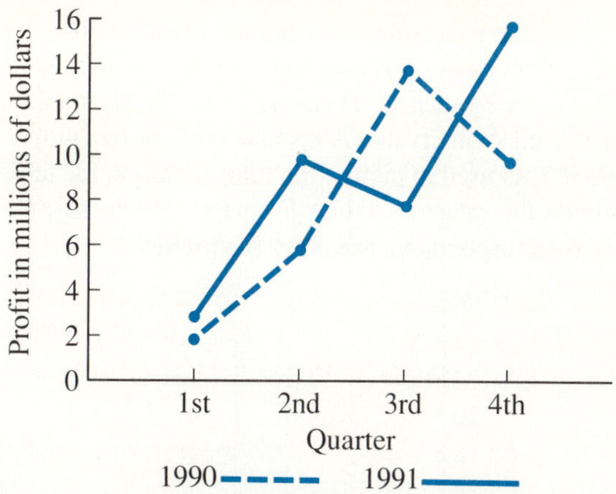

Figure 0.6 Double-Line Graph

EXAMPLE 5 Reading a Double-Line Graph

Refer to the graph in Figure 0.6.

(a) What was the profit made during the first quarter of 1991?

Since the dot on the solid line is halfway between the 2 and 4 million marks, the profit this quarter was 3 million dollars.

(b) Between which two consecutive quarters in 1990 was there a decrease in profit?

The dashed line graph goes down from 14 million dollars to 10 million dollars between the 3rd and 4th quarters of 1990.

(c) In which year did the company show the greater profit?

Adding up the quarterly profits in 1990 we obtain

$$2 + 6 + 14 + 10 = 32 \text{ million dollars.}$$

Adding up the quarterly profits in 1991 we obtain

$$3 + 10 + 8 + 16 = 37 \text{ million dollars.}$$

Thus, the greater profit was made in 1991.

Practice Exercise 5

Refer to the graph in Figure 0.6.

(a) What was the profit made during the 3rd quarter of 1990?

(b) Between which two consecutive quarters in 1991 did the profit increase the most?

(c) Which year had the better profit change between the 3rd and 4th quarters?

Answers: (a) 14 million dollars (b) between the 3rd and 4th quarters (c) 1991

5 HISTOGRAMS

Suppose sixty men in an organization list their heights in inches. The resulting data might be somewhat difficult to understand due to the large number of heights. To make the data easier to interpret, the heights could be divided into groups called **class intervals.** We might use 66 inches–68 inches as one interval, 68 inches–70 inches as another, and so forth. The number of heights in each class interval is called the class frequency. The table below summarizes the data collected.

Height in inches (class intervals)	*Number of men (class frequency)*
66–68	4
68–70	12
70–72	24
72–74	18
74–76	2

Suppose we assume that a person whose height is exactly 68 inches is in the class interval 68–70 and not in 66–68, and similarly for persons whose height is 70 inches, 72 inches, and 74 inches. Then each of the sixty men has his height in exactly one of the class intervals. A special type of bar graph, a **histogram,** shown in Figure 0.7, is used to picture the information in the table. The width of each bar represents the range of values in that class interval, and the length or height of the bar corresponds to the **class frequency.**

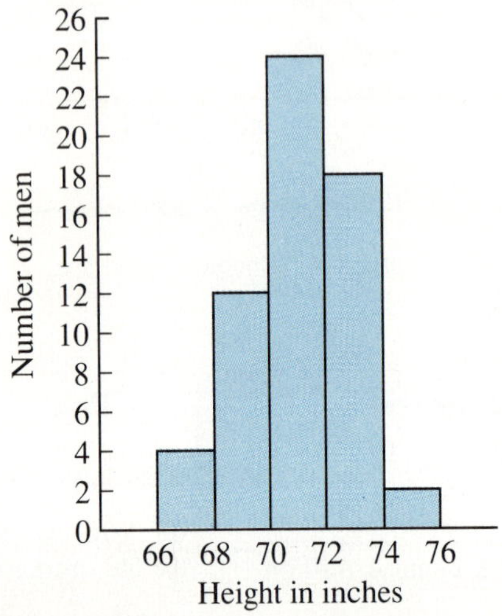

Figure 0.7 Histogram

EXAMPLE 6 Reading a Histogram

Use the histogram in Figure 0.7 to answer the following.

(a) How many men are 70 inches or taller?

We must add the class frequencies for the class intervals 70–72, 72–74, and 74–76.

$$24 + 18 + 2 = 44$$

There are 44 men 70 inches or taller.

(b) Find the ratio of men 68 to 70 inches tall to those 72 to 74 inches tall.

Since there are 12 men with heights between 68 and 70 inches and 18 men with heights between 72 and 74 inches, the ratio is

$$\frac{12}{18} = \frac{2}{3} \text{ or 2 to 3.}$$

Practice Exercise 6

Use the histogram in Figure 0.7 to answer the following.

(a) How many men are shorter than 72 inches?

(b) Find the ratio of men with heights between 66 and 68 inches to those between 74 and 76 inches.

Answers: (a) 40 (b) $\frac{2}{1}$ or 2 to 1

6 FREQUENCY POLYGONS

Sometimes data that is divided into class intervals is displayed using a special kind of broken-line graph. For instance, suppose we find the midpoint of each class interval, called the **class midpoint,** in the example above. The class midpoint of the class interval 66–68 inches is 67 inches, and the remaining class midpoints are 69 inches, 71 inches, 73 inches, and 75 inches. Locate the class midpoints in the histogram in Figure 0.7, and place a dot at the corresponding point on the top of each bar. When these dots are connected, as shown in Figure 0.8, the resulting broken-line graph is called a **frequency polygon.**

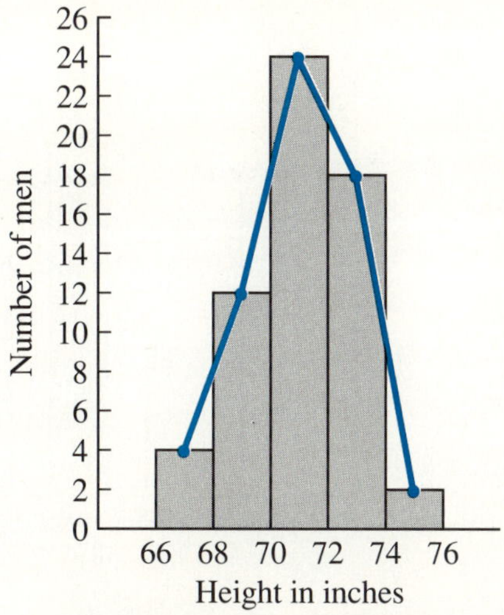

Figure 0.8 Frequency Polygon

HINT

Histograms and frequency polygons are used when data collected consists of a wide range of values such as numbers obtained by measurements.

0.5 EXERCISES A

Exercises 1–8 refer to the circle graph at the right which shows a breakdown of the amounts spent by the Warrens on an addition to their house.

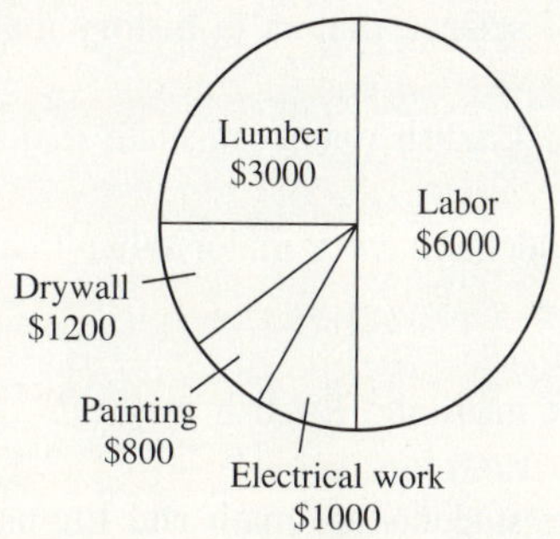

1. What was the total cost of the addition?

2. What was the largest expense item on the project?

3. What was the ratio of the cost of labor to the cost of lumber?

4. What was the ratio of the cost of painting to the cost of drywall?

5. What was the ratio of the cost of lumber to the cost of electrical work?

6. Lumber accounted for what percent of the total cost?

7. Labor accounted for what percent of the total cost?

8. Electrical work accounted for what percent of the total cost?

Exercises 9–20 refer to the bar graph at the right which shows the number of students by major course of study in the College of Arts and Science.

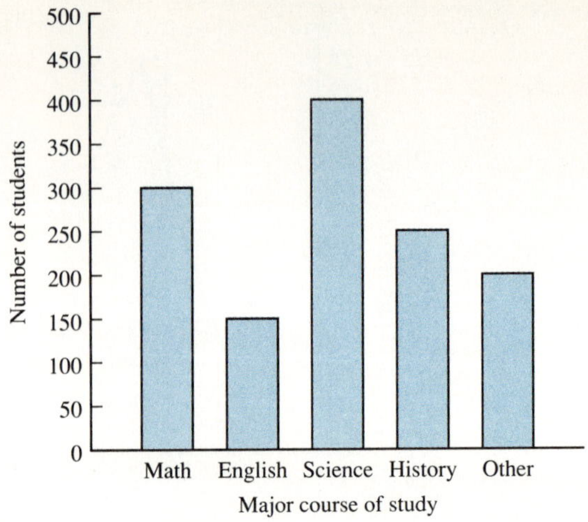

9. How many students are in the College of Arts and Science?

10. What is the ratio of math majors to the total number of students in the college?

11. What percent of students in the college are math majors (to the nearest tenth of a percent)?

12. What is the ratio of science majors to the total number of students in the college?

13. What percent of students in the college are science majors (to the nearest tenth of a percent)?

14. What percent of students are math and science majors (to the nearest tenth of a percent)?

15. What is the ratio of science majors to history majors?

16. What is the ratio of English majors to math majors?

17. What percent of students have a major other than math, science, English, or history (to the nearest tenth of a percent)?

18. What is the total of math and English majors?

19. What percent of the students are math and English majors (to the nearest tenth of a percent)?

20. What is the ratio of math and science majors to English and history majors?

Exercises 21–28 refer to the double-bar graph at the right which shows the number of male and female athletes at a university over a four-year period.

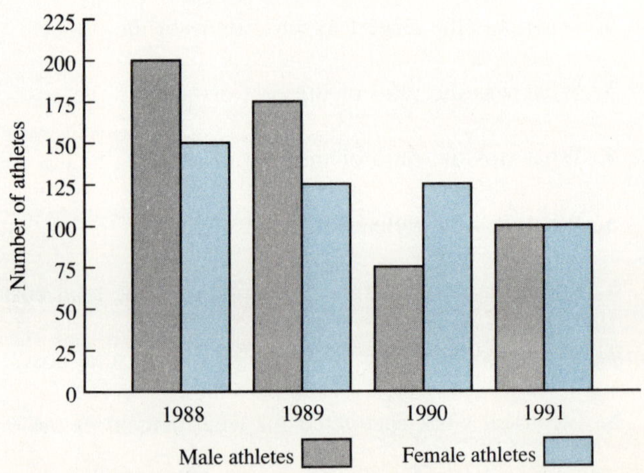

21. Which year had the greatest number of male athletes?

22. Which year had the smallest number of male athletes?

23. Which year had the greatest number of female athletes?

24. Which year had the smallest number of female athletes?

25. What was the ratio of male athletes to female athletes in 1988?

26. What was the ratio of female athletes to male athletes in 1990?

27. Which year had the same number of male and female athletes?

28. Would it be fair to conclude from the graph that some athletics programs were forced to be dropped? Explain.

The number of personal computers sold by a dealer in a six-week period is shown in the broken-line graph at the right. Use this graph in Exercises 29–38.

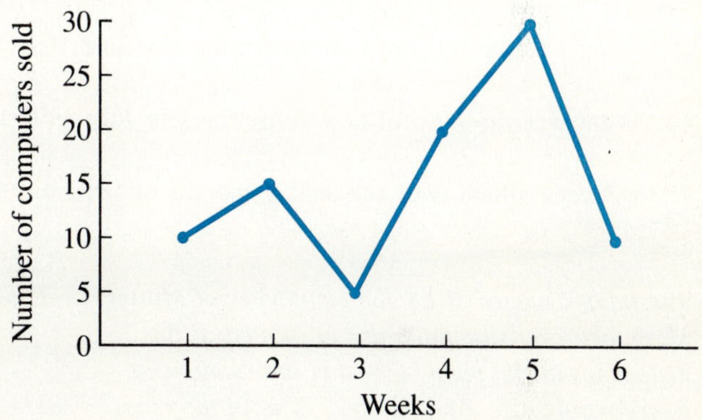

29. How many computers were sold in week 2?

30. During which week(s) were 10 computers sold?

31. How many computers were sold during the six-week period?

32. What was the ratio of computers sold in week 5 to week 3?

33. During which week did the dealer sell the most computers?

34. During which week did the dealer sell the least computers?

35. Between which two consecutive weeks did the sales increase the most?

36. Between which two consecutive weeks did the sales decrease the most?

37. What percent of the total sales were made during week 5?

38. What percent of the total sales were made during week 1?

The double-line graph at the right shows the comparison between new subscribers signed up for cable television during a four-month period in 1990 and 1991. Use this graph in Exercises 39–45.

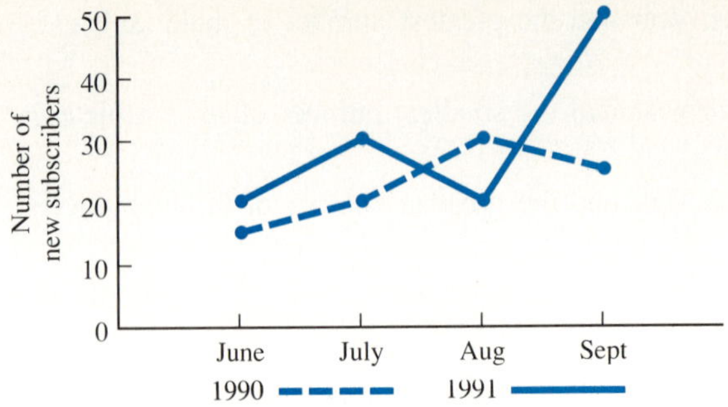

39. How many new subscribers were there in June of 1990?

40. How many new subscribers were there in August of 1991?

41. How many new subscribers were there in the four months of 1990?

42. How many new subscribers were there in the four months of 1991?

43. What was the ratio of new subscribers in June 1990 to June 1991?

44. What was the ratio of new subscribers in July 1991 to August 1991?

45. Between which two consecutive months of 1991 did the company have the greatest increase in new subscribers?

The hourly wages of all the employees of Mutter Manufacturing Company are presented in the histogram at the right. Use this information in Exercises 46–55. Assume that $5 is in the class interval $5–$7 not in $3–$5, similarly for $7, $9, and $11.

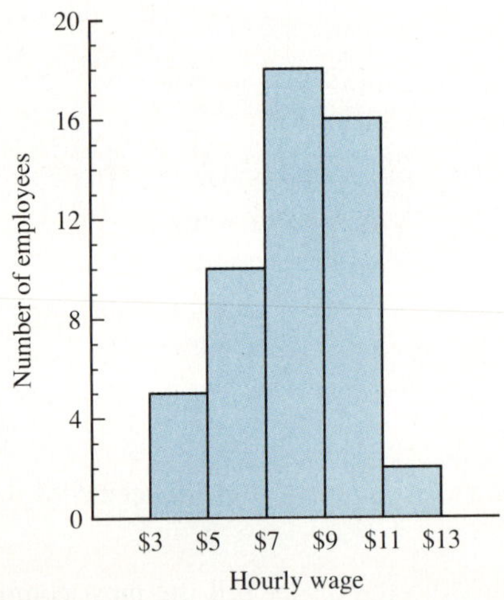

46. How many employees earn at least $3 and less than $5 an hour?

47. How many employees earn at least $7 and less than $9 an hour?

48. How many employees earn $7 an hour or more?

49 How many employees earn less than $9 an hour?

50. How many employees are there in Mutter Manufacturing Company?

51. What is the ratio of employees earning $9 or more an hour to the total number of employees?

52 What percent of the employees earn $9 or more an hour (to the nearest tenth of a percent)?

53. What is the ratio of employees earning less than \$7 an hour to those who earn \$9 or more an hour?

54. What values would be used as the class midpoints of the class intervals to construct the frequency polygon for this data?

55. Draw the frequency polygon in the figure above.

FOR REVIEW

56. Convert 3.7 to a fraction.

57. Convert $\frac{5}{9}$ to a decimal.

58. Convert 0.085 to a percent.

59. Convert $\frac{5}{11}$ to a percent.

60. Convert $7\frac{3}{4}\%$ to a decimal.

61. A chemist has a solution that is 0.6% salt. If the solution weighs 650 g, what is the weight of the salt?

ANSWERS: 1. \$12,000 2. labor 3. $\frac{2}{1}$ or 2 to 1 4. $\frac{2}{3}$ or 2 to 3 5. $\frac{3}{1}$ or 3 to 1 6. 25% 7. 50% 8. $8.\overline{3}\%$ or $8\frac{1}{3}\%$ 9. 1300 10. $\frac{3}{13}$ or 3 to 13 11. 23.1% 12. $\frac{4}{13}$ or 4 to 13 13. 30.8% 14. 53.8% 15. $\frac{8}{5}$ or 8 to 5 16. $\frac{1}{2}$ or 1 to 2 17. 15.4% 18. 450 19. 34.6% 20. $\frac{7}{4}$ or 7 to 4 21. 1988 22. 1990 23. 1988 24. 1991 25. $\frac{4}{3}$ or 4 to 3 26. $\frac{5}{3}$ or 5 to 3 27. 1991 28. Yes; the number of males and females seemed to decline over the period. 29. 15 30. weeks 1 and 6 31. 90 32. $\frac{6}{1}$ or 6 to 1 33. week 5 34. week 3 35. between weeks 3 and 4 36. between weeks 5 and 6 37. $33.\overline{3}\%$ or $33\frac{1}{3}\%$ 38. $11.\overline{1}\%$ or $11\frac{1}{9}\%$ 39. 15 40. 20 41. 90 42. 120 43. $\frac{3}{4}$ or 3 to 4 44. $\frac{3}{2}$ or 3 to 2 45. August and September 46. 5 47. 18 48. 36 49. 33 50. 51 51. $\frac{6}{17}$ or 6 to 17 52. 35.3% 53. $\frac{5}{6}$ or 5 to 6 54. \$4, \$6, \$8, \$10, \$12 55.

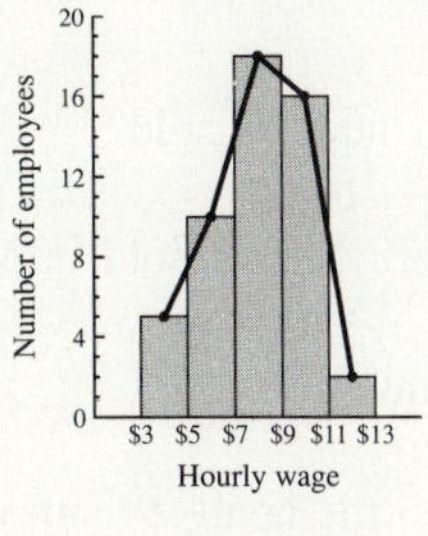

56. $\frac{37}{10}$ 57. $0.\overline{5}$ 58. 8.5% 59. $45.\overline{45}\%$ 60. 0.0775 61. 3.9 g

0.5 EXERCISES B

Exercises 1–8 refer to the circle graph at the right which gives the approximate land area of the seven continents in millions of square miles.

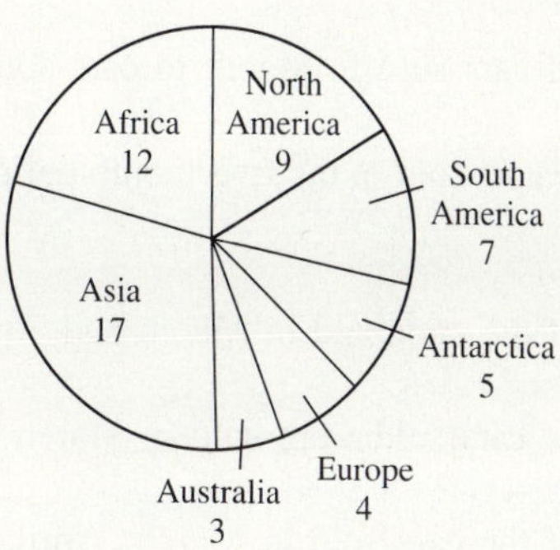

1. What is the approximate total land area of the seven continents?

2. What is the largest continent by area?

3. What is the ratio of the area of Africa to the area of North America?

4. What is the ratio of the area of Australia to the area of South America?

5. What is the ratio of the area of Europe to the area of Asia?

6. To the nearest tenth of a percent, the area of Asia is what percent of the total land area?

7. To the nearest tenth of a percent, the area of North America is what percent of the total land area?

8. North and South America together account for what percent of the total land area (to the nearest tenth of a percent)?

Exercises 9–20 refer to the bar graph at the right which shows the number of new cars sold by a dealer during a five-month period.

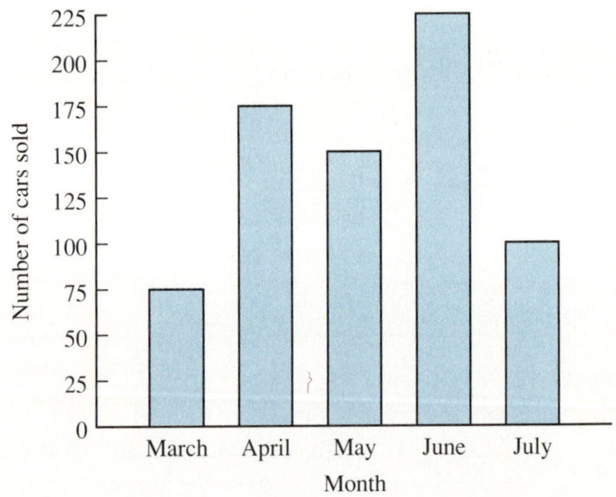

9. How many cars were sold in this five-month period?

10. What is the ratio of cars sold in March to the total number sold?

11. What percent of the cars sold in the five-month period were sold in March (to the nearest tenth of a percent)?

12. What is the ratio of cars sold in June to the total number sold?

13. What percent of the cars sold were sold in June (to the nearest tenth of a percent)?

14. What percent of the cars sold were sold in June and July (to the nearest tenth of a percent)?

15. What is the ratio of cars sold in June to cars sold in May?

16. What is the ratio of cars sold in March to cars sold in July?

17. What percent of the cars sold in the five-month period were sold in a month other than July (to the nearest tenth of a percent)?

18. What is the total of the cars sold in March and April?

19. What percent of the cars sold were sold in March and April (to the nearest tenth of a percent)?

20. What is the ratio of the cars sold in March, April, and May to the cars sold in June and July?

Exercises 21–28 refer to the double-bar graph at the right which shows the quarterly profits for a large corporation in millions of dollars for the years 1990 and 1991.

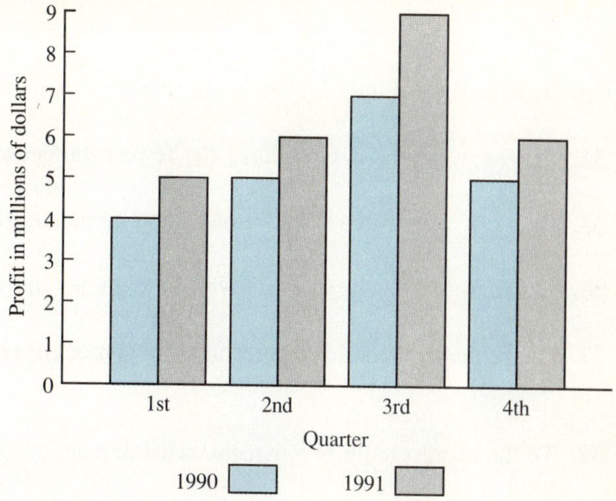

21. Which quarter showed the greatest profit in 1990?

22. Which quarter showed the smallest profit in 1990?

23. Which quarter showed the greatest profit in 1991?

24. Which quarter showed the smallest profit in 1991?

25. What was the profit in the 2nd quarter of 1990?

26. What was the profit in the 3rd quarter of 1991?

27. What was the total profit in all four quarters in 1991?

28. Would it be fair to say that during both years there was a steady increase in profits through the first three quarters and a decline during the fourth quarter? Explain.

The number of yards gained passing by quarterback Greg Wyatt in five home games of the 1991 season is shown in the broken-line graph at the right. Use this graph in Exercises 29–38.

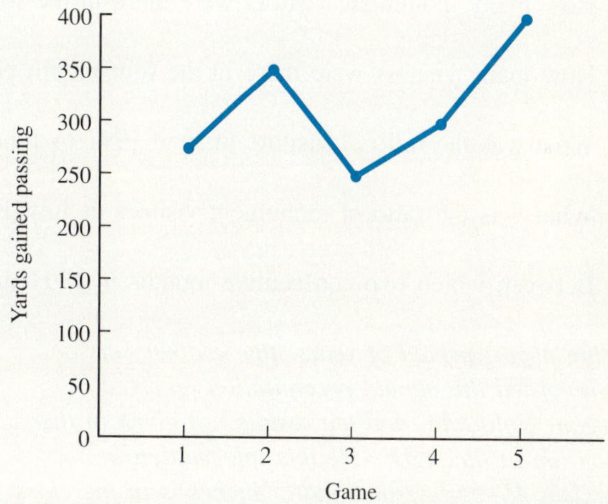

29. How many yards passing did Greg have in game 1?

30. During which game did Greg have 400 yards passing?

31. How many total yards passing did Greg have in the five home games?

32. What was the ratio of yards passing in game 2 to game 3?

33. In which game did Greg have the most yards passing?

34. In which game did Greg have the fewest yards passing?

35. Between which two consecutive home games did his yards passing increase the most?

36. Between which two consecutive home games did his yards passing decrease?

37. To the nearest tenth of a percent, what percent of Greg's total passing yards in the five home games was gained in week 5?

38. To the nearest tenth of a percent, what percent of Greg's total passing yards in the five home games was gained in week 1?

The double-line graph at the right shows the number of visitors (in thousands) to Sunset Crater National Monument during four months in the summers of 1990 and 1991. Use this graph in Exercises 39–45.

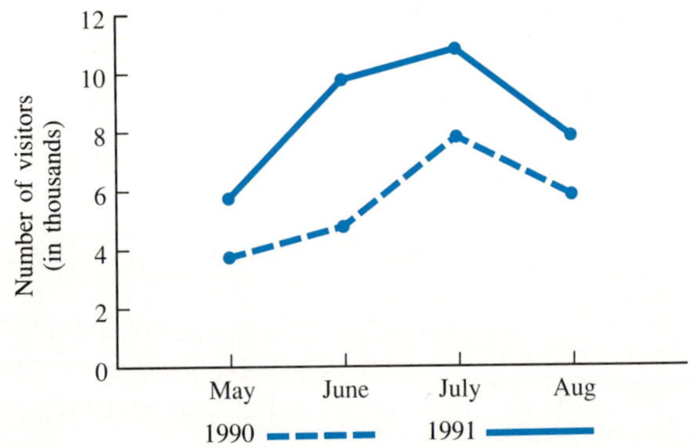

39. How many monument visitors were there in June of 1991?

40. How many visitors were there in June of 1990?

41. How many monument visitors were there in the four-month period in 1990?

42. How many visitors were there in the four-month period in 1991?

43. What was the ratio of visitors in June 1991 to June 1990?

44. What was the ratio of monument visitors in July 1990 to June 1990?

45. Between which two consecutive months in 1991 did the number of visitors increase the most?

During a past period of years, the weather service has recorded the annual precipitation received by Morgan, Colorado, and the results are given in the histogram at the right. Use this information in Exercises 46–55. Assume that 5 inches is in the class interval 5 inches–10 inches, not in 0 inches–5 inches, similarly for 10 inches, 15 inches, and 20 inches.

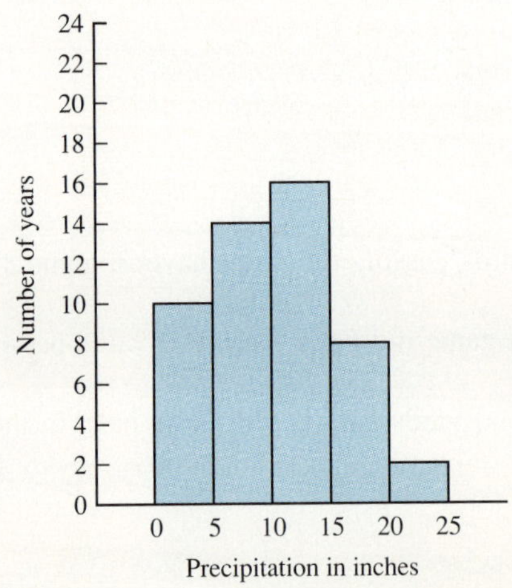

46. In how many years was the precipitation total less than 5 inches?

47. In how many years was the precipitation total at least 10 inches and less than 15 inches?

48. In how many years did Morgan receive 10 inches or more of precipitation?

49. In how many years did Morgan receive less than 15 inches of precipitation?

50. What was the total number of years in the recording period?

51. What is the ratio of the number of years of 15 to 20 inches of precipitation to the number of years of 20 to 25 inches?

52. In what percent of the total number of years did Morgan receive 20 inches or more of precipitation?

53. What is the ratio of the number of years of less than 5 inches of precipitation to the number of years of 10 inches or more of precipitation?

54. What values would be used as the class midpoints of the class intervals to construct the frequency polygon for this data?

55. Draw the frequency polygon in the figure above.

FOR REVIEW

56. Convert 8.3 to a fraction.

57. Convert $\frac{6}{11}$ to a decimal.

58. Convert 0.017 to a percent.

59. Convert $\frac{8}{9}$ to a percent.

60. Convert $10\frac{1}{4}\%$ to a decimal.

61. It has been estimated that 32% of the population of Northern, Minnesota is of Danish ancestry. If 19,270 people live in Northern, approximately how many of them are of Danish background?

0.5 EXERCISES C

A die was rolled 600 times and the number of times that each number (1, 2, 3, 4, 5, or 6) came up was recorded. The results of this experiment are shown in the bar graph at the right. The statistical probability *of obtaining a particular number is defined as the ratio of the number of times that number came up on the die to the total number of times the die is rolled. Use this information in Exercises 1–6 to give the statistical probability of rolling the particular number.*

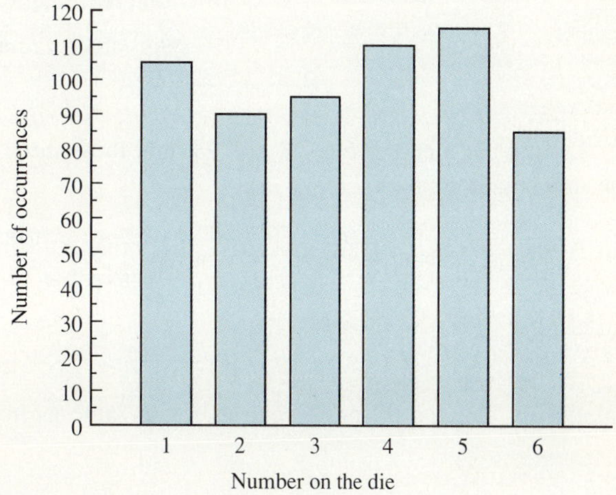

1. 1 [Answer: $\frac{7}{40}$]

2. 2

3. 3

4. 4

5. 5

6. 6

The letters in our alphabet occur with different frequencies in the words that we use. The letter "e" occurs most frequently and the letter "z" occurs least frequently. In fact, it has been shown statistically that in a large sample of words, the words on a page in a text for example, approximately 13% of the letters will be "e", 7% will be "a", 6% will be "s", and 8% will be "n". Select a page in one of your textbooks (a page with many words), and count the number of times e, a, s, *and* n *occur. Also, count the total number of letters on the page.*

7. Make a table showing the number of times each of the four letters occurs and transfer the information to a bar graph.

8. Which of these letters occurred most frequently?

9. Which of these letters occurred least frequently?

10. How do the percents of occurrence for the letters e, a, s, and n in your data compare to the percents given above?

0.6 MEAN, MEDIAN, AND MODE

STUDENT GUIDEPOSTS

1. Mean (Average)
2. Median
3. Mode

1 MEAN (AVERAGE)

In Section 0.5, numerical data were presented using circle graphs, bar graphs, broken-line graphs, and histograms. Often it is necessary to look for one number or statistic to characterize a whole collection of numbers. For example, suppose Jenny has scores of 69, 79, and 98 on three chemistry tests. To find out how she is doing in the course, Jenny might calculate the *mean* or *average* of her scores.

Mean (Average)

The **mean** or **average** of several values is the ratio of the sum of the values to the number of values.

$$\text{mean (average)} = \frac{\text{sum of values}}{\text{number of values}}$$

To find the mean of 69, 79, and 98, add these three scores and divide by 3 (the number of scores).

$$\text{mean} = \frac{69 + 79 + 98}{3} \begin{array}{l} \longleftarrow \text{Sum of the scores} \\ \longleftarrow \text{Number of scores} \end{array}$$

$$= \frac{246}{3} = 82$$

Making 69, 79, and 98 on three tests is the same as making

82, 82, and 82

on three tests. The mean or average gives Jenny a better idea of how well she is doing in chemistry. Her professor may use the average to compute her course grade.

EXAMPLE 1 Finding a Mean in a Business

The daily sales at George's Pie Place were \$285.85, \$192.50, \$222.82, \$315.16, \$427.45, and \$396.80 for a six-day week. To get a better idea of how well his business is doing, George would like to know the average daily sales for the week. What is the average?

Add the sales for the 6 days and divide by 6.

$$\begin{array}{r} \$285.85 \\ 192.50 \\ 222.82 \\ 315.16 \\ 427.45 \\ \underline{396.80} \\ \$1840.58 \end{array} \qquad \frac{\$1840.58}{6} \approx \$306.76 \quad \text{To the nearest cent}$$

George's shop is averaging about \$306.76 per day in sales.

Practice Exercise 1

City College is playing State College in a basketball game. The heights in inches of the starting five players for City are 80, 72, 82, 79, and 85. The starting five players for State have heights in inches of 81, 79, 76, 76, and 82. On the average, which team is taller?

Answer: City, with an average height of 79.6 inches, is taller than State, with an average of 78.8 inches.

In Example 1, if George had known that the total sales for the 6-day week were \$1840.58, he would only have had to divide by 6. The next example illustrates this type of problem.

EXAMPLE 2 Finding a Mean in Industry

Suppose tests by General Motors show that one of its cars will go 420 miles on a tank of gasoline. What is the average miles per gallon (mpg) of the car if the tank holds 12.5 gal?

In this problem the total miles traveled on the 12.5 gal is given. Thus, all that is needed to find the average is to divide 420 mi by 12.5 gal.

$$\frac{420 \text{ mi}}{12.5 \text{ gal}} = 33.6 \frac{\text{mi}}{\text{gal}} = 33.6 \text{ mpg}$$

Practice Exercise 2

The manufacturer of wood-burning stoves produced 78 stoves during a period of 12 days. What was the average daily production?

Answer: 6.5 stoves per day

2 MEDIAN

The mean of a collection of numbers can often be influenced by one very large or very small number (compared to the others). As a result, a mean is often a poor measure of the characteristics of a set of data. For example, the mean of the numbers

$$1, \quad 2, \quad 3, \quad 4, \quad 90$$

is

$$\frac{1+2+3+4+90}{5} = \frac{100}{5} = 20.$$

Since four of the five numbers are very small compared to the fifth number, to use 20 to describe this set of data might be somewhat misleading. Actually, a number closer to 1, 2, 3, or 4 than 20 would provide a better description. Another way to describe a set of data is to look at the value in the middle after the numbers are arranged by size. For example, consider the following.

$$1, \quad 2, \quad \underset{\substack{\uparrow \\ \textbf{middle}}}{3}, \quad 4, \quad 90$$

The value in the middle is 3, and certainly 3 gives a better description of the data set than the mean 20. The middle value of a collection of numbers that is arranged in order of size is the *median* of the data.

Suppose Russell made scores of 72, 65, and 88 on his first three tests. To find the middle score, or median score, arrange them in order of increasing value.

$$65, \quad \underset{\substack{\uparrow \\ \textbf{middle}}}{72}, \quad 88$$

The median score is 72.

But what about the middle score if Russell had four test scores, 72, 65, 88, and 95? Once again arrange them in order.

$$65, \quad 72, \underset{\substack{\uparrow \\ \textbf{middle}}}{\quad} 88, \quad 95$$

In this case, there is no middle score. To approximate a middle score, average the *two* scores in the middle.

$$\text{median} = \frac{72 + 88}{2} = 80$$

Median

The middle value of several values is called the **median.** To find the median:

1. Arrange the values in increasing order from left to right.
2. If there is an odd number of values, the median is the middle one.
3. If there is an even number of values, the median is the average of the *two* values in the middle.

EXAMPLE 3 FINDING A MEDIAN

Find the median of the given numbers.

(a) 5, 12, 2, 14, 8, 7, 9

First arrange the numbers in order.

2, 5, 7, **8,** 9, 12, 14

There are seven numbers, and the middle one is the fourth from the left. The median is 8.

(b) 125, 105, 216, 205, 110, 175

Arrange the numbers in order.

105, 110, **125, 175,** 205, 216

The two numbers in the middle are 125 and 175. Find their average.

$$\frac{125 + 175}{2} = 150$$

Thus, 150 is the median of the numbers.

PRACTICE EXERCISE 3

Find the median of the given numbers.

(a) 9, 20, 2, 18, 27

(b) 206, 140, 180, 375

Answers: (a) 18 (b) 193

EXAMPLE 4 An Application of Mean and Median

In Greenborough, the number of traffic accidents per day is recorded. The following are the numbers for one week.

12, 18, 22, 10, 8, 15, 27

Find the average number of accidents per day and the median number per day.

First find the average.

$$\frac{12 + 18 + 22 + 10 + 8 + 15 + 27}{7} = \frac{112}{7} = 16$$

To find the median, arrange the numbers in order.

8, 10, 12, **15,** 18, 22, 27

The median is 15 which is close to the mean of 16.

Practice Exercise 4

Jeanine made scores of 22, 24, 15, 27, 19, and 30 on six quizzes in geology lab. Find the average score and the median score on these quizzes.

Answer: average score: 22.8, median score: 23

HINT

Remember to arrange the numbers by size before selecting the middle one as the median. For example, the median of

5, 1, 2, 3, 8

is 3, not 2. This is easy to see after rewriting the numbers as 1, 2, 3, 5, 8.

EXAMPLE 5 Finding the Mean and Median Using a Table

The Lopez Engineering Company has 30 employees. Six of them make \$18,500 per year, 5 make \$19,200 per year, 2 make \$22,800 per year, 10 make \$24,500 per year, 4 make \$32,800 per year, 2 make \$38,000 per year, and 1 makes \$52,400 per year. What is the mean or average salary of the 30 employees? What is the median salary?

A good way to find the sum of the salaries is to set up a table.

Salary	*Number of employees*	*Salary times number of employees*
18,500	6	111,000
19,200	5	96,000
22,800	2	45,600
24,500	10	245,000
32,800	4	131,200
38,000	2	76,000
52,400	1	52,400
Totals	30	\$757,200

When each salary is multiplied by the number of employees making that salary and the results are added, the total is \$757,200. This is divided by 30, the total number of employees, to get the mean or average salary.

Practice Exercise 5

A new car dealer has 40 cars in stock. Five are valued at \$18,000, 12 at \$15,000, 10 at \$12,000, 7 at \$9000, and 6 at \$8000. What is the mean or average value of a car on his lot? What is the median value?

$$\frac{\$757{,}200}{30} = \$25{,}240$$

Since there are 30 employees, the median salary is the average of the middle two salaries once the salaries are arranged by size. This will be the average of the 15th and 16th salaries in the arrangement. Without actually listing all the salaries, we can find the 15th and 16th ones by looking at the table. Use the column for the number of employees and count down 15. Since the 15th and 16th positions are included in the 10 employees in the fourth row from the top, the 15th and 16th salary values are both $24,500, making the average of the two also $24,500. Thus, the median salary is $24,500.

Answer: mean = $12,525; median = $12,000

3 MODE

The final measure used to describe a set of data is called the *mode*. Consider the numbers

$$1, \quad 5, \quad 101, \quad 101.$$

The mean is

$$\frac{1 + 5 + 101 + 101}{4} = \frac{208}{4} = 52,$$

and the median is

$$\frac{5 + 101}{2} = \frac{106}{2} = 53.$$

But does 52 or 53 give an accurate reflection of the data set? Since two of the four numbers are 101, much greater than 52 or 53, it might be better to describe this data set using 101, the number that occurs most frequently in the set.

Mode

The value that occurs most frequently in a set of values is the **mode** of the data. If there are two values that occur more frequently than all the others, the data has two modes and is called **bimodal.** If no value occurs more frequently than the others in the set, the data has **no mode.**

EXAMPLE 6 Finding a Mode

Find the mode of the given numbers.

(a) 5, 8, 1, 8, 7, 9, 8, 5

Since 8 occurs three times, 5 occurs twice, and all the rest of the numbers occur only once, the mode is 8.

(b) 2, 7, 3, 7, 8, 8, 1, 5

Since 7 and 8 both occur twice and all other numbers occur only once, the set of numbers has two modes, 7 and 8. It is a bimodal set.

(c) 3, 9, 7, 5, 2, 1, 11

Since every number occurs exactly one time, there is no mode.

Practice Exercise 6

Find the mode of the given numbers.

(a) 1, 1, 1, 5, 2, 2, 9, 8

(b) 6, 6, 2, 1, 5, 2, 9, 10

(c) 2, 8, 5, 12, 9, 7

Answers: (a) 1 (b) 6 and 2 (bimodal) (c) no mode

0.6 EXERCISES A

Find the mean of the given numbers. Give answers to the nearest tenth when appropriate.

1. 7, 19, 13
2. 7, 9, 8, 12
3. 12, 22, 18, 25, 9
4. 102, 95, 116, 118, 125, 106
5. 7.2, 6.1, 8.6, 9.2, 5.5, 7.2, 8.8
6. 42, 43, 58, 55, 61, 41, 48, 48

Find the median.

7. 7, 19, 13
8. 7, 9, 8, 12
9. 12, 22, 25, 18, 9
10. 102, 95, 116, 118, 125, 106
11. 7.2, 6.1, 8.6, 9.2, 5.5, 7.2, 8.8
12. 42, 43, 58, 55, 61, 41, 48, 48

Find the mode.

13. 5, 6, 1, 6, 3, 2
14. 2, 3, 9, 2, 9, 7, 9, 11
15. 8, 3, 8, 5, 7, 3, 2
16. 2, 11, 17, 9, 6, 41
17. 4.3, 2.7, 4.3, 12.8, 15.1
18. 149, 150, 211, 150, 435, 147

Solve.

19. Suppose the Ford Motor Company's tests show that one of their cars will go 480 miles on a tank of gasoline. If the tank holds 13.6 gal, what is the average miles per gallon (mpg)?
20. A basketball player scored 252 points in 18 games. What is his average per game?
21. During a six-day week, the sales at Joseph's Shop for Men were $482.16, $316.58, $299.81, $686.47, $819.72, and $992.16. What were the average daily sales?
22. The heights of the starting five players on the Upstate University basketball team are 78 inches, 82 inches, 72 inches, 84 inches, and 73 inches. What is the average height? What is the median height?

23 The table at the right gives information on the yearly salaries of the employees at one plant of Mutter Manufacturing. Find the average salary. Find the median salary. What is the mode of the salaries?

Salary	*Number of Employees*	*Salary Times Number of Employees*
16,200	8	
20,800	10	
28,600	7	
32,000	2	
46,000	1	
Totals		

FOR REVIEW

The profit (in millions of dollars) made by the ABC Corporation over a six-year interval is shown in the broken-line graph at the right. Use this graph in Exercises 24–30.

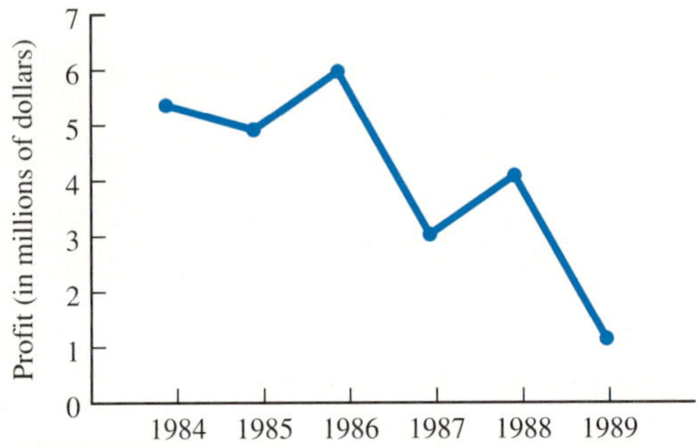

24. What profit was made in 1987?

25. What profit was made in 1986?

26. What was the total profit made in 1988 and 1989?

27. What was the total profit made during the six-year interval?

28. What was the ratio of the profit in 1988 to that in 1989?

29. What percent (to the nearest tenth) of the total profit in the six-year interval was made in 1987?

30. What percent (to the nearest tenth) of the total profit in the six-year interval was made in the years 1986 and 1987?

Exercises 31–34 refer to the circle graph at the right which shows the distribution of Chuck's gross monthly income of $2500.

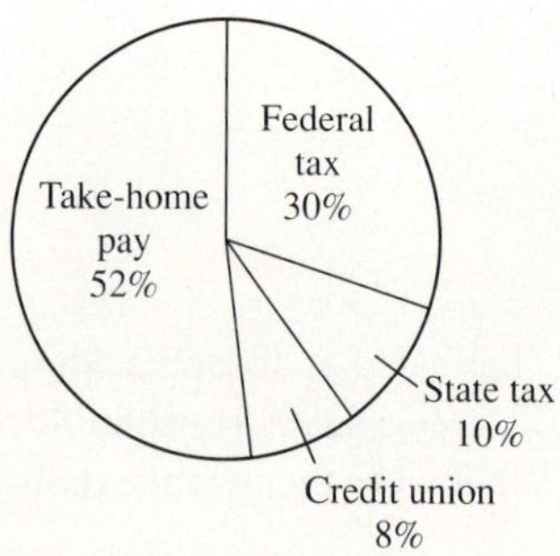

31. What is Chuck's monthly take-home pay?

32. How much federal tax does Chuck pay each month?

33. How much state tax does Chuck pay each month?

34. What is the ratio of federal tax to take-home pay?

ANSWERS: 1. 13 2. 9 3. 17.2 4. 110.3 5. 7.5 6. 49.5 7. 13 8. 8.5 9. 18 10. 111 11. 7.2 12. 48 13. 6 14. 9 15. 8 and 3 (bimodal) 16. no mode 17. 4.3 18. 150 19. 35.3 mpg 20. 14 21. $599.48 22. 77.8; 78 23. $23,135.71; $20,800; $20,800 24. 5 million dollars 25. 5.5 million dollars 26. 9 million dollars 27. 24.5 million dollars 28. $\frac{2}{1}$ or 2 to 1 29. 20.4% 30. 42.9% 31. $1300 32. $750 33. $250 34. $\frac{15}{26}$ or 15 to 26

0.6 EXERCISES B

Find the mean of the given numbers. Give answers to the nearest tenth when appropriate.

1. 11, 21, 19

2. 8, 12, 14, 22

3. 14, 35, 16, 19, 43

4. 130, 87, 102, 96, 145, 110

5. 6.3, 4.7, 8.2, 9.6, 7.1, 8.3, 2.5

6. 51, 57, 62, 65, 71, 83, 90, 97

Find the median.

7. 5, 12, 37

8. 6, 9, 10, 14

9. 5, 21, 27, 17, 19

10. 107, 93, 110, 84, 87, 130

11. 6.3, 4.7, 8.2, 9.6, 7.1, 8.3, 2.5

12. 51, 57, 62, 65, 90, 83, 71, 97

Find the mode.

13. 11, 17, 5, 17, 6, 21, 17

14. 6, 3, 6, 8, 11, 3, 21, 3

15. 2, 12, 5, 2, 8, 5, 7, 1

16. 3, 9, 21, 8, 7, 26, 5

17. 4.8, 7.6, 8.1, 9.3, 7.6, 2.4

18. 159, 276, 423, 159, 342, 177

Solve.

19. Tests show that a motorhome will travel 535.5 miles on a tank of gasoline. If the tank holds 42.5 gallons, what is the average miles per gallon (mpg)?

20. Quarterback Craig Austin completed 438 passes in 11 games. What is the average number of completions per game (to the nearest tenth)?

21. The number of TV sets sold per day by a dealer over a 12-day period is 2, 4, 6, 2, 10, 3, 6, 8, 12, 8, 22, and 12. What is the median number of sets sold per day? What is the average number of sets sold per day (to the nearest tenth)?

22. Maria scored 73, 85, 65, 92, 78, and 87 on six tests in math. What is her average score? What is her median score?

23. The Mutter Manufacturing record of sick-leave days available to employees is recorded below. What is the average number of days available? What is the median number? What is the mode?

Number of days sick leave available	*Number of employees*	*Number of days times number of employees*
14	4	
21	12	
28	8	
35	4	
Totals		

FOR REVIEW

The number of years of teaching experience by the faculty at CSU is shown in the histogram below. Use this graph in Exercises 24–30.

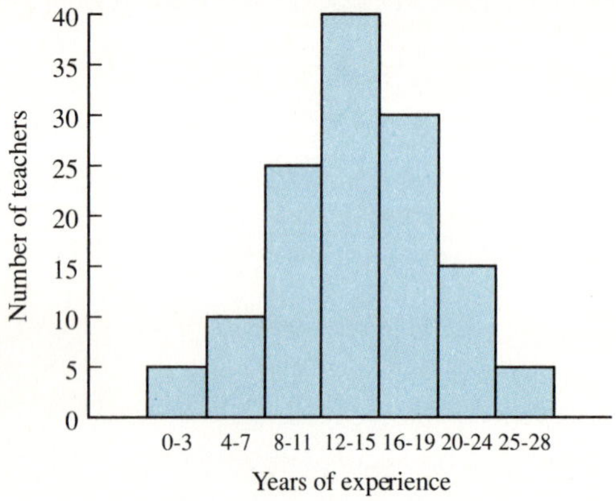

24. How many teachers have 0–3 years of experience?

25. How many teachers have 8–15 years of experience?

26. How many teachers have 12 or more years of experience?

27. How many teachers have 7 or fewer years of experience?

28. How many teachers in total are represented by this data?

29. What percent (to the nearest tenth) of the total number of teachers have 16 or more years of experience?

30. What is the ratio of the number of teachers with 16–19 years of experience to those with 7 or fewer years of experience?

Exercises 31–34 refer to the bar graph below which shows the number of students who preferred various cable television networks in a recent survey.

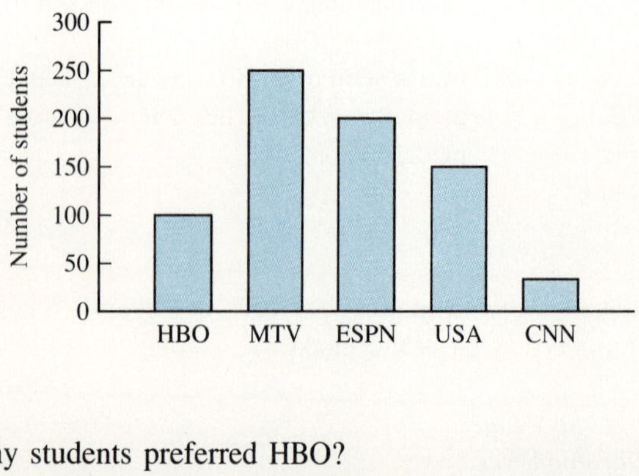

31. How many students preferred HBO?

32. How many students were in the survey?

33. To the nearest tenth of a percent, what percent of the students preferred ESPN?

34. What was the ratio of students preferring MTV to CNN?

0.6 EXERCISES C

Solve.

1. Suppose Marni made scores of 65, 72, and 88 on the first three tests in math. What score did she make on the fourth test if the average score on the four was 80?
[Answer: 95]

2. Suppose Burford made scores of 32, 38, 51, and 82 on four of five exams. If the median score of the five exams is 51, could the score on Burford's fifth test have been 35?

Which of the mean, median, or mode do you think is the best measure to describe each set of numbers in Exercises 3–6?

3. 1, 1, 1, 1, 20, 100

4. 1, 1, 99, 99
[Answer: mode (bimodal)]

5. 1, 2, 3, 1000

6. 1, 2, 3, 4, 5

0.7 MEASUREMENT

STUDENT GUIDEPOSTS

1. English Units of Measure
2. Unit Fraction
3. Household Units
4. Compound Units
5. Simplifying Measurement Units
6. Operations on Measurement Units

1 ENGLISH UNITS OF MEASURE

A standard system of units of measure is a useful tool for people living and working together. While many countries throughout the world use the metric system, discussed in the next section, the United States uses the English system.

Units of Measure: English System	
Length	1 foot (ft) = 12 inches (in) 1 yard (yd) = 3 feet (ft) 1 mile (mi) = 5280 feet (ft)
Volume	1 pint (pt) = 16 fluid ounces (fl oz) 1 quart (qt) = 2 pints (pt) 1 gallon (gal) = 4 quarts (qt)
Weight	1 pound (lb) = 16 ounces (oz) 1 ton = 2000 pounds (lb)
Time	1 minute (min) = 60 seconds (sec) 1 hour (hr) = 60 minutes (min) 1 day (da) = 24 hours (hr) 1 week (wk) = 7 days (da) 1 year (yr) = 365 days (da) = 52 weeks (wk) = 12 months (mo)

As the table shows, each type of measurement (length, volume, and so on) has several units in the English system. The weight of this book might be given in ounces, while the weight of a student would usually be given in pounds.

2 UNIT FRACTION

To work with units of measure, we need to know how to convert from one unit to another. A good way is to use **unit fractions,** fractions which have value 1. Some examples of unit fractions are

$$\frac{12 \text{ in}}{1 \text{ ft}}, \quad \frac{1 \text{ ft}}{12 \text{ in}}, \quad \frac{4 \text{ qt}}{1 \text{ gal}}, \quad \frac{1 \text{ gal}}{4 \text{ qt}}, \quad \frac{16 \text{ oz}}{1 \text{ lb}}, \quad \frac{1 \text{ lb}}{16 \text{ oz}}.$$

Each of the fractions has value 1 because the numerator and denominator are equal. It is important to show the unit of measure as well as the number. Obviously, $\frac{12}{1}$ *is not* 1, but $\frac{12 \text{ in}}{1 \text{ ft}}$ *is* 1 since

$$1 \text{ ft} = 12 \text{ in}.$$

Thus, to change 4 yd to feet, write

$$\begin{aligned} 4 \text{ yd} &= 4 \text{ yd} \times 1 && \text{Multiply by 1} \\ &= \frac{4 \text{ yd}}{1} \times \frac{3 \text{ ft}}{1 \text{ yd}} && \text{Divide out yards} \\ &= 4 \times 3 \text{ ft} \\ &= 12 \text{ ft}. \end{aligned}$$

With this method, the yards are divided out as if they were numbers.

HINT

The fact that the yards do divide out tells us that we were right to multiply by 3 ft instead of dividing by 3 ft. If we had tried to multiply

$$\frac{4 \text{ yd}}{1} \times \frac{1 \text{ yd}}{3 \text{ ft}},$$

the yards would not have divided out and the change of units could not be made.

EXAMPLE 1 Converting Feet to Inches

Convert 16 feet to inches (16 ft = ____ in).

Since

$$1 \text{ ft} = 12 \text{ in},$$

we need to use one of the unit fractions

$$\frac{12 \text{ in}}{1 \text{ ft}} \quad \text{or} \quad \frac{1 \text{ ft}}{12 \text{ in}}.$$

To divide out the ft in 16 ft, use $\frac{12 \text{ in}}{1 \text{ ft}}$.

$$\begin{aligned} 16 \text{ ft} &= 16 \text{ ft} \times 1 && \text{Multiply by 1} \\ &= \frac{16 \text{ ft}}{1} \times \frac{12 \text{ in}}{1 \text{ ft}} && \text{Divide out ft} \\ &= 16 \times 12 \text{ in} \\ &= 192 \text{ in} \end{aligned}$$

PRACTICE EXERCISE 1

Convert 23 feet to inches.

Answer: 276 in

EXAMPLE 2 Converting Inches to Feet

Convert 72 inches to feet (72 in = _____ ft).

For this problem use the unit fraction $\frac{1\text{ ft}}{12\text{ in}}$.

$$\begin{aligned} 72\text{ in} &= 72\text{ in} \times 1 && \text{Multiply by 1} \\ &= \frac{72\text{ in}}{1} \times \frac{1\text{ ft}}{12\text{ in}} && \text{Divide out inches} \\ &= \frac{72}{12}\text{ ft} \\ &= 6\text{ ft} \end{aligned}$$

Practice Exercise 2

Convert 108 inches to feet.

Answer: 9 ft

EXAMPLE 3 Conversion in Weight and Volume

Complete the conversion in units.

(a) 22 lb = _____ oz.

The unit fraction is $\frac{16\text{ oz}}{1\text{ lb}}$.

$$\begin{aligned} 22\text{ lb} &= 22\text{ lb} \times 1 && \text{Multiply by 1} \\ &= \frac{22\text{ lb}}{1} \times \frac{16\text{ oz}}{1\text{ lb}} && \text{Divide out lb} \\ &= 22 \times 16\text{ oz} \\ &= 352\text{ oz} \end{aligned}$$

(b) 6 pt = _____ qt.

$$\begin{aligned} 6\text{ pt} &= 6\text{ pt} \times 1 \\ &= \frac{6\text{ pt}}{1} \times \frac{1\text{ qt}}{2\text{ pt}} \\ &= \frac{6}{2}\text{ qt} \\ &= 3\text{ qt} \end{aligned}$$

Practice Exercise 3

Complete the conversion in units.

(a) 15 lb = _____ oz.

(b) 9 pt = _____ qt.

Answers: (a) 240 (b) 4.5

Some unit conversions may need two or more unit fractions.

EXAMPLE 4 Using Several Unit Fractions

Complete the change in units.

5 wk = _____ min.

$$\begin{aligned} 5\text{ wk} &= 5\text{ wk} \times 1 \\ &= \frac{5\text{ wk}}{1} \times \frac{7\text{ da}}{1\text{ wk}} && \text{Convert 5 wk to 35 da} \\ &= \frac{35\text{ da}}{1} \times \frac{24\text{ hr}}{1\text{ da}} && \text{Convert 35 da to 840 hr} \\ &= \frac{840\text{ hr}}{1} \times \frac{60\text{ min}}{1\text{ hr}} && \text{Convert 840 hr to 50,400 min} \\ &= 840 \times 60\text{ min} \\ &= 50{,}400\text{ min} \end{aligned}$$

Practice Exercise 4

Complete the change in units.

8 wk = _____ min

The steps can be combined.

$$5 \text{ wk} = \frac{5 \not{\text{wk}}}{1} \times \frac{7 \not{\text{da}}}{1 \not{\text{wk}}} \times \frac{24 \not{\text{hr}}}{1 \not{\text{da}}} \times \frac{60 \text{ min}}{1 \not{\text{hr}}}$$
$$= 5 \times 7 \times 24 \times 60 \text{ min}$$
$$= 50{,}400 \text{ min.}$$

Answer: 80,640

3 HOUSEHOLD UNITS

Some units of measure used in the kitchen are given in this table.

Household Units
1 tablespoon (tbsp) = 3 teaspoons (tsp)
1 cup = 16 tablespoons (tbsp)
= 8 fluid ounces (fl oz)
1 pint (pt) = 2 cups

EXAMPLE 5 Converting Household Units

Complete the change in units.

(a) 7 tbsp = ____ tsp.

$$7 \text{ tbsp} = 7 \text{ tbsp} \times 1$$
$$= \frac{7 \not{\text{tbsp}}}{1} \times \frac{3 \text{ tsp}}{1 \not{\text{tbsp}}} \qquad \frac{3 \text{ tsp}}{1 \text{ tbsp}} \text{ is a unit fraction}$$
$$= 7 \times 3 \text{ tsp} = 21 \text{ tsp}$$

(b) 40 tbsp = ____ pt.

$$40 \text{ tbsp} = \frac{40 \not{\text{tbsp}}}{1} \times \frac{1 \not{\text{cup}}}{16 \not{\text{tbsp}}} \times \frac{1 \text{ pt}}{2 \not{\text{cups}}}$$
$$= 40 \times \frac{1}{16} \times \frac{1}{2} \text{ pt} = \frac{40}{(16)(2)} \text{ pt}$$
$$= \frac{\not{2} \cdot \not{2} \cdot \not{2} \cdot 5}{\not{2} \cdot \not{2} \cdot \not{2} \cdot 2 \cdot 2} \text{ pt} = \frac{5}{4} \text{ pt} = 1\frac{1}{4} \text{ pt}$$

PRACTICE EXERCISE 5

Complete the change in units.

(a) 15 tbsp = ____ tsp.

(b) 120 tbsp = ____ pt.

Answers: (a) 45 (b) 3.75

4 COMPOUND UNITS

Remember that *percent* means "per hundred." We can also say that

$$\frac{12 \text{ in}}{1 \text{ ft}} \quad \text{means} \quad 12 \text{ inches per foot.}$$

Thus, 50 miles per hour can be written as

$$\frac{50 \text{ mi}}{1 \text{ hr}}.$$

Most of the time the 1 in the denominator is not written. 50 miles per hour may be written in any of the following ways.

$$\frac{50 \text{ mi}}{\text{hr}} \qquad 50\frac{\text{mi}}{\text{hr}} \qquad 50 \text{ mi/hr} \qquad 50 \text{ mph}$$

Other units can be written these ways.

$$12 \text{ feet per second} \quad \text{means} \quad \frac{12 \text{ ft}}{\text{sec}} = 12\frac{\text{ft}}{\text{sec}} = 12 \text{ ft/sec}$$

$$8 \text{ gallons per minute} \quad \text{means} \quad \frac{8 \text{ gal}}{\text{min}} = 8\frac{\text{gal}}{\text{min}} = 8 \text{ gal/min}$$

EXAMPLE 6 Converting Units of Rate

Complete the change in units.

$$30\frac{\text{mi}}{\text{hr}} = \_\_\_\_ \frac{\text{ft}}{\text{sec}}.$$

Combine the steps.

$$30\frac{\text{mi}}{\text{hr}} = \frac{30 \text{ mi}}{\text{hr}} = \frac{30 \cancel{\text{mi}}}{\cancel{\text{hr}}} \times \frac{5280 \text{ ft}}{1 \cancel{\text{mi}}} \times \frac{1 \cancel{\text{hr}}}{60 \cancel{\text{min}}} \times \frac{1 \cancel{\text{min}}}{60 \text{ sec}}$$

Divide out mi, hr, and min

$$= \frac{30 \times 5280 \text{ ft}}{60 \times 60 \text{ sec}}$$

$$= \frac{44 \text{ ft}}{\text{sec}} = 44\frac{\text{ft}}{\text{sec}}$$

Practice Exercise 6

Complete the change in units.

$$90\frac{\text{mi}}{\text{hr}} = \_\_\_\_ \frac{\text{ft}}{\text{sec}}$$

Answer: 132

EXAMPLE 7 Application of Conversion

Claudia wants to buy 15 inches of fringe which costs \$1.08 per yard. How much must she pay?

First convert dollars per yard to dollars per foot, then convert dollars per foot to dollars per inch.

$$\frac{1.08 \text{ dollars}}{\text{yd}} = \frac{1.08 \text{ dollars}}{1 \cancel{\text{yd}}} \times \frac{1 \cancel{\text{yd}}}{3 \cancel{\text{ft}}} \times \frac{1 \cancel{\text{ft}}}{12 \text{ in}}$$

Unit fractions

$$= \frac{1.08 \text{ dollars}}{3 \times 12 \text{ in}}$$

$$= \frac{0.03 \text{ dollars}}{\text{in}}$$

Since Claudia wants 15 in of the fringe, multiply 15 in by $\frac{0.03 \text{ dollars}}{\text{in}}$.

$$\frac{15 \cancel{\text{in}}}{1} \times \frac{0.03 \text{ dollars}}{\cancel{\text{in}}} = \$0.45$$

She must pay \$0.45 for the fringe.

Practice Exercise 7

Metal tubing costs \$3.60 per yard. How much must Larry pay for 28 inches?

Answer: \$2.80

5 SIMPLIFYING MEASUREMENT UNITS

Suppose the answer to a time-conversion problem is 64 minutes. When a time answer is more than 1 hour (60 min), the result is usually written in hours and minutes.

$$64 \text{ min} = \mathbf{60 \text{ min}} + 4 \text{ min}$$
$$= \mathbf{1 \text{ hr}}\ 4 \text{ min}$$

Plus sign left out

This is read "one hour four minutes." This form of the answer gives a better idea of the size of some measurements, especially when the numbers are large. For example, we have a better idea of the time if we are told a job will take 8 hr rather than 480 min.

Other measurement units can also be simplified. For example, to change 32 inches to feet and inches, write 32 as the sum of a multiple of 12 and a number less than 12.

$$\begin{aligned} 32 \text{ in} &= \mathbf{24 \text{ in}} + 8 \text{ in} \\ &= \mathbf{2 \text{ ft}}\ 8 \text{ in} \qquad 24 \text{ in} = 2 \text{ ft} \end{aligned}$$

This is read "two feet eight inches."

EXAMPLE 8 Simplifying Units

Simplify the following measurements.

(a) 3 hr 92 min

To change 92 minutes to hours and minutes, write 92 as the sum of a multiple of 60 and a number less than 60.

$$\begin{aligned} 3 \text{ hr } \mathbf{92 \text{ min}} &= 3 \text{ hr} + \mathbf{60 \text{ min}} + \mathbf{32 \text{ min}} && 92 = 60 + 32 \\ &= 3 \text{ hr} + 1 \text{ hr} + 32 \text{ min} && 60 \text{ min} = 1 \text{ hr} \\ &= 4 \text{ hr } 32 \text{ min} && \text{Add 3 hr and 1 hr} \end{aligned}$$

(b) 3 yd 8 ft

$$\begin{aligned} 3 \text{ yd } \mathbf{8 \text{ ft}} &= 3 \text{ yd} + \mathbf{6 \text{ ft}} + \mathbf{2 \text{ ft}} && 8 = 6 + 2 \\ &= 3 \text{ yd} + 2 \text{ yd} + 2 \text{ ft} && 6 \text{ ft} = 2 \text{ yd} \\ &= 5 \text{ yd } 2 \text{ ft} \end{aligned}$$

(c) 9 gal 18 qt 9 pt.

First change 9 pints to quarts and pints.

$$\begin{aligned} 9 \text{ pt} &= \mathbf{8 \text{ pt}} + 1 \text{ pt} \\ &= \mathbf{4 \text{ qt}}\ 1 \text{ pt} && 2 \text{ pt} = 1 \text{ qt} \end{aligned}$$

Adding 4 qt to 18 qt gives 22 qt. To change 22 qt to gallons and quarts, write 22 as the sum of a multiple of 4 and a number less than 4.

$$\begin{aligned} 22 \text{ qt } 1 \text{ pt} &= 20 \text{ qt} + 2 \text{ qt} + 1 \text{ pt} \\ &= 5(4 \text{ qt}) + 2 \text{ qt} + 1 \text{ pt} \\ &= 5 \text{ gal } 2 \text{ qt } 1 \text{ pt} && 4 \text{ qt} = 1 \text{ gal} \end{aligned}$$

Since

$$9 \text{ gal} + 5 \text{ gal} = 14 \text{ gal},$$

the final answer is 14 gal 2 qt 1 pt.

Practice Exercise 8

Simplify the following measurements.

(a) 7 hr 116 min

(b) 5 yd 14 ft

(c) 4 gal 34 qt 23 pt

Answers: (a) 8 hr 56 min (b) 9 yd 2 ft (c) 15 gal 1 qt 1 pt

6 OPERATIONS ON MEASUREMENT UNITS

In these simplifications, we have been adding measurements. In Example 8(c) we wrote

$$9 \text{ gal} + 5 \text{ gal} = 14 \text{ gal}.$$

Only measurements in the same unit can be added. Thus, 9 gal can be added to 5 gal, but 14 gal cannot be added to 2 qt without first converting one of the units to the other.

In some cases, we need to add measurements such as 8 ft 5 in and 6 ft 4 in. This problem is like adding mixed numbers. We can change both to inches and add or we can add them as mixed measurement units.

EXAMPLE 9 Adding Units

Add.

(a) 8 ft 5 in and 6 ft 4 in

Add each unit separately.

$$\begin{array}{rr} 8\text{ ft} & 5\text{ in} \\ +\ 6\text{ ft} & 4\text{ in} \\ \hline 14\text{ ft} & 9\text{ in} \end{array}$$

sum of feet ⟶ 14 ft; 9 in ⟵ sum of inches

The sum is 14 ft 9 in.

(b) 3 hr 28 min and 7 hr 54 min

$$\begin{array}{rr} 3\text{ hr} & 28\text{ min} \\ +\ 7\text{ hr} & 54\text{ min} \\ \hline 10\text{ hr} & 82\text{ min} \end{array}$$

sum of hours ⟶ 10 hr; 82 min ⟵ sum of minutes

Since 82 min is more than 1 hr, simplify the sum.

$$\begin{aligned} 10\text{ hr } \mathbf{82\text{ min}} &= 10\text{ hr} + \mathbf{60\text{ min} + 22\text{ min}} \\ &= 10\text{ hr} + 1\text{ hr} + 22\text{ min} \qquad 60\text{ min} = 1\text{ hr} \\ &= 11\text{ hr } 22\text{ min} \end{aligned}$$

The sum is 11 hr 22 min.

Practice Exercise 9

Add.

(a) 17 ft 7 in and 5 ft 2 in

(b) 14 hr 42 min and 5 hr 37 min

Answers: (a) 22 ft 9 in (b) 20 hr 19 min

To subtract measurement units, subtract only like units. As with subtraction of mixed numbers, sometimes we need to borrow. With measurement units, this means changing one unit of the larger measure to units of the smaller.

EXAMPLE 10 Subtracting Units

Subtract.

(a)

$$\begin{array}{rr} 8\text{ ft} & 5\text{ in} \\ -\ 6\text{ ft} & 4\text{ in} \\ \hline 2\text{ ft} & 1\text{ in} \end{array}$$

difference of feet ⟶ 2 ft; 1 in ⟵ difference of inches

The difference is 2 ft 1 in.

(b)

$$\begin{array}{rr} 17\text{ ft} & 2\text{ in} \\ -\ 9\text{ ft} & 8\text{ in} \\ \hline \end{array}$$

Since 2 inches is less than the 8 inches to be subtracted, borrow 1 ft (12 in) from 17 ft.

$$\begin{aligned} 17\text{ ft } 2\text{ in} &= 16\text{ ft} + 1\text{ ft} + 2\text{ in} \\ &= 16\text{ ft} + 12\text{ in} + 2\text{ in} \\ &= 16\text{ ft } 14\text{ in} \end{aligned}$$

Practice Exercise 10

Subtract.

(a)

$$\begin{array}{rr} 14\text{ ft} & 11\text{ in} \\ -10\text{ ft} & 4\text{ in} \\ \hline \end{array}$$

(b)

$$\begin{array}{rr} 25\text{ ft} & 1\text{ in} \\ -11\text{ ft} & 9\text{ in} \\ \hline \end{array}$$

Now subtract.

$$\begin{array}{rcrr} 17 \text{ ft } 2 \text{ in} = & & 16 \text{ ft} & 14 \text{ in} \\ -\ \ 9 \text{ ft } 8 \text{ in} = & - & 9 \text{ ft} & 8 \text{ in} \\ \hline & & 7 \text{ ft} & 6 \text{ in} \end{array}$$

The difference is 7 ft 6 in.

Answers: (a) 4 ft 7 in (b) 13 ft 4 in

Some multiplication problems involve measurement units. For example, suppose it takes 2 yd 1 ft 7 in of material to make one suit. To learn how much material is needed for 4 suits, multiply 2 yd 1 ft 7 in by 4. As with addition and subtraction, all units could be converted to inches and multiplied by 4, but the same result can be reached by multiplying 2 yd by 4, 1 ft by 4, and 7 in by 4, and then simplifying the result.

$$\begin{aligned} 4 \times (2 \text{ yd } 1 \text{ ft } 7 \text{ in}) &= 4(2 \text{ yd} + 1 \text{ ft} + 7 \text{ in}) \\ &= 8 \text{ yd} + 4 \text{ ft} + 28 \text{ in} \qquad \text{Distributive property} \end{aligned}$$

Now simplify. First change 28 in to feet: 28 in = 24 in + 4 in = 2 ft + 4 in. Then, rewrite the product.

$$\begin{aligned} 8 \text{ yd} + 4 \text{ ft} + \mathbf{28 \text{ in}} &= 8 \text{ yd} + 4 \text{ ft} + \mathbf{2 \text{ ft} + 4 \text{ in}} \qquad \text{Replace 28 in by 2 ft + 4 in} \\ &= 8 \text{ yd} + \mathbf{6 \text{ ft}} + 4 \text{ in} \qquad \text{Add 4 ft and 2 ft} \\ &= 8 \text{ yd} + \mathbf{2 \text{ yd}} + 4 \text{ in} \qquad 3 \text{ ft} = 1 \text{ yd} \\ &= 10 \text{ yd } 4 \text{ in} \end{aligned}$$

Thus, 10 yd 4 in is needed for the 4 suits. This could be written as 10 yd 0 ft 4 in.

EXAMPLE 11 Multiplying by a Non-Measurement Number

If a tank holds 7 gal 3 qt 1 pt of liquid, how much would be used to fill 6 tanks of the same size?

Multiply 7 gal 3 qt 1 pt by 6.

$$\begin{aligned} 6 \times (7 \text{ gal } 3 \text{ qt } 1 \text{ pt}) &= 6(7 \text{ gal} + 3 \text{ qt} + 1 \text{ pt}) \\ &= 42 \text{ gal} + 18 \text{ qt} + \mathbf{6 \text{ pt}} \\ &= 42 \text{ gal} + 18 \text{ qt} + \mathbf{3 \text{ qt}} \qquad 2 \text{ pt} = 1 \text{ qt} \\ &= 42 \text{ gal} + 21 \text{ qt} \qquad 18 \text{ qt} + 3 \text{ qt} = 21 \text{ qt} \\ &= 42 \text{ gal} + \mathbf{20 \text{ qt}} + 1 \text{ qt} \qquad 20 \text{ qt} = 5(4 \text{ qt}) \\ &= 42 \text{ gal} + 5 \text{ gal} + 1 \text{ qt} \qquad 4 \text{ qt} = 1 \text{ gal} \\ &= 47 \text{ gal } 1 \text{ qt} \qquad 42 \text{ gal} + 5 \text{ gal} = 47 \text{ gal} \end{aligned}$$

Practice Exercise 11

For a survival test each person was given 1 gal 2 qt 1 pt of water. How much water was needed to supply the 15 people taking the test?

Answer: 24 gal 1 qt 1 pt

It is possible to multiply measurement units by measurement units.

EXAMPLE 12 MULTIPLYING MEASUREMENT UNITS

Find the area of a rectangle which is 4 ft wide and 6 ft long.

Remember that the area of a rectangle is given by

$$\begin{aligned}\text{area} &= \text{length} \times \text{width}\\ &= 6 \text{ ft} \times 4 \text{ ft}.\end{aligned}$$

To work the problem find not only 6×4 but also ft $\times$ ft.

$$\begin{aligned}\text{area} &= 6 \text{ ft} \times 4 \text{ ft}\\ &= (6 \times 4)(\text{ft} \times \text{ft})\\ &= 24 \text{ ft}^2\\ &= 24 \text{ sq ft}\end{aligned}$$

The area is 24 sq ft. (The unit sq ft, or ft^2, used for area, is discussed in more detail in Chapter 12.)

PRACTICE EXERCISE 12

Find the area of a rectangular garden which is 10 yd wide and 17 yd long.

Answer: 170 yd^2

To divide a measurement by an ordinary number, first divide the largest unit by the number. If there is a remainder, add it to the next unit and divide again.

EXAMPLE 13 DIVIDING BY A NON-MEASUREMENT NUMBER

Divide 17 lb 6 oz by 3.

First divide 17 lb by 3.

$$\begin{array}{r} 5 \text{ lb} \\ 3\overline{)17 \text{ lb}} \\ \underline{15}\phantom{\text{ lb}} \\ 2 \text{ lb} \end{array}$$

Thus, 17 lb $\div$ 3 = 5 lb with a remainder of 2 lb. Change 2 lb to ounces.

$$2 \text{ lb} = 2(16) \text{ oz} = 32 \text{ oz}$$

Add the 6 oz from the original measurement to 32 oz.

$$32 \text{ oz} + 6 \text{ oz} = 38 \text{ oz}$$

Now divide 38 oz by 3.

$$\begin{array}{r} 12 \text{ oz} \\ 3\overline{)38 \text{ oz}} \\ \underline{3}\phantom{8 \text{ oz}} \\ 8\phantom{\text{ oz}} \\ \underline{6}\phantom{\text{ oz}} \\ 2 \text{ oz} \end{array}$$

The answer to this division is 12 oz with remainder 2 or $12\frac{2}{3}$ oz. The final answer to the problem (17 lb 6 oz) $\div$ 3 is

$$5 \text{ lb } 12\frac{2}{3} \text{ oz}.$$

PRACTICE EXERCISE 13

Divide 42 lb 13 oz by 5.

Answer: 8 lb 9 oz

Some problems involve division of a measurement by a measurement. Suppose you want to know how many shirts costing \$16 each can be bought for \$48.

$$\frac{48 \text{ dollars}}{16 \text{ dollars}} = \frac{3 \times \cancel{16 \text{ dollars}}}{\cancel{16 \text{ dollars}}} = 3$$

The dollars cancel and the answer is the whole number (with no measurement unit) 3. You can buy 3 shirts for \$48.

EXAMPLE 14 Dividing Measurement Units

How many 4-ft shelves can be cut from a board 16 ft long?

Divide 16 ft by 4 ft.

$$\frac{16 \text{ ft}}{4 \text{ ft}} = \frac{4 \times \cancel{4 \text{ ft}}}{\cancel{4 \text{ ft}}} = 4$$

Thus, 4 shelves can be cut from the board.

Practice Exercise 14

One uniform can be made from 3 yd of material. How many uniforms can be made from 51 yd of the material?

Answer: 17 uniforms

0.7 EXERCISES A

Study the lists of equal units before completing the following.

1. 5 ft = ________ in

2. 72 in = ________ ft

3. 72 in = ________ yd

4. 21 da = ________ wk

5. 64 oz = ________ lb

6. 36 hr = ________ da

7. 72 fl oz = ________ pt

8. 4500 lb = ________ ton

9. 6 wk = ________ da

10. 3 mi = ________ ft

11. 18 qt = ________ gal

12. 16 mo = ________ yr

13. 1 hr = ________ sec

14. 5400 sec = ________ hr

15. 96 in = ________ yd

16. 7 yd = ________ in

17. 3 ton = ________ oz

18. 96 fl oz = ________ qt

19. 2 mi = ________ in

20. 3 wk = ________ min

21. 7200 sec = ________ da

22. $15\frac{\text{mi}}{\text{hr}}$ = ________ $\frac{\text{ft}}{\text{sec}}$

23. $88\frac{\text{ft}}{\text{sec}}$ = ________ $\frac{\text{mi}}{\text{hr}}$

24. $5\frac{\text{cents}}{\text{in}}$ = ________ $\frac{\text{dollars}}{\text{yd}}$

25. $2\frac{\text{gal}}{\text{min}}$ = ________ $\frac{\text{qt}}{\text{sec}}$

26. $200\frac{\text{lb}}{\text{da}}$ = ________ $\frac{\text{ton}}{\text{wk}}$

27. $32\frac{\text{oz}}{\text{hr}}$ = ________ $\frac{\text{lb}}{\text{min}}$

Solve.

28. George buys 30 in of material which costs \$2.16 per yard. How much does he have to pay?

29. The speed limit is 55 miles per hour and Maria is driving at 5280 feet per minute. Is she driving over the speed limit? If so, by how much?

30. Richard is filling a tank at the rate of 4 pints per minute. How many gallons per hour is this?

31. A farmer needs to buy 3 miles of wire. What is the total cost if the wire costs 20¢ per foot?

Simplify.

32. 83 min

33. 3 hr 92 min

34. 5 yd 8 ft

35. 8 gal 17 qt

36. 14 lb 73 oz

37. 4 yd 7 ft 23 in

Perform the indicated operations and simplify.

38. 8 lb 3 oz + 4 lb 10 oz

39. 5 hr 52 min + 13 hr 41 min

40. 3 gal 3 qt + 4 gal 2 qt

41. 6 yd 2 ft 11 in + 7 yd 1 ft 9 in

42. 7 yd 2 ft − 3 yd 1 ft

43. 8 hr 30 min − 2 hr 26 min

44. 8 gal 1 qt − 4 gal 3 qt

45. 5 da 3 hr 8 min − 4 da 6 hr 40 min

46. 9 yd 2 ft 3 in − 2 yd 2 ft 10 in

47. 3 × (5 hr 16 min)

48. 5 × (4 gal 3 qt)

49. 4 × (2 yd 2 ft 11 in)

50. 6 × (2 hr 18 min 14 sec)

51. 3 in × 12 in

52. 7 yd × 12 yd

53. (5 lb 12 oz) ÷ 2

54. (7 hr 33 min) ÷ 3

55. 18 ft ÷ 3 ft

Solve.

56. How many $6 ski caps can be bought for $24?

57. A chemist has 5 gal 3 qt of one liquid to be mixed with 4 gal 2 qt of another. How much will he have after he mixes them?

58. Henry took 6 gal 2 qt of water on a camping trip. After using 2 gal 4 qt, how much water did he have left?

59. If it takes 1 lb 12 oz of flour to make one loaf of bread, how much will it take to make 7 loaves?

60. Find the area of a rectangle which is 16 inches long and 5 inches wide.

61. If it takes 1 lb 12 oz of flour to make one loaf of bread, how much will it take to make a loaf one half the size?

FOR REVIEW

Find the mean, median, and mode of the numbers given in Exercises 62–63.

62. 8, 15, 15, 22

63. 2, 4, 10, 3, 1, 8, 7

64. While typing a manuscript last week, Katie McCormick typed 23 pages on Monday, 8 pages on Tuesday, 31 pages on Wednesday, 16 pages on Thursday, and 12 pages on Friday. What was the average number of pages that she typed per day during that week?

ANSWERS: 1. 60 2. 6 3. 2 4. 3 5. 4 6. 1.5 7. 4.5 8. 2.25 9. 42 10. 15,840 11. 4.5 12. $1\frac{1}{3}$ 13. 3600 14. 1.5 15. $2\frac{2}{3}$ 16. 252 17. 96,000 18. 3 19. 126,720 20. 30,240 21. $\frac{1}{12}$ 22. 22 23. 60 24. 1.80 25. $\frac{2}{15}$ 26. 0.7 27. $\frac{1}{30}$ 28. \$1.80 29. yes; $5\frac{\text{mi}}{\text{hr}}$ 30. $30\frac{\text{gal}}{\text{hr}}$ 31. \$3168 32. 1 hr 23 min 33. 4 hr 32 min 34. 7 yd 2 ft 35. 12 gal 1 qt 36. 18 lb 9 oz 37. 6 yd 2 ft 11 in 38. 12 lb 13 oz 39. 19 hr 33 min 40. 8 gal 1 qt 41. 14 yd 1 ft 8 in 42. 4 yd 1 ft 43. 6 hr 4 min 44. 3 gal 2 qt 45. 20 hr 28 min 46. 6 yd 2 ft 5 in 47. 15 hr 48 min 48. 23 gal 3 qt 49. 11 yd 2 ft 8 in 50. 13 hr 49 min 24 sec 51. 36 sq in 52. 84 sq yd 53. 2 lb 14 oz 54. 2 hr 31 min 55. 6 56. 4 57. 10 gal 1 qt 58. 3 gal 2 qt 59. 12 lb 4 oz 60. 80 sq in 61. 14 oz 62. mean = 15; median = 15; mode = 15 63. mean = 5; median = 4; no mode 64. 18 pages per day

0.7 EXERCISES B

Complete the change in units.

1. 9 ft = ________ in

2. 132 in = ________ ft

3. 132 in = ________ yd

4. 63 da = ________ wk

5. 88 oz = ________ lb

6. 96 hr = ________ da

7. 108 fl oz = ________ pt

8. 15,000 lb = ________ ton

9. 13 wk = ________ da

10. 10 mi = ________ ft

11. 27 qt = ________ gal

12. 37 mo = ________ yr

13. $\frac{1}{2}$ hr = ________ sec

14. 400 sec = ________ hr

15. 144 in = ________ yd

16. 12 yd = ________ in

17. 0.1 ton = ________ oz

18. 160 fl oz = ________ qt

19. 0.4 mi = ________ in

20. $\frac{1}{2}$ wk = ________ min

21. 86,400 sec = ________ da

22. $75\frac{\text{mi}}{\text{hr}}$ = ________ $\frac{\text{ft}}{\text{sec}}$

23. $11\frac{\text{ft}}{\text{sec}}$ = ________ $\frac{\text{mi}}{\text{hr}}$

24. $20\frac{\text{cents}}{\text{in}}$ = ________ $\frac{\text{dollars}}{\text{yd}}$

25. $30\frac{\text{gal}}{\text{min}}$ = ________ $\frac{\text{qt}}{\text{sec}}$

26. $700\frac{\text{lb}}{\text{da}}$ = ________ $\frac{\text{ton}}{\text{wk}}$

27. $640\frac{\text{oz}}{\text{hr}}$ = ________ $\frac{\text{lb}}{\text{min}}$

Solve.

28. Sirloin steak is sale-priced at \$2.79 per pound. How much would the Moores have to pay for 35 lb?

29. Burt is driving at the rate of 55 feet per second. If he is driving where the speed limit is 40 miles per hour, how does his speed compare to the limit?

30. A water pump will deliver 210 gallons per minute. How many quarts per second is this?

31. Electrical wiring costs 5¢ per inch. How much would 25 yards of this wire cost?

Simplify.

32. 137 min

33. 7 hr 72 min

34. 4 yd 19 ft

35. 2 gal 25 qt

36. 6 lb 105 oz

37. 5 yd 14 ft 42 in

Perform the indicated operations and simplify.

38. $\begin{array}{r} 9 \text{ lb } 11 \text{ oz} \\ +\ 5 \text{ lb } \ \ 1 \text{ oz} \\ \hline \end{array}$

39. $\begin{array}{r} 12 \text{ hr } 39 \text{ min} \\ +\ 10 \text{ hr } 32 \text{ min} \\ \hline \end{array}$

40. $\begin{array}{r} 9 \text{ gal } 3 \text{ qt} \\ +\ 7 \text{ gal } 3 \text{ qt} \\ \hline \end{array}$

41. $\begin{array}{r} 12 \text{ yd } 1 \text{ ft } \ \ 5 \text{ in} \\ +\ 18 \text{ yd } 2 \text{ ft } 10 \text{ in} \\ \hline \end{array}$

42. $\begin{array}{r} 4 \text{ yd } 2 \text{ ft} \\ -\ 1 \text{ yd } 1 \text{ ft} \\ \hline \end{array}$

43. $\begin{array}{r} 19 \text{ hr } 49 \text{ min} \\ -\ 15 \text{ hr } 32 \text{ min} \\ \hline \end{array}$

44. $\begin{array}{r} 17 \text{ gal } 1 \text{ qt} \\ -\ 13 \text{ gal } 2 \text{ qt} \\ \hline \end{array}$

45. $\begin{array}{r} 22 \text{ da } \ \ 2 \text{ hr } 36 \text{ min} \\ -\ 21 \text{ da } 22 \text{ hr } 48 \text{ min} \\ \hline \end{array}$

46. $\begin{array}{r} 6 \text{ yd } 1 \text{ ft } 1 \text{ in} \\ -\ 3 \text{ yd } 2 \text{ ft } 2 \text{ in} \\ \hline \end{array}$

47. 4 × (3 hr 10 min)

48. 10 × (3 gal 2 qt)

49. 6 × (7 yd 1 ft 8 in)

50. 3 × (5 hr 40 min 25 sec)

51. 14 in × 22 in

52. 5 yd × 9 yd

53. (7 lb 8 oz) ÷ 3

54. (10 hr 20 min) ÷ 4

55. 33 ft ÷ 11 ft

Solve.

56. Steak costs \$3 per pound. How many pounds can Ralph buy with \$18?

57. For a company picnic, Ward brought 7 gal 2 qt of iced tea. If Sarah brought 10 gal 3 qt, how much tea did they have?

58. If the people at the picnic in Exercise 57 drank 15 gal 3 qt of the tea, how much was left?

59. Billy Joe has a chili recipe that takes 2 lb 6 oz of ground beef. How much meat should he buy to make 5 times his recipe?

60. Vicki needs to paint a wall which is 8 feet high and 12 feet wide. How many square feet of wall is this to paint?

61. A cookie recipe takes 1 lb 2 oz of flour. How much flour would be used if only one-third of the recipe is to be made?

FOR REVIEW

Find the mean, median, and mode of the numbers given in Exercises 62–63.

62. 60, 12, 17, 23

63. 6, 3, 9, 7, 1, 2, 7

64. Greg Odjakjian had three separate investment accounts last year that earned \$4235, \$6219, and \$3214, respectively. What was the average return on his investments?

0.7 EXERCISES C

Use a calculator to find the following changes in units. Give answers correct to the nearest tenth.

1. 627 in = ________ yd

2. $55 \dfrac{\text{mi}}{\text{hr}} = \_\_\_\_ \dfrac{\text{ft}}{\text{sec}}$
[Answer: 80.7]

3. 143 da = ________ yr

4. Mount McKinley in Alaska is the highest mountain in the United States at 20,320 feet above sea level. Give the height in miles.
[Answer: 3.8]

5. If material costs $2.70 per yard, how much does 8 feet of the material cost?
[Answer: $7.20]

6. Wire costs $13.85 per yard. How much does the wire cost per foot (to the nearest cent)?

0.8 THE METRIC SYSTEM AND CONVERSION

STUDENT GUIDEPOSTS

1 Length in Meters	4 Units of Measure
2 Volume in Liters	5 Converting Metric Units
3 Weight in Grams	6 Conversion Between Systems

Most of the world uses the metric system of measurement. It is a decimal system, and changes in units can be made by moving the decimal point. Thus, it is much easier to convert from one unit to another in the metric system than it is in the English system (just as easy as changing $1.27 to 127¢).

1 LENGTH IN METERS

The standard of length in the metric system is the **meter.** The opening in a normal doorway is about 2 meters high. Thus, a meter is about half the height of an opening in a doorway. A twin bed is about a meter wide and most newborn babies are about $\frac{1}{2}$ meter tall.

The actual length of the scale in Figure 0.9 is 0.1 meter. The distance between each mark is 0.01 meter.

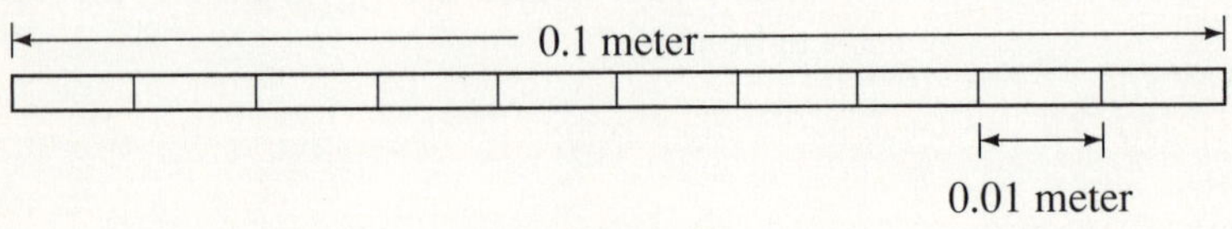

Figure 0.9

2 VOLUME IN LITERS

If we make a box as shown in Figure 0.10 (not actual size), the box holds one **liter.** This is the standard unit for capacity in the metric system. If you are thirsty on a summer day, you can probably drink a liter of liquid. It is a little less than three cans of soft drink.

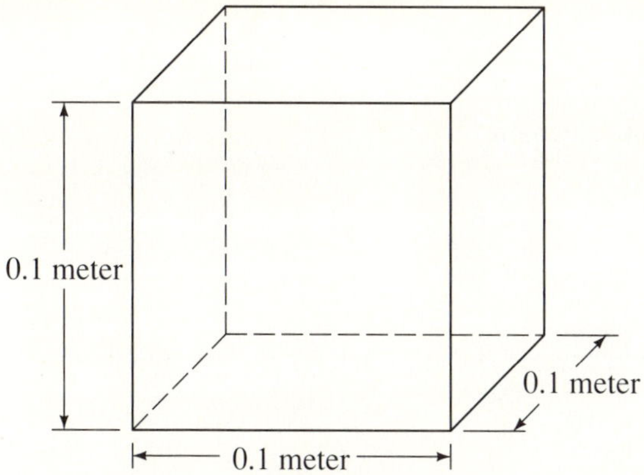

Figure 0.10

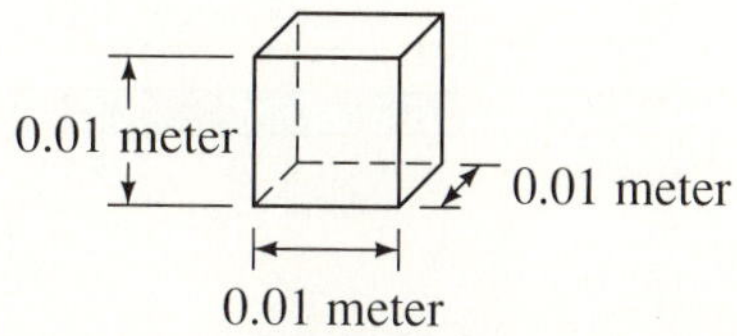

Figure 0.11

3 WEIGHT IN GRAMS

The standard for mass or weight at sea level is the **gram.** Figure 0.11 shows a cube of side 0.01 meter. If this cube were filled with water, the weight of the water would be 1 gram. One large paper clip (or two small ones) weighs about one gram and a large steak might weigh 1000 grams.

4 UNITS OF MEASURE

As with the English system, the various size units such as 1000 grams have names, and the naming is standard. For example, the names for 1000 meters, 1000 liters, and 1000 grams all start with the same prefix, *kilo*. Thus,

1000 meters = 1 *kilo*meter,
1000 liters = 1 *kilo*liter,
1000 grams = 1 *kilo*gram.

The following table lists the prefixes and symbols used in the metric system.

Number of units	1000	100	10	**1**	$\frac{1}{10}$ or 0.1	$\frac{1}{100}$ or 0.01	$\frac{1}{1000}$ or 0.001
Prefix	kilo	hecto	deka	**unit**	deci	centi	milli
Length	km	hm	dam	**meter(m)**	dm	cm	mm
Volume	kl	hl	dal	**liter(L)**	dl	cl	ml
Weight	kg	hg	dag	**gram(g)**	dg	cg	mg

The different size units are used in different applications. For example, the size of a room is given in meters, the distance between Miami and Washington, D.C., in kilometers, and bolt sizes in millimeters. In Figure 0.12, the paper clip and bolt are shown actual size.

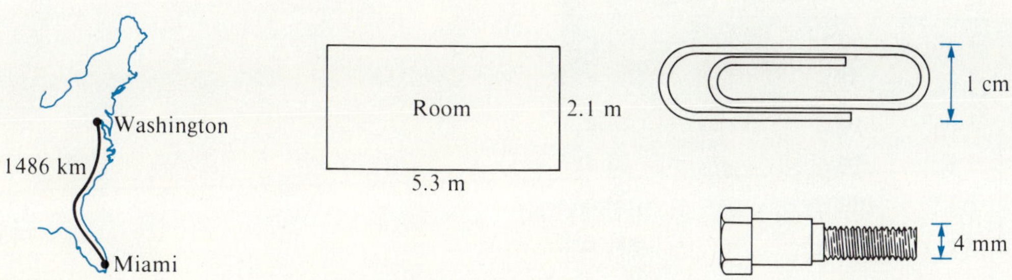

Figure 0.12

In Figure 0.13 (drawn to actual size), Figure 0.9 has been repeated using the new names we have learned and a few more details. This should give you a good idea of the sizes of some units of length and their relationship. A meter is 10 times the total length of the scale in Figure 0.13. A kilometer is about the length of five city blocks.

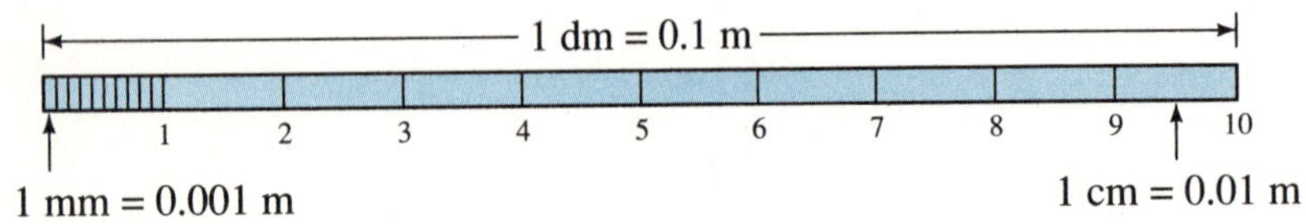

Figure 0.13

The most commonly used measures of length are the kilometer, meter, centimeter, and millimeter. These same prefixes—kilo, centi, and milli—are used for volume and weight also.

In volume units, the liter is used to measure gasoline and beverages. The kiloliter measures the capacity of large tanks, and the milliliter measures liquids in a laboratory.

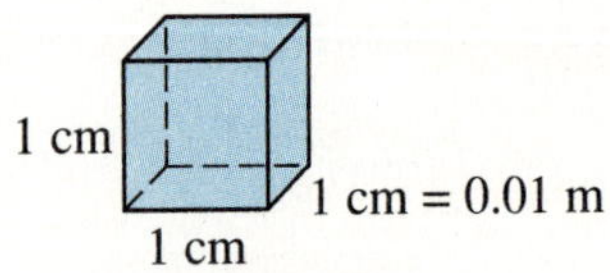

Figure 0.14

The cube in Figure 0.14, which is 1 cm on a side, has volume 1 ml and this much water weighs 1 g. Thus, the gram is a small unit of measurement. It is used in the laboratory where an even smaller unit, the milligram, is also needed. The unit that would suit many common weight applications is the kilogram. As has been mentioned, a large steak weighs about 1 kg (1000 g).

5 CONVERTING METRIC UNITS

Now that we have some idea of the size of metric units, we turn to conversion problems. Instead of using decimal conversion units such as 1 mm = 0.001 m, it will be easier to write all unit fractions in terms of 10, 100, or 1000. The following table gives the equal units.

Length	*Volume*	*Weight*
1 km = 1000 m	1 kl = 1000 L	1 kg = 1000 g
1 hm = 100 m	1 hl = 100 L	1 hg = 100 g
1 dam = 10 m	1 dal = 10 L	1 dag = 10 g
1 m = 10 dm	1 L = 10 dl	1 g = 10 dg
1 m = 100 cm	1 L = 100 cl	1 g = 100 cg
1 m = 1000 mm	1 L = 1000 ml	1 g = 1000 mg

HINT

Remember that multiplying and dividing by powers of ten can be done by moving the decimal point. Consider these examples.

$16.2 \times 100 = 1620$ Move two places to the right

$16.2 \div 100 = 0.162$ Move two places to the left

EXAMPLE 1 Converting by Using Unit Fractions

(a) 5.2 m = _____ cm.

Since 1 m = 100 cm, the unit fraction is $\frac{100 \text{ cm}}{1 \text{ m}}$.

$$\begin{aligned} 5.2 \text{ m} &= 5.2 \text{ m} \times 1 \\ &= \frac{5.2 \cancel{\text{m}}}{1} \times \frac{100 \text{ cm}}{1 \cancel{\text{m}}} \\ &= 5.2 \times 100 \text{ cm} \\ &= 520 \text{ cm} \end{aligned}$$

Notice that 5.2 is multiplied by 100, which moves the decimal point 2 places to the right. That is,

$$5.2 \text{ m} = 520. \text{ cm.}$$

2

(b) 362 ml = _____ dl.

The unit fractions are $\frac{1 \text{ L}}{1000 \text{ ml}}$ and $\frac{10 \text{ dl}}{1 \text{ L}}$.

$$\begin{aligned} 362 \text{ ml} &= \frac{362 \cancel{\text{ml}}}{1} \times \frac{1 \cancel{\text{L}}}{1000 \cancel{\text{ml}}} \times \frac{10 \text{ dl}}{1 \cancel{\text{L}}} \\ &= 362 \times \frac{1}{1000} \times \frac{10}{1} \text{ dl} \\ &= 362 \times \frac{10}{1000} \text{ dl} \\ &= 362 \times \frac{1}{100} \text{ dl} \\ &= 3.62 \text{ dl} \end{aligned}$$

Notice that 362 is divided by 100, which moves the decimal point 2 places to the left.

Practice Exercise 1

(a) 13.25 m = _____ cm.

(b) 64,800 ml = _____ dl.

Answers: (a) 1325 (b) 648

In Example 1(b), multiplying by one unit fraction moved the decimal point 3 places to the left and multiplying by the other moved it 1 place to the right. The same change of units could have been made by moving the decimal point 2 places to the left. This method of conversion is shown in the next example.

EXAMPLE 2 Converting by Moving Decimal Places

(a) 362 ml = _____ dl.

First show all the units in a diagram.

1000 L	100 L	10 L	1 L	0.1 L	0.01 L	0.001 L
(kl)	**(hl)**	**(dal)**		**(dl)**	**(cl)**	**(ml)**
				2	1	

Start at **ml** and make 2 moves to the *left* to get to **dl.** Thus, to change ml to dl move the decimal point 2 places to the *left*.

$$362 \text{ ml} = 3.\underset{2}{62} \text{ dl} = 3.62 \text{ dl}$$

(b) 0.0256 hg = _____ cg.

1000 g	100 g	10 g	1 g	0.1 g	0.01 g	0.001 g
(kg)	**(hg)**	**(dag)**		**(dg)**	**(cg)**	**(mg)**
	1	2	3	4		

Start at hg and move 4 times to the *right* to get to cg. Thus, to change hg to cg, move the decimal point 4 places to the *right*.

$$0.0256 \text{ hg} = \underset{4}{00256.} \text{ cg} = 256 \text{ cg}$$

(c) 3280 mm = _____ hm.

1000 m	100 m	10 m	1 m	0.1 m	0.01 m	0.001 m
(km)	**(hm)**	**(dam)**		**(dm)**	**(cm)**	**(mm)**
	5	4	3	2	1	

Start at mm and make 5 moves to the *left* to get to hm. Thus, to change mm to hm, move the decimal point 5 places to the *left*.

$$3280 \text{ mm} = 0.\underset{5}{03280} \text{ hm}$$

Practice Exercise 2

(a) 0.057 ml = _____ dl.

(b) 0.000882 hg = _____ cg.

(c) 426,000 mm = _____ hm.

Answers: (a) 0.00057 (b) 8.82 (c) 4.26

6 CONVERSION BETWEEN SYSTEMS

If the United States ever changes to the metric system, it will not be necessary to change units from one system to another. But while both systems are in use it is important to know the relationship between the two.

In the table below are some useful conversion units of measure. Notice that these units are all approximations.

Length, Volume, and Weight

	Metric to English	*English to Metric*
Length units	1 km ≈ 0.621 mi	1 mi ≈ 1.61 km
	1 m ≈ 39.37 in	1 in ≈ 2.54 cm
Volume units	1 L ≈ 1.06 qt	1 qt ≈ 0.946 L
Weight units	1 kg ≈ 2.20 lb	1 lb ≈ 454 g

Since we are familiar with the English units, these conversions will help us get a better idea of the size of the metric units. For example, on a quarter-mile track, a kilometer is about 2.5 times around. If two towns are 10 miles apart, they are about 16 km apart. A meter (39.37 in) is a little more than a yard (36 in). Figure 0.15 shows the relationship between centimeters and inches.

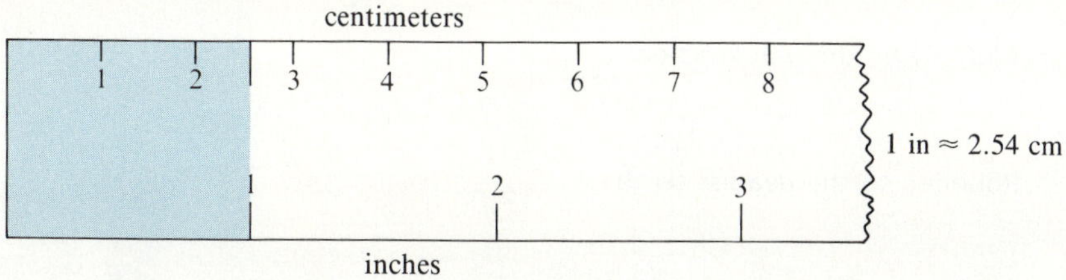

Figure 0.15

There is a little more milk in a liter container than there is in a quart, as shown in Figure 0.16.

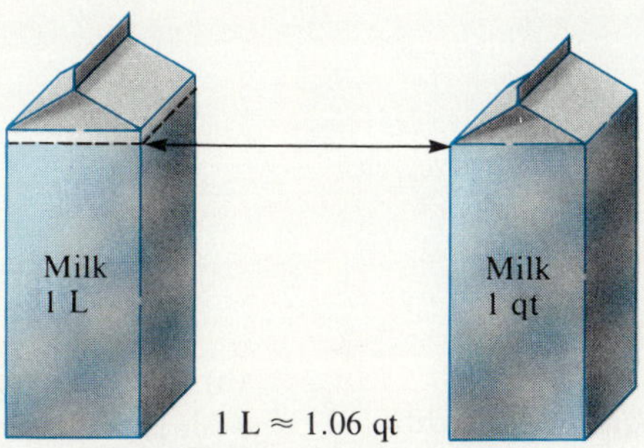

Figure 0.16

We now know that the 1-kg steak weighs 2.2 lb. Figure 0.17 shows the relative sizes of 1 lb and 1 kg of the same material.

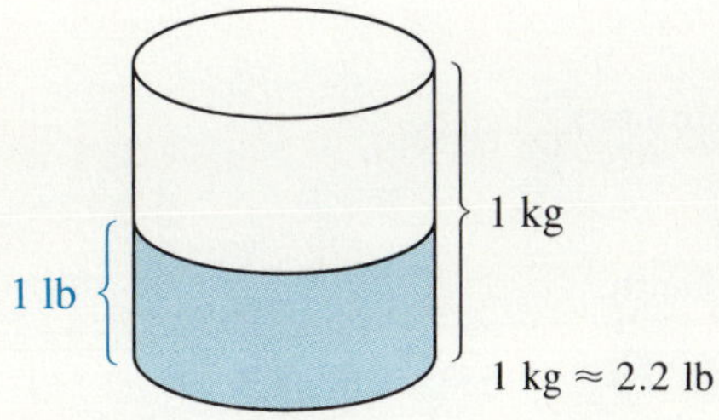

Figure 0.17

The following examples show conversions between the two systems. In most of these problems a calculator is helpful.

EXAMPLE 3 Converting Measures of Length

(a) 20 km = _____ mi.
Since 0.621 mi ≈ 1 km, the unit fraction is $\frac{0.621 \text{ mi}}{1 \text{ km}}$.

$$20 \text{ km} \approx \frac{20 \cancel{\text{km}}}{1} \times \frac{0.621 \text{ mi}}{1 \cancel{\text{km}}}$$ Multiply by the unit fraction

$$= 20 \times (0.621) \text{ mi}$$

$$\approx 12.4 \text{ mi}$$ Rounded to the nearest tenth

(b) 35 mi = _____ km.
Multiply by the unit fraction $\frac{1.61 \text{ km}}{1 \text{ mi}}$.

$$35 \text{ mi} \approx \frac{35 \cancel{\text{mi}}}{1} \times \frac{1.61 \text{ km}}{1 \cancel{\text{mi}}}$$ Multiply by the unit fraction

$$= 35 \times 1.61 \text{ km}$$

$$\approx 56.4 \text{ km}$$ Rounded to the nearest tenth

Practice Exercise 3

(a) 44 km = _____ mi.

(b) 120 mi = _____ km.

Answers: (a) 27.3 (b) 193.2

EXAMPLE 4 Converting Measures of Volume

(a) 520 ml = _____ qt.
Since the unit fraction that we know is $\frac{1.06 \text{ qt}}{1 \text{ L}}$, first change 520 ml to liters.

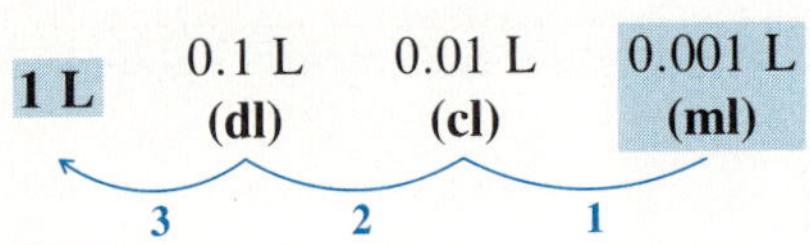

$$520 \text{ ml} = 0.520 \text{ L}$$

$$0.520 \text{ L} \approx \frac{0.52 \cancel{\text{L}}}{1} \times \frac{1.06 \text{ qt}}{1 \cancel{\text{L}}}$$

$$= (0.52)(1.06) \text{ qt}$$

$$\approx 0.55 \text{ qt}$$ Rounded to the nearest hundredth

(b) The gasoline tank on Roberto's car holds 12.6 gal. How many liters will it hold?

$$12.6 \text{ gal} = \frac{12.6 \cancel{\text{gal}}}{1} \times \frac{4 \text{ qt}}{1 \cancel{\text{gal}}}$$ Change 12.6 gal to quarts

$$= (12.6)(4) \text{ qt}$$

$$= 50.4 \text{ qt}$$

$$50.4 \text{ qt} \approx \frac{50.4 \cancel{\text{qt}}}{1} \times \frac{0.946 \text{ L}}{1 \cancel{\text{qt}}}$$ Change quarts to liters

$$= (50.4)(0.946) \text{ L}$$

$$\approx 47.7 \text{ L}$$ To the nearest tenth

Practice Exercise 4

(a) 1020 ml = _____ qt.

(b) Deb has a foreign car with a 40-L gas tank. How many gallons does the tank hold?

Answers: (a) 1.08 (b) 10.6 gal

EXAMPLE 5 Converting Measures of Weight

(a) 68.2 hg = _____ lb.

The unit fraction is $\frac{2.20 \text{ lb}}{1 \text{ kg}}$. First, change 62.8 hg to kilograms.

1000 g **(1 kg)**	100 g **(1 hg)**

1

68.2 hg = 6.82 kg

$$6.82 \text{ kg} \approx \frac{6.82 \text{ kg}}{1} \times \frac{2.20 \text{ lb}}{1 \text{ kg}}$$ Change kilograms to pounds

$$= (6.82)(2.20) \text{ lb}$$

$$= 15.0 \text{ lb}$$ To the nearest tenth

(b) A football team's starting fullback weighs 200 lb. How many kilograms does he weigh?

$$200 \text{ lb} \approx \frac{200 \text{ lb}}{1} \times \frac{454 \text{ g}}{1 \text{ lb}}$$

$$= (200)(454) \text{ g}$$

$$= 90{,}800 \text{ g}$$

Change 90,800 g to kilograms.

1000 g **(1 kg)**	100 g **(1 hg)**	10 g **(1 dag)**	**1 g**

3 2 1

90,800 g = 90.8 kg

Practice Exercise 5

(a) 2240 dg = _____ lb.

(b) A foreign basketball player trying to break into the NBA reported his weight as 97 kg. How many pounds does he weigh?

Answers: (a) 0.49 (b) 213 lb

Figure 0.18 shows a comparison of some common speed limits. Approximations are to the nearest kilometer per hour. A speed of $55\frac{\text{mi}}{\text{hr}}$ is about $89\frac{\text{km}}{\text{hr}}$.

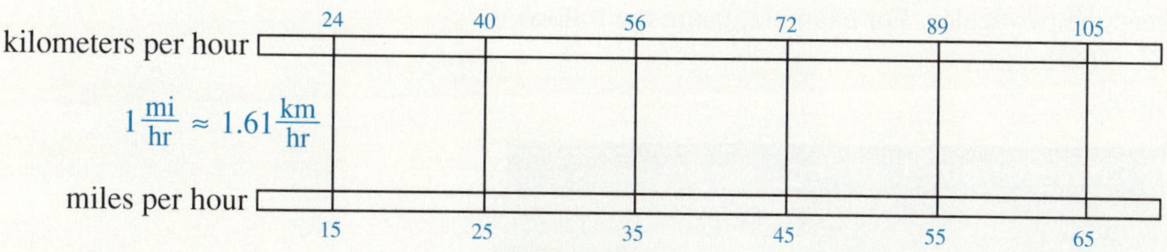

Figure 0.18

EXAMPLE 6 Conversion of Speed

Joe is driving at a rate of $95\frac{\text{km}}{\text{hr}}$. What is his speed in miles per hour?

$$95\frac{\text{km}}{\text{hr}} \approx \frac{95 \text{ km}}{\text{hr}} \times \frac{0.621 \text{ mi}}{1 \text{ km}}$$

$$\approx 59\frac{\text{mi}}{\text{hr}}$$ Rounded to the nearest mile per hour

Practice Exercise 6

Vera is driving $50\frac{\text{km}}{\text{hr}}$ in a $30\frac{\text{mi}}{\text{hr}}$ zone. Approximately how many miles per hour is she traveling over the speed limit?

Answer: She is about $1\frac{\text{mi}}{\text{hr}}$ over the speed limit.

The temperature scale used in the metric system is the **Celsius** (°C) scale. Water freezes at 0°C (32°F) and boils at 100°C (212°F). Figure 0.19 shows a comparison between degrees **Fahrenheit** (°F) and degrees Celsius. (In the past Celsius was called *centigrade*.)

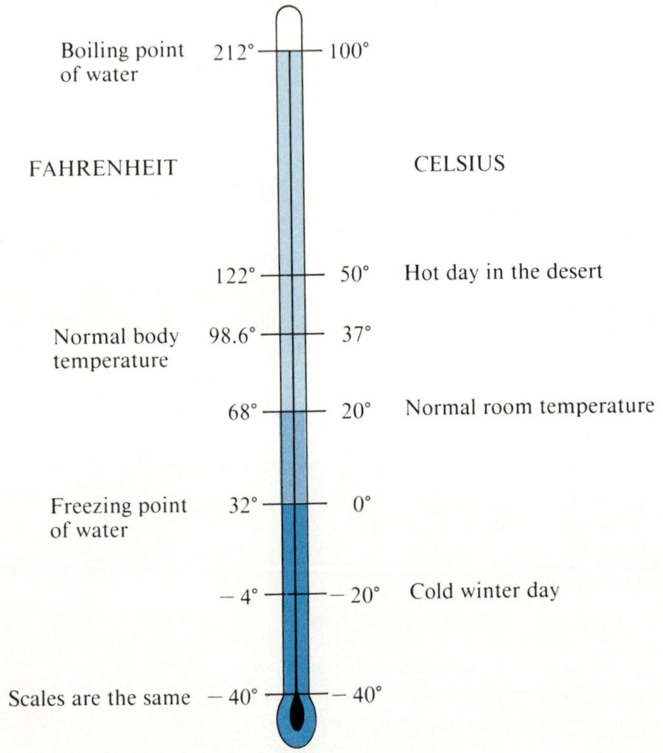

Figure 0.19

The equal temperatures on the scale in Figure 0.19 can be found once we know the relationship between the two scales. For example, using the following rules we can find that 20°C = 68°F.

To Convert from °C to °F

Multiply C, the temperature in °C, by $\frac{9}{5} = 1.8$ and add 32.

$$F = \frac{9}{5}C + 32 \quad \text{or} \quad F = 1.8C + 32$$

EXAMPLE 7 Converting Celsius to Fahrenheit

20°C = ____ °F.

$$F = \frac{9}{5}C + 32$$

Practice Exercise 7

35°C = ____ °F.

$$F = \frac{9}{5}\mathbf{20} + 32 \qquad C = 20$$

$$= \frac{9 \cdot 20}{5} + 32 \qquad \text{Multiply before adding}$$

$$= \frac{9 \cdot 4 \cdot 5}{5} + 32$$

$$= 36 + 32 = 68$$

Thus, 20°C is 68°F.

Answer: 95

To Convert from °F to °C

First subtract 32 from F, the temperature in °F, then multiply by $\frac{5}{9}$.

$$C = \frac{5}{9}(F - 32)$$

EXAMPLE 8 CONVERTING FAHRENHEIT TO CELSIUS

122°F = _____ °C.

$$C = \frac{5}{9}(F - 32)$$

$$= \frac{5}{9}(\mathbf{122} - 32) \qquad F = 122$$

$$= \frac{5}{9}(90) \qquad \text{Subtract inside parentheses first}$$

$$= \frac{5 \cdot \cancel{9} \cdot 10}{\cancel{9}} = 50 \qquad \text{Then multiply}$$

Thus, 122°F is 50°C.

PRACTICE EXERCISE 8

86°F = _____ °C.

Answer: 30

0.8 EXERCISES A

Answer true *or* false. *If the statement is false, tell why.*

1. The system of measurement used in most of the world is the metric system.
2. The opening in a normal doorway is about 10 meters high.
3. A liter is the space in a box which is 0.1 meter or one decimeter on a side.
4. The weight of water that can be put in a box which is 0.01 meter (1 centimeter) on a side is one gram.
5. A large steak might weigh about one gram.
6. To change units of length in the metric system, just move the decimal point.

Complete the following by moving the decimal point as shown in the examples.

7. 1 km = ________ m

8. 1 hl = ________ L

9. 1 dag = ________ g

10. 1 m = ________ dm

11. 1 mg = ________ g

12. 1 g = ________ mg

13. 62 g = ________ kg

14. 3.2 cm = ________ m

15. 16 dal = ________ L

16. 10.3 cm = ________ hm

17. 0.00721 kg = ________ dag

18. 63,000 ml = ________ kl

19. 1700 dm = ________ hm

20. 16.25 cg = ________ mg

21. 0.000212 hl = ________ ml

22. 92.6 mm = ________ cm

23. 4728 dg = ________ hg

24. 5,000,000 cl = ________ dal

25. 0.0052 hm = ________ mm

26. 0.07 dag = ________ mg

27. 3200 dl = ________ kl

28. 0.0953 km = ________ dm

29. 42 kg = ________ hg

30. 7.2 hl = ________ cl

31. $30\,\frac{\text{cm}}{\text{sec}} = \_\_\_\_\_\_\_\_ \frac{\text{m}}{\text{hr}}$

32. $20\,\frac{\text{cg}}{\text{da}} = \_\_\_\_\_\_\_\_ \frac{\text{g}}{\text{wk}}$

33. $360\,\frac{\text{kl}}{\text{hr}} = \_\_\_\_\_\_\_\_ \frac{\text{ml}}{\text{min}}$

Perform the indicated operations.

34. $\begin{array}{r} 2.05 \text{ m} \\ +\ 7.97 \text{ m} \\ \hline \end{array}$

35. $\begin{array}{r} 382.6 \text{ ml} \\ +\ \ 23.8 \text{ ml} \\ \hline \end{array}$

36. $\begin{array}{r} 983 \text{ kg} \\ -\ 724 \text{ kg} \\ \hline \end{array}$

37. $\begin{array}{r} 75.36 \text{ cm} \\ -\ 28.71 \text{ cm} \\ \hline \end{array}$

38. $\begin{array}{l} 16.3 \text{ m} \\ \times\ 5 \\ \hline \end{array}$

39. $\begin{array}{l} 16.3 \text{ m} \\ \times\ 5 \text{ m} \\ \hline \end{array}$

Solve.

40. How many 22-ml samples can be obtained from 330 ml of a chemical solution?

41. If steak costs \$6 per kilogram, how much would 600 grams of steak cost?

42. Gabi is traveling at $60\frac{\text{km}}{\text{hr}}$. How many meters per minute is this?

43. A rancher needs 10 km of wire. How much will it cost him if the wire is 40¢ per meter?

44. Which is longer, 1 mi or 1 km?

45. Which is longer, 3 cm or 1 in?

46. Which is more, 1 qt or 1 L?

47. Which is more, 1 kg or 3 lb?

Complete.

48. 16 km = ________ mi

49. 128 mi = ________ km

50. 2.6 m = ________ in

51. 72 in = ________ cm

52. 3 km = ________ ft

53. 3.2 ft = ________ cm

54. $80\frac{\text{km}}{\text{hr}} = \_\_\_\_\_\_\_\_ \frac{\text{mi}}{\text{hr}}$

55. $20\frac{\text{mi}}{\text{hr}} = \_\_\_\_\_\_\_\_ \frac{\text{km}}{\text{hr}}$

56. 20 qt = ________ L

57. 40 L = ________ gal

58. 50 kg = ________ lb

59. 5 lb = ________ g

60. 0.25 kg = ________ oz

61. 625 lb = ________ kg

62. 40°C = ________ °F

63. 50°F = ________ °C

64. 200°C = ________ °F

65. 200°F = ________ °C

Solve.

66. Patricia bought a French car with the capacity of the gas tank given as 48 L. How many gallons of gasoline will the tank hold?

67. Winston was driving in Mexico where the speed limit was $100\frac{\text{km}}{\text{hr}}$. What is this in miles per hour?

68. The average weight of the starting line of a football team is 258 lb. What is this in kilograms?

69. Would you put your hand in water at 95°C?

FOR REVIEW

Perform the indicated operations and simplify.

70.
```
  8 ft 11 in
+ 1 ft  3 in
```

71.
```
  8 lb 12 oz
- 3 lb 15 oz
```

72.
```
  12 gal 2 qt 1 pt
-  6 gal 3 qt 1 pt
```

73. 5 × (7 yd 2 ft 10 in)

74. (11 hr 40 min) ÷ 5

75. 42 ft ÷ 7 ft

ANSWERS: (Some answers are rounded.) 1. true 2. false (about 2 m) 3. true 4. true 5. false (about one kilogram) 6. true 7. 1000 8. 100 9. 10 10. 10 11. 0.001 12. 1000 13. 0.062 14. 0.032 15. 160 16. 0.00103 17. 0.721 18. 0.063 19. 1.7 20. 162.5 21. 21.2 22. 9.26 23. 4.728 24. 5000 25. 520 26. 700 27. 0.32 28. 953 29. 420 30. 72,000 31. 1080 32. 1.4 33. 6,000,000 34. 10.02 m 35. 406.4 ml 36. 259 kg 37. 46.65 cm 38. 81.5 m 39. 81.5 m^2 ($m \times m = m^2$) 40. 15 41. $3.60 42. $1000\frac{\text{m}}{\text{min}}$ 43. $4000 44. 1 mile 45. 3 cm 46. 1 L 47. 3 lb 48. 9.9 49. 206 50. 102.4 51. 183 52. 9837 53. 97.5 54. 50 55. 32 56. 18.9 57. 10.6 58. 110 59. 2270 60. 8.8 61. 284 62. 104 63. 10 64. 392 65. 93 66. 12.7 gal 67. $62\frac{\text{mi}}{\text{hr}}$ 68. 117 kg 69. no (95°C = 203°F) 70. 10 ft 2 in 71. 4 lb 13 oz 72. 5 gal 3 qt 73. 39 yd 2 ft 2 in 74. 2 hr 20 min 75. 6

0.8 EXERCISES B

Answer true *or* false. *If the statement is false, tell why.*

1. The United States is one of the few countries in the world that uses the English system.

2. The opening in a normal doorway is about 2 meters high.

3. The liter is used to measure gasoline in the metric system.

4. One gram of meat is more than one pound of meat.

5. A large steak might weigh about one kilogram.

6. Units of weight in the metric system can be changed by moving the decimal point.

Complete the following by moving the decimal point as shown in the examples.

7. 40 km = ________ m

8. 6 hl = ________ L

9. 0.05 dag = ________ g

10. 50 m = ________ dm

11. 0.061 mg = ________ g

12. 0.061 g = ________ mg

13. 420 g = ________ kg

14. 2300 cm = ________ m

15. 2.5 dal = ________ L

16. 6320 cm = ________ hm

17. 8.7 kg = ________ dag

18. 493,000 ml = ________ kl

19. 0.75 dm = ________ hm

20. 0.88 cg = ________ mg

21. 5.5 hl = ________ ml

22. 250 mm = ________ cm

23. 98,750 dg = ________ hg

24. 0.44 cl = ________ dal

25. 927 hm = ________ mm

26. 0.000082 dag = ________ mg

27. 200 dl = ________ kl

28. 0.000024 km = ________ dm

29. 0.081 kg = ________ hg

30. 4400 hl = ________ cl

31. $2 \frac{\text{cm}}{\text{sec}} = \underline{\qquad\qquad} \frac{\text{m}}{\text{hr}}$

32. $500 \frac{\text{cg}}{\text{da}} = \underline{\qquad\qquad} \frac{\text{g}}{\text{wk}}$

33. $0.06 \frac{\text{kl}}{\text{hr}} = \underline{\qquad\qquad} \frac{\text{ml}}{\text{min}}$

Perform the indicated operations.

34. $\begin{array}{r} 8.09 \text{ km} \\ + \ 3.66 \text{ km} \\ \hline \end{array}$

35. $\begin{array}{r} 24.89 \text{ dl} \\ + \ 13.77 \text{ dl} \\ \hline \end{array}$

36. $\begin{array}{r} 402 \text{ g} \\ - \ 389 \text{ g} \\ \hline \end{array}$

37. $\begin{array}{r} 4.735 \text{ m} \\ - \ 1.908 \text{ m} \\ \hline \end{array}$

38. $\begin{array}{r} 213.2 \text{ cm} \\ \times \ 8 \\ \hline \end{array}$

39. $\begin{array}{r} 213.2 \text{ cm} \\ \times \ 8 \text{ cm} \\ \hline \end{array}$

Solve.

40. Dr. Warren has 42 g of salt. How many 7-dg samples can she get from her supply?

41. Peaches are marked at $1.20 per kilogram. How much will 4000 g of peaches cost?

42. A tank is being filled at the rate of 25 liters per minute. What is this in kiloliters per hour?

43. A perfume is worth $2.60 per milliliter. How much would 5 liters cost?

44. Which is longer, 1 dm or 1 in?

45. Which is longer, 10 mi or 17 km?

46. Which is more, 1 gal or 4 L?

47. Which is more, 500 g or 1 lb?

Complete.

48. 125 km = ________ mi

49. 48 mi = ________ km

50. 0.056 m = ________ in

51. 9.5 in = ________ cm

52. 0.05 km = ________ ft

53. 0.15 ft = ________ cm

54. $25\frac{\text{km}}{\text{hr}} =$ ________ $\frac{\text{mi}}{\text{hr}}$

55. $100\frac{\text{mi}}{\text{hr}} =$ ________ $\frac{\text{km}}{\text{hr}}$

56. 0.5 qt = ________ L

57. 8.5 L = ________ gal

58. 0.2 kg = ________ lb

59. $\frac{1}{2}$ lb = ________ g

60. 0.0046 kg = ________ oz

61. 85 lb = ________ kg

62. 15°C = ________ °F

63. 77°F = ________ °C

64. 1000°C = ________ °F

65. 1000°F = ________ °C

Solve.

66. How many liters will a 19-gal gas tank hold?

67. Patricia's speedometer reads in kilometers per hour. How fast can she legally drive where the speed limit is $45\frac{\text{mi}}{\text{hr}}$?

68. Willy must have his weight down to 82 kg to compete in Europe. What is this in pounds?

69. Instructions call for a temperature control to be at 65°C. What is this in °F?

FOR REVIEW

Perform the indicated operations and simplify.

70. 29 ft 7 in + 18 ft 6 in

71. 2 lb 3 oz − 1 lb 7 oz

72. 8 gal 1 qt − 2 gal 3 qt 1 pt

73. 8 × (4 yd 1 ft 6 in)

74. (15 hr 36 min) ÷ 6

75. 132 ft ÷ 11 ft

0.8 EXERCISES C

Complete the following using an appropriate diagram.

1. $50\frac{\text{dl}}{\text{min}} =$ ____ $\frac{\text{hl}}{\text{hr}}$
[Answer: 3]

2. $140\frac{\text{kg}}{\text{wk}} =$ ____ $\frac{\text{dag}}{\text{hr}}$

3. $120\frac{\text{cm}}{\text{hr}} =$ ____ $\frac{\text{km}}{\text{sec}}$

4. If 1 mi = 1.61 km and a river is flowing at a rate of $4.2\frac{\text{mi}}{\text{hr}}$, how fast is it flowing in kilometers per hour?
[Answer: $6.762\frac{\text{km}}{\text{hr}}$]

Use a calculator to complete the following, and give answers rounded to the nearest hundredth.

5. 17.65 pt = ________ L
[Answer: 8.35]

6. $3\frac{1}{9}$ m = ________ in

7. 4.65 kg = ________ oz

8. $75\frac{\text{km}}{\text{hr}} =$ ________ $\frac{\text{ft}}{\text{sec}}$

9. 77.2°C = ________ °F
[Answer: 170.96]

10. 15.5°F = ________ °C

CHAPTER 0 REVIEW

KEY WORDS

0.1 A **numeral** is the name or symbol for a number.

The **natural** or **counting numbers** are 1, 2, 3, 4, ...

The **whole numbers** are 0, 1, 2, 3, ...

Fractions that are names for the same number are called **equivalent fractions.**

A fraction is **reduced to lowest terms** when 1 is the only natural number that divides both numerator and denominator.

A **prime number** is a natural number greater than 1 whose only divisors are 1 and itself.

0.2 The **reciprocal** of a fraction is the fraction formed by interchanging its numerator and denominator.

The **least common denominator (LCD)** of two fractions is the smallest number that has both denominators as factors.

0.3 A **place-value system** is a number system in which a digit has a particular value determined by its place in the numeral.

The **digits** are 0, 1, 2, 3, 4, 5, 6, 7, 8, and 9.

0.4 A **rational number** has a decimal representation that terminates or repeats.

An **irrational number** has a decimal representation that neither terminates nor repeats.

0.5 **Statistics** is a branch of mathematics that deals with collecting, organizing, and summarizing numerical data in order to describe or interpret it.

A **circle graph** uses pie-shaped sectors of a circle to represent different parts of a quantity.

A **bar graph** is used to compare various parts of a data set.

A **broken-line graph** can be used to show trends in data collected over a period of time.

A **histogram** is a special type of bar graph used to display data that is divided into groups called **class intervals.** The **class frequency** of an interval is represented by the length or height of each bar.

0.6 The **mean** or **average** of several values is the sum of the values divided by the number of values.

The **median** of a data set is the middle value or the average of the two middle values when the values are arranged in order of size.

The **mode** of a set of values is the value that occurs most frequently in the set.

0.7 The **English system** is the measurement system used in the United States.

A **unit fraction** is a fraction which has value 1 and is used to make measurement conversions.

0.8 The **metric system** is the system of measurement used in most of the countries of the world. The **meter** is the basic unit for length, the **liter** is the basic unit for volume, and the **gram** is the basic unit for weight.

The **Celsius** scale is the temperature scale used in the metric system.

The **Fahrenheit** scale is the temperature scale used in the English system.

KEY CONCEPTS

0.1 **1.** When reducing fractions by canceling common factors, do not make the numerator 0 when it is actually 1. For example,

$$\frac{5}{5\cdot 3} = \frac{\cancel{5}}{\cancel{5}\cdot 3} = \frac{1}{3}, \textit{ not } \frac{0}{3}.$$

2. Factors common to the numerator and denominator of a fraction may be divided or canceled. However, do not cancel numbers that are not factors. For example,

$$\frac{2\cdot 3}{2} = \frac{\cancel{2}\cdot 3}{\cancel{2}} = 3,$$

$$\text{but } \frac{2+3}{2} \text{ is } \textit{not} \text{ equal to } \frac{\cancel{2}+3}{\cancel{2}}.$$

0.2 **1.** To add or subtract fractions with different denominators, find the LCD. However, *do not* find the LCD to multiply or divide fractions.

2. For any nonzero number a, $\frac{0}{a} = 0$, but $\frac{a}{0}$ is undefined.

0.4 **1.** To convert **a percent to a decimal,** remove the % symbol and multiply by 0.01 (move the decimal point two places to the *left*). For example,

$$48\% = 48(0.01) = 0.48.$$

2. To convert **a decimal to a percent,** multiply by 100 (move the decimal point two places to the *right*) and attach the % symbol. For example,

$$0.172 = 0.172(100\%) = 17.2\%.$$

0.5 Numerical data can be displayed or pictured using a circle graph or a bar graph.
A histogram is often used when a large number of data items can best be described by dividing the data into groups.

0.6 **1.** Before finding the median of a collection of numbers, be sure to arrange the numbers in order of size.

2. If there is an even number of values, the median is the mean (average) of the two middle values.

3. A bimodal set of data has two values with the greatest frequency of occurrence. If no numbers in a collection occur more frequently than the others, the data set has no mode.

0.8 **1.** In the metric system, conversions in units can be made by moving the decimal point.

2. A meter is about half the height of a doorway opening. A liter is the volume of a cube 0.1 m (1 dm) on a side. A gram is the weight of water that can be put in a cube 0.01 m (1 cm) on a side.

3.

Prefix	*Symbol*	*Multiply standard unit by*
kilo	k	1000
hecto	h	100
deka	da	10
deci	d	$\frac{1}{10} = 0.1$
centi	c	$\frac{1}{100} = 0.01$
milli	m	$\frac{1}{1000} = 0.001$

4. To change °C to °F, use $F = \frac{9}{5}C + 32$.

5. To change °F to °C, use $C = \frac{5}{9}(F - 32)$.

REVIEW EXERCISES

Part I

0.1

1. Express 490 as a product of primes.

2. Reduce $\frac{110}{385}$ to lowest terms.

3. Write an equivalent fraction for $\frac{3}{11}$ having 44 for the denominator.

4. Supply the missing numerator. $\frac{9}{13} = \frac{}{39}$

0.2 *Perform the indicated operations.*

5. $\frac{3}{14} \cdot \frac{21}{9}$

6. $3\frac{1}{3} \div \frac{20}{9}$

7. $\frac{5}{9} + \frac{7}{12}$

8. $\frac{17}{18} - \frac{2}{3}$

9. Change $\frac{219}{4}$ to a mixed number.

10. Change $5\frac{2}{7}$ to an improper fraction.

11. On a map, one inch represents $\frac{3}{4}$ mi. How many miles are represented by $2\frac{1}{3}$ in?

12. Julie Simpson bought two pieces of fabric, one $3\frac{1}{8}$ yd long and the other $4\frac{3}{4}$ yd long. How many yards of fabric did she purchase?

0.3

13. Express 29.34 in expanded notation.

14. Express $2000 + 4 + \frac{4}{10} + \frac{2}{1000}$ in decimal notation.

15. Add.

$$\begin{array}{r} 14.5 \\ 0.007 \\ +\ 325.16 \\ \hline \end{array}$$

16. Subtract.

$$\begin{array}{r} 307.2 \\ -\ \ 12.009 \\ \hline \end{array}$$

17. Multiply.

$$\begin{array}{r} 1.31 \\ \times\ 0.02 \\ \hline \end{array}$$

18. Divide.

$0.005\overline{)36.15}$

0.4 **19.** Convert $\frac{3}{8}$ to a decimal.

20. Convert 27.235 to a fraction.

21. Convert 35.2% to a decimal.

22. Convert $\frac{4}{3}$ to a percent.

0.5 *Exercises 23–26 refer to the circle graph at the right which shows preference in vacation locations of 300 adults surveyed in a random poll. Give answers rounded to the nearest tenth.*

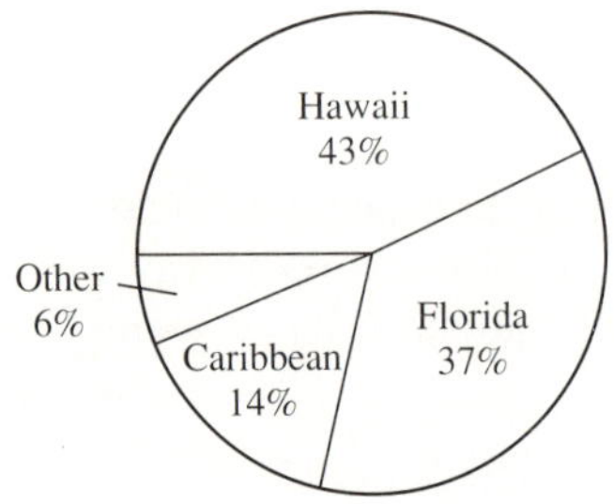

23. How many of those surveyed preferred Hawaii?

24. How many preferred the Caribbean?

25. What is the ratio of those preferring Hawaii to those preferring Florida?

26. How many preferred a vacation location other than Hawaii?

Exercises 27–32 refer to the broken-line graph at the right which shows the number of new homes sold in a subdivision over a six-month period.

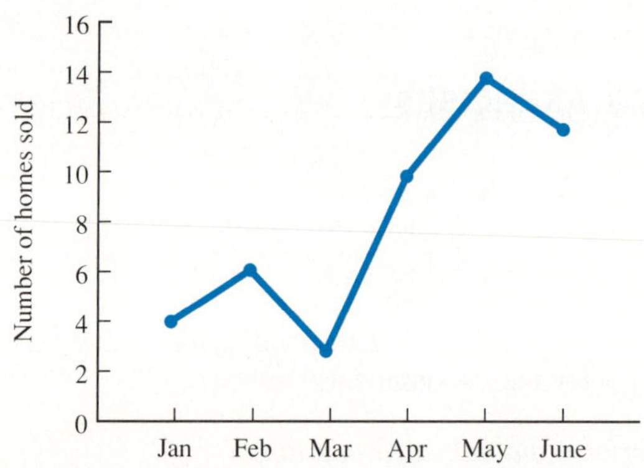

27. How many homes were sold in April?

28. How many homes were sold in May and June together?

29. Between which two consecutive months did sales of new homes show the greatest increase?

30. How many homes were sold in the six-month period?

31. What was the ratio of homes sold in April to those sold in March?

32. What percent (to the nearest tenth) of the homes sold in the six-month period were sold in January?

Exercises 33–38 refer to the histogram at the right which relates the number of cars to miles per gallon in an EPA survey. Assume that a car making 5 mpg, for example, is included in the range 5–10 mpg and not 0–5 mpg, similarly for 10 mpg, 15 mpg, and so forth.

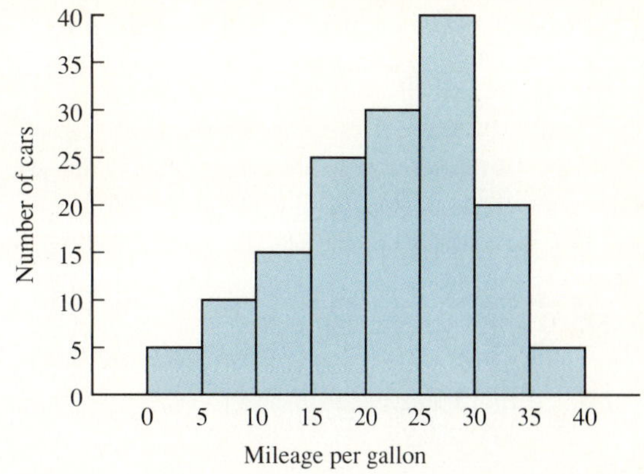

33. How many cars got at least 5 mpg and less than 10 mpg?

34. How many cars got 30 mpg or better?

35. How many cars got less than 20 mpg?

36. How many cars were represented in the survey?

37. What percent of the total number of cars got 35 mpg or better?

38. What is the ratio of cars that got 30 mpg or better to those that got less than 5 mpg?

0.6 *Find the mean and median. Give answers to the nearest tenth.*

39. 16, 12, 19, 22, 8

40. 375, 206, 311, 222, 320, 312

41. Maxine's car will go 589 miles on 15.5 gallons of gasoline. What is Maxine's average miles per gallon?

Find the mode.

42. 6, 5, 1, 3, 5, 7, 12, 5

43. 2, 9, 3, 2, 7, 3, 4, 12

44. The Thai Villa Restaurant had sales of \$1020, \$926, \$820, \$772, \$1428, and \$1362 over a six-day period. What were the average daily sales to the nearest dollar? What was the median?

45. The following table gives information of the monthly salaries of the employees of Lloyd's Shoppe. Find the average salary. Find the median salary. What is the mode of the salaries?

Salary	*Number of employees*	*Salary times number of employees*
\$ 820	4	
\$ 860	8	
\$ 920	5	
\$ 980	2	
\$1620	1	
Totals		

0.7 *Complete.*

46. 7 ft = ________ in

47. 30 in = ________ ft

48. 48 fl oz = ________ pt

49. 90 min = ________ hr

50. 3 ton = ________ lb

51. 18 pt = ________ gal

Perform the indicated operations and simplify.

52. $\begin{array}{r} 10 \text{ ft } 3 \text{ in} \\ + \ 7 \text{ ft } 2 \text{ in} \\ \hline \end{array}$

53. $\begin{array}{r} 10 \text{ ft } 3 \text{ in} \\ - \ 7 \text{ ft } 2 \text{ in} \\ \hline \end{array}$

54. $\begin{array}{r} 18 \text{ hr } 22 \text{ min} \\ + \ 5 \text{ hr } 56 \text{ min} \\ \hline \end{array}$

55. $\begin{array}{r} 18 \text{ hr } 22 \text{ min} \\ - \ 5 \text{ hr } 56 \text{ min} \\ \hline \end{array}$

56. $\begin{array}{r} 12 \text{ yd } 2 \text{ ft } 2 \text{ in} \\ + \ 6 \text{ yd } 2 \text{ ft } 8 \text{ in} \\ \hline \end{array}$

57. $\begin{array}{r} 12 \text{ yd } 2 \text{ ft } 2 \text{ in} \\ - \ 6 \text{ yd } 2 \text{ ft } 8 \text{ in} \\ \hline \end{array}$

58. $3 \times (5 \text{ gal } 3 \text{ qt } 1 \text{ pt})$

59. $8 \text{ ft} \times 3 \text{ ft}$

60. $(8 \text{ lb } 13 \text{ oz}) \div 3$

61. $18 \text{ dollars} \div 3 \text{ dollars}$

0.8 *Complete.*

62. 860 m = ________ km

63. 0.582 m = ________ cm

64. 483 ml = ________ dl

65. 48,300 mg = ________ kg

66. 0.00035 km = ________ dm

67. 5.6 L = ________ cl

Perform the indicated operations.

68. $\begin{array}{r} 15.2 \text{ m} \\ + \ 75.3 \text{ m} \\ \hline \end{array}$

69. $\begin{array}{r} 831.6 \text{ L} \\ - \ 416.9 \text{ L} \\ \hline \end{array}$

70. $\begin{array}{r} 42.8 \text{ g} \\ \times \quad 7 \\ \hline \end{array}$

Complete.

71. 12 km = ________ mi

72. 420 in = ________ m

73. $50 \dfrac{\text{mi}}{\text{hr}} =$ ________ $\dfrac{\text{km}}{\text{hr}}$

74. 22 qt = ________ L

75. 18 L = ________ gal

76. 16 lb = ________ g

77. Would water at 65°C feel cool to your hand?

78. If your Italian car has a 54-liter gas tank, how many gallons will it hold?

Part II

Exercises 79–82 refer to the bar graph at the right, which gives the number of adults choosing a particular sport as their favorite to view on television.

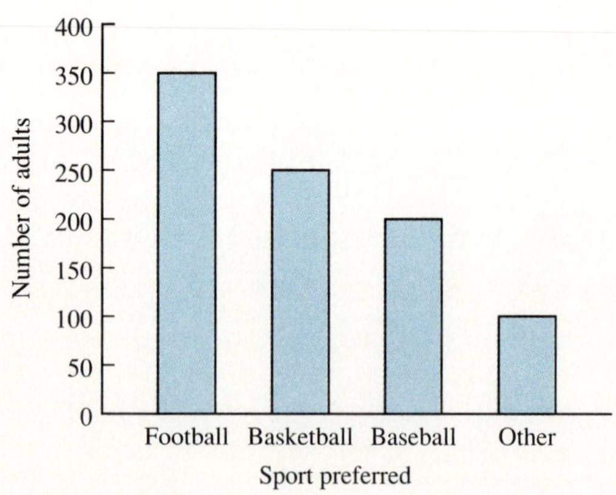

79. How many adults were in the survey?

80. What is the ratio of those preferring basketball to the total number surveyed?

81. What percent of those surveyed preferred basketball (to the nearest tenth of a percent)?

82. What is the ratio of those who prefer football to those who prefer baseball?

83. In a test of 560 video games, 420 were working. What fractional part were working? Reduce the fraction.

84. Graydon Bell works in Perko's Delicatessen as a registered cheese cutter. If he is paid \$9.85 per hour and worked a total of 35.6 hr last week, what did he earn?

Perform the indicated operations.

85. $0 \div \dfrac{5}{4}$

86. $4.7 - 0.001$

87. $1\dfrac{1}{5} + \dfrac{9}{10} - \dfrac{1}{3}$

88. $(1.04)(1.04)(1.04)$

89. Convert 0.5% to a decimal.

90. Convert $\frac{3}{11}$ to a decimal.

91. Convert 0.03 to a percent.

92. Convert 25 to a percent.

93. Change $\frac{31}{9}$ to a mixed number.

94. Convert 1.85 to a fraction.

95. Change $5\frac{23}{27}$ to an improper fraction.

96. Write an equivalent fraction for $\frac{2}{13}$ with 52 as the denominator.

Complete.

97. 0.032 kg = ________ oz

98. 7.8 ft = ________ cm

99. 114°F = ________°C

100. 6.35 dg = ________ dag

101. 9900 ml = ________ hl

102. 0.0008 g = ________ cg

103. 1 hr = ________ sec

104. 3200 oz = ________ ton

105. $22\ \dfrac{\text{ft}}{\text{sec}} = \underline{\qquad\qquad}\ \dfrac{\text{mi}}{\text{hr}}$

106. If it takes 2 lb 4 oz of flour to make a loaf of bread, how much will it take to make 5 loaves?

107. An electrician needs 60 yd of wire. What will be the total cost if the wire is 50¢ per foot?

Find the mean, median, and mode.

108. 12, 3, 9, 5, 5

109. 6.1, 2.5, 3.8, 9.2, 100.7

110. 18, 12, 17, 29, 40, 12

111. 4, 7, 4, 9, 7, 6, 8, 10

Exercises 112–120 refer to the double-bar graph at the right which compares the number of students enrolled in four math courses at State University in 1990 and 1991.

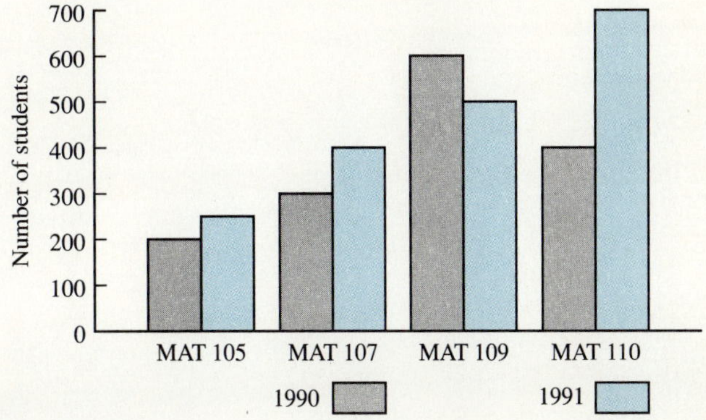

112. How many students were in MAT 107 in 1991?

113. How many students were in MAT 109 in 1990?

114. How many students were in these four math courses in 1990?

115. How many students were in these four math courses in 1991?

116. What was the ratio of students in MAT 110 in 1991 to those in MAT 110 in 1990?

117. What percent of students in all four classes in 1990 were in MAT 109?

118. During which year were there more students in MAT 109?

119. During which year were there fewer students in MAT 105?

120. How many students in total took MAT 105 in these two years?

ANSWERS: (Some answers are rounded.) 1. $2 \cdot 5 \cdot 7 \cdot 7$ 2. $\frac{2}{7}$ 3. $\frac{12}{44}$ 4. 27 5. $\frac{1}{2}$ 6. $\frac{3}{2}$ 7. $\frac{41}{36}$ 8. $\frac{5}{18}$ 9. $54\frac{3}{4}$ 10. $\frac{37}{7}$ 11. $1\frac{3}{4}$ mi 12. $7\frac{7}{8}$ yd 13. $20 + 9 + \frac{3}{10} + \frac{4}{100}$ 14. 2004.402 15. 339.667 16. 295.191 17. 0.0262 18. 7230 19. 0.375 20. $\frac{5447}{200}$ 21. 0.352 22. $133\frac{1}{3}\%$ 23. 129 24. 42 25. $\frac{43}{37}$ or 43 to 37 26. 171 27. 10 28. 26 29. between March and April 30. 49 31. $\frac{10}{3}$ or 10 to 3 32. 8.2% 33. 10 34. 25 35. 55 36. 150 37. $3.\overline{3}\%$ or $3\frac{1}{3}\%$ 38. $\frac{5}{1}$ or 5 to 1 39. 15.4; 16 40. 291; 311.5 41. 38 mpg 42. 5 43. 2 and 3 (bimodal) 44. $1055; $973 45. $917; $860; $860 46. 84 47. 2.5 48. 3 49. 1.5 50. 6000 51. 2.25 52. 17 ft 5 in 53. 3 ft 1 in 54. 24 hr 18 min 55. 12 hr 26 min 56. 19 yd 1 ft 10 in 57. 5 yd 2 ft 6 in 58. 17 gal 2 qt 1 pt 59. 24 ft^2 60. 2 lb 15 oz 61. 6 62. 0.86 63. 58.2 64. 4.83 65. 0.0483 66. 3.5 67. 560 68. 90.5 m 69. 414.7 L 70. 299.6 g 71. 7.452 72. 10.668 73. 80.5 74. 20.8 75. 4.77 76. 7264 77. no (it would feel hot: 65°C = 149°F.) 78. 14.31 gal 79. 900 80. $\frac{5}{18}$ or 5 to 18 81. 27.8% 82. $\frac{7}{4}$ or 7 to 4 83. $\frac{3}{4}$ 84. $350.66 85. 0 86. 4.699 87. $1\frac{23}{30}$ 88. 1.124864 89. 0.005 90. $0.\overline{27}$ 91. 3% 92. 2500% 93. $3\frac{4}{9}$ 94. $\frac{37}{20}$ 95. $\frac{158}{27}$ 96. $\frac{8}{52}$ 97. 1.13 98. 237.7 99. 46 100. 0.0635 101. 0.099 102. $\underline{0}.08$ 103. 3600 104. 0.1 105. 15 106. 11 lb 4 oz 107. $90 108. 6.8; 5; 5 109. 24.46; 6.1; no mode 110. $21.\overline{3}$; 17.5; 12 111. 6.875; 7; 4 and 7 112. 400 113. 600 114. 1500 115. 1850 116. $\frac{7}{4}$ or 7 to 4 117. 40% 118. 1990 119. 1990 120. 450

CHAPTER 0 TEST

1. Supply the missing term. $\frac{4}{15} = \frac{\quad}{90}$ 1. ________

2. In a contest, Michelle received 385 of the 525 possible points. What fractional part of the points did Michelle get? Reduce the fraction. 2. ________

Perform the indicated operation.

3. $\frac{2}{5} \cdot \frac{15}{22}$ 3. ________

4. $\frac{9}{14} \div \frac{2}{7}$ 4. ________

5. $\frac{5}{6} + \frac{7}{9}$ 5. ________

6. $\frac{17}{24} - \frac{2}{3}$ 6. ________

7. Change $\frac{28}{5}$ to a mixed number. 7. ________

8. Convert $\frac{1}{5}$ to a percent. 8. ________

9. Convert 3.15 to a fraction. 9. ________

10. The tax rate on purchases is 4%. What is the tax on a \$27 purchase? 10. ________

11. There were 720 tires in a shipment. If $\frac{3}{5}$ of these were whitewalls, how many whitewall tires were in the shipment? 11. ________

12. A bottle contains $7\frac{2}{5}$ liters of solution. If $2\frac{7}{10}$ liters are used, how many liters are left in the bottle? 12. ________

CHAPTER 0 TEST

CONTINUED

Problems 13–17 refer to the bar graph below which shows the preferences of 150 adults in type of restaurant food.

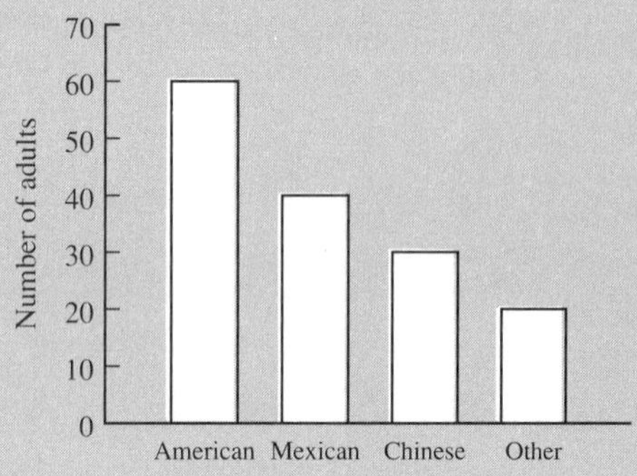

13. How many of those questioned preferred Mexican food? **13.** ______

14. How many preferred food other than American food? **14.** ______

15. What is the ratio of those preferring American food to those preferring Chinese food? **15.** ______

16. What percent of those interviewed preferred Mexican food? **16.** ______

17. What percent preferred some kind of food besides American? **17.** ______

The weights of all the members of a football team are presented in the histogram below. Use this information in problems 18–21. Assume that 175 pounds is in the class 175–200 pounds, not in 150–175 pounds, similarly for 200 pounds, 225 pounds, and 250 pounds.

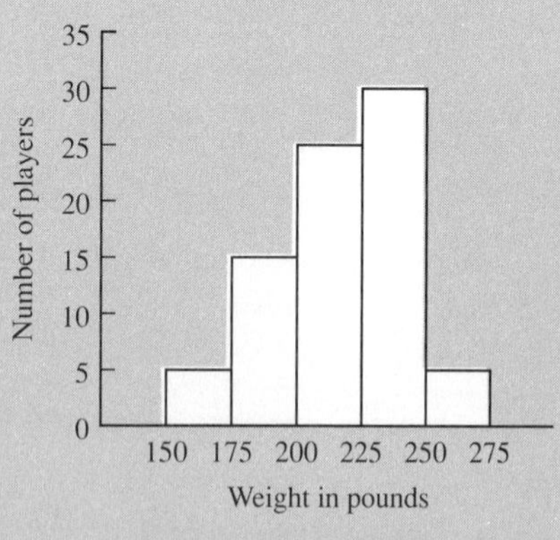

18. How many players weigh at least 200 and less than 225 pounds? **18.** ______

19. How many players are on the team? **19.** ______

20. What is the ratio of players weighing 225 pounds or more to the total number of players on the team? **20.** ______

21. What percent of the players weigh less than 200 pounds? **21.** ______

22. Find the mean of the given numbers. 22. ______

11, 22, 32, 17, 18

23. Find the median of the given numbers. 23. ______

11, 22, 32, 17, 18

24. Find the mode of the given numbers. 24. ______

14, 8, 9, 17, 8, 17, 23, 8, 5

25. The weights of the starting four defensive linemen for PSU are 246 pounds, 253 pounds, 257 pounds, and 262 pounds.
(a) What is the average weight? **(b)** What is the median weight?
(c) What is the mode of the weights?

25. (a) ______
(b) ______
(c) ______

Complete.

26. 40 oz = ______ lb 26. ______

27. 2800 ml = ______ hl 27. ______

28. 6.5 kg = ______ lb 28. ______

29. 65°C = ______°F 29. ______

Perform the indicated operations and simplify.

30.

$$\begin{array}{r} 12 \text{ yd } 2 \text{ ft } 3 \text{ in} \\ -\ \ 3 \text{ yd } 2 \text{ ft } 8 \text{ in} \\ \hline \end{array}$$

30. ______

31.

$$\begin{array}{r} 10 \text{ yd } 2 \text{ ft } \ \ 3 \text{ in} \\ +\ 12 \text{ yd } 2 \text{ ft } 11 \text{ in} \\ \hline \end{array}$$

31. ______

32. 28 lb ÷ 7 lb 32. ______

33. A small storage tank has a capacity of 4.3 gal. How many liters of a chemical will it hold? 33. ______

34. For an experiment to work, the temperature must be above 20°C. If the temperature is 59°F, will the experiment work? 34. ______

CHAPTER 1

Introduction to Algebra

1.1 VARIABLES, EXPONENTS, AND ORDER OF OPERATIONS

STUDENT GUIDEPOSTS

1. Variables
2. Exponential Notation
3. Order of Operations
4. Symbols of Grouping
5. Evaluating Algebraic Expressions
6. Formulas

1 VARIABLES

In algebra, we often use letters such as a, b, x, y, A and B to represent numbers. A letter that can be replaced by various numbers is called a **variable.** Lengthy verbal expressions can often be symbolized by brief algebraic expressions using variables, as in the following examples.

Three	times	a natural number	becomes $3 \cdot n$
↓	↓	↓	
3	$\cdot$	n	

Seven	plus	twice	a natural number	becomes $7 + 2 \cdot n$
↓	↓	↓	↓	
7	+	$2 \cdot$	n	

In each example, the variable n represents a natural number. Other notations for $3 \cdot n$ are $3(n)$, $(3)(n)$, and $3n$, with the last form preferred. The product of two natural numbers m and n would most likely be represented by mn, but could also be expressed by $m \cdot n$, $m(n)$, and $(m)(n)$. We usually avoid using the symbol $\times$ for multiplication since it can be confused with the letter x used as a variable.

2 EXPONENTIAL NOTATION

When a variable or number is multiplied by itself several times, as in

$$2 \cdot 2 \cdot 2 \cdot 2 \text{ or } x \cdot x \cdot x,$$

we can use **exponential notation** to avoid writing long strings of **factors** (the individual numbers in the expressed product). For example, we write $2 \cdot 2 \cdot 2 \cdot 2$ as 2^4,

$$\underbrace{2 \cdot 2 \cdot 2 \cdot 2}_{\textbf{4 factors}} = \underset{\uparrow \atop \textbf{base}}{2}^{4}, \leftarrow \textbf{exponent}$$

where 2 is the **base,** 4 the **exponent,** and 2^4 the **exponential expression** (read 2 to the **fourth power**). Similarly, $x \cdot x \cdot x = x^3$ is the **third power** or **cube** of x. The **square** or **second power** of a is a^2. The **first power** of a is a^1, which we write simply as a.

Exponential Notation

If a is any number and n is a natural number,

$$a^n = \underbrace{a \cdot a \cdot a \cdots a}_{n \text{ factors}}.$$

EXAMPLE 1 Using Exponents

Write in exponential notation.

(a) $\underbrace{7 \cdot 7 \cdot 7}_{3 \text{ factors}} = 7^3$

(b) $\underbrace{a \cdot a \cdot a \cdot a \cdot a \cdot a}_{6 \text{ factors}} = a^6$

(c) $\underbrace{3 \cdot 3}_{2 \text{ factors}} \cdot \underbrace{x \cdot x \cdot x \cdot x \cdot x}_{5 \text{ factors}} = 3^2x^5$

Practice Exercise 1

Write in exponential notation.

(a) $4 \cdot 4 \cdot 4 \cdot 4 \cdot 4 \cdot 4 \cdot 4$

(b) $z \cdot z \cdot z \cdot z$

(c) $6 \cdot 6 \cdot 6 \cdot y \cdot y \cdot y \cdot y$

Answers: (a) 4^7 (b) z^4 (c) 6^3y^4

3 ORDER OF OPERATIONS

When exponents or powers are used together with the operations of addition, subtraction, multiplication, and division, the resulting numerical expressions can be confusing unless we agree to an order of operations. For example,

$2 \cdot 5^2$ *could* equal 10^2 or 100, — If we first multiply then square

or $2 \cdot 5^2$ *could* equal $2 \cdot 25$ or 50. — If we first square then multiply

Similarly,

$2 \cdot 3 + 4$ *could* equal $2 \cdot 7$ or 14, — If we first add then multiply

or $2 \cdot 3 + 4$ *could* equal $6 + 4$ or 10. — If we first multiply then add

According to the following rule, we see that the second procedure in each of these examples is the correct one to use.

To Evaluate a Numerical Expression

1. Evaluate all powers, in any order, first.
2. Do all multiplications and divisions in order from left to right.
3. Do all additions and subtractions in order from left to right.

EXAMPLE 2 EVALUATING NUMERICAL EXPRESSIONS

Evaluate each numerical expression.

(a) $2 + 3 \cdot 4 = 2 + 12 = 14$ Multiply first, then add

(b) $20 - 3^2 = 20 - 9 = 11$ Square first, then subtract

(c) $5 \cdot 6 - 12 \div 3 = 30 - 4 = 26$ Multiply and divide first, then subtract

(d) $25 \div 5 + 3 \cdot 2^3 = 25 \div 5 + 3 \cdot 8$ Cube first

$= 5 + 24$ Divide and multiply second

$= 29$ Add last

PRACTICE EXERCISE 2

Evaluate each numerical expression.

(a) $5 \cdot 2 + 1$

(b) $6 + 2^3$

(c) $18 \div 9 + 3 \cdot 2$

(d) $16 \cdot 2 - 3 \cdot 2^2$

Answers: (a) 11 (b) 14 (c) 8 (d) 20

4 SYMBOLS OF GROUPING

Suppose that we want to evaluate three times the sum of 2 and 5. If we write $3 \cdot 2 + 5$ and use the above rule, we will obtain $6 + 5$ or 11. However, it is clear from the first sentence, that we want 3 times 7 or 21 for the result. **Symbols of grouping,** such as parentheses ( ), square brackets [ ], or braces { }, can help us symbolize the problem correctly as $3 \cdot (2 + 5)$. The grouping symbols contain the expression that must be evaluated first. In this case, we must add before multiplying.

$$3 \cdot (2 + 5) = 3 \cdot (7) = 21$$

Generally, we omit the dot and write $3(2 + 5)$ for $3 \cdot (2 + 5)$. Also, instead of writing $3 \cdot (7)$, we will write $(3)(7)$.

One other method of grouping is to use a fraction bar. For example, in

$$\frac{4 \cdot 5}{7 + 3},$$

the fraction bar acts like parentheses since the expression is the same as

$$(4 \cdot 5) \div (7 + 3).$$

We first multiply 4 and 5, then add 7 and 3, and finally we divide the results. Thus,

$$\frac{4 \cdot 5}{7 + 3} = \frac{20}{10} = 2.$$

To Evaluate a Numerical Expression Involving Grouping Symbols

Evaluate all expressions within the grouping symbols first, and begin with the innermost symbols of grouping if more than one set of symbols is present.

EXAMPLE 3 EVALUATING NUMERICAL EXPRESSIONS

Evaluate each numerical expression.

(a) $3 + (2 \cdot 5) = 3 + 10 = 13$ Evaluate inside the parentheses first

(b) $(4 + 5)2 + 3 = (9)2 + 3$ Work inside the parentheses first

$= 18 + 3$ Multiply before adding

$= 21$

PRACTICE EXERCISE 3

Evaluate each numerical expression.

(a) $3 \cdot (2 + 5)$

(b) $(6 - 1)3 - 2$

(c) $[3(8+2)+1]4 = [3(10)+1]4$ Innermost grouping symbol first
$= [30+1]4$ Multiply before adding inside brackets
$= [31]4$ Combine numbers inside brackets before multiplying
$= 124$

(c) $2[2(7+2)-1]$

Answers: (a) 21 (b) 13 (c) 34

We now summarize the order to follow when evaluating a numerical expression.

Order of Operations

1. Evaluate within grouping symbols first, beginning with the innermost set if more than one set is used.
2. Evaluate all powers.
3. Perform all multiplications and divisions in order from left to right.
4. Perform all additions and subtractions in order from left to right.

EXAMPLE 4 Evaluating Using All Rules

Evaluate each expression.

(a) $2^3 + 4^3 = 8 + 64 = 72$ Cube first, then add

(b) $(2+4)^3 = 6^3 = 216$ Add first inside parentheses, then cube

(c) $(3 \cdot 5)^2 = 15^2 = 225$ Multiply first, then square

(d) $3 \cdot 5^2 = 3 \cdot 25 = 75$ Square first, then multiply

(e) $5^2 - 4^2 = 25 - 16 = 9$ Square first, then subtract

(f) $(5-4)^2 = 1^2 = 1$ Subtract first, then square

(g) $3^2 - 15 \div 5 + 5 \cdot 2 = 9 - 15 \div 5 + 5 \cdot 2$ Square first
$= 9 - 3 + 10$ Divide and multiply in order
$= 16$ Subtract and add in order

(h) The square of seven minus three
$(7-3)^2 = 4^2 = 16$ Subtract first, then square

(i) Seven squared minus three squared
$7^2 - 3^2 = 49 - 9 = 40$ Square first, then subtract

PRACTICE EXERCISE 4

Evaluate each expression.

(a) $1^2 + 6^2$

(b) $(1+6)^2$

(c) $(2 \cdot 7)^2$

(d) $2 \cdot 7^2$

(e) $8^2 - 6^2$

(f) $(8-6)^2$

(g) $2^3 - 20 \div 4 + 4 \cdot 3$

(h) The square of nine plus two

(i) Nine squared plus two squared

Answers: (a) 37 (b) 49 (c) 196 (d) 98 (e) 28 (f) 4 (g) 15 (h) 121 (i) 85

CAUTION

Example 4 points out common errors to avoid when working with exponents. For example,

$$2^3 + 4^3 \neq (2+4)^3, \quad (3 \cdot 5)^2 \neq 3 \cdot 5^2, \quad 5^2 - 4^2 \neq (5-4)^2.$$

The symbol $\neq$ means "is *not* equal to." In general, if a, b, and n are numbers,

$$a^n + b^n \neq (a+b)^n, \quad (ab)^n \neq ab^n, \quad a^n - b^n \neq (a-b)^n.$$

5 EVALUATING ALGEBRAIC EXPRESSIONS

An **algebraic expression** contains variables as well as numbers. We can use the order of operations for evaluating numerical expressions to evaluate algebraic expressions when specific values for the variables are given.

To Evaluate an Algebraic Expression

1. Replace each variable (letter) with the specified value.
2. Proceed as in evaluating numerical expressions.

EXAMPLE 5 Evaluating an Algebraic Expression

Evaluate $2(a+b)-c$ when $a=3$, $b=4$, and $c=5$.

$$2(a+b)-c = 2(3+4)-5 \quad \text{Replace each letter with given value}$$
$$= 2(7)-5 \quad \text{Evaluate inside parentheses first}$$
$$= 14-5 = 9$$

Practice Exercise 5

Evaluate $3(x-y)+z$, when $x=4$, $y=2$, and $z=1$.

Answer: 7

EXAMPLE 6 Evaluating an Algebraic Expression

Evaluate $5[12-3(a+1)+b]-c$ when $a=2$, $b=7$, and $c=4$.

$$5[12-3(a+1)+b]-c = 5[12-3(2+1)+7]-4 \quad \text{Replace variables with numbers}$$
$$= 5[12-3(3)+7]-4$$
$$= 5[12-9+7]-4 \quad \text{Multiply before adding or subtracting}$$
$$= 5[10]-4 = 50-4 = 46$$

Practice Exercise 6

Evaluate $w-2[3(u-1)+v]$, when $u=3$, $v=1$, and $w=14$.

Answer: 0

EXAMPLE 7 Evaluating an Algebraic Expression

Evaluate $\dfrac{ab-1}{c}$ for the given values.

(a) $a=2$, $b=3$, and $c=0$

$$\frac{ab-1}{c} = \frac{(2)(3)-1}{0} \quad \text{Division by zero undefined}$$

That is, this expression is undefined when $c=0$.

(b) $a=3$, $b=\dfrac{1}{3}$, and $c=5$

$$\frac{ab-1}{c} = \frac{(3)\left(\frac{1}{3}\right)-1}{5} = \frac{1-1}{5} = \frac{0}{5} = 0 \quad \tfrac{0}{5}=0$$

Practice Exercise 7

Evaluate $\dfrac{3x}{yz-x}$ for the given values.

(a) $x=0$, $y=5$, $z=8$

(b) $x=6$, $y=2$, $z=3$

Answers: (a) 0 (b) undefined

EXAMPLE 8 Evaluating an Algebraic Expression

Evaluate the following when $a = 2$, $b = 1$, $c = 3$.

(a) $3a^2 = 3(2)^2$ Not $(3 \cdot 2)^2 = 6^2 = 36$

$= 3 \cdot 4 = 12$

(b) $2ab^2c^3 = 2(2)(1)^2(3)^3$

$= 2(2)(1)(27) = 4 \cdot 27 = 108$

(c) $(2c)^2 - 2c^2 = (2 \cdot 3)^2 - 2(3)^2$ Watch the substitution

$= 6^2 - 2 \cdot 9$

$= 36 - 18 = 18$

(d) $a^a + c^c = 2^2 + 3^3$

$= 4 + 27 = 31$

Practice Exercise 8

Evaluate the following when $u = 5$, $v = 3$, and $w = 1$.

(a) $2u^2$

(b) $3uv^2w$

(c) $(3u)^2 - 3u^2$

(d) $v^v - w^w$

Answers: (a) 50 (b) 135 (c) 150 (d) 26

6 FORMULAS

A **formula** is an algebraic statement that relates two or more quantities. For example, the formula

$$d = rt$$

relates the distance, d, traveled by an object moving at an average rate, r, for a period of time, t. Other familiar formulas related to geometric figures, such as

$A = lw$ Area of a rectangle in terms of its length and width

and $P = 2l + 2w$ Perimeter of a rectangle in terms of its length and width

are summarized on the inside front cover of the text. To use a particular formula in an applied problem, we must first understand the meaning of the variables, and next, evaluate the algebraic expression (formula) for the given values.

EXAMPLE 9 Evaluating Formulas

If we invest P dollars (called the principal) at an interest rate, r, compounded annually for a period of t years, it will grow to an amount, A, given by

$$A = P(1 + r)^t.$$

If a principal of \$1000 is invested at 12% interest, compounded annually, how much will be in the account at the end of two years?

$A = P(1 + r)^t$ Start with the formula

$= 1000(1 + 0.12)^2$ Substitute 1000 for P, 0.12 for r, and 2 for t

$= 1000(1.12)^2$

$= 1000(1.2544)$

$= 1254.4$

Thus, there will be \$1254.40 in the account. Notice that 12% was converted to the decimal 0.12, in order to solve this problem.

Practice Exercise 9

The perimeter of a rectangle, P, is given by $P = 2l + 2w$, where l is its length and w is its width. What is the perimeter of a rectangle with length 15 in and width 8 in?

Answer: 46 in

1.1 EXERCISES A

1. What is a letter used to represent a number called?

2. Given the exponential expression x^7, **(a)** what is the base? **(b)** what is the exponent?

3. In the expression $3x$, what is the exponent on x?

Write in exponential notation.

4. $8 \cdot 8 \cdot 8 \cdot 8 \cdot 8$

5. $(2x)(2x)$

6. $2 \cdot x \cdot x$

7. $(b + c)(b + c)(b + c)$

8. $7aaaaaaa$

9. $6 \cdot 6 \cdot 6 \cdot y \cdot y \cdot z \cdot z \cdot z \cdot z$

Write without using exponents.

10. $a^2b^3c^4$

11. $3y^3$

12. $(3y)^3$

13. 1^{51}

14. $a^3 - c^3$

15. $\left(\frac{2}{3}\right)^3 x^2$

Evaluate.

16. $3 \cdot 2^2$

17. $(3 \cdot 2)^2$

18. $(3 + 2)^2$

19. $3^2 + 2^2$

20. $(5 - 2)^3$

21. $5^3 - 2^3$

22. $8 - 2^3$

23. $(8 - 2)^3$

24. $(7 - 7)^3$

25. $9 - 2 \cdot 3 + 1$

26. $(9 - 2) \cdot 3 + 4 \cdot 0$

27. $\dfrac{2(3 - 1) - 4}{5}$

28. $\dfrac{4(3 + 5) - 10}{0}$

29. $12 \div 4 + 2 \cdot 3 - 1$

30. $2[8 - 2(4 - 1) + 5]$

31. $15 - 3\{2(5 - 4) + 3\}$

32. $10 - 2[8 - 2(7 - 3)]$

33. $16 \div 4 \cdot 2 \div 4 - 2$

34. The cube of the difference five minus two

35. Five cubed minus two cubed

36. Ten minus three squared

37. The square of the difference ten minus three

Evaluate when $a = 2$, $b = 3$, $c = 5$, $d = 0$, and $x = 12$.

38. $6a + b$

39. $6(a + b)$

40. $3ab \div d$

41. $2b^2$

42. $(2b)^2$

43. $(c - b)^2$

44. $c^2 - b^2$

45. $(a + b)^3$

46. $a^3 + b^3$

47. $3a^3 + 2$

48. $(3a)^3 + 2$

49. $a^a + b^b$

50. $\dfrac{2abcd - 3d}{x}$

51. $x - [a(b + 1) - c]$

52. $5a + 2[c + x \div b]$

Solve.

53. The perimeter of a rectangle, P, is given by $P = 2l + 2w$, where l is its length and w is its width. What is the perimeter of a rectangle of length 20 ft and width 13 ft?

54. The distance an automobile travels, d, at an average rate of speed, r, for a period of time, t, is given by $d = rt$. How far does a car travel in 11 hr at an average speed of 55 mph?

55. Use $A = P(1 + r)^t$ to find the amount of money in an account at the end of two years if a principal of \$500 is invested at 14% interest, compounded annually.

56. The area of a square, A, is given by $A = s^2$, where s is the length of a side. What is the area of a square with side 3.5 cm?

57. The area of a triangle, A, is given by $A = \frac{1}{2}bh$, where b is the length of its base and h is its height. Find the area of a triangle with height 15 m and base 8 m.

58. The temperature measured in degrees Celsius, °C, can be obtained from degrees Fahrenheit, °F, by using $C = \frac{5}{9}(F - 32)$. Find C when F is 200°.

59. The surface area of a cube, A, is given by $A = 6e^2$ where e is the length of an edge. Find the surface area of a cube with an edge of $2\frac{1}{2}$ in.

60 The surface area of a cylinder, A, with height, h, and base radius, r, is given by $A = 2\pi rh + 2\pi r^2$. Use 3.14 for π and find the surface area of a cylinder with a radius of 2 cm and a height of 10 cm.

ANSWERS: 1. variable 2. (a) x (b) 7 3. 1 4. 8^5 5. $(2x)^2$ 6. $2x^2$ 7. $(b + c)^3$, *not* $b^3 + c^3$ 8. $7a^7$ 9. $6^3y^2z^4$ 10. *aabbbcccc* 11. $3yyy$ 12. $(3y)(3y)(3y)$ or $27yyy$ 13. 1 14. $aaa - ccc$ 15. $\left(\frac{2}{3}\right)\left(\frac{2}{3}\right)\left(\frac{2}{3}\right)xx$ or $\frac{8}{27}xx$ 16. 12 17. 36 18. 25 19. 13 20. 27 21. 117 22. 0 23. 216 24. 0 25. 4 26. 21 27. 0 28. undefined 29. 8 30. 14 31. 0 32. 10 33. 0 34. 27 35. 117 36. 1 37. 49 38. 15 39. 30 40. undefined 41. 18 42. 36 43. 4 44. 16 45. 125 46. 35 47. 26 48. 218 49. 31 50. 0 51. 9 52. 28 53. 66 ft 54. 605 mi 55. $649.80 56. 12.25 cm^2 57. 60 m^2 58. $93.\overline{3}°$ 59. 37.5 in^2 60. 150.72 cm^2

1.1 EXERCISES B

1. When evaluating a numerical expression with more than one set of grouping symbols, always evaluate within which set of symbols first?
2. In the exponential expression a^6, **(a)** what is the base? **(b)** what is the exponent?
3. In the expression $8y$, what is the exponent on y?

Write in exponential notation.

4. $c \cdot c \cdot c \cdot c \cdot c$ **5.** $3 \cdot 3 \cdot y \cdot y \cdot y \cdot y \cdot y \cdot y$ **6.** $(3a)(3a)$

7. $3 \cdot a \cdot a$ **8.** $(x - y)(x - y)(x - y)$ **9.** $5 \cdot 5 \cdot 5 \cdot a \cdot a \cdot c \cdot c \cdot c$

Write without using exponents.

10. $a^3b^4c^2$ **11.** $2z^3$ **12.** $(2z)^3$

13. $x^2 - y^2$ **14.** $a^3 + b^3$ **15.** $\left(\frac{1}{2}\right)^3 a^2$

Evaluate.

16. $2 \cdot 7^2$ **17.** $(2 \cdot 7)^2$ **18.** $(2 + 7)^2$

19. $2^2 + 7^2$ **20.** $(4 - 3)^2$ **21.** $4^2 - 3^2$

22. $4 - 2^2$ **23.** $(4 - 2)^2$ **24.** $(3 - 3)^3$

25. $14 - 5 + 2 \cdot 6$ **26.** $0 \cdot (5 - 2) + 0$ **27.** $\dfrac{2(5 + 1) - 3}{3}$

28. $\dfrac{5(2 + 8) - 19}{0}$ **29.** $25 \div 5 + 2 \cdot 4 - 1$ **30.** $3[10 - 2(3 - 1) + 6]$

31. $20 - 2[3(6 - 5) + 4]$ **32.** $30 - 5[7 - 2(6 - 3)]$ **33.** $20 \div 5 \cdot 2 \div 4 - 2$

34. The square of the sum of eleven and three

35. Eleven squared plus three squared

36. Twelve minus two cubed

37. The cube of the difference twelve minus two

Evaluate when $x = 3$, $y = 2$, $z = 4$, $w = 0$, and $a = 24$.

38. $5x + y$

39. $5(x + y)$

40. $4xy \div w$

41. $3y^2$

42. $(3y)^2$

43. $(z - y)^2$

44. $z^2 - y^2$

45. $(x + z)^3$

46. $x^3 + z^3$

47. $2z^3 + 1$

48. $(2z)^3 + 1$

49. $x^y + y^x$

50. $\dfrac{3wxyz - 5w}{a}$

51. $a - [x(y + 3) - z]$

52. $3x + 4[x + z \div y]$

Solve.

53. The perimeter of a square, P, with side of length s is given by $P = 4s$. What is the perimeter of a square with side of length 15 ft?

54. The area of a rectangle, A, is given by $A = lw$, where l is its length and w its width. If a rectangle is 2 cm wide and 7 cm long, what is its area?

55. Use the formula $A = P(1 + r)^t$ to find the amount of money in an account at the end of two years if a principal of \$2000 is invested at 8% interest, compounded annually.

56. The area of a circle, A, with radius r, is given by $A = \pi r^2$. Use 3.14 for π and find the area of a circle with radius 20 m.

57. The volume, V, of a box with length l, width w, and height h, is given by $V = lwh$. If the box is 9 in long, 6 in wide, and 11 in high, what is its volume?

58. The temperature measured in degrees Fahrenheit, °F, can be obtained from degrees Celsius, °C, by using $F = \frac{9}{5}C + 32$. Find F when C is 20°.

59. The surface area of a sphere, A, with radius r, is given by $A = 4\pi r^2$. Use 3.14 for π and find the surface area of a sphere with radius 50 cm.

60. The surface area of a cylinder, A, is given by $A = 2\pi rh + 2\pi r^2$, where r is the radius of its base and h is its height. Find A if $r = 10$ in, $h = 8$ in, and $\pi = 3.14$.

1.1 EXERCISES C

Evaluate.

1. $2(5 - 3)^3 - [7 - (8 - 6)^2]$

2. $13 - [11 - (7 - 4)^2]^3$

3. $[(6 \div 3)^2 + 9 \cdot (3 - 1)^2]^2$
[Answer: 1600]

Evaluate when $a = 1.35$, $b = 2.75$, and $c = 6.2$. A calculator would be helpful for these problems.

4. $abc + b^2$

5. $c^2 - a^2 + b^2$

6. $2c^2 - (a + b)$

1.2 INTEGERS, RATIONAL NUMBERS, AND THE NUMBER LINE

STUDENT GUIDEPOSTS

1. Number Line
2. Less Than and Greater Than
3. Integers
4. Equal and Unequal Numbers
5. Rational Numbers
6. Absolute Value

1 NUMBER LINE

Many ideas in algebra can be better understood if we ''picture'' them in some way. Numbers are often pictured using a *number line*. Consider the set of whole numbers. We draw a line like the one in Figure 1.1, and select some unit of length. Then, starting at an arbitrary point that we label 0 and call the **origin,** we mark off unit lengths to the right, labeling the points 1, 2, 3, ... The result is a **number line** displaying the whole numbers, shown in Figure 1.1. Every whole number is paired with a point on this line. The arrowhead points to the direction in which the whole numbers continue.

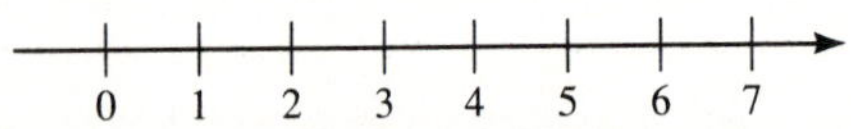

Figure 1.1

EXAMPLE 1 PAIRING NUMBERS WITH POINTS

What whole numbers are paired with a, b, and c on the number line in Figure 1.2?

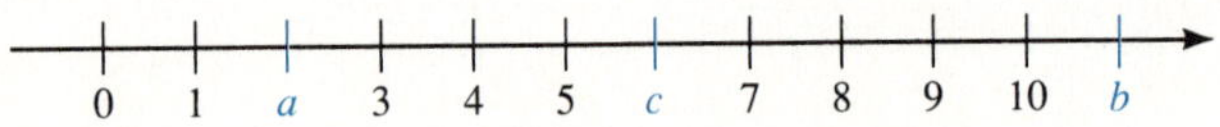

Figure 1.2

Point a is paired with 2, point b with 11, and point c with 6.

PRACTICE EXERCISE 1

What whole numbers are paired with x, y, and z on the following number line?

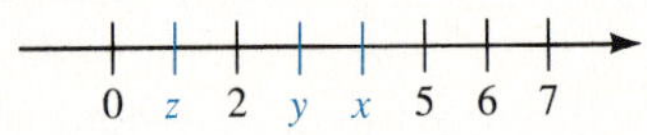

Answer: x **paired with 4,** y **paired with 3,** z **paired with 1**

2 LESS THAN AND GREATER THAN

The whole numbers occur in a natural order. For example, we know that 3 is less than (has a smaller value than) 8, and that 8 is greater than (has a larger value than) 3. Note on the number line in Figure 1.2, that 3 is to the left of 8, while 8 is to the right of 3. The symbol $<$ means **''is less than.''** We write ''3 is less than 8'' as

$$3 < 8.$$

Similarly, the symbol $>$ means **''is greater than.''** We write ''8 is greater than 3'' as

$$8 > 3.$$

Notice that $3 < 8$ and $8 > 3$ have the same meaning even though they are read differently.

Order of Whole Numbers

Suppose a and b are any two whole numbers.

1. If a is to the left of b on a number line, then $a < b$.
2. If a is to the right of b on a number line, then $a > b$.

EXAMPLE 2 Ordering Whole Numbers

Place the correct symbol ($<$ or $>$) between the numbers in the given pairs. Use part of a number line if necessary.

(a) 2 7

$2 < 7$. Notice that 2 is to the left of 7 on the number line in Figure 1.3.

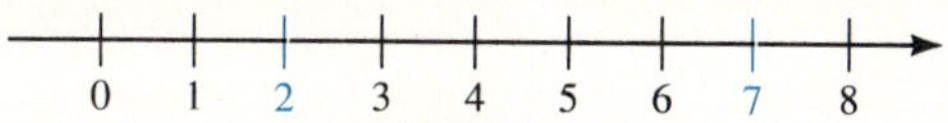

Figure 1.3

(b) 12 9

$12 > 9$. Notice that 12 is to the right of 9 on the number line in Figure 1.4.

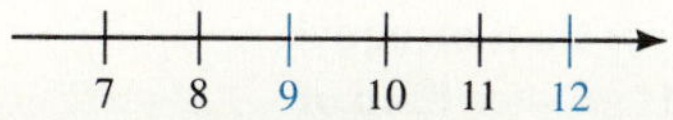

Figure 1.4

Practice Exercise 2

Place the correct symbol ($<$ or $>$) between the given pairs of numbers.

(a) 3 11

(b) 15 7

Answers: **(a)** $3 < 11$ **(b)** $15 > 7$

Notice in Example 2 that when $<$ and $>$ are used, the symbol *always points to the smaller of the two numbers*.

The number lines we have looked at so far display no numbers to the left of 0. What meaning could we give to a number to the left of 0? One example of such a number is a temperature of 5° below zero on a cold day, which is sometimes represented as −5° (read "negative 5°"). Others are shown in the following table.

Measurement	*Number*
5° above zero	5° (or +5°)
5° below zero	−5°
100 ft above sea level	100 ft (or +100 ft)
100 ft below sea level	−100 ft
$16 deposit into an account	$16 (or +$16)
$16 check written on an account	−$16

3 INTEGERS

When we put a negative sign in front of each of the counting numbers, we obtain the **negative integers.** The collection of negative integers together with the counting numbers (sometimes called **positive integers**) and zero is called the set of **integers.**

$I = \{. . . , -3, -2, -1, 0, 1, 2, 3, . . .\}$ Integers

The whole numbers are often called **nonnegative integers.** The number line in Figure 1.5 shows negative integers, positive integers, and zero. Note that we can write 1 as +1, 2 as +2, and so on. Also, we read −1 as "negative 1," for example.

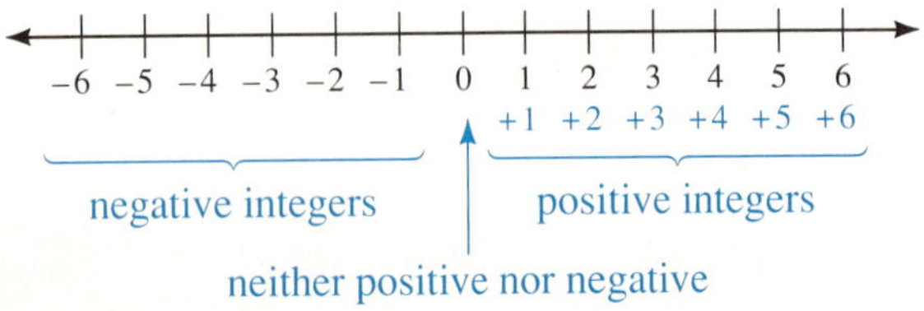

Figure 1.5

4 EQUAL AND UNEQUAL NUMBERS

We say that two numbers are **equal** (for example, $3 + 2 = 5$) if they correspond to the same point on a number line. If two numbers correspond to different points on a number line, they are **unequal.** We use $\neq$ to represent "is not equal to." We can expand our definitions of *less than* and *greater than* for whole numbers to include the integers. That is, one integer is **less than** a second if it is to the left of the second on a number line, and one integer is **greater than** a second if it is to the right of the second on a number line. If one number is **less than or equal to** another number, we use the symbol $\leq$ (similarly $\geq$ represents **greater than or equal to**). For example,

$$-4 \leq 4, \quad -2 \leq 0, \quad 0 \leq 3, \quad 2 \leq 2, \quad \text{and} \quad -1 \leq 1.$$

All negative integers are less than zero (left of zero), all positive integers are greater than zero (right of zero), and any positive integer is greater than any negative integer.

EXAMPLE 3 ORDER RELATIONS BETWEEN NUMBERS

Some relationships between numbers on the number line in Figure 1.6 follow.

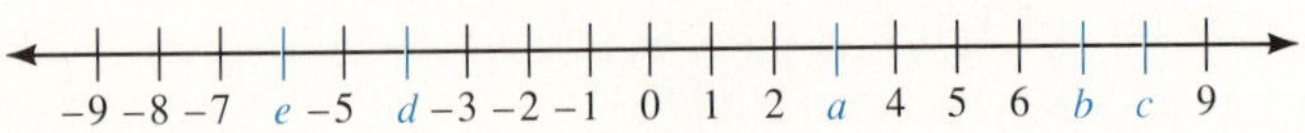

Figure 1.6

	In Words	*In Symbols*
(a)	1 is less than a	$1 < a$
(b)	c is greater than d	$c > d$
(c)	e is unequal to −5	$e \neq -5$
(d)	d is equal to −4	$d = -4$
(e)	b is less than or equal to b	$b \leq b$
(f)	a is greater than or equal to 2	$a \geq 2$

PRACTICE EXERCISE 3

Use Figure 1.6 and write the following in symbols.

(a) d is less than 0

(b) a is greater than or equal to −1

(c) b is equal to 7

(d) e is less than or equal to e

(e) a is greater than d

(f) c is unequal to −8

Answers: (a) $d < 0$ (b) $a \geq -1$ (c) $b = 7$ (d) $e \leq e$ (e) $a > d$ (f) $c \neq -8$

5 RATIONAL NUMBERS

In addition to the integers, the fractions we studied in Chapter 0, together with their negatives, can be shown or **plotted** on a number line. All of the fractions to the right of zero are called **positive rational numbers.** Those to the left of zero are **negative rational numbers,** and together, along with zero, they form the set of **rational numbers.** We can think of the rational numbers as

$Q = \{$all numbers that can be written as a quotient of two integers$\}$.

Any integer can be written as a quotient of two integers, for example $-2 = -\frac{2}{1}$ and $1 = \frac{1}{1}$, so all integers are also rational numbers. Some examples of rational numbers are plotted in Figure 1.7.

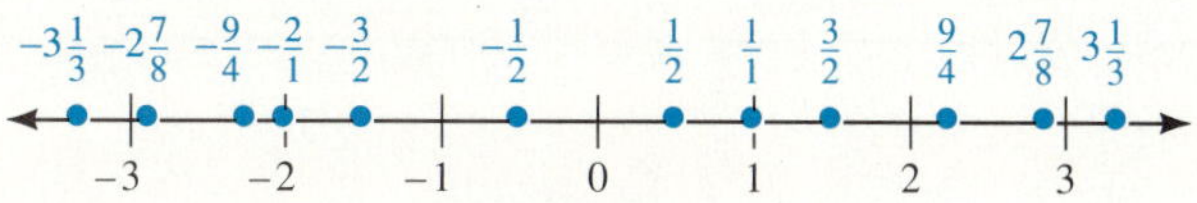

Figure 1.7

Just as with integers, we can define the order relationships of *less than* and *greater than* for rational numbers. For example, we see that

$$-\frac{1}{2} < \frac{1}{2}, \quad \frac{3}{2} < \frac{9}{4}, \quad 2\frac{7}{8} > -\frac{3}{2}, \quad \text{and} \quad -\frac{9}{4} > -3\frac{1}{3}$$

by looking at the number line in Figure 1.7.

A very important property of our number system states that if two numbers are unequal, then one of them must be less than the other. With two integers, it is easy to see if they are equal and, if not, which is less than the other. But this is not quite so easy for two rational numbers.

Ordering Rational Numbers

Given two positive rational numbers $\frac{a}{b}$ and $\frac{c}{d}$,

1. $\frac{a}{b} = \frac{c}{d}$ **whenever** $ad = bc$.
2. $\frac{a}{b} > \frac{c}{d}$ **whenever** $ad > bc$.
3. $\frac{a}{b} < \frac{c}{d}$ **whenever** $ad < bc$.

The product ad is called the **first cross product,** and bc is the **second cross product.** Thus, two positive fractions are equal if the cross products are equal, and the order of two unequal fractions is the same as the order of the first and second cross products. Similar results apply to negative rational numbers.

EXAMPLE 4 Ordering Rational Numbers

Place the correct symbol, =, >, or <, between the given pairs of fractions.

(a) $\frac{4}{6} \quad \frac{28}{42}$

Since $4 \cdot 42 = 168$ and $6 \cdot 28 = 168$, the fractions are equal.

$$\frac{4}{6} = \frac{28}{42}.$$

Practice Exercise 4

Place the correct symbol, =, >, or <, between the given pairs of fractions.

(a) $\frac{14}{3} \quad \frac{23}{5}$

(b) $\frac{24}{7}$ $\frac{39}{11}$

Since $24 \cdot 11 = 264$ and $7 \cdot 39 = 273$, and $264 < 273$,

$$\frac{24}{7} < \frac{39}{11}.$$

(c) $\frac{15}{9}$ $\frac{31}{20}$

Since $15 \cdot 20 = 300$ and $9 \cdot 31 = 279$, and $300 > 279$,

$$\frac{15}{9} > \frac{31}{20}.$$

(b) $\frac{21}{13}$ $\frac{18}{11}$

(c) $\frac{9}{28}$ $\frac{36}{112}$

Answers: (a) > (b) < (c) =

6 ABSOLUTE VALUE

In Chapter 0 we reviewed the four basic operations of addition, subtraction, multiplication, and division on the nonnegative rational numbers. When operating on *all* rational numbers we use the idea of *absolute value*.

Absolute Value of a Number

The **absolute value** of a number x is the distance from zero to x on a number line. We symbolize the absolute value of x by $|x|$.

EXAMPLE 5 Absolute Value Using a Number Line

(a) As shown in Figure 1.8, 3 is 3 units from 0. Thus, $|3| = 3$.

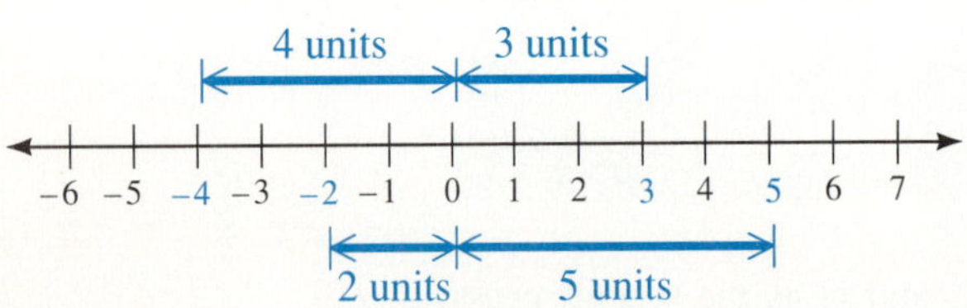

Figure 1.8

(b) 5 is 5 units from 0. Thus, $|5| = 5$.

(c) 0 is 0 units from 0. Thus, $|0| = 0$.

(d) -2 is 2 units from 0. Thus, $|-2| = 2$.

(e) -4 is 4 units from 0. Thus, $|-4| = 4$.

PRACTICE EXERCISE 5

Find the following absolute values.

(a) $|-8|$

(b) $|8|$

(c) $|0.05|$

(d) $|-0.05|$

(e) $\left|-\frac{3}{8}\right|$

Answers: (a) 8 (b) 8 (c) 0.05 (d) 0.05 (e) $\frac{3}{8}$

Remember that the absolute value of a positive number or zero is the number itself. The absolute value of a negative number is the positive number formed by removing the minus sign. Thus, *the absolute value of a number is always greater than or equal to zero, never negative.*

1.2 EXERCISES A

Answer true *or* false. *If the statement is false, tell why.*

1. $\{0, 1, 2, 3, \ldots\}$ is the set of nonnegative integers.

2. If a number x is located to the left of a number y on a number line, then $x > y$.

3. The numeral -7 is read "negative seven."

4. A temperature of 32° below zero could be written as $+32°$.

5. The distance from zero to a number y on a number line is called the absolute value of y.

6. An elevation of 4500 feet above sea level would be denoted by -4500.

7. A withdrawal of \$200 from a savings account could be denoted by $-\$200$.

Place the correct symbol, $=$, $>$, or $<$, between the given pairs of numbers.

8. $7 \quad 1$
9. $0 \quad 3$
10. $-2 \quad 0$
11. $2 \quad 9$
12. $-3 \quad 2$
13. $\frac{1}{2} \quad -\frac{1}{2}$
14. $-\frac{3}{4} \quad \frac{1}{4}$
15. $-5 \quad -5$

Is the given statement true or false? If the statement is false, tell why.

16. $-12 \le 0$
17. $6 \le 6$
18. $0 \le -3$
19. $-5 \le -5$
20. $-2 \le -7$
21. $\frac{2}{3} \le -\frac{2}{3}$
22. $4\frac{1}{2} \ge 4.5$
23. $-0.5 \le -\frac{1}{2}$

24. Plot the numbers $\frac{2}{3}$, $-\frac{7}{8}$, $\frac{5}{4}$, $-\frac{7}{4}$, $\frac{5}{2}$, and $-\frac{10}{3}$ on the given number line.

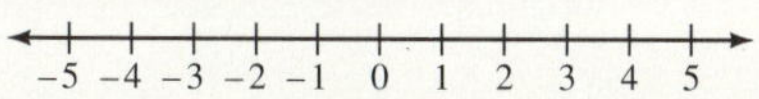

25. Insert the correct symbol, $=$, $>$, or $<$, between the two positive fractions.

(a) If $ad = bc$, then $\frac{a}{b} \quad \frac{c}{d}$.
(b) If $ad < bc$, then $\frac{a}{b} \quad \frac{c}{d}$.
(c) If $ad > bc$, then $\frac{a}{b} \quad \frac{c}{d}$.

(d) The products ad and bc are called ______________________.

Place the correct symbol, $=$, $>$, or $<$, between the given pairs of fractions.

26. $\frac{17}{43} \quad \frac{28}{79}$
27. $\frac{35}{11} \quad \frac{41}{13}$
28. $\frac{4}{11} \quad \frac{7}{19}$
29. $\frac{13}{50} \quad \frac{39}{150}$
30. $\frac{5}{7} \quad \frac{60}{83}$
31. $\frac{2}{9} \quad \frac{21}{91}$

Evaluate the absolute values.

32. $|17|$ **33.** $|-1|$ **34.** $|0|$ **35.** $\left|-\frac{3}{4}\right|$

36. $|3.1|$ **37.** $\left|-\frac{5}{2}\right|$ **38.** $|45|$ **39.** $|-0.8|$

FOR REVIEW

Evaluate when $x = 2$, $y = 4$, and $z = 5$.

40. $2z^2$ **41.** $(2z)^2$ **42.** $(y - x)^3$

43. $y^3 - x^3$ **44.** $z^2 - [xy - (x + y)]$ **45.** $\frac{(x + y)^2 - z^2}{x + y}$

ANSWERS: 1. true 2. false ($x < y$) 3. true 4. false ($-32°$) 5. true 6. false (4500) 7. true 8. $>$ 9. $<$ 10. $<$ 11. $<$ 12. $<$ 13. $>$ 14. $<$ 15. $=$ 16. true 17. true 18. false ($0 > -3$) 19. true 20. false ($-2 > -7$) 21. false $\left(\frac{2}{3} > -\frac{2}{3}\right)$ 22. true 23. true 24.

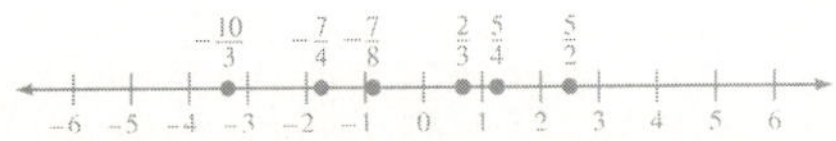

25. (a) $=$ (b) $<$ (c) $>$ (d) cross products 26. $>$ 27. $>$ 28. $<$ 29. $=$ 30. $<$ 31. $<$ 32. 17 33. 1 34. 0 35. $\frac{3}{4}$ 36. 3.1 37. $\frac{5}{2}$ 38. 45 39. 0.8 40. 50 41. 100 42. 8 43. 56 44. 23 45. $\frac{11}{6}$

1.2 EXERCISES B

Answer true *or* false. *If the statement is false, tell why.*

1. $\{\ldots, -3, -2, -1\}$ is the set of negative integers.

2. If a number w is located to the right of a number v on a number line, then $w < v$.

3. The numeral -11 is read "negative eleven."

4. The low temperature last winter in Twin Falls, Idaho of 38° below zero could be written $-38°$.

5. The absolute value of z, denoted by $|z|$, is the distance from zero to z on a number line.

6. An airplane flying at an altitude of 25,000 feet could be described as flying at $-25{,}000$ feet.

7. A check in the amount of \$35.00 written on an account could be denoted by $-\$35.00$.

Place the correct symbol, $=$, $>$, or $<$, between the given pairs of numbers.

8. 4 9 **9.** 5 0 **10.** 0 -7 **11.** 3 12

12. -5 1 **13.** $\frac{2}{3}$ $-\frac{2}{3}$ **14.** -0.2 0.2 **15.** -6 -6

Is the given statement true or false? If the statement is false, tell why.

16. $-3 \geq -4$ **17.** $-1 \geq -1$ **18.** $0 \leq -1$ **19.** $-8 \leq -8$

20. $-5 \leq -40$ **21.** $\frac{1}{4} \geq -\frac{1}{4}$ **22.** $3\frac{1}{5} \leq 3.2$ **23.** $-0.1 \leq -\frac{1}{10}$

24. Plot the numbers $\frac{1}{3}$, $-\frac{3}{4}$, $\frac{7}{4}$, $-\frac{9}{4}$, $\frac{7}{2}$, and $-\frac{11}{3}$ on a number line.

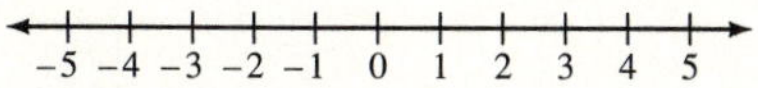

25. Insert the correct symbol, =, >, or <, between the two positive fractions.

(a) If $xw < yv$ then $\frac{x}{y}$ $\frac{v}{w}$. **(b)** If $xw = yv$ then $\frac{x}{y}$ $\frac{v}{w}$. **(c)** If $xw > yv$ then $\frac{x}{y}$ $\frac{v}{w}$.

(d) The cross products used to determine order on the two rational numbers $\frac{x}{y}$ and $\frac{v}{w}$ are ______ and ______.

Place the correct symbol, =, >, or <, between the given pairs of fractions.

26. $\frac{3}{5}$ $\frac{7}{10}$ **27.** $\frac{36}{11}$ $\frac{77}{20}$ **28.** $\frac{19}{43}$ $\frac{57}{129}$

29. $\frac{6}{7}$ $\frac{47}{55}$ **30.** $\frac{121}{9}$ $\frac{55}{3}$ **31.** $\frac{27}{201}$ $\frac{13}{121}$

Evaluate the absolute values.

32. $|25|$ **33.** $|-2|$ **34.** $\left|-\frac{1}{2}\right|$ **35.** $|-0|$

36. $|8.7|$ **37.** $\left|-\frac{9}{4}\right|$ **38.** $|112|$ **39.** $|-0.03|$

FOR REVIEW

Evaluate when $a = 10$, $b = 20$, and $c = 3$.

40. $4a^2$ **41.** $(4a)^2$ **42.** $(b - a)^2$

43. $b^2 - a^2$ **44.** $(ac - b)^2 - [b - (a - c)]$ **45.** $\frac{(b - a)^2 - (a - c)^2}{(b - a)^2}$

1.2 EXERCISES C

Place the correct symbol, =, >, or <, between the pairs of fractions.

1. $-\frac{14}{19}$ $-\frac{2}{3}$ **2.** $-\frac{21}{8}$ $-\frac{31}{11}$ **3.** $-\frac{77}{14}$ $-\frac{11}{2}$

1.3 ADDITION AND SUBTRACTION OF RATIONAL NUMBERS

STUDENT GUIDEPOSTS

1. Adding Numbers with Like Signs
2. Adding Numbers with Unlike Signs
3. Additive Identity
4. Additive Inverse (Negative)
5. Commutative Law of Addition
6. Associative Law of Addition
7. Adding More Than Two Numbers
8. Subtracting Numbers

In Chapter 0 we reviewed the four basic operations on the nonnegative rational numbers. Now we extend our treatment to addition and subtraction of all rational numbers. Consider the addition problem $3 + 2$ on the number line in Figure 1.9.

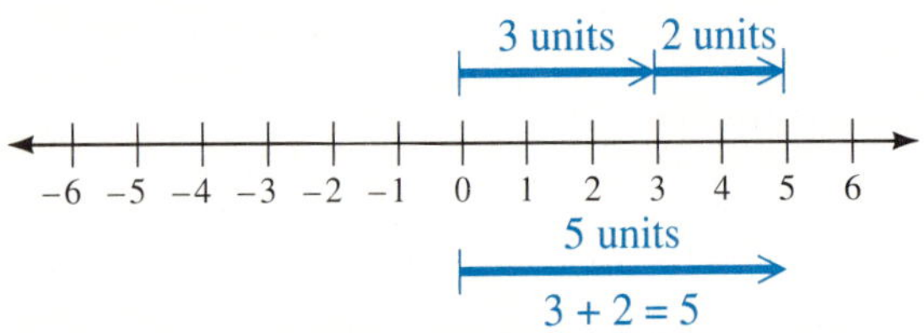

Figure 1.9

To find $3 + 2$, we start at 0, move 3 units to the *right,* and then move 2 more units to the *right*. We are then 5 units to the *right* of zero. Thus, $3 + 2 = 5$.

We can also find $3 + (-2)$ on a number line. We first move 3 units to the *right* and then 2 units to the *left*. This puts us 1 unit to the *right* of 0, as shown in Figure 1.10. Thus, $3 + (-2) = 1$.

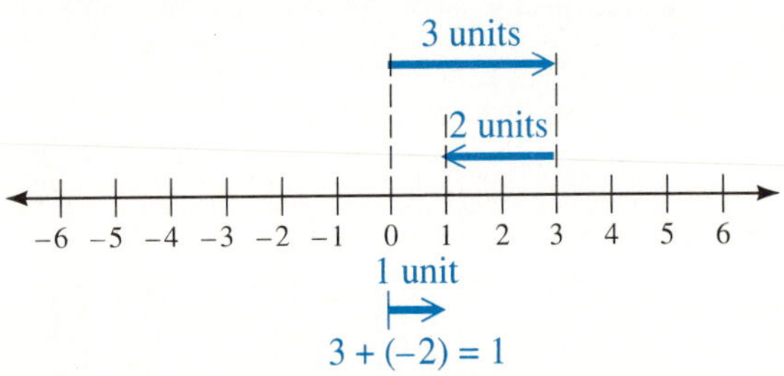

Figure 1.10

If we think of numbers as temperatures, an increase of 3° (+3) followed by a second increase of 2° (+2) results in a total increase of 5° (+5). Similarly, an increase of 3° (+3) followed by a decrease of 2° (−2) results in a net increase of 1° (+1).

Notice that when adding numbers using a number line, we move to the *right when a number is positive* and to the *left when it is negative*.

EXAMPLE 1 ADDING NUMBERS WITH LIKE SIGNS

Find $(-3) + (-2)$.

Draw a number line. Move 3 units to the left for -3 and then 2 more units to the left for -2. As seen in Figure 1.11, we end up 5 units to the left of 0, at -5. Thus, $(-3) + (-2) = -5$.

PRACTICE EXERCISE 1

Find $(-11) + (-8)$.

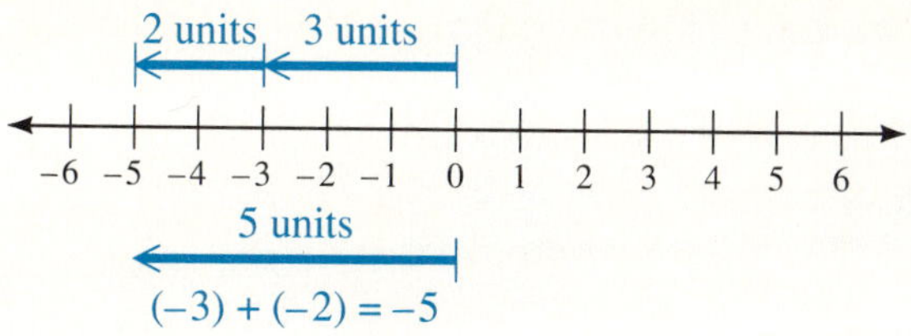

Figure 1.11

Answer: −19

EXAMPLE 2 ADDING NUMBERS WITH UNLIKE SIGNS

Find $(-3) + 2$.

Draw a number line. Move 3 units to the left and then 2 units to the right. As Figure 1.12 shows, we are then at -1. Thus, $(-3) + 2 = -1$.

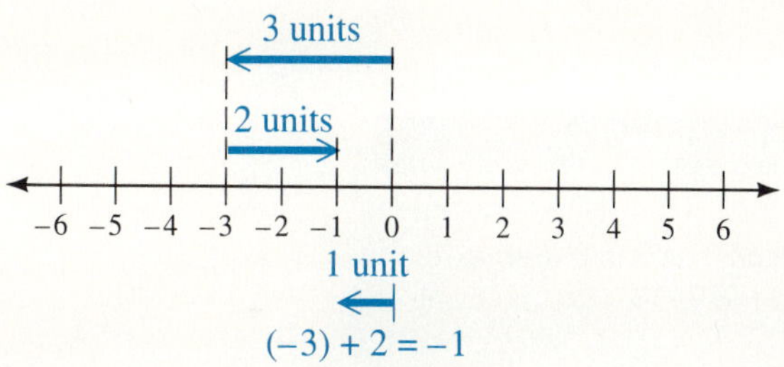

Figure 1.12

PRACTICE EXERCISE 2

Find $(-11) + 8$.

Answer: −3

EXAMPLE 3 ADDING NUMBERS WITH UNLIKE SIGNS

Find $2 + (-3)$.

Draw a number line. Move 2 units to the right and then 3 units to the left. (See Figure 1.13.) This puts us at -1. Thus, $2 + (-3) = -1$.

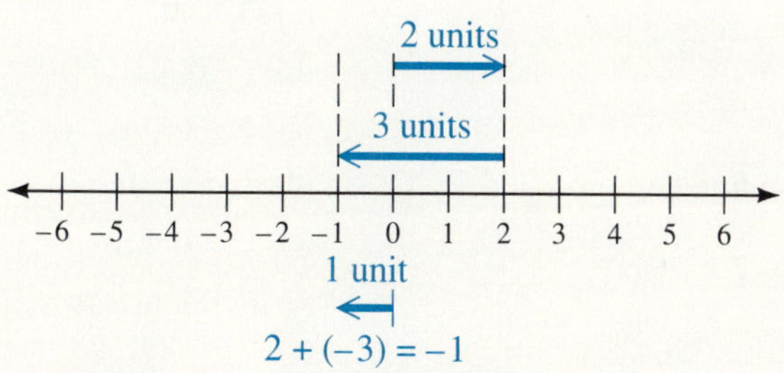

Figure 1.13

PRACTICE EXERCISE 3

Find $8 + (-11)$.

Answer: −3

1 ADDING NUMBERS WITH LIKE SIGNS

The number-line method of adding numbers shows us what addition means, but it takes too much time. For example, we found $(-3) + (-2)$ to be -5 using a number line (Figure 1.11). However, we could have added $3 + 2$ and attached a minus sign:

$$(-3) + (-2) = -(3 + 2) = -5.$$

This is an example of the following rule.

Adding Numbers Having Like Signs

To add two numbers having **like signs,** add their absolute values. The sum has the same sign as the numbers being added.

2 ADDING NUMBERS WITH UNLIKE SIGNS

When the signs of the numbers being added are not the same, we need another rule. For $3 + (-2) = 1$ (Figure 1.10), we could have found the difference between 3 and 2 and attached a plus sign:

$$3 + (-2) = +(3 - 2) = +1 = 1.$$

Also, for $(-3) + 2 = 2 + (-3) = -1$ (Figures 1.12 and 1.13), we could have subtracted 2 from 3 and attached a minus sign:

$$(-3) + 2 = 2 + (-3) = -(3 - 2) = -1.$$

These are examples of the following rule.

Adding Numbers Having Unlike Signs

To add two numbers having **unlike signs,** subtract the smaller absolute value from the larger absolute value. The result has the same sign as the number with the larger absolute value. If the absolute values are the same, the sum is zero.

Remember that if a number (other than zero) has no sign, this is the same as having a plus sign attached. For example, 5 and +5 are the same.

EXAMPLE 4 Adding Rational Numbers

Add using the preceding rules.

(a) $6 + 7 = 13$

(b) $(-6) + (-7) = -(6 + 7) = -13$

(c) $6 + (-7) = -(7 - 6) = -1$ $\quad |-7| = 7 > 6 = |6|$

(d) $(-6) + 7 = +(7 - 6) = +1 = 1$ $\quad |-6| = 6 < 7 = |7|$

(e) $12 + (-12) = 0$ $\quad |12| = 12 = |-12|$

Practice Exercise 4

Add.

(a) $5 + 14$

(b) $(-5) + (-14)$

(c) $5 + (-14)$

(d) $(-5) + 14$

(e) $(-14) + 14$

Answers: (a) 19 (b) −19 (c) −9 (d) 9 (e) 0

3 ADDITIVE IDENTITY

The number 0 plays a special role in our number system since for any number a,

$$a + 0 = 0 + a = a.$$

Intuitively, we see any number a remains identically the same when added to 0. For this reason, 0 is called the **additive identity.**

4 ADDITIVE INVERSE (NEGATIVE)

Example 4(e) illustrates another property of our number system. If a is any number, we call $-a$ the **additive inverse,** or **negative** of a, which means

$$a + (-a) = (-a) + a = 0.$$

EXAMPLE 5 Additive Identity and Inverses

Add.

(a) $0 + 7 = 7 + 0 = 7$ 0 is the additive identity

(b) $0 + (-7) = (-7) + 0 = -7$ 0 is the additive identity

(c) $(-2.3) + (-4.1) = -(2.3 + 4.1) = -6.4$

(d) $(-3.5) + 1.7 = -(3.5 - 1.7) = -1.8$

(e) $\frac{1}{2} + \left(-\frac{3}{4}\right) = -\left(\frac{3}{4} - \frac{1}{2}\right) = -\left(\frac{3}{4} - \frac{2}{4}\right) = -\frac{1}{4}$

Practice Exercise 5

Add.

(a) $13 + 0$

(b) $(-13) + 0$

(c) $(-4.02) + (-1.35)$

(d) $(-6.35) + 2.11$

(e) $\left(-\frac{2}{9}\right) + \frac{4}{11}$

Answers: (a) 13 (b) −13 (c) −5.37 (d) −4.24 (e) $\frac{14}{99}$

5 COMMUTATIVE LAW OF ADDITION

With two more properties of addition, we can change the order of addition, and rearrange grouping symbols in sums of three (or more) numbers. If a and b are two numbers, the **commutative law of addition** states that

$$a + b = b + a.$$

That is, the order of addition can be changed by the commutative law. For example, $16 + 5 = 5 + 16$.

6 ASSOCIATIVE LAW OF ADDITION

Given another number c, the **associative law of addition** states that

$$(a + b) + c = a + (b + c).$$

That is, the grouping symbols can be rearranged by the associative law. For example, $(5 + 2) + 3 = 5 + (2 + 3)$.

EXAMPLE 6 Commutative and Associative Laws

(a) Use the commutative law to complete the equation $5 + \frac{1}{4} =$ _____.

The commutative law says the order of addition can be changed, so

$$5 + \frac{1}{4} = \frac{1}{4} + 5.$$

(b) Use the associative law to complete the equation $(7 + \frac{1}{2}) + (-2) =$ _____.

The associative law says the grouping symbols can be rearranged, so

$$\left(7 + \frac{1}{2}\right) + (-2) = 7 + \left[\frac{1}{2} + (-2)\right].$$

Practice Exercise 6

(a) Complete the following using the commutative law.
$\left(-\frac{1}{2}\right) + 9 =$ _____

(b) Complete the following using the associative law.
$(7 + 0.5) + 2 =$ _____

Answers: (a) $9 + \left(-\frac{1}{2}\right)$
(b) $7 + (0.5 + 2)$

7 ADDING MORE THAN TWO NUMBERS

Taken together, the commutative and associative laws of addition allow us to add several signed numbers by reordering and regrouping the numbers to obtain the sum in the most convenient manner.

Adding More Than Two Numbers

1. Add all positive numbers.
2. Add all negative numbers.
3. Now use the rule for adding numbers with unlike signs.

EXAMPLE 7 Adding Several Numbers

Add.

$$\begin{aligned}&(-2) + (-3) + 2 + 7 + (-5) + 6 + 9 + (-1)\\&= [2 + 7 + 6 + 9] + [(-2) + (-3) + (-5) + (-1)]\\&= [2 + 7 + 6 + 9] + [-(2 + 3 + 5 + 1)]\\&= [24] + [-11] = +[24 - 11] = 13\end{aligned}$$

Practice Exercise 7

Add.

$5 + (-2) + (-1) + 7 + (-9) + 6$

Answer: 6

With practice we should be able to skip some intermediate steps in the preceding examples and make computations mentally.

8 SUBTRACTING NUMBERS

Subtraction of numbers can be defined as addition of an additive inverse or a negative. For example, we can think of $5 - 3 = 2$ as $5 + (-3) = 2$.

Subtracting Numbers

To subtract one number from another, change the sign of the number being subtracted and use the rule for adding numbers. Thus, for rational numbers a and b,

$$a - b = a + (-b).$$

EXAMPLE 8 Subtracting Numbers

Subtract.

(a) $5 - (+3) = 5 + (-3)$ Change sign and add
$= 5 - 3 = 2$

(b) $5 - (-3) = 5 + 3 = 8$ Change sign and add

(c) $(-5) - 3 = (-5) + (-3)$ Change sign and add
$= -(5 + 3) = -8$

Practice Exercise 8

Subtract.

(a) $17 - (+2)$

(b) $17 - (-2)$

(c) $(-17) - 2$

(d) $(-5) - (-3) = (-5) + 3$ Change sign and add
$= -(5 - 3) = -2$

(e) $(-5) - 0 = (-5) + (-0)$ Change sign and add
$= (-5) + 0$ The negative of zero is zero: $-0 = 0$
$= -5$

(d) $(-17) - (-2)$

(e) $0 - (-17)$

Answers: (a) 15 (b) 19 (c) −19 (d) −15 (e) 17

EXAMPLE 9 Subtracting Numbers

Subtract.

(a) $5 - (+5) = 5 + (-5) = 0$

(b) $5 - (-5) = 5 + 5 = 10$

(c) $(-5) - 5 = (-5) + (-5) = -(5 + 5) = -10$

(d) $(-5) - (-5) = (-5) + 5 = 0$

(e) $(-5.8) - (-3.2) = (-5.8) + 3.2 = -(5.8 - 3.2) = -2.6$

(f) $\left(-\frac{2}{3}\right) - \frac{1}{6} = \left(-\frac{2}{3}\right) + \left(-\frac{1}{6}\right) = \left(-\frac{4}{6}\right) + \left(-\frac{1}{6}\right) = -\frac{5}{6}$

Practice Exercise 9

Subtract.

(a) $12 - (+12)$

(b) $12 - (-12)$

(c) $(-12) - 12$

(d) $(-12) - (-12)$

(e) $(-1.2) - (-5.7)$

(f) $\left(-\frac{1}{8}\right) - \frac{3}{5}$

Answers: (a) 0 (b) 24 (c) −24 (d) 0 (e) 4.5 (f) $-\frac{29}{40}$

CAUTION

Although addition satisfies both the commutative and associative laws, subtraction does not. For example,

$3 = 5 - 2 \neq 2 - 5 = -3.$ Subtraction is *not* commutative

Likewise,

$(8 - 4) - 3 = 4 - 3 = 1,$

but $8 - (4 - 3) = 8 - 1 = 7.$ Subtraction is *not* associative

EXAMPLE 10 Application to Personal Banking

John had \$932 in his bank account on May 1. He deposited \$326 and \$791 and wrote checks for \$816, \$315, and \$940 during the month. What was his balance at the end of May?

We can consider the deposits as positive numbers and the checks written as negative numbers. We need to add 932, 326, 791, −816, −315, and −940.

$$\begin{array}{r} 932 \\ 326 \\ +\ 791 \\ \hline 2049 \end{array} \text{ Sum of deposits} \qquad \begin{array}{r} -816 \\ -315 \\ -940 \\ \hline -\ 2071 \end{array} \text{ Sum of checks} \qquad \begin{array}{r} -2071 \\ 2049 \\ \hline -\ 22 \end{array} \text{ Final sum}$$

Thus, John is overdrawn (in the red) \$22 on his account.

Practice Exercise 10

Marlo had a temperature of 100.1° at noon. It rose 2.3° before dropping 3.5° by 3:00 P.M. What was her temperature at 3:00 P.M.?

Answer: 98.9°

1.3 EXERCISES A

Perform the indicated operations.

1. $4 + 3$

2. $(-4) + (-3)$

3. $4 + (-3)$

4. $(-4) + 3$

5. $4 + 0$

6. $0 + (-4)$

7. $7 + (-7)$

8. $(-7) + 7$

9. $(-7) + (-7)$

10. $7 + 7$

11. $(-8) + (-5)$

12. $9 + (-1)$

13. $19 + (-25)$

14. $(-16) + 14$

15. $4 - 3$

16. $(-4) - (-3)$

17. $4 - (-3)$

18. $(-4) - 3$

19. $0 - 4$

20. $(-4) - 0$

21. $7 - (-7)$

22. $(-7) - 7$

23. $(-7) - (-7)$

24. $7 - 7$

25. $(-8) - (-5)$

26. $9 - (-1)$

27. $(-1) - 9$

28. $14 - 21$

29. $19 - (-25)$

30. $(-16) - 14$

31. $1.3 + (-2.5)$

32. $(-1.3) + 2.5$

33. $(-1.3) + (-2.5)$

34. $1.3 - (-2.5)$

35. $(-1.3) - 2.5$

36. $(-1.3) - (-2.5)$

37. $\frac{1}{2} + \left(-\frac{2}{5}\right)$

38. $\left(-\frac{1}{2}\right) + \frac{2}{5}$

39. $\left(-\frac{1}{2}\right) + \left(-\frac{2}{5}\right)$

40. $\frac{1}{2} - \left(-\frac{2}{5}\right)$

41. $\left(-\frac{1}{2}\right) - \frac{2}{5}$

42. $\left(-\frac{1}{2}\right) - \left(-\frac{2}{5}\right)$

43. $(-3) + 2 + (-7) + (-8) + 9$

44. $5 + 8 + 9 + (-3) + (-7)$

45. $(-8) + (-6) + 9 + (-3) + 5 + (-6)$

46. $(-4) + (-2) + 6 + 7 + (-1)$

47. Show by example that subtraction does not satisfy the commutative law. Can you find numbers a and b for which $a - b$ does equal $b - a$?

Add.

48. $\begin{array}{r} 23 \\ -16 \\ -36 \\ 5 \\ \underline{81} \end{array}$

49. $\begin{array}{r} -16 \\ -81 \\ -14 \\ -70 \\ \underline{-93} \end{array}$

50. $\begin{array}{r} -64 \\ 25 \\ 38 \\ -17 \\ \underline{-83} \end{array}$

51. When a cold front came through Cut Bank, Montana, the temperature dropped from 35° above zero to 13° below zero ($-13°$). What was the total change in temperature?

52. On March 1, Allen had \$64 in his bank account. During the month, he wrote checks for \$81, \$108, and \$192 and made deposits of \$123 and \$140. What was the value of his account at the end of the month?

53. A helicopter is 600 ft above sea level and a submarine directly below it is 250 ft below sea level (-250 ft). How far apart are they?

54. At 5:30 A.M., it was -42°F. By 2:00 P.M. the same day, the temperature had risen 57° to the recorded high temperature. Shortly thereafter, a cold front passed through dropping the temperature 41° by 5:00 P.M. What was the temperature at 5:00 P.M.?

FOR REVIEW

Place the correct symbol, =, <, or >, between the given pairs of numbers.

55. $-5 \quad -4$

56. $\frac{2}{3} \quad \frac{77}{110}$

57. $\frac{5}{12} \quad \frac{30}{72}$

58. $-4.7 \quad -1.2$

Evaluate the absolute values.

59. $|-81|$

60. $\left|-\frac{5}{4}\right|$

61. $|17|$

62. $|-7.25|$

ANSWERS: 1. 7 2. -7 3. 1 4. -1 5. 4 6. -4 7. 0 8. 0 9. -14 10. 14 11. -13 12. 8 13. -6 14. -2 15. 1 16. -1 17. 7 18. -7 19. -4 20. -4 21. 14 22. -14 23. 0 24. 0 25. -3 26. 10 27. -10 28. -7 29. 44 30. -30 31. -1.2 32. 1.2 33. -3.8 34. 3.8 35. -3.8 36. 1.2 37. $\frac{1}{10}$ 38. $-\frac{1}{10}$ 39. $-\frac{9}{10}$ 40. $\frac{9}{10}$ 41. $-\frac{9}{10}$ 42. $-\frac{1}{10}$ 43. -7 44. 12 45. -9 46. 6 47. $2 - 1 = 1 \neq -1 = 1 - 2$; if $a = b$ then $a - b = 0 = b - a$ 48. 57 49. -274 50. -101 51. $-48°$ 52. $-\$54$ 53. 850 ft 54. $-26°$ 55. < 56. < 57. = 58. < 59. 81 60. $\frac{5}{4}$ 61. 17 62. 7.25

1.3 EXERCISES B

Perform the indicated operations.

1. $5 + 2$
2. $(-5) + (-2)$
3. $5 + (-2)$
4. $(-5) + 2$
5. $0 + 5$
6. $(-5) + 0$
7. $5 + (-5)$
8. $(-5) + 5$
9. $(-5) + (-5)$
10. $5 + 5$
11. $(-9) + (-3)$
12. $(-2) + 11$
13. $18 + (-23)$
14. $(-15) + 19$
15. $8 - 7$
16. $(-8) - (-7)$
17. $8 - (-7)$
18. $(-8) - 7$
19. $10 - 0$
20. $0 - (-10)$
21. $6 - (-6)$

22. $(-6) - 6$ **23.** $(-6) - (-6)$ **24.** $6 - 6$

25. $(-11) - (-4)$ **26.** $(-13) - (-1)$ **27.** $(-1) - 7$

28. $(-12) - 24$ **29.** $18 - (-27)$ **30.** $(-17) - 11$

31. $1.5 + (-3.8)$ **32.** $(-1.5) + 3.8$ **33.** $(-1.5) + (-3.8)$

34. $1.5 - (-3.8)$ **35.** $(-1.5) - 3.8$ **36.** $(-1.5) - (-3.8)$

37. $\frac{1}{4} + \left(-\frac{3}{8}\right)$ **38.** $\left(-\frac{1}{4}\right) + \frac{3}{8}$ **39.** $\left(-\frac{1}{4}\right) + \left(-\frac{3}{8}\right)$

40. $\frac{1}{4} - \left(-\frac{3}{8}\right)$ **41.** $\left(-\frac{1}{4}\right) - \frac{3}{8}$ **42.** $\left(-\frac{1}{4}\right) - \left(-\frac{3}{8}\right)$

43. $(-5) + 4 + (-9) + (-10) + 6$ **44.** $5 + 9 + (-3) + 1 + (-10)$

45. $(-5) + (-7) + 10 + (-3) + 4 + (-7)$ **46.** $(-6) + (-3) + 6 + 10 + (-2)$

47. Show by example that subtraction does not satisfy the associative law. Can you find numbers, a, b, and c for which $(a - b) - c$ does equal $a - (b - c)$?

Add.

48.	**49.**	**50.**
26	−17	−78
−14	−20	32
−51	−43	51
3	−5	−60
92	−110	−92

51. In a five-hour period of time, the temperature in Empire, N.Y. dropped from 41° to 12° below zero. What was the total temperature change?

52. On July 4, Uncle Sam had \$97 in his checking account. During the next week he wrote checks for \$15, \$102, and \$312, and made deposits of \$145 and \$170. What was the value of his account at the end of the week?

53. A balloon is 1200 ft above sea level, and a diving bell directly below it is 340 ft beneath the surface of the water. How far apart are the two?

54. At 8:00 A.M., Leah Johnson had a temperature of 99.8°. It rose another 3.1° before falling 4.3° by noon. What was her temperature at noon?

FOR REVIEW

Place the correct symbol, =, <, or >, between the given pairs of numbers.

55. $-6 \quad -7$ **56.** $\frac{3}{4} \quad \frac{96}{120}$ **57.** $\frac{7}{8} \quad \frac{77}{88}$ **58.** $0 \quad -11.5$

Evaluate the absolute values.

59. $\left|-\frac{9}{10}\right|$ **60.** $|-210|$ **61.** $|42|$ **62.** $|-0.01|$

1.3 EXERCISES C

Perform the indicated operations.

1. $(-962) - (-508)$

2. $-48.62 - 40.002$

3. $64\frac{7}{24} - 98\frac{13}{16}$

[Answer: $-34\frac{25}{48}$]

1.4 MULTIPLICATION AND DIVISION OF RATIONAL NUMBERS

STUDENT GUIDEPOSTS

1. Multiplying Numbers
2. Multiplicative Identity
3. Multiplicative Inverse (Reciprocal)
4. Commutative Law of Multiplication
5. Associative Law of Multiplication
6. Dividing Numbers
7. Double Sign Properties

1 MULTIPLYING NUMBERS

In multiplying and dividing numbers, we need to decide what sign to give the product or quotient. To help us, we look for patterns in several products.

Decreases by 1 each time → (first factors); Decreases by 2 each time → (products)

$$\begin{array}{rcrl} 4\cdot 2 &=& 8 & \\ 3\cdot 2 &=& 6 & \text{6 is 2 less than 8} \\ 2\cdot 2 &=& 4 & \\ 1\cdot 2 &=& 2 & \\ 0\cdot 2 &=& 0 & \\ (-1)\cdot 2 &=& -2 & \text{-2 is 2 less than 0} \\ (-2)\cdot 2 &=& -4 & \text{-4 is 2 less than -2} \\ (-3)\cdot 2 &=& -6 & \\ (-4)\cdot 2 &=& -8 & \end{array}$$

We know the products $4 \cdot 2$, $3 \cdot 2$, $2 \cdot 2$, $1 \cdot 2$, and $0 \cdot 2$, and we can see that these products decrease by 2 each time. For this pattern to continue, a negative number times a positive number must be negative. For example,

$$(-1)\cdot 2 = -2, \quad (-2)\cdot 2 = -4, \quad (-3)\cdot 2 = -6, \quad (-4)\cdot 2 = -8.$$

Similarly, any positive number times any negative number must be negative. We use this to find a pattern in the following products.

Decreases by 1 each time → (first factors); Increases by 2 each time → (products)

$$\begin{array}{rcrl} 4\cdot(-2) &=& -8 & \\ 3\cdot(-2) &=& -6 & \text{-6 is 2 more than -8} \\ 2\cdot(-2) &=& -4 & \\ 1\cdot(-2) &=& -2 & \\ 0\cdot(-2) &=& 0 & \text{0 is 2 more than -2} \\ (-1)\cdot(-2) &=& 2 & \\ (-2)\cdot(-2) &=& 4 & \\ (-3)\cdot(-2) &=& 6 & \\ (-4)\cdot(-2) &=& 8 & \end{array}$$

For this pattern to continue, a negative number times a negative number must be a positive number. For example,

$$(-1)\cdot(-2) = 2, \quad (-2)\cdot(-2) = 4, \quad (-3)\cdot(-2) = 6, \quad (-4)\cdot(-2) = 8.$$

Our results are summarized in the following rule.

Multiplying Numbers

1. Multiply the absolute values of the numbers.
2. If both signs are positive or both negative, the product is positive.
3. If one sign is positive and one is negative, the product is negative.
4. If one number (or both numbers) is zero, the product is zero.

When multiplying numbers, remember that *like signs have a positive product and unlike signs have a negative product.*

EXAMPLE 1 MULTIPLYING NUMBERS

Multiply.

(a) $5 \cdot 2 = 10$ Like signs

(b) $(-5)(-2) = (5 \cdot 2) = 10$ Like signs

(c) $5 \cdot (-2) = -(5 \cdot 2) = -10$ Unlike signs

(d) $(-5) \cdot 2 = -(5 \cdot 2) = -10$ Unlike signs

(e) $0 \cdot (-8) = 0$

(f) $0 \cdot 0 = 0$

(g) $(3)\left(\frac{1}{3}\right) = 1$

PRACTICE EXERCISE 1

Multiply.

(a) $4 \cdot 11$

(b) $(-4)(-11)$

(c) $4 \cdot (-11)$

(d) $(-4) \cdot 11$

(e) $(12) \cdot 0$

(f) $0 \cdot (-12)$

(g) $\left(\frac{3}{7}\right)\left(\frac{7}{3}\right)$

Answers: (a) 44 (b) 44 (c) −44 (d) −44 (e) 0 (f) 0 (g) 1

2 MULTIPLICATIVE IDENTITY

The number 1 plays a role in multiplication similar to the one 0 plays in addition. Since for any number a,

$$\mathbf{a \cdot 1 = 1 \cdot a = a,}$$

we see that a number remains identically the same when multiplied by 1. For this reason, 1 is called the **multiplicative identity.**

3 MULTIPLICATIVE INVERSE (RECIPROCAL)

Example 1(g) illustrates a property of reciprocals that we have seen before. If a is any number except 0, we call $\frac{1}{a}$ the **multiplicative inverse** (or **reciprocal**) of a, which means that

$$a \cdot \frac{1}{a} = \frac{1}{a} \cdot a = 1.$$

EXAMPLE 2 Multiplicative Identity and Inverses

Multiply.

(a) $(-5)\left(-\frac{1}{5}\right) = 1$ — -5 and $-\frac{1}{5}$ are reciprocals or multiplicative inverses

(b) $1 \cdot (-7) = (-7) \cdot 1 = -7$ — 1 is the multiplicative identity

(c) $(-2.1) \cdot 3.5 = -(2.1)(3.5) = -7.35$

(d) $(-2.1) \cdot (-3.5) = (2.1)(3.5) = 7.35$

(e) $\left(\frac{1}{4}\right)\left(-\frac{3}{5}\right) = -\left(\frac{1}{4}\right)\left(\frac{3}{5}\right) = -\frac{3}{20}$

(f) $\left(-\frac{1}{4}\right)\left(-\frac{3}{5}\right) = \left(\frac{1}{4}\right)\left(\frac{3}{5}\right) = \frac{3}{20}$

Practice Exercise 2

Multiply.

(a) $\left(-\frac{1}{9}\right)(-9)$

(b) $1 \cdot (-13)$

(c) $(-4.1) \cdot (-2.9)$

(d) $(-4.1) \cdot (2.9)$

(e) $\left(\frac{1}{7}\right)\left(-\frac{3}{8}\right)$

(f) $\left(-\frac{1}{7}\right)\left(-\frac{3}{8}\right)$

Answers: (a) 1 (b) -13 (c) 11.89 (d) -11.89 (e) $-\frac{3}{56}$ (f) $\frac{3}{56}$

4 COMMUTATIVE LAW OF MULTIPLICATION

With two more properties of multiplication, like the ones for addition, we can change the order of multiplication and rearrange grouping symbols in products of three (or more) numbers. If a and b are two numbers, the **commutative law of multiplication** states that

$$a \cdot b = b \cdot a.$$

That is, the order of multiplication can be changed by the commutative law. For example, $4 \cdot 7 = 7 \cdot 4$.

5 ASSOCIATIVE LAW OF MULTIPLICATION

Given another number c, the **associative law of multiplication** states that

$$(a \cdot b) \cdot c = a \cdot (b \cdot c).$$

That is, the grouping symbols can be rearranged by the associative law. For example, $(2 \cdot 5) \cdot 3 = 2 \cdot (5 \cdot 3)$.

EXAMPLE 3 Commutative and Associative Laws

(a) Use the commutative law to complete the equation $7 \cdot \frac{3}{4} =$ _____.
The commutative law says the order of multiplication can be changed, so

$$7 \cdot \frac{3}{4} = \frac{3}{4} \cdot 7.$$

(b) Use the associative law to complete the equation $\left(3 \cdot \frac{1}{5}\right) \cdot (-5) =$ _____.

The associative law says the grouping symbols can be rearranged, so

$$\left(3 \cdot \frac{1}{5}\right) \cdot (-5) = 3 \cdot \left(\frac{1}{5} \cdot (-5)\right).$$

Practice Exercise 3

(a) Use the commutative law to complete the following.
$(0.09) \cdot (1000) =$ _____

(b) Use the associative law to complete the following.
$\left(\frac{1}{2} \cdot \frac{3}{10}\right) \cdot (-6) =$ _____

Answers: (a) $(1000) \cdot (0.09)$ (b) $\left(\frac{1}{2}\right) \cdot \left(\frac{3}{10} \cdot (-6)\right)$

Often we will need to multiply more than two numbers. To do this, multiply in pairs and keep track of the sign. Notice in Example 4 that *products involving an odd number of minus signs are negative while those with an even number are positive*.

EXAMPLE 4 Multiplying More Than Two Numbers

Multiply.

(a) $(3)(-2)(-4) = (-6)(-4) = 24$ Multiply 3 times −2 first

(b) $(-1)(-1)(-2)(-3)(-4) = (1)(-2)(-3)(-4)$ $(-1)(-1) = 1$

$= (-2)(-3)(-4)$ $(1)(-2) = -2$

$= (6)(-4)$ $(-2)(-3) = 6$

$= -24$

(c) $(-2)^3 = (-2)(-2)(-2) = (4)(-2)$ $(-2)(-2) = 4$

$= -8$

Practice Exercise 4

Multiply.

(a) $(2)(-1)(-4)(3)$

(b) $(-6)(-1)(-5)(-2)(-1)$

(c) $(-3)^4$

Answers: (a) 24 (b) −60 (c) 81

6 DIVIDING NUMBERS

We have seen that the only difference between multiplying positive numbers and multiplying rational numbers is finding the sign of the product. Recall that division can be thought of as multiplication by a reciprocal. For example,

$$15 \div 3 \text{ or } \frac{15}{3} \quad \text{is the same as} \quad 15 \cdot \frac{1}{3}.$$

Thus, we can use the same rules for signs in division that we used for multiplication.

Dividing Numbers

1. Divide the absolute values of the numbers.
2. If both signs are positive or both negative, the quotient is positive.
3. If one sign is positive and one negative, the quotient is negative.
4. Zero divided by any number (except zero) is zero, and division of any number by 0 is undefined.

Remember that the rule of signs for division is the same as for multiplication: *like signs have a positive quotient and unlike signs have a negative quotient*.

EXAMPLE 5 Dividing Numbers

Divide.

(a) $6 \div 3 = 2$ Like signs

(b) $(-6) \div (-3) = 6 \div 3 = 2$ Like signs

(c) $6 \div (-3) = -(6 \div 3) = -2$ Unlike signs

(d) $(-6) \div 3 = -(6 \div 3) = -2$ Unlike signs

Practice Exercise 5

Divide.

(a) $12 \div 4$

(b) $(-12) \div (-4)$

(c) $12 \div (-4)$

(d) $(-12) \div 4$

(e) $0 \div (-3) = 0$

(f) $(-3) \div 0$ is undefined

(g) $\frac{1}{2} \div \left(-\frac{3}{5}\right) = \frac{1}{2} \cdot \left(-\frac{5}{3}\right) = -\left(\frac{1}{2} \cdot \frac{5}{3}\right) = -\frac{5}{6}$

(e) $(-0.3) \div 0$

(f) $0 \div (-0.3)$

(g) $\left(-\frac{3}{5}\right) \div \frac{9}{10}$

Answers: (a) 3 (b) 3 (c) −3 (d) −3 (e) undefined (f) 0 (g) $-\frac{2}{3}$

CAUTION

Although multiplication satisfies both the commutative and associative laws, division does not. For example,

$$3 = 6 \div 2 \neq 2 \div 6 = \frac{2}{6} = \frac{1}{3}. \quad \text{Division is not commutative}$$

Likewise,

$$(8 \div 4) \div 2 = 2 \div 2 = 1$$

but $8 \div (4 \div 2) = 8 \div 2 = 4.$ Division is not associative

EXAMPLE 6 Application to Personal Banking

Willie wrote 3 checks for \$75 each and 4 checks for \$105 each. By how much did this change his bank balance?

The 3 checks for \$75 could be calculated as follows:

$$(3)(-75) = -225.$$

There are 4 checks for \$105, so

$$(4)(-105) = -420.$$

Thus, his account is changed by

$$-225 + (-420) = -225 - 420 = -645 \text{ dollars.}$$

His balance is \$645 less.

Practice Exercise 6

A diver descended below the surface of the ocean by diving 25 ft each minute for 11 minutes. How far below sea level was he at this time?

Answer: 275 ft (−275)

EXAMPLE 7 Application to Personal Finance

Mary, Sue, and Maria share an apartment with total rent of \$720. By how much is each woman's bank account changed if they each write a check for an equal share of the rent?

Since the rent is a decrease in their accounts, we use −720. To find the shares, we divide −720 by 3:

$$-720 \div 3 = -240.$$

Thus, each woman's account is decreased by \$240.

Practice Exercise 7

During a 5-hr period, the temperature decreased 42°. What was the average decrease per hour?

Answer: 8.4° (−8.4°)

When evaluating numerical expressions involving rational numbers, follow the order of operations and rules of grouping given in Section 1.1. That is, evaluate within grouping symbols first, beginning with the innermost, evaluate powers next, and then perform multiplication and division left to right followed by addition and subtraction.

EXAMPLE 8 ORDER OF OPERATIONS

Evaluate each numerical expression.

(a) $(-3) + 5 \cdot (-1) - (-4)$

$= (-3) + (-5) - (-4)$ Multiply first

$= (-8) - (-4)$ Add first from left

$= (-8) + 4$ Then subtract

$= -4$

(b) $(-2)^3 + 1 = (-2)(-2)(-2) + 1$ $a^3 = aaa$

$= (4)(-2) + 1$ $(-2)(-2) = 4$

$= (-8) + 1$ Multiply first

$= -7$ Then add

(c) $\frac{1}{2} - \left\{\frac{1}{4} - \left[\left(-\frac{3}{4}\right) - \left(-\frac{1}{4}\right)\right]\right\}$

$= \frac{1}{2} - \left\{\frac{1}{4} - \left[\left(-\frac{3}{4}\right) + \frac{1}{4}\right]\right\}$

$= \frac{1}{2} - \left\{\frac{1}{4} - \left[-\frac{2}{4}\right]\right\}$ Innermost parentheses first

$= \frac{1}{2} - \left\{\frac{1}{4} + \frac{2}{4}\right\}$

$= \frac{1}{2} - \left(\frac{3}{4}\right)$

$= \frac{2}{4} - \frac{3}{4} = -\frac{1}{4}$

PRACTICE EXERCISE 8

Evaluate each numerical expression.

(a) $5 + (-3) \cdot (-2) - 6$

(b) $(-4)^2 + (-1)^3$

(c) $-0.15 + [3.2 - (5.65 - 1.6)]$

Answers: (a) 5 (b) 15 (c) −1

7 DOUBLE SIGN PROPERTIES

Notice in Example 8 that we use parentheses to avoid writing plus and minus signs right next to each other. The parentheses may be removed and the two signs replaced by one according to the following rule, part 4 of which is called the **double negative property**.

Double Sign Properties

If a is any number,

1. $+(+a) = +a$ **2.** $+(-a) = -a$

3. $-(+a) = -a$ **4.** $-(-a) = +a.$

Notice that when a plus sign is before parentheses, we *do not* change the sign inside when the parentheses and plus sign are removed. However, when a minus sign is before parentheses, we *do* change the sign inside when the parentheses and minus sign are removed.

EXAMPLE 9 Double Sign Properties

Write without parentheses.

(a) $+(+6) = +6 = 6$

(b) $+(-8) = -8$

(c) $-(+7) = -7$

(d) $-(-5) = +5 = 5$

(e) $-[-(-2)] = -[+2] = -2$ Remove innermost parentheses first

Practice Exercise 9

Write without parentheses.

(a) $+(+11)$

(b) $+(-19)$

(c) $-(+17)$

(d) $-(-15)$

(e) $-[+(-12)]$

Answers: (a) 11 (b) −19 (c) −17 (d) 15 (e) 12

EXAMPLE 10 Evaluating Expressions

Evaluate when $a = -3$ and $b = -2$.

(a) $a - b = (-3) - (-2) = (-3) + 2 = -1$ Use parentheses to substitute

(b) $b - [-a] = (-2) - [-(-3)] = (-2) - [3]$
$= -2 - 3 = -5$

(c) $2a - 3b = 2(-3) - 3(-2) = -6 - (-6) = -6 + 6 = 0$

(d) $|a - 2b| = |(-3) - 2(-2)| = |(-3) - (-4)|$
$= |(-3) + 4| = |1| = 1$

Practice Exercise 10

Evaluate when $x = -5$ and $y = -1$.

(a) $x - y$

(b) $y - (-x)$

(c) $2x - 6y$

(d) $|y - 3x|$

Answers: (a) −4 (b) −6 (c) −4 (d) 14

EXAMPLE 11 Evaluating Expressions

Evaluate $3[2(a + 2) - 3(b + c)]$ when $a = -2$, $b = 4$, and $c = -8$.

$3[2(a + 2) - 3(b + c)] = 3[2(-2 + 2) - 3(4 + (-8))]$ Replace variables with numbers

$= 3[2(0) - 3(-4)]$ Add inside parentheses first

$= 3[0 + 12]$

$= 3[12]$

$= 36$

Practice Exercise 11

Evaluate $-4[(u - v) - (w - u)]$ when $u = -3$, $v = 5$, and $w = -1$.

Answer: 40

1.4 EXERCISES A

Perform the indicated operations.

1. $(2)(6)$ **2.** $(-2)(-6)$ **3.** $(2)(-6)$ **4.** $(-2)(6)$

5. $0 \cdot (-6)$ **6.** $(-5) \cdot 7$ **7.** $(-3)(-9)$ **8.** $(-10)(8)$

9. $(9)(-12)$

10. $(12)(-12)$

11. $(-12)(-12)$

12. $(-15) \cdot 20$

13. $6 \div 2$

14. $(-6) \div (-2)$

15. $6 \div (-2)$

16. $(-6) \div 2$

17. $0 \div (-6)$

18. $(-6) \div 0$

19. $\dfrac{99}{-11}$

20. $\dfrac{-36}{-4}$

21. $\dfrac{-12}{12}$

22. $\dfrac{0}{-38}$

23. $\dfrac{-48}{-16}$

24. $\dfrac{52}{-13}$

25. $(1.2)(-2.3)$

26. $(-1.2)(-2.3)$

27. $(-1.2)(2.3)$

28. $\left(\frac{2}{5}\right)\left(-\frac{15}{4}\right)$

29. $\left(-\frac{2}{5}\right)\left(-\frac{15}{4}\right)$

30. $\left(-\frac{2}{5}\right)\left(\frac{15}{4}\right)$

31. $6.4 \div (-0.8)$

32. $(-6.4) \div (-0.8)$

33. $\dfrac{-6.4}{0.8}$

34. $\frac{3}{5} \div \left(-\frac{9}{15}\right)$

35. $\left(-\frac{3}{5}\right) \div \left(-\frac{9}{15}\right)$

36. $\dfrac{-\frac{3}{5}}{\frac{9}{15}}$

37. $(-1)(-1)(-1)$

38. $(-1)(2)(-3)$

39. $(-1)(-2)(-3)(-4)$

40. $(-1)(-2)(-3)(4)(-5)(6)(-1)$

41. $(-1)(-1)(-1)(-1)(-1)(-1)(-1)$

42. Show by example that division does not satisfy the associative law. Can you find numbers a, b, and c for which $(a \div b) \div c$ does equal $a \div (b \div c)$?

43. Tony had to write four $12 checks during the month of June. By how much did this change his bank balance?

44 Martha had 423 points in a contest, but then she received 20 penalty points 3 times. What was her total after the 3 penalties?

Evaluate each numerical expression.

45. $(-3)^3 + 27$

46. $(-1)^5 + (-1)^3$

47. $2 - [3 - (4 - 1)]$

48. $4 + 2[(-1) - (-5)]$

49. $(3 - 8) \cdot 4 + 1$

50 $3 - \dfrac{2 + (-4)}{1 - 5}$

Write without parentheses using only one sign (or no sign).

51. $-(-3)$

52. $-(+4)$

53. $+(-x)$

54. $-(-x)$

55. $-[+(-2)]$

56. $-[-(+3)]$

57. $-[-(-4)]$

58. $+[-(-10)]$

Evaluate the following expressions when $a = -1$, $b = 2$, *and* $c = -4$.

59. $a^2 - 1$

60. $abc + b^2$

61. $3(a + b) + c$

62. $2(b - a) - c$

63 $2[3(a + 1) + 2(4 + c)]$

64. $a^2 + b^2 + c$

FOR REVIEW

Perform the following operations.

65. $(-12) + 9$

66. $(-12) + (-9)$

67. $15 + (-8)$

68. $(-12) - 9$

69. $(-12) - (-9)$

70. $15 - (-8)$

Add.

71. $\begin{array}{r} -13 \\ 12 \\ 26 \\ \underline{-18} \end{array}$

72. $\begin{array}{r} -28 \\ -29 \\ -42 \\ \underline{-36} \end{array}$

73. $\begin{array}{r} -92 \\ 81 \\ 27 \\ \underline{-61} \end{array}$

74. $\begin{array}{r} -102 \\ 321 \\ -421 \\ \underline{-\ 60} \end{array}$

75 The Brisebois family owed the credit union \$3000. They repaid \$1200 and later borrowed \$650. What number represents the status of their account?

ANSWERS: 1. 12 2. 12 3. −12 4. −12 5. 0 6. −35 7. 27 8. −80 9. −108 10. −144 11. 144 12. −300 13. 3 14. 3 15. −3 16. −3 17. 0 18. undefined 19. −9 20. 9 21. −1 22. 0 23. 3 24. −4 25. −2.76 26. 2.76 27. −2.76 28. $-\frac{3}{2}$ 29. $\frac{3}{2}$ 30. $-\frac{3}{2}$ 31. −8 32. 8 33. −8 34. −1 35. 1 36. −1 37. −1 38. 6 39. 24 40. −720 41. −1 42. $(4 \div 2) \div 2 = 2 \div 2 = 1 \neq 4 = 4 \div 1 = 4 \div (2 \div 2)$; if $a = b = c = 1$, then $(1 \div 1) \div 1 = 1 \div 1 = 1 = 1 \div 1 = 1 \div (1 \div 1)$ 43. −\$48 44. 363 points 45. 0 46. −2 47. 2 48. 12 49. −19 50. $\frac{5}{2}$ 51. 3 52. −4 53. $-x$ 54. x 55. 2 56. 3 57. −4 58. 10 59. 0 60. 12 61. −1 62. 10 63. 0 64. 1 65. −3 66. −21 67. 7 68. −21 69. −3 70. 23 71. 7 72. −135 73. −45 74. −262 75. −\$2450

1.4 EXERCISES B

Perform the indicated operations.

1. $(5)(9)$

2. $(-5)(-9)$

3. $(5)(-9)$

4. $(-5)(9)$

5. $(-7)(0)$

6. $(-2)(-7)$

7. $(12)(-2)$

8. $(-12)(2)$

9. $(-6)(-13)$ **10.** $(-6)(13)$ **11.** $(-11)(-11)$ **12.** $(-15)(30)$

13. $9 \div 3$ **14.** $(-9) \div (-3)$ **15.** $9 \div (-3)$ **16.** $(-9) \div 3$

17. $0 \div (-5)$ **18.** $(-5) \div 0$ **19.** $\dfrac{88}{-11}$ **20.** $\dfrac{-24}{-8}$

21. $\dfrac{-15}{-15}$ **22.** $\dfrac{0}{-25}$ **23.** $\dfrac{-100}{10}$ **24.** $\dfrac{52}{-26}$

25. $(1.5)(-2.2)$ **26.** $(-1.5)(-2.2)$ **27.** $(-1.5)(2.2)$ **28.** $\left(\dfrac{3}{5}\right)\left(-\dfrac{10}{9}\right)$

29. $\left(-\dfrac{3}{5}\right)\left(-\dfrac{10}{9}\right)$ **30.** $\left(-\dfrac{3}{5}\right)\left(\dfrac{10}{9}\right)$ **31.** $4.8 \div (-0.6)$ **32.** $(-4.8) \div (-0.6)$

33. $\dfrac{-4.8}{-0.6}$ **34.** $\dfrac{2}{5} \div \left(-\dfrac{6}{15}\right)$ **35.** $\left(-\dfrac{2}{5}\right) \div \left(-\dfrac{6}{15}\right)$ **36.** $\dfrac{-\frac{2}{5}}{\frac{6}{15}}$

37. $(-1)(-1)(-1)(-1)$ **38.** $(-1)(-2)(-3)$ **39.** $(-1)(2)(-3)(4)$

40. $(-1)(-2)(3)(-1)(5)(-4)$ **41.** $(-1)(-1)(-1)(-1)(-1)(-1)(-1)(-1)(-1)$

42. Show by example that division does not satisfy the commutative law. Can you find numbers a and b for which $a \div b$ does equal $b \div a$?

43. During each of 5 consecutive hours, the temperature dropped 7°. By how much did this change the temperature?

44. Hank Anderson received 330 points for a performance, but then was penalized 25 points each for four rule infractions. What was his point total after the penalties were imposed?

Evaluate each numerical expression.

45. $(-3) + 2 \cdot (-4)$ **46.** $(-2)^2 + 3$ **47.** $4 - [2 - (5 - 3)]$

48. $6 + 3[(-2) - (-4)]$ **49.** $(4 - 7) \cdot 5 + 2$ **50.** $2 - \dfrac{4 + (-5)}{3 - 5}$

Write without parentheses using only one sign (or no sign).

51. $-(-5)$ **52.** $+(-2)$ **53.** $-(+a)$ **54.** $-(-a)$

55. $-[+(-3)]$ **56.** $-[-(+2)]$ **57.** $-[-(-7)]$ **58.** $+[-(-8)]$

Evaluate the following expressions when $x = -2$, $y = 1$, and $z = -3$.

59. $3xy + z$ **60.** $xyz + z^2$ **61.** $3(y - x) - z$

62. $x + \dfrac{y}{z}$ **63.** $2[4(x + 2) + 7(z + 3)]$ **64.** $x^2 + y^2 + z$

FOR REVIEW

Perform the following operations.

65. $(-17) + 6$ **66.** $(-17) + (-6)$ **67.** $17 + (-6)$

68. $(-17) - 6$ **69.** $(-17) - (-6)$ **70.** $17 - (-6)$

Add.

71. -16, 15, 3, -27

72. -40, -27, -19, -52

73. -97, 27, 13, -47

74. -205, 420, -318, -65

75. A football team lost 6 yards on first down, gained 11 yards on second down, and lost 4 yards on third down. If the series started on their 25 yard line, where was the ball placed on fourth down?

1.4 EXERCISES C

Evaluate each numerical expression.

1. $-2 - [3 - (4 - 2)^2]^2$ **2.** $[17 - (-5 + 8)^2 - 6 \div 2]^2$

3. $-[2^3 - (2 - 4)]^2 - \dfrac{16 - (-6)}{15 - 4}$

[Answer: -102]

A calculator would be helpful in Exercises 4–6.

4. $(3.65)^2 - (2.57 - 1.22)^2$ **5.** $0.0123[7.2 - (5.5 - 4.6)]$

6. $\dfrac{1.63 - 0.7}{4.8} + \dfrac{(0.76)^2}{1 - (2.71 - 2.11)}$

1.5 THE DISTRIBUTIVE LAWS AND SIMPLIFYING EXPRESSIONS

STUDENT GUIDEPOSTS

1. Distributive Laws
2. Terms, Factors, and Coefficients
3. Collecting Like Terms
4. Simplified Expressions

1 DISTRIBUTIVE LAWS

The operations of addition (or subtraction) and multiplication are related by two important properties called the **distributive laws.** Consider the numerical expression $2(3 + 5)$. Since the order of operations tells us to work within the parentheses first, we evaluate it as follows:

$$2(3 + 5) = 2(8) = 16.$$

However, in this case if we were to "distribute" the product of 2 over the sum of $3 + 5$,

$$2(3 + 5) = 2 \cdot 3 + 2 \cdot 5 = 6 + 10 = 16,$$

we would obtain the same result. Similarly,

$$4(8-3) = 4(5) = 20,$$

$$\text{and}\quad 4(8-3) = 4\cdot 8 - 4\cdot 3 = 32 - 12 = 20.$$

Thus, multiplication can be "distributed over" addition and subtraction. These examples illustrate the following laws.

Distributive Laws of Multiplication Over Addition and Subtraction

If a, b, and c are numbers,

1. $a(b + c) = ab + ac$, and **2.** $a(b - c) = ab - ac$.

Since multiplication is commutative, products are not affected by changing the order of multiplication. Thus we also have

$$(b + c)a = ba + ca \quad\text{and}\quad (b - c)a = ba - ca.$$

Also, multiplication distributes over sums with more than two terms.

$$a(b + c + d) = ab + ac + ad$$

EXAMPLE 1 EVALUATING USING THE DISTRIBUTIVE LAW

Evaluate each numerical expression in two ways.

(a) $7(3 + 2) = 7\cdot 3 + 7\cdot 2 = 21 + 14 = 35$. Also, $7(3 + 2) = 7(5) = 35$.

(b) $5(8 - 3) = 5\cdot 8 - 5\cdot 3 = 40 - 15 = 25$. Also, $5(8 - 3) = 5(5) = 25$.

(c) $6(7 + 3 - 5) = 6\cdot 7 + 6\cdot 3 - 6\cdot 5 = 42 + 18 - 30 = 30$. Also, $6(7 + 3 - 5) = 6(5) = 30$.

(d) $-7(3 + 2) = (-7)(3) + (-7)(2) = (-21) + (-14) = -35$. Also, $-7(3 + 2) = (-7)(5) = -35$.

(e) $-2(-3 - 6) = (-2)(-3) - (-2)(6) = 6 - (-12) = 6 + 12 = 18$. Also, $-2(-3 - 6) = -2(-9) = (2)(9) = 18$.

PRACTICE EXERCISE 1

Evaluate each numerical expression in two ways.

(a) $4(1 + 9)$

(b) $3(2 - 7)$

(c) $5(4 - 6 + 3)$

(d) $-3(5 - 1)$

(e) $-9(-2 - 8)$

Answers: (a) 40 (b) −15 (c) 5 (d) −12 (e) 90

The distributive laws are useful in computation, but they are more useful when we work with algebraic expressions. In Section 1.1, we introduced algebraic expressions involving sums, differences, products, or quotients of numbers and variables.

2 TERMS, FACTORS, AND COEFFICIENTS

A **term** is a part of an expression that is a product of numbers and variables and that is separated from the rest of the expression by plus or minus signs. The numbers and letters that are multiplied in a term are **factors** of the term. The numerical factor is the **(numerical) coefficient** of the term. For example,

$$3x, \quad 4a + 7, \quad 3x + 7y - z, \quad 4x + 5a + 3 - 8x$$

are algebraic expressions with one, two, three, and four terms, respectively. In $4a + 7$, the term $4a$ has factors 4 and a, and 4 is the coefficient of the term.

Two terms are **similar** or **like terms** if they contain the same variables to the same powers. In $4x + 5a + 3 - 8x$, the terms $4x$ and $-8x$ are like terms. Note that the minus sign goes with the term and thus the coefficient of $-8x$ is -8.

If the terms of an expression have a common factor, the distributive laws can be used in reverse to **remove the common factor** by a process called **factoring.** We use the distributive law in the following way.

$$ab + ac = a(b + c)$$

EXAMPLE 2 Factoring Using the Distributive Law

Use the distributive laws to factor.

(a) $3x + 3y = 3(x + y)$ — The distributive law in reverse order

(b) $4a - 4b = 4(a - b)$ — 4 and $(a - b)$ are factors of $4a - 4b$

(c) $5u + 5v - 5w = 5(u + v - w)$

(d) $3x + 6 = 3 \cdot x + 3 \cdot 2$ — 6 is $3 \cdot 2$

$= 3(x + 2)$ — Factor out 3

(e) $8a - 8 = 8 \cdot a - 8 \cdot 1$ — Express 8 as $8 \cdot 1$

$= 8(a - 1)$ — Factor out 8

(f) $-3x - 3y = (-3)x + (-3)y$ — Factor out -3

$= (-3)(x + y)$

Note that $-3x - 3y$ is not $-3(x - y)$ since $-3(x - y) = -3x + 3y$.

PRACTICE EXERCISE 2

Use the distributive laws to factor.

(a) $7a + 7b$

(b) $3u - 3w$

(c) $11c - 11d + 11v$

(d) $6u + 6$

(e) $3w - 15$

(f) $-12a - 12b$

Answers: (a) $7(a + b)$ (b) $3(u - w)$ (c) $11(c - d + v)$ (d) $6(u + 1)$ (e) $3(w - 5)$ (f) $-12(a + b)$

In any factoring problem, we check by multiplying. For example, since

$$3(x + y) = 3x + 3y \quad \text{and} \quad 4(a - b) = 4a - 4b,$$

our factoring in the first two parts of Example 2 is correct.

3 COLLECTING LIKE TERMS

When an expression contains like terms, it can be simplified by **collecting like terms.** This process is illustrated in the next example.

EXAMPLE 3 Collecting Like Terms

Use the distributive laws to collect like terms.

(a) $5x + 7x = (5 + 7)x = 12x$ — Factor out x

(b) $-8x + 2x = (-8 + 2)x = -6x$ — Factor out x

(c) $-2a + 7a - 9a = (-2 + 7 - 9)a = -4a$ — Factor out a

(d) $6y + 2y - y + 5 = 6 \cdot y + 2 \cdot y - 1 \cdot y + 5$ — $-y = -1 \cdot y$

$= (6 + 2 - 1)y + 5$ — Factor y out of first three terms

$= 7y + 5$

The terms $7y$ and 5 *cannot* be collected since they are not like terms. Thus, $7y + 5$ *is not* $12y$. You can see this more easily if you replace y by some number. For example, when $y = 2$,

$$7y + 5 = 7(2) + 5 = 14 + 5 = 19, \quad \text{but} \quad 12y = 12(2) = 24.$$

PRACTICE EXERCISE 3

Use the distributive laws to collect like terms.

(a) $3u + 10u$

(b) $-9v + 5v$

(c) $-w + 2w - 5w$

(d) $3z - z + 2z + 13$

(e) $4a + 7b - a + 6b = 4a - a + 7b + 6b$ Commutative law

$= 4 \cdot a - 1 \cdot a + 7 \cdot b + 6 \cdot b$ $-a = -1 \cdot a$

$= (4 - 1)a + (7 + 6)b$ Distributive law

$= 3a + 13b$

With practice some steps can be left out. We should be able to see, for example, that $4a - a = 3a$ and $7b + 6b = 13b$.

(f) $0.07x + x = (0.07)x + 1 \cdot x$ $x = 1 \cdot x$

$= (0.07 + 1)x$ Distributive law

$= 1.07x$

(e) $6x + 6y - x - 7y$

(f) $p + (0.14)p$

Answers: (a) $13u$ (b) $-4v$ (c) $-4w$ (d) $4z + 13$ (e) $5x - y$ (f) $(1.14)p$

4 SIMPLIFIED EXPRESSIONS

When all like terms of an expression have been collected, we say that the expression has been **simplified.**

Using the distributive law and the fact that $-x = (-1) \cdot x$, we get

$$-(a + b) = (-1)(a + b) = (-1)(a) + (-1)(b) = -a - b,$$
$$-(a - b) = (-1)(a - b) = (-1)(a) - (-1)(b) = -a + b,$$
$$-(-a - b) = (-1)(-a - b) = (-1)(-a) - (-1)(b) = a + b.$$

These observations lead us to the next rule.

To Simplify an Expression by Removing Parentheses

1. When a negative sign is before parentheses, remove the parentheses (and the negative sign in front of the parentheses) by changing the sign of every term within the parentheses.
2. When a plus sign is before parentheses, remove the parentheses without changing any of the signs of the terms.

EXAMPLE 4 Removing Parentheses

Simplify by removing parentheses.

(a) $-(x + 1) = -x - 1$ Change all signs

(b) $-(x - 1) = -x + 1$ Change all signs

(c) $-(-x + y + 5) = +x - y - 5$

$= x - y - 5$ Change all signs

(d) $+(-x + y + 5) = -x + y + 5$ Change *no* signs

(e) $x - (y - 3) = x - y + 3$ Change all signs within parentheses

Practice Exercise 4

Simplify by removing parentheses.

(a) $-(2 + y)$

(b) $-(-2 - y)$

(c) $-(5 - u - w)$

(d) $+(5 - u - w)$

(e) $a - (1 - b)$

Answers: (a) $-2 - y$ (b) $2 + y$ (c) $-5 + u + w$ (d) $5 - u - w$ (e) $a - 1 + b$

The process of removing parentheses can be called **clearing parentheses.**

EXAMPLE 5 Clearing Parentheses and Collecting Like Terms

Clear parentheses and collect like terms.

(a) $3x + (2x - 7) = 3x + 2x - 7$ Do not change signs
$= (3 + 2)x - 7$ Collect like terms
$= 5x - 7$

(b) $y - (4y - 4) = y - 4y + 4$ Change all signs within parentheses
$= (1 - 4)y + 4$ Collect like terms
$= -3y + 4$

(c) $3 - (5a + 2) + 7a = 3 - 5a - 2 + 7a$ Change all signs within parentheses
$= 7a - 5a + 3 - 2$ Commutative law
$= 2a + 1$ Collect like terms

(d) $3x - (-2x - 7) + 5 = 3x + 2x + 7 + 5$ Change all signs within parentheses
$= (3 + 2)x + 7 + 5$
$= 5x + 12$

Practice Exercise 5

Clear parentheses and collect like terms.

(a) $a + (2 - 3a)$

(b) $b - (1 - b)$

(c) $2x - (3x + 4) - x$

(d) $5 - (-8 - 2w) + 8$

Answers: (a) $-2a + 2$ (b) $2b - 1$ (c) $-2x - 4$ (d) $2w + 21$

CAUTION

Change *all* signs within the parentheses.

$$-(-2x - 7) = 2x + 7, \quad \textit{not} \quad 2x - 7.$$

EXAMPLE 6 Evaluating Expressions

Evaluate the expressions when $a = -2$ and $b = -1$.

(a) $3a + 7 = 3(-2) + 7 = -6 + 7 = 1.$

Note how using parentheses at the substitution step helps us avoid ambiguous statements. Without parentheses we would have $3 \cdot -2 + 7$, which is confusing.

(b) $4a^2 = 4(-2)^2 = 4 \cdot 4 = 16$ Only -2 is squared. Compare this example with the next one

(c) $(4a)^2 = (4 \cdot (-2))^2 = (-8)^2 = 64$ From (b) and (c) we see that $4a^2 \neq (4a)^2$

(d) $-a^2 = -(-2)^2 = -(4) = -4$ Compare this with the next example

(e) $(-a)^2 = (-(-2))^2 = (2)^2 = 4$ From (d) and (e) we see that $-a^2 \neq (-a)^2$

(f) $-a - b = -(-2) - (-1) = 2 + 1 = 3$

(g) $-a^3 = -(-2)^3 = -(-8) = +8 = 8$

Practice Exercise 6

Evaluate the expressions when $u = -3$ and $w = -1$.

(a) $2u - 3$

(b) $3w^2$

(c) $(3w)^2$

(d) $-u^3$

(e) $(-u)^3$

(f) $-u - w$

(g) $-(-u + w^2)$

Answers: (a) -9 (b) 3 (c) 9 (d) 27 (e) 27 (f) 4 (g) -4

CAUTION

Two common mistakes to avoid are: (1) forgetting to change *all* signs when a minus sign appears in front of a set of parentheses (for example, $-(x-2)$ is not $-x-2$), and (2) making sign errors when evaluating expressions such as $-a^2$ (for example, $-a^2$ is not the same as $(-a)^2$).

1.5 EXERCISES A

1. **(a)** Compute $-2(3-8)$ **(b)** Compute $(-2)(3)-(-2)(8)$ **(c)** Why are these two equal?

2. **(a)** Compute $-3(4+2)$ **(b)** Compute $(-3)(4)+(-3)(2)$ **(c)** Why are these two equal?

3. In $2x-3y+7-9x$ the terms $2x$ and $-9x$ are called similar or ______ terms.

How many terms does each expression have?

4. $2a+b-3$ **5.** $-2x-3y+4-z$ **6.** $3w$

Multiply.

7. $4(x+y)$ **8.** $5(x-y)$ **9.** $10(x+2y)$

10. $10(x+2y+3z)$ **11.** $2(2a+1+6b)$ **12.** $-5(x+y)$

Use the distributive laws to factor.

13. $4x+4y$ **14.** $5x-5y$ **15.** $10x+20y$

16. $10x+20y+30z$ **17.** $4a+2+12b$ **18.** $-5x-5y$

Simplify by removing parentheses.

19. $-(y+2)$ **20.** $-(-a-b)$ **21.** $-(-x+2)$

22. $-(2y-2)$ **23.** $x-(3-b)$ **24.** $-(x-y-4)$

25. $+(-x-y-z)$ **26.** $-(1+x)+(a-b)$ **27.** $+(u-v)-(w-3)$

Use the distributive laws and collect like terms.

28. $3x-8x$ **29.** $-4z-9z$ **30.** $1-2x+x$

31. $y - 3 - 5y$

32. $2x - y - 5x + y$

33. $a - b + a - b$

34. $1 - x + 2y - 3x - y$

35. $-a + 3 + b - 2a + b$

36. $u - 3 + 3v - 2u + 1$

37. $\frac{1}{2}x - \frac{2}{3}y - \frac{5}{2}x - \frac{1}{3}y$

38. $-\frac{3}{4}a + \frac{1}{4} + \frac{3}{4}a - \frac{1}{4}b$

39. $2.1u - 3 - 5.8u$

40. $x - 3(x + 1)$

41. $2a - (1 - 3a)$

42. $2(1 - 2u) + u$

43. $2x - (-x + 1) + 3$

44. $a + (2 - 5a) - 3$

45. $7u - (-3u - 1) + 5$

46. $-2[x - 3(x + 1)]$

47. $a - [3a - (1 - 2a)]$

48. $2 - [3u - (-2 + 3u)]$

49. $3(x - 2) - 2[5y - (2x - y)]$

50. $2(3a - 4b) - (-4a + b) - (-4b - 5a)$

51. $-[2u - (u - v)] - 3[(u - 2v) - 3v]$

Evaluate the following expressions when $x = -3$, $y = -1$ and $z = 2$.

52. $-2x - y$

53. $2x^2$

54. $(2x)^2$

55. $-2x^2$

56. $(-2x)^2$

57. $-x - y + z$

58. $x^2 - y^2 - z^2$

59. $-3y^2 - (x + z)$

60. $z^3 - x^3$

61. $x^2 - 4yz$

62. $(x - y)^3$

63. $x^3 - y^3$

64. $x^2y^2z^2$

65. $2x + 3z + y + 1$

66. $-(-x)$

67. $|x + y|$

68. $|x - y|$

69. $|x^2 + y^2|$

FOR REVIEW

Perform the indicated operations.

70. $(3)(-11)$

71. $(-3)(-11)$

72. $(-12) \div 6$

73. $(-12) \div (-6)$

74. $\left(-\frac{1}{3}\right)\left(-\frac{3}{4}\right)$

75. $(-6.3) \div (-0.9)$

ANSWERS: 1. (a) 10 (b) 10 (c) distributive law 2. (a) −18 (b) −18 (c) distributive law 3. like 4. 3 5. 4 6. 1 7–12. answers given in exercises 13–18 13–18. answers given in exercises 7–12 19. $-y-2$ 20. $a+b$ 21. $x-2$ 22. $-2y+2$ 23. $x-3+b$ 24. $-x+y+4$ 25. $-x-y-z$ 26. $-1-x+a-b$ 27. $u-v-w+3$ 28. $-5x$ 29. $-13z$ 30. $1-x$ 31. $-4y-3$ 32. $-3x$ 33. $2a-2b$ 34. $1-4x+y$ 35. $-3a+3+2b$ 36. $-u-2+3v$ 37. $-2x-y$ 38. $\frac{1}{4}-\frac{1}{4}b$ 39. $-3-3.7u$ 40. $-2x-3$ 41. $5a-1$ 42. $2-3u$ 43. $3x+2$ 44. $-4a-1$ 45. $10u+6$ 46. $4x+6$ 47. $-4a+1$ 48. 0 49. $7x-12y-6$ 50. $15a-5b$ 51. $-4u+14v$ 52. 7 53. 18 54. 36 55. −18 56. 36 57. 6 58. 4 59. −2 60. 35 61. 17 62. −8 63. −26 64. 36 65. 0 66. −3 67. 4 68. 2 69. 10 70. −33 71. 33 72. −2 73. 2 74. $\frac{1}{4}$ 75. 7

1.5 EXERCISES B

1. (a) Compute $-3(5-2)$. (b) Compute $(-3)(5)-(-3)(2)$. (c) Why are these two equal?
2. (a) Compute $-4(6+2)$. (b) Compute $(-4)(6)+(-4)(2)$. (c) Why are these two equal?
3. In $2a+b+8-10a$ the terms $2a$ and $-10a$ are called like or ______ terms.

How many terms does each expression have?

4. $2a+7+6w-3$
5. $4a+8b$
6. $2w-3+7b+a+c$

Multiply.

7. $4(x-y)$
8. $a(x+y)$
9. $10(a+4b)$
10. $11(x+2y+3z)$
11. $2(3x+1+4y)$
12. $-6(a+w)$

Use the distributive laws to factor.

13. $4x-4y$
14. $ax+ay$
15. $10a+40b$
16. $11x+22y+33z$
17. $6x+2+8y$
18. $-6a-6w$

Simplify by removing parentheses.

19. $-(x+3)$
20. $-(-x-y)$
21. $-(-w+5)$
22. $-(4a-4)$
23. $a-(2-y)$
24. $-(a-b-9)$
25. $+(-a-b-w)$
26. $-(1+a)+(w-c)$
27. $+(x-y)-(a-8)$

Use the distributive laws and collect like terms.

28. $-3x+8x$
29. $-3w-10w$
30. $1-3x+2x$
31. $5y+2-3y$
32. $x+y-x-y$
33. $2y-u+y-3u$
34. $1-2a+b-a-3b$
35. $w+3-2w+b-4$
36. $z-4+3z-a+2$
37. $\frac{1}{4}a-\frac{1}{5}w+\frac{3}{4}a+\frac{2}{5}w$
38. $-\frac{1}{3}x+\frac{1}{2}-\frac{2}{3}x+\frac{1}{3}y$
39. $2.5w-3+4.1w$
40. $a-2(a+5)$
41. $2w-(4-5w)$
42. $3(1-5x)+x$

43. $2b - (-b + 2) + 5$

44. $y + (3 - 2y) - 5$

45. $2y - (-5y - 2) + 7$

46. $-3[a - 2(a + 2)]$

47. $x - [4x - (3 - 4x)]$

48. $4 - [2w - (-3 - 2w)]$

49. $-6(x + 2y) - 5[8x - (-2x + y)]$

50. $-3(-a + 3b) + 5(-2a - 2b) - 2(3a - b)$

51. $-[(6u - v) - 5v] - 4[-u - (2u - v)]$

Evaluate the following expressions when $a = -2$, $b = -1$, and $c = 3$.

52. $-3a - b$

53. $3a^2$

54. $(3a)^2$

55. $-3a^2$

56. $(-3a)^2$

57. $-a - b + c$

58. $a^2 - b^2 - c^2$

59. $-2b^2 - (a + c)$

60. $c^3 - a^3$

61. $a^2 - 3bc$

62. $(a - b)^3$

63. $a^3 - b^3$

64. $a^2b^2c^2$

65. $2a + b - 3c + 5$

66. $-(-a)$

67. $|a + c|$

68. $|a - c|$

69. $|b^2 + c^2|$

FOR REVIEW

Perform the indicated operations.

70. $(4)(-10)$

71. $(-4)(-10)$

72. $(-18) \div 9$

73. $(-18) \div (-9)$

74. $\left(-\frac{1}{5}\right)\left(-\frac{5}{6}\right)$

75. $(-5.4) \div (-0.9)$

1.5 EXERCISES C

Use the distributive laws and collect like terms.

1. $3\{x - 2[x - (y - x)] - (2x - y)\}$

2. $4\{[5(a - 2b) - 3b] - 2[6a - (3b - a)]\}$

3. $7x - \{[6x - 2y - 3(2x - y)] - 3[-x - (3y - x)]\}$

4. $-5\{[-4(a + 5b) - 2b] - 7[a - (-a - b)]\}$

1.6 IRRATIONAL AND REAL NUMBERS

STUDENT GUIDEPOSTS

1. Irrational and Real Numbers
2. Perfect Squares and Square Roots
3. Principal Square Roots, Radicals, and Radicands
4. Summary of Basic Properties
5. More Formal Definitions

So far we have concentrated on working with the rational numbers. Remember that a **ratio**nal number is the quotient or **ratio** of two integers. Equivalently a rational number is a number with a decimal form that either terminates or repeats a block of digits.

1 IRRATIONAL AND REAL NUMBERS

There are many numbers that cannot be expressed this way. One of the most familiar of these is π, the number equal to the ratio of the circumference of any circle to its diameter. Such numbers are called *irrational numbers*. In decimal notation **irrational numbers** do not terminate nor do they have a repeating block of digits. The set of **real numbers** includes both the rational numbers and the irrational numbers.

2 PERFECT SQUARES AND SQUARE ROOTS

Before identifying some of the more familiar irrational numbers, we look at the idea of perfect squares and their *square roots*.

Perfect Square and Square Root

When an integer is squared, the result is a **perfect square.** Either of the identical factors of a perfect square number is a **square root** of the number.

EXAMPLE 1 Square Roots

(a) 4 is a perfect square since $4 = 2^2$. We call 2 a square root of 4. Since $4 = (-2)^2$, -2 is also a square root of 4.

(b) 25 is a perfect square since $25 = 5^2$. The two square roots of 25 are 5 and -5.

(c) 0 is a perfect square since $0 = 0^2$. Unlike other perfect squares, 0 has only one square root, namely itself, 0.

Practice Exercise 1

(a) Is 1 a perfect square?

(b) Is 81 a perfect square?

(c) Is 40 a perfect square?

Answers: (a) yes; $1 = 1^2$ (b) yes; $81 = 9^2$ (c) No; no integer times itself is equal to 40.

CAUTION

Do not confuse the terms *square* and *square root!*

5 is a **square** root of 25 → ← 25 is the **square** of 5

$$5^2 = 5 \cdot 5 = 25$$

3 PRINCIPAL SQUARE ROOTS, RADICALS, AND RADICANDS

All perfect squares other than 0 have two square roots, one positive and one negative. When we refer to the square root of a number, do we mean the positive root or the negative root? To avoid confusion, we refer to the positive root as the **principal square root** and we use the symbol $\sqrt{\ }$, called a **radical,** to represent it. The principal or positive square root of 4 is $\sqrt{4}$, or 2. The number under the radical (4 in this case) is the **radicand.** Since the radical symbol only designates the principal or positive (possibly zero) square root of a number, we designate the negative square root as $-\sqrt{\ }$. For example, $-\sqrt{4} = -2$.

The first sixteen perfect squares and their square roots are listed in the following table.

Perfect square	*Positive square root of N*	*Negative square root of N*
N	$\sqrt{N}$	$-\sqrt{N}$
0	0	$-0 = 0$
1	1	-1
4	2	-2
9	3	-3
16	4	-4
25	5	-5
36	6	-6
49	7	-7
64	8	-8
81	9	-9
100	10	-10
121	11	-11
144	12	-12
169	13	-13
196	14	-14
225	15	-15

In addition to whole number perfect squares, there are also fractional perfect squares. A fraction that can be factored into the product of two identical fractional factors is a **perfect square,** and each factor is a **square root** of the fraction.

EXAMPLE 2 Fractional Perfect Squares

(a) $\frac{4}{9}$ is a perfect square, with $\sqrt{\frac{4}{9}} = \frac{2}{3}$ and $-\sqrt{\frac{4}{9}} = -\frac{2}{3}$.

(b) $\frac{25}{81}$ is a perfect square, with $\sqrt{\frac{25}{81}} = \frac{5}{9}$ and $-\sqrt{\frac{25}{81}} = -\frac{5}{9}$.

Practice Exercise 2

(a) Is $\frac{36}{121}$ a perfect square?

(b) Is $\frac{50}{200}$ a perfect square?

Answers: (a) yes; $\frac{36}{121} = \left(\frac{6}{11}\right)^2$
(b) yes; first reduce the fraction to $\frac{1}{4}$, and $\frac{1}{4} = \left(\frac{1}{2}\right)^2$.

Example 2 shows that a fraction is a perfect square if its numerator and denominator are whole number perfect squares. However, a fraction can be a perfect square without this feature. For example, if we reduce $\frac{8}{18}$ to lowest terms, we change it to a ratio of whole number perfect squares:

$$\sqrt{\frac{8}{18}} = \sqrt{\frac{\not{2} \cdot 4}{\not{2} \cdot 9}} = \sqrt{\frac{4}{9}} = \frac{2}{3}.$$

Thus far we have discussed only square roots of perfect squares. What about square roots of other numbers? This problem can be divided into two parts, square roots of positive numbers and square roots of negative numbers. Square roots of negative numbers are not real numbers and will not be discussed in this text until Chapter 7. However, square roots of positive numbers that are not perfect squares supply us with many examples of irrational numbers. These include $\sqrt{2}$, $\sqrt{3}$, $\sqrt{5}$, $\sqrt{6}$, and $\sqrt{7}$ among many, many more.

Suppose we consider $\sqrt{2}$, the number which when squared is 2. We know that

$$\sqrt{1} = 1 \text{ and } \sqrt{4} = 2.$$

Since $1 < 2 < 4$, we would expect $\sqrt{1} < \sqrt{2} < \sqrt{4}$ so that

$$1 < \sqrt{2} < 2.$$

Since there is no integer between 1 and 2, $\sqrt{2}$ cannot be an integer. In more advanced work it can be shown that $\sqrt{2}$ is not a rational number either.

Most of our work in this text will be with rational numbers until Chapter 7 when we consider operations with radicals.

EXAMPLE 3 Number Systems

Find the numbers in the set $\{-5, 0, 4, \sqrt{7}, 5, -\pi, \frac{3}{2}, -1.8\}$ that belong to the specified set.

(a) Natural numbers: 4 and 5

(b) Whole numbers: 0, 4, and 5

(c) Integers: 0, 4, 5, and -5

(d) Rational numbers: 0, 4, 5, -5, $\frac{3}{2}$, and -1.8

(e) Irrational numbers: $\sqrt{7}$, $-\pi$

(f) Real numbers: all the numbers in the set are real numbers

Practice Exercise 3

Find the numbers in the set $\{-7, -\frac{3}{5}, -\sqrt{17}, 0, 11, \pi, \frac{11}{2}, 7.22\}$ that belong to the specified set.

(a) Natural numbers

(b) Whole numbers

(c) Integers

(d) Rational numbers

(e) Irrational numbers

(f) Real numbers

Answers: (a) 11 (b) 0, 11 (c) -7, 0, 11 (d) -7, $-\frac{3}{5}$, 0, 11, $\frac{11}{2}$, 7.22 (e) $-\sqrt{17}$, π (f) all the numbers

The diagram in Figure 1.14 shows the relationships among the number systems we have discussed.

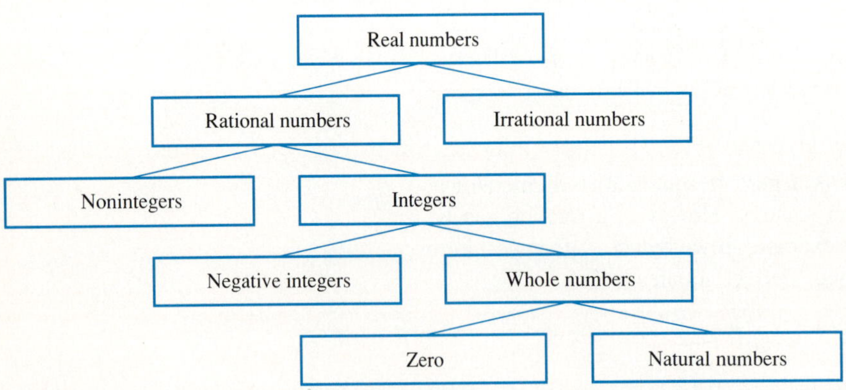

Figure 1.14

4 SUMMARY OF BASIC PROPERTIES

The properties of rational numbers that we have discussed in this chapter are also true for all real numbers. We now give a summary of the basic properties of real numbers a, b, and c.

Basic Properties of Real Numbers

Let a, b, and c be real numbers.

Closure laws	$a + b$ is a real number ab is a real number
Commutative laws	$a + b = b + a$ $ab = ba$
Associative laws	$a + (b + c) = (a + b) + c$ $a(bc) = (ab)c$
Additive identity	$a + 0 = a = 0 + a$
Multiplicative identity	$a \cdot 1 = a = 1 \cdot a$
Negatives (additive inverses)	$a + (-a) = 0 = (-a) + a$
Reciprocals (multiplicative inverses)	$a\left(\frac{1}{a}\right) = 1 = \left(\frac{1}{a}\right)a,\ a \neq 0$
Distributive law	$a(b + c) = ab + ac$

A summary of the properties 0, 1, and -1 is now given for a any real number.

Properties of 0, 1, and −1

Let a be a real number.

Properties of 0 under subtraction	$a - 0 = a$ $0 - a = -a$
Properties of 0 under multiplication	$a \cdot 0 = 0$ $0 \cdot a = 0$
Properties of 0 under division	$0 \div a = 0, \quad a \neq 0$ $a \div 0$ is undefined
Properties of 1 under division	$1 \div a = \frac{1}{a}, \quad a \neq 0$ $a \div 1 = a$
Property of -1 under multiplication	$(-1)a = -a$

EXAMPLE 4 Recognizing Properties of 0, 1, and −1

State the property that verifies the truth of each statement.

(a) $0 \cdot (4.5) = 0$ Property of 0 under multiplication

(b) $5 \div 1 = 5$ Property of 1 under division

(c) $0 \div \frac{3}{4} = 0$ Property of 0 under division

(d) $(-1) \cdot (-3) = -(-3) = 3$ Property of -1 under multiplication

Practice Exercise 4

State the property that verifies the truth of each statement.

(a) $8 - 0 = 8$

(b) $2 \div 1 = 2$

(c) $4.8 \div 0$ is undefined.

(d) $\frac{1}{7}(0) = 0$

Answers: (a) property of 0 under subtraction (b) property of 1 under division (c) property of 0 under division (d) property of 0 under multiplication

5 MORE FORMAL DEFINITIONS

We have used intuitive definitions to discuss equality, inequalities, and absolute value in previous sections. Now that we have the real number system we will give more formal definitions of these concepts.

Two numbers a and b that correspond to the same point on a number line are **equal,** written $a = b$. For example,

$$\frac{4}{2} = 2, \quad 1.5 = \frac{3}{2}, \quad \text{and} \quad \frac{3}{11} = 0.\overline{27}.$$

Numbers that are not equal (written $a \neq b$) correspond to different points.

The **properties of equality,** assumed to be true without proof and listed below, are used for solving equations and simplifying algebraic expressions.

Properties of Equality

Let a, b, and c be real numbers.

Reflexive law	$a = a$
Symmetric law	If $a = b$, then $b = a$.
Transitive law	If $a = b$ and $b = c$, then $a = c$.
Substitution law	If $a = b$, then either may replace the other in any statement without affecting the truth of the statement.

EXAMPLE 5 Recognizing Properties of Equality

State the property of equality illustrated by each statement.

(a) $4 = 4, \quad -3 = -3, \quad \frac{1}{2} = \frac{1}{2}$ Reflexive law

(b) If $x = 8$, then $8 = x$. Symmetric law

Practice Exercise 5

State the property of equality illustrated by each statement.

(a) If $y = 7$ and $7 = w$, then $y = w$.

(b) If $x = 2$ and $z + x = 8$, then $z + 2 = 8$.

Answers: (a) transitive law (b) substitution law

In Section 1.2, we discussed the order relationships "less than" and "greater than" from an intuitive standpoint using the terms *left* and *right* when related to points on a number line. We now present a formal definition of these relationships.

Inequalities

Let a and b be real numbers.

1. We say that a **is less than** b, denoted $a < b$, if there is a positive number c such that $a + c = b$.
2. We say that a **is greater than** b, denoted $a > b$, if there is a positive number c such that $a - c = b$.

For example,

$5 < 12$ since 7 is a positive number such that $5 + 7 = 12$,
and $7 > 3$ since 4 is a positive number such that $7 - 4 = 3$.

Two more inequality properties can now be added to the list of properties of the set of real numbers.

Properties of Inequality

Let a, b, and c be real numbers.

Trichotomy law Exactly one of the following holds: $a > b$, $a = b$, or $a < b$.

Transitive law If $a < b$ and $b < c$, then $a < c$.

By the trichotomy law, either

$$6 < z \text{ or } 6 = z \text{ or } 6 > z.$$

By the transitive law,

$$\text{if } x < 9 \text{ and } 9 < y, \text{ then } x < y.$$

Absolute value was defined in Section 1.2 as the distance between points on a number line. The following more formal definition is useful in our further work in algebra.

Absolute Value

Let a be a real number. The **absolute value** of a is denoted by $|a|$ and

$$|a| = \begin{cases} a \text{ if } a \geq 0 \\ -a \text{ if } a < 0. \end{cases}$$

If $a = -3$, then

$$|a| = |-3| = -(-3) \qquad a = -3 < 0, \text{ so } |a| = -a$$
$$= 3. \qquad \text{Double-negation property}$$

The double-negation property ensures that the absolute value of a number is *never* a negative number.

1.6 EXERCISES A

Evaluate the square roots.

1. $\sqrt{36}$
2. $-\sqrt{49}$
3. $\sqrt{225}$
4. $-\sqrt{196}$
5. $\sqrt{0}$
6. $-\sqrt{1}$
7. $-\sqrt{169}$
8. $\sqrt{81}$
9. $\sqrt{\frac{9}{25}}$
10. $\sqrt{\frac{64}{196}}$
11. $-\sqrt{\frac{49}{81}}$
12. $-\sqrt{\frac{36}{225}}$

13. $\sqrt{\frac{1}{144}}$ **14.** $\sqrt{\frac{4}{169}}$ **15.** $\sqrt{\frac{49}{196}}$ **16.** $\sqrt{\frac{9}{81}}$

17 $\sqrt{\frac{8}{18}}$ **18.** $-\sqrt{\frac{2}{50}}$ **19.** $\sqrt{\frac{12}{48}}$ **20.** $\sqrt{\frac{75}{363}}$

21. Compute **(a)** $(16)^2$ **(b)** $\sqrt{256}$

22. Compute **(a)** $(17)^2$ **(b)** $\sqrt{289}$

23. Compute **(a)** $(18)^2$ **(b)** $\sqrt{324}$

24. Compute **(a)** $(19)^2$ **(b)** $\sqrt{361}$

25. If x is any whole number, what is $\sqrt{x^2}$?

26. If x is any natural number, what is $-\sqrt{x^2}$?

27. **(a)** What is the square of 9? **(b)** What is the principal square root of 9?

28. **(a)** What is the square of $\frac{1}{4}$? **(b)** What is the principal square root of $\frac{1}{4}$?

29 Between what two positive integers is $\sqrt{15}$ located?

30. Between what two positive integers is $\sqrt{150}$ located?

In Exercises 31–36, list the numbers in the set $\left\{-\frac{7}{2}, -1, 0, -\sqrt{11}, -3, 2.66, 2\pi, \frac{\sqrt{2}}{2}\right\}$ *that belong to the specified set.*

31. Natural numbers **32.** Whole numbers **33.** Integers

34. Rational numbers **35.** Irrational numbers **36.** Real numbers

Of the natural numbers, the whole numbers, the integers, the rational numbers, and the real numbers, which is the smallest set that contains each of the following numbers?

37. -3 **38.** 0 **39.** $\sqrt{13}$ **40.** -10

41. 22 **42.** -1.326 **43.** $-\frac{11}{5}$ **44.** $3\sqrt{3}$

Name the property that justifies each statement. Assume that all letters represent real numbers.

45. $3 \cdot x = x \cdot 3$ **46.** $x + 0 = x$ **47.** $8(xy) = (8x)y$

48. $2 + y = y + 2$ **49.** $1 \cdot 5 = 5$ **50.** $7 + (-7) = 0$

51. $\left(-\frac{1}{4}\right)(-4) = 1$

52. $(4 + x) + 1 = 4 + (x + 1)$

53. $3 + 9$ is a real number

54. $3(a + b) = 3a + 3b$

55. $(2 + 3) + 9 = 9 + (2 + 3)$

56. $0 + (a + b) = (0 + a) + b$

57. $3(a + 7) = (a + 7)3$

58. $-(-3) = 3$

59. $7 \div 0$ is undefined

60. $0 \cdot (-8) = 0$

61. $-\left(-\frac{2}{3}\right) = \frac{2}{3}$

62. $(-1)(10) = -10$

63. If $a < b$ and $b < 7$ then $a < 7$.

64. $2 < 5$ since $3 > 0$ and $2 + 3 = 5$.

Decide whether each statement is true *or* false.

65. The fact that $7 = 7$ illustrates the symmetric law of equality.

66. If $a = -3$ then $-3 = a$ by the symmetric law of equality.

67. If $w = 2$ and $2 = c$, then we know that $w = c$ by the transitive law of equality.

68. If $x = 4$ and $a = 1 + x$, then $a = 5$ by the substitution law of equality.

Evaluate the absolute value expressions.

69. $|3 - 8|$

70. $|3| - |8|$

71. $|(3)(-8)|$

72. $|3||-8|$

73. $\left|\frac{8}{-2}\right|$

74. $\frac{|8|}{|-2|}$

75. $|(-2) + 1|$

76. $|-2| + |1|$

Answer each of the following.

77. For what real number a will $|a| = |-a|$?

78. Give an example to show that $|a + b| \neq |a| + |b|$.

79. If $a < 0$, is $-a$ positive or negative?

80. If $a < 0$, is $-(-a)$ positive or negative?

FOR REVIEW

Remove parentheses and collect like terms.

81. $2a - (4a - 1)$

82. $-(1 - a) + (a + 1)$

83. $-[-(1 - a)]$

84. $2[y - (3y - 2)]$

85. $-3[-y + (1 - y)]$

86. $-[-(-y)] + y$

ANSWERS: 1. 6 2. −7 3. 15 4. −14 5. 0 6. −1 7. −13 8. 9 9. $\frac{3}{5}$ 10. $\frac{4}{7}$ 11. $-\frac{7}{9}$ 12. $-\frac{2}{5}$ 13. $\frac{1}{12}$ 14. $\frac{2}{13}$ 15. $\frac{1}{2}$ 16. $\frac{1}{3}$ 17. $\frac{2}{3}$ 18. $-\frac{1}{5}$ 19. $\frac{1}{2}$ 20. $\frac{5}{11}$ 21. (a) 256 (b) 16 22. (a) 289 (b) 17 23. (a) 324 (b) 18 24. (a) 361 (b) 19 25. x 26. $-x$ 27. (a) 81 (b) 3 28. (a) $\frac{1}{16}$ (b) $\frac{1}{2}$ 29. 3 and 4 30. 12 and 13 31. none 32. 0 33. −1, 0, −3 34. $-\frac{7}{2}$, −1, 0, −3, 2.66 35. $-\sqrt{11}$, 2π, $\frac{\sqrt{2}}{2}$ 36. all the numbers 37. integers 38. whole numbers 39. real numbers 40. integers 41. natural numbers 42. rational numbers 43. rational numbers 44. real numbers 45. commutative law of multiplication 46. existence of additive identity 47. associative law of multiplication 48. commutative law of addition 49. existence of multiplicative identity 50. existence of negatives 51. existence of reciprocals 52. associative law of addition 53. closure law of addition 54. distributive law 55. commutative law of addition 56. associative law of addition 57. commutative law of multiplication 58. double-negation property 59. property of zero under division 60. property of zero under multiplication 61. double-negation property 62. property of −1 under multiplication 63. transitive law of inequality 64. definition of *less than* 65. false (reflexive law) 66. true 67. true 68. true 69. 5 70. −5 71. 24 72. 24 73. 4 74. 4 75. 1 76. 3 77. every real number 78. Answers will vary; one set that will work is $a = 1$ and $b = -1$. 79. positive 80. negative 81. $-2a + 1$ 82. $2a$ 83. $1 - a$ 84. $-4y + 4$ 85. $6y - 3$ 86. 0

1.6 EXERCISES B

Evaluate the square roots.

1. $\sqrt{196}$
2. $-\sqrt{36}$
3. $\sqrt{49}$
4. $-\sqrt{121}$
5. $\sqrt{144}$
6. $-\sqrt{64}$
7. $-\sqrt{100}$
8. $\sqrt{169}$
9. $\sqrt{\frac{4}{9}}$
10. $\sqrt{\frac{25}{196}}$
11. $-\sqrt{\frac{81}{121}}$
12. $-\sqrt{\frac{36}{49}}$
13. $\sqrt{\frac{1}{225}}$
14. $\sqrt{\frac{4}{36}}$
15. $\sqrt{\frac{49}{121}}$
16. $\sqrt{\frac{4}{16}}$
17. $\sqrt{\frac{8}{32}}$
18. $-\sqrt{\frac{50}{98}}$
19. $\sqrt{\frac{12}{75}}$
20. $\sqrt{\frac{147}{363}}$
21. Compute (a) $(20)^2$ (b) $\sqrt{400}$
22. Compute (a) $(21)^2$ (b) $\sqrt{441}$
23. Compute (a) $(24)^2$ (b) $\sqrt{576}$
24. Compute (a) $(25)^2$ (b) $\sqrt{625}$
25. If a is any whole number, what is $\sqrt{a^2}$?
26. If a is any natural number, what is $-\sqrt{a^2}$?
27. (a) What is the square of 16? (b) What is the principal square root of 16?
28. (a) What is the square of $\frac{1}{9}$? (b) What is the principal square root of $\frac{1}{9}$?
29. Between what two positive integers is $\sqrt{30}$ located?
30. Between what two positive integers is $\sqrt{200}$ located?

In Exercises 31–36, list the numbers in the set $\left\{6, \frac{5}{4}, -0.235, 0, \sqrt{22}, -4, \frac{\pi}{2}\right\}$ *that belong to the specified set.*

31. Natural numbers
32. Whole numbers
33. Integers
34. Rational numbers
35. Irrational numbers
36. Real numbers

Of the natural numbers, the whole numbers, the integers, the rational numbers, and the real numbers, which is the smallest set that contains each of the following numbers?

37. 0 **38.** -16 **39.** 45 **40.** $\sqrt{23}$

41. 63.21 **42.** 4π **43.** $6\sqrt{5}$ **44.** $-\frac{21}{5}$

Name the property that justifies each statement. Assume that all letters represent real numbers.

45. $3(y + 2) = 3y + 3 \cdot 2$ **46.** $5 \cdot c = c \cdot 5$ **47.** $0 \div (-13) = 0$

48. $7\left(\frac{1}{7}\right) = 1$ **49.** $3 + z = z + 3$ **50.** $(3 + 5)a = 3a + 5a$

51. $1 \div 15 = \frac{1}{15}$ **52.** $8 \div 0$ is undefined **53.** $(2 + a) + 5 = 2 + (a + 5)$

54. $4(y + 2) = (y + 2)4$ **55.** $3 \cdot 14$ is a real number **56.** $(4 + 1) + 8 = 8 + (4 + 1)$

57. $0 + (3 + d) = (0 + 3) + d$ **58.** $7 \div 1 = 7$ **59.** $(13)(0) = 0$

60. $(-1)(5) = -5$ **61.** $0 \div 2 = 0$ **62.** $-(-2.5) = 2.5$

63. If $x < 2$ and $2 < w$ then $x < w$. **64.** $4 > -1$ since $5 > 0$ and $4 - 5 = -1$.

Decide whether each statement is true *or* false.

65. If $x = 2$ then $2 = x$ by the transitive law of equality.

66. If $a = 3$ and $y = a + 3$, then $y = 6$ by the substitution law of equality.

67. The fact that $\sqrt{2} = \sqrt{2}$ illustrates the reflexive law of equality.

68. If $y = -3$ and $-3 = w$ then $y = w$ by the transitive law of equality.

Evaluate the absolute value expressions.

69. $|2 - 9|$ **70.** $|2| - |9|$ **71.** $|(2)(-9)|$ **72.** $|2||-9|$

73. $\left|\frac{9}{-3}\right|$ **74.** $\frac{|9|}{|-3|}$ **75.** $|(-3) + 1|$ **76.** $|-3| + |1|$

Answer each of the following.

77. For what real number a will $|a| = |a + 1|$?

78. Give an example to show that $|a - b| \neq |a| - |b|$.

79. If $a > 0$, is $-a$ positive or negative?

80. If $a > 0$, is $-(-a)$ positive or negative?

FOR REVIEW

Remove parentheses and collect like terms.

81. $4y - (6y - 3)$ **82.** $-(2 - y) + (2 + y)$ **83.** $-[-(3 - y)]$

84. $3[w - (2w - 4)]$ **85.** $-4[-w + (2 - w)]$ **86.** $-[-(-w)]-w$

1.6 EXERCISES C

Evaluate the square roots where a is a natural number.

1. $\sqrt{25a^2}$ **2.** $\sqrt{\frac{16}{a^4}}$ **3.** $-\sqrt{\frac{625}{576}}$ **4.** $-\sqrt{\frac{847}{1008}}$

[Answer: $-\frac{11}{12}$]

Use a calculator to find the square root of each number to two decimal places.

5. $\sqrt{2}$ **6.** $\sqrt{7}$ **7.** $\sqrt{26}$ **8.** $\sqrt{191}$

[Answer: 13.82]

1.7 INTEGER EXPONENTS

STUDENT GUIDEPOSTS

1. Exponential Notation
2. Product Rule
3. Quotient Rule
4. Power Rule
5. Powers of Products and Quotients
6. Zero Exponents
7. Negative Exponents
8. Summary of Rules for Exponents

1 EXPONENTIAL NOTATION

In Section 1.1 we introduced exponents and exponential notation.

Exponential Notation

If a is any number and n is a positive integer,

$$a^n = \underbrace{a \cdot a \cdot a \cdots a}_{n \text{ factors}},$$

where a is the base, n the exponent, and a^n the exponential expression.

EXAMPLE 1 Using Exponential Notation

Write without using exponents.

(a) $4^5 = \underbrace{4 \cdot 4 \cdot 4 \cdot 4 \cdot 4}_{5 \text{ factors}}$ The product is 1024

(b) $3y^2 = 3\underbrace{yy}_{2\ ys \text{ as factors}}$ 3 *is not* squared

(c) $(3y)^2 = \underbrace{(3y)(3y)}_{2\ (3y)s \text{ as factors}}$ 3 *is* squared

(d) $2^2 + 3^2 = \underbrace{2 \cdot 2}_{2\ 2s} + \underbrace{3 \cdot 3}_{2\ 3s}$ This simplifies to 13, *not* $(2+3)^2 = 5^2 = 25$

(e) $1^{32} = \underbrace{1 \cdot 1 \cdot 1 \cdots 1}_{32 \text{ factors}} = 1$ 1 to any power is always 1

PRACTICE EXERCISE 1

Write without using exponents.

(a) 9^7

(b) $6w^3$

(c) $(6w)^3$

(d) $4^2 + 8^2$

(e) 1^{27}

(f) $(-1)^5$

(g) $(-1)^{14}$

(f) $(-1)^3 = (-1)(-1)(-1)$

$= (+1)(-1) = -1$ — -1 to an *odd* power is -1

(g) $(-1)^4 = (-1)(-1)(-1)(-1)$

$= (-1)(-1) = 1$ — -1 to an *even* power is $+1$ or 1

Answers: (a) $9 \cdot 9 \cdot 9 \cdot 9 \cdot 9 \cdot 9 \cdot 9$ (b) $6 \cdot w \cdot w \cdot w$ (c) $(6w)(6w)(6w)$ (d) $4 \cdot 4 + 8 \cdot 8$ (e) 1 (f) -1 (g) 1

CAUTION

Two of the most common errors made with exponents have been shown in the above examples:

$$3y^2 \neq (3y)^2 \quad \text{and} \quad 2^2 + 3^2 \neq (2 + 3)^2.$$

Parts (f) and (g) of Example 1 illustrate the following general rule: An odd power of a negative number is negative, and an even power of a negative number is positive.

2 PRODUCT RULE

When we combine terms containing exponential expressions by multiplication, division, or taking powers, our work can be simplified by using the basic properties of exponents. For example,

$$a^3 \cdot a^2 = \underbrace{(a \cdot a \cdot a)}_{\substack{3\\ \textbf{factors}}}\underbrace{(a \cdot a)}_{\substack{2\\ \textbf{factors}}} = \underbrace{a \cdot a \cdot a \cdot a \cdot a}_{\substack{5\\ \textbf{factors}}} = a^5.$$

When two exponential expressions *with the same base* are multiplied, the product is that base raised to the sum of the exponents on the original expressions.

Product Rule for Exponents

If a is any number, and m and n are positive integers.

$$a^m a^n = a^{m+n}.$$

(To multiply powers with the same base, add exponents.)

EXAMPLE 2 Using the Product Rule

Find the product.

(a) $a^3 \cdot a^4 = a^{3+4} = a^7$

(b) $5^3 \cdot 5^7 = 5^{3+7} = 5^{10}$ — Not 25^{10}

(c) $2^3 \cdot 2^2 \cdot 2^6 = 2^{3+2+6} = 2^{11}$ — The rule applies to more than two factors

(d) $3x^3 \cdot x^2 = 3x^{3+2} = 3x^5$ — The 3 is not raised to the powers

Practice Exercise 2

Find the product.

(a) $w^5 \cdot w^8$

(b) $8^2 \cdot 8^8$

(c) $3^2 \cdot 3^3 \cdot 3^5 \cdot 3^{11}$

(d) $7a^5 \cdot a^{12}$

Answers: (a) w^{13} (b) 8^{10} (c) 3^{21} (d) $7a^{17}$

3 QUOTIENT RULE

When two powers with the same base are divided, for example,

$$\frac{a^5}{a^2} = \frac{\overbrace{a \cdot a \cdot a \cdot a \cdot a}^{5 \text{ factors}}}{\underbrace{a \cdot a}_{2 \text{ factors}}} = \frac{a \cdot a \cdot a \cdot \not{a} \cdot \not{a}}{\not{a} \cdot \not{a}} = \underbrace{a \cdot a \cdot a}_{3 \text{ factors}} = a^3,$$

the quotient can be found by raising the base to the difference of the exponents $(5 - 2 = 3)$.

Quotient Rule for Exponents

If a is any number except zero, and m, n, and $m - n$ are positive integers, then

$$\frac{a^m}{a^n} = a^{m-n}.$$

(To divide powers with the same base, subtract exponents.)

EXAMPLE 3 Using the Quotient Rule

Find the quotient.

(a) $\frac{a^7}{a^3} = a^{7-3} = a^4$

(b) $\frac{5^8}{5^5} = 5^{8-5} = 5^3$

(c) $\frac{2^3}{3^4}$ Cannot be simplified using the rule of exponents since the bases are different

(d) $\frac{3x^3}{x^2} = 3x^{3-2} = 3x^1 = 3x$ $x^1 = x$

(e) $\frac{2yy^3}{y^2} = \frac{2y^1y^3}{y^2} = \frac{2y^{1+3}}{y^2}$

$= \frac{2y^4}{y^2} = 2y^{4-2} = 2y^2$

Practice Exercise 3

Find the quotient.

(a) $\frac{x^{10}}{x^4}$

(b) $\frac{7^{12}}{7^9}$

(c) $\frac{8^3}{5^2}$

(d) $\frac{2a^7}{a}$

(e) $\frac{7z^2z^8}{z^7}$

Answers: (a) x^6 (b) 7^3 (c) cannot be simplified using the quotient rule (d) $2a^6$ (e) $7z^3$

4 POWER RULE

When a power is raised to a power, for example,

$$(a^2)^3 = \underbrace{(a^2)(a^2)(a^2)}_{3 \text{ factors}} = (a \cdot a)(a \cdot a)(a \cdot a)$$
$$= \underbrace{a \cdot a \cdot a \cdot a \cdot a \cdot a}_{6 \text{ factors}} = a^6,$$

the resulting exponential expression can be found by raising the base to the product of the exponents $(2 \cdot 3 = 6)$.

Power Rule

If a is any number, and m and n are positive integers,

$$(a^m)^n = a^{mn}.$$

(To raise a power to a power, multiply exponents.)

CAUTION

Do not confuse the power rule with the product rule. For example,

$$(a^5)^2 = a^{5\cdot 2} = a^{10},$$

$$\text{but} \quad a^5a^2 = a^{5+2} = a^7.$$

EXAMPLE 4 Using the Power Rule

Find the powers.

(a) $(a^3)^8 = a^{3\cdot 8} = a^{24}$

(b) $2(x^3)^2 = 2x^{3\cdot 2} = 2x^6$

PRACTICE EXERCISE 4

Find the powers.

(a) $(w^5)^7$

(b) $8(a^4)^3$

Answers: (a) w^{35} (b) $8a^{12}$

5 POWERS OF PRODUCTS AND QUOTIENTS

A product or quotient of expressions is often raised to a power. For example,

$$(3x^2)^4 = \underbrace{(3x^2)\cdot(3x^2)\cdot(3x^2)\cdot(3x^2)}_{\text{4 factors}}$$

$$= \underbrace{3\cdot 3\cdot 3\cdot 3}_{\text{4 factors}}\cdot\underbrace{x^2\cdot x^2\cdot x^2\cdot x^2}_{\text{4 factors}} = 3^4(x^2)^4,$$

$$\text{and} \quad \left(\frac{2}{y^2}\right)^3 = \underbrace{\frac{2}{y^2}\cdot\frac{2}{y^2}\cdot\frac{2}{y^2}}_{\text{3 factors}} = \frac{\overbrace{2\cdot 2\cdot 2}^{\text{3 factors}}}{\underbrace{y^2\cdot y^2\cdot y^2}_{\text{3 factors}}} = \frac{2^3}{(y^2)^3}.$$

These illustrate the next rule.

Powers of Products and Quotients

If a and b are any numbers, and n is a positive integer, then

$$(ab)^n = a^nb^n \quad \text{and} \quad \left(\frac{a}{b}\right)^n = \frac{a^n}{b^n} \quad (b \text{ not zero}).$$

EXAMPLE 5 Powers of Products and Quotients

Simplify.

(a) $(2y)^5 = 2^5\cdot y^5 = 2^5y^5 = 32y^5$

PRACTICE EXERCISE 5

Simplify.

(a) $(5a)^3$

(b) $(3a^2b^3)^4 = 3^4 \cdot (a^2)^4 \cdot (b^3)^4$ Raise each factor to the fourth power

$= 3^4a^8b^{12}$

$= 81a^8b^{12}$

(c) $\left(\frac{2}{x^3}\right)^4 = \frac{2^4}{(x^3)^4}$ $\left(\frac{a}{b}\right)^n = \frac{a^n}{b^n}$

$= \frac{16}{x^{3\cdot 4}}$ $(a^m)^n = a^{mn}$

$= \frac{16}{x^{12}}$

(d) $\left(\frac{3a^2}{b}\right)^3 = \frac{(3a^2)^3}{b^3}$ $\left(\frac{a}{b}\right)^n = \frac{a^n}{b^n}$

$= \frac{3^3(a^2)^3}{b^3}$ $(ab)^n = a^nb^n$

$= \frac{27a^6}{b^3}$

(b) $(2u^3w^5)^2$

(c) $\left(\frac{3}{z^2}\right)^3$

(d) $\left(\frac{4x}{y^3}\right)^2$

Answers: (a) $125a^3$ (b) $4u^6w^{10}$ (c) $\frac{27}{z^6}$ (d) $\frac{16x^2}{y^6}$

CAUTION

Rules similar to the product and quotient rules above do not exist for sums and differences. For example,

$$(2^2 + 3^2)^3 \text{ is } \textit{not } (2^2)^3 + (3^2)^3,$$

$$\text{and } (1^3 - 4^3)^2 \text{ is } \textit{not } (1^3)^2 - (4^3)^2.$$

Also, exponential expressions with different bases cannot be combined by adding exponents. For example, in general,

$$a^2 \cdot b^5 \text{ is } \textit{not } (ab)^7.$$

6 ZERO EXPONENTS

We know that if a is not zero,

$$\frac{a^m}{a^n} = a^{m-n}.$$

Suppose we extend the quotient rule to include $m = n$. Then

$$\frac{a^m}{a^m} = a^{m-m} = a^0 \quad \text{and also} \quad \frac{a^m}{a^m} = 1.$$

(Any number divided by itself is 1.) This suggests the following definition.

Zero Exponent

If a is any number except zero,

$$a^0 = 1.$$

EXAMPLE 6 Using Zero Exponents

Simplify.

(a) $5^0 = 1$

(b) $(2a^2b^3)^0 = 1$ (assuming $a \neq 0$ and $b \neq 0$)

Practice Exercise 6

Simplify.

(a) $21^0 =$ ______.

(b) If $x \neq 0$ and $y \neq 0$, $(7xy^5)^0 =$ ______.

Answers: (a) 1 (b) 1

7 NEGATIVE EXPONENTS

Again, consider

$$\frac{a^m}{a^n} = a^{m-n} \quad (a \neq 0).$$

What happens when $n > m$? For example, if we let $n = 5$ and $m = 2$ and we extend the quotient rule to include $m - n < 0$, we have

$$\frac{a^m}{a^n} = \frac{a^2}{a^5} = a^{2-5} = a^{-3}.$$

If we look at the problem another way, we have

$$\frac{a^2}{a^5} = \frac{\cancel{a} \cdot \cancel{a}}{\cancel{a} \cdot \cancel{a} \cdot a \cdot a \cdot a} = \frac{1}{a \cdot a \cdot a} = \frac{1}{a^3}.$$

Thus, we conclude that $a^{-3} = \frac{1}{a^3}$. This suggests a way to define exponential expressions with negative integer exponents.

Negative Exponents

If $a \neq 0$ and n is a positive integer ($-n$ is a negative integer), then

$$a^{-n} = \frac{1}{a^n}.$$

EXAMPLE 7 Using Negative Exponents

Simplify and write without negative exponents.

(a) $5^{-3} = \frac{1}{5^3} = \frac{1}{125}$ $\quad 5^{-3}$ is *not* -5^3 nor $(-3)(5)$

(b) $4^{-2} = \frac{1}{4^2} = \frac{1}{16}$

(c) $\frac{1}{3^{-2}} = \frac{1}{\frac{1}{3^2}} = \frac{1}{\frac{1}{9}} = 1 \cdot \frac{9}{1} = 9 = 3^2$

(d) $(-2)^{-5} = \frac{1}{(-2)^5} = \frac{1}{-32} = -\frac{1}{32}$

Practice Exercise 7

Simplify and write without negative exponents.

(a) 7^{-2}

(b) 2^{-5}

(c) $\frac{1}{5^{-1}}$

(d) $(-6)^{-2}$

Answers: (a) $\frac{1}{49}$ (b) $\frac{1}{32}$ (c) 5 (d) $\frac{1}{36}$

Example 7 shows that we can "remove" negative exponents simply by moving an exponential expression with a negative exponent from numerator to denominator (or denominator to numerator) and changing the sign of the exponent.

CAUTION

a^{-n} is not equal to $-a^n$ nor to $(-n)a$. As shown in Example 7 (a), 5^{-3} is $\frac{1}{125}$ and not -5^3, which is -125, nor $(-3)(5)$, which is -15.

8 SUMMARY OF RULES FOR EXPONENTS

All the rules of exponents stated for positive integer exponents apply to all integer exponents: positive, negative, and zero. These are summarized as follows.

Rules for Exponents

Let a and b be any two numbers, and m and n any two integers.

1. $a^m a^n = a^{m+n}$ **2.** $\dfrac{a^m}{a^n} = a^{m-n} \quad (a \neq 0)$

3. $(a^m)^n = a^{mn}$ **4.** $(ab)^n = a^n b^n$

5. $\left(\dfrac{a}{b}\right)^n = \dfrac{a^n}{b^n} \quad (b \neq 0)$ **6.** $a^0 = 1 \quad (a \neq 0)$

7. $a^{-n} = \dfrac{1}{a^n} \quad (a \neq 0)$ **8.** $\dfrac{1}{a^{-n}} = a^n \quad (a \neq 0)$

EXAMPLE 8 Using All Rules

Simplify and write without negative exponents.

(a) $(2a)^{-1} = \dfrac{1}{(2a)^1} = \dfrac{1}{2a}$ $(2a)^{-1}$ is *not* $-2a$

(b) $2a^{-1} = 2\dfrac{1}{a} = \dfrac{2}{a}$ The exponent -1 is only on a, not on 2

(c) $y^3y^{-5} = y^{3+(-5)} = y^{-2} = \dfrac{1}{y^2}$

(d) $\dfrac{a^2b^{-3}}{a^{-1}b^5} = a^{2-(-1)}b^{-3-5} = a^3b^{-8} = a^3 \cdot \dfrac{1}{b^8} = \dfrac{a^3}{b^8}$

(e) $\dfrac{1}{x^{-3}} = \dfrac{1}{\frac{1}{x^3}} = 1 \cdot \dfrac{x^3}{1} = x^3$

(f) $(2a^{-2}b)^{-3} = 2^{-3}(a^{-2})^{-3}b^{-3} = \dfrac{1}{2^3}a^{(-2)(-3)}\dfrac{1}{b^3}$

$= \dfrac{1}{8} \cdot a^6 \dfrac{1}{b^3} = \dfrac{a^6}{8b^3}$

PRACTICE EXERCISE 8

Simplify and write without negative exponents.

(a) $(5w)^{-2}$

(b) $5w^{-2}$

(c) $u^{-7}u^5$

(d) $\dfrac{x^4y^{-3}}{x^{-1}y^2}$

(e) $\dfrac{1}{m^{-7}}$

(f) $(6y^{-3}z^{-1})^{-2}$

(g) $\left(\frac{a^2}{2y^{-3}}\right)^{-2} = \frac{(a^2)^{-2}}{2^{-2}(y^{-3})^{-2}} = \frac{a^{-4}}{\frac{1}{2^2}y^6} = \frac{\frac{1}{a^4}}{\frac{y^6}{2^2}}$

$= \frac{1}{a^4} \cdot \frac{2^2}{y^6} = \frac{4}{a^4y^6}$

(g) $\left(\frac{3w^{-1}}{x^2}\right)^{-3}$

Answers: (a) $\frac{1}{25w^2}$ (b) $\frac{5}{w^2}$ (c) $\frac{1}{u^2}$ (d) $\frac{x^5}{y^5}$ (e) m^7 (f) $\frac{y^6z^2}{36}$ (g) $\frac{w^3x^6}{27}$

EXAMPLE 9 Evaluation of Exponential Expressions

Evaluate the following when $a = -2$ and $b = 3$.

(a) $a^{-1} = (-2)^{-1} = \frac{1}{-2} = -\frac{1}{2}$ $\quad (-2)^{-1}$ is *not* $+2$

(b) $\frac{a^{-2}}{b} = \frac{(-2)^{-2}}{3} = \frac{\frac{1}{(-2)^2}}{3} = \frac{\frac{1}{4}}{3} = \frac{1}{4} \cdot \frac{1}{3} = \frac{1}{12}$

(c) $(a + b)^{-1} = (-2 + 3)^{-1} = 1^{-1} = \frac{1}{1} = 1$

(d) $a^{-1} + b^{-1} = (-2)^{-1} + (3)^{-1}$

$= \frac{1}{-2} + \frac{1}{3} = -\frac{3}{6} + \frac{2}{6} = -\frac{1}{6}$

Practice Exercise 9

Evaluate the following when $x = -3$ and $y = 5$.

(a) x^{-1}

(b) $\frac{x^2}{y^{-1}}$

(c) $(y - x)^{-1}$

(d) $y^{-1} - x^{-1}$

Answers: (a) $\frac{1}{-3}$ (b) 45 (c) $\frac{1}{8}$ (d) $\frac{8}{15}$

1.7 EXERCISES A

Write in exponential notation.

1. $8 \cdot 8 \cdot 8 \cdot 8$ **2.** $2 \cdot 2 \cdot y \cdot y \cdot y$ **3.** $(2x)(2x)(2x)(2x)$

Write without using exponents.

4. $2y^4$ **5.** $(2y)^4$ **6.** $a^2 + b^2$

Simplify and write without negative exponents.

7. $x^2 \cdot x^5$ **8.** $a^3 \cdot a^2 \cdot a^4$ **9.** $2y^2 \cdot y^8$

10. $\frac{a^4}{a^3}$ **11.** $\frac{2y^5}{y^3}$ **12.** $(a^3)^4$

13. $(2x^3)^4$ **14.** $2(x^3)^4$ **15.** $\left(\frac{2}{x^3}\right)^4$

16. $\frac{a^3}{b^5}$

17. 5^0

18. 0^0

19. $(2a)^0 \quad (a \neq 0)$

20. $(2x)^{-1}$

21. $2x^{-1}$

22. $\frac{2x^7}{x^9}$

23. $3y^4y^{-7}$

24. $(3y)^{-2}$

25. $3y^{-2}$

26. $(3y^{-2})^3$

27 $\left(\frac{2y}{x^3}\right)^{-2}$

28. $\frac{b^{-2}}{a^{-4}}$

29. $\frac{a^{-4}b^2}{b^{-2}}$

30 $\frac{a^{-2}b^2}{a^4b^{-3}}$

31. $(3xy^2)(4x^2y^3)$

32. $(2x^2y^2)^2(3x^2y)^3$

33. $(-xy)(2x^3y)(4xy^3)$

34. $\frac{x^3y^{-5}}{x^4y^{-6}}$

35. $\frac{3^0a^3b^{-8}}{ab^4}$

36. $\frac{3^{-1}x^{-1}y^{-5}}{x^{-6}y^2}$

37. $(x^{-2}y^{-1})^{-2}$

38. $(x^2y^{-3})^{-4}$

39. $(3x^{-1}y)^{-2}$

40. $\left(\frac{a^{-5}}{b^{-1}}\right)^{-1}$

41. $\left(\frac{2a^{-3}}{b^3}\right)^{-2}$

42. $\left(\frac{2a^3b^{-2}}{a^{-5}b}\right)^{-3}$

Evaluate when $a = -2$ and $b = 3$.

43. $3a^2$

44. $(3a)^2$

45. $-3a^2$

46. $(-3a)^2$

47. $-b^2$

48. $(-b)^2$

49. $a^2 - b^2$

50. $(a - b)^2$

51 a^{-2}

52 $-2a$

53 $-a^2$

54 $a^{-2} + b^{-2}$

55 $(a + b)^{-2}$

56. $\frac{a^{-1}}{b^{-2}}$

57. a^{-3}

58. $(-a)^{-3}$

59 $a^{-1}b^{-1}$

60 $(ab)^{-1}$

FOR REVIEW

Evaluate.

61. $-\sqrt{\dfrac{25}{121}}$ **62.** $\sqrt{\dfrac{18}{50}}$ **63.** $|4 \div 2 + 3 \cdot 5|$ **64.** $-|-[-(2-3)]|$

Complete each statement using the specified law.

65. Commutative law of addition: $8 + a =$ _________

66. Associative law of multiplication: $4(y \cdot 7) =$ _________

67. Existence of a multiplicative identity: $1 \cdot 28 =$ _________

68. Distributive law: $2(a + y) =$ _________

69. Associative law of addition: $x + (3 + 7) =$ _________

70. Distributive law: $3a + 3w =$ _________

ANSWERS: 1. 8^4 2. 2^2y^3 3. $(2x)^4$ 4. $2yyyy$ 5. $(2y)(2y)(2y)(2y)$ 6. $aa + bb$ 7. x^7 8. a^9 9. $2y^{10}$ 10. $a^1 = a$ 11. $2y^2$ 12. a^{12} 13. $16x^{12}$ 14. $2x^{12}$ 15. $\frac{16}{x^{12}}$ 16. cannot be simplified further 17. 1 18. undefined 19. 1 20. $\frac{1}{2x}$ 21. $\frac{2}{x}$ 22. $\frac{2}{x^2}$ 23. $\frac{3}{y^3}$ 24. $\frac{1}{9y^2}$ 25. $\frac{3}{y^2}$ 26. $\frac{27}{y^6}$ 27. $\frac{x^6}{4y^2}$ 28. $\frac{a^4}{b^2}$ 29. $\frac{b^4}{a^4}$ 30. $\frac{b^5}{a^6}$ 31. $12x^3y^5$ 32. $108x^{10}y^7$ 33. $-8x^5y^5$ 34. $\frac{y}{x}$ 35. $\frac{a^2}{b^{12}}$ 36. $\frac{x^5}{3y^7}$ 37. x^4y^2 38. $\frac{y^{12}}{x^8}$ 39. $\frac{x^2}{9y^2}$ 40. $\frac{a^5}{b}$ 41. $\frac{a^6b^6}{4}$ 42. $\frac{b^9}{8a^{24}}$ 43. 12 44. 36 45. −12 46. 36 47. −9 48. 9 49. −5 50. 25 51. $\frac{1}{4}$ 52. 4 53. −4 54. $\frac{13}{36}$ 55. 1 56. $-\frac{9}{2}$ 57. $-\frac{1}{8}$ 58. $\frac{1}{8}$ 59. $-\frac{1}{6}$ 60. $-\frac{1}{6}$ 61. $-\frac{5}{11}$ 62. $\frac{3}{5}$ 63. 17 64. −1 65. $a + 8$ 66. $(4y) \cdot 7$ 67. 28 68. $2a + 2y$ 69. $(x + 3) + 7$ 70. $3(a + w)$

1.7 EXERCISES B

Write in exponential notation.

1. $4 \cdot 4 \cdot z \cdot z \cdot z$ **2.** $(3w)(3w)(3w)$ **3.** $3www$

Write without using exponents.

4. $4y^3$ **5.** $(4y)^3$ **6.** $x^2 + y^2$

Simplify and write without negative exponents.

7. $y^2 \cdot y^7$ **8.** $x^3 \cdot x^2 \cdot x^6$ **9.** $2z^2 \cdot z^5$

10. $\dfrac{b^5}{b^2}$ **11.** $\dfrac{2y^7}{y^4}$ **12.** $(w^3)^5$

13. $(2c^2)^4$ **14.** $2(y^3)^5$ **15.** $\left(\dfrac{2}{a^2}\right)^3$

16. $\dfrac{x^3}{y^4}$ **17.** 7^0 **18.** -0^0

19. $(4x)^0 \quad (x \neq 0)$

20. $(5y)^{-1}$

21. $5y^{-1}$

22. $\dfrac{3a^3}{a^7}$

23. $4z^4z^{-9}$

24. $(5w)^{-2}$

25. $5w^{-2}$

26. $(2a^{-2})^3$

27. $\left(\dfrac{2y}{x^3}\right)^2$

28. $\dfrac{b^{-5}}{a^{-3}}$

29. $\dfrac{a^{-3}b^{-3}}{b^{-4}}$

30. $\dfrac{x^{-2}y^4}{x^3y^{-3}}$

31. $(-5x^2y^2)(3x^3y)$

32. $(x^3y^2)^3(2x^3y)^2$

33. $(-2x)(5x^5y)(-3xy^2)$

34. $\dfrac{a^4b^{-3}}{a^{-2}b^2}$

35. $\dfrac{5^0a^{-5}b^{-2}}{a^4b^{-1}}$

36. $\dfrac{2^{-2}x^3y^{-4}}{x^{-2}y^{-2}}$

37. $(a^{-1}b^{-5})^{-1}$

38. $(a^{-6}b^3)^{-3}$

39. $(4a^{-1}b^{-1})^{-3}$

40. $\left(\dfrac{x^{-4}}{y^{-2}}\right)^{-2}$

41. $\left(\dfrac{3x^4}{y^3}\right)^{-1}$

42. $\left(\dfrac{3x^{-2}y^{-1}}{x^3y^{-5}}\right)^{-4}$

Evaluate when $x = -3$ and $y = 2$.

43. $4x^2$

44. $(4x)^2$

45. $-4x^2$

46. $(-4x)^2$

47. $-x^2$

48. $(-x)^2$

49. $y^2 - x^2$

50. $(y - x)^2$

51. x^{-2}

52. $-3y$

53. $-y^3$

54. $x^{-2} + y^{-2}$

55. $(x + y)^{-2}$

56. $\dfrac{x^{-1}}{y^{-2}}$

57. x^{-3}

58. $(-x)^{-3}$

59. $x^{-1}y^{-1}$

60. $(xy)^{-1}$

FOR REVIEW

Evaluate.

61. $-\sqrt{\dfrac{64}{49}}$

62. $\sqrt{\dfrac{8}{162}}$

63. $|9 \div 3 - 2 \cdot 4|$

64. $-|-[-(4 - 9)]|$

Complete each statement using the specified law.

65. Existence of an additive identity: $6 + 0 =$ _____

66. Closure law of multiplication: $4 \cdot 8$ is _____

67. Distributive law: $3(x + w) =$ _____

68. Commutative law of multiplication: $3x =$ _____

69. Existence of negatives: $(-12) +$ _____ $= 0$

70. Distributive law: $4a + 4m =$ _____

1.7 EXERCISES C

Simplify and write without negative exponents.

1. $\dfrac{a^{-2}b^3c^2}{a^{-3}b^{-2}c^{-2}}$

2. $\dfrac{3^0x^{-6}(y^{-1})^{-2}}{x^{-2}y^{-3}}$

3. $\dfrac{2^{-2}(x^2)^{-3}y^3z^{-2}}{3^{-1}x^{-1}(yz)^{-1}}$ $\left[\text{Answer: } \dfrac{3y^4}{4x^5z}\right]$

4. $\left(\dfrac{5^0a^{-6}b^2c^3}{a^2b^{-1}c^{-2}}\right)^{-1}$

5. $\left(\dfrac{2^{-1}x^{-5}y^{-8}}{4^{-1}x^3y^{-4}}\right)^{-2}$

6. $\left[\left(\dfrac{a^2b^{-2}}{a^{-5}b^{-1}}\right)^{-1}\right]^{-2}$ $\left[\text{Answer: } \dfrac{a^{14}}{b^2}\right]$

1.8 SCIENTIFIC NOTATION

STUDENT GUIDEPOSTS

1. Scientific Notation
2. Calculations Using Scientific Notation
3. Significant Digits

When we compute with very large or very small numbers on a calculator, problems can arise because the display is limited (usually 8 digits). For example, if we use a calculator to multiply

$$(290{,}000)(15{,}000),$$

the product might be given as in Figure 1.15.

Figure 1.15

The number 4.35×10^9 is *scientific notation* for 4,350,000,000, which has ten digits, too many for the display. This shorthand notation depends on the use of integer exponents.

1 SCIENTIFIC NOTATION

A scientist might use the number

$$235{,}000{,}000{,}000{,}000{,}000{,}000$$

but, instead of writing out all the zeros, he or she would write

$$2.35 \times 10^{20}.$$

This short form is easier to use in computations. Likewise, the number

$$0.000000000057$$

could be written as 5.7×10^{-11}.

A number is written in **scientific notation** if it is the product of a power of 10 and a number that is greater than or equal to 1 and less than 10.

To Write a Number in Scientific Notation

1. Move the decimal point to the position immediately to the right of the first nonzero digit.
2. Multiply by a power of ten that is equal to the number of decimal places moved. The exponent on 10 is positive if the original number is greater than 10 and negative if the number is less than 1.

EXAMPLE 1 Converting to Scientific Notation

Write in scientific notation.

(a) $2{,}500{,}000 = 2.5 \times 10^6$ (first nonzero digit; 6 places) Count 6 decimal places

(b) $0.0000025 = 2.5 \times 10^{-6}$ (first nonzero digit; 6 places) Count 6 decimal places

(c) $4{,}321{,}000{,}000 = 4.321 \times 10^9$ (9 places)

(d) $0.00000000001 = 1 \times 10^{-11}$ (11 places)

(e) $0.1 = 1 \times 10^{-1}$ (1 place)

(f) $4.8 = 4.8 \times 10^0$

PRACTICE EXERCISE 1

Write in scientific notation.

(a) 18,300,000

(b) 0.000087

(c) 65,240,000,000,000

(d) 0.0000000001

(e) 0.5

(f) 9.7

Answers: (a) 1.83×10^7 (b) 8.7×10^{-5} (c) 6.524×10^{13} (d) 1×10^{-10} (e) 5×10^{-1} (f) 9.7×10^0

EXAMPLE 2 Converting to Standard Notation

Write in standard notation.

(a) $5.4 \times 10^5 = 540{,}000$ Count 5 decimal places

(b) $5.4 \times 10^{-5} = 0.000054$ Count 5 decimal places

PRACTICE EXERCISE 2

Write in standard notation.

(a) 6.1×10^7

(b) 6.1×10^{-7}

(c) 5.35×10^{12}

(d) 9.03×10^{-6}

(c) $8.94 \times 10^{13} = 89,400,000,000,000$

(d) $2.113 \times 10^{-8} = 0.00000002113$

Answers: (a) 61,000,000
(b) 0.00000061
(c) 5,350,000,000,000
(d) 0.00000903

2 CALCULATIONS USING SCIENTIFIC NOTATION

Scientific notation not only shortens the notation for certain numbers but also simplifies calculations involving very large or very small numbers.

EXAMPLE 3 Performing Operations with Scientific Notation

Perform the indicated operations using scientific notation.

(a) $(20,000)(3,000,000) = (2 \times 10^4)(3 \times 10^6)$

$= (2 \cdot 3) \times (10^4 \cdot 10^6)$ — Use the commutative law

$= 6 \times 10^{10}$ — $10^4 \cdot 10^6 = 10^{4+6} = 10^{10}$

(b) $(2.4 \times 10^{-13})(5.0 \times 10^6) = 12 \times 10^{-7}$

$= \mathbf{1.2} \times 10^1 \times 10^{-7}$ — Obtain a number between 1 and 10

$= 1.2 \times 10^{-6}$

(c) $\dfrac{8.2 \times 10^{-4}}{4.1 \times 10^8} = \left(\dfrac{8.2}{4.1}\right) \times \left(\dfrac{10^{-4}}{10^8}\right)$

$= 2.0 \times (10^{-4-8})$ — $\dfrac{a^m}{a^n} = a^{m-n}$

$= 2.0 \times 10^{-12}$

(d) $\dfrac{(2.65 \times 10^{-3})(4.18 \times 10^{-8})}{3.21 \times 10^6}$

We will use a scientific calculator to carry out these operations. If your calculator works in algebraic logic, these are the steps to follow.*

2.65 [EE] 3 [+/−] [×] 4.18 [EE] 8 [+/−] [÷] 3.21 [EE] 6 [=] → [3.4508 − 17]

The result, rounded to two decimal places, is 3.45×10^{-17}.

Practice Exercise 3

Perform the indicated operations using scientific notation.

(a) (45,000)(6,000,000)

(b) $(5.6 \times 10^{-8})(3.2 \times 10^{15})$

(c) $\dfrac{9.9 \times 10^{-3}}{1.1 \times 10^{10}}$

(d) $\dfrac{(4.38 \times 10^4)(6.15 \times 10^{-7})}{1.06 \times 10^{-3}}$

Answers: (a) 2.7×10^{11}
(b) 1.792×10^8 (c) 9.0×10^{-13}
(d) 2.54×10^1

3 SIGNIFICANT DIGITS

Notice that we rounded the answer in Example 3(d) to the same number of decimal places as the numbers in the original problem. Rounding is necessary when approximate numbers are used in calculations, since it would be inappropriate to give an answer with a higher degree of accuracy than the values used to compute it. The idea of *significant digits* is often used to describe approximate values. The number of **significant digits** is always one more than the number of

*If you have a calculator that uses Reverse Polish Notation (RPN), consult your operator's manual.

decimal places given when the number is written in scientific notation. Thus, there are three significant digits in 3.45×10^{-17}, the answer to Example 3(d).

To find the number of significant digits, write the number in scientific notation, count the decimal places, and add 1. The following table gives several examples.

Number	*Scientific Notation*	*Significant Digits*
315.6	3.156×10^2	4
0.000 012	1.2×10^{-5}	2
7,860,300	7.8603×10^6	5
0.000 920	9.20×10^{-4}	3

The last entry in the table indicates that when one or more zeros *follow* nonzero digits to the *right* of the decimal point, they are significant digits.

1.8 EXERCISES A

Write in scientific notation.

1. 370 **2.** 0.0037 **3.** 98,000

4. 0.00012 **5.** 8360 **6.** 0.00279

7. 0.0000000000756 **8.** 2,650,000,000 **9.** 0.01

Write in standard notation.

10. 2.3×10^2 **11.** 2.3×10^{-2} **12.** 8.7×10^{-5}

13. 4.58×10^6 **14.** 7.51×10^{-8} **15.** 6.64×10^{10}

Perform the indicated operations using scientific notation.

16. $(4 \times 10^5)(1 \times 10^6)$ **17.** (40,000,000)(20,000) **18.** $(1 \times 10^{-10})(5 \times 10^7)$

19 (0.0000022)(300) **20.** $\dfrac{3.3 \times 10^{12}}{1.1 \times 10^{-2}}$ **21.** $\dfrac{0.0000006}{0.03}$

22. $\dfrac{(4 \times 10^7)(6 \times 10^{-5})}{(8 \times 10^{10})}$ **23.** $\dfrac{(0.0231)(0.000572)}{866{,}000}$ **24.** $\dfrac{(3200)(0.000081)}{0.0000012}$

Write the number in each statement using scientific notation.

25. A light-year, the distance that light will travel in one year, is approximately 5,870,000,000,000 mi.

26. The thickness of a particular surface is 0.0000415 cm.

27. The national debt of a country is \$86,500,000,000 and the population is 27,300,000. Use scientific notation and a calculator to find the amount of debt per person.

State the number of significant digits in each number.

28. 6805

29. 0.00000802

30. 0.300

FOR REVIEW

Evaluate when $a = -2$ and $b = 5$.

31. a^{-1}

32. a^{-2}

33. $-2a$

34. b^{-2}

35. $-2b$

36. $a^{-1} + b^{-1}$

37. $(a + b)^{-1}$

38. $(a - b)^{-2}$

39. $a^{-2} - b^{-2}$

ANSWERS: **1.** 3.7×10^2 **2.** 3.7×10^{-3} **3.** 9.8×10^4 **4.** 1.2×10^{-4} **5.** 8.36×10^3 **6.** 2.79×10^{-3} **7.** 7.56×10^{-11} **8.** 2.65×10^9 **9.** 1×10^{-2} **10.** 230 **11.** 0.023 **12.** 0.000087 **13.** 4,580,000 **14.** 0.0000000751 **15.** 66,400,000,000 **16.** 4×10^{11} **17.** 8×10^{11} **18.** 5×10^{-3} **19.** 6.6×10^{-4} **20.** 3×10^{14} **21.** 2×10^{-5} **22.** 3×10^{-8} **23.** 1.53×10^{-11} **24.** 2.2×10^5 **25.** 5.87×10^{12} **26.** 4.15×10^{-5} **27.** \$3170 **28.** 4 **29.** 3 **30.** 3 **31.** $-\frac{1}{2}$ **32.** $\frac{1}{4}$ **33.** 4 **34.** $\frac{1}{25}$ **35.** -10 **36.** $-\frac{3}{10}$ **37.** $\frac{1}{3}$ **38.** $\frac{1}{49}$ **39.** $\frac{21}{100}$

1.8 EXERCISES B

Write in scientific notation.

1. 5400

2. 0.0054

3. 386

4. 0.000028

5. 33,000

6. 0.1

7. 0.000000000254

8. 4,620,000,000

9. 0.001

Write in standard notation.

10. 8.4×10^3

11. 8.4×10^{-3}

12. 5.42×10^{-7}

13. 7.25×10^7

14. 2.06×10^{-6}

15. 4.99×10^{12}

Perform the indicated operations using scientific notation.

16. $(5 \times 10^3)(1 \times 10^4)$

17. (60,000,000)(50,000)

18. $(1 \times 10^{-12})(3 \times 10^8)$

19. (0.00000044)(500)

20. $\dfrac{4.4 \times 10^{15}}{1.1 \times 10^{-3}}$

21. $\dfrac{0.00000008}{0.04}$

22. $\dfrac{(4 \times 10^5)(6 \times 10^{-3})}{(8 \times 10^{-7})}$

23. $\dfrac{(0.00541)(0.0635)}{1{,}250{,}000}$

24. $\dfrac{(64{,}100)(0.00498)}{0.000000451}$

Write the number in each statement using scientific notation.

25. The earth is approximately 93,000,000 miles from the sun.

26. The measure of one calorie is equal to 0.000000278 kilowatt-hours.

27. The sun is 93,000,000 miles from the earth and light travels at 186,000 miles per second. Use scientific notation and a calculator to find how long it takes light to reach the earth from the sun.

State the number of significant digits in each number.

28. 9,720,300

29. 0.0000003005

30. 0.0005020

FOR REVIEW

Evaluate when $x = -3$ and $y = 2$.

31. x^{-1}

32. x^{-2}

33. $-2x$

34. y^{-2}

35. $-2y$

36. $x^{-1} + y^{-1}$

37. $(x + y)^{-1}$

38. $(x - y)^{-2}$

39. $x^{-2} - y^{-2}$

1.8 EXERCISES C

Perform the indicated operations using scientific notation.

1. $\dfrac{(2.5 \times 10^{-3})(4.2 \times 10^{-8})}{(5.0 \times 10^{7})(8.4 \times 10^{-10})}$

2. $\dfrac{(0.000036)(0.0001)^{-1}}{(1{,}200{,}000)(1 \times 10^{7})^{-2}}$ [Answer: 3×10^7]

CHAPTER 1 REVIEW

KEY WORDS

1.1 A **variable** is a letter that can be replaced by various numbers.

An **exponent** is a power on a number or variable indicating how many times the number or variable is used as a factor.

The **base** of an exponential expression is the number or variable that is raised to a power.

1.2 The **integers** are

$\ldots, -3, -2, -1, 0, 1, 2, 3, \ldots.$

The **rational numbers** are the numbers that can be written as the ratio of two integers.

1.3 The **additive identity** is 0.

The **additive inverse** or **negative** of a is $-a$.

1.4 The **multiplicative identity** is 1.

The **multiplicative inverse** or **reciprocal** of a ($a \neq 0$) is $\frac{1}{a}$.

1.5 A **term** is part of an expression that is a product of numbers and variables and that is separated from the rest of the expression by plus or minus signs.

The **factors** of a term are numbers or letters that are multiplied in the term.

The **coefficient** of a term is the numerical factor.

1.6 An **irrational number** is a number that is not rational and has a decimal representation that does not terminate or repeat.

The **real numbers** are the rational numbers together with the irrational numbers.

A **perfect square** results when an integer or a rational number is squared.

Either of the identical factors of a perfect square number is a **square root** of the number.

1.7 An **exponential expression** is of the form a^n where a is the **base** and n is the **exponent.**

1.8 A number is written in **scientific notation** if it is the product of a power of 10 and a number that is greater than or equal to 1 and less than 10.

KEY CONCEPTS

1.1
1. In an expression such as $2y^3$, only y is cubed—not $2y$. That is, $2y^3$ is not the same as $(2y)^3$.
2. When simplifying a numerical expression,
   First, evaluate within grouping symbols.
   Second, evaluate all powers.
   Third, perform all multiplications and divisions in order from left to right.
   Fourth, perform all additions and subtractions in order from left to right.

1.2
1. If a and b are any numbers such that a is to the left of b on a number line, then $a < b$ and $b > a$.
2. The absolute value of a number x, denoted by $|x|$, is the distance from x to zero on a number line and, as a result, is never negative.

1.3 If a, b, and c are any numbers,

(a) $a + 0 = 0 + a = a$, Additive identity

(b) $a + (-a) = (-a) + a = 0$, Additive inverses (negatives)

(c) $a + b = b + a$, Commutative law of addition

(d) $(a + b) + c = a + (b + c)$. Associative law of addition

1.4 If a, b, and c are any numbers,

(a) $a \cdot 1 = 1 \cdot a = a$, Multiplicative identity

(b) $a \cdot \frac{1}{a} = \frac{1}{a} \cdot a = 1 \quad (a \neq 0)$, Multiplicative inverses (reciprocals)

(c) $ab = ba$, Commutative law of multiplication

(d) $(ab)c = a(bc)$. Associative law of multiplication

1.5
1. The distributive laws,
$$a(b + c) = ab + ac$$
$$\text{and} \quad a(b - c) = ab - ac,$$
are used to factor and to collect like terms in algebraic expressions.
2. Only like terms can be combined. For example, $4y + 5 \neq 9y$ since $4y$ and 5 are *not* like terms.
3. A minus sign before a set of parentheses changes the sign of *every* term inside when the parentheses are removed. For example, $-(2 - a) = -2 + a$, *not* $-2 - a$.

1.6
1. The radical symbol by itself only designates the principal (positive or zero) square root of a number. For example, $\sqrt{9} = 3$, not -3. We must write $-\sqrt{9}$ to represent -3.
2. Absolute value of a is defined as
$$|a| = \begin{cases} a \text{ if } a \geq 0 \\ -a \text{ if } a < 0. \end{cases}$$

1.7
1. If a and b are any numbers, and m and n are integers, the following rules for exponents hold.

(a) $a^m a^n = a^{m+n}$

(b) $\frac{a^m}{a^n} = a^{m-n}$, if $a \neq 0$

(c) $(a^m)^n = a^{mn}$

(d) $(ab)^n = a^n b^n$

(e) $\left(\frac{a}{b}\right)^n = \frac{a^n}{b^n}$, if $b \neq 0$

(f) $a^0 = 1$, if $a \neq 0$, and 0^0 is undefined.

REVIEW EXERCISES

Part I

1.1 *Write in exponential notation.*

1. $aaaaa$ **2.** $(3z)(3z)(3z)$ **3.** $(a + w)(a + w)$

Write without using exponents.

4. b^7 **5.** $2x^3$ **6.** $(2x)^3$

Evaluate.

7. $2 \cdot 5 - 9 \div 3 + 1$ **8.** $2 \cdot 3^3 - 3 \cdot 2^3$

Evaluate when $a = 2$, $b = 5$, and $x = 4$.

9. $3a^2$ **10.** $(3a)^2$ **11.** 3^2a

12. $2 + 5[x + (b - a)]$ **13.** $2a + 3[x \div a \cdot b - 4]$

1.2 *Find the absolute values.*

14. $|12|$ **15.** $|-7|$ **16.** $|-2.51|$

Are the following true *or* false*? If the statement is false, tell why.*

17. $-11 < -1$ **18.** $0 \leq -4$ **19.** $5 \leq 5$

Place the correct symbol, =, >, or <, between the pairs of fractions.

20. $\frac{2}{17} \quad \frac{10}{85}$ **21.** $\frac{13}{5} \quad \frac{28}{11}$ **22.** $\frac{6}{7} \quad \frac{11}{12}$

23. The lowest temperature ever recorded in Hawley Lake, Arizona, was 41° below zero. How could this be written?

1.3 *Perform the indicated operations.*

24. $(-2) + (-5)$ **25.** $2 + (-5)$ **26.** $(-2) - (-5)$

27. $(-2) - 5$ **28.** $2 - (-5)$ **29.** $0 - (-2)$

30. $(-2.3) + (-1.5)$ **31.** $\left(\frac{1}{5}\right) - \left(-\frac{3}{10}\right)$ **32.** $\left(-\frac{1}{5}\right) - \left(-\frac{3}{10}\right)$

33. $(-5) + (-3) + 5 + (-8) + (-4) + (-6)$

34. On first down the LSU football team made 7 yards. It lost 9 yards on second down, then picked up 11 yards on third down. If the series started at the LSU 30 yard line, where was the ball spotted on fourth down?

1.4 *Perform the indicated operations.*

35. $(-4)(2)$ **36.** $(-14) \div 7$ **37.** $(-14) \div (-7)$

38. $\frac{0}{-12}$ **39.** $\left(-\frac{1}{3}\right)\left(\frac{15}{8}\right)$ **40.** $(-2.5)(-1.6)$

41. $(-2)(-2)(-1)(2)(-3)(-1)(-1)(-1)$

42. When Bertha was on a crash diet, she lost 7 pounds in each of 6 consecutive weeks. Use a number to express her total weight loss.

Evaluate the following:

43. $(-2) + (-8) \div (-4)$

44. $(-1)^3 + 5$

45. $3 - 2[1 - (4 - 3)]$

46. $(6 - 8) \cdot 2 - (-4)$

Write without parentheses using only one sign (or no sign).

47. $-(-9)$

48. $-(-y)$

49. $-[-(-8)]$

50. $-[+(-x)]$

1.5 *Factor.*

51. $-4x + 12$

52. $-2 - 6x - 10y$

53. $-5x - 25y$

Collect like terms.

54. $3a - 2 - 7a + 4 - a$

55. $-x + 2 - 3y + 4x + y$

Clear parentheses and collect like terms.

56. $y - (2y - 3) + y - (-y + 2)$

57. $-3[2a - (4 - a)] - (-a - 3)$

Evaluate when $b = -2$ and $c = 4$.

58. $7b^2$

59. $-7b^2$

60. $(-7b)^2$

61. $(c - b)^2$

62. $-b^3$

63. $|b^2 - c|$

1.6 *Evaluate the square roots.*

64. $-\sqrt{169}$

65. $\sqrt{\frac{48}{75}}$

66. $-\sqrt{\frac{27}{3}}$

67. **(a)** What is the square of 25? **(b)** What is the principal square root of 25?

Of the natural numbers, the whole numbers, the integers, the rational numbers, and the real numbers, which is the smallest set that contains the given number?

68. 11

69. -13

70. $\frac{4}{5}$

71. $-\sqrt{17}$

72. -3.2

73. 0

Name the property that justifies each statement. Assume that all variables represent real numbers.

74. $-(-4.5) = 4.5$

75. $5x + 5y = 5(x + y)$

76. $7\left(\frac{1}{7}\right) = 1$

77. $0 + (x + y) = x + y$

78. $9 + (-9) = 0$

79. $x < 3$, $x = 3$, or $x > 3$

Evaluate.

80. $|(-3) + 2 + (-5) + (-8)|$

81. $|(-2)(-5) + (-3)(3)|$

1.7 *Simplify and write without negative exponents.*

82. $2y^7y^2$

83. $(3x^2)^3$

84. $(x^2y^3)^5$

85. $2a^0 \quad (a \neq 0)$

86. $(2a)^0 \quad (a \neq 0)$

87. $\dfrac{2b^2b^{-5}}{b^7}$

88. $5a^{-1}$

89. $(5a)^{-1}$

90. $\dfrac{a^{-1}}{b^{-1}}$

91. $(a^2b^{-3})^{-2}$

92. $\dfrac{3^0a^{-5}b^7}{a^4b^{-1}}$

93. $\left(\dfrac{2^{-1}x^2y^{-4}}{x^{-2}y^{-2}}\right)^{-1}$

1.8 *Write in scientific notation.*

94. 0.000000411

95. 549,000,000,000

Write without using scientific notation.

96. 6.15×10^{12}

97. 4×10^{-8}

98. Use a calculator to simplify and write the following in scientific notation.

$$\frac{(0.000000081)(21{,}000{,}000)}{(0.000055)}$$

99. State the number of significant digits in the number 0.500.

Part II

Perform the indicated operations.

100. $\left(\dfrac{3}{4}\right) \div \left(-\dfrac{15}{8}\right)$

101. $(0)(24)$

102. $(-1)(-2)(-4)(-6)$

103. $\left(\dfrac{1}{8}\right) + \left(-\dfrac{3}{4}\right)$

104. $(6.8) - (9.2)$

105. $\left(-\dfrac{2}{7}\right) - \left(-\dfrac{3}{14}\right)$

Evaluate.

106. $(-3) - 14 \div 7$

107. $(2 - 8) \cdot 5 + 4 \div 2$

108. $|-1.7|$

109. $|6 - 8|$

110. $2^3 - (6 - 3) \cdot 5$

111. $-[-(-8)]$

112. $-\sqrt{64}$

113. $\sqrt{\dfrac{32}{50}}$

114. $-\sqrt{\dfrac{169}{144}}$

Evaluate when $x = -2$ and $y = -5$.

115. $9x^2$

116. $-y^3$

117. $|x - y^2|$

Factor.

118. $-8x + 2$ **119.** $-6x - 12$ **120.** $3a - 6b + 9c$

Place the correct symbol, =, >, or <, between the pairs of fractions.

121. $\frac{6}{7} \quad \frac{4}{5}$ **122.** $\frac{21}{11} \quad \frac{19}{10}$ **123.** $\frac{36}{26} \quad \frac{18}{13}$

Clear parentheses and collect like terms.

124. $x + (x - 3) - (2x + 1)$ **125.** $-2[5 - (y - 2)] - 3(y + 1)$

Simplify and write without negative exponents.

126. $(-2x^2)^3$ **127.** $-2(x^2)^3$ **128.** $\left(\frac{x^3}{y^2}\right)^{-2}$

129. $(a^2b^{-4})^{-2}$ **130.** $\frac{2^0x^{-4}y^3}{x^2y^{-3}}$ **131.** $\left(\frac{ab^{-2}}{a^{-3}b^{-3}}\right)^{-1}$

132. Use $A = P(1 + r)^t$ to find the amount of money in an account at the end of two years if a principal of \$5000 is invested at 15% interest, compounded annually.

ANSWERS: 1. a^5 2. $(3z)^3$ 3. $(a + w)^2$ 4. $bbbbbbb$ 5. $2xxx$ 6. $(2x)(2x)(2x)$ or $8xxx$ 7. 8 8. 30 9. 12 10. 36 11. 18 12. 37 13. 22 14. 12 15. 7 16. 2.51 17. true 18. false $(0 > -4)$ 19. true 20. = 21. > 22. < 23. −41° 24. −7 25. −3 26. 3 27. −7 28. 7 29. 2 30. −3.8 31. $\frac{1}{2}$ 32. $\frac{1}{10}$ 33. −21 34. 39 yard line 35. −8 36. −2 37. 2 38. 0 39. $-\frac{5}{8}$ 40. 4 41. −24 42. −42 pounds 43. 0 44. 4 45. 3 46. 0 47. 9 48. y 49. −8 50. x 51. $-4(x - 3)$ 52. $-2(1 + 3x + 5y)$ 53. $-5(x + 5y)$ 54. $-5a + 2$ 55. $3x - 2y + 2$ 56. $y + 1$ 57. $-8a + 15$ 58. 28 59. −28 60. 196 61. 36 62. 8 63. 0 64. −13 65. $\frac{4}{5}$ 66. −3 67. (a) 625 (b) 5 68. natural numbers 69. integers 70. rational numbers 71. real numbers 72. rational numbers 73. whole numbers 74. double-negation property 75. distributive law 76. existence of reciprocals 77. existence of an additive identity 78. existence of negatives 79. trichotomy law 80. 14 81. 1 82. $2y^9$ 83. $27x^6$ 84. $x^{10}y^{15}$ 85. 2 86. 1 87. $\frac{2}{b^{10}}$ 88. $\frac{5}{a}$ 89. $\frac{1}{5a}$ 90. $\frac{b}{a}$ 91. $\frac{b^6}{a^4}$ 92. $\frac{b^8}{a^9}$ 93. $\frac{2y^2}{x^4}$ 94. 4.11×10^{-7} 95. 5.49×10^{11} 96. 6,150,000,000,000 97. 0.00000004 98. 3.1×10^4 99. 3 100. $-\frac{2}{5}$ 101. 0 102. 48 103. $-\frac{5}{8}$ 104. −2.4 105. $-\frac{1}{14}$ 106. −5 107. −28 108. 1.7 109. 2 110. −7 111. −8 112. −8 113. $\frac{4}{5}$ 114. $-\frac{13}{12}$ 115. 36 116. 125 117. 27 118. $-2(4x - 1)$ 119. $-6(x + 2)$ 120. $3(a - 2b + 3c)$ 121. > 122. > 123. = 124. −4 125. $-y - 17$ 126. $-8x^6$ 127. $-2x^6$ 128. $\frac{y^4}{x^6}$ 129. $\frac{b^8}{a^4}$ 130. $\frac{y^6}{x^6}$ 131. $\frac{1}{a^4b}$ 132. \$6612.50

CHAPTER 1 TEST

1. Write $a \cdot a \cdot a$ in exponential notation. 1. ______

2. Evaluate. $42 \div 6 - (4 - 2) \cdot 3$ 2. ______

Evaluate when $x = 3$, $y = 2$, *and* $z = 5$.

3. $3z^2$ 3. ______

4. $(3z)^2$ 4. ______

5. $(x + y)^3$ 5. ______

6. $x^3 + y^3$ 6. ______

7. $z + 2[(x + y) - z]$ 7. ______

8. Use $A = P(1 + r)^t$ to find the amount of money in an account at the end of two years if a principal of \$500 is invested at 9% interest, compounded annually. 8. ______

9. True or false: the number $\frac{2}{7}$ is the negative of $\frac{7}{2}$. 9. ______

10. Find the absolute value. $|-13|$ 10. ______

11. Is the following true or false? $-15 < -5$ 11. ______

12. Place the correct symbol, $=$, $<$, or $>$, between the pair of fractions. $\frac{6}{13} \quad \frac{3}{5}$ 12. ______

13. The lowest temperature recorded in Polar, Idaho was 49° below zero. If the highest temperature was 92°, what is the difference between these extremes? 13. ______

CHAPTER 1 TEST

CONTINUED

Perform the indicated operations.

14. $(-8) + (-14)$ | **14.** ______

15. $5 - (-3)$ | **15.** ______

16. $\left(\frac{1}{3}\right)\left(-\frac{3}{8}\right)$ | **16.** ______

17. $(-4.5) \div (-1.5)$ | **17.** ______

18. $(-3) + 8 - (-7) - 9 - (-2)$ | **18.** ______

19. $(-2)(-1)(-1)(-1)(3)(-3)$ | **19.** ______

20. Evaluate. $(2 - 7) \cdot 5 - (-6)$ | **20.** ______

21. Write $-[-(-6)]$ without parentheses using only one sign (or no sign). | **21.** ______

22. Factor. $-3y + 15$ | **22.** ______

23. Collect like terms. $3y - 2b + 5 + b - 5y$ | **23.** ______

24. Clear parentheses and collect like terms.
$y - (2y - 3) + y$ | **24.** ______

25. Evaluate $x^2 - y^2$ when $x = 4$ and $y = 1$. | **25.** ______

26. Multiply. $-3(4y - 2)$ | **26.** ______

27. Evaluate. $\sqrt{121}$ | **27.** ______

28. Evaluate. $-\sqrt{\frac{50}{32}}$ | **28.** ______

CHAPTER 1 TEST

CONTINUED

Are the following true or false?

29. $\sqrt{81}$ is an integer. 29. ______________

30. 4π is a rational number. 30. ______________

Simplify and write without negative exponents.

31. $2y^5y^2$ 31. ______________

32. $\dfrac{a^7}{a^3}$ 32. ______________

33. $(-2x^2)^3$ 33. ______________

34. z^{-3} 34. ______________

35. $\dfrac{x^{-6}y^3}{x^2y^{-4}}$ 35. ______________

36. Evaluate y^{-3} when $y = -4$. 36. ______________

37. Evaluate. $\dfrac{3.9 \times 10^{-4}}{1.3 \times 10^{-2}}$ 37. ______________

CHAPTER 2

Linear Equations and Inequalities

2.1 LINEAR EQUATIONS AND THE ADDITION-SUBTRACTION RULE

STUDENT GUIDEPOSTS

1 Equations	3 Equivalent Equations
2 Linear Equations	4 Addition-Subtraction Rule

1 EQUATIONS

An **equation** is a statement that two quantities are equal. The two quantities are written with an equal sign (=) between them. Some equations are true, some are false, and for some the truth value cannot be determined. For example,

$1 + 2 = 3$	is true,
$2 + 5 = 3 - 7$	is false, and
$x + 2 = 9$	is neither true nor false since the value of x is not known.

Equations such as $x + 2 = 9$, which involve variables, are our primary concern in this chapter. If the variable can be replaced by a number that makes the resulting equation true, that number is a **solution** of the equation. The process of finding all solutions is **solving the equation.**

2 LINEAR EQUATIONS

In this chapter we concentrate on **linear equations** in one variable in which the exponent on the variable is 1. For this reason, a linear equation is often called a **first-degree equation.** Every linear equation in one variable x can be written in the form

$$ax + b = 0$$

where a and b are known real numbers and $a \neq 0$. Some simple linear equations can be solved directly by inspection or observation. Others require more sophisticated techniques which are discussed in the material that follows. For example, it is easy to see that 5 is a solution to the equation $x = 5$, whereas finding the solution of $2x + 3 = 8 - x$ is not so obvious and will require something beyond simple inspection. The equation

$$x + 2 = 9$$

has 7 as a solution because when x is replaced by 7,

$$7 + 2 = 9 \quad \text{True}$$

is a true equation. Note that 3 is not a solution since

$$3 + 2 = 9 \quad \text{False}$$

is false. An equation such as this, which is true for some replacements of the variable and false for others, is a **conditional equation.** The equation

$$x + 1 = 1 + x$$

has many solutions. In fact every number is a solution of this equation. An equation like this, called an **identity,** has the entire set of real numbers for its solution set. The equation

$$x + 1 = x$$

has no solutions. An equation like this is a **contradiction.**

3 EQUIVALENT EQUATIONS

Consider the two equations

$$x + 1 = 3 \quad \text{and} \quad x = 2.$$

Both equations have 2 as a solution. Although it is easier to see that 2 is a solution to the second, both can be solved by inspection. When two equations have exactly the same solutions, they are **equivalent equations.** To solve an equation, we try to change it into an equivalent equation that can be solved by direct inspection. This usually means that the variable is isolated on one side, as in $x = 2$.

4 ADDITION-SUBTRACTION RULE

Following is the first rule for solving equations.

Addition-Subtraction Rule

An equivalent equation is obtained if the same quantity is added to or subtracted from both sides of an equation.

Suppose a, b, and c are real numbers.

If $a = b$, then $a + c = b + c$ and $a - c = b - c$.

For example, if we start with the true equation $5 = 5$ and then add 2 to both sides or subtract 3 from both sides,

$$5 + 2 = 5 + 2 \quad \text{and} \quad 5 - 3 = 5 - 3,$$

the resulting equations are also true since $7 = 7$ and $2 = 2$.

EXAMPLE 1 Using the Addition-Subtraction Rule

Solve. $x - 2 = 7$

$x - 2 + 2 = 7 + 2$ Add 2 to both sides to isolate x on the left side

$x + 0 = 9$

$x = 9$

The solution is 9.

PRACTICE EXERCISE 1

Solve. $y - 8 = -1$

Answer: 7

EXAMPLE 2 Using the Addition-Subtraction Rule

Solve. $y + 3 = 17$

$y + 3 - 3 = 17 - 3$ Subtract 3 from both sides to isolate y on the left side

$y + 0 = 14$

$y = 14$

The solution is 14.

Practice Exercise 2

Solve. $z + 9 = 4$

Answer: -5

Get in the habit of checking all solutions you get to an equation. To do this, replace the variable throughout the original equation with the solution you get, to see whether the resulting equation is true.

EXAMPLE 3 Using the Addition-Subtraction Rule

Solve. $3 = x + 5$

$3 - 5 = x + 5 - 5$ Subtract 5 from both sides to isolate x on the right side

$-2 = x + 0$

$-2 = x$

The solution is -2.

Check: $3 \stackrel{?}{=} -2 + 5$ Replace x with -2 in the original equation

$3 = 3$

The solution -2 does check.

Practice Exercise 3

Solve. $21 = z + 15$

Answer: 6

In Example 3, we isolated the variable on the right side of the equation instead of the left side as in the first two examples. Since $-2 = x$ and $x = -2$ are equivalent, we see that it makes no difference whether the isolated variable is on the left or the right.

Equations involving fractions or decimals are solved in the same way.

EXAMPLE 4 Equation with Fractions

Solve. $\frac{1}{2} + y = -3$

$\frac{1}{2} - \frac{1}{2} + y = -3 - \frac{1}{2}$ Subtract $\frac{1}{2}$ from both sides

$0 + y = -\frac{6}{2} - \frac{1}{2}$ The LCD of -3 and $-\frac{1}{2}$ is 2

$y = \frac{-6 - 1}{2}$

$y = -\frac{7}{2}$ Subtract fractions

The solution is $-\frac{7}{2}$. To check this, substitute $-\frac{7}{2}$ for y in $\frac{1}{2} + y = -3$.

Practice Exercise 4

Solve. $\frac{3}{4} + x = \frac{1}{8}$

$$\frac{1}{2} + \left(-\frac{7}{2}\right) \stackrel{?}{=} -3$$
$$\frac{1}{2} - \frac{7}{2} \stackrel{?}{=} -3$$
$$\frac{1-7}{2} \stackrel{?}{=} -3$$
$$-\frac{6}{2} \stackrel{?}{=} -3$$
$$-3 = -3$$

Answer: $-\frac{5}{8}$

2.1 EXERCISES A

Give an example of each of the following.

1. A true equation
2. A contradiction
3. A linear equation

Solve the following equations by direct observation or inspection.

4. $x + 3 = 5$
5. $2y = 8$
6. $2 + z = 2 - z$
7. $2 + x = 2 + x$
8. $y + 5 = y + 1$
9. $2z = 5z$

Use the equation $y - 1 = 5$ to answer Exercises 10–12.

10. What is the variable?
11. What is the left side of the equation?
12. Is it an identity?

Use the addition-subtraction rule to solve. Check all solutions.

13. $x + 3 = 7$
14. $y - 4 = 9$
15. $z + 12 = -4$
16. $x - 9 = -6$
17. $y + 16 = 15$
18. $z + 11 = 13$
19. $4 = x + 1$
20. $-7 = y - 3$
21. $-3 = z + 2$
22. $\frac{1}{3} + x = 5$
23. $\frac{3}{2} + y = -2$
24. $-\frac{1}{4} + z = \frac{3}{4}$
25. $x - 2.1 = 3.4$
26. $-4.2 = y + 1.7$
27. $-2.6 = z - 3.9$

28. $x + 2\frac{1}{2} = 5$

29. $y - 3\frac{2}{3} = 4$

30. $1\frac{3}{8} + z = \frac{1}{4}$

In Exercises 31–34, is the statement true *or* false*? If the statement is false, tell why.*

31. -2 is a solution to $x + 2 = 0$.

32. $y + 3 = 5$ is a linear equation.

33. Zero is a solution of $x = 2x$.

34. $x + 2 = x$ is an identity.

FOR REVIEW

Exercises 35–38 review material from Chapters 0 and 1 to help prepare for the next section. Simplify each expression.

35. $2(5x)$

36. $(-1)(-z)$

37. $(-4)\left(-\frac{1}{4}y\right)$

38. $\left(-\frac{2}{5}\right)\left(-\frac{5}{2}x\right)$

ANSWERS: 1–3. Answers will vary. 4. 2 5. 4 6. 0 7. any number is a solution. 8. no solution 9. 0 10. y 11. $y - 1$ 12. no 13. 4 14. 13 15. -16 16. 3 17. -1 18. 2 19. 3 20. -4 21. -5 22. $\frac{14}{3}$ 23. $-\frac{7}{2}$ 24. 1 25. 5.5 26. -5.9 27. 1.3 28. $2\frac{1}{2}$ 29. $7\frac{2}{3}$ 30. $-1\frac{1}{8}$ 31. true 32. true 33. true 34. false (contradiction) 35. $10x$ 36. z 37. y 38. x

2.1 EXERCISES B

Give an example of each of the following.

1. A false equation

2. An identity

3. An equation that is neither true nor false

Solve the following equations by direct observation or inspection.

4. $x + 5 = 12$

5. $3y = 15$

6. $4 + z = 4 - z$

7. $1 + x = 1 + x$

8. $y + 2 = y + 8$

9. $8z = 3z$

Use the equation $x + 3 = 19$ to answer Exercises 10–12.

10. What is the right side of the equation?

11. What is the solution?

12. Is this a contradiction?

Solve and check.

13. $x + 2 = 12$

14. $y - 3 = 10$

15. $z + 13 = -2$

16. $x - 11 = -9$

17. $y + 17 = 18$

18. $z + 9 = 18$

19. $5 = x + 7$

20. $-3 = y - 3$

21. $-4 = z + 3$

22. $\frac{1}{3} + x = 8$

23. $\frac{1}{3} + y = \frac{7}{3}$

24. $-\frac{1}{4} + z = 1$

25. $x - 2.3 = 5.7$

26. $-5.7 = y - 2.3$

27. $-3.8 = z - 1.2$

28. $x + 1\frac{3}{4} = 3$

29. $y - 2\frac{1}{5} = 7$

30. $3\frac{2}{3} + z = \frac{1}{9}$

In Exercises 31–34, is the statement true *or* false? *If the statement is false, tell why.*

31. -4 is a solution to $4 + x = 0$.

32. $z + 2 = 8$ is a linear equation.

33. A solution of $3y = y$ is 0.

34. $y + 1 = 1 - y$ is a contradiction.

FOR REVIEW

Exercises 35–38 review material from Chapters 0 and 1 to help prepare for the next section. Simplify each expression.

35. $8(4x)$

36. $\frac{1}{3}(3y)$

37. $\left(-\frac{1}{2}\right)(-2x)$

38. $\left(-\frac{2}{3}\right)\left(-\frac{3}{2}z\right)$

2.1 EXERCISES C

Solve. A calculator would be helpful for these exercises.

1. $3.75 - x = 7\frac{2}{5}$

2. $0.00125 + y = \frac{1}{1000}$

3. $125\frac{7}{8} = 95.2 - z$
[Answer: -30.675]

2.2 THE MULTIPLICATION-DIVISION RULE

In Section 2.1 we were careful to choose equations for which the coefficient of the variable was 1. When the coefficient of the variable is a number other than 1, we need to carry out an additional step using the following rule.

Multiplication-Division Rule

An equivalent equation is obtained if both sides of an equation are multiplied or divided by the same nonzero quantity.

Suppose a, b, and c are real numbers with $c \neq 0$.

$$\text{If } a = b, \text{ then } \quad ac = bc \quad \text{and} \quad \frac{a}{c} = \frac{b}{c}.$$

For example, if we start with the true equation $6 = 6$ and then multiply both sides by 2 or divide both sides by 2,

$$6 \cdot 2 = 6 \cdot 2 \quad \text{and} \quad \frac{6}{2} = \frac{6}{2},$$

the resulting equations are also true since $12 = 12$ and $3 = 3$.

EXAMPLE 1 Using the Multiplication-Division Rule

Solve. $2x = 10$

$\frac{1}{2} \cdot 2x = \frac{1}{2} \cdot 10$ Multiply both sides by the reciprocal of 2, which is $\frac{1}{2}$

$1 \cdot x = 5$

$x = 5$

Practice Exercise 1

Solve. $9y = -27$

The solution is 5.

Check: $2 \cdot 5 \stackrel{?}{=} 10$ Replace x with 5 in the original equation

$10 = 10$

Answer: -3

Equivalently, in Example 1 we could divide both sides of the equation by 2.

$$2x = 10$$

$$\frac{2x}{2} = \frac{10}{2} \quad \text{Divide both sides by 2}$$

$$\frac{2}{2} \cdot x = 5$$

$$1 \cdot x = 5$$

$$x = 5$$

EXAMPLE 2 Using the Multiplication-Division Rule

Solve. $\frac{1}{3}x = 18$

$$3 \cdot \frac{1}{3}x = 3 \cdot 18 \quad \text{Multiply both sides by the reciprocal of } \tfrac{1}{3}\text{, which is 3}$$

$$1 \cdot x = 54$$

$$x = 54$$

The solution is 54. Check.

Practice Exercise 2

Solve. $\frac{1}{8}y = -1$

Answer: -8

Notice in Examples 1 and 2 that we used the multiplication-division rule to make the coefficient of the variable equal to 1. This is the necessary final step in solving an equation. With practice, many of the steps may be performed mentally.

EXAMPLE 3 Using the Multiplication-Division Rule

Solve. $\frac{x}{\frac{4}{5}} = 20$

In an equation of this type, simplify the left side first.

$$\frac{x}{\frac{4}{5}} = \frac{\frac{x}{1}}{\frac{4}{5}} = \frac{x}{1} \div \frac{4}{5} = \frac{x}{1} \cdot \frac{5}{4} = \frac{5}{4} \cdot x$$

Thus, we are actually solving

$$\frac{5}{4} \cdot x = 20.$$

$$\frac{4}{5} \cdot \frac{5}{4} \cdot x = \frac{4}{5} \cdot 20 \quad \text{Multiply both sides by } \tfrac{4}{5}$$

$$1 \cdot x = 16 \quad \tfrac{4}{5} \cdot 20 = \tfrac{4 \cdot 20}{5} = \tfrac{4 \cdot 4}{1} = 16$$

$$x = 16$$

Practice Exercise 3

Solve. $\frac{z}{\frac{1}{7}} = -35$

The solution is 16.

Check: $\dfrac{16}{\frac{4}{5}} \stackrel{?}{=} 20$

$$\frac{5}{4} \cdot 16 \stackrel{?}{=} 20$$

$$20 = 20$$

Answer: -5

EXAMPLE 4 Equation with Decimals

Solve. $-2.7x = 54$

$$\frac{-2.7x}{-2.7} = \frac{54}{-2.7} \quad \text{Divide both sides by } -2.7$$

$$\frac{-2.7}{-2.7} \cdot x = \frac{54}{-2.7}$$

$$1 \cdot x = -20$$

$$x = -20$$

The solution is -20.

Check: $-2.7(-20) \stackrel{?}{=} 54$

$$54 = 54$$

Practice Exercise 4

Solve. $0.6y = -7.2$

Answer: -12

We can remember the multiplication-division rule with the statement, "multiply or divide both sides of the equation by the same quantity."

2.2 EXERCISES A

Use the multiplication-division rule to solve. Check all solutions.

1. $5x = 25$
2. $56 = 8y$
3. $-11z = 77$
4. $12x = -84$
5. $-72 = -y$
6. $-2z = 2$
7. $-5x = -50$
8. $-60y = 360$
9. $\dfrac{z}{1} = 10$
10. $\dfrac{2}{3}x = 10$
11. $\dfrac{y}{4} = -3$
12. $-\dfrac{2}{5}z = 8$
13. $\dfrac{x}{\frac{1}{3}} = 27$
14. $\dfrac{y}{\frac{7}{2}} = -6$
15. $\dfrac{z}{-\frac{1}{8}} = 16$
16. $\dfrac{2x}{3} = 4$
17. $\dfrac{3y}{2} = -6$
18. $\dfrac{4z}{-3} = -8$

19. $4x = -24.8$ **20.** $1.2y = 7.2$ **21.** $-0.8z = -6.4$

22. $2\frac{1}{2}x = 10$ **23.** $-3\frac{1}{3}y = 20$ **24.** $-4\frac{1}{5}z = -63$

FOR REVIEW

Solve.

25. $-8 + x = -12$ **26.** $-\frac{3}{4} + y = -\frac{7}{4}$ **27.** $7.4 + z = -3.2$

28. Is $2 + x = x + 2$ an identity? **29.** Is 0 a solution of $x - 3 = -3 - x$?

ANSWERS: 1. 5 2. 7 3. −7 4. −7 5. 72 6. −1 7. 10 8. −6 9. 10 10. 15 11. −12 12. −20 13. 9 14. −21 15. −2 16. 6 17. −4 18. 6 19. −6.2 20. 6 21. 8 22. 4 23. −6 24. 15 25. −4 26. −1 27. −10.6 28. yes 29. yes

2.2 EXERCISES B

Use the multiplication-division rule to solve. Check.

1. $7x = 49$ **2.** $72 = 6y$ **3.** $-11z = 44$

4. $12x = -96$ **5.** $-63 = -y$ **6.** $-3z = 48$

7. $-20x = -300$ **8.** $30y = -240$ **9.** $\frac{z}{1} = -30$

10. $\frac{2}{3}x = 14$ **11.** $\frac{y}{5} = -3$ **12.** $-\frac{2}{7}z = 8$

13. $\frac{x}{\frac{1}{2}} = 10$ **14.** $\frac{y}{\frac{1}{9}} = 27$ **15.** $\frac{z}{-\frac{2}{3}} = 12$

16. $\frac{3x}{5} = 6$ **17.** $\frac{4y}{3} = -2$ **18.** $\frac{8z}{-5} = -16$

19. $3x = -25.5$ **20.** $1.2y = 8.4$ **21.** $-1.2z = -4.8$

22. $3\frac{2}{3}x = 22$ **23.** $-4\frac{1}{4}y = 17$ **24.** $-1\frac{3}{5}z = -32$

FOR REVIEW

Solve.

25. $-21 + x = -25$ **26.** $-\frac{5}{6} + y = \frac{7}{12}$ **27.** $1.9 + z = -6.7$

28. Is $x + 1 = x - 1$ a contradiction? **29.** Is -6 a solution of $x - 6 = 0$?

2.2 EXERCISES C

Solve. A calculator might be helpful in these exercises.

1. $0.0568x = 0.006816$ **2.** $9\frac{7}{12}y = -7\frac{3}{16}$ **3.** $-15\frac{3}{4}z = 0.525$

[Answer: $-0.0\overline{3}$]

2.3 SOLVING EQUATIONS BY COMBINING RULES

STUDENT GUIDEPOSTS

1. Equations with Similar Terms
2. Equations with the Variable on Both Sides
3. Equations with Parentheses
4. Equations with Fractional or Decimal Coefficients
5. Summary of Methods

When we solve an equation, we isolate the variable on one side so we can find the solution by inspection. If we need to use both rules to do this, we generally use the addition-subtraction rule before the multiplication-division rule, as in the following example.

EXAMPLE 1 Using Both Rules

Solve. $3x + 5 = 11$

$3x + 5 - 5 = 11 - 5$ Use addition-subtraction rule first to subtract 5 from both sides

$3x + 0 = 6$

$3x = 6$

$\frac{1}{3} \cdot 3x = \frac{1}{3} \cdot 6$ Now use multiplication-division rule to multiply both sides by $\frac{1}{3}$

$1 \cdot x = 2$

$x = 2$

The solution is 2.

Check: $3 \cdot 2 + 5 \stackrel{?}{=} 11$

$6 + 5 \stackrel{?}{=} 11$

$11 = 11$

PRACTICE EXERCISE 1

Solve. $4y + 10 = 2$

Answer: -2

1 EQUATIONS WITH SIMILAR TERMS

Sometimes the sides of an equation have similar terms. These terms should be combined before attempting to apply the rules.

EXAMPLE 2 Equation with Similar Terms

Solve. $15 = 4y + 3y$

$15 = 7y$ Combine similar terms

$\frac{1}{7} \cdot 15 = \frac{1}{7} \cdot 7y$ Multiply both sides by $\frac{1}{7}$

$\frac{15}{7} = 1 \cdot y$

$\frac{15}{7} = y$

The solution is $\frac{15}{7}$. Check by substituting in the original equation.

Practice Exercise 2

Solve. $9z - 3z = 2$

Answer: $\frac{1}{3}$

2 EQUATIONS WITH THE VARIABLE ON BOTH SIDES

When an equation has terms involving the variable on both sides, we use the addition-subtraction rule to get them together on the same side so they can then be combined.

EXAMPLE 3 Variable on Both Sides

Solve. $7x - 2 = 2x + 3$

$7x - 2 + 2 = 2x + 3 + 2$ Add 2 to both sides

$7x + 0 = 2x + 5$

$7x = 2x + 5$

$7x - 2x = 2x - 2x + 5$ Subtract $2x$ from both sides

$5x = 0 + 5$

$5x = 5$

$\frac{1}{5} \cdot 5x = \frac{1}{5} \cdot 5$ Multiply both sides by $\frac{1}{5}$

$1 \cdot x = 1$

$x = 1$

The solution is 1. Check.

Practice Exercise 3

Solve. $4 + 3y = -8 - y$

Answer: -3

To solve an equation, use the addition-subtraction rule to isolate all terms involving the variable on one side of the equation and then combine them. Next apply, if necessary, the multiplication-division rule. Always remember to combine similar terms before applying the rules.

EXAMPLE 4 VARIABLE IN SEVERAL TERMS

Solve. $2 - x + 10 = 3x + 2x + 24$

$-x + 12 = 5x + 24$ Combine similar terms

$-x + 12 - 12 = 5x + 24 - 12$ Subtract 12 from both sides

$-x = 5x + 12$

$-x - 5x = 5x - 5x + 12$ Subtract $5x$ from both sides

$-6x = 12$

$\left(-\frac{1}{6}\right)(-6x) = \left(-\frac{1}{6}\right)12$ Multiply both sides by $-\frac{1}{6}$

$1 \cdot x = -2$

$x = -2$

The solution is -2.

Check: $2 - (-2) + 10 \stackrel{?}{=} 3(-2) + 2(-2) + 24$

$2 + 2 + 10 \stackrel{?}{=} -6 - 4 + 24$

$14 = 14$

PRACTICE EXERCISE 4

Solve. $4y + 2 - y = 3 + 5y - 9$

Answer: 4

3 EQUATIONS WITH PARENTHESES

When solving an equation containing parentheses, remove the parentheses using the distributive property. The resulting equation can then be solved using previous methods.

EXAMPLE 5 EQUATION WITH PARENTHESES

Solve. $3(x + 6) = 21$

$3 \cdot x + 3 \cdot 6 = 21$ To remove parentheses, multiply both 6 and x by 3

$3x + 18 = 21$

$3x + 18 - 18 = 21 - 18$ Subtract 18

$3x = 3$

$\frac{1}{3} \cdot 3x = \frac{1}{3} \cdot 3$ Multiply by $\frac{1}{3}$

$1 \cdot x = 1$

$x = 1$

The solution is 1. Check.

PRACTICE EXERCISE 5

Solve. $5(z + 2) = 10$

Answer: 0

EXAMPLE 6 EQUATION WITH PARENTHESES

Solve. $6z - (3z - 4) = 14$

$6z - 3z + 4 = 14$ Remove parentheses by changing signs

$3z + 4 = 14$

$3z + 4 - 4 = 14 - 4$ Subtract 4

$3z = 10$

$z = \frac{10}{3}$

PRACTICE EXERCISE 6

Solve. $4y - (y - 8) = -1$

The solution is $\frac{10}{3}$.

Check: $6\left(\frac{10}{3}\right) - \left(3\left(\frac{10}{3}\right) - 4\right) \stackrel{?}{=} 14$

$$20 - (10 - 4) \stackrel{?}{=} 14$$
$$20 - 6 \stackrel{?}{=} 14$$
$$14 = 14$$

Answer: -3

EXAMPLE 7 Equation with Parentheses

Solve. $2x - 4(2x - 4) = 22 - 3(x - 5)$

$$2x - 8x + 16 = 22 - 3x + 15 \quad \text{Remove parentheses}$$
$$-6x + 16 = 37 - 3x \quad \text{Combine like terms}$$
$$-6x + 16 - 16 = 37 - 3x - 16 \quad \text{Subtract 16}$$
$$-6x = 21 - 3x$$
$$-6x + 3x = 21 - 3x + 3x \quad \text{Add } 3x$$
$$-3x = 21$$
$$\left(-\frac{1}{3}\right)(-3x) = \left(-\frac{1}{3}\right)21 \quad \text{Multiply by } -\tfrac{1}{3}$$
$$1 \cdot x = -7$$
$$x = -7$$

The solution is -7. Check.

PRACTICE EXERCISE 7

Solve.

$3z - 2(1 - z) = 7 - 5(z + 3)$

Answer: $-\frac{3}{5}$

4 EQUATIONS WITH FRACTIONAL OR DECIMAL COEFFICIENTS

If an equation has fractional coefficients, it helps to simplify first by multiplying both sides by the least common denominator (LCD) of all fractions. For example, to solve

$$\frac{1}{2}x + \frac{3}{4} = \frac{5}{6},$$

first multiply both sides by the least common denominator 12.

$$\mathbf{12}\left(\frac{1}{2}x + \frac{3}{4}\right) = \mathbf{12} \cdot \frac{5}{6}$$
$$\mathbf{12} \cdot \frac{1}{2}x + \mathbf{12} \cdot \frac{3}{4} = 12 \cdot \frac{5}{6} \quad \text{Use distributive law}$$
$$6x + 9 = 10$$

We have less chance of making an error solving this equation than the original. The same remarks apply to equations with decimal coefficients. First eliminate the decimals by multiplying both sides by an appropriate power of 10. For example, to solve

$$1.5y + 3.25 = 6$$

multiply both sides by 100 to clear all decimals.

$$\begin{aligned} 100(1.5y + 3.25) &= 100 \cdot 6 \\ 150y + 325 &= 600 \\ 150y &= 275 \\ y &= \frac{275}{150} = \frac{11}{6} \end{aligned}$$

We have combined several steps and performed some of the calculations mentally in the above example. With practice, you should be able to do the same.

5 SUMMARY OF METHODS

We conclude this section by summarizing how to solve linear equations.

To Solve a Linear Equation

1. Simplify both sides by clearing parentheses, fractions, or decimals. Collect like terms.
2. Use the addition-subtraction rule to isolate all variable terms on one side and all constant terms on the other side. Collect like terms when possible.
3. Use the multiplication-division rule to obtain a variable with coefficient of 1.
4. Check the solution by substituting in the original equation.
5. If an identity results, the original equation has every real number as a solution. If a contradiction results, there is no solution.

2.3 EXERCISES A

Solve each equation. Check all solutions.

1. $2x + 1 = 5$

2. $3y - 1 = 8$

3. $-4z + 3 = 5$

4. $\frac{3}{4}x + \frac{1}{4} = 1$

5. $\frac{2}{5} - y = \frac{3}{10}$

6. $12 = -1.2z + 24$

7. $2x + 3x = 10$

8. $3y - y = 7$

9. $\frac{1}{4}z + \frac{1}{2}z = -9$

10. $9.2x - 3.1x = 12.2$

11 $-\frac{1}{7}y + y = 18$

12. $3z = 5 - 7z$

13. $32 - 8x = 3 - 9x$

14 $-7 + 21y + 23 = 16$

15. $z + 3z = 8 - 2z + 10$

16. $x + 9 + 6x = 2 + x + 1$

17 $6y - 4y + 1 = 12 + 2y - 11$

18. $2.1x - 3.2 = -8.4x - 45.2$

19 $3y + \frac{5}{2}y + \frac{3}{2} = \frac{1}{2}y + \frac{5}{2}y$

Remove parentheses and solve.

20. $3(2x + 1) = 21$

21. $10 = 5(y - 20)$

22. $2(5z + 1) = 42$

23. $3x - 4(x + 2) = 5$

24. $2y - (13 - 2y) = 59$

25. $3z - (z + 4) = 0$

26. $5(x + 4) - 4(x + 3) = 0$

27. $4(y - 3) - 6(y + 1) = 0$

28. $9z - (3z - 18) = 36$

29 $5(2 - 3x) = 15 - (x + 7)$

30. $8y - (2y + 5) = 0$

31. $9z - (5z - 2) = 8$

32. $7(x - 5) = 10 - (x + 1)$

33. $(y - 9) - (y + 6) = 4y$

34. $8(2z + 1) = 4(7z + 7)$

35. $5 + 3(x + 2) = 3 - (x + 2)$

36 $\frac{1}{3}(6x - 9) = \frac{1}{2}(8x - 4)$

37 $5(x + 1) - 4x = x - 5$

38 $3(2 - 4x) = 4(2x - 1) - 2(1 + x)$

39. $2 + y = 3 - 2[1 - 2(y + 1)]$

40. $-2(z - 1) - 3(2z + 1) = -9$

41 $-4(-x + 1) + 3(2x + 3) - 7x = -10$

FOR REVIEW

Solve.

42. $\dfrac{x}{-\frac{1}{5}} = 20$

43. $\dfrac{y}{-5} = -2$

44. $\dfrac{3z}{7} = -6$

45. $-2\dfrac{3}{8}x = -19$

46. $y - \dfrac{2}{3} = \dfrac{1}{9}$

47. $3.6 - z = 7.2$

ANSWERS: 1. 2 2. 3 3. $-\frac{1}{2}$ 4. 1 5. $\frac{1}{10}$ 6. 10 7. 2 8. $\frac{7}{2}$ 9. −12 10. 2 11. 21 12. $\frac{1}{2}$ 13. −29 14. 0 15. 3 16. −1 17. every real number (identity) 18. −4 19. $-\frac{3}{5}$ 20. 3 21. 22 22. 4 23. −13 24. 18 25. 2 26. −8 27. −9 28. 3 29. $\frac{1}{7}$ 30. $\frac{5}{6}$ 31. $\frac{3}{2}$ 32. $\frac{11}{2}$ 33. $-\frac{15}{4}$ 34. $-\frac{5}{3}$ 35. $-\frac{5}{2}$ 36. $-\frac{1}{2}$ 37. no solution 38. $\frac{2}{3}$ 39. −1 40. 1 41. −5 42. −4 43. 10 44. −14 45. 8 46. $\frac{7}{9}$ 47. −3.6

2.3 EXERCISES B

Solve and check.

1. $3x + 1 = 7$

2. $4y - 2 = 10$

3. $-2z + 5 = 9$

4. $\dfrac{2}{3}x + \dfrac{1}{3} = 1$

5. $\dfrac{3}{5} - y = \dfrac{7}{10}$

6. $10 = -0.5z + 5$

7. $3x + 7x = 40$

8. $4y - y = 27$

9. $\dfrac{1}{8}z + \dfrac{5}{8}z = -6$

10. $1.3x + 2.7x = 4.4$

11. $-\dfrac{1}{5}y + y = 8$

12. $5z = 2 - 9z$

13. $25 - 7x = 4 - 10x$

14. $-2 + 8y + 27 = 1$

15. $z + 2z = 3 - 2z + 17$

16. $x + 3 + 4x = 2 + x + 1$

17. $3y - y + 10 = 14 + 2y - 4$

18. $6.4 - 4.2x = 16.8x + 90.4$

19. $2x - \dfrac{1}{3} + \dfrac{2}{3}x = \dfrac{4}{9} + \dfrac{1}{3}x$

Clear parentheses and solve.

20. $2(3x + 1) = 14$

21. $9 = 3(4y - 1)$

22. $7(2z - 1) = -35$

23. $4x - 2(x + 1) = -1$

24. $7y - (4 - 3y) = 11$

25. $2z - (z + 1) = 0$

26. $6(x + 2) - 2(2x + 1) = 0$

27. $4(2y - 1) - 7(y + 2) = 0$

28. $6z - (2z - 5) = 0$

29. $3(2 - 4x) = 13 - (x + 1)$

30. $7y - (2y + 15) = 0$

31. $8z - (3z - 1) = 0$

32. $4(x - 2) = 12 - (x + 3)$

33. $(y - 2) - (y + 2) = 4y$

34. $9(2z - 1) = 3(z + 2)$

35. $7 + 3(x + 1) = 5 - (x + 1)$

36. $\frac{1}{2}(2y - 4) = \frac{2}{3}(9y + 3)$

37. $6(z - 1) - 3z = 3z + 8$

38. $2(1 - 3x) = 5(2x + 1) - 3(1 + x)$

39. $4 + y = 8 - 3[2 - (y + 2)]$

40. $-3(z - 4) - 2(3z + 1) = -8$

41. $-7(-x + 2) + 5(2x + 1) - 11x = 3$

FOR REVIEW

Solve.

42. $-\frac{3}{5}x = 6$

43. $\frac{y}{-\frac{1}{7}} = 14$

44. $\frac{2z}{9} = -12$

45. $-3\frac{1}{7}x = -88$

46. $\frac{3}{4} - y = \frac{1}{2}$

47. $8.4 + z = -9.2$

2.3 EXERCISES C

Solve and check.

1. $0.02x - (0.56x - 7.33) = 2.15 - (0.7x - 0.11)$

2. $3\frac{2}{3}x + 7\frac{1}{9} - \left(\frac{x}{6} - 4\frac{1}{9}\right) = -\left(2\frac{1}{6} - 6\frac{2}{9}x\right)$
[Answer: $\frac{241}{49}$]

3. $x(x - 3) - 3x[1 - (x - 1)] = 4x(x - 5) + 22$

4. $-3[1 - (x^2 - 2)] - 2x(1 - x) = -5[4 - x(x - 2)]$
[Answer: $-\frac{11}{8}$]

2.4 TECHNIQUES OF SOLVING APPLICATION PROBLEMS

STUDENT GUIDEPOSTS

1. Translating Words into Symbols
2. Solving Applied Problems
3. Consecutive Integer Problems
4. Geometry Problems

1 TRANSLATING WORDS INTO SYMBOLS

Solving a word problem with algebra involves two basic steps. First, we translate the words of the problem into an algebraic equation, and second, we solve the equation. We learned how to solve several types of equations in the first three sections of this chapter, and now we concentrate on translating words into equations. Some common words and their symbolic translations are presented on the following page.

Symbol	*Stands for*
$+$	and, sum, sum of, added to, increased by, more than
$-$	minus, less, subtracted from, less than, diminished by, difference between, difference, decreased by
$\cdot$	times, product, product of, multiplied by, of
$\div$	divided by, quotient of, ratio
$=$	equals, is equal to, is as much as, is, is the same as, gives, yields

Any letter (we often use x) can stand for the unknown or desired quantity. Some examples of translations of these phrases follow.

A number increased by 5 is 13.
$$x + 5 = 13$$

Two subtracted from a number is 5.
$$x - 2 = 5$$

Three percent of a number is 15.
$$0.03 \cdot x = 15$$

Twice my age in two years will be 20.
$$2 \cdot (x + 2) = 20$$

Four times a number, diminished by 12, is as much as twice the number.
$$4 \cdot x - 12 = 2x$$

It is sometimes helpful to use a letter as the variable that is indicative of the quantity it represents. For example, we might use n for a natural number, t for time, s for salary, V for volume, or T for Ted's age.

EXAMPLE 1 Translating Phrases into Symbols

Select a variable to represent each quantity and translate the phrase into symbols.

Word phrase	*Symbolic translation*
(a) Twice a natural number	$2n$
(b) The time in 6 hours	$t + 6$
(c) My salary less \$200 for taxes	$S - 200$
(d) The price less 10% of the price	$p - 0.10p$
(e) One-third the volume	$\frac{1}{3}V$

Practice Exercise 1

Select a variable and translate each phrase into symbols.

(a) One-half a whole number

(b) The rate, increased by 5 mph

(c) The price less \$15

(d) The population, diminished by 20%

(e) Three times the area

Answers: (a) $\frac{1}{2}w$ (b) $r + 5$ (c) $p - 15$ (d) $p - 0.20p$ (e) $3A$

EXAMPLE 2 Translating Sentences into Symbols

Use x for the variable and translate each word sentence into symbols.

	Word expression	*Symbolic translation*
(a)	The product of a number and 3 is 22.	$3x = 22$
(b)	Four times a number is 8.	$4x = 8$
(c)	Twice a number, increased by 8, is 11.	$2x + 8 = 11$
(d)	Seven is 4 less than three times a number.	$7 = 3x - 4$
(e)	Seven is 4 less three times a number.	$7 = 4 - 3x$
(f)	One-tenth of a number is 6.	$\frac{1}{10}x = 6$
(g)	Twice a number, less five times the number, is the same as 12.	$2x - 5x = 12$
(h)	A number is 15% of 28.	$x = (0.15)(28)$
(i)	A number subtracted from 2 is 5.	$2 - x = 5$
(j)	Two subtracted from a number is 8.	$x - 2 = 8$
(k)	Twice the sum of a number and 3 is four times the square of the number.	$2(x + 3) = 4x^2$
(l)	Eight more than a number is 12.	$8 + x = 12$
(m)	My age 3 years ago was 35. (Let x be my present age.)	$x - 3 = 35$

Practice Exercise 2

Use x for the variable and translate into symbols.

(a) Three times a number is 5 less the number.

(b) Thirty is 12% of the principal.

(c) Ann's age in 8 years will be 24.

(d) Five times the price, less \$20, is \$45.

(e) The cube of a number is 24 more than 10 times the number.

Answers: (a) $3x = 5 - x$ (b) $30 = 0.12x$ (c) $x + 8 = 24$ (d) $5x - 20 = 45$ (e) $x^3 = 24 + 10x$

2 SOLVING APPLIED PROBLEMS

Because there is a wide variety of applied problems, no single technique works for all problems. But following these steps will help.

To Solve an Applied Problem

1. Read the problem carefully (perhaps several times) and decide what quantity (or quantities) must be found.
2. Represent the unknown quantity (or quantities) with a letter.
3. Make a sketch or diagram whenever possible.
4. Determine which expressions are equal and write an equation.
5. Solve the equation and state the answer to the problem.
6. Check to see if the answer (or answers) satisfies the conditions of the original problem. This can often be done mentally.

In the remainder of this section we solve several types of applied problems.

EXAMPLE 3 Number Problem

If three times a number is subtracted from 20, the result is twice the number. Find the number.

Let $x =$ the desired number,

$20 - 3x =$ three times the number subtracted from 20.

Practice Exercise 3

Twenty less a number is 3 times the number. Find the number.

$$20 - 3x = 2x \quad \text{Symbolic translation of the problem}$$
$$20 = 5x$$
$$4 = x$$

The number is 4. Check: 20 minus $3 \cdot 4$ is indeed twice 4.

Answer: 5

3 CONSECUTIVE INTEGER PROBLEMS

Consecutive integers are integers that are next to each other in counting order. For example, 3, 4, and 5 are consecutive integers. If x is an integer, the next consecutive integer is $x + 1$, and the next after that is $x + 2$. We can also consider consecutive even or odd integers such as 4, 6, 8, or 9, 11, 13. In these cases, if x is an integer, the next consecutive even (or odd) integer is $x + 2$, and the next after that is $x + 4$.

EXAMPLE 4 Consecutive Integer Problem

The sum of two consecutive integers is 47. Find the integers.

Let x = the first integer,

$x + 1$ = the next consecutive integer.

$$x + (x + 1) = 47$$
$$2x + 1 = 47$$
$$2x = 46$$
$$x = 23 \quad \text{and} \quad x + 1 = 24$$

The integers are 23 and 24. Check: 23 and 24 are indeed consecutive integers that add up to 47.

Practice Exercise 4

Find two consecutive odd integers whose sum is 56.

Answer: 27 and 29

EXAMPLE 5 Number Problem

A 20-ft section of rope is cut into two pieces, one 7 ft longer than the other. How long is each piece?

Let x = the length of the shorter piece,

$x + 7$ = the length of the longer piece. (Why?)

Make a sketch as in Figure 2.1.

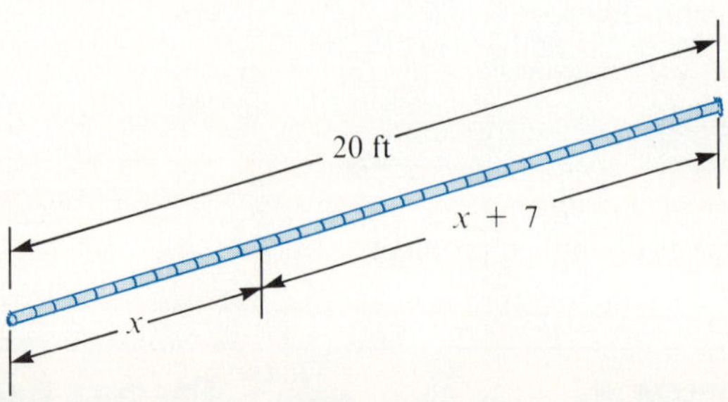

Figure 2.1

$$x + (x + 7) = 20$$
$$2x + 7 = 20$$
$$2x = 13$$

Practice Exercise 5

A 48-in gold chain is cut into two lengths, one 12 in shorter than the other. How long is each piece?

$$x = \frac{13}{2}$$

$$\text{and} \quad x + 7 = \frac{27}{2}$$

The pieces are $\frac{13}{2}$ ft and $\frac{27}{2}$ ft. Check: $\frac{27}{2}$ is indeed 7 more than $\frac{13}{2}$, and $\frac{13}{2} + \frac{27}{2} = 20$.

Answer: One is 18 in and the other is 30 in.

4 GEOMETRY PROBLEMS

Many applied problems require certain basic formulas. In particular, problems involving geometric figures might need one of the formulas given inside the front cover. As you read through each problem, decide which formula is needed. Remember to make a sketch.

EXAMPLE 6 Geometry Problem

The second angle of a triangle is six times as large as the first angle. If the third angle is 45° more than twice the first, find the measure of each angle.

Remember that the sum of the measures of the angles of any triangle is 180°. Make a sketch as in Figure 2.2.

Let x = the measure of the first angle,
$6x$ = the measure of the second angle,
$2x + 45$ = the measure of the third angle.

$$\begin{aligned} x + 6x + (2x + 45) &= 180 \\ 9x + 45 &= 180 \\ 9x &= 135 \\ x &= 15 \\ \text{so} \quad 6x &= 90 \\ \text{and} \quad 2x + 45 &= 75 \end{aligned}$$

Figure 2.2

The measures are 15°, 90°, and 75°. Check: Since 15° + 90° + 75° = 180°, 90° = 6(15°), and 75° = 45° + 2(15°), the measures do check.

PRACTICE EXERCISE 6

The second angle of a triangle is four times the first, and the third is 20° less than three times the first. Find the measures of the angles.

Answer: 25°, 100°, and 55°

EXAMPLE 7 Construction Problem

A farmer plans to build a cylindrical storage silo with radius 5.2 m. How high will it need to be to store 724 m^3 of grain? [Use 3.14 for π.]

The volume of a cylinder is given by $V = \pi r^2 h$. See Figure 2.3.

$$V = \pi r^2 h$$

$$724 = (3.14)(5.2)^2 h \qquad \text{Substitution}$$

PRACTICE EXERCISE 7

Find the height (to the nearest tenth) of a cylindrical tank with surface area 500 ft^2 and radius 3.5 ft. The surface area of a cylinder is $A = 2\pi rh + 2\pi r^2$.

$$\frac{724}{(3.14)(5.2)^2} = h$$

$$8.53 \approx h \quad \text{(to the nearest hundredth)}$$

Thus, the silo must be about 8.5 m high.

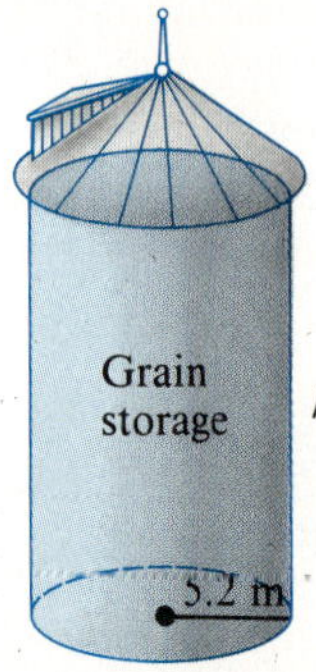

Figure 2.3

Answer: 19.2 ft

In Example 7 we used the symbol ≈ to represent "*is approximately equal to*." Use ≈ for approximate or rounded numbers.

2.4 EXERCISES A

Select a variable to represent each quantity and translate the phrase into symbols.

1. The sum of a number and 4

2. The time in 2 hours

3. Her salary, increased by \$400

4. Twice the volume

5. \$4000 less the cost

6. The reciprocal of a number

7. Twice the number of votes, increased by 500

8. 5% of the price

Letting x represent the unknown number, translate each statement into symbols.

9. Eight times a number is 72.

10. A number subtracted from 3 is 10.

11. A number increased by 7 is 21.

12. A number divided by 3 is equal to 4.

13. My age in 8 years will be 26.

14. Twice a number, increased by 3, is 14.

15. Twice the sum of a number and 5 is the same as the number.

16. Seven times a number, less 3, is as much as twice the number, plus 5.

17. A number less 8 is as much as the number squared.

18. Eleven is 5 less than eight times a number.

19. Two-thirds of a number is 36.

20. A number is 12% of 25.

21. When 9 is added to three times a number and the result tripled, the resulting quantity is the same as 13 more than 6 times the number.

22. A number increased by its reciprocal is the same as five times the number, decreased by 7.

Solve each of the following applied problems. (Some problems have been started.)

23. If 3 is added to five times a number, the result is 38. Find the number.

Let x = the number,
$5x$ = five times the number.

24. The sum of two numbers is 24. The larger number is five times the smaller number. Find the two numbers.

Let x = one number,
$5x$ = the other number.

25. Sam made \$10 more than twice what Pete earned in one month. If together they earned \$760, how much did each earn that month?

Let x = amount Pete earned in one month,
$2x + 10$ = amount Sam earned in one month.

26. A woman burns up three times as many calories running as she does when walking the same distance. If she runs 2 mi and walks 5 mi and burns up a total of 770 cal, how many calories does she burn up while running 1 mi?

Let x = calories burned walking 1 mi,
$3x$ = calories burned running 1 mi.

27. The sum of two consecutive odd integers is 76. Find the two integers. [*Hint:* If one integer is x, the other is $x + 2$. Why?]

28. Two-fifths of a man's income each month goes to taxes. If he pays \$848 in taxes each month, what is his monthly income?

29. The sum of three consecutive integers is 126. Find the three integers.

30. The sum of Jan's age and Juan's age is 78 years. If Jan is 6 years younger than Juan, how old is each?

31. A board is 11 ft long. It is to be cut into three pieces in such a way that the second piece is twice as long as the first piece and the third piece is 3 ft longer than the first piece. Find the length of each piece.

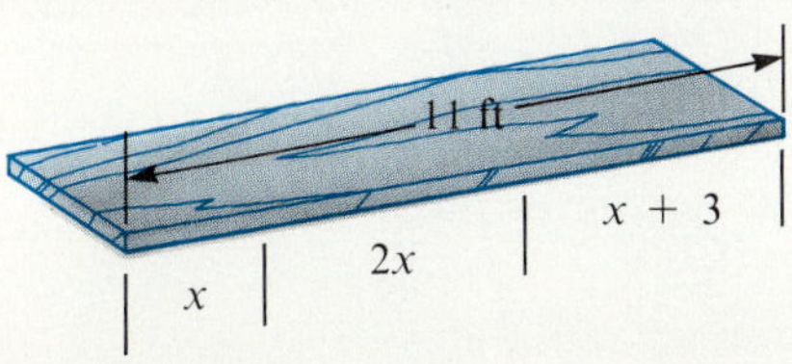

32. A pole is standing in a small lake. If one-sixth of the length of the pole is in the sand at the bottom of the lake, 25 ft are in the water, and two-thirds of the total length is in the air above the water, what is the length of the pole?

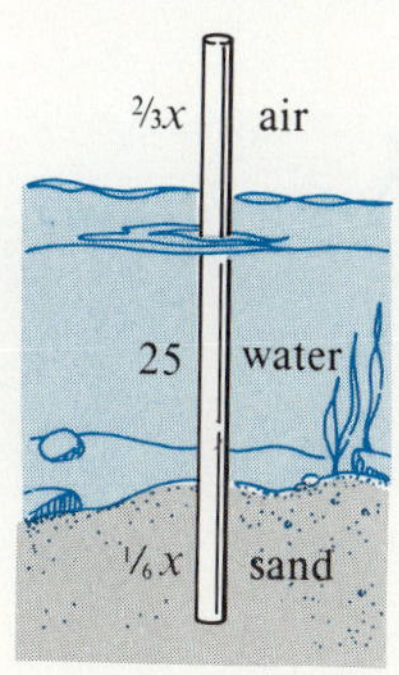

Let x = length of pole.

33 Becky must have an average of 90 on four tests in geology to get an A in the course. What is the lowest score she can get on the fourth test if her first three scores are 96, 78, and 91?

34 International Car Rental will rent a compact car for \$12.50 per day plus 10¢ per mile driven. If Charlene rented a compact car for two days and was billed a total of \$73.00, how many miles did she drive?

35. The perimeter of a rectangle is 86 in. If the length is 19 in more than the width, find the dimensions.

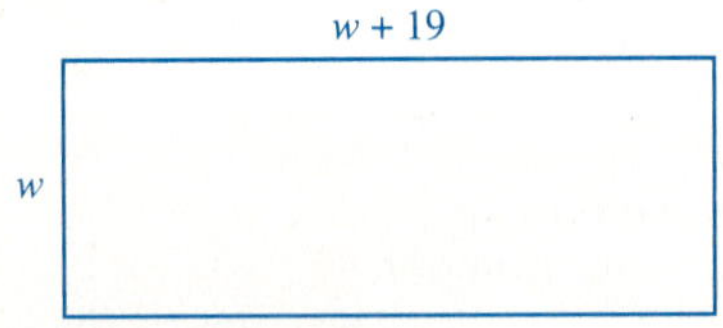

Let w = width,
$w + 19$ = length.
$P = 2l + 2w$
$86 = 2(w + 19) + 2w$

36. If the second angle of a triangle is 50° more than the first angle, and the third angle is eleven times the first, find the measure of each angle.

Let x = the measure of the first angle.
$x + 50 =$
$11x =$

37. If two angles are supplementary (they have measures totaling 180°) and the second is 15° less than twice the first, find the measure of each angle.

38. An isosceles triangle has two equal sides called legs and a third side called the base. If each leg of an isosceles triangle is five times the base and the perimeter is 77 m, find the lengths of the legs and base.

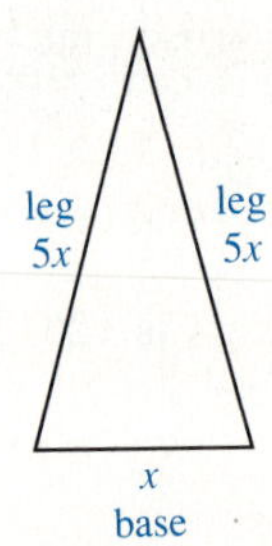

Let x = base.
$5x =$

39. The circumference of a circular garden is 182 ft. How many feet of pipe are required to reach from the edge of the garden to a fountain in the center of the garden? [Use 3.14 for π.]

Fountain

40. Find the height of a parallelogram whose base is 19 cm and whose area is 456 cm^2.

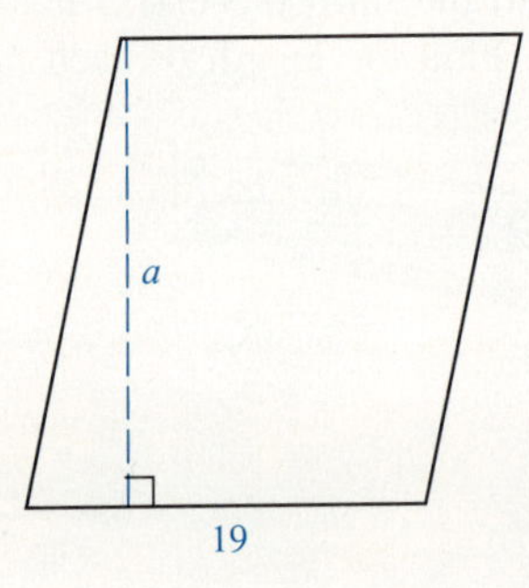

Let a = height

41 A sphere of radius 2 ft is dropped into a rectangular tank that is 10 ft long and 8 ft wide. To the nearest tenth, how much does this raise the level of water in the tank?

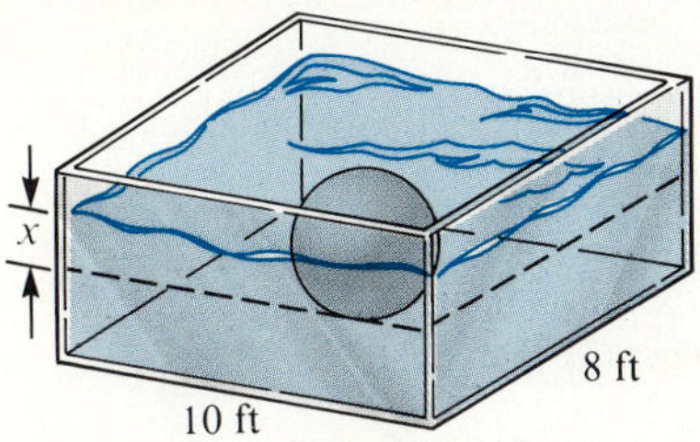

Let x = rise in level.

42. Find the height of a cylindrical tank with volume 3200 in^3 and radius 8 in. [Use 3.14 for π.]

43. A cube with edge 10 in is submerged in a rectangular tank of water. If the tank is 40 in by 50 in, how much does the level of water in the tank rise?

44. A rancher wishes to enclose his north pasture with fence selling for 28¢ per linear foot. If the pasture is rectangular in shape, 3 mi long and 2 mi wide, how much will the project cost?

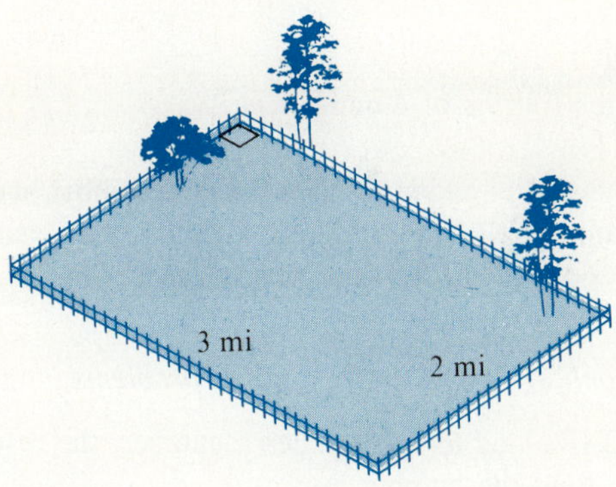

FOR REVIEW

Solve.

45. $\frac{1}{2}y + \frac{3}{4}y = -40$

46. $0.2(y + 1) - 0.3(y - 1) = 0.9$

47. $7(2y - 1) = 3 + 14y$

48. $-3[z - (4z + 1)] = 5z - 1$

49. $z^2 - 3 = (3 + z)z$

50. $\frac{z + 2}{5} - \frac{2z - 1}{2} = 1$

ANSWERS: Some answers are rounded. **1.** $n + 4$ **2.** $t + 2$ **3.** $s + 400$ **4.** $2V$ **5.** $4000 - c$ **6.** $\frac{1}{n}$ **7.** $2v + 500$ **8.** $0.05p$ **9.** $8x = 72$ **10.** $3 - x = 10$ **11.** $x + 7 = 21$ **12.** $\frac{x}{3} = 4$ **13.** $x + 8 = 26$ **14.** $2x + 3 = 14$ **15.** $2(x + 5) = x$ **16.** $7x - 3 = 2x + 5$ **17.** $x - 8 = x^2$ **18.** $11 = 8x - 5$ **19.** $\frac{2}{3}x = 36$ **20.** $x = (0.12)(25)$ **21.** $3(3x + 9) = 6x + 13$ **22.** $x + \frac{1}{x} = 5x - 7$ **23.** 7 **24.** 4, 20 **25.** Pete: \$250; Sam: \$510 **26.** 210 calories **27.** 37, 39 **28.** \$2120 **29.** 41, 42, 43 **30.** Jan is 36 and Juan is 42. **31.** 2 ft, 4 ft, 5 ft **32.** 150 ft **33.** 95 **34.** 480 mi **35.** 12 in, 31 in **36.** 10°, 60°, 110° **37.** 65°, 115° **38.** base: 7 m; legs: 35 m **39.** about 29 ft **40.** 24 cm **41.** 0.4 ft **42.** 15.9 in **43.** 0.5 in **44.** \$14,784 **45.** -32 **46.** -4 **47.** no solution **48.** -1 **49.** -1 **50.** $-\frac{1}{8}$

2.4 EXERCISES B

Select a variable to represent each quantity and translate the phrase into symbols.

1. A number subtracted from 20
2. The time 6 hours ago
3. My wages less \$60
4. Five times the area
5. Triple the cost
6. A number added to its reciprocal
7. 25% of a number
8. Cost plus 20% of the cost

Letting x represent the unknown number, translate the following into symbols.

9. Seven times a number is 42.
10. A number subtracted from 8 is 1.
11. A number increased by 5 is 37.
12. A number divided by 7 is equal to 3.
13. Jeff's age in 3 years will be 39.
14. Twice a number, increased by 4, is 20.
15. Three times the sum of a number and 2 is the same as the number.
16. Six times a number, less 9, is as much as twice the number, plus 7.
17. A number less 15 is as much as the number squared.
18. The quotient of 4 more than a number and 2 less the number is equal to -2.
19. Three-fourths of a number is 28.
20. A number is 8% of 300.
21. When 11 is added to three times a number and the result doubled, the resulting quantity is the same as 15 more than four times the number.
22. A number decreased by its reciprocal is the same as ten times the number, increased by 5.

Solve each of the following applied problems.

23. If 2 is added to seven times a number, the result is 79. Find the number.
24. The sum of two numbers is 52. The larger is three times the smaller number. Find the two numbers.
25. Carlos made \$3000 more than twice what Ernie earned one year. If together they earned \$31,500, how much did each earn that year?
26. A man burns up one-third as many calories walking as he does when running the same distance. If he walks 3 km and runs 4 km and burns up a total of 1125 cal, how many calories does he burn up while walking 1 km?
27. The sum of two consecutive even integers is 170. Find the two integers.
28. Three-fourths of the operating budget of a university goes for faculty salaries. If faculty salaries total \$6,000,000, what is the operating budget?
29. The sum of three consecutive odd integers is 87. Find the integers.
30. In an election with only two candidates, the winner received 40 votes more than twice the total for the loser. If a total of 1480 votes were cast, how many did each candidate receive?
31. A wire is 40 in long. It is to be cut into three pieces so that the second piece is three times as long as the first piece and the third piece is 5 in longer than the first piece. Find the length of each piece.
32. On a recent trip, Bob Packard traveled one-fifth the total distance by car, 300 miles by boat, and one-half the total distance by air. What was the total distance traveled on the trip?

33. To get a B in history, Bernie must have an average of 80 on four tests. What is the lowest score he can get on the fourth test if his first three scores are 87, 81, and 68?

34. The cost of renting a carpet shampooer is \$4.50 per hour plus \$3.15 per quart of shampoo. If the Pearsons used 3 quarts of shampoo and were billed \$49.95, how many hours did they use the machine?

35. The perimeter of a rectangle is 120 ft. If the length is 22 ft more than the width, find the dimensions.

36. If the second angle of a triangle is 36° more than the first angle, and the third is six times the first, find the measure of each angle.

37. Two angles are supplementary and the second is 30° less than twice the first. Find the measure of each angle.

38. Each leg of an isosceles triangle is four times the base, and the perimeter of the triangle is 99 in. Find the lengths of the legs and base.

39. The circumference of a circular patio is 125.6 yd. How far is it from the edge of the patio to a flagpole in its center? [Use 3.14 for π.]

40. Find the height of a parallelogram whose base is 12.5 cm and whose area is 255 cm^2.

41. A sphere of radius 9 inches is dropped into a rectangular tank 20 in long and 18 in wide. To the nearest tenth, how much does this raise the level of the water in the tank?

42. Find the height of a cylindrical tank with volume 301.44 ft^3 and radius 4 ft. [Use 3.14 for π.]

43. A cube with edge 15 cm is submerged in a rectangular tank of water. If the tank is 30 cm by 40 cm, how much does the level of the water in the tank rise?

44. Mr. Williams wishes to put a fence around his property. If his land is rectangular in shape, 1 mi wide and 2 mi long, and fencing costs 32¢ per linear foot, how much will the project cost?

FOR REVIEW

Solve.

45. $\frac{1}{3}y + \frac{5}{3}y = -9$

46. $0.2(x + 1) - 0.3(x - 2) = 0.5$

47. $6(2z - 1) = 4 + 12z$

48. $-2[x - (2x + 1)] = 5 - x$

49. $y^2 + 2 = y(y - 1)$

50. $\frac{z + 1}{3} - \frac{z + 2}{2} = 1$

2.4 EXERCISES C

Solve.

1. Three integers are consecutive multiples of 5. If their sum is 90, find the integers.

2. One side of a triangle is 14 cm. Another side is one-fourth the perimeter, and the third side is 4 less than one-third the perimeter. Find the perimeter of the triangle. [Answer: 24 cm]

2.5 APPLICATIONS USING LINEAR EQUATIONS

STUDENT GUIDEPOSTS

1. Tax Problems
2. Commission Problems
3. Discount Problems
4. Simple Interest Problems
5. Rate Problems

Applications of percent are common. We may pay a 5% sales tax, read that Mr. Levy won an election with 58% of the vote, or hear that a basketball player shot 82% from the free throw line.

When working with percent, the word "of" translates to *multiplied by* or *times*. There are three basic questions to answer with percent problems.

(1) What is 30% of 40?
$$x = (0.30) \cdot 40 \qquad \text{Thus, } x = 12$$

(2) Twelve is 30% of what number?
$$12 = (0.30) \cdot x \qquad \text{Thus, } x = 40$$

(3) What percent of 40 is 12?
$$x \cdot 40 = 12 \qquad \text{Thus, } x = 0.3 = 30\%$$

In each case percent is expressed as a decimal (see Section 0.4), and the resulting translation is a linear equation.

1 TAX PROBLEMS

Many types of tax problems involve percent. For example, in most states a tax is charged on purchases made in retail stores. The rate of this tax, called a **sales tax,** varies from location to location. In general, however, we have

sales tax = (tax rate) · (selling price)

and **total cost = (selling price) + (sales tax).**

EXAMPLE 1 Sales Tax Problem

The sales tax rate in Maryville, Texas is 4%. How much tax would be charged on a purchase of $42? What is the total cost?

In effect, we are asked: What is 4% of 42?

Let x = the tax. We obtain the following equation.

$$x = (0.04)(42) = 1.68$$

The tax is $1.68, and the total cost is $43.68.

Practice Exercise 1

How much tax would be charged on a purchase totaling $145 if the tax rate is 6%? What is the total cost including tax?

Answer: $8.70; $153.70

2 COMMISSION PROBLEMS

Some salespersons receive all or part of their income as a percent of their sales. This is the **commission,** and the **commission rate** is the percent involved. In general,

commission = (commission rate) · (total sales).

For example, if a person works on a 20% commission and has total sales of \$8360, her commission is 20% of \$8360, which amounts to

$$(0.20)(\$8360) = \$1672.$$

EXAMPLE 2 Commission Problem

What are the total sales on which Gino received a commission of \$1140 if his commission rate is 12%?

We are asked: 12% of what is \$1140?

Let x = Gino's total sales. The question then translates to this equation.

$$(0.12) \cdot x = 1140$$

$$x = \frac{1140}{0.12} = 9500$$

Gino's total sales amounted to \$9500.

PRACTICE EXERCISE 2

If Terri earned a commission of \$1455 during July and her commission rate is 15%, what were her total sales that month?

Answer: \$9700

3 DISCOUNT PROBLEMS

A **discount** is a reduction in the regular price of an item put on sale. The **discount rate** is the percent of the regular price that the item is reduced, and the **sale price** is the price after the discount is taken. In general,

discount = (discount rate) · (regular price)

and **sale price = (regular price) − (discount).**

EXAMPLE 3 Discount Problem

A sport coat is discounted 30% and the sale price is \$87.50. What was the original price?

Let x = the original price of the sport coat,

$(0.30)x$ = the amount of the discount.

Then we need to solve the following equation.

$$x - (0.30)x = 87.50 \quad \text{Regular price} - \text{discount} = \text{sale price}$$

$$x(1 - 0.30) = 87.50 \quad \text{Distributive law}$$

$$x(0.70) = 87.50$$

$$x = 125$$

Thus, the sport coat originally sold for \$125.

PRACTICE EXERCISE 3

The sale price on a new car is \$13,416. If the car has been discounted 14%, what was the original price?

Answer: \$15,600

4 SIMPLE INTEREST PROBLEMS

When money is invested or deposited in a bank and earns **simple interest,** the formula for the interest earned, I, is

$$I = Prt,$$

where P is the **principal** or amount invested, r is the **yearly interest rate,** and t is the **time** in years. The amount in the account A at the end of t years is

$$A = P + I = P + Prt.$$

EXAMPLE 4 Simple Interest Problem

In three years, Pat Polinski wishes to have \$10,850 available to purchase a car. How much should he invest today at 12% simple interest to have this amount available when he needs it?

Let x = the amount of money Pat should invest today. In three years the interest earned will be

$$I = Prt$$
$$I = x \cdot (0.12)(3) = (0.36)x.$$

We need to solve the following equation.

$$x + (0.36)x = 10{,}850 \qquad A = P + I$$
$$(1 + 0.36)x = 10{,}850 \qquad \text{Distributive law}$$
$$x = \frac{10{,}850}{1.36}$$
$$x \approx 7977.94$$

Pat must invest approximately \$7977.94.

Practice Exercise 4

In five years the Raffertys wish to have \$15,000 available for their daughter's education. How much should they invest today at 11% simple interest to have this amount when needed?

Answer: approximately \$9677.42

5 RATE PROBLEMS

Another important formula relates the distance (d) that an object travels in a given time (t) at a constant or uniform rate (r).

$$d = rt \quad \text{distance} = \text{(rate)(time)}$$

An automobile traveling at a rate of 55 mph for 3 hr will travel a distance of

$$d = rt = 55 \cdot 3 = 165 \text{ miles.}$$

Many rate or distance problems depend in some way on this basic formula.

EXAMPLE 5 Rate-Distance Problem

A hiker crossed a canyon by walking 4 mph the first 2 hr and 3 mph the next 3 hr. What was the total distance that she hiked?

A sketch like Figure 2.4 is helpful.

Practice Exercise 5

On a trip a woman drove her car for 6 hr at 55 mph and then rode a boat for 3 hr at 30 mph. What was the total distance traveled?

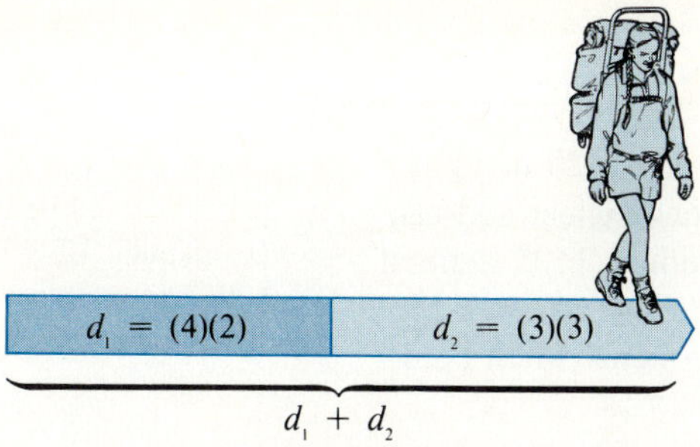

Figure 2.4

Let d_1 = distance hiked the first 2 hr. Then $d_1 = (4)(2) = 8$ mi.
Let d_2 = distance hiked the second 3 hr. Then $d_2 = (3)(3) = 9$ mi.
The total distance hiked is $d_1 + d_2 = 8 + 9 = 17$ mi.

Answer: 420 mi

EXAMPLE 6 Rate-Distance Problem

Two cars leave the same point traveling in opposite directions. The second car travels 10 mph faster than the first and after 3 hr they are 300 mi apart. How fast is each car traveling?

Make a sketch like Figure 2.5.

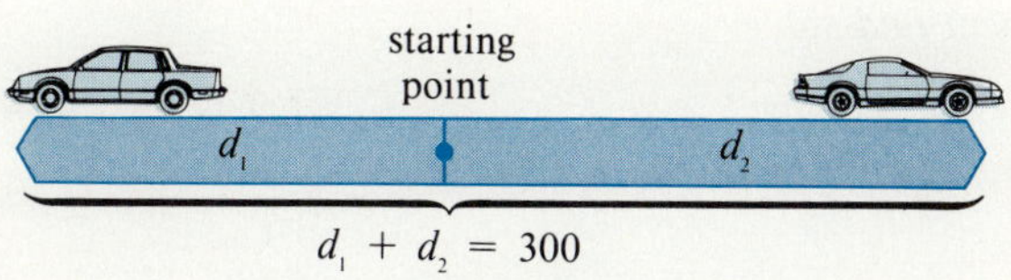

Figure 2.5

Let d_1 = the distance the first car travels,
r = rate of the first car.
Then $d_1 = 3r$.
Let d_2 = the distance the second car travels,
$r + 10$ = rate of the second car. (Why?)
Then $d_2 = 3(r + 10)$.
We need to solve the following equation.

$$d_1 + d_2 = 300$$
$$3r + 3(r + 10) = 300 \quad \text{Substitution}$$
$$3r + 3r + 30 = 300$$
$$6r = 270$$
$$r = 45$$
$$\text{and} \quad r + 10 = 55$$

The first car is traveling 45 mph and the second, 55 mph.

Practice Exercise 6

A jet plane leaves Miami and travels toward Chicago at 425 mph. At the same time a small plane leaves Chicago and travels 150 mph toward Miami. If the cities are 1150 mi apart, how long will it take the planes to meet and how far will they be from Miami at that time?

Answer: 2 hr and 850 mi

Other rate-time problems are solved in a similar manner.

EXAMPLE 7 Rate-Time Production Problem

A machine that was manufactured 15 years ago can produce 20 items per day. A newer machine, manufactured last year, is more efficient and can produce 45 items per day. How long will it take to produce 1105 items if both machines are turned on together?

Let x = number of days it takes to produce 1105 items when both machines work together,

$n_1 = 20x$ = number of items produced by the older machine in x days,

$n_2 = 45x$ = number of items produced by the newer machine in x days.

We need to solve the following equation.

$$n_1 + n_2 = 1105$$

$$20x + 45x = 1105 \quad \text{Substitution}$$

$$65x = 1105$$

$$x = 17$$

It will take 17 days for the two machines to produce 1105 items.

Practice Exercise 7

One printer can print 5 form letters per minute while a second can print 8 per minute. How long will it take both printers together to print 468 form letters?

Answer: 36 min

2.5 EXERCISES A

Translate each question into a mathematical statement and solve.

1. 36 is 15% of what?

2. 12% of 240 is what?

3. What is $\frac{1}{2}$% of 400?

Solve.

4. A family spent \$180 a month for food. This was 15% of their monthly income. What was their monthly income?

   Let x = monthly income,
   $0.15x$ = amount spent for food.
   $0.15x = 180$

5. If \$0.43 sales tax is charged on a coffeepot that sells for \$21.50, what is the tax rate?

   Let x = tax rate.
   $21.50x =$

6. An automobile salesperson received a 2% commission on the price of each vehicle sold. What total commission will be earned on the sale of a car priced at \$9850 and a truck priced at \$13,500?

7. A retailer discounted all merchandise in her shop by 40%. What are the discount and sale price on a suit regularly priced at \$135.50?

8 A tank contained 985 gal of fuel. An additional quantity of fuel was added making a total of 1182 gal. What was the percent increase?

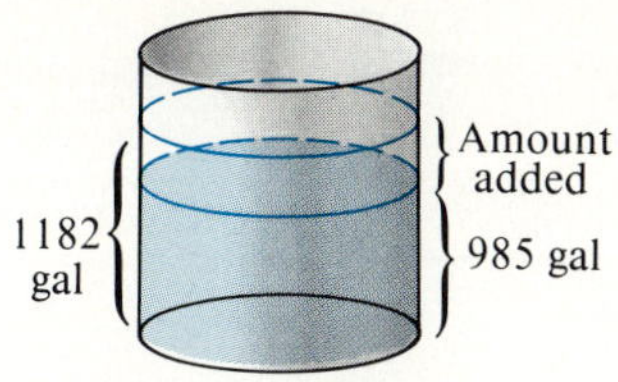

9 The population in Deserted, Arizona, was 567 in 1990. This was a 35% increase over the population in 1980. What was the population in 1980?

10. The retail selling price of a sofa is $806.25. What is the cost to the dealer if she sells at a 25% markup on the cost?

11 Paul Pendergast must have $11,470 available to pay off a loan in full five years from today. How much should he invest at 11% simple interest in order to obtain this amount?

12 Lucky Lucia won $10,000 in the New York Lottery. She invested part in a savings account that earned 12% simple interest and the rest in a mutual fund that paid 15% annually. If at the end of one year she earned an income of $1275 from the two, how much was invested at each rate?

13. Troy Hudson has $17,000 invested in bonds that pay 13% annually. How much more money must he invest in a savings account paying 17% so that the average return on the two investments amounts to 14%?

14. A floor diagram of the Perkos' family room is given below. Suppose they purchased enough carpeting to cover the floor wall-to-wall, and the total bill for the job came to $1719.90, including a 5% sales tax. Assuming there was no waste, how much did the carpet cost per square yard?

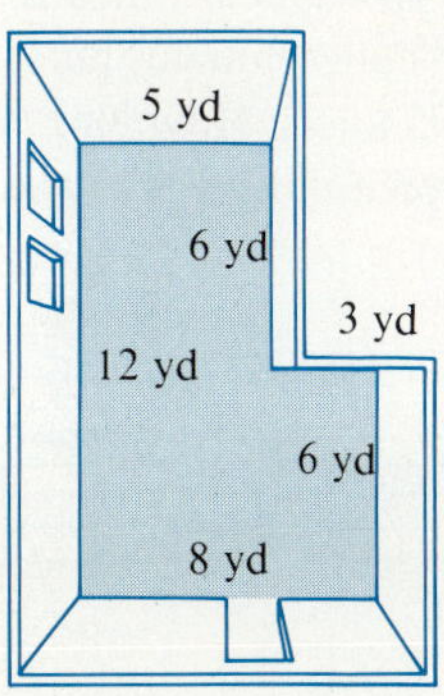

15 The We-Try-Harder Rental Company charges $13 per day and 8¢ per mile to rent an Oldsmobile Ciera. If Jeff rented a Ciera for 4 days and the total bill was $199.68, including a 4% sales tax, how many miles did he drive?

16. Two trains leave the same city, one traveling north and the other traveling south. If the first train is moving 15 mph faster than the second, and if after 3 hr they are 405 mi apart, how fast is each train traveling?

17. Two hikers head towards a mountain 25 mi away. The first travels 4 mph and the second travels 2.5 mph. How far apart will they be after 6 hr?

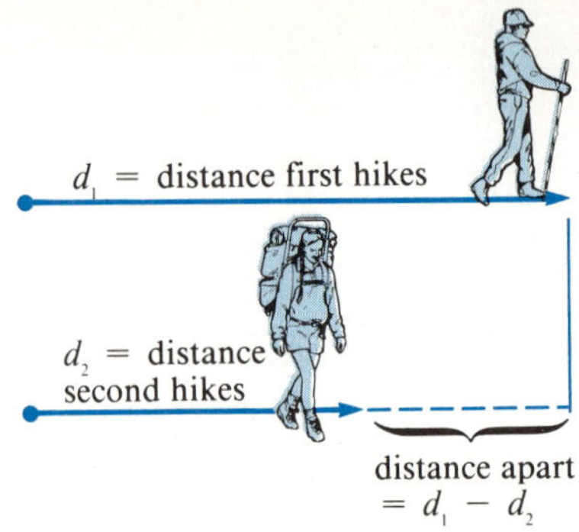

18 A woman hikes to the bottom of the Grand Canyon at a rate of 4.5 mph and returns to the top by mule traveling at a rate of 2.5 mph. If the total time of the trip is 7 hr, how long did she walk? What was the total length of the trip?

19 Two families leave their homes at 9:30 A.M., planning to meet for a picnic at a point between them. If one travels at a rate of 55 mph, the other travels at 50 mph, and they live 315 mi apart, at what time do they meet?

20. Two machines are turned on at 8:00 A.M. If one can produce 36 items each hour and the other can produce 50 items each hour, at what time will they have produced a total of 645 items?

21. Two distance runners, Bob and Murph, run the same course with Bob running 1.5 mph slower than Murph. If Murph finishes in 4 hr and Bob takes 5 hr to finish, what is the rate of each? How far do they run?

22 Two printing presses print 14,850 fliers. Their rates of production differ by 15 fliers per minute. What is the rate of each if together they complete the job in 150 min?

23 A private boat traveling at 20 knots (a knot is 1 nautical mile per hour) is 10 nautical miles from a harbor when a Coast Guard cutter initiates pursuit on the same course traveling 35 knots. How long will it take for the cutter to overtake the private boat?

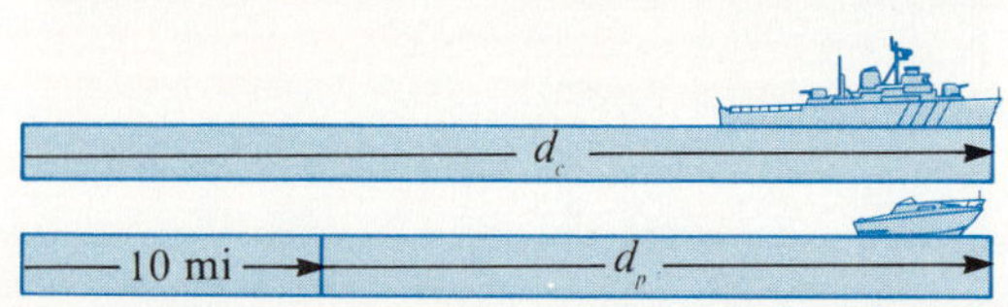

FOR REVIEW

Select a variable to represent each quantity and translate the phrase into symbols.

24. Price plus 6% of the price

25. Twice a number, less its reciprocal

Solve.

26. On a recent football booster trip, it was noted that the bus was completely full and that there were 12 more men than women. If the bus has a capacity of 56 passengers, how many men were on the trip?

27. The sum of two consecutive even integers is 114. Find the integers.

28. Two angles are complementary if their sum is 90°. One of two complementary angles is 30° more than twice the other. Find the measure of each angle.

29. A farmer wishes to enclose a rectangular pen with 168 yd of fencing in such a way that the length is three times the width. What will be the dimensions of the pen?

ANSWERS: 1. 240 2. 28.8 3. 2 4. $1200 5. 2% 6. $467 7. $54.20, $81.30 8. 20% 9. 420 10. $645 11. $7400 12. $7500 at 12%, $2500 at 15% 13. $5666.67 14. $21.00 15. 1750 miles 16. 60 mph, 75 mph 17. 9 mi 18. 2.5 hr, 22.5 mi 19. 12:30 P.M. 20. 3:30 P.M. 21. Murph: 7.5 mph; Bob: 6 mph; 30 mi 22. 42 fliers/min, 57 fliers/min 23. 40 min 24. $p + 0.06p$ 25. $2n - \frac{1}{n}$ 26. 34 men 27. 56, 58 28. 20°, 70° 29. 21 yd by 63 yd

2.5 EXERCISES B

Translate each question into a mathematical statement and solve.

1. 9 is 2% of what?

2. 3% of 180 is what?

3. 8% of what is 68?

Solve.

4. Larry spent $3600 for a used car. This was 30% of his annual income. What was his annual income?

5. What is the tax rate in Southern, Georgia, if a tax of $6.25 is charged on an item selling for $250?

6. The Blue Carpet Realty Co. receives a 6% commission on all real estate sales. During August the company sold a house for $89,500 and a lot for $22,300. What was the total commission on these sales?

7. Bjorn was making $24,500 per year. The next year he earned $27,440. What was the percent increase from one year to the next?

8. Marvin received a 14% raise and now makes $14,592. What was his former salary?

9. The retail selling price of a video recorder is $1022. What is the cost to the dealer if he sells at a 40% markup on the cost?

10. What amount of money must be invested today at 14% simple interest to reach $9672 in four years?

11. A fuel tank contained 125 gal of heating oil. An additional quantity of oil was added, making a total of 165 gal. What was the percent increase?

12. If \$4200 is invested in stocks at 10%, how much additional money must be invested in a savings account that earns 14% to make the total earnings equivalent to an investment of 13%?

13. When his rich aunt passed away, Martin received an inheritance of \$25,000. He put part of the money in an account that paid 14% simple interest, and the rest into a stock that paid a 17% annual dividend. If at the end of the year he had a total income of \$3920 from the two, how much was invested in each?

14. A floor diagram of the Fancher's recreation room is given below. Suppose they purchased enough carpeting to cover the floor wall-to-wall, and the total bill for the job came to \$962.00, including a 4% sales tax. Assuming there is no waste, what was the price per square yard of the carpet?

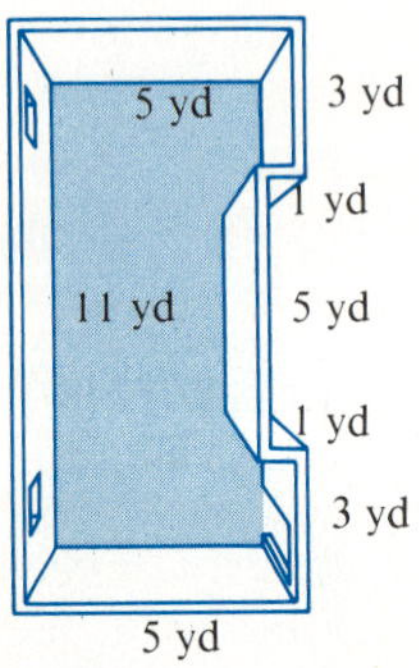

15. Luxury Limo Rentals charges \$50 per day and 21¢ per mile to rent a Continental. If Beth rented a Continental for 3 days and the total bill was \$307.44, including a 5% sales tax, how many miles did she drive?

16. Two cars leave the same park, one traveling east and the other traveling west. If the first is moving 10 mph faster than the second, and if after 4 hr they are 424 mi apart, how fast is each traveling?

17. Two truckers leave San Francisco at the same time heading east. The first is traveling at a rate of 60 mph and the second at 54 mph. How far apart will they be in 7 hr?

18. Jerry hikes to a waterfall at a rate of 3.5 mph and returns at a rate of 3 mph. If the total time of the trip is 6.5 hr, how long did it take to reach the waterfall? What was the total distance traveled?

19. Two families leave their homes at 10:00 A.M., planning to meet for a reunion at a point between them. If one travels at a rate of 60 mph, the other travels at a rate of 54 mph, and they live 285 mi apart, at what time do they meet?

20. Two machines are turned on at 9:00 A.M. If one can produce 40 items each hour and the other can produce 32 items each hour, at what time will they have produced a total of 360 items?

21. Two hikers, Jenny and Christine, hike the same trail with Jenny hiking 1 mph slower than Christine. If Jenny finishes in 4 hr and Christine takes 3 hr to finish, what is the rate of each? How long is the trail?

22. A 1977-model machine, which can produce 30 units an hour, has made 15 units when a 1987-model machine is turned on. If the 1987-model can produce 45 units an hour, how long will it be before it has produced the same number of units as the older model?

23. A fighter plane flying at a speed of 750 mph is to overtake a transport plane flying at a speed of 500 mph. If the transport took off at 8:30 A.M. and the fighter left 1.5 hr later, at what time will it overtake the transport?

FOR REVIEW

Select a variable to represent each quantity and translate the phrase into symbols.

24. My weight before gaining 10 lb

25. Twice the reciprocal of a number

Solve.

26. A support beam is standing in a river. If one-third of the height of the beam is in the sand at the bottom of the river, two-ninths of the total height is in the air above the water, and 16 ft are in the water, what is the length of the beam?

27. Sean, Gary, and Troy were born in consecutive years. If the sum of their ages is 54, how old is each?

28. Marv and Burford leave their campsite and walk in opposite directions around a lake. If the shoreline is 15 miles long, Marv walks 0.5 mph faster than Burford, and they meet in 2 hr, how fast does each walk?

29. A rancher has 700 ft of fencing for enclosing a small rectangular pasture with length four times the width. What will be the dimensions of the pasture?

2.5 EXERCISES C

Solve. A calculator may be helpful in some of these exercises.

1. On Tuesday Galen Benning bought ten shares of stock. On Wednesday the value of the shares went up 5%, and on Thursday the value fell 5%. How much did Galen pay for the ten shares on Tuesday if he sold them on Thursday for $99.75? [Answer: $100]

2. The Brinegars plan to sell their home. They want to receive $65,000 after deducting the sales commission of 6% on the selling price. Rounded to the nearest dollar, at what selling price should the house be listed?

3. During a consecutive three-year period, Scott Williams earned a 10%, 20%, and 30% raise in salary, respectively. What was the total percent change from the beginning of the first year to the end of the third year?

4. A container holds 100 qt of a solution that is 12% alcohol and 88% water. How much water must be added to get a solution that is 10% alcohol? [Answer: 20 quarts]

5. Aprile Kyle rents a boat at 9:00 A.M. She is told it will travel 20 mph upstream and 25 mph downstream. If she decides to travel as far downstream as possible and be back by 6:00 P.M., at what time should she turn back, and how far downriver will she be at that time?

6. Assume that sound travels at 0.34 km/sec and that a bullet travels 0.68 km/sec. If the gun is fired at a target and 3 sec later the sound of impact is heard, how far away is the target? [Answer: 0.68 km]

7. Six seconds after Kristen fires a gun at a rock cliff she hears the sound of impact. If sound travels at 1095 ft/sec and the bullet at 2190 ft/sec, how far is the cliff from Kristen?

8. A ship located 1 mi from an island fires one of its guns. The sound reaches the island through the water 3.73 sec before it arrives through the air. If sound travels 4.5 times faster through water than through air, approximately how fast does sound travel in water? [Answer: 4954 ft/sec]

9. Distance between ships can be determined when one ship sends signals that travel through the water at different rates of speed and the second ship measures the time between the signals. Two signals traveling at 1.2 mi/sec and 2.5 mi/sec are received 39 seconds apart. How far apart are the ships? [Answer: 90 mi]

10. Repeat Exercise 9 if the signals are received 15 seconds apart. Give answer correct to the nearest tenth of a mile.

11. An object that is thrown downward with an initial velocity of 120 ft/sec satisfies the equation $v = 32t + 120$ where v is the velocity in ft/sec and t is in seconds. What is the velocity of the object after 3 seconds? After how many seconds will the velocity be 280 ft/sec? [Answer: 216 ft/sec; 5 sec]

12. An object that is thrown upward with an initial velocity of 48 ft/sec satisfies the equation $v = -32t + 48$ where v is velocity in ft/sec and t is in seconds. What is the velocity of the object after 2.5 seconds? After how many seconds will the velocity be 0?

A type of problem with many applications in physics is a lever problem. *A* ***lever*** *is a rigid rod or beam rotating about a fixed point, called the* ***fulcrum*** *of the lever. A lever is in* ***equilibrium*** *if the forces on the lever are balanced according to a fundamental law of physics, the* **law of the lever.** *Suppose three forces, F_1, F_2, and F_3, are applied to a lever at distances d_1, d_2, and d_3 as shown in the figure. For equilibrium, the law of the lever requires that*

$$F_1d_1 + F_2d_2 = F_3d_3.$$

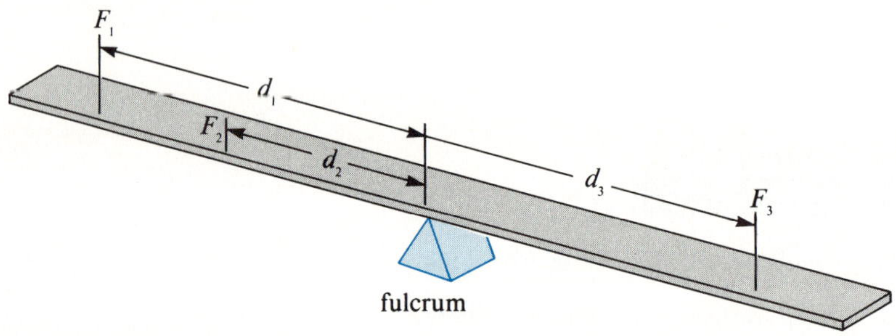

Use this information in the following problems.

13. Jeff, who weighs 160 pounds, and Annie, who weighs 110 pounds, are sitting at opposite ends of a seesaw. Each is 8 feet from the fulcrum. Where should their 80-lb daughter Cindy sit, to balance the system? [Answer: 5 ft from the fulcrum on Annie's side]

14. Repeat Exercise 13 but this time use their son Steve, who weighs 64 pounds.

15. Repeat Exercise 13 but this time use their dog, Chuck, who weighs 32 pounds.

SUPPLEMENTARY APPLIED PROBLEMS

The following exercises use the techniques presented in the previous three sections.

1. Seven more than twice Herman's age is equal to 73. How old is Herman?

2. The perimeter of Ivan Danielson's yard is 528 yd. If the yard is rectangular in shape with length 24 yd more than the width, what are the dimensions?

3. On a recent field trip, Cindy noticed that the bus was completely full and that there were 14 more girls than boys on the bus. If the bus has a capacity of 54 children, how many boys were on the bus?

4. Bob, Mary, and Sharon were born in consecutive years. If the sum of their ages is 36, how old is each?

5 A pole is standing in a pond. If one-fourth of the height of the pole is in the sand at the bottom of the pond, 9 ft are in the water, and three-eighths of the total height is in the air above the water, what is the length of the pole?

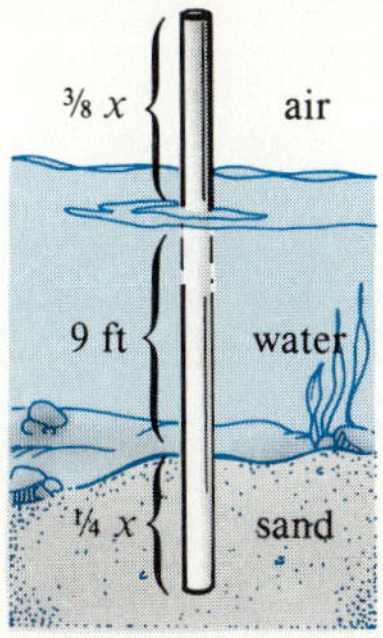

6. In an election with only two candidates, the winner received 50 votes more than twice the total for the loser. If a total of 1310 votes were cast, how many did each candidate receive?

7. Four-fifths of the operating budget of a community college goes for faculty salaries. If the total of all faculty salaries is $4,000,000, what is the total operating budget?

8. If $16.80 is charged as tax on the sale of a washing machine priced at $420, what is the tax rate?

9. A fuel tank contained 840 gallons of fuel. An additional quantity of fuel was added making a total of 1092 gallons. What was the percent increase?

10 The measure of the second angle of a triangle is one-fourth the measure of the third and 30° more than the first. Find the measure of each.

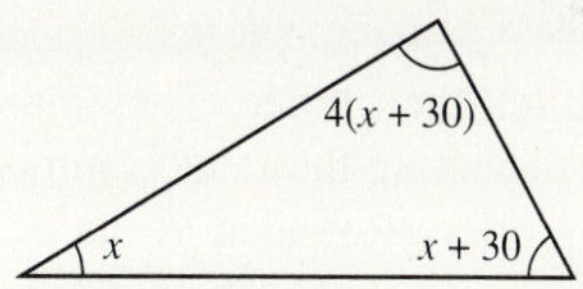

11. International Car Rental will rent a car for $15.50 per day plus 10¢ per mile driven. If Karin rented a car for two days and was billed a total of $93.00, how many miles did she drive?

12. Tim Chandler bought a pair of running shoes on sale for $33.60. If the sale price was discounted 30% from the original price, what was the original price?

13 A rectangular patio is twice as long as it is wide. If its length were decreased 8 ft and the width were increased 8 ft, it would be square. What are its dimensions?

14. The relationship between °F and °C is given by the formula $F = \frac{9}{5}C + 32$. Determine °C for a temperature of 50°F.

15. How much simple interest will be earned on a deposit of \$800 for one year at an interest rate of 12% per year?

16. Barbara purchased a Ford Bronco for \$16,500. If the options totaled \$5800, what percent of the purchase price was the base price of the vehicle?

17. Leah was able to do 40 more situps than Lynn on a recent fitness test. If the total for both girls was 290, how many situps did Lynn do?

18. The perimeter of a square is eleven cm less than five times the length of a side. Find the length of a side.

19 The population of Red Rock, Utah, was 740 in 1990. This was a 20% decrease in the population of 1980. What was the population in 1980?

20 Gene won \$10,000 in the Illinois Lottery. He invested part in a savings account that earned 14% simple interest, and the rest in a fund that paid 11% simple interest. If at the end of one year he earned an income of \$1310 from the two, how much was invested at each rate?

21. The Arnotes plan to sell their home. They must receive \$82,000 after deducting the sales commission of 6% on the selling price. Rounded to the nearest dollar, at what selling price should the house be listed?

22. Dr. Cotera wishes to enclose a rectangular plot of land with 140 yards of fencing in such a way that the length is four times the width. What will be the dimensions of the plot?

23. On a recent vacation, Leona traveled one-third the total distance by car, 400 miles by boat, and one-half the total distance by air. What was the total distance traveled on the vacation?

24 Becky must have an average of 90 on four tests in geology to get an A in the course. What is the lowest score she can make on the fourth test if her first three grades are 96, 78, and 91?

25. When ice floats in water, approximately $\frac{8}{9}$ of the height of the ice is below the surface of the water. If the tip of an iceberg is 35 feet above the surface of the water, what is the approximate depth of the iceberg?

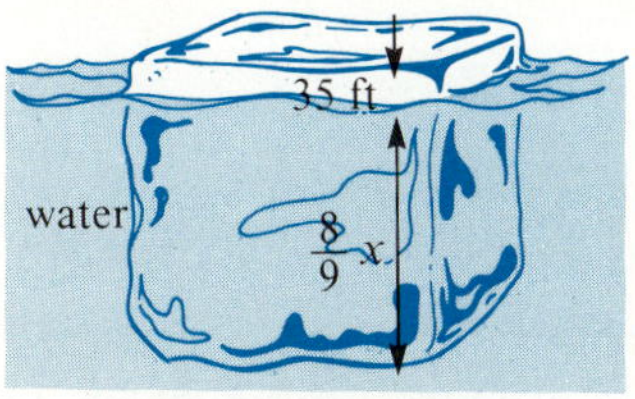

26. A boat traveled from Puerto Vallarta to an island and back again in a total time of 10.5 hours. How far is the island from Puerto Vallarta if the boat averaged 15 mph going to the island and 20 mph returning?

27. In a school election, Morton beats his two opponents by 230 and 175 votes, respectively. If there were a total of 1635 votes cast, how many did Morton receive?

28. A meteorite weighing 4.5 tons was discovered near Meteor Crater, Arizona, in 1918. If the meteorite contains about 87% iron, how many pounds of iron does it contain?

29. After draining their swimming pool to make several minor repairs, the Bonnetts plan to refill it using their garden hose. Suppose the hose lets water in at a rate of 20 gallons per minute and the pool holds 36,000 gallons. Will the pool be filled by noon on Sunday if the water is turned on at 9:00 A.M. Saturday?

30. To estimate the trout population in Lake Louise, 400 banded trout were released on June 1, and the lake was closed to all fishing. On June 10, a sample of 100 trout were caught, 6 of which were banded. As a result, the ranger estimated that 6% of all the trout in Lake Louise were banded. Approximately how many trout were in the lake?

ANSWERS: 1. 33 yr 2. 120 yd, 144 yd 3. 20 4. Bob is 11, Mary is 12, Sharon is 13. 5. 24 ft 6. 420, 890 7. $5,000,000 8. 4% 9. 30% 10. 5°, 35°, 140° 11. 620 mi 12. $48.00 13. 32 ft, 16 ft 14. 10°C 15. $96 16. 64.8% 17. 125 18. 11 cm 19. 925 20. $7000 in savings, $3000 in the fund 21. $87,234 22. 14 yd, 56 yd 23. 2400 mi 24. 95 25. 315 ft 26. 90 mi 27. 680 votes 28. 7830 lb 29. No. It will take 30 hours to fill the pool. 30. 6667 trout

2.6 LITERAL EQUATIONS

STUDENT GUIDEPOSTS

1. Equations with the Unknown in One Term
2. Equations with the Unknown in Several Terms

1 EQUATIONS WITH THE UNKNOWN IN ONE TERM

Thus far the equations we have solved have contained only one variable. A **literal equation** (or **letter equation**) is an equation that has two or more variables. Many formulas familiar to us are literal equations, such as

$$A = lw, \quad P = 2l + 2w, \quad d = rt, \quad A = \pi r^2, \quad I = Prt.$$

At times we want to solve a literal equation for a particular variable. For example, the equation $d = rt$ can be solved for t and for r,

$$t = \frac{d}{r} \quad \text{and} \quad r = \frac{d}{t}.$$

Remember that the letters in a literal equation, that is, the variables, simply represent numbers. Any of the rules for solving equations with numerical values also apply to literal equations. It can be helpful to solve first a numerical equation similar in form to the literal equation. We show this in the examples.

EXAMPLE 1 Solving a Literal Equation

Solve $T = cn$ for n.

Literal equation		*Similar numerical equation*	
$T = cn$		$5 = 3n$	
$\frac{1}{c} \cdot T = \frac{1}{c} \cdot cn$	Multiply by $\frac{1}{c}$	$\frac{1}{3} \cdot 5 = \frac{1}{3} \cdot 3n$	Multiply by $\frac{1}{3}$
$\frac{T}{c} = n$		$\frac{5}{3} = n$	

Practice Exercise 1

Solve $P = RT$ for R.

Answer: $R = \frac{P}{T}$

Keep in mind that solving a literal equation (or any other equation) depends on our ability to isolate the variable on one side of the equation.

EXAMPLE 2 Solving a Literal Equation

Solve $F = v + w$ for v.

Literal equation		*Similar numerical equation*	
$F = v + w$		$5 = v + 7$	
$F - w = v + w - w$	Subtract w	$5 - 7 = v + 7 - 7$	Subtract 7
$F - w = v$		$-2 = v$	

Practice Exercise 2

Solve $h = u + z$ for z.

Answer: $z = h - u$

Geometry problems often require the solution of a literal equation. In the following example we work with the formula for perimeter of a rectangle.

EXAMPLE 3 Geometry Problem

Solve the formula for the perimeter of a rectangle, $P = 2l + 2w$, for w.

Literal equation	*Similar numerical equation*
$P = 2l + 2w$	$12 = 2 \cdot 3 + 2w$
$P - 2l = 2l - 2l + 2w$	$12 - 2 \cdot 3 = 2 \cdot 3 - 2 \cdot 3 + 2w$
$P - 2l = 2w$	$6 = 2w$
$\frac{1}{2}(P - 2l) = \frac{1}{2} \cdot 2w$	$\frac{1}{2} \cdot 6 = \frac{1}{2} \cdot 2w$
$\frac{P - 2l}{2} = w$	$3 = w$

Practice Exercise 3

Solve $m = 3n + 5k$ for k.

Answer: $k = \frac{m - 3n}{5}$

2 EQUATIONS WITH THE UNKNOWN IN SEVERAL TERMS

When the variable for which we want to solve occurs in more than one term, we collect these terms on one side of the equation first. This is illustrated in the next example.

EXAMPLE 4 Solving a Literal Equation

Solve $By = 5p - by$ for y.

Literal equation		*Similar numerical equation*	
$By = 5p - by$		$3y = 5 \cdot 7 - 2y$	
$By + by = 5p$	Add by	$3y + 2y = 5 \cdot 7$	Add $2y$
$y(B + b) = 5p$	Factor out y using the distributive law	$y(3 + 2) = 35$	Factor out y
		$5y = 35$	Simplify
$y = \dfrac{5p}{B + b}$	Divide by $(B + b)$	$y = \dfrac{35}{5} = 7$	Divide by 5

PRACTICE EXERCISE 4

Solve $mx = nx + ky$ for x.

Answer: $x = \frac{ky}{m - n}$

CAUTION

The step at which y is factored out in the numerical equation in Example 4 is usually omitted and performed mentally, resulting in the simplified equation $5y = 35$. In the literal equation, however, we need this step to obtain $y(B + b)$ so we can divide both sides by the coefficient of y, $B + b$. Also, *do not* try to solve $By = 5p - by$ for y by writing

$$y = \frac{5p - by}{B}.$$

Notice that y is still a factor of a term on the right, so the equation has not been solved for y.

Many formulas used by scientists use several variables and need to be solved as literal equations. The following is an example from physics.

EXAMPLE 5 Physics Problem

The distance above ground level of an object under the influence of gravity is given by the equation

$$h = \frac{1}{2}gt^2 + vt + s.$$

Solve this equation for v, then find v when $h = 650$ ft, $g = 32$ ft/sec^2, $t = 2$ sec, and $s = 1250$ ft.

To isolate the vt term, subtract $\frac{1}{2}gt^2$ and s from both sides of the equation.

$$h = \frac{1}{2}gt^2 + vt + s$$

$$h - \frac{1}{2}gt^2 - s = \frac{1}{2}gt^2 - \frac{1}{2}gt^2 + vt + s - s$$

PRACTICE EXERCISE 5

Solve $h = \frac{1}{2}gt^2 + vt + s$ for s, then find s when $h = 1000$, $g = 32$ ft/sec^2, $t = 4$ sec, and $v = -50$ ft/sec.

$$h - \frac{1}{2}gt^2 - s = vt$$

$$\frac{h - \frac{1}{2}gt^2 - s}{t} = \frac{vt}{t} \quad \text{Divide by } t$$

$$\frac{h - \frac{1}{2}gt^2 - s}{t} = v$$

Now find v, which is initial velocity, when $h = 650$ ft, $g = 32$ ft/sec^2, $t = 2$ sec, and $s = 1250$ ft.

$$v = \frac{h - \frac{1}{2}gt^2 - s}{t} = \frac{650 - \frac{1}{2}(32)(2)^2 - 1250}{(2)}$$

$$= \frac{650 - 64 - 1250}{2} = -332$$

Thus, the initial velocity is -332 ft/sec, which means 332 ft/sec downward.

Answer: $s = h - \frac{1}{2}gt^2 - vt$; $s = 944$ ft

The following summarizes the steps for solving literal equations.

To Solve a Literal Equation

1. Simplify each side of the equation.
2. Using the additive property, change the equation into an equivalent equation having all terms with the specified variable on one side and all other terms on the other side.
3. Factor the specified variable from all terms (if necessary) and divide both sides by the coefficient of the variable, combining all like terms when appropriate.

2.6 EXERCISES A

Solve.

1. $a = b + c$ for c

2. $a = bc$ for b

3. $a = bc + d$ for b

4. $a + b = c + d$ for c

5. $u = 2v + 2w$ for v

6. $x + z + y = w$ for z

7. $abc = w$ for a

8. $aA = a + A$ for a

9. $P = 20a + b$ for a

10. $a = 3(b + c)d$ for d

11. $a = 3(b + c)d$ for b

12. $F = \frac{9}{5}C + 32$ for C

13. $V = lwh$ for w

14 $A = \frac{1}{2}(b_1 + b_2)h$ for b_1

15. $P = a + b + c$ for c

16. $A = P + Prt$ for r

17. $A = P + Prt$ for P

18. $I = Prt$ for r

19. $a_n = a_1 + (n - 1)d$ for d

20. $a_n = a_1 + (n - 1)d$ for n

21. $A = \frac{1}{2}bh$ for b

22. $E = IR + Ir$ for r

23. $S = 2\pi rh + 2\pi r^2$ for h

24. $bx - c = ax$ for x

25. $3by - c = 2ay$ for y

26 $2(x - b) + 5x = 4c$ for x

27. $x(b + 2) = 3b + 4x$ for x

For each of the following, write an appropriate formula for the specified variable, and then find the required value of that variable.

28. If the distance traveled d and the rate of travel r are given, find a formula for the time of travel t. What is the value of t when d is 495 mi and r is 45 mph?

29. If the area A and length l of a rectangle are given, find a formula for the width w. What is the value of w when A is 3.75 cm^2 and l is 2.5 cm?

30. If the rate r, time t, and simple interest I are given, find a formula for the principal P. What is the value of P when r is 15%, t is 4 yr, and I is \$1500?

31. If two sides of a triangle, a and b, and the perimeter P are given, find a formula for the third side c. What is the value of c when a is 4.3 in, b is 5.2 in, and P is 17.0 in?

32. If the perimeter P and width w of a rectangle are given, find a formula for the length l. What is the value of l when P is 38 yd and w is 6 yd?

33. If the total cost T of a number of items n is given, find a formula for the cost per item c. What is the value of c when T is \$52.50 and n is 35?

34. If the two bases b_1 and b_2 and the area A of a trapezoid are given, find a formula for the height h of the trapezoid. What is the value of h if b_1 is 7 ft, b_2 is 19 ft, and A is 110.5 ft^2?

35. If the volume V and two edges a and b of a rectangular solid are given, find a formula for the third edge c. What is the value of c if V is 31 yd^3, a is 3.1 yd, and b is 2.5 yd?

FOR REVIEW

36. After an 8% raise, Mick's new salary is \$14,472. What was his former salary?

37. What amount of money invested at $7\frac{1}{2}$% simple interest will increase to \$1612.50 in one year?

38. Jackson Black received a commission of \$396 on sales amounting to \$5280. What was his commission rate?

39. After receiving a 20% discount on the selling price, Dyane paid \$74.76, including 5% sales tax, for a watch. What was the price of the watch before the discount?

40. Becky can ride her bike to school in 10 min, but it takes her an hour if she walks. Her speed walking is 15 mph slower than her speed when riding. Find each rate.

41. One secretary can fold and stuff 12 envelopes per minute while another can do only 9 per minute. How long will it take them to complete the job of folding and stuffing 6000 envelopes if the first works for 45 min before being joined by the other?

ANSWERS: **1.** $c = a - b$ **2.** $b = \frac{a}{c}$ **3.** $b = \frac{a-d}{c}$ **4.** $c = a + b - d$ **5.** $v = \frac{u-2w}{2}$ **6.** $z = w - y - x$ **7.** $a = \frac{w}{bc}$ **8.** $a = \frac{A}{A-1}$ **9.** $a = \frac{P-b}{20}$ **10.** $d = \frac{a}{3b+3c}$ **11.** $b = \frac{a-3cd}{3d}$ **12.** $C = \frac{5}{9}(F - 32)$ **13.** $w = \frac{V}{lh}$ **14.** $b_1 = \frac{2A - b_2h}{h}$ **15.** $c = P - a - b$ **16.** $r = \frac{A-P}{Pt}$ **17.** $P = \frac{A}{1+rt}$ **18.** $r = \frac{I}{Pt}$ **19.** $d = \frac{a_n - a_1}{n-1}$ **20.** $n = \frac{a_n - a_1 + d}{d}$ **21.** $b = \frac{2A}{h}$ **22.** $r = \frac{E-IR}{I}$ **23.** $h = \frac{S - 2\pi r^2}{2\pi r}$ **24.** $x = \frac{c}{b-a}$ **25.** $y = \frac{c}{3b-2a}$ **26.** $x = \frac{4c+2b}{7}$ **27.** $x = \frac{3b}{b-2}$ **28.** $t = \frac{d}{r}$; 11 hr **29.** $w = \frac{A}{l}$; 1.5 cm **30.** $P = \frac{I}{rt}$; \$2500 **31.** $c = P - a - b$; 7.5 in **32.** $l = \frac{P-2w}{2}$; 13 yd **33.** $c = \frac{T}{n}$; \$1.50 **34.** $h = \frac{2A}{b_1+b_2}$; 8.5 ft **35.** $c = \frac{V}{ab}$; 4.0 yd **36.** \$13,400 **37.** \$1500 **38.** 7.5% **39.** \$89.00 **40.** riding: 18 mph, walking: 3 mph **41.** 4 hr and 20 min

2.6 EXERCISES B

Solve.

1. $a = b + c$ for b

2. $a = bc$ for c

3. $a = bc + d$ for c

4. $a + b = c + d$ for d

5. $u = 2v + 2w$ for w

6. $x + z + y = w$ for y

7. $abc = w$ for c

8. $Bb = B + b$ for b

9. $P = 20a + b$ for b

10. $a = 3(b + c)d$ for c

11. $C = \frac{5}{9}(F - 32)$ for F

12. $V = lwh$ for h

13. $A = \frac{1}{2}(b_1 + b_2)h$ for h

14. $A = \frac{1}{2}(b_1 + b_2)h$ for b_2

15. $P = a + b + c$ for c

16. $A = P + Prt$ for t

17. $U = A + Acw$ for A

18. $I = Prt$ for t

19. $a_n = a_1 + (n - 1)d$ for a_1

20. $A = \frac{1}{2}bh$ for h

21. $E = IR + Ir$ for R

22. $C = 2\pi r$ for r

23. $ax = c - 3bx$ for x

24. $4ay = c + 2by$ for y

25. $dx - 3(x - a) = 2b$ for x

26. $a(y + b) + c = dy$ for y

27. $v = k + gt$ for g

For each of the following, write an appropriate formula for the specified variable, and find the required value of that variable.

28. If the distance traveled d and the time of travel t are given, find a formula for the rate of travel r. What is the value of r when d is 350 km and t is 5 hr?

29. If the area A and the width w of a rectangle are given, find a formula for the length l. What is the value of l when A is 18.9 ft^2 and w is 3.5 ft?

30. If the principal P, time t, and simple interest I are given, find a formula for the rate r. What is the value of r when P is \$400, t is 3 yr, and I is \$144?

31. If two sides of a triangle, b and c, and the perimeter P are given, find a formula for the third side a. What is the value of a when b is 2.8 cm, c is 4.3 cm, and P is 15.0 cm?

32. If the perimeter P and length l of a rectangle are given, find a formula for the width w. What is the value of w when P is 40 m and l is 13 m?

33. If the total cost T of a number of items n is given, find a formula for the cost per item c. What is the value of c when T is \$74.00 and n is 40?

34. If the volume V and two edges, b and c, of a rectangular solid are given, find a formula for the third edge a. What is the value of a if V is 10.78 cm^3, b is 3.5 cm, and c is 1.4 cm?

35. If the surface area S and two edges, b and c, of a rectangular solid are given, find a formula for the third edge a. What is the value of a if S is 184 ft^2, b is 4 ft, and c is 8 ft?

FOR REVIEW

36. Paulette received a 6% raise, making her new salary \$11,130. What was her former salary?

37. What amount of money invested at 14% simple interest will increase to \$2125 at the end of five years?

38. Eric Bradley borrowed a sum of money from a savings and loan office at 7% simple interest. At the end of 3 years he repaid the loan in the amount of \$1663.75. How much did he originally borrow?

39. After receiving a 30% discount on the selling price, Verna paid \$29.40, including 5% sales tax, for a pair of shoes. What was the price of the shoes before the discount?

40. Two cars, that are 436 mi apart and whose speeds differ by 5 mph, are traveling towards each other. What is the speed of each if they meet in 4 hr?

41. Two airplanes leave Chicago at 12:00 noon and fly in opposite directions. If one flies at 410 mph and the other 120 mph faster, how long will it take them to be 3290 mi apart?

2.6 EXERCISES C

Solve.

1. $S_n = \frac{n}{2}[2a_1 + (n-1)d]$ for d

2. $S_n = \frac{n}{2}[2a_1 + (n-1)d]$ for a_1

3. $b(x + y) - a = Ax$ for x $\left[\text{Answer: } x = \frac{a - by}{b - A}\right]$

4. $y(a + 1) = 3a + 2y$ for y

For each of the following, write an appropriate formula for the specified variable, and then find the required value of that variable.

5. If the length of one base a, the height h, and the area A of a trapezoid are given, find a formula for the length of the other base b. What is the value of b when a is 5 in, h is 8 in, and A is 44 in^2?

6. If the surface area S and two edges, a and b, of a rectangular solid are given, find a formula for the third edge c. What is the value of c if S is 62 m^2, a is 3 m, and b is 2 m? $\left[\text{Answer: } c = \frac{S - 2ab}{2b + 2a}; \text{ 5 m}\right]$

2.7 SOLVING LINEAR INEQUALITIES

STUDENT GUIDEPOSTS

1. Graphing Inequalities
2. Linear Inequalities
3. Interval Notation
4. Solving Linear Inequalities

1 GRAPHING INEQUALITIES

An inequality involving a variable such as

$$x > 1$$

means all real numbers greater than 1. This can be graphed on a number line as in Figure 2.6. The open dot at the point 1 indicates that 1 is not included in the graph. Since the inequality $x \le 1$ includes the point 1, it has a solid dot at 1 in Figure 2.7.

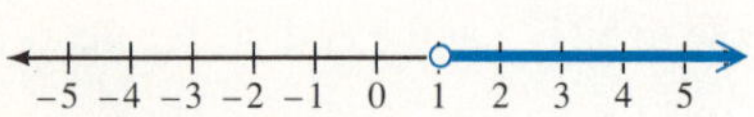

Figure 2.6 $x > 1$

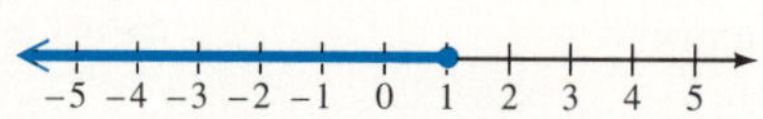

Figure 2.7 $x \le 1$

Inequalities such as $-2 < x < 1$ indicate the points between -2 and 1; the graph is shown in Figure 2.8. The graph of $-3 \le x < 2$ is in Figure 2.9.

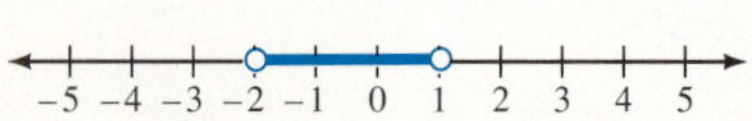

Figure 2.8 $-2 < x < 1$

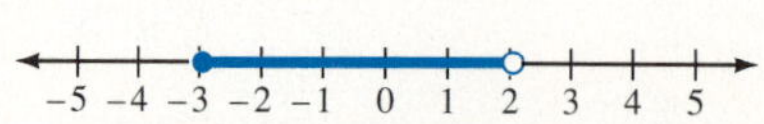

Figure 2.9 $-3 \le x < 2$

2 LINEAR INEQUALITIES

In this section, we will solve and graph statements, such as

$$x + 7 < 2, \quad 3x \geq 9, \quad 3x - 1 > 11, \quad \text{and} \quad 2x + 1 \leq 5 - 7x,$$

which are called **linear inequalities.** A linear inequality can be written as

$$\boldsymbol{ax + b < 0}, \quad \boldsymbol{ax + b > 0}, \quad \boldsymbol{ax + b \leq 0}, \quad \text{or} \quad \boldsymbol{ax + b \geq 0},$$

where a and b are real numbers, $a \neq 0$.

3 INTERVAL NOTATION

Before solving linear inequalities like these, we introduce a new way to write and graph inequalities, *interval notation*. The set described by $-2 < x < 1$ is an **open interval** because it contains neither end point, and is denoted $(-2, 1)$. The set $-2 \leq x \leq 1$ is a **closed interval** because it contains both end points, and is denoted $[-2, 1]$. In general, if $a < b$ we make the following definitions.

$(a, b) = \{x|a < x < b\}$ Open interval

$[a, b] = \{x|a \leq x \leq b\}$ Closed interval

$(a, b] = \{x|a < x \leq b\}$ Half-open interval

$[a, b) = \{x|a \leq x < b\}$ Half-open interval

Note that $(a, b]$ and $[a, b)$ are both called **half-open intervals** because each contains only one end point.

For the statement $x > 1$, we write the **infinite interval,** $(1, \infty)$. The infinity symbol ∞ is not a number but simply indicates that all numbers greater than 1 are included in the set. The symbol $-\infty$, read "minus infinity" or "negative infinity," is also used in the following definitions.

$$(-\infty, a) = \{x|x < a\}$$
$$(-\infty, a] = \{x|x \leq a\}$$
$$(a, \infty) = \{x|x > a\}$$
$$[a, \infty) = \{x|x \geq a\}$$
$$(-\infty, \infty) = \{x|x \text{ is any real number}\}$$

EXAMPLE 1 Using Interval Notation

Write each inequality in interval notation and then graph.

(a) $-3 \leq x < 0$

In the notation for a half-open interval, $-3 \leq x < 0$ becomes $[-3, 0)$. Figure 2.10 shows two methods of graphing the interval, the dot notation we used to introduce graphing, as well as interval notation. Note that [replaces the closed dot on the left and) replaces the open dot on the right.

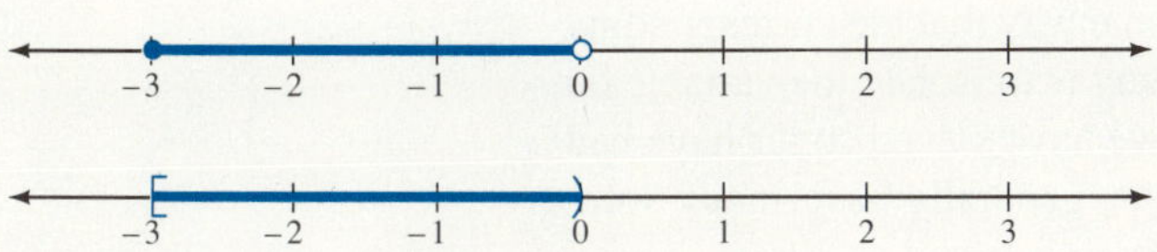

Figure 2.10 $[-3, 0)$

Practice Exercise 1

Write each inequality in interval notation and then graph.

(a) $-1 < x \leq 2$

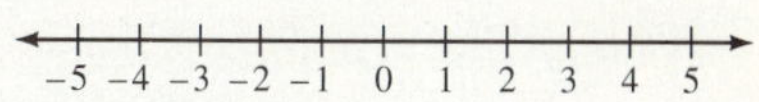

(b) $x < 2$

This infinite interval can be written as $(-\infty, 2)$. Figure 2.11 again shows the interval graphed in both dot notation and interval notation.

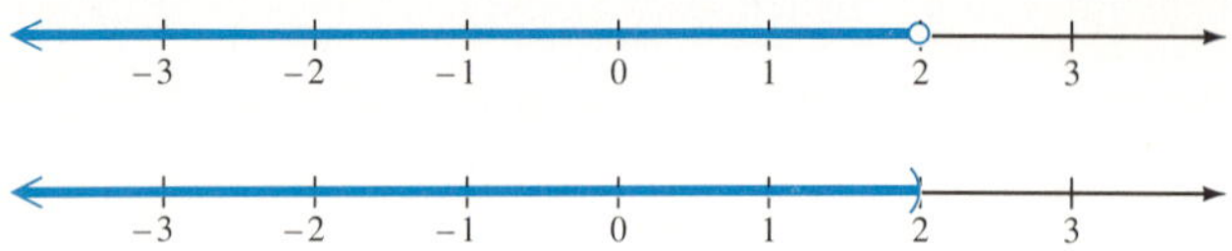

Figure 2.11 $(-\infty, 2)$

(b) $x \geq -2$

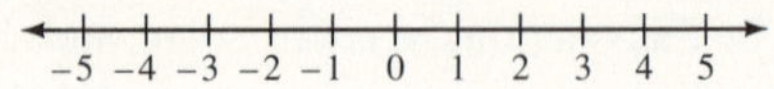

Answers: (a) **Graph** $(-1, 2]$. (b) **Graph** $[-2, \infty)$.

4 SOLVING LINEAR INEQUALITIES

We are now ready to solve linear inequalities such as

$$x + 2 < 7, \quad 3x \geq 9, \quad 3x - 1 > 11, \quad \text{and} \quad 2x + 1 \leq 5 - 7x.$$

A **solution** to an inequality is a number that, when substituted for the variable, makes the inequality true.

If the inequality symbols ($<$, $\geq$, $>$, and $\leq$) in the above statements were replaced with the equality symbol ($=$), we would know how to solve the resulting equations. Fortunately, solving an inequality is much like solving an equation since most of the same basic rules apply. There is one exception that will be discussed shortly. If we add 4 to both sides of the true inequality $5 < 8$, we have another true inequality.

$$5 + 4 < 8 + 4$$
$$9 < 12 \quad \text{True}$$

Similarly, if we subtract 11 from both sides, we again obtain a true inequality.

$$5 - 11 < 8 - 11$$
$$-6 < -3 \quad \text{True}$$

These observations illustrate the property which states that the same number (positive or negative) can be added to both sides of an inequality.

Additive Property of Inequalities

Let a, b, and c be numbers.

If $a < b$, then $a + c < b + c$.

That is, $a < b$ and $a + c < b + c$ are **equivalent inequalities** (they have the same solutions).

This rule is used to solve inequalities in the same way that the corresponding additive property is used to solve equations. The key is to isolate the variable on one side of the inequality. Most of the equations we have solved so far have had only one solution (identities excepted). Inequalities generally have many solutions.

EXAMPLE 2 Using the Additive Property

Solve and graph.

$$x + 7 < 2$$
$$x + 7 - 7 < 2 - 7 \quad \text{Subtract 7 from both sides}$$
$$x < -5$$

The solutions are *all* numbers less than -5, $(-\infty, -5)$, and the graph is given in Figure 2.12.

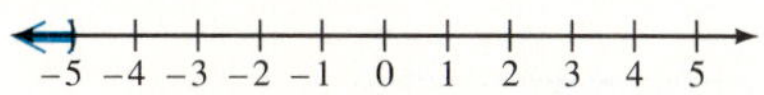

Figure 2.12 $(-\infty, -5)$

Practice Exercise 2

Solve and graph. $y + 4 \geq -2$

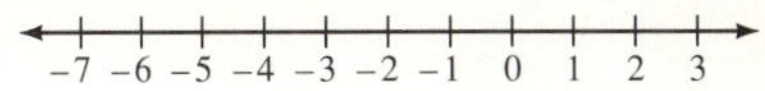

Answer: Graph $y \geq -6$, $[-6, \infty)$

If we start again with the true inequality

$$5 < 8$$

and multiply both sides by 2, we obtain the true inequality

$$10 < 16.$$

However, if we multiply both sides by -2 we obtain the false inequality

$$-10 < -16. \quad -10 \text{ is really greater than } -16$$

To obtain a true inequality when we multiply by the negative number -2, we must reverse the symbol of inequality, that is, change from $<$ to $>$ or from $>$ to $<$. The next property gives the rules for multiplying (or dividing) both sides of an inequality by positive or negative numbers.

Multiplicative Property of Inequalities

Let a, b, and c be numbers.

If $c > 0$ and $a < b$, then $ac < bc$.

That is, if $c > 0$, then $a < b$ and $ac < bc$ are equivalent inequalities.

If $c < 0$ and $a < b$, then $ac > bc$.

That is, if $c < 0$, then $a < b$ and $ac > bc$ are equivalent inequalities.

The only substantial difference between solving an equation and solving an inequality concerns multiplying or dividing both sides by a negative number.

CAUTION

Always reverse the symbol of inequality when multiplying (or dividing) both sides of an inequality by a negative number.

EXAMPLE 3 Using the Multiplicative Property

Solve and graph.

(a) $3x \geq 9$

$\frac{1}{3}(3x) \geq \frac{1}{3} \cdot 9$ $\quad$ $\frac{1}{3}$ is positive, so the inequality remains the same

$x \geq 3$

The solutions are *all* numbers greater than or equal to 3. Generally, we will simply write $x \geq 3$ or $[3, \infty)$. The graph is in Figure 2.13.

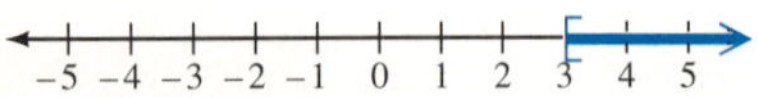

Figure 2.13 $[3, \infty)$

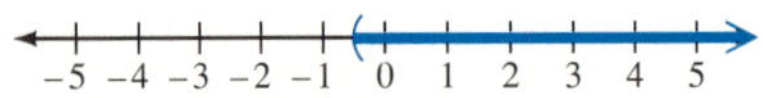

Figure 2.14 $\left(-\frac{1}{2}, \infty\right)$

(b) $-\frac{1}{2}x < \frac{1}{4}$

$(-2)\left(-\frac{1}{2}x\right) > (-2)\left(\frac{1}{4}\right)$ $\quad$ -2 is negative, so the inequality symbol is reversed

$x > -\frac{1}{2}$

The solution is $x > -\frac{1}{2}$, or $\left(-\frac{1}{2}, \infty\right)$. The graph is given in Figure 2.14.

PRACTICE EXERCISE 3

Solve and graph.

(a) $7y < 35$

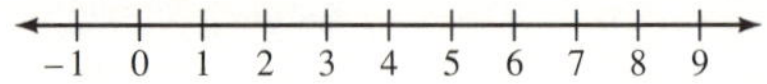

(b) $-3x \geq 4$

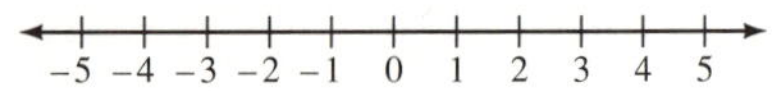

Answers: (a) Graph $y < 5$, $(-\infty, 5)$. (b) Graph $x \leq -\frac{4}{3}$, $\left(-\infty, -\frac{4}{3}\right]$.

As with solving equations, many times we need a combination of the additive and multiplicative properties. We usually use the additive property first, as in the next example.

EXAMPLE 4 Using a Combination of Rules

Solve and graph.

(a) $3x - 1 > 11$

$3x - 1 + 1 > 11 + 1$ $\quad$ Add 1

$3x > 12$

$\frac{1}{3} \cdot 3x > \frac{1}{3} \cdot 12$ $\quad$ Multiply by *positive* $\frac{1}{3}$

$x > 4$

The solution, $x > 4$ or $(4, \infty)$, is graphed in Figure 2.15.

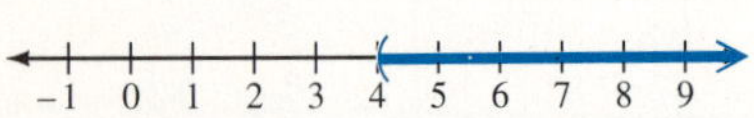

Figure 2.15 $(4, \infty)$

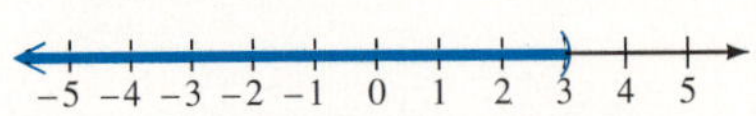

Figure 2.16 $(-\infty, 3)$

(b) $5 - 4x > 2 - 3x$

$5 - 5 - 4x > 2 - 5 - 3x$ $\quad$ Subtract 5

$-4x > -3 - 3x$

$-4x + 3x > -3 - 3x + 3x$ $\quad$ Add $3x$

$-x > -3$

PRACTICE EXERCISE 4

Solve and graph.

(a) $9y + 1 \leq 19$

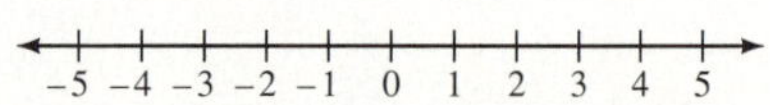

(b) $9 + 7y \leq 4 - 3y$

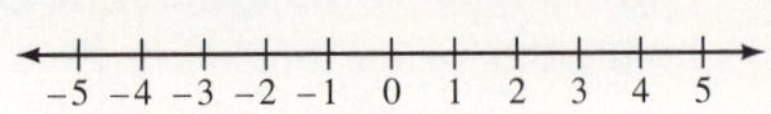

$$(-1)(-x) < (-1)(-3) \quad \text{Multiply by } -1 \text{ and } \textit{reverse} \text{ inequality}$$
$$x < 3$$

The solution, $x < 3$ or $(-\infty, 3)$, is graphed in Figure 2.16.

Answers: (a) Graph $y \le 2$, $(-\infty, 2]$. (b) Graph $y \le -\frac{1}{2}$, $\left(-\infty, -\frac{1}{2}\right]$.

When an inequality contains parentheses, remove all parentheses, collect like terms (if such exist) on each side, and proceed as in previous cases.

EXAMPLE 5 Solving an Inequality Containing Parentheses

Solve and graph.

$$x - 3(2 + x) > 2(3x - 1)$$
$$x - 6 - 3x > 6x - 2 \quad \text{Remove parentheses}$$
$$-2x - 6 > 6x - 2 \quad \text{Collect like terms}$$
$$-2x - 6 + 6 > 6x - 2 + 6 \quad \text{Add 6}$$
$$-2x > 6x + 4$$
$$-2x - 6x > 6x - 6x + 4 \quad \text{Subtract } 6x$$
$$-8x > 4$$
$$\left(-\frac{1}{8}\right)(-8x) < \left(-\frac{1}{8}\right)(4) \quad \text{Multiply by } -\tfrac{1}{8} \text{ and reverse the inequality}$$
$$x < -\frac{1}{2}$$

The solution, $x < -\frac{1}{2}$ or $\left(-\infty, -\frac{1}{2}\right)$ is graphed in Figure 2.17.

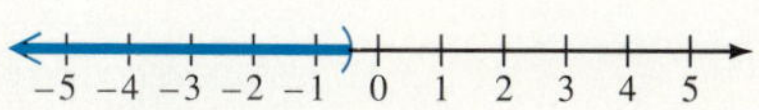

Figure 2.17 $\left(-\infty, -\frac{1}{2}\right)$

PRACTICE EXERCISE 5

Solve and graph.

$$2y - 4(y - 3) \le 3(2y + 1)$$

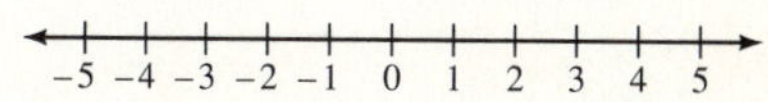

Answer: Graph $y \ge \frac{9}{8}$, $\left[\frac{9}{8}, \infty\right)$.

As is the case with linear equations, it is best to clear all fractions or decimals first.

EXAMPLE 6 Solving an Inequality Involving Fractions

Solve. $\dfrac{5x - 1}{7} \le 2$

$$7\left(\frac{5x - 1}{7}\right) \le 7(2) \quad \text{Multiply both sides by 7 to clear the fraction}$$
$$5x - 1 \le 14$$
$$5x \le 15 \quad \text{Add 1 to both sides}$$
$$x \le 3 \quad \text{Divide both sides by 5}$$

In interval notation, the solution is $(-\infty, 3]$.

PRACTICE EXERCISE 6

Solve. $-0.2y + 1.8 > 3.2$

[*Hint:* Multiply both sides by 10 first to clear the decimals.]

Answer: $(-\infty, -7)$

We now give a summary of inequality-solving techniques.

To Solve a Linear Inequality

1. To simplify both sides, remove grouping symbols and combine like terms. To eliminate fractions or decimals, multiply both sides by an appropriate positive number (the LCD of the fractions or a power of ten for decimals).
2. Use the addition-subtraction rule to isolate all variable terms on one side and all constant terms on the other side. Collect like terms when possible.
3. Use the multiplication-division rule to obtain a variable with coefficient 1. Remember to reverse the symbol of the inequality whenever multiplying or dividing both sides by a negative number.

We conclude this section with an application.

EXAMPLE 7 GRADE-COMPUTING PROBLEM

Krista must have at least 90% of the points in her math class to get an A. The grade is determined by three 100-point tests and a 150-point final exam. If she has 85, 92, and 96 on her tests, what range of scores on the final will give her an A?

Let x = Krista's total points on final exam,
450 = total points possible,
$(0.9)450$ = 90% of the total points.

Thus, x must satisfy the following inequality.

$$85 + 92 + 96 + x \geq 0.9(450)$$
$$273 + x \geq 405$$
$$x \geq 132$$

Krista's score must be greater than or equal to 132 to get an A.

PRACTICE EXERCISE 7

What range of scores on the final will give Krista (see Example 7) an A or B in the course if at least 80% of the points are necessary?

Answer: Krista's score must be greater than or equal to 87 to get an A or B.

2.7 EXERCISES A

Solve the following inequalities. Give answers in interval notation.

1. $x + 3 < 7$

2. $7 < z - 3$

3. $y + 2.1 \geq -3.4$

4. $-2z > 10$

5. $\frac{1}{3}y \leq -5$

6. $-\frac{1}{4}x \geq \frac{3}{2}$

7. $3z - 1 < 8$

8. $1 - 3y < 8$

9. $2z + 3 > 5z - 3$

10. $\frac{2y-1}{3} < 5$

11. $\frac{x+3}{-2} < 17$

12. $-y \leq 4$

13. $-1.5x < 4.5$

14. $0.4z + 1.6 < 2.4$

15. $-8y - 21 \geq 7 - 15y$

16. $5(3 - x) - 10 \leq 25$

17. $3(2z + 8) \leq 4(z - 3)$

18. $5(y + 3) + 1 \leq y - 4$

Solve and graph the following inequalities. Give answers in interval notation.

19. $x - 4 > 0$

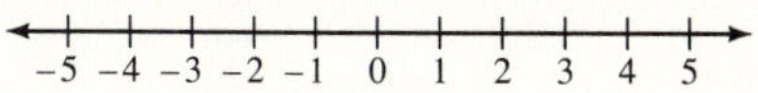

20. $z + 2 \leq 0$

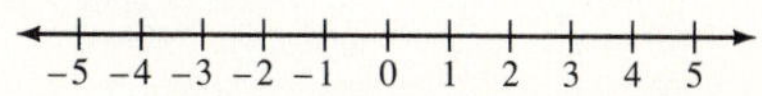

21. $2x - 5 < -1$

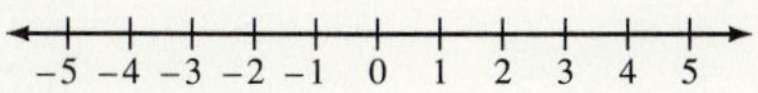

22. $3 - 5x \geq -7$

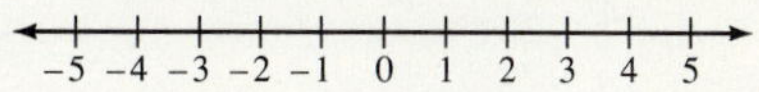

23. $3(z + 7) \leq 2z + 18$

24. $4(x - 2) + 2 \geq 2(x - 6)$

Letting x represent the unknown number, translate the following into symbols and solve.

25. Twice a number is greater than 8.

26. If 3 is subtracted from a number, the result is more than 20.

27. If twice a number is diminished by 7, the result is at least 12.

28. One-third of a number is no more than 15.

Translate to an inequality and solve.

29 Walt has scores of 86, 83, and 97 on three history tests. What range of scores can he make on test four to have an average of 90 or better?

30 To make a profit on the sale of television sets, a store owner knows that the total income on sales, S, must exceed the total of all costs involved, C. If n represents the number of televisions sold, $S = 150n$ and $C = 125n + 350$. How many sets must be sold for the owner to make a profit?

Are the following statements true or false? If false, change the second inequality to make the statement true.

31. If $x > 9$, then $-3x > -27$.

32. If $x < 4$, then $3x < 12$.

33. If $x \geq 8$, then $x - 1 \geq 7$.

34. If $x < -2$, then $-x > 2$.

FOR REVIEW

Solve.

35. $c = \dfrac{wrt}{1000}$ for w

36. $ax + by + c = 0$ for x

37. $e = mc^2$ for m

38. $ax + 2b = 3(x + c)$ for x

39. $3dy = d + 2ay$ for y

40. $3dy = d + 2ay$ for d

ANSWERS: 1. $x < 4$; $(-\infty, 4)$ 2. $z > 10$; $(10, \infty)$ 3. $y \geq -5.5$; $[-5.5, \infty)$ 4. $z < -5$; $(-\infty, -5)$ 5. $y \leq -15$; $(-\infty, -15]$ 6. $x \leq -6$; $(-\infty, -6]$ 7. $z < 3$; $(-\infty, 3)$ 8. $y > -\frac{7}{3}$; $\left(-\frac{7}{3}, \infty\right)$ 9. $z < 2$; $(-\infty, 2)$ 10. $y < 8$; $(-\infty, 8)$ 11. $x > -37$; $(-37, \infty)$ 12. $y \geq -4$; $[-4, \infty)$ 13. $x > -3$; $(-3, \infty)$ 14. $z < 2$; $(-\infty, 2)$ 15. $y \geq 4$; $[4, \infty)$ 16. $x \geq -4$; $[-4, \infty)$ 17. $z \leq -18$; $(-\infty, -18]$ 18. $y \leq -5$; $(-\infty, -5]$ 19. $(4, \infty)$ 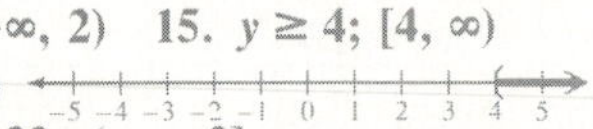 20. $(-\infty, -2]$ 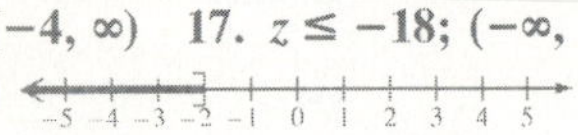 21. $(-\infty, 2)$ 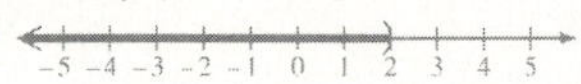 22. $(-\infty, 2]$ 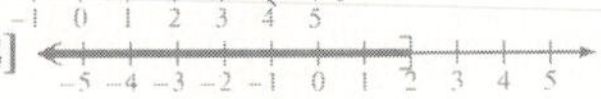 23. $(-\infty, -3]$ 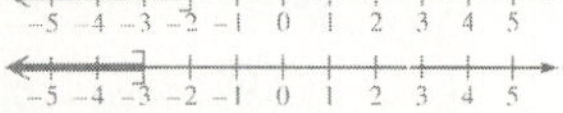 24. $[-3, \infty)$ 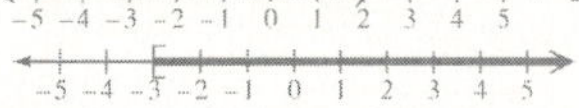 25. $2x > 8$; $x > 4$, $(4, \infty)$ 26. $x - 3 > 20$; $x > 23$, $(23, \infty)$ 27. $2x - 7 \geq 12$; $x \geq \frac{19}{2}$, $\left[\frac{19}{2}, \infty\right)$ 28. $\frac{1}{3}x \leq 15$; $x \leq 45$, $(-\infty, 45]$ 29. Walt's score, s, must satisfy $s \geq 94$. 30. $n > 14$, that is, 15 or more sets must be sold. 31. false; $-3x < -27$ 32. true 33. true 34. true 35. $w = \frac{1000c}{rt}$ 36. $x = \frac{-by - c}{a}$ 37. $m = \frac{e}{c^2}$ 38. $x = \frac{3c - 2b}{a - 3}$ 39. $y = \frac{d}{3d - 2a}$ 40. $d = \frac{2ay}{3y - 1}$

2.7 EXERCISES B

Solve the following inequalities. Give answers in interval notation.

1. $y + 5 \geq 3$

2. $x - 5 > 2$

3. $z - \dfrac{1}{2} < \dfrac{3}{4}$

4. $-9 \leq -3x$

5. $-3 \geq \dfrac{1}{4}z$

6. $\dfrac{7}{10} < -\dfrac{1}{5}y$

7. $5 - 2x > 3$

8. $2z - 5 > 3$

9. $4x - 5 \leq 2x + 13$

10. $\frac{3z-2}{5} \geq 2$

11. $\frac{4-y}{-3} \geq 13$

12. $-z > 17$

13. $-1.2y \geq -7.2$

14. $3.2 - 0.8x \geq 4.8$

15. $-3z - 39 \leq 13 - 2z$

16. $2(1 - y) - 4 > 6$

17. $2(3x + 1) > 5(x - 4)$

18. $-4z - 3 + z < 2(z + 12)$

Solve and graph the following inequalities. Give answers in interval notation.

19. $y - 3 < 0$

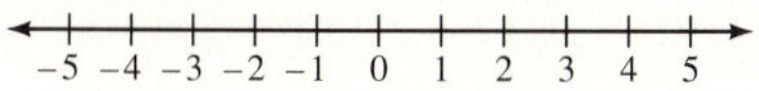

20. $x + 1 \geq 0$

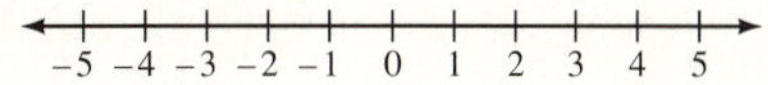

21. $5z - 2 > -7$

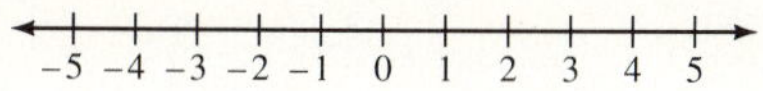

22. $7 - 3y \leq -2$

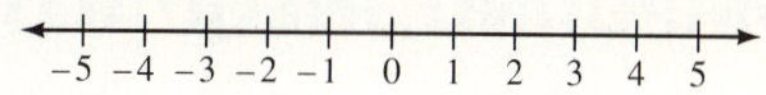

23. $5(x + 1) < 3x + 1$

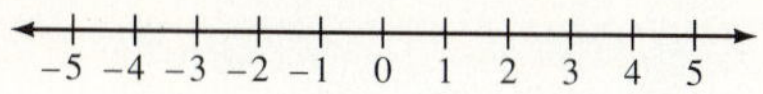

24. $4 - (x - 2) < 8 - 2x$

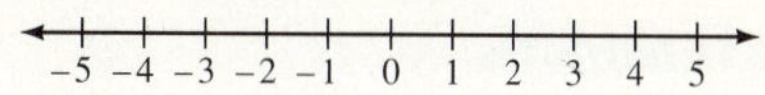

Letting x represent the unknown number, translate the following into symbols and solve.

25. My age in 3 years will be less than 25.

26. If 5 is subtracted from a number, the result is more than 3.

27. If three times a number is increased by 9, the result is at least 21.

28. Three-fourths of a number is no less than 81.

Translate to an inequality and solve.

29. Missy has scores of 80, 86, and 97 on three geology tests. What range of scores can she make on test four to have an average of 90 or better?

30. To make a profit on the sale of clocks, a dealer knows that the total income on sales, S, must exceed the total of all costs involved, C. If n is the number of clocks sold, $S = 40n$ and $C = 25n + 150$. How many clocks must be sold to make a profit?

Are the following statements true or false? If false, change the second inequality to make the statement true.

31. If $y > 5$, then $-2y > -10$.

32. If $y < 3$, then $8y < 24$.

33. If $y \geq 12$, then $y - 1 \geq 11$.

34. If $y < -4$, then $-y > 4$.

FOR REVIEW

Solve.

35. $c = \frac{wrt}{1000}$ for t

36. $ax + by + c = 0$ for y

37. $g = mt^2$ for m

38. $ax + 3b = 7(x - c)$ for x

39. $4dz = d + 3bz$ for z

40. $4dz = d + 3bz$ for d

2.7 EXERCISES C

Solve. Give answers in interval notation.

1. $\frac{2x}{3} + \frac{1}{4}(x - 4) \le \frac{x}{12} - \frac{1}{3}(x + 1)$
[*Hint:* Clear all fractions first.]

2. $(x + 1)(x - 2) > (x - 3)(x - 5)$
[Answer: $x > \frac{17}{7}$; $\left(\frac{17}{7}, \infty\right)$]

Are the following statements true or false? If the statement is false, explain why.

3. If $x > -1$, then $3 - 2x < 5$.

4. If $x < 5$, then $x^2 < 5x$.

5. If $x < -2$ and $y < 0$, then $xy > -2y$.

6. If $y < 2$, then $y^2 < 4$. [Answer: false]

2.8 COMPOUND INEQUALITIES

STUDENT GUIDEPOSTS

1. *And* Statements
2. *Or* Statements

1 *AND* STATEMENTS

We often encounter problems that combine two linear inequalities. When two inequalities are combined by using the word *and* or *or*, the result is a **compound statement.** Consider the compound statement

$$x + 2 > 0 \quad \textit{and} \quad 2(x - 1) < 4.$$

Solving each inequality gives the equivalent compound statement

$$x > -2 \quad \textit{and} \quad x < 3,$$

which could also be given by

$$-2 < x \quad \textit{and} \quad x < 3,$$

by the **chain of inequalities**

$$-2 < x < 3,$$

or by the open interval $(-2, 3)$. The numbers that satisfy this chain of inequalities are solutions to both inequalities: numbers that are *both* greater than -2 *and* less than 3. The graphs of $x > -2$, $x < 3$, and $-2 < x < 3$ are all shown in Figure 2.18.

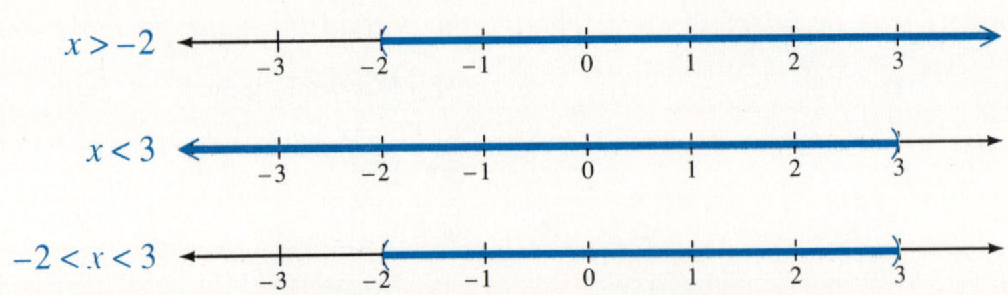

Figure 2.18 $-2 < x < 3$ or $(-2, 3)$

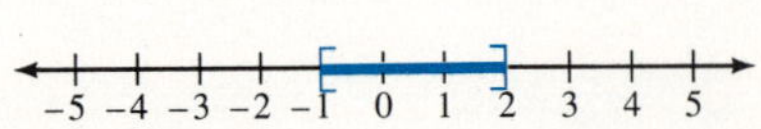

Figure 2.19 $-1 \le x \le 2$ or $[-1, 2]$

Notice that the graph of $-2 < x < 3$ includes all points that $x > -2$ and $x < 3$ have in common. Similarly, the graph of $-1 \le x \le 2$ $(-1 \le x$ and $x \le 2)$ is shown in Figure 2.19, the graph of $-3 \le x < 0$ $(-3 \le x$ and $x < 0)$ in Figure 2.20, and the graph of $1 < x \le 2$ $(1 < x$ and $x \le 2)$ in Figure 2.21.

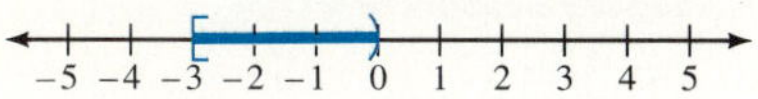

Figure 2.20 $-3 \le x < 0$ or $[-3, 0)$

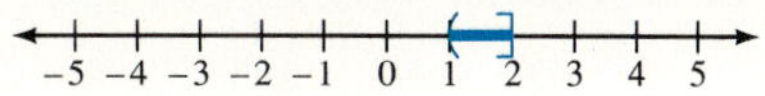

Figure 2.21 $1 < x \le 2$ or $(1, 2]$

The compound statement

$$2 < x \quad \text{and} \quad x < -1$$

has no solution, since there are no numbers that are *both* less than -1 *and* greater than 2. If we try to merge the two inequalities into a single chain as before, we obtain

$$2 < x < -1, \qquad \text{This is wrong}$$

which states that 2 is less than -1, a contradiction. Such chains should always be avoided.

A chain of inequalities may be more complex. For example,

$$-1 < 2x + 1 < 3$$

is

$$-1 < 2x + 1 \quad \textit{and} \quad 2x + 1 < 3.$$

Note that -1 is indeed less than 3, so the chain makes sense.

EXAMPLE 1 Solving an *And* Inequality

Solve and graph $-1 < 2x + 1 < 3$.

We need to solve two inequalities.

$$-1 < 2x + 1 \quad \text{and} \quad 2x + 1 < 3$$
$$-1 - 1 < 2x + 1 - 1 \quad \text{and} \quad 2x + 1 - 1 < 3 - 1 \qquad \text{Subtract 1}$$
$$-2 < 2x \quad \text{and} \quad 2x < 2$$
$$-1 < x \quad \text{and} \quad x < 1 \qquad \text{Divide by 2}$$

Rewritten as a single chain, the solution is

$$-1 < x < 1,$$

and the graph is shown in Figure 2.22.

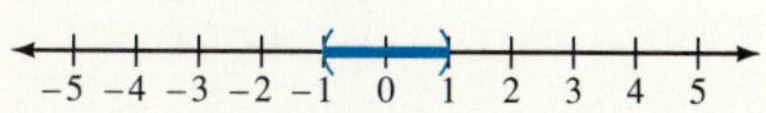

Figure 2.22 $(-1, 1)$

Practice Exercise 1

Solve and graph $-7 \le 3x - 4 \le 2$.

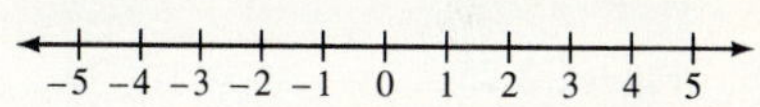

Answer: Graph $-1 \le x \le 2$, $[-1, 2]$.

Alternatively and more compactly, we can solve the original chain by the following method.

$$-1 < 2x + 1 < 3$$
$$-1 - 1 < 2x + 1 - 1 < 3 - 1 \qquad \text{Subtract 1 throughout}$$
$$-2 < 2x < 2$$
$$\frac{1}{2}(-2) < \frac{1}{2}(2x) < \frac{1}{2}(2) \qquad \text{Multiply throughout by } \tfrac{1}{2}$$
$$-1 < x < 1$$

2 *OR* STATEMENTS

The second type of compound statement uses the connective *or*. For example,

$$x + 1 < 0 \quad \textit{or} \quad 3(x - 1) > 3$$

simplifies to

$$x < -1 \quad \textit{or} \quad x > 2.$$

A solution to a compound statement using the word *or* solves *either* (or both) inequalities.

CAUTION

You *cannot* form a single chain when the word *or* is used. Notice that $2 < x < -1$ is a senseless statement (2 is not less than -1) and should *not* be used to describe the compound statement $x < -1$ *or* $x > 2$.

The graph of an *or* combination is usually *two* intervals on a number line, such as the graph of $x < -1$ or $x > 2$ shown in Figure 2.23. Similarly, $x < -2$ or $x \geq 1$ is graphed in Figure 2.24. Also, the graph of $x \leq 0$ or $x > 3$ is shown in Figure 2.25, and the graph of $x \leq -1$ or $x \geq 0$ is in Figure 2.26. Finally, the graph of $x \leq 2$ or $x \geq -1$ in Figure 2.27 is the entire number line.

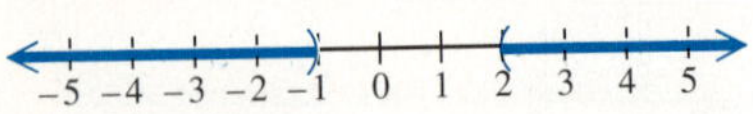

Figure 2.23 $(-\infty, -1)$ or $(2, \infty)$

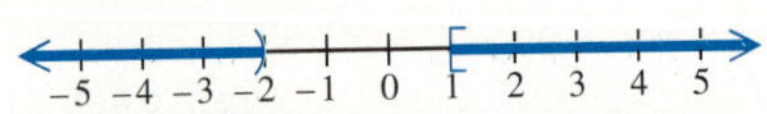

Figure 2.24 $(-\infty, -2)$ or $[1, \infty)$

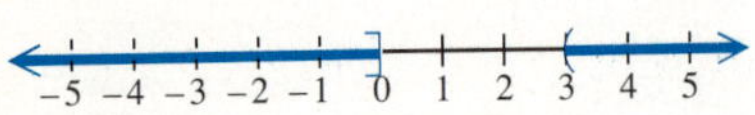

Figure 2.25 $(-\infty, 0]$ or $(3, \infty)$

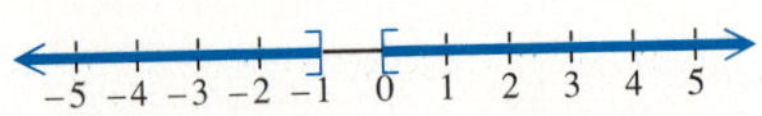

Figure 2.26 $(-\infty, -1]$ or $[0, \infty)$

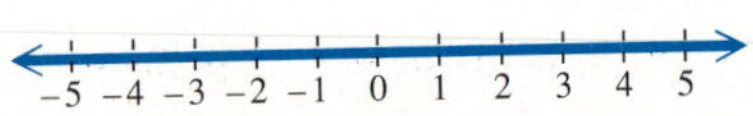

Figure 2.27 $(-\infty, \infty)$

EXAMPLE 2 SOLVING AN *OR* INEQUALITY

Solve and graph the following compound statements.

(a)

$2x + 1 \leq -1$	or	$2x + 1 > 3$	
$2x + 1 - 1 \leq -1 - 1$	or	$2x + 1 - 1 > 3 - 1$	Subtract 1
$2x \leq -2$	or	$2x > 2$	
$x \leq -1$	or	$x > 1$	Divide by 2

The solution is $x \leq -1$ or $x > 1$,

or, in interval notation, $(-\infty, -1]$ or $(1, \infty)$. The graph is shown in Figure 2.28.

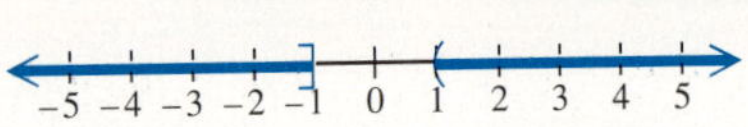

Figure 2.28 $(-\infty, -1]$ or $(1, \infty)$

PRACTICE EXERCISE 2

Solve and graph.

(a) $4x - 3 < -5$ or $4x - 3 \geq 5$

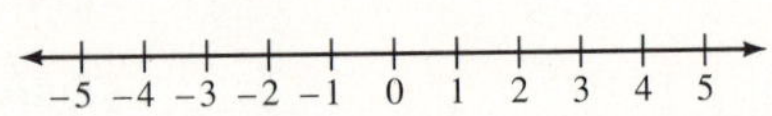

(b) Solve and graph the following compound statement.

$$3 - 4x < -1 \quad \text{or} \quad 3 - 4x \geq 9$$
$$-4x < -4 \quad \text{or} \quad -4x \geq 6$$
$$x > 1 \quad \text{or} \quad x \leq -\frac{6}{4} \qquad \text{Reverse}$$
$$x \leq -\frac{3}{2}$$

The solution is $x > 1 \text{ or } x \leq -\frac{3}{2}$,

or, in interval notation, $(1, \infty)$ or $\left(-\infty, -\frac{3}{2}\right]$ and the graph is given in Figure 2.29.

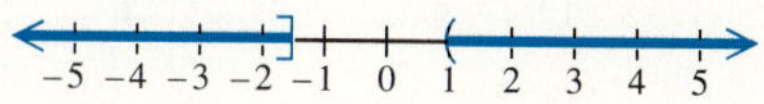

Figure 2.29 $(1, \infty)$ or $\left(-\infty, -\frac{3}{2}\right]$

(b) $2 - 5x \leq -3$ or $2 - 5x > 7$

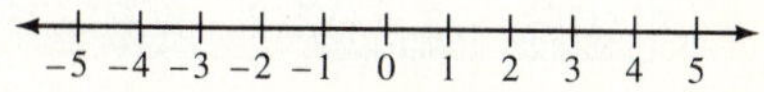

Answers: (a) Graph $x < -\frac{1}{2}$ or $x \geq 2$; $\left(-\infty, -\frac{1}{2}\right)$ or $[2, \infty)$.
(b) Graph $x < -1$ or $x \geq 1$; $(-\infty, -1)$ or $[1, \infty)$.

There are numerous applications of compound inequalities. The following is one example.

EXAMPLE 3 Temperature Problem

Celsius (C) and Fahrenheit (F) temperatures are related by $F = \frac{9}{5}C + 32$. If a hot tub is advertised to have a temperature range of 95°F to 104°F, inclusive, within what range does the Celsius temperature fall?

First write F using a chain of inequalities.

$$95 \leq F \leq 104$$

Now substitute $F = \frac{9}{5}C + 32$ and solve the chain.

$$95 \leq \frac{9}{5}C + 32 \leq 104$$
$$95 - 32 \leq \frac{9}{5}C \leq 104 - 32 \qquad \text{Subtract 32 throughout}$$
$$63 \leq \frac{9}{5}C \leq 72$$
$$(63)(5) \leq 9C \leq (72)(5) \qquad \text{Multiply by 5}$$
$$\frac{(63)(5)}{9} \leq C \leq \frac{(72)(5)}{9} \qquad \text{Divide by 9}$$
$$35 \leq C \leq 40$$

The temperature is between 35°C and 40°C, inclusive.

Practice Exercise 3

A student must have an average mark between 70% and 80%, inclusive, to receive a grade of C. If Bobbi has 69%, 82%, and 79% on the first three tests this semester, what range of scores on the fourth test would give her a C?

Answer: Bobbi must score between 50 and 90, inclusive.

Remember that an *and* compound inequality (such as $-4 < x$ and $x \leq 3$) can be expressed as a single chain of inequalities ($-4 < x \leq 3$) and generally has as its graph a single interval on a number line. An *or* compound inequality (such as $x < -1$ or $x > 2$) *cannot* be expressed as a single chain of inequalities (*never write* $2 < x < -1$) and generally has as its graph two separate intervals on a number line.

2.8 EXERCISES A

Solve and graph each compound inequality. Give answer in interval notation.

1. $x + 2 < 2$ and $x + 2 > -1$

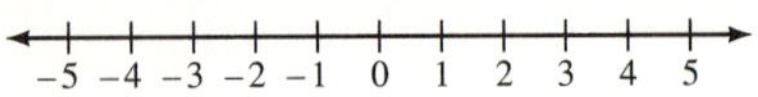

2. $z + 3 < 0$ or $z > 0$

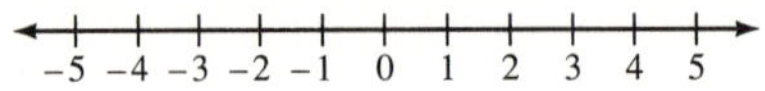

3. $y \geq 0$ and $y - 2 \leq 2$

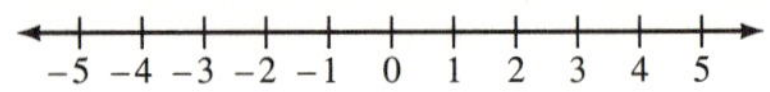

4. $x - 2 < 0$ or $x - 4 \geq 0$

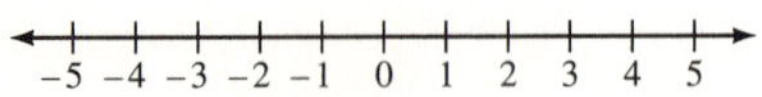

5. $-5 \leq 4z + 3 < 5$

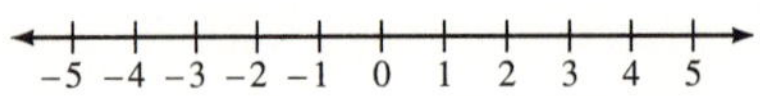

6. $4y + 3 < -5$ or $4y + 3 \geq 5$

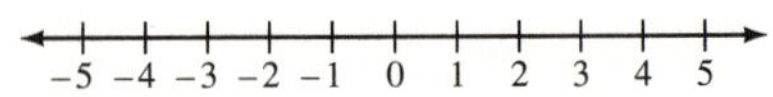

7. $0 > 3 - x$ or $3 - x \geq 1$

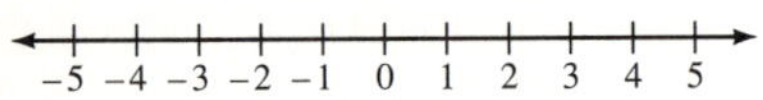

8. $0 < -z < 2$

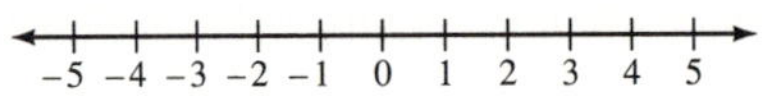

9. $0 \leq 3 - x < 1$

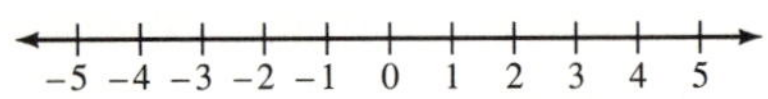

Letting x represent the unknown number, translate the following into symbols and solve.

10. A number is less than 8 and greater than -1.

11. A number is at least 2 and at most 9.

12. When twice a number is increased by 1, the result is between 7 and 13.

13. One less a number is greater than 10 or less than -3.

Solve.

14. Fahrenheit (F) and Celsius (C) temperatures are related by $C = \frac{5}{9}(F - 32)$. If the Celsius temperature is between -15°C and 30°C, inclusive, within what range will the Fahrenheit temperature fall?

15 A student must have an average percent between 80% and 90%, inclusive, in order to receive a grade of B. If Jake has 69%, 82%, and 79% on the first three tests this semester, what range of scores on the fourth test will give him a B?

FOR REVIEW

Solve.

16. $\dfrac{z - 3}{-4} < 3$

17. $3(2 - y) + y > 14$

18. $3(2z - 1) \geq -2(1 - z)$

19. The force in pounds F required to stretch a spring x inches beyond its normal length is given by $F = 5.5x$. What are the corresponding values for x if F is greater than 6.6 pounds?

Exercises 20–22 review material from Sections 1.2 and 1.6 to help you prepare for the next section. Evaluate the absolute values.

20. $|(-2) - (-1)(-2)|$

21. $\dfrac{|(-5)(-1) - 7|}{|-2|}$

22. $\dfrac{|(-2)(-7) - 20|}{|-18 + (-24)|}$

ANSWERS:

1. $(-3, 0)$;

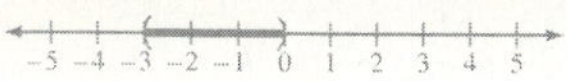

2. $(-\infty, -3)$ or $(0, \infty)$;

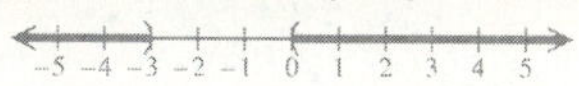

3. $[0, 4]$;

4. $(-\infty, 2)$ or $[4, \infty)$;

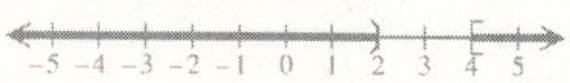

5. $\left[-2, \frac{1}{2}\right)$;

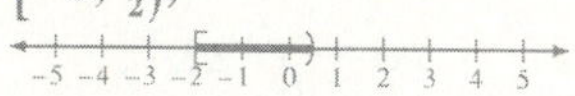

6. $(-\infty, -2)$ or $\left[\frac{1}{2}, \infty\right)$;

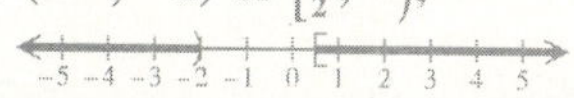

7. $(-\infty, 2]$ or $(3, \infty)$;

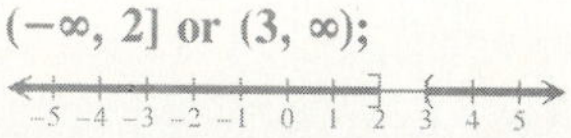

8. $(-2, 0)$;

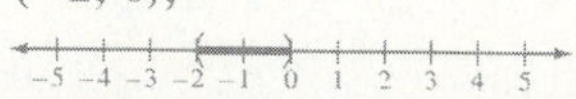

9. $(2, 3]$;

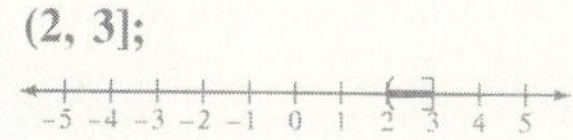

10. $-1 < x < 8$; $(-1, 8)$ **11.** $2 \le x \le 9$; $[2, 9]$ **12.** $7 < 2x + 1 < 13$; $(3, 6)$ **13.** $1 - x > 10$ or $1 - x < -3$; $(-\infty, -9)$ or $(4, \infty)$ **14.** $5° \le F° \le 86°$ **15.** His score must be between 90% and 100%, inclusive. **16.** $z > -9$ **17.** $y < -4$ **18.** $z \ge \frac{1}{4}$ **19.** $x > 1.2$ inches **20.** 4 **21.** 1 **22.** $\frac{1}{7}$

2.8 EXERCISES B

Solve and graph each compound inequality. Give answer in interval notation.

1. $y + 3 > 2$ and $y + 3 \le 5$

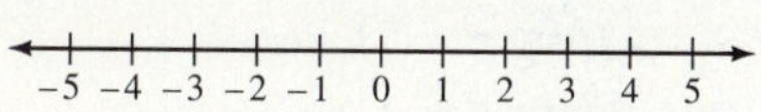

2. $x + 2 < 0$ or $x > 0$

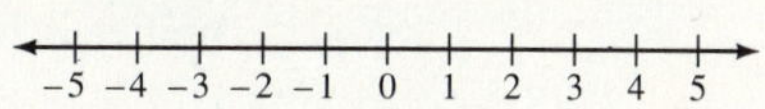

3. $z + 1 \ge 0$ and $z - 1 \le 0$

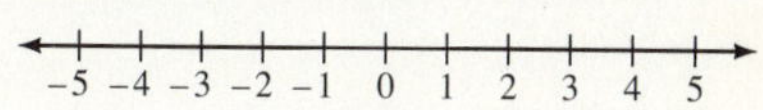

4. $y - 1 < 0$ or $y - 3 \ge 0$

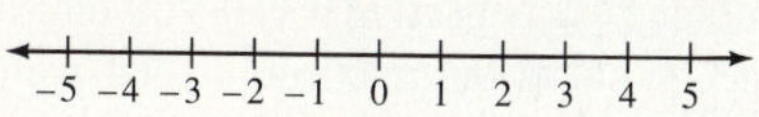

5. $-4 \le 3x + 2 < 3$

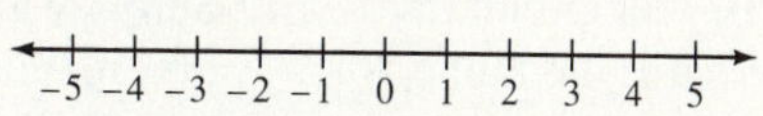

6. $2z + 1 \le -5$ or $2z + 1 > 5$

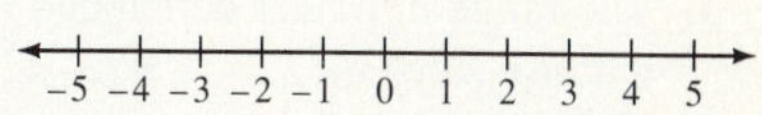

7. $0 \ge 2 - y$ or $2 - y > 1$

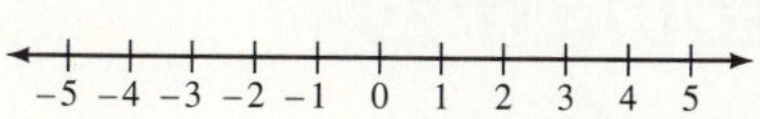

8. $-2 \le -y < 3$

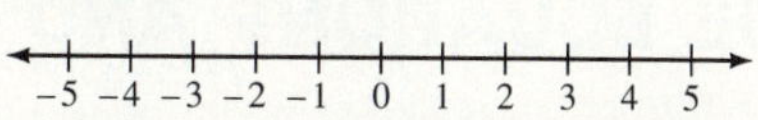

9. $4 > 1 - y \ge 0$

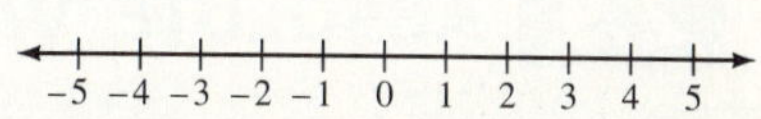

Letting x represent the unknown number, translate the following into symbols and solve.

10. A number is greater than or equal to 3 and less than 7.

11. A number is at most 0 and at least -1.

12. When three times a number is decreased by 5, the result is a number between 4 and 22.

13. Seven less a number is more than 5 or less than -5.

Solve.

14. Fahrenheit (F) and Celsius (C) temperatures are related by $C = \frac{5}{9}(F - 32)$. If the Celsius temperature is between -5°C and 30°C, inclusive, within what range will the Fahrenheit temperature fall?

15. A student must have an average percent between 70% and 80%, inclusive, to earn a grade of C. If Syd has 45%, 51%, and 62% on the first three tests this semester, what range of scores on the fourth test will give him a C?

FOR REVIEW

Solve.

16. $\frac{z-2}{-3} > 5$

17. $2(3 - y) + y < 8$

18. $5(z - 1) < -2(z - 15)$

19. The force in pounds F required to stretch a spring x inches beyond its normal length is given by $F = 6.2x$. What are the corresponding values for x if F is greater than 15.5 pounds?

Exercises 20–22 review material from Sections 1.2 and 1.6 to help you prepare for the next section. Evaluate the absolute values.

20. $|(-3) - (4)(-2)|$

21. $\frac{|(-6)(-2) - 12|}{|-5|}$

22. $\frac{|(-8)(4) - (-16)|}{|-20 + (-12)|}$

2.8 EXERCISES C

Solve.

1. The perimeter of a rectangular enclosure must be less than or equal to 100 yd. If one side must be 20 yd long, what is the maximum length for the other side?

2. Mr. James manufactures novelty items. The cost in dollars, c, of producing n items in a day is given by $c = 100 + 3n$. The number of dollars, s, generated by daily sales is given by $s = 5n$. In order to realize a profit, $s > c$. What is the minimum number of items that must be produced each day for Mr. James to realize a profit? [Answer: 51 items]

The formula $W = EI$ is used to describe the state of an electrical circuit, where W represents the power measured in watts, E represents the pressure in volts, and I represents the current in amperes.

3. The range in power demanded in a 110-volt circuit in a house satisfies $220 \le W \le 2200$. What is the range in current in this circuit?

4. Suppose a 20-ampere fuse is used in a 110-volt circuit in a mobile home. What is the maximum total wattage of all items on the circuit before "blowing" the fuse? [Answer: 2200 watts]

2.9 ABSOLUTE VALUE EQUATIONS AND INEQUALITIES

STUDENT GUIDEPOSTS

1. Absolute Value Equations
2. Absolute Value Inequalities

1 ABSOLUTE VALUE EQUATIONS

In Chapter 1 we defined absolute value as

$$|a| = \begin{cases} a & \text{if } a \ge 0 \\ -a & \text{if } a < 0. \end{cases}$$

To solve an absolute value equation, such as

$$|x| = 5,$$

we need to solve two equations, since

$$|x| = \begin{cases} x & \text{if } x \geq 0 \\ -x & \text{if } x < 0. \end{cases}$$

The two equations to solve are

$$x = 5 \quad \text{and} \quad -x = 5.$$

The second equation becomes $x = -5$, giving us the two equations

$$x = 5 \quad \text{and} \quad x = -5.$$

Therefore, $|x| = 5$ implies two solutions, $x = 5$ and $x = -5$.

EXAMPLE 1 Solving an Absolute Value Equation

Solve $|x| = 12$.

From the above discussion, $|x| = 12$ gives us the two solutions,

$$x = 12 \quad \text{and} \quad x = -12.$$

Check: $|12| = 12$ and $|-12| = 12$. Thus, the solutions 12 and -12 do check.

Practice Exercise 1

Solve. $|y| = 9$

Answer: 9, -9

An absolute value equation in which the expression inside the absolute value bars is more complex, is solved similarly using the following technique.

Absolute Value Equations

To solve an absolute value equation of the form

$$|Expression| = a$$

when $a \geq 0$, solve the two related equations

$$Expression = a \quad \text{and} \quad Expression = -a.$$

If $a < 0$, $|Expression| = a$ has no solution.

EXAMPLE 2 Solving an Absolute Value Equation

Solve $|2x + 1| = 3$.

We need to solve

$$\begin{aligned} 2x + 1 &= 3 \\ 2x + 1 - 1 &= 3 - 1 \\ 2x &= 2 \\ \frac{1}{2}(2x) &= \frac{1}{2}(2) \\ x &= 1 \end{aligned} \qquad \text{and} \qquad \begin{aligned} 2x + 1 &= -3. \\ 2x + 1 - 1 &= -3 - 1 \\ 2x &= -4 \\ \frac{1}{2}(2x) &= \frac{1}{2}(-4) \\ x &= -2 \end{aligned}$$

The solutions are 1 and -2. Check by substituting each solution into the original equation.

Practice Exercise 2

Solve. $|4x - 2| = 6$

Answer: -1, 2

EXAMPLE 3 Solving an Absolute Value Equation

Solve $|1 - 3x| = 0$.

We must solve $1 - 3x = 0$ and $1 - 3x = -0$. But since $-0 = 0$, we really have only one equation to solve.

$$1 - 3x = 0$$
$$-3x = -1 \quad \text{Subtract 1 from both sides}$$
$$x = \frac{1}{3} \quad \text{Divide both sides by } -3$$

The solution is $\frac{1}{3}$. Check in the original equation.

Practice Exercise 3

Solve. $|4 - 5y| = 0$

Answer: $\frac{4}{5}$

CAUTION

$|5x - 1| = -2$, has no solution. Since the absolute value of a number is always ≥ 0, it cannot equal -2.

Other types of absolute value equations are solved in a similar way, as in the next example.

EXAMPLE 4 Solving an Absolute Value Equation

Solve $|3x + 1| = |2 - x|$.

For this equation to be true, $3x + 1$ must equal $2 - x$ or its negative. Thus, we need to solve

$$3x + 1 = 2 - x \quad \text{and} \quad 3x + 1 = -(2 - x).$$
$$4x = 1 \qquad\qquad 3x + 1 = -2 + x$$
$$x = \frac{1}{4} \qquad\qquad 2x = -3$$
$$x = -\frac{3}{2}$$

The two solutions are $\frac{1}{4}$ and $-\frac{3}{2}$. Check these in the original equation.

Practice Exercise 4

Solve. $|2x - 3| = |4 - 3x|$

Answer: 1, $\frac{7}{5}$

2 ABSOLUTE VALUE INEQUALITIES

Solutions to absolute value equations are usually found by solving two related equations. Similarly, solutions to **absolute value inequalities** are solutions to two related inequalities. Consider the inequality

$$|x| < 2.$$

Geometrically, since $|x|$ represents the distance that x is located from the origin, $|x| < 2$ means that x cannot be further from zero than 2 units in either direction; that is, it must be located between -2 and 2. This is the compound statement

$$-2 < x < 2 \quad \text{Written as a chain of inequalities}$$

$$\text{or} \quad x > -2 \quad \textit{and} \quad x < 2. \quad \text{Written using the connective } \textit{and}$$

Thus, to solve $|x| < 2$, we simply give the solution using the related compound statement $-2 < x < 2$ ($x > -2$ *and* $x < 2$).

To solve the inequality

$$|x| > 2$$

we are interested in those numbers x whose distance from zero is more than 2 units in either direction; that is, numbers that are less than -2 or greater than 2. This is the compound statement

$$x < -2 \quad or \quad x > 2.$$

Thus, the solution to $|x| > 2$ is given using the compound statement $x < -2$ *or* $x > 2$.

Absolute value inequalities with more complex expressions within the absolute value bars are solved in a similar way.

Absolute Value Inequalities

1. To solve an absolute value inequality of the form $|Expression| > a$ where $a > 0$, solve the compound statement

   ***Expression* < −*a* or *Expression* > *a*.**

2. To solve an absolute value inequality of the form $|Expression| < a$ where $a > 0$, solve either the compound statement

   Expression* < *a* and *Expression* > −*a

   or the chain of inequalities

   −*a* < *Expression* < *a*.

3. Use similar procedures for $|Expression| \geq a$ and $|Expression| \leq a$.

Since $|Expression|$ represents the distance that *Expression* is located from the origin, $|Expression| < a$ means that *Expression* cannot be farther from zero than a units in either direction; that is, it must be located between $-a$ and a. Similarly, $|Expression| > a$ means that *Expression* must be more than a units from zero in either direction; that is, it must be less than $-a$ or greater than a. These relationships are shown in Figure 2.30.

Figure 2.30 Absolute Value Inequalities

EXAMPLE 5 Solving an Absolute Value Inequality

Solve $|x| > 3$.

The solution is

$$x < -3 \quad or \quad x > 3, \qquad \text{Expression is } x \text{ in this case}$$

or, in interval notation, $(-\infty, -3)$ or $(3, \infty)$.

PRACTICE EXERCISE 5

Solve. $|y| \geq 10$

Answer: $y \leq -10$ or $y \geq 10$; $(-\infty, -10]$ or $[10, \infty)$

EXAMPLE 6 Solving an Absolute Value Inequality

Solve $|x| \le 5$.

The solution is $x \le 5$ *and* $x \ge -5$, which is equivalent to

$$-5 \le x \le 5, \quad \text{Expression is } x$$

or, in interval notation, $[-5, 5]$.

Practice Exercise 6

Solve. $|y| < 9$

Answer: $-9 < y < 9$; $(-9, 9)$

EXAMPLE 7 Solving an Absolute Value Inequality

Solve $|x - 2| > 7$. Graph the solution.

We need to solve

$$x - 2 < -7 \quad \textit{or} \quad x - 2 > 7. \qquad \text{Expression is } x - 2$$

$$x - 2 + 2 < -7 + 2 \quad \text{or} \quad x - 2 + 2 > 7 + 2$$

$$x < -5 \quad \text{or} \quad x > 9$$

The solution is $x < -5$ or $x > 9$. (*Do not write* $9 < x < -5$.) In interval notation, the solution is $(-\infty, -5)$ or $(9, \infty)$, which is graphed in Figure 2.31.

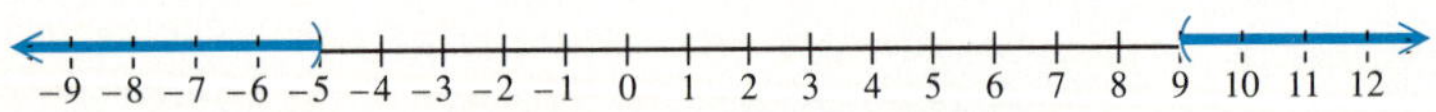

Figure 2.31 $(-\infty, -5)$ or $(9, \infty)$

Practice Exercise 7

Solve and graph. $|y + 8| \ge 2$

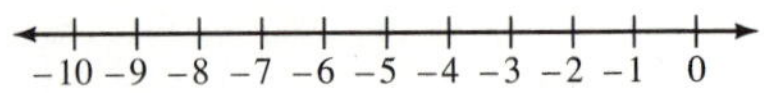

Answer: Graph $y \le -10$ or $y \ge -6$; $(-\infty, -10]$ or $[-6, \infty)$.

EXAMPLE 8 Solving Absolute Value Inequalities

(a) Solve $|x + 3| \le 4$. Graph the solution.

We need to solve $x + 3 \le 4$ *and* $x + 3 \ge -4$, which is equivalent to

$$-4 \le x + 3 \le 4. \qquad \text{Expression is } x + 3$$

$$-4 - 3 \le x + 3 - 3 \le 4 - 3 \qquad \text{Subtract 3 throughout}$$

$$-7 \le x \le 1$$

The solution is $-7 \le x \le 1$. (Equivalently, $x \le 1$ and $x \ge -7$.) In interval notation, the solution is $[-7, 1]$, which is graphed in Figure 2.32.

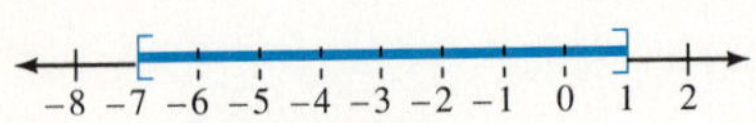

Figure 2.32 $[-7, 1]$

(b) Solve $|1 - 3x| \ge 7$. Graph the solution.

We need to solve

$$1 - 3x \le -7 \quad \textit{or} \quad 1 - 3x \ge 7. \qquad \text{Expression is } 1 - 3x$$

$$-3x \le -8 \quad \text{or} \quad -3x \ge 6 \qquad \text{Subtract 1}$$

$$x \ge \frac{8}{3} \quad \text{or} \quad x \le -2 \qquad \text{Reverse inequalities}$$

Practice Exercise 8

Solve and graph.

(a) $|y - 5| < 5$

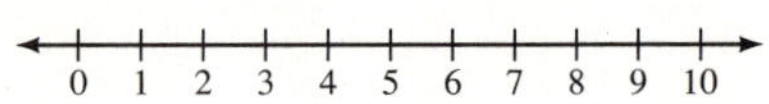

(b) $|2 - 4y| \ge 6$

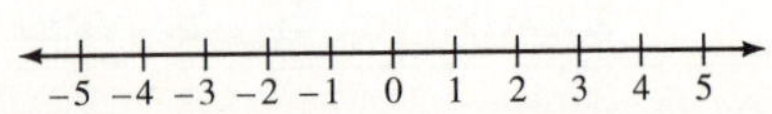

The solution is $x \ge \frac{8}{3}$ or $x \le -2$. (*Do not use* $\frac{8}{3} \le x \le -2$ to describe the solution.) In interval notation, the solution is $(-\infty, -2]$ or $\left[\frac{8}{3}, \infty\right)$, which is graphed in Figure 2.33.

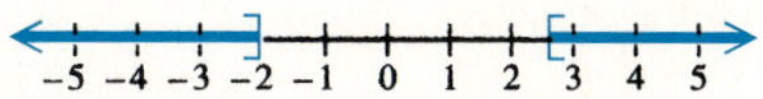

Figure 2.33 $(-\infty, -2]$ or $\left[\frac{8}{3}, \infty\right)$

Answers: (a) Graph $0 < y < 10$; $(0, 10)$. (b) Graph $y \le -1$ or $y \ge 2$; $(-\infty, -1]$ or $[2, \infty)$.

CAUTION

To solve an absolute value inequality, *memorize the rules in the box on page 243,* since the very first step is translating to a compound statement. It is not enough to solve $2x + 1 < 5$ when $|2x + 1| < 5$ is given. This is a common mistake.

Can we solve

$$|1 - x| < -2 \quad ?$$

Notice that -2 is negative. This inequality has no solutions since $|Expression|$ is always nonnegative. Similarly, to solve

$$|1 - x| \le 0$$

we would solve $|1 - x| = 0$ or $1 - x = 0$ which has 1 as the only solution. Finally, to solve

$$|1 - x| \ge -2$$

we would observe that since $|1 - x|$ is always greater than or equal to 0, hence greater than or equal to -2, every real number is a solution.

2.9 EXERCISES A

Solve.

1. $|x| = 7$

2. $|z| = 0$

3. $|y| = -7$

4. $|x + 1| = 4$

5. $|z + 1| + 8 = 4$

6. $|y + 1| = 0$

7. $|2x - 5| = 9$

8. $|5 - 2z| - 1 = 8$

9. $|2y - 5| = 0$

10. $|5 - 2y| = 0$

11. $\left|\frac{1}{2}z - \frac{3}{4}\right| = \frac{1}{4}$

12. $|2.2y - 1.1| = 5.5$

13. $|x + 2| = |3x - 4|$

14. $\left|\frac{1}{2}z + 1\right| = \left|\frac{3}{4}z - 1\right|$

15. $|y - 1| = |7 + y|$

Letting x represent the unknown number, translate the following into symbols and solve.

16. The absolute value of a number is 15.

17. If 2 is added to a number, the absolute value of the result is 5.

18. The absolute value of 3 more than one-half of a number is 8.

19. The absolute value of the sum of a number and 1 is equal to the absolute value of the number itself.

Solve. Give answers using interval notation.

20. $|x| < 9$

21. $|x| \geq 10$

22. $|x| < -1$

23. $|x| < 0$

24. $|x| \leq 0$

25. $|x + 1| < 3$

26. $|x - 1| < 3$

27. $|x + 4| > 3$

28. $|x + 4| \geq 3$

29. $|x - 5| \leq 7$

30. $|2x + 1| < 9$

31. $|2x + 1| \geq 9$

32. $|1 - 2x| > 7$

33. $|1 - 2x| \leq 7$

34. $\left|\dfrac{x + 1}{2}\right| < 3$

35. $\left|\dfrac{x + 1}{2}\right| > 3$

36. $\left|\dfrac{x - 3}{-4}\right| \leq 5$

37. $\left|\dfrac{x - 3}{-4}\right| \geq 5$

38. $|2x + 5| < -1$

39. $|2x + 5| \geq 0$

Letting x represent the unknown number, translate the following into symbols and solve.

40. The absolute value of a number is less than 20.

41. If 3 is added to twice a number, the absolute value of the result is at least 7.

Graph the following absolute value inequalities.

42. $|x - 1| < 2$

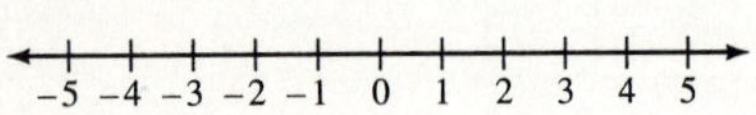

43. $|x - 1| \geq 2$

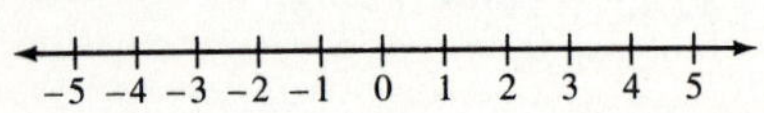

44. $|2x + 3| \geq 3$

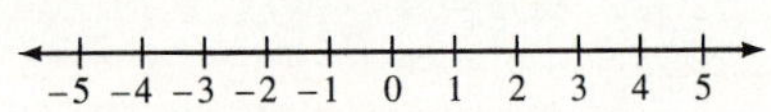

45. $|2x + 3| < 3$

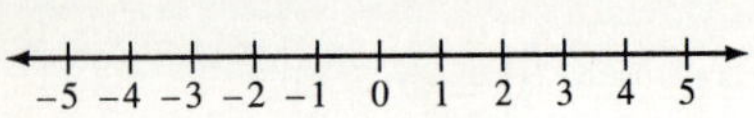

46. $|2x + 3| \leq 0$

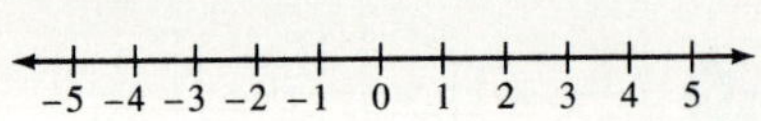

47. $|2x + 3| < 0$

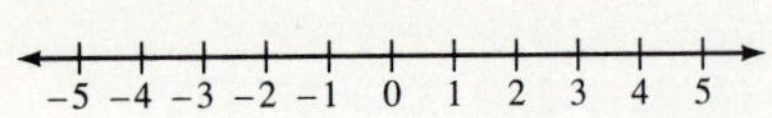

FOR REVIEW

Graph the following on a number line.

48. $(2x + 3) - (4x - 1) \le 0$

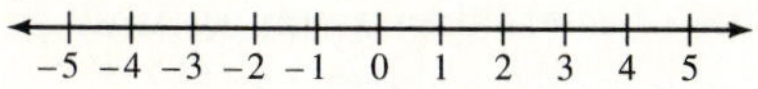

49. $4(2x - 3) > 7x - 10$

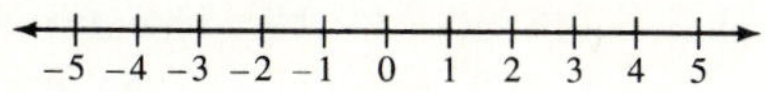

50. $0 > 2 - x$ *or* $2 - x \ge 2$

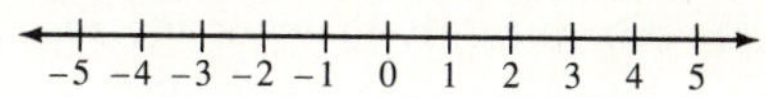

51. $-1 \le x \le 3$

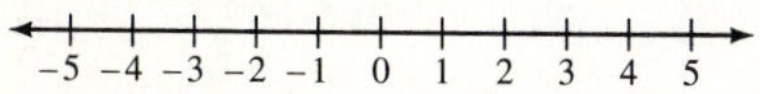

52. $x > \frac{1}{2}$ *or* $2x < 1$

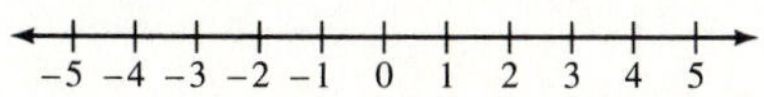

53. $-3 < x + 1 < 3$

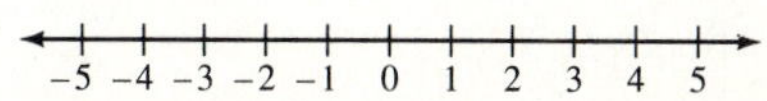

Solve.

54. Three times a number, decreased by 5, is at least -17 and no more than 1. What numbers satisfy this statement?

55. Five less a number is at least 4 or at most -1. What numbers satisfy this statement?

ANSWERS: **1.** 7, −7 **2.** 0 **3.** no solution **4.** 3, −5 **5.** no solution **6.** −1 **7.** 7, −2 **8.** 7, −2 **9.** $\frac{5}{2}$ **10.** $\frac{5}{2}$ **11.** 2, 1 **12.** 3, −2 **13.** 3, $\frac{1}{2}$ **14.** 8, 0 **15.** −3 **16.** $|x| = 15$; 15, −15 **17.** $|x + 2| = 5$; 3, −7 **18.** $|3 + \frac{1}{2}x| = 8$; 10, −22 **19.** $|x + 1| = |x|$; $-\frac{1}{2}$ **20.** $-9 < x < 9$ ($x > -9$ *and* $x < 9$); $(-9, 9)$ **21.** $x \le -10$ or $x \ge 10$ (do not write $10 \le x \le -10$); $(-\infty, -10]$ or $[10, \infty)$ **22.** no solution **23.** no solution **24.** 0 **25.** $-4 < x < 2$; $(-4, 2)$ **26.** $-2 < x < 4$; $(-2, 4)$ **27.** $x < -7$ or $x > -1$; $(-\infty, -7)$ or $(-1, \infty)$ **28.** $x \le -7$ or $x \ge -1$; $(-\infty, -7]$ or $[-1, \infty)$ **29.** $-2 \le x \le 12$; $[-2, 12]$ **30.** $-5 < x < 4$; $(-5, 4)$ **31.** $x \le -5$ or $x \ge 4$; $(-\infty, -5]$ or $[4, \infty)$ **32.** $x > 4$ or $x < -3$; $(-\infty, -3)$ or $(4, \infty)$ **33.** $-3 \le x \le 4$; $[-3, 4]$ **34.** $-7 < x < 5$; $(-7, 5)$ **35.** $x < -7$ or $x > 5$; $(-\infty, -7)$ or $(5, \infty)$ **36.** $-17 \le x \le 23$; $[-17, 23]$ **37.** $x \le -17$ or $x \ge 23$; $(-\infty, -17]$ or $[23, \infty)$ **38.** no solution **39.** Every real number is a solution. **40.** $|x| < 20$; $-20 < x < 20$ **41.** $|2x + 3| \ge 7$; $x \le -5$ or $x \ge 2$

42. 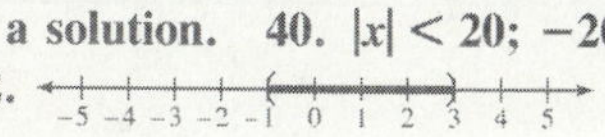 **43.** 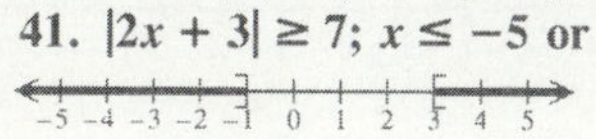 **44.** 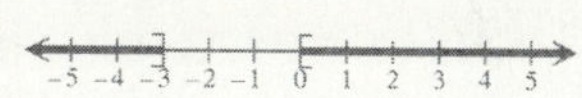

45. 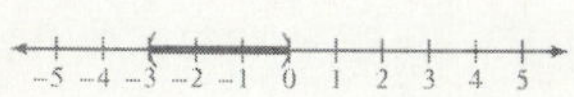 **46.** 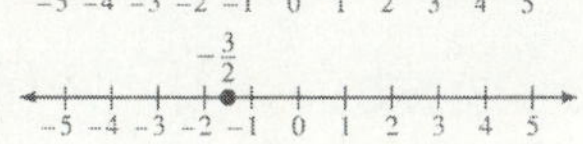 **47.** 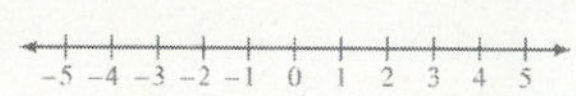

48. 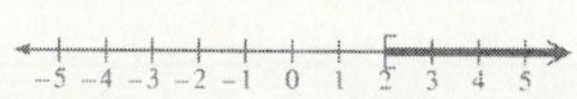 **49.** 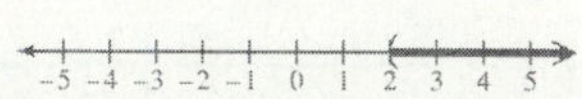 **50.** 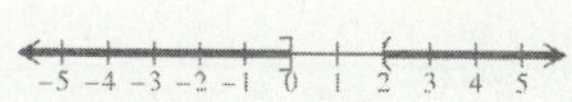

51. 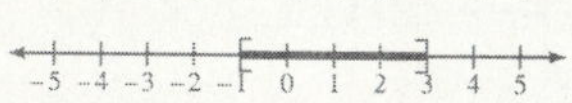 **52.** 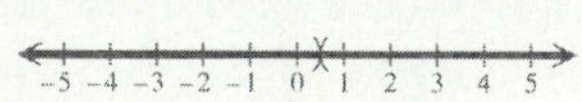 **53.** 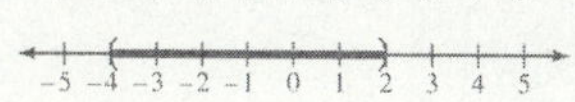

54. $-4 \le x \le 2$; $[-4, 2]$ **55.** $x \le 1$ *or* $x \ge 6$; $(-\infty, 1]$ or $[6, \infty)$

2.9 EXERCISES B

Solve.

1. $|y| = 13$

2. $|x| = 0$

3. $|z| = -2$

4. $|y + 5| = 3$

5. $|3x + 4| + 3 = 1$

6. $|2z + 5| = 0$

7. $|3y - 1| = 14$

8. $|4 - 3x| - 2 = 3$

9. $|7z - 3| = 0$

10. $|8 - 3y| = 0$

11. $\left|\frac{1}{5}x - \frac{3}{10}\right| = \frac{1}{2}$

12. $|1.1z - 4.4| = 7.7$

13. $|3y - 1| = |2y + 5|$

14. $\left|\frac{7}{8}x + \frac{1}{4}\right| = \left|\frac{1}{2}x - \frac{5}{8}\right|$

15. $|2z + 5| = |3 + 2z|$

Letting x represent the unknown number, translate the following into symbols and solve.

16. The absolute value of a number is 0.

17. If 5 is subtracted from a number, the absolute value of the result is 13.

18. The result is 17 when the absolute value of the difference between a number and 4 is obtained.

19. The absolute value of 1 less than a number is equal to the absolute value of the number itself.

Solve. Give answers using interval notation.

20. $|y| < 7$

21. $|x| \geq 3$

22. $|x| < -8$

23. $|z| \geq 0$

24. $|y| > 0$

25. $|x + 2| < 11$

26. $|z - 3| < 7$

27. $|y - 3| > 17$

28. $|x - 3| \leq 7$

29. $|3z + 2| < 11$

30. $|3y + 2| \geq 11$

31. $|2 - x| \leq 3$

32. $|2 - z| > 3$

33. $\left|\frac{y + 5}{3}\right| \leq 5$

34. $\left|\frac{x + 5}{3}\right| \geq 5$

35. $\left|\frac{x - 6}{-2}\right| < 3$

36. $\left|\frac{x - 6}{-2}\right| > 3$

37. $|3z + 2| < -1$

38. $|3z + 2| \leq 0$

39. $|3z + 2| \geq 0$

Letting x represent the unknown number, translate the following into symbols and solve.

40. The absolute value of a number is more than 20.

41. Add -5 to twice a number and take the absolute value of the result. This quantity is no more than 17.

Graph the following absolute value inequalities.

42. $|x - 2| < 3$

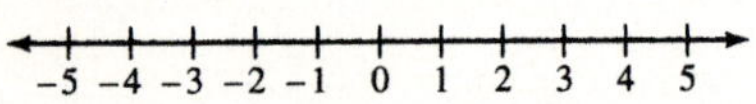

43. $|x - 2| \geq 3$

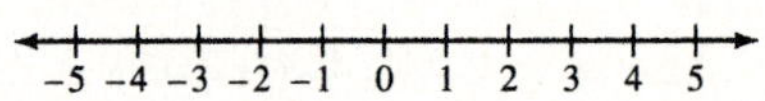

44. $|4x - 8| \geq 8$

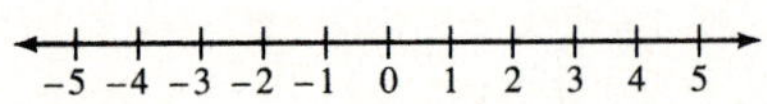

45. $|4x - 8| < 8$

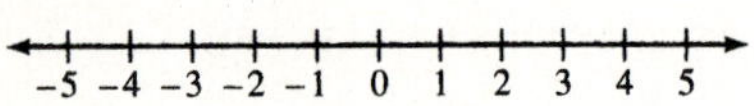

46. $|4x - 8| \geq 0$

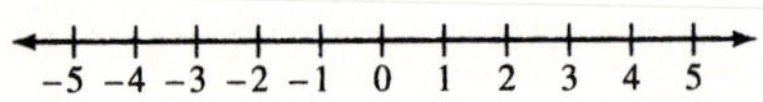

47. $|4x - 8| < 0$

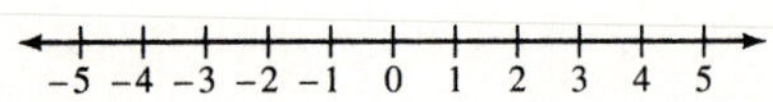

FOR REVIEW

Graph the following on a number line.

48. $(x + 5) - (2x - 1) \geq 0$

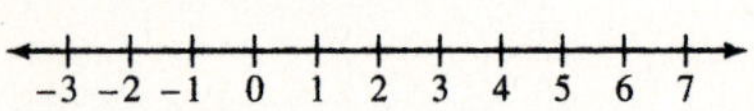

49. $3(2x - 1) < 5x - 2$

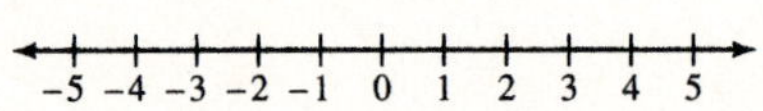

50. $0 > 3 - x$ *or* $3 - x \geq 1$

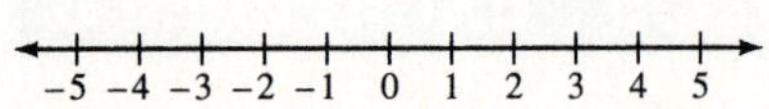

51. $-2 \leq x \leq 1$

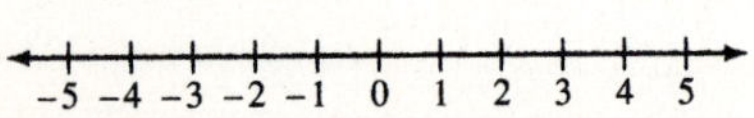

52. $x > \frac{1}{3}$ *or* $3x < 1$

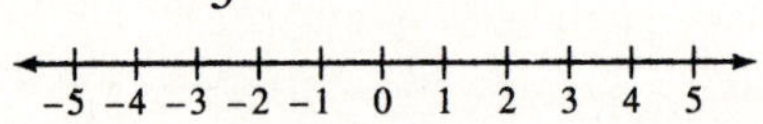

53. $-2 < x + 1 < 2$

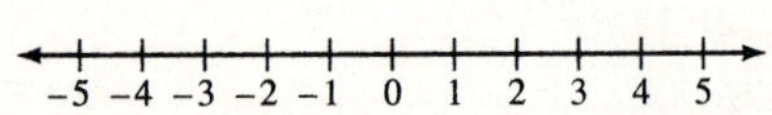

Solve.

54. Two less a number is at most 17 and at least 3. What numbers satisfy this statement?

55. Three less a number is at most -2 or at least 1. What numbers satisfy this statement?

2.9 EXERCISES C

Solve.

1. $|x+1| = x+1$ [*Hint:* When is the absolute value of a number equal to the number itself?]
2. $|x+1| = -x-1$
3. $|x-1| = 1-x$
4. $|x-1| = x-1$
5. $|x+1| > x$ [Answer: all real numbers]
6. $|x+1| < x$ [Answer: no solution]

CHAPTER 2 REVIEW

KEY WORDS

2.1 An **equation** is a statement that two quantities are equal.

A **solution** to an equation is a number that makes the equation true.

A **linear equation** can be written in the form $ax + b = 0$.

A **conditional equation** is true for some replacements of the variable and false for others.

An **identity** has the entire set of real numbers for its solution set.

A **contradiction** has no solutions.

Equivalent equations have exactly the same solutions.

2.4 **Consecutive integers** are integers that are next to each other in the regular counting order.

2.6 A **literal equation** is an equation that involves two or more variables.

2.7 A **linear inequality** can always be written in one of the forms:

$ax + b < 0$, $ax + b > 0$, $ax + b \le 0$, or $ax + b \ge 0$.

$(a, b) = \{x|a < x < b\}$ is an **open interval.**

$[a, b] = \{x|a \le x \le b\}$ is a **closed interval.**

$(a, b] = \{x|a < x \le b\}$ and $[a, b) = \{x|a \le x < b\}$ are **half-open intervals.**

$(-\infty, a) = \{x|x < a\}$, $(-\infty, a] = \{x|x \le a\}$, $(a, \infty) = \{x|x > a\}$, $[a, \infty) = \{x|x \ge a\}$, and $(-\infty, \infty) = \{x|x$ is any real number$\}$ are **infinite intervals.**

A **solution** to an inequality is any number that, when substituted for the variable, makes the inequality true.

Two inequalities are **equivalent inequalities** if they have exactly the same solutions.

2.8 A **compound statement** is an inequality formed by joining two inequalities with the words *and* or *or*.

When an *and* statement such as $a < x$ *and* $x < b$ is written as $a < x < b$, this second form is called a **chain of inequalities.**

KEY CONCEPTS

2.1 When using the addition-subtraction rule to solve an equation, add or subtract the same expression on both sides.

2.2 When using the multiplication-division rule to solve an equation, multiply by a number that makes the coefficient of x equal to 1. For example, to solve

$$\frac{x}{\frac{4}{5}} = 20 \qquad \text{multiply by } \tfrac{4}{5}, \textit{ not } \text{by } \tfrac{5}{4}.$$

2.3
1. Use the addition-subtraction rule before the multiplication-division rule when a combination is necessary to solve an equation.
2. When solving an equation involving parentheses, first use the distributive law to clear all parentheses.

2.4 When solving an applied problem, read the problem several times and describe the variable completely. Also, make a sketch showing the information given, if appropriate.

2.5 1. Sales tax = (tax rate) · (selling price) and total cost = (selling price) + (sales tax)

2. Discount = (discount rate) · (regular price) and sale price = (regular price) − (discount).

3. Commission = (commission rate) · (total sales)

4. The formula for simple interest, I, is $I = Prt$, where P is the principal, r is the yearly interest rate, and t is the time in years. The amount in the account, A, after t years is $A = P + I = P + Prt$.

5. Many distance-rate problems use the formula $d = rt$ [distance = rate · time].

2.6 It may help to solve a similar numerical equation before solving a given literal equation.

2.7 1. If the same number or quantity is added to both sides of an inequality, the result is an equivalent inequality.

2. When both sides of an inequality are multiplied (or divided) by a negative number, be sure to *reverse* the symbol of inequality to obtain an equivalent inequality.

2.8 Inequalities connected by the word *and* can generally be written as a single chain of inequalities; those connected by the word *or* cannot be written as a chain.

2.9 1. If $a > 0$ and $|expression| = a$, then solve

$$expression = a \quad \text{and}$$
$$expression = -a.$$

2. If $a > 0$ and $|expression| > a$, then solve

$$expression < -a \quad \text{or}$$
$$expression > a.$$

3. If $a > 0$ and $|expression| < a$, then solve

$$-a < expression < a.$$

4. An inequality of the form $|expression| > b$, where $b < 0$, has every number as a solution, while $|expression| < b$, where $b < 0$, has no solutions.

5. To solve $|expression| \le 0$, solve the equation

$$expression = 0.$$

6. $|expression| \ge 0$ has every real number as a solution, $|expression| < 0$ has no solution, and $|expression| > 0$ has every real number as a solution except those for which $expression = 0$.

REVIEW EXERCISES

Part I

2.1, 2.2, 2.3 *Solve.*

1. $4 - 3x = -5$
2. $\frac{1}{3}y = -2$
3. $3 - 2(x - 1) = x - 10$
4. $4(2y - 4) + 6 = 6y - 2(y - 1)$
5. $(2z + 1) - (3z - 7) = 0$
6. $\frac{x}{2} - \frac{x}{7} = 1$
7. $y(y + 2) = y^2 - 6$
8. $2[x - (1 - 3x)] = 3(x + 1)$

2.4 *Select a variable to represent each quantity and translate the phrase into symbols.*

9. the time 4 hours ago
10. twice the surface area

Solve.

11. If three times a number is increased by 2, the result is the same as 20 less than five times the number. Find the number.

12. The perimeter of a rectangle is 20.8 cm. If the length is 3.5 cm more than twice the width, find the dimensions.

2.5 **13.** 60.5 is 11% of what?

14. 7% of 320 is what?

15. After a 7% raise, Andy Woodring's new salary is $14,445. What was his former salary?

16. What amount of money invested at 15% simple interest will increase to $1140 at the end of six years?

17. A fuel tank contained 620 gal of fuel. More fuel was added, making the total 775 gal. What was the percent increase?

18. A man leaves Flagstaff at 8:00 A.M., traveling by car to Lake Powell, a distance of 135 miles, at a rate of 45 mph. He then drives a boat to Rainbow Bridge, a distance of 50 miles, at a rate of 20 mph. At what time does he reach the bridge?

2.6 **19.** Solve $M = N + Nat$ for N.

20. Solve $b(x + 2) = x + a$ for x.

2.7 *Solve and graph on a number line.*

21. $x - 9 \leq 2(x - 6)$

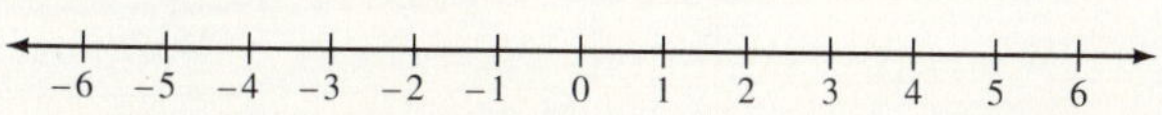

22. $-3y + 1 + 2y < 5(y - 3)$

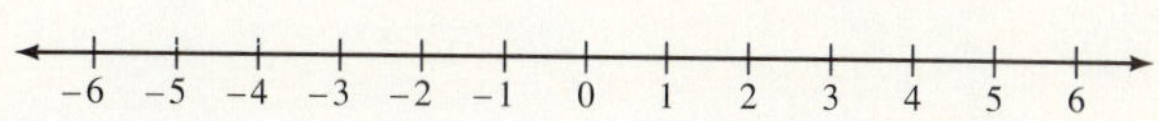

2.8 *Solve and graph on a number line. Give the answer in interval notation.*

23. $y + 2 \leq -1$ or $y + 2 \geq 1$

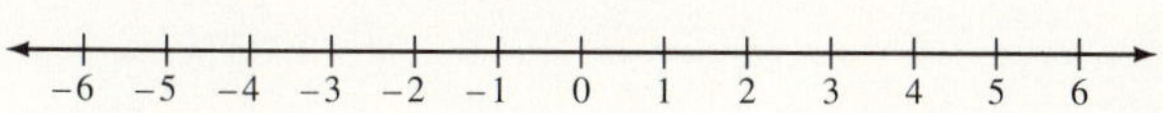

24. $-3 \leq 2x + 1 < 7$

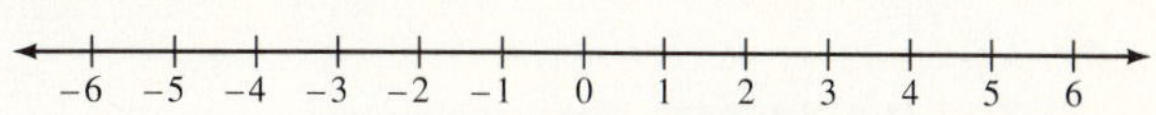

25. A student must have an average between 80% and 90%, inclusive, to receive a grade of B. Ted has 78%, 64%, and 92% on the first three tests. What range of scores on the fourth test would give him a B?

2.9 *Solve.*

26. $|1 - 7x| = 6$

27. $|1 - 7x| = 0$

28. $|x - 3| = |2 + x|$

Solve. Give answer in interval notation.

29. $|1 - 3x| < 4$ **30.** $|3y + 4| \geq 28$ **31.** $|2z + 11| < -3$

Graph each inequality on a number line.

32. $|x - 2| \leq 2$

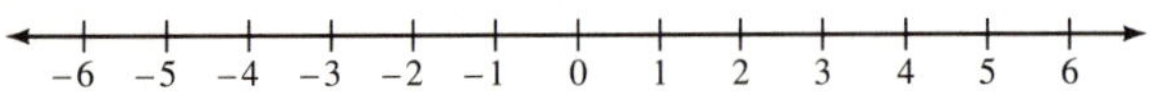

33. $|2 - y| > 1$

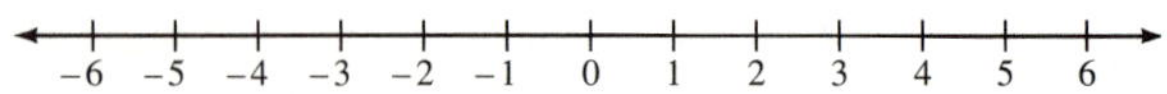

Part II

Solve. When appropriate, give answer in interval notation.

34. Two trains leave the same town, one traveling east and the other west. If one train is moving 10 mph faster than the other, and if after 2 hr they are 260 mi apart, how fast is each traveling?

35. The population of Appleville, Washington, was 740 in 1990. This was a 20% decrease in the population of 1980. What was the population in 1980?

36. $|2x - 3| \leq 7$

37. $|1 - 7x| = -3$

38. $\dfrac{z+1}{3} - \dfrac{z-1}{2} = 1$

39. $3(x - 4) - (x + 7) = 5$

40. A basketball player made 12 shots in 20 attempts. What was her shooting percentage?

41. After receiving a 30% discount on the selling price, Dan Busch paid $35.28, including 5% sales tax, for a running suit. What was the price of the suit before the discount?

42. Peter is 9 years older than Rhoda. If the sum of their ages is 87, how old is each?

43. A steel rod is 17 m long. It is to be cut into two pieces in such a way that one piece is 7 m longer than the other. How long is each piece?

44. $|2x - 5| \leq 15$

45. $|1 - x| > 3$

46. Solve $a = \frac{1}{2}uv + t$ for u.

47. Find the measures of the angles of a triangle if the second is three times the first and the third is 30° more than twice the first.

Solve and graph on a number line. Give answer in interval notation.

48. $x - 1 < -5$ or $x - 1 > 2$

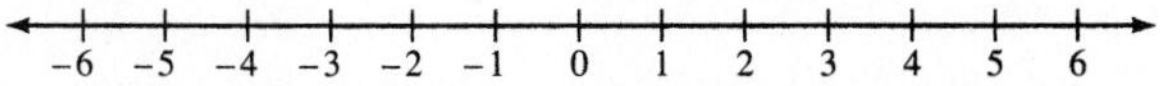

49. $-1 \le 1 - 2x \le 3$

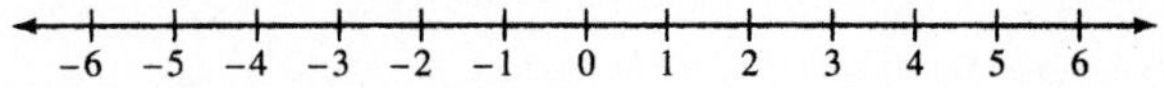

50. $|3x + 2| \le 5$

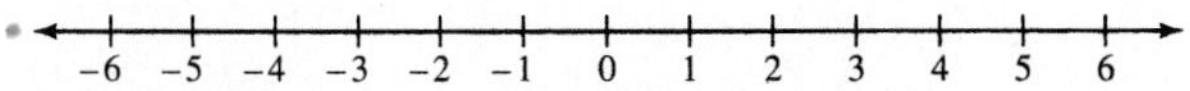

51. $|4 - x| > 1$

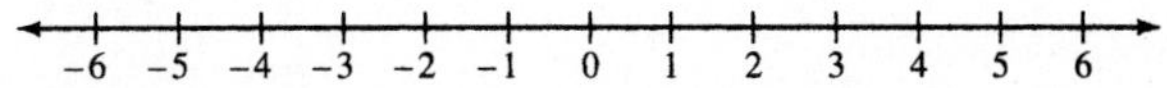

52. Solve. $B = 40a + w$ for a.

53. Solve. $|2y - 3| = |6 - y|$

54. Solve. $3z - (1 - 2z) = 4 - (z - 1)$

55. Solve. $|4x - 5| = 0$

Select a variable to represent each quantity and translate the phrase into symbols.

56. His wages plus 10% of his wages

57. \$1000 less the cost

58. The power in watts W on an electrical circuit is related to the current I in amperes by $W = 110I$. If the power demand on an appliance is between 220 watts and 550 watts, what is the range of current in the circuit?

ANSWERS: **1.** 3 **2.** −6 **3.** 5 **4.** 3 **5.** 8 **6.** $\frac{14}{5}$ **7.** −3 **8.** 1 **9.** $t - 4$ **10.** $2S$ **11.** 11 **12.** 8.1 cm by 2.3 cm **13.** 550 **14.** 22.4 **15.** \$13,500 **16.** \$600 **17.** 25% **18.** 1:30 P.M. **19.** $N = \frac{M}{1 + at}$ **20.** $x = \frac{a - 2b}{b - 1}$
21. $x \ge 3$

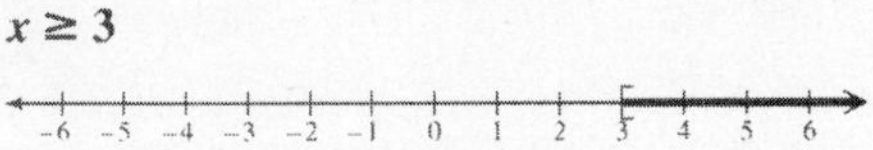

22. $y > \frac{8}{3}$

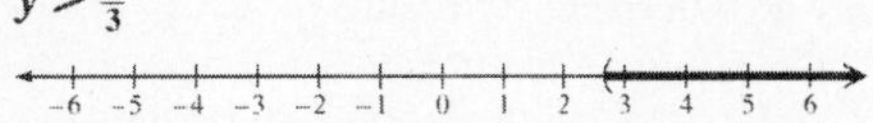

23. $(-\infty, -3]$ or $[-1, \infty)$

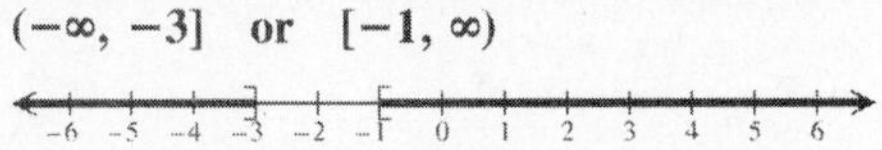

24. $[-2, 3)$

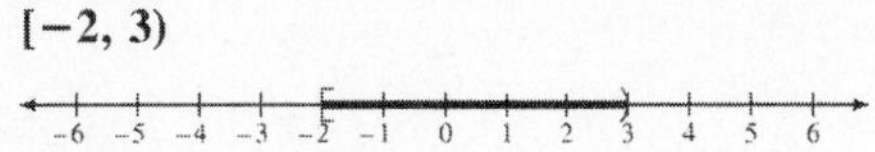

25. His score must be between 86% and 100% (the maximum score possible), inclusive. **26.** $-\frac{5}{7}, 1$ **27.** $\frac{1}{7}$ **28.** $\frac{1}{2}$ **29.** $\left(-1, \frac{5}{3}\right)$ **30.** $\left(-\infty, -\frac{32}{3}\right]$ or $[8, \infty)$ **31.** no solution **32.**

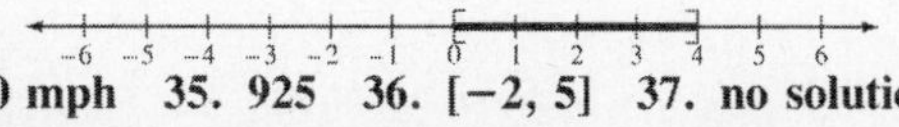

33.

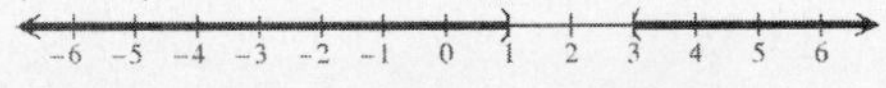

34. 60 mph, 70 mph **35.** 925 **36.** $[-2, 5]$ **37.** no solution **38.** −1 **39.** 12 **40.** 60% **41.** \$48.00 **42.** Peter is 48, Rhoda is 39. **43.** 5 m, 12 m **44.** $[-5, 10]$ **45.** $(-\infty, -2)$ or $(4, \infty)$ **46.** $u = \frac{2a - 2t}{v}$ **47.** 25°, 75°, 80° **48.** $(-\infty, -4)$ or $(3, \infty)$

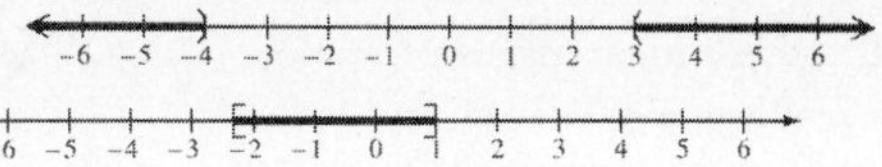

49. $[-1, 1]$

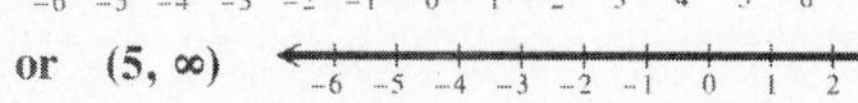

50. $\left[-\frac{7}{3}, 1\right]$

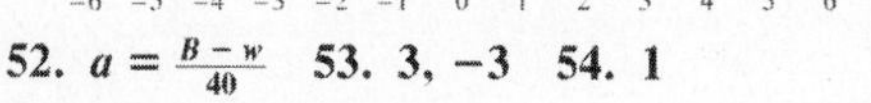

51. $(-\infty, 3)$ or $(5, \infty)$

52. $a = \frac{B - w}{40}$ **53.** 3, −3 **54.** 1 **55.** $\frac{5}{4}$ **56.** $w + 0.10w$ **57.** $1000 - c$ **58.** between 2 amperes and 5 amperes

CHAPTER 2 TEST

Solve.

1. $6 - 5x = 31$ 1. ____________

2. $(4z + 1) - (3z - 5) = 0$ 2. ____________

3. $2(3y - 1) - 4 = 2y - (6 - y)$ 3. ____________

4. $a - (2a - 1) = \frac{2}{3}(5 - a)$ 4. ____________

5. $\frac{x}{6} - \frac{x}{5} = 2$ 5. ____________

6. A wire that is 33 inches long is cut into two pieces in such a way that one piece is 5 inches shorter than the other. How long is each piece? 6. ____________

7. Find the measures of the angles of a triangle if the second angle is three times the first, and the third is 15° more than seven times the first. 7. ____________

8. Two cars leave Denver at 8:00 A.M., one traveling north and the other traveling south. If one is going 55 mph and the other 60 mph, at what time will they be 575 mi apart? 8. ____________

9. 42 is 12% of what? 9. ____________

10. After receiving an 8% raise, Mary has a new salary of $16,848. What was her former salary? 10. ____________

CHAPTER 2 TEST

CONTINUED

11. The population of South Keys, Florida, was 828 in 1990 and 720 in 1980. What was the percent increase in population over this period of time?

11. ______________

12. Solve $W = 30a + c$ for a.

12. ______________

13. Solve $u(v - w) + 2 = vw$ for w.

13. ______________

14. Solve and graph. $x - 8 \geq 2(x - 5)$

14.

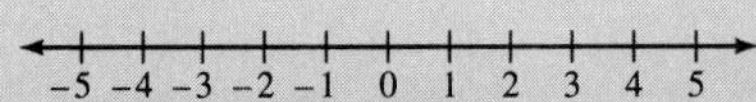

15. Solve and graph. $y - 1 < -3$ or $y - 1 > 3$

15.

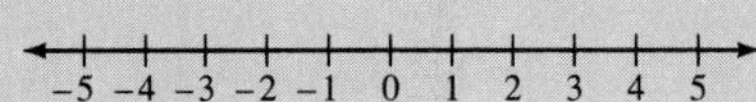

16. Solve and graph. $|z - 3| < 2$

16.

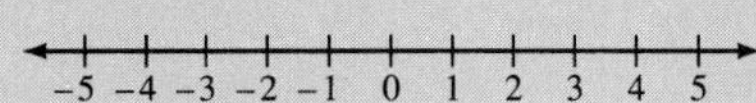

Solve.

17. $|4 - 3x| = 7$

17. ______________

18. $|5 - 2y| \leq 9$ (Give answer in interval notation.)

18. ______________

19. $|x - 5| = |2x + 1|$

19. ______________

20. One-third of a number x is greater than -2.

20. ______________

CHAPTER 3

Graphs, Relations, and Functions

3.1 GRAPHING EQUATIONS IN TWO VARIABLES

STUDENT GUIDEPOSTS

1. The Coordinate System
2. Plotting Points
3. Solutions to Equations in Two Variables
4. Graphing Equations

1 THE COORDINATE SYSTEM

Solutions to inequalities in one variable were graphed on number lines in Chapter 2. We will now be graphing equations and inequalities in two variables. When two number lines are placed together as in Figure 3.1, so that the two origins coincide and the lines are perpendicular, the result is a **Cartesian** or **rectangular coordinate system** (named after French mathematician René Descartes) or a **coordinate plane.**

The horizontal number line is the ***x*-axis,** the vertical number line is the ***y*-axis,** and the point of intersection of the axes is the **origin.**

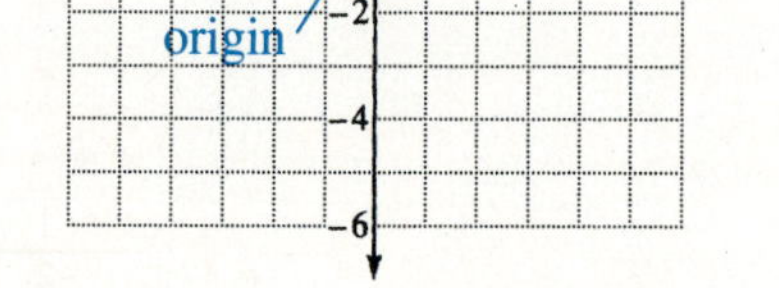

Figure 3.1 Rectangular or Cartesian Coordinate System

2 PLOTTING POINTS

Recall that there is one and only one point on a number line associated with each real number. A similar situation exists for points in a plane and **ordered pairs** of numbers. For example, the ordered pair (2, 3) can be identified with a point in a coordinate plane as follows:

The first number, 2, called the ***x*-coordinate** or **abscissa** of the point, is associated with a value on the horizontal or x-axis.

The second number, 3, called the ***y*-coordinate** or **ordinate** of the point, is associated with a value on the vertical or y-axis.

The pair (2, 3) is associated with the point where the vertical line through 2 on the x-axis and the horizontal line through 3 on the y-axis intersect. See Figure 3.2.

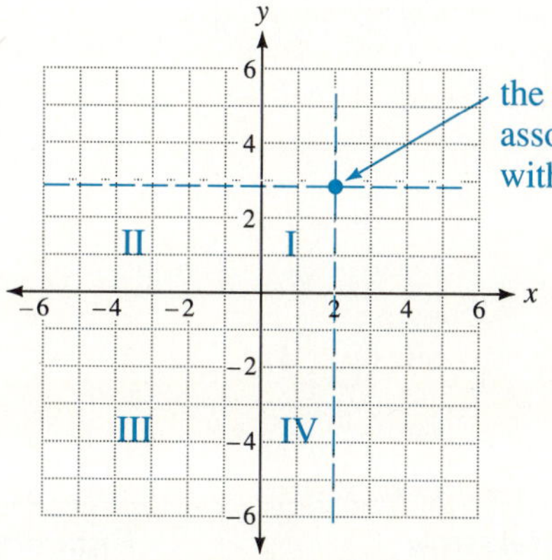

Figure 3.2 Plotting the Point (2, 3)

The axes in a coordinate system separate the plane into four sections called **quadrants.** The first, second, third, and fourth quadrants are identified by the Roman numerals I, II, III, and IV, respectively, in the coordinate plane in Figure 3.2. The signs of the x-coordinate (first) and y-coordinate (second) in the various quadrants are as follows.

I: (+, +) **II:** (−, +) **III:** (−, −) **IV:** (+, −)

We often use (x, y) to refer to an ordered pair of numbers. Thus, the point in the plane associated with the pair (x, y) has x as its x-coordinate and y as its y-coordinate. When we identify the point P in a plane corresponding to the ordered pair (x, y), sometimes called "the point (x, y)," we are plotting the point $P(x, y)$.

EXAMPLE 1 Plotting Points

The points associated with the ordered pairs (3, 2), (−1, 3), (−3, −2), and (2, −2) are plotted in the coordinate plane in Figure 3.3.

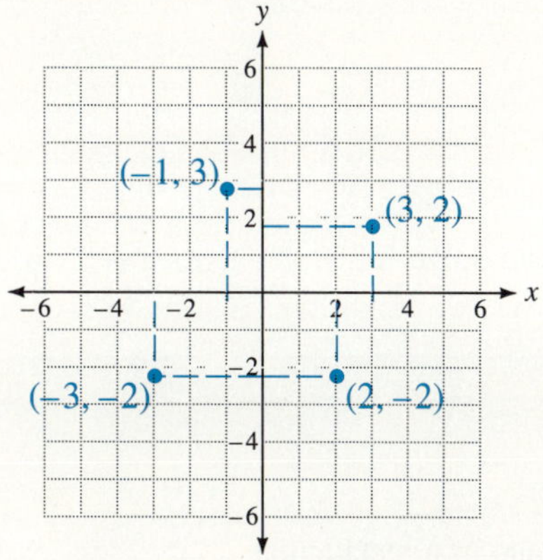

Figure 3.3 Plotting Points

Practice Exercise 1

Give the ordered pair associated with each point and the quadrant in which each is located.

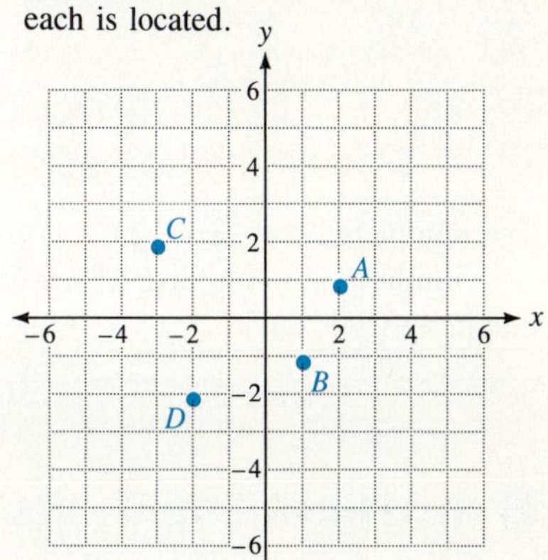

Answers: $A(2, 1)$, I; $B(1, -1)$, IV; $C(-3, 2)$, II; $D(-2, -2)$, III

3 SOLUTIONS TO EQUATIONS IN TWO VARIABLES

Graphing equations with two variables requires a rectangular coordinate system since **solutions** to these equations are ordered pairs of numbers that when substituted for the variables result in a true equation.

For example, the equation

$$y = 3x - 2$$

has infinitely many solutions. One solution is (2, 4), since if x is replaced with 2 and y with 4, the resulting equation is true as shown below.

$$y = 3x - 2$$
$$4 = 3(2) - 2 \qquad x = 2 \text{ and } y = 4$$
$$4 = 6 - 2$$
$$4 = 4$$

You can verify that $(0, -2)$ and $(-1, -5)$ are also solutions, while $(-2, 4)$ is not.

EXAMPLE 2 Finding Solutions to an Equation

Given the equation $2x + 3y = 12$, complete the following ordered pairs so that they are solutions to the equation.

$$(0, \quad), \qquad (\quad, 0), \qquad (3, \quad), \qquad (\quad, -2)$$

To complete the ordered pair (0,), substitute 0 for x in $2x + 3y = 12$ and solve for y.

$$2(0) + 3y = 12$$
$$3y = 12$$
$$y = 4$$

The completed ordered pair is (0, 4).

To complete the ordered pair (, 0), substitute 0 for y in $2x + 3y = 12$ and solve for x.

$$2x + 3(0) = 12$$
$$2x = 12$$
$$x = 6$$

The completed ordered pair is (6, 0).

To complete the pair (3,), substitute 3 for x and solve for y.

$$2(3) + 3y = 12$$
$$6 + 3y = 12$$
$$3y = 6$$
$$y = 2$$

The completed pair is (3, 2).

Similarly, to complete (, −2), substitute −2 for y and solve for x, obtaining $(9, -2)$.

Practice Exercise 2

Given $3x - 4y = 24$, complete the following ordered pairs so that they are solutions to the equation.

$$(0, \quad), \ (\quad, 0), \ (-4, \quad), \ (\quad, 1)$$

Answers: $(0, -6)$, $(8, 0)$, $(-4, -9)$, $\left(\frac{28}{3}, 1\right)$

4 GRAPHING EQUATIONS

The **graph** of an equation in two variables is the set of points in a rectangular coordinate system that corresponds to solutions of the equation. Since there are infinitely many solutions, we cannot find and plot each possible pair. Usually we plot enough points to see a pattern and then connect these points with a line or curve, the graph of the equation.

Consider again the equation $y = 3x - 2$. One way to display some of its solutions is to make a **table of values.** We choose several values for x, substitute these values into the equation, and compute the corresponding values for y. We begin by making a table such as the one below. Each y-value we find is placed beside the x-value used to calculate it. (Calculations are usually done mentally or as scratch work.)

Substitution	*Result in* $y = 3x - 2$
$x = 0$	$y = 3(0) - 2 = -2$
$x = 1$	$y = 3(1) - 2 = 1$
$x = -1$	$y = 3(-1) - 2 = -5$
$x = 2$	$y = 3(2) - 2 = 4$
$x = -2$	$y = 3(-2) - 2 = -8$
$x = 3$	$y = 3(3) - 2 = 7$
$x = -3$	$y = 3(-3) - 2 = -11$

x	y	*Solutions*
0	−2	(0, −2)
1	1	(1, 1)
−1	−5	(−1, −5)
2	4	(2, 4)
−2	−8	(−2, −8)
3	7	(3, 7)
−3	−11	(−3, −11)

Now we plot the points that correspond to these ordered-pair solutions in a rectangular coordinate system, as in Figure 3.4. It appears that all seven points lie on a straight line. Thus, it is reasonable to assume that the graph of this equation is the straight line passing through these seven points in Figure 3.5.

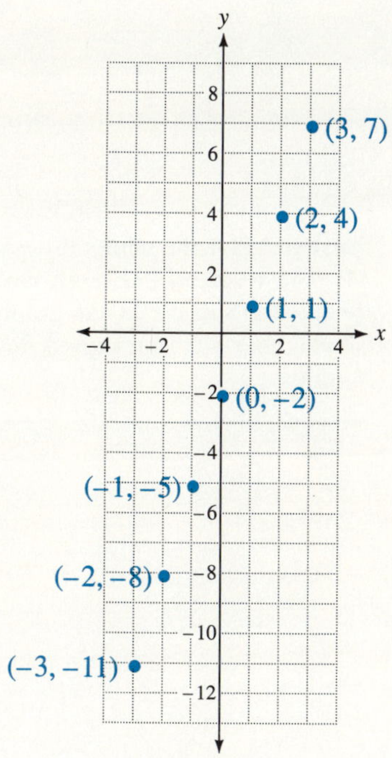

Figure 3.4
Plotting Solutions to $y = 3x - 2$

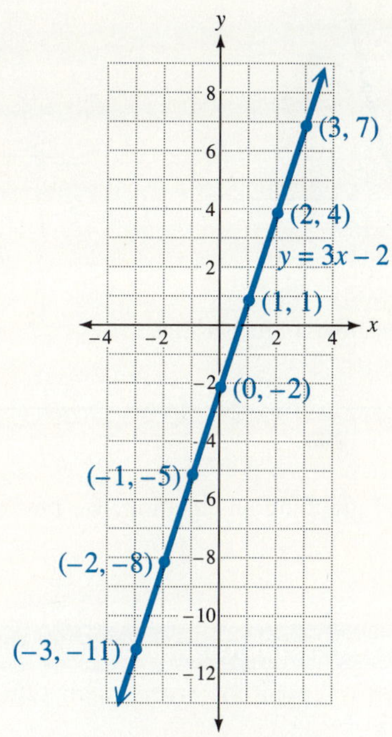

Figure 3.5
Connecting Points in Straight Line

We are now ready to graph other equations in two variables. In Example 3 on the following page we select values of x which give points that are easily plotted.

EXAMPLE 3 Graphing an Equation

Graph $y = x^2$ in a rectangular coordinate system.

We begin by making a table of values.

Substitution	*Result in* $y = x^2$
$x = 0$	$y = (0)^2 = 0$
$x = 1$	$y = (1)^2 = 1$
$x = -1$	$y = (-1)^2 = 1$
$x = 2$	$y = (2)^2 = 4$
$x = -2$	$y = (-2)^2 = 4$
$x = 3$	$y = (3)^2 = 9$
$x = -3$	$y = (-3)^2 = 9$

x	y	*Solutions*
0	0	(0, 0)
1	1	(1, 1)
−1	1	(−1, 1)
2	4	(2, 4)
−2	4	(−2, 4)
3	9	(3, 9)
−3	9	(−3, 9)

Plot the solution points and connect them with a smooth curve to obtain the graph shown in Figure 3.6.

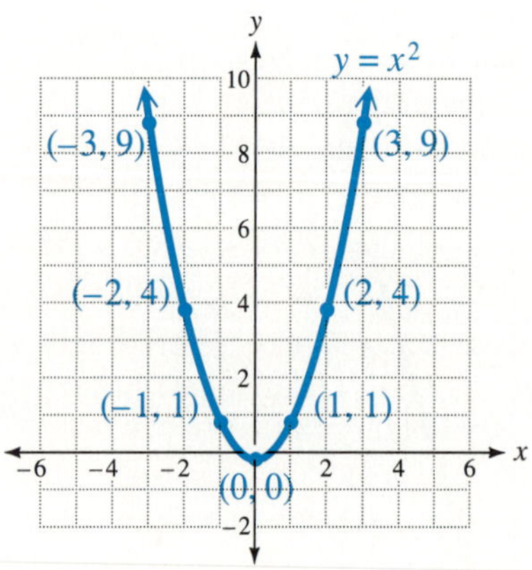

Figure 3.6 Graph Different from a Straight Line

Practice Exercise 3

Graph $y = x^2 - 5$ in a rectangular coordinate system.

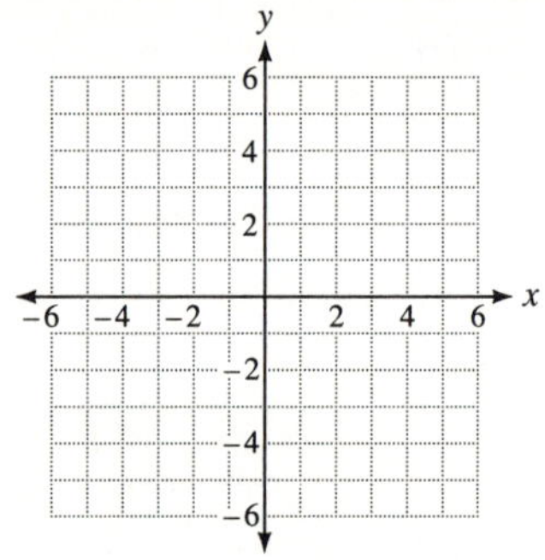

Answer: Plot the points (0, −5), (1, −4), (−1, −4), (2, −1), and (−2, −1). Notice that the graph is similar to $y = x^2$ but moved down 5 units.

Before considering another example of graphing an equation, we present a summary of the method to use.

To Graph an Equation in Two Variables x and y

1. Make a table of values. These values represent the ordered-pair solutions of the equation.
2. Plot the points that correspond to these solutions in a rectangular coordinate system.
3. Connect the points with a line or curve.

EXAMPLE 4 Graphing an Equation

Graph $y = |x|$ in a rectangular coordinate system.

Make a table of values.

Substitution	*Result in* $y = \|x\|$
$x = 0$	$y = \|0\| = 0$
$x = 1$	$y = \|1\| = 1$
$x = -1$	$y = \|-1\| = 1$
$x = 2$	$y = \|2\| = 2$
$x = -2$	$y = \|-2\| = 2$
$x = 3$	$y = \|3\| = 3$
$x = -3$	$y = \|-3\| = 3$

x	y	*Solutions*
0	0	(0, 0)
1	1	(1, 1)
−1	1	(−1, 1)
2	2	(2, 2)
−2	2	(−2, 2)
3	3	(3, 3)
−3	3	(−3, 3)

Plot the points associated with the ordered-pair solutions in the table to obtain the graph shown in Figure 3.7.

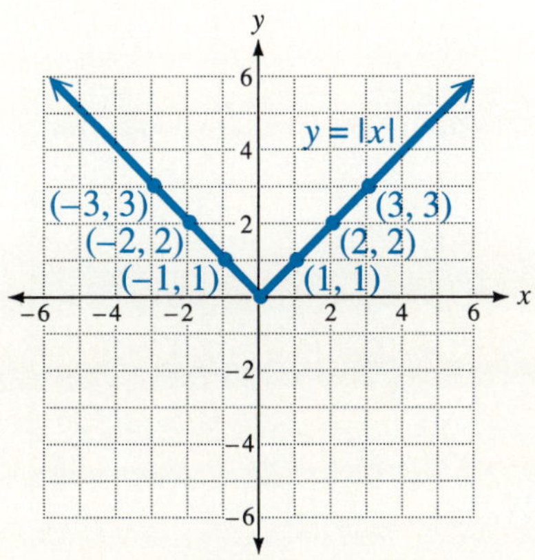

Figure 3.7
Graph with Straight-Line Segments

Practice Exercise 4

Graph $y = |x| + 1$ in a rectangular coordinate system.

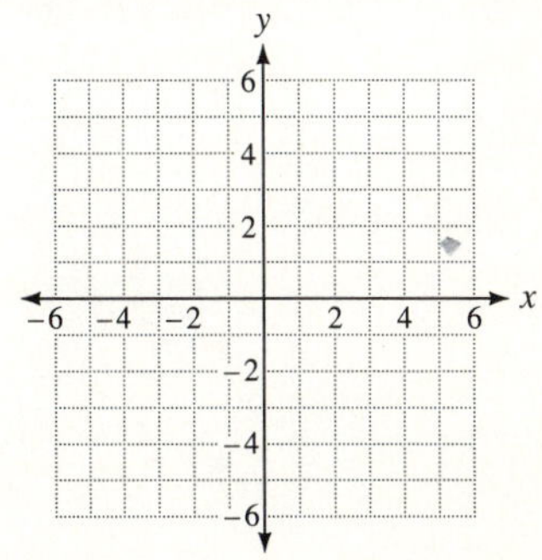

Answer: Plot the points (0, 1), (2, 3), (−2, 3) and notice that the graph is similar to $y = |x|$ but moved up 1 unit.

3.1 EXERCISES A

1. Plot the points associated with the given pairs $A(1, 4)$, $B(4, -2)$, $C(-3, 2)$, $D(-3, 0)$, $E(3, 0)$, $F(0, 0)$, $G(-3, -3)$, and $H(0, -2)$.

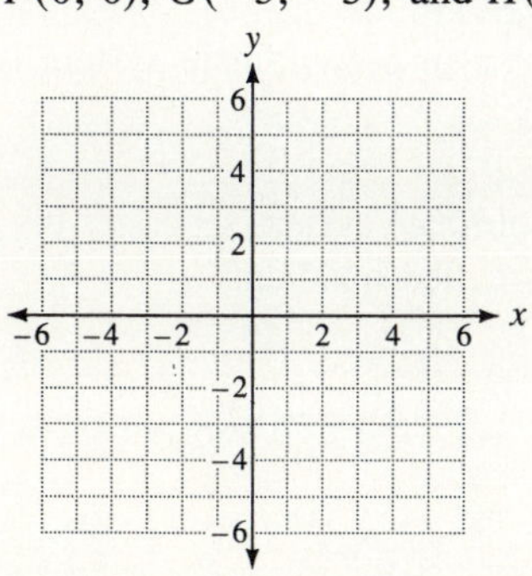

2. Give the coordinates of the points A, B, C, D, E, F, G, and H.

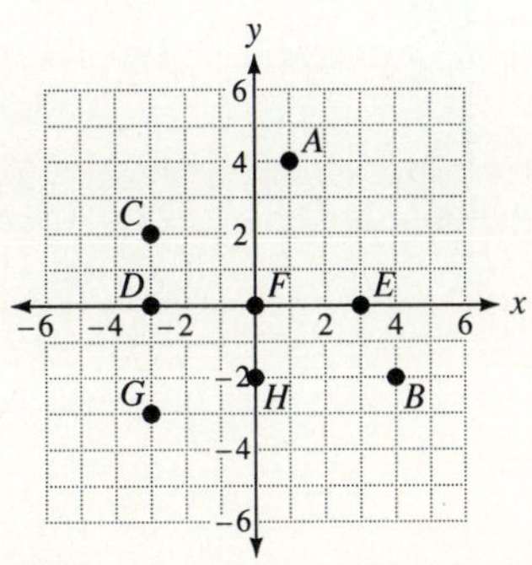

3. Plot the points associated with the pairs $J\left(\frac{1}{2}, 2\right)$, $K\left(-\frac{5}{2}, 1\right)$, $L\left(-2, -\frac{7}{4}\right)$, and $M\left(3, -\frac{3}{4}\right)$.

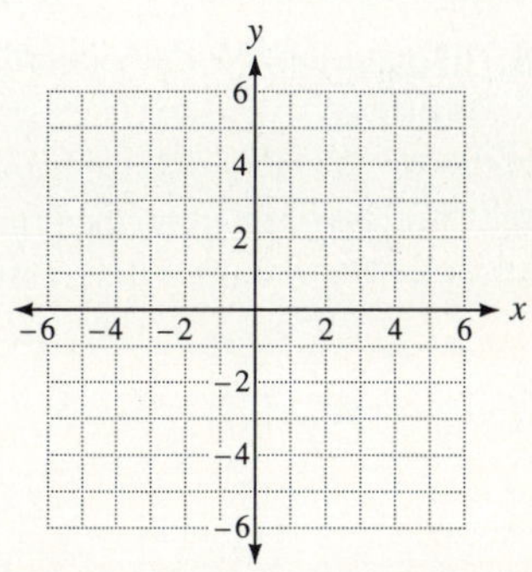

4. In which quadrants are the points J, K, L, and M of Exercise 3 located?

Give the quadrant that contains each point.

5. $(-2, -3)$

6. $(1, 7)$

7. $(-2, 5)$

8. $\left(\sqrt{2}, -\frac{1}{2}\right)$

9. Give the coordinates of each point. What do these points have in common? What do their coordinates have in common?

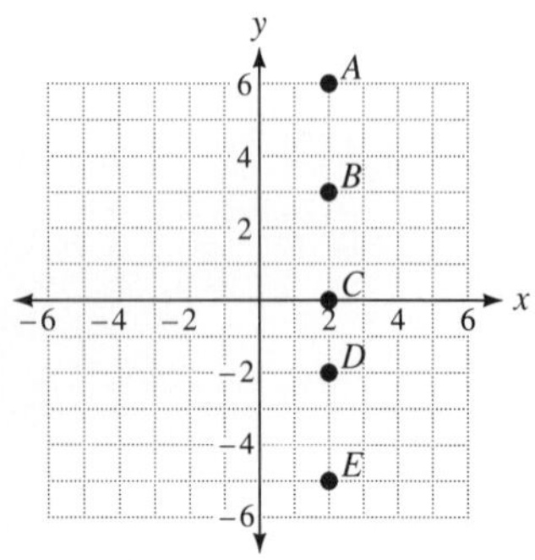

10. Give the coordinates of each point. What do these points have in common? What do their coordinates have in common?

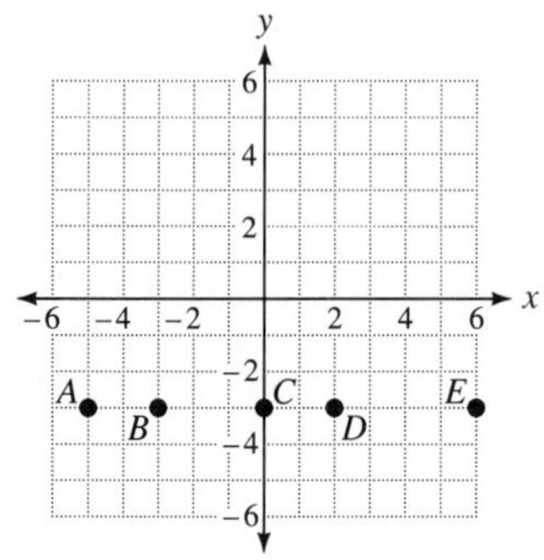

11. Find the coordinates of each point described below.

(a) Point A

(b) Point B, that is 2 units down from A

(c) Point C that is 5 units to the left of A

(d) Point D that is 1 unit up and 2 units to the right of A

(e) Point E that is 4 units down and 1 unit to the left of A

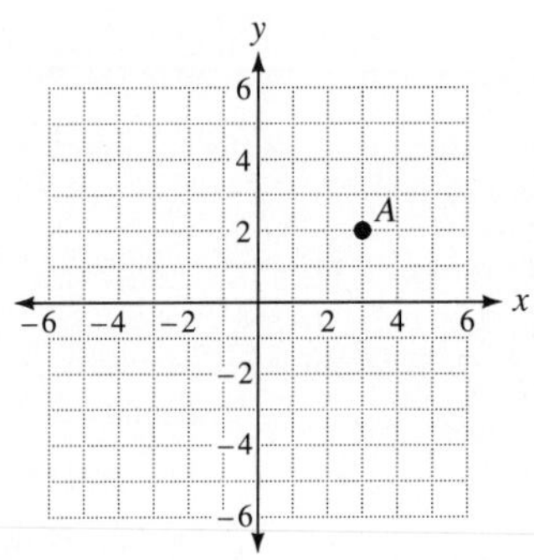

Write true *or* false *for Exercises* 12–21. *If the statement is false, explain why.*

12. A configuration in which points corresponding to ordered pairs of real numbers are plotted in a plane is a rectangular or Cartesian coordinate system.

13. The ordered pair $(3, -4)$ determines a point in quadrant IV with x-coordinate 3 and y-coordinate -4.

14. The ordered pair $(-8, -5)$ determines a point in quadrant III with abscissa -8 and ordinate -5.

15. In quadrant I, the x-coordinate of a point is always negative.

16. In quadrant II, the y-coordinate of a point is always positive.

17. In quadrant III, the x-coordinate of a point is always negative.

18. In quadrant IV, the y-coordinate of a point is always positive.

19. The horizontal axis in a coordinate system is also called the y-axis.

20. The perpendicular lines in a rectangular coordinate system are called the axes.

21. A rectangular coordinate system separates the plane into four regions called quadrants.

Answer each question below.

22. In which quadrants do points have coordinates with opposite signs?

23. The x-coordinate of a point is positive in which quadrants?

24. The y-coordinate of a point is negative in which quadrants?

25. What is the x-coordinate of any point on the y-axis?

26. What is the distance between $A(5, 2)$ and $B(8, 2)$?

27. What is the distance between $C(4, 2)$ and $D(4, -3)$?

In each of the following, find the missing coordinate so that the ordered pair is a solution to the equation.

28. $x + y + 2 = 0$

(a) (0,) **(b)** (, 0)

(c) (1,) **(d)** (, −2)

29. $x - y = 0$

(a) (0,) **(b)** (, 0)

(c) (5,) **(d)** (, −3)

Graph the following equations.

30. $x + y + 2 = 0$

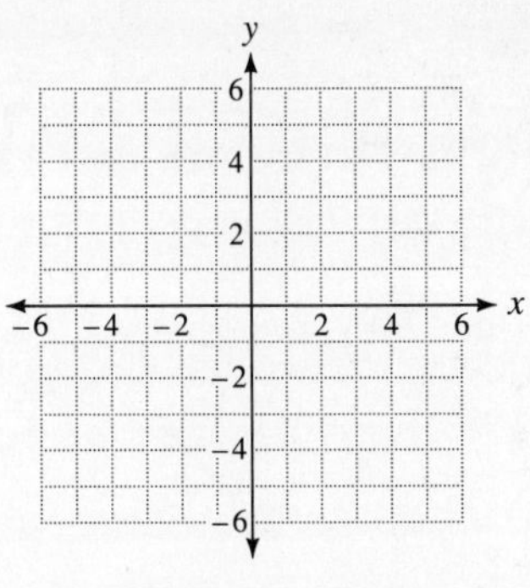

31. $3x + y = 0$

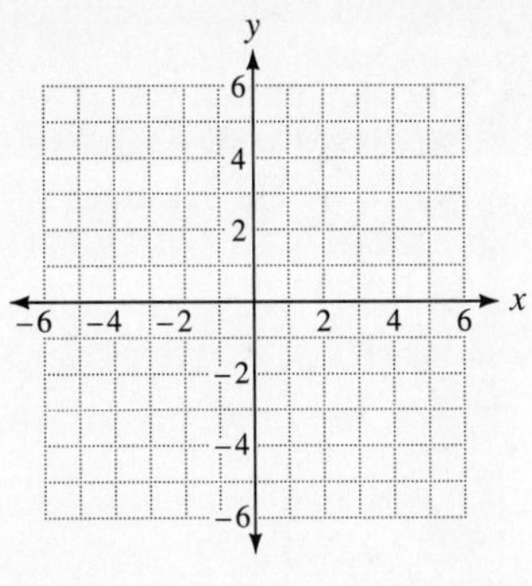

32. $x - y = 2$

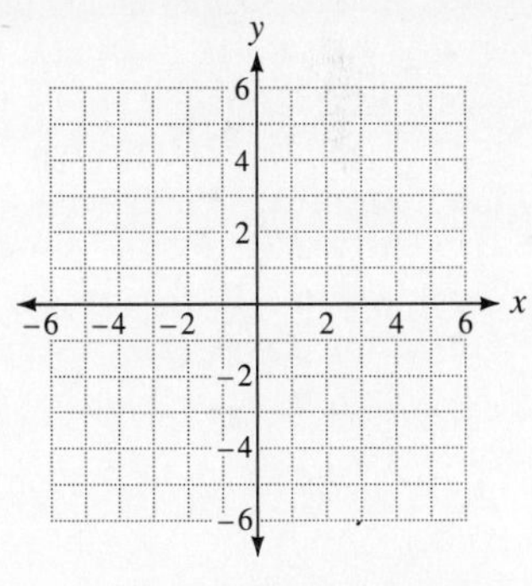

33. $x - y = 0$

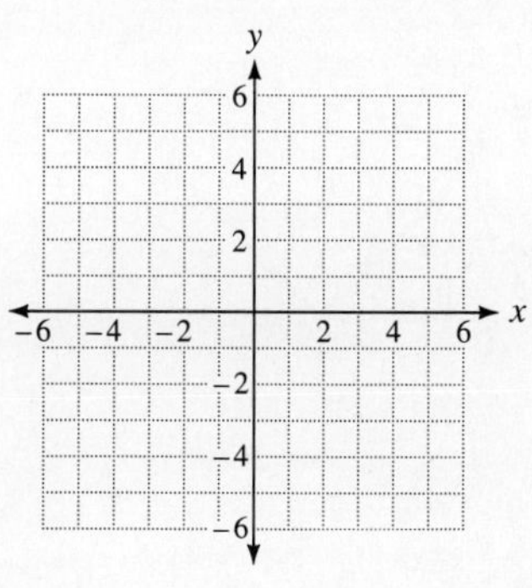

34. $y = x^2 - 1$

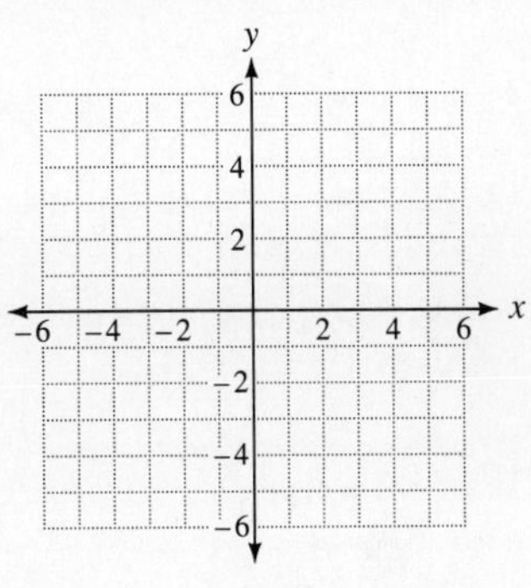

35. $y = x^2 + 1$

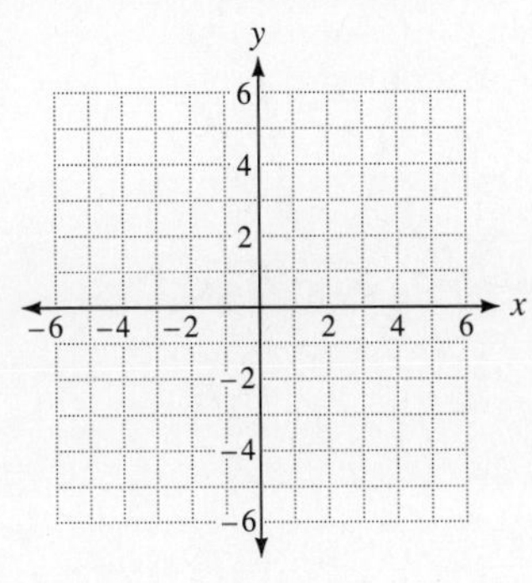

36. $y = |x| + 2$

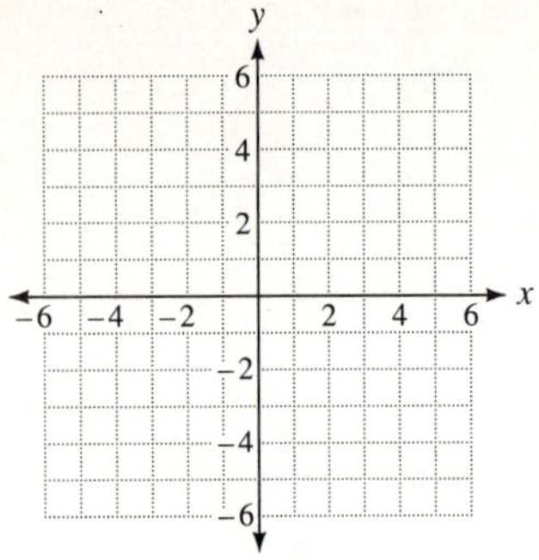

37. $y = |x| - 3$

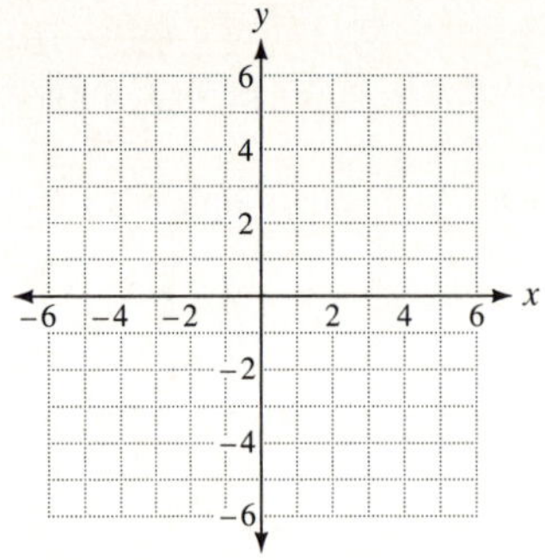

38. $y = |x - 1|$

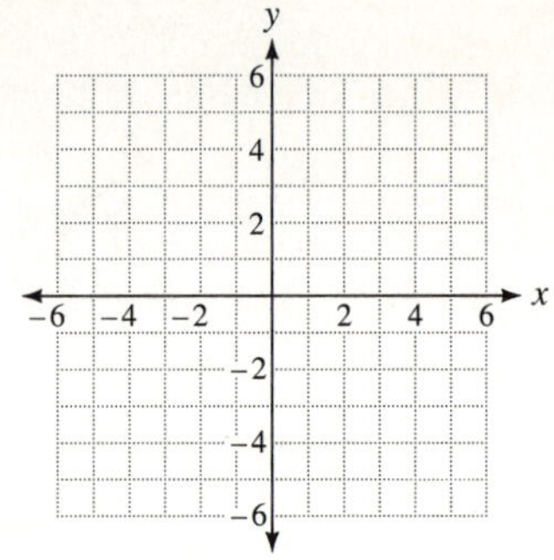

39 Mr. Goldstein has estimated that the cost (in dollars), y, of producing a number, x, of machine parts in his plant is given by the equation $y = 200x + 30$. Complete the following ordered pairs: (1,), (5,), (10,). What is the cost of producing **(a)** 1 part? **(b)** 5 parts? **(c)** 10 parts?

FOR REVIEW

Solve the absolute value equation or inequality.

40. $|2x - 5| = 7$ **41.** $|3x + 1| < 4$ **42.** $|2x - 6| \geq 2$

ANSWERS: 1. Answers are given in the coordinate system in Exercise 2. 2. See Exercise 1. 3.

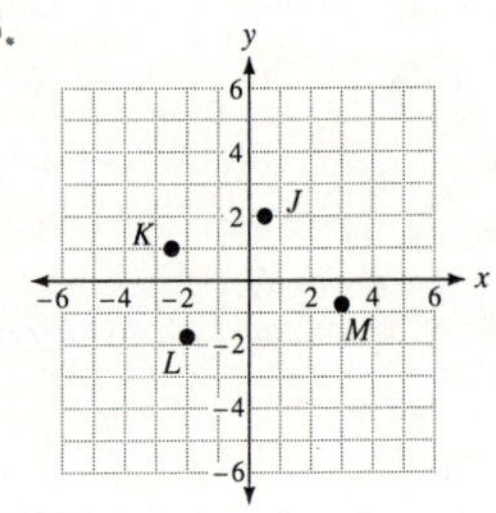

4. J: I; K: II; L: III; M: IV 5. III 6. I 7. II 8. IV 9. $A(2, 6)$, $B(2, 3)$, $C(2, 0)$, $D(2, -2)$, $E(2, -5)$; the points all lie on the same vertical line; all points have x-coordinate 2 10. $A(-5, -3)$, $B(-3, -3)$, $C(0, -3)$, $D(2, -3)$, $E(6, -3)$; the points all lie on the same horizontal line; all points have y-coordinate -3 11. (a) (3, 2) (b) (3, 0) (c) $(-2, 2)$ (d) (5, 3) (e) $(2, -2)$ 12. true 13. true 14. true 15. false; positive 16. true 17. true 18. false; negative 19. false; x-axis 20. true 21. true 22. II and IV 23. I and IV 24. III and IV 25. 0 26. 3 units 27. 5 units 28. (a) -2 (b) -2 (c) -3 (d) 0 29. (a) 0 (b) 0 (c) 5 (d) -3

30.

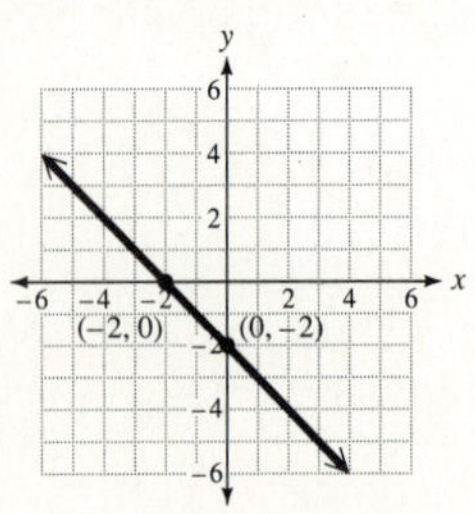

31.

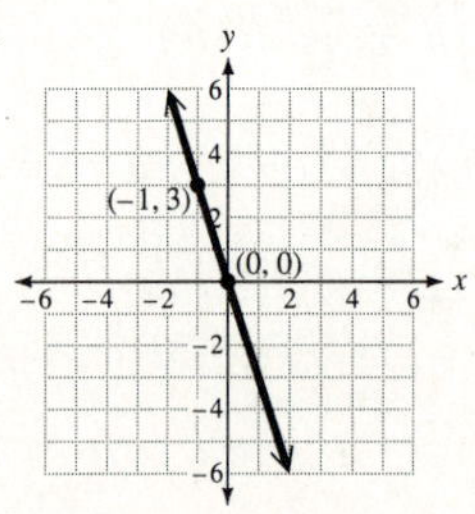

32.

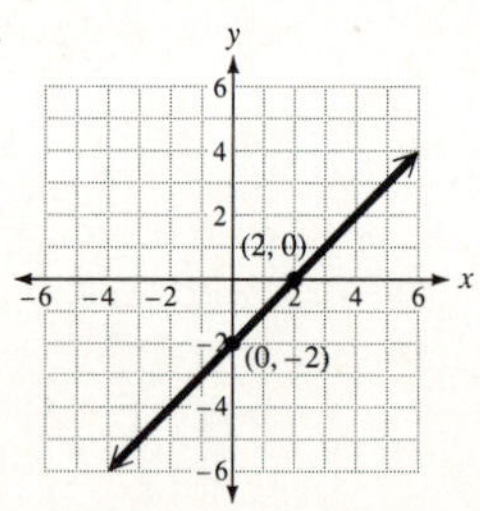

33.

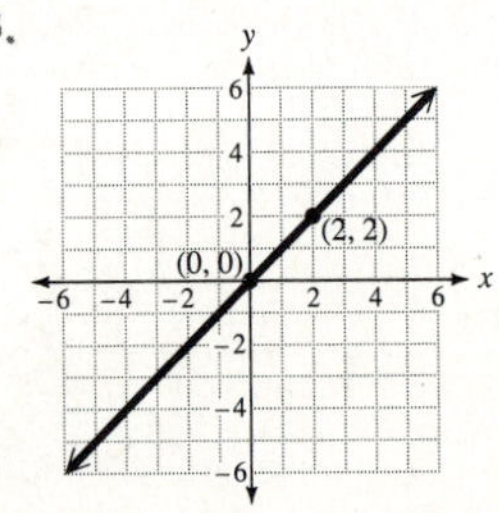

34.

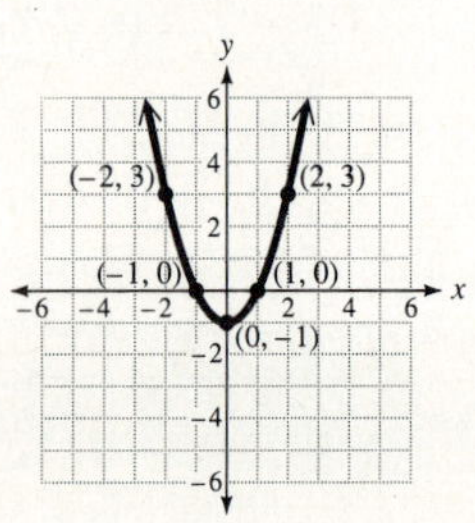

35.

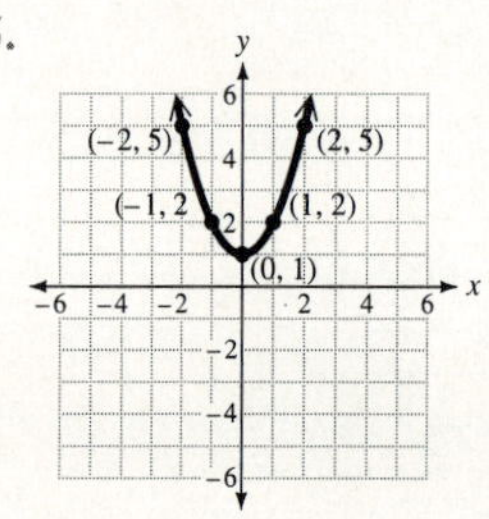

36.

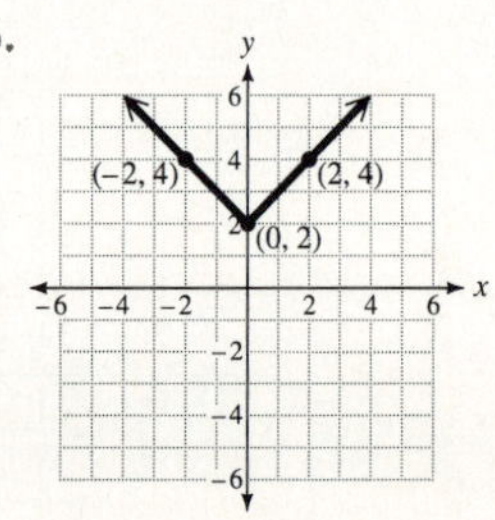

37.

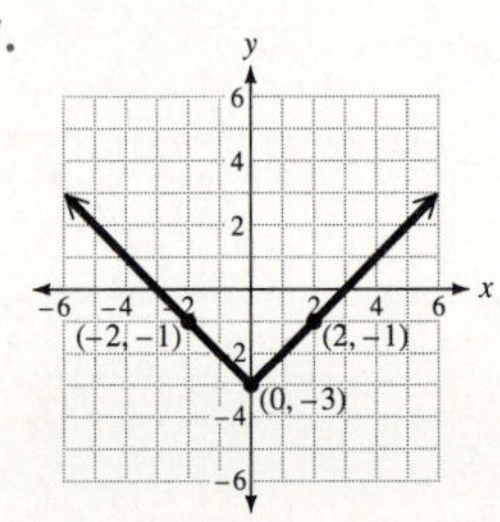

38. 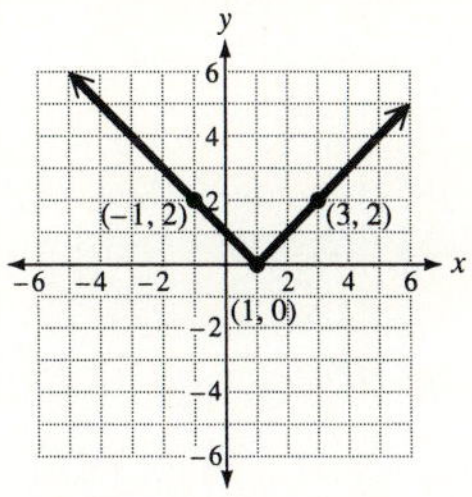

39. (1, 230), (5, 1030), (10, 2030) (a) \$230 (b) \$1030 (c) \$2030 **40.** 6, −1
41. $-\frac{5}{3} < x < 1$ **42.** $x \geq 4$ or $x \leq 2$

3.1 EXERCISES B

1. Plot the points corresponding to the ordered pairs $A(1, 5)$, $B(-4, 3)$, $C(5, 2)$, $D(0, 1)$, $E(-6, 0)$, $F(0, -3)$, $G(-6, -6)$, $H(3, -4)$.

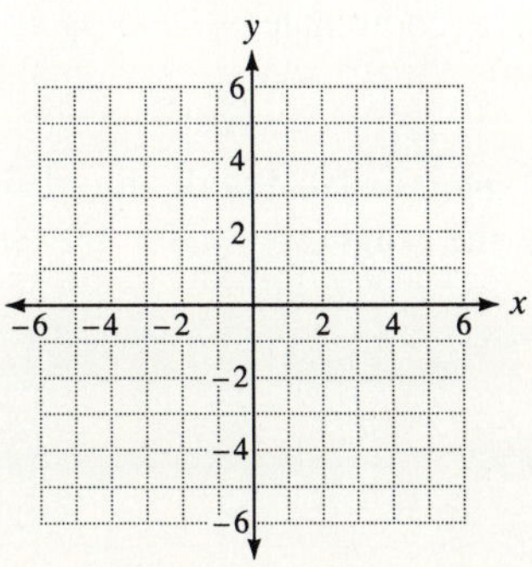

2. Give the coordinates of the points A, B, C, D, E, F, G, and H.

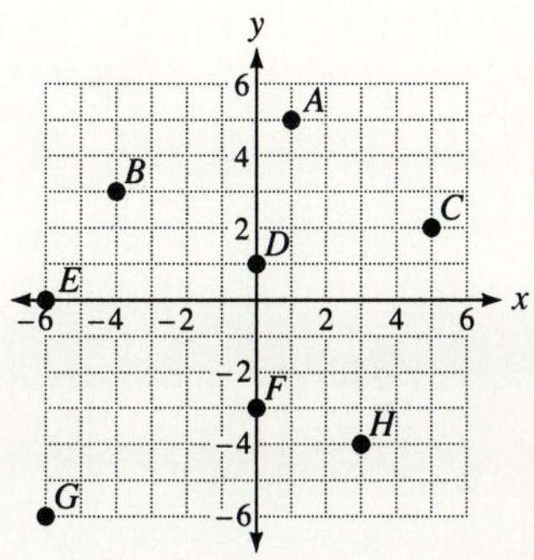

3. Plot the points associated with the ordered pairs $P\left(-\frac{1}{2}, 3\right)$, $Q\left(\frac{7}{2}, 4\right)$, $R\left(-3, -\frac{3}{4}\right)$, and $S\left(2, -\frac{7}{3}\right)$.

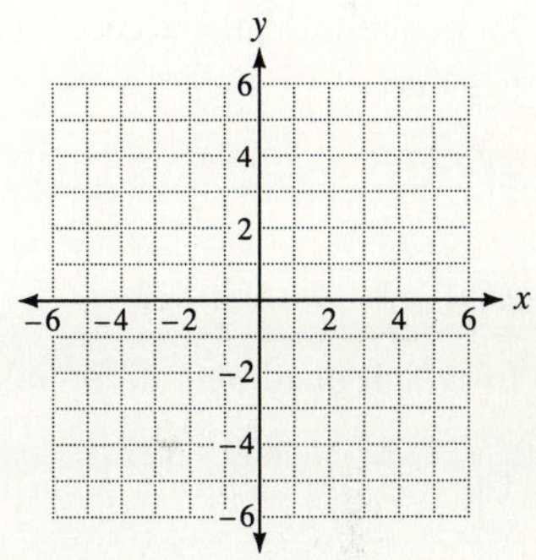

4. In what quadrants are the points P, Q, R, and S of Exercise 3 located?

Give the quadrant that contains each point.

5. $(-4, 5)$ **6.** $(6, -2)$ **7.** $(-5, -1)$ **8.** $\left(\frac{1}{3}, \sqrt{2}\right)$

9. Give the coordinates of each point. What do these points have in common? What do their coordinates have in common?

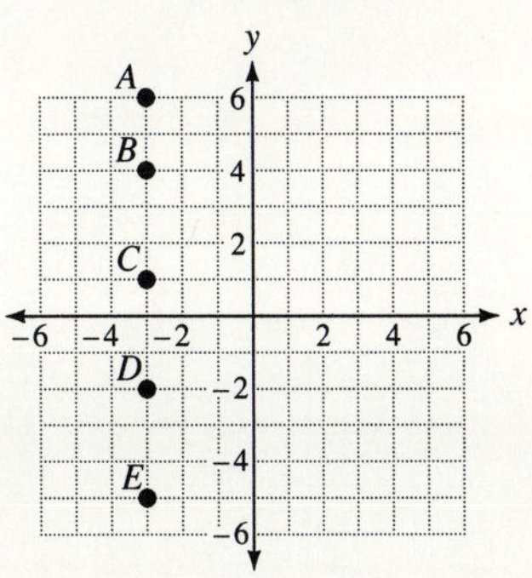

10. Give the coordinates of each point. What do these points have in common? What do their coordinates have in common?

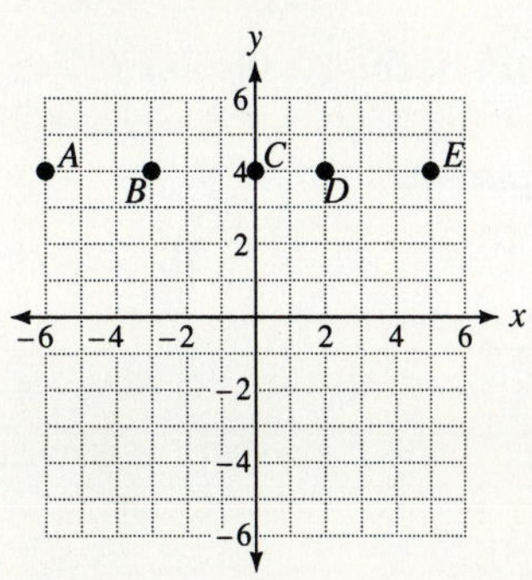

11. Find the coordinates of each point described below.
  (a) Point A
  (b) Point B that is 3 units up from A
  (c) Point C that is 4 units to the right of A
  (d) Point D that is 2 units up and 5 units to the left of A
  (e) Point E that is 1 unit down and 3 units to the right of A

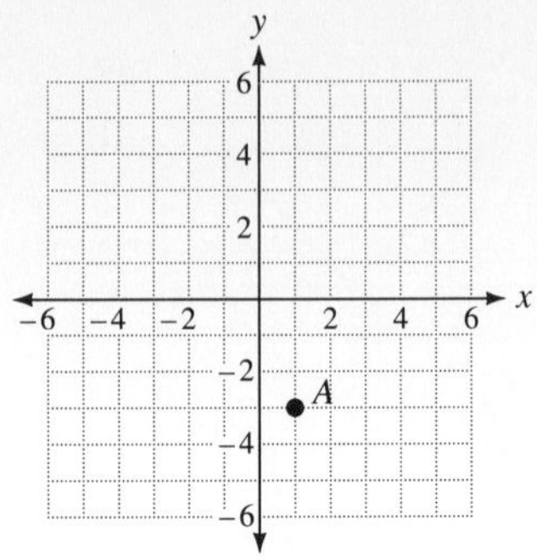

Write true *or* false *for Exercises* 12–21. *If the statement is false, explain why.*

12. Another name for a rectangular or Cartesian coordinate system is a coordinate plane.

13. The ordered pair $(-2, 8)$ determines a point in quadrant I with x-coordinate -2 and y-coordinate 8.

14. The ordered pair $(7, 2)$ determines a point in quadrant I with abscissa 2 and ordinate 7.

15. In quadrant I, the y-coordinate of a point is always positive.

16. In quadrant II, the x-coordinate of a point is always positive.

17. In quadrant III, the y-coordinate of a point is always negative.

18. In quadrant IV, the x-coordinate of a point is always negative.

19. The vertical axis in a coordinate system is also called the y-axis.

20. The coordinates of the origin are $(0, 0)$.

21. When we find the point associated with an ordered pair, we are plotting the point.

Answer each question below.

22. In which quadrants do points have coordinates with the same sign?

23. The x-coordinate of a point is negative in which quadrants?

24. The y-coordinate of a point is positive in which quadrants?

25. What is the y-coordinate of any point on the x-axis?

26. What is the distance between $A(-1, 3)$ and $B(-1, -4)$?

27. What is the distance between $C(6, -2)$ and $D(-2, -2)$?

In each of the following, find the missing coordinates so that the ordered pair is a solution to the equation.

28. $2x - y + 1 = 0$; **(a)** $(0,\ )$ **(b)** $(\ , 0)$ **(c)** $(-1,\ )$ **(d)** $(\ , 3)$

29. $x + y = 0$; **(a)** $(0,\ )$ **(b)** $(\ , 0)$ **(c)** $(4,\ )$ **(d)** $(\ , -5)$

Graph the following equations.

30. $x + y - 2 = 0$

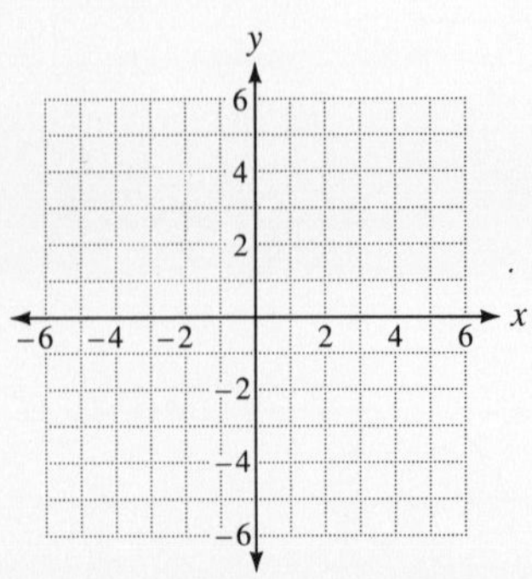

31. $2y - x = 0$

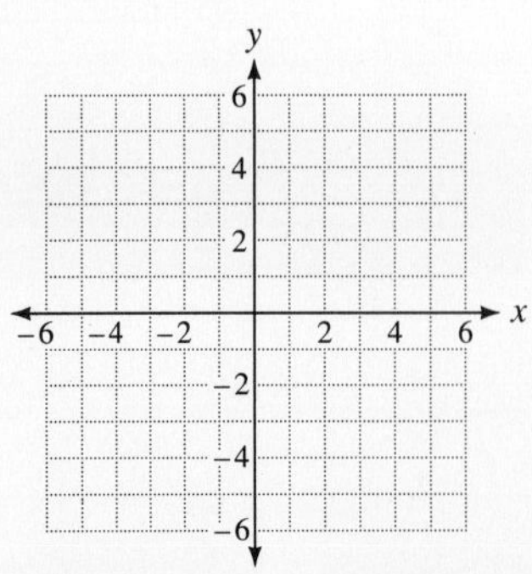

32. $y - x = 1$

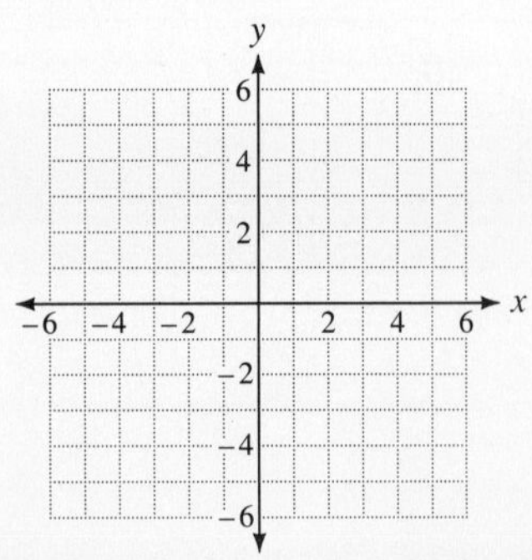

33. $y - x = 0$

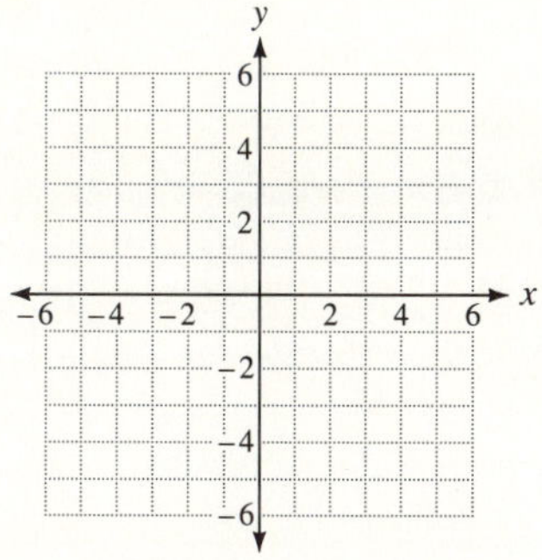

34. $y = x^2 + 2$

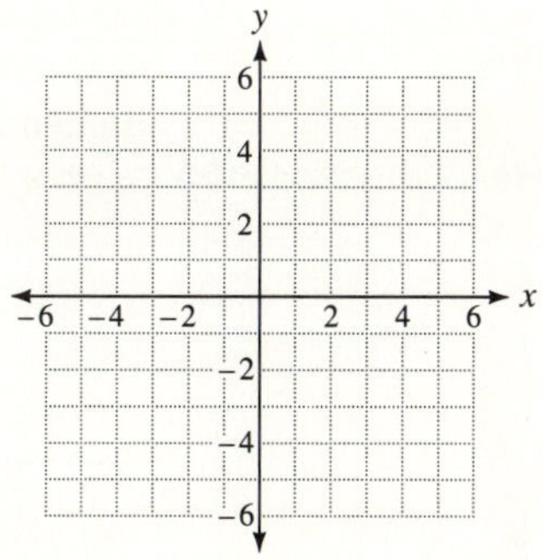

35. $y = x^2 - 3$

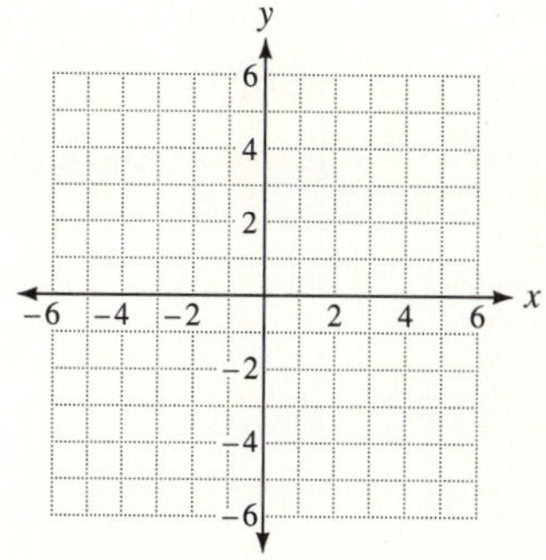

36. $y = |x| - 1$

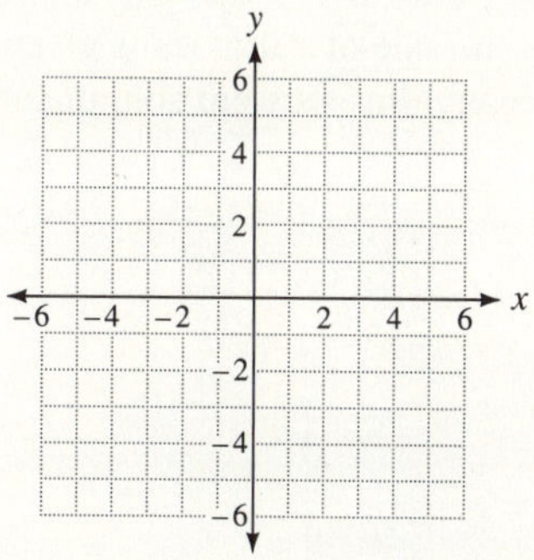

37. $y = |x| + 3$

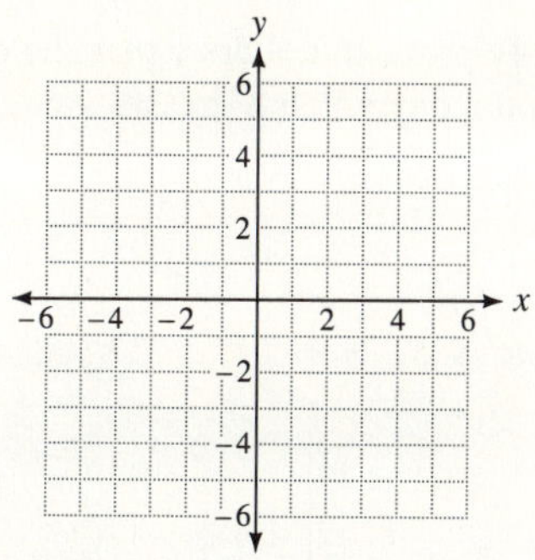

38. $y = |x + 1|$

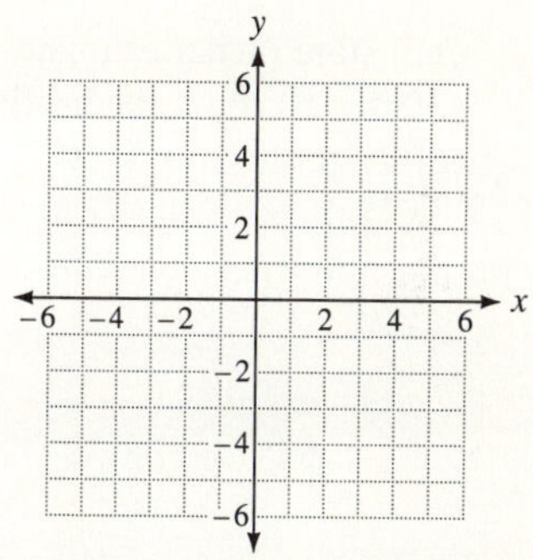

39. The annual salary of Sam Passamante, a salesman, is given in terms of his total sales x by the equation $y = 15{,}000 + (0.2)x$. Complete the ordered pairs (10,000,), (20,000,) and (50,000,). How much will Sam earn if he sells **(a)** 10,000 items? **(b)** 20,000 items? **(c)** 50,000 items?

FOR REVIEW

Solve the absolute value equation or inequality.

40. $|3x + 5| = 4$

41. $|2x - 4| \le 2$

42. $|2x + 8| > 3$

3.1 EXERCISES C

1. Give the coordinates of the vertices (corners) of a rectangle that is 7 units long and 3 units wide if the rectangle is in the first quadrant, its longest side lies along the x-axis, and one vertex is at the origin. [*Hint:* Make a sketch.]

2. Repeat Exercise 1 for a rectangle that is in the third quadrant with all other conditions the same.

3. Give the equation in two variables x and y in which the value of y is 3 more than twice the value of x. Graph the equation.

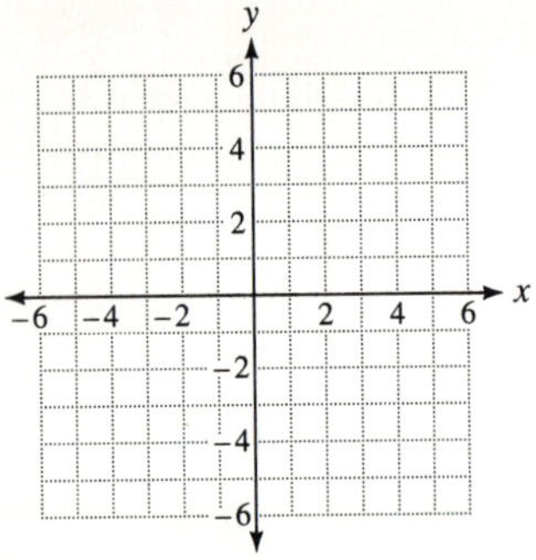

4. Graph $x + y = 2$ and $2x - y = 1$ in the same coordinate system. What are the coordinates of the point of intersection of the two graphs? Is this ordered pair a solution to both equations? [Answer: (1, 1); yes]

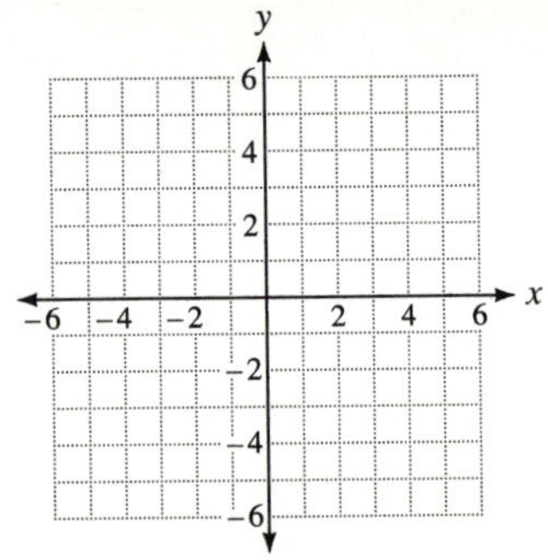

5. A retail store owner estimates that the daily profit in dollars y that she can make on the sale of x dresses is given by $y = 30x - 50$. Find the profit she made on a day when 8 dresses were sold. How many dresses were sold on a day when her profit was \$1000?

Graph the following equations.

6. $y = x^3$

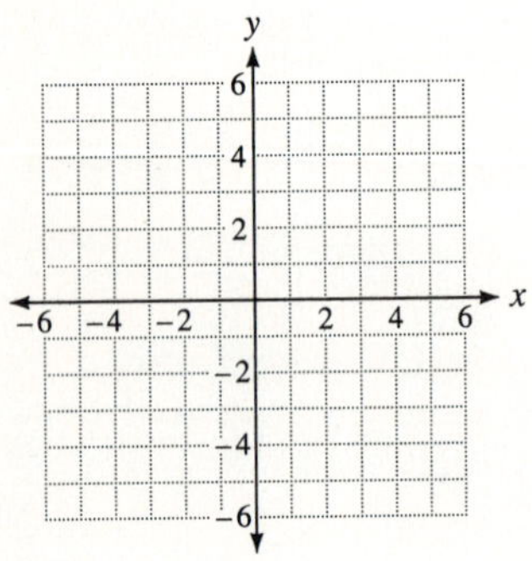

7. $x^2 + y^2 = 4$

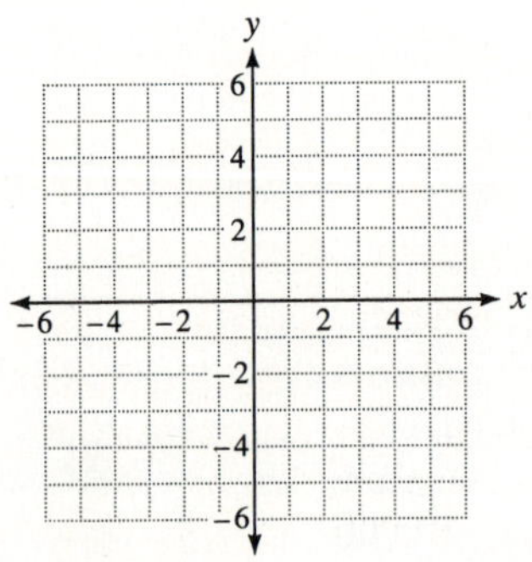

8. $y = -x^2$

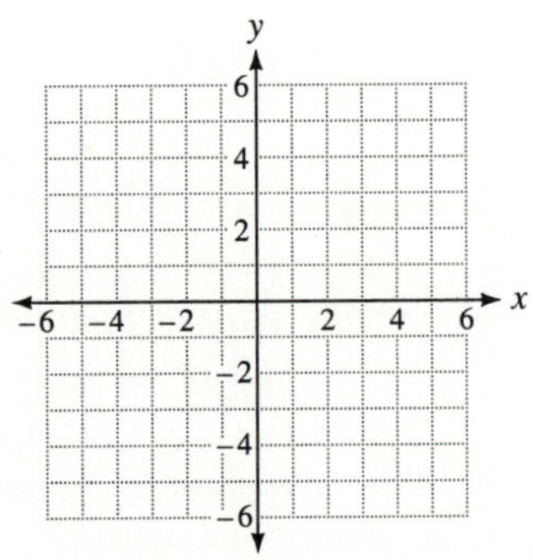

9. Suppose that $(-2, 3)$, $(6, 3)$, and $(4, -2)$ are three of the four vertices of a parallelogram. Find the coordinates of the fourth vertex.

10. What is the area of the parallelogram found in Exercise 9? [Answer: 40 square units]

3.2 LINEAR EQUATIONS

STUDENT GUIDEPOSTS

1. Linear Equations in Two Variables
2. The Intercept Method of Graphing
3. Graphing Special Linear Equations

1 LINEAR EQUATIONS IN TWO VARIABLES

A linear equation in one variable like those we studied in Chapter 2 can always be written in the form

$$ax + b = 0. \quad a, b \text{ constants}$$

A **linear equation in two variables** x and y can be written in the **general form**

$$ax + by + c = 0. \quad a, b, c \text{ constants with } a \text{ and } b \text{ not both zero}$$

Linear equations are also called **first-degree equations** since the variables occur to the first power only. Following are examples of both linear equations and those that are not linear.

EXAMPLE 1 IDENTIFYING LINEAR AND NONLINEAR EQUATIONS

Equation	*Nature*	*General form*	*Constants*
(a) $3x - y = 4$	linear	$3x - y - 4 = 0$	$a = 3, b = -1, c = -4$
(b) $x = y + 3$	linear	$x - y - 3 = 0$	$a = 1, b = -1, c = -3$
(c) $x + 2 = 0$	linear	$x + 0 \cdot y + 2 = 0$	$a = 1, b = 0, c = 2$
(d) $3y - 8 = 0$	linear	$0 \cdot x + 3y - 8 = 0$	$a = 0, b = 3, c = -8$
(e) $y = x^2 + 1$	not linear	(x to second power)	
(f) $2x + 3y^3 = 1$	not linear	(y to third power)	
(g) $xy = -1$	not linear	(cannot be written in form $ax + by + c = 0$)	
(h) $y = 3 - \frac{2}{x}$	not linear	(cannot be written in form $ax + by + c = 0$)	

PRACTICE EXERCISE 1

Decide whether or not each equation is linear.

(a) $8x + 2y = 7$

(b) $y = 2x - 5$

(c) $x^2 - y^2 = 1$

(d) $2 = x - 7y$

(e) $3xy + 2 = 0$

(f) $y = \frac{1}{x}$

(g) $x = y$

(h) $y + 5 = 0$

Answers: (a) yes (b) yes (c) no (d) yes (e) no (f) no (g) yes (h) yes

2 THE INTERCEPT METHOD OF GRAPHING

In Section 3.1 we graphed equations, some of which were linear, by making a table of values of solutions to the equation, plotting these points, and drawing a smooth curve through them. This method can be improved upon if we know ahead of time that an equation has a particular geometric shape as its graph.

Every **line**ar equation in two variables has a straight **line** for its graph. Since two points determine a line, we need to find only two solutions to graph a linear equation in two variables. In most cases, the two solutions easiest to find correspond to the **intercepts.** The points where a line crosses the x-axis and y-axis, respectively, are the ***x*-intercept** and the ***y*-intercept.** Since the y-intercept is a point on the y-axis, it has x-coordinate 0 as shown in the following table. Similarly, the x-intercept is a point on the x-axis, and has y-coordinate 0.

x	y
0	
	0

Complete this to find *y*-intercept (row with $x = 0$)

Complete this to find *x*-intercept (row with $y = 0$)

EXAMPLE 2 GRAPHING BY THE INTERCEPT METHOD

(a) Graph $3x - 5y - 15 = 0$.
First find the x- and y-intercepts by completing the following table.

x	y
0	
	0

PRACTICE EXERCISE 2

(a) Graph $2x - 3y = 6$. Use axes on next page.

Substitute 0 for x and solve for y.

$$3\,(0) - 5y - 15 = 0 \qquad x = 0 \text{ determines the } y\text{-intercept}$$
$$-5y - 15 = 0$$
$$-5y = 15$$
$$y = -3$$

Then substitute 0 for y and solve for x.

$$3x - 5\,(0) - 15 = 0 \qquad y = 0 \text{ determines the } x\text{-intercept}$$
$$3x - 15 = 0$$
$$3x = 15$$
$$x = 5$$

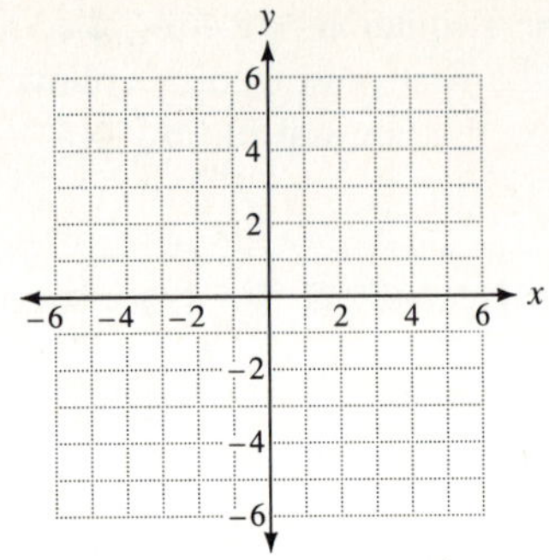

The completed table

x	y
0	−3
5	0

shows the y-intercept $(0, -3)$ and the x-intercept $(5, 0)$. If we plot these two points and draw the line between them, we obtain the graph of $3x - 5y - 15 = 0$ in Figure 3.8. *Although only two points are necessary, it is wise to calculate a third solution as a check.* For example, if $x = 1$, we calculate y to be $-\frac{12}{5}$ or -2.4, and it certainly appears that $(1, -2.4)$ is on the line $3x - 5y - 15 = 0$.

(b) Graph $x - 3y = 0$.

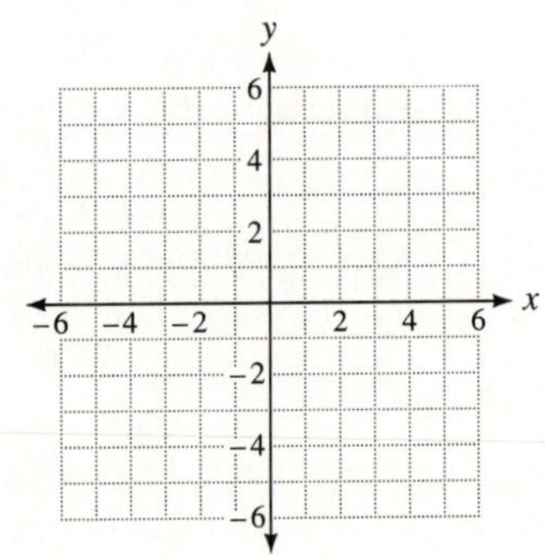

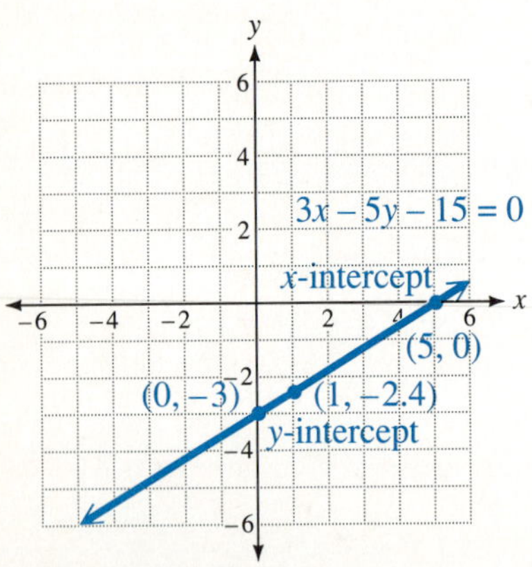

Figure 3.8 Graphing Using Intercepts

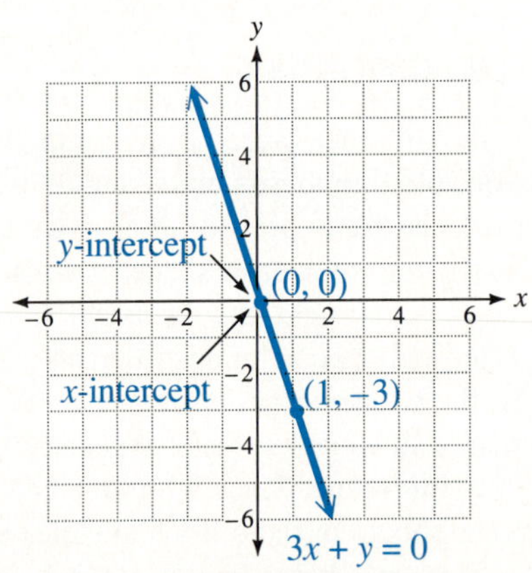

Figure 3.9

(b) Graph $3x + y = 0$.
Here, the constant term c is 0, and the table of intercepts looks like this.

x	y
0	0
0	0

That is, the x-intercept and y-intercept are the same point, the origin $(0, 0)$. When this occurs, we need to find another point on the line. If $x = 1$, $y = -3$, so $(1, -3)$ is also on the line. Plot $(0, 0)$ and $(1, -3)$ and connect the points to obtain the graph in Figure 3.9.

Answers: (a) line with intercepts (3, 0) and (0, −2) (b) line through (0, 0) and (3, 1)

3 GRAPHING SPECIAL LINEAR EQUATIONS

Two special cases of linear equations are presented in the next example which shows what happens to the graph when $a = 0$ or $b = 0$ in an equation $ax + by + c = 0$. When $a = 0$, the graph is a horizontal line, and when $b = 0$, the graph is a vertical line.

EXAMPLE 3 Graphing Horizontal and Vertical Lines

(a) Graph $3y + 5 = 0$.
The x-term is missing so $a = 0$. The equation, $3y + 5 = 0$, can be simplified to

$$y = -\frac{5}{3}$$

so that solutions are of the form

$$\left(x, -\frac{5}{3}\right) \text{ for } x \text{ any real number.}$$

Since y cannot be zero, there is no x-intercept. However, when $x = 0$, $y = -\frac{5}{3}$ so that $\left(0, -\frac{5}{3}\right)$ is the y-intercept. The graph is the horizontal line passing through y-intercept $\left(0, -\frac{5}{3}\right)$ parallel to the x-axis, as shown in Figure 3.10.

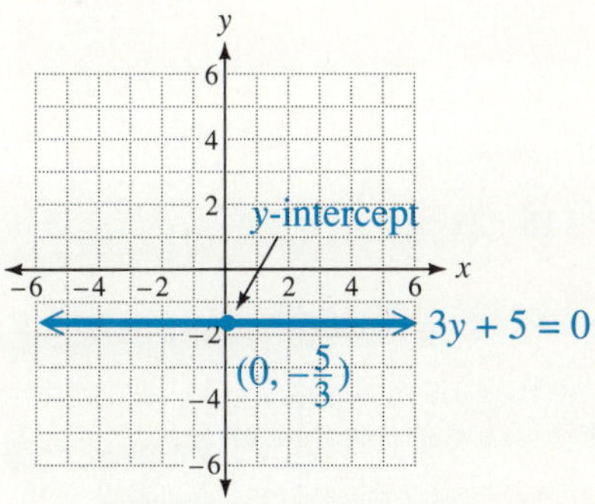

Figure 3.10 Horizontal Line

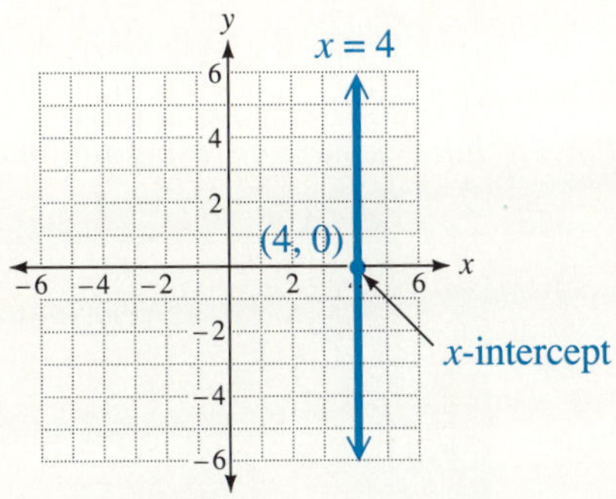

Figure 3.11 Vertical Line

(b) Graph $x = 4$.
This time $b = 0$, so the y-term is missing and the solutions are of the form

$$(4, y) \text{ for } y \text{ any real number.}$$

Since x cannot be zero, there is no y-intercept. However, if $y = 0$, x is 4 so that $(4, 0)$ is the x-intercept. The graph is the vertical line passing through x-intercept $(4, 0)$ parallel to the y-axis, as shown in Figure 3.11.

Practice Exercise 3

(a) Graph $2y - 6 = 0$.

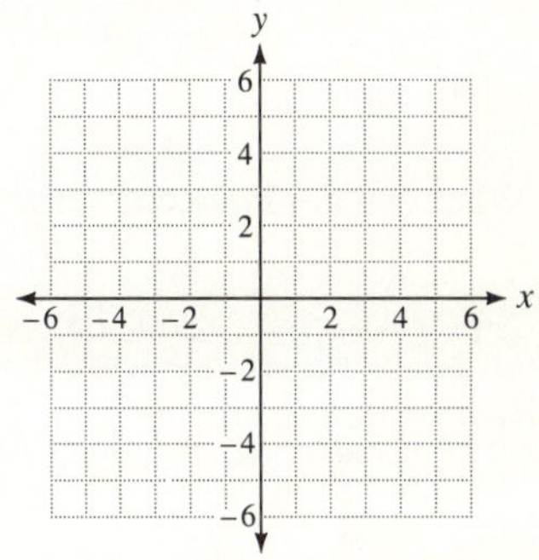

(b) Graph $3x + 9 = 0$.

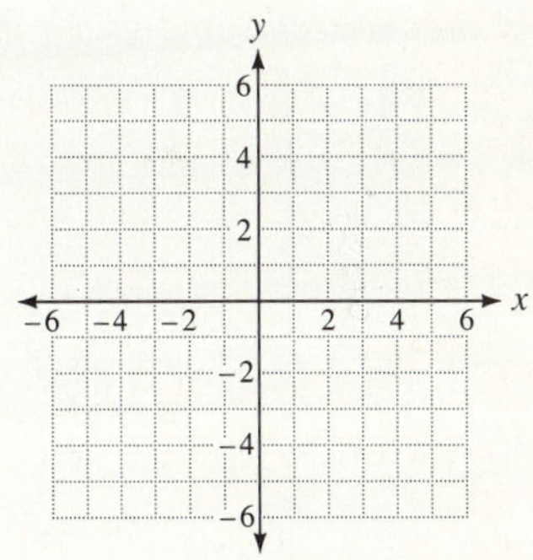

Answers: (a) horizontal line through $(0, 3)$ (b) vertical line through $(-3, 0)$

We conclude this section with a summary of the method for graphing a linear equation using intercepts.

To Graph a Linear Equation $ax + by + c = 0$

1. If $a \neq 0$ and $b \neq 0$: Plot the intercepts. If both intercepts are $(0, 0)$, find an additional point. Draw the line through the two points.
2. If $a = 0$, the equation can be written as $y =$ a constant. The graph is a horizontal line (parallel to the x-axis) through the y-intercept.
3. If $b = 0$, the equation can be written as $x =$ a constant. The graph is a vertical line (parallel to the y-axis) through the x-intercept.

3.2 EXERCISES A

Write true *or* false. *If the statement is false, explain why.*

1. A linear equation has as its graph a straight line.

2. The general form of a linear equation is $ax + by + c = 0$.

3. An equation of the form $ax + c = 0$, $a \neq 0$, has as its graph a straight line parallel to the x-axis.

4. An equation of the form $ax + by = 0$, $a \neq 0$ and $b \neq 0$, has x-intercept and y-intercept $(0, 0)$.

5. A point at which a graph crosses the x-axis is called a y-intercept.

6. Usually, the best two points to use when graphing a linear equation are the intercepts.

7. What are the intercepts of the line $x = 0$? What is its graph?

Is the given equation linear? Explain.

8. $x + y + 1 = 0$

9. $x^2 + xy + y^2 = 7$

10. $x - xy = 0$

11. $\dfrac{2}{x} + \dfrac{3}{y} = 1$

12. $x = 5$

13. $3y - 7 = 0$

Find the intercepts and graph the following linear equations using the intercept method.

14. $3x + 4y - 12 = 0$

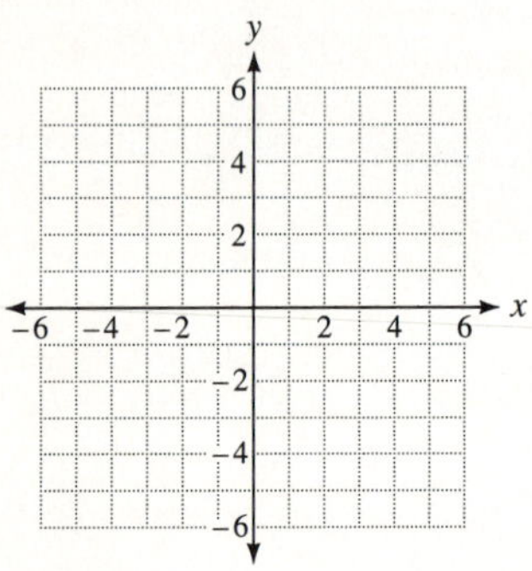

15. $2x - y = 4$

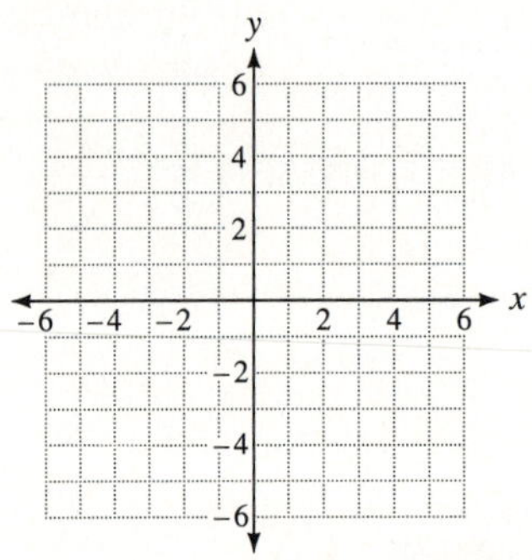

16. $2x + y = 1$

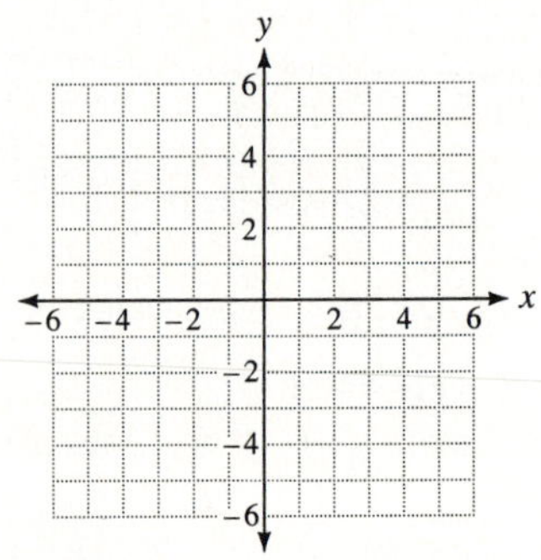

17. $2x - y = 0$

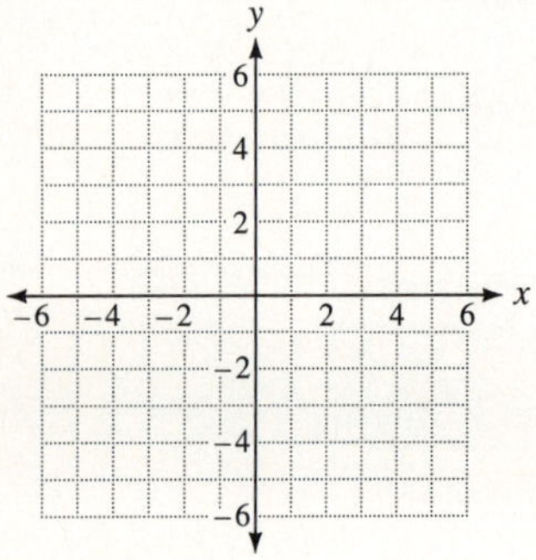

18. $x + 3y = 0$

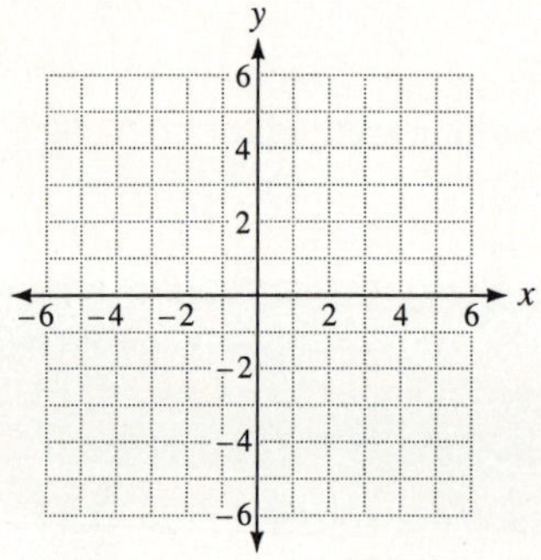

19. $y = -5$

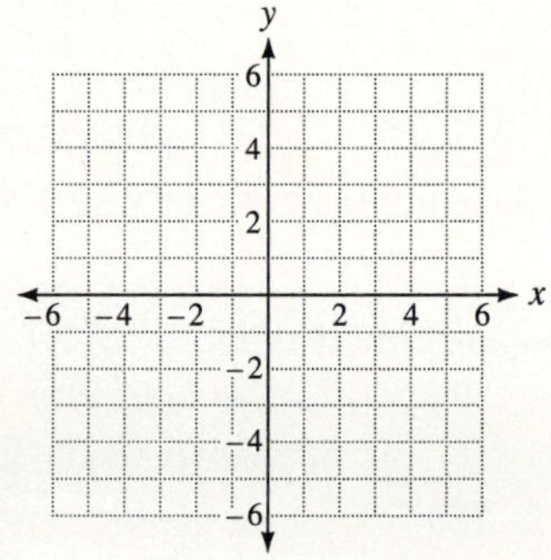

20. $2y - 6 = 0$

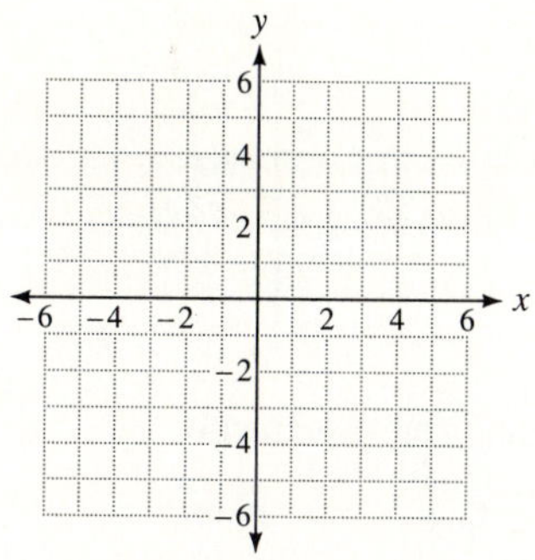

21. $x = 2$

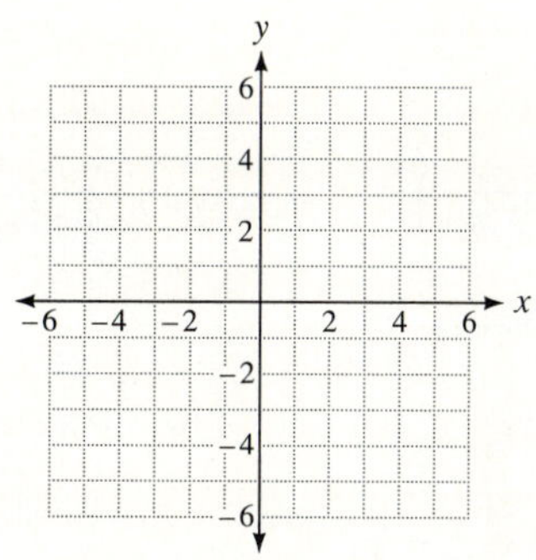

22. $2x + 5 = 0$

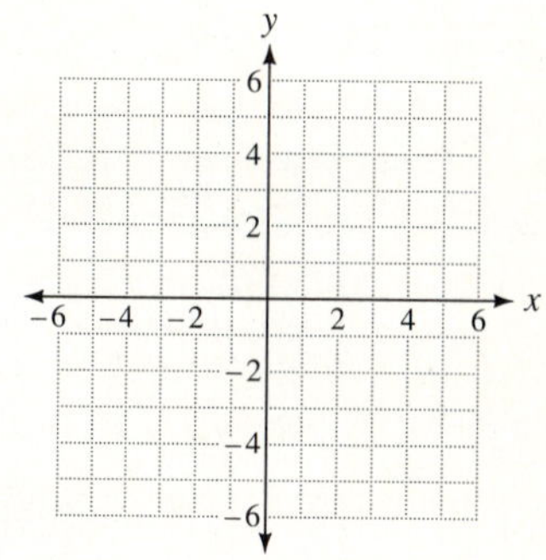

FOR REVIEW

Write true *or* false. *If the statement is false, explain why.*

23. The coordinates of the origin are (0, 0).

24. Every point on the y-axis has y-coordinate equal to zero.

25. Every point above the x-axis has x-coordinate equal to a positive number.

26. Every point to the left of the y-axis has x-coordinate equal to a negative number.

27. Points for which both coordinates are positive lie in quadrant III.

28. Points that have x-coordinate negative and y-coordinate positive lie in quadrant II.

29. In Jefferson City, Missouri, it costs 90¢ plus 10¢ for every tenth of a mile for a taxi ride. The cost in cents, y, of riding a taxi is given by $y = 10x + 90$, where x is the number of tenths of a mile driven. Complete the ordered pairs (5,   ), (50,   ), and (100,   ). How much does it cost to ride **(a)** 5 tenths of a mile? **(b)** 50 tenths of a mile? **(c)** 100 tenths of a mile?

The following exercises review material from Section 1.1 that will help you prepare for the next section. Evaluate each expression.

30. $\dfrac{-2-(-4)}{3-(-1)}$ **31.** $\dfrac{6-(-5)}{-2-(-2)}$ **32.** $\dfrac{-1-(-1)}{2-(-3)}$

ANSWERS: 1. true 2. true 3. false; to the y-axis 4. true 5. false; x-intercept 6. true 7. x-intercept (0, 0); every point on the y-axis is a y-intercept; the graph is the y-axis 8. yes 9. no 10. no 11. no 12. yes 13. yes

14.

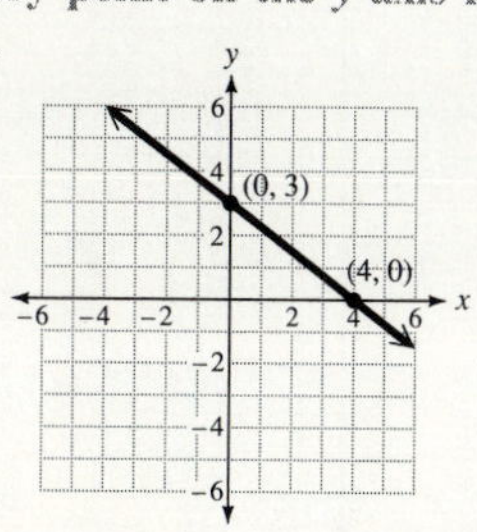

15.

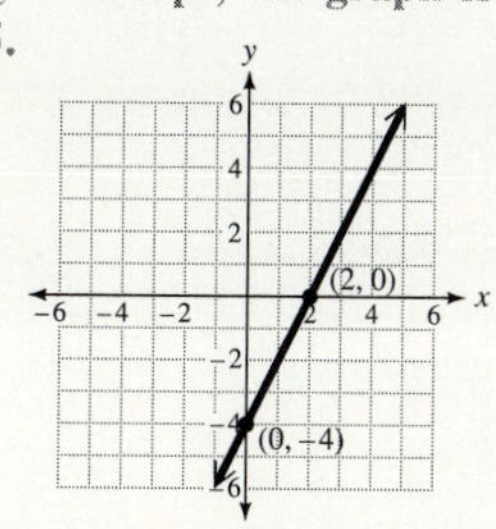

16.

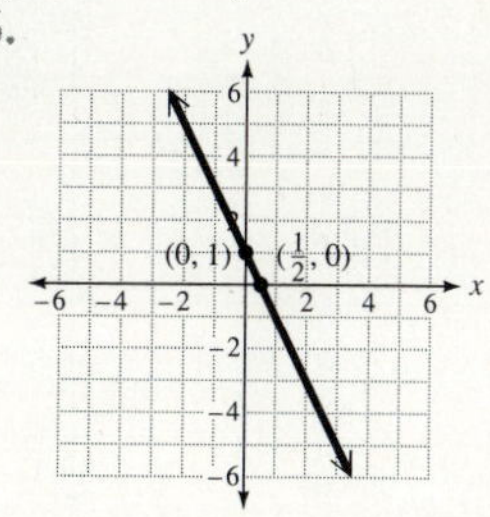

17.

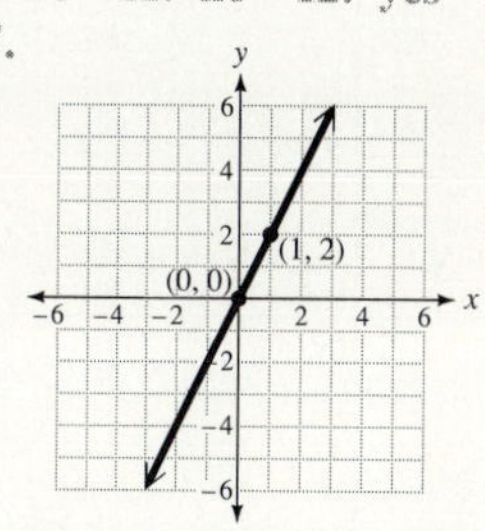

18.

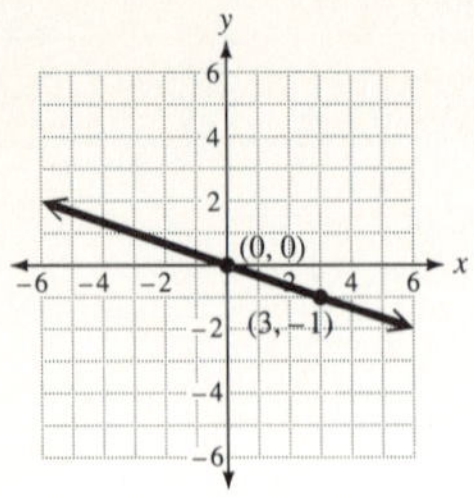

19.

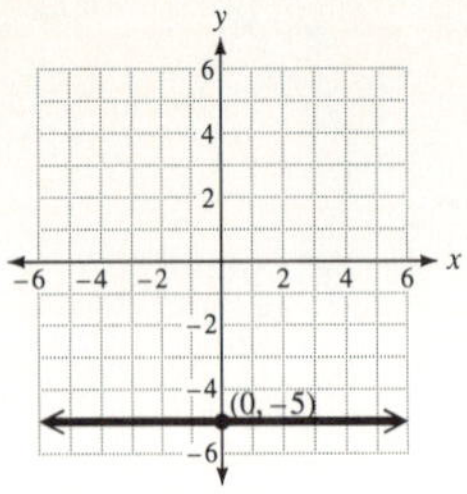

(no x-intercept)

20.

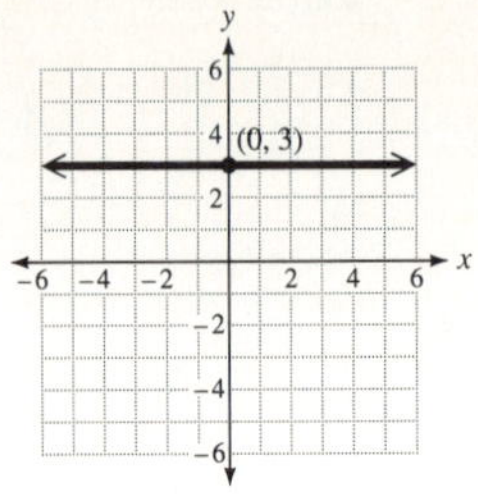

(no x-intercept)

21.

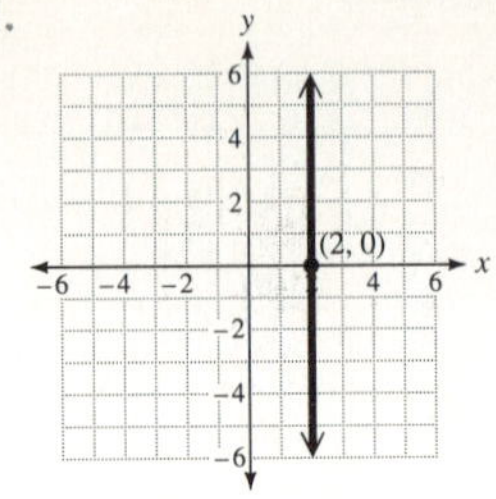

(no y-intercept)

22.

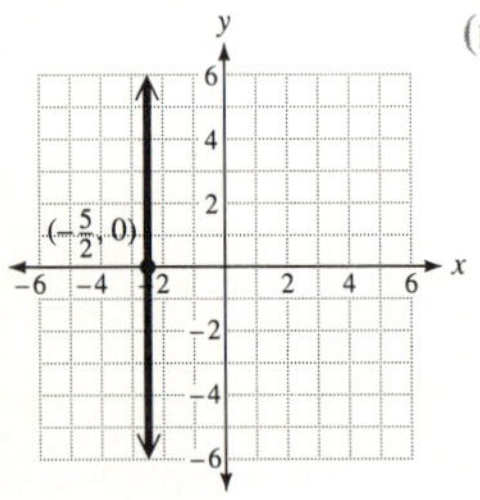

(no y-intercept) **23.** true **24.** false; x-coordinate is zero **25.** false; y-coordinates are positive **26.** true **27.** false; quadrant I **28.** true **29.** (5, 140), (50, 590), (100, 1090); (a) \$1.40 (b) \$5.90 (c) \$10.90 **30.** $\frac{1}{2}$ **31.** undefined **32.** 0

3.2 EXERCISES B

Write true *or* false. *If the statement is false, explain why.*

1. Linear equations are also called first-degree equations.

2. The equation $ax + by + c = 0$, $a \neq 0$, or $b \neq 0$, is called the general form of a linear equation.

3. An equation of the form $by + c = 0$ has as its graph a straight line parallel to the y-axis.

4. In an equation $ax + by + c = 0$, if $c = 0$, $a \neq 0$, $b \neq 0$, the point (0, 0) is both the x-intercept and the y-intercept.

5. A point at which a graph crosses the y-axis is called an x-intercept.

6. To graph a linear equation, only two points are needed.

7. What are the intercepts of the line $y = 0$? What is its graph?

Is the given equation linear? Explain.

8. $3x - y + 2 = 0$

9. $2x^2 - xy - y^2 = 3$

10. $xy = 3$

11. $x - \dfrac{1}{y} = 8$

12. $3x + 7 = 0$

13. $y = \dfrac{1}{2}$

Find the intercepts and graph the following linear equations using the intercept method.

14. $4x - 3y - 12 = 0$

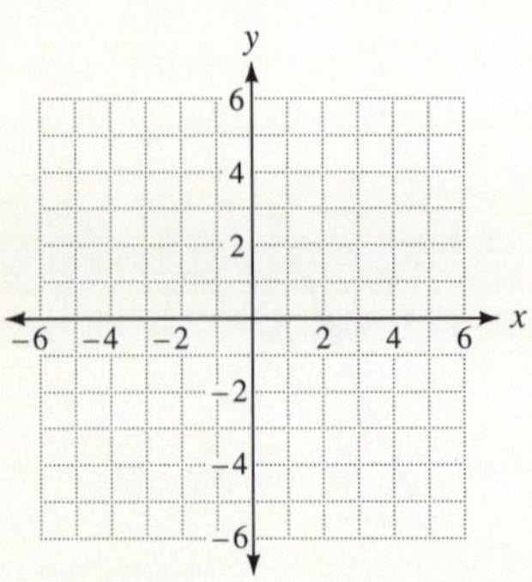

15. $2x + y = 4$

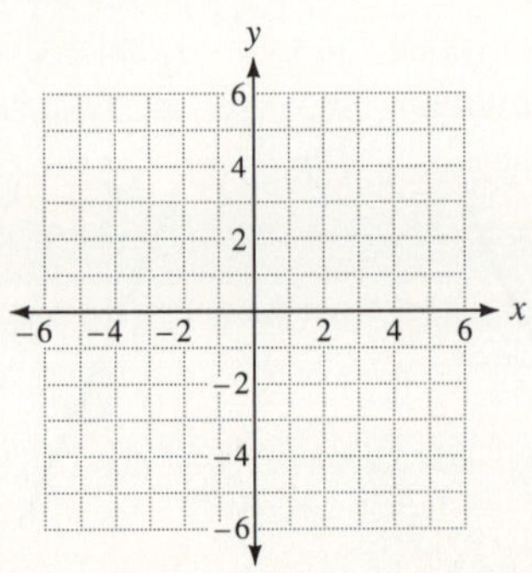

16. $2x - y = 1$

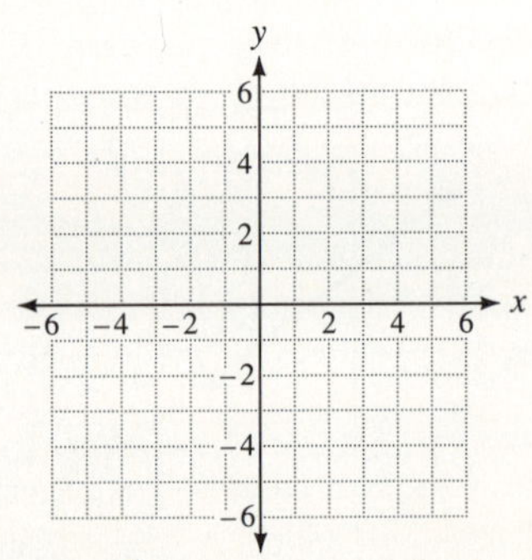

17. $2x + y = 0$

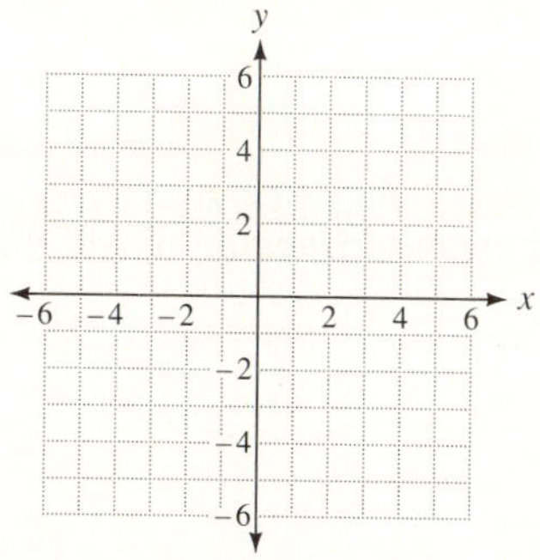

18. $x - 3y = 0$

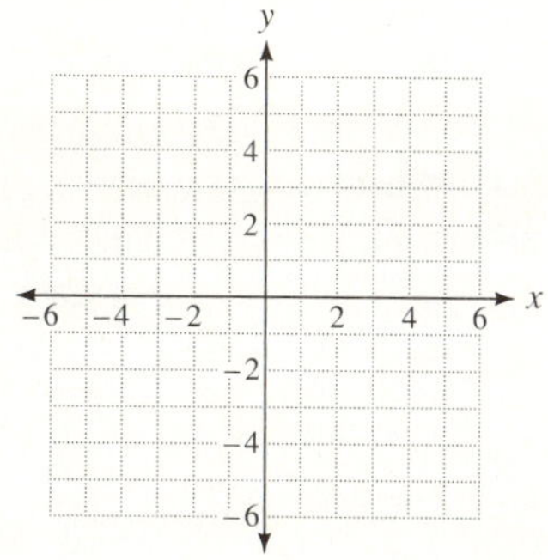

19. $y = 3$

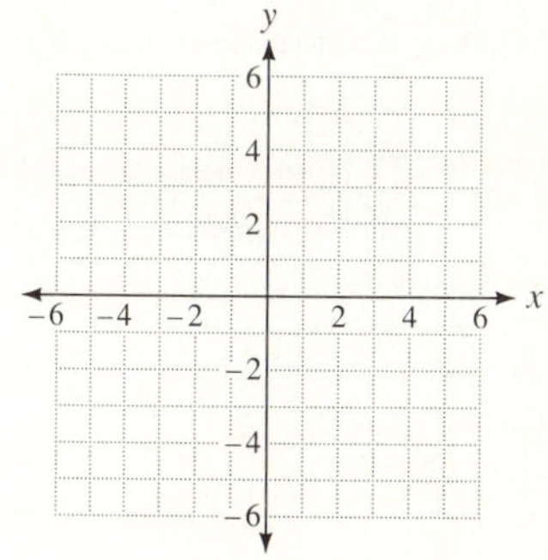

20. $3y + 15 = 0$

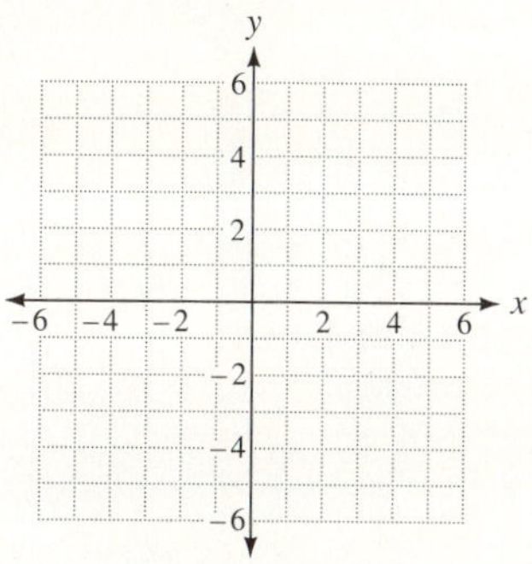

21. $x = -5$

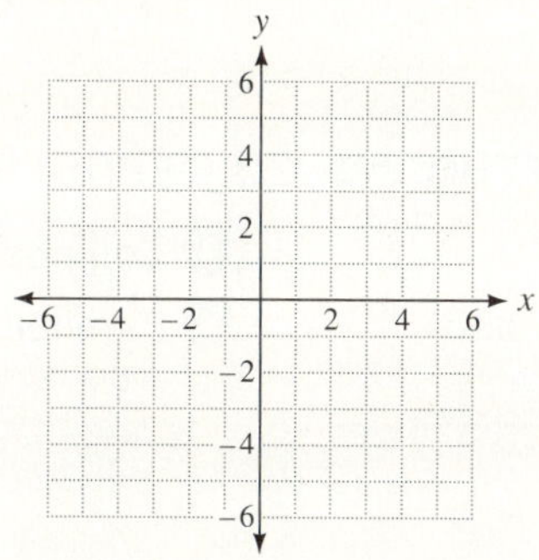

22. $2x - 7 = 0$

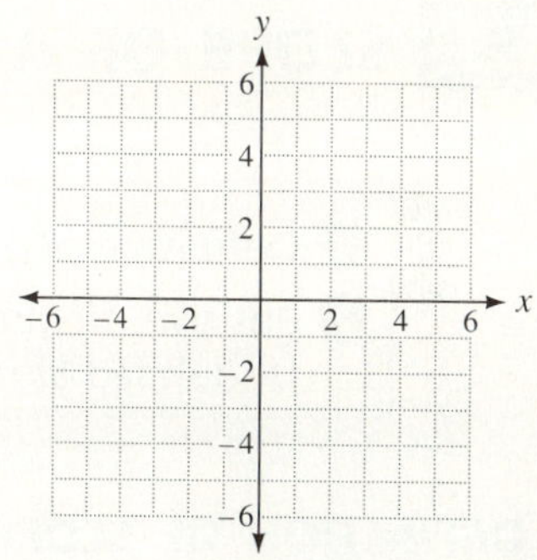

FOR REVIEW

Write true *or* false. *If the statement is false, explain why.*

23. The point with coordinates (0, 0) is called the origin.

24. Every point on the x-axis has x-coordinate equal to zero.

25. Every point below the x-axis has x-coordinate equal to a negative number.

26. Every point to the right of the y-axis has x-coordinate equal to a positive number.

27. Points for which both coordinates are negative lie in quadrant II.

28. Points that have x-coordinate positive and y-coordinate negative lie in quadrant IV.

29. The number, y, of antelope that can live on a prairie is related to the number of acres, x, in the prairie. If this relationship is given by the equation $y = 2x + 10$, complete the ordered pairs (5,), (100,), and (1000,). How many antelope can the prairie sustain if it contains **(a)** 5 acres? **(b)** 100 acres? **(c)** 1000 acres?

The following exercises review material from Section 1.1 that will help you prepare for the next section. Evaluate each expression.

30. $\dfrac{-3-(-5)}{5-(-1)}$

31. $\dfrac{-4-(-4)}{-2-(-3)}$

32. $\dfrac{4-(-7)}{-3-(-3)}$

3.2 EXERCISES C

1. Graph $y = 3x + 2$, $y = \frac{1}{3}x + 2$, $y = 0x + 2$, $y = -\frac{1}{3}x + 2$, and $y = -3x + 2$ in the same coordinate system. What common characteristics do all the lines possess?

2. Graph $y = 2x + 3$, $y = 2x + 1$, $y = 2x + 0$, $y = 2x - 1$, and $y = 2x - 3$ in the same coordinate system. What common characteristics do all the lines possess?

3.3 SLOPE OF A LINE

STUDENT GUIDEPOSTS

1. Definition of Slope
2. Positive, Negative, Zero, and Undefined Slopes
3. Collinear Points
4. Parallel and Perpendicular Lines

1 DEFINITION OF SLOPE

We have seen that the graph of a linear equation may be horizontal (parallel to the x-axis), vertical (parallel to the y-axis), may "slope" upward from lower left to upper right, or may "slope" downward from upper left to lower right. The *slope* of a line (or lack of slope) can be defined precisely.

To define the slope of a line, suppose that $P(x_1, y_1)$ and $Q(x_2, y_2)$ are two points on a line. (The subscript notation distinguishes between the points while identifying the x- and y-coordinates.) The *slope* of the line is the ratio of the vertical change, *rise,* to the horizontal change, *run,* as we move from (x_1, y_1) to (x_2, y_2) along the line. The rise is $y_2 - y_1$, the change in y-coordinates, and the run is $x_2 - x_1$, the change in x-coordinates. Refer to Figure 3.12.

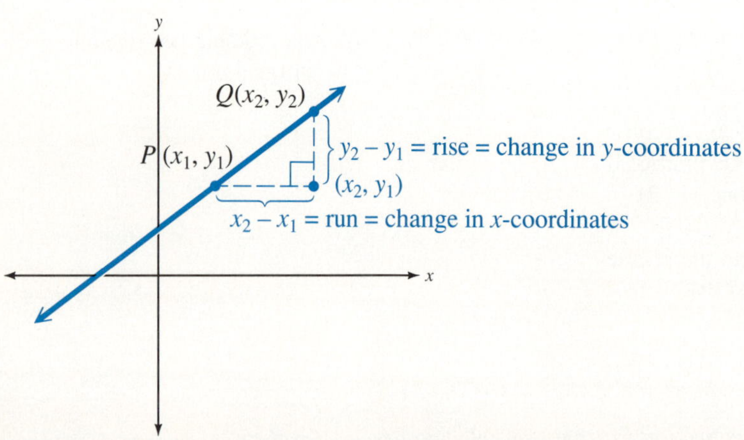

Figure 3.12 Slope $= \dfrac{\text{rise}}{\text{run}} = \dfrac{y_2 - y_1}{x_2 - x_1}$

The Slope of a Line

Let $P(x_1, y_1)$ and $Q(x_2, y_2)$ be two points on a *nonvertical* line. The **slope** m of the line is given by

$$m = \frac{y_2 - y_1}{x_2 - x_1} = \frac{\text{rise}}{\text{run}} = \frac{\text{change in } y\text{-coordinates}}{\text{change in } x\text{-coordinates}}.$$

EXAMPLE 1 Finding the Slope of a Line

Find the slope of the line passing through the given points.

(a) (5, 4) and (2, 1)

Let $P(x_1, y_1)$ be identified with (5, 4) and $Q(x_2, y_2)$ with (2, 1), as shown in Figure 3.13. The slope is given by

$$m = \frac{y_2 - y_1}{x_2 - x_1} = \frac{1-4}{2-5} = \frac{-3}{-3} = 1.$$

Thus, the line goes up one unit for each unit to the right, and it has *positive slope*. The same slope would have been obtained had we let (2, 1) be $P(x_1, y_1)$ and (5, 4) be $Q(x_2, y_2)$. You should calculate m with the different names of the points to check this. *The slope is the same regardless of how the two points are chosen.*

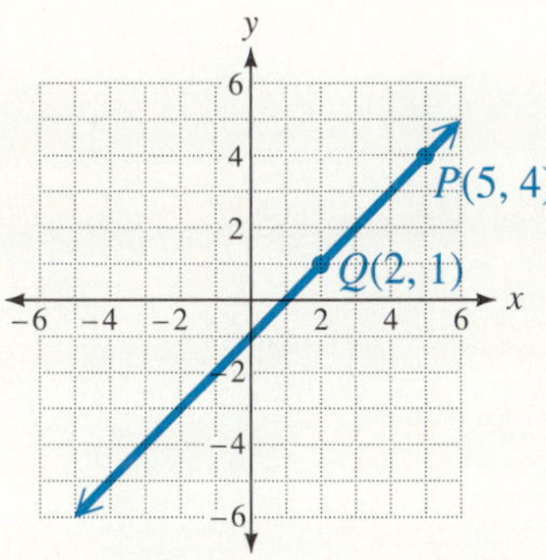

Figure 3.13 Positive Slope

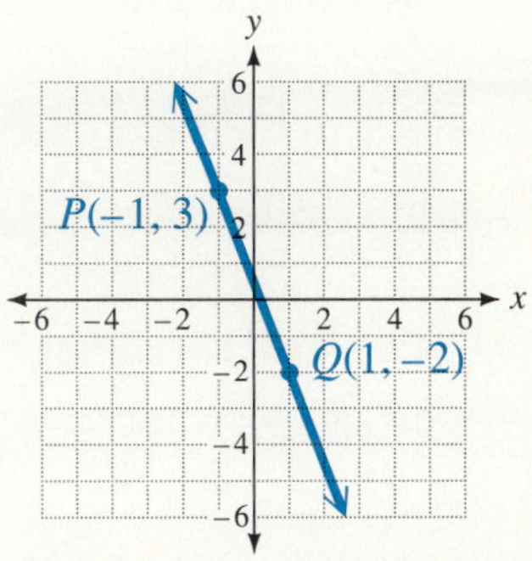

Figure 3.14 Negative Slope

(b) (−1, 3) and (1, −2)

Let $P(x_1, y_1) = P(-1, 3)$ and $Q(x_2, y_2) = Q(1, -2)$, as in Figure 3.14.

$$m = \frac{y_2 - y_1}{x_2 - x_1} = \frac{-2-3}{1-(-1)} = \frac{-5}{1+1} = -\frac{5}{2}$$

Thus, the line goes down 5 units for each 2 units to the right, and it has *negative slope*.

Practice Exercise 1

Find the slope of the line passing through the given points.

(a) (2, 3) and (4, 6)

(b) (1, −3) and (−2, 4)

Answers: (a) $\frac{3}{2}$ (b) $-\frac{7}{3}$

CAUTION

Are the coordinates subtracted in the same order? To find the slope of the line through $P(1, 2)$ and $Q(4, 3)$, you cannot compute

$$\frac{3-2}{1-4} = \frac{y_2 - y_1}{x_1 - x_2}. \quad \text{This is wrong}$$

EXAMPLE 2 Finding Zero Slope and Undefined Slope

Find the slope of the line passing through the given points.

(a) $(-2, -2)$ and $(4, -2)$

Identifying $P(x_1, y_1)$ with $(-2, -2)$ and $Q(x_2, y_2)$ with $(4, -2)$ as in Figure 3.15, we obtain

$$m = \frac{y_2 - y_1}{x_2 - x_1} = \frac{-2 - (-2)}{4 - (-2)} = \frac{-2 + 2}{4 + 2} = \frac{0}{6} = 0.$$

The horizontal line has *zero slope*.

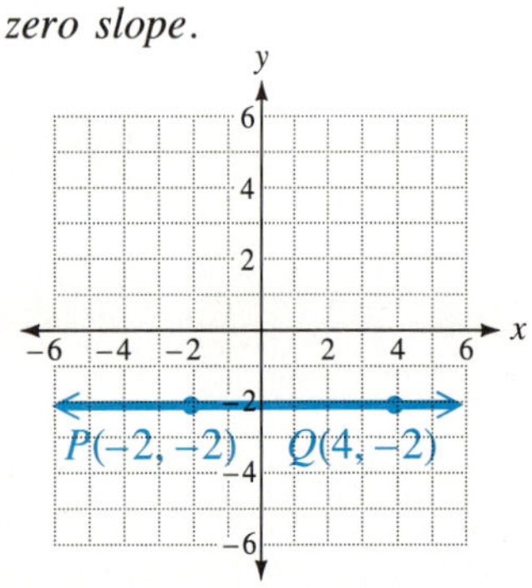

Figure 3.15 Horizontal Line: Zero Slope

(b) $(3, 2)$ and $(3, -1)$

Let $P(x_1, y_1) = (3, 2)$ and $Q(x_2, y_2) = (3, -1)$, as in Figure 3.16.

$$m = \frac{y_2 - y_1}{x_2 - x_1} = \frac{-1 - 2}{3 - 3} = \frac{-3}{0} \quad \text{This is undefined}$$

A vertical line has an *undefined slope*.

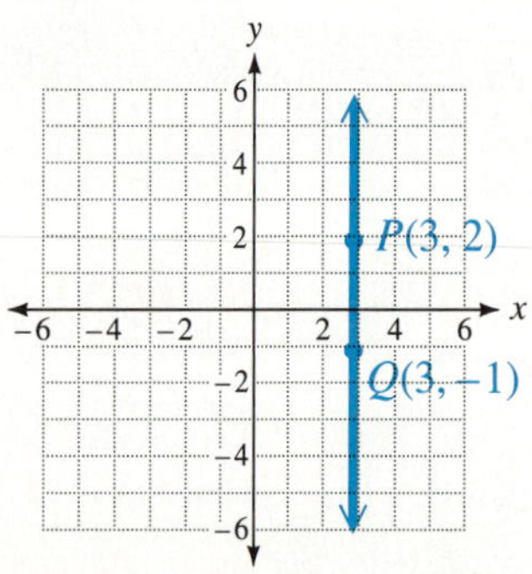

Figure 3.16 Vertical Line: Undefined Slope

Practice Exercise 2

Find the slope of the line passing through the given points.

(a) $(5, 3)$ and $(-4, 3)$

(b) $(-1, 3)$ and $(-1, -2)$

Answers: (a) 0 (b) undefined slope

2 POSITIVE, NEGATIVE, ZERO, AND UNDEFINED SLOPES

Examples 1 and 2 suggest the following summary.

The Slope of a Line

1. A line that "slopes" from lower left to upper right has **positive slope.**
2. A line that "slopes" from upper left to lower right has **negative slope.**
3. A horizontal line (parallel to the x-axis) has **zero slope.**
4. A vertical line (parallel to the y-axis) has an **undefined slope.**

The graphs in Figure 3.17 show the four possibilities listed above.

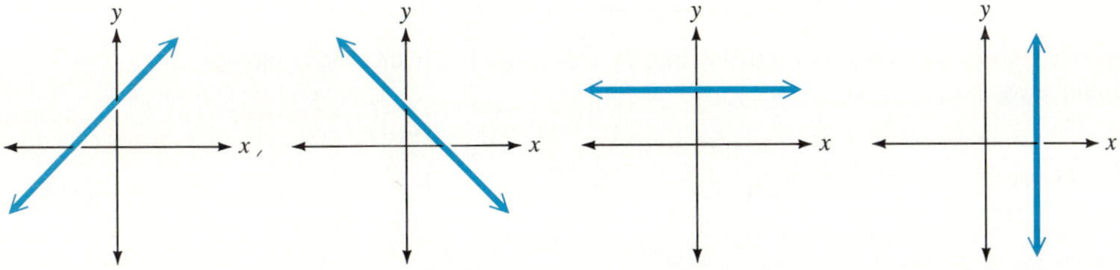

Figure 3.17 Types of Slope

3 COLLINEAR POINTS

Slope can be useful in solving a variety of problems. For example, suppose we are given three points P, Q, and R, and we wish to know if the three points are **collinear** (lie on the same straight line). We find the slope of the line through P and Q, then the slope of the line through Q and R. The points are on the same straight line through P, Q, and R provided the two slopes are the same.

EXAMPLE 3 Showing Points Collinear

Tell whether $P(5, 3)$, $Q(1, 1)$, and $R(-3, -1)$ are collinear.

The slope of the line through P and Q is

$$m_1 = \frac{1-3}{1-5} = \frac{-2}{-4} = \frac{1}{2}.$$

The slope of the line through Q and R is

$$m_2 = \frac{-1-1}{-3-1} = \frac{-2}{-4} = \frac{1}{2}.$$

Since $m_1 = m_2$, the three points lie on the same straight line (that is, they are collinear).

Practice Exercise 3

Tell whether the points $P(2, 3)$, $Q(-1, -2)$, and $R(4, 6)$ are collinear.

Answer: no

4 PARALLEL AND PERPENDICULAR LINES

The slope of a line can also be used to tell when two lines are **parallel** (never intersect) or **perpendicular** (intersect at right angles).

Slopes of Parallel and Perpendicular Lines

Given two distinct nonvertical lines with slopes m_1 and m_2:

1. If $m_1 = m_2$, the lines are parallel (equal slopes determine parallel lines).
2. If $m_1 m_2 = -1$, the lines are perpendicular (slopes that are negative reciprocals determine perpendicular lines).

Also, two distinct vertical lines (both with undefined slope) are parallel, and a vertical line and a horizontal line (with slopes undefined and 0, respectively) are perpendicular.

EXAMPLE 4 Showing Lines Parallel

Verify that the line l_1 through $(2, -1)$ and $(-3, 3)$ and the line l_2 through $(4, 1)$ and $(-1, 5)$ are parallel. See Figure 3.18.

$$\text{The slope of } l_1 \text{ is } m_1 = \frac{3-(-1)}{-3-2} = \frac{3+1}{-5} = -\frac{4}{5}.$$

$$\text{The slope of } l_2 \text{ is } m_2 = \frac{5-1}{-1-4} = \frac{4}{-5} = -\frac{4}{5}.$$

Since $m_1 = m_2$, the lines are parallel.

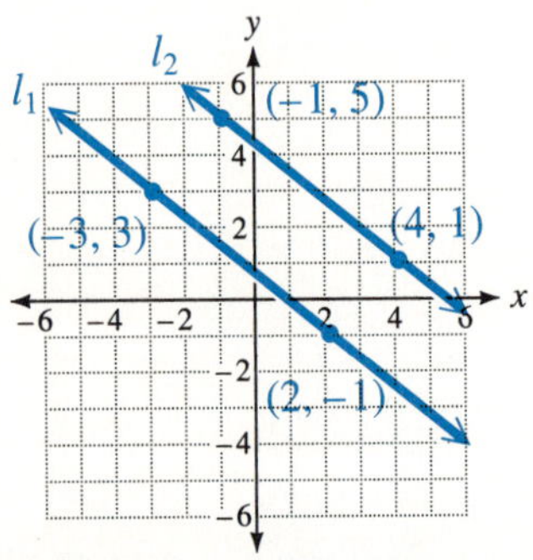

Figure 3.18 Parallel Lines

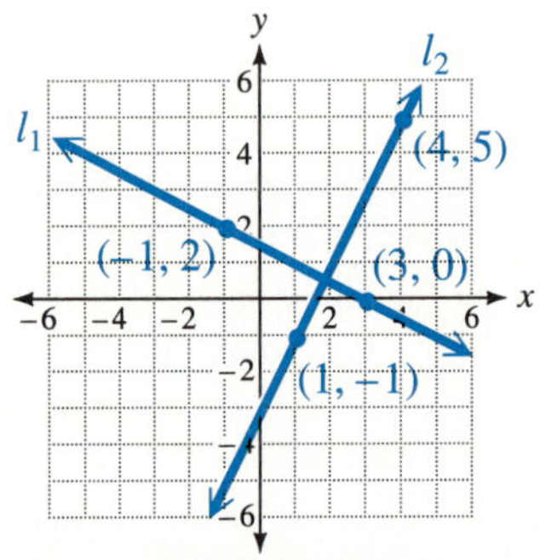

Figure 3.19 Perpendicular Lines

Practice Exercise 4

Is the line l_1 through $(-1, 3)$ and $(2, 4)$ parallel to l_2 through $(5, 2)$ and $(2, 1)$?

Answer: yes

EXAMPLE 5 Showing Lines Perpendicular

Verify that the line l_1 through $(-1, 2)$ and $(3, 0)$ is perpendicular to the line l_2 through $(4, 5)$ and $(1, -1)$. See Figure 3.19.

$$\text{The slope of } l_1 \text{ is } m_1 = \frac{0-2}{3-(-1)} = \frac{-2}{3+1} = \frac{-2}{4} = -\frac{1}{2}.$$

$$\text{The slope of } l_2 \text{ is } m_2 = \frac{-1-5}{1-4} = \frac{-6}{-3} = 2.$$

Since $m_1 m_2 = \left(-\frac{1}{2}\right)(2) = -1$, the lines are perpendicular. Note that $m_1 = -\frac{1}{2} = -\frac{1}{m_2}$.

Practice Exercise 5

Is l_1 through $(-1, 4)$ and $(6, 3)$ perpendicular to l_2 through $(-1, 2)$ and $(-2, -5)$?

Answer: yes

3.3 EXERCISES A

Answer Exercises 1–5 with one of the following phrases: positive slope, negative slope, zero slope, undefined slope.

1. A line parallel to the y-axis has ________________.
2. A line that "slopes" from lower left to upper right has ________________.
3. A line that "slopes" from upper left to lower right has ________________.
4. A line parallel to the x-axis has ________________.
5. The x-axis has ________________.

Find the slope of the line passing through the given pair of points.

6. $(7, -1)$ and $(3, 3)$

7. $(-5, 2)$ and $(-1, -6)$

8. $(-4, 1)$ and $(-1, 3)$

9. $(4, 2)$ and $(4, -2)$

10. $(1, 7)$ and $(3, 3)$

11. $(-2, 7)$ and $(-5, 7)$

12. **(a)** What is the slope of a line through points A and B?
(b) A line through A with slope -2 passes through which other marked point?
(c) Does the line through C and E have positive or negative slope?

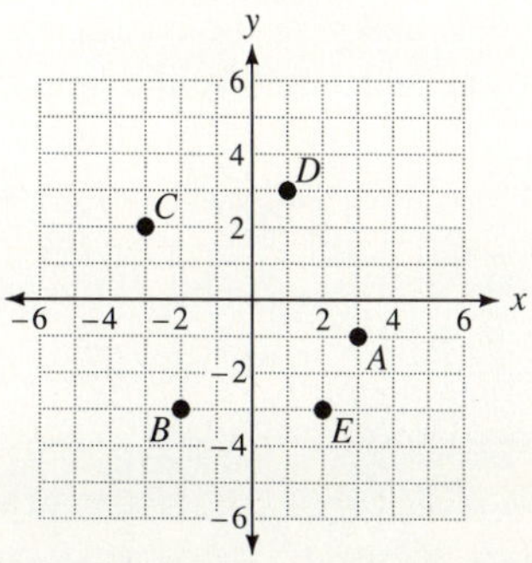

13. **(a)** What is the slope of a line through points H and F?
(b) A line through I with slope $\frac{2}{3}$ passes through which other marked point?
(c) Does the line through G and F have positive or negative slope?

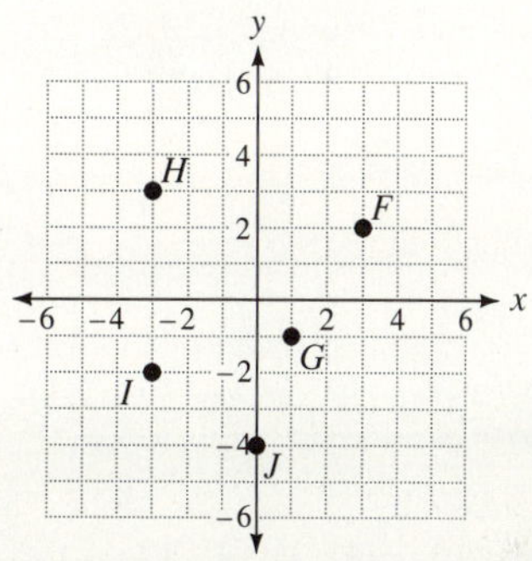

14. Find the slope of a line that is parallel to the line through $(5, -1)$ and $(3, 3)$. Explain.

15. Find the slope of a line that is perpendicular to the line through $(5, -1)$ and $(3, 3)$. Explain.

16. Verify that the line l_1 through $(1, 5)$ and $(-2, -1)$ and the line l_2 through $(-1, -7)$ and $(4, 3)$ are parallel.

17 Verify that the line l_1 through $(0, 2)$ and $(1, -1)$ and the line l_2 through $(0, -1)$ and $(3, 0)$ are perpendicular.

18. Find the slope of the line with equation $x - 3y + 6 = 0$ by first finding two points on the line. Solve the equation for y and compare the coefficient of x with the slope you obtained.

19. Are the points $P(-2, 1)$, $Q(2, 3)$, and $R(0, 2)$ collinear? Explain.

FOR REVIEW

20. Find the intercepts and graph the line with equation $3x - 2y - 6 = 0$.

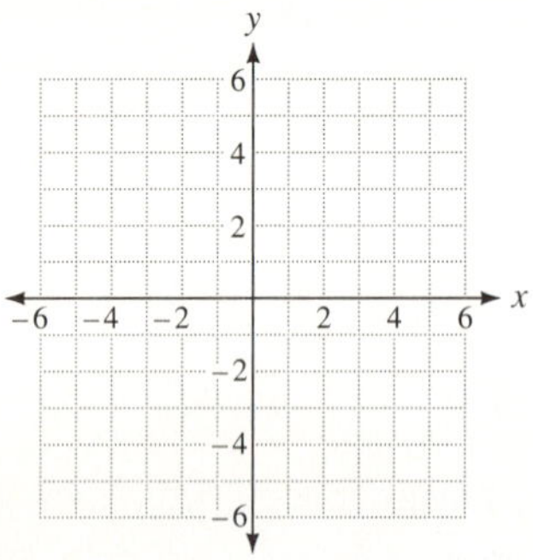

21. Find the intercepts and graph the line with equation $3y - 4 = 0$.

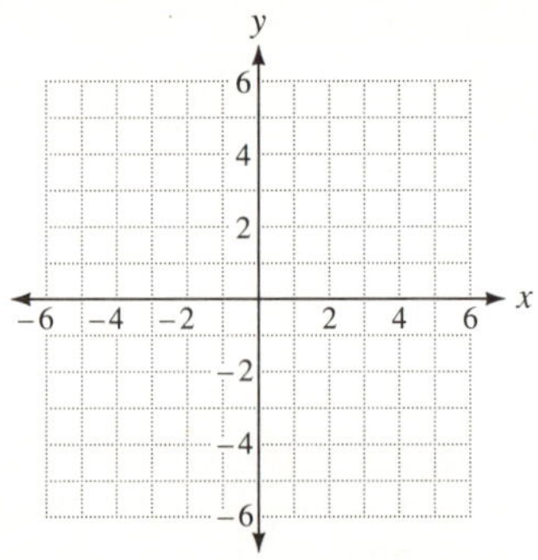

The following exercises will help you prepare for the topics presented in the next section.

22. Remove parentheses and write $y - 2 = \frac{3}{5}(x - (-4))$ in the form $ax + by + c = 0$.

23. Start at point $A(0, -4)$ and give the coordinates of point B if B is 3 units up and 2 units to the right of A.

ANSWERS: 1. undefined slope 2. positive slope 3. negative slope 4. zero slope 5. zero slope 6. −1 7. −2 8. $\frac{2}{3}$ 9. undefined slope 10. −2 11. 0 12. (a) $\frac{2}{5}$ (b) D (c) negative slope 13. (a) $-\frac{1}{6}$ (b) F (c) positive slope 14. −2; parallel lines have the same slope, and the given line has slope −2 15. $\frac{1}{2}$; perpendicular lines must have slopes that are negative reciprocals, and the given line has slope −2, the negative reciprocal of which is $\frac{1}{2}$ 16. Both have slope 2. 17. $m_1 = -3$ and $m_2 = \frac{1}{3}$ (negative reciprocals) 18. $\frac{1}{3}$; the coefficient of x is also $\frac{1}{3}$. 19. Yes. The slope of the line through P and Q is $\frac{1}{2}$, and the slope of the line through Q and R is $\frac{1}{2}$. 20. x-intercept $(2, 0)$; y-intercept $(0, -3)$ 21. no x-intercept; y-intercept $(0, \frac{4}{3})$ 22. $3x - 5y + 22 = 0$ 23. $B(2, -1)$

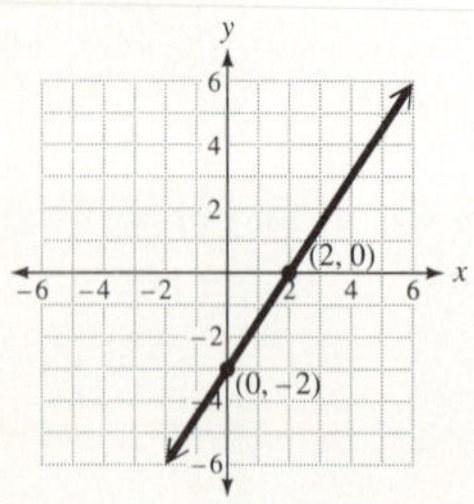

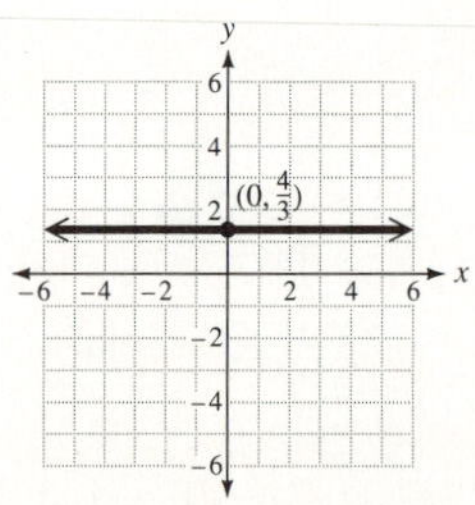

3.3 EXERCISES B

1. A line with zero slope is parallel to which axis?

2. A line with undefined slope is parallel to which axis?

3. Is it true that a line with positive slope "slopes" from lower left to upper right?

4. Is it true that a line with negative slope "slopes" from upper left to lower right?

5. Which coordinate axis has undefined slope? Which coordinate axis has zero slope?

Find the slope of the line passing through the given pair of points.

6. $(4, -2)$ and $(2, 2)$

7. $(-3, -1)$ and $(1, -5)$

8. $(-5, 2)$ and $(-1, 3)$

9. $(-1, 5)$ and $(9, 5)$

10. $(-2, 5)$ and $(4, -1)$

11. $(-3, 5)$ and $(-3, 4)$

12. **(a)** What is the slope of a line through points B and E?
(b) A line through C with slope 3 passes through which other marked point?
(c) Does the line through A and D have positive or negative slope?

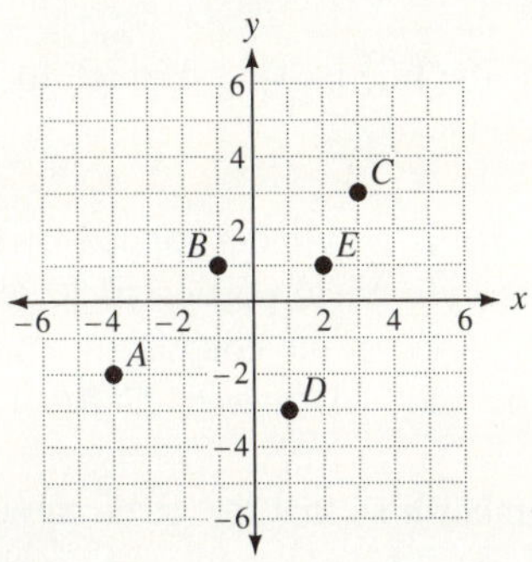

13. **(a)** What is the slope of a line through points F and H?
(b) A line through H with slope $-\frac{1}{2}$ passes through which other marked point?
(c) Does the line through G and F have positive or negative slope?

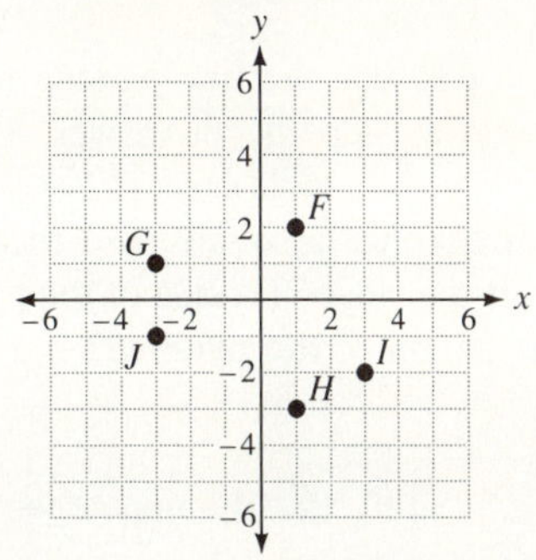

14. What is the slope of a line parallel to the line through $(3, -4)$ and $(2, 5)$?

15. What is the slope of a line perpendicular to the line through $(3, -4)$ and $(2, 5)$?

16. Verify that the line l_1 through $(3, -1)$ and $(4, 2)$ and the line l_2 through $(-2, 1)$ and $(-3, -2)$ are parallel.

17. Verify that the line l_1 through $(4, 0)$ and $(-2, 1)$ and the line l_2 through $(3, 5)$ and $(2, -1)$ are perpendicular.

18. Find the slope of the line with equation $3x + y = 6$ by first finding two points on the line. Solve the equation for y and compare the coefficient of x with the slope you obtained.

19. Are the points $P(2, 2)$, $Q(-1, 0)$, and $R(-3, -1)$ collinear? Explain.

FOR REVIEW

20. Find the intercepts and graph the line with equation $3x + y - 6 = 0$.

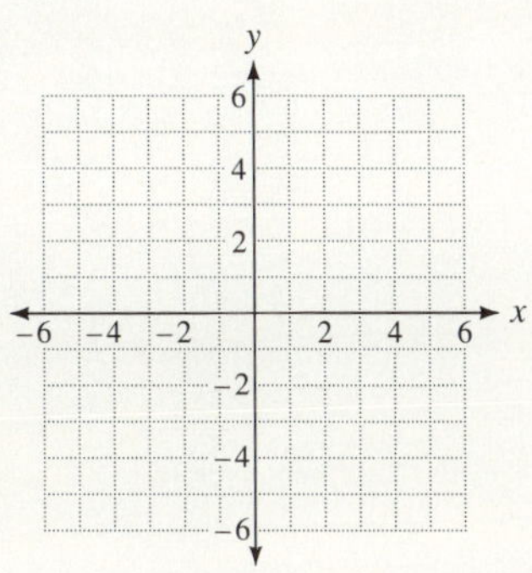

21. Find the intercepts and graph the line with equation $3x + 6 = 0$.

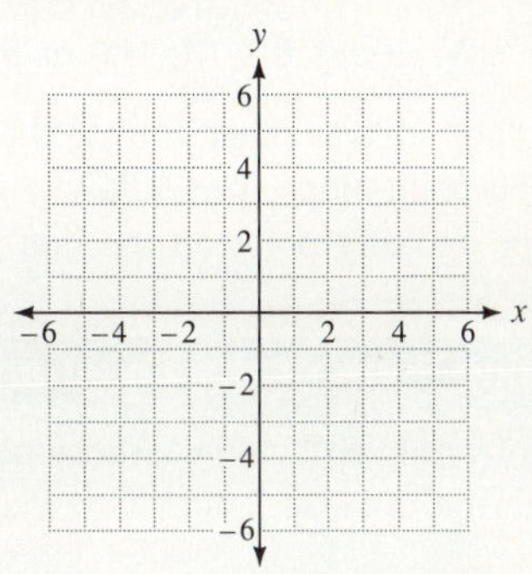

The following exercises will help you prepare for the topics presented in the next section.

22. Remove parentheses and write $y - (-3) = -\frac{4}{3}(x + 2)$ in the form $ax + by + c = 0$.

23. Start at point $A(0, 2)$ and give the coordinates of point B if B is 5 units down and 3 units to the left of A.

3.3 EXERCISES C

1. Is the triangle with vertices $A(-2, 7)$, $B(-3, 3)$, and $C(-6, 8)$ a right triangle? Explain. [*Hint:* Are two adjacent sides perpendicular?]

2. Is the parallelogram with vertices $A(2, 1)$, $B(4, 1)$, $C(-1, -1)$, and $D(1, -1)$ a rectangle? Explain.

3. Find the value of x so that the line passing through $(x, 1)$ and $(3, -x)$ has slope 5. [Answer: 4]

4. Find the slope of the line passing through $(u, 2 - v)$ and $(u + 1, 5 - v)$.

5. A pyramid in Mexico has a square base measuring 400 ft on a side. If the slope of a side of the pyramid is $\frac{6}{5}$, what is the height of the pyramid?

6. A grain storage silo is a cylinder topped by a cone. The height of the cylindrical portion is 20 feet with radius 8 ft. If the slope of the cone is $\frac{1}{4}$, what is the total height of the silo? [Answer: 22 ft]

7. What is the equation of the line through $(2, -1)$ with slope 0? With undefined slope?

8. Find the real number n so that the given three points are collinear: $(6, 3)$, $(n, -1)$, and $(-3, -3)$

3.4 FORMS OF THE EQUATION OF A LINE

STUDENT GUIDEPOSTS

1. Point-Slope Form
2. Slope-Intercept Form
3. Equation of the Line through Two Points
4. Graphing Equations Using Slope

1 POINT-SLOPE FORM

In Section 3.2 we defined the general form of a linear equation in two variables to be

$$ax + by + c = 0 \qquad (a \neq 0 \text{ or } b \neq 0).$$

In this section we will look at two other forms, the *point-slope form* and the *slope-intercept form*. The point-slope form allows us to find the equation of a line when given its slope and any point on the line.

Point-Slope Form of the Equation of a Line

To find the equation of a nonvertical line with slope m and passing through the point (x_1, y_1), substitute these values into the **point-slope form**

$$y - y_1 = m(x - x_1).$$

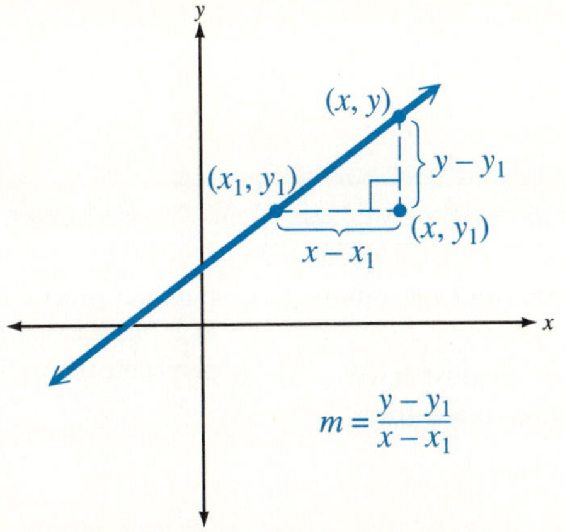

Figure 3.20 Point-Slope Form

To see how this formula is obtained, refer to Figure 3.20 and think of a general point on the line passing through (x_1, y_1) with slope m as having coordinates (x, y). The slope m is given by

$$m = \frac{y - y_1}{x - x_1},$$

where (x, y) plays the role of the point (x_2, y_2) in the formula for m. Multiplying both sides by $(x - x_1)$, we obtain

$$(x - x_1) \cdot m = (x - x_1) \cdot \frac{(y - y_1)}{(x - x_1)},$$

which is equivalent to the point-slope form $y - y_1 = m(x - x_1)$.

EXAMPLE 1 Using Point-Slope Form

Find the point-slope form of the equation of the line with slope $\frac{1}{2}$ passing through the point $(-3, 1)$. Also write the general form of the equation of this line.

We have $(x_1, y_1) = (-3, 1)$ and $m = \frac{1}{2}$, so by substituting we obtain the point-slope form

$$y - y_1 = m(x - x_1)$$

$$y - 1 = \frac{1}{2}[x - (-3)]. \qquad \text{Watch the sign}$$

To obtain the general form of the equation, eliminate fractions, clear parentheses, and then collect all terms on the left side of the equation.

$$y - 1 = \frac{1}{2}(x + 3)$$

$$2(y - 1) = 2 \cdot \frac{1}{2}(x + 3) \qquad \text{Multiply by LCD} = 2$$

$$2y - 2 = x + 3 \qquad \text{Remove parentheses}$$

$$2y - 2 - x - 3 = x + 3 - x - 3 \qquad \text{Subtract } x \text{ and } 3$$

$$-x + 2y - 5 = 0$$

$$x - 2y + 5 = 0 \qquad \text{Multiply by } -1$$

Practice Exercise 1

Give the general form of the equation of a line with slope $\frac{1}{4}$ passing through $(2, -3)$.

Answer: $x - 4y - 14 = 0$

To change $-x + 2y - 5 = 0$ to $x - 2y + 5 = 0$ in Example 1, we multiplied both sides by -1. Either form is acceptable as the general form. However, a positive leading coefficient on x is usually preferred.

EXAMPLE 2 Finding the Equation of a Perpendicular Line

Find the general form of the equation of a line passing through the point $(-1, 2)$ and perpendicular to a line with slope 3.

Since the line is perpendicular to a line with slope 3, it must have slope $-\frac{1}{3}$ (negative reciprocal of 3). Substitute in the point-slope form.

$$y - y_1 = m(x - x_1)$$

$$y - 2 = -\frac{1}{3}(x - (-1))$$

$$y - 2 = -\frac{1}{3}(x + 1) \quad \text{Watch signs}$$

$$3(y - 2) = 3\left(-\frac{1}{3}\right)(x + 1) \quad \text{Eliminate fractions}$$

$$3(y - 2) = -(x + 1)$$

$$3y - 6 = -x - 1 \quad \text{Remove parentheses}$$

$$x + 3y - 5 = 0 \quad \text{Add } x \text{ and 1 to both sides}$$

Practice Exercise 2

Find the general form of the equation of a line passing through $(3, -5)$ and perpendicular to a line with slope $\frac{5}{6}$.

Answer: $6x + 5y + 7 = 0$

2 SLOPE-INTERCEPT FORM

The point-slope form lets us find the equation of a line when we are given a point on the line and the slope of the line. A special case occurs when the given point is the y-intercept.

Slope-Intercept Form of the Equation of a Line

If the equation of a nonvertical line is solved for y, the resulting equation is in **slope-intercept** form

$$y = mx + b$$

where m is the slope and $(0, b)$ is the y-intercept.

To see how this formula is obtained, consider the y-intercept $(0, b)$ as a given point on a line with slope m. Substitute in the point-slope form to obtain the slope-intercept form.

$$y - y_1 = m(x - x_1)$$

$$y - b = m(x - 0)$$

$$y - b = mx$$

$$y = mx + b$$

An important use of the slope-intercept form is that it allows us to find quickly the slope (and y-intercept) of a line without first finding two points on the line. To find the slope-intercept form of a line, simply solve the equation for y. The coefficient of x is the slope, and the constant term is the y-coordinate of the y-intercept.

EXAMPLE 3 Finding Slope and y-Intercept

Find the slope and y-intercept of the line with the following equation.

$$y = \frac{1}{2}x - 3$$

Since the equation is already solved for y (that is, already in slope-intercept form), the slope is $\frac{1}{2}$ and the y-intercept is $(0, -3)$. (Note that $b = -3$, not 3.) The following diagram might be helpful for remembering these facts.

$$y = \frac{1}{2}x - 3 = \frac{1}{2}x + (-3) = mx + b$$

$m = \frac{1}{2} = \text{slope}$ $\quad$ $(0, -3) = y\text{-intercept}$ $\quad$ $[-3 = b]$

Practice Exercise 3

Find the slope and y-intercept of the line $y = -4x - 5$.

Answer: -4, $(0, -5)$

EXAMPLE 4 Finding Slope and y-Intercept

Find the slope and the y-intercept of the line with equation

$$2x - 3y + 5 = 0.$$

First solve for y to obtain the slope-intercept form.

$$-3y = -2x - 5 \qquad \text{Subtract } 2x \text{ and } 5$$

$$\left(-\frac{1}{3}\right)(-3y) = \left(-\frac{1}{3}\right)(-2x - 5) \qquad \text{Multiply by } -\tfrac{1}{3}$$

$$y = \frac{2}{3}x + \frac{5}{3} \qquad \text{Remove parentheses}$$

$m = \frac{2}{3} = \text{slope}$ $\quad$ $\left(0, \frac{5}{3}\right) = y\text{-intercept}$ $\quad$ $\left[\frac{5}{3} = b\right]$

The slope is $\frac{2}{3}$ and the y-intercept is $\left(0, \frac{5}{3}\right)$.

Practice Exercise 4

Find the slope and y-intercept of $4x + 3y - 2 = 0$.

Answer: $-\frac{4}{3}$, $\left(0, \frac{2}{3}\right)$

In Exercise 18 of 3.3 Exercises A we found the slope of the line with equation $x - 3y + 6 = 0$ by first finding two points on the line and then using the definition of the slope. We might use the two points $(0, 2)$ and $(3, 3)$, and substitute to obtain

$$m = \frac{y_2 - y_1}{x_2 - x_1} = \frac{3 - 2}{3 - 0} = \frac{1}{3}.$$

With what we have just learned, we can solve the same problem in a much simpler way by solving for y.

$$-3y = -x - 6 \qquad \text{Subtract } x \text{ and } 6$$

$$y = \frac{1}{3}x + 2 \qquad \text{Divide through by } -3$$

$m = \text{slope} = \frac{1}{3}$

The ability to find the slope of a line more quickly by solving for y is also helpful when considering parallel and perpendicular lines.

EXAMPLE 5 Finding if Lines Are Parallel or Perpendicular

Tell whether the lines with equations $3x - 2y + 7 = 0$ and $2x + 3y - 6 = 0$ are parallel, perpendicular, or neither.

Solve each equation for y (write each in slope-intercept form) to find the slope.

$$\begin{aligned} 3x - 2y + 7 &= 0 \\ -2y &= -3x - 7 \\ y &= \frac{3}{2}x + \frac{7}{2} \\ m_1 &= \frac{3}{2} \end{aligned} \qquad \begin{aligned} 2x + 3y - 6 &= 0 \\ 3y &= -2x + 6 \\ y &= -\frac{2}{3}x + 2 \\ m_2 &= -\frac{2}{3} \end{aligned}$$

Since $m_1 = \frac{3}{2}$ and $m_2 = -\frac{2}{3}$ are negative reciprocals, the two lines are perpendicular.

Practice Exercise 5

Tell whether the lines with equations $5x - 2y + 8 = 0$ and $2x + 5y - 17 = 0$ are parallel, perpendicular, or neither.

Answer: perpendicular

EXAMPLE 6 Recognizing Coinciding Lines

Tell whether the lines with equations $x - 4y + 2 = 0$ and $3x - 12y + 6 = 0$ are parallel, perpendicular, or neither.

Solve each equation for y.

$$\begin{aligned} x - 4y + 2 &= 0 \\ -4y &= -x - 2 \\ y &= \frac{1}{4}x + \frac{2}{4} \\ y &= \frac{1}{4}x + \frac{1}{2} \\ m_1 &= \frac{1}{4} \end{aligned} \qquad \begin{aligned} 3x - 12y + 6 &= 0 \\ -12y &= -3x - 6 \\ y &= \frac{3}{12}x + \frac{6}{12} \\ y &= \frac{1}{4}x + \frac{1}{2} \\ m_2 &= \frac{1}{4} \end{aligned}$$

Since $m_1 = m_2$ we are tempted to conclude that the two lines are parallel. However, the y-intercepts are also equal [both are $(0, \frac{1}{2})$] so the equations determine the same line, that is, the two lines coincide.

Practice Exercise 6

Tell whether the lines with equations $9x + 3y - 6 = 0$ and $3x + y - 2 = 0$ are parallel, perpendicular, or coincide.

Answer: The lines coincide.

Once we have used the slope-intercept form to identify the slope of a line, we can then find the equation of a line parallel or perpendicular to it.

EXAMPLE 7 The Equation of a Line Parallel to a Line

Find the equation of the line passing through the point $(-2, 5)$ parallel to the line with equation $4x + 2y - 9 = 0$.

Since we want a line parallel to the line with equation $4x + 2y - 9 = 0$, we must find the slope of this line. Solving for y,

$$\begin{aligned} 2y &= -4x + 9 \\ y &= -2x + \frac{9}{2}. \end{aligned} \qquad \text{Slope-intercept form}$$

Practice Exercise 7

Find the equation of the line passing through $(4, -2)$ perpendicular to the line with equation $5x - 3y + 10 = 0$.

The slope is -2, and this is also the slope of the desired line (parallel lines have equal slopes) which must pass through $(-2, 5)$. To find the desired line, substitute in the point-slope form.

$$\begin{aligned} y - y_1 &= m(x - x_1) \\ y - 5 &= -2(x - (-2)) \\ y - 5 &= -2(x + 2) \\ y - 5 &= -2x - 4 \\ 2x + y - 1 &= 0 \end{aligned}$$

The general form of the equation of the line through $(-2, 5)$ parallel to the line $4x + 2y - 9 = 0$ is $2x + y - 1 = 0$.

Answer: $3x + 5y - 2 = 0$

CAUTION

The point-slope form lets us find the equation of a line when we are given a point and the slope. It does not work, of course, when the line is vertical with undefined slope.

3 EQUATION OF THE LINE THROUGH TWO POINTS

To find the equation of a line passing through two points, use the slope formula with the point-slope form; as in the next example.

EXAMPLE 8 The Equation of the Line through Two Points

Find the general form of the equation of a line passing through the points $(4, -1)$ and $(2, 7)$.

Let $(x_1, y_1) = (4, -1)$ and $(x_2, y_2) = (2, 7)$ and find the slope.

$$m = \frac{y_2 - y_1}{x_2 - x_1} = \frac{7 - (-1)}{2 - 4} = \frac{8}{-2} = -4$$

Now use $m = -4$ and either point, say $(4, -1)$, in the point-slope form.

$$\begin{aligned} y - y_1 &= m(x - x_1) \\ y + 1 &= -4(x - 4) \\ y + 1 &= -4x + 16 \\ 4x + y - 15 &= 0 \end{aligned}$$

The general form of the equation of the line passing through $(4, -1)$ and $(2, 7)$ is $4x + y - 15 = 0$.

PRACTICE EXERCISE 8

Find the general form of the equation of a line passing through $(-1, 5)$ and $(6, -3)$.

Answer: $8x + 7y - 27 = 0$

The name of the form of the equation tells us what we need to know to use that form.

Which Form to Use

1. If we know the *slope* and the *y-intercept,* we use the *slope-intercept* form.
2. If we know the *slope* and *any point,* we use the *point-slope* form.
3. If we know *two points,* we use the *slope formula* followed by the *point-slope* form.

Many applied problems can be described by a linear equation in two variables. The next example illustrates this.

EXAMPLE 9 Sales Projection Problem

In 1982 the total sales of a newly formed manufacturing company amounted to \$55,000. During the year 1989, the total sales were \$111,000. Assuming that the total sales y in year x can be approximated by a linear equation, find this equation and approximate the total sales expected in 1996.

Suppose we identify 1982 as year 1, $x = 1$. Then 1989 corresponds to year 8 and $x = 8$. When $x = 1$, $y = 55{,}000$ and when $x = 8$, $y = 111{,}000$, giving us the two points $(x_1, y_1) = (1, 55{,}000)$ and $(x_2, y_2) = (8, 111{,}000)$ on the line describing total sales. First find the slope of the line.

$$m = \frac{y_2 - y_1}{x_2 - x_1} = \frac{111{,}000 - 55{,}000}{8 - 1} = \frac{56{,}000}{7} = 8000$$

Now use 8000 for m and $(1, 55{,}000)$ for (x_1, y_1) in the point-slope form.

$$y - 55{,}000 = 8000(x - 1)$$
$$y - 55{,}000 = 8000x - 8000$$
$$y = 8000x + 47{,}000 \quad \text{Slope-intercept form}$$

We leave the equation in slope-intercept form since we want the value of y when x is 15 (in year 1996).

$$y = 8000(15) + 47{,}000 \quad x = 15$$
$$= 120{,}000 + 47{,}000$$
$$= 167{,}000$$

In 1996, the company can expect total sales of about \$167,000.

Practice Exercise 9

Kopy Kat Video Services can duplicate 5 videos at a cost of \$40. With volume purchasing of blank tapes, the owner discovers that 80 videos can be duplicated for \$565. If y is the cost of duplicating x videos, find a linear equation describing this relationship. Use the equation to determine the cost of duplicating 100 videos.

Answer: $y = 7x + 5$; \$705

4 GRAPHING EQUATIONS USING SLOPE

In Section 3.2 we learned how to graph linear equations using the intercepts. We conclude this section by showing how the graph of a linear equation given in slope-intercept form can be easily obtained using the rise-over-run concept of slope. The method is illustrated in the following example.

EXAMPLE 10 Graphing Using Slope

Graph $y = \frac{4}{3}x - 2$.

Since the y-intercept is $(0, -2)$, we know that the line passes through this point. We must find the one line of the infinitely many through $(0, -2)$ that has slope $\frac{4}{3}$. To do this, we will find a second point using

Practice Exercise 10

Graph $y = 2x + 1$. [*Hint:* The slope is $2 = \frac{2}{1}$, so the rise is 2 and the run is 1.]

the rise of 4 units and the run of 3 units from slope $\frac{4}{3}$. Starting at the y-intercept $(0, -2)$, we move up 4 units (a rise of 4) and then move right 3 units (a run of 3), ending up at the point $(3, 2)$. The line passes through $(0, -2)$ and $(3, 2)$ as shown in Figure 3.21.

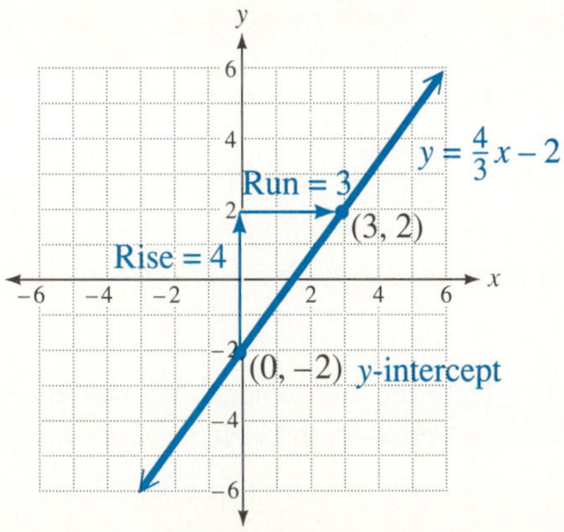

Figure 3.21

Answer: The line passes through $(0, 1)$ and $(1, 3)$.

To graph using slope, we do not need the equation of the line. All that is needed is one point on the line and the slope of the line.

CAUTION

If the slope is $-\frac{2}{3}$, we can say the rise is -2 and the run is 3. A rise of -2 means move *down* two units. Alternatively, if we say the rise is 2 and the run is -3, a run of -3 means move *left* 3 units. Either interpretation will result in the same straight line.

3.4 EXERCISES A

Supply the required information for Exercises 1–21.

1. Find the general form of the equation of the line passing through the point $(3, -1)$ with slope -2. [*Hint:* First find the point-slope form and then transform it to the general form.]

2. Find the general form of the equation of the line passing through the point $(2, -4)$ parallel to a line with slope $\frac{1}{3}$. Also give the x-intercept and y-intercept.

3. Find the general form of the equation of the line passing through the point $(-1, -3)$ perpendicular to a line with slope $\frac{1}{5}$.

4. Find the general form of the equation of the line passing through the points $(7, -1)$ and $(5, 3)$.

5. Find the slope-intercept form of the equation of the line passing through the points $(-2, 3)$ and $(1, -3)$. What is the slope? The y-intercept?

6. Find the slope and y-intercept of the line with equation $5x - 2y + 4 = 0$.

7. Find the slope and y-intercept of the line with equation $5x + 4 = 0$.

8. Find the slope and y-intercept of the line with equation $-2y + 4 = 0$.

9. Tell whether the lines with equations $5x - 3y + 8 = 0$ and $3x + 5y - 7 = 0$ are parallel, perpendicular, or neither. Justify your answer.

10. Tell whether the lines with equations $2x - y + 7 = 0$ and $-6x + 3y - 1 = 0$ are parallel, perpendicular, or neither. Justify your answer.

11. Tell whether the lines with equations $2x + 1 = 0$ and $-3y - 4 = 0$ are parallel, perpendicular, or neither. Justify your answer.

12. Find the general form of the equation of the line passing through $(-1, 4)$ parallel to the line with equation $-3x - y + 4 = 0$.

13 Find the general form of the equation of the line passing through $(-1, 4)$ perpendicular to the line with equation $-3x - y + 4 = 0$.

14. Find the general form of the equation of the line through $(-2, -4)$ with $m = -3$.

15. Find the general form of the equation of the line with y-intercept $(0, 2)$ and slope -5.

16 Find the general form of the equation of the line with x-intercept $(3, 0)$ and slope $\frac{1}{2}$.

17. Find the slope and y-intercept of the line with equation $6x + 2y - 10 = 0$.

18. Find the slope and y-intercept of the line with equation $3y + 9 = 0$.

19. Find the slope and y-intercept of the line with equation $x = 7$.

20 Find the general form of the equation of a line through $(2, 3)$ with undefined slope.

21. Consider the line that passes through the points $(1, 6)$ and $(5, -2)$.
(a) Find the slope of a line parallel to this line. **(b)** Find the slope of a line perpendicular to this line.

Some application problems involve information that can be described by a straight-line graph. If two pieces of information are given, and they are interpreted as points on the graph of a line, an equation describing the situation can be found. Use this idea in the following two exercises.

22 John makes wood-burning stoves. He discovers that 10 stoves can be constructed for \$2800, and 25 stoves can be made for \$6125. If y is the cost required to make x stoves, find a linear equation describing this relationship. Use the equation to calculate the cost of making 31 stoves.

23. In 1987 (call this year 1), the Mutter Manufacturing Company had sales of \$40,000. Three years later, in 1990 (year 4), total sales were \$90,000. If y represents total sales in year x, find a linear equation describing this relationship. Use the equation to approximate total sales for the year 1991 (year 5).

Graph the following equations using the slope method illustrated in Example 10.

24. $y = \frac{2}{3}x - 4$

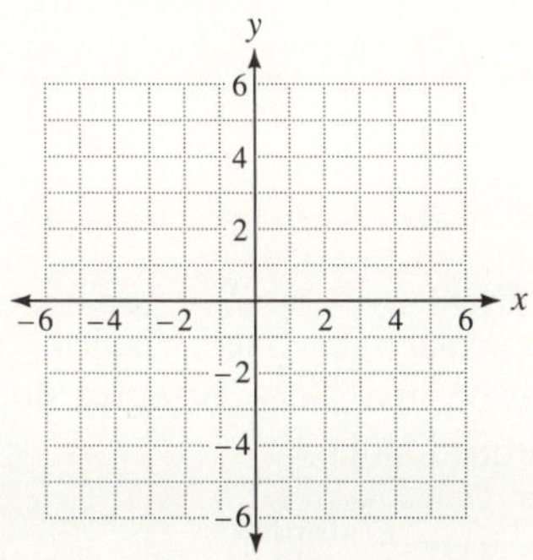

25. $y = 4x + 1$

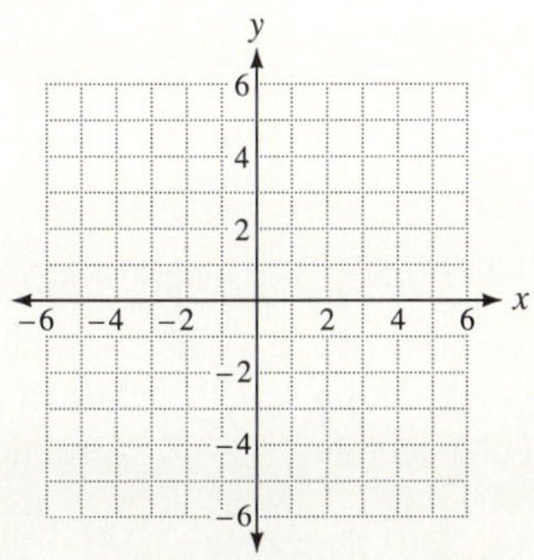

26. $y = -\frac{2}{3}x + 3$

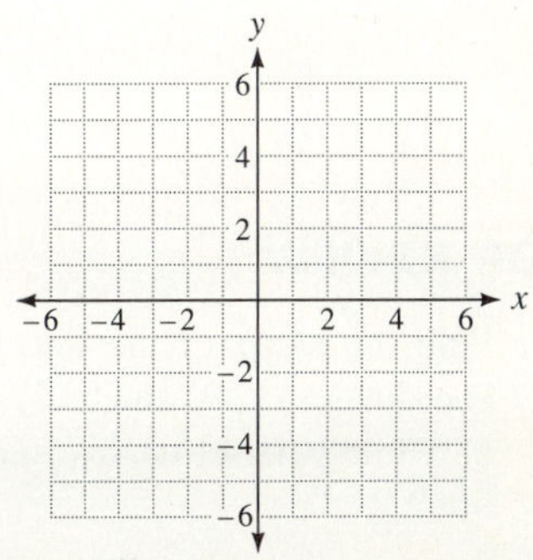

27. $y = -3x + 2$

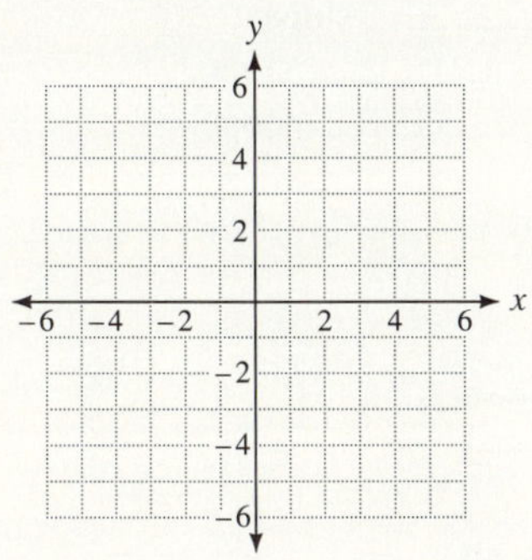

28. $x + 3y - 6 = 0$

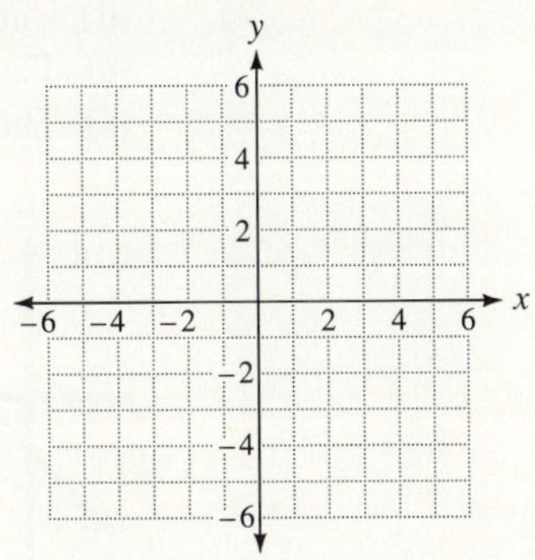

29. $3x - 4y + 8 = 0$

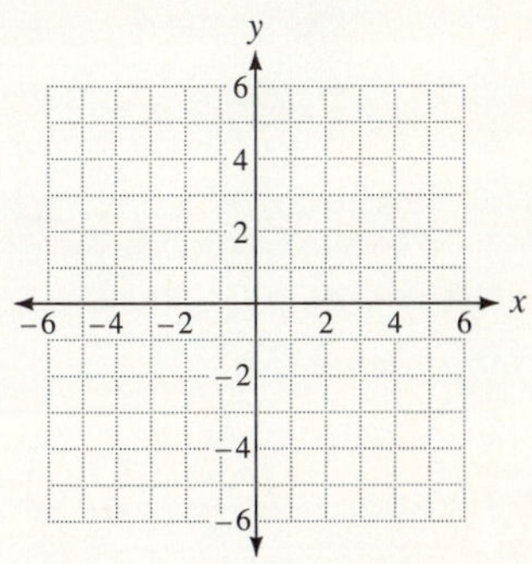

Without finding the equation of the line, graph the line using the given information and the slope method.

30. Through $(1, 2)$ with slope -3

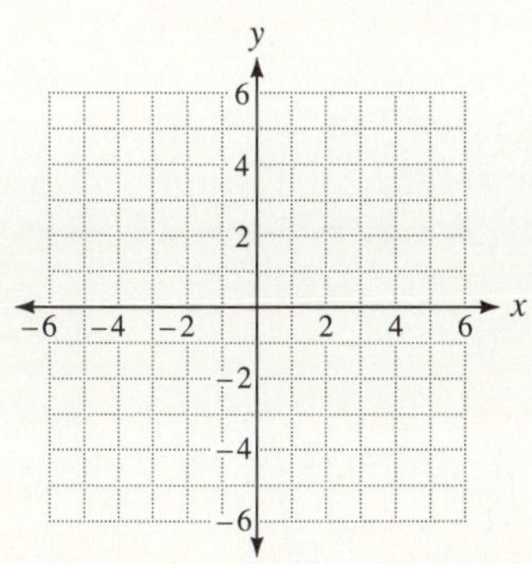

31. Through $(-3, 2)$ with slope $\frac{3}{5}$

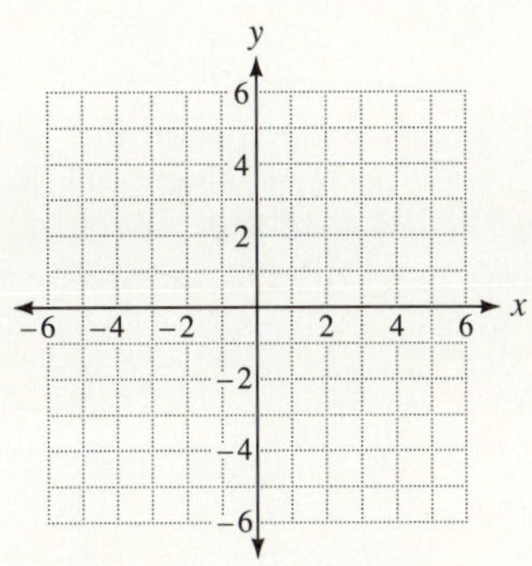

32. Through $(-4, 2)$ parallel to the line $y = 3x + 5$

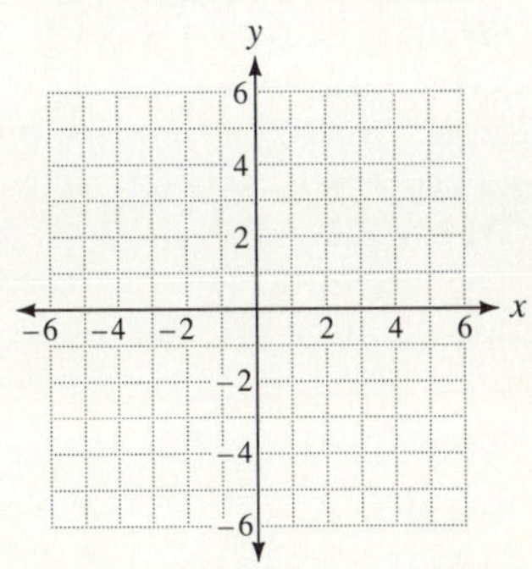

33. The graph shows the relationship between the distance y traveled by a car and the time of travel x. Find the equation that describes this relationship and use it to find the distance traveled after 10 hours.

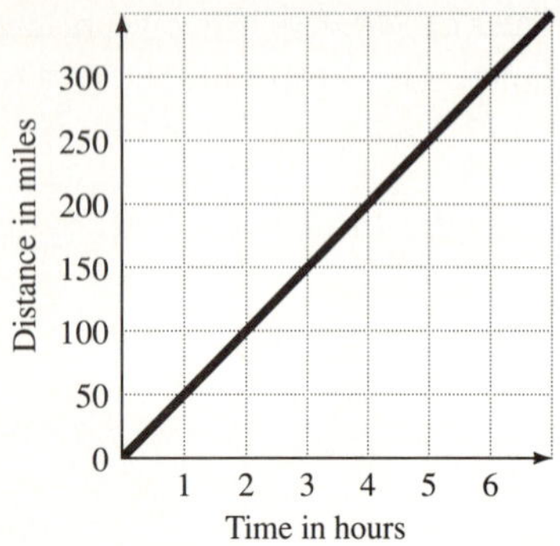

34. The graph shows the relationship between the cost of a rental mobile home y and the depreciation allowed for tax purposes x. Find the equation that describes this relationship and use it to find the value of the home after five and one-half years.

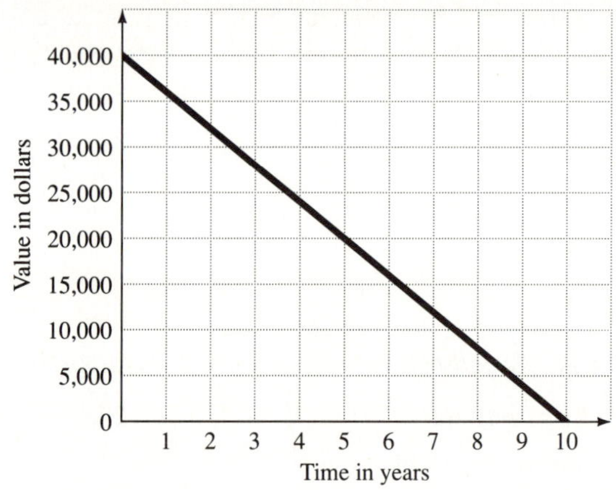

FOR REVIEW

35. Find the slope of the line joining the points with coordinates $(1, 4)$ and $(-7, -12)$. What is the slope of a line parallel to this line? perpendicular to this line?

36. Without finding the actual values, tell whether each line in the figure has positive slope, negative slope, zero slope, or undefined slope.

(a) Line l_1 has ________ slope.
(b) Line l_2 has ________ slope.
(c) Line l_3 has ________ slope.
(d) Line l_4 has ________ slope.
(e) The x-axis has ________ slope.
(f) The y-axis has ________ slope.

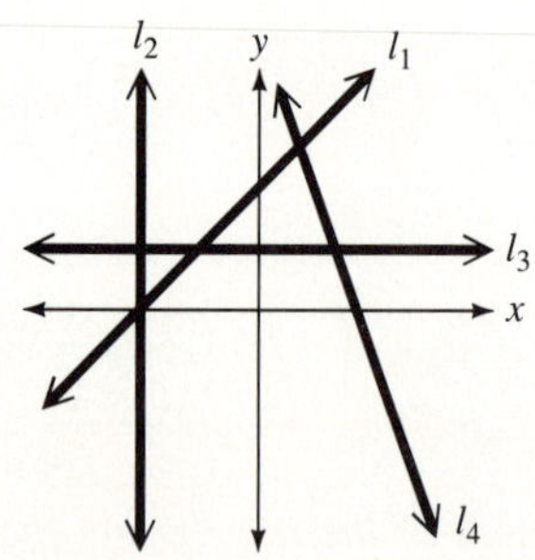

ANSWERS: **1.** $2x + y - 5 = 0$ **2.** $x - 3y - 14 = 0$; x-intercept $(14, 0)$; y-intercept $\left(0, -\frac{14}{3}\right)$ **3.** $5x + y + 8 = 0$ **4.** $2x + y - 13 = 0$ **5.** $y = -2x - 1$; slope $= -2$; y-intercept $(0, -1)$ **6.** slope $= \frac{5}{2}$; y-intercept $(0, 2)$ **7.** undefined slope; no y-intercept **8.** slope $= 0$; y-intercept $(0, 2)$ **9.** perpendicular $\left(\text{slopes are } \frac{5}{3} \text{ and } -\frac{3}{5}\right)$ **10.** parallel (both slopes are 2 and the y-intercepts are different) **11.** perpendicular (each is parallel to one of the axes) **12.** $3x + y - 1 = 0$ **13.** $x - 3y + 13 = 0$ **14.** $3x + y + 10 = 0$ **15.** $5x + y - 2 = 0$ **16.** $x - 2y - 3 = 0$ **17.** slope $= -3$; y-intercept $(0, 5)$ **18.** slope $= 0$; y-intercept $(0, -3)$ **19.** undefined slope; no y-intercept **20.** $x - 2 = 0$ **21.** (a) -2 (b) $\frac{1}{2}$ **22.** $665x - 3y + 1750 = 0$; \$7455 **23.** $50{,}000x - 3y + 70{,}000 = 0$; \$106,666.67

24.
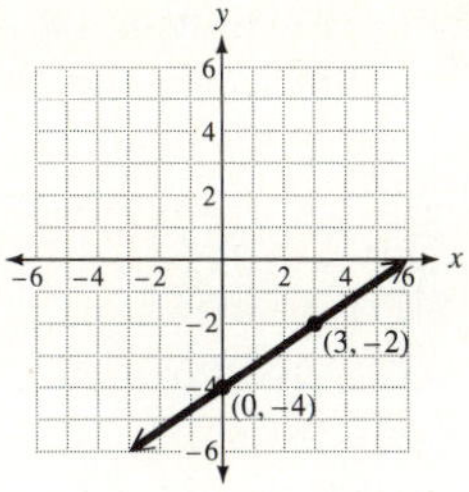

25.
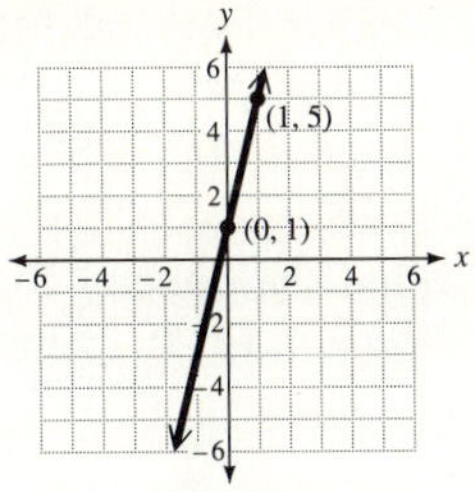

26.
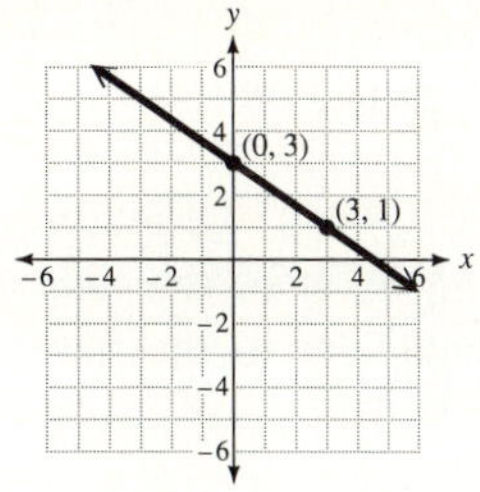

27.
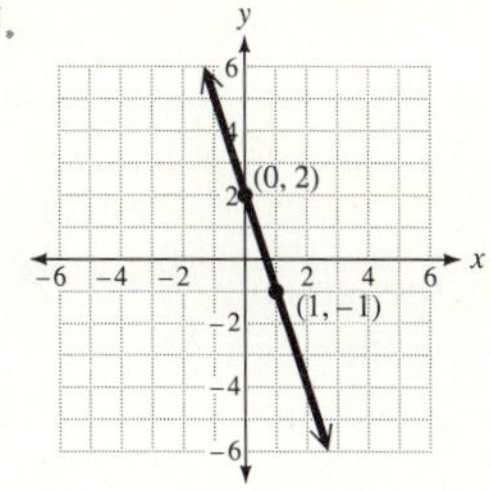

28.
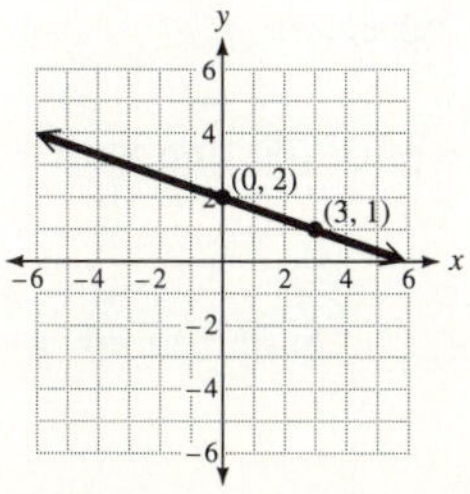

29.
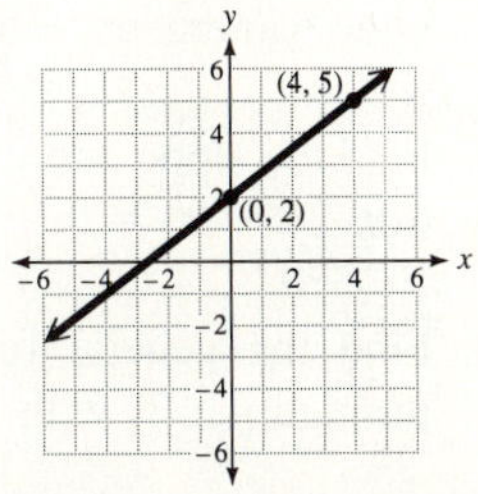

30.
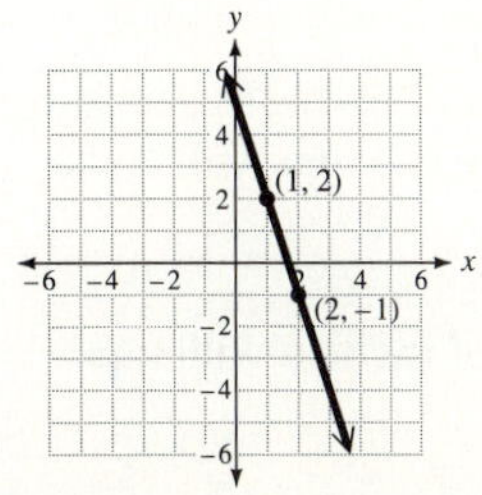

31.
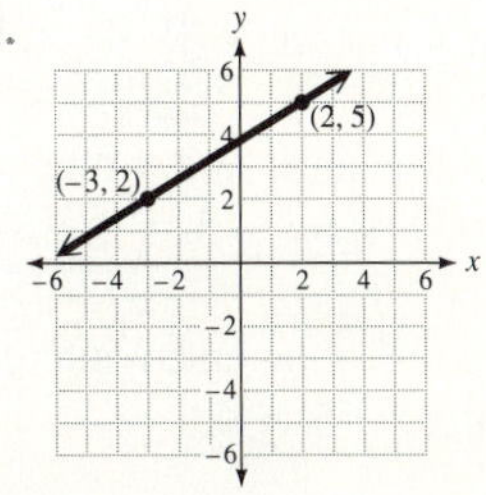

32.
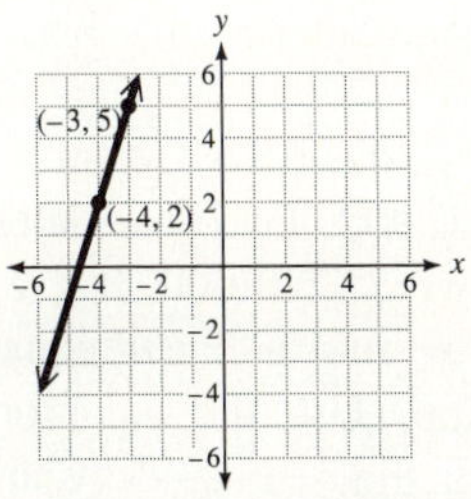

33. $y = 50x$; 500 miles **34.** $y = -4000x + 40{,}000$; \$18,000 **35.** 2; 2; $-\frac{1}{2}$
36. (a) positive (b) undefined (c) zero (d) negative (e) zero (f) undefined

3.4 EXERCISES B

Supply the required information for Exercises 1–23.

1. Find the general form of the equation of the line passing through the point $(-2, -4)$, with slope -3.

2. Find the general form of the equation of the line passing through the point $(-3, 5)$ parallel to a line with slope -4. Also give the x-intercept and y-intercept.

3. Find the general form of the equation of the line passing through the point $(3, -5)$ perpendicular to a line with slope 7.

4. Find the general form of the equation of the line passing through the points $(-7, 1)$ and $(3, -5)$.

5. Find the slope-intercept form of the equation of the line passing through the points $(4, -5)$ and $(2, 8)$. What is the slope? The y-intercept?

6. Find the slope and y-intercept of the line with equation $6x + 2y - 10 = 0$.

7. Find the slope and y-intercept of the line with equation $x + 7 = 0$.

8. Find the slope and y-intercept of the line with equation $3y + 9 = 0$.

9. Tell whether the lines with equations $2x - y + 3 = 0$ and $x + 2y - 5 = 0$ are parallel, perpendicular, or neither. Justify your answer.

10. Tell whether the lines with equations $5x - 2y + 3 = 0$ and $-10x + 4y + 3 = 0$ are parallel, perpendicular, or neither. Justify your answer.

11. Tell whether the lines with equations $4x - y + 7 = 0$ and $3x - 1 = 0$ are parallel, perpendicular, or neither. Justify your answer.

12. Find the general form of the equation of the line passing through $(-2, 5)$ parallel to the line with equation $4x + 2y - 9 = 0$.

13. Find the general form of the equation of the line passing through $(4, -7)$ perpendicular to the line with equation $x - 5y + 7 = 0$.

14. Find the general form of the equation of the line through $(3, 0)$ with $m = -\frac{1}{2}$.

15. Find the general form of the equation of the line with y-intercept $(0, 4)$ and slope -3.

16. Find the general form of the equation of the line with x-intercept $(-5, 0)$ and slope 2.

17. Find the slope and y-intercept of the line with equation $3x + 6y - 5 = 0$.

18. Find the slope and y-intercept of the line with equation $2y + 10 = 0$.

19. Find the slope and y-intercept of the line with equation $2x - 5 = 0$.

20. Find the general form of the equation of a line through $(2, 3)$ with slope zero.

21. Consider the line that passes through the points $(2, -1)$ and $(-3, 9)$.
(a) Find the slope of a line parallel to this line. **(b)** Find the slope of a line perpendicular to this line.

22. Willie makes leather bags. He knows that 5 bags can be made for \$45, and 20 bags can be made for \$140. If y is the cost of making x bags, find a linear equation describing this relationship. Use the equation to calculate the cost of making 65 bags.

23. In 1988 (call this year 1), the BRS Construction Company had sales of \$100,000. Two years later, in 1990 (year 3), total sales were \$220,000. If y represents total sales in year x, find a linear equation describing this relationship. Use the equation to approximate total sales for the year 1993 (year 6).

Graph the following equations using the slope method illustrated in Example 10.

24. $y = \dfrac{3}{2}x + 1$

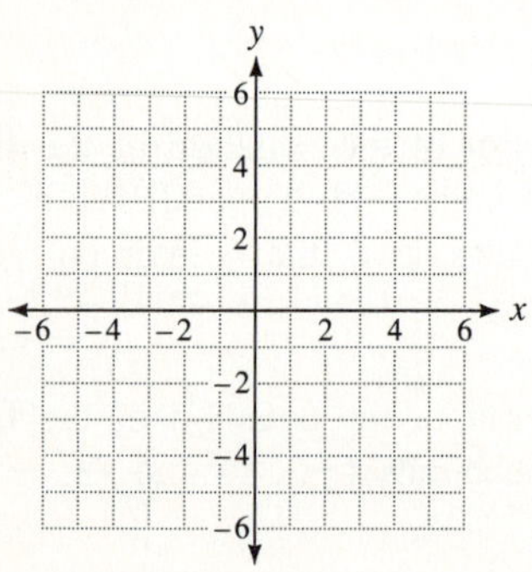

25. $y = 4x - 1$

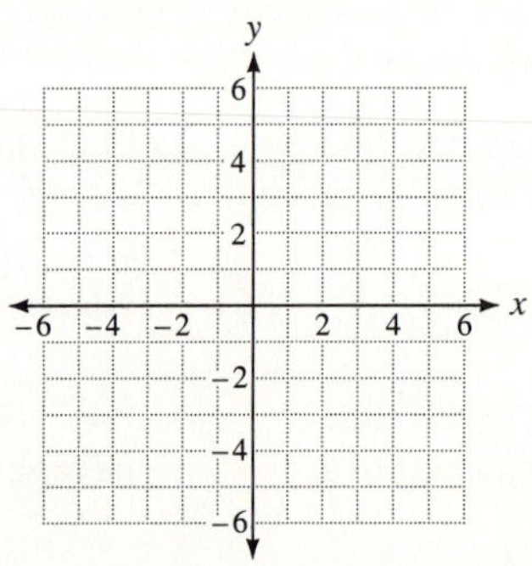

26. $y = -\dfrac{3}{2}x + 5$

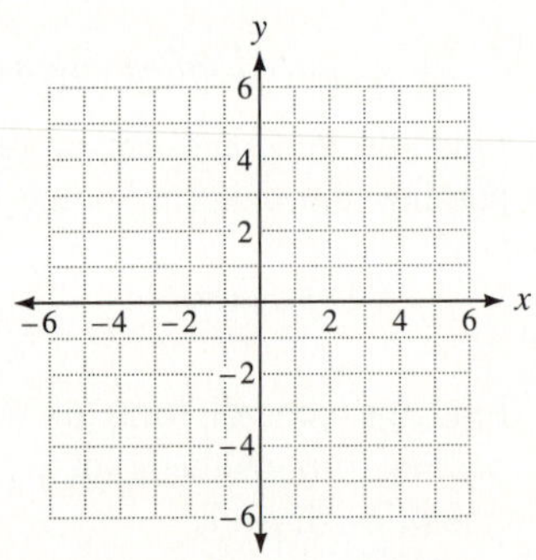

27. $y = -5x + 3$

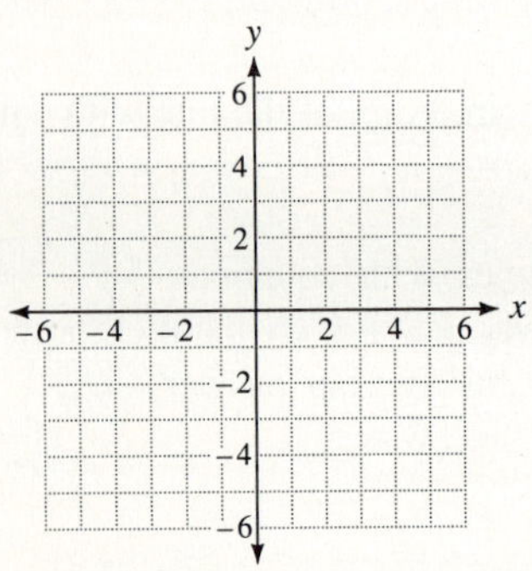

28. $x - 3y + 9 = 0$

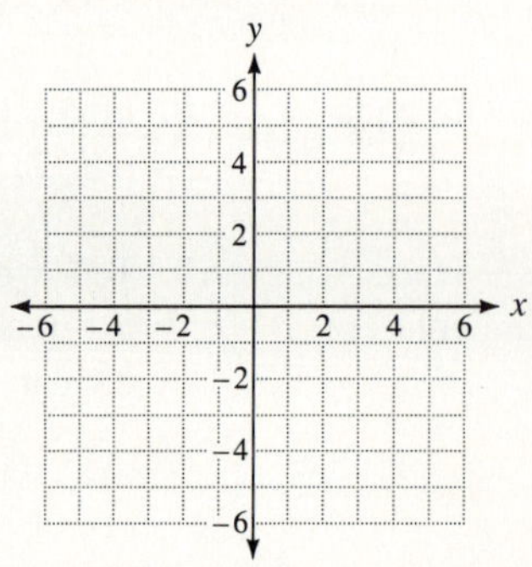

29. $3x + 4y - 8 = 0$

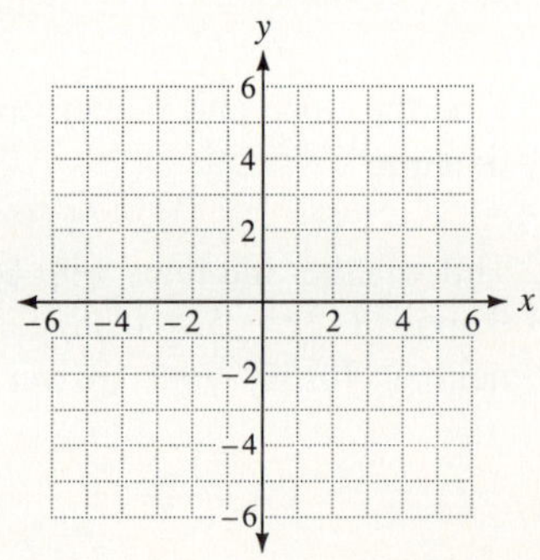

Without finding the equation of the line, graph the line using the given information and the slope method.

30. Through (2, 3) with slope $\frac{1}{2}$

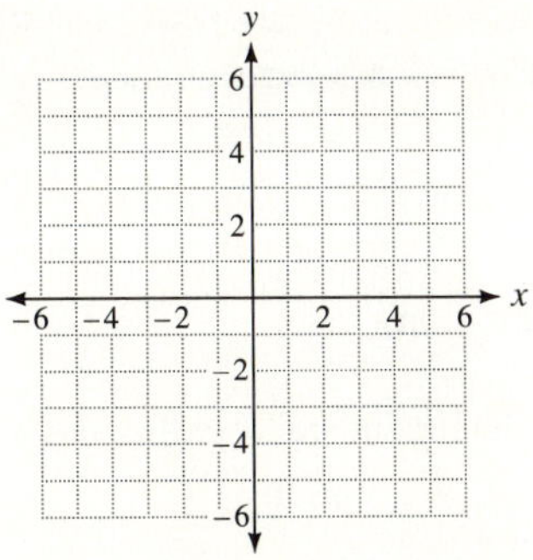

31. Through (−3, −4) with slope $\frac{4}{5}$

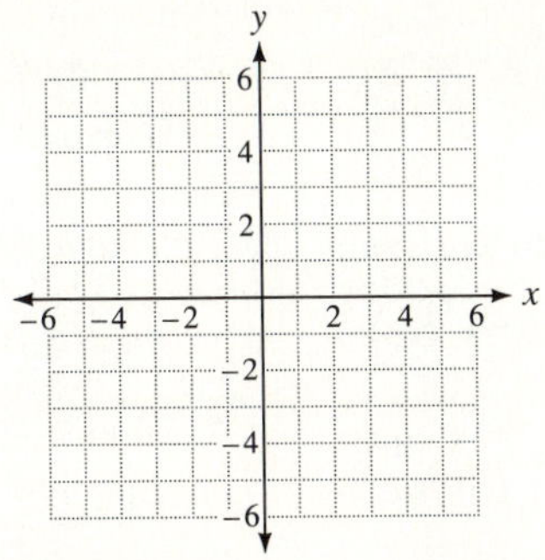

32. Through (3, −5) parallel to the line $y = 4x + 7$

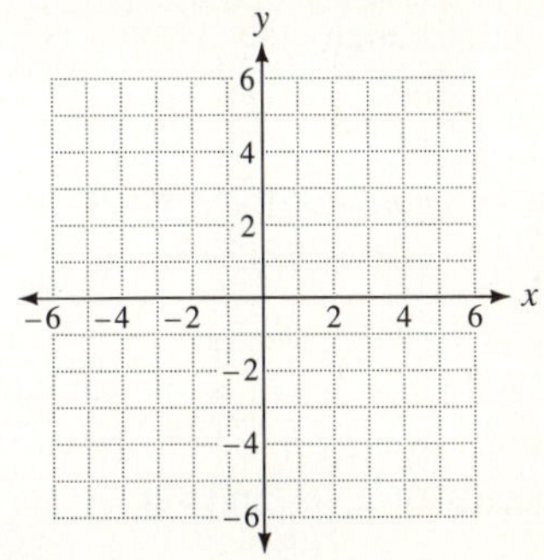

33. The graph shows the relationship between the monthly income y and total sales x of an insurance salesperson. Find the equation that describes this relationship and use it to find the monthly income when monthly sales are $32,000.

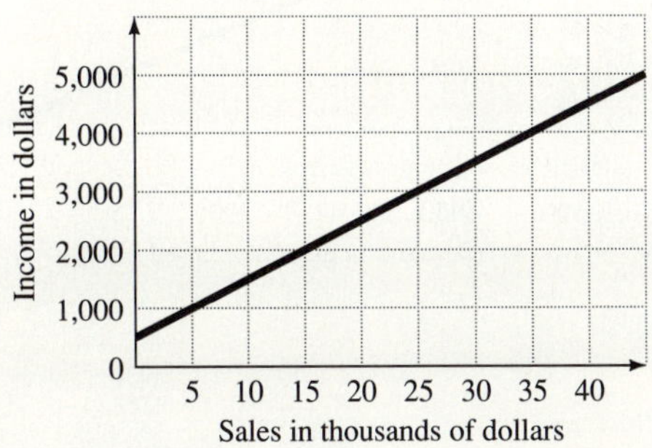

34. The graph shows the relationship between the cost of a delivery truck y and the depreciation allowed for tax purposes x. Find the equation that describes this relationship and use it to find the value of the truck after four and one-half years.

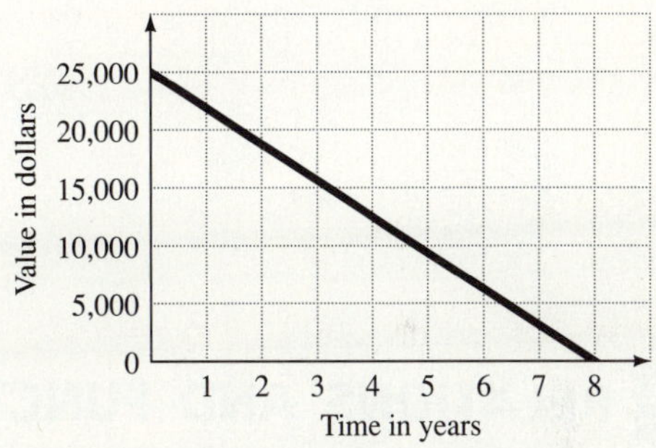

FOR REVIEW

35. Find the slope of the line joining (2, −5) and (−8, −7). What is the slope of a line parallel to this line? perpendicular to this line?

36. Without finding the actual values, tell whether each line in the figure has positive slope, negative slope, zero slope, or undefined slope.

(a) Line l_1 has ________ slope.
(b) Line l_2 has ________ slope.
(c) Line l_3 has ________ slope.
(d) Line l_4 has ________ slope.
(e) The y-axis has ________ slope.
(f) The x-axis has ________ slope.

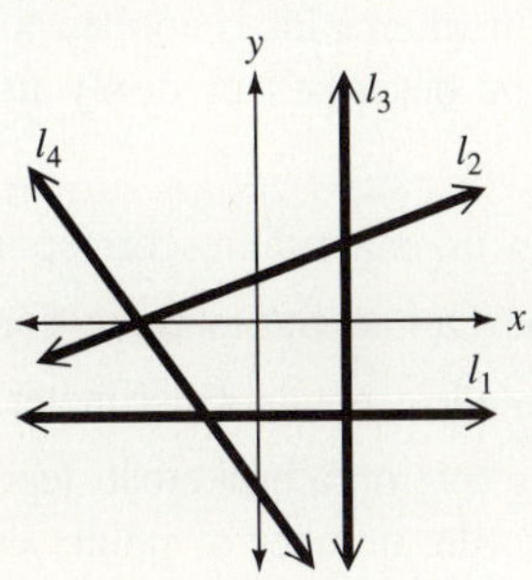

3.4 EXERCISES C

1. Give the general form of the equation of the line passing through the points $\left(\frac{1}{2}, -\frac{3}{4}\right)$ and $\left(-\frac{3}{8}, \frac{5}{12}\right)$.

2. Find the value of a so that the lines with equations $x + 2y = 3$ and $ax + y = 7$ are perpendicular. [Answer: -2]

Find the equations of the (a) vertical and (b) horizontal lines through each point.

3. $(-1, 5)$

4. $(-\sqrt{3}, \sqrt{2})$

5. $(\sqrt{5}, \pi)$

6. Find the slope of a line parallel to the line with equation $ax + by + c = 0$. [*Hint:* The answer will be in terms of a, b, and c.]

7. Find the slope of a line perpendicular to the line with equation $ax + by + c = 0$.

8. An economist uses a demand curve to show the relationship between the price of a product y and the demand by consumers for the product, x. Find the equation of the demand curve graphed here and use it to find the demand for the product when the price is \$2.50 per unit.

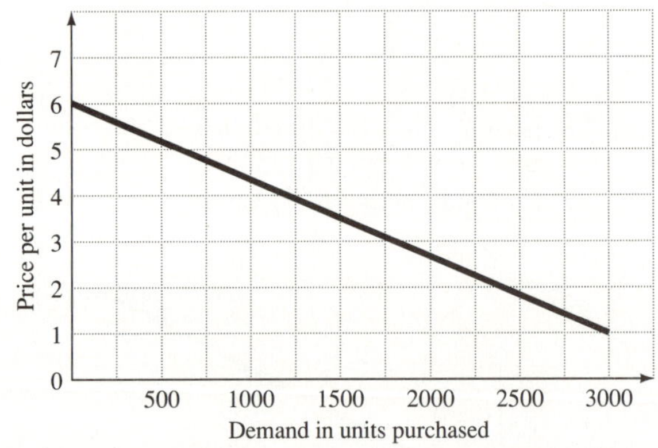

3.5 RELATIONS AND FUNCTIONS

STUDENT GUIDEPOSTS

1. Relations
2. Functions
3. Dependent and Independent Variables
4. Graphing Relations and Functions
5. Vertical Line Test
6. Functions Described by Equations

1 RELATIONS

In this section we introduce two central concepts in mathematics: relations and functions. The idea of a relation is already familiar since relationships or correspondences between two sets of objects play an important role in everyday life. When the terminology of mathematics is applied to these concepts, we discover that many relations can be described precisely using equations. Consider the following *relations:*

1. To each automobile in the state there corresponds a license number.
2. To each person in the United States there corresponds a legal last name.
3. To each triangle there corresponds a perimeter.
4. To each of the starting five on a basketball team in a particular game, there corresponds a number, the number of points each scores in that game.
5. To each parent in the United States there corresponds one or more children.

Definition of a Relation (by Rule)

A correspondence between a first set of objects, the **domain,** and a second set of objects, the **range,** is called a **relation** if to each element of the domain there corresponds *one or more* elements in the range.

2 FUNCTIONS

Each of the above five correspondences is a relation. In fact, the first four share an even more specific property in that they are *functions*.

Definition of a Function (by Rule)

A **function** is a relation with the additional property that to each element in the domain there corresponds *one and only one* element in the range.

CAUTION

By definition, all functions are relations but not all relations are functions. The fifth relation above is not a function since a parent may correspond to more than one child.

EXAMPLE 1 Identifying Relations and Functions

Which of the following relations are functions?

(a) Relation: item to its price
Domain: items in a discount store
Range: the set of all prices

This relation *is* a function since each item has one and only one selling price. Notice that the same price (an element of the range) may correspond to several different items (elements of the domain). This does not change the fact that the relation of an item to its price is a function.

(b) Relation: number to its square
Domain: the set of integers
Range: the set of whole number perfect squares

This relation *is* a function since each integer has one and only one square.

(c) Relation: number to its square root
Domain: the set of whole number perfect squares
Range: the set of integers

This *is not* a function since each whole number perfect square, other than 0, has two integer square roots. For example, 2 and -2 are both square roots of 4. Thus, two numbers in the range (in this case 2 and -2) correspond to the same number in the domain (in this case 4).

(d) Relation: number to its principal square root
Domain: the set of whole number perfect squares
Range: the set of whole numbers

This relation *is* a function since each whole number perfect square has exactly one principal (nonnegative) square root.

Practice Exercise 1

Which of the following relations are functions?

(a) Relation: car to its price
Domain: cars in stock
Range: the set of all prices

(b) Relation: number to its cube root
Domain: integer perfect cubes
Range: integers

(c) Relation: "is the brother of"
Domain: males with siblings
Range: people with brothers

(d) Relation: six times a number
Domain: real numbers
Range: real number multiples of 6

Answers: (a) function (b) function (c) not a function (d) function

One way to represent a function is to use a table. For example, consider the starting five on a basketball team and the points each scored in a given game. The domain of this function is the set of players while the range is the set of scores. We often use arrows to indicate the object in the range (the arrow points *to* it) which corresponds to the object in the domain (the arrow points *from* it).

Domain		*Range*
Troy	⟶	20
Rick	⟶	19
Eric	⟶	14
Jeff	⟶	14
Andy	⟶	9

Notice that one number in the range, 14, is identified with two members in the domain. The definition of a function allows for this (which is possible in reality), but it does not allow a player, an element of the domain, to be identified with two different numbers in the range (which *is impossible* in reality). We usually do not list an object in the range more than once. Thus, the function would take the following form.

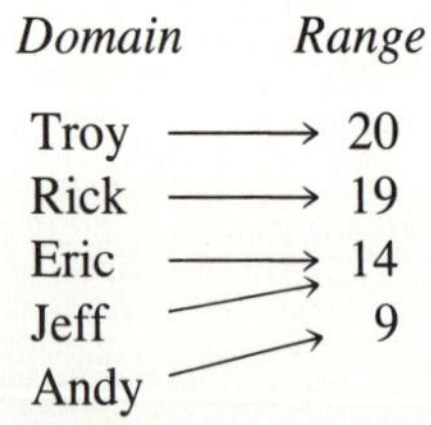

EXAMPLE 2 Using Tables to Describe Functions

Which of the following relations are functions?

(a)

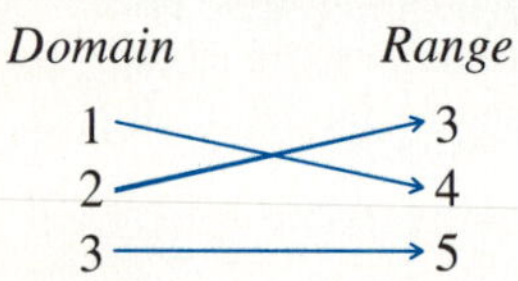

This relation *is* a function since each object in the domain corresponds to exactly one object in the range.

(b)

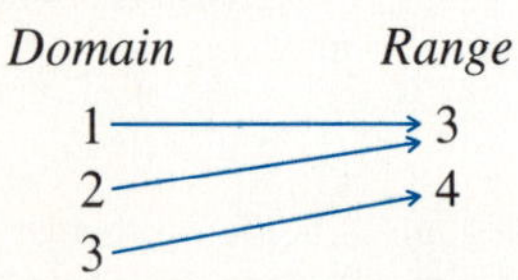

This relation *is* a function since 3 (in the range) can correspond to the two objects 1 and 2 (in the domain).

(c)

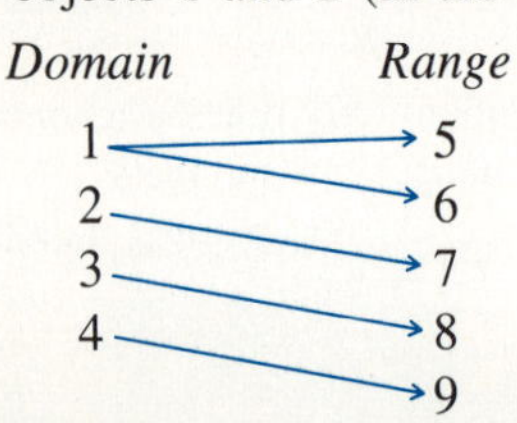

This relation *is not* a function since 1 (in the domain) corresponds to two objects, 5 and 6 (in the range).

Practice Exercise 2

Tell whether each relation is a function.

(a)

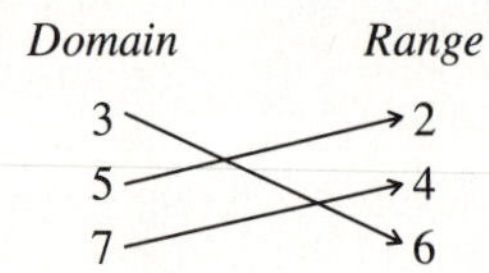

(b)

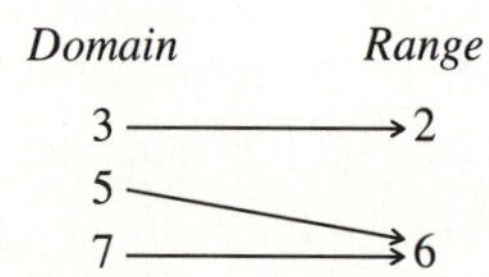

(c)

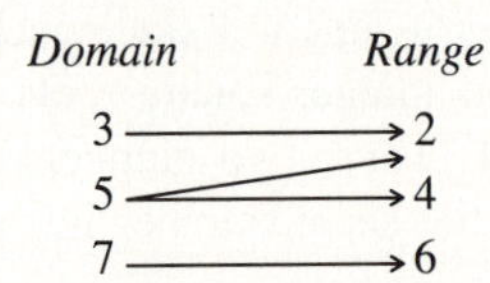

Answers: (a) function (b) function (c) not a function

3 DEPENDENT AND INDEPENDENT VARIABLES

When the objects in the domain and range of a relation are real numbers, a set of ordered pairs of numbers (x, y) can be used to define the relation. If $x \rightarrow y$, that is, if x is related to y, then the pair (x, y) is included in the relation. We think of the value of y as depending on the value of x and call y the **dependent variable** and x the **independent variable.** The set of all values of the independent variable is the domain of the relation, and the set of all values of the dependent variable is the range.

Consider again the relations given in Example 2. Using ordered pairs, the relation in (a) is $\{(1, 4), (2, 3), (3, 5)\}$, the relation in (b) is $\{(1, 3), (2, 3), (3, 4)\}$, and the relation in (c) is $\{(1, 5), (1, 6), (2, 7), (3, 8), (4, 9)\}$.

The observations above are restated in the following alternative definition of relation and function.

Definition of Relation and Function (Using Ordered Pairs)

Any set of ordered pairs of numbers is called a **relation.** A **function** is a relation with the additional restriction that no two ordered pairs can have the same x-coordinate and different y-coordinates.

Notice that the relations in Example 2 (a) and (b) are indeed functions, but the relation in (c) is not since (1, 5) and (1, 6) have the same x-coordinate and different y-coordinates.

4 GRAPHING RELATIONS AND FUNCTIONS

The ordered-pair definition of a relation provides a useful way to graph relations and functions since ordered pairs can be plotted in a rectangular coordinate system.

EXAMPLE 3 Graphing Relations and Functions

List the ordered pairs in each relation, plot its graph, and tell whether it is a function.

(a) *Relation*

Domain *Range*

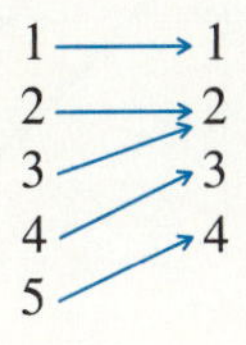

Ordered pairs in the relation

(1, 1)
(2, 2)
(3, 2)
(4, 3)
(5, 4)

Graph of the relation

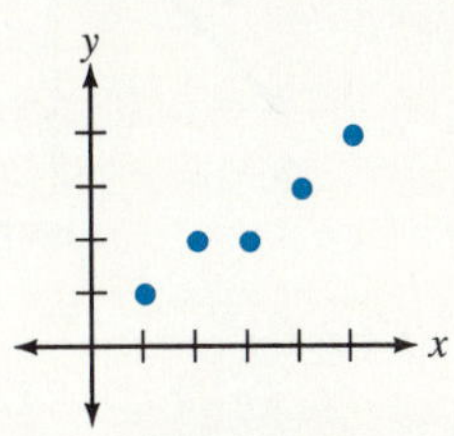

Figure 3.22

Since every number in the domain is paired with exactly one number in the range, this relation is a function.

Practice Exercise 3

List the ordered pairs and tell whether the relation is a function.

(a) *Domain* *Range*

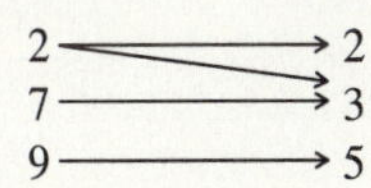

(b) *Relation*

Domain *Range*

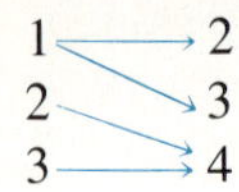

Ordered pairs in the relation

(1, 2)
(1, 3)
(2, 4)
(3, 4)

Graph of the relation

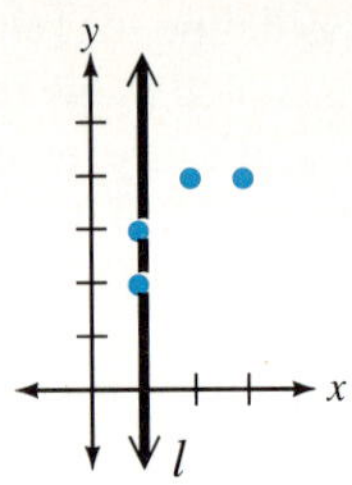

Figure 3.23

Since the number 1 in the domain is paired with two numbers, 2 and 3, in the range, this relation is not a function.

(b) *Domain* *Range*

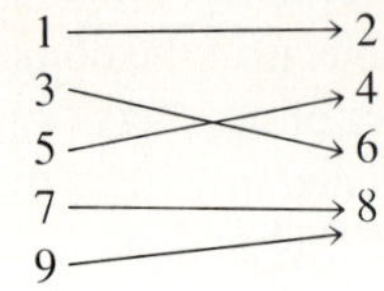

Answers: (a) (2, 2), (2, 3), (7, 3), (9, 5); not a function (b) (1, 2), (3, 6), (5, 4), (7, 8), (9, 8); function

5 VERTICAL LINE TEST

Notice in Figure 3.23 that two points on the graph, (1, 2) and (1, 3), lie on the same vertical line l. This property is true of all relations that are not functions. On the other hand, any collection of points that has the property that no vertical line passes through two or more points, is the graph of a function. This allows us to use the following test.

Vertical Line Test

If each vertical line intersects the graph of a relation in no more than one point, the relation is a function.

EXAMPLE 4 Using the Vertical Line Test

Use the vertical line test to tell whether the graph is the graph of a function.

(a) Figure 3.24 *is* the graph of a function since no vertical line crosses the graph twice.

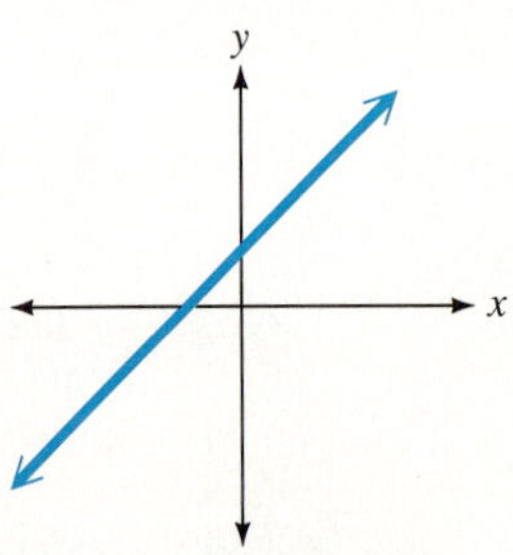

Figure 3.24

(b) Figure 3.25 *is not* the graph of a function since the vertical line l crosses the graph twice. Note that two values of y (y_1 and y_2) correspond to the same value of x (x_1).

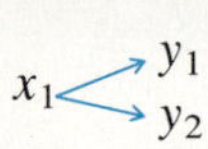

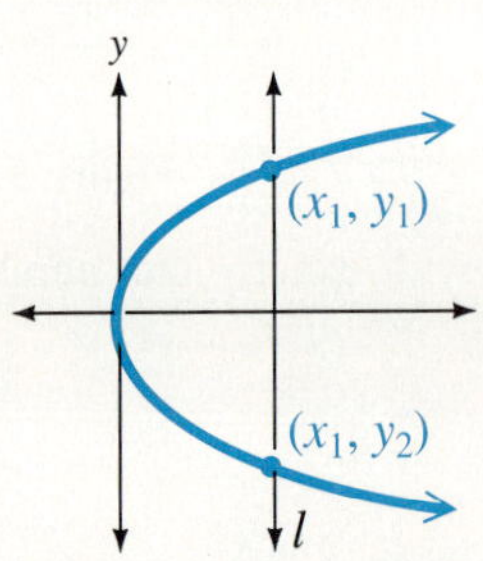

Figure 3.25

Practice Exercise 4

Use the vertical line test to tell whether or not each graph is the graph of a function.

(a)

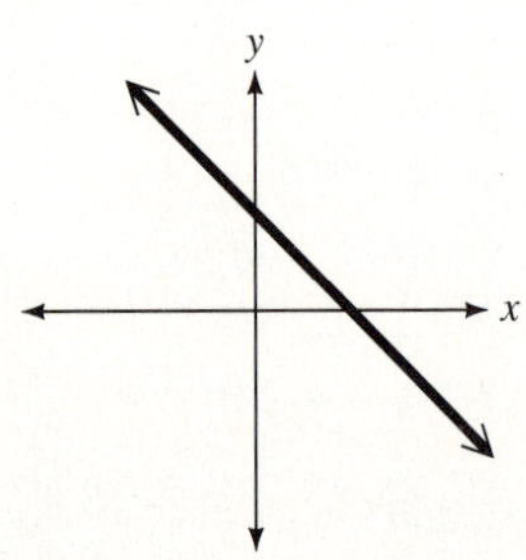

(b)

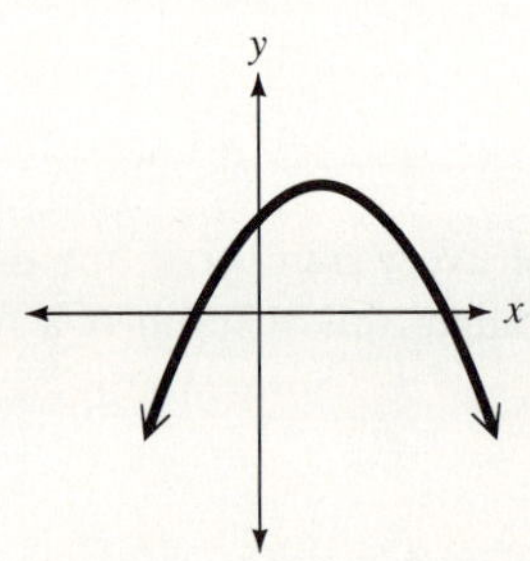

(c) Figure 3.26 *is not* the graph of a function, as shown by the vertical line l. Here a given value of x (x_1) corresponds to three values of y (y_1, y_2, and y_3).

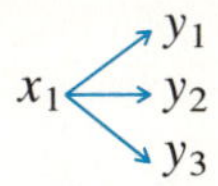

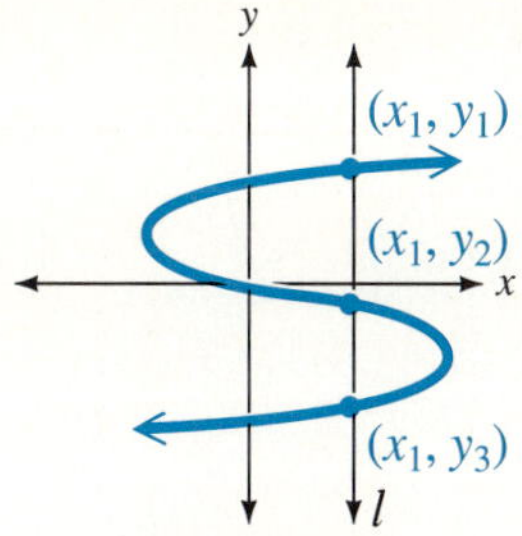

Figure 3.26

(c)

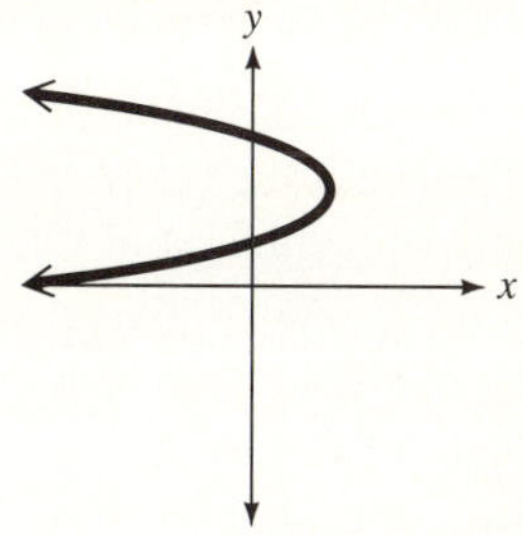

Answers: (a) function (b) function (c) not a function

6 FUNCTIONS DESCRIBED BY EQUATIONS

Up to now we have discussed functions described by diagram, by ordered pairs, and by graphs. Often, an equation is used to describe a relation for which the ordered pair solutions (x, y) fit the definition of a function. In fact, nearly every linear equation

$$ax + by + c = 0$$

describes a function. The only linear equations that do not, have $b = 0$ and a vertical line for a graph. For example, $2x - 3 = 0$ does not describe a function, as is clear from the graph of $2x - 3 = 0$ or $2x = 3$ or $x = \frac{3}{2}$ in Figure 3.27. This cannot be the graph of a function since for one value of x, $\frac{3}{2}$, there are more than one (in fact infinitely many) values of y. However, any other type of line in the plane is the graph of a function. We will consider functions defined by equations in more detail in the following sections.

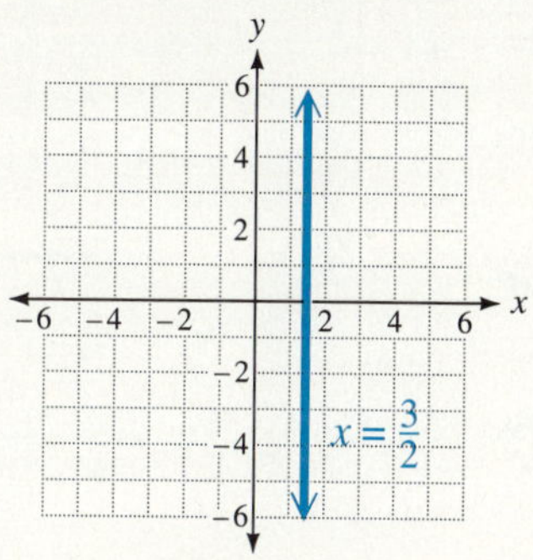

Figure 3.27

3.5 EXERCISES A

Write true *or* false *for Exercises 1 and 2. If the statement is false, explain why.*

1. A relation is a correspondence between a first set called the domain and a second set called the range of the relation.

2. If x corresponds to y ($x \to y$) in a function, the ordered pair (x, y) belongs to the function, and the point in the plane that corresponds to this ordered pair is a point on the graph of the function.

Tell which of the relations in Exercises 3–4 are functions.

3. Relation: a number to its fourth roots
Domain: the set of whole numbers
Range: the set of fourth roots of whole numbers

4. Relation: person to his/her annual salary
Domain: all salaried people
Range: the set of salaries

Which of the following are functions? Explain.

5.
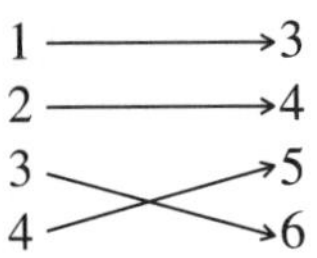

6.
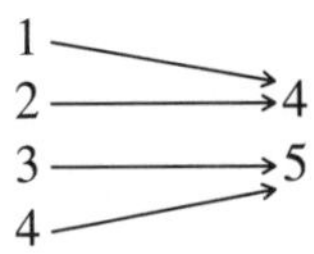

7.
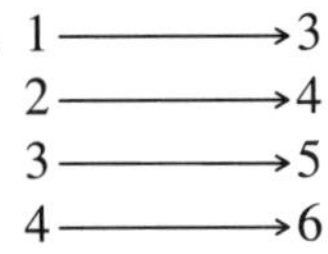

8.
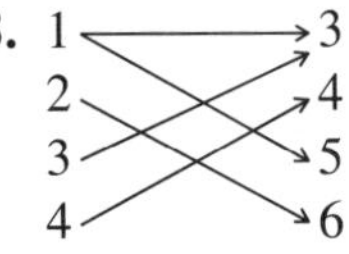

9.
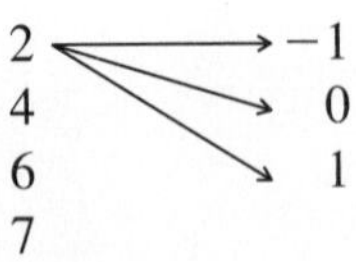

10.
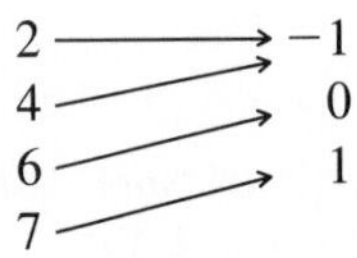

11.
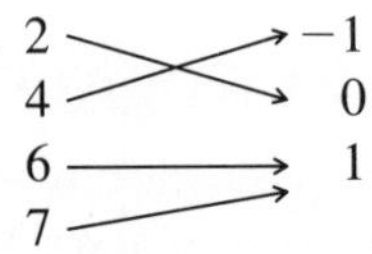

12.
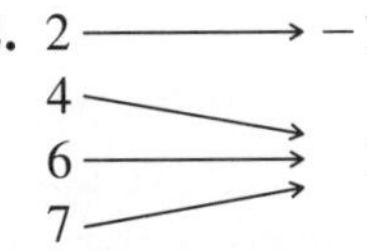

Which of the following are graphs of functions? Explain.

13.
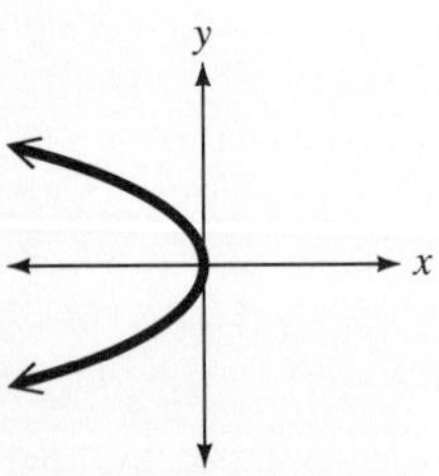

14.
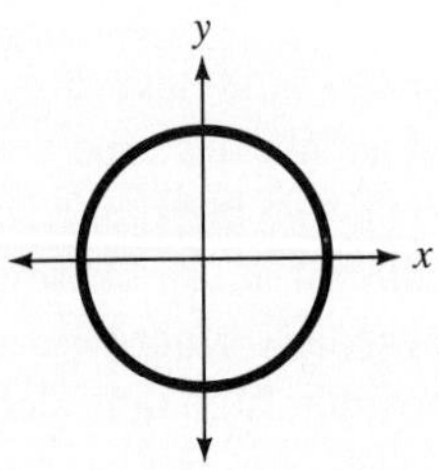

15.
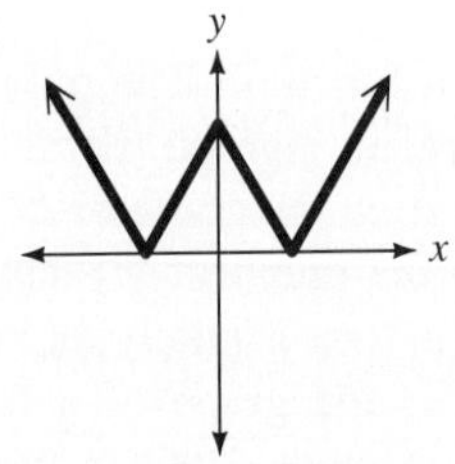

16.
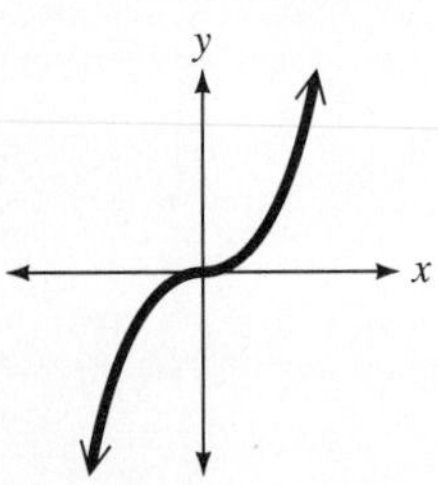

17.
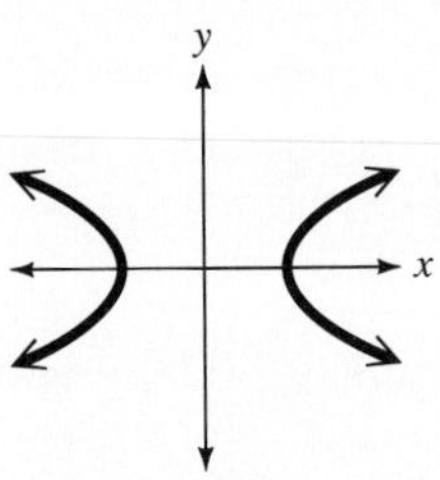

18.
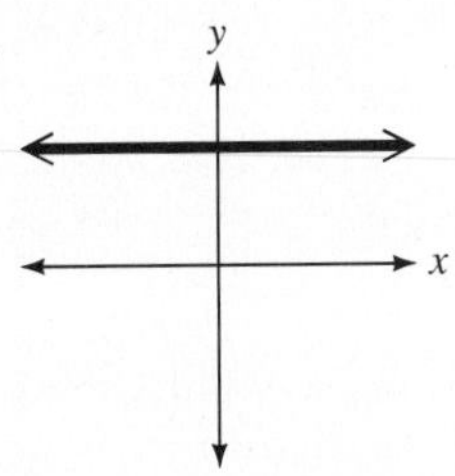

19.
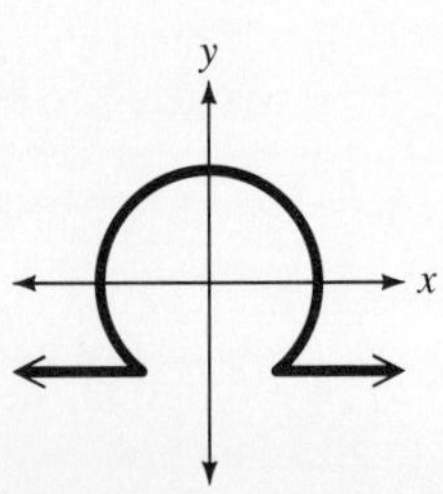

20.
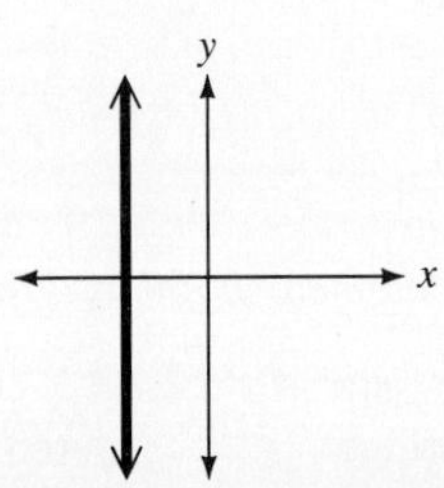

21.
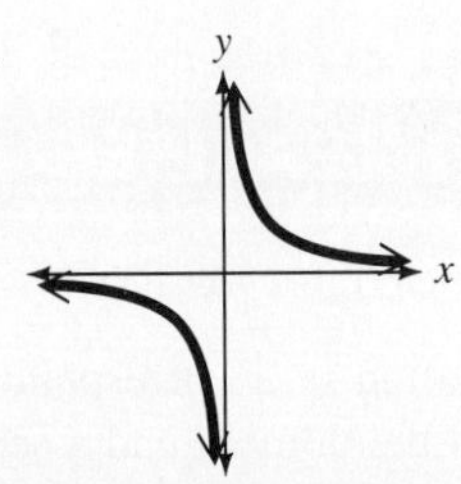

Graph the equation and use your graph to tell whether the equation describes a function.

22. $3x + 5y - 15 = 0$

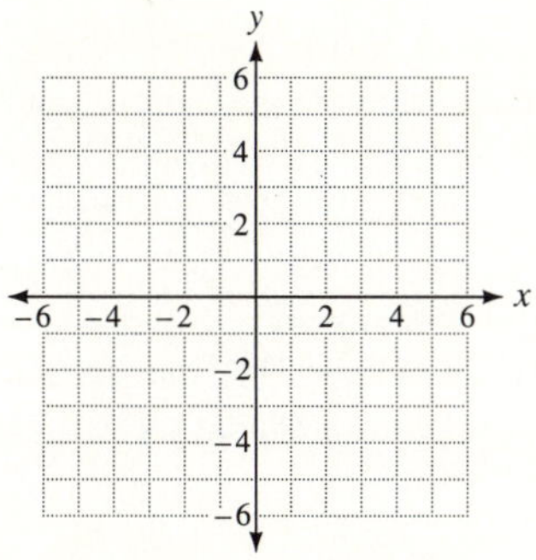

23. $3y - 12 = 0$

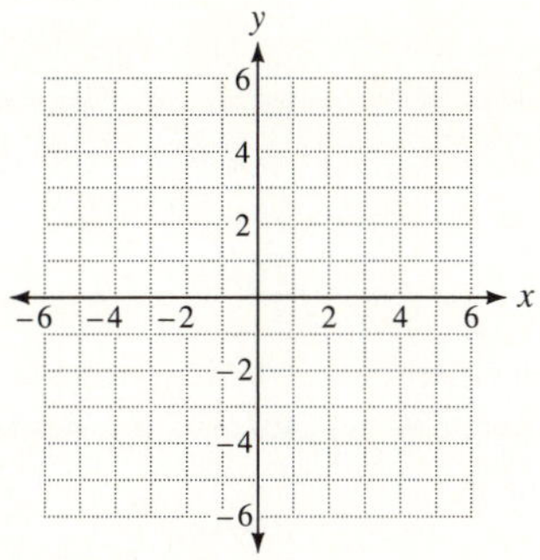

24. $5x - 15 = 0$

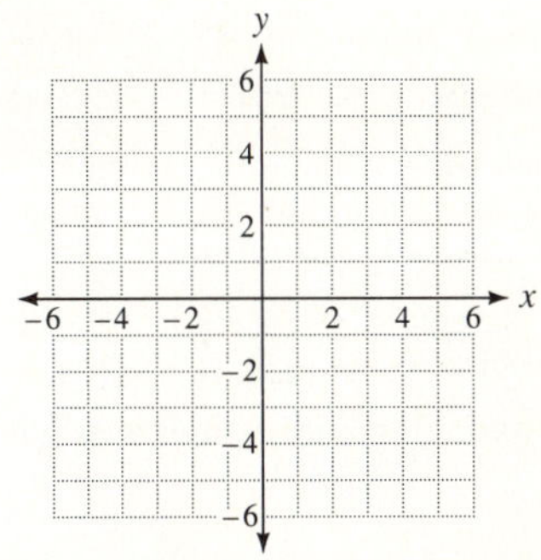

25. Consider the relation described by the set of ordered pairs $\{(1, 3), (2, 5), (3, 4), (4, 2), (5, 2)\}$.

(a) Write the relation using arrows as in Example 3.

(b) Graph the relation in the given rectangular coordinate system.

(c) Is the relation a function? Explain.

(a)

(b)

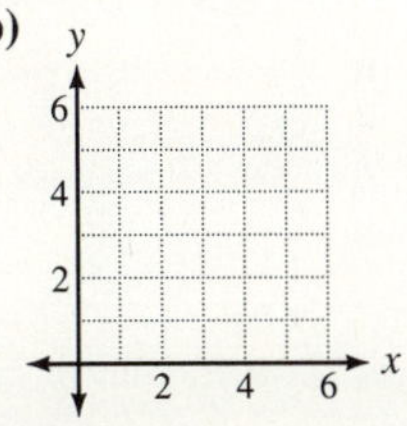

(c)

In Exercises 26–27, list the ordered pairs in each relation. What is the domain of the relation? What is the range of the relation? Is the relation a function?

26. $\{(x, y)$ where $x = y^2$ and $y = 1, 2, 3\}$

27. $\{(x, y)$ where $x^2 + y^2 = 1$ and $x = -1, 0, 1\}$

FOR REVIEW

28. Give the point-slope form of the equation of a line passing through (x_1, y_1) with slope m.

29. Give the slope-intercept form of the equation of a line with slope m and y-intercept $(0, b)$.

30. Give the slope of a line parallel to a line with slope m.

31. Find the slope-intercept form of the line with general equation $6x + 3y - 9 = 0$. What is the slope? The y-intercept?

32. Find the general form of the equation of the line passing through $(-5, 2)$ with slope $m = -3$.

33. Terry Larsen manufactures a special part for farm equipment. He can make 15 parts for \$70 and 60 parts for \$192. If y is the cost of making x parts, find a linear equation describing this relationship and use it to calculate the cost of making 100 parts, to the nearest dollar.

The following exercises review material from Chapter 1 and will help you prepare for the next section.

34. Will the expression $\frac{1}{x-2}$ be a real number for every real number x?

35. Evaluate $2x^2 - x + 1$ when $x = -1$.

ANSWERS: 1. true 2. true 3. not a function 4. function 5. function 6. function 7. function 8. not a function 9. not a function 10. function 11. function 12. function 13. not the graph of a function 14. not the graph of a function 15. graph of a function 16. graph of a function 17. not the graph of a function 18. graph of a function 19. not the graph of a function 20. not the graph of a function 21. graph of a function

22. function

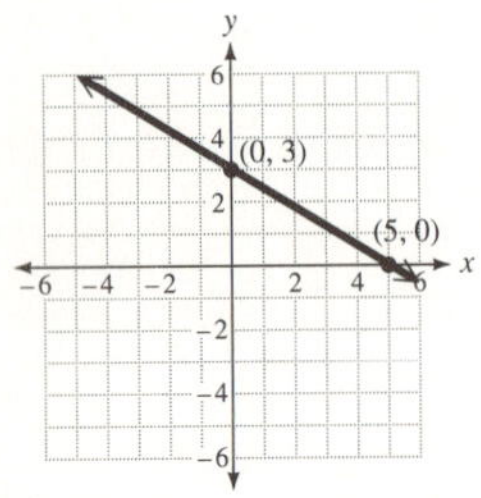

23. function

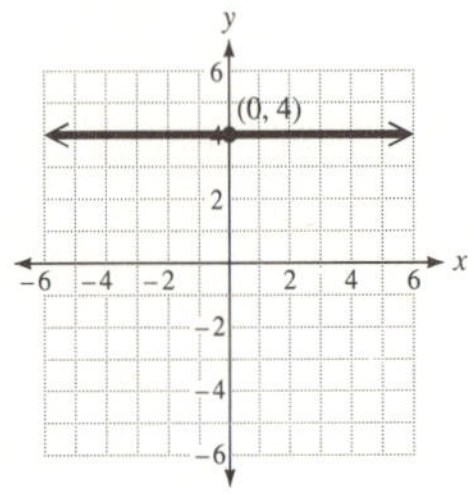

24. not a function

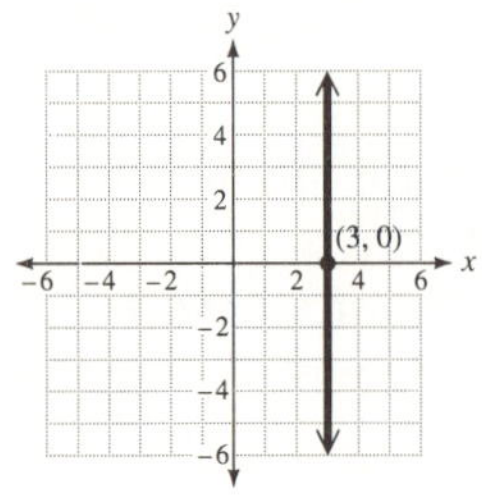

25. (a)

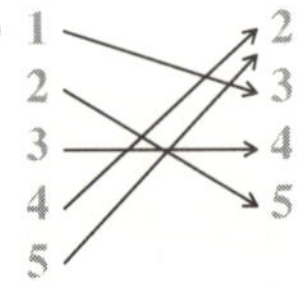

(b)

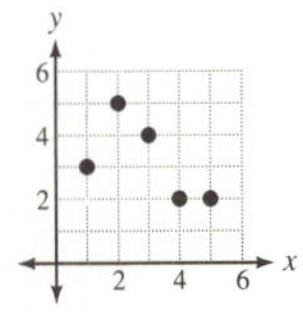

(c) Yes. The graph passes the vertical line test.

26. $\{(1, 1), (4, 2), (9, 3)\}$; $\{1, 4, 9\}$; $\{1, 2, 3\}$; yes 27. $\{(-1, 0), (0, 1), (0, -1), (1, 0)\}$; $\{-1, 0, 1\}$; $\{-1, 0, 1\}$; no 28. $y - y_1 = m(x - x_1)$ 29. $y = mx + b$ 30. m 31. $y = -2x + 3$; $m = -2$; $(0, 3)$ 32. $3x + y + 13 = 0$ 33. $122x - 45y + 1320 = 0$; \$300 34. for every real number except 2 35. 4

3.5 EXERCISES B

Write true *or* false *for Exercises 1 and 2. If the statement is false, explain why.*

1. A relation with the property that every object in its domain corresponds to exactly one object in its range is called a function.

2. If (x, y) belongs to a function, then x corresponds to y.

Tell which of the relations in Exercises 3–4 are functions.

3. Relation: a number to its cube
Domain: the set of all integers
Range: the set of all perfect cubes

4. Relation: person to his/her age
Domain: all living people
Range: the set of all ages

Which of the following are functions? Explain.

5.

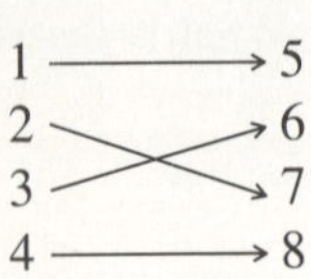

6.

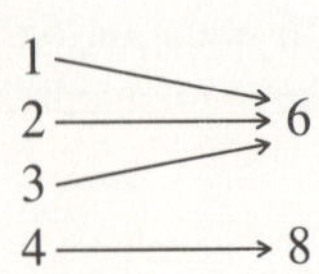

7.

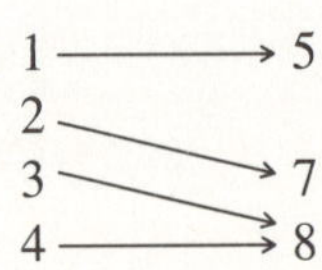

8.

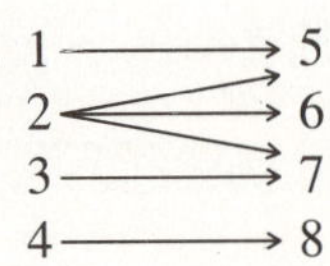

9.

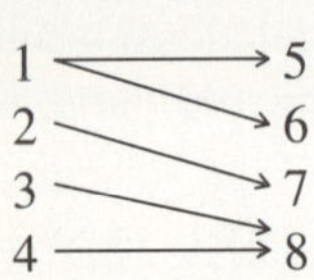

10.

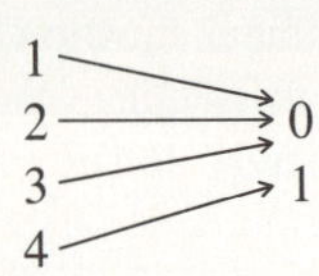

11.

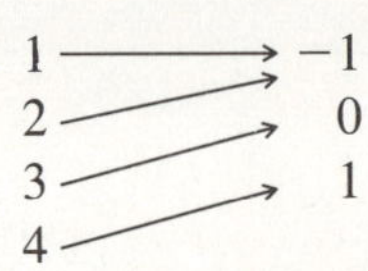

12.

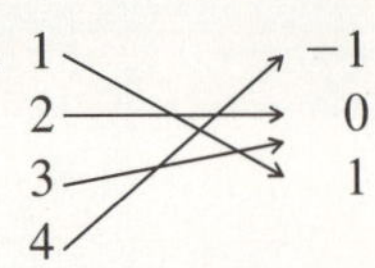

Which of the following are graphs of functions? Explain.

13.

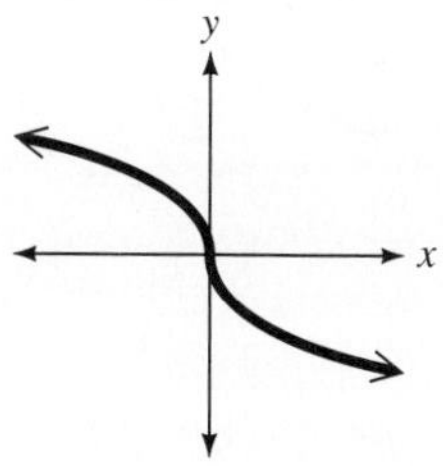

14.

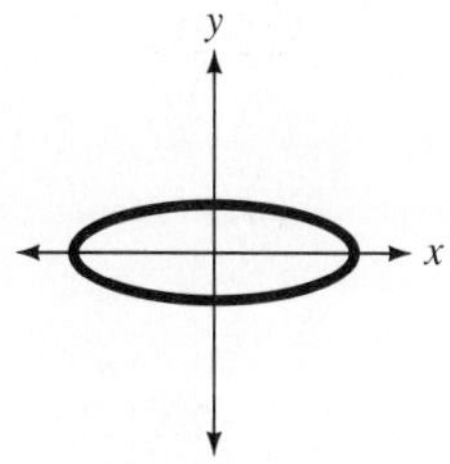

15.

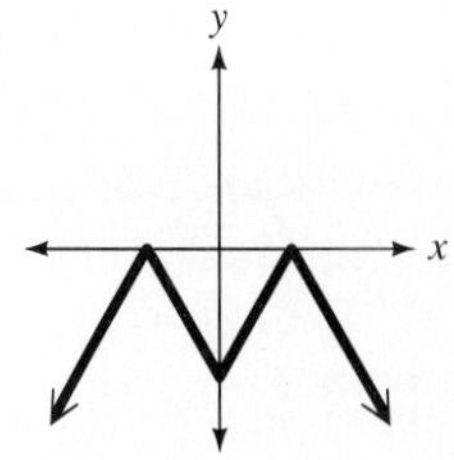

16.

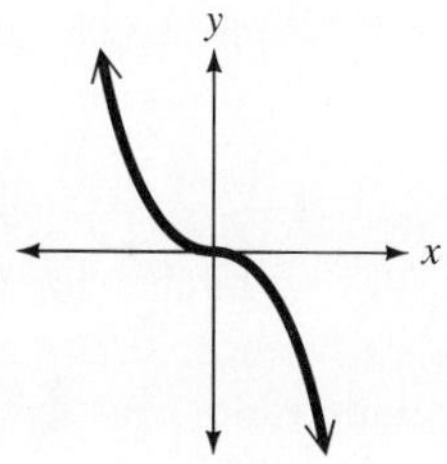

17.

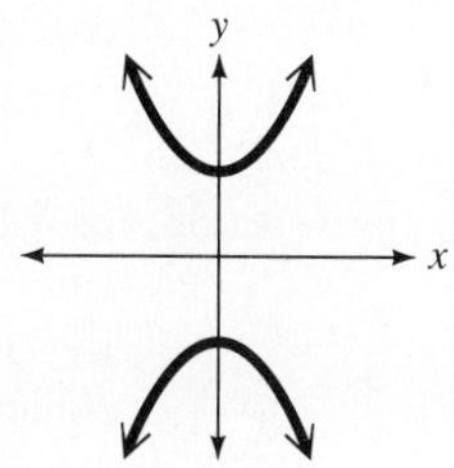

18.

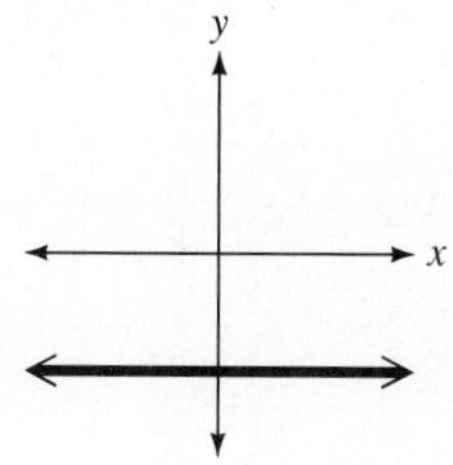

19.

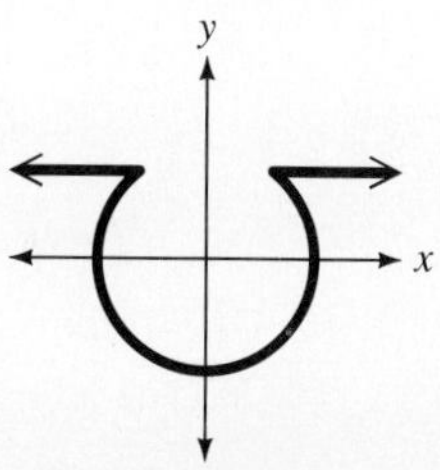

20.

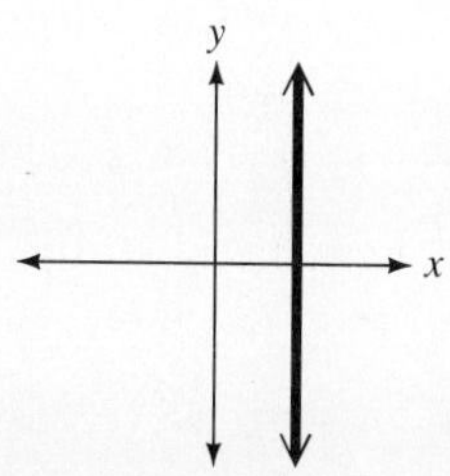

21.

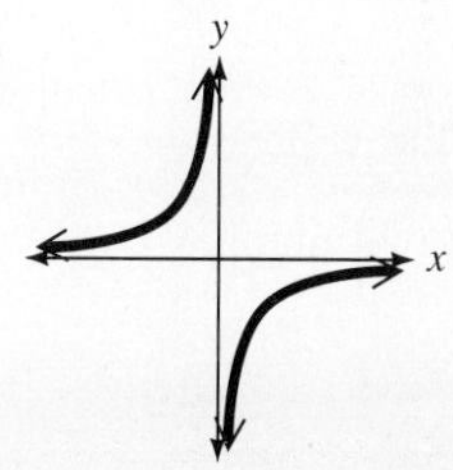

Graph the equation and use your graph to tell whether the equation describes a function.

22. $4y + 12 = 0$

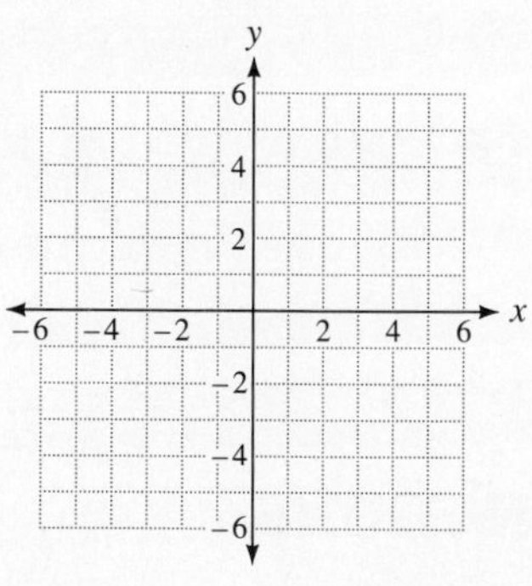

23. $x + y = 0$

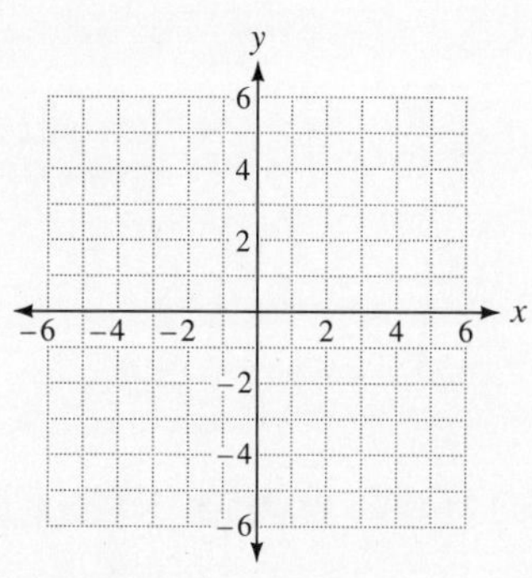

24. $5x + 20 = 0$

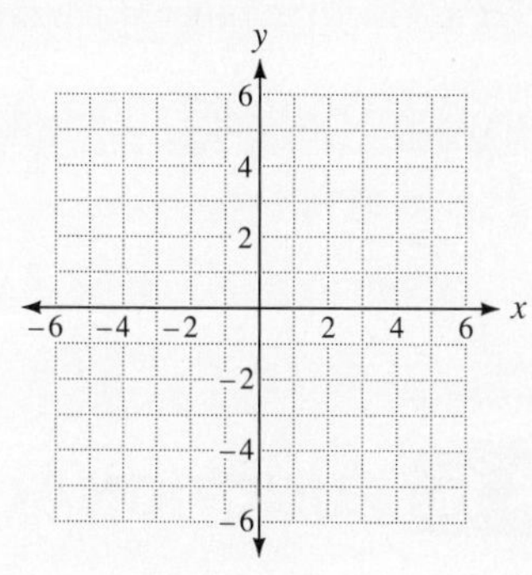

25. Consider the relation described by the set of ordered pairs $\{(1, 2), (1, 3), (2, 1), (3, 5), (4, 5)\}$.
(a) Write the relation using arrows as in Example 3.
(b) Graph the relation in a Cartesian coordinate system.
(c) Is the relation a function? Explain.

(a)

(b)

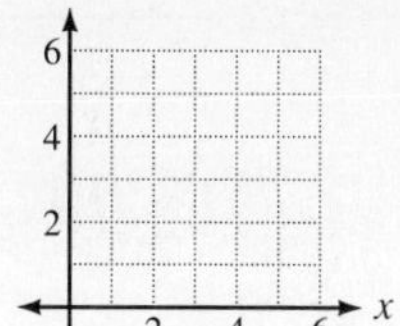

(c)

In Exercises 26–27, list the ordered pairs in each relation. What is the domain of the relation? What is the range of the relation? Is the relation a function?

26. $\{(x, y)$ where $y = x^2$ and $y = 0, 1, 4\}$

27. $\{(x, y)$ where $x^2 + y^2 = 4$ and $y = -2, 0, 2\}$

FOR REVIEW

28. Give the type of the equation $y - y_1 = m(x - x_1)$.

29. Give the type of the equation $y = mx + b$.

30. Give the slope of a line perpendicular to a line with slope m.

31. Find the slope-intercept form of the line with general form $3x - 6y - 24 = 0$. What is the slope? The y-intercept?

32. Find the general form of the equation of the line passing through $(-3, 7)$ with slope $m = \frac{1}{2}$.

33. Phil Mortensen builds a special type of hitch. He can make 3 hitches for \$170, or 7 hitches for \$342. If y is the cost of making, x hitches, find a linear equation describing this relationship, and use it to calculate the cost of making 10 hitches.

The following exercises review material from Chapter 1 and will help you prepare for the next section.

34. Will the expression $\dfrac{1}{x^2 + 1}$ be a real number for every real number x?

35. Evaluate $-3x^2 + x - 2$ when $x = -1$.

3.5 EXERCISES C

Which of the following relations are functions?

1. $\{(x, y) | y = \sqrt{x + 1} \quad \text{and} \quad x = 1, 2, 3\}$
[*Hint:* List the ordered pairs.]

2. $\{(x, y) | y^2 = x + 1 \quad \text{and} \quad x = 1, 2, 3\}$

3. $\{(x, y) | y = |x| \quad \text{and} \quad x = -1, 0, 1\}$

4. $\{(x, y) | x = |y| \quad \text{and} \quad x = 0, 1\}$
[Answer: not a function]

3.6 FUNCTIONAL NOTATION AND SPECIAL FUNCTIONS

STUDENT GUIDEPOSTS

1. Specifying the Domain of a Function
2. Functional Notation $f(x)$
3. Linear Functions
4. Constant Functions
5. The Identity Function

1 SPECIFYING THE DOMAIN OF A FUNCTION

When a function consists of a set of ordered pairs (x, y) where x is the independent variable and y is the dependent variable, we think of y *as a function of* x. That is, if a value of x is given, one and only one value of y can be found.

Equations that are solved for y in terms of x often describe a function. For example,

$$y = 5, \quad y = x, \quad y = 2x + 3, \quad y = x^2 + 1, \quad \text{and} \quad y = \frac{1}{x}$$

all describe functions consisting of ordered pairs (x, y). That is, for a given real number x, there is one and only one corresponding value for y making the equation true. Unless we specify otherwise, *the domain of a function described by an equation is the set of all real numbers for which the expression in x is defined*. The domain for each of the functions

$$y = 5, \quad y = x, \quad y = 2x + 3, \quad \text{and} \quad y = x^2 + 1,$$

is the entire set of real numbers. However, the domain of

$$y = \frac{1}{x}$$

is the set of all real numbers except 0. Remember, division by zero is not defined, so no value can be found for y when x is 0. In general, the domain of a function is restricted so that *numbers that make a denominator zero are not permitted*.

EXAMPLE 1 Specifying Domains

Specify the domain of each function.

(a) The function $y = \dfrac{1}{x - 2}$ is defined for all real numbers except $x - 2 = 0$. Thus, the domain is the set of all real numbers except 2 $(x \neq 2)$.

(b) The function $y = \dfrac{1}{x + 1}$ is defined for all real numbers except $x + 1 = 0$. The domain is the set of all real numbers except -1 $(x \neq -1)$.

(c) The function $y = 3x^2 + x + 1$ is defined for all real numbers. The domain is the set of all real numbers.

Practice Exercise 1

Specify the domain of each function.

(a) $y = \dfrac{1}{x + 6}$

(b) $y = 2x - 8$

(c) $y = \dfrac{x + 2}{x - 3}$

Answers: (a) $x \neq -6$ (b) all real numbers (c) $x \neq 3$

2 FUNCTIONAL NOTATION $f(x)$

Many times we use a letter to name a function (f and g are common choices). Suppose that a function f is described by the equation

$$y = x^2 + 1.$$

To show that y depends on choices of x, we use $f(x)$ for y and write

$$f(x) = x^2 + 1, \qquad y \text{ is } f(x)$$

where $f(x)$ is read "f of x." To determine the value of the function f for a given value of x, we substitute the value into the formula $f(x) = x^2 + 1$. For example, to compute $f(2)$, we replace each x with 2 and simplify. It is sometimes helpful to place parentheses around the number we are substituting for x.

$$f(2) = (2)^2 + 1 = 4 + 1 = 5$$

Remember that $f(2) = 5$ is the y-value that corresponds to the x-value 2. That is, $2 \to f(2)$, and the ordered pair $(2, f(2))$ or $(2, 5)$ belongs to the function. Also, when

$$x = -3, \quad y = f(-3) = (-3)^2 + 1 = 9 + 1 = 10;$$
$$x = 0, \quad y = f(0) = (0)^2 + 1 = 0 + 1 = 1.$$

EXAMPLE 2 Functional Notation

Consider the function g defined by

$$g(x) = 2x + 3.$$

(a) $g(0) = 2(0) + 3 = 0 + 3 = 3$ $\quad 0 \to 3$; $(0, 3)$ belongs to g

(b) $g(4) = 2(4) + 3 = 8 + 3 = 11$ $\quad 4 \to 11$; $(4, 11)$ belongs to g

(c) $g(-2) = 2(-2) + 3 = -4 + 3 = -1$ $\quad -2 \to -1$; $(-2, -1)$ belongs to g

(d) $g(a) = 2(a) + 3 = 2a + 3$ $\quad a \to 2a + 3$; $(a, 2a + 3)$ belongs to g

(e) $g(b + 1) = 2(b + 1) + 3 = 2b + 2 + 3 = 2b + 5$

PRACTICE EXERCISE 2

$h(x) = 4x - 5$

(a) $h(0) =$

(b) $h(2) =$

(c) $h(-2) =$

(d) $h(a) =$

(e) $h(b - 1) =$

Answers: (a) -5 (b) 3 (c) -13 (d) $4a - 5$ (e) $4b - 9$

3 LINEAR FUNCTIONS

Since we know from Section 3.4 that a linear equation in slope-intercept form is $y = mx + b$, it follows that a **linear function** can be described by an equation of the form

$$f(x) = mx + b \qquad (m \text{ and } b \text{ constants}).$$

The domain of every linear function is the set of real numbers. To graph a linear function, we graph the linear equation that defines it.

EXAMPLE 3 A Linear Function

The linear function f defined by

$$f(x) = 2x + 1$$

has the following values.

$$f(1) = 2(1) + 1 = 3$$

$f(0) = 2(0) + 1 = 1$

$f(1) = 2(1) + 1 = 3$

$f(2) = 2(2) + 1 = 5$

$f(-1) = 2(-1) + 1 = -1$

$f(a + 1) = 2(a + 1) + 1$
$= 2a + 2 + 1 = 2a + 3$

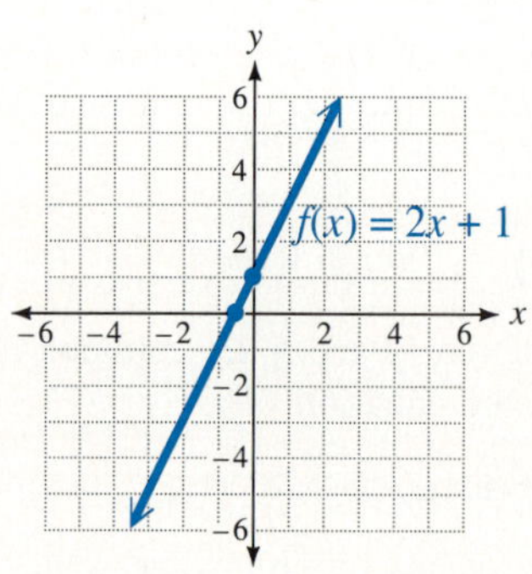

Figure 3.28 Linear Function

To graph the linear function $f(x) = 2x + 1$, graph $y = 2x + 1$ as in Figure 3.28 by finding the intercepts $(0, 1)$ and $\left(-\frac{1}{2}, 0\right)$ or by using the slope method discussed in Section 3.3.

PRACTICE EXERCISE 3

$f(x) = -3x + 2$

(a) $f(0) =$

(b) $f(-2) =$

(c) $f(1) =$

(d) $f(a) =$

(e) $f(b + 1) =$

Answers: (a) 2 (b) 8 (c) -1 (d) $-3a + 2$ (e) $-3b - 1$

4 CONSTANT FUNCTIONS

If $m = 0$ in a linear function $f(x) = mx + b$, we obtain a **constant function**

$$f(x) = b. \qquad (b \text{ a constant})$$

The domain of a constant function is the entire set of real numbers, and no matter what value x assumes, the corresponding y-value, $f(x)$, is the same constant b. Since $f(x)$ is another name for y, and since every equation of the form $y = b$ describes a horizontal line, every constant function has as its graph a straight line parallel to the x-axis.

EXAMPLE 4 A Constant Function

The constant function f defined by

$$f(x) = 3$$

has the following values and is graphed in Figure 3.29.

$$f(0) = 3$$
$$f(1) = 3$$
$$f(2) = 3$$
$$f(-1) = 3$$
$$f(a + 1) = 3$$

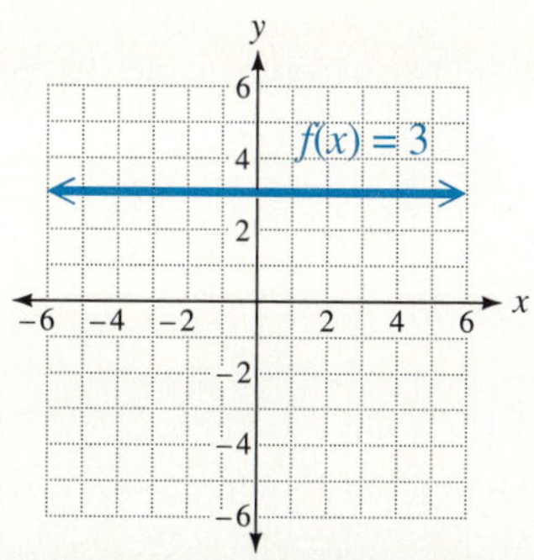

Figure 3.29 Constant Function

Practice Exercise 4

$g(x) = -5$

(a) $g(0) =$

(b) $g(3) =$

(c) $g(-6) =$

(d) $g(b) =$

(e) $g(a - 1) =$

Answers: (a) -5 (b) -5 (c) -5 (d) -5 (e) -5

5 THE IDENTITY FUNCTION

Another special linear function results when $b = 0$ and $m = 1$ in $f(x) = mx + b$. The identity function

$$f(x) = x$$

has domain the entire set of real numbers, and identifies a real number with itself. For example, $f(4) = 4$, $f(100) = 100$, and $f(-1) = -1$. The identity function is graphed in Figure 3.30.

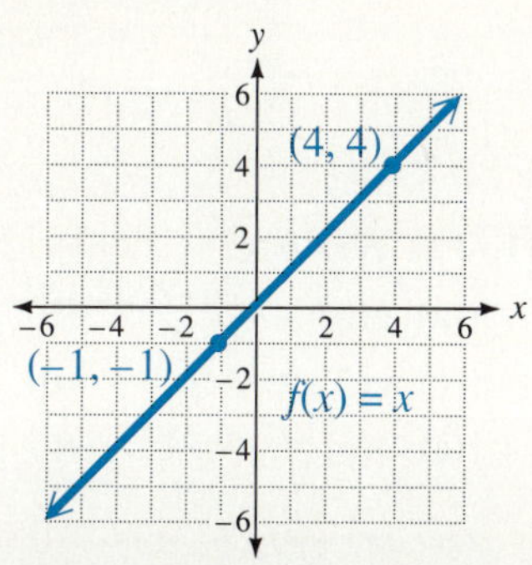

Figure 3.30 Identity Function

The identity function will be considered in more detail in Chapter 9 when additional properties of functions are discussed.

Many applied problems can be simplified using the idea of a function. One quantity or variable may be given in terms of another in an equation that describes a functional relationship. This is illustrated in the next example.

EXAMPLE 5 Sales Problem

A sales representative earns an annual salary of \$12,000 plus a 10% commission on total sales. Express her annual earnings, y, as a function, f, of her total sales, x. What are her annual earnings when her yearly sales total \$85,000?

The function defined by this information is

$$y = f(x) = 12{,}000 + 0.10x.$$

When her yearly sales total \$85,000, her annual earnings amount to $f(85{,}000)$.

$$\begin{aligned} f(85{,}000) &= 12{,}000 + (0.10)(85{,}000) \\ &= 12{,}000 + 8500 = 20{,}500 \end{aligned}$$

Thus, she earned \$20,500 that year.

Practice Exercise 5

A sales representative earns \$9000 plus 8% commission on total sales. Express his earnings as a function of total sales, x, and find the earnings on sales of \$120,000.

Answer: $f(x) = 9000 + 0.08x$; \$18,600

3.6 EXERCISES A

Specify the domain of each function defined by the given equation.

1. $y = x + 1$

2. $y = 5x - 2$

3. $y = \dfrac{5}{x + 5}$

4. $y = \dfrac{1}{4 - x}$

5. $y = x^2 + 3$

6. $y = 9x^2 + 5$

7. $y = \dfrac{x}{x - 1}$

8. $y = \dfrac{5}{x^2 + 5}$

9. $y = 3x^2 + 4x + 1$

Evaluate each of the following.

10. Let $f(x) = 2x + 5$.

(a) $f(0) =$ **(b)** $f(1) =$

(c) $f(3) =$ **(d)** $f(-2) =$

(e) $f(a) =$ **(f)** $f(b - 1) =$

11. Let $g(x) = x^2 - 3$.

(a) $g(0) =$ **(b)** $g(-1) =$

(c) $g(3) =$ **(d)** $g(-2) =$

(e) $g(b) =$ **(f)** $g(-a) =$

12. Let $f(x) = 2x^3 - 1$.

(a) $f(0) =$ **(b)** $f(-1) =$

(c) $f(2) =$ **(d)** $f(-3) =$

(e) $f(a) =$ **(f)** $f(-b) =$

13. Let $g(x) = -7$.

(a) $g(0) =$ **(b)** $g(-1) =$

(c) $g(2) =$ **(d)** $g(-3) =$

(e) $g(a) =$ **(f)** $g(b - 1) =$

14. Let $f(x) = -3x + 1$.

(a) $f(0) =$ **(b)** $f(-1) =$

(c) $f(2) =$ **(d)** $f(-3) =$

(e) $f(b) =$ **(f)** $f(a - 1) =$

15. Let $g(x) = 4x^2 - 3x + 1$.

(a) $g(0) =$ **(b)** $g(-1) =$

(c) $g(2) =$ **(d)** $g(-3) =$

(e) $g(a) =$ **(f)** $g(-b) =$

16. Let $f(x) = 1 - x$.

(a) $f(0) =$ **(b)** $f(-1) =$

(c) $f(2) =$ **(d)** $f\left(-\frac{3}{2}\right) =$

(e) $f(b) =$ **(f)** $f(a - 1) =$

17. Let $f(x) = -x$.

(a) $f(0) =$ **(b)** $f(-1) =$

(c) $f(2) =$ **(d)** $f\left(-\frac{3}{2}\right) =$

(e) $f(b) =$ **(f)** $f(a - 1) =$

State whether the given function is a constant or linear function, and sketch its graph.

18. $f(x) = -2x + 1$

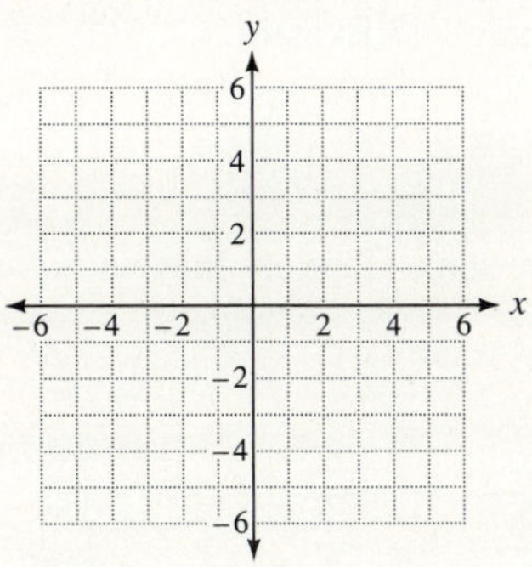

19. $f(x) = 2x + 1$

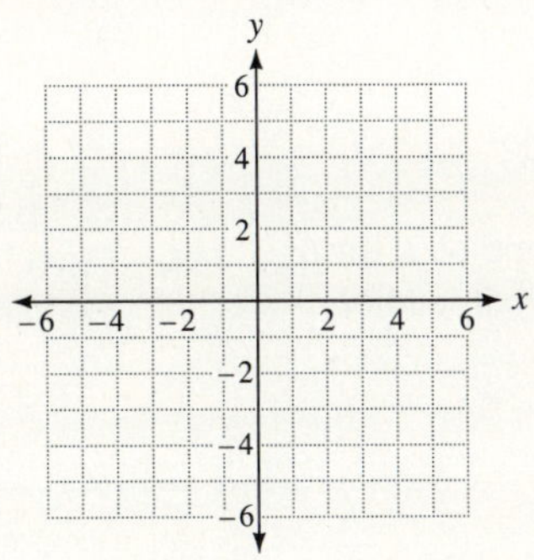

20. $f(x) = -2$

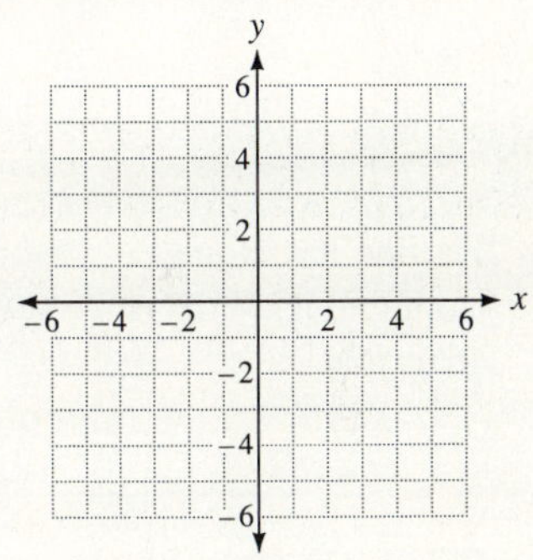

21. Write an equation for the cost function, f, described by selling x books at \$11.95 a book. How much do 15 books cost?

22. The distance that a car travels at 55 mph can be found by multiplying the total number of hours driven, x, by the hourly rate. Express this functional relationship using the letter g, and use it to find the distance traveled in 8 hr.

23 A salesman earns an annual salary of \$15,000 plus a 12% commission on sales. Express his annual earnings as a function, f, of his total sales, x. What are his annual earnings when his yearly sales total \$100,000?

24 An object is thrown upward with initial velocity of 128 ft per sec. If its height after x sec is given by $f(x) = -16x^2 + 128x$, at what height will it be in 2 sec?

FOR REVIEW

Which of the following are functions? Explain.

25. 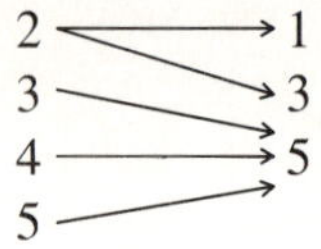

26. 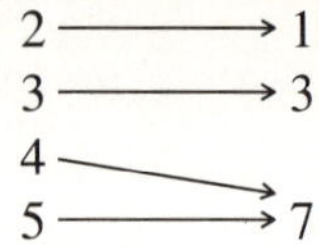

27. 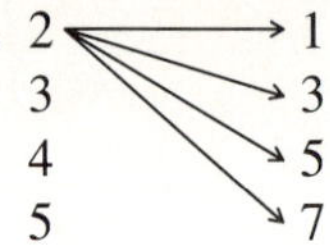

Which of the following are graphs of functions? Explain.

28. 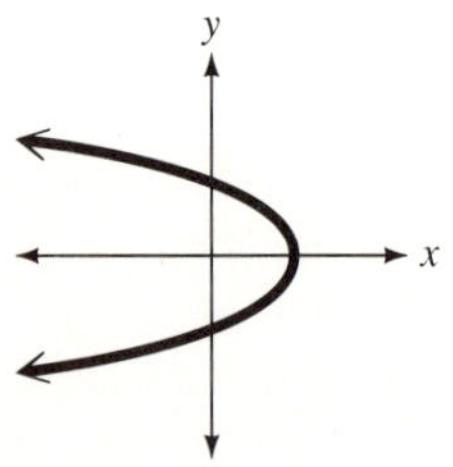

29. 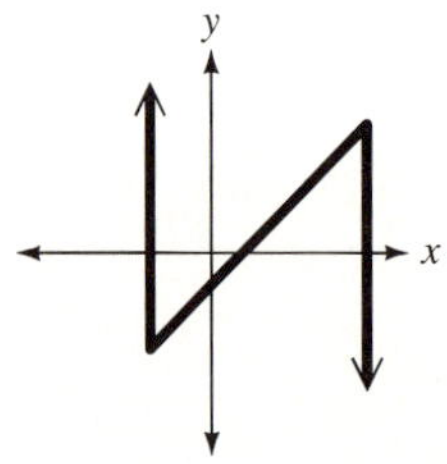

30. 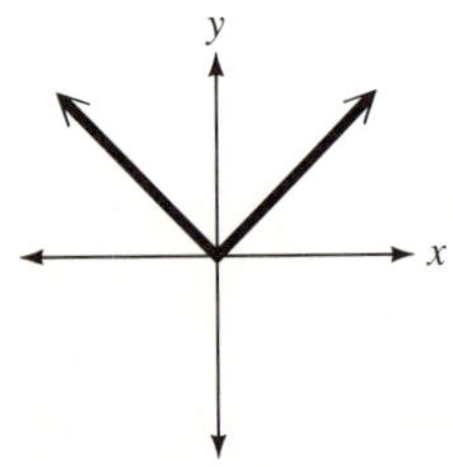

The following exercises, primarily from Section 3.3, will help you prepare for the material in Chapter 4.

31. Find the value of y if x is 2 in $3x - 2y + 8 = 0$.

32. Are the lines with equations $5x - 4y - 8 = 0$ and $5x - 4y + 8 = 0$ parallel? Explain.

33. If a line has slope $-\frac{1}{3}$, what is the slope of a line parallel to it? What is the slope of a line perpendicular to it?

34. If two lines have the same slope are they necessarily parallel lines?

ANSWERS: **1.** all real numbers **2.** all real numbers **3.** all real numbers except -5 **4.** all real numbers except 4 **5.** all real numbers **6.** all real numbers **7.** all real numbers except 1 **8.** all real numbers **9.** all real numbers **10.** (a) 5 (b) 7 (c) 11 (d) 1 (e) $2a + 5$ (f) $2b + 3$ **11.** (a) -3 (b) -2 (c) 6 (d) 1 (e) $b^2 - 3$ (f) $a^2 - 3$ **12.** (a) -1 (b) -3 (c) 15 (d) -55 (e) $2a^3 - 1$ (f) $-2b^3 - 1$ **13.** (a) -7 (b) -7 (c) -7 (d) -7 (e) -7 (f) -7 **14.** (a) 1 (b) 4 (c) -5 (d) 10 (e) $-3b + 1$ (f) $-3a + 4$ **15.** (a) 1 (b) 8 (c) 11 (d) 46 (e) $4a^2 - 3a + 1$ (f) $4b^2 + 3b + 1$ **16.** (a) 1 (b) 2 (c) -1 (d) $\frac{5}{2}$ (e) $1 - b$ (f) $2 - a$ **17.** (a) 0 (b) 1 (c) -2 (d) $\frac{3}{2}$ (e) $-b$ (f) $1 - a$ **18.** linear function **19.** linear function **20.** constant function

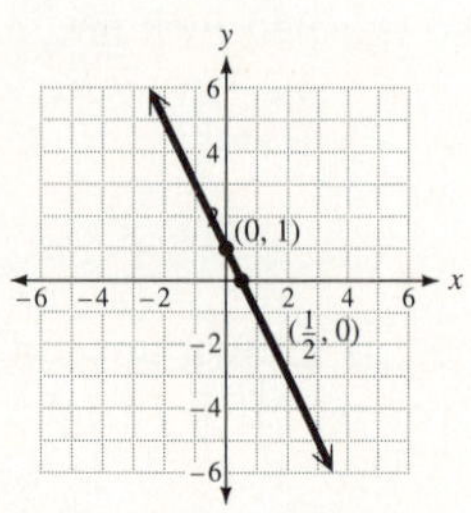

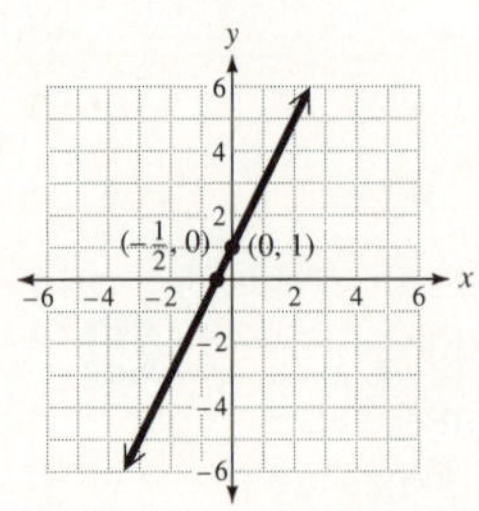

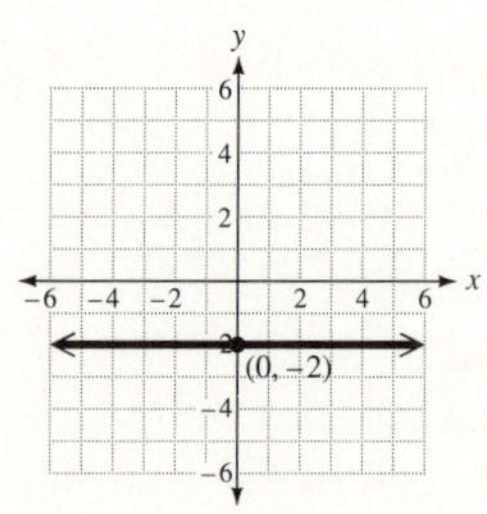

21. $f(x) = 11.95x$; \$179.25 **22.** $g(x) = 55x$; 440 miles **23.** $f(x) = 15{,}000 + 0.12x$; \$27,000 **24.** 192 feet **25.** not a function **26.** a function **27.** not a function **28.** not the graph of a function **29.** not the graph of a function **30.** graph of a function **31.** $y = 7$ **32.** Yes. Both have slope $\frac{5}{4}$ but different y-intercepts. **33.** $-\frac{1}{3}$; 3 **34.** No. They could be the same line.

3.6 EXERCISES B

Specify the domain of each function defined by the given equation.

1. $y = x - 1$

2. $y = 9x - 5$

3. $y = \dfrac{2}{x + 2}$

4. $y = \dfrac{1}{3 - x}$

5. $y = x^2 + 1$

6. $y = 8x^2 + 1$

7. $y = \dfrac{x}{1 - x}$

8. $y = \dfrac{1}{x^2 + 1}$

9. $y = 2x^2 + x - 3$

For each function, evaluate ***(a)*** $f(0)$ ***(b)*** $f(1)$ ***(c)*** $f(2)$ ***(d)*** $f(-3)$ ***(e)*** $f(b)$ ***(f)*** $f(-a)$.

10. $f(x) = 3x - 1$
(a) **(b)**
(c) **(d)**
(e) **(f)**

11. $f(x) = x^2 + 2$
(a) **(b)**
(c) **(d)**
(e) **(f)**

12. $f(x) = 3x^3 - 2$
(a) **(b)**
(c) **(d)**
(e) **(f)**

13. $f(x) = 4x^2 - x + 3$
(a) **(b)**
(c) **(d)**
(e) **(f)**

For each function, evaluate ***(a)*** $f(0)$ ***(b)*** $f(-1)$ ***(c)*** $f\left(-\frac{1}{2}\right)$ ***(d)*** $f(3)$ ***(e)*** $f(a)$ ***(f)*** $f(b - 1)$.

14. $f(x) = -3$
(a) **(b)**
(c) **(d)**
(e) **(f)**

15. $f(x) = -2x + 5$
(a) **(b)**
(c) **(d)**
(e) **(f)**

16. $f(x) = 3 - x$
(a) **(b)**
(c) **(d)**
(e) **(f)**

17. $f(x) = x$
(a) **(b)**
(c) **(d)**
(e) **(f)**

State whether the given function is a constant or linear function, and sketch its graph.

18. $f(x) = 2x - 1$

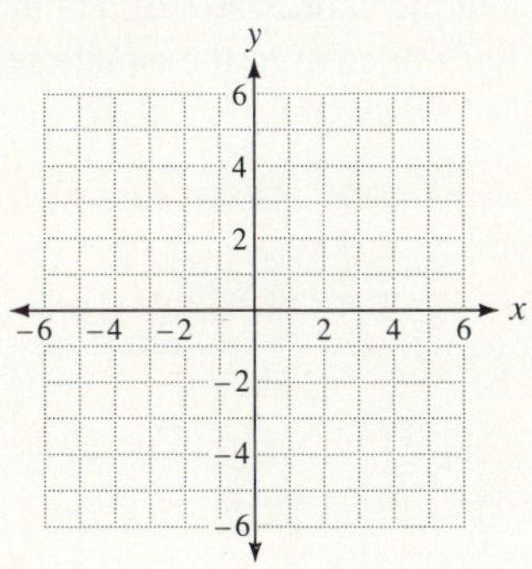

19. $f(x) = -2x - 1$

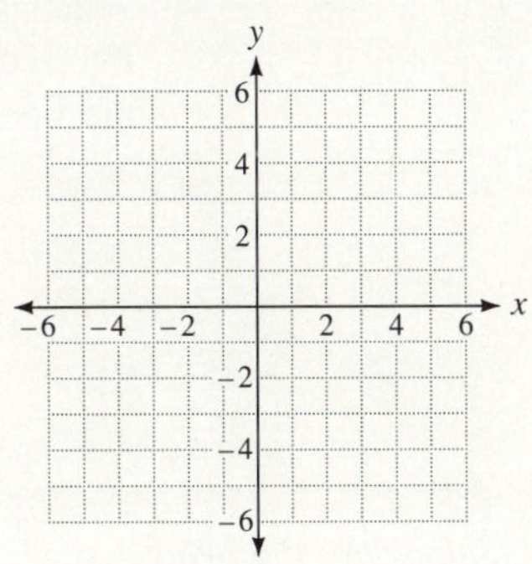

20. $f(x) = 5$

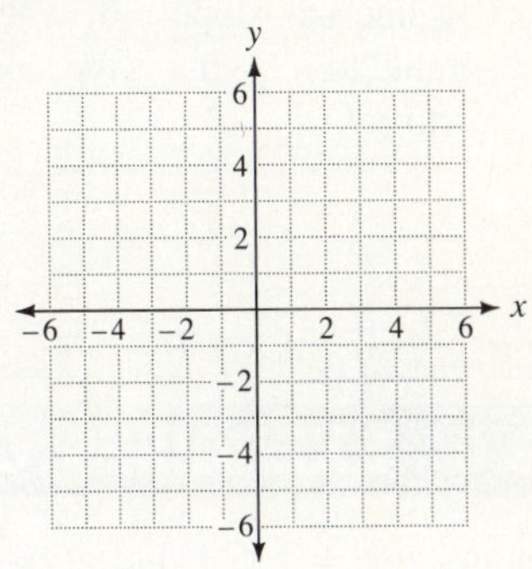

21. Write an equation for the cost function, f, described by selling x tickets at \$8.50 a ticket. How much do 50 tickets cost?

22. The distance that a motor home travels at 50 mph can be found by multiplying the number of hours driven, x, by the hourly rate. Express this functional relationship using the letter g, and find the distance traveled in 14 hours.

23. A realtor earns an annual salary of \$10,000 plus a 6% commission on her sales. Express her annual earnings as a function, f, of her sales, x. What does she earn in a year when she sells \$450,000 worth of real estate?

24. An object is thrown upward with initial velocity of 128 ft per sec. If its height after x sec is given by $f(x) = -16x^2 + 128x$, at what height will it be in 5 sec?

FOR REVIEW

Which of the following are functions? Explain.

25. 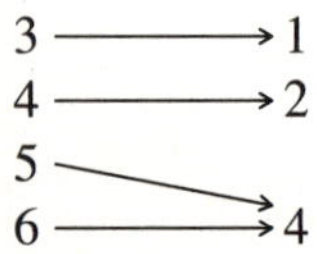

26. 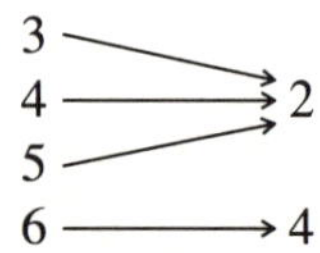

27. 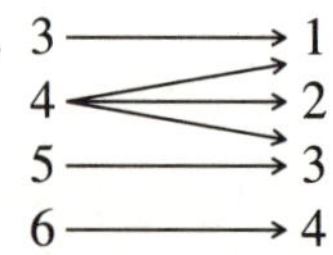

Which of the following are graphs of functions? Explain.

28. 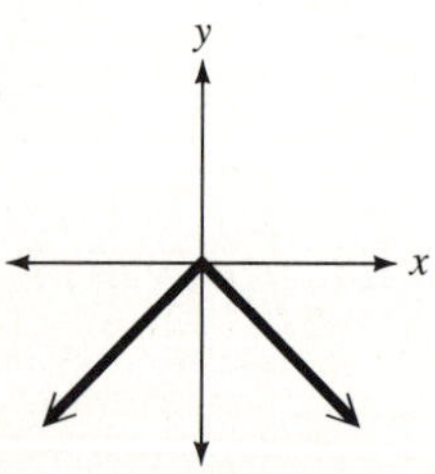

29. 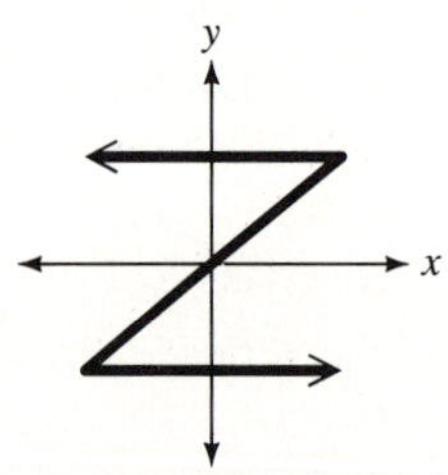

30. 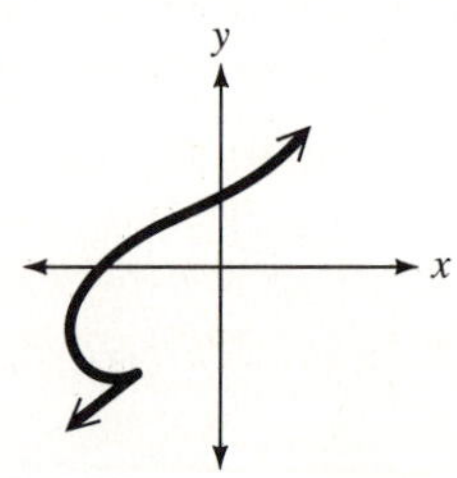

The following exercises, primarily from Section 3.3, will help you prepare for the material in Chapter 4.

31. Find the value of x if y is -2 in $3x - 2y + 8 = 0$.

32. Are the lines with equations $2x - 4y + 10 = 0$ and $x - 2y + 5 = 0$ parallel? Explain.

33. If a line has slope -6, what is the slope of a line parallel to it? What is the slope of a line perpendicular to it?

34. What can be said about the coordinates of a point of intersection of two lines relative to the equations of the lines?

3.6 EXERCISES C

Suppose $f(x) = x + 1$ and $g(x) = x^2 - 5$. Evaluate the following.

1. $f(g(1))$ [Answer: -3]

2. $g(f(1))$

3. $f(g(x))$

4. $f(x + h)$

5. $\dfrac{f(x + h) - f(x)}{h}$ [Answer: 1]

6. $\dfrac{g(x + h) - g(x)}{h}$

CHAPTER 3 REVIEW

KEY WORDS

3.1 A **rectangular (Cartesian) coordinate system** is a configuration used to graph equations in two variables.

The point of intersection of the axes in a rectangular coordinate system is called the **origin.**

A point with coordinates given by the ordered pair (x, y) has ***x*-coordinate (abscissa)** x and ***y*-coordinate (ordinate)** y.

The axes in a coordinate system divide the plane into four regions called **quadrants.**

A **solution** to an equation in two variables is an ordered pair of numbers that when substituted makes the equation true.

The **graph** of an equation in two variables is the set of points in a coordinate system that corresponds to solutions of the equation.

3.2 A **linear** (or **first-degree**) **equation in two variables** can be written in the form $ax + by + c = 0$, a and b not both zero.

Points where the graph of a linear equation (line) crosses the x-axis and y-axis, respectively, are called the ***x*-intercept** and ***y*-intercept.**

3.3 The **slope** of a line is a number which is the ratio of the vertical change (rise) to the horizontal change (run) as we move from one point to another on the line.

Points are **collinear** if they lie on the same line.

Two lines are **parallel** if they never intersect.

Two lines are **perpendicular** if they intersect at right angles.

3.5 A correspondence between a first set of objects, the **domain,** and a second set of objects, the **range,** is a **relation** if to each element in the domain there corresponds one or more elements in the range.

A **function** is a relation with the additional property that to each element in the domain there corresponds one and only one element in the range.

If (x, y) is an element of a function, x is the **independent variable** and y is the **dependent variable.**

KEY CONCEPTS

3.1 To graph an equation, construct a table of values, plot the points corresponding to these solutions in a rectangular coordinate system, and join these points with a smooth curve.

3.2
1. The general form of the equation of a line is $ax + by + c = 0$. If $a = 0$, the line is a horizontal line, parallel to the x-axis, with y-intercept $\left(0, -\frac{c}{b}\right)$. If $b = 0$, the line is a vertical line, parallel to the y-axis, with x-intercept $\left(-\frac{c}{a}, 0\right)$.
2. To graph a linear equation using the intercept method, plot the intercepts and draw the line through them. If both intercepts are $(0, 0)$, plot another solution and draw the line through it and $(0, 0)$.

3.3
1. The slope of a line passing through points with coordinates (x_1, y_1) and (x_2, y_2) is $m = \frac{y_2 - y_1}{x_2 - x_1}$.
2. If m_1 and m_2 are the slopes of two lines and if $m_1 = m_2$, then the lines are parallel; if $m_1 = -\frac{1}{m_2}$, then the lines are perpendicular.

3.4
1. Point-slope form: $y - y_1 = m(x - x_1)$.
2. Slope-intercept form: $y = mx + b$.
3. To find the equation of a line through two points, find the slope of the line first, then use the point-slope form with either of the given points.
4. To graph a line using the slope method, plot the y-intercept and use the slope to determine the rise and run of the graph which will give a second point on the line.

3.5
1. A relation can be thought of as a set of ordered pairs of numbers. If no two pairs have the same x-coordinate and different y-coordinate, the relation is a function.
2. Vertical line test: If every vertical line intersects the graph of a relation in no more than one point, the relation is a function.

3.6
1. Unless specified otherwise, the domain of a function described by an equation is the set of all real numbers for which the expression in x is defined.

2. If a function is written in functional notation, for example, $f(x) = 3x + 5$, $f(x)$ represents y. To find y when x is 2, we calculate $f(2) = 3(2) + 5 = 11$.

3. Linear function: $f(x) = mx + b$.
4. Constant function: $f(x) = c$.
5. Identity function: $f(x) = x$.

REVIEW EXERCISES

Part I

3.1 **1.** Give the coordinates of the points A, B, C, and D. In which quadrant is each point located?

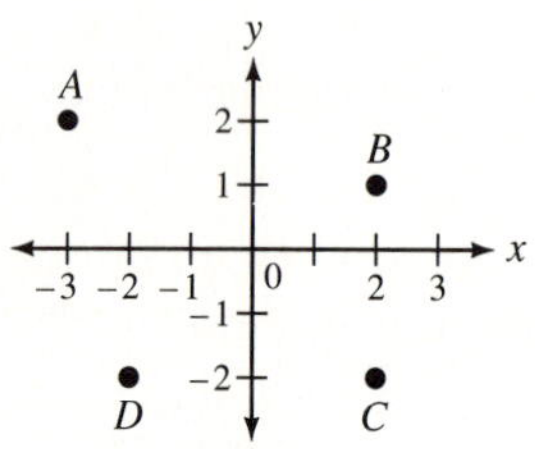

2. Give the coordinates of the vertices of a rectangle 4 units long and 3 units wide if the rectangle is in the first quadrant, its longest side is parallel to the x-axis, and one vertex is at $(1, 2)$.

Graph the following equations.

3. $y = x^2 - 2$

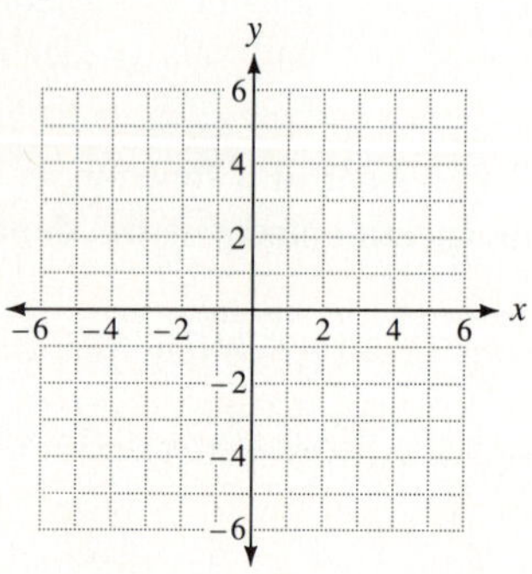

4. $y = |x| - 4$

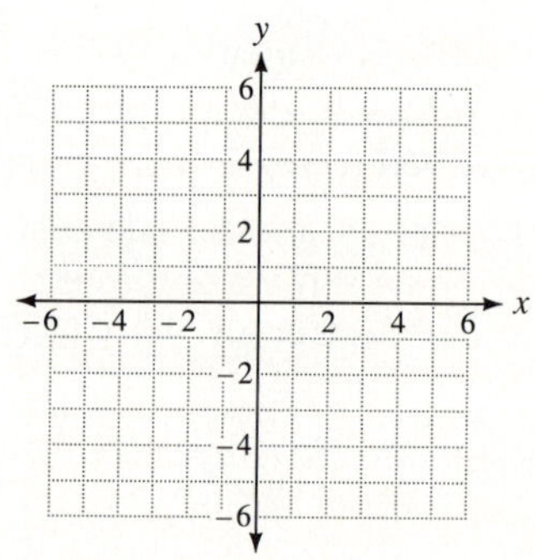

5. Mike Ratliff has estimated that the cost (in dollars), y, of producing a number, x, of computer parts in his plant is given by the equation $y = 3x + 40$. Complete the ordered pairs (1,   ), (100,   ), (1000,   ). What is the cost of producing **(a)** 1 part? **(b)** 100 parts? **(c)** 1000 parts?

3.2–3.3 *Give the intercepts, slope, and graph of the following equations.*

6. $x + 2y - 4 = 0$

x-intercept =

y-intercept =

slope =

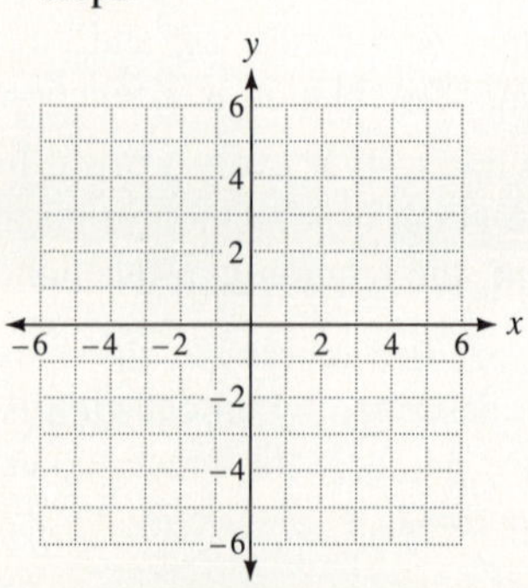

7. $2x + 1 = 0$

x-intercept =

y-intercept =

slope =

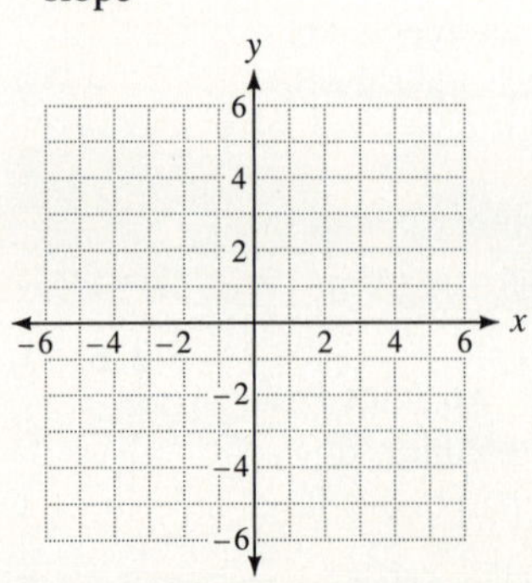

8. $y - 3 = 0$

x-intercept =

y-intercept =

slope =

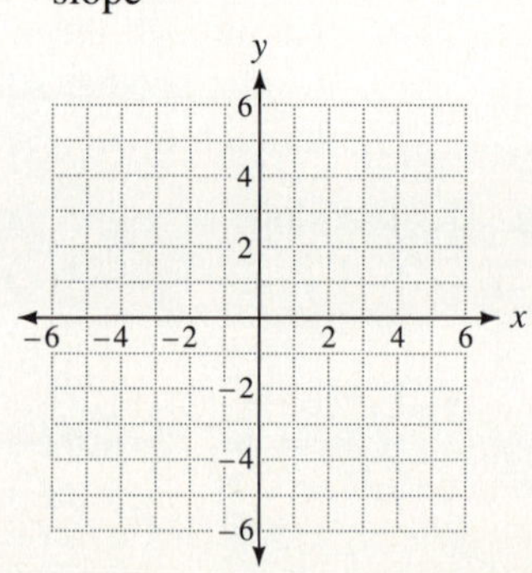

9. Does a line that slopes from upper left to lower right have positive slope or negative slope?

10. Which coordinate axis has undefined slope?

11. What is the slope of the line passing through $(3, -8)$ and $(-1, -2)$? What is the slope of a line that is perpendicular to this line?

12. Are the points $P(2, 5)$, $Q(-4, 1)$, and $R(5, 7)$ collinear? Explain.

3.4 *Are the given pairs of lines* parallel, perpendicular, *or* neither? *Explain.*

13. $3x - y + 2 = 0$
$x + 3y - 7 = 0$

14. $x + 2 = 0$
$y - 2 = 0$

15. $2x - 5y + 2 = 0$
$-4x + 10y - 2 = 0$

16. Find the general form of the equation of the line that has slope -4 and passes through $(-1, 3)$.

17. Find the slope-intercept form of the equation of the line with equation $4x + 2y - 10 = 0$. What is the slope? the y-intercept?

18. Find the general form of the equation of the line passing through the points $(-2, 5)$ and $(6, 9)$.

19. Find the general form of the equation of the line that passes through the point $(-3, 1)$ and is perpendicular to the line $2x + y - 3 = 0$.

20. What are the intercepts and the slope of the line with equation $3y - 15 = 0$?

21. What is the equation of the line passing through $(-4, 5)$ parallel to the x-axis? What is the equation of the line through $(-4, 5)$ parallel to the y-axis?

3.5 *Which of the following are functions?*

22. 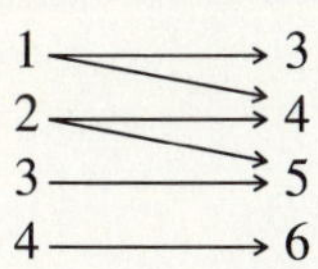

23. 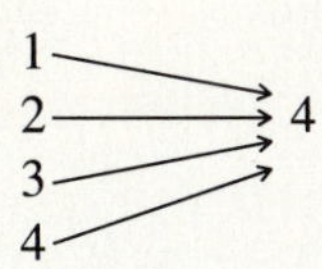

24. 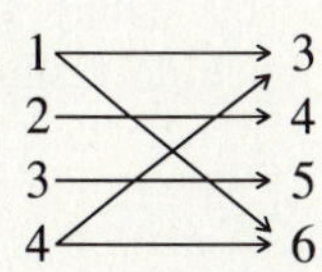

Which of the following are graphs of functions?

25. 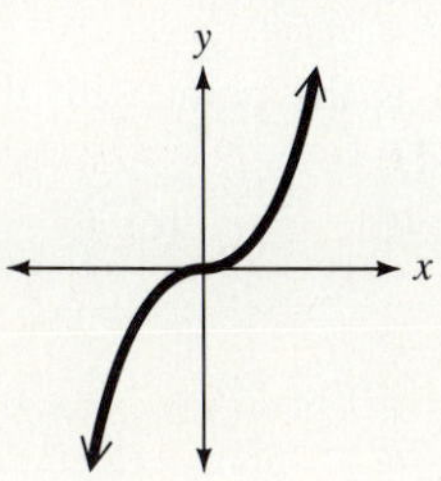

26. 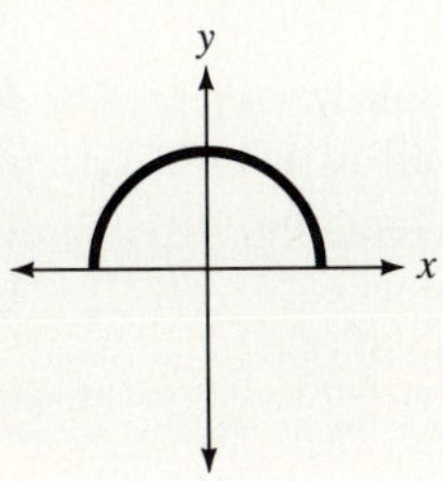

27. 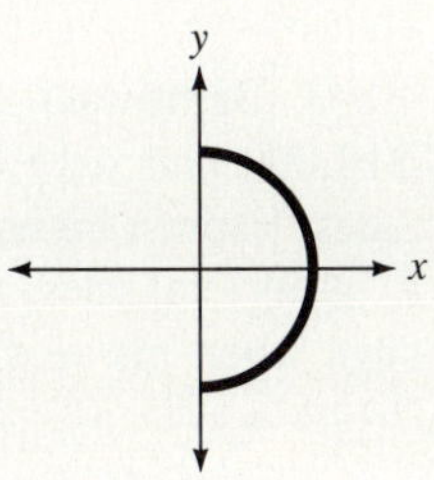

28. Consider the relation described by the set of ordered pairs $\{(2, 3), (1, 3), (3, 3), (4, 4), (4, 1)\}$.

(a) Write this relation using arrows.

(b) Graph the relation in the given Cartesian coordinate system.

(c) Is the relation a function? Explain.

(a)

(b)

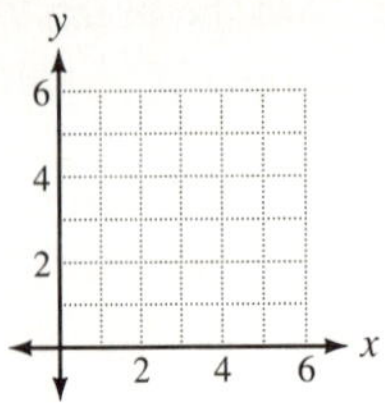

(c)

3.6 *For the given function, supply the required information for parts* **(a)** *through* **(f)**. *Graph the equation for part* **(g)**.

29. $f(x) = -2$

(a) $f(0) =$

(b) $f(1) =$

(c) $f(-2) =$

(d) $f(b) =$

(e) $f(-a) =$

(f) Is this a constant or linear function?

(g)

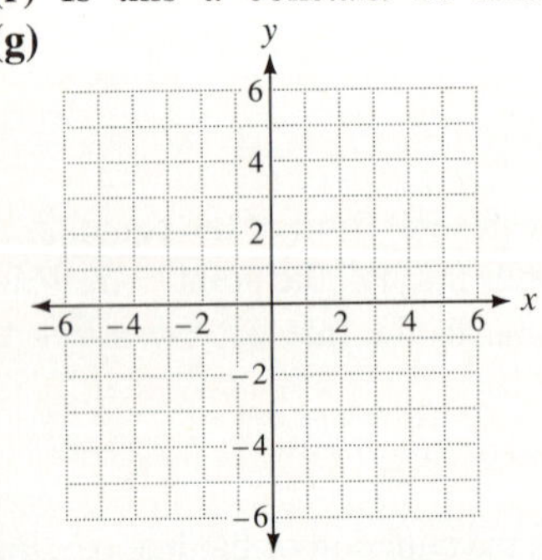

30. $f(x) = -3x + 6$

(a) $f(0) =$

(b) $f(1) =$

(c) $f(-2) =$

(d) $f(b) =$

(e) $f(-a) =$

(f) Is this a constant or linear function?

(g)

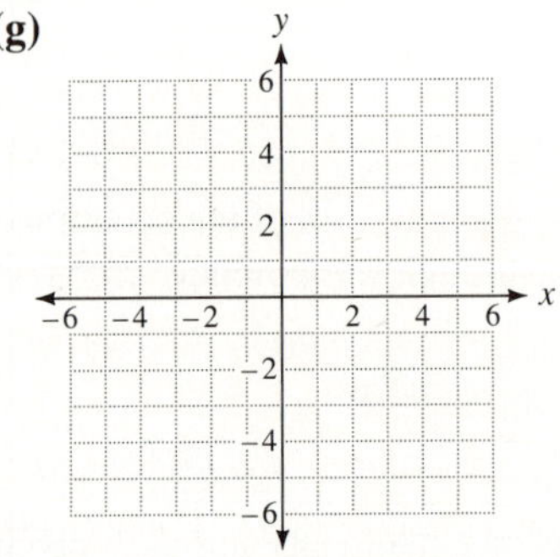

Specify the domain of each function defined by the given equation.

31. $y = 8x + 5$

32. $y = x^2 + 10$

33. $y = \dfrac{7}{7 - x}$

34. $y = \dfrac{1}{x^2 + 10}$

35. Write an equation for the cost function, g, described by selling x shirts at \$16.95 a shirt. How much do 8 shirts cost?

36. What is the equation describing the identity function?

Part II

37. Which coordinate axis has zero slope?

38. Give the slope and y-intercept of the line with equation $3x - 2y + 12 = 0$.

39. Barry Bertrowski earns an annual salary of \$12,000 plus a 15% commission on all of his sales. Express his annual earnings as a function, f, of his total sales, x. What are his annual earnings when his yearly sales total is \$230,000?

40. In which quadrant is the point with the given coordinates located?

(a) $(-2, 8)$ **(b)** $(6, -3)$ **(c)** $(4, \pi)$

(d) $(-\sqrt{2}, -3)$

Graph the following equations.

41. $y = |x - 2|$

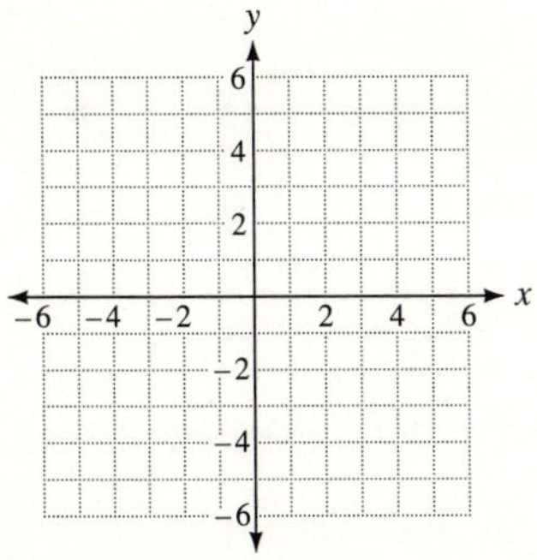

42. $6x + y - 6 = 0$

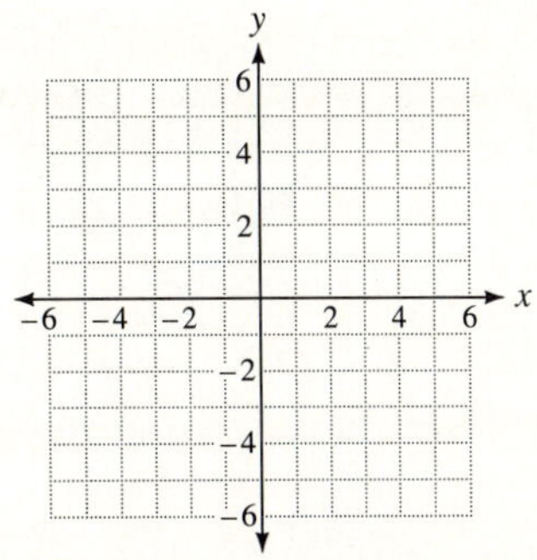

43. What is the slope of the line passing through $(0, -2)$ and $(7, -2)$?

44. Find the general form of the equation of the line that passes through $(-2, 1)$ and is parallel to the line $3x - y + 2 = 0$.

45. $f(x) = 4x - 2$

(a) $f(0) =$ **(b)** $f(1) =$

(c) $f(-2) =$ **(d)** $f(b) =$

(e) $f(-a) =$

(f) Is this a constant or linear function?

(g) Graph the function using the slope method.

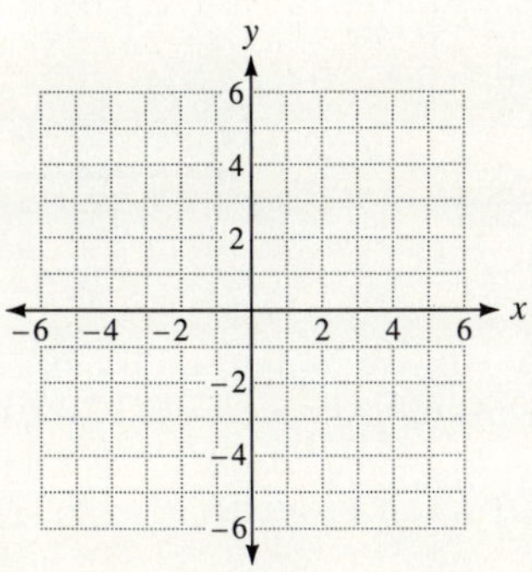

46. Is the following a function?

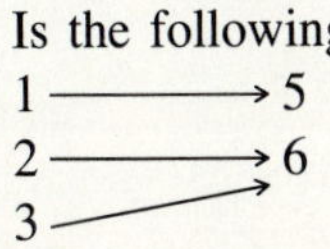

47. Are the lines with equations

$$2x - y + 3 = 0$$
$$-4x + 2y - 6 = 0$$

parallel, perpendicular, or neither?

48. What are the slope and y-intercept of the line with equation $2x + 10 = 0$?

49. If $g(x) = -x^2 + 6$, find **(a)** $g(-1)$, and **(b)** $g(c)$.

50. Terry McGinnis has found that the cost in dollars, y, of typing x pages of manuscript is given by $y = 8x + 25$. Complete the ordered pair (500,) and use it to find the cost of typing 500 pages.

ANSWERS: 1. $A(-3, 2)$ in II; $B(2, 1)$ in I; $C(2, -2)$ in IV; $D(-2, -2)$ in III 2. (1, 2), (5, 2), (5, 5), (1, 5)

3.

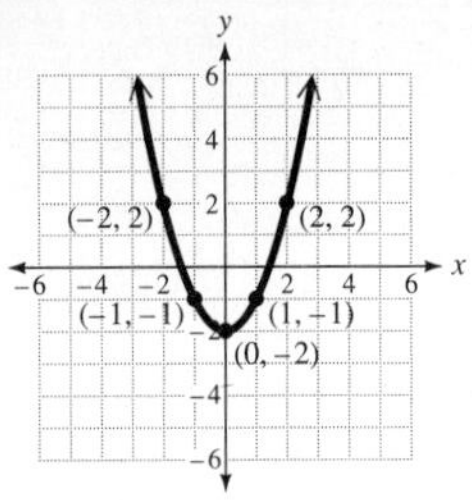

4.

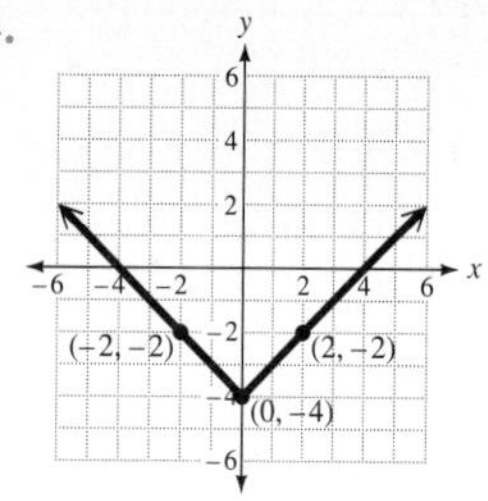

5. (1, 43), (100, 340), (1000, 3040); (a) \$43 (b) \$340 (c) \$3040

6. x-intercept (4, 0); y-intercept (0, 2); slope $-\frac{1}{2}$

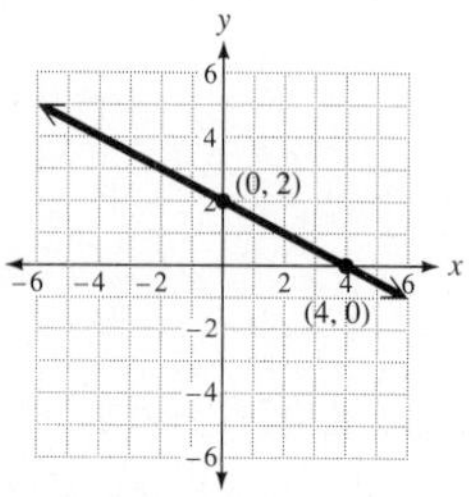

7. x-intercept $\left(-\frac{1}{2}, 0\right)$; no y-intercept; undefined slope

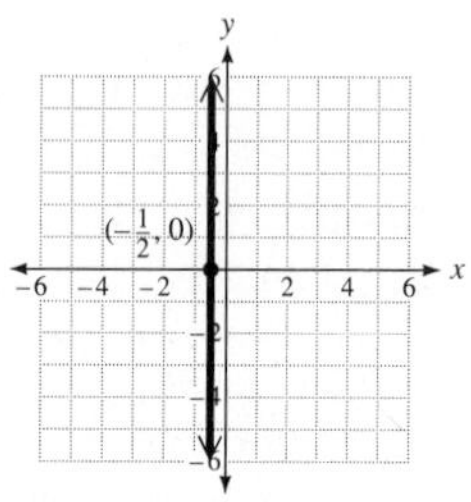

8. no x-intercept; y-intercept (0, 3); slope 0

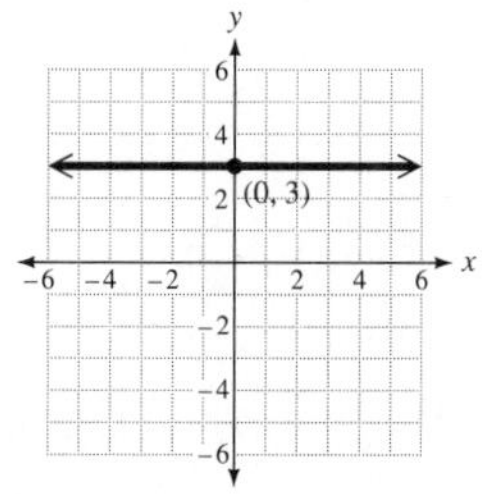

9. negative slope 10. y-axis 11. $-\frac{3}{2}$; $\frac{2}{3}$ 12. Yes. The slope of the line through P and Q is $\frac{2}{3}$, as is the slope of the line through Q and R. 13. perpendicular (slopes are 3 and $-\frac{1}{3}$) 14. perpendicular (each is parallel to an axis) 15. parallel (both have slope $\frac{2}{5}$) 16. $4x + y + 1 = 0$ 17. $y = -2x + 5$; slope -2; (0, 5) 18. $x - 2y + 12 = 0$ 19. $x - 2y + 5 = 0$

20. no x-intercept; y-intercept (0, 5); slope 0 21. $y = 5$; $x = -4$ 22. not a function 23. function 24. not a function 25. function 26. function 27. not a function

28. (a)

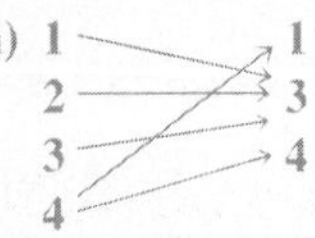

(b)

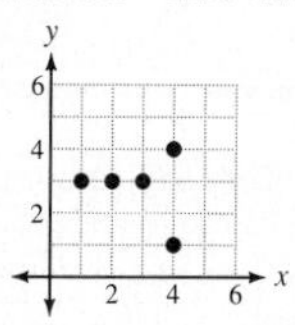

(c) No. The graph does not pass the vertical line test.

29. (a) $f(0) = -2$
(b) $f(1) = -2$
(c) $f(-2) = -2$
(d) $f(b) = -2$
(e) $f(-a) = -2$
(f) constant
(g)

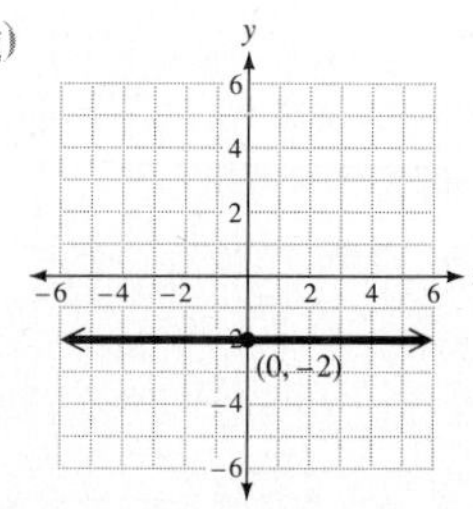

30. (a) $f(0) = 6$
(b) $f(1) = 3$
(c) $f(-2) = 12$
(d) $f(b) = -3b + 6$
(e) $f(-a) = 3a + 6$
(f) linear
(g)

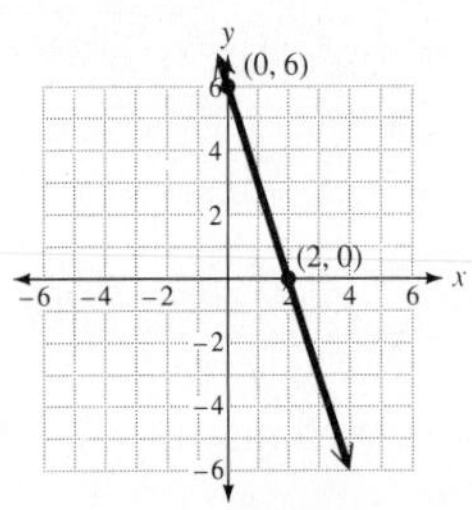

31. all real numbers 32. all real numbers 33. all real numbers except 7 34. all real numbers 35. $g(x) = 16.95x$; \$135.60 36. $f(x) = x$ 37. x-axis 38. $\frac{3}{2}$; (0, 6) 39. $f(x) = 12{,}000 + 0.15x$; \$46,500 40. (a) II (b) IV (c) I (d) III

41.

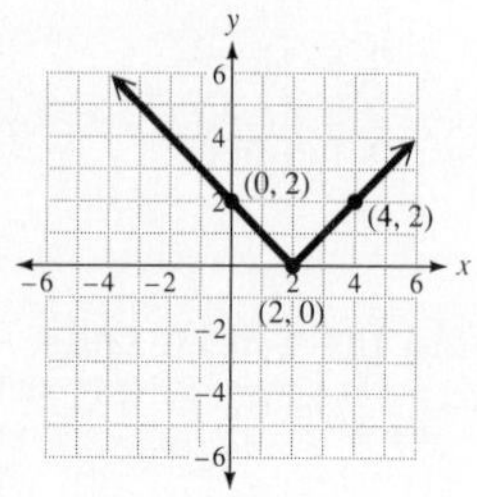

42.

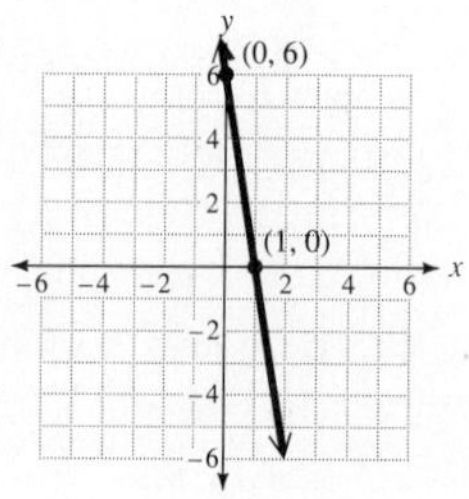

43. 0 44. $3x - y + 7 = 0$

45. (a) $f(0) = -2$
(b) $f(1) = 2$
(c) $f(-2) = -10$
(d) $f(b) = 4b - 2$
(e) $f(-a) = -4a - 2$
(f) linear
(g)

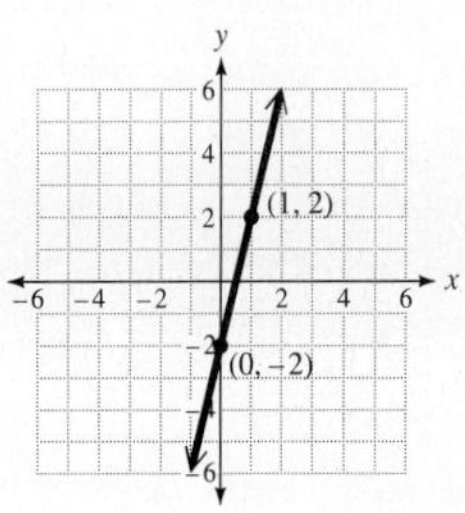

46. yes 47. Neither. The lines are the same. 48. undefined slope; no y-intercept 49. (a) 5 (b) $-c^2 + 6$ 50. (500, 4025); \$4025

CHAPTER 3 TEST

Exercises 1–5 refer to the equation $3x + 2y = 12$.

1. Write the equation in slope-intercept form. **1.** ______________

2. What is the x-intercept? **2.** ______________

3. What is the y-intercept? **3.** ______________

4. What is the slope? **4.** ______________

5. Graph the equation. **5.**

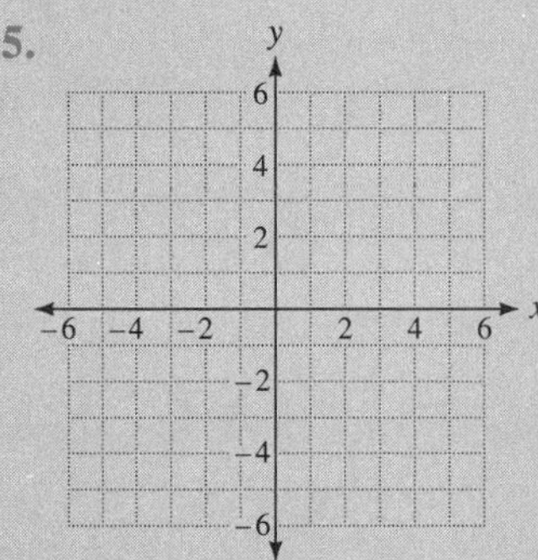

6. Ricardo estimates that the cost (in dollars), y, of repairing a number, x, of electronic detectors is given by the equation $y = 4x + 30$. Complete the ordered pair (10,   ) and use it to find the cost of repairing 10 detectors. **6.** ______________

7. What is the slope of the line passing through $(-2, 7)$ and $(4, -5)$? What is the slope of a line perpendicular to this line? **7.** ______________

8. Are the lines represented by the given equations parallel, perpendicular, or neither? **8.** ______________

$$2x - 5y + 7 = 0$$
$$-2x + 5y + 7 = 0$$

9. Find the general form of the equation of the line that has slope -3 and passes through $(-2, 4)$. **9.** ______________

CHAPTER 3 TEST

(CONTINUED)

10. Find the general form of the equation of the line passing through $(-1, 7)$ and $(4, -3)$. **10.** ____________

11. What is the equation of the line through $(-2, 8)$ parallel to the x-axis? **11.** ____________

12. Is 1 ⟶ 4 a function? **12.** ____________
2 ⟶ 5
3 ↗ 5

13. Is the curve at right the graph of a function?

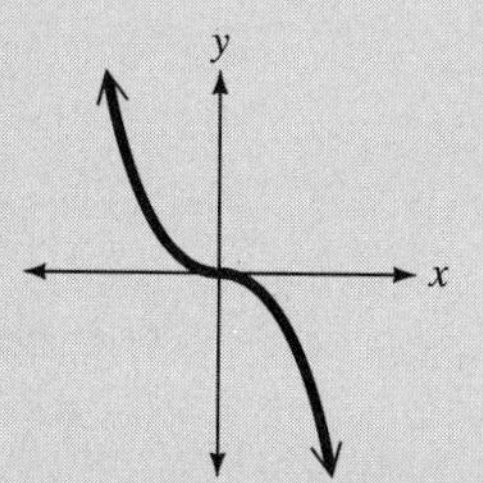

13. ____________

14. Is the relation $\{(2, 1), (1, 3), (1, 4), (3, 5)\}$ a function? **14.** ____________

Exercises 15–18 refer to the function $f(x) = 3x + 3$.

15. Find $f(-2)$. **15.** ____________

16. Find $f(a - 1)$. **16.** ____________

17. Is this a constant or linear function? **17.** ____________

18. Graph the function. **18.**

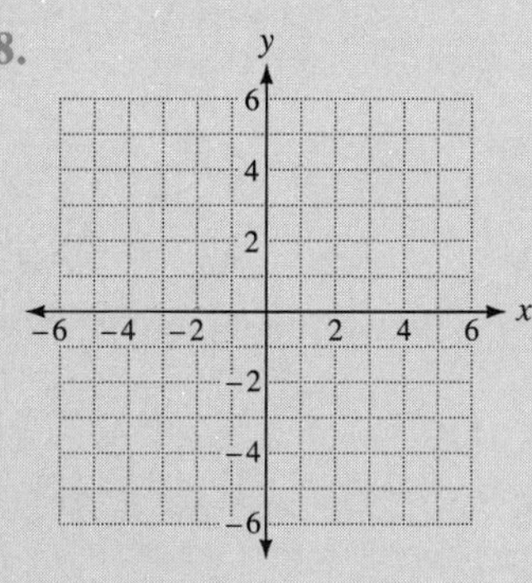

19. Specify the domain of the function $y = \dfrac{1}{x + 3}$. **19.** ____________

20. Write an equation for the cost function, g, described by selling x pairs of shoes at \$20 a pair. **20.** ____________

CHAPTER 4

Systems of Linear Equations and Inequalities

4.1 SYSTEMS OF TWO LINEAR EQUATIONS IN TWO VARIABLES

STUDENT GUIDEPOSTS

1 Systems of Equations
2 The Graphing Method
3 Types of Linear Systems
4 Number of Solutions to a System

In Chapter 3 we saw that a linear equation in the two variables x and y, such as

$$x - y - 1 = 0,$$

has many solutions that are ordered pairs of numbers that make the equation true. Some of the solutions to this equation are

$$(0, -1), \quad (1, 0), \quad (-1, -2), \quad (2, 1), \quad \text{and} \quad (5, 4).$$

The graph of the equation is a straight line, every point of which corresponds to a solution to the equation. Suppose that we have a second linear equation,

$$2x + y - 5 = 0.$$

This equation too has infinitely many solutions, some of which are

$$(0, 5), \quad (1, 3), \quad (-1, 7), \quad (2, 1), \quad \text{and} \quad (-3, 11).$$

1 SYSTEMS OF EQUATIONS

A **system of two linear equations in two variables** is a pair of linear equations such as the following.

$$x - y - 1 = 0$$
$$2x + y - 5 = 0$$

A **solution** to a system of equations is an ordered pair of numbers that is a solution to *both* equations. For our particular example, because (2, 1) is listed as a solution to both of the equations, (2 1) is a solution to the system of equations. We can check this by substituting (2, 1) into each of the equations.

$$2 - 1 - 1 \stackrel{?}{=} 0 \qquad\qquad 2(2) + 1 - 5 \stackrel{?}{=} 0$$
$$0 = 0 \qquad\qquad 4 + 1 - 5 \stackrel{?}{=} 0$$
$$0 = 0$$

2 THE GRAPHING METHOD

If we write both equations above in slope-intercept form,

$$x - y - 1 = 0 \qquad\qquad 2x + y - 5 = 0$$
$$-y = -x + 1 \qquad\qquad y = -2x + 5$$
$$y = x - 1$$

we can see that the slopes are unequal ($m_1 = 1$ and $m_2 = -2$) and that the graphs of the equations are intersecting lines. Graphing both equations in the same rectangular coordinate system in Figure 4.1, we see that the point of intersection corresponds to the solution of the system, (2, 1).

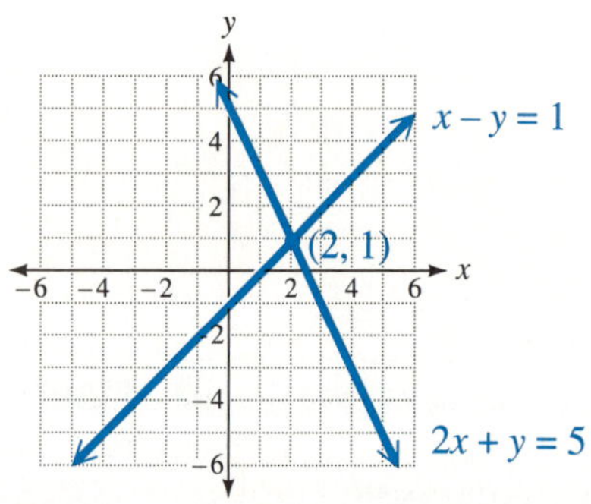

Figure 4.1

This procedure illustrates the **graphing method** for solving a system: We plot the graphs of the equations, estimate the point of intersection (if such a point exists), and verify by substitution that the ordered pair corresponding to the point does indeed solve each equation.

EXAMPLE 1 SOLVING SYSTEMS BY GRAPHING

Solve the system using the graphing method.

$$x + y = 2$$
$$2x - y = -5$$

We graph each equation by finding its intercepts.

$x + y = 2$

x	y
0	2
2	0

$2x - y = -5$

x	y
0	5
$-\frac{5}{2}$	0

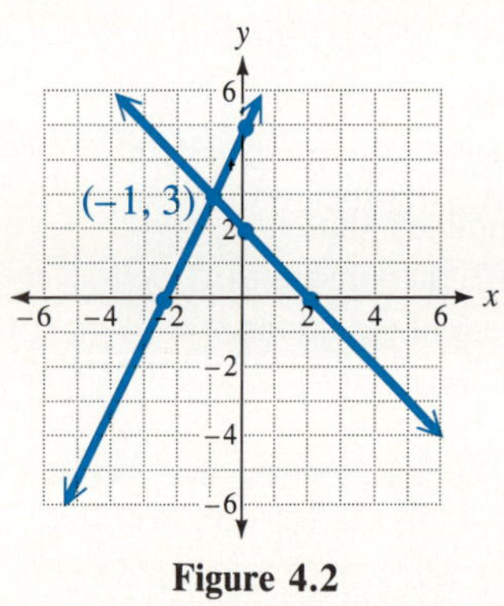

Figure 4.2

PRACTICE EXERCISE 1

Solve the system using the graphing method.

$$x + y = 0$$
$$2x - y = 6$$

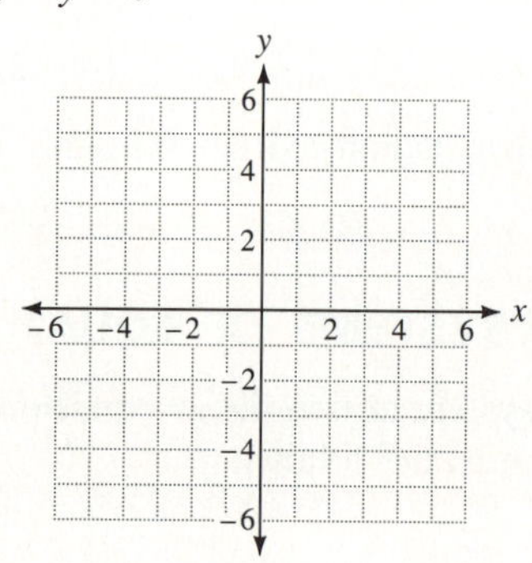

The point of intersection of the two lines, graphed in Figure 4.2, appears to have coordinates $(-1, 3)$. We check by substitution in both equations.

$$\begin{aligned} (-1) + (3) &\stackrel{?}{=} 2 \\ 2 &= 2 \end{aligned} \qquad \begin{aligned} 2(-1) - (3) &\stackrel{?}{=} -5 \\ -2 - 3 &\stackrel{?}{=} -5 \\ -5 &= -5 \end{aligned}$$

The solution to the system is $(-1, 3)$.

Answer: $(2, -2)$

3 TYPES OF LINEAR SYSTEMS

Before we consider other methods of finding solutions to systems of equations, it is helpful to study the nature of these solutions. When each of two linear equations in a system is graphed in the same rectangular coordinate system, one of three possibilities occurs:

1. The lines coincide, and the system of equations has infinitely many solutions. The system is said to be **dependent.**
2. The lines are parallel, and the system of equations has no solution. The system is said to be **inconsistent.**
3. The lines intersect in exactly one point, and the system of equations has exactly one solution. The system is said to be **independent** and **consistent.**

As examples of these three cases consider the following three systems of equations and their corresponding graphs in Figure 4.3.

(A) $2x - y = -4$	(C) $2x - y = -4$	(E) $2x - y = -4$
(B) $6x - 3y = -12$	(D) $2x - y = 2$	(F) $x + y = 1$

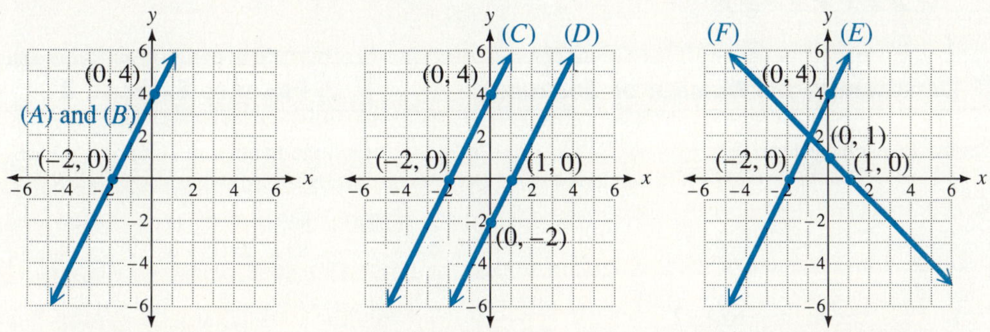

Lines coincide — system dependent

Lines are parallel — system inconsistent

Lines intersect in exactly one point — system independent and consistent

Figure 4.3 Types of Linear Systems

In the first system, lines (A) and (B) coincide; in the second system, lines (C) and (D) are parallel; and in the third system, lines (E) and (F) intersect in one point. Let us write each equation in slope-intercept form (solve each equation for y).

(A) $y = 2x + 4$	(C) $y = 2x + 4$	(E) $y = 2x + 4$
(B) $y = 2x + 4$	(D) $y = 2x - 2$	(F) $y = -x + 1$
same slope	**same slope**	**different slopes**
same intercept	**different intercept**	

Remember that in slope-intercept form, the coefficient of x is the slope and the constant term determines the y-intercept. Also, parallel or coinciding lines have the same slope (or both have undefined slope).

4 NUMBER OF SOLUTIONS TO A SYSTEM

Recognizing different types of linear systems allows us to know the number of solutions to a particular system.

To Find the Number of Solutions to a System of Equations

1. Write each equation in slope-intercept form and identify the slope and y-intercept of each.
2. If the slopes are equal and the y-intercepts are equal, the graphs coincide and the system has *infinitely many solutions*.
3. If the slopes are equal and the y-intercepts are unequal, the graphs are parallel and the system has *no solution*.
4. If the slopes are unequal, the graphs intersect in exactly one point and the system has *exactly one solution*.

CAUTION

The rule above does not hold when one or both of the equations in a system is of the form $ax = c$, since such equations cannot be solved for y. However, since the graph of an equation of this form is always a line parallel to the y-axis, we know parallel or coinciding lines result only if both equations in the system are of this type.

EXAMPLE 2 DETERMINING THE NUMBER OF SOLUTIONS

Find the number of solutions to each system.

(a) $3x - 2y + 1 = 0$
$-4y + 6x = -2$

Solve each equation for y.

$$-2y = -3x - 1 \qquad -4y = -6x - 2$$

$$y = \frac{3}{2}x + \frac{1}{2} \qquad y = \frac{6}{4}x + \frac{2}{4} = \frac{3}{2}x + \frac{1}{2}$$

Since the slopes and y-intercepts are equal, the lines coincide, and the system has infinitely many solutions. That is, any solution to one equation also solves the other and is a solution to the system.

(b) $2x - 5y + 1 = 0$
$15y - 6x = 1$

Solve each equation for y.

$$-5y = -2x - 1 \qquad 15y = 6x + 1$$

$$y = \frac{2}{5}x + \frac{1}{5} \qquad y = \frac{6}{15}x + \frac{1}{15} = \frac{2}{5}x + \frac{1}{15}$$

Since the slopes are equal and the y-intercepts are unequal, the lines are parallel. Thus, the system has no solution.

PRACTICE EXERCISE 2

Find the number of solutions to each system.

(a) $4x + y = -1$
$-8x - 2y = 1$

(b) $x - 3y + 5 = 0$
$-2x + 6y = 10$

(c) $x - 5y = 6$
$4 + y = 0$

(c) $2x - 5 = 0$
$2x + y = 5$

Since the first equation has as its graph a line with undefined slope (parallel to the y-axis), and the second equation has as its graph a line with slope -2, the lines intersect and the system has exactly one solution.

Answers: (a) no solution (b) infinitely many solutions (c) exactly one solution

4.1 EXERCISES A

Answer true *or* false *for each statement. If the statement is false, explain why.*

1. In a system of equations, if the lines coincide, the system has infinitely many solutions and is said to be dependent.

2. In a system of equations, if the lines are parallel, the system has no solution and is said to be consistent.

3. When the slopes of the lines in a system of equations are unequal, the graphs are intersecting lines.

4. When the slopes of the lines in a system of equations are unequal, the system is said to be consistent and independent.

5. Is (3, 0) a solution to the given systems?

(a) $3x - 2y - 9 = 0$
$x + 8y - 3 = 0$

(b) $x + y + 1 = 4$
$2x - y + 2 = 4$

(c) $x = 3$
$2x - 4y - 6 = 0$

6. Is $\left(\frac{1}{2}, -5\right)$ a solution to the given systems?

(a) $2x + y + 4 = 0$
$8x - 3y - 11 = 0$

(b) $2x - 1 = 0$
$y + 5 = 0$

(c) $3y + 15 = 0$
$4x - y - 7 = 0$

Without solving, tell the number of solutions to the given system.

7. $5x - 3y + 7 = 0$
$-5x + 3y - 7 = 0$

8. $5x - 3y + 7 = 0$
$3x - 5y - 7 = 0$

9. $2x + 7 = 0$
$2x + 7y = 5$

10. $2x + 7 = 0$
$7x + 2 = 0$

11. $x + y + 1 = 0$
$y + x - 1 = 0$

12. $3x + 3y + 3 = 0$
$5x + 5y + 5 = 0$

13. $y - 6 = 0$
$x + 6 = 0$

14. $x + 2y + 3 = 0$
$3x + 2y + 1 = 0$

15. $3x + y = 0$
$x + 3y = 0$

Solve the following systems using the graphing method.

16. $x + y = 4$
$3x - y = 0$

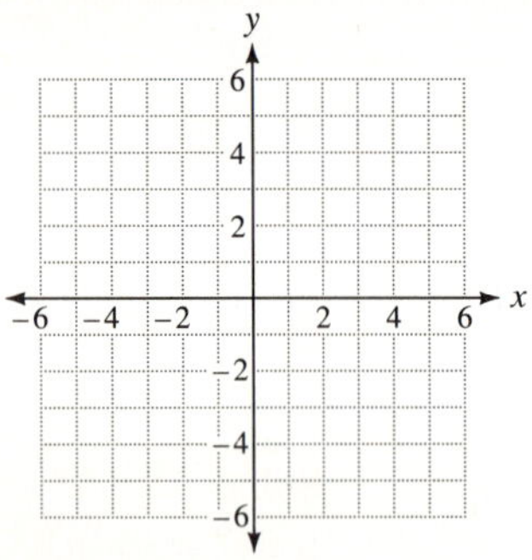

17. $2x - y = 1$
$x + 5y = 6$

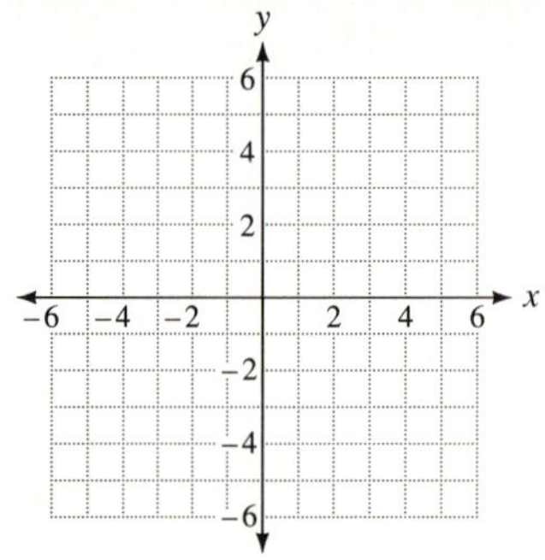

18. $3x + 2y = 5$
$-6x - 4y = 5$

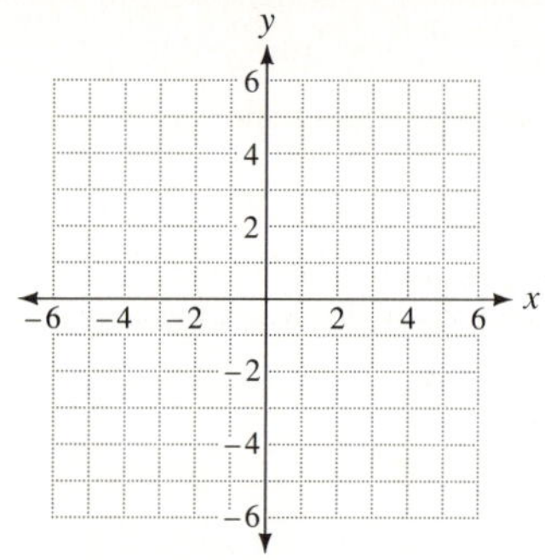

19. $5x - y = 0$
$x + 4y = 0$

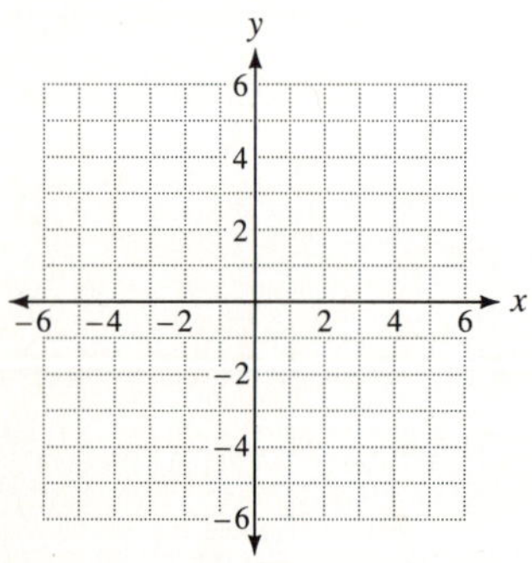

20 $2x + 3y = -2$
$-4x + y = -3$

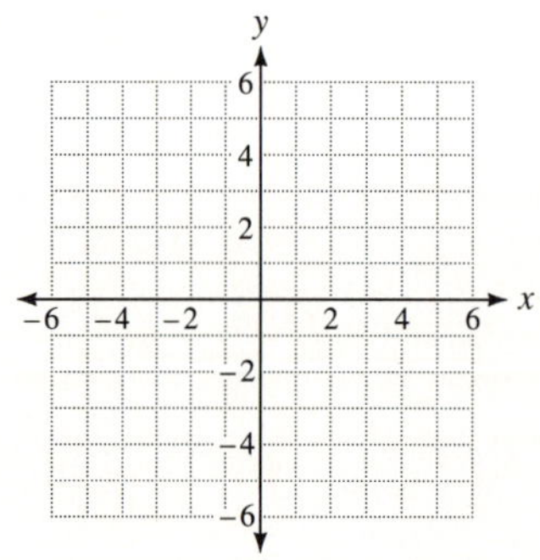

21. $x - 4 = 0$
$2x + y = 7$

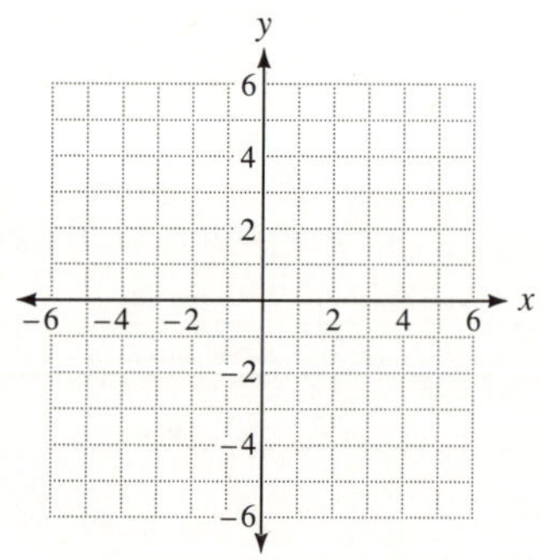

FOR REVIEW

The following exercises will help you prepare for the next section.

22. Solve $6x + 3y - 15 = 0$ for y. Substitute this expression for y into $x - 2y + 5 = 0$. Solve the resulting equation for the numerical value of x.

23. Consider the ordered pairs $(x, x + 2)$, for $x = -2, -1, 0, 1, 2$. Show that each of these ordered pairs is a solution to $3x - 3y = -6$.

24. Find the value of y if $x = 3$ and $y = 2x - 5$.

25. Find the value of x if $y = -2$ and $x = \frac{1}{2}y + 4$.

ANSWERS: 1. true 2. false (inconsistent) 3. true 4. true 5. (a) yes (b) no (c) yes 6. (a) no (b) yes (c) yes 7. infinitely many 8. exactly one 9. exactly one 10. none 11. none 12. infinitely many 13. exactly one 14. exactly one 15. exactly one 16. (1, 3) 17. (1, 1) 18. no solution (graphs are parallel) 19. (0, 0) 20. $\left(\frac{1}{2}, -1\right)$ 21. (4, −1) 22. $y = -2x + 5$; $x = 1$ 23. Substituting the values in each pair shows it is a solution. 24. $y = 1$ 25. $x = 3$

4.1 EXERCISES B

Answer true *or* false *for each statement. If the statement is false, explain why.*

1. If a system of equations has no solution, the graphs of the system are intersecting lines.
2. If a system of equations has exactly one solution, the graphs of the system are parallel lines.
3. When the slopes of the lines in a system are equal but the y-intercepts are unequal, the graphs of the system are parallel lines.
4. A system has infinitely many solutions whenever the graphs coincide.
5. Is (0, 2) a solution to the given systems?

(a) $3x - y = -2$, $x + 7y = 14$ **(b)** $y + 2 = 0$, $3x - y = -2$ **(c)** $4x - 2y = -4$, $x + 6y = 12$

6. Is $\left(-1, \frac{1}{3}\right)$ a solution to the given systems?

(a) $x + 3y = 0$, $2x - 6y = 0$ **(b)** $3x - 6y = -5$, $x + 9y = 2$ **(c)** $5x + 9y = -2$, $-x - 3y = 0$

Without solving, tell the number of solutions to the given system.

7. $-5x + 2y + 3 = 0$, $-5x - 2y + 3 = 0$
8. $-5x + 2y + 3 = 0$, $5x - 2y - 3 = 0$
9. $2x + y = 7$, $2x - 3 = 0$
10. $3y + 5 = 0$, $y - 7 = 0$
11. $x + y - 3 = 0$, $x - y + 3 = 0$
12. $2x + 2y - 2 = 0$, $5x + 5y - 5 = 0$
13. $2x = 8$, $8y = 2$
14. $-5x + 2y + 3 = 0$, $10x - 4y + 3 = 0$
15. $x - 7y = 0$, $7x - y = 0$

Solve the following systems using the graphing method.

16. $3x + y = -1$, $x - y = -3$

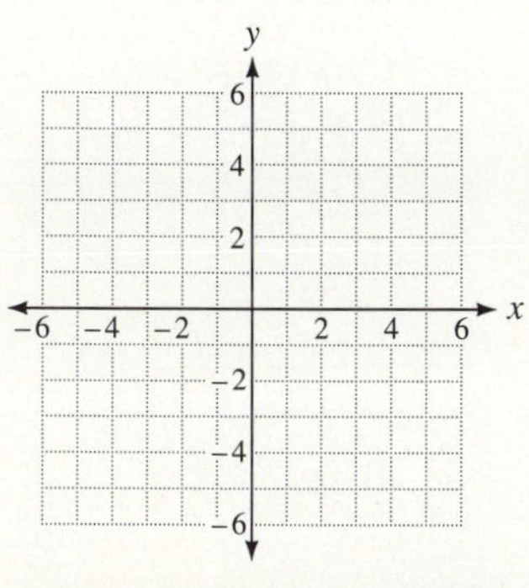

17. $2x - 3y = -2$, $x + y = 4$

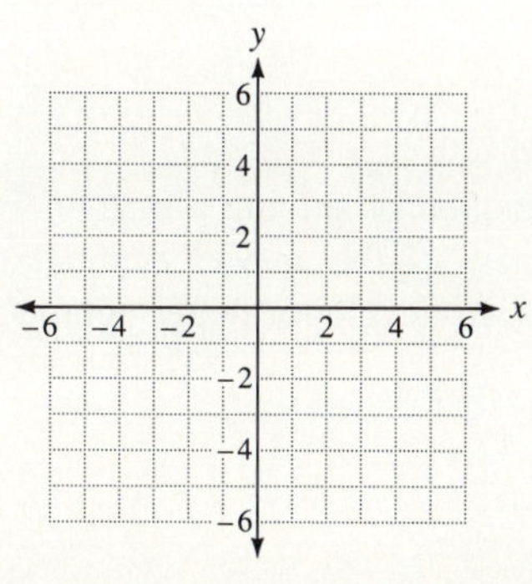

18. $5x - 2y = 7$, $-5x + 2y = 7$

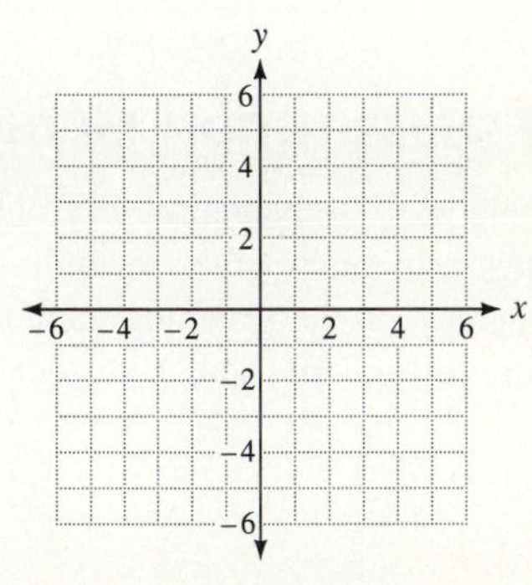

19. $x + 3y = -1$
$3x + 5y = 5$

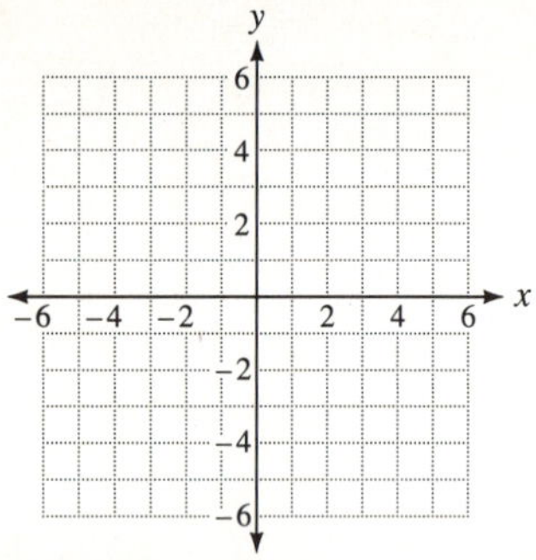

20. $3x - y = -1$
$-6x + 5y = 8$

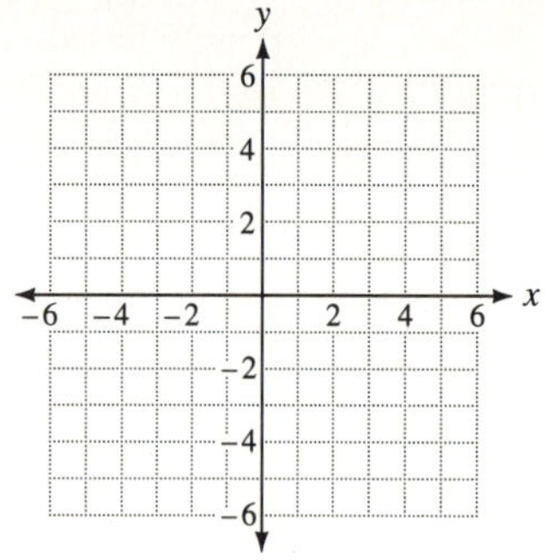

21. $y + 3 = 0$
$3x - 2y = 0$

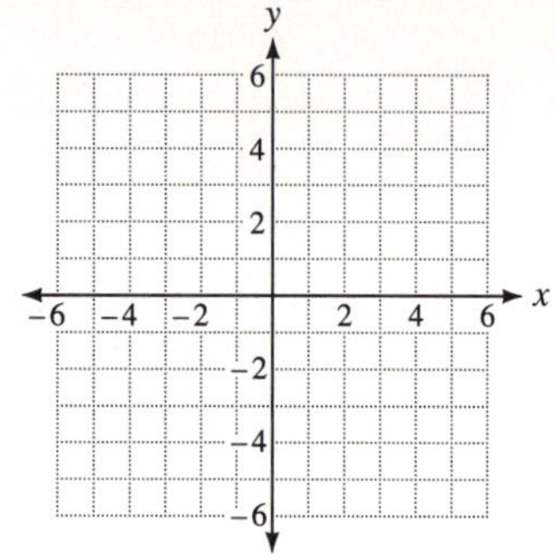

FOR REVIEW

The following exercises will help you prepare for the next section.

22. Solve $x + y + 3 = 0$ for x. Substitute this expression for x into $2x - y - 6 = 0$. Solve the resulting equation for the numerical value of y.

23. Consider the ordered pairs $(x, 2 - x)$, for $x = -2, -1, 0, 1, 2$. Show that each of these pairs is a solution to $2x + 2y - 4 = 0$.

24. Find the value of x if $y = -2$ and $x = 3y + 1$.

25. Find the value of y if $x = 6$ and $y = \frac{1}{3}x - 5$.

4.1 EXERCISES C

Tell whether $(0.25, 1.78)$ *is a solution to the given system.*

1. $0.3x + 1.5y = 2.745$
$2.8x - 0.7y = -0.546$

2. $3.4x + 1.6y - 3.698 = 0$
$-0.7x + 2.1y - 3.913 = 0$

4.2 SOLVING SYSTEMS OF TWO EQUATIONS

STUDENT GUIDEPOSTS

1 Substitution Method
2 Elimination Method

1 SUBSTITUTION METHOD

There are disadvantages to graphing as a method of solving a system. First, it takes too much time; second, the correctness of the answer depends on the accuracy of the graphing techniques. Here we develop a better approach, the **substitution method.** Consider the system

$$x + y = 4$$
$$3x - y = 0.$$

This system can be changed into one equation in one unknown. Solve the first equation for x,

$$x = -y + 4,$$

and substitute this value of x into the second equation.

$$3(-y + 4) - y = 0$$

The result is an equation in the single variable y which can be solved for y.

$$-3y + 12 - y = 0$$
$$-4y = -12$$
$$y = 3$$

Substitute this value of y into either of the original equations, for example, the second,

$$3x - (3) = 0,$$

and solve for x.

$$3x = 3$$
$$x = 1$$

The solution to the system is (1, 3). Check by substituting (1, 3) into each of the original equations. Compare your work here with the results of Exercise 16 in the A exercises for Section 4.1. Although we used two different methods to solve the same system, we obtained the same solution each time.

To Solve a System of Equations Using the Substitution Method

1. Solve one of the equations for one of the variables.
2. Substitute that value of the variable in the *remaining* equation.
3. Solve this new equation and substitute the numerical solution into either of the two *original* equations to find the numerical value of the second variable.
4. Check your solution in both original equations.

EXAMPLE 1 Using the Substitution Method

Solve by the substitution method.

$$2y - x - 2 = 0$$
$$2x + y - 6 = 0$$

We can solve either equation for either variable, but one way to avoid fractions is to solve the second equation for y.

$$y = -2x + 6$$

Substitute $-2x + 6$ for y in the first equation.

$$2(-2x + 6) - x - 2 = 0$$
$$-4x + 12 - x - 2 = 0$$
$$-5x + 10 = 0$$
$$-5x = -10$$
$$x = 2$$

Practice Exercise 1

Solve by the substitution method.

$$2x - y = -11$$
$$x + 5y = 0$$

Now substitute 2 for x in

$$\begin{aligned} y &= -2x + 6. \\ &= -2(2) + 6 \\ &= -4 + 6 \\ &= 2 \end{aligned}$$

The solution is (2, 2). Check in both equations.

Answer: $(-5, 1)$

EXAMPLE 2 A System with Infinitely Many Solutions

Solve by the substitution method.

$$\begin{aligned} 2x - y &= 1 \\ -4x + 2y &= -2 \end{aligned}$$

To avoid fractions, solve the first equation for y.

$$y = 2x - 1$$

Substitute this expression into the second equation.

$$\begin{aligned} -4x + 2(2x - 1) &= -2 \\ -4x + 4x - 2 &= -2 \\ -2 &= -2 \quad \text{An identity} \end{aligned}$$

When we obtain an identity, return to the original system and examine the nature of the solutions by solving each equation for y.

$$\begin{aligned} 2x - y &= 1 & \qquad -4x + 2y &= -2 \\ -y &= -2x + 1 & 2y &= 4x - 2 \\ y &= 2x - 1 & y &= 2x - 1 \end{aligned}$$

The slopes and y-intercepts are equal, so the lines coincide, the system is dependent, and there are infinitely many solutions. In fact, any pair of numbers

$$(x, 2x - 1) \quad \text{Since } y = 2x - 1$$

for x any real number, is a solution to the system. Several representative solutions are $(0, -1)$, $(1, 1)$, $(2, 3)$, and $(-1, -3)$, obtained by letting $x = 0, 1, 2,$ and -1, respectively.

Practice Exercise 2

Solve by the substitution method.

$$\begin{aligned} 5x + y &= -1 \\ -10x - 2y &= 2 \end{aligned}$$

Answer: $(x, -5x - 1)$ for any real number x

EXAMPLE 3 A System with No Solution

Solve by the substitution method.

$$\begin{aligned} 2x + 4y &= 1 \\ x + 2y &= 8 \end{aligned}$$

To avoid fractions, solve the second equation for x.

$$x = -2y + 8$$

Substitute this value into the first equation.

$$\begin{aligned} 2(-2y + 8) + 4y &= 1 \\ -4y + 16 + 4y &= 1 \\ 16 &= 1 \quad \text{A contradiction} \end{aligned}$$

Practice Exercise 3

Solve by the substitution method.

$$\begin{aligned} x - 3y &= 1 \\ 3x - 9y &= -1 \end{aligned}$$

When we obtain a contradiction, return to the original system and examine the nature of the solutions by solving each equation for y.

$$2x + 4y = 1 \qquad\qquad x + 2y = 8$$
$$4y = -2x + 1 \qquad\qquad 2y = -x + 8$$
$$y = -\frac{1}{2}x + \frac{1}{4} \qquad\qquad y = -\frac{1}{2}x + 4$$

The slopes are equal and the y-intercepts are unequal, so the lines are parallel, the system is inconsistent, and there are no solutions to the system.

Answer: no solution

Examples 2 and 3 lead us to the following result that will apply to the next method we study as well as to the substitution method.

Solving Dependent or Inconsistent Systems

In solving a system of equations,

1. if an **identity** results, the lines coincide, the system is dependent, and there are **infinitely many solutions;**
2. if a **contradiction** results, the lines are parallel, the system is inconsistent, and there is **no solution.**

2 ELIMINATION METHOD

An alternative for solving systems of equations, known as the **elimination method,** is perhaps superior to substitution for systems such as

$$3x + 5y = -2$$
$$5x + 3y = 2.$$

Solving this system by the substitution method would involve fractions regardless of our choice of variable and choice of equation. The elimination method lets us avoid computations with fractions.

To help in our approach to the above system, we first consider a somewhat simpler system.

$$x + y = 4$$
$$2x - y = 5$$

Observe that $2x - y$ and 5 are both names or expressions for the same number. Remember that if the same expression is added to or subtracted from both sides of an equation the results are equal. Therefore, we can add $2x - y$ to the left side of the first equation and 5 to the right to obtain another equation with the same solution.

$$(x + y) + (2x - y) = 4 + 5$$
$$3x = 9$$

Generally, we add vertically and avoid the intermediate steps.

$$\begin{array}{r} x + y = 4 \\ \underline{2x - y = 5} \\ 3x \quad\; = 9 \\ x = 3 \end{array}$$

By adding the two equations, we "eliminate" the y term and obtain one equation in one variable. Once we know that x is 3, we substitute this value into either one of the two original equations to find the value of y. Substitute in the first equation.

$$\begin{aligned} 3 + y &= 4 \\ y &= 1 \end{aligned}$$

The solution is (3, 1). Check this by substituting in both original equations.

At times, addition or subtraction alone will not eliminate a variable and yield one equation in only one variable. When this happens, we may have to multiply one or both equations by a number that makes the coefficients of one of the variables negatives of each other. Then by adding the two equations we eliminate a variable and proceed as before. This is illustrated in the next example.

EXAMPLE 4 Using the Elimination Method

Solve by the elimination method.

$$\begin{aligned} 5x - 2y &= -13 \\ 2x + y &= 11 \end{aligned}$$

By adding or subtracting the two equations as given, we do not eliminate a variable. However, if we multiply both sides of the second equation by 2 (using the multiplication rule), the resulting equation is

$$4x + 2y = 22.$$

Add this equation to the first equation.

$$\begin{aligned} 5x - 2y &= -13 \\ 4x + 2y &= 22 \\ \hline 9x \quad &= 9 \\ x &= 1 \end{aligned}$$

Substitute 1 for x in the second equation.

$$\begin{aligned} 2(1) + y &= 11 \\ 2 + y &= 11 \\ y &= 9 \end{aligned}$$

The solution is (1, 9). Check by substitution.

Practice Exercise 4

Solve by the elimination method.

$$\begin{aligned} x - 4y &= -14 \\ 3x + y &= -3 \end{aligned}$$

Answer: $(-2, 3)$

At times the multiplication rule must be applied to both equations before adding or subtracting. Let us solve our original system.

EXAMPLE 5 Using the Elimination Method

Solve by the elimination method.

$$\begin{aligned} 3x + 5y &= -2 \\ 5x + 3y &= 2 \end{aligned}$$

Multiply the first equation by -5 and the second equation by 3, making the coefficients of the x terms -15 and 15, respectively. Add to eliminate x.

Practice Exercise 5

Solve by the elimination method.

$$\begin{aligned} 2x + 7y &= 8 \\ 3x - 5y &= 12 \end{aligned}$$

$$\begin{array}{rll} -15x - 25y = & 10 & \text{−5 times first equation} \\ 15x + 9y = & 6 & \text{3 times second equation} \\ \hline -16y = & 16 & \\ y = & -1 & \end{array}$$

Substitute -1 for y in the first equation.

$$3x + 5(-1) = -2$$
$$3x - 5 = -2$$
$$3x = 3$$
$$x = 1$$

The solution is $(1, -1)$. Check by substitution.

Answer: (4, 0)

To Solve a System of Equations Using the Elimination Method

1. Apply the multiplication rule to one or both equations (if necessary) to transform them so that addition or subtraction will eliminate a variable.
2. Eliminate the variable by adding or subtracting.
3. Solve the resulting single-variable equation and substitute this value into one of the original equations and solve.
4. Check your answer by substitution in both original equations.

You may ask, ''Should I use substitution or elimination for this problem?'' If one of the variables has a coefficient of 1 or -1, it is usually easier to solve for it and use substitution. Otherwise, you probably should use elimination.

Suppose we try to solve the following system by elimination.

$$3x - 2y = 5$$
$$-3x + 2y = -5$$

Notice that we can eliminate the variable x by addition.

$$\begin{array}{rr} 3x - 2y = & 5 \\ -3x + 2y = & -5 \\ \hline 0 = & 0 \end{array}$$

However, we have actually eliminated both variables and obtained an identity. Just as with the substitution method, this indicates that there are infinitely many solutions. (Verify that the lines do coincide.) If we solve the first equation for y, we obtain

$$-2y = -3x + 5$$
$$y = \frac{3}{2}x - \frac{5}{2}.$$

The solutions to the system take the form $\left(x, \frac{3}{2}x - \frac{5}{2}\right)$ where x is any real number.

We conclude this section with an example of a system that is not linear but that can be solved by the techniques for linear systems, after changing variables.

EXAMPLE 6 Solving a Nonlinear System

Solve the system.

$$\frac{3}{x} - \frac{5}{y} = -2$$
$$\frac{6}{x} + \frac{15}{y} = 16$$

This system is not linear, but if we rewrite it as

$$3\left(\frac{1}{x}\right) - 5\left(\frac{1}{y}\right) = -2$$
$$6\left(\frac{1}{x}\right) + 15\left(\frac{1}{y}\right) = 16$$

and make the substitution $u = \frac{1}{x}$ and $v = \frac{1}{y}$, we obtain

$$3u - 5v = -2$$
$$6u + 15v = 16$$

which is linear in u and v. To solve for u and v, we multiply the first equation by 3 and add.

$$\begin{aligned} 9u - 15v &= -6 \\ 6u + 15v &= 16 \\ \hline 15u \qquad &= 10 \\ u &= \frac{2}{3} \end{aligned}$$

Substituting u into one of the equations, we find $v = \frac{4}{5}$. Since $u = \frac{1}{x}$ and $v = \frac{1}{y}$,

$$\frac{2}{3} = \frac{1}{x} \quad \text{and} \quad \frac{4}{5} = \frac{1}{y}$$
$$x = \frac{3}{2} \quad \text{and} \quad y = \frac{5}{4}.$$

Thus, the solution is $\left(\frac{3}{2}, \frac{5}{4}\right)$. Check.

Practice Exercise 6

Solve the system.

$$\frac{2}{x} + \frac{4}{y} = -4$$
$$\frac{3}{x} - \frac{1}{y} = 8$$

Answer: $\left(\frac{1}{2}, -\frac{1}{2}\right)$

CAUTION

When you have found the values of substitute variables such as u and v in Example 6, you're not finished with the problem! Be sure you back-substitute and find the values for the variables x and y in the original problem.

4.2 EXERCISES A

Solve using the substitution method.

1. $x + 2y = -1$
$2x - 3y = 12$

2. $4x - y = -3$
$2x - 3y = 1$

3. $2x + 4y = -12$
$3x + 5y = -14$

4. $2x - 3y = 2$
$-4x + 6y = 2$

5 $5x - 5y = 10$
$x - y = 2$

6. $x - 4 = 0$
$x + y = 1$
[*Hint:* x is always 4 in the first equation so substitute to get y.]

Solve using the elimination method.

7. $3x + 2y = 4$
$3x - 3y = 9$

8. $2x - 5y = 1$
$4x + 2y = 14$

9. $2x - 7y = 24$
$3x + 2y = -14$

10 $5x - 7y = 3$
$-10x + 14y = -1$

11. $5x - 3y = 4$
$2x + 4y = -1$

12. $3x + 3y = 12$
$7x + 7y = 28$

Solve using the substitution or elimination method (whichever seems appropriate).

13. $x + 2y = -9$
$3x + 4y = -17$

14. $x + 2 = 0$
$3x + 2y = 10$

15. $5x - 11y = -6$
$11x - 5y = 6$

16. $3y - 15 = 0$
$3x - 15y = 0$

17. $3x + y = -1$
$-4x + 2y = -7$

18. $7x - 7y = 7$
$-13x + 13y = -13$

19 $\frac{3}{4}x - \frac{2}{3}y = 1$
$\frac{3}{8}x - \frac{1}{6}y = 1$

20 $0.6x - 0.7y = 1.3$
$-1.2x + 1.4y = 7.3$

21 $\frac{4}{x} - \frac{2}{y} = -2$
$\frac{2}{x} + \frac{1}{y} = 3$

22. $\frac{1}{u} + \frac{1}{v} = 5$
$\frac{2}{u} - \frac{1}{v} = 4$

23. Suppose that in the process of solving a system of equations we obtain $0 = 2$. What does this tell us about the system?

24. Suppose that in the process of solving a system of equations we obtain $0 = 0$. What does this tell us about the system?

FOR REVIEW

25. If the two lines in a system coincide, how many solutions does the system have? What is the system called?

26. If the two lines in a system are parallel, how many solutions does the system have? What is the system called?

27. If the two lines in a system intersect, how many solutions does the system have? What is the system called?

28. Solve by the graphing method.
$x + y = 0$
$2x + y = -3$

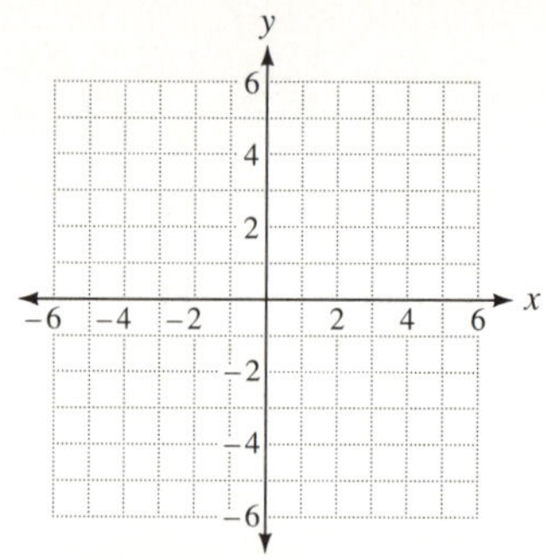

Exercises 29–30 review material from Section 2.4. They will help you prepare for the next section. Let x represent the unknown variable and translate each sentence into symbols. Do not solve.

29. Three times what number, less 8, is 22?

30. Two liters of acid are in how many liters of a 5% acid solution?

ANSWERS: **1.** (3, −2) **2.** (−1, −1) **3.** (2, −4) **4.** no solution **5.** infinitely many solutions of the form $(x, x - 2)$ for x any real number **6.** (4, −3) **7.** (2, −1) **8.** (3, 1) **9.** (−2, −4) **10.** no solution **11.** $\left(\frac{1}{2}, -\frac{1}{2}\right)$ **12.** infinitely many solutions of the form $(x, 4 - x)$ for x any real number **13.** (1, −5) **14.** (−2, 8) **15.** (1, 1) **16.** (25, 5) **17.** $\left(\frac{1}{2}, -\frac{5}{2}\right)$ **18.** infinitely many solutions of the form $(x, x - 1)$ for x any real number **19.** (4, 3) **20.** no solution **21.** $\left(2, \frac{1}{2}\right)$ **22.** $\left(\frac{1}{3}, \frac{1}{2}\right)$ **23.** The system has no solution. **24.** The system has infinitely many solutions. **25.** infinitely many; dependent **26.** no solution; inconsistent **27.** exactly one; consistent and independent **28.** (−3, 3) **29.** $3x - 8 = 22$ **30.** $2 = 0.05x$

4.2 EXERCISES B

Solve using the substitution method.

1. $3x - y = 7$
$x + 4y = -2$

2. $5x - y = 0$
$3x - 2y = -7$

3. $4x + 2y = 3$
$-2x - y = -7$

4. $2x + y = 3$
$-10x - 5y = -15$

5. $4x + y = -1$
$-6x + 5y = -18$

6. $4y - 12 = 0$
$8 - 2x = 0$

Solve using the elimination method.

7. $4x - 2y = -18$
$4x + 2y = 10$

8. $3x - 2y = -4$
$-5x + 4y = 10$

9. $2x - 5y = -14$
$3x + 4y = 2$

10. $4x - 9y = 3$
$-8x + 18y = 3$

11. $3x + y = 3$
$-12x - 4y = -12$

12. $7x - 2y = -1$
$5x + 4y = 2$

Solve using the substitution or elimination method (whichever seems appropriate).

13. $7x + 5y = 2$
$5x + 7y = -2$

14. $4x + 16 = 0$
$2x + y = -1$

15. $x + 5y = -2$
$3x - y = 10$

16. $3x - y = -9$
$2x + 4y = 8$

17. $3x - 4y = 2$
$-6x + 4y = -3$

18. $6x - 6y = 6$
$-15x + 15y = -15$

19. $\frac{5}{2}x + \frac{3}{4}y = 2$
$\frac{3}{4}x - \frac{1}{16}y = -4$

20. $1.5x - 2.2y = 3.8$
$-4.5x + 6.6y = -7.6$

21. $\frac{5}{u} - \frac{3}{v} = -\frac{1}{2}$
$\frac{10}{u} + \frac{9}{v} = 4$

22. $\frac{4}{3u} + \frac{1}{9v} = 3$
$\frac{5}{2u} - \frac{2}{3v} = 3$

23. If an identity is obtained when solving a system of equations, what can be said about the solution to the system?

24. If a contradiction is obtained when solving a system of equations, what can be said about the solution to the system?

FOR REVIEW

25. When the two lines in a system have different slopes, how many solutions does the system have? What is the system called?

26. When the two lines in a system have the same slope but different y-intercepts, how many solutions does the system have? What is the system called?

27. When the two lines in a system have the same slope and the same y-intercepts, how many solutions does the system have? What is the system called?

28. Solve by the graphing method.
$2x - y = -3$
$x + 4y = -15$

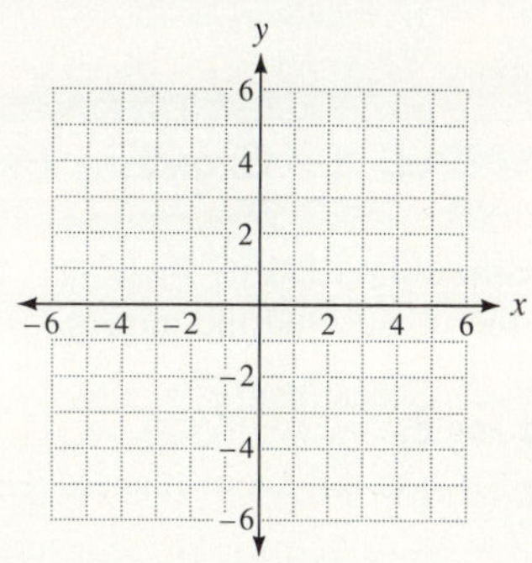

Exercises 29–30 review material from Section 2.4. They will help you prepare for the next section. Let x represent the unknown variable and translate each sentence into symbols. Do not solve.

29. A distance of 150 miles was traveled at what rate of speed if the time of travel was 3 hours?

30. Seven coats were purchased at what price each if the total amount spent was $280?

4.2 EXERCISES C

1. Find the values for a and b so that $(x, y) = (-3, 2)$ is a solution to the following system.
$ax + by = 4$
$-2ax - by = 2$
[*Hint:* Substitute -3 for x and 2 for y in both equations, and solve the resulting system for a and b.]

2. Solve the system for x and y in terms of the nonzero constants a and b.
$ax + y = b$
$ax - y = 2b$

3. Suppose that $\left(\frac{1}{2}, 3\right)$ and $(-4, 0)$ are two of the solutions to the equation $ax + by = -8$. Find a and b.
[Answer: $a = 2$, $b = -3$]

4. Solve for x and y.

$$\frac{x+y}{2} + \frac{x-y}{4} = 3$$
$$\frac{x+y}{2} - \frac{x-y}{4} = 1$$

5. Find the value(s) of a that will make the following system dependent.

$$\begin{aligned} 2x + 5y &= -1 \\ 6x + 15y &= a \end{aligned}$$

6. Find the value(s) of a that will make the following system inconsistent.

$$\begin{aligned} 2x + 5y &= -1 \\ 6x + 15y &= a \end{aligned}$$

4.3 APPLICATIONS USING SYSTEMS OF TWO EQUATIONS

STUDENT GUIDEPOSTS

1. Combination Problems
2. Mixture Problems

In previous chapters we solved several kinds of applied problems. Some of those, as well as other problems, can be simplified by using two variables (instead of one) and translating them into a system of equations. Several examples illustrate this idea.

EXAMPLE 1 Geometry Problem

Two angles are **complementary** (that is, their sum is 90°). If one is 10° more than seven times the other, find the angles.

Let x = one angle,
y = the second angle.

Then $x + y = 90$. Why? If x is 10° more than seven times y, then

$$x = 7y + 10.$$

The system

$$\begin{aligned} x + y &= 90 \\ x - 7y &= 10 \end{aligned}$$

can be solved by substituting $7y + 10$ for x in $x + y = 90$.

$$\begin{aligned} (7y + 10) + y &= 90 \\ 8y + 10 &= 90 \\ 8y &= 80 \\ y &= 10 \end{aligned}$$

Now substitute 10 for y in $x = 7y + 10$.

$$x = 7(10) + 10 = 70 + 10 = 80$$

The angles are 10° and 80°. Since $10° + 80° = 90°$ and $80° = 7(10°) + 10°$, the answer checks.

PRACTICE EXERCISE 1

Two angles are **supplementary** (their sum is 180°). If one is 30° less than four times the other, find the angles.

Answer: 138° and 42°

EXAMPLE 2 DISTANCE-RATE PROBLEM

By going 20 mph for one period of time and then 30 mph for another, a boater traveled from Glen Canyon Dam to the end of Lake Powell, a distance of 180 miles. If she had gone 21 mph throughout the same period of time, she would only have reached Hite Marina, a distance of 147 miles. How many hours did she travel at each speed?

Let x = the number of hours that she traveled at 20 mph,
y = the number of hours that she traveled at 30 mph.

Since

$$(\text{distance}) = (\text{rate})(\text{time}),$$

it follows that

$$\text{distance traveled at 20 mph} = 20x$$
$$\text{distance traveled at 30 mph} = 30y.$$

The total distance traveled is the sum of the two distances traveled at the two rates, so the first equation is

$$20x + 30y = 180.$$

She would have traveled 147 miles at a rate of 21 mph for the total time $x + y$. Thus the second equation is

$$21(x + y) = 147 \quad \text{or} \quad 21x + 21y = 147.$$

It is usually wise to simplify equations before solving. Dividing the first equation by 10 and the second by 7, we obtain the following system.

$$2x + 3y = 18$$
$$3x + 3y = 21$$

Subtract the first from the second.

$$x = 3$$

Then

$$2(3) + 3y = 18$$
$$3y = 12$$
$$y = 4.$$

She traveled 3 hours at 20 mph and 4 hours at 30 mph. Check this in the words of the problem.

PRACTICE EXERCISE 2

A train travels 330 miles by going 60 mph for one period of time and 70 mph for another. Had it gone 62 mph throughout the trip it would have traveled only 310 miles. How many hours did it travel at each speed?

Answer: 2 hr at 60 mph and 3 hr at 70 mph

1 COMBINATION PROBLEMS

In another kind of word problem, a **combination problem,** two quantities are combined in two different ways, resulting in two equations. Many combination problems involve *total value*. The **total value** of something composed of a number of units, each having the same value, is given by

(total value) = (value per unit)(number of units).

For example, the total value of 8 pounds of chicken worth $1.10 per pound is

$$(\text{total value}) = (\text{value per pound})(\text{number of pounds})$$
$$= (\$1.10)(8)$$
$$= \$8.80.$$

Similarly, the value of a collection of 12 quarters and four dimes is

$$\begin{aligned}(\text{total value}) &= (\text{value per quarter})(\text{number of quarters}) \\ &\quad + (\text{value per dime})(\text{number of dimes}) \\ &= (25¢)(12) + (10¢)(4) \\ &= 340¢ = \$3.40.\end{aligned}$$

The next example applies total value equations to a combination problem.

EXAMPLE 3 Combination Problem

Bill Poole bought 5 shirts (of the same value) and 4 pairs of socks (of the same value) for \$87. He returned to the same store a week later and purchased (at the same prices) 2 more shirts and 6 more pairs of socks for \$48. Find the price of each.

Let x = price of one shirt,
y = price of one pair of socks.
$5x$ = value of 5 shirts, Value = (price per shirt)(number of shirts)
$4y$ = value of 4 pairs of socks.

Thus, the system to solve is

$$5x + 4y = 87$$
$$2x + 6y = 48.$$

Divide the second equation by 2 to simplify.

$$x + 3y = 24$$

Solve for x,

$$x = 24 - 3y,$$

and substitute into the first equation.

$$\begin{aligned}5(24 - 3y) + 4y &= 87 \\ 120 - 15y + 4y &= 87 \\ -11y &= -33 \\ y &= 3\end{aligned}$$

Now substitute 3 for y in $x = 24 - 3y$.

$$\begin{aligned}x &= 24 - 3y \\ &= 24 - 3(3) = 24 - 9 = 15\end{aligned}$$

Each shirt cost \$15 and each pair of socks cost \$3.

Practice Exercise 3

Three 5.25-inch computer disks and seven 3.5-inch computer disks cost \$16.25. Ten 5.25-inch disks and four 3.5-inch disks cost \$15.50. Find the price of each type of disk.

Answer: 5.25-inch: \$0.75; 3.5-inch: \$2.00

2 MIXTURE PROBLEMS

In a certain type of combination problem called a **mixture problem,** the mixing together of two quantities results in two equations we call the *quantity equation* and the *value equation*.

EXAMPLE 4 Mixture Problem

The owner of a candy store wishes to mix candy selling for \$1.50 per lb with nuts selling for \$1.00 per lb to obtain a party mix to be sold for \$1.20 per lb. How many pounds of each must be used to obtain 50 lb of the mixture?

Practice Exercise 4

A shipping crate contains 95 boxes, some weighing 1 lb each and the others 3 lb each. How many of each

Let c = the number of pounds of candy,
n = the number of pounds of nuts.

It sometimes helps to place all given information into a summary table. To avoid decimals, we convert all monetary units to cents.

	Number of pounds	*Value per pound*	*Value*
Candy	c	150	$150c$
Nuts	n	100	$100n$
Total	50	120	120(50)

First we obtain the quantity equation

(number lb of candy) + (number lb of nuts) = (number lb of mixture),

which translates to

$$c + n = 50.$$

Next we obtain the value equation.

(value of candy) + (value of nuts) = (total value of mixture)

That is, the value of a mixture is equal to the sum of the values of its parts. Thus, the value equation is

$$150c + 100n = 120 \cdot 50$$

or $$3c + 2n = 120.$$ Divide through by 50

The system we have to solve is

$$\begin{aligned} c + n &= 50 \\ 3c + 2n &= 120. \end{aligned}$$

Solve the first equation for c.

$$c = 50 - n$$

Then substitute into the second.

$$\begin{aligned} 3(50 - n) + 2n &= 120 \\ 150 - 3n + 2n &= 120 \\ -n &= -30 \\ n &= 30 \end{aligned}$$

Now substitute 30 for n in $c = 50 - n$.

$$c = 50 - 30 = 20$$

The store owner must use 20 lb of candy and 30 lb of nuts to obtain the desired mixture.

type are in the crate if the total weight of the contents of the crate is 135 lb?

Answer: 75 1-lb boxes and 20 3-lb boxes

EXAMPLE 5 Coin Problem

Burford has 30 coins consisting of dimes and quarters. If the total value is \$4.20, find the number of dimes and the number of quarters.

Let d = the number of dimes,
q = the number of quarters.

Once again, a table might be helpful.

Practice Exercise 5

Matt found a coin purse containing 40 coins with a total value of \$2.90. If there are only nickels and dimes in the purse, how many of each are there?

	Number of coins	*Value per coin*	*Value*
Dimes	d	10	$10d$
Quarters	q	25	$25q$
Total	30	—	420

The quantity equation

(number of dimes) + (number of quarters) = (number of coins)

translates to $d + q = 30.$

The value equation

(value of dimes) + (value of quarters) = (value of coins)

becomes

$$10d + 25q = 420, \quad \text{which reduces to} \quad 2d + 5q = 84$$

after dividing both sides by 5. Solve the first equation for d, $d = 30 - q$, and substitute.

$$\begin{aligned} 2(30 - q) + 5q &= 84 \\ 60 - 2q + 5q &= 84 \\ 3q &= 24 \\ q &= 8 \end{aligned}$$

Now substitute 8 for q in $d = 30 - q$.

$$d = 30 - 8 = 22$$

Thus, Burford has 22 dimes and 8 quarters.

Answer: 18 dimes and 22 nickels

EXAMPLE 6 Mixture Problem in Chemistry

A chemist has two solutions, each containing a certain percentage of acid. If solution A is 5% acid and solution B is 15% acid, how much of each should be mixed to obtain 20 liters of a solution that is 12% acid?

Let x = number of liters of 5% acid solution,
y = number of liters of 15% acid solution.

The quantity equation is

$$x + y = 20.$$

The value equation this time equates the amount of acid in the solutions.

$$\begin{aligned} 0.05x &= \text{amount of acid in } x \text{ liters of 5\% solution} \\ 0.15y &= \text{amount of acid in } y \text{ liters of 15\% solution} \\ (0.12)(20) &= \text{amount of acid in 20 liters of 12\% mixture} \end{aligned}$$

So we have

$$\begin{aligned} 0.05x + 0.15y &= (0.12)(20) \\ 5x + 15y &= 240 && \text{Clear all decimals} \\ x + 3y &= 48. && \text{Divide by 5} \end{aligned}$$

Solving $x + y = 20$ for x, $x = 20 - y$. Substitute.

$$\begin{aligned} (20 - y) + 3y &= 48 \\ 20 + 2y &= 48 \\ 2y &= 28 \\ y &= 14 \end{aligned}$$

Practice Exercise 6

A 15% salt solution is to be mixed with a 40% salt solution to obtain 5 quarts of a 25% salt solution. How many quarts of each should be used?

Now substitute 14 for y in $x = 20 - y$.

$$x = 20 - y = 20 - 14 = 6$$

The chemist should mix 6 liters of 5% solution with 14 liters of 15% solution to get the 12% acid solution.

Answer: 15% solution: 3 qt; 40% solution: 2 qt

4.3 EXERCISES A

Solve using a system of equations.

1. The sum of two numbers is 50 and their difference is 16. Find the numbers.

   Let x = first number,
   y = second number.
   $x + y =$ ____
   $x - y =$ ____

2. Pete is 7 years younger than Jim. Three years from now the sum of their ages will be 33. How old is each now?

   Let x = Jim's age,
   y = Pete's age,
   $x + 3$ = Jim's age in 3 yr,
   $y + 3$ = Pete's age in 3 yr.
   $y = x -$ ____
   $(x + 3) + (y + 3) =$ ____

3. Two angles are supplementary (their sum is 180°), and one is 18° more than five times the other. Find the angles.

   Let x = measure of first angle,
   y = measure of second angle.
   $x + y =$ ____

4. By traveling 30 mph for one period of time and 40 mph for another, Mandy traveled 230 miles. Had she gone 10 mph faster throughout, she would have traveled 300 miles. How many hours did she travel at each rate?

   Let x = time traveling 30 mph,
   y = time traveling 40 mph.

5. Four books (of the same kind) and 6 pens (of the same kind) cost $9.00. Three books and 9 pens also cost $9.00. Find the cost of one book and of one pen.

6. A candy mix sells for $1.10 per lb. If the mix has two kinds of candy, one worth 90¢ per lb and the other worth $1.50 per lb, how many pounds of each would be in a 30-lb mixture?

7. A collection of dimes and nickels is worth $3.75. If there are 55 coins in the collection, how many of each are there?

8. If there were 450 people at a play, the total receipts were $600, and the admission price was $2.00 for adults and 75¢ for children, how many adults and how many children were in attendance?

9. Mr. Smith is 43 years old. If the sum of Mr. Smith's two daughters' ages is 23, and if three times the age of the younger plus the age of the older daughter is equal to the age of Mr. Smith, how old is each daughter?

10. A 40-ft rope is cut in two pieces. One piece is 1 ft more than twice the other. How long is each piece?

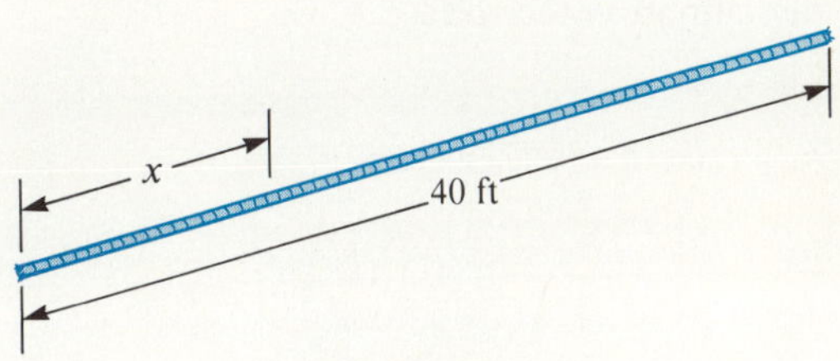

11. In a right triangle, one acute angle is 6° less than three times the other. Find each acute angle.

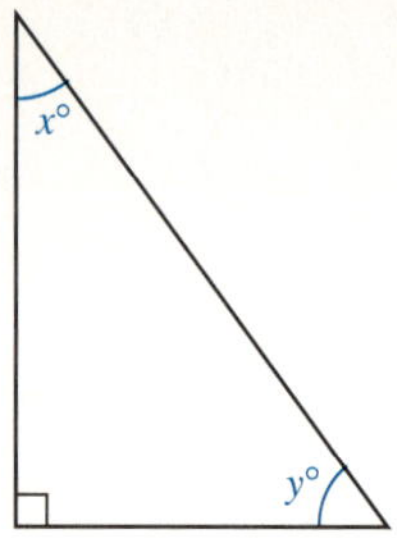

12. A collection of quarters and nickels is worth \$3.50. If there are 30 coins in the collection, how many of each are there?

13. Abby invested \$2000 for one year in two separate savings accounts. One account paid 8% simple interest and the other 9% simple interest. How much was invested in each account if the total interest from both was \$175?

14 The total interest earned on a sum of money invested for one year was \$183. If \$500 more was invested at 10% simple interest than was invested at 9% simple interest, how much was invested at each rate?

15. The owner of Carrie's Coffee House wishes to mix two blends of coffee, one selling for \$3.60 per pound and the other for \$4.80 per pound, to obtain a 20-lb mixture selling for \$4.20 per pound. How many pounds of each must she use?

16 A chemist has one solution which is 50% acid and a second which is 25% acid. How much of each should he use to make 10 L of a 40%-acid solution?

17. The perimeter of a rectangular field is 192 yards. If the length is 8 yards more than the width, find the dimensions.

18 The Small Car Rental Agency charges a daily fee plus a mileage fee. Mr. Green was charged \$143.12 for 3 days and 316 miles, and Mr. Brown was charged \$147.44 for 5 days and 242 miles. What is the company's daily rate, and what is its mileage rate?

19 A houseboat sailed 30 miles downstream in 3 hr and returned to the same marina upstream in 5 hr. Find the rate of the houseboat in still water and the rate of the current in the river.

20. Two trucks leave Kansas City, one traveling east and the other traveling west. After 4 hours they are 492 miles apart. If one truck is traveling 7 mph faster than the other, what is the speed of each?

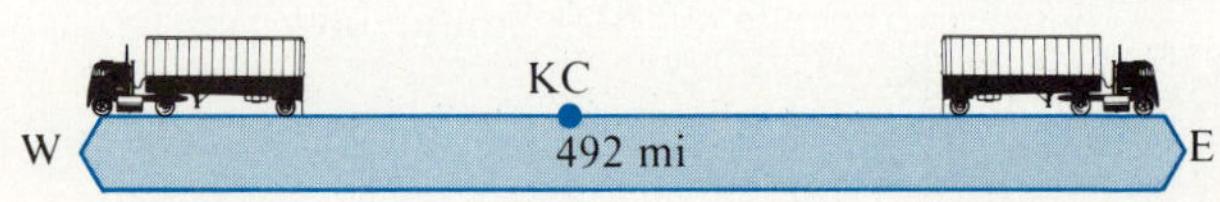

FOR REVIEW

Solve by any method.

21. $3x + 8y = 13$
$2x - 3y = 17$

22. $-x + y = 7$
$12x - 12y = 7$

23. $\frac{2}{3}x + \frac{1}{9}y = 6$
$\frac{1}{4}x + \frac{3}{4}y = 15$

The following exercises from Chapter 1 will help you prepare for the next section. In Exercises 24–26, evaluate each expression when $x = -1$, $y = 3$, and $z = -5$.

24. $2x + 4y - z$

25. $-3x - y + 2z$

26. $-10x + 12y - 3z$

Find the numerical value of y in each equation if $x = 2$ and $z = -4$.

27. $3x + y - 2z = 13$

28. $7x - 2y - z = 4$

29. $-8x + 2y - 7z = 10$

ANSWERS: 1. 17, 33 2. Pete is 10, Jim is 17. 3. 27°, 153° 4. 5 hr at 30 mph, 2 hr at 40 mph 5. book is \$1.50, pen is 50¢ 6. 20 lb of 90¢ candy, 10 lb of \$1.50 candy 7. 20 dimes; 35 nickels 8. 210 adults, 240 children 9. 10 yr, 13 yr 10. 13 ft, 27 ft 11. 24°, 66° 12. 10 quarters, 20 nickels 13. \$500 at 8%, \$1500 at 9% 14. \$1200 at 10%, \$700 at 9% 15. 10 lb of each 16. 4 L of 25% solution, 6 L of 50% solution 17. 52 yd, 44 yd 18. \$14 per day, \$0.32 per mile 19. houseboat: 8 mph; river: 2 mph 20. 58 mph, 65 mph 21. (7, −1) 22. no solution 23. (6, 18) 24. 15 25. −10 26. 61 27. $y = -1$ 28. $y = 7$ 29. $y = -1$

4.3 EXERCISES B

Solve using a system of equations.

1. Twice one number plus a second number is 42. The first minus the second is −6. Find the numbers.

2. The sum of Jan's and Sue's ages is 15. In 3 years, Jan will be twice as old as Sue. What are their present ages?

3. Two angles are complementary (their sum is 90°). If one is 10° more than seven times the other, find the angles.

4. Maria rode her bike at 15 mph for a time but had mechanical problems and had to continue her trip walking at 4 mph. She covered 42 miles. Had she been able to ride all the way, she would have covered 75 miles in the same time. How long did she ride, and how long did she walk?

5. Seven pads of paper and 5 pencils cost \$5.20. Two pads and 18 pencils cost \$4.80. Find the price of each.

6. Nuts worth \$2.25 per lb are mixed with candy worth \$1.50 per lb. The mixture sells for \$1.90 per lb. How many pounds of nuts and how many pounds of candy should be used to make a 60-lb mixture?

7. There are 38 coins in a collection of nickels and quarters. If the total value of the collection is \$5.10, how many of each are there?

8. A total of 8340 people paid to see a college basketball game. Tickets were \$1.50 for children, \$2.50 for adults, and the total receipts were \$19,620. How many children and how many adults paid to see the game?

9. Mari is 26 years older than Larry. In 10 years Mari will be twice as old as Larry is then. How old is each now?

10. Phillip bought 6 pounds of hamburger and 4 pounds of steak for \$22.50. If steak is three times the price of hamburger, what is the price per pound of each?

11. In a basketball game, Troy scored 4 more than twice as many points as Shawn. If one of them had scored 2 more points, their total would have been 60. How many points did each score?

12. Wanda invested \$6000 in bonds, \$7500 in savings certificates, and received \$1515 in interest for the year. If the interest rate on the certificates was 4% higher than the rate on the bonds, what were the respective interest rates?

13. Michele invested \$5000 for one year in two separate savings accounts. One account paid 7% simple interest and the other 9% simple interest. How much was invested in each account if the total interest from both was \$410?

14. The total interest earned on a sum of money invested for one year was \$80. If \$300 more was invested at 8% simple interest than at 6% simple interest, how much was invested at each rate?

15. A shop owner mixes candy worth \$1.40 per lb with candy worth \$1.95 per lb to get 110 lb of a party mix worth \$1.75 per lb. How many pounds of each did she use?

16. A lab technician is diluting a 70% alcohol solution with water. How much water and how much alcohol must be mixed to obtain 21 L of a 30%-alcohol solution?

17. In an endurance race Fred ran 3 times as far as Sam. If their total distance was 12.8 miles, how far did each run?

18. A new machine can produce 120 copies per minute while an older model produces only 30 copies per minute. The new machine had been operating 15 minutes when the older one was started. What was the total time each was operating if together they produced 8850 copies?

19. Grayden Bell sails his boat upstream from a marina 210 miles at constant speed in 7 hr. He returns downstream to the same marina at the same speed in 6 hr. What are the speed of the boat in still water and the speed of the current?

20. A plane flying with the wind flew 600 miles in 4 hr. The return trip against the wind took 5 hr. Find the rate of the plane in calm air and the rate of the wind.

FOR REVIEW

Solve by any method.

21. $3x + y = 10$
$5x + 7y = 6$

22. $-x + y = -7$
$12x - 12y = 84$

23. $\frac{1}{2}x + \frac{3}{4}y = 5$
$\frac{3}{8}x - \frac{3}{2}y = -21$

The following exercises from Chapter 1 will help you prepare for the next section. In Exercises 24–26, evaluate each expression when $x = -2$, $y = 1$, and $z = -3$.

24. $3x - y + 2z$

25. $-2x + y - 4z$

26. $5x - 7y + 8z$

Find the numerical value of x in each equation if $y = -3$ and $z = 0$.

27. $4x - 8y + 7z = 0$

28. $3x + 7y - 9z = 0$

29. $-2x + 8y - 13z = 5$

4.3 EXERCISES C

Solve.

1. The ones digit of a whole number is 1 more than twice the tens digit. When the two digits are reversed, the new number is 4 less than twice the original number. Find the original number. [Answer: 49]

2. A shipping crate contains two types of items, one weighing 3 lb each and the other $\frac{1}{2}$ lb each. Suppose we know there are 90 items in the crate. If the crate weighs 30 lb when empty, and now weighs 200 lb, how many of each type of item are in the crate?

3. The radiator in an automobile holds 14 quarts. How much antifreeze should be mixed with a 20% antifreeze solution to obtain a 40% antifreeze mixture that will fill the radiator?

4. In a laboratory experiment, a biologist must keep a control group of animals on a strict diet. Each animal must receive 19.5 grams of protein and 2.2 grams of fat every day. The biologist has purchased two food blends, HealthChow and GroMeal. If HealthChow contains 30% protein and 4% fat and GroMeal contains 25% protein and 2% fat, how many grams of each should be mixed to provide the correct diet for one animal?

4.4 SYSTEMS OF THREE LINEAR EQUATIONS IN THREE VARIABLES

STUDENT GUIDEPOSTS

1. Linear Equations in Three Variables
2. Systems of Three Linear Equations
3. Inconsistent and Dependent Systems
4. Applications of Systems of Three Equations

1 LINEAR EQUATIONS IN THREE VARIABLES

An equation of the form

$$ax + by + cz = d$$

when a, b, c, and d are constant real numbers and x, y, and z are variables, is called a **linear equation in three variables.** The term *linear equation* in this context is a bit misleading (perhaps *first-degree equation* is better) since the graph of a linear equation in three variables is actually a plane in space, not a line. However, our solutions to linear equations in three variables will be purely algebraic, patterned after similar situations with linear equations in two variables.

A solution to a linear equation such as

$$2x + y - 3z = 3$$

is an **ordered triple** of numbers. For example, $(1, -2, -1)$ is a solution to the above equation since if x, y, and z are replaced with 1, -2, and -1, in that order, the equation is true.

$$2(1) + (-2) - 3(-1) \stackrel{?}{=} 3$$
$$2 - 2 + 3 \stackrel{?}{=} 3$$
$$3 = 3$$

The *order* in the ordered triple is always the x-value first, the y-value second, and the z-value third. If we substitute -1 for x, -2 for y, and 1 for z,

$$2(-1) + (-2) - 3(1) \stackrel{?}{=} 3$$
$$-2 - 2 - 3 \stackrel{?}{=} 3$$
$$-7 \neq 3,$$

we discover that $(-1, -2, 1)$ is not a solution to the equation.

2 SYSTEMS OF THREE LINEAR EQUATIONS

A **system of three linear equations in three variables** is a trio of linear equations such as the following.

$$x + y + z = 2$$
$$-2x - y + z = 1$$
$$x - 2y - z = 1$$

A solution to such a system is an ordered triple of numbers that is a solution to *all three* equations. It is easy to verify by substitution that $(1, -1, 2)$ is a solution to each of the above equations, and hence is a solution to the system.

For a system of three equations in three variables there may be exactly one solution, no solution, or infinitely many solutions. Graphically, these possibilities correspond to three planes, A, B and C, intersecting in exactly one point as in Figure 4.4(a), having no single point found in all three planes as in Figure 4.4(b), and having infinitely many points in common as in Figure 4.4(c).

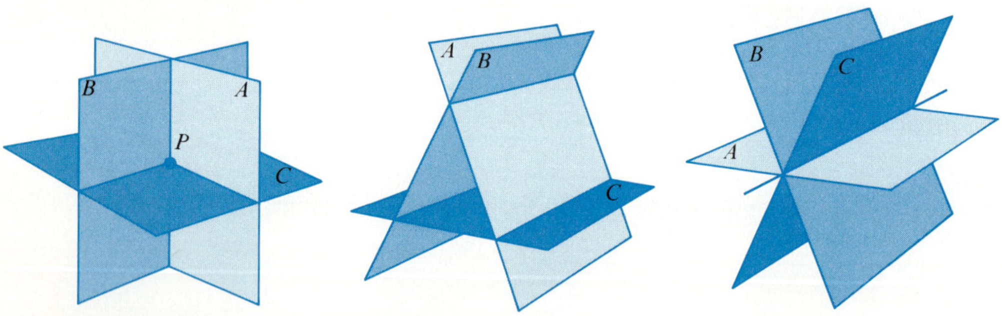

(a) Point P is common to all three planes

(b) No points are common to all three planes

(c) All points on the line of intersection are common to all three planes

Figure 4.4 Number of Solutions

To solve a system we combine methods of elimination and substitution similar to those in Section 4.2 to reduce to a system of two equations in two variables. The basic steps of this technique follow.

To Solve a System of Three Equations in Three Variables

1. Select any two of the three equations and eliminate one of the variables by addition or subtraction.
2. Use the remaining equation from the original system together with either equation used in Step 1, and eliminate the *same* variable as in Step 1.
3. Solve the resulting pair of equations in two variables.
4. Substitute the values obtained in Step 3 into any of the three original equations to obtain the value of the third variable.
5. Check the solution in all three of the original equations.
6. If, at any step, a contradiction is obtained, the system has no solution. If an identity is obtained after using only two equations, the system has either infinitely many solutions or no solution, depending on what happens when the third equation is involved.

In the process of solving a system, letter labels help us keep track of equations and steps. This is shown in the next example.

EXAMPLE 1 Solving a System of Three Equations

Solve the system.

$$\begin{aligned} &\textbf{(A)} \quad 3x + 2y + 3z = 3 \\ &\textbf{(B)} \quad 4x - 5y + 7z = 1 \\ &\textbf{(C)} \quad 2x + 3y - 2z = 6 \end{aligned}$$

Eliminate x from (A) and (B) by adding 4 times (A) to -3 times (B).

$$\begin{array}{lrcl} \textbf{4(A)} & 12x + 8y + 12z &=& 12 \\ \textbf{-3(B)} & -12x + 15y - 21z &=& -3 \\ \hline \textbf{(D)} & 23y - 9z &=& 9 \end{array} \qquad 4(A) - 3(B) = (D)$$

Eliminate x from (B) and (C) by adding (B) to -2 times (C).

$$\begin{array}{lrcl} \textbf{(B)} & 4x - 5y + 7z &=& 1 \\ \textbf{-2(C)} & -4x - 6y + 4z &=& -12 \\ \hline \textbf{(E)} & -11y + 11z &=& -11 \end{array} \qquad (B) - 2(C) = (E)$$

Simplify (E) by multiplying through by $-\frac{1}{11}$ (or dividing by -11).

$$\textbf{(F)} \quad y - z = 1 \qquad (F) = -\frac{1}{11}(E)$$

We have the following system.

$$\begin{aligned} &\textbf{(D)} \quad 23y - 9z = 9 \\ &\textbf{(F)} \quad y - z = 1 \end{aligned}$$

Solve (F) for y, obtaining $y = 1 + z$, and substitute into (D).

$$\begin{aligned} 23(1 + z) - 9z &= 9 \\ 23 + 23z - 9z &= 9 \\ 14z &= -14 \\ z &= -1 \end{aligned}$$

Substitute -1 for z in (F).

$$\begin{aligned} y - (-1) &= 1 \\ y + 1 &= 1 \\ y &= 0 \end{aligned}$$

Substitute 0 for y and -1 for z in (A).

$$\begin{aligned} 3x + 2(0) + 3(-1) &= 3 \\ 3x - 3 &= 3 \\ 3x &= 6 \\ x &= 2 \end{aligned}$$

The solution is $(2, 0, -1)$. Check in all three equations.

PRACTICE EXERCISE 1

Solve the system using the steps given.

$$\begin{aligned} &\text{(A)} \quad 2x - y + 3z = 5 \\ &\text{(B)} \quad 3x + 5y - z = 1 \\ &\text{(C)} \quad -x + 3y + 2z = -5 \end{aligned}$$

Step I: Use (A) and (C) and eliminate x to obtain equation (D).

Step II: Use (B) and (C) and eliminate x to obtain equation (E).

Step III: Solve the system of two equations in the two variables y and z, equations (D) and (E), to obtain the numerical values of y and z.

Step IV: Substitute the values for y and z obtained in III back into (C) to find the value of x.

Answer: $(2, -1, 0)$

EXAMPLE 2 Solving a System with Some Coefficients Zero

Solve the system.

$$\begin{aligned} &\textbf{(A)} \quad x + y + z = 6 \\ &\textbf{(B)} \quad x - z = -2 \\ &\textbf{(C)} \quad y + 3z = 11 \end{aligned}$$

PRACTICE EXERCISE 2

Solve the system.

$$\begin{aligned} &\text{(A)} \quad x + y - 2z = -1 \\ &\text{(B)} \quad y + z = 5 \\ &\text{(C)} \quad 2x - y + 4z = -6 \end{aligned}$$

When a variable is missing from one or more equations, we can shorten our work a bit. Since x does not appear in (C), if we subtract (B) from (A) to eliminate x, the resulting equation can be paired with (C) immediately.

$$\begin{array}{lrcl} \textbf{(A)} & x + y + z & = & 6 \\ \textbf{(B)} & x \quad\;\; - z & = & -2 \\ \hline \textbf{(D)} & y + 2z & = & 8 \end{array} \qquad \text{(A) − (B) = (D)}$$

Thus, we have the following system.

$$\textbf{(C)} \quad y + 3z = 11$$
$$\textbf{(D)} \quad y + 2z = 8$$

Subtracting (D) from (C), we eliminate y.

$$\textbf{(E)} \quad z = 3 \qquad \text{(C) − (D) = (E)}$$

Substitute 3 for z in (C).

$$y + 3(3) = 11$$
$$y + 9 = 11$$
$$y = 2$$

Substitute 3 for z in (B).

$$x - 3 = -2$$
$$x = 1$$

The solution is (1, 2, 3). Check in all three equations.

[*Hint:* Use equations (A) and (C) and eliminate x to obtain equation (D) in y and z. Pair (D) with (B) and solve.]

Answer: $(-3, 4, 1)$

3 INCONSISTENT AND DEPENDENT SYSTEMS

As with systems of two equations, when a linear system of three equations has no solution, it is inconsistent, and a contradiction will result when we try to solve it.

EXAMPLE 3 Working with an Inconsistent System

Solve the system.

$$\begin{array}{lr} \textbf{(A)} & 3x - y + 2z = 4 \\ \textbf{(B)} & -6x + 2y - 4z = 1 \\ \textbf{(C)} & 5x - 3y + 8z = 0 \end{array}$$

If we try to eliminate y by adding 2 times (A) to (B)

$$\begin{array}{lr} \textbf{2(A)} & 6x - 2y + 4z = 8 \\ \textbf{(B)} & -6x + 2y - 4z = 1 \\ \hline & 0 = 9 \end{array}$$

we obtain a contradiction, so the system is inconsistent and has no solution.

Practice Exercise 3

Solve the system.

(A) $x - 3y + z = 2$
(B) $2x + y - 3z = 1$
(C) $3x - 9y + 3z = -1$

Answer: no solution

If we obtain an identity such as $0 = 0$ when solving a linear system of three equations, and only two equations have been used to this point, we know that either the system has infinitely many solutions and is dependent, or else it has no solution and is inconsistent. Then we need to use the third equation to know the nature of the system. Notice this differs from systems of two equations where an identity at *any* step *always* means there are infinitely many solutions.

EXAMPLE 4 Working with a Dependent System

Solve the following system.

$$\begin{aligned} &\textbf{(A)} & x - 3y + z &= 4 \\ &\textbf{(B)} & x + 5y - z &= 2 \\ &\textbf{(C)} & -2x + 2y - z &= -7 \end{aligned}$$

If we add (A) and (B) we obtain

$$\textbf{(D)}\quad 2x + 2y = 6$$

which reduces to (E) by multiplying both sides by $\frac{1}{2}$.

$$\textbf{(E)}\quad x + y = 3$$

If we add (A) and (C) we obtain

$$\textbf{(F)}\quad -x - y = -3.$$

If we add (E) and (F),

$$\begin{aligned} &\textbf{(E)} & x + y &= 3 \\ &\textbf{(F)} & -x - y &= -3 \\ \hline & & 0 &= 0 \end{aligned}$$

we obtain an identity. Since all three equations have been used in the process of elimination, we can conclude that the system has infinitely many solutions and is dependent.

PRACTICE EXERCISE 4

Solve the system using the steps given.

$$\begin{aligned} &\textbf{(A)} & x - y + z &= 3 \\ &\textbf{(B)} & 2x - 2y + 2z &= -1 \\ &\textbf{(C)} & -x + y - z &= -3 \end{aligned}$$

Step 1: Use (A) and (C) and add to eliminate x. An identity results. Equation (B) must now be used to determine if the system is inconsistent or dependent.

Step 2: Multiply (A) by -2 and add to (B) to again eliminate x. The result is a contradiction, so the system is inconsistent.

Answer: no solution

4 APPLICATIONS OF SYSTEMS OF THREE EQUATIONS

Many applied problems can be translated into a system of three equations in three variables. As with problems in two equations, be careful to write out all details such as complete definitions.

EXAMPLE 5 Geometry Problem

In a triangle, the largest angle is 70° more than the smallest angle, and the remaining angle is 10° more than three times the smallest angle. Find the measure of each angle.

Let x = the measure of the smallest angle,
y = the measure of the middle angle,
z = the measure of the largest angle.

Since z is 70° more than x, we have

$$x + 70 = z$$

or

$$\textbf{(A)}\quad x - z = -70.$$

Since y is 10° more than 3 times x,

$$3x + 10 = y$$

or

$$\textbf{(B)}\quad 3x - y = -10.$$

Finally, the sum of the measures of the angles of a triangle is 180°.

$$\textbf{(C)}\quad x + y + z = 180$$

PRACTICE EXERCISE 5

The smallest angle in a triangle is one-third the measure of the middle-sized angle, and the largest angle is twice the sum of the other two. Find the measure of each angle.

We now have the following system.

$$\begin{aligned} &\textbf{(A)} \quad x \qquad\quad - z = -70 \\ &\textbf{(B)} \quad 3x - y \qquad = -10 \\ &\textbf{(C)} \quad x + y + z = 180 \end{aligned}$$

y is missing in (A), and if we add (B) and (C), we get another equation with y missing.

$$\textbf{(D)} \quad 4x + z = 170 \qquad \text{(B) + (C) = (D)}$$

We need to solve

$$\begin{aligned} &\textbf{(A)} \quad x - z = -70 \\ &\textbf{(D)} \quad 4x + z = 170. \end{aligned}$$

Add (A) and (D).

$$\begin{aligned} 5x &= 100 \\ x &= 20 \end{aligned}$$

Substitute 20 for x in (A).

$$\begin{aligned} 20 - z &= -70 \\ -z &= -90 \\ z &= 90 \end{aligned}$$

Substitute 20 for x in (B).

$$\begin{aligned} 3(20) - y &= -10 \\ 60 - y &= -10 \\ -y &= -70 \\ y &= 70 \end{aligned}$$

The angles have measures 20°, 70°, and 90°. Check.

Answer: 15°, 45°, 120°

EXAMPLE 6 Interest Problem

Joe has \$10,000 divided into three separate investments. Part of the money is invested in certificates at 7%, part in bonds at 8%, and the rest invested in his brother Bob's private business. If the business earns 6%, the total earnings from all the investments will amount to \$660. However, if the business slumps, Joe loses 6% on his investment, and the total earnings from the three will amount to only \$60. How much is invested in each category?

Let x = the number of dollars invested in certificates,
y = the number of dollars invested in bonds,
z = the number of dollars invested in Bob's business.

Then we must solve

$$\begin{aligned} x + \quad y + \quad z &= 10{,}000 && \text{Total invested is \$10,000} \\ 0.07x + 0.08y + 0.06z &= 660 && \text{Amount earned if business earns 6\%} \\ 0.07x + 0.08y - 0.06z &= 60. && \text{Amount earned if business loses; note the minus } 0.06z \end{aligned}$$

It is best to clear all decimals and then solve.

$$\begin{aligned} &\textbf{(A)} \quad x + y + z = 10{,}000 \\ &\textbf{(B)} \quad 7x + 8y + 6z = 66{,}000 && \text{Multiply by 100} \\ &\textbf{(C)} \quad 7x + 8y - 6z = 6000 && \text{Multiply by 100} \end{aligned}$$

Practice Exercise 6

Vanessa borrowed \$80,000 in three different loans to start a small business. She borrowed two amounts, totaling \$70,000, at 11% and 10%, respectively. She borrowed the rest at 13%. How much was borrowed at each rate if the total annual simple interest was \$8500?

Adding (B) and (C) eliminates z.

$$\textbf{(D)}\quad 14x + 16y = 72{,}000 \qquad \text{(B) + (C) = (D)}$$

Adding (B) to -6 times (A) will also eliminate z.

$$\textbf{(E)}\quad x + 2y = 6000 \qquad \text{(B) − 6(A) = (E)}$$

Solve (E) for x, $x = 6000 - 2y$, and substitute into (D).

$$\begin{aligned} 14(6000 - 2y) + 16y &= 72{,}000 \\ 84{,}000 - 28y + 16y &= 72{,}000 \\ -12y &= -12{,}000 \\ y &= 1000 \end{aligned}$$

Substitute 1000 for y in (E).

$$\begin{aligned} x + 2(1000) &= 6000 \\ x + 2000 &= 6000 \\ x &= 4000 \end{aligned}$$

Substitute 4000 for x and 1000 for y in (A).

$$\begin{aligned} 4000 + 1000 + z &= 10{,}000 \\ 5000 + z &= 10{,}000 \\ z &= 5000 \end{aligned}$$

Joe has invested \$4000 in certificates, \$1000 in bonds, and \$5000 in his brother's business. Check.

Answer: \$20,000 at 11%, \$50,000 at 10%, and \$10,000 at 13%

4.4 EXERCISES A

Solve the systems using the procedures shown in the examples.

1. $\begin{aligned} x + y + z &= 2 \\ -x + y - 2z &= -5 \\ 2x - y + z &= 1 \end{aligned}$

2. $\begin{aligned} x + 2y + 3z &= 6 \\ -2x + y - z &= -2 \\ -x + 3y - 2z &= 0 \end{aligned}$

3. $\begin{aligned} x - 2y - z &= 2 \\ 2x - y + z &= 7 \\ 3x + 2y + z &= 2 \end{aligned}$

4. $\begin{aligned} x + y + z &= 8 \\ x - y - z &= 0 \\ x + 2y + z &= 9 \end{aligned}$

5 $\begin{aligned} 3x + y - z &= 4 \\ y - 2z &= 5 \\ 2x + z &= -1 \end{aligned}$

6. $\begin{aligned} 2x + 3y &= 13 \\ 5x - 2z &= -2 \\ 4y - z &= 6 \end{aligned}$

7. $\begin{aligned} 4x - y + 3z &= 3 \\ -8x + 2y - 6z &= -7 \\ x + 5y - z &= 2 \end{aligned}$

8. $\begin{aligned} 2x - 3y + 4z &= 3 \\ 5x + 4y - 2z &= 7 \\ -3x + 2y - 5z &= -6 \end{aligned}$

9.
$$\begin{aligned} 2x - 5y + 7z &= 6 \\ x + 4y - 5z &= -2 \\ -6x + 15y - 21z &= 5 \end{aligned}$$

10.
$$\begin{aligned} 3x + y &= -1 \\ 5y - 8z &= 17 \\ -2x + z &= 5 \end{aligned}$$

11.
$$\begin{aligned} 4x &= 8 \\ -5y &= 3 \\ 2z &= 7 \end{aligned}$$

12.
$$\begin{aligned} 0.3x - 2.1y + 1.2z &= -3 \\ 5.7x + 4.2y - 4.8z &= -3 \\ 1.1x - 6.9y + 3.6z &= -19 \end{aligned}$$
[*Hint:* Clear decimals first.]

Each system in Exercises 13 and 14 is either inconsistent or dependent.

13.
$$\begin{aligned} 3x - 2y + 5z &= 8 \\ 6x - 4y + 10z &= 4 \\ 2x + 3y + 4z &= 5 \end{aligned}$$

14.
$$\begin{aligned} 2x - y + z &= 1 \\ -x + 2y - z &= 1 \\ -3x - z &= -3 \end{aligned}$$

Solve using a system of three equations.

15. The sum of three numbers is 6. The first, plus twice the second, minus the third is -3. Three times the first, minus the second, plus twice the third is 18. Find the three numbers.

16. A collection of 90 coins consisting of nickels, dimes, and quarters has a value of \$11.50. If the number of quarters is twice the number of nickels, find the number of each type of coin.

17. The smallest angle of a triangle is one-third the middle-sized angle, and the largest angle is 5° more than the middle-sized angle. Find the measure of each.

18. Anita has \$10,000 divided into three separate investments. Part is invested in a mutual fund which earns 8%, part is in time certificates which earn 7%, and the rest is invested in a business. If the business does well, it will earn 10% and her total earnings will amount to \$840. If the business loses 2%, her total earnings will amount to only \$360. How much is invested in each category?

19. The average of Manny's three scores in geometry is 78. If the first is 10 points less than the second and the third is 4 points more than the second, find all three scores. [*Hint:* the first sentence translates to $\frac{x+y+z}{3} = 78$.]

20. Find three numbers such that the sum of the first and second is -1, the sum of the second and third is -4, and the sum of the first and third is 7.

21. The average price of oranges, apples, and limes is 50¢ per pound. Six pounds of oranges, 5 pounds of apples, and 2 pounds of limes cost \$6.00. If limes cost 25¢ per pound more than apples, what is the price per pound of each?

22. At a recent rock concert, a total of 1000 tickets were sold. The prices of the three different types of tickets were \$4, \$5, and \$10, and the total receipts were \$4900. If twice as many \$4 tickets were sold as \$5 tickets, how many tickets at each price were sold?

FOR REVIEW

Solve using a system of equations.

23. By traveling 50 mph for one period of time and then 55 mph for another, Pam traveled 310 miles. If she had gone 5 mph faster throughout, she would have traveled 340 miles. How many hours did she travel at each rate?

24. If 2 lb of nuts and 3 lb of candy cost a total of \$5.40, and 3 lb of nuts and 4 lb of candy cost a total of \$7.50, find the cost of 1 lb of nuts and of 1 lb of candy.

25. A chemist has one solution which is 10% acid and a second which is 5% acid. How much of each should be used to make 20 L of an 8%-acid solution?

26. A boat travels 48 miles downstream in 2 hours and returns the 48 miles upstream in 3 hours. Find the speed of the boat and the speed of the stream.

The following exercises review material from Chapter 1 and will help you prepare for the material in the next section. Find the value of each expression.

27. $(-2)(3) - 4(-1) + (-2)(-3)$

28. $(-4)(-3) - (-2)(5) + (-4)(1)$

ANSWERS: 1. $(-1, 0, 3)$ 2. $(1, 1, 1)$ 3. $(1, -2, 3)$ 4. $(4, 1, 3)$ 5. $(0, 3, -1)$ 6. $(2, 3, 6)$ 7. inconsistent; no solution 8. $(1, 1, 1)$ 9. inconsistent; no solution 10. $(-2, 5, 1)$ 11. $\left(2, -\frac{3}{5}, \frac{7}{2}\right)$ 12. $(10, 20, 30)$ 13. inconsistent; no solution 14. dependent; infinitely many solutions 15. $(3, -1, 4)$ 16. 10 nickels, 60 dimes, 20 quarters 17. 25°, 75°, 80° 18. \$2000 in a mutual fund, \$4000 in certificates, \$4000 in a business 19. 70, 80, 84 20. $(5, -6, 2)$ 21. oranges cost 45¢ per pound, apples cost 40¢ per pound, limes cost 65¢ per pound 22. 600 \$4 tickets, 300 \$5 tickets, and 100 \$10 tickets 23. 4 hours at 50 mph; 2 hours at 55 mph 24. nuts are 90¢ per lb, candy is \$1.20 per lb 25. 8 L of 5% solution, 12 L of 10% solution 26. boat: 20 mph; stream: 4 mph 27. 4 28. 18

4.4 EXERCISES B

Solve the systems using the procedures shown in the examples.

1.
$$\begin{aligned} x + y + z &= 1 \\ -x + y - 3z &= -5 \\ 4x - y + 2z &= 0 \end{aligned}$$

2.
$$\begin{aligned} x - 2y + z &= 7 \\ 2x + y - z &= 3 \\ -3x + 2y + z &= -9 \end{aligned}$$

3.
$$\begin{aligned} 3x - y + 3z &= 2 \\ 2x + 3y + z &= 7 \\ x + y + 2z &= 0 \end{aligned}$$

4.
$$\begin{aligned} 2x + y - z &= 2 \\ x - y + z &= 1 \\ x + 2y + 3z &= 6 \end{aligned}$$

5.
$$\begin{aligned} -x + 4y + 3z &= 3 \\ 5x + y - z &= 2 \\ 2x - 8y - 6z &= -7 \end{aligned}$$

6.
$$\begin{aligned} x + 2y &= -2 \\ 2x + z &= 8 \\ -3y - z &= 9 \end{aligned}$$

7.
$$\begin{aligned} 3x - 2y - z &= -2 \\ 5x + y - 5z &= 3 \\ x - 3y + 3z &= -1 \end{aligned}$$

8.
$$\begin{aligned} x + y + z &= -3 \\ 4x - 4y + 2z &= -2 \\ -5x + 4y - 2z &= 3 \end{aligned}$$

9.
$$\begin{aligned} 2x - 12y + z &= 17 \\ 3x + 12y - 4z &= -2 \\ 4x - 8y - 6z &= -6 \end{aligned}$$

10.
$$\begin{aligned} 2x - 3y &= -7 \\ 3y - 2z &= -3 \\ 4x - 4z &= 5 \end{aligned}$$

11.
$$\begin{aligned} -3z &= 9 \\ 5x &= 4 \\ 9y &= -12 \end{aligned}$$

12.
$$\begin{aligned} \frac{1}{2}x + \frac{1}{3}y - \frac{1}{4}z &= 2 \\ \frac{1}{4}x - \frac{1}{2}y + \frac{1}{2}z &= 2 \\ -\frac{1}{2}x + \frac{1}{2}y - \frac{1}{4}z &= -1 \end{aligned}$$

Each system in Exercises 13 and 14 is either inconsistent or dependent.

13.
$$\begin{aligned} 2x - 5y + 7z &= 6 \\ x + 4y - 5z &= -2 \\ -6x + 15y - 21z &= 5 \end{aligned}$$

14.
$$\begin{aligned} 2x - y + z &= 5 \\ x - y + 2z &= 1 \\ -3x + y &= -9 \end{aligned}$$

Solve using a system of three equations.

15. The sum of twice a first number, three times a second, and four times a third is 20. The sum of the first and the third is 10 more than the second. The first is the sum of the second and third. Find the numbers.

16. Connie has a total of 27 bills, consisting of ones, fives, and tens, in a cash box. The total value of these bills is \$127. If there is one more ten than there are fives, how many of each are in the box?

17. The largest angle of a triangle is 20° more than twice the middle-sized angle. The sum of the smallest and middle-sized is one-half the largest. Find the measure of each.

18. Howard has a total of \$50,000 invested in stocks, bonds, and certificates. The amount in stocks returned 12% for the year, each of the other accounts returned 9%, and his total income from the investments for the year was \$4860. Had he reversed the amounts invested in stocks and bonds, he would have earned \$5100. How much was invested in each category?

19. Alberta has an average of 88 on three math tests. If her first test score was 13 points higher than the second, and the third was 5 points more than the second, find her three scores.

20. Find three numbers such that the first is five times the third, the second minus the third is -10, and four times the third is 3 less than the first.

21. A chemist has three acid solutions A, B, and C with an average concentration of 25% acid. If he mixes 5 L of A, 4 L of B, and 11 L of C, he obtains 20 L of 24% acid. If the percent acid in A is twice that in C, find the concentration of each solution.

22. A total of 5000 tickets were sold for the conference championship game. The prices of the three different types of tickets were \$2, \$3, and \$5. If three times as many \$2 tickets were sold as \$5 tickets, and the total receipts were \$14,000, how many tickets at each price were sold?

FOR REVIEW

Solve using a system of equations.

23. A 60-ft piece of rope is cut into two pieces. One is 12 ft more than three times the other. How long is each piece?

24. In a right triangle, one acute angle is 15° more than four times the other. Find each acute angle.

25. If there are 26 coins in a collection of dimes and quarters, and the value of the collection is \$4.70, how many dimes and how many quarters are there?

26. A chemist has one solution which is 15% acid and another which is 20% acid. How many gallons of each should be mixed together to obtain 30 gallons of a solution that is 18% acid?

The following exercises review material from Chapter 1 and will help you prepare for the next section. Find the value of each expression.

27. $(-1)(-3) - 5(-2) + (-4)(-2)$

28. $(2)(-5) - (-3)(-2) + (-6)(-1)$

4.4 EXERCISES C

Solve.

1. $$\begin{aligned} w + x + y + z &= 2 \\ 2w + x + y + z &= 3 \\ w + x + y \quad\;\; &= 0 \\ w + x \qquad\quad\;\; &= 1 \end{aligned}$$

2. $$\begin{aligned} w + 2z &= 5 \\ 2w - 3y &= -5 \\ 4x + z &= 11 \\ x - 5y &= -3 \end{aligned}$$

[Answer: $w = -1$, $x = 2$, $y = 1$, $z = 3$]

3. $\dfrac{1}{u} + \dfrac{1}{v} + \dfrac{3}{w} = 4$

$\dfrac{2}{u} - \dfrac{1}{v} + \dfrac{4}{w} = -2$

$\dfrac{3}{u} - \dfrac{2}{v} - \dfrac{1}{w} = 2$

4. $\dfrac{1}{u} + \dfrac{1}{v} = -1$

$\dfrac{1}{v} + \dfrac{1}{w} = 1$

$\dfrac{1}{u} + \dfrac{1}{w} = 6$

5. A three-digit whole number has the property that the sum of the three digits is 9, and the ones digit is 3 more than the tens digit. Also, if the hundreds digit and the tens digit are interchanged, the value of the number remains the same. Find the number.

6. The general form of the equation of a circle in a rectangular coordinate system is given by $x^2 + y^2 + Ax + By + C = 0$, where A, B, and C are constants. Find the general form of the equation of a circle passing through $(-1, -2)$, $(1, 0)$, and $(3, -2)$.
[Answer: $x^2 + y^2 - 2x + 4y + 1 = 0$]

7. Ponderosa Products manufactures three models of wood-burning stoves, the Aspen, the Birch, and the Sierra. Each stove is made in three stages requiring the specified times in hours listed in the table below. Due to vacation schedules and equipment maintenance, the cutting, welding, and finishing departments have available a maximum of 75, 47, and 36 workhours per week, respectively. Determine the number of stoves of each model that should be made weekly for Ponderosa Products to be at full capacity operation.

	Aspen	Birch	Sierra
Cutting	4	5	7
Welding	2	3	5
Finishing	2	2	4

[Answer: 3 Aspen stoves, 7 Birch stoves, 4 Sierra stoves]

8. The owner of a chemical plant received an order for a special blend of fertilizer that consists of, among other things, 760 pounds of nitrate, 540 pounds of phosphate, and 106 pounds of iron. He has available three mixes with the compositions shown in the table below. How many pounds of each should he blend to fill the order?

	Nitrate (in %)	Phosphate (in %)	Iron (in %)
GroWell	20	15	4
YieldHi	25	20	2
FastGro	30	18	5

4.5 DETERMINANTS AND CRAMER'S RULE

STUDENT GUIDEPOSTS

1 Second-order Determinants
2 Cramer's Rule for Two Equations
3 Third-order Determinants
4 Cramer's Rule for Three Equations

In this section we define a determinant and use it to solve systems of linear equations. The method we develop is a numerical technique that can easily be programmed in a computer or a hand calculator. However, without a computing device, this method is probably less desirable than either the substitution method or the elimination method.

1 SECOND-ORDER DETERMINANTS

As motivation for our basic definition, suppose we solve a general system of two linear equations in x and y,

$$a_1x + b_1y = c_1$$
$$a_2x + b_2y = c_2,$$

where a_1, b_1, c_1, a_2, b_2, and c_2 are real numbers. Multiply the first equation by b_2 and the second by b_1.

$$a_1 b_2 x + b_1 b_2 y = c_1 b_2 \quad \text{b_2 times the first equation}$$
$$a_2 b_1 x + b_2 b_1 y = c_2 b_1 \quad \text{b_1 times the second equation}$$

Subtract the second from the first to eliminate y.

$$a_1b_2x - a_2b_1x = c_1b_2 - c_2b_1 \quad b_1b_2y - b_2b_1y = 0 \cdot y = 0$$
$$(a_1b_2 - a_2b_1)x = c_1b_2 - c_2b_1 \quad \text{Factor out } x$$
$$x = \frac{c_1b_2 - c_2b_1}{a_1b_2 - a_2b_1} \quad \text{Divide by } a_1b_2 - a_2b_1, \text{ if } a_1b_2 - a_2b_1 \neq 0$$

Similarly, we may first eliminate x and then solve for y.

$$y = \frac{a_1c_2 - a_2c_1}{a_1b_2 - a_2b_1}$$

We can remember the numbers $a_1b_2 - a_2b_1$, $c_1b_2 - c_2b_1$, and $a_1c_2 - a_2c_1$ if we look at the arrangements of coefficients and constants in the original system, and relate these numbers to the following definition.

Second-Order Determinants

If the four numbers a_1, b_1, a_2, b_2 are written in a square array within vertical bars,

$$\begin{vmatrix} a_1 & b_1 \\ a_2 & b_2 \end{vmatrix}$$

the result is **a second-order determinant** and has **value**

$$\begin{vmatrix} a_1 & b_1 \\ a_2 & b_2 \end{vmatrix} = a_1b_2 - a_2b_1. \quad \text{Difference of cross-products}$$

EXAMPLE 1 EVALUATING SECOND-ORDER DETERMINANTS

Evaluate the second-order determinants.

(a) $\begin{vmatrix} 1 & 2 \\ 3 & 4 \end{vmatrix} = (1)(4) - (3)(2) = 4 - 6 = -2$

(b) $\begin{vmatrix} c_1 & b_1 \\ c_2 & b_2 \end{vmatrix} = c_1b_2 - c_2b_1$

PRACTICE EXERCISE 1

Evaluate the second-order determinants.

(a) $\begin{vmatrix} 2 & 3 \\ 5 & -1 \end{vmatrix}$

(b) $\begin{vmatrix} a_1 & c_1 \\ a_2 & c_2 \end{vmatrix}$

Answers: (a) -17
(b) $a_1c_2 - a_2c_1$

2 CRAMER'S RULE FOR TWO EQUATIONS

Return to the original system of equations at the beginning of this section with the solutions

$$x = \frac{c_1b_2 - c_2b_1}{a_1b_2 - a_2b_1} \quad \text{and} \quad y = \frac{a_1c_2 - a_2c_1}{a_1b_2 - a_2b_1}.$$

We can use the definition of second-order determinant to rewrite the solutions

$$x = \frac{\begin{vmatrix} c_1 & b_1 \\ c_2 & b_2 \end{vmatrix}}{\begin{vmatrix} a_1 & b_1 \\ a_2 & b_2 \end{vmatrix}} \quad \text{and} \quad y = \frac{\begin{vmatrix} a_1 & c_1 \\ a_2 & c_2 \end{vmatrix}}{\begin{vmatrix} a_1 & b_1 \\ a_2 & b_2 \end{vmatrix}}$$

The determinant

$$A = \begin{vmatrix} a_1 & b_1 \\ a_2 & b_2 \end{vmatrix}$$

which appears in both denominators of the solutions is often called the **determinant of the coefficients of the system of equations.** The determinant in the numerator of the solution for x is denoted by

$$A_x = \begin{vmatrix} c_1 & b_1 \\ c_2 & b_2 \end{vmatrix}$$

and is formed from A by replacing the coefficients of the x terms (the a's) with the constants (the c's). Similarly, the determinant in the numerator of the solution for y is denoted by

$$A_y = \begin{vmatrix} a_1 & c_1 \\ a_2 & c_2 \end{vmatrix}$$

and is formed from A by replacing the coefficients of y (the b's) with the constants (the c's). Using this notation, we have **Cramer's rule for solving a system of two linear equations in two variables.**

Cramer's Rule

The system

$$a_1x + b_1y = c_1$$
$$a_2x + b_2y = c_2$$

has solutions $x = \dfrac{A_x}{A}$ and $y = \dfrac{A_y}{A}$, if $A \neq 0$,

where $A = \begin{vmatrix} a_1 & b_1 \\ a_2 & b_2 \end{vmatrix}$, $A_x = \begin{vmatrix} c_1 & b_1 \\ c_2 & b_2 \end{vmatrix}$, $A_y = \begin{vmatrix} a_1 & c_1 \\ a_2 & c_2 \end{vmatrix}$.

EXAMPLE 2 Using Cramer's Rule to Solve Two Equations

Solve using Cramer's rule.

$$3x + 5y = -2$$
$$5x + 3y = 2$$

Practice Exercise 2

Solve using Cramer's rule.

$$3x + 2y = 0$$
$$5x - y = -13$$

The determinant of the coefficients of the system is

$$A = \begin{vmatrix} 3 & 5 \\ 5 & 3 \end{vmatrix},$$

the determinant in the numerator for x is

$$A_x = \begin{vmatrix} -2 & 5 \\ 2 & 3 \end{vmatrix}$$

(we have replaced 3 and 5, the coefficients of x, with the constants -2 and 2), and the determinant in the numerator for y is

$$A_y = \begin{vmatrix} 3 & -2 \\ 5 & 2 \end{vmatrix}$$

(we have replaced 5 and 3, the coefficients of y, with the constants -2 and 2). Then, by Cramer's rule,

$$x = \frac{A_x}{A} = \frac{\begin{vmatrix} -2 & 5 \\ 2 & 3 \end{vmatrix}}{\begin{vmatrix} 3 & 5 \\ 5 & 3 \end{vmatrix}} = \frac{(-2)(3) - (2)(5)}{(3)(3) - (5)(5)} = \frac{-6 - 10}{9 - 25} = \frac{-16}{-16} = 1$$

$$y = \frac{A_y}{A} = \frac{\begin{vmatrix} 3 & -2 \\ 5 & 2 \end{vmatrix}}{\begin{vmatrix} 3 & 5 \\ 5 & 3 \end{vmatrix}} = \frac{(3)(2) - (5)(-2)}{(3)(3) - (5)(5)} = \frac{6 - (-10)}{9 - 25} = \frac{16}{-16} = -1.$$

The solution to the system is $(1, -1)$.

$$A = \begin{vmatrix} 3 & 2 \\ 5 & -1 \end{vmatrix} =$$

$$A_x = \begin{vmatrix} 0 & 2 \\ -13 & -1 \end{vmatrix} =$$

$$A_y = \begin{vmatrix} & \\ & \end{vmatrix} =$$

$$x = \frac{A_x}{A} =$$

$$y =$$

Answer: $A = -13$, $A_x = 26$, $A_y = -39$, $x = -2$, $y = 3$; the solution is $(-2, 3)$

If we try to solve a system such as

$$\begin{aligned} 2x - y &= 3 \\ -4x + 2y &= 1, \end{aligned}$$

we have $\quad x = \dfrac{A_x}{A} = \dfrac{7}{0} \quad$ and $\quad y = \dfrac{A_y}{A} = \dfrac{14}{0}.$

Since the determinant of the coefficients of the system A has value 0, both denominators are zero. Upon closer examination of the system, we see that the lines are parallel, the system is inconsistent, and there is no solution. Likewise, if we solve

$$\begin{aligned} 2x - y &= 3 \\ -4x + 2y &= -6, \end{aligned}$$

we have

$$x = \frac{A_x}{A} = \frac{0}{0} \quad \text{and} \quad y = \frac{A_y}{A} = \frac{0}{0}.$$

Again, the determinant of the coefficients, A, has value 0, resulting in zero denominators, but this time both numerators are also 0. Upon closer examination of the system, we see that the lines coincide, the system is dependent, and there are infinitely many solutions of the form $(x, 2x - 3)$ for any real number x.

Inconsistent and Dependent Systems

When a system is solved using Cramer's rule,

1. if $A = 0$, and either $A_x \neq 0$ or $A_y \neq 0$, then the system is inconsistent and has **no solution;**
2. if $A = 0$, and both $A_x = 0$ and $A_y = 0$, then the system is dependent and has **infinitely many solutions.**

3 THIRD-ORDER DETERMINANTS

Cramer's rule can be applied to systems of three linear equations in three variables (in fact, to systems of n linear equations in n variables, for $n \geq 2$) by suitably defining **third-order determinant** (and ***n*th-order determinant**).

Third-Order Determinants

If the nine numbers $a_1, b_1, c_1, a_2, b_2, c_2, a_3, b_3, c_3$ are written in a square array within vertical bars,

$$\begin{vmatrix} a_1 & b_1 & c_1 \\ a_2 & b_2 & c_2 \\ a_3 & b_3 & c_3 \end{vmatrix}$$

The result is a **third-order determinant** and has **value**

$$\begin{vmatrix} a_1 & b_1 & c_1 \\ a_2 & b_2 & c_2 \\ a_3 & b_3 & c_3 \end{vmatrix} = a_1 \begin{vmatrix} b_2 & c_2 \\ b_3 & c_3 \end{vmatrix} - a_2 \begin{vmatrix} b_1 & c_1 \\ b_3 & c_3 \end{vmatrix} + a_3 \begin{vmatrix} b_1 & c_1 \\ b_2 & c_2 \end{vmatrix}.$$

Notice that the third-order determinant is evaluated by evaluating three second-order determinants. Remember the definition by concentrating on the first column of numbers: each number in the first column multiplies the second-order determinant which remains if the row and column which contain the number are deleted from the third-order determinant.

$$a_1 \begin{vmatrix} b_2 & c_2 \\ b_3 & c_3 \end{vmatrix} \qquad -a_2 \begin{vmatrix} b_1 & c_1 \\ b_3 & c_3 \end{vmatrix} \qquad + a_3 \begin{vmatrix} b_1 & c_1 \\ b_2 & c_2 \end{vmatrix}$$

$$\begin{vmatrix} a_1 & b_1 & c_1 \\ a_2 & b_2 & c_2 \\ a_3 & b_3 & c_3 \end{vmatrix} \qquad \begin{vmatrix} a_1 & b_1 & c_1 \\ a_2 & b_2 & c_2 \\ a_3 & b_3 & c_3 \end{vmatrix} \qquad \begin{vmatrix} a_1 & b_1 & c_1 \\ a_2 & b_2 & c_2 \\ a_3 & b_3 & c_3 \end{vmatrix}$$

Be sure to *subtract* a_2 times the second-order determinant.

EXAMPLE 3 EVALUATING A THIRD-ORDER DETERMINANT

Evaluate the third-order determinant.

$$\begin{vmatrix} 5 & -1 & -2 \\ 4 & -1 & 0 \\ 1 & 2 & 3 \end{vmatrix} = 5 \begin{vmatrix} -1 & 0 \\ 2 & 3 \end{vmatrix} - 4 \begin{vmatrix} -1 & -2 \\ 2 & 3 \end{vmatrix} + 1 \begin{vmatrix} -1 & -2 \\ -1 & 0 \end{vmatrix}$$

$$= 5[(-1)(3) - (2)(0)] - 4[(-1)(3) - (2)(-2)] + 1[(-1)(0) - (-1)(-2)]$$
$$= 5[-3 - 0] - 4[-3 + 4] + 1[0 - 2]$$
$$= 5[-3] - 4[1] + 1[-2] = -15 - 4 - 2 = -21$$

PRACTICE EXERCISE 3

Evaluate the third-order determinant.

$$\begin{vmatrix} 2 & 0 & 5 \\ 3 & 4 & -2 \\ -1 & 1 & 2 \end{vmatrix}$$

Answer: 55

4 CRAMER'S RULE FOR THREE EQUATIONS

The definition we have given for a third-order determinant would seem reasonable if we were to solve a general system of three linear equations.

$$a_1x + b_1y + c_1z = d_1$$
$$a_2x + b_2y + c_2z = d_2$$
$$a_3x + b_3y + c_3z = d_3$$

The solutions to this system have the following form.

$$x = \frac{\begin{vmatrix} d_1 & b_1 & c_1 \\ d_2 & b_2 & c_2 \\ d_3 & b_3 & c_3 \end{vmatrix}}{\begin{vmatrix} a_1 & b_1 & c_1 \\ a_2 & b_2 & c_2 \\ a_3 & b_3 & c_3 \end{vmatrix}}, \quad y = \frac{\begin{vmatrix} a_1 & d_1 & c_1 \\ a_2 & d_2 & c_2 \\ a_3 & d_3 & c_3 \end{vmatrix}}{\begin{vmatrix} a_1 & b_1 & c_1 \\ a_2 & b_2 & c_2 \\ a_3 & b_3 & c_3 \end{vmatrix}}, \quad z = \frac{\begin{vmatrix} a_1 & b_1 & d_1 \\ a_2 & b_2 & d_2 \\ a_3 & b_3 & d_3 \end{vmatrix}}{\begin{vmatrix} a_1 & b_1 & c_1 \\ a_2 & b_2 & c_2 \\ a_3 & b_3 & c_3 \end{vmatrix}}$$

This result is also known as **Cramer's rule for a system of three linear equations in three variables.** The determinant of the coefficient appears in the denominator of each fraction, and the determinant in each numerator is formed by replacing the coefficients of the respective variable with the constants. As with systems of two equations in two variables, we use the following notation.

$$A = \begin{vmatrix} a_1 & b_1 & c_1 \\ a_2 & b_2 & c_2 \\ a_3 & b_3 & c_3 \end{vmatrix} \quad A_x = \begin{vmatrix} d_1 & b_1 & c_1 \\ d_2 & b_2 & c_2 \\ d_3 & b_3 & c_3 \end{vmatrix} \quad A_y = \begin{vmatrix} a_1 & d_1 & c_1 \\ a_2 & d_2 & c_2 \\ a_3 & d_3 & c_3 \end{vmatrix} \quad A_z = \begin{vmatrix} a_1 & b_1 & d_1 \\ a_2 & b_2 & d_2 \\ a_3 & b_3 & d_3 \end{vmatrix}$$

The solutions to the system, using Cramer's rule, are

$$x = \frac{A_x}{A}, \quad y = \frac{A_y}{A}, \quad z = \frac{A_z}{A}, \quad \text{if } A \neq 0.$$

EXAMPLE 4 Using Cramer's Rule to Solve Three Equations

Solve using Cramer's rule.

$$\begin{aligned} x + y + z &= 2 \\ -2x - y + z &= 1 \\ x - 2y - z &= 1 \end{aligned}$$

The three solutions are given by

$$x = \frac{A_x}{A} = \frac{\begin{vmatrix} 2 & 1 & 1 \\ 1 & -1 & 1 \\ 1 & -2 & -1 \end{vmatrix}}{\begin{vmatrix} 1 & 1 & 1 \\ -2 & -1 & 1 \\ 1 & -2 & -1 \end{vmatrix}} \qquad y = \frac{A_y}{A} = \frac{\begin{vmatrix} 1 & 2 & 1 \\ -2 & 1 & 1 \\ 1 & 1 & -1 \end{vmatrix}}{\begin{vmatrix} 1 & 1 & 1 \\ -2 & -1 & 1 \\ 1 & -2 & -1 \end{vmatrix}}$$

$$z = \frac{A_z}{A} = \frac{\begin{vmatrix} 1 & 1 & 2 \\ -2 & -1 & 1 \\ 1 & -2 & 1 \end{vmatrix}}{\begin{vmatrix} 1 & 1 & 1 \\ -2 & -1 & 1 \\ 1 & -2 & -1 \end{vmatrix}}$$

PRACTICE EXERCISE 4

Solve using Cramer's rule.

$$\begin{aligned} x - y + z &= -2 \\ x + 2y - 3z &= -6 \\ 4x + y + 5z &= -7 \end{aligned}$$

In each fraction, the denominator is the determinant of the coefficients, and the numerator has the corresponding column of coefficients of each variable replaced with the column of constants. Next we evaluate each of the four different third-order determinants.

$$A = \begin{vmatrix} 1 & 1 & 1 \\ -2 & -1 & 1 \\ 1 & -2 & -1 \end{vmatrix} = (1)\begin{vmatrix} -1 & 1 \\ -2 & -1 \end{vmatrix} - (-2)\begin{vmatrix} 1 & 1 \\ -2 & -1 \end{vmatrix} + (1)\begin{vmatrix} 1 & 1 \\ -1 & 1 \end{vmatrix}$$

$$= (1)(3) + (2)(1) + (1)(2) = 3 + 2 + 2 = 7$$

$$A_x = \begin{vmatrix} 2 & 1 & 1 \\ 1 & -1 & 1 \\ 1 & -2 & -1 \end{vmatrix} = (2)\begin{vmatrix} -1 & 1 \\ -2 & -1 \end{vmatrix} - (1)\begin{vmatrix} 1 & 1 \\ -2 & -1 \end{vmatrix} + (1)\begin{vmatrix} 1 & 1 \\ -1 & 1 \end{vmatrix}$$

$$= (2)(3) - (1)(1) + (1)(2) = 6 - 1 + 2 = 7$$

$$A_y = \begin{vmatrix} 1 & 2 & 1 \\ -2 & 1 & 1 \\ 1 & 1 & -1 \end{vmatrix} = (1)\begin{vmatrix} 1 & 1 \\ 1 & -1 \end{vmatrix} - (-2)\begin{vmatrix} 2 & 1 \\ 1 & -1 \end{vmatrix} + (1)\begin{vmatrix} 2 & 1 \\ 1 & 1 \end{vmatrix}$$

$$= (1)(-2) + (2)(-3) + (1)(1) = -2 - 6 + 1 = -7$$

$$A_z = \begin{vmatrix} 1 & 1 & 2 \\ -2 & -1 & 1 \\ 1 & -2 & 1 \end{vmatrix} = (1)\begin{vmatrix} -1 & 1 \\ -2 & 1 \end{vmatrix} - (-2)\begin{vmatrix} 1 & 2 \\ -2 & 1 \end{vmatrix} + (1)\begin{vmatrix} 1 & 2 \\ -1 & 1 \end{vmatrix}$$

$$= (1)(1) + (2)(5) + (1)(3) = 1 + 10 + 3 = 14$$

Thus, substituting the values of A, A_x, A_y, and A_z, we have

$$x = \frac{A_x}{A} = \frac{7}{7} = 1, \qquad y = \frac{A_y}{A} = \frac{-7}{7} = -1, \qquad z = \frac{A_z}{A} = \frac{14}{7} = 2,$$

giving the solution $(1, -1, 2)$.

Answer: $A = 23$, $A_x = -69$, $A_y = 0$, and $A_z = 23$, the solution is $(-3, 0, 1)$

As is the case with Cramer's rule for systems of two equations, when the determinant of the coefficients is zero ($A = 0$), there is no solution if at least one of A_x, A_y, or A_z is not zero, and there are infinitely many solutions if all of A_x, A_y, and A_z are zero.

4.5 EXERCISES A

Evaluate the determinants.

1. $\begin{vmatrix} 3 & 5 \\ 2 & 4 \end{vmatrix}$

2. $\begin{vmatrix} 2 & 5 \\ -3 & -1 \end{vmatrix}$

3. $\begin{vmatrix} -3 & -7 \\ -6 & -8 \end{vmatrix}$

4. $\begin{vmatrix} 2 & 5 \\ 0 & 0 \end{vmatrix}$

5. $\begin{vmatrix} 2 & -1 \\ 2 & -1 \end{vmatrix}$

6. $\begin{vmatrix} 10 & 20 \\ 30 & 40 \end{vmatrix}$

7. $\begin{vmatrix} 3 & 1 & 2 \\ 4 & 0 & -2 \\ -2 & 1 & 3 \end{vmatrix}$

8. $\begin{vmatrix} 1 & 1 & 1 \\ -1 & 2 & -2 \\ -3 & 4 & -3 \end{vmatrix}$

9. $\begin{vmatrix} 0 & -4 & -2 \\ 4 & 2 & -1 \\ 0 & 6 & 3 \end{vmatrix}$

10. $\begin{vmatrix} 3 & 0 & 0 \\ 0 & -2 & 0 \\ 0 & 0 & 4 \end{vmatrix}$

Solve the systems using Cramer's rule.

11. $\begin{aligned} x + 2y &= -1 \\ 2x - 3y &= 12 \end{aligned}$

12. $\begin{aligned} 4x - y &= -3 \\ 2x - 3y &= 1 \end{aligned}$

13. $\begin{aligned} 5x + 5y &= 10 \\ x + y &= 2 \end{aligned}$

14 $\begin{aligned} 3x + 2y &= 7 \\ -x + 4y &= 0 \end{aligned}$

15. $\begin{aligned} 3x - y &= -11 \\ 2x + 4z &= -2 \\ -3y + 5z &= -1 \end{aligned}$

16. $\begin{aligned} 2x + 3y &= -1 \\ 3y + z &= 6 \\ 3y + 4z &= 15 \end{aligned}$

17. $\begin{aligned} 2x - y &= 3 \\ 2y - z &= 3 \\ x + y + z &= 9 \end{aligned}$

18 $\begin{aligned} 4x - 3y + 8z &= 12 \\ 2x - \frac{3}{2}y + 4z &= 11 \\ x - 5z &= -10 \end{aligned}$

FOR REVIEW

Solve.

19. $\begin{aligned} 3x - 5y + z &= 4 \\ 5x + 4y &= 5 \\ 2x - 3y - 3z &= -1 \end{aligned}$

20. $\begin{aligned} x + y &= 1 \\ -y + z &= -1 \\ x + z &= 2 \end{aligned}$

21. The average of a student's first three scores on math tests is 84. If the second score was 5 points higher than the first, and the third was 2 points higher than the second, what were the three scores?

22. Jill has \$20,000 invested in stocks, bonds, and securities. She has twice as much in stocks as she does in bonds. The return on her investments was 12% for the stocks, 9% for the bonds, and 8% for the securities. If her income for the year was \$1960, what amount does she have invested in each?

Exercises 23–26, reviewing material covered in Chapter 2, will help you prepare for the next section. Solve each inequality.

23. $2x - 1 \le 3x + 4$

24. $3(x - 4) - (x - 1) > 5$

25. $2(x - 5) < 1 - (x - 1)$

26. $-3x \ge 2 - 4(1 - 2x)$

ANSWERS: 1. 2 2. 13 3. −18 4. 0 5. 0 6. −200 7. 6 8. 7 9. 0 10. −24 11. (3, −2) 12. (−1, −1) 13. infinitely many solutions of the form $(x, 2-x)$ for x any real number 14. $\left(2, \frac{1}{2}\right)$ 15. (−3, 2, 1) 16. (−2, 1, 3) 17. (3, 3, 3) 18. no solution 19. (1, 0, 1) 20. inconsistent; no solution 21. 80, 85, 87 22. \$8000 in stocks, \$4000 in bonds, \$8000 in securities 23. $x \geq -5$ 24. $x > 8$ 25. $x < 4$ 26. $x \leq \frac{2}{11}$

4.5 EXERCISES B

Evaluate the determinants.

1. $\begin{vmatrix} -1 & 0 \\ 3 & 4 \end{vmatrix}$
2. $\begin{vmatrix} 5 & 0 \\ -3 & 2 \end{vmatrix}$
3. $\begin{vmatrix} -1 & -1 \\ -2 & -4 \end{vmatrix}$
4. $\begin{vmatrix} 2 & 0 \\ 5 & 0 \end{vmatrix}$
5. $\begin{vmatrix} 5 & 5 \\ 6 & 6 \end{vmatrix}$
6. $\begin{vmatrix} -2 & -7 \\ -3 & -4 \end{vmatrix}$
7. $\begin{vmatrix} 1 & -2 & 1 \\ 2 & 3 & 2 \\ 3 & 1 & 3 \end{vmatrix}$
8. $\begin{vmatrix} -2 & 0 & 6 \\ 1 & 2 & -1 \\ 4 & -2 & -5 \end{vmatrix}$
9. $\begin{vmatrix} 3 & -2 & -4 \\ 0 & 0 & 6 \\ 2 & 0 & 8 \end{vmatrix}$
10. $\begin{vmatrix} 0 & 0 & 7 \\ 0 & 3 & 0 \\ -1 & 0 & 0 \end{vmatrix}$

Solve the systems using Cramer's rule.

11. $\begin{aligned} 3x + 2y &= -3 \\ 6x - 3y &= 8 \end{aligned}$
12. $\begin{aligned} 2x \phantom{{}+y} &= 4 \\ -3x + y &= 7 \end{aligned}$
13. $\begin{aligned} -3x + 3y &= 5 \\ 2x - 2y &= -1 \end{aligned}$
14. $\begin{aligned} 2x + 2y &= 0 \\ x - 5y &= -18 \end{aligned}$
15. $\begin{aligned} 2x + y - z &= -3 \\ x - y + z &= 0 \\ -3x + 2y - 4z &= -3 \end{aligned}$
16. $\begin{aligned} -x + 5y + z &= 2 \\ 3x - y + 4z &= 3 \\ 6x - 2y + 8z &= 7 \end{aligned}$
17. $\begin{aligned} 4x \phantom{{}+3y} + z &= 5 \\ 3y - 6z &= 0 \\ 5x \phantom{{}+3y} + 5z &= 10 \end{aligned}$
18. $\begin{aligned} 5x + 3y - 2z &= -4 \\ 2y - z &= 8 \\ 3x \phantom{{}+3y} + 4z &= -1 \end{aligned}$

FOR REVIEW

Solve.

19. $\begin{aligned} 3x - 3y + z &= 0 \\ 4x - 2y - z &= 3 \\ 2x + 3y - 2z &= -2 \end{aligned}$
20. $\begin{aligned} x + y \phantom{{}+z} &= 2 \\ -y + z &= -4 \\ x \phantom{{}+y} + z &= 3 \end{aligned}$
21. The smallest angle of a triangle is one-third the middle-sized angle, and the middle-sized angle is one-half the largest. Find the measure of each angle.
22. The average of the weights of the three interior linemen for the Lumberjacks is 257 pounds. If the sum of the first two is 500 pounds, and the third is 21 pounds heavier than the second, what is the weight of each?

Exercises 23–26, reviewing material covered in Chapter 2, will help you prepare for the next section. Solve each inequality.

23. $3x - 4 \geq 4x + 2$
24. $4(x - 1) - 2(x - 3) < 6$
25. $5(x - 3) > 2 - (x - 1)$
26. $-2x \leq 6 - 3(4 - x)$

4.5 EXERCISES C

Solve for x.

1. $\begin{vmatrix} x & 2 \\ 3 & 1 \end{vmatrix} = 3$
2. $\begin{vmatrix} x & 0 & 0 \\ 0 & 1 & 2 \\ 0 & -1 & 3 \end{vmatrix} = 10$
   [Answer: $x = 2$]
3. $\begin{vmatrix} x & 1 \\ 1 & x \end{vmatrix} = 3$

4.6 LINEAR INEQUALITIES IN TWO VARIABLES

STUDENT GUIDEPOSTS

1. Linear Inequalities in Two Variables
2. Graphing Linear Inequalities
3. Systems of Linear Inequalities
4. Graphing Systems of Linear Inequalities

In Chapter 2 we solved inequalities in one variable, such as

$$2x + 1 < 3 \quad \text{and} \quad x - 2 \geq 2(x - 4),$$

and graphed their solutions on a number line. For example, if we solve

$$2x + 1 < 3$$

the solution $x < 1$, or $(-\infty, 1)$, is graphed in Figure 4.5.

Figure 4.5 $x < 1$ or $(-\infty, 1)$

1 LINEAR INEQUALITIES IN TWO VARIABLES

We now consider inequalities in two variables in which the variables are raised only to the first power. Such inequalities, called **linear inequalities in two variables,** can always be written in one of the forms

$$\mathbf{ax + by > c,} \quad \mathbf{ax + by < c,} \quad \mathbf{ax + by \geq c,} \quad \text{or} \quad \mathbf{ax + by \leq c,}$$

where a, b, and c are real number constants, a and b not both zero. For example,

$$2x + 3y > 6$$

is a linear inequality in two variables x and y. As with linear equations, a solution to such an inequality is an ordered pair (x, y) which makes the inequality true. Thus $(-1, 4)$ is a solution to $2x + 3y > 6$ since

$$2(-1) + 3(4) > 6$$
$$10 > 6 \quad \text{is true.}$$

However, $(1, -4)$ is not a solution since

$$2(1) + 3(-4) > 6$$
$$-10 > 6 \quad \text{is false.}$$

Remember that the graph of a linear equation such as $2x + 3y = 6$ is a line in the plane. Notice in Figure 4.6 that the graph of $2x + 3y = 6$ divides the plane into a region above the line, the line itself, and a region below the line.

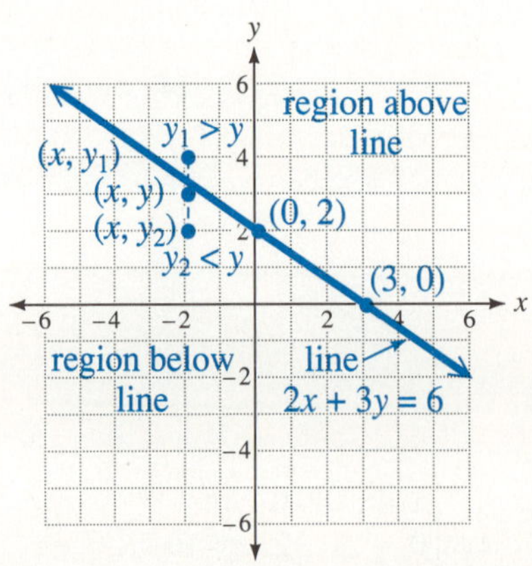

Figure 4.6 Regions Formed by a Line

2 GRAPHING LINEAR INEQUALITIES

The graph of an inequality such as $2x + 3y > 6$ consists of all the points on one side of the boundary line $2x + 3y = 6$, while the graph of $2x + 3y < 6$ is all the points on the other side. To decide which side of the line to graph for an inequality, consider the point (x, y) on the line and the points (x, y_1) and (x, y_2) above and below the line in Figure 4.6. Since $y_1 > y$, the point (x, y_1) would satisfy $2x + 3y > 6$. Similarly, since $y_2 < y$, the point (x, y_2) would satisfy $2x + 3y < 6$.

Suppose we graph $2x + 3y > 6$. We can select any point *not* on the line $2x + 3y = 6$ and use it as a **test point.** Since (0, 0) is not on the line, we choose it for the test point to make the arithmetic easy.

$$\begin{aligned} 2x + 3y &> 6 \\ 2(0) + 3(0) &> 6 \\ 0 &> 6 \quad \text{This is false} \end{aligned}$$

Since this is a false inequality we graph the points on the side of the line not containing (0, 0). If the point (0, 0) is used as a test point for the inequality $2x + 3y < 6$, a true inequality is obtained.

$$\begin{aligned} 2x + 3y &< 6 \\ 2(0) + 3(0) &< 6 \\ 0 &< 6 \quad \text{This is true} \end{aligned}$$

To graph $2x + 3y < 6$ we shade the region containing the test point (0, 0). The graphs of both $2x + 3y > 6$ and $2x + 3y < 6$ are shown in Figure 4.7 with the appropriate region for each inequality shaded.

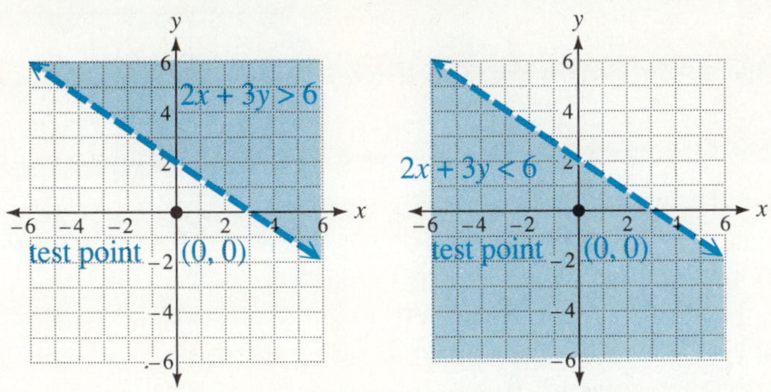

Figure 4.7 Graphing Linear Inequalities

Notice that the lines in Figure 4.7 are dashed. This is done to show that the points on the line are *not* part of the graph. To graph $2x + 3y \geq 6$ and $2x + 3y \leq 6$ we would make the line solid to indicate that points on the line *are* part of the graph.

To Graph a Linear Inequality in Two Variables

1. Graph the boundary line, using a dashed line if the inequality is $<$ or $>$ and a solid line if it is $\leq$ or $\geq$.
2. Choose a test point that is not on the boundary line and substitute it into the inequality.
3. Shade the region that includes the test point if a true inequality is obtained, and shade the region that does not contain the test point if a false inequality results.

EXAMPLE 1 Graphing a Linear Inequality

(a) Graph $2x - 5y \le 10$.

First graph the line $2x - 5y = 10$ using a solid line because the inequality $\le$ includes the points on the line. To graph the line use the intercepts $(0, -2)$ and $(5, 0)$. Now select a test point which is not on the line. The point $(0, 0)$ is again easy to use.

$$2x - 5y \le 10$$
$$2(0) - 5(0) \le 10$$
$$0 \le 10 \quad \text{This is true}$$

Since $0 \le 10$ is true we shade the side of the graph that *contains* the test point $(0, 0)$. See Figure 4.8.

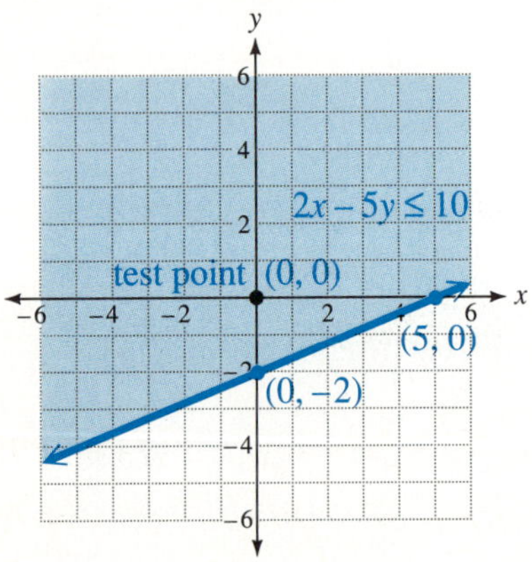

Figure 4.8 Graph Using $\le$

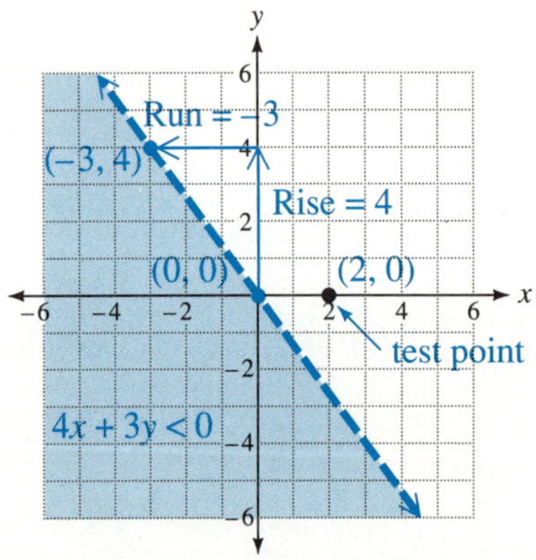

Figure 4.9 Graph Using $<$

(b) Graph $4x + 3y < 0$.
We graph $4x + 3y = 0$ with a dashed line since the inequality is $<$. This time we graph the line using the slope and y-intercept. Since $3y = -4x$, we have $y = -\frac{4}{3}x + 0$ as the slope-intercept form. Starting at the y-intercept (also the x-intercept) $(0, 0)$, we move up 4 units (a rise of 4) and left 3 units (a run of -3) to obtain a second point $(-3, 4)$ on the line. Since $(0, 0)$ is on the line we choose a different test point such as $(2, 0)$.

$$4x + 3y < 0$$
$$4(2) + 3(0) < 0$$
$$8 < 0 \quad \text{This is false}$$

Since $8 < 0$ is false, we shade the region in Figure 4.9 that does not contain the test point $(2, 0)$.

Practice Exercise 1

(a) Graph $3x + y \ge 6$.

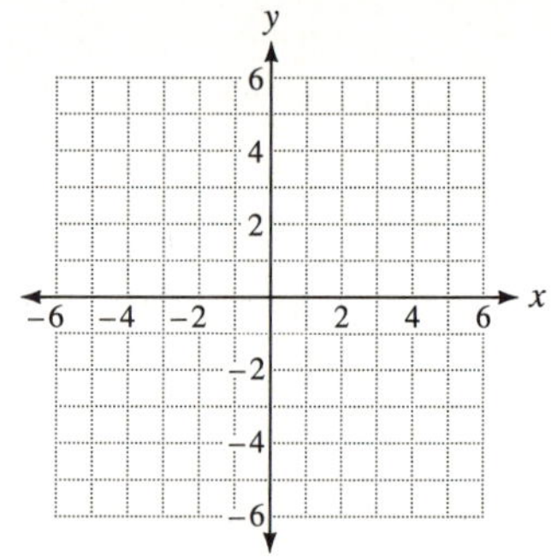

(b) Graph $2x - 5y > 0$.

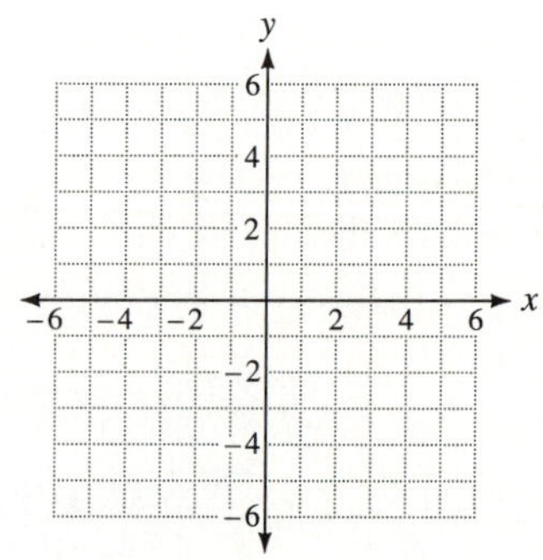

Answers: (a) The graph is the region above and including the line with intercepts $(0, 6)$ and $(2, 0)$. (b) The graph is the region below (not including) the line with intercepts $(0, 0)$ and passing through the point $(5, 2)$.

An inequality in two variables such as $2x - 6 > 0$ has the y-term missing (the coefficient of y is zero). To graph this type of inequality the procedure is the same, but the work can be simplified if we first solve for x, $x > 3$.

We now recognize the graph of the boundary line $x = 3$ as a vertical line through the point $(3, 0)$. Since the inequality is $>$, we dash the line. Next, the test point $(0, 0)$ gives a false inequality.

$$x > 3$$
$$0 > 3 \quad \text{This is false}$$

Shade the region that does not contain (0, 0) as in Figure 4.10.

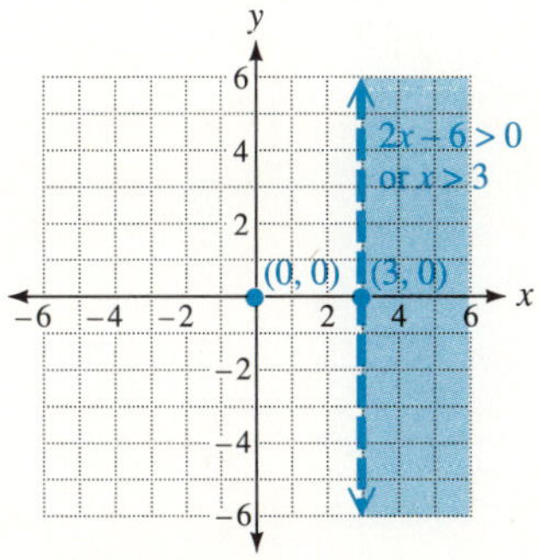

Figure 4.10 Regions Determined by Vertical Line

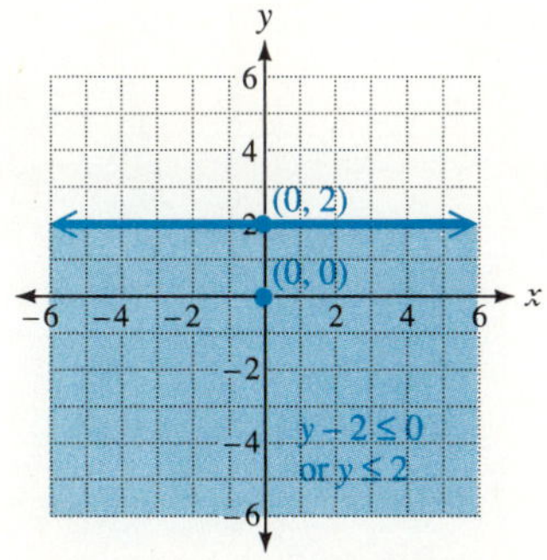

Figure 4.11 Regions Determined by Horizontal Line

Suppose we graph $y - 2 \leq 0$. This inequality can be written $y \leq 2$; the line to plot is the horizontal line $y = 2$. Use the test point (0, 0) to obtain the true inequality $0 \leq 2$ and shade the region that contains the test point. See Figure 4.11.

3 SYSTEMS OF LINEAR INEQUALITIES

When we graph a linear inequality in two variables, the points on the graph correspond to ordered pairs of numbers that are solutions to the inequality. A **system of linear inequalities in two variables** is simply two or more linear inequalities in two variables. A solution to such a system is an ordered pair of numbers that solves all inequalities in the system.

4 GRAPHING SYSTEMS OF LINEAR INEQUALITIES

Graphing a system of inequalities is outlined in the following.

To Graph a System of Linear Inequalities in Two Variables

1. Draw the graph of each inequality in the system.
2. The graph consists of those points common to the graphs of the individual inequalities.

EXAMPLE 2 Graphing a System of Inequalities

(a) Graph the systems of inequalities.

$$3x + 4y < 12$$
$$y - x \geq 1$$

Graph $3x + 4y = 12$ with a dashed line (the inequality is $<$) and $y - x = 1$ with a solid line (the inequality is $\geq$). The test point (0, 0) can be used for each inequality.

$3x + 4y < 12$		$y - x \geq 1$	
$3(0) + 4(0) < 12$		$0 - 0 \geq 1$	
$0 < 12$	True	$0 \geq 1$	False

Thus, the graph of the system includes all points both below the line $3x + 4y = 12$ *and* above and on the line $y - x = 1$. The graph is shaded in Figure 4.12 on the following page.

Practice Exercise 2

(a) Graph the systems of inequalities.

$$3x + 4y > 12$$
$$y - x \leq 1$$

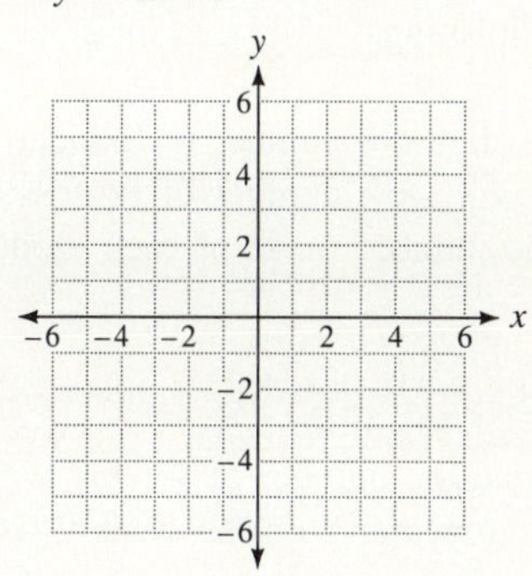

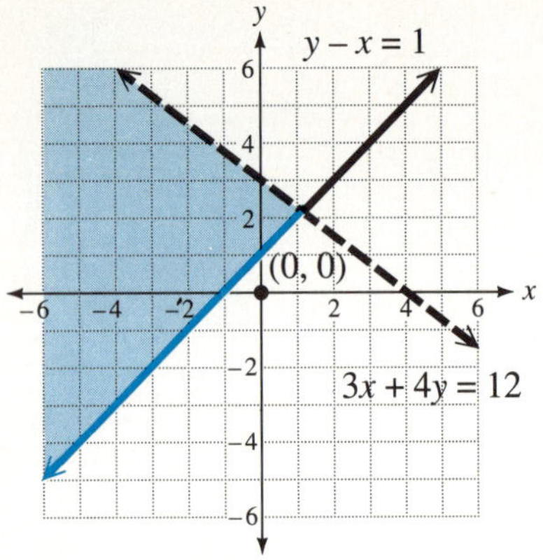

Figure 4.12

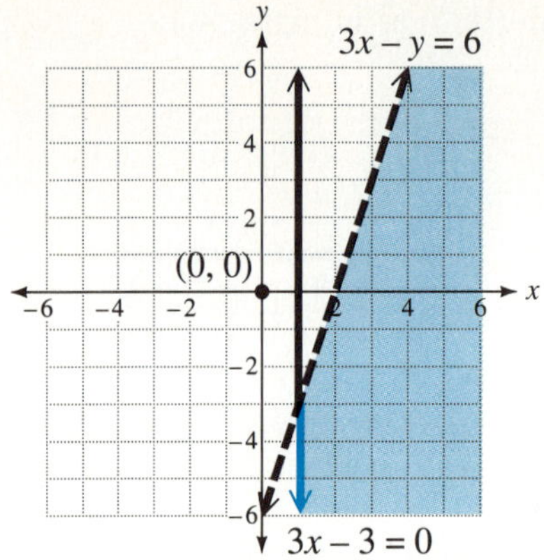

Figure 4.13

(b) $3x - y < 6$
$3x - 3 \le 0$

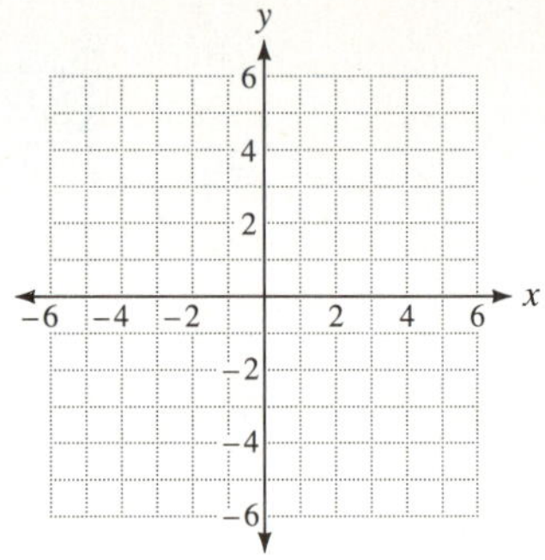

(b) $3x - y > 6$
$3x - 3 \ge 0$

Use the test point (0, 0) after the boundary lines $3x - y = 6$ and $3x - 3 = 0$ are graphed.

$3x - y > 6$		$3x - 3 \ge 0$	
$3(0) - 0 > 6$		$3(0) - 3 \ge 0$	
$0 > 6$	False	$-3 \ge 0$	False

Thus, the graph consists of all points both below the line $3x - y = 6$ *and* to the right of and on the line $3x - 3 = 0$ ($x = 1$). The graph is shaded in Figure 4.13.

Answers: (a) Your graph should be the region above the line with intercepts (0, 3) and (4, 0) but on and below the line with intercepts (0, 1) and (−1, 0). (b) Your graph should be the region above the line with intercepts (0, −6) and (2, 0) but on and to the left of the vertical line $x = 1$.

Many applied problems, especially in business, can be described and solved using a system of inequalities.

EXAMPLE 3 Production Problem

Bikes Inc. manufactures two types of racing bicycles, the Standard and the Professional models. The table below summarizes the weekly production data. Determine the various combinations of bicycles that can be made each week under the limitations on the workhours available in each stage of production.

	Workhours for Standard model	*Workhours for Professional model*	*Maximum workhours per week*
Construction	8	12	216
Finishing	2	2	48

Let x = number of Standard models made per week,
y = number of Professional models made per week.

The number made of each model must not be negative, so

$$x \ge 0$$
$$y \ge 0.$$

Since the number of weekly workhours for constructing Standard models, $8x$, together with the weekly workhours for constructing Professional

Practice Exercise 3

Graph the system.

$$x \ge 0$$
$$y \ge 0$$
$$x + y \le 4$$
$$x + 2y \le 6$$

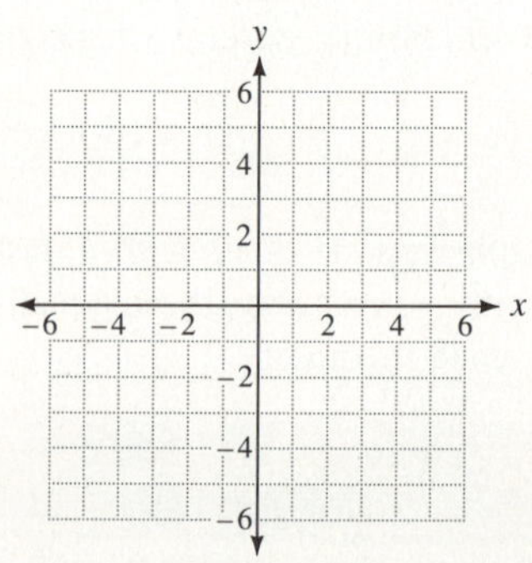

models, $12y$, cannot exceed 216, a third inequality is

$$8x + 12y \leq 216.$$

Similarly, the weekly workhours allocated to finishing define a fourth inequality

$$2x + 2y \leq 48.$$

Any solution to the system

$$\begin{aligned} x &\geq 0 \\ y &\geq 0 \\ 8x + 12y &\leq 216 \\ 2x + 2y &\leq 48 \end{aligned}$$

provides a possible weekly production schedule. The graph of these solutions is shaded in Figure 4.14. Notice that the inequalities $x \geq 0$ and $y \geq 0$ restrict the graph to points in the first quadrant while the other two inequalities restrict the graph to points *below* both boundary lines. For example, one possible production schedule would be 18 Standard models and 6 Professional models, corresponding to the point of intersection of the two lines (18, 6). A point outside the graph, such as (15, 12), corresponding to 15 Standard and 12 Professional models, is impossible.

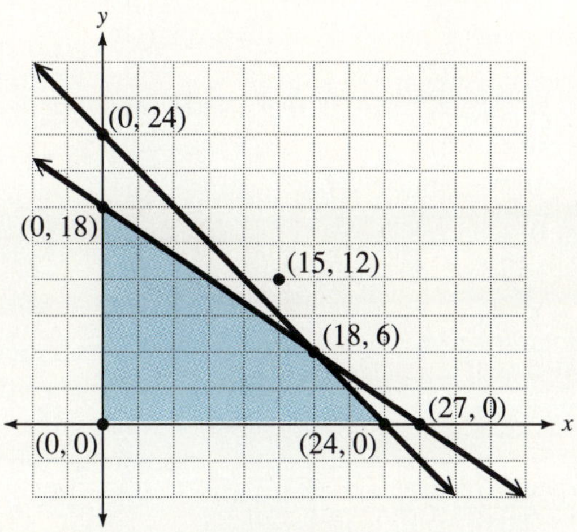

Figure 4.14

Answer: The graph consists of all points in quadrant I below the lines $x + y = 4$ and $x + 2y = 6$ together with points on these boundary lines and the positive axes.

4.6 EXERCISES A

Graph the linear inequalities in two variables.

1. $x + y > 2$

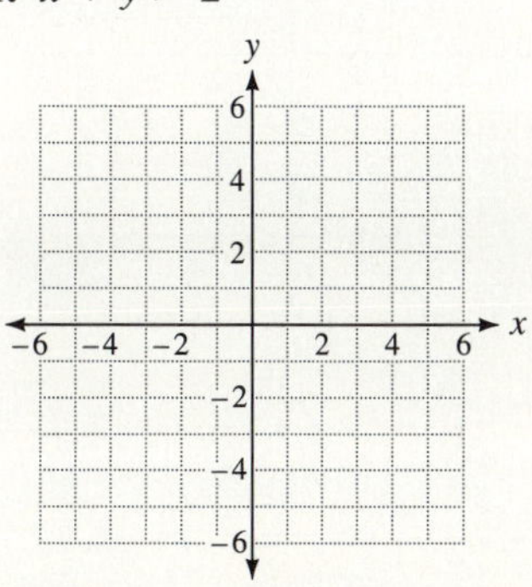

2. $x + y \leq 2$

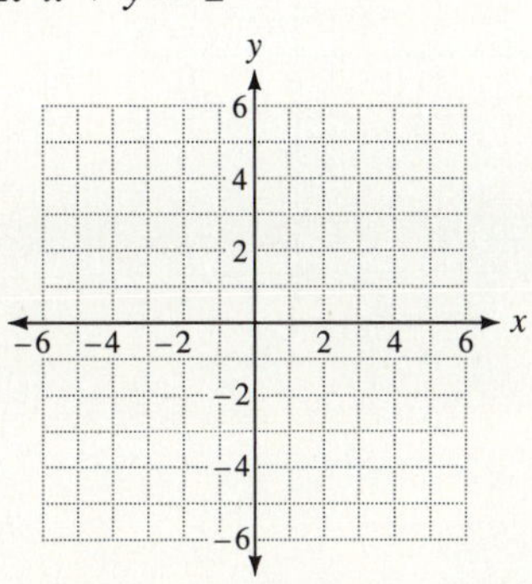

3. $x - y \geq -1$

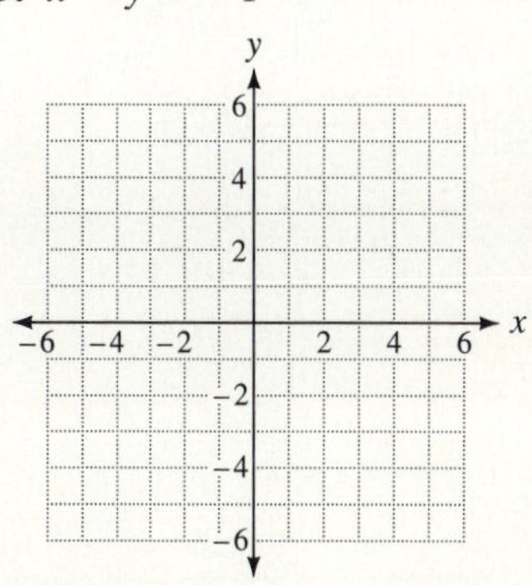

4. $x - y < -1$

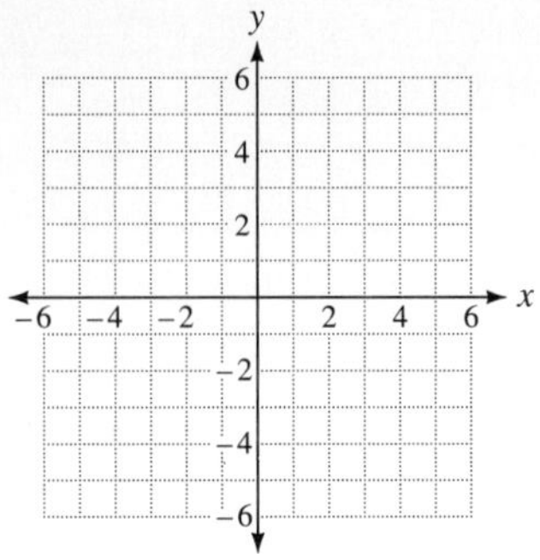

5. $3x + 4y < 12$

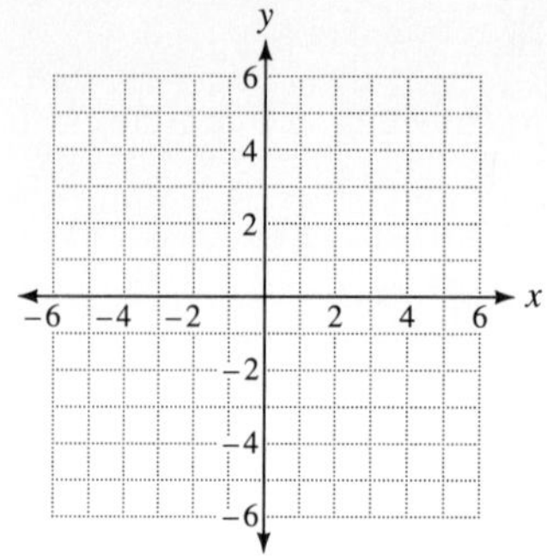

6. $3x + 4y \geq 12$

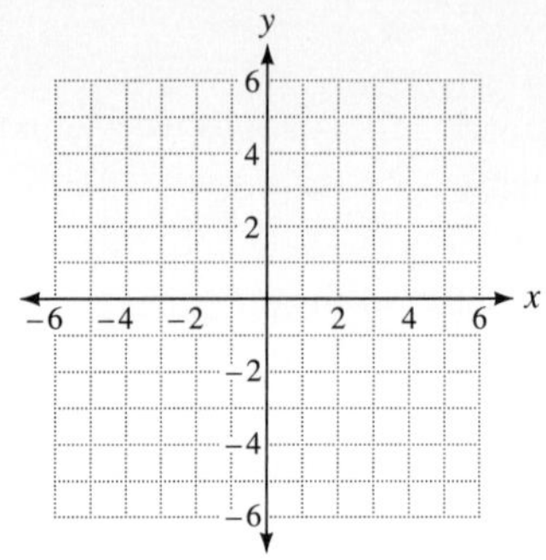

7. $4x + 8 > 0$

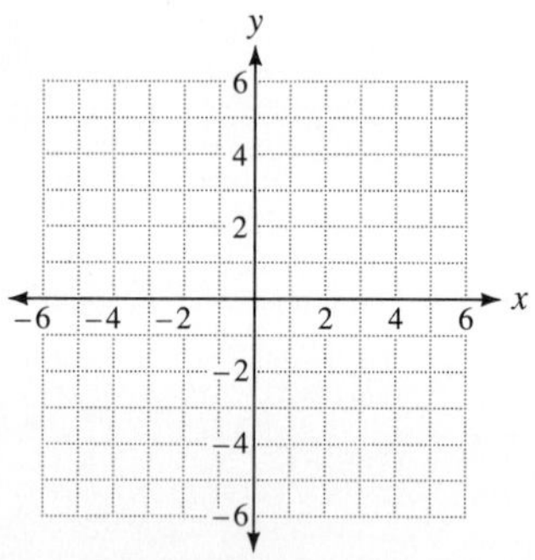

8. $4x + 8 \leq 0$

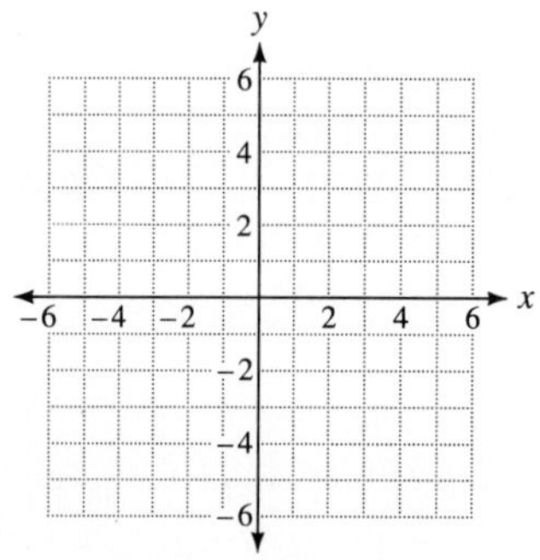

9. $2y - 6 \leq 0$

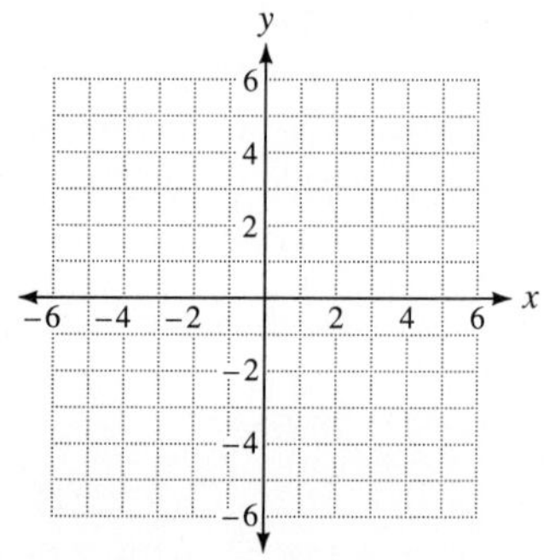

10. $2y - 6 > 0$

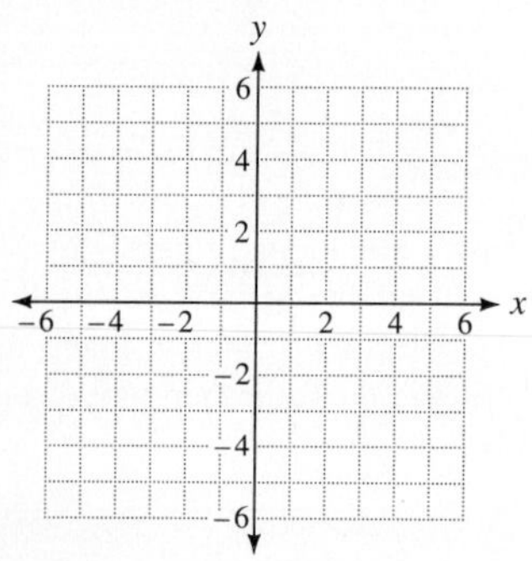

11. $2x + y < 0$

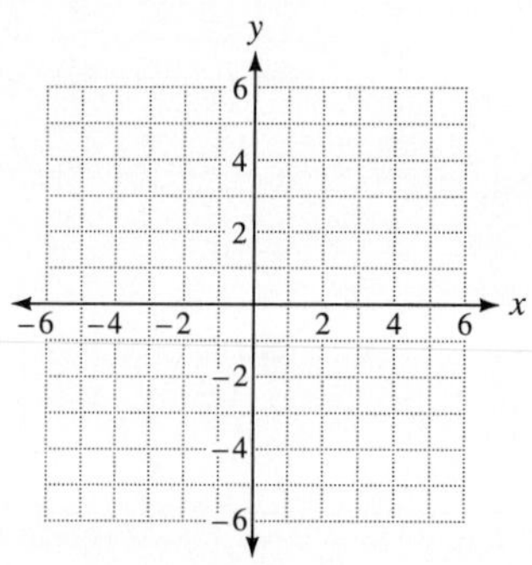

12. $2x + y \geq 0$

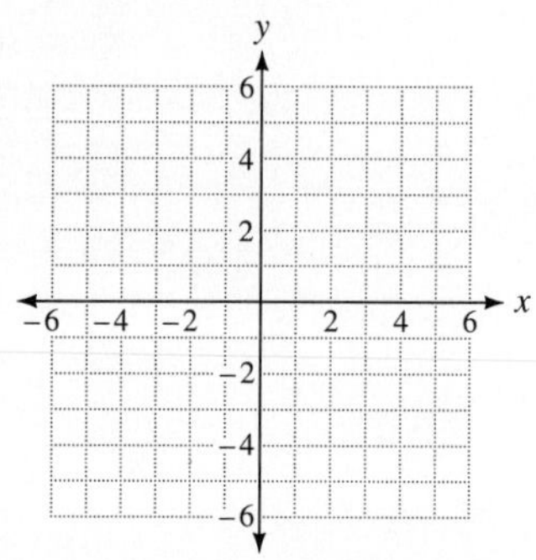

13. $3x + 2 > 5x$

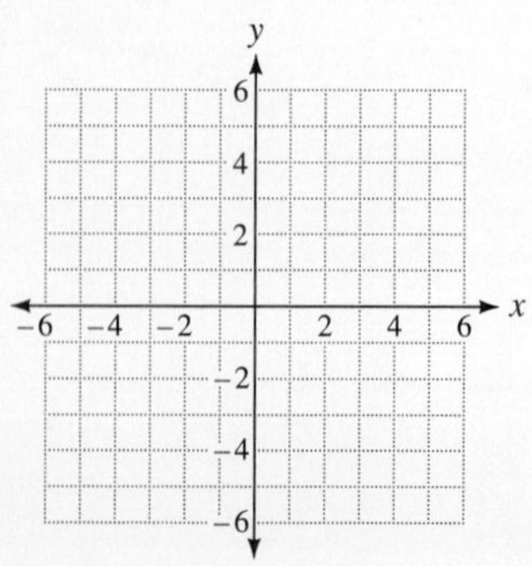

14. $3(x + y) \geq 2(x - y)$

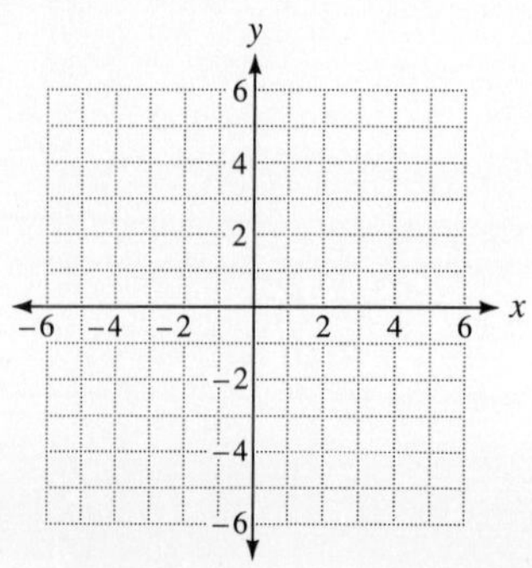

15. $5x + 6 < -3y - 9$

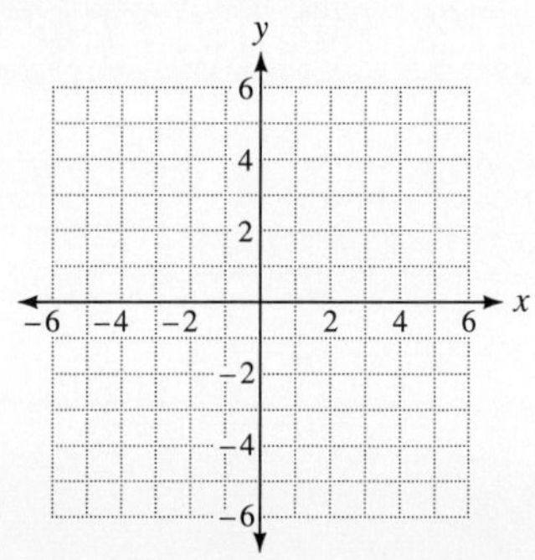

Graph each system of inequalities.

16. $3x - 2y \geq 6$
$x + 2y > -2$

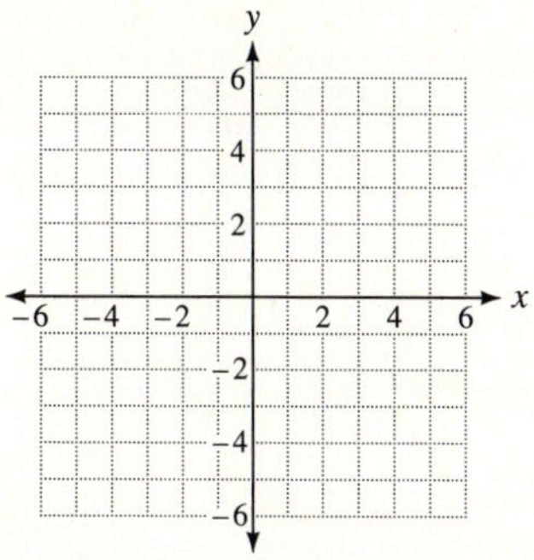

17. $3x - 2y < 6$
$x + 2y \geq -2$

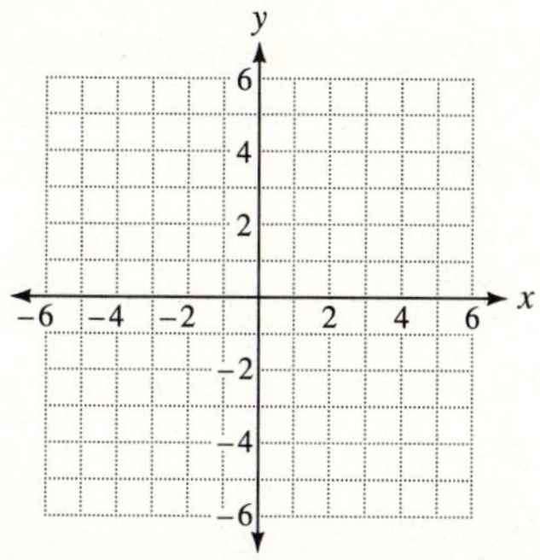

18. $3x - 2y \leq 6$
$x + 2y < -2$

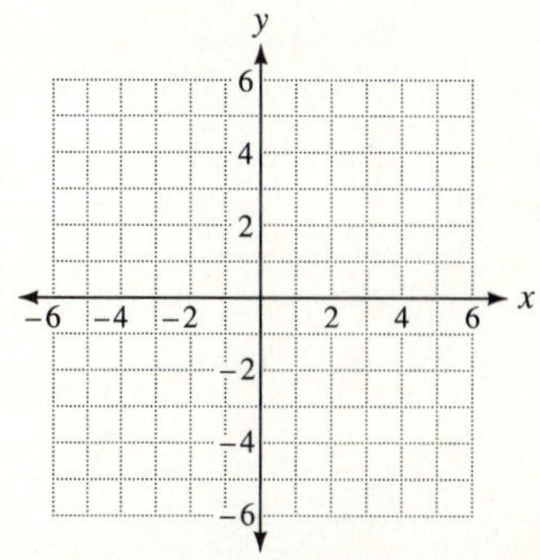

19. $x - y < 1$
$2x - 2y \geq -4$

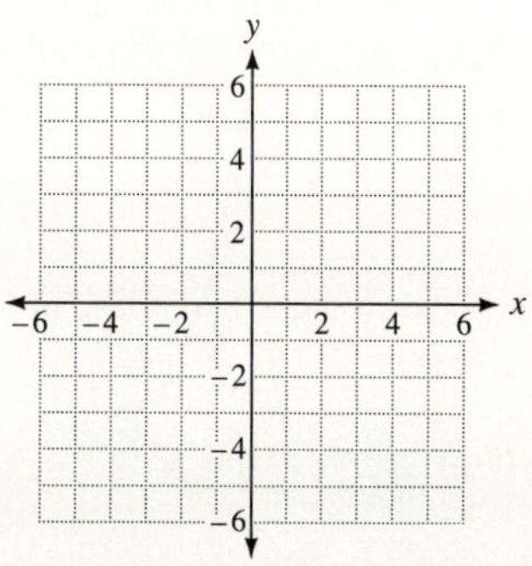

20. $x + 1 \leq 0$
$3y + 6 > 0$

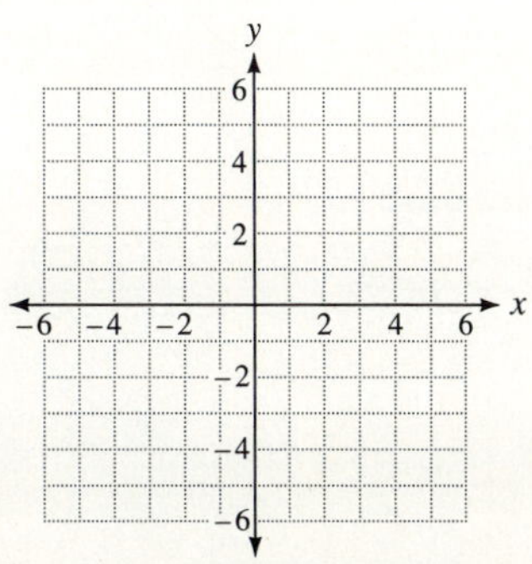

21. $3y - x \leq 3$
$x + 2 > 0$

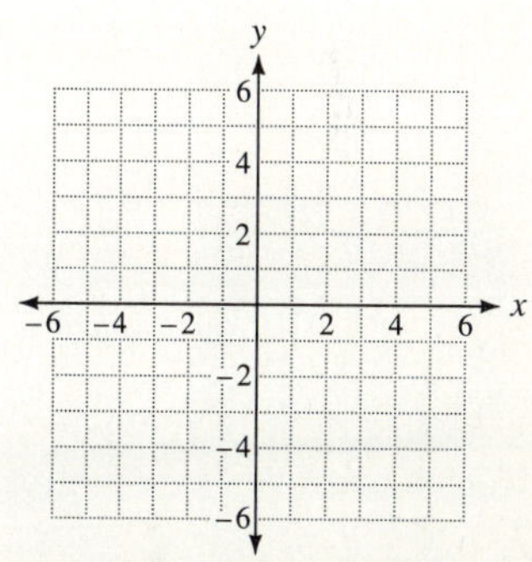

22. $y \geq x$
$y \geq -x$

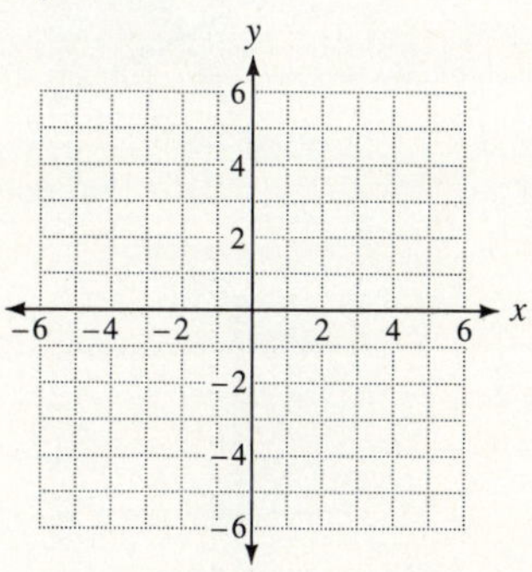

23. $x \geq 0$
$y \geq 0$

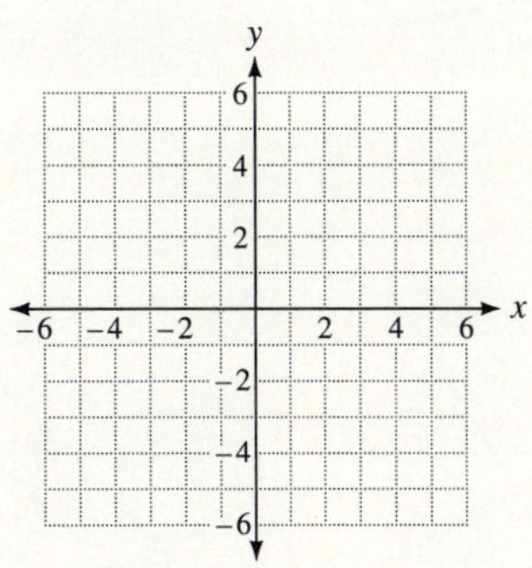

24. $x \geq 0$
$y \geq 0$
$x + y < 1$

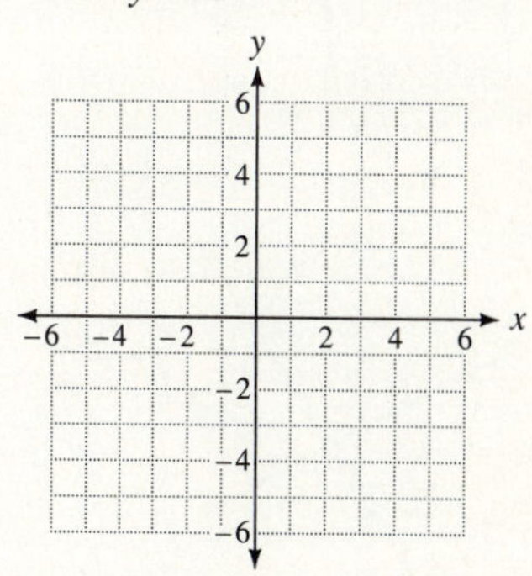

FOR REVIEW

Evaluate the following determinants.

25. $\begin{vmatrix} -3 & 4 \\ 2 & -8 \end{vmatrix}$

26. $\begin{vmatrix} 0 & 0 & 7 \\ 0 & 3 & 0 \\ -1 & 0 & 0 \end{vmatrix}$

Solve using Cramer's rule.

27. $\begin{aligned} 4x + y &= -3 \\ -3x - 2y &= -4 \end{aligned}$

28. $\begin{aligned} x + y - z &= -6 \\ 4x + 3y + 2z &= 6 \\ -2x - 5y + 2z &= 12 \end{aligned}$

ANSWERS:

1. 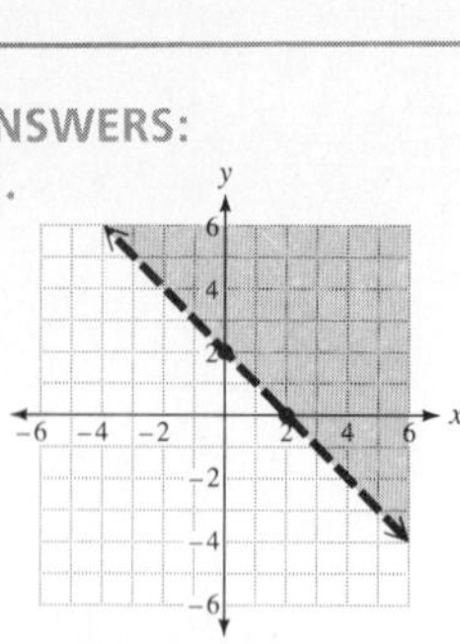
2. 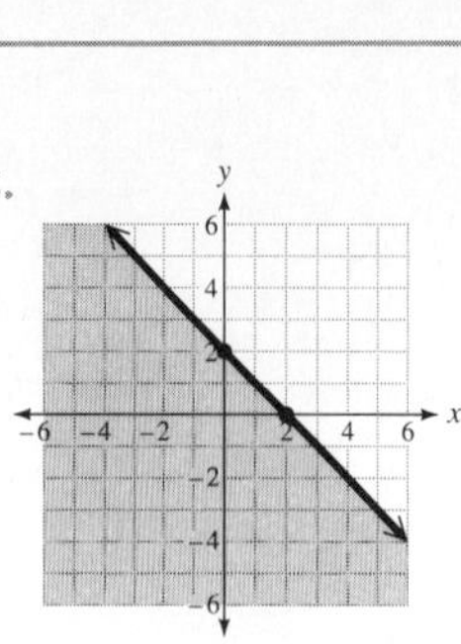
3. 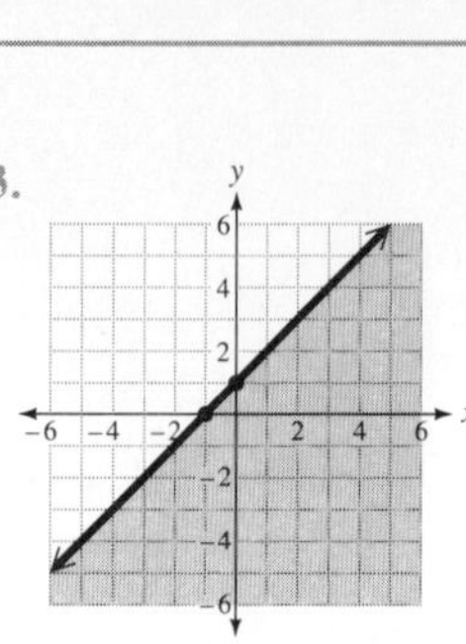
4. 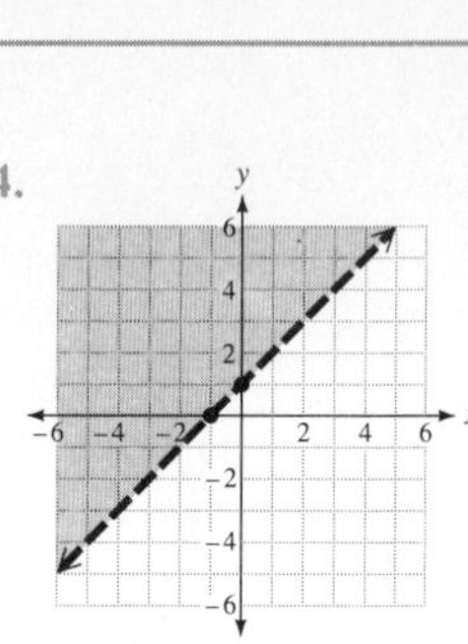
5. 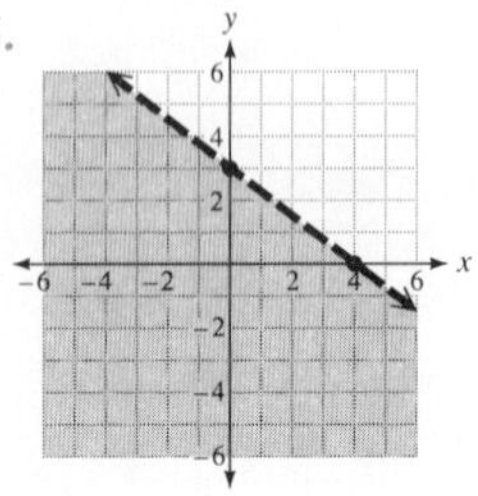
6. 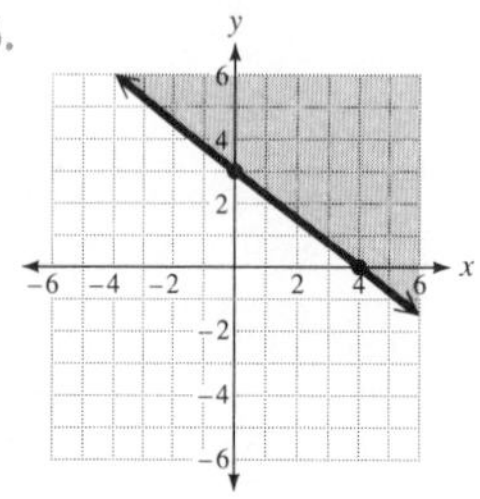
7. 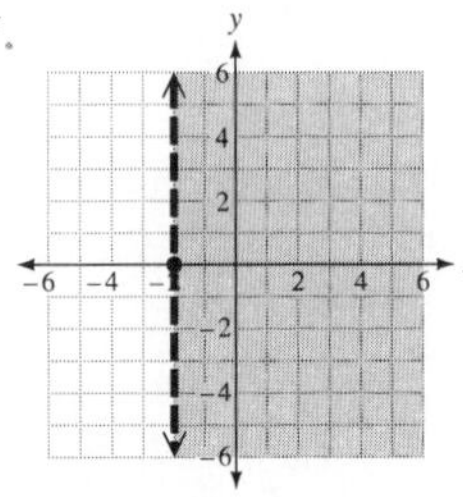
8. 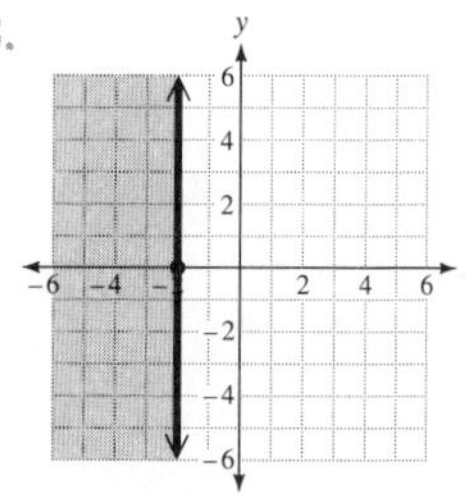
9. 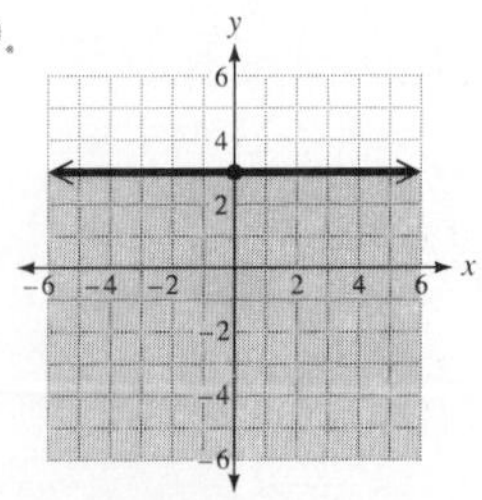
10. 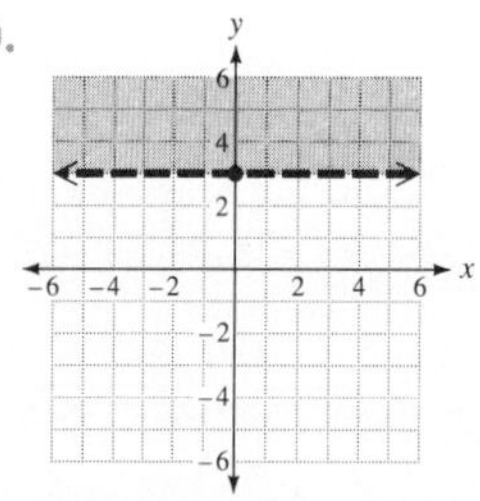
11. 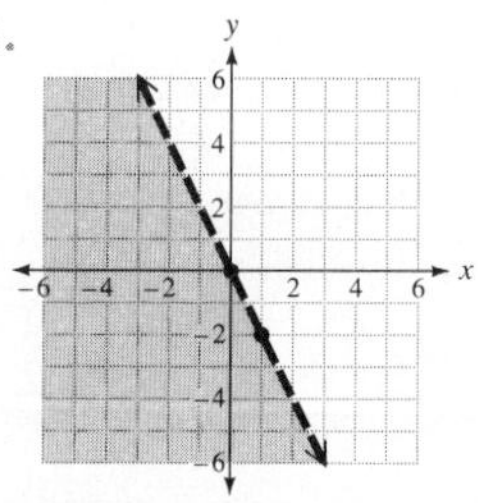
12. 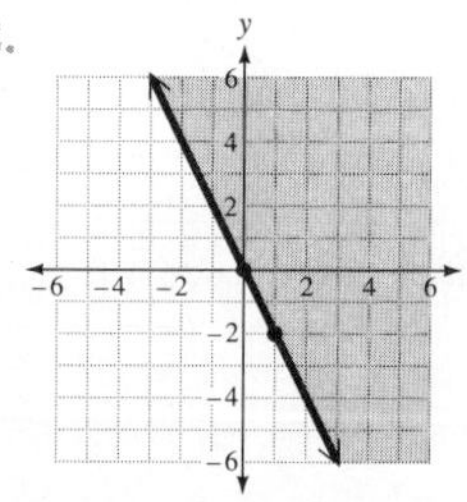
13. 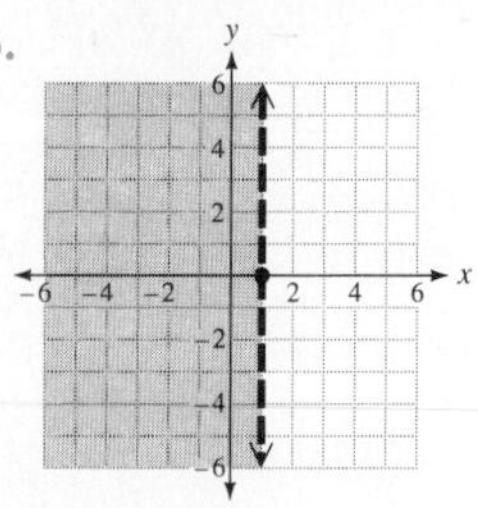
14. 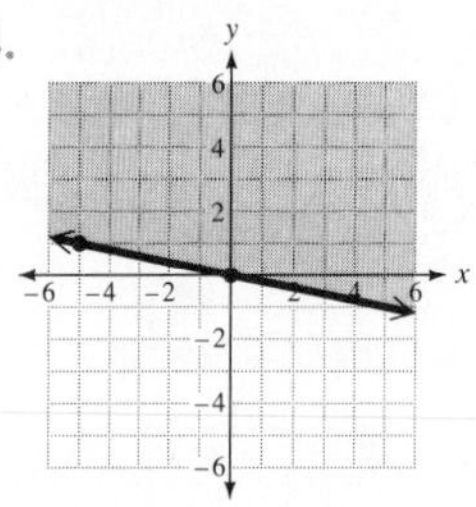
15. 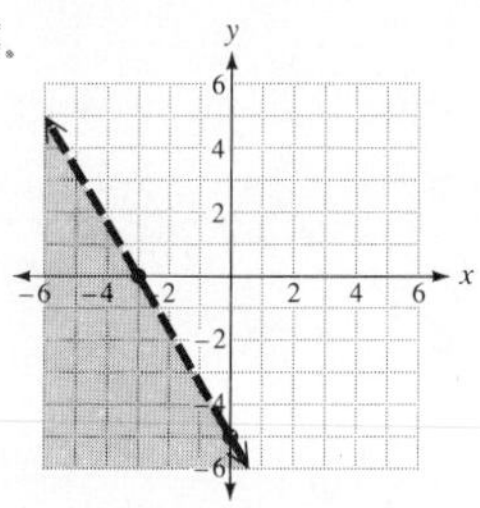
16. 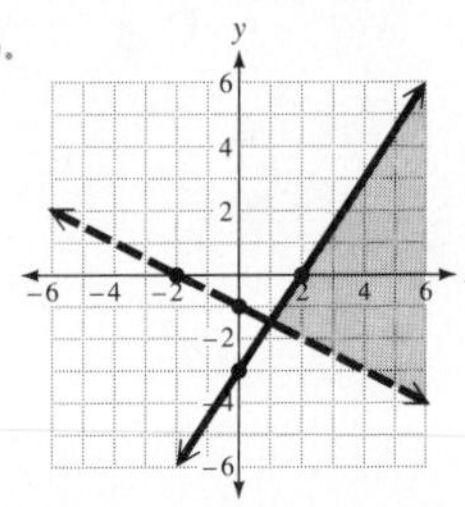
17. 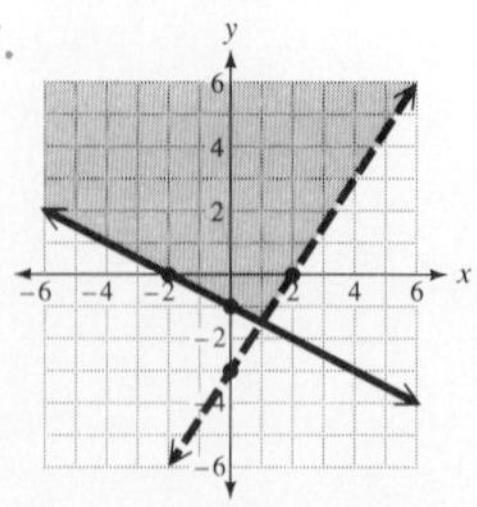
18. 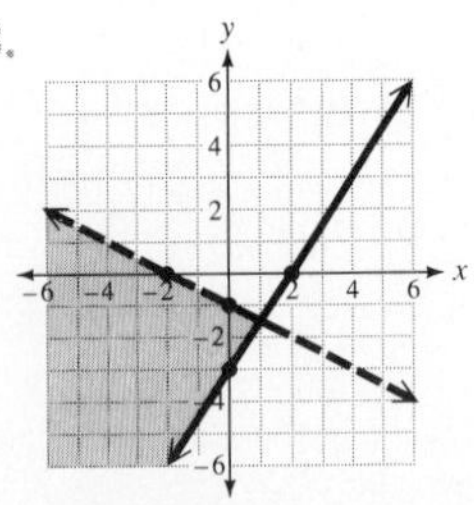
19. 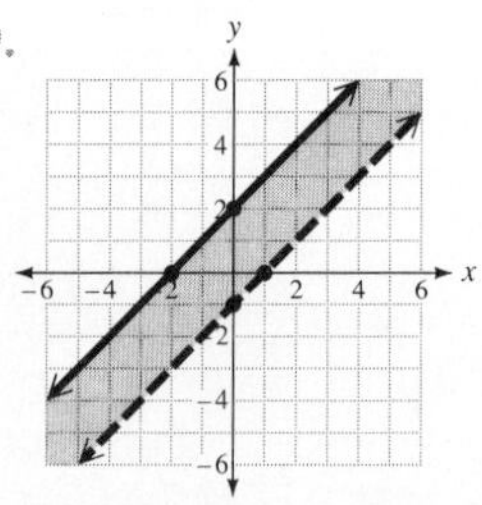
20. 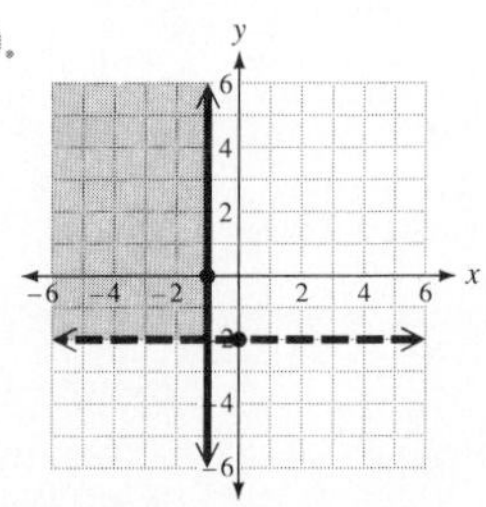
21. 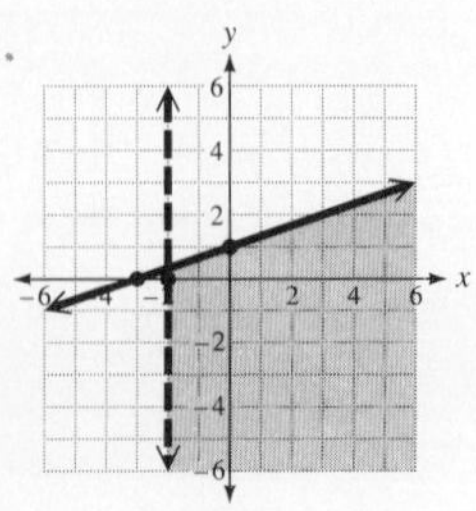
22. 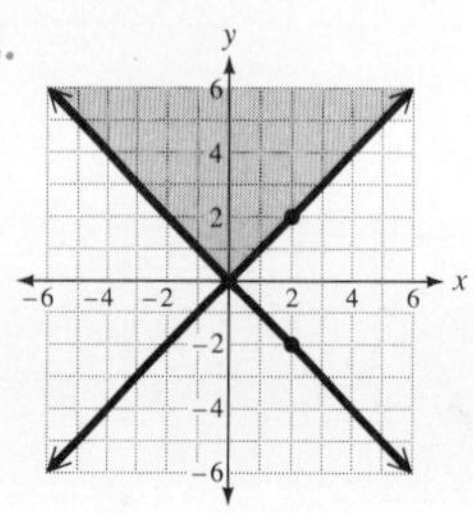
23. 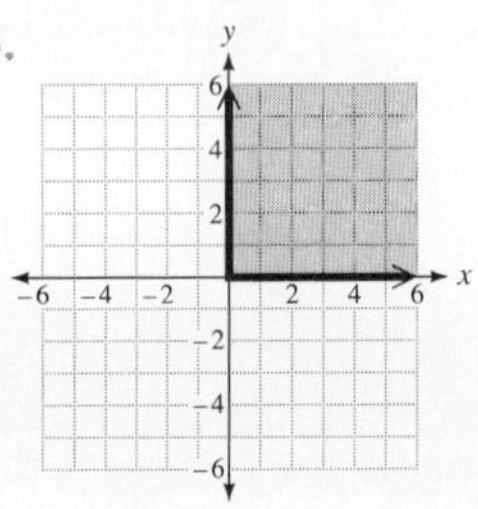
24. 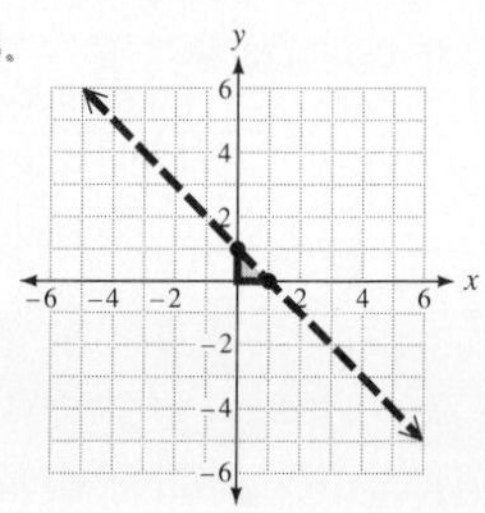

25. 16 26. 21 27. $(-2, 5)$ 28. $(-1, 0, 5)$

4.6 EXERCISES B

Graph the linear inequalities in two variables.

1. $x + y < 1$

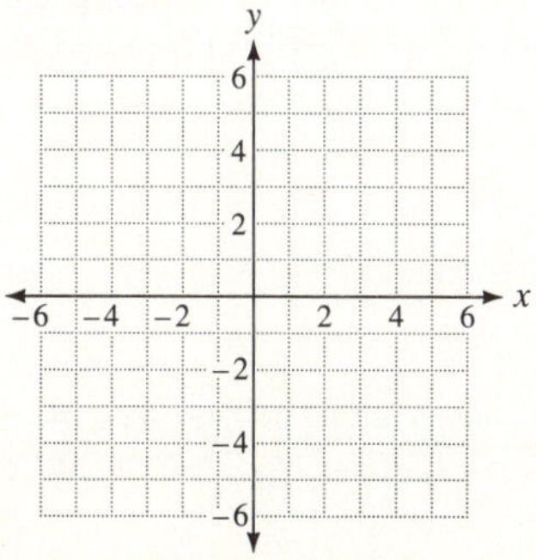

2. $x + y \geq 1$

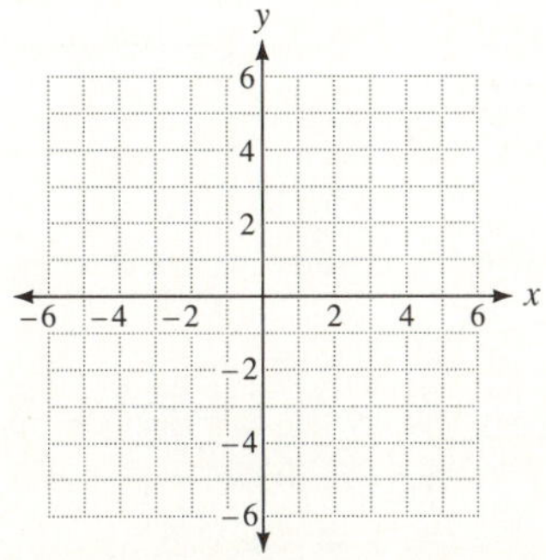

3. $x - y \geq -2$

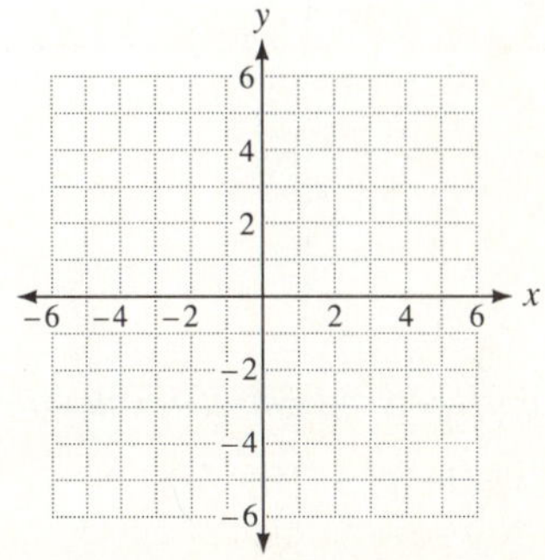

4. $x - y < -2$

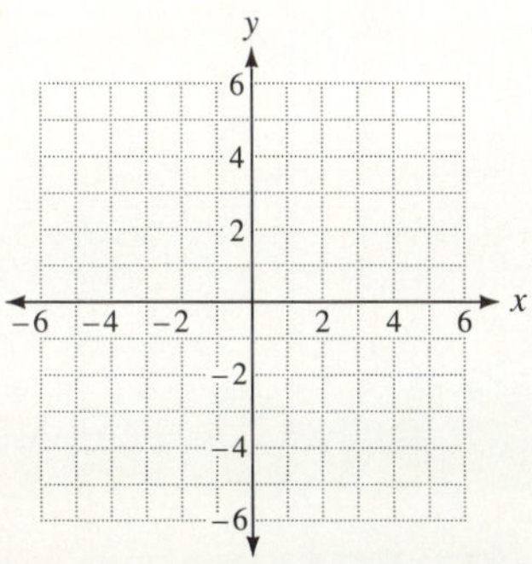

5. $4x - 3y > 12$

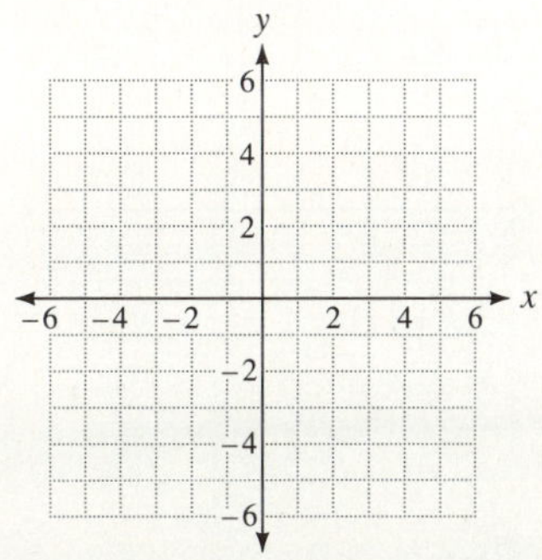

6. $4x - 3y \leq 12$

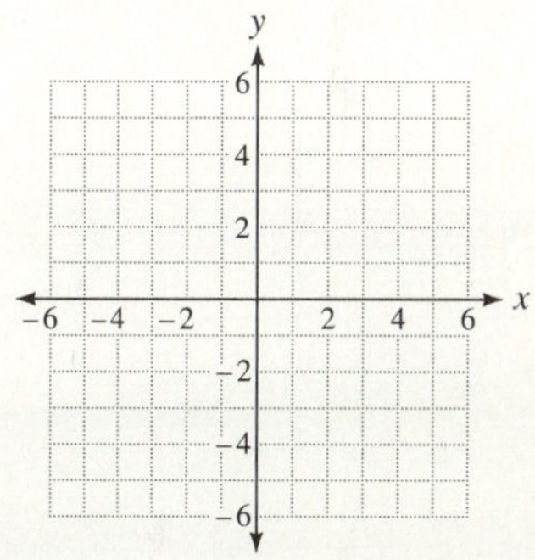

7. $3x - 9 \geq 0$

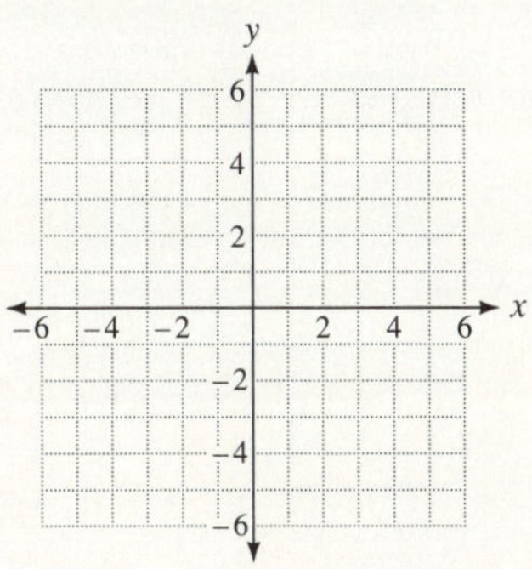

8. $3x - 9 < 0$

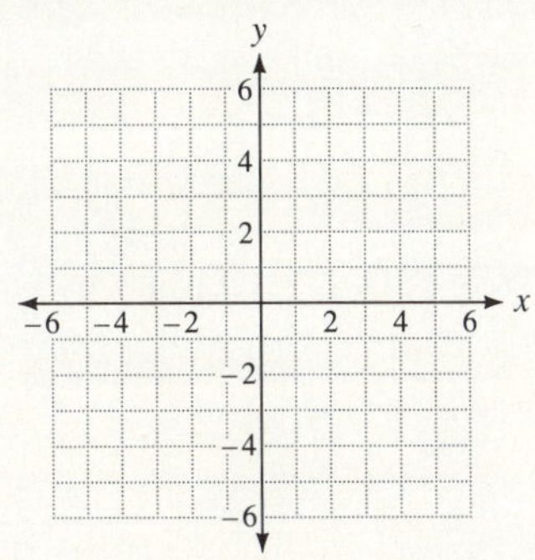

9. $2y - 8 < 0$

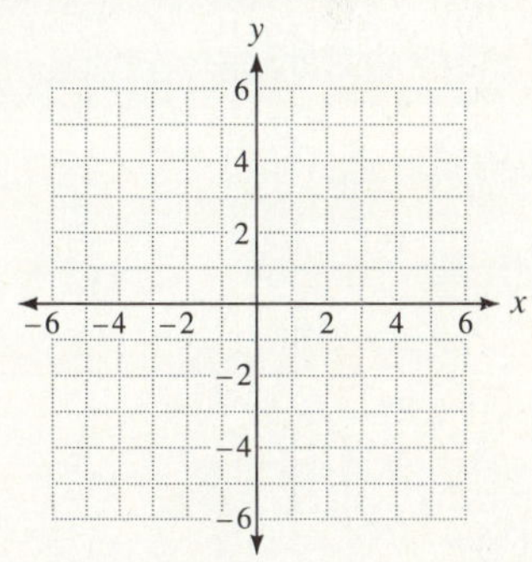

10. $2y - 8 \geq 0$

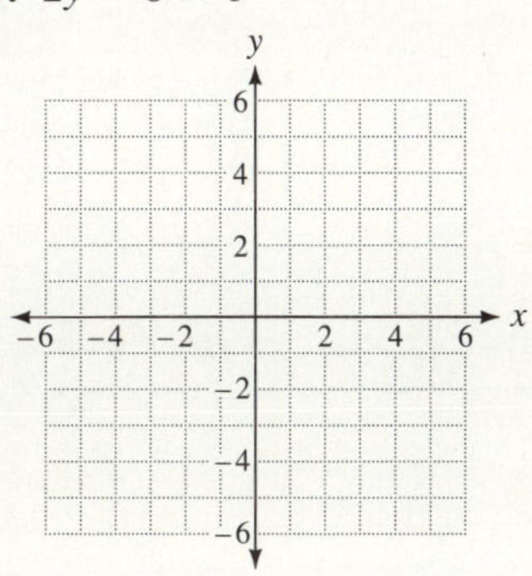

11. $4y + x \leq 0$

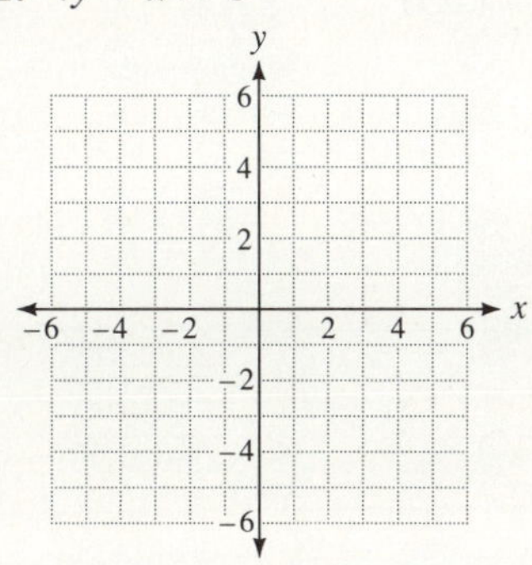

12. $4y + x > 0$

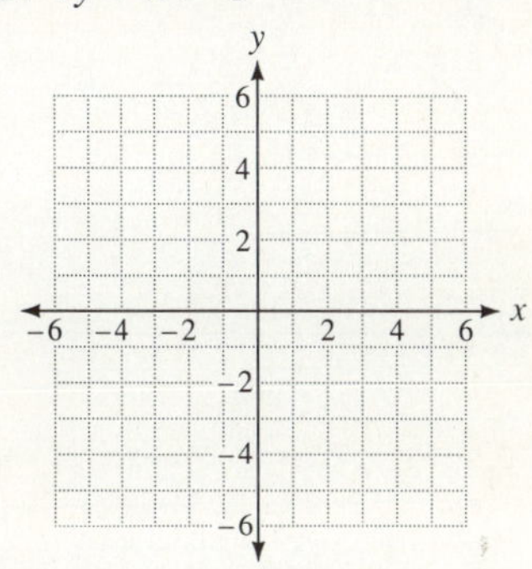

13. $3x + 2 \le 5x$

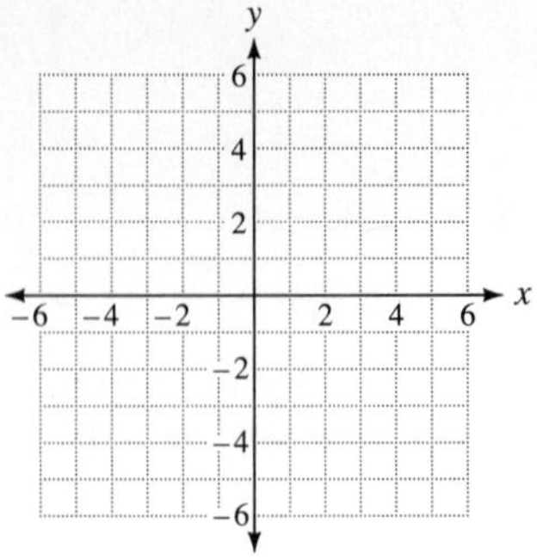

14. $3(x + y) < 2(x - y)$

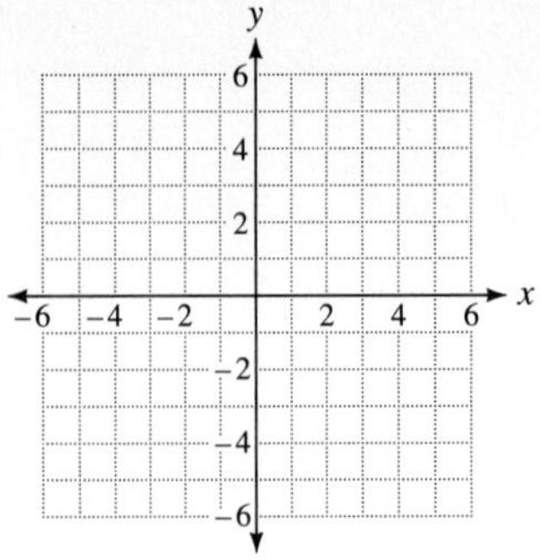

15. $5x + 6 \ge -3y - 9$

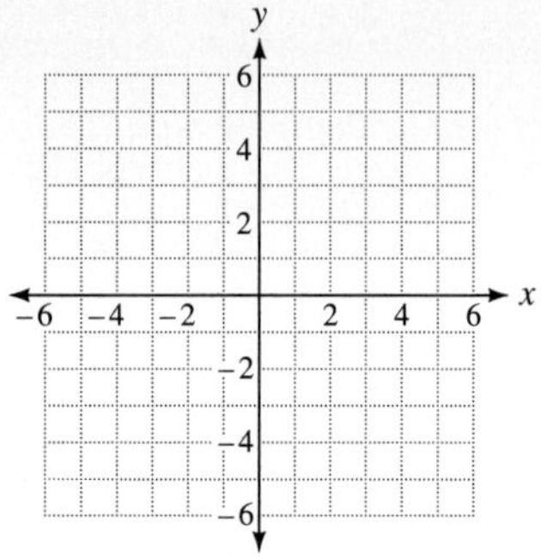

Graph each system of inequalities.

16. $2x - 5y \ge 10$
$\frac{1}{2}x + \ \ y > -1$

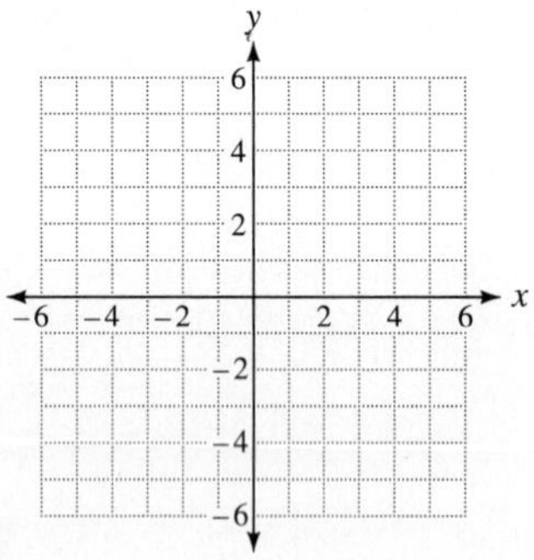

17. $2x - 5y < 10$
$\frac{1}{2}x + \ \ y \ge -1$

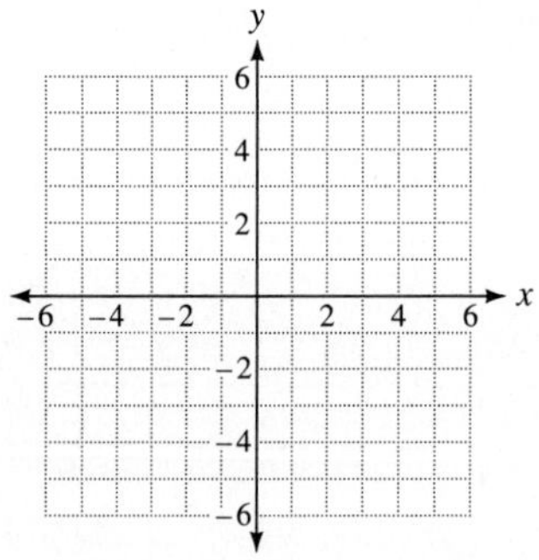

18. $2x - 5y \le 10$
$\frac{1}{2}x + \ \ y < -1$

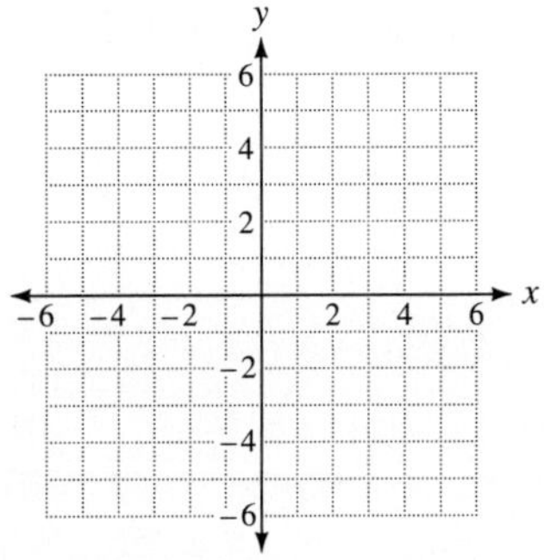

19. $x + \ \ y \ge -3$
$3x + 3y < 6$

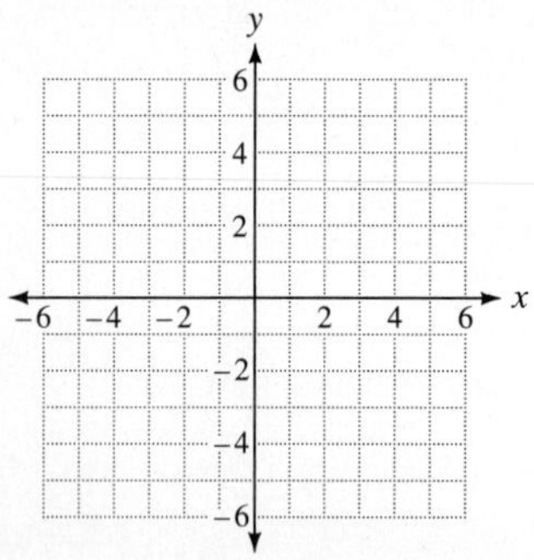

20. $2x + 4 > 0$
$y - 1 \le 0$

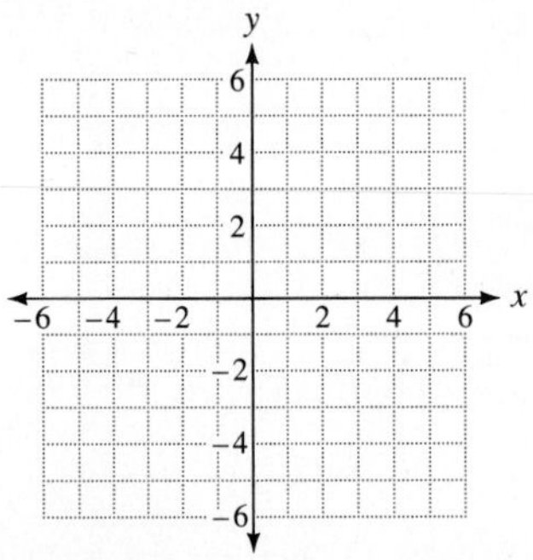

21. $4y - x \ge 4$
$y - 2 < 0$

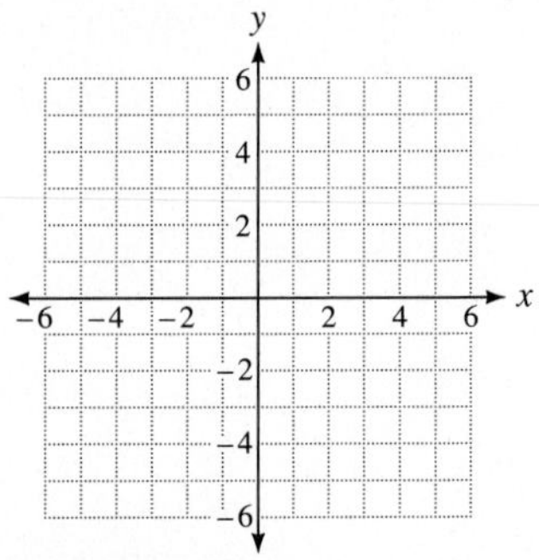

22. $2x + \ \ y \le 0$
$x - 2y \ge 0$

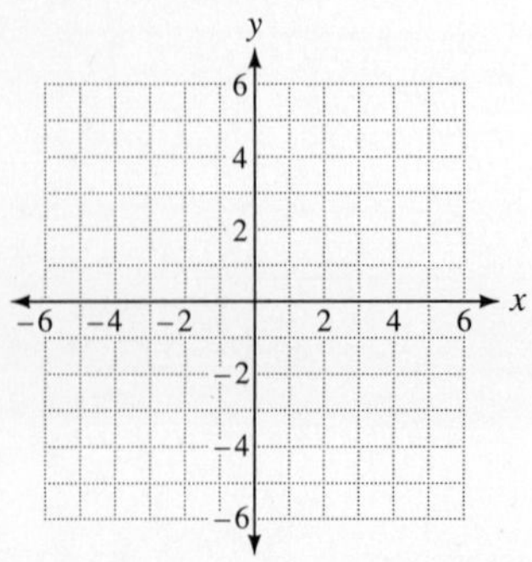

23. $x \le 0$
$y \ge 0$

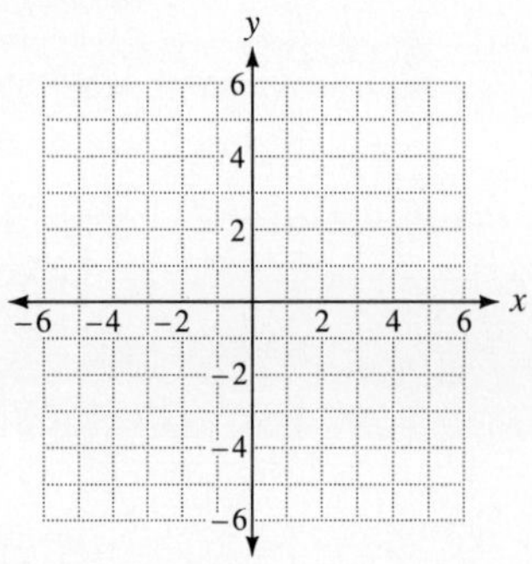

24. $x \le 0$
$y \ge 0$
$y - x \le 3$

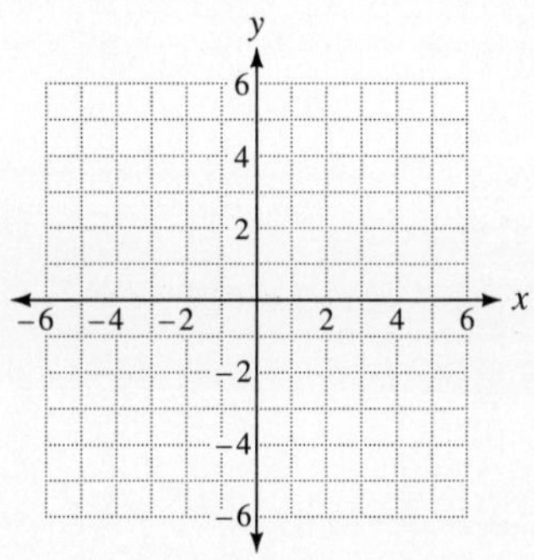

FOR REVIEW

Evaluate the following determinants.

25. $\begin{vmatrix} -4 & 5 \\ 2 & -3 \end{vmatrix}$

26. $\begin{vmatrix} 3 & 0 & 0 \\ 0 & -5 & 0 \\ 0 & 0 & 2 \end{vmatrix}$

Solve using Cramer's rule.

27. $2x - 9y = 15$
$4x + 7y = 5$

28. $2x + 2y - 3z = 1$
$5x + 5y + 4z = -9$
$3x \quad + 2z = 7$

4.6 EXERCISES C

1. Evaluate the determinant and graph the resulting inequality.

$\begin{vmatrix} x & y \\ 5 & 1 \end{vmatrix} < 5$

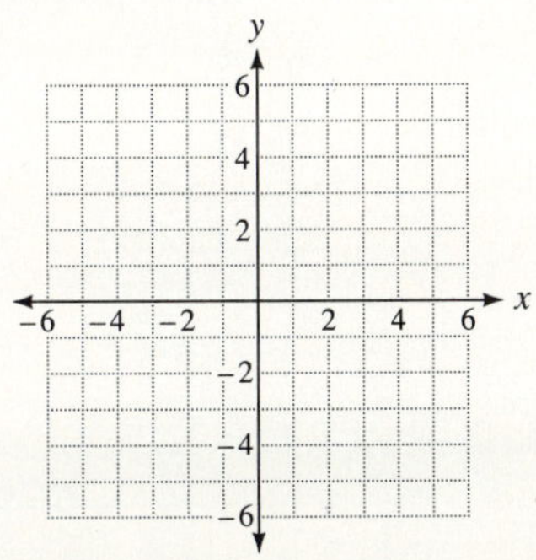

2. Graph the relation $\{(x, y) | x \geq 0, \ y \geq 0, \text{ and } 3x + 2y \leq 6\}$.

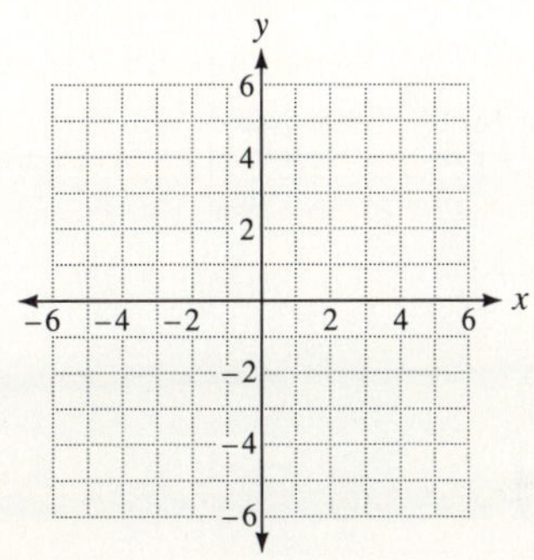

Graph each system of inequalities.

3. $x - y > 1$
$2x - 2y \leq -4$

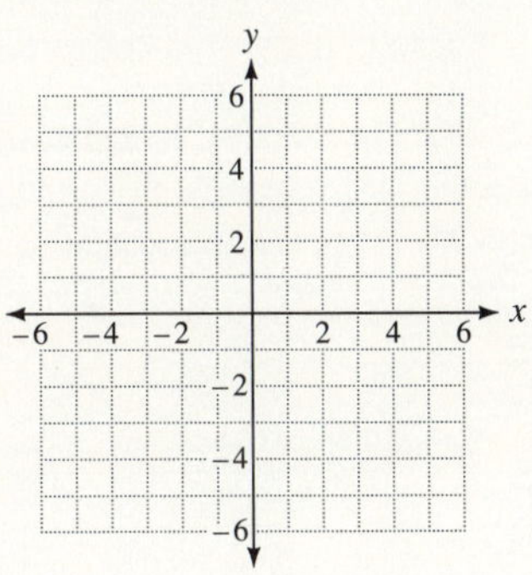

4. $0 \leq x \leq 3$
$y \geq 0$
$2y - x \leq 4$

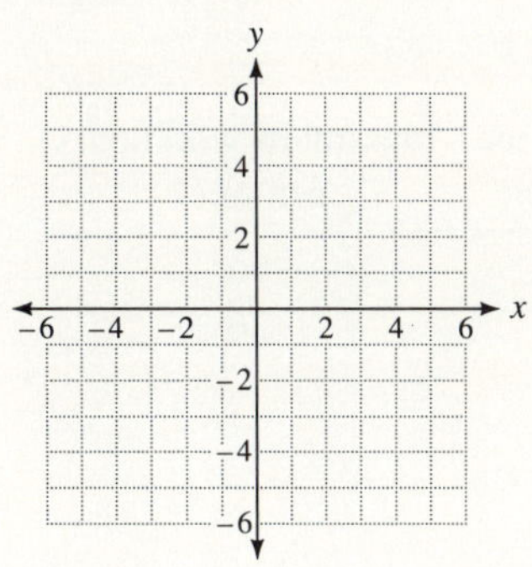

5. $x - 2 > 0$
$2 - x > 0$

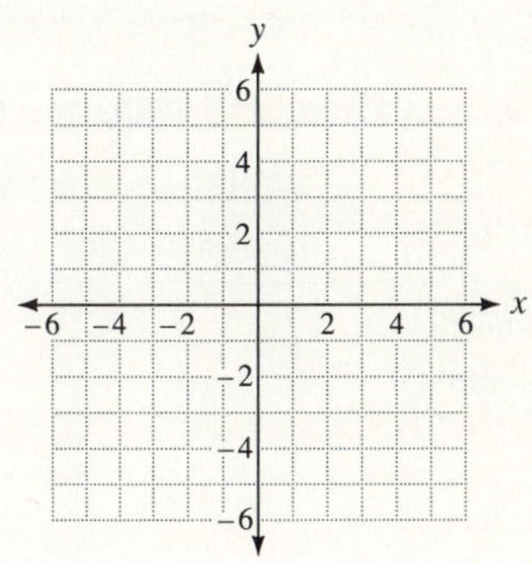

6. $y + 3 > 0$
$y + 4 \leq 0$

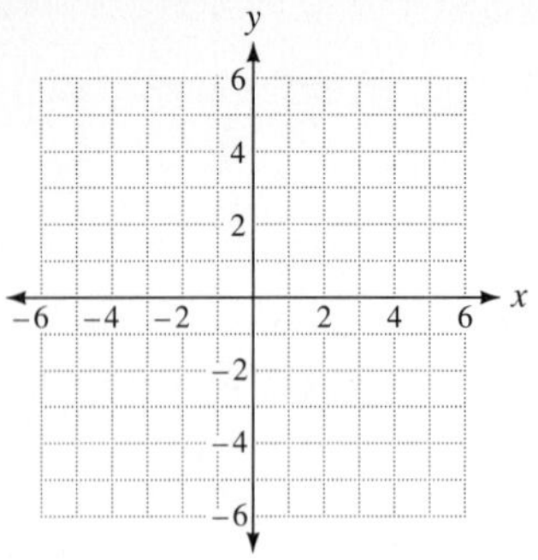

7. $x \leq 0$
$y \geq 0$
$2x - 5y \geq -10$
$x - y \geq -3$

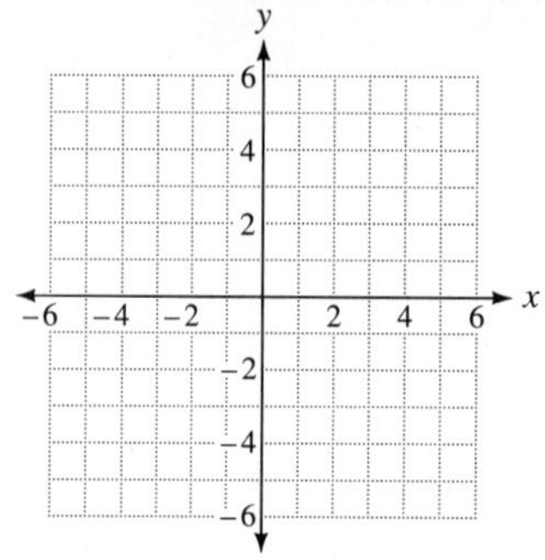

8. $x \leq 0$
$y \geq 0$
$2x - y \geq -6$
$x + 2y \leq 2$

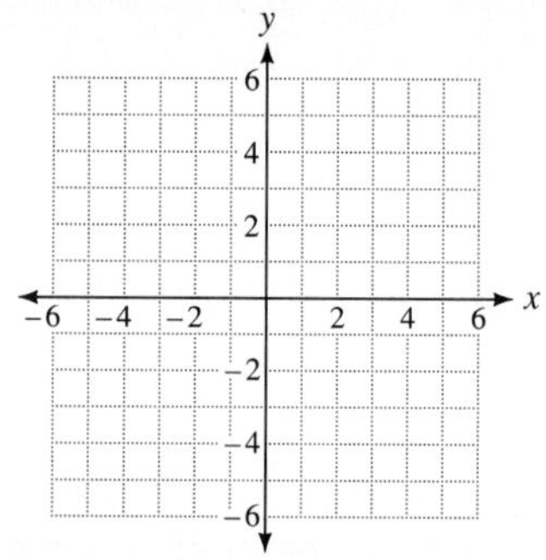

9. Recreation Furniture manufactures two types of picnic tables, one that seats four and another that seats six. The table below shows the production information. The number made weekly of each type of table should not exceed the workhours available in each manufacturing stage. Form the system of inequalities described by this information and graph the system.

	Workhours for four-seat table	Workhours for six-seat table	Maximum workhours per week
Construction	6	8	120
Finishing	1	3	30

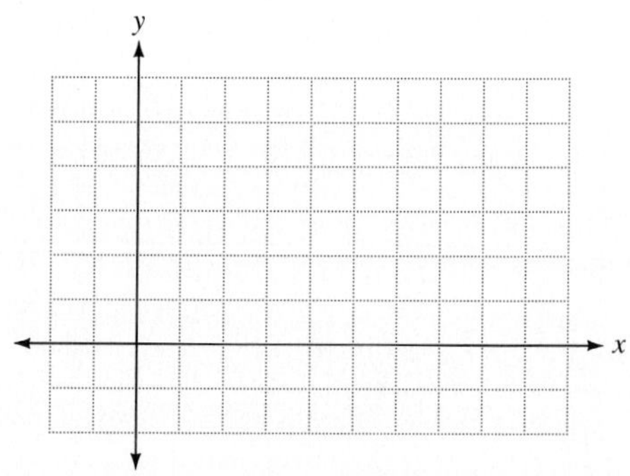

10. Repeat Exercise 9 using the table below.

	Workhours for four-seat table	Workhours for six-seat table	Maximum workhours per week
Construction	3	6	96
Finishing	1	4	36

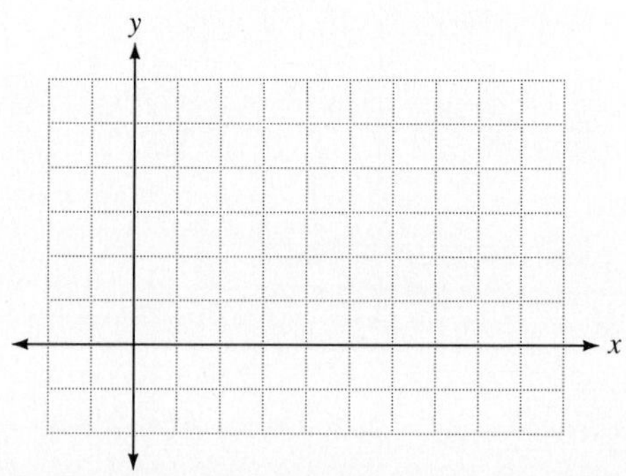

CHAPTER 4 REVIEW

KEY WORDS

4.1 A **system of equations** is a pair of linear equations in two variables.

A **solution** to a system of two linear equations in two variables is an ordered pair that solves each equation.

A system of equations is **dependent** if it has infinitely many solutions (the lines coincide).

A system of equations is **inconsistent** if it has no solution (the lines are parallel).

A system of equations is **independent** and **consistent** if it has exactly one solution (the lines intersect).

4.3 Two angles are **complementary** if their sum is 90°.

Two angles are **supplementary** if their sum is 180°.

4.4 An equation of the form $ax + by + cz = d$, where a, b, c, and d are real numbers, is a **linear equation in three variables.**

A **solution** to a system of three linear equations in three variables is an ordered triple of numbers that solves each equation in the system.

4.5 A **determinant** is a square array of numbers written within vertical bars whose value is a real number.

4.6 A **linear inequality in two variables** is an inequality that can be written in one of the forms $ax + by > c$, $ax + by < c$, $ax + by \geq c$, or $ax + by \leq c$.

A **system of inequalities** is two or more linear inequalities in two variables.

KEY CONCEPTS

4.1 **1.** A system of equations can be solved by the graphing method, but this method has limitations since it depends on the accuracy of the graphs.

2. By writing the equations in a system in slope-intercept form, it is easy to determine the relationship of the graphs and hence the number of solutions to the system.

4.2 **1.** To solve a system of two equations, use either the substitution method or the elimination method.

2. If an identity (such as $0 = 0$) results when solving a system of two equations, the system has infinitely many solutions.

3. If a contradiction (such as $5 = 0$) results when solving a system of two equations, the system has no solution.

4.3 To solve an applied problem that results in a system, make sure you define the variables accurately when you translate the words to equations. Sometimes a summary table can be useful.

4.4 When solving a system of three equations in three variables, if a contradiction results at any step, the system is inconsistent and has no solution. If an identity results after using only two equations, the system has either no solution *or* infinitely many solutions (is dependent); the third equation must be used to determine which case occurs.

4.5 **1.** Second-order determinant:

$$\begin{vmatrix} a_1 & b_1 \\ a_2 & b_2 \end{vmatrix} = a_1b_2 - a_2b_1$$

2. Third-order determinant:

$$\begin{vmatrix} a_1 & b_1 & c_1 \\ a_2 & b_2 & c_2 \\ a_3 & b_3 & c_3 \end{vmatrix} = a_1\begin{vmatrix} b_2 & c_2 \\ b_3 & c_3 \end{vmatrix} - a_2\begin{vmatrix} b_1 & c_1 \\ b_3 & c_3 \end{vmatrix} + a_3\begin{vmatrix} b_1 & c_1 \\ b_2 & c_2 \end{vmatrix}$$

3. Cramer's rule:

$$x = \frac{A_x}{A}, \quad y = \frac{A_y}{A}, \quad z = \frac{A_z}{A}$$

4.6 **1.** To graph a linear inequality in two variables, graph the equation using a solid line if the inequality is either $\leq$ or $\geq$ and a dashed line if it is either $<$ or $>$. Use a test point to determine which region should be shaded.

2. To graph a system of linear inequalities, graph each inequality in the system first. The graph of the system consists of all points in the region common to the graphs of the individual inequalities.

REVIEW EXERCISES

Part I

4.1 *Tell whether (1, −2) is a solution to the system.*

1. $2x + 3y + 4 = 0$
$-3x + y - 5 = 0$

2. $x - 2y = 5$
$4x + 2y = 0$

Without solving, give the number of solutions to the given system.

3. $2x - y = 7$
$x - 2y = 7$

4. $3x + 4y - 12 = 0$
$-6x - 8y + 12 = 0$

5. $2x + 2y = 2$
$5x + 5y = 5$

Solve using the graphing method.

6. $x + y = 3$
$2x - y = 3$

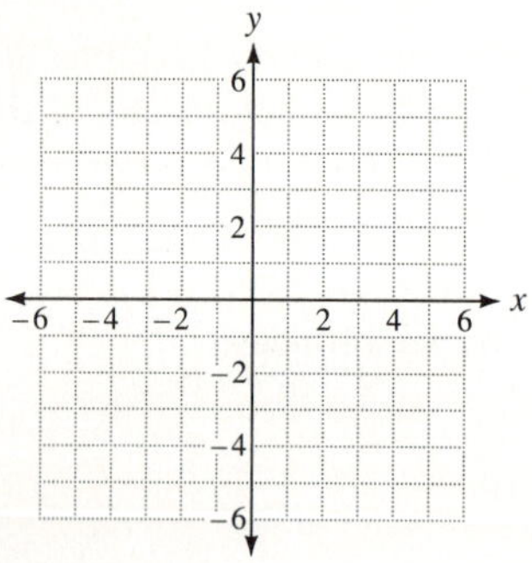

4.2 *Solve using either the substitution method or the elimination method, whichever seems appropriate.*

7. $3x - y = 2$
$-6x + 2y = 2$

8. $5x + 3y = -1$
$2x - 4y = -16$

9. $x + 7y = -1$
$-2x - 14y = 2$

4.3 *Solve using a system of equations.*

10. The sum of two numbers is 13 and their difference is 3. Find the numbers.

11. Janet McShane bought 2 shirts of the same value and 5 pairs of pants of the same value for \$94. Later she returned and bought 3 shirts and 1 pair of pants for \$50. How much does one shirt cost and how much does one pair of pants cost?

12. Chuck had \$5000 divided into two investments. The part invested in bonds at 8% together with the other part invested in certificates at 7% earned \$370 one year. How much was invested in each category?

13. An airplane flies 900 miles in 2 hours with the wind and 1050 miles in 3 hours against the wind. Find the speed of the plane in still air and the speed of the wind.

4.4 *Solve.*

14. $x + 2y = 3$
$y - 2z = -5$
$2x + z = 5$

15. $4x - 6y + 10z = -8$
$-2x + 3y - 5z = -4$
$7x - 6y + 3z = -5$

16. Determine whether $(-1, 2, 3)$ is a solution to the system.

$$\begin{aligned} x + y + z &= 4 \\ 3x - y + 2z &= 1 \\ -2x - y + z &= 3 \end{aligned}$$

Solve using a system of equations.

17. The average of a student's three scores is 80. The sum of the first and second scores is 150, and the third score is 5 more than the second. Find all three scores.

18. Joey has a collection of 75 coins consisting of pennies, nickels, and dimes with a value of \$4.78. If there are 5 fewer dimes than pennies and nickels combined, how many of each coin are there?

19. Find three numbers such that the sum of the first and second is twice the third, the first plus the third is 2, and the second minus the third is -4.

20. The largest angle of a triangle is twice the smallest, and the remaining angle is 20° less than the largest. Find the measure of each angle.

4.5 *Evaluate the determinants.*

21. $\begin{vmatrix} 3 & 7 \\ 12 & -2 \end{vmatrix}$

22. $\begin{vmatrix} 2 & -5 & 4 \\ 3 & -6 & -1 \\ -4 & 10 & -8 \end{vmatrix}$

Solve using Cramer's rule.

23. $\begin{aligned} 2x + 3y &= 6 \\ -x - 4y &= -3 \end{aligned}$

24. $\begin{aligned} x + 5y + 4z &= 7 \\ -2x - 8y + 7z &= -14 \\ 4y + 5z &= 0 \end{aligned}$

4.6 *Graph in a rectangular coordinate system.*

25. $x - 2y \le -2$

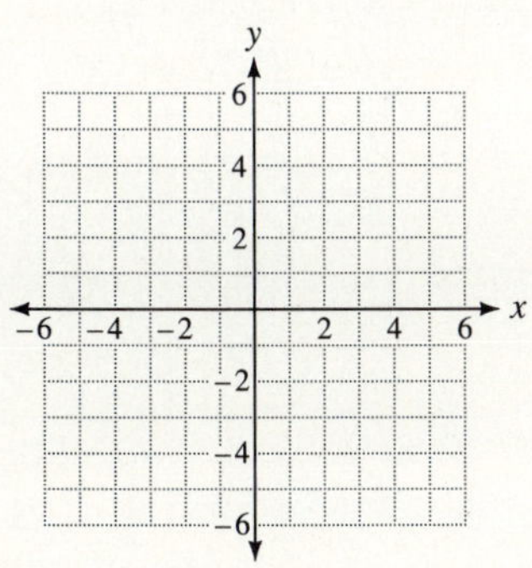

26. $2y + 4 > 0$

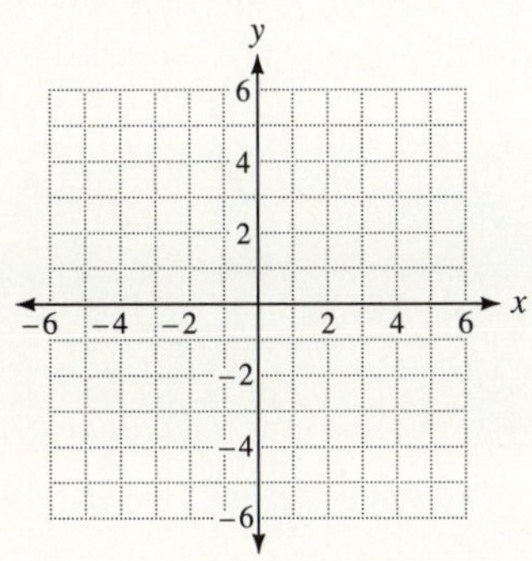

27. $3 - 3x \ge 0$

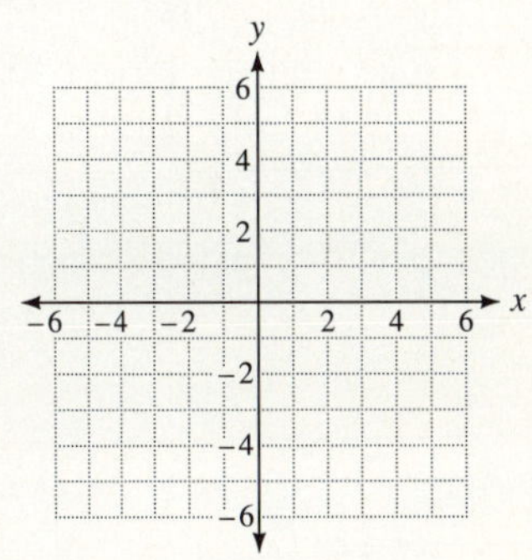

28. $3x + 2y < 6$

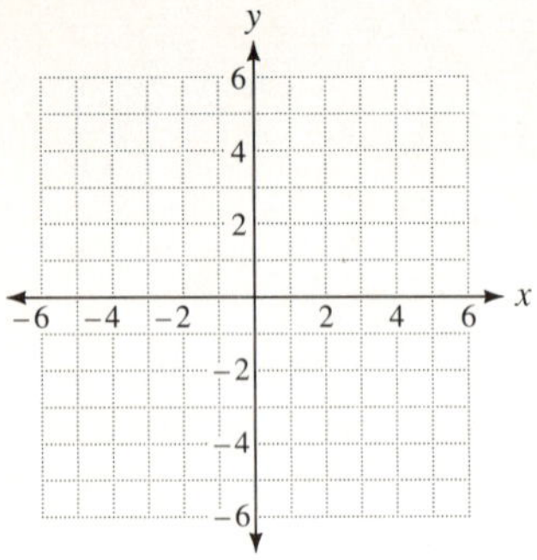

29. $3x + 2y < 6$
$2y + 4 > 0$

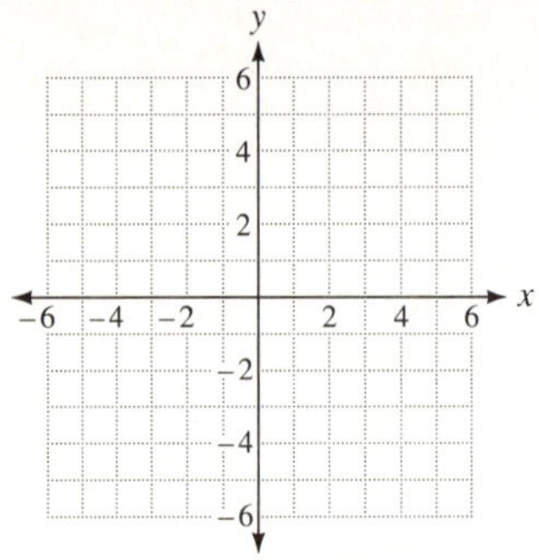

30. $y \leq x + 1$
$y \geq x - 1$

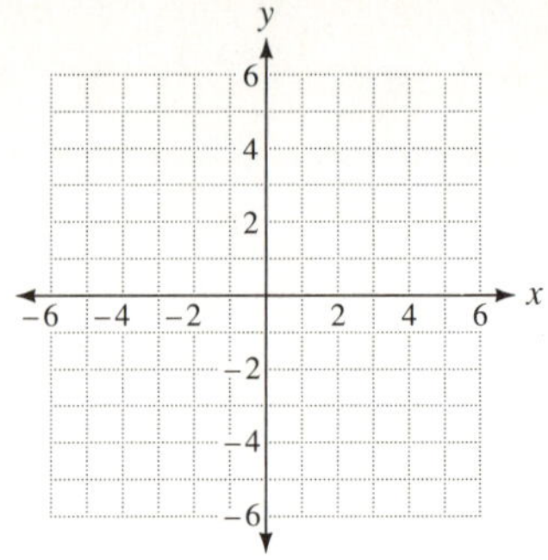

Part II

Solve.

31. $3x + y = -2$
$3x - y = -10$

32. $4x - 2y = 7$
$-2x + y = -1$

33. Dr. Gordon Johnson has one solution that is 15% salt and another that is 20% salt. How many gallons of each should he mix together to have 50 gallons of a solution that is 18% salt?

34. Gail has a collection of 80 coins made up of nickels and dimes. If the value of the collection is \$6.75, how many of each are there?

35. $x + y + z = 4$
$2x - y + z = -3$
$-x + 2y - 3z = 1$

36. $x - 3y + 2z = 6$
$4x - 2y + 3z = 14$
$2x + 4y - z = 2$

37. The sum of the ages of Mark, Larry, and Jane is 53. Jane is 5 years younger than Larry, and in two years, Mark will be the same age as Larry is now. How old is each?

38. $3x - 2y = 5$
$-3x + 2y = -5$

Graph in a rectangular coordinate system.

39. $2x + 3y \leq -12$

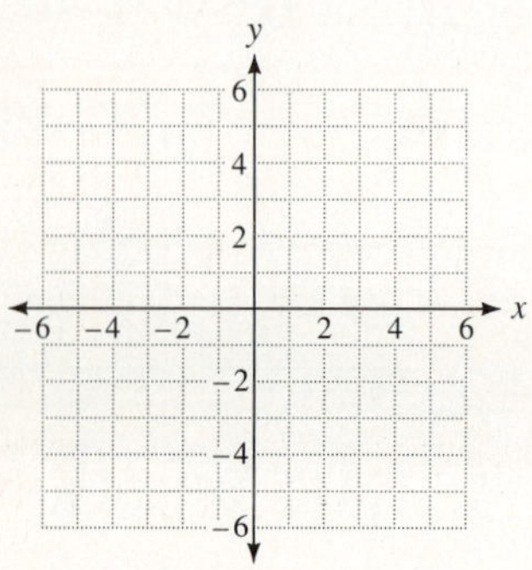

40. $x - y < 5$
$2x + y \geq 4$

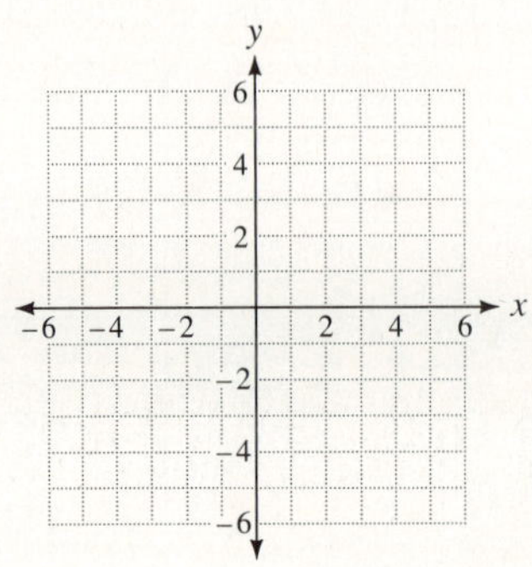

41. $x \geq 0$
$y \geq 0$
$x + y \leq 4$

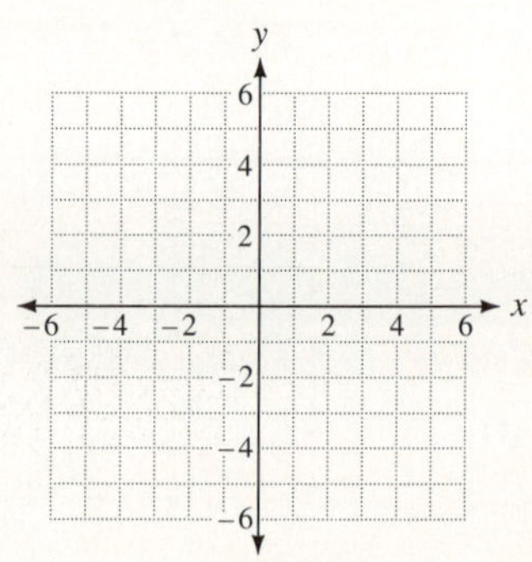

42. Without solving, tell the number of solutions to:

$$\begin{aligned} 2x + y &= 5 \\ 2x - y &= -5. \end{aligned}$$

43. Is $(3, -4)$ a solution to the following system?

$$\begin{aligned} 5x + y &= 11 \\ -3x + 2y &= -1 \end{aligned}$$

44. Solve using Cramer's rule.

$$\begin{aligned} 2x - 3y &= -5 \\ 6x + y &= 35 \end{aligned}$$

45. Evaluate the determinant.

$$\begin{vmatrix} 3 & -1 & 2 \\ 4 & 0 & 3 \\ -2 & -2 & 1 \end{vmatrix}$$

ANSWERS: 1. no 2. yes 3. exactly one (lines intersect) 4. no solution (lines are parallel) 5. infinitely many (lines coincide) 6. (2, 1) 7. no solution 8. $(-2, 3)$ 9. infinitely many solutions of the form $\left(x, -\frac{1}{7}x - \frac{1}{7}\right)$ for x any real number 10. 8, 5 11. shirt is \$12, pants are \$14 12. \$2000 in bonds, \$3000 in certificates 13. plane: 400 mph; wind: 50 mph 14. (1, 1, 3) 15. no solution 16. yes 17. 65, 85, 90 18. 18 pennies, 22 nickels, 35 dimes 19. 3, -5, -1 20. 40°, 60°, 80° 21. -90 22. 0 23. (3, 0) 24. (7, 0, 0)

25.
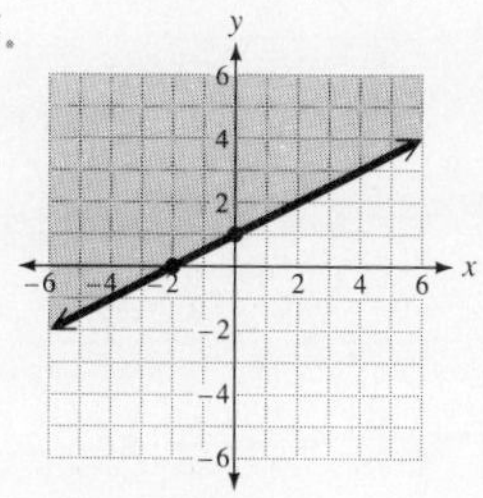

26.
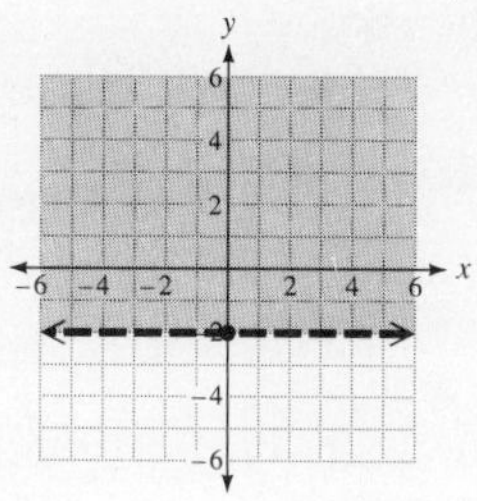

27.
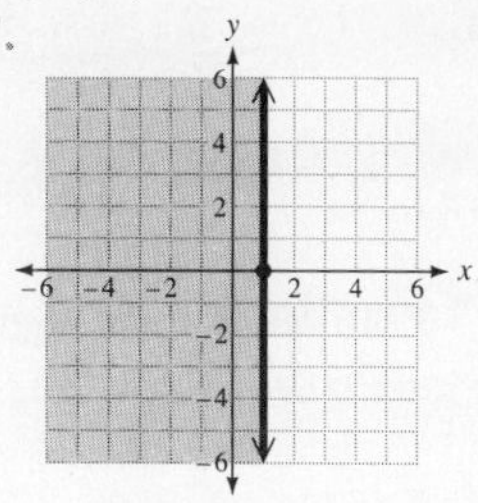

28.
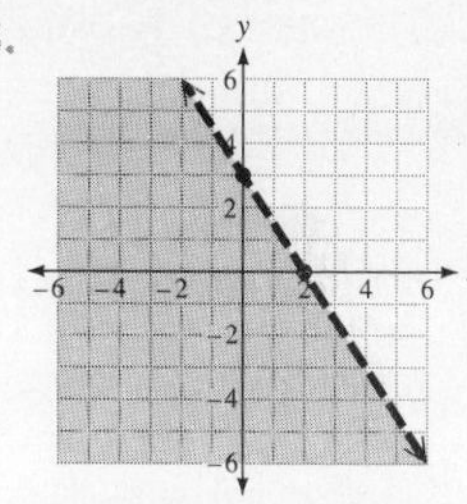

29.
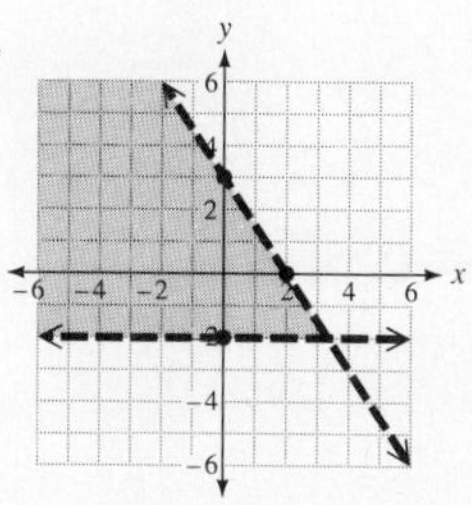

30.
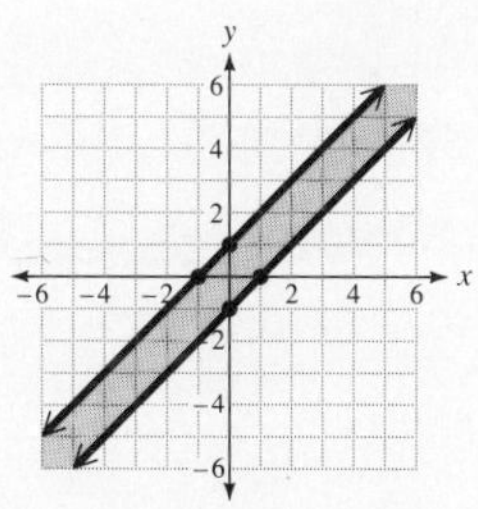

31. $(-2, 4)$ 32. no solution 33. 20 gal of 15% solution, 30 gal of 20% solution 34. 55 dimes, 25 nickels 35. $(-1, 3, 2)$ 36. infinitely many solutions; the system is dependent 37. Mark is 18, Larry is 20, Jane is 15. 38. $\left(x, \frac{3}{2}x - \frac{5}{2}\right)$ for x any real number

39.
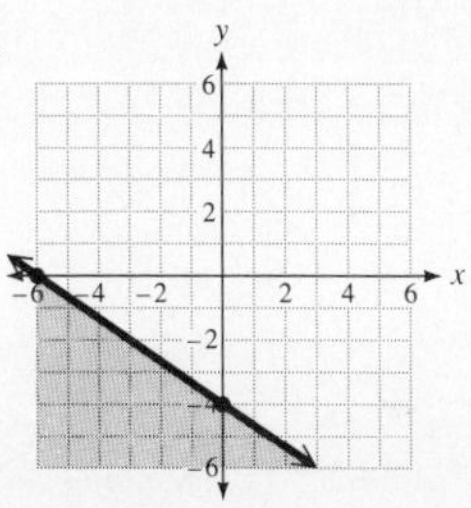

40.
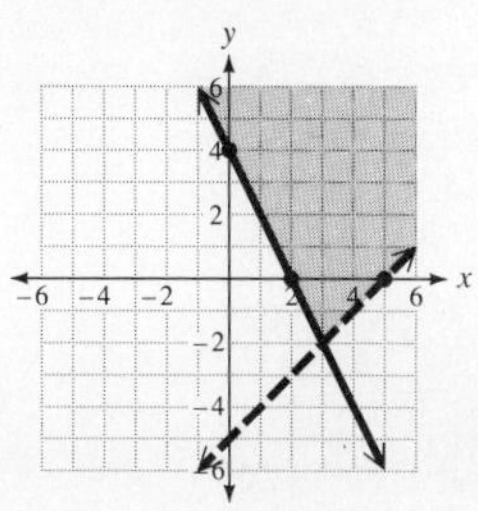

41.
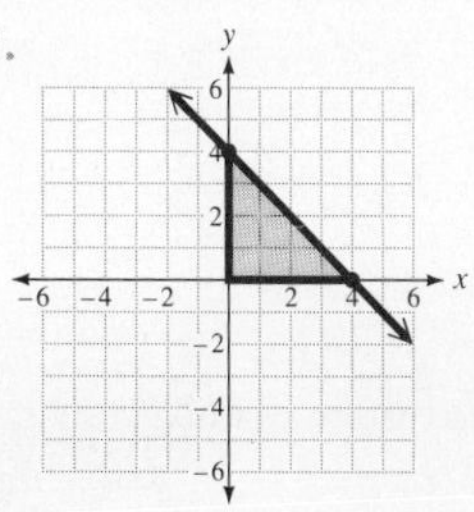

42. exactly one (lines intersect) 43. no 44. (5, 5) 45. 12

CHAPTER 4 TEST

1. Tell whether $(-2, 4)$ is a solution of the system. **1.** ______

$$\begin{aligned} 2x + 3y &= 8 \\ 3x - y &= 10 \end{aligned}$$

2. Without solving, give the number of solutions. **2.** ______

$$\begin{aligned} -4x + 6y &= 10 \\ 2x - 3y &= -5 \end{aligned}$$

Solve.

3. **3.** ______

$$\begin{aligned} 2x + 5y &= 5 \\ -4x - 8y &= -12 \end{aligned}$$

4. **4.** ______

$$\begin{aligned} 2x - y &= 3 \\ -2x + y &= -3 \end{aligned}$$

5. The sum of two numbers is 17. Twice the first minus the second is 1. Find the numbers. **5.** ______

6. A mixture of nuts sells for \$2.50 per pound. If the mix is composed of two kinds of nuts, one worth \$2.00 per pound and the other worth \$2.80 per pound, how many pounds of each are in 48 pounds of the mix? **6.** ______

7. **7.** ______

$$\begin{aligned} x + y - z &= 4 \\ x - y - z &= 2 \\ 3x - y - 3z &= -4 \end{aligned}$$

8. **8.** ______

$$\begin{aligned} x + y - z &= 3 \\ 2x - y + 3z &= -4 \\ 3x + 2y - 2z &= 7 \end{aligned}$$

9. Bill has \$10,000 divided into three separate investments. Part is invested in stocks earning 8%, part is in a savings account earning 7%, and the rest is invested in a business. If the business does well, it will earn 10% and Bill's total earnings on the three investments will amount to \$810. If the business does poorly, he will lose 4% and his total earnings will amount to \$390. Set up the system of equations but do not solve. **9.** ______

CHAPTER 4 TEST

CONTINUED

Evaluate.

10. $\begin{vmatrix} 6 & -2 \\ 5 & 1 \end{vmatrix}$

10. ______________________

11. $\begin{vmatrix} 2 & 1 & 0 \\ -1 & 3 & 2 \\ 1 & 4 & 0 \end{vmatrix}$

11. ______________________

Solve for ***x only*** *using Cramer's rule.*

12. $3x - y = 5$
$2x + 3y = -4$

12. ______________________

13. $x + y + 3z = 5$
$y - z = 2$
$x + z = 3$

13. ______________________

Graph the system.

14. $x - 3y < 6$
$2x + y \geq 4$

14.

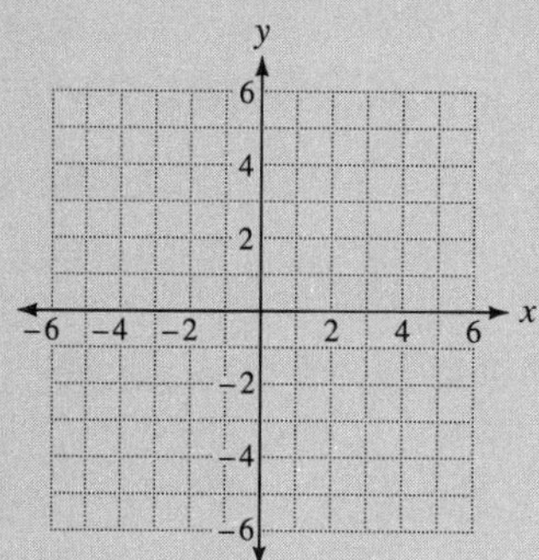

15. $x + y < 5$
$2x - 4 \geq 0$

15.

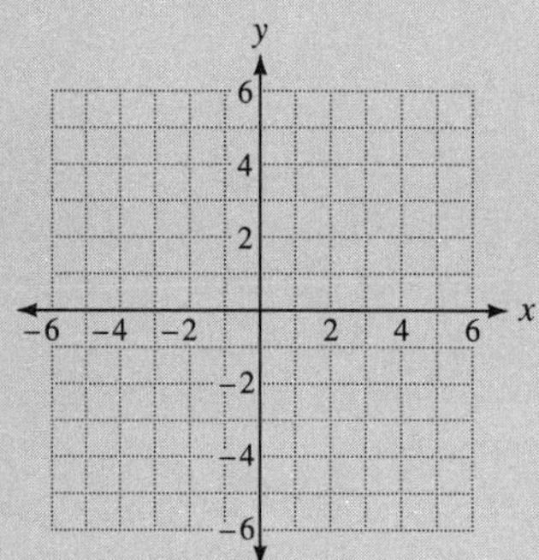

CHAPTER 5

Polynomials and Factoring

5.1 ADDITION AND SUBTRACTION OF POLYNOMIALS

STUDENT GUIDEPOSTS

1 Polynomials	4 Adding Polynomials
2 Collecting Like Terms	5 Subtracting Polynomials
3 Descending and Ascending Order	6 Evaluating Polynomials

1 POLYNOMIALS

In Chapter 1 we said that an algebraic expression involves sums, differences, products, or quotients of numbers and variables raised to various powers. Remember that within an algebraic expression, a term is a product of a number, called the numerical coefficient, and one or more variables raised to a power. A **polynomial** is an algebraic expression having as terms products of numbers and variables with whole-number exponents.

These Are Polynomials

$$-11x^2 + x^3 - x + 3, \quad x + y, \quad 3x^2y, \quad 2, \quad -8a^5b^4 + 6ab - 2, \quad \frac{3}{7}xy - \frac{4}{5}$$

The terms in the first polynomial are $-11x^2$, x^3, $-x$, and 3. The coefficients of these terms are -11, 1, -1, and 3, respectively.

These Are Not Polynomials

$$2\sqrt{x} - 1, \quad \frac{x+y}{x-y}, \quad 2^a, \quad 5|b| + 3$$

These examples illustrate algebraic expressions that are *not* polynomials. In a polynomial, a variable cannot appear within a radical sign, in the denominator, as an exponent, or within an absolute value.

A **monomial** is a polynomial with one term, while a **binomial** has two terms and a **trinomial** has three terms. A polynomial with more than three terms has no special name; we simply call it a polynomial.

A polynomial in one variable has the same variable in each of its terms, while a polynomial in several variables has two or more variables. We define the **degree of a term** to be the power on the variable, if there is only one variable in the term, and the sum of the powers on the variables for terms with more than one variable. The **degree of a polynomial** is equal to the degree of the term with the highest degree.

A polynomial that is a real number with no variables, such as 5, is said to have degree zero since 5 can be written $5x^0$ (remember $x^0 = 1$). However, the real number zero has no degree. We classify several polynomials by the number of terms and give the degree of each one in the following table.

Polynomial	*Type*	*Degree*	*Reason for degree*
$3x^4 + 2x^2 - 1$	Trinomial	4	Highest degree term is $3x^4$.
$5x$	Monomial	1	Highest degree term is $5x^1$.
-6	Monomial	0	$-6 = -6x^0$
0	Monomial	No degree	Zero has no degree.
$-2a^2b^3 + 8ab^5$	Binomial	6	Highest degree term is $8a^1b^5$ and $1 + 5 = 6$.
$2x^2 + xy + y^2 - 3$	Polynomial	2	Highest degree terms are $2x^2$, xy, and y^2.
$-3xyz + 2xy - 5z$	Trinomial	3	Highest degree term is $-3x^1y^1z^1$ and $1 + 1 + 1 = 3$.

As we saw in Chapter 1, like terms in a polynomial contain the same variables raised to the same powers. In the following table one term is like the original and the other one is unlike the original.

Term	*Like term*	*Unlike term*
$3x^4$	$-9x^4$	$3x^5$
5	-8	$-8x$
$7x^3y$	$-5x^3y$	$-5xy^3$
$-6a^2b^2c$	$15a^2b^2c$	$15a^2b^2c^2$
x^2b^3	$12x^2b^3$	$12x^2y^3$

2 COLLECTING LIKE TERMS

We can collect the like terms in a polynomial using the distributive law, as we saw in Section 1.5.

EXAMPLE 1 Collecting Like Terms

Collect like terms in each of the polynomials.

(a) $6y^2 - 2y^3 + 3y - 4y^2 + 1 - 3y - 7$

$= 6y^2 - 4y^2 - 2y^3 + 3y - 3y + 1 - 7$ — Commute so like terms are adjacent

$= (6 - 4)y^2 - 2y^3 + (3 - 3)y + (1 - 7)$ — Use distributive law

$= 2y^2 - 2y^3 + 0 \cdot y - 6$

$= 2y^2 - 2y^3 - 6$

PRACTICE EXERCISE 1

Collect like terms.

(a) $2a^3 + 3a^4 - a^3 + a - a^4 + 2a$

(b) $7a^2b^3 + 3a^3b^2 - a^2b^2 - 4a^2b^3 + 9a^3b^2$

(b) $9x^4y^2 + x^4y^2 - 4xy^3 + 3 + 6xy^3 - 5$
$= 9x^4y^2 + x^4y^2 - 4xy^3 + 6xy^3 + 3 - 5$ Commute
$= (9 + 1)x^4y^2 + (-4 + 6)xy^3 + (3 - 5)$ Distributive law
$= 10x^4y^2 + 2xy^3 - 2$

Answers: (a) $a^3 + 2a^4 + 3a$
(b) $3a^2b^3 + 12a^3b^2 - a^2b^2$

3 DESCENDING AND ASCENDING ORDER

To add polynomials, it is helpful to arrange terms in order. Arrange the polynomial

$$3x^2 - 5x^3 + 4x^6 - 2 + 3x$$

with the highest degree term first, followed by the next highest, and so forth.

$4x^6 - 5x^3 + 3x^2 + 3x - 2$ Exponents get smaller from left to right

The polynomial is now in **descending order.** Note that any constant term will be last in this order since $-2 = -2x^0$. To write the polynomial in **ascending order** put the smallest power first, followed by the next higher power, and so forth.

$-2 + 3x + 3x^2 - 5x^3 + 4x^6$ Exponents get larger from left to right

For polynomials in several variables we need to choose the variable to be ordered. For example, for descending powers of x we write

$3x^3y + 7x^2y^3 - 5xy^2 + 12$ Descending powers of x

and for descending powers of y

$7x^2y^3 - 5xy^2 + 3x^3y + 12.$ Descending powers of y

The descending order of one of the variables is often used for adding or subtracting polynomials.

EXAMPLE 2 Descending Order

Collect like terms and write in descending powers of a.

(a) $-3a^3 + a^7 - 7a - 4a^7 + a^3 + 1$
$= a^7 - 4a^7 - 3a^3 + a^3 - 7a + 1$
$= (1 - 4)a^7 + (-3 + 1)a^3 - 7a + 1$
$= -3a^7 - 2a^3 - 7a + 1$

(b) $2ab - 7a^2b^3 - b^2 + 4ab - 8a^3 + 2b^2 + 2a^2b^3$
$= -8a^3 - 7a^2b^3 + 2a^2b^3 + 2ab + 4ab - b^2 + 2b^2$
$= -8a^3 + (-7 + 2)a^2b^3 + (2 + 4)ab + (-1 + 2)b^2$
$= -8a^3 - 5a^2b^3 + 6ab + b^2$

PRACTICE EXERCISE 2

Collect like terms and write in descending powers of x.

(a) $4x - 3x^4 + 5x - 3x^5 + x^4 + 9x^5$

(b) $3x^2y^2 - 4x^3y + 7xy^3 - x^3y - 2x^2y^2$

Answers: (a) $6x^5 - 2x^4 + 9x$
(b) $-5x^3y + x^2y^2 + 7xy^3$

4 ADDING POLYNOMIALS

To add polynomials, we write the problem with descending powers of a variable, remove parentheses, and collect like terms.

EXAMPLE 3 Adding Polynomials

Add the polynomials.

(a) $-3x + x^4 - 5x^3 + 7$ and $3 + 3x - 5x^4 + 8x^2$

$(x^4 - 5x^3 - 3x + 7) + (-5x^4 + 8x^2 + 3x + 3)$ Arrange in descending order

$= x^4 - 5x^3 - 3x + 7 - 5x^4 + 8x^2 + 3x + 3$ Remove parentheses

$= (1 - 5)x^4 - 5x^3 + 8x^2 + (-3 + 3)x + (7 + 3)$ Collect like terms

$= -4x^4 - 5x^3 + 8x^2 + 0 \cdot x + 10$

$= -4x^4 - 5x^3 + 8x^2 + 10$

Although we usually add polynomials horizontally, we may also add vertically writing like terms in columns and leaving spaces for terms that are missing.

$$\begin{array}{rrrrr} x^4 - 5x^3 & & - 3x & + 7 \\ -5x^4 & + 8x^2 & + 3x & + 3 \\ \hline -4x^4 - 5x^3 & + 8x^2 & + 0x & + 10 \end{array} = -4x^4 - 5x^3 + 8x^2 + 10$$

Leave spaces where terms are missing

(b) $x^4y - 5x^3y^2 - 3xy^3 + 7y$ and $-5x^4y + 8x^2y + 4xy^3 + 3$

$(x^4y - 5x^3y^2 - 3xy^3 + 7y) + (-5x^4y + 8x^2y + 4xy^3 + 3)$

$= x^4y - 5x^3y^2 - 3xy^3 + 7y - 5x^4y + 8x^2y + 4xy^3 + 3$ Remove parentheses

$= (1 - 5)x^4y - 5x^3y^2 + 8x^2y + (-3 + 4)xy^3 + 7y + 3$ Collect like terms

$= -4x^4y - 5x^3y^2 + 8x^2y + xy^3 + 7y + 3$

PRACTICE EXERCISE 3

Add the polynomials.

(a) $9y - y^5 - 7y^2 + 2$ and $10y^2 + y^5 - 7y + 5$

(b) $x^2y^2 - 9xy + 8x + 3$ and $-4y + 2xy - 5 + 6x^2y^2$

Answers: (a) $3y^2 + 2y + 7$ (b) $7x^2y^2 - 7xy + 8x - 4y - 2$

With practice some steps shown in Example 3 can be omitted. We can add more than two polynomials the same way.

EXAMPLE 4 Adding Three Polynomials

Add $-3x^3y + 4x^2y^2 - 8x + y$, $5x^2y^2 - 7x + y + 5$, and $9x^3y + 4x - 3y$ using columns.

$$\begin{array}{rrrrr} -3x^3y & + 4x^2y^2 & - 8x & + y & \\ & 5x^2y^2 & - 7x & + y & + 5 \\ 9x^3y & & + 4x & - 3y & \\ \hline 6x^3y & + 9x^2y^2 & - 11x & - y & + 5 \end{array}$$

PRACTICE EXERCISE 4

Add $8ab - 2a^2b + 5ab^2$, $9a^2b - 4ab - 8ab^2$, and $-3ab^2 - 2ab + a^2b$.

Answer: $8a^2b + 2ab - 6ab^2$

5 SUBTRACTING POLYNOMIALS

Subtracting polynomials is the same as adding except that signs require more thought. We have to be careful to write the problem correctly and to change all signs on the subtrahend (polynomial being subtracted). For example, if we want to subtract $3x^2 - 5x + 7$ from $9x^2 - 4x - 8$, we write

$$(9x^2 - 4x - 8) - (3x^2 - 5x + 7).$$

To subtract, we change all the signs in $3x^2 - 5x + 7$ and then add.

$-(3x^2 - 5x + 7) = -3x^2 + 5x - 7$ Remove parentheses and change all three signs

Note that $3x^2$ becomes $-3x^2$, $-5x$ becomes $+5x$, and $+7$ becomes -7.

$(9x^2 - 4x - 8) - (3x^2 - 5x + 7)$

$= 9x^2 - 4x - 8 - 3x^2 + 5x - 7$ Remove parentheses and change all signs in the subtrahend

$= 9x^2 - 3x^2 - 4x + 5x - 8 - 7$ Collect like terms

$= (9 - 3)x^2 + (-4 + 5)x + (-8 - 7)$ Distributive law

$= 6x^2 + x - 15$

To do the same problem using columns, we add $-(3x^2 - 5x + 7) = -3x^2 + 5x - 7$ to $9x^2 - 4x - 8$. That is, we change all signs on the subtrahend and add.

$$\begin{array}{r} 9x^2 - 4x - 8 \\ -3x^2 + 5x - 7 \\ \hline 6x^2 + x - 15 \end{array}$$

Arrange like terms in vertical columns
Change all signs in the subtrahend and add

EXAMPLE 5 SUBTRACTING POLYNOMIALS

(a) Subtract $7x - 3x^2$ from $4 - 2x - 4x^2$.

$(-4x^2 - 2x + 4) - (-3x^2 + 7x)$

$= -4x^2 - 2x + 4 + 3x^2 - 7x$ Change signs

$= -4x^2 + 3x^2 - 2x - 7x + 4$ Collect like terms

$= (-4 + 3)x^2 + (-2 - 7)x + 4$ Distributive law

$= -x^2 - 9x + 4$

(b) Subtract $-3a^2b^3 + 9ab^2 - 5$ from $4a^2b^3 + 7ab^2 - 3b^2 + 8$.

$(4a^2b^3 + 7ab^2 - 3b^2 + 8) - (-3a^2b^3 + 9ab^2 - 5)$

$= 4a^2b^3 + 7ab^2 - 3b^2 + 8 + 3a^2b^3 - 9ab^2 + 5$ Change signs

$= (4 + 3)a^2b^3 + (7 - 9)ab^2 - 3b^2 + (8 + 5)$ Collect like terms

$= 7a^2b^3 - 2ab^2 - 3b^2 + 13$

To subtract $-3a^2b^3 + 9ab^2 - 5$ from $4a^2b^3 + 7ab^2 - 3b^2 + 8$ using columns, change the signs in $-3a^2b^3 + 9ab^2 - 5$ and add.

$$\begin{array}{r} 4a^2b^3 + 7ab^2 - 3b^2 + 8 \\ +3a^2b^3 - 9ab^2 \qquad + 5 \\ \hline 7a^2b^3 - 2ab^2 - 3b^2 + 13 \end{array}$$

Change signs
Add

PRACTICE EXERCISE 5

(a) Subtract $-4x^2 + 5x^3$ from $2x + x^3 - 9x^2$.

(b) Subtract $6x^2y - 9xy^2 + 3xy$ from $5x^2y + 7xy^2 - 2xy$.

Answers: (a) $-5x^2 - 4x^3 + 2x$ (b) $-x^2y + 16xy^2 - 5xy$

Addition and subtraction of polynomials can be combined, as in the next example.

EXAMPLE 6 ADDING AND SUBTRACTING POLYNOMIALS

Perform the indicated operations.

(a) $(5a^3 + a^2 - 6) + (7a^2 - 3) - (-4a^3 + 7a - 5)$

$= 5a^3 + a^2 - 6 + 7a^2 - 3 + 4a^3 - 7a + 5$

$= 9a^3 + 8a^2 - 7a - 4$ Some steps deleted

By columns we proceed as follows.

$$\begin{array}{r} 5a^3 + a^2 \qquad - 6 \\ 7a^2 \qquad - 3 \\ +4a^3 \qquad - 7a + 5 \\ \hline 9a^3 + 8a^2 - 7a - 4 \end{array}$$

Change signs

PRACTICE EXERCISE 6

Perform the indicated operations.

(a) $(4b^4 - 3b^2 + 1) + (-3b^4 + 2) - (b^4 - 9b^2 - 5)$

(b) $(9a^2b^2 - 2ab) + (-4ab + 8) - (-a^2b^2 + ab)$

(b) $(5x^3y^3 - 3x^2y) + (-7x^3y^3 + 4xy) - (9x^2y + 8xy)$
$= 5x^3y^3 - 3x^2y - 7x^3y^3 + 4xy - 9x^2y - 8xy$
$= (5 - 7)x^3y^3 + (-3 - 9)x^2y + (4 - 8)xy$
$= -2x^3y^3 - 12x^2y - 4xy$

Answers: (a) $6b^2 + 8$ (b) $10a^2b^2 - 7ab + 8$

6 EVALUATING POLYNOMIALS

In applied problems it is often necessary to evaluate polynomials for given values of the variable. Example 7 shows how this evaluation can be done.

EXAMPLE 7 Evaluating Polynomials

Evaluate the polynomials for the given values of the variables.

(a) $2x^2 - 7x + 1$ for $x = -3$

$2x^2 - 7x + 1 = 2(-3)^2 - 7(-3) + 1$ Substitute -3 for x using parentheses
$= 2(9) + 21 + 1$
$= 18 + 21 + 1 = 40$

(b) $5a^2b^3 - 2ab + 5$ for $a = 2$ and $b = -1$

$5a^2b^3 - 2ab + 5$
$= 5(2)^2(-1)^3 - 2(2)(-1) + 5$ Substitute $a = 2$ and $b = -1$ using parentheses
$= 5(4)(-1) + 4 + 5$
$= -20 + 4 + 5 = -11$

PRACTICE EXERCISE 7

Evaluate the polynomials.

(a) $-3b^3 + 2b^2 - 5$ for $b = 2$

(b) $7x^3y - 4x^2y^2 - 6$ for $x = -1$ and $y = -3$

Answers: (a) -21 (b) -21

The following application to business involves addition, subtraction, and evaluation of polynomials.

EXAMPLE 8 Business Problem

In a small business the cost of manufacturing is $M(u) = 2u^2 - 3u + 5$ and the cost of the wholesale operation is $W(u) = u^2 + u + 6$ where u is the number of units involved.

(a) Find the polynomial which represents the total cost.

The total cost $C(u)$ is $M(u) + W(u)$.

$C(u) = M(u) + W(u) = (2u^2 - 3u + 5) + (u^2 + u + 6)$
$= 2u^2 + u^2 - 3u + u + 5 + 6$ Collect like terms
$= 3u^2 - 2u + 11$ Total cost equation

(b) Find the total cost when 10 units are produced and sold.

$C(u) = 3u^2 - 2u + 11$ Total cost equation from (a)
$C(10) = 3(10)^2 - 2(10) + 11$ Substitute $u = 10$
$= 3(100) - 2(10) + 11$
$= 300 - 20 + 11 = 291$

The cost for 10 units is \$291.

PRACTICE EXERCISE 8

In a manufacturing operation,

$M(u) = 3u^2 - 3u + 2$ and
$W(u) = u^2 - u + 4$.

(a) Find $C(u) = M(u) + W(u)$.

(b) Find $C(u)$ when $u = 20$.

(c) If the total income from selling u units is $I(u) = 4u^2 - 3u$, find the profit $P(u)$.

The profit $P(u)$ is income minus cost.

$$P(u) = I(u) - C(u) = (4u^2 - 3u) - (3u^2 - 2u + 11)$$
$$= 4u^2 - 3u - 3u^2 + 2u - 11 \quad \text{Change signs}$$
$$= 4u^2 - 3u^2 - 3u + 2u - 11$$
$$= u^2 - u - 11 \quad \text{Profit equation}$$

(d) Find the profit when 15 units are produced and sold.

$$P(u) = u^2 - u - 11 \quad \text{Profit equation from (c)}$$
$$P(15) = (15)^2 - 15 - 11$$
$$= 225 - 15 - 11 = 199$$

The profit is \$199.

(c) If $I(u) = 5u^2 - 4u$, find $P(u) = I(u) - C(u)$.

(d) Find $P(u)$ when $u = 20$.

Answers: (a) $4u^2 - 4u + 6$ (b) \$1526 (c) $u^2 - 6$ (d) \$394

5.1 EXERCISES A

Tell whether the following are monomials, binomials, or trinomials and give the degree of each.

1. $-4x^2 - 3$ **2.** 10 **3.** $2x^4 - 3x^3$ **4.** $18a^5b^4$

5. $2xy + 3$ **6.** 0 **7.** $5xy^2 - 8x^2y - 17$ **8.** $9a^2b^4 - \frac{1}{3}$

9. $-14a^5 + b^4 - 7ab$ **10.** $-x - y$ **11.** $a^5b^6c^9 - 3a^2b^{20} - 3$ **12.** $27x^2y^2z^2$

Collect like terms and write in descending order.

13. $3y - 6 + 8y + 5$

14. $3x^2 - 2x + x^2 - 5x + 1$

15. $-6x^3 + 7x - 13x^5 - 7x + x^3$

16. $5y^8 - 6y^4 + 4y^3 - 6 + 7y^8 + 5y^4 + 1$

17. $-7 - 3a - 2a^2 + a^4 + 4a^2 + 3$

18. $5y^3 - 6y^7 - 12y^3 - 8y - 3y^7 + 1$

19. $0.5a^2 - 0.2 - 1.3a^2 - 0.7$

20 $\frac{1}{2}x^3 - \frac{1}{3}x - \frac{1}{4}x^3 - \frac{1}{9}x + 1$

Collect like terms and write in descending powers of the indicated variable.

21. $3xy - 2y - 5xy;\ x$

22. $7x^2y^3 - 2xy + 5xy - 5x^2y^3;\ x$

23. $5a^2b^2 - 2ab + 3 - 7a^2b^2 + 3ab - 1;\quad a$

24. $8a^3b^2 - 2a^2b^3 + 3ab - 4a^2b^3 - ab;\quad b$

25. $3x^4y^8 - 2xy^9 - 4x^4y^8 + 5xy^9 - 4x^{10}$; y

26. $-2abc^2 - 5a + bc + 3abc^2$; c

Add the polynomials.

27. $5x - 2$ and $-7x - 8$

28. $3xy + 4$ and $-7xy + 5$

29. $3y^2 - 5y + 2$ and $-7y^2 + y - 1$

30. $8a^2b - 3ab + 2$ and $9a^2b + 7ab - 5$

31. $-3x^4 - 7x^2 + 2$ and $x^3 + 5x - 5$

32. $10a^3b^3 - 7a^2b - 4ab^2$ and $-9a^3b^3 - 7a^2b + 5ab$

Subtract the polynomials.

33. $3y - 8$ from $2y + 5$

34. $-2xy + 3y$ from $9xy - 7y$

35. $-7a^2 + 2a - 5$ from $4a^2 - 5a - 3$

36. $7a^2b + ab - 5$ from $-10a^2b + ab + 8$

37. $-5y^4 - 6y^2 + 2y - 8$ from $y^4 + y^3 - y^2 + 1$

38. $-8a^3b^3 + 5a^2b + 10ab^2$ from $6a^3b^3 + 3a^2b - 4ab$

Perform the indicated operations and simplify.

39. $(3a^3 - 9a^4 + a^5 - 2) + (-6a + 7a^2 - a^3 + 2a^5)$

40. $(1.3a^2 - 5.2a + 3.7) + (2.3a^2 + 4.8a - 2.8)$

41. $\left(\frac{2}{3}z^2 - \frac{1}{5}z + 2\right) + \left(\frac{1}{2}z^2 + z - \frac{3}{4}\right)$

42. $(4y - 5y^3) + (-7 + 2y - y^2) - (7y^3 + 2 - 3y)$

43. $(7.3x^2y^2 - 8.7xy + 14.2) + (-6.7x^2y^2 + 2.3xy + 5.9)$

44 $(6x^3y^3 + 5x^2y^2) + (11x^2y^2 - 4xy) - (2x^3y^3 - 2x^2y^2 + 7xy)$

45. $3x - [4y - 7(5x - 2y) - (x - y) - 3(-4y - 2x)]$

46 $5a^2 - 4b(3a - [2b - b(a - 1) - (ab + b)])$

Add.

47.
$$\begin{array}{r} 5x^4 - 6x^3 - 2x^2 \qquad + 5 \\ -3x^4 \qquad + 8x^2 + 3x + 2 \\ \underline{-7x^4 + 8x^3 - 5x^2 \qquad + 9} \end{array}$$

48.
$$\begin{array}{r} 3.7a^3 - 8.6a + 9.3 \\ -5.8a^3 + 3.7a + 1.8 \\ \underline{8.5a^3 + 9.2a - 2.2} \end{array}$$

Evaluate the polynomial for the given values of the variables.

49. $10x^2 - 20x + 5;\ x = 2$

50. $5a^2b^2 - 2ab + 4;\ a = -1,\ b = 3$

51 $3uvw - 4u^2v;\ u = 5,\ v = -2,\ w = 3$

52. An index in a wholesale operation is given by the polynomial $x^2 + 8y^2 - 3xy + 2$. Find the index when $x = 5$ and $y = 2$.

In a business the manufacturing costs are given by the equation $M(u) = 3u^2 - 4u + 4$, and the cost of the wholesale operation is $W(u) = u^2 - u + 3$ where u is the number of units produced. See Example 8.

53 Find the polynomial $C(u)$ that represents the total cost of manufacturing and wholesale and use it in Exercises 54–55.

54. Find the total cost when 20 units are produced and sold.

55. Find the total cost when 5 units are produced and sold.

56 If the total income from the operation is $I(u) = 5u^2 - 2u$, find the polynomial that gives the profit $P(u)$ and use it to answer Exercises 57–58.

57. What is the profit when 20 units are produced and sold?

58. What is the profit when 5 units are produced and sold?

FOR REVIEW

Exercises 59–64 review material from Chapter 1 to help you prepare for the next section. Perform the indicated multiplication.

59. a^2a^4

60. $(a^4b^3)(ab^5)$

61. $-3(2x - y)$

62. $(x^3)^4$

63. $(x^2y^2)^3$

64. $12(x^2 + y^2)$

ANSWERS: 1. binomial; 2 2. monomial; 0 3. binomial; 4 4. monomial; 9 5. binomial; 2 6. monomial; no degree 7. trinomial; 3 8. binomial; 6 9. trinomial; 5 10. binomial; 1 11. trinomial; 22 12. monomial; 6 13. $11y - 1$ 14. $4x^2 - 7x + 1$ 15. $-13x^5 - 5x^3$ 16. $12y^8 - y^4 + 4y^3 - 5$ 17. $a^4 + 2a^2 - 3a - 4$ 18. $-9y^7 - 7y^3 - 8y + 1$ 19. $-0.8a^2 - 0.9$ 20. $\frac{1}{4}x^3 - \frac{4}{9}x + 1$ 21. $-2xy - 2y$ 22. $2x^2y^3 + 3xy$ 23. $-2a^2b^2 + ab + 2$ 24. $-6a^2b^3 + 8a^3b^2 + 2ab$ 25. $3xy^9 - x^4y^8 - 4x^{10}$ 26. $abc^2 + bc - 5a$ 27. $-2x - 10$ 28. $-4xy + 9$ 29. $-4y^2 - 4y + 1$ 30. $17a^2b + 4ab - 3$ 31. $-3x^4 + x^3 - 7x^2 + 5x - 3$ 32. $a^3b^3 - 14a^2b - 4ab^2 + 5ab$ 33. $-y + 13$ 34. $11xy - 10y$ 35. $11a^2 - 7a + 2$ 36. $-17a^2b + 13$ 37. $6y^4 + y^3 + 5y^2 - 2y + 9$ 38. $14a^3b^3 - 2a^2b - 10ab^2 - 4ab$ 39. $3a^5 - 9a^4 + 2a^3 + 7a^2 - 6a - 2$ 40. $3.6a^2 - 0.4a + 0.9$ 41. $\frac{7}{6}z^2 + \frac{4}{5}z + \frac{5}{4}$ 42. $-12y^3 - y^2 + 9y - 9$ 43. $0.6x^2y^2 - 6.4xy + 20.1$ 44. $4x^3y^3 + 18x^2y^2 - 11xy$ 45. $33x - 31y$ 46. $5a^2 - 12ab + 8b^2 - 8ab^2$ 47. $-5x^4 + 2x^3 + x^2 + 3x - 2$ 48. $6.4a^3 + 4.3a + 8.9$ 49. 5 50. 55 51. 110 52. 29 53. $4u^2 - 5u + 7$ 54. \$1507 55. \$82 56. $u^2 + 3u - 7$ 57. \$453 58. \$33 59. a^6 60. a^5b^8 61. $-6x + 3y$ 62. x^{12} 63. x^6y^6 64. $12x^2 + 12y^2$

5.1 EXERCISES B

Tell whether the following are monomials, binomials, or trinomials and give the degree of each.

1. $8x^2 - 2x + 1$ **2.** $3x - 1$ **3.** $3a^2b^2 - 2$ **4.** $6x^3y^7$

5. 42 **6.** $-abc + 2b + 3c$ **7.** 0 **8.** $\frac{2}{3}x^2b^3$

9. $x + 2y$ **10.** $a^2b^3c^4 + 2bc - 5a^{10}$ **11.** $5y^3 + x^2y^2$ **12.** $x^2 + y^2 + z^2$

Collect like terms and write in descending order.

13. $2y - 5 - 7y - 8$

14. $2x^3 - x^2 + 5x^3 - 2x + 3x^2$

15. $10x - 5x^3 + 3x + 8x^3 - x$

16. $-2y^3 + 4y^8 - 3y^3 + 7 - 1 + 2y^8$

17. $2a - 5 + 8a^2 - a^4 + 4 - 2a + 4a^4$

18. $-10y + 8y + 9y^{10} - y + 4y^{10}$

19. $0.5a^3 - 1.2a + 0.7a^3 + 0.6a$

20. $\frac{2}{3}x^2 - \frac{1}{3} + \frac{1}{9}x - \frac{1}{3}x^2 + \frac{2}{3}x - 1$

Collect like terms and write in descending powers of the indicated variable.

21. $2x^2y - xy + 3x^2y$; x

22. $8x^2y^2 - 3xy^2 - 5x^2y^2 + 6xy^2$; x

23. $a^3b^2 - 5ab + 3 - 2a^3b^2 - 7ab + 2$; a

24. $3a + 5a^2b - 3b^3 + ab^2 - 3a^2b + 2a$; b

25. $-xy^5 + 2x^9y^9 - 3y^{10} + xy^5 - 5x^9y^9 + 8y^{10}$; y

26. $7a^2 - 3abc + 4a^2 + 9abc + 8c^2$; c

Add the polynomials.

27. $3y - 5$ and $7y + 8$

28. $-8xy - 4$ and $xy - 1$

29. $5a^3 - 3a + 7$ and $7a^3 - 10a - 8$

30. $-6x^3y + x^2y^2 - 5xy$ and $8x^3y - 10x^2y^2 + 6$

31. $x^2 - 2x + 1$ and $x^4 + x^3 - x^2 + 5$

32. $a^3b^2 - a^2b^3$ and $a^3b^2 - 5a^2b^3 + 7$

Subtract the polynomials.

33. $-8x - 7$ from $9x + 2$

34. $6xy - 3$ from $-2xy + 7$

35. $9x^3 - 7x^2 - 2x$ from $-7x^2 + 7x - 8$

36. $12x^3y - 7x^2y^2 + 14xy$ from $7x^3y + 6x^2y^2 - 5$

37. $x^4 - 3x^2 + 2x + 1$ from $3x^4 - 5x^3 + 7$

38. $-6x^3y^3 + x^2y^2 - xy$ from $x^3y^3 - x^2y^2 - xy$

Perform the indicated operations and simplify.

39. $(3a^3 - 9a^4 + a^5 - 2) - (-6a + 7a^2 - a^3 + 2a^5)$

40. $\left(\frac{3}{4}x^2 - \frac{1}{3}x - \frac{4}{5}\right) - \left(-\frac{1}{2}x^2 + \frac{4}{9}x - \frac{3}{25}\right)$

41. $(-5.8c^2 + 6.2c - 4.1) - (2.3c^2 + 4.8c - 1.3)$

42. $(2 - 2a^4 + a^3) - (a^3 - 3a^4 + 7a + 5) - (14a + a^3 - 3a^2)$

43. $\left(\frac{2}{9}a^3b^2 + \frac{3}{4}ab - \frac{2}{5}b\right) - \left(\frac{1}{3}a^3b^2 - \frac{1}{8}ab + \frac{3}{10}b\right)$

44. $(-8a^3b - 5a^2b^2) - (-3a^2b^2 - 10a) - (-7a^3b - 9a^2b^2 + 4a)$

45. $7a - [2b - 5(9a - 5b - 3) - 4 - 7(a - b)]$

46. $7x^2 - 2x\{3y - [-6y - x(y + 3) - y(x - 2)]\}$

Add.

47.
$$\begin{array}{r} 180y^3 - 24y^2 - 85y \phantom{{}-240} \\ -35y^3 + 128y^2 + 99y - 240 \\ 67y^2 - 108y + 560 \\ \hline \end{array}$$

48.
$$\begin{array}{r} \frac{2}{7}x^3 - 5x^2 + \frac{3}{5}x + 2 \\ \frac{3}{14}x^3 - \frac{5}{6}x^2 - \frac{7}{10}x + \frac{3}{2} \\ \hline \end{array}$$

Evaluate the polynomial for the given values of the variables.

49. $3x^2 - 4x - 5;\ x = 3$

50. $2ab^2 - 3a^2b + ab;\ a = 2,\ b = -1$

51. $-6uvw + 7u - 5w;\quad u = 2,\ v = -3,\ w = 7$

52. A manufacturing index is given by the polynomial $2x^2 - 5xy + y^2 - 5$. Find the index when $x = 3$ and $y = 4$.

In a sales operation the wholesale cost is given by $W(u) = u^2 - 2u + 4$, and the retail cost is $R(u) = u^2 - u + 8$ where u is the number of units sold.

53. Find the polynomial $C(u)$ which represents the total cost of both operations and use it in Exercises 54–55.

54. Find the total cost when 2 units are sold.

55. Find the total cost when 30 units are sold.

56. If the total income from the operation is $I(u) = 3u^2 - 2u$, find the polynomial which gives the profit $P(u)$ and use it to answer Exercises 57–58.

57. What is the profit when 2 units are sold?

58. What is the profit when 30 units are sold?

FOR REVIEW

Exercises 59–64 review material from Chapter 1 to help you prepare for the next section.

59. x^3x^7

60. $(u^2v^5)(u^3v^4)$

61. $-4(a - 3c)$

62. $(y^4)^2$

63. $(x^5y)^4$

64. $8(xy + z^2)$

5.1 EXERCISES C

Evaluate the polynomial for the given values of the variables.

1. $xy - 3x + 2y^2;\ x = \frac{1}{2},\ y = -\frac{1}{3}$

2. $0.0005a^3 + 0.001a^2 + 0.01a - 0.007;\ a = 0.3$

3. $x^2y - x^3 + xy^3 - 5;\ x = 2a,\ y = -a$

4. $\frac{4}{3}w^3 - \frac{4}{5}w^2 + w - \frac{1}{6};\ w = \frac{1}{2}$
[Answer: $\frac{3}{10}$]

Perform the indicated operations.

5. $(-10x^4y^3 + 3x^3y^4 - 7x^2z + 5z) + (10x^4y^3 + 9x^3y^4 - 6z) - (12x^3y^4 - z)$

6. $\frac{1}{4}(16a^3b^2 - 8ab^3 + 12ab) - \frac{1}{5}(25a^3b^2 + 5ab^3 - 10ab) + \frac{2}{3}(9ab - 6ab^3)$
[Answer: $-a^3b^2 - 7ab^3 + 11ab$]

5.2 MULTIPLICATION OF POLYNOMIALS

STUDENT GUIDEPOSTS

1. Multiplying by Monomials
2. Multiplying Binomials
3. The FOIL Method
4. Special Products

1 MULTIPLYING BY MONOMIALS

To multiply two monomials, we use the commutative and associative laws with the law of exponents that we learned in Chapter 1,

$$a^m a^n = a^{m+n}.$$

EXAMPLE 1 Multiplying Monomials

(a) Multiply $6ab^2$ by $-a^3$.

$$(6ab^2)(-a^3) = (6)(-1)(aa^3)b^2 \qquad -a^3 = (-1)a^3$$
$$= -6a^{1+3}b^2 \qquad a = a^1$$
$$= -6a^4b^2$$

PRACTICE EXERCISE 1

(a) Multiply $-x^4y^6$ by $9x^6y$.

(b) Multiply $2a^3b^4$ by $-8ab^5$.

(b) Multiply $-3x^2y^3$ by $5x^5y^2$.

$$(-3x^2y^3)(5x^2y^2) = (-3)(5)(x^2x^5)(y^3y^2)$$ Use the commutative law to change the order of factors

$$= -15x^{2+5}y^{3+2}$$ Use $a^m a^n = a^{m+n}$

$$= -15x^7y^5$$

Answers: (a) $-9x^{10}y^7$ (b) $-16a^4b^9$

The product of a monomial and a binomial or any other polynomial uses the distributive law as well as the methods for multiplying monomials.

EXAMPLE 2 Multiplying a Binomial by a Monomial

Multiply.

$$-5x^2y(4x^2y^2 - 6y)$$

$$= (-5x^2y)(4x^2y^2) - (-5x^2y)(6y)$$ Distributive property

$$= (-5)(4)(x^2x^2)(yy^2) - (-5)(6)x^2(yy)$$

$$= -20x^{2+2}y^{1+2} - (-30)x^2y^{1+1}$$ $a^m a^n = a^{m+n}$

$$= -20x^4y^3 + 30x^2y^2$$

Practice Exercise 2

Multiply $-3ab^3(2a^2b^2 - 7ab^4)$.

Answer: $-6a^3b^5 + 21a^2b^7$

2 MULTIPLYING BINOMIALS

We can multiply two binomials the way we multiply a monomial times a binomial if we think of one binomial as a single term and use the distributive law twice.

$$(x + 2y)(5x - y) = (x + 2y)(5x) + (x + 2y)(-y)$$ Distributive law

$$= (x)(5x) + (2y)(5x) + (x)(-y) + (2y)(-y)$$ Distributive law again, twice

$$= 5x^2 + 10xy - xy - 2y^2$$

$$= 5x^2 + 9xy - 2y^2$$

EXAMPLE 3 Multiplying Binomials

Multiply.

(a) $(2x - 1)(x + 3) = (2x - 1)(x) + (2x - 1)(3)$ Distributive law

$$= (2x)(x) - (1)(x) + (2x)(3) - (1)(3)$$ Distributive law

$$= 2x^2 - x + 6x - 3$$

$$= 2x^2 + 5x - 3$$ Collect like terms

(b) $(3a^2 + 2b)(2a^3 - 5b)$

$$= (3a^2 + 2b)(2a^3) + (3a^2 + 2b)(-5b)$$

$$= (3a^2)(2a^3) + (2b)(2a^3) + (3a^2)(-5b) + (2b)(-5b)$$

$$= 6a^5 + 4a^3b - 15a^2b - 10b^2$$

Practice Exercise 3

Multiply.

(a) $(4x + 3)(x - 2)$

(b) $(4u^3 - 3v)(2u^2 + 7v)$

Answers: (a) $4x^2 - 5x - 6$ (b) $8u^5 + 28u^3v - 6u^2v - 21v^2$

3 THE FOIL METHOD

Notice that when we multiplied $(x + 2y)(5x - y)$ above, each term in the first binomial was multiplied by each term in the second and then like terms were collected. A shorter way to find this product is to use the **FOIL method.** The letters FOIL stand for first terms Ⓕ, outside terms Ⓞ, inside terms Ⓘ, and last terms Ⓛ. If we multiply each of these pairs of terms, we obtain the same product as we did using the distributive law twice.

$$\begin{aligned}(x + 2y)(5x - y) &= \overset{F}{(x)(5x)} + \overset{O}{(x)(-y)} + \overset{I}{(2y)(5x)} + \overset{L}{(2y)(-y)}\\ &= 5x^2 - xy + 10xy - 2y^2\\ &= 5x^2 + 9xy - 2y^2\end{aligned}$$

Practice with the FOIL method will let us find products of binomials mentally. It is a time-saving technique.

EXAMPLE 4 Using the FOIL Method

Use the FOIL method to multiply the binomials.

$$\begin{aligned}(5a + 2b)(3a - 7b) &= \overset{F}{(5a)(3a)} + \overset{O}{(5a)(-7b)} + \overset{I}{(2b)(3a)} + \overset{L}{(2b)(-7b)}\\ &= 15a^2 - 35ab + 6ab - 14b^2\\ &= 15a^2 - 29ab - 14b^2\end{aligned}$$

Practice Exercise 4

Use the FOIL method to multiply.

$(9x - y)(6x - 5y)$

Answer: $54x^2 - 51xy + 5y^2$

Another method of multiplying involves writing one polynomial above the other and arranging like terms of the product in vertical columns. The column method is especially useful with trinomials or larger polynomials.

EXAMPLE 5 Multiplying Using Columns

Multiply using columns.

(a)

$$\begin{array}{lll} & 3x^2 - 5xy + 2y^2 & \\ & 7x - 8y & \\ \hline & 21x^3 - 35x^2y + 14xy^2 & 7x \text{ times the top polynomial}\\ & \quad - 24x^2y + 40xy^2 - 16y^3 & -8y \text{ times the top polynomial}\\ \hline & 21x^3 - 59x^2y + 54xy^2 - 16y^3 & \end{array}$$

(b)

$$\begin{array}{ll} 6x^2 - 7xy + 5y^2 & \\ 4x^2 + 3xy - 2y^2 & \\ \hline 24x^4 - 28x^3y + 20x^2y^2 & 4x^2 \text{ times the top polynomial}\\ \quad 18x^3y - 21x^2y^2 + 15xy^3 & 3xy \text{ times the top polynomial}\\ \quad\quad - 12x^2y^2 + 14xy^3 - 10y^4 & -2y^2 \text{ times the top polynomial}\\ \hline 24x^4 - 10x^3y - 13x^2y^2 + 29xy^3 - 10y^4 & \end{array}$$

Practice Exercise 5

Multiply.

(a)

$$\begin{array}{l} 2x^2 + 3xy - 4y^2\\ 6x - y\\ \hline \end{array}$$

(b)

$$\begin{array}{l} 2x^2 - 5xy - 6y^2\\ 9x^2 - xy + 4y^2\\ \hline \end{array}$$

Answers: (a) $12x^3 + 16x^2y - 27xy^2 + 4y^3$ (b) $18x^4 - 47x^3y - 41x^2y^2 - 14xy^3 - 24y^4$

4 SPECIAL PRODUCTS

Products of certain binomials occur frequently and deserve special consideration. Look at the following three products.

(1) $(2x + 3y)(2x - 3y) = (2x)^2 - 6xy + 6xy - (3y)^2 = 4x^2 - 9y^2$

(2) $(2x + 3y)(2x + 3y) = (2x)^2 + 6xy + 6xy + (3y)^2 = 4x^2 + 12xy + 9y^2$

(3) $(2x - 3y)(2x - 3y) = (2x)^2 - 6xy - 6xy + (3y)^2 = 4x^2 - 12xy + 9y^2$

In (1) we have an example of the general formula

$(a + b)(a - b) = a^2 - ab + ab - b^2$ Using FOIL method

$\mathbf{(a + b)(a - b) = a^2 - b^2}$.   Difference of squares

We call this the difference of squares since the middle terms add to zero and the difference of squares is left. We can also write general formulas for both (2) and (3) above.

$(a + b)^2 = a^2 + ab + ab + b^2$ Using FOIL method

$(a + b)^2 = a^2 + 2ab + b^2$ Perfect square binomial

(sum)² = (first term)² + 2(product) + (last term)²

$(a - b)^2 = a^2 - ab - ab + b^2$ Using FOIL method

$(a - b)^2 = a^2 - 2ab + b^2$ Perfect square binomial

(difference)² = (first term)² + 2(product) + (last term)²

The products (2) and (3) fall into the pattern: the square of the first term, plus or minus twice the product of the first and last terms, plus the square of the last. These three formulas are special products that will be helpful when we factor polynomials in Section 5.4.

Special Product Formulas

The product of a sum and difference of two terms is the square of the first minus the square of the second.

1. $(a + b)(a - b) = a^2 - b^2$

The square of a binomial is the square of the first term plus twice the product of the terms plus the square of the second term.

2. $(a + b)(a + b) = (a + b)^2 = a^2 + 2ab + b^2$

3. $(a - b)(a - b) = (a - b)^2 = a^2 - 2ab + b^2$

CAUTION

Pay special attention to the difference between the first and third formulas: $(a - b)^2$ is not $a^2 - b^2$. Also note that $(a + b)^2 \neq a^2 + b^2$.

Now we can find certain products using the FOIL method or the special product formulas.

FOIL *Method* | *Using* $(a + b)(a - b) = a^2 - b^2$

$$(x + 6)(x - 6) = \overset{F}{x^2} - \overset{O}{6x} + \overset{I}{6x} - \overset{L}{36}$$
$$= x^2 - 36$$

$$(x + 6)(x - 6) = x^2 - 6^2 \qquad a = x,\ b = 6$$
$$= x^2 - 36$$

FOIL *Method* | *Using* $(a + b)^2 = a^2 + 2ab + b^2$

$$(x + 6)(x + 6) = x^2 + 6x + 6x + 36$$
$$= x^2 + 12x + 36$$

$$(x + 6)^2 = x^2 + 2(x)(6) + 6^2 \qquad a = x,\ b = 6$$
$$= x^2 + 12x + 36$$

FOIL *Method* | *Using* $(a - b)^2 = a^2 - 2ab + b^2$

$$(x - 6)(x - 6) = x^2 - 6x - 6x + 36$$
$$= x^2 - 12x + 36$$

$$(x - 6)^2 = x^2 - 2(x)(6) + 6^2 \qquad a = x,\ b = 6$$
$$= x^2 - 12x + 36$$

In the following examples we use the rules of exponents $(ab)^n = a^n b^n$ and $(a^m)^n = a^{mn}$.

EXAMPLE 6 Using Special Product Formulas

Find each product using the special product formulas.

(a) $(3x - 2y)(3x + 2y) = (3x)^2 - (2y)^2$ Use $(a - b)(a + b) = a^2 - b^2$, with $a = 3x$ and $b = 2y$

$= 3^2x^2 - 2^2y^2$ $(ab)^n = a^n b^n$

$= 9x^2 - 4y^2$

(b) $(3x + 2y)^2$

$= (3x)^2 + 2(3x)(2y) + (2y)^2$ Use $(a + b)^2 = a^2 + 2ab + b^2$, with $a = 3x$ and $b = 2y$

$= 9x^2 + 12xy + 4y^2$ $(ab)^n = a^n b^n$

(c) $(3x - 2y)^2$

$= (3x)^2 - 2(3x)(2y) + (2y)^2$ Use $(a - b)^2 = a^2 - 2ab + b^2$, with $a = 3x$ and $b = 2y$

$= 9x^2 - 12xy + 4y^2$ $(ab)^n = a^n b^n$

Practice Exercise 6

Find each product using the special product formulas.

(a) $(7a - 4b)(7a + 4b)$

(b) $(7a + 4b)^2$

(c) $(7a - 4b)^2$

Answers: (a) $49a^2 - 16b^2$ (b) $49a^2 + 56ab + 16b^2$ (c) $49a^2 - 56ab + 16b^2$

EXAMPLE 7 Using Special Product Formulas

Find each product using the special product formulas.

(a) $(5a^3 - 4b^2)(5a^3 + 4b^2) = (5a^3)^2 - (4b^2)^2$

Use $(a - b)(a + b) = a^2 - b^2$, with $a = 5a^3$ and $b = 4b^2$

$= 5^2(a^3)^2 - 4^2(b^2)^2$

$= 25a^6 - 16b^4$ $(a^m)^n = a^{mn}$

(b) $(5a^3 + 4b^2)^2 = (5a^3)^2 + 2(5a^3)(4b^2) + (4b^2)^2$

Use $(a + b)^2 = a^2 + 2ab + b^2$ with $a = 5a^3$ and $b = 4b^2$

$= 5^2(a^3)^2 + 40a^3b^2 + 4^2(b^2)^2$ $(ab)^n = a^n b^n$

$= 25a^6 + 40a^3b^2 + 16b^4$ $(a^m)^n = a^{mn}$

Practice Exercise 7

Find each product using the special product formulas.

(a) $(9x^4 - y^3)(9x^4 + y^3)$

(b) $(9x^4 - y^3)^2$

Answers: (a) $81x^8 - y^6$ (b) $81x^8 - 18x^4y^3 + y^6$

You should try to multiply polynomials mentally. Don't start multiplying, however, until you have examined the polynomials to determine the easiest method to use. At times, as in the next example, a special product formula might be applied more than once.

EXAMPLE 8 Multiplying Three Polynomials

Multiply $(x + 2y)(x - 2y)(x^2 + 4y^2)$.

Notice that $(x + 2y)(x - 2y)$ can be found using the special product $(a + b)(a - b)$. This product, $x^2 - 4y^2$, can be multiplied by $x^2 + 4y^2$ using the same formula.

$$\begin{aligned}&(x + 2y)(x - 2y)(x^2 + 4y^2)\\ &= (x^2 - 4y^2)(x^2 + 4y^2)\\ &= x^4 - 16y^4\end{aligned}$$

Practice Exercise 8

Multiply.

$(9u^2 + v^2)(3u - v)(3u + v)$

Answer: $81u^4 - v^4$

5.2 EXERCISES A

Multiply the polynomials.

1. $(3x)(7y)$

2. $(-5x^2)(-6xy)$

3. $(8a^2b)(-9ab^2)$

4. $-4x^3y(2xy - 5)$

5. $8a^2b^2(3ab - 2a + 5b)$

6. $-7a^3b(3a^2b^2 + 5a^2 - 6b^2)$

Multiply using the FOIL *method.*

7. $(x + y)(x + 2y)$

8. $(x - y)(x - 2y)$

9. $(x - y)(x + 2y)$

10. $(x + y)(x - 2y)$

11. $(2a + b)(a + 3b)$

12. $(2a - b)(a - 3b)$

13. $(2a + b)(a - 3b)$

14. $(2a - b)(a + 3b)$

15. $(5x - 2y)(x + 3y)$

16. $(10a + 3b)(7a - 4b)$

17. $(4x - 9y)(3x + 5y)$

18. $(4x + 9y)(3x - 5y)$

Multiply using columns.

19. $\begin{array}{r}7a^2 - 3ab + 4b^2\\ \underline{3a - 2b}\end{array}$

20. $\begin{array}{r}2x^2 + 4xy - 5y^2\\ \underline{9x + 2y}\end{array}$

21. $\begin{array}{r}5x^2y^2 - 3xy - 2\\ \underline{8xy - 7}\end{array}$

22. $\begin{array}{r}2a - ab + 3b\\ \underline{4a + 2ab - 3b}\end{array}$

23. $\begin{array}{r}7a - 2b + 5\\ \underline{3a - b + 2}\end{array}$

24. $\begin{array}{r}a^2 - 3ab + 2b^2\\ \underline{2a^2 + 4ab - 3b^2}\end{array}$

Multiply the polynomials.

25. $3(x - 5y)(2x + y)$

26. $a^2b(a + 2b)(a - 2b)$

27. $(x + 3y)^3$

28 $(a - 5b)(2a + b)(2a - 3b)$

29. $(x + y)(x^2 - xy + y^2)$

30 $(a - b)[3a^2 - (a + b)(a - 2b)]$

In an environmental study the final result $F(t)$ is given by $F(t) = N(t)S(t)$. Find $F(t)$ for the given $N(t)$ and $S(t)$.

31. $N(t) = 2t,\ S(t) = t^3 - t^2$

32. $N(t) = 2t - 5,\ S(t) = 3t + 2$

33. $N(t) = 2t^2 - 3t + 1,\ S(t) = 3t - 1$

34 $N(t) = t^2 - 2t - 5,\ S(t) = 2t^2 + t - 6$

35. Find $F(2)$ for $F(t)$ in Exercise 31.

36. Find $F(-2)$ for $F(t)$ in Exercise 32.

37. Find $F(3)$ for $F(t)$ in Exercise 33.

38. Find $F(-1)$ for $F(t)$ in Exercise 34.

Use the special product formulas to find the products.

39. $(x + 3)(x - 3)$

40. $(a - 4)^2$

41. $(b + 8)^2$

42. $(3x + 5)(3x - 5)$

43. $(3x + 5y)(3x - 5y)$

44. $(5a + 7b)^2$

45. $(8a - 5b)(8a - 5b)$

46. $(5x + 9y)^2$

47. $\left(\frac{1}{2}x - y\right)\left(\frac{1}{2}x + y\right)$

48 $(0.5x + 0.1y)^2$

49. $(x^2 + 1)(x^2 - 1)$

50. $(x^2 - 1)^2$

51. $(2a^2 + b^2)(2a^2 - b^2)$

52. $(3a^2 + b^2)^2$

53. $(3a^2 - 2b^4)(3a^2 + 2b^4)$

54. $(a^3 + 5b^2)^2$

55. $(x - y)^2 + (x + y)^2$

56 $[a - (b + c)][a + (b + c)]$

57. $[(2x - y)(2x + y)]^2$

58. $[2x^2 - (x - 2y)][2x^2 + (x - 2y)]$

59. $(u - 3v)(u + 3v)(u^2 + 9v^2)$

60. $(4x + 3y)(4x - 3y)(16x^2 + 9y^2)$

If Roy Troutman deposits A dollars in a savings account at i interest rate, the amount in his account at the end of 2 years is given by the equation $S = A(1 + i)^2$.

61. Obtain an equivalent equation by performing all the indicated multiplications.

62 Check your work in Exercise 61 by using both formulas to find S when $A = \$1000$ and $i = 0.12$.

FOR REVIEW

Tell whether the following are monomials, binomials, or trinomials and give the degree of each.

63. $2x^4 + 3y^3$ **64.** $a^2b^2 - c^2 + ac$ **65.** -6 **66.** $5x^2y^2z^2 - 2x^7y - 5z^4$

Perform the indicated operations.

67. $(x^2 + y^2 - 3) - (7x^2 - 4y^2 + 5) - (x^2 + 5y^2 - 3)$

68. $(6x^2y^2 - 2xy + 5) + (3x^2y^2 - 10) - (-10x^2y^2 + 5xy - 2)$

Exercises 69–72 review material from Section 1.5 to help you prepare for the next section. Use the distributive law to factor each expression.

69. $2x + 8y$ **70.** $uv + 3v$ **71.** $3x - 6y + 15z$ **72.** $aw - 2w + zw$

ANSWERS: 1. $21xy$ 2. $30x^3y$ 3. $-72a^3b^3$ 4. $-8x^4y^2 + 20x^3y$ 5. $24a^3b^3 - 16a^3b^2 + 40a^2b^3$ 6. $-21a^5b^3 - 35a^5b + 42a^3b^3$ 7. $x^2 + 3xy + 2y^2$ 8. $x^2 - 3xy + 2y^2$ 9. $x^2 + xy - 2y^2$ 10. $x^2 - xy - 2y^2$ 11. $2a^2 + 7ab + 3b^2$ 12. $2a^2 - 7ab + 3b^2$ 13. $2a^2 - 5ab - 3b^2$ 14. $2a^2 + 5ab - 3b^2$ 15. $5x^2 + 13xy - 6y^2$ 16. $70a^2 - 19ab - 12b^2$ 17. $12x^2 - 7xy - 45y^2$ 18. $12x^2 + 7xy - 45y^2$ 19. $21a^3 - 23a^2b + 18ab^2 - 8b^3$ 20. $18x^3 + 40x^2y - 37xy^2 - 10y^3$ 21. $40x^3y^3 - 59x^2y^2 + 5xy + 14$ 22. $8a^2 - 2a^2b^2 + 6ab + 9ab^2 - 9b^2$ 23. $21a^2 - 13ab + 29a + 2b^2 - 9b + 10$ 24. $2a^4 - 2a^3b - 11a^2b^2 + 17ab^3 - 6b^4$ 25. $6x^2 - 27xy - 15y^2$ 26. $a^4b - 4a^2b^3$ 27. $x^3 + 9x^2y + 27xy^2 + 27y^3$ 28. $4a^3 - 24a^2b + 17ab^2 + 15b^3$ 29. $x^3 + y^3$ 30. $2a^3 - a^2b + ab^2 - 2b^3$ 31. $F(t) = 2t^4 - 2t^3$ 32. $F(t) = 6t^2 - 11t - 10$ 33. $F(t) = 6t^3 - 11t^2 + 6t - 1$ 34. $F(t) = 2t^4 - 3t^3 - 18t^2 + 7t + 30$ 35. 16 36. 36 37. 80 38. 10 39. $x^2 - 9$ 40. $a^2 - 8a + 16$ 41. $b^2 + 16b + 64$ 42. $9x^2 - 25$ 43. $9x^2 - 25y^2$ 44. $25a^2 + 70ab + 49b^2$ 45. $64a^2 - 80ab + 25b^2$ 46. $25x^2 + 90xy + 81y^2$ 47. $\frac{1}{4}x^2 - y^2$ 48. $0.25x^2 + 0.1xy + 0.01y^2$ 49. $x^4 - 1$ 50. $x^4 - 2x^2 + 1$ 51. $4a^4 - b^4$ 52. $9a^4 + 6a^2b^2 + b^4$ 53. $9a^4 - 4b^8$ 54. $a^6 + 10a^3b^2 + 25b^4$ 55. $2x^2 + 2y^2$ 56. $a^2 - b^2 - 2bc - c^2$ 57. $16x^4 - 8x^2y^2 + y^4$ 58. $4x^4 - x^2 + 4xy - 4y^2$ 59. $u^4 - 81v^4$ 60. $256x^4 - 81y^4$ 61. $S = A + 2Ai + Ai^2$ 62. \$1254.40 63. binomial; 4 64. trinomial; 4 65. monomial; 0 66. trinomial; 8 67. $-7x^2 - 5$ 68. $19x^2y^2 - 7xy - 3$ 69. $2(x + 4y)$ 70. $v(u + 3)$ 71. $3(x - 2y + 5z)$ 72. $w(a - 2 + z)$

5.2 EXERCISES B

Multiply the polynomials.

1. $(5x)(8y)$ **2.** $(-8x^2)(-2xy^2)$ **3.** $(5ab^2)(-7a^2b)$

4. $-9xy^3(3xy - 2)$ **5.** $4a^3b^2(ab + 5a - 7b)$ **6.** $-10ab^3(5a^2b^2 + 2a^2 - 9b^2)$

Multiply using the FOIL *method.*

7. $(x + 3y)(x + y)$
8. $(x - 3y)(x - y)$
9. $(x - 3y)(x + y)$
10. $(x + 3y)(x - y)$
11. $(3a + b)(a + 2b)$
12. $(3a - b)(a - 2b)$
13. $(3a + b)(a - 2b)$
14. $(3a - b)(a + 2b)$
15. $(2x - 3y)(4x + y)$
16. $(12a + 5b)(3a - 7b)$
17. $(3x - 7y)(5x + 2y)$
18. $(3x + 7y)(5x - 2y)$

Multiply using columns.

19. $\begin{array}{l} 4a^2 - 5ab + 6b^2 \\ \underline{2a \quad - 3b \qquad\quad} \end{array}$
20. $\begin{array}{l} 3x^2 + 6xy - 2y^2 \\ \underline{7x \quad + 3y \qquad\quad} \end{array}$
21. $\begin{array}{l} 3x^2y^2 - 4xy - 5 \\ \underline{6xy \quad - 5 \qquad\quad} \end{array}$
22. $\begin{array}{l} 3a - 2ab + 5b \\ \underline{2a + 3ab - 4b} \end{array}$
23. $\begin{array}{l} 5a - 3b + 2 \\ \underline{2a - 2b + 1} \end{array}$
24. $\begin{array}{l} 3a^2 + \ ab - 5b^2 \\ \underline{\ a^2 - 3ab + 2b^2} \end{array}$

Multiply the polynomials.

25. $5(3x + 7y)(x - y)$
26. $ab^3(a - 3b)(a + 3b)$
27. $(2x + y)^3$
28. $(2a - 7b)(a + b)(5a - b)$
29. $(x - y)(x^2 + xy + y^2)$
30. $(2a - b)[a^2 - (2a - b)(a - b)]$

In an ecological study the final result $F(t)$ is given by $F(t) = E(t)H(t)$. Find $F(t)$ for the given $E(t)$ and $H(t)$.

31. $E(t) = t^2 - t,\ H(t) = 3t^2$
32. $E(t) = 3t + 4,\ H(t) = 2t - 3$
33. $E(t) = 3t^2 + 2t - 1,\ H(t) = 2t + 1$
34. $E(t) = -t^2 + 3t - 2,\ H(t) = t^2 + 3t - 4$
35. Find $F(3)$ for $F(t)$ in Exercise 31.
36. Find $F(-3)$ for $F(t)$ in Exercise 32.
37. Find $F(-1)$ for $F(t)$ in Exercise 33.
38. Find $F(2)$ for $F(t)$ in Exercise 34.

Use the special product formulas to find the products.

39. $(x + 7)(x - 7)$
40. $(a - 10)^2$
41. $(b + 9)^2$
42. $(2x + 7)(2x - 7)$
43. $(2x + 7y)(2x - 7y)$
44. $(9a + b)^2$
45. $(4a - 9b)^2$
46. $(10x + 3y)^2$
47. $\left(\frac{1}{3}x - y\right)\left(\frac{1}{3}x + y\right)$
48. $(0.2x + 0.5y)^2$
49. $(x^2 + 3)(x^2 - 3)$
50. $(x^2 - 3)^2$
51. $(a^2 - 2b^2)(a^2 + 2b^2)$
52. $(a^2 - 3b^2)^2$
53. $(5a^4 - 2b^2)(5a^4 + 2b^2)$
54. $(3a^2 + b^3)^2$
55. $(x + y)^2 - (x - y)^2$
56. $[a - (b - c)][a + (b - c)]$
57. $[(3x + 2y)(3x - 2y)]^2$
58. $[2x^2 + (x + 2y)][2x^2 - (x + 2y)]$
59. $(x + 5y)(x - 5y)(x^2 + 25y^2)$
60. $(2u + 3w)(2u - 3w)(4u^2 + 9w^2)$

If Susie Wong deposits A dollars in a savings account at i interest rate, the amount in her account at the end of 2 years is given by the equation $S = A(1 + i)^2$.

61. Obtain an equivalent equation by performing all the indicated multiplications.

62. Check your work in Exercise 61 by using both formulas to find S when $A = \$2000$ and $i = 0.14$.

FOR REVIEW

Tell whether the following are monomials, binomials, or trinomials and give the degree of each.

63. $7x^4 + 2x + 1$

64. $a^3b^2 - 5a^5b^3$

65. 0

66. 25

Perform the indicated operations.

67. $(3a^2 + 2b^2 - 5) - (-4a^2 + 3b^2 + 7) - (2a^2 - 8)$

68. $(3a^3b^3 + 5a^2b^2 - 4) - (-5a^3b^3 + 10ab + 7) - (6a^2b^2 - 3ab + 4)$

Exercises 69–72 review material from Section 1.5, to help you prepare for the next section. Use the distributive law to factor each expression.

69. $3x + 6y$

70. $ax + 7x$

71. $2x + 6y - 12z$

72. $xa - 6a + ya$

5.2 EXERCISES C

Multiply. Assume that m and n represent positive integers.

1. $[(x^n)^{2n}]^2$
[Answer: x^{4n^2}]

2. $(a^n)^{n^2}(a^{2n})^3$

3. $x^2y^n(x^ny^3 - xy^n)$

4. $(x^n + y^m)(x^n - 2y^m)$

5. $a^2(a^{m+n})^{m-n}$

6. $(a^2)^{mn}(a^{m-n})^{m-n}$
[Answer: $a^{m^2+n^2}$]

7. $(2x^n + 1)^2$
[Answer: $4x^{2n} + 4x^n + 1$]

8. $(2x^n - 1)^2$

9. $(2x^n - 1)(2x^n + 1)$

10. $(x^{2n} + y^{2n})(x^n + y^n)(x^n - y^n)$
[Answer: $x^{4n} - y^{4n}$]

11. $(x^n - y^n)^2(x^n + y^n)^2$

12. $(x^{n+1} + y^{n+1})(x^{n+1} - y^{n+1})$
[Answer: $x^{2n+2} - y^{2n+2}$]

13. $(a^n + b^n)^2$

14. $(2a^n - 3b^n)(a^n + b^n)$

5.3 COMMON FACTORS AND GROUPING

STUDENT GUIDEPOSTS

1. Greatest Common Factor (GCF)
2. Factoring by Grouping

1 GREATEST COMMON FACTOR (GCF)

We begin our study of factoring with polynomials having a common factor in each term. Remember that a factor of a term is a number or power of a variable that is a divisor of the term. For example,

$2, x, x^2$ are factors of $2x^2$

$2, 3, 6, x, y$ are factors of $6xy$.

From the above we see that $2x^2$ and $6xy$ have *common factors* of 2 and x, while

$2x$ is the *greatest common factor*. The **greatest common factor (GCF)** of a polynomial is the common factor of each term containing the largest numerical factor and the highest power of each variable. We factor an expression like $2x^2 - 6xy$ by factoring out the GCF.

$$2x^2 - 6xy = 2x(x - 3y)$$

Notice that factoring uses the distributive law in reverse of the order used to multiply polynomials. Thus, factoring can be checked by multiplying. We shall see later that the process of factoring will be useful in such operations as simplifying fractions and solving equations.

To factor a polynomial such as $3x^2y + 6xy^2$, we could first factor each term into prime factors,

$$3x^2y + 6xy^2 = 3 \cdot x \cdot x \cdot y + 2 \cdot 3 \cdot x \cdot y \cdot y,$$

and then note that $3xy$ is the greatest common factor. To complete the factoring, we write

$$3x^2y + 6xy^2 = 3xy(x + 2y).$$

The step of factoring into primes may be omitted once we gain experience at recognizing common factors.

The expression $3x^2y + 6xy^2$ could also have been factored as $3(x^2y + 2xy^2)$ or $3x(xy + 2y^2)$, but the remaining binomial in each case still has a common factor in each term. When the instruction is given to factor an expression we will always factor completely by removing the greatest common factor.

EXAMPLE 1 Finding the Greatest Common Factor

Find the GCF of each polynomial.

(a) $12x^2y^3 + 20xy^2 = 2 \cdot 2 \cdot 3 \cdot x \cdot x \cdot y^2 \cdot y + 2 \cdot 2 \cdot 5 \cdot x \cdot y^2$

The GCF is $2 \cdot 2 \cdot x \cdot y^2 = 4xy^2$.

(b) $27a^3b^3 - 18a^4b^2 = 3 \cdot 3 \cdot 3 \cdot a^3 \cdot b^2 \cdot b - 2 \cdot 3 \cdot 3 \cdot a^3 \cdot a \cdot b^2$

The GCF is $3 \cdot 3 \cdot a^3 \cdot b^2 = 9a^3b^2$.

Practice Exercise 1

Find the GCF.

(a) $18x^4y + 24x^3y^2$

(b) $35a^5b^3 - 42a^2b^5$

Answers: (a) $6x^3y$ (b) $7a^2b^3$

EXAMPLE 2 Factoring By Removing the GCF

Factor each polynomial.

(a) $5x^2 + 10xy = 5 \cdot x \cdot x + 2 \cdot 5 \cdot x \cdot y$ GCF is $5x$

$= 5x(x + 2y)$ Distributive law

(b) $3x^2y + 6xy^2 = 3 \cdot x \cdot x \cdot y + 2 \cdot 3 \cdot x \cdot y \cdot y$ GCF is $3xy$

$= 3xy(x + 2y)$ Distributive law

(c) $12x^2y^3 + 20xy^2$

$= 2 \cdot 2 \cdot 3 \cdot x \cdot x \cdot y^2 \cdot y + 2 \cdot 2 \cdot 5 \cdot x \cdot y^2$ GCF is $4xy^2$

$= 4xy^2(3xy + 5)$ Distributive law

With practice we can shorten our work.

(d) $27a^3b^3 - 18a^4b^2 = (9a^3b^2) \cdot 3b - (9a^3b^2) \cdot 2a$ GCF is $9a^3b^2$

$= 9a^3b^2(3b - 2a)$ Distributive law

(e) $2a^2b - ab = (ab) \cdot 2a - (ab) \cdot 1$ GCF is ab

$= ab(2a - 1)$ The 1 helps us remember what remains after ab is factored out

Practice Exercise 2

Factor.

(a) $7x^3 + 14x^2y$

(b) $4x^3y + 12x^2y^2$

(c) $32x^4y^4 - 12x^2y^2$

(d) $6a^5b^3 - 24a^4b^5$

(e) $5x^2y^2 - 5xy$

(f) $7x^2 - y^2 = 7x^2 - y^2$ GCF is 1; cannot be factored

(g) $4a^2b^3 - 6a^2b^2 + 8ab^2$

$= (2ab^2) \cdot 2ab - (2ab^2) \cdot 3a + (2ab^2) \cdot 4$ GCF is $2ab^2$

$= 2ab^2(2ab - 3a + 4)$

(h) $-6x^3y^3 - 12x^2y^2 - 18xy$

$= -(6xy) \cdot x^2y^2 - (6xy) \cdot 2xy - (6xy) \cdot 3$ GCF is $6xy$

$= 6xy(-x^2y^2 - 2xy - 3)$

$= -6xy(x^2y^2 + 2xy + 3)$ -1 is factored out

(f) $3a^2b^2 + c^2$

(g) $2a^4b^2 - 10a^3b^3 - 6a^2b^4$

(h) $-5xy^2 - 15x^2y + 10x^2y^2$

Answers: (a) $7x^2(x + 2y)$
(b) $4x^2y(x + 3y)$
(c) $4x^2y^2(8x^2y^2 - 3)$
(d) $6a^4b^3(a - 4b^2)$
(e) $5xy(xy - 1)$
(f) cannot be factored
(g) $2a^2b^2(a^2 - 5ab - 3b^2)$
(h) $-5xy(y + 3x - 2xy)$

In Example 2(h) two answers are correct. Most of the time we factor out negative numbers, if necessary, to leave a plus sign in the first term of the remaining polynomial.

In some cases polynomials may contain symbols of grouping, or we may introduce grouping symbols to aid in factoring. For example, the polynomial

$$5x(x + y) + 7(x + y)$$

has the common factor $x + y$ in each of its terms. If we let the variable u equal $x + y$, the expression becomes $5xu + 7u$ and can be factored as follows.

$5xu + 7u = 5x \cdot u + 7 \cdot u$ GCF is u

$= (5x + 7)u$

$= (5x + 7)(x + y)$ Since $u = x + y$

We may omit the substitution of the variable u when we see that a grouped expression like $(x + y)$ can be considered as a single factor.

EXAMPLE 3 Common Binomial Factors

Factor each polynomial.

(a) $2a(a + 2b) + 3b(a + 2b) = 2au + 3bu$ Let $u = a + 2b$

$= (2a + 3b)u$ The GCF is u

$= (2a + 3b)(a + 2b)$ Since $u = a + 2b$

(b) $7x(2x - 5y) - y(2x - 5y) = (7x - y)(2x - 5y)$ GCF is $2x - 5y$

(c) $5a(b - 1) - (b - 1) = 5a(b - 1) - (1)(b - 1)$ GCF is $b - 1$

$= (5a - 1)(b - 1)$ Distributive law

Practice Exercise 3

Factor.

(a) $5x(2x - 7y) + 4y(2x - 7y)$

(b) $9a(a - 3b) - b(a - 3b)$

(c) $6x(y - 2) - (y - 2)$

Answers: (a) $(5x + 4y)(2x - 7y)$
(b) $(9a - b)(a - 3b)$
(c) $(6x - 1)(y - 2)$

2 FACTORING BY GROUPING

The ability to factor out a common binomial factor, as shown in Example 3, allows us to **factor by grouping.** With this process we group terms, factor the individual groups, and then factor the complete expression.

EXAMPLE 4 Factoring By Grouping

Factor by grouping.

(a) $3x^2 + 6xy + 5x + 10y$

$= (3x^2 + 6xy) + (5x + 10y)$

$= 3x(x + 2y) + 5(x + 2y)$ Factor $3x$ from first parentheses and 5 from second using the distributive law

$= (3x + 5)(x + 2y)$ Since $x + 2y$ is the GCF

(b) $6ab^2 - 3ab - 14b + 7$

$= (6ab^2 - 3ab) - (14b - 7)$ Note the -7, not 7

$= 3ab(2b - 1) - 7(2b - 1)$ $3ab$ from first parentheses and 7 from second

$= (3ab - 7)(2b - 1)$ Since $2b - 1$ is the GCF

PRACTICE EXERCISE 4

Factor by grouping.

(a) $7x^3 + 14x^2y + 3xy + 6y^2$

(b) $12a^2b - 4ab - 15a + 5$

Answers: (a) $(7x^2 + 3y)(x + 2y)$ (b) $(4ab - 5)(3a - 1)$

Sometimes an expression can be grouped in more than one way. Consider the expression in Example 4(b).

$6ab^2 - 3ab - 14b + 7$

$= 6ab^2 - 14b - 3ab + 7$ Commutative law

$= 2b(3ab - 7) - 1(3ab - 7)$ Watch the sign here

$= (2b - 1)(3ab - 7)$

Notice that we obtain the same pair of factors as before; they are simply written in a different order.

Generally, we first try to find any common factors in all the terms of an expression. Once we have done this, if the expression has four terms, we try to factor by grouping, as illustrated in Example 5.

EXAMPLE 5 Factoring the GCF and by Grouping

Factor.

$4x^3y + 2x^2y^2 - 2x^2y - xy^2$

$= xy[4x^2 + 2xy - 2x - y]$ Remove greatest common factor

$= xy[(4x^2 + 2xy) - (2x + y)]$ $2x + y$, not $2x - y$

$= xy[2x(2x + y) - (1)(2x + y)]$ $2x$ from first parentheses and 1 from second

$= xy[(2x - 1)(2x + y)]$ Since $2x + y$ is the GCF

$= xy(2x - 1)(2x + y)$

PRACTICE EXERCISE 5

Factor.

$6x^4y + 3x^3y^2 - 14x^3 - 7x^2y$

Answer: $x^2(3xy - 7)(2x + y)$

Factoring can simplify calculations in some applications, as in Example 6.

EXAMPLE 6 Physics Problem

If an object is propelled upward with initial velocity of 96 ft/sec from an initial height of 240 ft, ignoring air resistance, the height h of the object in feet after t seconds is given by $h = -16t^2 + 96t + 240$. Factor the

PRACTICE EXERCISE 6

Refer to Example 6 to find the height of an object after 3 seconds.

greatest common factor from the right side of this equation and use the factored form to find the height of an object 5 seconds after it is propelled upward.

Factoring, we obtain

$$\begin{aligned} h &= -16t^2 + 96t + 240 \\ &= -16(t^2 - 6t - 15). \end{aligned}$$

Substitute 5 for t to obtain the height after 5 seconds.

$$\begin{aligned} h &= -16(5^2 - 6(5) - 15) \\ &= -16(25 - 30 - 15) \\ &= -16(-20) = 320 \end{aligned}$$

Thus, the object is 320 ft above the ground after 5 seconds.

Answer: 384 ft

5.3 EXERCISES A

Factor. First try to remove the GCF from each polynomial. If the expression has four terms, try to factor by grouping.

1. $5x + 10y$

2. $12a - 3b$

3. $14x + 7$

4. $5a^2b - 10ab^2$

5. $6x^2y^2 + 12xy$

6. $2ab^2 - 2ab$

7. $3a^2b^4 - 5$

8. $24x^8y^4 + 9x^3y^2$

9. $7a^2b - 14ab^2 + 28ab$

10. $-3x^5y^5 - 6x^4y^4 - 9x^6y^6$

11. $42a^5b^4 - 66a^4b^5 - 36a^4b^4$

12. $15a^3b^2 - 45a^2b^3 - 60a^2b$

13 $64xy^5z^2 - 128x^2y^3z^3 - 144xy^2z^4$

14 $3a(a + b) - 2b(a + b)$

15. $7x(5x - y) + 2y(5x - y)$

16. $2a^2(a + 4b) - 9b(a + 4b)$

17 $5x^2 + 2xy + 10x + 4y$

18. $a^2b - ab^2 + 5a - 5b$

19. $9x^2y - 18xy^2 - xy + 2y^2$

20 $2a^2b + 6ab - 3a - 9$

21. $x^3y^2 - y^3 + x^3y - y^2$

22. $a^3b^2c^3 - a^3c^2 - b^2c^4 + c^3$

23. $5x^2y - 5xy + 20x - 20$

24 $7x^3y^3 + 21x^2y^2 - 10x^3y^2 - 30x^2y$

An economist finds that certain economic conditions are related to the equation $E = 25c^2e^3 - 10c^3e^2$.

25 Factor the expression for E.

26. Check your factoring in Exercise 25 by evaluating both expressions for $c = 2$ and $e = 3$.

If an object is propelled upward from a tower with initial velocity of v ft/sec from an initial height of s ft, ignoring air resistance, the height h of the object in feet after t seconds is given by $h = -16t^2 + vt + s$. *Use the given values of v and s in Exercises 27–28, and express h as a product of factors. Use the factored form to find the height of an object that is propelled upward after* ***(a)*** *3 seconds and* ***(b)*** *10 seconds.*

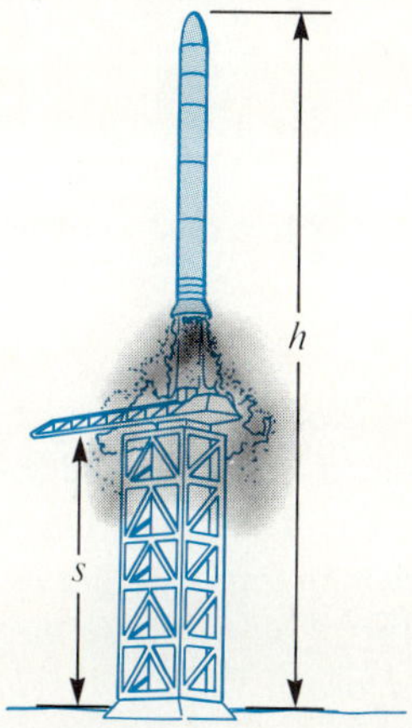

27. $v = 144$ ft/sec and $s = 160$ ft

28. $v = 112$ ft/sec and $s = 1920$ ft

FOR REVIEW

Multiply.

29. $6x^2y(5x^2 - 4xy + 10y^2)$

30. $(2x - y)(4x - 3y)$

31. $(5a - 3b)(10a + 7b)$

32. $(7x + 3y)(8x^2 - 4xy + 5y^2)$

33. $(3a + 2b)(3a - 2b)$

34. $(-4x + 3y)^2$

ANSWERS: 1. $5(x + 2y)$ 2. $3(4a - b)$ 3. $7(2x + 1)$ 4. $5ab(a - 2b)$ 5. $6xy(xy + 2)$ 6. $2ab(b - 1)$ 7. cannot be factored 8. $3x^3y^2(8x^5y^2 + 3)$ 9. $7ab(a - 2b + 4)$ 10. $-3x^4y^4(xy + 2 + 3x^2y^2)$ 11. $6a^4b^4(7a - 11b - 6)$ 12. $15a^2b(ab - 3b^2 - 4)$ 13. $16xy^2z^2(4y^3 - 8xyz - 9z^2)$ 14. $(3a - 2b)(a + b)$ 15. $(7x + 2y)(5x - y)$ 16. $(2a^2 - 9b)(a + 4b)$ 17. $(x + 2)(5x + 2y)$ 18. $(ab + 5)(a - b)$ 19. $y(9x - 1)(x - 2y)$ 20. $(2ab - 3)(a + 3)$ 21. $y(y + 1)(x^3 - y)$ 22. $c^2(a^3 - c)(b^2c - 1)$ 23. $5(xy + 4)(x - 1)$ 24. $x^2y(7y - 10)(xy + 3)$ 25. $E = 5c^2e^2(5e - 2c)$ 26. 1980 27. (a) 448 ft (b) 0 ft (the object is on the ground) 28. (a) 2112 ft (b) 1440 ft 29. $30x^4y - 24x^3y^2 + 60x^2y^3$ 30. $8x^2 - 10xy + 3y^2$ 31. $50a^2 + 5ab - 21b^2$ 32. $56x^3 - 4x^2y + 23xy^2 + 15y^3$ 33. $9a^2 - 4b^2$ 34. $16x^2 - 24xy + 9y^2$

5.3 EXERCISES B

Factor. First try to remove the GCF from each polynomial. If the expression has four terms, try to factor by grouping.

1. $6x - 9y$

2. $15a + 10b$

3. $18y - 9$

4. $14a^2b^2 + 21ab$

5. $10x^2y - 5xy^2$

6. $16a^2b^2 - 6ab$

7. $7a^3b^3 + 6$

8. $20x^9y^2 - 15x^6y^6$

9. $-6a^5b^5 - 8a^4b^4 - 4a^3b^2$

10. $11xy^2 - 22x^2y - 44xy$

11. $96a^4b^4 + 72a^9b^6 + 108a^3b^6$

12. $1000a^3b^3 + 10ab - 100a^2b^2$

13. $35x^9y^9z^9 - 49x^7y^7z^7 - 77x^5y^5z^5$

14. $5ab(2a - b) + 4b(2a - b)$

15. $9x(3x + 2y) - 5y(3x + 2y)$

16. $5a^2(3a - 4b) + 12b^2(3a - 4b)$

17. $7x^2 - 14xy + 4x - 8y$

18. $a^3 + a^2 + ab^2 + b^2$

19. $5x^2y^2 + 10x^2 - 7y^2 - 14$

20. $3a^2b^2 - 9ab^2 - 5a + 15$

21. $x^3y^3 + y^4 - x^3y^2 - y^3$

22. $a^2b^2c^2 - a^2b^2c - b^2c^3 + b^2c^2$

23. $6x^2y^2 + 6xy - 24xy^2 - 24y$

24. $6x^4y^4 - 6x^3y^2 + 8xy^2 - 8$

An environmentalist finds that $E = 20r^4s^2 - 35r^2s$.

25. Factor the expression for E.

26. Check your factoring in Exercise 25 by evaluating both expressions for $r = 3$ and $s = -1$.

If an object is propelled upward from a tower with initial velocity of v ft/sec from an initial height of s ft, ignoring air resistance, the height h of the object in feet after t seconds is given by $h = -16t^2 + vt + s$. *Use the given values of v and s in Exercises 27–28, and express h as a product of factors. Use the factored form to find the height of an object that is propelled upward after* **(a)** 3 *seconds and* **(b)** 10 *seconds.*

27. $v = 80$ ft/sec and $s = 1600$ ft

28. $v = 160$ ft/sec and $s = 2800$ ft

FOR REVIEW

Multiply.

29. $-8xy^2(4x^2y^2 + 3xy - 5)$

30. $(3x + y)(5x - 7y)$

31. $(4a - 5b)(9a - 2b)$

32. $(2x - 3y)\left(\frac{1}{3}x^2 + \frac{1}{2}xy + \frac{2}{3}y^2\right)$

33. $(5a + 2b)^2$

34. $(2x^2 - y)(2x^2 + y)$

5.3 EXERCISES C

Factor. Assume that n represents a positive integer.

1. $x^{n+2}y^{n+2} - x^ny^n$

2. $x^{3n}y^{n+1} + 5x^{2n}y^n$
[*Hint:* $x^{2n}y^n$ is a common factor of both terms.]

3. $x^ny + y + 2x^n + 2$
[Answer: $(x^n + 1)(y + 2)$]

4. $x^ny^n - y^n + 2x^n - 2$

5.4 FACTORING TRINOMIALS

STUDENT GUIDEPOSTS

1. Factoring $x^2 + bx + c$
2. Factoring $ax^2 + bx + c$, $a \neq 1$
3. Factoring Trinomials by Grouping
4. Factoring Using Substitution

1 FACTORING $x^2 + bx + c$

We begin our study of factoring trinomials of the form $ax^2 + bx + c$ by considering the case where a is 1. Suppose we use the FOIL method to obtain a trinomial product.

$$\begin{aligned}(x + 5)(x - 7) &= \underset{(F)}{x \cdot x} + \underset{(O)}{x \cdot (-7)} + \underset{(I)}{5 \cdot x} + \underset{(L)}{5 \cdot (-7)} \\ &= x^2 - 7x + 5x - 35 \\ &= \underset{(F)}{x^2} \quad \underset{(O)+(I)}{- 2x} \quad \underset{(L)}{- 35}\end{aligned}$$

The product of the first terms, (F), gives x^2, the product of the last terms, (L), gives -35, and the sum of (O) and (I) gives the middle term, $-2x$. To factor $x^2 - 2x - 35$, we reverse the multiplication steps above and fill in the blanks in

$$x^2 - 2x - 35 = (x + \_\_)(x + \_\_).$$

The numbers in the blanks must multiply to give -35 and must add to give -2.

$$x^2 - 2x - 35 = (x + 5)(x - 7)$$

(sum: $5 + (-7) = -2$; product: $5 \cdot (-7) = -35$)

To Factor $x^2 + bx + c$

1. Write $x^2 + bx + c = (x + \_\_)(x + \_\_)$.
2. List all pairs of integers whose product is c.
3. Fill in the blanks with the pair from the list whose sum is b.
4. Check your work by multiplying the factors to see if their product equals the original trinomial.

EXAMPLE 1 FACTORING TRINOMIALS OF THE FORM $x^2 + bx + c$

Factor each trinomial.

(a) $x^2 + 7x + 10$ ($b = 7$ and $c = 10$)

$x^2 + 7x + 10 = (x + \_\_)(x + \_\_)$

Factors of $c = 10$	*Sum of factors*
10, 1	$10 + 1 = 11$
2, 5	**$2 + 5 = 7$**
$-10, -1$	$-10 + (-1) = -11$
$-2, -5$	$-2 + (-5) = -7$

PRACTICE EXERCISE 1

Factor.

(a) $x^2 + 7x + 12$

Since the product of 2 and 5 is 10, which equals c, and the sum of 2 and 5 is 7, which equals b, we write

$$x^2 + 7x + 10 = (x + 2)(x + 5).$$

Note that since $c = 10 > 0$, the factors in each pair must have the same sign. Also, since $b = 7 > 0$, negative factors will not work. As a final check of our work, we use FOIL to multiply.

$$\begin{aligned}(x + 2)(x + 5) &= x^2 + 5x + 2x + 10\\ &= x^2 + 7x + 10\end{aligned}$$

(b) $x^2 - 7x + 10$ ($b = -7$ and $c = 10$)

$x^2 - 7x + 10 = (x + \_\_)(x + \_\_)$

Factors of $c = 10$	*Sum of factors*
10, 1	$10 + 1 = 11$
2, 5	$2 + 5 = 7$
−10, −1	$-10 + (-1) = -11$
−2, −5	$-2 + (-5) = -7$

$x^2 - 7x + 10 = (x - 2)(x - 5)$ Check using FOIL

Since $b = -7$, positive factors will not work.

(c) $x^2 + 3x - 10$ ($b = 3$ and $c = -10$)

$x^2 + 3x - 10 = (x + \_\_)(x + \_\_)$

Factors of $c = -10$	*Sum of factors*
10, −1	$10 + (-1) = 9$
−10, 1	$-10 + 1 = -9$
2, −5	$2 + (-5) = -3$
−2, 5	$-2 + 5 = 3$

$x^2 + 3x - 10 = (x - 2)(x + 5)$ Check using FOIL

Since $c = -10 < 0$, the factors of c must have opposite signs.

(d) $x^2 - 3x - 10$ ($b = -3$ and $c = -10$)

$x^2 - 3x - 10 = (x + \_\_)(x + \_\_)$

Factors of $c = -10$	*Sum of factors*
10, −1	$10 + (-1) = 9$
−10, 1	$-10 + 1 = -9$
2, −5	$2 + (-5) = -3$
−2, 5	$-2 + 5 = 3$

$x^2 - 3x - 10 = (x + 2)(x - 5)$ Check using FOIL

(b) $x^2 - 7x + 12$

(c) $x^2 + x - 12$

(d) $x^2 - x - 12$

Answers: (a) $(x + 3)(x + 4)$ (b) $(x - 3)(x - 4)$ (c) $(x - 3)(x + 4)$ (d) $(x + 3)(x - 4)$

Our results in Example 1 would have been similar had the trinomials been in two variables. Compare Example 1 with the following.

$$\begin{aligned}x^2 + 7xy + 10y^2 &= (x + 2y)(x + 5y)\\ x^2 - 7xy + 10y^2 &= (x - 2y)(x - 5y)\\ x^2 + 3xy - 10y^2 &= (x - 2y)(x + 5y)\\ x^2 - 3xy - 10y^2 &= (x + 2y)(x - 5y)\end{aligned}$$

Thus, factoring with two or more variables can be as easy as with one variable. The only difference is that we are filling in the blanks in

$$x^2 + bxy + cy^2 = (x + \_\_y)(x + \_\_y).$$

EXAMPLE 2 Factoring Trinomials in Two Variables

Factor $x^2 - 8xy + 12y^2$.

$$x^2 - 8xy + 12y^2 = (x + \_\_y)(x + \_\_y)$$

Factors of $c = 12$	*Sum of factors*
12, 1	$12 + 1 = 13$
6, 2	$6 + 2 = 8$
4, 3	$4 + 3 = 7$
−12, −1	$-12 + (-1) = -13$
−6, −2	$-6 + (-2) = -8$
−4, −3	$-4 + (-3) = -7$

$$x^2 - 8xy + 12y^2 = (x - 6y)(x - 2y)$$

To check we multiply.

$$(x - 6y)(x - 2y) = x^2 - 2xy - 6xy + 12y^2 = x^2 - 8xy + 12y^2$$

Practice Exercise 2

Factor $x^2 - 11xy + 28y^2$.

Answer: $(x - 4y)(x - 7y)$

When factoring trinomials, remember to remove any common factors first, as in the next example.

EXAMPLE 3 Removing the GCF First

Factor $2a^2 - 10ab - 28b^2$.

First factor out the common factor 2.

$$2a^2 - 10ab - 28b^2 = 2(a^2 - 5ab - 14b^2)$$

Now factor the trinomial.

Factors of $c = -14$	*Sum of factors*
14, −1	$14 + (-1) = 13$
−14, 1	$-14 + 1 = -13$
7, −2	$7 + (-2) = 5$
−7, 2	$-7 + 2 = -5$

Thus we know $a^2 - 5ab - 14b^2 = (a - 7b)(a + 2b)$ and $2a^2 - 10ab - 28b^2 = 2(a - 7b)(a + 2b)$.

Practice Exercise 3

Factor $3x^2 + 12x - 36$.

Answer: $3(x - 2)(x + 6)$

CAUTION

When factoring a trinomial with a common factor, don't forget to include that factor in the final result. In Example 3 the common factor 2 is included with the factors $(a - 7b)$ and $(a + 2b)$.

2 FACTORING $ax^2 + bx + c$, $a \neq 1$

When factoring $ax^2 + bx + c$ with $a \neq 1$, we use the following trial and error method.

To Factor $ax^2 + bx + c$

1. Write $ax^2 + bx + c = (\_x + \_)(\_x + \_)$.
2. List all pairs of integers whose product is a and use these in the first blanks in each binomial factor.
3. List all pairs of integers whose product is c and use these in the second blanks.
4. Use trial and error to determine which pairs give the correct value for b.
5. Check your work by multiplying the factors.

EXAMPLE 4 Factoring $ax^2 + bx + c$ when $a \neq 1$

Factor $3x^2 + 17x + 10$. $(a = 3,\ b = 17,\ c = 10)$

Factors of 10

$$3x^2 + 17x + 10 = (\_x + \_)(\_x + \_)$$

Factors of 3

Since all terms of the trinomial are positive, list only the positive factors of a and c. List the factors of c in both orders as a reminder to try all possibilities.

Factors of a	*Factors of c*
3, 1	10, 1
	1, 10
	5, 2
	2, 5

$3x^2 + 17x + 10$

$= (\_x + \_)(\_x + \_)$

$= (3x + \_)(x + \_)$ The only factors of 3 are 3 and 1

$\stackrel{?}{=} (3x + 10)(x + 1)$ Does not work since $3x + 10x = 13x$

$\stackrel{?}{=} (3x + 1)(x + 10)$ Does not work since $30x + x = 31x$

$\stackrel{?}{=} (3x + 5)(x + 2)$ Does not work since $6x + 5x = 11x$

$\stackrel{?}{=} (3x + 2)(x + 5)$ This works since $15x + 2x = 17x$

$$3x^2 + 17x + 10 = (3x + 2)(x + 5)$$

To check we multiply.

$$(3x + 2)(x + 5) = 3x^2 + 15x + 2x + 10 = 3x^2 + 17x + 10$$

Practice Exercise 4

Factor $3x^2 + 13x + 14$.

Answer: $(3x + 7)(x + 2)$

When factoring $ax^2 + bxy + cy^2$ with $a \neq 1$ we fill in the blanks in

$$ax^2 + bxy + cy^2 = (\_x + \_y)(\_x + \_y).$$

EXAMPLE 5 Factoring $ax^2 + bxy + cy^2$ when $a \neq 1$

Factor $8x^2 - 13xy + 5y^2$. ($a = 8$, $b = -13$, $c = 5$)

Factors of 5

$$8x^2 - 13xy + 5y^2 = (\_\_x + \_\_y)(\_\_x + \_\_y)$$

Factors of 8

Since $c = 5 > 0$ and $b = -13 < 0$, the factors of c must both be negative.

Factors of a	*Factors of c*
8, 1	−1, −5
1, 8	
4, 2	
2, 4	

$8x^2 - 13xy + 5y^2$

$= (\_\_x + \_\_y)(\_\_x + \_\_y)$

$= (\_\_x - y)(\_\_x - 5y)$ The only factors of c are -1 and -5

$\stackrel{?}{=} (8x - y)(x - 5y)$ Does not work since $-40xy - xy = -41xy$

$\stackrel{?}{=} (x - y)(8x - 5y)$ This works because $-5xy - 8xy = -13xy$

$$8x^2 - 13xy + 5y^2 = (x - y)(8x - 5y)$$

To check we multiply.

$$(x - y)(8x - 5y) = 8x^2 - 5xy - 8xy + 5y^2$$
$$= 8x^2 - 13xy + 5y^2$$

Practice Exercise 5

Factor $6x^2 - 31xy + 35y^2$.

Answer: $(3x - 5y)(2x - 7y)$

In general when factoring we insist that all factors be polynomials with integer coefficients. This means that there are many polynomials that cannot be factored. For example, $x^2 + x + 1$, $x^2 + 2xy + 2y^2$, $3u^2 - 4uv + 8v^2$, and $5x^2 - 2xy - y^2$ cannot be factored into binomials with integer coefficients. To show that $x^2 + 2xy + 2y^2$ cannot be factored, we can list the factors of $c = 2$ and see that they do not add to give $b = 2$.

Factors of c = 2	*Sum of factors*
1, 2	1 + 2 = 3

Since 1 and 2 are the only positive factors of $c = 2$, and $1 + 2 = 3 \neq 2$, the trinomial cannot be factored.

Always remove common factors first, as in the next example.

EXAMPLE 6 Removing the GCF First

Factor $-6x^2 - 21xy + 45y^2$.

We first factor out the common factor and make $a > 0$.

$$-6x^2 - 21xy + 45y^2 = -3(2x^2 + 7xy - 15y^2)$$

Next, we list the factors of a and c for the trinomial $2x^2 + 7xy - 15y^2$ ($a = 2$, $b = 7$, $c = -15$).

Practice Exercise 6

Factor $-10x^2 - 95xy + 50y^2$.

Factors of a	*Factors of c*		
2, 1	15, −1	and	−15, 1
	1, −15	and	−1, 15
	5, −3	and	−5, 3
	3, −5	and	−3, 5

$-6x^2 - 21xy + 45y^2 = -3(2x^2 + 7xy - 15y^2)$ First factor out the common factor

$= -3(2x + \_\_y)(x + \_\_y)$

$\stackrel{?}{=} -3(2x + 5y)(x - 3y)$ Does not work

$\stackrel{?}{=} -3(2x - 5y)(x + 3y)$ Does not work

$\stackrel{?}{=} -3(2x + 3y)(x - 5y)$ Does not work

$\stackrel{?}{=} -3(2x - 3y)(x + 5y)$ Works

$$-6x^2 - 21xy + 45y^2 = -3(2x - 3y)(x + 5y)$$

Check this by multiplying.

Answer: $-5(2x - y)(x + 10y)$

3 FACTORING TRINOMIALS BY GROUPING

Another method for factoring trinomials involves factoring by grouping. Consider the following.

$2x^2 + 11xy + 12y^2 = 2x^2 + 8xy + 3xy + 12y^2$ $11xy = 8xy + 3xy$

$= (2x^2 + 8xy) + (3xy + 12y^2)$ Group terms

$= 2x(x + 4y) + 3y(x + 4y)$ Factor the groups

$= (2x + 3y)(x + 4y)$ Factor out the common factor $x + 4y$

We have factored $2x^2 + 11xy + 12y^2$ by grouping the appropriate terms. But how do we decide to write $11xy = 8xy + 3xy$? Notice that 8 and 3 are factors of $2 \cdot 12 = 24$. That is, 8 and 3 are factors of the product ac in $ax^2 + bxy + cy^2$.

To Factor $ax^2 + bxy + cy^2$ by Grouping

1. Find the product ac.
2. List the factors of ac until a pair is found which add to give b.
3. Write bxy as a sum using these factors as coefficients of xy.
4. Factor the result by grouping.

EXAMPLE 7 Factoring a Trinomial by Grouping

Factor $3x^2 - 8xy + 5y^2$ by grouping.

Factors of $ac = 3 \cdot 5 = 15$	*Sum of factors*
15, 1	$15 + 1 = 16$
−15, −1	$-15 + (-1) = -16$
5, 3	$5 + 3 = 8$
−5, −3	$-5 + (-3) = -8$

PRACTICE EXERCISE 7

Factor $2x^2 - 11xy + 9y^2$ by grouping.

$$3x^2 - 8xy + 5y^2 = 3x^2 - 5xy - 3xy + 5y^2$$
$$= x(3x - 5y) - y(3x - 5y) \quad \text{Factor out } x \text{ and } -y$$
$$= (x - y)(3x - 5y) \quad \text{Common factor is } 3x - 5y$$

Check this by multiplying.

Answer: $(2x - 9y)(x - y)$

EXAMPLE 8 FACTORING A TRINOMIAL BY GROUPING

Factor $7u^2 - 33uv - 10v^2$ by grouping.

We try those factors of $ac = 7(-10) = -70$ which seem most likely to add to $b = -33$.

Factors of $ac = -70$	*Sum of factors*
$35, -2$	$35 + (-2) = 33$
$-35, 2$	$-35 + 2 = -33$

$$7u^2 - 33uv - 10v^2 = 7u^2 - 35uv + 2uv - 10v^2 \quad 7u^2 + 2uv - 35uv - 10v^2 \text{ also works}$$
$$= 7u(u - 5v) + 2v(u - 5v) \quad \text{Factor groups}$$
$$= (7u + 2v)(u - 5v) \quad \text{Common factor is } u - 5v$$

Check this by multiplying.

PRACTICE EXERCISE 8

Factor $5x^2 + 4xy - 33y^2$ by grouping.

Answer: $(5x - 11y)(x + 3y)$

4 FACTORING USING SUBSTITUTION

The factors of some polynomials can be more easily found if we make a simplifying substitution. For example, $2(x - y)^2 - 3(x - y) + 1$ can be factored by letting $u = x - y$. Thus,

$$2(x - y)^2 - 3(x - y) + 1$$
$$= 2u^2 - 3u + 1 \quad \text{Let } u = x - y$$
$$= (u - 1)(2u - 1)$$
$$= [(x - y) - 1][2(x - y) - 1] \quad \text{Return to the original variables}$$
$$= (x - y - 1)(2x - 2y - 1).$$

EXAMPLE 9 FACTORING A TRINOMIAL BY SUBSTITUTION

Factor $x^4 + 5x^2 + 6$.

First we note that $x^4 = (x^2)^2$ and then let $u = x^2$.

$$x^4 + 5x^2 + 6 = (x^2)^2 + 5x^2 + 6$$
$$= u^2 + 5u + 6 \quad u = x^2$$
$$= (u + 2)(u + 3)$$
$$= (x^2 + 2)(x^2 + 3) \quad \text{Return to the original variable}$$

PRACTICE EXERCISE 9

Factor $x^4 - 5x^2 + 6$.

Answer: $(x^2 - 2)(x^2 - 3)$

5.4 EXERCISES A

Factor completely.

1. $x^2 + 6x + 5$
2. $x^2 - 6x + 5$
3. $x^2 + 4x - 5$
4. $x^2 - 4x - 5$
5. $y^2 - 4y - 21$
6. $y^2 - 10y + 21$
7. $y^2 + 4y - 21$
8. $y^2 + 10y + 21$
9. $w^2 + 4w - 45$
10. $w^2 - 2w - 63$
11. $x^2 - x - 56$
12. $x^2 + 15x + 56$
13. $x^2 + 2x + 4$
14. $y^2 + 12y + 36$
15. $u^2 + 21u + 110$
16. $u^2 - u - 110$
17. $x^2 + 6xy + 5y^2$
18. $x^2 - 6xy + 5y^2$
19. $x^2 + 4xy - 5y^2$
20. $x^2 - 4xy - 5y^2$
21. $u^2 - 2uv - 35v^2$
22. $2w^2 + 19w + 24$
23. $2w^2 - 13w - 24$
24. $-4w^2 - 34w - 70$
25. $7w^2 - 14w + 7$
26. $2x^2 + 13xy - 24y^2$
27. $2x^2 - 19xy + 24y^2$
28. $2x^2 + 4xy + y^2$
29. $2u^2v^2 - 7uv - 30$
30. $5x^2 + 17xy - 12y^2$
31. $6x^2 + 23xy + 20y^2$
32. $-45u^2 + 150uv - 125v^2$
33. $6x^2 - 19xy - 20y^2$
34. $4x^2 - 32xy + 63y^2$
35. $2u^2 - 7uv + 6v^2$
36. $21u^2 - 23uv - 20v^2$

Factor by substitution.

37. $(x + 2)^2 - 3(x + 2) + 2$

38. $(x + y)^2 + 6(x + y) + 8$

39. $y^4 + y^2 - 6$

40. $u^6 - 7u^3 + 12$

A laboratory assistant finds that her work in evaluation of lab equations is made easy if she factors the equations. Factor each of the equations and evaluate for $u = 5$ and $v = -2$.

41. $K = u^2 + 6uv + 8v^2$

42. $L = u^2 - 2uv - 8v^2$

43. $M = 3u^2 - 5uv - 2v^2$

44. $N = 10u^2 - 23uv + 12v^2$

45. When a surface-to-air missile is fired with a velocity of 656 ft/sec from a submarine at a depth of -640 ft, the height h in feet of the missile is given by $h = -16t^2 + 656t - 640$. When h is negative, the missile is still under water; when h is zero, the missile is at water level; and when h is positive, the missile is above the surface of the water. Factor the right side of the equation giving h and use the factored form to determine the location of the missile **(a)** 0.5 seconds, **(b)** 1 second, and **(c)** 20 seconds after it has been fired.

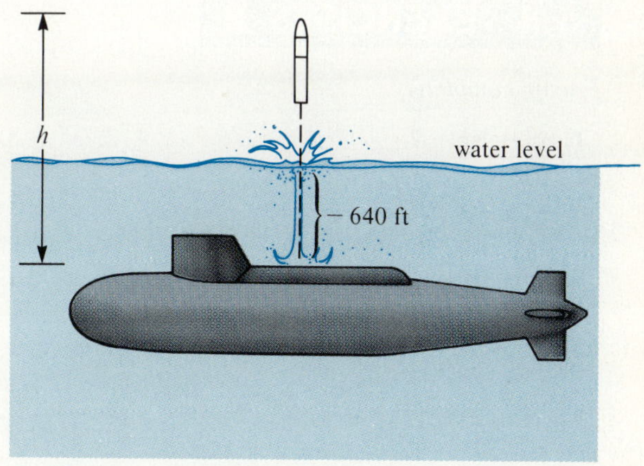

FOR REVIEW

Factor completely.

46. $20x^4y^2 + 28x^2y^3$

47. $25x^3y^3 - 35x^2y^3 + 45xy^3$

48. $2u^2 - 3uv - 4uv + 6v^2$

49. $21u^2 + 12uv - 35uv - 20v^2$

50. $4x^2 - 14xy + 10xy - 35y^2$

51. $12x^2 - 20xy - 9xy + 15y^2$

Exercises 52–55 review material from Section 5.2 to help you prepare for topics in the next section. Find each product.

52. $(y + 5)(y - 5)$ **53.** $(2x - 7y)(2x + 7y)$ **54.** $(3a + 4)^2$ **55.** $(5x - 2y)^2$

ANSWERS: 1. $(x + 1)(x + 5)$ 2. $(x - 1)(x - 5)$ 3. $(x - 1)(x + 5)$ 4. $(x + 1)(x - 5)$ 5. $(y + 3)(y - 7)$ 6. $(y - 3)(y - 7)$ 7. $(y - 3)(y + 7)$ 8. $(y + 3)(y + 7)$ 9. $(w - 5)(w + 9)$ 10. $(w + 7)(w - 9)$ 11. $(x + 7)(x - 8)$ 12. $(x + 7)(x + 8)$ 13. cannot be factored 14. $(y + 6)(y + 6)$ 15. $(u + 10)(u + 11)$ 16. $(u + 10)(u - 11)$ 17. $(x + y)(x + 5y)$ 18. $(x - y)(x - 5y)$ 19. $(x - y)(x + 5y)$ 20. $(x + y)(x - 5y)$ 21. $(u + 5v)(u - 7v)$ 22. $(w + 8)(2w + 3)$ 23. $(w - 8)(2w + 3)$ 24. $-2(w + 5)(2w + 7)$ 25. $7(w - 1)(w - 1)$ 26. $(x + 8y)(2x - 3y)$ 27. $(x - 8y)(2x - 3y)$ 28. cannot be factored 29. $(uv - 6)(2uv + 5)$ 30. $(x + 4y)(5x - 3y)$ 31. $(2x + 5y)(3x + 4y)$ 32. $-5(3u - 5v)(3u - 5v)$ 33. $(x - 4y)(6x + 5y)$ 34. $(2x - 7y)(2x - 9y)$ 35. $(u - 2v)(2u - 3v)$ 36. $(3u - 5v)(7u + 4v)$ 37. $x(x + 1)$ 38. $(x + y + 2)(x + y + 4)$ 39. $(y^2 - 2)(y^2 + 3)$ 40. $(u^3 - 3)(u^3 - 4)$ 41. $(u + 2v)(u + 4v)$; -3 42. $(u + 2v)(u - 4v)$; 13 43. $(3u + v)(u - 2v)$; 117 44. $(5u - 4v)(2u - 3v)$; 528 45. (a) -316 ft (b) 0 ft (c) 6080 ft 46. $4x^2y^2(5x^2 + 7y)$ 47. $5xy^3(5x^2 - 7x + 9)$ 48. $(u - 2v)(2u - 3v)$ 49. $(3u - 5v)(7u + 4v)$ 50. $(2x + 5y)(2x - 7y)$ 51. $(3x - 5y)(4x - 3y)$ 52. $y^2 - 25$ 53. $4x^2 - 49y^2$ 54. $9a^2 + 24a + 16$ 55. $25x^2 - 20xy + 4y^2$

5.4 EXERCISES B

Factor completely.

1. $x^2 + 8x + 7$ **2.** $x^2 - 8x + 7$ **3.** $x^2 + 6x - 7$

4. $x^2 - 6x - 7$ **5.** $y^2 + 7y - 18$ **6.** $y^2 - 11y + 18$

7. $y^2 - 7y - 18$ **8.** $y^2 + 11y + 18$ **9.** $w^2 + 5w - 50$

10. $w^2 + 2w - 63$ **11.** $x^2 + x - 56$ **12.** $x^2 - 15x + 56$

13. $x^2 - 2x + 4$ **14.** $y^2 + 14y + 48$ **15.** $u^2 - 18u + 72$

16. $u^2 - 6u - 72$ **17.** $x^2 + 8xy + 7y^2$ **18.** $x^2 - 8xy + 7y^2$

19. $x^2 + 6xy - 7y^2$ **20.** $x^2 - 6xy - 7y^2$ **21.** $u^2 + 2uv - 35v^2$

22. $2w^2 + 19w + 45$ **23.** $2w^2 - 19w + 45$ **24.** $-6w^2 + 51w - 105$

25. $5w^2 - 20w + 20$ **26.** $2x^2 - xy - 45y^2$ **27.** $2x^2 + xy - 45y^2$

28. $2x^2 - 4xy + y^2$ **29.** $2u^2v^2 + 7uv - 30$ **30.** $5x^2 - 23xy - 10y^2$

31. $6x^2 - 7xy - 20y^2$ **32.** $-36u^2 - 120uv - 100v^2$ **33.** $6x^2 - 11xy - 21y^2$

34. $4x^2 - 4xy - 63y^2$ **35.** $15u^2 - 14uv - 8v^2$ **36.** $12u^2 - 52uv + 35v^2$

Factor by substitution.

37. $(x - 5)^2 + 4(x - 5) + 3$ **38.** $(x + y)^2 - 5(x + y) - 6$

39. $y^4 - 7y^2 + 10$ **40.** $u^6 - u^3 - 30$

An assistant in an investment office finds that his work in evaluation of stock market equations is made easy by factoring the equations. Factor each of the equations and evaluate for $m = -3$ and $n = 2$.

41. $P = m^2 - 8mn + 15n^2$

42. $Q = m^2 + 2mn - 15n^2$

43. $R = 3m^2 + mn - 2n^2$

44. $S = 6m^2 - 19mn + 10n^2$

45. When a surface-to-air missile is fired with a velocity of 576 ft/sec from a submarine at a depth of -560 ft, the height h in feet of the missile is given by $h = -16t^2 + 576t - 560$. When h is negative, the missile is still under water; when h is zero, the missile is at water level; and when h is positive, the missile is above the surface of the water. Factor the right side of the equation giving h and use the factored form to determine the location of the missile **(a)** 0.5 seconds, **(b)** 1 second, and **(c)** 20 seconds after it has been fired.

FOR REVIEW

Factor completely.

46. $21x^2y^3 - 14xy^2$

47. $42x^4y^4 - 14x^4y^3 - 98x^4y^2$

48. $15u^2 + 6uv - 20uv - 8v^2$

49. $12u^2 - 42uv - 10uv + 35v^2$

50. $6x^2 + 20xy - 27xy - 90y^2$

51. $10x^2 - 16xy - 35xy + 56y^2$

Exercises 52–55 review material from Section 5.2 to help you prepare for topics in the next section. Find each product.

52. $(x - 3)(x + 3)$

53. $(4u + 5v)(4u - 5v)$

54. $(2x + 3)^2$

55. $(5u - 3v)^2$

5.4 EXERCISES C

Factor. Assume that n represents a positive integer.

1. $24x^2 - 65xy - 50y^2$

2. $198a^2 - 393ab + 189b^2$
[*Hint:* A number has 3 as a factor if 3 is a factor of the sum of its digits.

3. $x^{2n} + 5x^n + 6$
[Answer: $(x^n + 2)(x^n + 3)$]

4. $(x + 1)^{2n} + 3(x + 1)^n + 2$
[*Hint:* Substitute u for $x + 1$.]

5. $y^{2n+2} - y^{n+1} - 2$

6. $3x^{6n} + 2x^{5n} - x^{4n}$

5.5 SPECIAL FORMULAS AND SUMMARY OF FACTORING TECHNIQUES

STUDENT GUIDEPOSTS

1. Difference of Squares
2. Perfect Squares
3. Sum and Difference of Cubes
4. Factoring by Using Substitution

1 DIFFERENCE OF SQUARES

In Section 5.2 we saw that

$$(2x + 3y)(2x - 3y) = (2x)^2 - (3y)^2 = 4x^2 - 9y^2.$$

By reversing this we can factor $4x^2 - 9y^2$ as $(2x + 3y)(2x - 3y)$. This is an example of the general formula

$$a^2 - b^2 = (a + b)(a - b) \qquad \text{Difference of squares}$$

which was proved by multiplying $(a + b)(a - b)$. Notice that in this formula we are factoring a binomial which is the difference of two perfect squares.

EXAMPLE 1 Factoring the Difference of Squares

Factor using $a^2 - b^2 = (a + b)(a - b)$.

(a) $x^2 - 49 = x^2 - 7^2$ $\quad$ $x = a$ and $7 = b$ in $a^2 - b^2 = (a + b)(a - b)$

$= (x + 7)(x - 7)$

(b) $16u^2 - 25v^2 = 4^2u^2 - 5^2v^2$ $\quad$ $4^2u^2 = (4u)^2$

$= (4u)^2 - (5v)^2$ $\quad$ $4u = a$ and $5v = b$ in $a^2 - b^2 = (a + b)(a - b)$

$= (4u + 5v)(4u - 5v)$

(c) $18x^6 - 8y^4 = 2(9x^6 - 4y^4)$ $\quad$ Factor out common factor

$= 2[3^2(x^3)^2 - 2^2(y^2)^2]$

$= 2[(3x^3)^2 - (2y^2)^2]$ $\quad$ $3x^3 = a$ and $2y^2 = b$

$= 2(3x^3 + 2y^2)(3x^3 - 2y^2)$

Practice Exercise 1

Factor using

$a^2 - b^2 = (a + b)(a - b)$.

(a) $y^2 - 81$

(b) $64u^2 - 49v^2$

(c) $75u^4 - 12v^6$

Answers: (a) $(y + 9)(y - 9)$
(b) $(8u + 7v)(8u - 7v)$
(c) $3(5u^2 + 2v^3)(5u^2 - 2v^3)$

CAUTION

Although $a^2 - b^2$ can be factored, $a^2 + b^2$ cannot. While $x^2 - 49$ was factored in Example 1, $x^2 + 49$ cannot be factored.

2 PERFECT SQUARES

We also considered in Section 5.2 the two results

$$(2x + 3y)^2 = (2x)^2 + 2(6xy) + (3y)^2 = 4x^2 + 12xy + 9y^2$$

and

$$(2x - 3y)^2 = (2x)^2 - 2(6xy) + (3y)^2 = 4x^2 - 12xy + 9y^2.$$

Again the process of multiplication can be reversed to factor the two expressions on the right. The general formulas to use are

$$a^2 + 2ab + b^2 = (a + b)^2 \qquad \text{Perfect square trinomial}$$

and

$$a^2 - 2ab + b^2 = (a - b)^2. \qquad \text{Perfect square trinomial}$$

Notice that these formulas apply only when the trinomial has two terms that are perfect squares and another that is plus or minus twice the product of the square roots of the squared terms. For example, in

$$x^2 + 8x + 16 = x^2 + 2(4x) + 4^2,$$

x^2 and 4^2 are perfect squares and $8x = 2 \cdot 4x = 2ab$ where $x = a$ and $4 = b$. Thus,

$$x^2 + 8x + 16 = (x + 4)^2.$$

EXAMPLE 2 Factoring Perfect Square Trinomials

Factor using the perfect square formulas.

(a) $u^2 + 10uv + 25v^2$

$= u^2 + 2(5uv) + 5^2v^2$ First and last terms are perfect squares

$= u^2 + 2u(5v) + (5v)^2$ $u = a$, $5v = b$

$= (u + 5v)^2$

(b) $9x^2 + 42xy + 49y^2 = 3^2x^2 + 2(3)(7xy) + 7^2y^2$

$= (3x)^2 + 2(3x)(7y) + (7y)^2$ $3x = a$, $7y = b$

$= (3x + 7y)^2$

(c) $u^2 - 10uv + 25v^2 = u^2 - 2(5uv) + 5^2v^2$

$= u^2 - 2u(5v) + (5v)^2$ $u = a$, $5v = b$

$= (u - 5v)^2$

(d) $36x^4 - 84x^2y + 49y^2 = 6^2(x^2)^2 - 2(6)(7x^2y) + 7^2y^2$

$= (6x^2)^2 - 2(6x^2)(7y) + (7y)^2$ $6x^2 = a$, $7y = b$

$= (6x^2 - 7y)^2$

PRACTICE EXERCISE 2

Factor using the perfect square formulas.

(a) $x^2 + 22x + 121$

(b) $16x^2 + 72xy + 81y^2$

(c) $x^2 - 22x + 121$

(d) $9x^4y^4 - 30x^2y^2 + 25$

Answers: (a) $(x + 11)^2$ (b) $(4x + 9y)^2$ (c) $(x - 11)^2$ (d) $(3x^2y^2 - 5)^2$

3 SUM AND DIFFERENCE OF CUBES

The next two formulas are used to factor binomials whose terms are perfect cubes.

$$a^3 + b^3 = (a + b)(a^2 - ab + b^2) \quad \text{Sum of cubes}$$

$$a^3 - b^3 = (a - b)(a^2 + ab + b^2) \quad \text{Difference of cubes}$$

To prove these we multiply.

$$\begin{array}{r} a^2 - ab + b^2 \\ a + b \\ \hline a^3 - a^2b + ab^2 \\ a^2b - ab^2 + b^3 \\ \hline a^3 + 0a^2b + 0ab^2 + b^3 \\ = a^3 + b^3 \end{array} \qquad \begin{array}{r} a^2 + ab + b^2 \\ a - b \\ \hline a^3 + a^2b + ab^2 \\ - a^2b - ab^2 - b^3 \\ \hline a^3 + 0a^2b + 0ab^2 - b^3 \\ = a^3 - b^3 \end{array}$$

Notice that $a^2 - ab + b^2$ and $a^2 + ab + b^2$ cannot be factored further.

EXAMPLE 3 Factoring the Sum of Cubes

Factor using the sum of cubes formula.

(a) $x^3 + 8 = x^3 + 2^3$ $x = a$ and $2 = b$ in $a^3 + b^3 = (a + b)(a^2 - ab + b^2)$

$= (x + 2)(x^2 - 2x + 4)$

(b) $u^3 + 27v^3 = u^3 + 3^3v^3$

$= u^3 + (3v)^3$

$= (u + 3v)[u^2 - u(3v) + (3v)^2]$

$= (u + 3v)(u^2 - 3uv + 9v^2)$

PRACTICE EXERCISE 3

Factor using the sum of cubes formula.

(a) $x^3 + 125$

(b) $64u^3 + v^3$

(c) $27x^9 + 8y^3$

(c) $8x^3 + 125y^6$
$= 2^3x^3 + 5^3(y^2)^3$
$= (2x)^3 + (5y^2)^3$ $\quad 2x = a,\ 5y^2 = b$
$= (2x + 5y^2)[(2x)^2 - (2x)(5y^2) + (5y^2)^2]$
$= (2x + 5y^2)(4x^2 - 10xy^2 + 25y^4)$

Answers: (a) $(x + 5)(x^2 - 5x + 25)$
(b) $(4u + v)(16u^2 - 4uv + v^2)$
(c) $(3x^3 + 2y)(9x^6 - 6x^3y + 4y^2)$

CAUTION

To factor a sum or difference of cubes, you must identify a and b and substitute these values into the appropriate formula. This means that *both formulas must be memorized.*

EXAMPLE 4 FACTORING THE DIFFERENCE OF CUBES

Factor using the difference of cubes formula.

(a) $y^3 - 64 = y^3 - 4^3$ $\quad y = a$ and $4 = b$ in $a^3 - b^3 = (a - b)(a^2 + ab + b^2)$
$= (y - 4)(y^2 + 4y + 16)$

(b) $u^3 - 27v^3 = u^3 - (3v)^3$ $\quad u = a,\ 3v = b$
$= (u - 3v)[u^2 + u(3v) + (3v)^2]$
$= (u - 3v)(u^2 + 3uv + 9v^2)$

(c) $125x^3y^3 - 64 = 5^3x^3y^3 - 4^3$
$= (5xy)^3 - 4^3$ $\quad 5xy = a,\ 4 = b$
$= (5xy - 4)[(5xy)^2 + (5xy)(4) + (4)^2]$
$= (5xy - 4)(25x^2y^2 + 20xy + 16)$

PRACTICE EXERCISE 4

Factor using the difference of cubes formula.

(a) $27 - x^3$

(b) $125u^3 - v^3$

(c) $8u^3 - 27v^3w^3$

Answers: (a) $(3 - x)(9 + 3x + x^2)$
(b) $(5u - v)(25u^2 + 5uv + v^2)$
(c) $(2u - 3vw)(4u^2 + 6uvw + 9v^2w^2)$

CAUTION

Do not write $a^2 - 2ab + b^2$ for $a^2 - ab + b^2$ or $a^2 + 2ab + b^2$ for $a^2 + ab + b^2$ when factoring the sum or difference of cubes.

In some factoring problems we need to use more than one of the formulas, as in the next example.

EXAMPLE 5 USING MORE THAN ONE FORMULA

Factor.

(a) $x^6 - 1$
$= (x^3)^2 - 1^2$ $\quad$ Factor difference of squares first with $x^3 = a$ and $1 = b$
$= (x^3 + 1)(x^3 - 1)$
$= [(x + 1)(x^2 - x + 1)][(x - 1)(x^2 + x + 1)]$ $\quad$ Sum and difference of cubes
$= (x + 1)(x - 1)(x^2 - x + 1)(x^2 + x + 1)$

PRACTICE EXERCISE 5

Factor.

(a) $y^6 - 64$

(b) $u^2 - (v^2 - 2v + 1)$

(c) $u^4 - 16v^4$

(b) $x^2 + 2x + 1 - y^2$
$= (x^2 + 2x + 1) - y^2$
$= (x + 1)^2 - y^2$ $\quad x + 1 = a$ and $y = b$ in $a^2 - b^2 = (a + b)(a - b)$
$= [(x + 1) + y][(x + 1) - y]$
$= (x + 1 + y)(x + 1 - y)$

(c) $x^4 - 1 = (x^2)^2 - 1^2$ $\quad x^2 = a$ and $1 = b$ in $a^2 - b^2 = (a - b)(a + b)$
$= (x^2 - 1)(x^2 + 1)$
$= (x - 1)(x + 1)(x^2 + 1)$ $\quad x^2 + 1$ cannot be factored

Answers: (a) $(y + 2)(y - 2)(y^2 - 2y + 4)(y^2 + 2y + 4)$ (b) $(u + v - 1)(u - v + 1)$ (c) $(u + 2v)(u - 2v)(u^2 + 4v^2)$

4 FACTORING BY USING SUBSTITUTION

The substitution method used in Section 5.4 could have been used in Example 5. Example 6 illustrates this method.

EXAMPLE 6 Factoring by Using Substitution

Factor $25(x + 2y)^2 - 10(x + 2y) + 1$ by letting $u = x + 2y$.

$25(x + 2y)^2 - 10(x + 2y) + 1$
$= 25u^2 - 10u + 1$ $\quad$ Let $u = x + 2y$
$= (5u)^2 - 2(5u)(1) + 1$ $\quad 5u = a$ and $1 = b$
$= (5u - 1)^2$
$= [5(x + 2y) - 1]^2$ $\quad u = x + 2y$
$= (5x + 10y - 1)^2$

Practice Exercise 6

Factor $(5x + y)^2 - 8(5x + y) + 16$.

Answer: $(5x + y - 4)^2$

Here we summarize the special formulas used in factoring.

Special Formulas

1. $a^2 - b^2 = (a + b)(a - b)$	Difference of squares
2. $a^2 + 2ab + b^2 = (a + b)^2$	Perfect square trinomial
3. $a^2 - 2ab + b^2 = (a - b)^2$	Perfect square trinomial
4. $a^3 + b^3 = (a + b)(a^2 - ab + b^2)$	Sum of cubes
5. $a^3 - b^3 = (a - b)(a^2 + ab + b^2)$	Difference of cubes

Some applications involve expressions that can be factored. This is illustrated in the next example.

EXAMPLE 7 Geometry Problem

A metal machine part in the shape of a cube with edge E has a rectangular solid removed from the center as shown in Figure 5.1. The volume of the remaining metal is given by $V = E^3 - Ee^2$. Find the factored form of this equation and use both formulas to compute the volume when $E = 12$ cm and $e = 3$ cm.

Practice Exercise 7

The current I in amperes varies according to time t in seconds in an electrical circuit. If $I = 2t^2 - 20t + 50$, find the factored form of this equation and use both formulas to compute the current when $t = 5$ sec.

Factoring, we have
$$V = E^3 - Ee^2$$
$$= E(E^2 - e^2)$$
$$= E(E + e)(E - e).$$

Substitute 12 for E and 3 for e in $V = E^3 - Ee^2$ first.
$$V = (12)^3 - (12)(3)^2$$
$$= 1728 - 108 = 1620$$

Now substitute in the factored formula.
$$V = 12(12 + 3)(12 - 3)$$
$$= 12(15)(9) = 1620$$

In both cases we obtain 1620 cm^3 for the volume of the machine part.

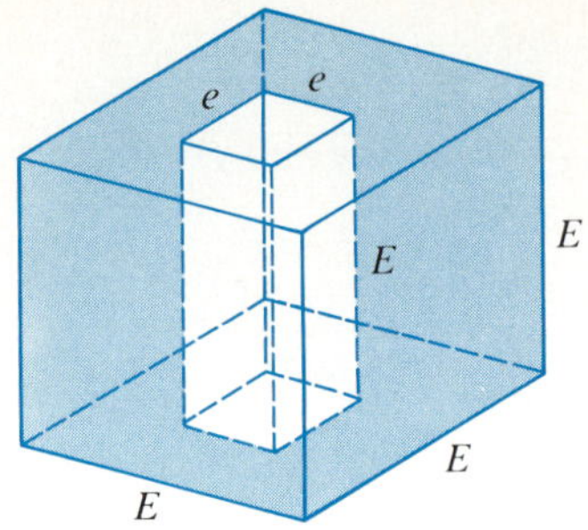

Figure 5.1

Answer: $I = 2(t - 5)^2$; 0 amperes

The exercises for this section start with applications of special formulas and conclude with a variety of factoring problems. As preparation we summarize the factoring techniques that we have learned in this chapter.

Factoring Techniques

1. Factor out any common factor. It is also helpful when factoring $ax^2 + bxy + cy^2$ to factor out (-1) if a is negative.
2. To factor a binomial, use

   $a^2 - b^2 = (a + b)(a - b)$ Difference of squares
   $a^3 + b^3 = (a + b)(a^2 - ab + b^2)$ Sum of cubes
   $a^3 - b^3 = (a - b)(a^2 + ab + b^2).$ Difference of cubes

3. To factor a trinomial, use the techniques of Section 5.4, but also consider perfect squares using the rules

   $a^2 + 2ab + b^2 = (a + b)^2$ Perfect square trinomial
   and $a^2 - 2ab + b^2 = (a - b)^2.$ Perfect square trinomial

   These may apply when the first and last terms of a trinomial are perfect squares.
4. To factor a polynomial with four terms, try to factor by grouping.

5.5 EXERCISES A

Factor using special formulas.

1. $x^2 - 9$ **2.** $y^2 - 6y + 9$ **3.** $u^3 - 8$

4. $y^2 - 64$ **5.** $u^2 - 12uv + 36v^2$ **6.** $7x^2 - 7y^2$

7. $8x^3 + 27y^3$ **8.** $4u^2 + 12uv + 9v^2$ **9.** $20x^2 - 45y^2$

10. $u^2 + 2uv + 2v^2$

11. $40x^3 + 5y^9$

12. $25u^2 - 9v^6$

13. $125u^3 + 27v^3$

14 $(x - y)^2 - 9$

15. $(x - 5)^2 - 4(x - 5) + 4$

Factor.

16. $x^2 - 14xy + 48y^2$

17. $x^2 - 2xy - 48y^2$

18. $6x^2 - 17xy + 10y^2$

19. $35x^2 - 2xy - y^2$

20. $-5x - 5y$

21. $27x^2y - 9x^2y^2 + 81xy^2$

22. $-10u^2 + 40v^2$

23 $7u^3 + 56v^3$

24. $2x^2 - 4xy + 2y^2$

25. $u^2 + 18uv + 80v^2$

26. $u^2 + 22uv + 121v^2$

27. $3x^2 - 36xy + 105y^2$

28. $4x^2y^2 - 4xy + 1$

29. $21u^2 - 4uv - v^2$

30. $250x^3 + 54y^3$

31. $3u^2 - 2uv - 16v^2$

32. $36u^2 + 60uv + 25v^2$

33. $x^2 + 5xy + 5y^2$

34. $-5x^2 + 70xy - 240y^2$

35. $u^6 - v^6$

36. $(3x + y)^2 - 9$

37 $(x + y)^2 + 2(x + y) + 1$

38. $27u^3v^3 - 8$

39 $3uv + u - 3v^2 - v$

40. $30x^2 + 31xy - 21y^2$

41. $x^3y^3 + z^3$

42. $9x^6 - 42x^3y + 49y^2$

43. $-16x^4 + 64y^2$

44 $-18u^2 - 69uv - 60v^2$

45. $5x^2 + 15xy + 15y^2$

46. $(2x + y)^2 - 6(2x + y) + 8$

47. $x^4 - x^2 - 6$

48 $(u + v)^2 + 3(u + v)(x + y) + 2(x + y)^2$

The area of a circular garden with a circular fountain in the center is given by the equation $A = \pi R^2 - \pi r^2$ where R is the radius of the garden and r is the radius of the fountain.

49. Find the factored form of the equation.

50. Use both formulas to find the area when $R = 10$ ft and $r = 4$ ft. (Use 3.14 for π.)

51. In an electrical circuit, the current I in amperes varies according to time t in seconds. If $I = 3t^2 - 36t + 108$, find the factored form of this equation and use both formulas to compute the current when $t = 6$ sec.

FOR REVIEW

The following exercises review material from Section 5.1 along with factoring to help you prepare for the next section. Evaluate each polynomial when $x = 2$. Then factor the polynomial and evaluate the factored form when $x = 2$. Which form is easier to evaluate?

52. $x^2 + x - 6$

53. $5x^2 - 20$

54. $3x^3 - 24$

ANSWERS: 1. $(x + 3)(x - 3)$ 2. $(y - 3)^2$ 3. $(u - 2)(u^2 + 2u + 4)$ 4. $(y + 8)(y - 8)$ 5. $(u - 6v)^2$ 6. $7(x + y)(x - y)$ 7. $(2x + 3y)(4x^2 - 6xy + 9y^2)$ 8. $(2u + 3v)^2$ 9. $5(2x + 3y)(2x - 3y)$ 10. cannot be factored 11. $5(2x + y^3)(4x^2 - 2xy^3 + y^6)$ 12. $(5u + 3v^3)(5u - 3v^3)$ 13. $(5u + 3v)(25u^2 - 15uv + 9v^2)$ 14. $(x - y + 3)(x - y - 3)$ 15. $(x - 7)^2$ 16. $(x - 6y)(x - 8y)$ 17. $(x + 6y)(x - 8y)$ 18. $(6x - 5y)(x - 2y)$ 19. $(7x + y)(5x - y)$ 20. $-5(x + y)$ 21. $9xy(3x - xy + 9y)$ 22. $-10(u + 2v)(u - 2v)$ 23. $7(u + 2v)(u^2 - 2uv + 4v^2)$ 24. $2(x - y)^2$ 25. $(u + 8v)(u + 10v)$ 26. $(u + 11v)^2$ 27. $3(x - 5y)(x - 7y)$ 28. $(2xy - 1)^2$ 29. $(3u - v)(7u + v)$ 30. $2(5x + 3y)(25x^2 - 15xy + 9y^2)$ 31. $(u + 2v)(3u - 8v)$ 32. $(6u + 5v)^2$ 33. cannot be factored 34. $-5(x - 6y)(x - 8y)$ 35. $(u + v)(u - v)(u^2 - uv + v^2)(u^2 + uv + v^2)$ 36. $(3x + y + 3)(3x + y - 3)$ 37. $(x + y + 1)^2$ 38. $(3uv - 2)(9u^2v^2 + 6uv + 4)$ 39. $(u - v)(3v + 1)$ 40. $(15x - 7y)(2x + 3y)$ 41. $(xy + z)(x^2y^2 - xyz + z^2)$ 42. $(3x^3 - 7y)^2$ 43. $-16(x^2 + 2y)(x^2 - 2y)$ 44. $-3(2u + 5v)(3u + 4v)$ 45. $5(x^2 + 3xy + 3y^2)$ 46. $(2x + y - 2)(2x + y - 4)$ 47. $(x^2 + 2)(x^2 - 3)$ 48. $(u + v + x + y)(u + v + 2x + 2y)$ 49. $A = \pi(R + r)(R - r)$ 50. 84π ft$^2 \approx 264$ ft^2 51. $I = 3(t - 6)^2$; 0 amperes 52. $(x - 2)(x + 3)$; 0 53. $5(x + 2)(x - 2)$; 0 54. $3(x - 2)(x^2 + 2x + 4)$; 0

5.5 EXERCISES B

Factor using special formulas.

1. $x^2 + 6x + 9$

2. $y^3 + 8$

3. $u^2 - 14u + 49$

4. $y^3 + 64$ **5.** $u^2 + 12uv + 36v^2$ **6.** $7x^3 + 7y^3$

7. $8x^3 - 27y^3$ **8.** $4u^2 - 12uv + 9v^2$ **9.** $x^2 + y^2$

10. $9u^2 + 6uv + v^2$ **11.** $49x^4 + 14x^2y^2 + y^4$ **12.** $9u^6 - 30u^3v^2 + 25v^4$

13. $5u^3 - 5000v^3$ **14.** $(x + y)^2 - 25$ **15.** $(x - y)^2 + 6(x - y) + 9$

Factor.

16. $x^2 + 13xy + 42y^2$ **17.** $x^2 - xy - 42y^2$ **18.** $5x^2 - 12xy - 32y^2$

19. $14x^2 + 25xy - 25y^2$ **20.** $4x^2y^2 - 2xy$ **21.** $21x^3y - 35x^2y^2 + 14x^2y$

22. $-108u^2 + 12v^2$ **23.** $40u^3 - 135v^3$ **24.** $8x^2 + 32xy + 32y^2$

25. $3u^2 - 8uv + 4v^2$ **26.** $144u^2 - 24uv + v^2$ **27.** $40x^2 - 28xy + 4y^2$

28. $9x^2y^2 - 6xy + 1$ **29.** $10u^2 + 99uv - 10v^2$ **30.** $24x^3 - 81y^3$

31. $30u^2 + 13uv - 3v^2$ **32.** $49u^2 - 42uv + 9v^2$ **33.** $8x^2 + 8xy + 3y^2$

34. $-28x^2 + 140xy - 175y^2$ **35.** $u^8 - v^8$ **36.** $9(x + 5y)^2 - 4$

37. $4(2x + y)^2 - 4(2x + y) + 1$ **38.** $u^3v^3 + x^3y^3$ **39.** $5u^2 - 3uv - 9v^2 + 15uv$

40. $36x^2 - 9xy - 10y^2$ **41.** $x^3y^3z^3 - 8$ **42.** $25x^4 + 30x^2y^2 + 9y^4$

43. $-288x^2 + 8y^4$ **44.** $25u^2 - 130uv + 25v^2$ **45.** $12x^2 - 12xy + 24y^2$

46. $(x^2 + y^2)^2 + 4(x^2 + y^2) + 4$ **47.** $x^4 - 25$ **48.** $(u^2v^2 + 1)^2 + 4(u^2v^2 + 1) + 3$

The volume of a hollow cylindrical column which is 10 m *high is given by the equation* $V = 10\pi R^2 - 10\pi r^2$ *where* R *is the outer radius and* r *is the inner radius.*

49. Find the factored form of the equation.

50. Use both formulas to find the volume when $R =$ 4 m and $r =$ 2 m. (Use 3.14 for π.)

51. In an electrical circuit, the current I in amperes varies according to time t in seconds. If $I = 4t^2 - 80t + 400$, find the factored form of this equation and use both formulas to compute the current when $t =$ 10 sec.

FOR REVIEW

The following exercises review material from Section 5.1 along with factoring to help you prepare for the next section. Evaluate each polynomial when $x = -3$. *Then factor the polynomial and evaluate the factored form when* $x = -3$. *Which form is easier to evaluate?*

52. $x^2 - 2x - 15$ **53.** $7x^2 - 63$ **54.** $2x^3 + 54$

5.5 EXERCISES C

Factor. Assume that n represents a positive integer.

1. $x^{4n} - y^{4n}$

2. $x^{3n} - y^{3n}$
[Answer: $(x^n - y^n)(x^{2n} + x^ny^n + y^{2n})$]

3. $x^{3n} + y^{3n}$

4. $x^{2n} + 2x^n y^n + y^{2n}$

5. $x^{2n} - 2x^n y^n + y^{2n}$

6. $x^n y^n + y^n + x^n + 1$
[Answer: $(x^n + 1)(y^n + 1)$]

7. $8x^{3n} - 27y^{3n}$
[Answer: $(2x^n - 3y^n)(4x^{2n} + 6x^n y^n + 9y^{2n})$]

8. $8x^{3n} + 27y^{3n}$

5.6 APPLICATIONS USING FACTORING

STUDENT GUIDEPOSTS

1. The Zero-Product Rule
2. Quadratic Equations
3. Applied Problems

1 THE ZERO-PRODUCT RULE

Factoring can be used to solve equations and applied problems. The polynomial in an equation such as

$$x^2 + 2x - 15 = 0$$

can be factored to give

$$(x - 3)(x + 5) = 0.$$

This equation can then be solved by using a property of the real number system called the **zero-product rule,** which states that if a product of two or more factors is zero, then at least one of the factors must be zero.

Zero-Product Rule

If a and b are numbers or expressions and

if $ab = 0$, then $a = 0$ or $b = 0$.

Using the zero-product rule, the solutions of an equation like

$$(x - 3)(x + 5) = 0$$

are found by setting each factor equal to zero and solving for x.

EXAMPLE 1 USING THE ZERO-PRODUCT RULE

Solve $(x - 3)(x + 5) = 0$.

By the zero-product rule,

$$\begin{aligned} x - 3 &= 0 \quad \text{or} \quad & x + 5 &= 0 \\ x &= 3 & x &= -5. \end{aligned}$$

The solutions are 3 and -5. Check these by substituting into the original equation.

PRACTICE EXERCISE 1

Solve $(x + 4)(x - 10) = 0$.

Answer: -4, 10

The zero-product rule can be extended to products of three or more factors, as shown in the next example.

EXAMPLE 2 Extending the Zero-Product Rule

Solve $x(x - 3)(2x + 3) = 0$.

$$x = 0 \quad \text{or} \quad x - 3 = 0 \quad \text{or} \quad 2x + 3 = 0 \qquad \text{Zero-product rule}$$
$$x = 0 \qquad x = 3 \qquad 2x = -3$$
$$x = -\frac{3}{2}$$

The solutions are 0, 3, and $-\frac{3}{2}$. Check by substituting in the original equation.

PRACTICE EXERCISE 2

Solve $(x - 2)(x + 9)(2x - 5) = 0$.

Answer: 2, -9, $\frac{5}{2}$

CAUTION

Note that the zero-product rule applies only to *zero-products*. It cannot be used to solve equations like $ab = 6$, $a + b = 0$, or $a - b = 0$. For example,

$$(x - 2) - (2x - 1) = 0$$

is solved the same way we solved linear equations in Chapter 2. We first clear parentheses and then combine like terms. Also, if $(x - 1)(x + 5) = 6$, this does not mean that each factor is 6.

2 QUADRATIC EQUATIONS

In Example 1, if we had started with

$$x^2 + 2x - 15 = 0,$$

we could have factored to get

$$(x - 3)(x + 5) = 0$$

and then continued as in the example. We have, in effect, solved a *quadratic equation*. A **quadratic equation** or **second-degree equation** is an equation that can be written in the form

$$ax^2 + bx + c = 0,$$

where x is the variable and a, b, and c are constants with $a \neq 0$. We will solve equations of this type using factoring in this section but will wait for a complete discussion of quadratic equations until Chapter 8.

EXAMPLE 3 Solving a Quadratic Equation

Solve the quadratic equation by factoring.

$$x^2 - x - 12 = 0$$
$$(x - 4)(x + 3) = 0 \qquad \text{Factor}$$
$$x - 4 = 0 \quad \text{or} \quad x + 3 = 0 \qquad \text{Zero-product rule}$$
$$x = 4 \qquad x = -3$$

PRACTICE EXERCISE 3

Solve $x^2 - 2x - 63 = 0$.

Check: $(4)^2 - 4 - 12 \stackrel{?}{=} 0$ $\quad$ $(-3)^2 - (-3) - 12 \stackrel{?}{=} 0$

$16 - 4 - 12 \stackrel{?}{=} 0$ $\quad$ $9 + 3 - 12 \stackrel{?}{=} 0$

$16 - 16 = 0$ $\quad$ $12 - 12 = 0$

The solutions are 4 and -3.

Answer: 9, -7

When we solve applied problems, some equations may involve fractions or may require collecting terms. The following examples illustrate these techniques.

EXAMPLE 4 Solving Quadratic Equations

Solve.

(a) $5x^2 - 2x + \frac{1}{5} = 0$

$25x^2 - 10x + 1 = 0$ Multiply by 5 to clear fractions

$(5x - 1)^2 = 0$

$5x - 1 = 0$ or $5x - 1 = 0$ Zero-product rule

$5x = 1$ $\quad$ $5x = 1$

$x = \frac{1}{5}$ $\quad$ $x = \frac{1}{5}$

There is only one solution, $\frac{1}{5}$. Check by substituting.

(b) $2x^2 - 2x = 12$

$2x^2 - 2x - 12 = 0$ Collect terms on left side

$2(x^2 - x - 6) = 0$ Factor out 2

$x^2 - x - 6 = 0$ Divide both sides by 2

$(x - 3)(x + 2) = 0$ Factor

$x - 3 = 0$ or $x + 2 = 0$ Zero-product rule

$x = 3$ $\quad$ $x = -2$

The solutions are 3 and -2. Check by substituting.

(c) $5x^2 = 3x$

$5x^2 - 3x = 0$ Collect terms on left side

$x(5x - 3) = 0$ Factor out x

$x = 0$ or $5x - 3 = 0$ Zero-product rule

$5x = 3$

$x = \frac{3}{5}$

The solutions are 0 and $\frac{3}{5}$. Check by substituting.

PRACTICE EXERCISE 4

Solve.

(a) $\frac{4}{3}x^2 + 4x + 3 = 0$

(b) $3x^2 - 18x = 48$

(c) $9x^2 = -6x$

Answers: (a) $-\frac{3}{2}$ (b) 8, -2 (c) 0, $-\frac{2}{3}$

CAUTION

When solving an equation such as $5x^2 = 3x$ in Example 4(c), students sometimes make the mistake of dividing both sides by x. This gives $5x = 3$, leading to $x = \frac{3}{5}$, but the solution $x = 0$ is lost. Never divide both sides of an equation by an expression containing the variable.

The steps for solving an equation using the zero-product rule are summarized below.

To Solve an Equation Using the Zero-Product Rule

1. If necessary, rewrite the equation making one side 0 by collecting like terms and clearing fractions.
2. Factor the nonzero side.
3. Use the zero-product rule to set each factor equal to 0.
4. Solve each equation obtained in step 3.
5. Check all solutions in the original equation.

3 APPLIED PROBLEMS

Factoring is used to solve many applied problems. In the following examples, write out detailed descriptions of the variables and draw and label any geometric figures.

EXAMPLE 5 Consecutive Odd Integer Problem

One-third the product of two consecutive positive odd integers is 85. Find the integers.

Let x = the first positive odd integer,

$x + 2$ = the next consecutive odd integer,

$\frac{1}{3}x(x + 2) = \frac{x(x + 2)}{3}$ = one-third their product.

The equation we must solve is

$$\frac{x(x + 2)}{3} = 85.$$

Multiply by 3 to clear fractions.

$$x(x + 2) = 255$$
$$x^2 + 2x = 255$$
$$x^2 + 2x - 255 = 0$$
$$(x - 15)(x + 17) = 0$$
$$x - 15 = 0 \quad \text{or} \quad x + 17 = 0$$
$$x = 15 \qquad\qquad x = -17$$

Since x must be a positive integer, we rule out -17. Thus, 15 is the first integer, and the next odd integer is 17 ($x + 2 = 17$). Since 15 and 17 are consecutive positive odd integers and $\frac{1}{3}(15)(17) = 85$, these solutions check.

Practice Exercise 5

Five times the product of two consecutive positive even integers is 840. Find the integers.

Let x = the first even integer,

$x + 2$ = the second even integer.

Answer: 12, 14

EXAMPLE 6 Geometry Problem

Find the length and width of a rectangle if the length is 5 cm more than the width and the area is 84 cm^2. See the sketch in Figure 5.2.

Let w = width of the rectangle,

$w + 5$ = length of the rectangle.

Practice Exercise 6

Find the length and width of a rectangle if the width is 8 ft less than the length and the area is 105 ft^2.

$$
\begin{aligned}
w(w+5) &= 84 \\
w^2 + 5w &= 84 \\
w^2 + 5w - 84 &= 0 \\
(w-7)(w+12) &= 0
\end{aligned}
$$

$$w - 7 = 0 \quad \text{or} \quad w + 12 = 0$$

$$w = 7 \qquad\qquad w = -12$$

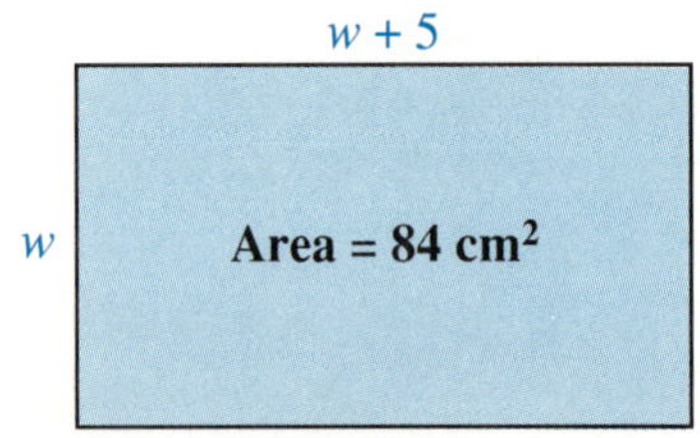

Figure 5.2

Since -12 cannot be a width of a rectangle, the width is 7 cm and the length is $w + 5 = 7 + 5 = 12$ cm.

Answer: 15 ft, 7 ft

EXAMPLE 7 Business Problem

A wholesale warehouse uses the equation $P(n) = 2n^2 - 7n - 15$ to calculate the profit P on the number n of refrigerators sold.

(a) Find the profit when 20 refrigerators are sold.
This is an evaluation problem. We calculate $P(20)$.

$$P(n) = 2n^2 - 7n - 15 \qquad \text{Profit equation}$$

$$P(20) = 2(20)^2 - 7(20) - 15 \qquad n = 20$$

$$= 2(400) - 140 - 15 = 645$$

The profit is \$645 when 20 are sold.

(b) Find the number sold when the profit was \$57.
Since $P(n) = \$57$ we need to solve the following equation.

$$P(n) = 2n^2 - 7n - 15 = 57 \qquad \text{Profit was \$57}$$

$$2n^2 - 7n - 72 = 0 \qquad \text{Collect like terms}$$

$$(2n + 9)(n - 8) = 0 \qquad \text{Factor}$$

$$2n + 9 = 0 \quad \text{or} \quad n - 8 = 0 \qquad \text{Zero-product rule}$$

$$n = -\frac{9}{2} \quad \text{or} \quad n = 8$$

Since n is a number of refrigerators, we rule out $-\frac{9}{2}$. Thus, 8 refrigerators were sold.

Practice Exercise 7

A retail store uses the equation $P(n) = 2n^2 - 5n - 50$ to calculate the profit P on the number n of dresses sold.

(a) Find the profit when 15 dresses are sold.

(b) Find the number sold when the profit was \$100.

Answers: (a) \$325 (b) 10

5.6 EXERCISES A

Solve.

1. $(x - 3)(x + 6) = 0$

2. $(2x - 1)(x + 7) = 0$

3. $(5y - 1)(2y - 3) = 0$

Exercises 54–56 review material from Section 1.7 to help you prepare for the next chapter. Simplify and write without negative exponents.

54. $\frac{x^2}{x^5}$ **55.** $\frac{3x^3y}{6xy^4}$ **56.** $\left(\frac{4^0xy^2}{2x^2y}\right)^{-2}$

ANSWERS: 1. 3, −6 2. $\frac{1}{2}$, −7 3. $\frac{1}{5}$, $\frac{3}{2}$ 4. −2.3, 5.7 5. $\frac{3}{2}$, $-\frac{7}{2}$ 6. −3, 9 7. 0, 1, 3 8. −1, 2, −3 9. 5, 8, $-\frac{1}{2}$ 10. 8 11. $-\frac{1}{3}$, 5 12. 2, 3 13. −5, 6 14. 7, −9 15. −3, −10 16. 0, $\frac{5}{3}$ 17. 5 18. $\frac{3}{2}$, −3 19. $\frac{1}{3}$, −1 20. $\frac{2}{5}$, 3 21. 2, 4 22. 1 23. −2, −3 24. 2, 3 25. 2, 8 26. 0, 1 27. 3, −3 28. 5 or −1 29. 17 yr 30. 20, 22 31. 6 or −6 32. Jose: 22 yr; Mike: 10 yr 33. 3 cm, 4 cm 34. 6 in 35. 12 yr 36. 26 yd, 20 yd 37. 6, 8 38. 3 yr 39. 11 ft, 5 ft 40. 16 cm, 9 cm 41. 6 cm, 14 cm 42. 125 amperes 43. 45 amperes 44. 5 seconds 45. 1 second and 9 seconds 46. \$215 47. \$1150 48. 0 49. 4 50. $(3x + 8y)(3x - 8y)$ 51. $(11x - 2y)^2$ 52. $2(3u - 2v)(9u^2 + 6uv + 4v^2)$ 53. $(5u + v^2)(25u^2 - 5uv^2 + v^4)$ 54. $\frac{1}{x^3}$ 55. $\frac{x^2}{2y^3}$ 56. $\frac{4x^2}{y^2}$

5.6 EXERCISES B

Solve.

1. $(x - 5)(x + 9) = 0$

2. $(3x - 2)(x + 5) = 0$

3. $(4y - 3)(y - 8) = 0$

4. $(x + 4.9)(x - 8.8) = 0$

5. $\left(y - \frac{8}{3}\right)\left(y + \frac{2}{3}\right) = 0$

6. $8(x - 2)(x + 10) = 0$

7. $y(y + 3)(y - 2) = 0$

8. $(x + 3)(x - 2)(x + 5) = 0$

9. $(y + 6)(y + 9)(2y + 3) = 0$

10. $(x + 4) - (3x - 4) = 0$

11. $(2y - 3)^2(y + 6)^2 = 0$

12. $x^2 - 4x + 3 = 0$

13. $x^2 - 3y - 18 = 0$

14. $y^2 + 4y - 32 = 0$

15. $\frac{1}{4}x^2 + x = \frac{21}{4}$

16. $7u = 5u^2$

17. $v^2 = 16v - 64$

18. $x^2 - \frac{3}{2}x = \frac{5}{2}$

19. $3x^2 = 10x - 8$

20. $5u^2 + 6 = -17u$

21. $\frac{1}{3}y^2 + \frac{4}{3}y + 1 = 0$

22. $0.1v^2 - 0.6v + 0.9 = 0$

23. $x(x - 6) = -8$

24. $u(u + 5) = -2u - 10$

25. $5y^2 = 40y - 75$

26. $y^3 + 4y^2 + 4y = 0$

27. $3x^2 - 8 = 2x^2 + 8$

Solve.

28. If the square of Waldo's age is decreased by 300, the result is twenty times his age. How old is Waldo?

29. The product of 1 less than a number and 7 more than the number is 105. Find the number.

30. Find two consecutive negative odd integers whose product is 143.

31. The product of Ed's age in two years and his age two years ago is 320. What is Ed's present age?

32. The length of a rectangle is 8 ft more than the width. Find the dimensions if its area is 384 ft^2.

33. Adding 4 to the square of Zeke's age is the same as subtracting 3 from eight times his age. How old is Zeke?

30. The product of two consecutive positive even integers is 440. Find the integers.

31. One more than a number times 1 less than the number is 35. Find the number.

32. Jose is 12 years older than Mike. The product of their ages is 220. How old is each?

33. The hypotenuse of a right triangle is 5 cm, and one leg is 1 cm longer than the other. Find the length of each leg.

Let x = length of one leg.

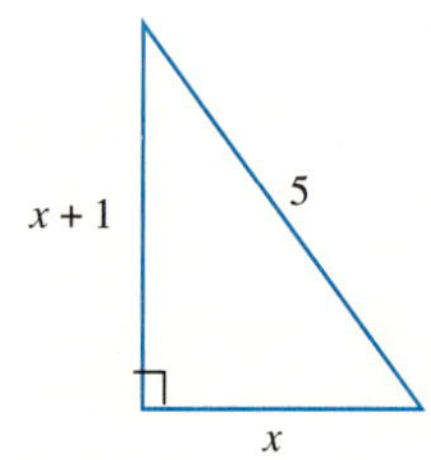

34 The number of square inches in the area of a square is 12 more than the number of inches in its perimeter. Find the length of each side.

35. Wanda's age 4 years ago times her age in 6 years is 144. How old is Wanda?

36. A rectangular garden is 6 yd longer than it is wide. What are the dimensions of the garden if its area is 520 yd^2?

Let w = width of garden.

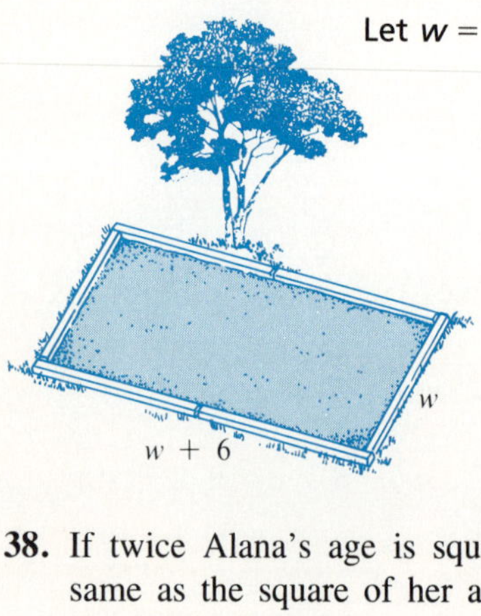

37 The sum of the squares of two positive consecutive even integers is 100. Find the integers.

38. If twice Alana's age is squared, the result is the same as the square of her age 3 years from now. How old is Alana?

39. The area of a parallelogram is 55 ft^2. If the base is 1 ft greater than twice the altitude, find the base and the altitude.

40 A box is 12 cm high. The length is 7 cm more than the width. If the volume is 1728 cm^3, find the length and width.

Let w = width of box.

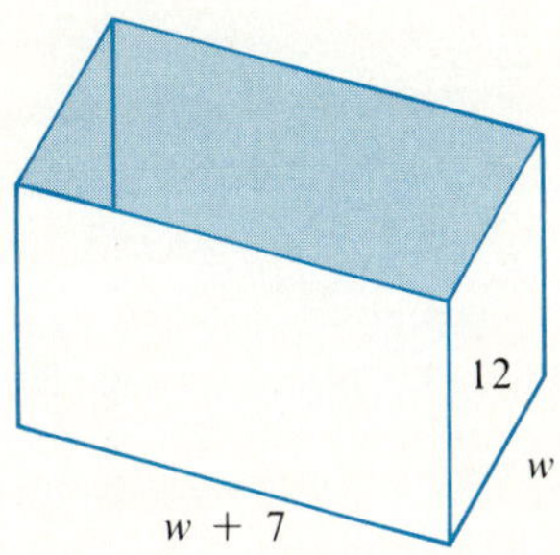

41 The area of a triangle is 42 cm^2. If the base is 8 cm less than the height, find the base and height.

The current I in amperes in an electrical circuit varies according to time t in seconds by the equation $I = 5t^2 - 50t + 125$.

42. What is the initial amperage (when $t = 0$)?

43. What is the amperage in 2 seconds (when $t = 2$)?

44. In how many seconds will the amperage be 0 amperes?

45. In how many seconds will the amperage be 80 amperes?

The profit P of a retail operation is given by the equation $P(n) = 3n^2 - 2n - 10$ *where n is the number of suits sold.*

46. Find the profit when $n = 9$.

47. Find the profit when 20 suits are sold.

48. How many suits were sold on a day when the profit was $-\$10$ (there was a \$10 loss)?

49 How many suits were sold on a day when the profit was \$30?

FOR REVIEW

Factor.

50. $9x^2 - 64y^2$

51. $121x^2 - 44xy + 4y^2$

52. $54u^3 - 16v^3$

53. $125u^3 + v^6$

Exercises 54–56 review material from Section 1.6 to help you prepare for the next chapter. Simplify and write without negative exponents.

54. $\dfrac{x^2}{x^5}$ **55.** $\dfrac{3x^3y}{6xy^4}$ **56.** $\left(\dfrac{4^0xy^2}{2x^2y}\right)^{-2}$

ANSWERS: **1.** 3, −6 **2.** $\frac{1}{2}$, −7 **3.** $\frac{1}{5}$, $\frac{3}{2}$ **4.** −2.3, 5.7 **5.** $\frac{3}{2}$, $-\frac{7}{2}$ **6.** −3, 9 **7.** 0, 1, 3 **8.** −1, 2, −3 **9.** 5, 8, $-\frac{1}{2}$ **10.** 8 **11.** $-\frac{1}{3}$, 5 **12.** 2, 3 **13.** −5, 6 **14.** 7, −9 **15.** −3, −10 **16.** 0, $\frac{5}{3}$ **17.** 5 **18.** $\frac{3}{2}$, −3 **19.** $\frac{1}{3}$, −1 **20.** $\frac{2}{5}$, 3 **21.** 2, 4 **22.** 1 **23.** −2, −3 **24.** 2, 3 **25.** 2, 8 **26.** 0, 1 **27.** 3, −3 **28.** 5 or −1 **29.** 17 yr **30.** 20, 22 **31.** 6 or −6 **32.** Jose: 22 yr; Mike: 10 yr **33.** 3 cm, 4 cm **34.** 6 in **35.** 12 yr **36.** 26 yd, 20 yd **37.** 6, 8 **38.** 3 yr **39.** 11 ft, 5 ft **40.** 16 cm, 9 cm **41.** 6 cm, 14 cm **42.** 125 amperes **43.** 45 amperes **44.** 5 seconds **45.** 1 second and 9 seconds **46.** \$215 **47.** \$1150 **48.** 0 **49.** 4 **50.** $(3x + 8y)(3x - 8y)$ **51.** $(11x - 2y)^2$ **52.** $2(3u - 2v)(9u^2 + 6uv + 4v^2)$ **53.** $(5u + v^2)(25u^2 - 5uv^2 + v^4)$ **54.** $\frac{1}{x^3}$ **55.** $\frac{x^2}{2y^3}$ **56.** $\frac{4x^2}{y^2}$

5.6 EXERCISES B

Solve.

1. $(x - 5)(x + 9) = 0$

2. $(3x - 2)(x + 5) = 0$

3. $(4y - 3)(y - 8) = 0$

4. $(x + 4.9)(x - 8.8) = 0$

5. $\left(y - \dfrac{8}{3}\right)\left(y + \dfrac{2}{3}\right) = 0$

6. $8(x - 2)(x + 10) = 0$

7. $y(y + 3)(y - 2) = 0$

8. $(x + 3)(x - 2)(x + 5) = 0$

9. $(y + 6)(y + 9)(2y + 3) = 0$

10. $(x + 4) - (3x - 4) = 0$

11. $(2y - 3)^2(y + 6)^2 = 0$

12. $x^2 - 4x + 3 = 0$

13. $x^2 - 3y - 18 = 0$

14. $y^2 + 4y - 32 = 0$

15. $\dfrac{1}{4}x^2 + x = \dfrac{21}{4}$

16. $7u = 5u^2$

17. $v^2 = 16v - 64$

18. $x^2 - \dfrac{3}{2}x = \dfrac{5}{2}$

19. $3x^2 = 10x - 8$

20. $5u^2 + 6 = -17u$

21. $\dfrac{1}{3}y^2 + \dfrac{4}{3}y + 1 = 0$

22. $0.1v^2 - 0.6v + 0.9 = 0$

23. $x(x - 6) = -8$

24. $u(u + 5) = -2u - 10$

25. $5y^2 = 40y - 75$

26. $y^3 + 4y^2 + 4y = 0$

27. $3x^2 - 8 = 2x^2 + 8$

Solve.

28. If the square of Waldo's age is decreased by 300, the result is twenty times his age. How old is Waldo?

29. The product of 1 less than a number and 7 more than the number is 105. Find the number.

30. Find two consecutive negative odd integers whose product is 143.

31. The product of Ed's age in two years and his age two years ago is 320. What is Ed's present age?

32. The length of a rectangle is 8 ft more than the width. Find the dimensions if its area is 384 ft^2.

33. Adding 4 to the square of Zeke's age is the same as subtracting 3 from eight times his age. How old is Zeke?

34. The length of a rectangle is 10 m more than the width. Find the length and width if the area is 264 m^2.

35. Five times the product of two positive consecutive integers is 550. Find the integers.

36. The square of a number is 48 more than 8 times the number. Find the number.

37. The area of a triangle is 114 ft^2 and the height is 7 ft more than the base. Find the height and base.

38. If the sides of a square are lengthened by 3 cm, the area will be 196 cm^2. Find the length of a side.

39. What number has the property that the number times itself is the same as the number plus itself?

40. A ladder 13 ft long leans against a wall. If the base of the ladder is 5 ft from the wall, how far would the lower end have to be pulled out so that the upper end is pulled down the same amount?

41. The sum of the squares of two positive consecutive integers is 113. Find the integers.

The current I in amperes in an electrical circuit varies according to time t in seconds by $I = 3t^2 - 48t + 192$.

42. What is the initial amperage (when $t = 0$)?

43. What is the amperage in 5 seconds (when $t = 5$)?

44. In how many seconds will the amperage be 0 amperes?

45. In how many seconds will the amperage be 108 amperes?

The profit P of a wholesale operation is given by the equation $P(n) = 2n^2 - 3n - 12$, where n is the number of appliances sold.

46. Find the profit when $n = 7$.

47. Find the profit when 15 appliances are sold.

48. How many appliances were sold on a day when the profit was $-\$13$ (a \$13 loss)?

49. How many appliances were sold on a day when the profit was \$123?

FOR REVIEW

Factor.

50. $125x^2 - 20y^2$

51. $25x^2 + 30xy + 9y^2$

52. $27u^3 - v^6$

53. $64u^3 + 125v^3$

Exercises 54–56 review material from Section 1.7 to help you prepare for the next chapter. Simplify and write without negative exponents.

54. $\dfrac{a^4}{a^{11}}$

55. $\dfrac{2u^4v^2}{8u^2v^5}$

56. $\left(\dfrac{3^0x^2y}{3xy^3}\right)^{-2}$

5.6 EXERCISES C

Solve for x. Assume that a and b are constant real numbers.

1. $x(x - 1)(x + 3)(x + 7) = 0$

2. $(x^2 - x - 2)(x^3 - x^2 - 20x) = 0$
[Answer: 2, −1, 0, 5, −4]

3. $(x - a)(x + 3b) = 0$

4. $2x^2 + ax - 2bx - ab = 0$

5. $(x - 3)^2 - a^2 = 0$

6. $(x - 3)(x + 2) - a^2 = (2x - 1)(x + 2) - 2a^2$
[Answer: $a - 2$, $-a - 2$]

CHAPTER REVIEW

KEY WORDS

5.1 A **polynomial** is an algebraic expression with terms that are products of numbers and variables raised to whole-number exponents.

A **monomial** is a polynomial with one term.

A **binomial** is a polynomial with two terms.

A **trinomial** is a polynomial with three terms.

The **degree of a term** in a polynomial is the exponent on the variable if only one variable is present. When more variables are present in a term, the **degree of the term** is the sum of the exponents on the variables.

The **degree of a polynomial** is equal to the degree of the term with the highest degree.

Terms in a polynomial that have the same variables raised to the same powers are called **like terms.**

When like terms in a polynomial are combined using the distributive law we are **collecting like terms.**

A polynomial is written in **descending (ascending) order** when the terms are arranged with the highest (lowest) degree term first, followed by the next highest (lowest), and so forth.

5.3 The **greatest common factor (GCF)** of a polynomial is the common factor of each term that contains the largest numerical factor and the highest power of each variable.

5.6 A **quadratic equation (second-degree equation)** is an equation that can be written in the form $ax^2 + bx + c = 0$ with $a \neq 0$.

KEY CONCEPTS

5.1 **1.** The degree of a nonzero monomial such as 5 is 0. The zero monomial 0 has no degree.

2. Only like terms can be collected. For example, $3x^2 - 2x$ cannot be simplified since $3x^2$ and $-2x$ are not like terms.

3. Change *all* signs when removing parentheses preceded by a minus sign in a subtraction problem. For example, $3x^3 - (2x^2 - 5x + 6) = 3x^3 - 2x^2 + 5x - 6$.

5.2 **1.** Use the FOIL method to multiply binomials or the column method to multiply polynomials with several terms.

2. Special product formulas:

(a) $(a + b)(a - b) = a^2 - b^2$

(b) $(a + b)(a + b) = (a + b)^2 = a^2 + 2ab + b^2$ (*not* $a^2 + b^2$)

(c) $(a - b)(a - b) = (a - b)^2 = a^2 - 2ab + b^2$ (*not* $a^2 - b^2$)

5.3 **1.** In any factoring problem always remove the greatest common factor (GCF) first.

2. Polynomials with four terms can often be factored by grouping.

5.4 **1.** To factor a trinomial of the form $x^2 + bx + c$, form all pairs of integers whose product is c. Use the pair whose sum is b in $x^2 + bx + c = (x + \_\_)(x + \_\_)$.

2. To factor $ax^2 + bx + c$, $a \neq 1$, list all pairs of integers whose product is a and list all pairs whose product is c. Use trial and error to determine which pairs give the correct value for b.

3. You should never give a wrong answer to a factoring problem because you can always check your work by multiplying the factors.

5.5 **1.** Difference of squares:

$$a^2 - b^2 = (a + b)(a - b)$$

2. Perfect square trinomials:

$$a^2 + 2ab + b^2 = (a + b)^2$$

and

$$a^2 - 2ab + b^2 = (a - b)^2$$

3. Sum of cubes:

$$a^3 + b^3 = (a + b)(a^2 - ab + b^2)$$

4. Difference of cubes:

$$a^3 - b^3 = (a - b)(a^2 + ab + b^2)$$

5.6 The zero-product rule states that if $ab = 0$, then $a = 0$ or $b = 0$. This rule is used to solve equations that can be factored.

REVIEW EXERCISES

Part I

5.1 *For each polynomial, (**a**) tell whether it is a monomial, binomial, or trinomial, (**b**) give the term of degree 2 and its coefficient, and (**c**) give the degree of the polynomial.*

1. $x^2 + 2x + 5$ **2.** $-4ab$ **3.** $6x^4y^5z^2 - 7x^4 + 9y^2$

Collect like terms and write in descending powers of x.

4. $5x^2y - 3xy - 4x^2y + 5$ **5.** $6x^2y^2 - 8x^3y + 4xy - 6x^3y - x^2y^2$

Evaluate the polynomial for the given values of the variables.

6. $-2x^2 - 5x + 10;\ x = -2$ **7.** $3a^2b^3 - 7ab^2 + 5b;\ a = -1,\ b = 2$

8. Add $4y^2 - 16y + y^3 - 2$ and $-3y + 7y^3 + 2 - y^2$.

9. Subtract $-10x^3y^2 + 6x^2y^2 - 5xy$ from $5x^3y^2 - 4x^2y^2 + 12$.

Perform the indicated operations.

10. $(7a^3b^2 - 3a^2b) + (6a^3b^2 + 5ab^2) - (-2a^3b^2 + 7ab^2)$

11. $(a^3b^3 - 5a + 2b) + (-2a^3b^3 + 4a - 8b) - (5a^3b^3 - 6b)$

Add.

12. $\begin{array}{r} 36x^2y^2 - 6xy + 5 \\ \underline{22x^2y^2 + 18xy - 17} \end{array}$ **13.** $\begin{array}{r} 105x^3y^2 + 71x^2y^3 - 8xy + 5 \\ \underline{-82x^3y^2 - 92x^2y^3 - 9xy + 18} \end{array}$

5.2 *Multiply.*

14. $(3u^2v)(-8uv^2)$ **15.** $-18uv^3(2u - 5v)$ **16.** $(x - 2y)(2x - y)$

17. $(u + 8)^2$ **18.** $(u - 8)^2$ **19.** $(3x - 5y)(7x + 4y)$

20. $(4u + 11v)(2u + 3v)$ **21.** $(2u - 5v)(2u + 5v)$ **22.** $(7u - v)^2$

23. $(3x^2 + 2y)^2$ **24.** $(2x^2 - 5y^2)(2x^2 + 5y^2)$ **25.** $(3u - v)(u - 5v)(u + 5v)$

26. $[3x^2 - (x - y)][3x^2 + (x - y)]$ **27.** $(2x + 3y)^3$ **28.** $\begin{array}{r} 6x^2y^2 - 7xy + 5 \\ 4xy + 3 \\ \hline \end{array}$

5.3–5.5 *Factor.*

29. $5u^2v^2 - 10uv$ **30.** $-11u^3v^2 - 77u^2v^2 - 88uv^2$ **31.** $15x^2 - 10xy + 3xy - 2y^2$

32. $x^2 - 4x - 21$ **33.** $3y^2 + 23y + 14$ **34.** $u^2 + 10uv + 21v^2$

35. $-5x^2 - 15xy + 50y^2$ **36.** $6x^2 - 54y^2$ **37.** $8u^3 + v^3$

38. $8u^3 - v^3$ **39.** $4x^2 - 12xy + 9y^2$ **40.** $9x^2 + 30xy + 25y^2$

41. $2x^2 + xy - 8x - 4y$ **42.** $(x - y)^2 - 4$ **43.** $(x^2 + y^2)^2 - 3(x^2 + y^2) + 2$

44. $(x + y)^2 - (u - v)^2$ **45.** $x^4 - 5x^2 + 4$ **46.** $(3x + 5)^2 - 7(3x + 5) + 12$

5.6 *Solve.*

47. $(x + 2)(x - 10) = 0$ **48.** $y\left(y - \frac{1}{2}\right)(y + 7) = 0$

49. $3x^2 - 2x = 5$ **50.** $3u(u + 1) = u^2 - 2u - 2$

Solve.

51. Robert is 6 years older than Wong. The product of their ages is 160. How old is each?

52. The area of a triangle is 20 m^2 and the base is 3 m more than the height. Find the base and height.

The profit P of a wholesale operation is given by the equation $P(n) = 2n^2 - 4n - 10$ *where n is the number of widgets sold.*

53. Find the profit when 20 widgets are sold.

54. How many widgets were sold on a day when the profit was $150?

Part II

Factor.

55. $16u^4 - v^4$

56. $6y^2 + y - 40$

57. $6x^2 - 14xy - 21xy + 49y^2$

58. $9x^2 + 6x + 1$

59. $27a^3 - w^3$

60. $4u^2 - 20u + 25$

Perform the indicated operations.

61. $(10 - 12y^2 + y^3 + 8y) - (6y + 5 - 7y^3 + 8y^2)$

62. $(x - 2y)(x + 2y)$

63. $(6a + 7b)(a + 2b)$

64. $(3u - 5)^2$

65. $(4x + 3y)^2$

66. $(5a^2b^2 - 5ab + 14) + (6a^2b^2 + 7ab - 5)$

In a business the manufacturing cost is $M(u) = 2u^2 - 3u + 5$, *the wholesale cost is* $W(u) = u^2 - 2u + 4$, *and the income is* $I(u) = 4u^2 - 2u$ *where u is the number of units manufactured and sold.*

67. Find the total cost $C(u) = M(u) + W(u)$.

68. Find the total cost when u is 20.

69. Find the profit $P(u) = I(u) - C(u)$.

70. Find the number of units manufactured and sold when the profit was $9.

71. Collect like terms and write in descending powers of x. $6x^4y^3 - 7x^6y - 3xy + 7x^6y$

72. Multiply.

$$\begin{array}{l} 5x^2 + 11xy - 6y^2 \\ \underline{3x - 7y} \end{array}$$

The area of seating in a circular auditorium with a circular stage in the center is $A = \pi R^2 - \pi r^2$ where R is the radius of the auditorium and r is the radius of the stage.

73. Find the factored form of this equation.

74. Use the factored form to find the area of seating when $R = 20$ m and $r = 4$ m. (Use 3.14 for π.)

Solve.

75. $(2x - 1)(5x - 3) = 0$

76. $3x^2 - 22x - 16 = 0$

77. $x^2 = \frac{7}{3}x$

78. $(3x - 1)(2x + 1)^2 = 0$

79. The product of 1 less than a number and 7 more than the number is 153. Find the number.

80. The current I in amperes in an electrical circuit varies according to time t in seconds by the equation $I = 6t^2 - 48t + 96$. **(a)** What is the initial amperage in the circuit? **(b)** In how many seconds will the amperage be 0 amperes?

ANSWERS: 1. (a) trinomial (b) x^2, 1 (c) 2 2. (a) monomial (b) $-4ab$, -4 (c) 2 3. (a) trinomial (b) $9y^2$, 9 (c) 11 4. $x^2y - 3xy + 5$ 5. $-14x^3y + 5x^2y^2 + 4xy$ 6. 12 7. 62 8. $8y^3 + 3y^2 - 19y$ 9. $15x^3y^2 - 10x^2y^2 + 5xy + 12$ 10. $15a^3b^2 - 3a^2b - 2ab^2$ 11. $-6a^3b^3 - a$ 12. $58x^2y^2 + 12xy - 12$ 13. $23x^3y^2 - 21x^2y^3 - 17xy + 23$ 14. $-24u^3v^3$ 15. $-36u^2v^3 + 90uv^4$ 16. $2x^2 - 5xy + 2y^2$ 17. $u^2 + 16u + 64$ 18. $u^2 - 16u + 64$ 19. $21x^2 - 23xy - 20y^2$ 20. $8u^2 + 34uv + 33v^2$ 21. $4u^2 - 25v^2$ 22. $49u^2 - 14uv + v^2$ 23. $9x^4 + 12x^2y + 4y^2$ 24. $4x^4 - 25y^4$ 25. $3u^3 - u^2v - 75uv^2 + 25v^3$ 26. $9x^4 - x^2 + 2xy - y^2$ 27. $8x^3 + 36x^2y + 54xy^2 + 27y^3$ 28. $24x^3y^3 - 10x^2y^2 - xy + 15$ 29. $5uv(uv - 2)$ 30. $-11uv^2(u^2 + 7u + 8)$ 31. $(5x + y)(3x - 2y)$ 32. $(x - 7)(x + 3)$ 33. $(3y + 2)(y + 7)$ 34. $(u + 3v)(u + 7v)$ 35. $-5(x + 5y)(x - 2y)$ 36. $6(x - 3y)(x + 3y)$ 37. $(2u + v)(4u^2 - 2uv + v^2)$ 38. $(2u - v)(4u^2 + 2uv + v^2)$ 39. $(2x - 3y)^2$ 40. $(3x + 5y)^2$ 41. $(2x + y)(x - 4)$ 42. $(x - y - 2)(x - y + 2)$ 43. $(x^2 + y^2 - 1)(x^2 + y^2 - 2)$ 44. $(x + y + u - v)(x + y - u + v)$ 45. $(x + 1)(x - 1)(x + 2)(x - 2)$ 46. $(3x + 2)(3x + 1)$ 47. -2, 10 48. 0, $\frac{1}{2}$, -7 49. $\frac{5}{3}$, -1 50. $-\frac{1}{2}$, -2 51. Robert is 16, Wong is 10. 52. 8 m, 5 m 53. \$710 54. 10 55. $(2u - v)(2u + v)(4u^2 + v^2)$ 56. $(2y - 5)(3y + 8)$ 57. $(2x - 7y)(3x - 7y)$ 58. $(3x + 1)^2$ 59. $(3a - w)(9a^2 + 3aw + w^2)$ 60. $(2u - 5)^2$ 61. $8y^3 - 20y^2 + 2y + 5$ 62. $x^2 - 4y^2$ 63. $6a^2 + 19ab + 14b^2$ 64. $9u^2 - 30u + 25$ 65. $16x^2 + 24xy + 9y^2$ 66. $11a^2b^2 + 2ab + 9$ 67. $C(u) = 3u^2 - 5u + 9$ 68. \$1109 69. $P(u) = u^2 + 3u - 9$ 70. 3 71. $6x^4y^3 - 3xy$ 72. $15x^3 - 2x^2y - 95xy^2 + 42y^3$ 73. $A = \pi(R + r)(R - r)$ 74. approximately 1206 m^2 75. $\frac{1}{2}$, $\frac{3}{5}$ 76. $-\frac{2}{3}$, 8 77. 0, $\frac{7}{3}$ 78. $-\frac{1}{2}$, $\frac{1}{3}$ 79. 10, -16 80. (a) 96 amperes (b) 4 seconds

CHAPTER 5 TEST

Given the polynomial $2x^4 - 5x^2y + y^2$:

1. Give the term which has degree 3. 1. ________

2. What is the degree of the polynomial? 2. ________

3. Is the polynomial a monomial, binomial, or trinomial? 3. ________

4. Evaluate the polynomial when $x = 2$ and $y = -1$. 4. ________

Perform the indicated operations and simplify.

5. $(3x^3 - 2x^2 + x - 7) + (2x^2 - x + 3)$ 5. ________

6. $(3a^3b^2 - 2ab + b^4) - (4a^3b^2 - 2ab - 3b^4)$ 6. ________

7. $(2x^2y^3 - xy) + (3x^2y^3 + 7) - (7 - xy)$ 7. ________

8. Add. $\begin{array}{l} 2ab - 3a^2b^2 + 8 \\ \underline{6ab + 3a^2b^2 - 7} \end{array}$ 8. ________

Multiply.

9. $-4u^2v(3u - 2v)$ 9. ________

10. $(a - 3b)(a + 3b)$ 10. ________

11. $(z + 5)^2$ 11. ________

12. $(z - 5)^2$ 12. ________

13. $(3x - 2y)(4x + 5y)$ 13. ________

14. $(a^2 + b^2)(a - b)(a + b)$ 14. ________

Factor.

15. $-11x^2y^3 + 66xy^2 - 44x^2y^2$ 15. ________

16. $6u^2 - uv - v^2$ 16. ________

17. $2x^2 - 8$ 17. ________

18. $a^3 + 8b^3$ 18. ________

19. $x^2 - 6xy + 9y^2$ 19. ________

20. $bx - 2b - xy + 2y$ 20. ________

21. $(x + y)^2 - 5(x + y) + 6$ 21. ________

Solve.

22. $(x + 1)(x - 7) = 0$ 22. ________

23. $2y^2 + 9y = 5$ 23. ________

24. Mack is 3 years older than Betty and the product of their ages is 180. How old is each? 24. ________

25. A retailer's profit is given by the equation $P(n) = 2n^2 - 59n - 5$ where n is the number of items sold. How many items were sold on Tuesday when the profit was $25? 25. ________

CHAPTER 6

Rational Expressions

6.1 BASIC CONCEPTS OF ALGEBRAIC FRACTIONS

STUDENT GUIDEPOSTS

1. Rational Expressions
2. Fundamental Principle of Fractions
3. Reducing to Lowest Terms

In Chapter 1 we saw that a rational number or fraction is a quotient of integers. In this chapter we define an **algebraic fraction** as a fraction that contains a variable in the numerator, the denominator, or both, such as

$$\frac{2x}{x-1}, \quad \frac{\sqrt{y}+1}{2y+1}, \quad \frac{(a+b)^2}{2}, \quad \frac{7}{(x-y)(y+1)}.$$

1 RATIONAL EXPRESSIONS

An algebraic fraction that is the ratio of two polynomials is called a **rational expression.** Of the algebraic fractions listed above, only

$$\frac{\sqrt{y}+1}{2y+1}$$

is *not* a rational expression since $\sqrt{y}+1$ is not a polynomial. Most of the algebraic fractions that we will consider are rational expressions.

Since division by zero is undefined, with rational expressions we must be careful to exclude replacements for the variable that make the denominator equal to zero.

To Find the Values for Which a Rational Expression Is Not Defined

1. Set the denominator equal to zero.
2. Solve the resulting equation; any solution must be excluded as a possible value of the variable.

EXAMPLE 1 FINDING VALUES TO EXCLUDE

Find the values for the variable that must be excluded.

(a) $\frac{3x}{x-5}$

Setting $x - 5 = 0$ and solving, we obtain $x = 5$. Then 5 must be excluded since if x is 5, the denominator is zero.

PRACTICE EXERCISE 1

Find the values for the variable that must be excluded.

(a) $\frac{w}{2w+1}$

(b) $\dfrac{3y}{y^2 - y - 6}$

$$y^2 - y - 6 = 0$$
$$(y - 3)(y + 2) = 0 \quad \text{Factor}$$
$$y - 3 = 0 \quad \text{or} \quad y + 2 = 0 \quad \text{Zero-product rule}$$
$$y = 3 \qquad\qquad y = -2$$

The values that must be excluded are 3 and -2.

(c) $\dfrac{x^2 - y^2}{5}$

Since the denominator can never be zero, there are no values that have to be excluded. The fraction is defined for all values of x and y.

(b) $\dfrac{a + 3}{27}$

(c) $\dfrac{z - 8}{z^2 - 1}$

Answers: (a) $-\frac{1}{2}$ (b) none (c) 1 and -1

2 FUNDAMENTAL PRINCIPLE OF FRACTIONS

In Chapter 0, we saw that multiplying or dividing the numerator and denominator of a fraction by the same nonzero number gives us equivalent fractions.

$$\frac{4}{10} = \frac{4 \cdot 5}{10 \cdot 5} = \frac{20}{50} \qquad \tfrac{4}{10} \text{ is equivalent to } \tfrac{20}{50}$$
$$\frac{4}{10} = \frac{4 \div 2}{10 \div 2} = \frac{2}{5} \qquad \tfrac{4}{10} \text{ is equivalent to } \tfrac{2}{5}$$

We can generalize this idea to include algebraic fractions.

Fundamental Principle of Fractions

If the numerator and denominator of an algebraic fraction are multiplied or divided by the same nonzero expression or number, the resulting algebraic fraction is **equivalent** to the original. That is,

$$\frac{a}{b} = \frac{ac}{bc} \quad \text{and} \quad \frac{ac}{bc} = \frac{a}{b} \quad (b \neq 0 \text{ and } c \neq 0).$$

EXAMPLE 2 DETERMINING EQUIVALENT FRACTIONS

Are the fractions equivalent?

(a) $\dfrac{x}{x - y}$ and $\dfrac{x(x + y)}{(x - y)(x + y)}$

These are equivalent (if $x + y \neq 0$). Note that $\frac{x(x + y)}{(x - y)(x + y)}$ has been built up from $\frac{x}{x - y}$ by multiplying numerator and denominator by $(x + y)$, while $\frac{x}{x - y}$ has been reduced from $\frac{x(x + y)}{(x - y)(x + y)}$ by dividing numerator and denominator by $(x + y)$.

(b) $\dfrac{3x}{x + 2}$ and $\dfrac{9}{x - 3}$

These are *not* equivalent since there is no expression that both the numerator and denominator of one fraction can be multiplied by to give the other.

PRACTICE EXERCISE 2

Are the fractions equivalent?

(a) $\dfrac{2a}{a + 1}$ and $\dfrac{2a^2}{a^2 + a}$

(b) $\dfrac{2w}{2w - 5}$ and $\dfrac{w}{w - 5}$.

Answers: (a) yes (b) no

3 REDUCING TO LOWEST TERMS

A fraction is **reduced to lowest terms** when 1 (or -1) is the only number or expression that divides both numerator and denominator. For example, we might reduce $\frac{-12}{42}$ to lowest terms as follows.

$$\frac{-12}{42} = \frac{(-1)\cdot\cancel{2}\cdot 2\cdot\cancel{3}}{\cancel{2}\cdot\cancel{3}\cdot 7} = \frac{-2}{7}$$ Factor completely; divide out common factors

When we factor completely, we factor into **primes,** numbers or expressions that can be divided only by themselves and 1. Sometimes we can reduce a fraction without factoring all the way to primes. This can happen when we recognize larger common factors in the numerator and denominator. For example, we might reduce $\frac{-12}{42}$ as follows.

$$\frac{-12}{42} = \frac{(-1)\cdot\cancel{6}\cdot 2}{\cancel{6}\cdot 7} = \frac{-2}{7}$$

Since we recognize 6 as a factor common to the numerator and denominator, it was not necessary to factor 6 into primes.

CAUTION

Divide out *factors* only, not terms. Only expressions *multiplied by* (not added to) every other expression in the numerator or denominator can be divided out. For example,

$$\frac{2\cdot 4}{2} = \frac{\cancel{2}\cdot 4}{\cancel{2}} \quad \text{but} \quad \frac{2+4}{2} \neq \frac{\cancel{2}+4}{\cancel{2}}.$$

The term "cancel" is sometimes used for "divide out." However, to avoid errors like the one shown in the caution statement above, it is probably better to use "divide out" since division is the inverse operation to multiplication.

To Reduce an Algebraic Fraction to Lowest Terms

1. Factor numerator and denominator completely.
2. Divide out all common factors.
3. Multiply the remaining factors.

EXAMPLE 3 Reducing Fractions to Lowest Terms

Reduce to lowest terms.

(a) $$\frac{20x^2y}{50x^3y^4} = \frac{\cancel{2}\cdot 2\cdot\cancel{5}\cdot\cancel{x}\cdot\cancel{x}\cdot\cancel{y}}{\cancel{2}\cdot\cancel{5}\cdot 5\cdot\cancel{x}\cdot\cancel{x}\cdot x\cdot\cancel{y}\cdot y\cdot y\cdot y} = \frac{2}{5x\cdot y\cdot y\cdot y}$$
$$= \frac{2}{5xy^3}$$

Compare this with the result obtained by applying the rules of exponents to x and y.

$$\frac{20x^2y}{50x^3y^4} = \frac{20}{50}x^{2-3}y^{1-4} = \frac{2\cdot\cancel{10}}{5\cdot\cancel{10}}x^{-1}y^{-3} = \frac{2}{5}\,\frac{1}{x}\,\frac{1}{y^3} = \frac{2}{5xy^3}$$

PRACTICE EXERCISE 3

Reduce to lowest terms.

(a) $\dfrac{15ab^2}{25a^3b^3}$

Here we reduced $\frac{20}{50}$ by dividing out the common factor, 10. With practice, our work could be shortened as follows:

$$\frac{20x^2y}{50x^3y^4} = \frac{2 \cdot \cancel{10} \cdot \cancel{x^2} \cdot \cancel{y}}{5 \cdot \cancel{10} \cdot \cancel{x^2} \cdot x \cdot \cancel{y} \cdot y^3} = \frac{2}{5xy^3}$$ Divide out common factor $10x^2y$

(b) $\frac{6x^3 + 2x}{2x} = \frac{\cancel{2x}(3x^2 + 1)}{\cancel{2x}} = \frac{3x^2 + 1}{1} = 3x^2 + 1$

Never divide out terms, as in $\frac{6x^3 + \cancel{2x}}{\cancel{2x}}$. This is wrong

(c) $\frac{x + xy}{y} = \frac{x(1 + y)}{y}$ Cannot be reduced; it is already in lowest terms

(d) $\frac{x^2 + x - 2}{x^2 - 1} = \frac{\cancel{(x - 1)}(x + 2)}{\cancel{(x - 1)}(x + 1)}$

$= \frac{x + 2}{x + 1}$

(e) $\frac{2a^3 + a^2b - ab^2}{a^2 + ab} = \frac{a(2a^2 + ab - b^2)}{a(a + b)}$

$= \frac{\cancel{a(a + b)}(2a - b)}{\cancel{a(a + b)}}$

$= 2a - b$

(b) $\frac{3w^2 + 9w}{3w}$

(c) $\frac{2y}{3y + 6}$

(d) $\frac{a^2 - 4}{a^2 + a - 6}$

(e) $\frac{x^3 - x^2y - 2xy^2}{x^3 - 3x^2y + 2xy^2}$

Answers: (a) $\frac{3}{5a^2b}$ (b) $w + 3$ (c) cannot be reduced (d) $\frac{a + 2}{a + 3}$ (e) $\frac{x + y}{x - y}$

CAUTION

When dividing out factors, remember that the quotient is 1 (not zero). For example,

$$\frac{x}{3x^2} = \frac{\cancel{x}}{3 \cdot \cancel{x} \cdot x} = \frac{\mathbf{1}}{3 \cdot \mathbf{1} \cdot x} = \frac{1}{3x},$$

not $\frac{0}{3x}$. We can write small 1's by factors that are divided out, as shown below, to remind us of this.

$$\frac{x}{3x^2} = \frac{\overset{1}{\cancel{x}}}{3\underset{1}{\cancel{x}}x} = \frac{1}{3x}$$

Suppose we wish to reduce $\frac{y - x}{x - y}$ to lowest terms. Notice that $y - x$ and $x - y$ are negatives of each other.

$\frac{y - x}{x - y} = \frac{(-1)(-y + x)}{(x - y)}$ Factor -1 from each term in the numerator

$= \frac{(-1)\cancel{(x - y)}}{\cancel{(x - y)}}$ $-y + x = x - y$ by the commutative law

$= -1$

In general we have shown that

$$\frac{a - b}{b - a} = -1, \quad \text{when } a \neq b.$$

EXAMPLE 4 Reducing a Fraction with Negative Factors

Reduce $\frac{y^2 - xy}{x^2 - xy}$ to lowest terms.

$$\frac{y^2 - xy}{x^2 - xy} = \frac{y(y-x)}{x(x-y)} = \frac{y}{x} \cdot \frac{y-x}{x-y} = \frac{y}{x} \cdot (-1) = -\frac{y}{x}$$

Practice Exercise 4

Reduce $\frac{m^2 - n^2}{mn - m^2}$ to lowest terms.

Answer: $\frac{m+n}{-m}$

Finding values to exclude in a rational expression in an applied problem can be slightly different. This is shown in the next example.

EXAMPLE 5 Interest Problem

The formula

$$P = \frac{A}{1 + 0.08t}$$

gives the relationship between a principal P that must be deposited in an account earning 8% simple interest to accumulate a specified amount A at the end of t years.

(a) Find the value of P if A is to be \$1160 after 2 years.
Substitute 1160 for A and 2 for t.

$$P = \frac{1160}{1 + 0.08(2)} = \frac{1160}{1 + 0.16} = \frac{1160}{1.16} = 1000$$

Thus, the principal must be \$1000.

(b) What values of t must be excluded in the formula for P?

Usually we only want to exclude values for the variable that make a denominator zero. For $1 + 0.08t$ to equal 0, t would be $\frac{-1}{0.08} = -12.5$. However, since a negative number for time in years would not make sense in this problem, every negative value for t must be excluded.

Practice Exercise 5

The base of a triangle with area A and height h is given by the formula $b = \frac{2A}{h}$.

(a) Find b if $A = 18.9\text{ cm}^2$ and $h = 5.4$ cm.

(b) What values of h must be excluded in the formula?

Answers: (a) 7.0 cm (b) all values of $h \le 0$

6.1 EXERCISES A

Find the values for the variable that must be excluded.

1. $\frac{x}{3x+1}$
2. $\frac{y^2+3}{y+3}$
3. $\frac{3x+1}{x}$
4. $\frac{a+1}{(a-1)(a+2)}$
5. $\frac{5}{y^2-2y+1}$
6. $\frac{a+1}{2a^3+7a^2+6a}$
7. $\frac{4x+5}{x^2-x}$
8. $\frac{5-w}{5+w^2}$

Are the given fractions equivalent? Explain. (Assume values that make the denominator zero are excluded.)

9. $\frac{2}{3}, \frac{6}{9}$
10. $\frac{4}{x}, \frac{x}{4}$
11. $\frac{5}{a}, \frac{-5}{-a}$

12. $\frac{-7}{x}, \frac{7}{-x}$

13. $\frac{x+1}{y}, \frac{2(x+1)}{2y}$

14. $\frac{a+b}{3}, \frac{5a+5b}{15}$

15. $\frac{2}{x^3}, \frac{2+y}{x^3+y}$

16. $\frac{a+b}{a^2-b^2}, \frac{1}{a-b}$

17. $\frac{a^2-b^2+c^2}{b-c}, \frac{-a^2+b^2-c^2}{c-b}$

Reduce to lowest terms. (Assume values that make the denominator zero are excluded.)

18. $\frac{45}{105}$

19. $\frac{91}{39}$

20. $\frac{9x^7y^4}{18x^3y^5}$

21. $\frac{7a^2b^3z^5}{21ab^5z^2}$

22. $\frac{2x}{2x^2+2x}$

23. $\frac{y+4}{y^2+4}$

24. $\frac{a^2-b^2}{a^2-2ab+b^2}$

25. $\frac{x^3-y^3}{x-y}$

26. $\frac{a-z}{z-a}$

27. $\frac{x-2y}{2y-x}$

28. $\frac{x+2y}{2y+x}$

29. $\frac{x-2y}{2y+x}$

30. $\frac{a^4-b^4}{a^2+b^2}$

31 $\frac{xz+xw+yz+yw}{z^2+zw}$

32. $\frac{15x+7x^2-2x^3}{x^2-8x+15}$

33. $\frac{a^2-2a}{2-a}$

34 $\frac{9a^2-15a+25}{27a^3+125}$

35. $\frac{3z^2-2z-1}{z^3-z^2+2z-2}$

36. $\frac{8u^3-27}{2u-3}$

37. $\frac{(x+2)(x^2-3x+2)}{(x-2)(x^2+x-2)}$

38. $\frac{y^2-x^2+2x-1}{y^2-x^2+y+x}$

Dr. Gordon Johnson's assistant, Burford, worked Exercises 39–40 as indicated. Why was Burford fired?

39. $\frac{2+4x}{4x} = \frac{2+\cancel{4x}}{\cancel{4x}} = 2$

40. $\frac{x+2}{x^2-4} = \frac{\cancel{x+2}}{\cancel{(x+2)}(x-2)} = \frac{0}{x-2} = 0$

41. Reduce $\frac{5-xy}{xy-5}$ to lowest terms.

42. Reduce $\frac{5+xy}{xy+5}$ to lowest terms.

43. Reduce $\dfrac{x^2 - xy - 2y^2}{x + y}$ to lowest terms. Substitute 2 for x and -1 for y in both the original and the reduced fraction to show that the two are equivalent.

In Exercises 44–45, use a calculator to find the value of P, to the nearest cent, if $P = \dfrac{A}{1 + 0.07t}$ *and A and t have the given values.*

44 $A = \$5000$ and $t = 6$ yr

45. $A = \$10{,}000$ and $t = 12$ yr

46. You have probably observed the change in the pitch of a train whistle as the train moves quickly past you. The normal frequency (number of vibrations per second) of the whistle appears to increase as the train approaches and decrease as the train moves away. Physicists call this the **Doppler effect,** named for Austrian physicist C. J. Doppler (1803–53). The formula that describes the Doppler effect is

$$F = \frac{1100f}{1100 - V},$$

where F is the observed frequency of a sound, f is the actual frequency of a sound, and V is the velocity of the train (a positive number if the train is approaching and a negative number if it is going away). Suppose a train whistle produces a sound with a frequency of 612.0 vibrations per second. What is the observed frequency **(a)** as the train is approaching at 80 mph? **(b)** as the train is going away at 80 mph?

FOR REVIEW

Exercises 47–49 review multiplication and division of fractions to help you prepare for the next section.

47. $\dfrac{3}{8} \cdot \dfrac{4}{27}$

48. $\dfrac{4}{5} \div \dfrac{8}{35}$

49. $\dfrac{9}{5} \div 3$

50. Factor 13,650 into a product of prime integers.

ANSWERS: 1. $-\frac{1}{3}$ 2. -3 3. 0 4. 1, -2 5. 1 6. 0, -2, $-\frac{3}{2}$ 7. 0, 1 8. none 9. yes 10. no 11. yes 12. yes 13. yes 14. yes 15. no 16. yes 17. yes 18. $\frac{3}{7}$ 19. $\frac{7}{3}$ 20. $\frac{x^4}{2y}$ 21. $\frac{az^3}{3b^2}$ 22. $\frac{1}{x+1}$ 23. $\frac{y+4}{y^2+4}$ 24. $\frac{a+b}{a-b}$ 25. $x^2 + xy + y^2$ 26. -1 27. -1 28. 1 29. $\frac{x-2y}{2y+x}$ 30. $a^2 - b^2$ 31. $\frac{x+y}{z}$ 32. $\frac{x(2x+3)}{3-x}$ 33. $-a$ 34. $\frac{1}{3a+5}$ 35. $\frac{3z+1}{z^2+2}$ 36. $4u^2 + 6u + 9$ 37. 1 38. $\frac{y+x-1}{y+x}$ 39. Terms have been divided out, *not* factors. 40. The numerator is 1, *not* 0. 41. -1 42. 1 43. $x - 2y$; both values are 4 44. \$3521.13 45. \$5434.78 46. (a) 660.0 vibrations/sec (b) 570.5 vibrations/sec 47. $\frac{1}{18}$ 48. $\frac{7}{2}$ 49. $\frac{3}{5}$ 50. $2 \cdot 3 \cdot 5 \cdot 5 \cdot 7 \cdot 13$

6.1 EXERCISES B

Find the values for the variable that must be excluded.

1. $\dfrac{y}{2y - 1}$

2. $\dfrac{a^2 + 7}{a - 2}$

3. $\dfrac{4z - 3}{z}$

4. $\dfrac{x + 5}{(x + 1)(x - 3)}$

5. $\dfrac{16}{a^2 - 4a + 4}$

6. $\dfrac{x - 1}{x^3 + x^2 - 6x}$

7. $\dfrac{3a + 2}{a^2 + a}$

8. $\dfrac{3c + 2}{3c^2 + 2}$

Are the given fractions equivalent? Explain. (Assume values that make the denominator zero are excluded.)

9. $\dfrac{3}{4}, \dfrac{15}{20}$

10. $\dfrac{5}{y}, \dfrac{y}{5}$

11. $\dfrac{-3}{x^2}, \dfrac{3}{-x^2}$

12. $\dfrac{-4}{-y^2}, \dfrac{4}{y^2}$

13. $\dfrac{a+3}{b}, \dfrac{4(a+3)}{4b}$

14. $\dfrac{x-y}{7}, \dfrac{3x-3y}{21}$

15. $\dfrac{5+x}{y^3+x}, \dfrac{5}{y^3}$

16. $\dfrac{x+2}{x^2-4}, \dfrac{1}{x-2}$

17. $\dfrac{x^2-5+y}{x-2y}, \dfrac{5-x^2-y}{2y-x}$

Reduce to lowest terms. (Assume values that make the denominator zero are excluded.)

18. $\dfrac{39}{52}$

19. $\dfrac{60}{135}$

20. $\dfrac{15a^3b^8}{10ab^5}$

21. $\dfrac{16x^3y^7u^5}{12xy^8u^{10}}$

22. $\dfrac{5u}{5u^2+5u}$

23. $\dfrac{a+2}{a^2+4}$

24. $\dfrac{x^2-y^2}{x^2+2xy+y^2}$

25. $\dfrac{u^3+v^3}{u+v}$

26. $\dfrac{2a-3w}{3w-2a}$

27. $\dfrac{2x-y}{y-2x}$

28. $\dfrac{2x+y}{y+2x}$

29. $\dfrac{2x-y}{y+2x}$

30. $\dfrac{x^2-y^2}{x^4-y^4}$

31. $\dfrac{ab-ac+db-dc}{ca+cd}$

32. $\dfrac{u^2-10u}{10-u}$

33. $\dfrac{2a^2-a-3}{2a^3-3a^2+2a-3}$

34. $\dfrac{8u^3-27}{2u-3}$

35. $\dfrac{(x+2)(x^2-3x+2)}{(x-2)(x^2+x-2)}$

36. $\dfrac{9a^2-15a+25}{27a^3+125}$

37. $\dfrac{a^2-2ab+b^2-4}{a^2-b^2-2a-2b}$

38. $\dfrac{u^3+1+2u^2+2u}{u^3-1}$

Dr. Gordon Johnson's assistant Burford, worked Exercises 39–40 as indicated. Why was Burford fired?

39. $\dfrac{x+3z}{3z} = \dfrac{x+\cancel{3z}}{\cancel{3z}} = x$

40. $\dfrac{y-1}{y^2-1} = \dfrac{\cancel{y-1}}{(y+1)\cancel{(y-1)}} = \dfrac{0}{y+1} = 0$

41. Reduce $\dfrac{2-3a}{3a-2}$ to lowest terms.

42. Reduce $\dfrac{2+3a}{3a+2}$ to lowest terms.

43. Substitute -1 for x and 3 for y in $x-y$ and $\dfrac{x^2-3xy+2y^2}{x-2y}$. Why are both results equal to -4?

In Exercises 44–45, use a calculator to find the value of P, to the nearest cent, if $P = \dfrac{A}{1+0.07t}$ *and A and t have the given values.*

44. $A = \$2500$ and $t = 5$ yr

45. $A = \$10{,}000$ and $t = 7$ yr

46. By the Doppler effect, the frequency of a sound such as a train whistle appears to increase as the train approaches and decrease as the train moves away. The formula that describes the Doppler effect is

$$F = \frac{1100f}{1100 - V},$$

where F is the observed frequency of a sound, f is the actual frequency of a sound, and V is the velocity of the train (a positive number if the train is approaching and a negative number if it is going away). Suppose a car horn produces a sound with a frequency of 414.0 vibrations per second. What is the observed frequency **(a)** as the car is approaching at 65 mph? **(b)** as the car is going away at 65 mph?

FOR REVIEW

Exercises 47–49 review multiplication and division of fractions to help you prepare for the next section.

47. $\frac{6}{7} \cdot 28$

48. $\frac{3}{4} \div \frac{5}{4}$

49. $12 \div \frac{4}{5}$

50. Factor 48,510 into a product of prime integers.

6.1 EXERCISES C

Find the values for the variable that must be excluded.

1. $\dfrac{ab}{ab + b - 5a - 5}$

[*Hint:* Factor the denominator by grouping terms.]

2. $\dfrac{x}{(x^2 - 4x + 4) - 25}$

[Answer: 7, −3]

Reduce to lowest terms. (Assume values that make the denominator zero are excluded.)

3. $\dfrac{y^2 + 6y + 5}{(y - 1)^2 - 4}$

4. $\dfrac{a^2 + ab + b^2}{a^4 - b^3a + a^3b - b^4}$

6.2 MULTIPLICATION AND DIVISION OF FRACTIONS

STUDENT GUIDEPOSTS

1 Multiplying Fractions
2 Dividing Fractions

1 MULTIPLYING FRACTIONS

Remember that the product of numerical fractions is equal to the product of all numerators divided by the product of all denominators. That is,

$$\frac{a}{b} \cdot \frac{c}{d} = \frac{ac}{bd}. \quad b \neq 0 \text{ and } d \neq 0$$

The same is true for algebraic fractions. In both cases, the resulting fraction should be reduced to lowest terms, which is easy to do if all common factors in the numerator and denominator are divided out before multiplying. For example,

$$\frac{6}{35} \cdot \frac{7}{3} = \frac{6 \cdot 7}{35 \cdot 3} = \frac{2 \cdot \cancel{3} \cdot \cancel{7}}{5 \cdot \cancel{7} \cdot \cancel{3}} = \frac{2}{5}$$

Had we multiplied first to obtain $\frac{42}{105}$, we would have had to factor (undo this multiplication) to reduce the product.

To Multiply Two or More Algebraic Fractions

1. Factor all numerators and denominators and place the indicated product of all numerator factors over the product of all denominator factors.
2. Divide out common factors to reduce to lowest terms.
3. Multiply the remaining numerator factors, then multiply the remaining denominator factors.

EXAMPLE 1 Multiplying Fractions

Multiply and simplify.

(a) $\dfrac{3a}{8} \cdot \dfrac{2}{9a^2} = \dfrac{\cancel{3} \cdot \cancel{a} \cdot \cancel{2}}{\cancel{2} \cdot 2 \cdot 2 \cdot \cancel{3} \cdot 3 \cdot \cancel{a} \cdot a} = \dfrac{1}{2 \cdot 2 \cdot 3 \cdot a} = \dfrac{1}{12a}$

In a problem like this, we can take several shortcuts without first factoring to primes. The work might appear as

$$\dfrac{\overset{1 \cdot 1 \cdot 1}{\cancel{3} \cdot \cancel{a} \cdot \cancel{2}}}{\underset{4 \cdot 3 \cdot a}{\cancel{8} \cdot \cancel{9}\,\cancel{a^2}}} = \dfrac{1}{12a},$$

in which 3 was divided into 9, 2 into 8, and a into a^2.

(b) $\dfrac{x^2 - 5x + 6}{x^2 + x - 6} \cdot \dfrac{x + 3}{x - 2} = \dfrac{\cancel{(x - 2)}(x - 3)\cancel{(x + 3)}}{\cancel{(x + 3)}\cancel{(x - 2)}(x - 2)} = \dfrac{x - 3}{x - 2}$

(c) $\dfrac{a^2 - 4b^2}{49a^3 - 4ab^2} \cdot \dfrac{7a^2 - 2ab}{a + 2b} = \dfrac{(a - 2b)\cancel{(a + 2b)} \cdot \cancel{a(7a - 2b)}}{\cancel{a(7a - 2b)}(7a + 2b)\cancel{(a + 2b)}}$

$= \dfrac{a - 2b}{7a + 2b}$

(d) $\dfrac{y^2 - 2y - 3}{y^2 - y - 2} \cdot \dfrac{y^2 + 3y - 10}{(5 + y)(1 - 2y)} \cdot \dfrac{2y^2 - 3y + 1}{y^2 - 4y + 3}$

$= \dfrac{\cancel{(y - 3)}\cancel{(y + 1)}\cancel{(y - 2)}\cancel{(y + 5)}(2y - 1)\cancel{(y - 1)}}{\cancel{(y + 1)}\cancel{(y - 2)}\cancel{(5 + y)}(1 - 2y)\cancel{(y - 1)}\cancel{(y - 3)}}$ Cancel common factors

$= \dfrac{2y - 1}{1 - 2y} = -1$ $\qquad \dfrac{a - b}{b - a} = -1$

Practice Exercise 1

Multiply and simplify.

(a) $\dfrac{5x^3}{7} \cdot \dfrac{14}{25x}$

(b) $\dfrac{a^2 + 3a - 4}{a^2 - 16} \cdot \dfrac{a - 4}{a + 1}$

(c) $\dfrac{u^2 - 3uv}{3u^2 - 8uv - 3v^2} \cdot \dfrac{3u^2 + 10uv + 3v^2}{u^2 + 3uv}$

(d) $\dfrac{a^2 - 1}{a^2 + a} \cdot \dfrac{a - 2}{a^2 - 3a + 2} \cdot \dfrac{a^2 + 5a}{2a + 1}$

Answers: (a) $\frac{2x^2}{5}$ (b) $\frac{a-1}{a+1}$ (c) 1 (d) $\frac{a+5}{2a+1}$

CAUTION

There will be a lot of extra work if we multiply numerators and denominators without factoring first. For example, in Exercise 1(b), if we multiply first we obtain

$$\frac{x^2 - 5x + 6}{x^2 + x - 6} \cdot \frac{x + 3}{x - 2} = \frac{x^3 - 2x^2 - 9x + 18}{x^3 - x^2 - 8x + 12}.$$

To reduce the product to lowest terms, we would now have to factor the two third-degree polynomials in the numerator and denominator. *Never multiply without first factoring and dividing out common factors.*

2 DIVIDING FRACTIONS

When one numerical fraction is divided by another, for example,

$$\frac{2}{3} \div \frac{2}{15} \quad \text{or} \quad \frac{\frac{2}{3}}{\frac{2}{15}},$$

the fraction $\frac{2}{15}$ is called the **divisor.** Recall that to calculate the quotient, we multiply $\frac{2}{3}$ by the reciprocal of $\frac{2}{15}$, $\frac{15}{2}$, the fraction formed by interchanging the numerator and the denominator.

$$\frac{2}{3} \div \frac{2}{15} = \frac{2}{3} \cdot \frac{15}{2} = \frac{2}{3} \cdot \frac{3 \cdot 5}{2} = \frac{\cancel{2} \cdot \cancel{3} \cdot 5}{\cancel{3} \cdot \cancel{2}} = 5$$

Thus, if $\frac{a}{b}$ and $\frac{c}{d}$ are two fractions with $\frac{c}{d} \neq 0$, then

$$\frac{a}{b} \div \frac{c}{d} = \frac{a}{b} \cdot \frac{d}{c}.$$

The following table shows several fractions and their reciprocals.

Fraction	$\frac{3}{y}$	$\frac{x+y}{(x-y)^2}$	$5 = \frac{5}{1}$	$y = \frac{y}{1}$	$\frac{1}{2}$	$a + b$
Reciprocal	$\frac{y}{3}$	$\frac{(x-y)^2}{x+y}$	$\frac{1}{5}$	$\frac{1}{y}$	$2 = \frac{2}{1}$	$\frac{1}{a+b}$ $\left(\text{not } \frac{1}{a} + \frac{1}{b}\right)$

Notice that in all cases the product of a fraction and its reciprocal is 1. Division of algebraic fractions is similar to division of numerical fractions.

To Divide One Algebraic Fraction by Another

1. Find the reciprocal of the divisor.
2. Multiply following the same procedure as for multiplying fractions.

EXAMPLE 2 Dividing Fractions

Divide and simplify.

(a) $\frac{9x^5}{20y^7} \div \frac{27x^3}{10y^4} = \frac{9x^5}{20y^7} \cdot \frac{10y^4}{27x^3}$ The reciprocal of $\frac{27x^3}{10y^4}$ is $\frac{10y^4}{27x^3}$

$$= \frac{\cancel{9x^3}x^2 \cdot \cancel{10y^4}}{2 \cdot \cancel{10y^4}y^3 \cdot 3 \cdot \cancel{9x^3}}$$

$$= \frac{x^2}{6y^3}$$

(b) $\frac{a}{a+2} \div \frac{a}{5} = \frac{a}{a+2} \cdot \frac{5}{a}$ The reciprocal of $\frac{a}{5}$ is $\frac{5}{a}$

$$= \frac{\cancel{a} \cdot 5}{(a+2) \cdot \cancel{a}} = \frac{5}{a+2}$$

(c) $\frac{x+y}{x} \div x = \frac{x+y}{x} \cdot \frac{1}{x}$ The reciprocal of x is $\frac{1}{x}$

$$= \frac{x+y}{x^2}$$

(d) $\frac{x^2-6x+9}{x^2-4} \div \frac{x-3}{x+2} = \frac{x^2-6x+9}{x^2-4} \cdot \frac{x+2}{x-3}$

$$= \frac{\cancel{(x-3)}(x-3)\cancel{(x+2)}}{(x-2)\cancel{(x+2)}\cancel{(x-3)}} = \frac{x-3}{x-2}$$

PRACTICE EXERCISE 2

Divide and simplify.

(a) $\frac{5a^3}{16b^5} \div \frac{25a^2}{12b^3}$

(b) $\frac{w}{6} \div \frac{w}{6+w}$

(c) $\frac{a}{a-b} \div \frac{a}{b}$

(d) $\frac{m^2-3mn+2n^2}{m^2-n^2} \div \frac{m+2n}{m+n}$

(e) $\dfrac{a^2 - b^2}{a^2 + a - 2} \cdot \dfrac{a^2 + 2a}{a^2 + 2ab + b^2} \div \dfrac{a^3 - b^3}{a^2 - a + ba - b}$

$$= \frac{(a + b)(a - b)}{(a - 1)(a + 2)} \cdot \frac{a(a + 2)}{(a + b)(a + b)} \cdot \frac{a(a - 1) + b(a - 1)}{(a - b)(a^2 + ab + b^2)}$$

$$= \frac{(a + b)(a - b)a(a + 2)(a + b)(a - 1)}{(a - 1)(a + 2)(a + b)(a + b)(a - b)(a^2 + ab + b^2)}$$

$$= \frac{a}{a^2 + ab + b^2}$$

(e) $\dfrac{x + y}{x - y} \cdot \dfrac{x^2 - xy}{x^2 - xy + y^2} \div \dfrac{x^2 + xy}{x^3 + y^3}$

Answers: (a) $\frac{3a}{20b^2}$ (b) $\frac{6 + w}{6}$ (c) $\frac{b}{a - b}$ (d) $\frac{m - 2n}{m + 2n}$ (e) $x + y$

6.2 EXERCISES A

Multiply and simplify.

1. $\dfrac{2ab}{5} \cdot \dfrac{25}{4a^3}$

2. $\dfrac{-4a^3b^2}{3ab^3} \cdot \dfrac{9a^5b^7}{8a^3b^3}$

3. $\dfrac{2c}{7} \cdot \dfrac{7b + 14}{4(b + 2)}$

4. $\dfrac{3x^2}{x + 2} \cdot \dfrac{x^2 - 4}{3x}$

5. $\dfrac{4x + 8}{9x + 9} \cdot \dfrac{x^2 - 1}{2x + 4}$

6. $\dfrac{y}{y^2 - yz - 12z^2} \cdot \dfrac{y^2 + 3yz}{y - 4z}$

7. $\dfrac{a^2 + ab}{a - 5b} \cdot \dfrac{5b - a}{b^2 - a^2}$

8. $\dfrac{9a^2 - 6a + 1}{28 + 7a} \cdot \dfrac{a^2 - 16}{3a^2 - 13a + 4}$

9. $\dfrac{a^2 - 6a - 16}{2a^2 - 128} \cdot \dfrac{a^2 + 16a + 64}{3a^2 + 30a + 48}$

10. $\dfrac{4 - a^2}{c^2 + 2cd + d^2} \cdot \dfrac{c^2 - d^2}{2 - a - a^2} \cdot \dfrac{c + d - ca - da}{2a - a^2}$

Find the reciprocal of each expression.

11. $\dfrac{4}{5}$

12. 8

13. $\dfrac{x}{3}$

14. $2z$

15. $\dfrac{u}{v + u}$

16. $x + 3$

17. $4x + 3$

18. 0.27

Divide and simplify.

19. $\dfrac{a}{a + 3} \div \dfrac{a}{5}$

20. $\dfrac{3(a + b)}{7} \div \dfrac{9(a + b)}{28}$

21. $\dfrac{(u^2v^3)^2}{3uv} \div \dfrac{(uv^2)^3}{12u^3v}$

22. $\dfrac{x^2 - x - 2}{y^2} \div \dfrac{x^2 - 1}{y^2}$

23. $\dfrac{4}{a^2 - 9} \div \dfrac{4a^2 - 4a - 24}{a^2 - 6a + 9}$

24. $\dfrac{a^3 - 27}{a^2 - 3a} \div \dfrac{a^2 + 3a + 9}{a}$

25. $\dfrac{2a^2 - 2a}{2a + 4} \div \dfrac{a^2 - 1}{a^2 + 3a + 2}$

26. $\dfrac{a^2 - 4a + 3}{a^2 - 2a - 15} \div \dfrac{2a^2 - 6a + 4}{a^2 + 3a - 10}$

Perform the indicated operations.

27. $\dfrac{u^2 - v^2}{u^2 - uv + v^2} \cdot \dfrac{2u^2 - 3uv - 2v^2}{u^2 + 2uv + v^2} \div \dfrac{2u^2 - uv - v^2}{u^3 + v^3}$

28 $\dfrac{3s^2 - 5st - 2t^2}{t^2 + st - 2s^2} \cdot \dfrac{t^2 + 4st - 5s^2}{12s^2 + 7st + t^2} \div \dfrac{5s^2 - 9st - 2t^2}{8s^2 - 2st - t^2}$

FOR REVIEW

29. Burford worked the following problem. What is wrong with it?

$$\frac{2 + (x + y)}{3(x + y)} = \frac{2 + \cancel{(x + y)}}{3\cancel{(x + y)}} = \frac{2}{3}$$

Find the values of the variable that must be excluded.

30. $\dfrac{x + 7}{x^2 + 5x + 6}$

31. $\dfrac{3}{x^2}$

Are the given fractions equivalent?

32. $\dfrac{x + y}{x^2 - y^2}, \dfrac{1}{x - y}$

33. $\dfrac{x^2}{4}, \dfrac{x^2 + 5}{9}$

Reduce to lowest terms.

34. $\dfrac{x^2 + 3x}{x^2 + 5x + 6}$

35. $\dfrac{a^3 - ab^2}{a^2 - ab}$

Exercises 36–41 review adding and subtracting fractions to help you prepare for the next section. Perform the indicated operations.

36. $\dfrac{7}{6} + \dfrac{1}{6}$

37. $\dfrac{7}{10} - \dfrac{3}{10}$

38. $\dfrac{2}{15} + \dfrac{2}{21}$

39. $\frac{3}{35} - \frac{4}{63}$

40. $8 + \frac{6}{7}$

41. $\frac{41}{3} - 3$

ANSWERS: 1. $\frac{5b}{2a^2}$ 2. $\frac{-3a^4b^3}{2}$ 3. $\frac{c}{2}$ 4. $x(x-2)$ 5. $\frac{2(x-1)}{9}$ 6. $\frac{y^2}{(y-4z)^2}$ 7. $\frac{-a}{b-a}$ or $\frac{a}{a-b}$ 8. $\frac{3a-1}{7}$ 9. $\frac{1}{6}$ 10. $\frac{c-d}{a}$ 11. $\frac{5}{4}$ 12. $\frac{1}{8}$ 13. $\frac{3}{x}$ 14. $\frac{1}{2z}$ 15. $\frac{v+u}{u}$ 16. $\frac{1}{x+3}$ 17. $\frac{1}{4x+3}$ 18. $\frac{100}{27}$ 19. $\frac{5}{a+3}$ 20. $\frac{4}{3}$ 21. $4u^3$ 22. $\frac{x-2}{x-1}$ 23. $\frac{1}{(a+3)(a+2)}$ 24. 1 25. a 26. $\frac{(a-3)(a+5)}{2(a-5)(a+3)}$ 27. $u - 2v$ 28. $\frac{2s-t}{2s+t}$ 29. Terms have been divided out, *not* factors. 30. $-2, -3$ 31. 0 32. yes 33. no 34. $\frac{x}{x+2}$ 35. $a + b$ 36. $\frac{4}{3}$ 37. $\frac{2}{5}$ 38. $\frac{8}{35}$ 39. $\frac{1}{45}$ 40. $\frac{62}{7}$ 41. $\frac{32}{3}$

6.2 EXERCISES B

Multiply and simplify.

1. $\frac{5x^3y^2}{x^2y^5} \cdot \frac{x^5y^7}{10x^3y}$

2. $\frac{-2x^2y}{7x} \cdot \frac{14xy^3}{6x^2y^5}$

3. $\frac{u^2v^2}{a-b} \cdot \frac{2b-2a}{4uv}$

4. $\frac{ab}{a-3b} \cdot \frac{(a-3b)^2}{a^2b-3ab^2}$

5. $\frac{a^3+a^2b}{5a} \cdot \frac{25}{3a+3b}$

6. $\frac{x^3-y^3}{2xy} \cdot \frac{4x^2y^2}{x^2+xy+y^2}$

7. $\frac{x^2-4}{x^2-4x+4} \cdot \frac{x^2-9x+14}{x^3+2x^2}$

8. $\frac{2x^2-5xy-3y^2}{4x^2-y^2} \cdot \frac{8x^2+10xy+3y^2}{4x^2-9xy-9y^2}$

9. $\frac{x^2+4x-5}{2x^2+16x} \cdot \frac{x^3-64x}{x^2-9x+8}$

10. $\frac{x^2+3xy+2y^2}{x^2-x} \cdot \frac{x^3-x^2y+xy^2}{x^2+xy-2y^2} \cdot \frac{x^2-x-xy+y}{x^3+y^3}$

Find the reciprocal of each expression.

11. $\frac{1}{7}$

12. $\frac{7}{y}$

13. y

14. $\frac{1}{y+z}$

15. $\frac{a-b}{b}$

16. $\frac{3x-5}{2y+1}$

17. $2\frac{2}{3}$

18. 0

Divide and simplify.

19. $\frac{3x}{x-1} \div \frac{6x}{x-1}$

20. $\frac{20a^2b}{18ab^3} \div \frac{5a^4b}{9a^2b^2}$

21. $\frac{(xy^2)^4}{5x^3y^5} \div \frac{x^7y^7}{(5xy^2)^3}$

22. $\frac{a^2-ab}{a+b} \div \frac{a-b}{a^2+2ab+b^2}$

23. $\frac{uv-uw+xv-xw}{v-w} \div \frac{v^2-2vw+w^2}{xv-xw}$

24. $\frac{x-y}{4x+4y} \div \frac{x^2-2xy+y^2}{x^2-y^2}$

25. $\frac{z^2-z-6}{z^2-81} \div \frac{z^2-9z+18}{4z+36}$

26. $\frac{x^2-y^2}{x^3-y^3} \div \frac{x^2+2xy+y^2}{x^2+xy+y^2}$

Perform the indicated operations.

27. $\frac{y^2+xy-y-x}{x^4-x^3+x^2-x} \cdot \frac{2x^4+2x^3+2x^2+2x}{x^3-x+x^2y-y} \div \frac{2y-2}{x^2-2x+1}$

28. $\dfrac{ax - ay + bx - by}{ax + ay - bx - by} \cdot \dfrac{ax + ay - 3bx - 3by}{ax - ay - 3bx - 3by} \div \dfrac{a + b}{b - a}$

FOR REVIEW

29. Burford worked the following problem. What is wrong with it?

$$\frac{a-5}{a^2-4a-5} = \frac{a-5}{(a-5)(a+1)} = \frac{\cancel{a-5}}{\cancel{(a-5)}(a+1)} = \frac{0}{a+1} = 0$$

Find the values of the variable that must be excluded.

30. $\dfrac{x - 8}{x^2 - 4x + 3}$

31. $\dfrac{3 + y}{y^2}$

Are the given fractions equivalent?

32. $\dfrac{x - y}{x^2 - y^2}, \dfrac{1}{x + y}$

33. $\dfrac{y^3}{2}, \dfrac{y^3 + 3}{5}$

Reduce to lowest terms.

34. $\dfrac{x^2 - 5x}{x^2 - 8x + 15}$

35. $\dfrac{a^2b - a^3}{a^3 - ab^2}$

Exercises 36–41 review adding and subtracting fractions to help you prepare for the next section. Perform the indicated operations.

36. $\dfrac{3}{10} + \dfrac{9}{10}$

37. $\dfrac{5}{6} - \dfrac{1}{6}$

38. $\dfrac{5}{6} + \dfrac{1}{10}$

39. $\dfrac{4}{15} - \dfrac{2}{21}$

40. $2 + \dfrac{3}{5}$

41. $\dfrac{32}{5} - 4$

6.2 EXERCISES C

Perform the indicated operations.

1. $\dfrac{x^3 + x^2 - 20x}{(x - 1)^2 - 9} \cdot \dfrac{x^3 + 8}{ax + 5a + x + 5} \div \dfrac{x^3 - 2x^2 + 4x}{a^2 - 6a - 7}$

2. $\dfrac{ax - bx + a - b}{b^2 - 3ab - 10a^2} \div \left[\dfrac{b^2 - a^2}{2b^2 + 7ab + 6a^2} \div \dfrac{b^2 - 4ab - 5a^2}{x^3 + 1}\right]$ $\left[\text{Answer: } -\dfrac{2b + 3a}{x^2 - x + 1}\right]$

6.3 ADDITION AND SUBTRACTION OF FRACTIONS

STUDENT GUIDEPOSTS

1. Adding and Subtracting Like Fractions
2. Least Common Denominator (LCD)
3. Rewriting Fractions
4. Adding and Subtracting Unlike Fractions

1 ADDING AND SUBTRACTING LIKE FRACTIONS

As with multiplication and division, addition and subtraction of algebraic fractions parallel the corresponding operations on numerical fractions. Fractions that have the same denominator are called **like fractions.** We add or subtract like fractions such as $\frac{a}{b}$ and $\frac{c}{b}$ as follows.

$$\frac{a}{b}+\frac{c}{b}=\frac{a+c}{b} \quad \text{and} \quad \frac{a}{b}-\frac{c}{b}=\frac{a-c}{b}$$

For example,

$$\frac{3}{5}+\frac{1}{5}=\frac{3+1}{5}=\frac{4}{5} \quad \text{and} \quad \frac{3}{5}-\frac{1}{5}=\frac{3-1}{5}=\frac{2}{5}.$$

The resulting sum or difference should always be reduced to lowest terms.

EXAMPLE 1 Adding and Subtracting Like Fractions

Perform the indicated operation.

(a) $\frac{4}{9}+\frac{2}{9}=\frac{4+2}{9}=\frac{6}{9}=\frac{\cancel{3}\cdot 2}{\cancel{3}\cdot 3}=\frac{2}{3}$ Reduce to lowest terms

(b) $\frac{a+b}{5a}+\frac{a^2-b}{5a}=\frac{(a+b)+(a^2-b)}{5a}$

$$=\frac{a+a^2}{5a}=\frac{\cancel{a}(1+a)}{5\cancel{a}}=\frac{1+a}{5}$$

(c) $\frac{2x+y}{x-1}-\frac{x-y}{x-1}=\frac{(2x+y)-(x-y)}{x-1}$ Use parentheses and watch signs

$$=\frac{2x+y-x+y}{x-1}=\frac{x+2y}{x-1}$$

PRACTICE EXERCISE 1

Perform the indicated operations.

(a) $\frac{1}{12}+\frac{7}{12}$

(b) $\frac{x-y}{xy}+\frac{y^2-x}{xy}$

(c) $\frac{b+3a}{a+2}-\frac{b-a}{a+2}$

Answers: (a) $\frac{2}{3}$ (b) $\frac{y-1}{x}$ (c) $\frac{4a}{a+2}$

CAUTION

Parentheses are extremely important, especially in a subtraction problem. In the above example, $-(x-y)$ becomes $-x+y$ when the parentheses are cleared. If you do not enclose the numerator in parentheses, it is easy to make a sign error.

Often, two fractions have denominators that are negatives of each other. When this occurs, multiply both the numerator and denominator of *one* of the fractions by -1. Since we are multiplying the fraction by the multiplicative identity, 1, in the form $\frac{-1}{-1}$, the result is equivalent to the original and has the same denominator as the other fraction.

EXAMPLE 2 Fractions whose Denominators Are Negatives

Perform the indicated operation.

(a) $$\frac{x}{3} + \frac{x-y}{-3} = \frac{x}{3} + \frac{(-1)(x-y)}{(-1)(-3)}$$
$$= \frac{x}{3} + \frac{-x+y}{3} = \frac{x+(-x+y)}{3} = \frac{y}{3}$$

(b) $$\frac{4y}{y-2} - \frac{y+4}{2-y} = \frac{4y}{y-2} - \frac{(-1)(y+4)}{(-1)(2-y)}$$
$$= \frac{4y}{y-2} - \frac{-y-4}{-2+y}$$
$$= \frac{4y}{y-2} - \frac{(-y-4)}{y-2}$$
$$= \frac{4y-(-y-4)}{y-2} = \frac{4y+y+4}{y-2} \quad \text{Watch signs}$$
$$= \frac{5y+4}{y-2}$$

Practice Exercise 2

Perform the indicated operation.

(a) $\frac{2-a}{-5} + \frac{2}{5}$

(b) $\frac{3u}{u-v} - \frac{v-2u}{v-u}$

Answers: (a) $\frac{a}{5}$ (b) $\frac{u+v}{u-v}$

2 LEAST COMMON DENOMINATOR (LCD)

To add or subtract fractions with different denominators, called **unlike fractions,** we need to convert them to equivalent fractions with a common denominator. For example, to add $\frac{4}{15}$ and $\frac{5}{6}$, one common denominator is $15 \cdot 6 = 90$.

$$\frac{4}{15} + \frac{5}{6} = \frac{4 \cdot 6}{15 \cdot 6} + \frac{5 \cdot 15}{6 \cdot 15} = \frac{24}{90} + \frac{75}{90} = \frac{99}{90} = \frac{\cancel{9} \cdot 11}{\cancel{9} \cdot 10} = \frac{11}{10}$$

If we use the **least common denominator,** it takes less effort to reduce the final sum or difference to lowest terms. For example, we could have used 30 as the least common denominator above.

$$\frac{4}{15} + \frac{5}{6} = \frac{4 \cdot 2}{15 \cdot 2} + \frac{5 \cdot 5}{6 \cdot 5} = \frac{8}{30} + \frac{25}{30} = \frac{33}{30} = \frac{\cancel{3} \cdot 11}{\cancel{3} \cdot 10} = \frac{11}{10}$$

The method for finding the least common denominator (LCD) of two or more fractions, including algebraic fractions, is summarized here.

To Find the LCD of Two or More Fractions

1. Factor all denominators and reduce all fractions to lowest terms.
2. When there are no common factors in any two denominators, the LCD is the product of *all* denominators.
3. When there are common factors in two or more denominators, each factor appears in the LCD as many times as it appears in the denominator where it is found the greatest number of times.

EXAMPLE 3 Finding the Least Common Denominator

Find the LCD of the fractions.

(a) $\frac{7}{150}$ and $\frac{2}{315}$

Factor the denominators.

$$150 = 2 \cdot 3 \cdot 5 \cdot 5 \quad \text{and} \quad 315 = 3 \cdot 3 \cdot 5 \cdot 7$$

The LCD must consist of one 2, two 3s, two 5s, and one 7. Thus, the LCD is $2 \cdot 3 \cdot 3 \cdot 5 \cdot 5 \cdot 7 = 3150$.

(b) $\frac{3}{x}$ and $\frac{5}{x+y}$

Since x and $x + y$ are already completely factored, and since there are no common factors in the two denominators, the LCD is $x(x + y)$.

(c) $\frac{3a+1}{a^2-b^2}$ and $\frac{7b}{2a-2b}$

Factor the denominators.

$$a^2 - b^2 = (a-b)(a+b) \quad \text{and} \quad 2a - 2b = 2(a-b)$$

The LCD must consist of one $(a - b)$, one $(a + b)$, and one 2. Thus, the LCD is $2(a - b)(a + b)$.

Practice Exercise 3

Find the LCD of the fractions.

(a) $\frac{2}{105}$ and $\frac{7}{60}$

(b) $\frac{4}{a+1}$ and $\frac{3}{a}$

(c) $\frac{2x}{x^2-4}$ and $\frac{x+2}{x^2-2x}$

Answers: (a) 420 (b) $a(a + 1)$ (c) $x(x + 2)(x - 2)$

Be sure the fractions are reduced to lowest terms before finding the LCD.

EXAMPLE 4 Reducing Fractions before Finding the LCD

Find the LCD of $\frac{a+1}{a^2+2a+1}$ and $\frac{5-a}{a^2-4a-5}$.

Factor the denominators.

$$a^2 + 2a + 1 = (a+1)(a+1) \quad \text{and}$$
$$a^2 - 4a - 5 = (a+1)(a-5)$$

We might almost conclude that the LCD is $(a + 1)^2(a - 5)$. However, we can reduce the given fractions to lowest terms as follows.

$$\frac{a+1}{a^2+2a+1} = \frac{\cancel{(a+1)}}{(a+1)\cancel{(a+1)}} = \frac{1}{a+1}$$

$$\frac{5-a}{a^2-4a-5} = \frac{5-a}{(a-5)(a+1)} = \frac{(-1)\cancel{(a-5)}}{\cancel{(a-5)}(a+1)} \qquad 5 - a \text{ and } a - 5 \text{ are negatives}$$

$$= \frac{-1}{a+1}$$

Thus, to find the LCD of the given fractions, all we need to do is find the LCD of $\frac{1}{a+1}$ and $\frac{-1}{a+1}$, which is simply $a + 1$.

Practice Exercise 4

Find the LCD of $\frac{2y-x}{x^2-xy-2y^2}$ and $\frac{x-y}{x^2-y^2}$.

Answer: $x + y$

3 REWRITING FRACTIONS

Before we add and subtract unlike fractions, we reverse the process of reducing a fraction to lowest terms and consider rewriting a fraction as an equivalent fraction with a specified denominator.

EXAMPLE 5 Rewriting Fractions

Write each fraction as an equivalent fraction with the specified denominator.

(a) $\frac{3y}{2x}$ with denominator $10x^2y$

Since $10x^2y = 2 \cdot 5 \cdot xxy = (2x)(5xy)$, we multiply numerator and denominator by $5xy$.

$$\frac{3y}{2x} = \frac{3y\,(5xy)}{2x\,(5xy)} = \frac{15xy^2}{10x^2y}$$

(b) $\frac{4a + b}{2a - b}$ with denominator $4a^2 - b^2$

Since $4a^2 - b^2 = (2a - b)(2a + b)$, we multiply numerator and denominator by $2a + b$.

$$\frac{4a + b}{2a - b} = \frac{(4a + b)\,(2a + b)}{(2a - b)\,(2a + b)} = \frac{8a^2 + 6ab + b^2}{4a^2 - b^2}$$

Practice Exercise 5

Write each fraction as an equivalent fraction with the specified denominator.

(a) $\frac{4a}{x^2}$ with denominator $6x^3y$

(b) $\frac{x + y}{x - y}$ with denominator $x^2 - y^2$

Answers: (a) $\frac{24axy}{6x^3y}$ (b) $\frac{x^2 + 2xy + y^2}{x^2 - y^2}$

4 ADDING AND SUBTRACTING UNLIKE FRACTIONS

To add or subtract unlike fractions, we first find their LCD and then convert each fraction to an equivalent fraction with the LCD as the denominator. The technique is illustrated by adding two numerical fractions.

$\frac{2}{21} + \frac{3}{35} = \frac{2}{3 \cdot 7} + \frac{3}{5 \cdot 7}$ — $21 = 3 \cdot 7$ and $35 = 5 \cdot 7$; the LCD is $3 \cdot 5 \cdot 7 = 105$

$= \frac{2 \cdot 5}{3 \cdot 7 \cdot 5} + \frac{3 \cdot 3}{5 \cdot 7 \cdot 3}$ — Multiply numerator and denominator of $\frac{2}{21}$ by 5, of $\frac{3}{35}$ by 3

$= \frac{2 \cdot 5 + 3 \cdot 3}{3 \cdot 5 \cdot 7}$ — Denominators are now the same; add numerators and place the sum over the LCD

$= \frac{10 + 9}{3 \cdot 5 \cdot 7}$ — Leave denominator factored and simplify the numerator

$= \frac{19}{105}$ — Since 19 has no factor of 3, 5, or 7, the resulting fraction is in lowest terms

The same procedure applies to algebraic fractions.

To Add or Subtract Algebraic Fractions

1. Express denominators in completely factored form and reduce all fractions.
2. Find the LCD of the fractions.
3. Multiply the numerator and denominator of each fraction by all factors present in the LCD but missing in the denominator of the particular fraction, so the fractions have the same denominator.
4. Indicate the sum or difference of all numerators, using parentheses if needed, and place the result over the LCD.
5. Simplify the numerator and reduce the resulting fraction.

EXAMPLE 6 Adding Unlike Fractions

Add.

$$\frac{5}{6a}+\frac{a+1}{15a^2}$$

$$=\frac{5}{2\cdot 3a}+\frac{a+1}{3\cdot 5\cdot a\cdot a}$$ Factor denominators

$$=\frac{5\cdot(5\cdot a)}{2\cdot 3\cdot a\cdot(5\cdot a)}+\frac{(a+1)(2)}{3\cdot 5\cdot a\cdot a\cdot(2)}$$ Supply missing factors: LCD is $2\cdot 3\cdot 5\cdot a\cdot a$

$$=\frac{5\cdot 5\cdot a+(a+1)\cdot 2}{2\cdot 3\cdot a\cdot 5\cdot a}$$ Add numerators over the LCD

$$=\frac{25a+2a+2}{2\cdot 3\cdot 5\cdot a\cdot a}$$

$$=\frac{27a+2}{30a^2}$$ No common factors; the sum is in lowest terms

Practice Exercise 6

Add.

$$\frac{y^2+1}{25y^3}+\frac{7}{10y}$$

Answer: $\frac{37y^2+2}{50y^3}$

EXAMPLE 7 Subtracting Unlike Fractions

Subtract.

(a) $$\frac{3}{a+b}-\frac{2}{a-b}$$ Denominators already factored; LCD = $(a+b)(a-b)$

$$=\frac{3(a-b)}{(a+b)(a-b)}-\frac{2(a+b)}{(a+b)(a-b)}$$ Supply missing factors

$$=\frac{3(a-b)-2(a+b)}{(a+b)(a-b)}$$ Subtract numerators over LCD

$$=\frac{3a-3b-2a-2b}{(a+b)(a-b)}$$ Watch signs

$$=\frac{a-5b}{(a+b)(a-b)}$$ Since no common factors exist, this is in lowest terms

(b) $$\frac{2xy}{x^2-y^2}-\frac{y}{x+y}$$

$$=\frac{2xy}{(x-y)(x+y)}-\frac{y}{x+y}$$ LCD = $(x-y)(x+y)$

$$=\frac{2xy}{(x-y)(x+y)}-\frac{y(x-y)}{(x+y)(x-y)}$$

$$=\frac{2xy-y(x-y)}{(x-y)(x+y)}$$

$$=\frac{2xy-yx+y^2}{(x-y)(x+y)}$$

$$=\frac{xy+y^2}{(x-y)(x+y)}$$ This one can be simplified

$$=\frac{y\cancel{(x+y)}}{(x-y)\cancel{(x+y)}}=\frac{y}{x-y}$$

Practice Exercise 7

Subtract.

(a) $$\frac{5}{x+1}-\frac{8}{x-1}$$

(b) $$\frac{3a}{a-b}-\frac{6ab}{a^2-b^2}$$

Answers: (a) $\frac{-3x-13}{(x+1)(x-1)}$ (b) $\frac{3a}{a+b}$

EXAMPLE 8 Reducing Fractions before Subtracting

Subtract.

$$\frac{y^2 - 1}{y^3 - 1} - \frac{y}{2y^2 + 2y + 2}$$

$$= \frac{\cancel{(y - 1)}(y + 1)}{\cancel{(y - 1)}(y^2 + y + 1)} - \frac{y}{2(y^2 + y + 1)}$$ Factor and reduce first fraction

$$= \frac{y + 1}{y^2 + y + 1} - \frac{y}{2(y^2 + y + 1)}$$ LCD $= 2(y^2 + y + 1)$

$$= \frac{2(y + 1)}{2(y^2 + y + 1)} - \frac{y}{2(y^2 + y + 1)}$$ Supply missing factor

$$= \frac{2(y + 1) - y}{2(y^2 + y + 1)}$$ Subtract numerators over LCD

$$= \frac{2y + 2 - y}{2(y^2 + y + 1)} = \frac{y + 2}{2(y^2 + y + 1)}$$ Already reduced to lowest terms

Practice Exercise 8

Add.

$$\frac{1 - x}{x} + \frac{x^2 - y^2}{x^2 - 2xy + y^2}$$

Answer: $\frac{2xy + x - y}{x(x - y)}$

Adding or subtracting three or more fractions uses the same technique. This is shown in the next example.

EXAMPLE 9 Adding and Subtracting Three Fractions

Perform the indicated operations.

$$\frac{2}{x^2 - 4} + \frac{1}{x + 2} - \frac{x}{x^2 - x - 2}$$

$$= \frac{2}{(x + 2)(x - 2)} + \frac{1}{x + 2} - \frac{x}{(x + 1)(x - 2)}$$ LCD $= (x + 2)(x - 2)(x + 1)$

$$= \frac{2(x + 1)}{(x + 2)(x - 2)(x + 1)} + \frac{(1)(x - 2)(x + 1)}{(x + 2)(x - 2)(x + 1)} - \frac{x(x + 2)}{(x + 1)(x - 2)(x + 2)}$$ Supply missing factors

$$= \frac{2(x + 1) + (1)(x - 2)(x + 1) - x(x + 2)}{(x + 2)(x - 2)(x + 1)}$$ Add and subtract numerators over LCD

$$= \frac{2x + 2 + x^2 - x - 2 - x^2 - 2x}{(x + 2)(x - 2)(x + 1)}$$ Multiply and remove parentheses

$$= \frac{-x}{(x + 2)(x - 2)(x + 1)}$$ Collect like terms

Practice Exercise 9

Perform the indicated operations.

$$\frac{2y}{y^2 + 3y - 4} - \frac{1}{y + 3} - \frac{y}{y^2 + 2y - 3}$$

Answer: $\frac{4 - y}{(y - 1)(y + 4)(y + 3)}$

6.3 EXERCISES A

Perform the indicated operation.

1. $\frac{3}{x} + \frac{8}{x}$

2. $\frac{3y + 2}{y - 4} + \frac{-2 - 3y}{y - 4}$

3. $\frac{a}{2} + \frac{2a - 1}{-2}$

4. $\frac{2a}{a - b} + \frac{3a}{b - a}$

5. $\frac{x}{8x - 6} - \frac{5}{2(3 - 4x)}$

6. $\frac{z}{1 - z^2} - \frac{1 - z}{(z - 1)(z + 1)}$

Find the LCD of each pair of fractions.

7. $\frac{1}{39}$ and $\frac{5}{33}$

8. $\frac{2}{y}$ and $\frac{3}{5y}$

9. $\frac{a + b}{a}$ and $\frac{3}{4a^5}$

10. $\frac{6}{5b}$ and $\frac{3}{5b + 5}$

11. $\frac{a}{3a + 3b}$ and $\frac{b}{a^2 - b^2}$

12. $\frac{3}{b^2 + 5b + 6}$ and $\frac{b}{b^2 - 5b - 6}$

13. $\frac{2}{x^2 - y^2}$ and $\frac{3}{x^2 - 2xy + y^2}$

14. $\frac{2}{a - b}$ and $\frac{a}{b - a}$

15. $\frac{z + 2}{z^2 + 4z + 4}$ and $\frac{z - 1}{z^2 + z - 2}$

16. $\frac{a + 3}{a^3 - 27}$ and $\frac{a}{a^2 + 3a + 9}$

17. $\frac{x + 3}{5x^4 - 15x^3 - 50x^2}$ and $\frac{x - 7}{10x^3 - 100x^2 + 250x}$

Write each fraction as an equivalent fraction with the specified denominator.

18. $\frac{8}{3y}$; $27y^3z$

19. $\frac{a}{a - 1}$; $4a - 4$

20. $\frac{y}{y + 4}$; $y^2 - 16$

Perform the indicated operation.

21. $\frac{3}{5x} + \frac{5 - 6x}{10x^2}$

22. $\frac{3}{a + 4} - \frac{2}{a - 4}$

23. $\frac{2}{y + 3} - \frac{7}{y - 3}$

24. $\frac{4}{3z} + \frac{1 - 12z}{9z^2}$

25. $\frac{2}{x - 2} + \frac{5}{x}$

26. $\frac{7}{a^2 - 25} + \frac{2}{a - 5}$

27. $\frac{2x}{9 - x^2} - \frac{1}{3 - x}$

28. $\frac{5}{3y - 3} + \frac{3}{1 - y}$

29. $\frac{xy}{3x - 3y} + \frac{xy}{5y - 5x}$

30. $\frac{3}{2a^2 - 2a} + \frac{5}{2 - 2a}$

31. $\frac{y^2}{y - 1} - 2$

32. $\frac{2x + 1}{x + 2} + x$

33 $a - b - \frac{-ab^2}{a^2 + ab}$

34. $\frac{y}{y^2 - y - 20} + \frac{2}{y + 4}$

35 $\frac{2a - 1}{a^2 - 4a + 4} + \frac{2a + 3}{4 - a^2}$

36. $\frac{-2z}{8z^3 + 27} + \frac{1}{4z^2 - 6z + 9}$

37. $\frac{2}{x^2 - x - 6} - \frac{1}{x^2 - 2x - 3}$

38. $\frac{2}{a^2 + ab - 2b^2} + \frac{2}{a^2 + 3ab + 2b^2}$

39. $\frac{x + 2}{x^2 + 5x + 6} - \frac{x + 1}{x^2 + 4x + 3}$

40. $\frac{4 - 3a}{3a^2 + 6a + 12} + \frac{a^2 - 4}{a^3 - 8}$

41. $\frac{u}{u^2 - v^2} + \frac{v}{u^2 - 2uv + v^2} - \frac{1}{u + v}$

42. $\frac{-x}{x^2 + x - 2} - \frac{1}{x^2 - 2x + 1} + \frac{1}{x + 2}$

43. $\frac{a}{a^2 - 5a + 6} - \frac{2}{a^2 - 4a + 3} - \frac{3}{a^2 - 3a + 2}$

44 $\frac{x + y}{x^2 + 2xy + y^2} - \frac{x}{x^2 - xy - 2y^2} + \frac{y}{x^2 - 4xy - 5y^2}$

The early Greeks were aware of the formula

$$M = 180 - \frac{360}{n}$$

which gives the degree measure M of each interior angle of a regular polygon with n sides. Express M as a single fraction and use the result in Exercises 45–46.

45 What is the measure in degrees of each interior angle of a regular pentagon (5 sides)?

46 What is the measure in degrees of each interior angle of a regular octagon (8 sides)?

FOR REVIEW

Perform the indicated operation.

47. $\dfrac{a^3 - b^3}{a^2 - b^2} \cdot \dfrac{a^2 + 2ab + b^2}{a^2 + ab + b^2}$

48. $\dfrac{x^2 - 2xy + y^2}{xy - y^2} \div \dfrac{x^3 - xy^2}{x^2 + 2xy + y^2}$

The following exercises review material covered in Chapter 2 to help you prepare for the next section. Solve each equation.

49. $2 - (2a - 1) = 3(a - 3)$

50. $\dfrac{x - 1}{3} = \dfrac{3x - 2}{12}$

ANSWERS: 1. $\frac{11}{x}$ 2. 0 3. $\frac{1-a}{2}$ 4. $\frac{-a}{a-b}$ or $\frac{a}{b-a}$ 5. $\frac{x+5}{8x-6}$ 6. $\frac{1}{1-z^2}$ 7. 429 8. $5y$ 9. $4a^5$ 10. $5b(b+1)$ 11. $3(a-b)(a+b)$ 12. $(b+2)(b+3)(b-6)(b+1)$ 13. $(x+y)(x-y)^2$ 14. $a-b$ 15. $z+2$ 16. $(a-3)(a^2+3a+9)$ 17. $10x^2(x-5)^2(x+2)$ 18. $\frac{72y^2z}{27y^3z}$ 19. $\frac{4a}{4a-4}$ 20. $\frac{y^2-4y}{y^2-16}$ 21. $\frac{1}{2x^2}$ 22. $\frac{a-20}{(a+4)(a-4)}$ 23. $\frac{-5y-27}{(y+3)(y-3)}$ 24. $\frac{1}{9z^2}$ 25. $\frac{7x-10}{x(x-2)}$ 26. $\frac{2a+17}{(a-5)(a+5)}$ 27. $\frac{-1}{x+3}$ 28. $\frac{-4}{3(y-1)}$ 29. $\frac{2xy}{15(x-y)}$ 30. $\frac{3-5a}{2a(a-1)}$ 31. $\frac{y^2-2y+2}{y-1}$ 32. $\frac{x^2+4x+1}{x+2}$ 33. $\frac{a^2}{a+b}$ 34. $\frac{3y-10}{(y-5)(y+4)}$ 35. $\frac{4(a+1)}{(a-2)^2(a+2)}$ 36. $\frac{3}{8z^3+27}$ 37. $\frac{x}{(x-3)(x+2)(x+1)}$ 38. $\frac{4a}{(a-b)(a+b)(a+2b)}$ 39. 0 40. $\frac{10}{3(a^2+2a+4)}$ 41. $\frac{2uv}{(u-v)^2(u+v)}$ 42. $\frac{-2x-1}{(x-1)^2(x+2)}$ 43. $\frac{a^2-6a+13}{(a-1)(a-2)(a-3)}$ 44. $\frac{y(8y-x)}{(x+y)(x-2y)(x-5y)}$ 45. 108° 46. 135° 47. $a+b$ 48. $\frac{x+y}{xy}$ 49. $\frac{12}{5}$ 50. 2

6.3 EXERCISES B

Perform the indicated operation.

1. $\dfrac{4}{2y} - \dfrac{5}{2y}$

2. $\dfrac{3b}{b+1} - \dfrac{2b-1}{b+1}$

3. $\dfrac{3b}{4} + \dfrac{4b-3}{-4}$

4. $\dfrac{4x}{3-x} - \dfrac{x+1}{x-3}$

5. $\dfrac{2x}{10-5x} - \dfrac{3}{5(x-2)}$

6. $\dfrac{z-1}{(1-z)(1+z)} - \dfrac{1+z}{z^2-1}$

Find the LCD of each pair of fractions.

7. $\frac{5}{14}$ and $\frac{7}{38}$

8. $\frac{1}{4}$ and $\frac{3}{x}$

9. $\frac{3}{a}$ and $\frac{2}{9a}$

10. $\frac{3}{2x}$ and $\frac{1}{2x + 2}$

11. $\frac{u}{2u - 2v}$ and $\frac{3}{u^2 - v^2}$

12. $\frac{x}{x^2 + 4x + 3}$ and $\frac{5}{x^2 + 6x - 7}$

13. $\frac{6}{a^2 - b^2}$ and $\frac{7}{a^2 + 2ab + b^2}$

14. $\frac{3}{5a - b}$ and $\frac{b}{b - 5a}$

15. $\frac{a + 3}{a^2 + 6a + 9}$ and $\frac{a - 4}{a^2 - a - 12}$

16. $\frac{x - 2}{x^3 + 8}$ and $\frac{2x}{x^2 - 2x + 4}$

17. $\frac{3y + 2}{3y^5 - 6y^4 + 3y^3}$ and $\frac{y + 1}{6y^4 + 24y^3 - 30y^2}$

Write each fraction as an equivalent fraction with the specified denominator.

18. $\frac{5}{4a}$; $12a^2b$

19. $\frac{a}{a - b}$; $a^2 - b^2$

20. $\frac{2xy}{3x - y}$; $3x^2 + 2xy - y^2$

Perform the indicated operation.

21. $\frac{7}{2x} + \frac{2 - 14x}{4x^2}$

22. $\frac{5}{a - 3} - \frac{4}{a + 3}$

23. $\frac{3}{y + 5} - \frac{6}{y - 5}$

24. $\frac{6}{7z} + \frac{3 - 18z}{21z^2}$

25. $\frac{5}{x + 4} + \frac{6}{x}$

26. $\frac{3}{a^2 - 16} + \frac{2}{a - 4}$

27. $\frac{6y}{4 - y^2} - \frac{1}{2 + y}$

28. $\frac{8}{2x - 2} + \frac{5}{1 - x}$

29. $\frac{a^2b^2}{4a - 4b} + \frac{a^2b^2}{7b - 7a}$

30. $\frac{4}{3x^2 - 3x} + \frac{5}{3 - 3x}$

31. $\frac{x^2}{x - 2} - 3$

32. $\frac{3y - 2}{y - 5} + y$

33. $x + y - \frac{-xy^2}{x^2 - xy}$

34. $\frac{2x}{x^2 - 3x - 4} + \frac{3}{x + 1}$

35. $\frac{5u - 3}{u^2 - 10u + 25} + \frac{u - 8}{25 - u^2}$

36. $\frac{-6x}{27x^3 - 8} + \frac{2}{9x^2 + 6x + 4}$

37. $\frac{-3}{a^2 + 4a - 5} - \frac{2}{a^2 + 6a + 5}$

38. $\dfrac{3}{u^2 + 5uv + 6v^2} + \dfrac{-1}{u^2 + 3uv + 2v^2}$

39. $\dfrac{a + 3}{a^2 + 10a + 21} - \dfrac{a - 2}{a^2 + 5a - 14}$

40. $\dfrac{2x - 5}{4x^2 - 12x + 36} + \dfrac{x^2 - 9}{x^3 + 27}$

41. $\dfrac{b}{a^2 - b^2} - \dfrac{1}{a - b} + \dfrac{a}{a^2 + 2ab + b^2}$

42. $\dfrac{5}{x^2 - 5x + 6} - \dfrac{x}{x^2 - 4x + 4} + \dfrac{1}{x - 2}$

43. $\dfrac{u}{u^2 - 2u - 3} - \dfrac{u}{u^2 + 3u + 2} - \dfrac{5}{u^2 - u - 6}$

44. $\dfrac{x - y}{x^2 - 2xy + y^2} - \dfrac{y}{x^2 + xy - 2y^2} + \dfrac{x}{x^2 + 2xy - 3y^2}$

The early Greeks were aware of the formula

$$M = 180 - \frac{360}{n}$$

which gives the degree measure M of each interior angle of a regular polygon with n sides. Express M as a single fraction and use the result in Exercises 45–46.

45. What is the measure in degrees of each interior angle of a regular hexagon (6 sides)?

46. What is the measure in degrees of each interior angle of a regular decagon (10 sides)?

FOR REVIEW

Perform the indicated operations.

47. $\dfrac{2x^2 + 9x - 5}{2x^2 - 3x + 1} \div \dfrac{x + 5}{x - 1}$

48. $\dfrac{3a + 6b}{a^2 + ab - 2b^2} \cdot \dfrac{a^3 - b^3}{a^2 + ab + b^2}$

The following exercises review material covered in Chapter 2 to help you prepare for the next section. Solve each equation.

49. $7 - 4x = -5$

50. $\dfrac{x + 2}{4} + \dfrac{x + 7}{8} = 1$

6.3 EXERCISES C

Perform the indicated operation.

1. $\dfrac{-5}{5 - x} - \dfrac{x}{x - 5}$

2. $\dfrac{a}{(a - b)(a - 2b)} + \dfrac{b}{(b - a)(2b - a)}$

Find the LCD of each group of fractions.

3. $\dfrac{1}{x^2+x}, \dfrac{2x}{x^2-x-2}$, and $\dfrac{3}{x^2-4x+4}$

4. $\dfrac{a-b}{a^2-b^2}, \dfrac{x+1}{ax+bx+a+b}$, and $\dfrac{a-3b}{a^2-2ab-3b^2}$
[Answer: $a+b$]

5. $\dfrac{2b}{a^2-b^2} - \dfrac{b}{a^2+ab} - \dfrac{a^2+2ab+b^2}{a^3-ab^2}$
$\left[\text{Answer: } \dfrac{1}{b-a}\right]$

6. $\dfrac{-x}{x^3-y^3} - \dfrac{1}{y^2-x^2} + \dfrac{1}{x^2+xy+y^2}$

6.4 SOLVING FRACTIONAL EQUATIONS

STUDENT GUIDEPOSTS

1. Fractional Equations
2. The Cross-Product Equation
3. Literal Fractional Equations

1 FRACTIONAL EQUATIONS

An equation that contains one or more algebraic fractions or rational expressions is called a **fractional equation.** To **clear the fractions,** we multiply both sides of the equation by the LCD of all fractions and then solve the equation using the techniques of Chapter 2. Every solution to this new equation must be checked in the *original* equation; if the solution makes one of the denominators equal to zero, it must be discarded. If all solutions are discarded, the equation has no solution.

EXAMPLE 1 SOLVING A FRACTIONAL EQUATION

Solve $\dfrac{5}{x} - \dfrac{7}{6} = \dfrac{3}{2x}$.

$$6x\left[\frac{5}{x} - \frac{7}{6}\right] = 6x\left[\frac{3}{2x}\right] \quad \text{Multiply both sides by the LCD, } 6x$$

$$6\cancel{x}\cdot\frac{5}{\cancel{x}} - \cancel{6}x\cdot\frac{7}{\cancel{6}} = 6\cancel{x}\cdot\frac{3}{2\cancel{x}} \quad \text{Distribute}$$

$$30 - 7x = 9 \quad \text{Clear fractions}$$

$$-7x = -21 \quad \text{Subtract 30 from both sides}$$

$$x = 3 \quad \text{Divide both sides by } -7$$

Check: $\dfrac{5}{3} - \dfrac{7}{6} \stackrel{?}{=} \dfrac{3}{2\cdot 3}$

$$\frac{2\cdot 5}{2\cdot 3} - \frac{7}{6} \stackrel{?}{=} \frac{3}{6}$$

$$\frac{10}{6} - \frac{7}{6} \stackrel{?}{=} \frac{3}{6}$$

$$\frac{3}{6} = \frac{3}{6}$$

The solution is 3.

PRACTICE EXERCISE 1

Solve $\dfrac{1}{x} - \dfrac{1}{3} = \dfrac{1}{3x}$.

Answer: 2

To Solve a Fractional Equation

1. Find the LCD of all fractions in the equation.
2. Multiply both sides of the equation by the LCD to clear fractions. Make sure that *all* terms are multiplied.
3. Solve this equation.
4. Check all possible solutions in the original equation. THIS IS IMPORTANT!

EXAMPLE 2 Solving a Fractional Equation

Solve $\frac{2}{y} = \frac{3}{4-y}$.

$$y(4-y)\left[\frac{2}{y}\right] = y(4-y)\left[\frac{3}{4-y}\right] \quad \text{Multiply both sides by the LCD, } y(4-y) \text{ to clear fractions}$$

$$2(4-y) = 3y$$

$$8 - 2y = 3y$$

$$8 = 5y \quad \text{Add } 2y \text{ to both sides}$$

$$\frac{8}{5} = y \quad \text{Divide both sides by 5}$$

Check: $\frac{2}{\frac{8}{5}} \stackrel{?}{=} \frac{3}{4-\frac{8}{5}}$

$$\frac{10}{8} \stackrel{?}{=} \frac{3}{\frac{12}{5}}$$

$$\frac{5}{4} \stackrel{?}{=} \frac{15}{12}$$

$$\frac{5}{4} \stackrel{?}{=} \frac{5}{4}$$

The solution is $\frac{8}{5}$.

Practice Exercise 2

Solve $\frac{3}{z} = \frac{6}{z-4}$.

Answer: -4

2 THE CROSS-PRODUCT EQUATION

Equations that consist of two equated fractions, such as the one in Example 2, can be solved using a shortcut. We know that $\frac{a}{b} = \frac{c}{d}$ if and only if the cross-products ad and bc are equal. As a result, to solve

$$\frac{2}{y} = \frac{3}{4-y}$$

we can set equal the cross-products $2(4-y)$ and $3y$, forming the equation

$$2(4-y) = 3y.$$

Notice that we immediately arrive at the second step in Example 2 above. However, this technique works only for these special equations of two fractions. Do not try to use it, for example, to solve an equation like the one in Example 1.

CAUTION

Check all possible solutions. Multiplying both sides of an equation by a variable can produce ''solutions'' which are values for the variable that make a denominator zero in the original equation. This is shown in the next example.

EXAMPLE 3 Solving a Fractional Equation with No Solution

Solve $\dfrac{a^2+9}{a^2-9} - \dfrac{3}{a+3} = \dfrac{-a}{3-a}$.

We first observe that $a^2 - 9 = (a-3)(a+3)$. If we multiply the numerator and denominator of $\frac{-a}{3-a}$ by -1, we obtain the equivalent fraction $\frac{a}{a-3}$. Then the LCD is $(a-3)(a+3)$, and we multiply both sides by it.

$$(a-3)(a+3)\left[\frac{a^2+9}{(a-3)(a+3)} - \frac{3}{(a+3)}\right] = (a-3)(a+3)\left[\frac{a}{(a-3)}\right]$$

$$\cancel{(a-3)}\cancel{(a+3)} \cdot \frac{a^2+9}{\cancel{(a-3)}\cancel{(a+3)}} - (a-3)\cancel{(a+3)} \cdot \frac{3}{\cancel{(a+3)}} = \cancel{(a-3)}(a+3) \cdot \frac{a}{\cancel{(a-3)}}$$

$$\begin{aligned} a^2 + 9 - (a-3)\cdot 3 &= (a+3)\cdot a \\ a^2 + 9 - (3a - 9) &= a^2 + 3a \\ a^2 + 9 - 3a + 9 &= a^2 + 3a \\ -3a + 18 &= 3a \\ 18 &= 6a \\ 3 &= a \end{aligned}$$

Check: $\dfrac{3^2+9}{3^2-9} - \dfrac{3}{3+3} \stackrel{?}{=} \dfrac{-3}{3-3}$

$$\frac{18}{0} - \frac{3}{6} = \frac{-3}{0}$$

Since division by 0 is undefined, 3 must be discarded as a possible value for a. There is no solution.

Practice Exercise 3

Solve $\dfrac{15}{w^2+5w} + \dfrac{w+4}{w+5} = \dfrac{w+3}{w}$.

Answer: no solution (0 does not check)

Some fractional equations reduce to a quadratic equation, as in the example on the following page.

EXAMPLE 4 Solving a Quadratic Fractional Equation

Solve $\frac{y}{3} - \frac{6}{y} = 1$.

$$3y\left[\frac{y}{3} - \frac{6}{y}\right] = 3y[1] \quad \text{Multiply by LCD, } 3y$$

$$\cancel{3}y \cdot \frac{y}{\cancel{3}} - 3\cancel{y} \cdot \frac{6}{\cancel{y}} = 3y \quad \text{Distribute}$$

$$y^2 - 18 = 3y \quad \text{This time we have a quadratic equation}$$

$$y^2 - 3y - 18 = 0$$

$$(y + 3)(y - 6) = 0 \quad \text{Factor}$$

$$y + 3 = 0 \quad \text{or} \quad y - 6 = 0$$

$$y = -3 \qquad\qquad y = 6$$

Check: $\frac{-3}{3} - \frac{6}{-3} \stackrel{?}{=} 1$ $\qquad$ $\frac{6}{3} - \frac{6}{6} \stackrel{?}{=} 1$

$-1 + 2 \stackrel{?}{=} 1$ $\qquad$ $2 - 1 \stackrel{?}{=} 1$

$1 = 1$ $\qquad$ $1 = 1$

-3 checks. $\qquad$ 6 checks.

The solutions are -3 and 6.

Practice Exercise 4

Solve $\frac{x}{4} + \frac{3}{x} = 2$.

Answer: 2 and 6

CAUTION

A common mistake is to confuse addition or subtraction of algebraic fractions with solving fractional equations. Compare the two problems:

Solve: $\frac{2}{y} + \frac{y}{y+1} = 1$

$$y(y+1)\left[\frac{2}{y} + \frac{y}{y+1}\right] = y(y+1)(1)$$

$$\cancel{y}(y+1)\frac{2}{\cancel{y}} + y\cancel{(y+1)}\frac{y}{\cancel{y+1}} = y^2 + y$$

$$2y + 2 + y^2 = y^2 + y$$

$$2y + 2 = y$$

$$y = -2$$

Add: $\frac{2}{y} + \frac{y}{y+1} + 1$

$$\frac{2(y+1)}{y(y+1)} + \frac{y \cdot y}{y(y+1)} + \frac{y(y+1)}{y(y+1)}$$

$$\frac{2(y+1) + y^2 + y(y+1)}{y(y+1)}$$

$$\frac{2y + 2 + y^2 + y^2 + y}{y(y+1)}$$

$$\frac{2y^2 + 3y + 2}{y(y+1)}$$

The LCD $y(y + 1)$ must be found in both cases, but it is used in different ways. In the equation, we multiply both sides by $y(y + 1)$ to clear the fractions and eventually find the numerical solution -2. In the addition problem, each term is written as an equivalent fraction with denominator $y(y + 1)$. We then find the sum of these fractions, obtaining an algebraic expression.

EXAMPLE 5 NUMBER PROBLEM

If the numerator of a fraction exceeds the denominator by 6, and the value of the fraction is $\frac{5}{3}$, find the fraction.

Let $x =$ the denominator of the fraction,
$x + 6 =$ the numerator of the fraction. (Why?)

The fraction is $\frac{x+6}{x}$, so the equation we must solve is

$$\frac{x+6}{x} = \frac{5}{3}. \qquad \text{The LCD is } 3x$$

$$3x \cdot \frac{x+6}{x} = 3x \cdot \frac{5}{3}$$

$$3(x+6) = x \cdot 5 \qquad \text{Use parentheses}$$

$$3x + 18 = 5x$$

$$18 = 2x$$

$$9 = x$$

$$x + 6 = 15$$

The fraction is $\frac{15}{9}$. Since the numerator 15 is 6 more than the denominator 9, and $\frac{15}{9}$ reduces to $\frac{5}{3}$, $\frac{15}{9}$ checks.

PRACTICE EXERCISE 5

If the denominator of a fraction exceeds the numerator by 9 and the value of the fraction is $\frac{4}{7}$, find the fraction.

Let $x =$ numerator,
$x + 9 =$ denominator.

Answer: $\frac{12}{21}$

3 LITERAL FRACTIONAL EQUATIONS

Many literal equations are fractional equations that can be solved using the method of this section.

EXAMPLE 6 LITERAL FRACTIONAL EQUATION

Solve $\frac{1}{a} + \frac{1}{b} = c$ for a.

$$\frac{1}{a} + \frac{1}{b} = c$$

$$ab\left(\frac{1}{a} + \frac{1}{b}\right) = abc \qquad \text{Multiply both sides by the LCD, } ab$$

$$ab \cdot \frac{1}{a} + ab \cdot \frac{1}{b} = abc \qquad \text{Distributive law}$$

$$b + a = abc$$

$$b = abc - a \qquad \text{Subtract } a \text{ to get all terms with the desired variable on the same side}$$

$$b = a(bc - 1) \qquad \text{Factor out } a \text{ using the distributive law}$$

$$\frac{b}{bc - 1} = a \qquad \text{Divide by } bc - 1\text{, the coefficient of } a$$

PRACTICE EXERCISE 6

Solve $\frac{1}{w} - \frac{1}{u} = \frac{1}{v}$ for w.

Answer: $w = \frac{uv}{u+v}$

EXAMPLE 7 Electricity Problem

The resistance R of an electrical circuit is given by $\frac{1}{R} = \frac{1}{R_1} + \frac{1}{R_2}$ where R_1 and R_2 are resistors in parallel, as shown in Figure 6.1. Solve this equation for R_1.

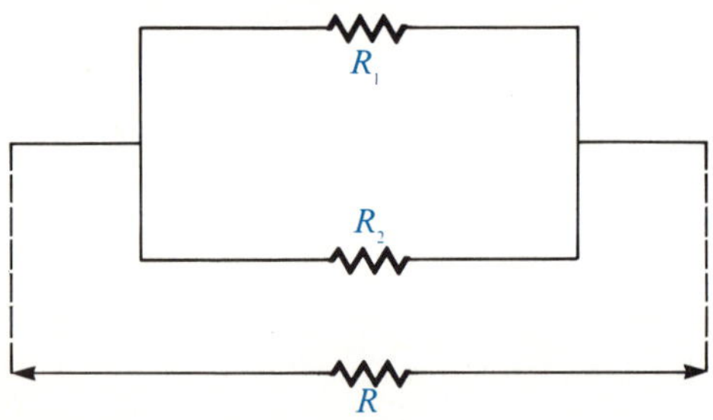

Figure 6.1

$$\frac{1}{R} = \frac{1}{R_1} + \frac{1}{R_2}$$

$$RR_1R_2\left[\frac{1}{R}\right] = RR_1R_2\left[\frac{1}{R_1} + \frac{1}{R_2}\right] \quad \text{Multiply both sides by the LCD, } RR_1R_2$$

$$R_1R_2 = RR_2 + RR_1$$

$$R_1R_2 - RR_1 = RR_2 \quad \text{Subtract to get both terms with } R_1 \text{ on the left}$$

$$R_1[R_2 - R] = RR_2 \quad \text{Factor out } R_1$$

$$R_1 = \frac{RR_2}{R_2 - R} \quad \text{Divide by } R_2 - R\text{, the coefficient of } R_1$$

Practice Exercise 7

Solve $\frac{1}{R} = \frac{1}{R_1} + \frac{1}{R_2}$ for R.

Answer: $R = \frac{R_1R_2}{R_1 + R_2}$

6.4 EXERCISES A

Solve.

1. $\frac{2}{x} - \frac{1}{x} = -6$

2. $\frac{7}{y} = \frac{5}{y-2}$

3. $\frac{1}{y-1} + \frac{2}{y+1} = 0$

4. $\frac{a-2}{a+3} = \frac{a+3}{a+2}$

5. $\frac{1}{a+2} + \frac{1}{a-2} = \frac{1}{a^2-4}$

6. $\frac{11}{c^2-25} - \frac{2}{c-5} = \frac{1}{c+5}$

7. $\frac{x}{x+4} = \frac{4}{x-4} + \frac{x^2+16}{x^2-16}$

8. $\frac{1}{z^2-2z+1} + \frac{1}{z-1} = 2$

9. $\frac{1}{a} = \frac{-6}{a^2+5}$

10. $\dfrac{2y}{y-1} - 2 = \dfrac{5}{2y}$

11. $\dfrac{x^2}{x+12} = 1$

12. $5 + \dfrac{x}{x-3} = \dfrac{-6}{x-3}$

13. $\dfrac{25}{y-5} - \dfrac{25}{y} = \dfrac{1}{4}$

14 $\dfrac{-1}{a^2-3a} = \dfrac{1}{a} + \dfrac{a}{a-3}$

In Exercises 15–18, either solve the given equation or perform the indicated operation.

15. $\dfrac{3x+5}{6} = \dfrac{4+3x}{5}$

16. $\dfrac{3x+5}{6} + \dfrac{4+3x}{5}$

17. $\dfrac{2}{y+1} - \dfrac{3}{y} = 0$

18. $\dfrac{2}{y+1} - \dfrac{3}{y}$

19. Find two numbers whose sum is 45 and whose quotient is $\frac{2}{3}$.

20 Harriet is three-fourths as old as Lorraine, and the difference of their ages is 10 years. How old is each?

21. The reciprocal of 5 less than a number is twice the reciprocal of the number. Find the number.

22. The numerator of a fraction is 4 less than the denominator. If both the numerator and denominator are increased by 2, the value of the fraction is $\frac{2}{3}$. Find the fraction.

23 A student made 75, 80, and 82 on the first three of four tests in a course. What must she make on the fourth test for an average score of 83?

24. If five times the reciprocal of a number is equal to the reciprocal of 4 more than the number, find the number.

25. The sum of the reciprocals of two consecutive integers is nine times the reciprocal of their product. Find the integers.

26. If Brenda is 5 years older than Jane, and the quotient of their ages is $\frac{5}{4}$, how old is each?

Solve each equation for the indicated variable.

27. $\dfrac{1}{a} + \dfrac{1}{b} = c$ for b

28. $\dfrac{1}{a} + \dfrac{1}{b} = \dfrac{1}{c}$ for c

29. $\dfrac{a^{-1}}{c} = \dfrac{d^{-1}}{b}$ for a
$\left[Hint: \text{Substitute } \frac{1}{a} \text{ for } a^{-1} \text{ and } \frac{1}{d} \text{ for } d^{-1}.\right]$

30. $a^{-1} = b + c^{-1}$ for c

31. Solve the simple interest formula
$$P = \frac{A}{1 + rt} \text{ for } r.$$

32. Solve Doppler's formula
$$F = \frac{1100f}{1100 - V} \text{ for } V.$$

FOR REVIEW

Perform the indicated operations.

33. $\dfrac{2y - 1}{y^2 - 4y + 4} + \dfrac{2y + 1}{4 - y^2}$

34. $\dfrac{12}{z^3 + 8} + \dfrac{z}{z^2 - 2z + 4} - \dfrac{1}{z + 2}$

ANSWERS: 1. $-\frac{1}{6}$ 2. 7 3. $\frac{1}{3}$ 4. $-\frac{13}{6}$ 5. $\frac{1}{2}$ 6. 2 7. no solution 8. $\frac{1}{2}$, 2 9. $-5, -1$ 10. 5 11. 4, -3 12. $\frac{3}{2}$ 13. 25, -20 14. -2, 1 15. $\frac{1}{3}$ 16. $\frac{33x + 49}{30}$ 17. -3 18. $\frac{-y - 3}{y(y + 1)}$ 19. 18, 27 20. Harriet is 30, Lorraine is 40. 21. 10 22. $\frac{6}{10}$ 23. 95 24. -5 25. 4, 5 26. Brenda is 25, Jane is 20. 27. $b = \frac{a}{ac - 1}$ 28. $c = \frac{ab}{a + b}$ 29. $a = \frac{bd}{c}$ 30. $c = \frac{a}{1 - ab}$ 31. $r = \frac{A - P}{Pt}$ 32. $V = \frac{1100(F - f)}{F}$ 33. $\frac{6y}{(y - 2)^2(y + 2)}$ 34. $\frac{4}{z^2 - 2z + 4}$

6.4 EXERCISES B

Solve.

1. $\dfrac{3}{a + 1} = \dfrac{5}{a}$

2. $\dfrac{3}{z} + \dfrac{1}{z} = -4$

3. $\dfrac{x + 1}{x + 2} = \dfrac{x + 2}{x - 1}$

4. $\dfrac{1}{z - 5} + \dfrac{7}{z + 3} = 0$

5. $\dfrac{6}{x + 3} + \dfrac{2}{x - 3} = \dfrac{20}{x^2 - 9}$

6. $\dfrac{12}{z^2 - 9} + \dfrac{3}{z + 3} = \dfrac{2}{z - 3}$

7. $\dfrac{9 + y}{16 + 2y} = \dfrac{7}{16 + 2y}$

8. $\dfrac{8}{x^2 + 8x + 16} + \dfrac{2}{x + 4} = 3$

9. $\dfrac{10}{y - 1} = \dfrac{16}{y - 1} - 3y$

10. $\dfrac{u^2}{35 - 2u} = 1$

11. $\dfrac{a + 2}{a - 1} + \dfrac{3}{a} = 1$

12. $7 + \dfrac{5}{z - 2} = \dfrac{z}{z - 2}$

13. $\dfrac{3x}{2x + 3} = \dfrac{8x^2 + 12}{4x^2 - 9} - \dfrac{x}{2x - 3}$

14. $\dfrac{-10}{y^2 + 3y - 10} = \dfrac{y}{y - 2} + \dfrac{1}{y + 5}$

In Exercises 15–18, either solve the given equation or perform the indicated operation.

15. $\frac{5x-2}{3}=\frac{2+7x}{4}$ **16.** $\frac{5x-2}{3}-\frac{2+7x}{4}$ **17.** $\frac{5}{y+2}+\frac{2}{y}=0$ **18.** $\frac{5}{y+2}+\frac{2}{y}$

19. If the numerator of a fraction exceeds the denominator by 6, and the value of the fraction is $\frac{5}{3}$, find the fraction.

20. Sandi is four-fifths as old as Richard, and the difference in their ages is 7 years. How old is each?

21. Twice the reciprocal of a number, increased by the number itself, is -3. Find the number.

22. The denominator of a fraction is 1 more than the numerator. If both the numerator and denominator are decreased by 3, the value of the fraction is $\frac{4}{5}$. Find the fraction.

23. If the average weight of the four linemen for the Rams is 243 lb, and the weights of three of them are 220 lb, 240 lb, and 260 lb, how much does the fourth one weigh?

24. When 5 is added to the reciprocal of a number, the result is the same as 16 times the reciprocal of the number. Find the number.

25. The reciprocal of the smaller of two consecutive even integers, less the reciprocal of the larger integer, is equal to the reciprocal of the larger integer. Find the integers.

26. Walter is 8 years younger than Lupe and the quotient of their ages is $\frac{21}{17}$. How old is each?

Solve each equation for the indicated variable.

27. $\frac{1}{a}+\frac{1}{b}=\frac{1}{c}$ for a

28. $\frac{ab}{c}=\frac{wx}{d}$ for d

29. $(ab)^{-1}=a^{-1}+b^{-1}$ for b

30. $b^{-1}+c^{-1}=2$ for b

31. Solve the temperature conversion formula

$$\frac{C}{F-32}=\frac{5}{9} \text{ for } F.$$

32. Solve the simple interest formula

$$t=\frac{A-P}{Pr} \textit{ for } P.$$

FOR REVIEW

Perform the indicated operations.

33. $\frac{3a+2}{a^2-6a+9}+\frac{3a-2}{9-a^2}$

34. $\frac{3}{z^3+1}+\frac{z}{z^2-z+1}-\frac{1}{z+1}$

6.4 EXERCISES C

Solve.

1. $\frac{1}{x^2-x-2}=\frac{3}{x^2-2x-3}+\frac{1}{x^2-5x+6}$

2. $\frac{x+1}{x+2}-\frac{x+3}{x+5}=\frac{x+2}{x+5}-\frac{x+1}{x+3}$

[Answer: $1, -\frac{5}{2}$]

3. Solve the focal length formula
$$\frac{1}{f} = \frac{1}{d_1} + \frac{1}{d_2} \text{ for } d_1.$$

4. Solve the gas law
$$\frac{PV}{T} = \frac{pv}{t} \text{ for } P.$$

In Exercises 5–6, either solve the equation or perform the indicated operation.

5. $\dfrac{3}{x^2 + x - 2} + \dfrac{2}{x^2 - 1} = \dfrac{-1}{x^2 + 3x + 2}$

6. $\dfrac{3}{x^2 + x - 2} + \dfrac{2}{x^2 - 1} - \dfrac{1}{x^2 + 3x + 2}$

6.5 RATIO, PROPORTION, AND VARIATION

STUDENT GUIDEPOSTS

1. Ratios and Proportions
2. Solving Proportions
3. Direct Variation
4. Inverse Variation
5. Joint Variation
6. Combined Variation

1 RATIOS AND PROPORTIONS

Numerous applied problems can be simplified using the concepts of ratio, proportion, and variation. The **ratio** of one number a to another number b is the quotient $\frac{a}{b}$, sometimes expressed by the notation $a{:}b$. Ratios occur in applications such as percent, rate of speed, and cost. Consider the following examples.

Applications	*Ratio involved*
6% sales tax	$\frac{\$6}{\$100} = \$6$ per \$100 tax
240 miles in 4 hours	$\frac{240 \text{ mi}}{4 \text{ hr}} = 60$ mph
\$80 for 4 shirts	$\frac{\$80}{4 \text{ shirts}} = \$20$ per shirt

Ratios are used to define *proportions*.

Proportions

An equation stating that two ratios are equal is called a **proportion.** That is, if $\frac{a}{b}$ and $\frac{c}{d}$ are two ratios, the equation

$$\frac{a}{b} = \frac{c}{d}$$

is a proportion. It is read "a is to b as c is to d."

2 SOLVING PROPORTIONS

In the proportion

$$\frac{3}{5} = \frac{18}{30},$$

the numbers 3 and 30 are the **extremes** of the proportion, and 5 and 18 are the **means.** Consider the proportion

$$\frac{a}{b} = \frac{c}{d}$$

with extremes a and d and means b and c. If we multiply both sides by the LCD, bd, we obtain

$$bd\frac{a}{b} = bd\frac{c}{d}$$
$$ad = bc,$$

the cross-product equation. We have just verified the following property of proportions.

Means-Extremes Property of Proportions

In any proportion, the product of the extremes equals the product of the means.

The means-extremes property allows us to solve for an unknown term in a proportion, called **solving the proportion.** Suppose we solve the proportion

$$\frac{x}{3} = \frac{8}{27}.$$
$$27x = 3 \cdot 8 \quad \text{Product of extremes = product of means}$$
$$27x = 24$$
$$x = \frac{24}{27} = \frac{8}{9}$$

Ratio and proportion are useful in many applied problems. Suppose that it takes 3 hours to grade 24 tests. How many hours would it take to grade 56 tests? Letting x be the number of hours to grade 56 tests, the proportion

$$\text{time to grade 24 tests} \longrightarrow \frac{3}{24} = \frac{x}{56} \longleftarrow \text{time to grade 56 tests}$$

describes the problem. That is, the ratios of the number of hours to the number of tests are equal. Using the means-extremes property,

$$3 \cdot 56 = 24x$$
$$7 = x.$$

It would take 7 hours to grade 56 tests.

EXAMPLE 1 Using Proportions

If a boat uses 14 gallons of gas to go 102 miles, how many gallons would be needed to go 510 miles?

Let x = the number of gallons needed to go 510 miles.

Practice Exercise 1

If a scout troop can pick up 7 bags of trash in 2 hr, how many bags can they pick up in 8 hr?

The equation to solve is

$$\frac{14}{102} = \frac{x}{510}$$

miles on 14 gallons → 102; 510 ← miles on x gallons

$$102x = 510 \cdot 14 \qquad \text{Cross-product equation}$$

$$x = \frac{510 \cdot 14}{102}$$

$$x = \frac{\cancel{102} \cdot 5 \cdot 14}{\cancel{102}} = 70$$

It would take 70 gallons to go 510 miles.

Answer: 28 bags

Two triangles are **similar** if the angles of one are equal respectively to the angles of the other. To show similarity, all that is necessary is to show that two angles of one triangle are equal to two angles of the other. (Why?) Similar triangles have the property that corresponding sides (sides opposite the equal angles) are proportional. If $\angle A = \angle A'$, $\angle B = \angle B'$, and $\angle C = \angle C'$, as in Figure 6.2 ($\angle$ is the symbol for *angle*), then $\triangle ABC$ is similar to $\triangle A'B'C'$ ($\triangle$ is the symbol for *triangle*). The fact that the sides are proportional is expressed by the following equations.

$$\frac{a}{a'} = \frac{b}{b'} = \frac{c}{c'}$$

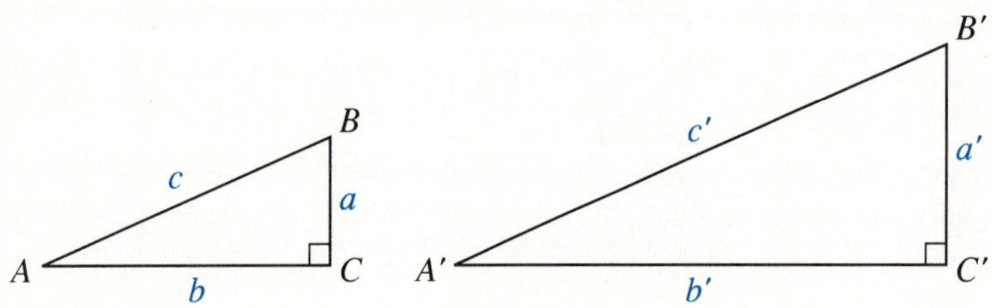

Figure 6.2

EXAMPLE 2 Similar Triangle Application

If a man who is 6 ft tall stands in the sunlight and casts a shadow 14 ft long, while a flagpole casts a shadow 35 ft long, how tall is the pole?

It is best to make a sketch, as in Figure 6.3.

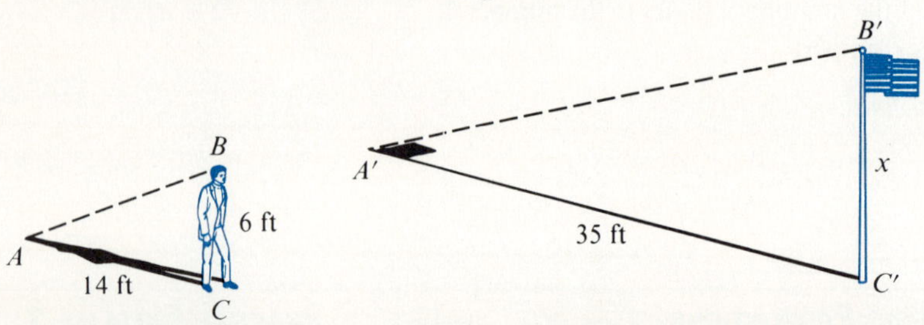

Figure 6.3

We assume that $\angle ACB = \angle A'C'B' = 90°$ (a right angle). Since both measurements are taken at the same time, the sun's rays make the same angle with level ground, that is, $\angle BAC = \angle B'A'C'$. Thus, $\triangle ABC$ is similar to $\triangle A'B'C$.

Practice Exercise 2

A tree casts a shadow 4 ft long at the same time a building, which is 85 ft tall, casts a shadow 17 ft in length. How tall is the tree?

$$\frac{AC}{A'C'} = \frac{BC}{B'C'}$$
$$\frac{14}{35} = \frac{6}{x}$$
$$14x = 6 \cdot 35$$
$$x = \frac{6 \cdot 35}{14} = 15$$

The pole is 15 ft tall.

Answer: 20 ft

EXAMPLE 3 Number Problem

What number must be added to each of 20, 8, 32, and 14 to give four numbers which are in proportion in that order?

Let x = the number to be added.

We need to solve

$$\frac{20 + x}{8 + x} = \frac{32 + x}{14 + x}.$$
$$(20 + x)(14 + x) = (8 + x)(32 + x)$$
$$280 + 34x + x^2 = 256 + 40x + x^2$$
$$280 + 34x = 256 + 40x$$
$$24 = 6x$$
$$4 = x$$

The desired number is 4. Check.

Practice Exercise 3

What number must be subtracted from each of 8, 12, 20, and 40 to give four numbers which are in proportion in that order?

Let x = number to be subtracted.

Answer: 5

3 DIRECT VARIATION

Three types of relationships identified with the term **variation** provide another way of looking at proportion problems. For example, the distance a car travels **varies** or changes in a given period of time as the average speed varies. Similarly, the strength of a wooden beam varies as we vary the length, width, and thickness of the beam, and the amount of illumination from a light source varies as we vary the distance from the source. The study of these and other types of variation is important to the scientist since many laws in engineering, chemistry, physics, and biology are stated using variation.

The first type of relationship, *direct variation,* can be written as a ratio equal to a constant.

Direct Variation

If two variables x and y have a constant ratio c,

$$\frac{y}{x} = c \quad \text{or} \quad y = cx,$$

we say that **y varies directly as x, y is proportional to x,** or simply **y varies as x.** The number c is called the **constant of variation** or **constant of proportionality.**

Perhaps the simplest example of direct variation is found in the formula for the circumference of a circle. The circumference is equal to π times the diameter.

$$C = \pi d$$

Thus, π is the constant of variation and, in fact, π is defined as the ratio of the circumference of any circle to its diameter. We are often interested in finding the constant of variation and the equation of variation that describe a given situation.

EXAMPLE 4 Using Direct Variation

Find the equation of variation if y varies directly as x, and $y = 16$ when $x = 4$.

The first step is to write the equation described by the phrase *y varies directly as x*.

$$y = cx$$

Next we substitute the given values for y and x.

$$16 = c(4)$$
$$4 = c$$

The equation of variation is $y = 4x$.

Practice Exercise 4

Find the equation of variation if y varies directly as x, and $y = 25$ when $x = 2$.

Answer: $y = 12.5x$

EXAMPLE 5 Electricity Problem

According to Ohm's law, the voltage V in an electrical circuit varies directly as the current I in the circuit. If the voltage is 20 volts when the current is 3 amps, find the variation equation and use it to determine the voltage when the current is 27 amps.

The equation described is $V = cI$. Substitute $V = 20$ and $I = 3$.

$$20 = c(3)$$
$$\frac{20}{3} = c$$

The equation of variation is $V = \frac{20}{3}I$. When $I = 27$, $V = \frac{20}{3} \cdot 27 = 20 \cdot 9 = 180$ volts.

Practice Exercise 5

If the voltage in a circuit is 110 volts when the current is 9 amps, use Ohm's law to determine the voltage when the current is 18 amps.

Answer: 220 volts

4 INVERSE VARIATION

The second kind of variation, *inverse variation,* can be written as a product equal to a constant.

Inverse Variation

If two variables x and y have a constant product c,

$$xy = c \quad \text{or} \quad y = \frac{c}{x},$$

we say that **y varies inversely as x** or **y is inversely proportional to x.** Again **c** is the **constant of variation.**

If $y = \frac{c}{x}$, then if y becomes large, x must get small, and if y becomes small, x must get large. Perhaps the simplest example of inverse variation is the relationship between the time t and the rate r it takes for a vehicle to travel a fixed distance d. If the rate is increased, the time decreases, and if the rate is decreased, the time increases. This variation is given by

$$d = rt, \quad t = \frac{d}{r}, \quad r = \frac{d}{t},$$

where, in all three equations, we assume that d is constant (the constant of variation).

As with direct variation, we must be able to calculate the constant of variation and obtain the variation equation in an inverse variation problem.

EXAMPLE 6 Physics Problem

The volume V of a gas kept at a constant temperature varies inversely as the pressure P. If the volume is 150 ft^3 when the pressure is 18 lb/ft^2, find the variation equation and use it to determine the volume when the pressure is 12 lb/ft^2.

The equation described is

$$V = \frac{c}{P}.$$

Substitute $V = 150$ and $P = 18$.

$$150 = \frac{c}{18}$$

$$2700 = c$$

The equation of variation is $V = \frac{2700}{P}$, and when $P = 12$, $V = \frac{2700}{12} = 225$ ft^3.

Practice Exercise 6

If the volume of a gas is 200 ft^3 when the pressure is 25 lb/ft^2, use the gas equation to find the pressure when the volume is 500 ft^3.

Answer: 10 lb/ft^2

5 JOINT VARIATION

Often, a quantity will vary directly as the product of two quantities. When this occurs we have a *joint variation* problem.

Joint Variation

If three variables x, y, and z are so related that

$$z = cxy$$

where c is a constant, we say that z varies directly as x and directly as y or that z **varies jointly as x and y.**

EXAMPLE 7 Using Joint Variation

Find the equation of variation if z varies jointly as x and y, and $z = 36$ when $x = 3$ and $y = 2$.

The first step is to write the equation described by the phrase *z varies jointly as x and y.*

$$z = cxy$$

Practice Exercise 7

Find the equation of variation if u varies jointly as v and w, and $u = 45$ when $v = 3$ and $w = 5$.

Next we substitute the given values for z, x, and y.

$$36 = c \cdot 3 \cdot 2$$
$$6 = c$$

The equation of variation is $z = 6xy$.

Answer: $u = 3vw$

6 COMBINED VARIATION

Any of the preceding types of variation can be combined to form more complex variation equations. These **combined variations** are best understood by considering the following:

Statement	*Translates to*
y varies directly as x and inversely as z (There is only one constant of variation, c, and only one equation.)	$y = \dfrac{cx}{z}$
a varies directly as the square of b and inversely as the cube of t	$a = \dfrac{cb^2}{t^3}$
u varies jointly as v and the square of w and inversely as the square root of x	$u = \dfrac{cvw^2}{\sqrt{x}}$

To Solve a Variation Problem

1. Translate the variation word description to an equation involving the constant of variation.
2. Using given values for the variables, substitute into the variation equation and determine the constant of variation.
3. If needed, using the variation equation with the calculated constant of variation, find the missing value of one of the variables.

6.5 EXERCISES A

Solve.

1. If a boat uses 20 gal of gas to go 75 mi, how many miles can the boat travel on 36 gal of gas?

2. In an election the winning candidate won by a 5 to 3 ratio. If he received 820 votes, how many votes did the loser receive?

3. If 150 ft of wire weighs 45 lb, what will 240 ft of the same wire weigh?

4. If on a map $\frac{3}{4}$ in represents 20 mi, how many miles will be represented by 9 in?

5. If 3 lb of steak cost $8.25, how much will 5 lb of the steak cost?

6. What number must be added to each of 25, 35, 10, and 15 to give four numbers that are in proportion in that order?

7 A rope 65 ft long is cut into two pieces which have the ratio of 8:5. How long is each piece?

8. A boy 5 ft tall casts a shadow 12 ft long at the same time that a tower casts a shadow 252 ft long. How tall is the tower?

Write the equation of variation for each of the following.

9. u varies directly as t.

10. x varies directly as y and inversely as z.

11. u varies jointly as x, y, and z.

12. a varies directly as b and inversely as the product of f and g.

13 a varies directly as u and inversely as the product of v and w, and $a = 4$ when $u = 3$, $v = 3$, and $w = 4$.

14. z varies directly as the cube of x and inversely as the square root of y, and $z = 4$ when $x = 2$ and $y = 9$.

15. The current I in an electrical circuit having constant voltage varies inversely as the resistance R of the circuit.

16. The frequency f of a guitar string of a given length varies directly as the square root of the tension t of the string.

Solve.

17 The gravitational attraction A between two masses varies inversely as the square of the distance d between them. If the force of attraction is 64 lb when the masses are 9 ft apart, find the attraction when the masses are 24 ft apart.

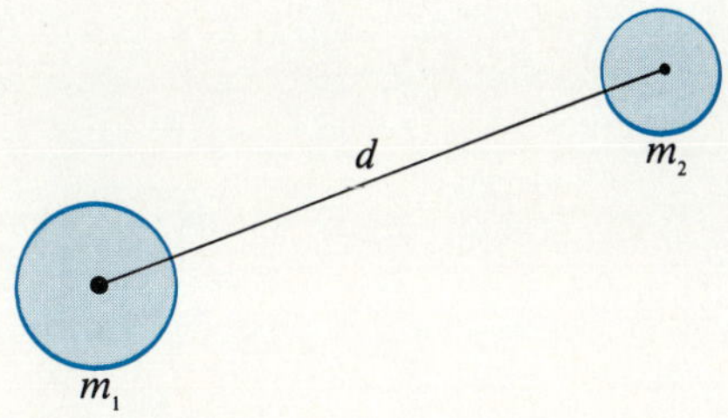

18. The period of vibration P of a pendulum varies directly as the square root of its length l. If the period of vibration of a pendulum of length 64 cm is $\frac{1}{3}$ sec, find the length of a pendulum with a period of $\frac{5}{24}$ sec.

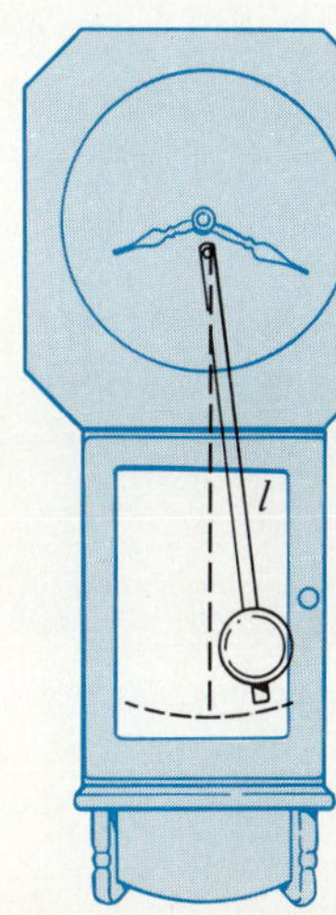

19. The intensity I of light varies inversely as the square of the distance d from the source. If the intensity of illumination on a screen 8 ft from a light is 5 foot-candles, find the intensity on a screen 16 ft from the light.

20 The weight w of a body above the surface of the earth varies inversely as the square of its distance d from the center of the earth. If an object weighs 150 lb on the surface of the earth, what would the object weigh 1000 mi above the surface of the earth? [*Hint:* Use 4000 mi for the radius of the earth.]

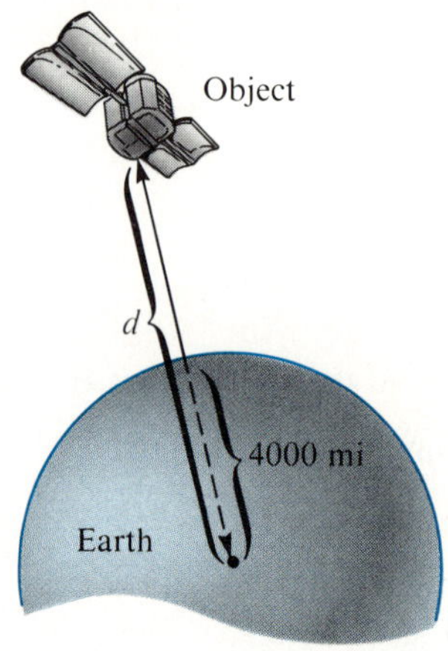

FOR REVIEW

Solve.

21. $\dfrac{1}{x^2 - 1} = \dfrac{1}{x - 1}$

22. $\dfrac{4z + 2}{z^2 - z} + \dfrac{5}{z - 1} = \dfrac{1}{z}$

Solve for the indicated variable.

23. $A = \dfrac{ra}{r + a}$ for r

24. $I = \dfrac{nm}{A + np}$ for p

Exercises 25–26 will help you prepare for the next section.

25. If Mike can do a job in 8 hours, how much of the job will he be able to do in **(a)** 1 hr, **(b)** 3 hr, **(c)** 8 hr?

26. Let x be the speed of a boat in miles per hour in still water. What would represent the actual speed of a boat traveling downstream if the speed of the current is 4 mph? What would represent the actual speed of the boat traveling upstream?

ANSWERS: 1. 135 mi 2. 492 votes 3. 72 lb 4. 240 mi 5. \$13.75 6. 5 7. 40 ft; 25 ft 8. 105 ft 9. $u = ct$ 10. $x = \frac{cy}{z}$ 11. $u = cxyz$ 12. $a = \frac{cb}{fg}$ 13. $a = \frac{16u}{vw}$ 14. $z = \frac{3x^3}{2\sqrt{y}}$ 15. $I = \frac{c}{R}$ 16. $f = c\sqrt{t}$ 17. 9 lb 18. 25 cm 19. $\frac{5}{4}$ foot-candles 20. 96 lb 21. 0 22. $-\frac{3}{8}$ 23. $r = \frac{aA}{a - A}$ 24. $p = \frac{nm - AI}{In}$ 25. (a) $\frac{1}{8}$ of the job (b) $\frac{3}{8}$ of the job (c) $\frac{8}{8}$ of the job, that is, the whole job 26. $x + 4$ mph; $x - 4$ mph

6.5 EXERCISES B

Solve.

1. If it takes 2 hours to type 10 pages, how many hours would it take to type 45 pages?

2. In a recent election, Representative Wettaw won by a 7 to 4 ratio. If his opponent got 6100 votes, how many votes did he get?

3. In a sample of 92 tires, 3 were defective. How many defective tires would you expect in a sample of 644 tires?

4. On a trail map, $\frac{1}{2}$ in represents 15 mi. How many miles are represented by $2\frac{1}{2}$ in?

5. If 4 lb of candy cost $5.40, how much will 7 lb of the candy cost?

6. What number when subtracted from 12, 22, 3, and 4 will make the resulting numbers, in that order, proportional?

7. A chemist wishes to make a mixture of concentrated hydrochloric acid and distilled water so that the ratio of acid to water is 7:3. If he starts with 21 L of acid, how much water should be mixed with the acid?

8. In two similar triangles, the sides of one are 6 in, 7 in, and 12 in. If the shortest side in the other is 18 in, find the other two sides.

Write the equation of variation for each of the following.

9. m varies jointly as n and p.

10. a varies jointly as b and the square root of d.

11. u varies directly with v and inversely with w.

12. y varies jointly as w and the square of x and inversely as the square root of z.

13. y varies jointly as x and z and inversely as the square of w, and $y = 18$ when $x = 2$, $z = 6$, and $w = 2$.

14. a varies jointly as s and the square of t and inversely as the cube of u, and $a = 1$ when $s = 2$, $t = 3$, and $u = 4$.

15. The volume V of a rectangular box with a fixed height varies jointly as its length l and width w.

16. The volume V of a cone varies jointly as its height h and the square of the radius r of its base.

Solve.

17. *The horsepower P* needed to move a ship varies directly as the cube of the speed s. If the horsepower needed for a speed of 15 mph is 23,625, find the horsepower needed for a speed of 20 mph.

18. The volume V of a given mass of gas varies directly as the temperature T and inversely as the pressure P. If $V = 360$ in^3 when $T = 450°$ and $P = 25$ lb/in^2, find the volume when $T = 350°$ and $P = 20$ lb/in^2.

19. The time T it takes to make an enlargement of a photo negative varies directly as the area A of the enlargement. If it takes 32 sec to make an 8-by-10 enlargement, find the time it takes to make a 5-by-7 enlargement.

20. The force of the wind blowing against the side of a building varies jointly as the surface area of the side and the square of the velocity of the wind. If a 4 mph wind exerts a force of 100 lb against the side of a building with surface area 100 ft^2, how much force will a wind of 20 mph exert against a surface with area 400 ft^2.

FOR REVIEW

Solve.

21. $\dfrac{a^2}{12 - a} = 1$

22. $\dfrac{14}{y^2 + 8y - 33} = \dfrac{y}{y - 3} - \dfrac{1}{y + 11}$

Solve for the indicated variable.

23. $\dfrac{A}{a} = \dfrac{B + b}{b}$ for b

24. $I = \dfrac{nm}{A + nP}$ for n

Exercises 25–26 will help you prepare for the next section.

25. If Stephanie can do a job in x hours, how much of the job will she be able to do in **(a)** 1 hr, **(b)** 4 hr, **(c)** x hr?

26. An airplane can fly 450 mph in calm air. What would represent the ground speed of the airplane if it is flying into a headwind with velocity x mph? What would represent the ground speed flying with the wind?

6.5 EXERCISES C

Solve.

1. To estimate the number of antelope in a game preserve, a ranger catches 40 antelope, tags their ears, and returns them to the preserve. Later on, he catches 25 antelope and finds 4 of them tagged. Use a proportion to estimate the number of antelope in the preserve.

2. The uniformly distributed safe load L that a horizontal beam can carry varies jointly as its width w and the square of its depth d, and inversely as its length l. If a 10-ft beam with width 2 ft and depth $\frac{1}{2}$ ft can carry a weight of 500 lb, how much can a 10-ft beam support if its width is 2 ft and its depth is 1 ft?
[Answer: 2000 lb]

3. When the means of a proportion are equal, as in $\frac{a}{b} = \frac{b}{c}$, b is called the **mean proportional** between a and c. The altitude h drawn to the hypotenuse c of a right triangle is the mean proportional between the segments of the hypotenuse (x and y). Find h if $x = 9$ inches and $y = 16$ inches. See the figure.

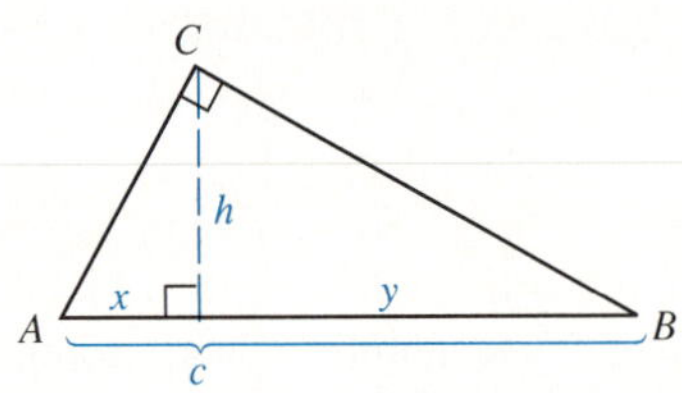

6.6 MORE APPLICATIONS USING RATIONAL EXPRESSIONS

STUDENT GUIDEPOSTS

1 Motion Problems
2 Work Problems

1 MOTION PROBLEMS

Some motion problems that use the formula

$$d = rt \quad \text{(distance = rate} \cdot \text{time)}$$

translate into fractional equations. When solving motion problems that involve two distances, two times, and two rates, it is helpful if we precisely describe each of these six quantities. Study the following examples carefully.

EXAMPLE 1 Motion Problem

The speed of a stream is 4 mph. A boat travels 48 mi upstream in the same time it takes to travel 72 mi downstream. What is the speed of the boat in still water?

Let x = speed of the boat in still water.

When the boat goes downstream, its speed relative to the stream bank is its speed in still water *increased* by the speed of the stream. Thus,

$x + 4$ = speed of the boat when traveling downstream.

When the boat travels upstream, its speed relative to the bank is its speed in still water *decreased* by the speed of the stream. Thus,

$x - 4$ = speed of the boat when traveling upstream,

72 = distance traveled downstream,

48 = distance traveled upstream.

Again we calculate the two times using $t = \frac{d}{r}$.

$$\frac{72}{x+4} = \text{time of travel downstream}$$

$$\frac{48}{x-4} = \text{time of travel upstream}$$

Since the two times are equal, we need to solve the equation

$$\frac{72}{x+4} = \frac{48}{x-4}.$$

$$72(x-4) = 48(x+4) \quad \text{Cross-product equation}$$

$$72x - 288 = 48x + 192$$

$$24x - 288 = 192 \quad \text{Subtract } 48x \text{ from both sides}$$

$$24x = 480 \quad \text{Add 288 to both sides}$$

$$x = 20$$

The speed of the boat in still water is 20 mph.

Practice Exercise 1

The speed of a plane in still air is 250 mph. If the plane can travel 846 mi with the wind in the same time that it can travel 654 mi against the wind, what is the wind speed?

Let x = wind speed,

$250 + x$ = rate with wind,

$250 - x$ = rate against wind.

Answer: 32 mph

CAUTION

In Example 1, a common mistake is to conclude that the speed upstream is $4 - x$ rather than $x - 4$. Which number, x or 4, must be larger if the boat actually makes progress upstream? Do you see why $x - 4$ is the correct rate?

EXAMPLE 2 Motion Problem

On a business trip, Terry McGinnis traveled 300 miles from Chicago. She averaged 65 mph for most of the trip but was slowed to 40 mph during a violent storm. If the total time of the trip was 5 hours, how many miles did she drive at 40 mph?

Let x = the number of miles driven at 40 mph,

$300 - x$ = the number of miles driven at 65 mph.

Practice Exercise 2

Kim Doane rode a bicycle 960 miles on a vacation. She averaged 50 miles per day most of the trip but was slowed to 35 miles per day after she pulled a muscle. If the total time of the trip was 21 days, how many days did she average 35 miles per day?

We can calculate the two different times driven at each rate using $t = \frac{d}{r}$. Then

$$\frac{x}{40} = \text{number of hours driven at 40 mph,}$$

$$\frac{300 - x}{65} = \text{number of hours driven at 65 mph.}$$

Since the total time of travel was 5 hours, we need to solve the following equation.

$$\frac{x}{40} + \frac{300 - x}{65} = 5$$

$$(5)(8)(13)\left[\frac{x}{40} + \frac{300 - x}{65}\right] = (5)(8)(13)(5) \qquad \text{The LCD is } (5)(8)(13)$$

$$13x + 8(300 - x) = 2600$$

$$13x + 2400 - 8x = 2600$$

$$5x = 200$$

$$x = 40$$

Thus, Terry drove 40 miles of the trip at the reduced rate of 40 mph.

Answer: 6 days

2 WORK PROBLEMS

Consider the following problem: Jim can do a job in 3 hr and Dave can do the same job in 7 hr. How long would it take them to do the job if they worked together? This type of problem is usually called a *work problem* and three principles must be kept in mind.

1. The time it takes to do a job when the individuals work together must be less than the time it takes for the fastest worker to complete the job alone. Thus the time together is *not* the average of the two times (which would be 5 hr in this case). Since Jim can do the job alone in 3 hr, with help the time must clearly be less than 3 hr.
2. If a job can be done in t hr, in 1 hr $\frac{1}{t}$ of the job will be completed. For example, since Jim can do the job in 3 hr, in 1 hr he would do $\frac{1}{3}$ of the job. Similarly, in 1 hr Dave would do $\frac{1}{7}$ of the job since he can do it all in 7 hr.
3. The work done by Jim in 1 hr added to the work done by Dave in 1 hr equals the amount of work done together in 1 hr. Thus, if t is the time it takes to complete the job working together,

   (amount Jim does in 1 hr) + (amount Dave does in 1 hr)
   = (amount done together in 1 hr)

   translates to

$$\frac{1}{3} + \frac{1}{7} = \frac{1}{t}.$$

We see that this word problem translates to a fractional equation. Multiply both sides by the LCD, $21t$.

$$21t\left[\frac{1}{3} + \frac{1}{7}\right] = 21t\left(\frac{1}{t}\right)$$

$$21t\left(\frac{1}{3}\right) + 21t\left(\frac{1}{7}\right) = 21$$
$$7t + 3t = 21$$
$$10t = 21$$
$$t = \frac{21}{10}$$

It would take Jim and Dave $\frac{21}{10}$ hr (2 hr 6 min) to complete the job working together. Does this seem reasonable in view of (1) above? Check.

A variation of this type of problem is to look for the time it takes one individual when given the time for the other and the time working together.

EXAMPLE 3 Work Problem

G & L Painters have the painting contract for all the new homes in a subdivision. When Gary and Lew work together, it takes 5 days to paint one house. If Gary paints one house by himself, it takes 7 days to complete the job. How long would it take Lew to paint one house if he worked alone?

We have the following information.

$5 =$ the number of days to paint one house together,
$\frac{1}{5} =$ the amount painted together in 1 day,
$7 =$ the number of days for Gary to paint one house,
$\frac{1}{7} =$ the amount painted by Gary in 1 day.
Let $x =$ the number of days for Lew to paint one house,
$\frac{1}{x} =$ the amount painted by Lew in 1 day.

(amount by Gary) + (amount by Lew) = (amount together)

$$\frac{1}{7} + \frac{1}{x} = \frac{1}{5}$$

msp,509Multiply both sides of this equation by the LCD, $35x$.

$$\mathbf{35x}\left(\frac{1}{7} + \frac{1}{x}\right) = \mathbf{35x}\left(\frac{1}{5}\right)$$
$$5x + 35 = 7x$$
$$35 = 2x$$
$$\frac{35}{2} = x$$

Thus, it would take Lew $17\frac{1}{2}$ days to paint one house by himself. Check.

Practice Exercise 3

Bob and Lennie can rebuild an engine in 3 days when they work together. If Bob can do the same job alone in 4 days, how long would it take Lennie to do it if he works alone?

Let $t =$ time for Lennie to do the job working alone

$$\frac{1}{4} + \frac{1}{t} = \_\_\_\_\_\_.$$

Answer: 12 days

CAUTION

When solving work problems (as with any word problem), be neat and complete. Do not take shortcuts, especially when writing down the pertinent information. Writing detailed descriptions of the variables can eliminate errors. Pattern your work after the examples.

The work problem on the next page results in a quadratic equation.

EXAMPLE 4 Work Problem

It takes pipe A 9 days longer to fill a reservoir than pipe B. If the two pipes are turned on together, they can fill the reservoir in 20 days. How long would it take each pipe to fill the reservoir alone?

Let t = the number of days required for B to fill the reservoir,

$t + 9$ = the number of days required for A to fill the reservoir,

20 = the number of days required to fill the reservoir together,

$\frac{1}{t}$ = the amount filled by B in 1 day,

$\frac{1}{t+9}$ = the amount filled by A in 1 day,

$\frac{1}{20}$ = the amount filled by A and B working together in 1 day.

We must solve $\frac{1}{t} + \frac{1}{t+9} = \frac{1}{20}$. The LCD = $20t(t + 9)$

$$20t(t+9)\frac{1}{t} + 20t(t+9)\frac{1}{t+9} = 20t(t+9)\frac{1}{20}$$

$$20(t+9) + 20t = t(t+9)$$

$$20t + 180 + 20t = t^2 + 9t$$

$$40t + 180 = t^2 + 9t$$

$$0 = t^2 - 31t - 180$$

This time we obtain a quadratic equation to solve.

$$(t+5)(t-36) = 0$$

$$t + 5 = 0 \quad \text{or} \quad t - 36 = 0$$

$$t = -5 \qquad\qquad t = 36$$

$$t + 9 = 45$$

Since -5 is meaningless for a time in a work problem, we have $t = 36$ as the only solution. Therefore it takes pipe B 36 days to fill the reservoir alone, and pipe A takes 45 days to fill it. Check.

Practice Exercise 4

When Matt and Wendy work together it takes 6 days to wash the windows in a building. If it takes Matt 9 days longer than Wendy to wash them by himself, how long would it take each to complete the job working alone?

Answer: Matt: 18 days; Wendy: 9 days

6.6 EXERCISES A

Solve.

1. A freight train travels 5 mph slower than a passenger train. If the freight train travels 260 mi in the same time that the passenger train travels 280 mi, find the speed of each.

Let x = speed of the passenger train,

$x - 5$ = speed of the freight train.

$260 =$

$280 =$

$\frac{280}{x} =$

$\frac{260}{x-5} =$

2. The speed of a stream is 5 mph. If a boat travels 50 mi downstream in the same time that it takes to travel 25 mi upstream, what is the speed of the boat in still water?

Let x = speed of the boat in still water,

$x - 5$ = speed of the boat going upstream.

$x + 5 =$

$50 =$

$25 =$

$\frac{50}{x+5} =$

$\frac{25}{x-5} =$

3. A plane flies 480 mi with the wind and 330 mi against the wind in the same length of time. If the speed of the wind is 25 mph, what is the speed of the plane in still air?

Let x = speed of the plane in still air.

$x + 25 =$

$x - 25 =$

$480 =$

$330 =$

4. The speed of a boat in still water is 24 mph. If the boat travels 54 mi upstream in the same time that it takes to travel 90 mi downstream, what is the speed of the stream?

5. A man walks a distance of 12 mi at a rate 8 mph slower than the rate he rides a bicycle for a distance of 24 mi. If the total time of the trip is 5 hr, how fast does he walk? How fast does he ride?
[*Hint:* $\frac{12}{x} + \frac{24}{x+8} = 5$]

6 A woman flies from Phoenix to Denver (a distance of 800 mi) at a rate 40 mph faster than on the return trip. If the total time of the trip is 9 hr, what was her rate going to Denver, and what was her rate returning to Phoenix?

7. The time of a plane trip of 450 mi, with the wind is three-fifths the time of the return trip against the wind. If the plane travels 120 mph in still air, what is the wind speed?

8 Dennis and Alberta rent a boat at the Riverside Campground at 9:00 A.M. They are told that due to the current of the river, the boat will travel 6 mph upstream and 12 mph downstream. If they must return the boat at 1:00 P.M. the same day, how far upstream can they travel, and what time must they turn back?

9. If Clyde can do a job in 8 days and Irv can do the same job in 3 days, how long would it take them to do the job together?

8 = the number of days for Clyde to do the job

$\frac{1}{8}$ = the amount Clyde does in 1 day

$3 =$

$\frac{1}{3} =$

Let t = the number of days to do the job together,

$\frac{1}{t}$ = the amount done in 1 day working together.

10. If pipe A can fill a tank in 20 hr and pipe B can fill the same tank in 15 hr, how long would it take to fill the tank if both pipes fill together?

$20 =$

$\frac{1}{20} =$

$15 =$

$\frac{1}{15} =$

Let $t =$ the number of hours to fill tank together.

$\frac{1}{t} =$

11. On the day that the Wisconsin cheese shipment arrives at Perko's Delicatessen, it takes Perko 3 hr to slice and display the cheese. When Perko and his assistant Graydon work together, it takes only 2 hr to process the same shipment. How long would it take Graydon to slice and display the cheese if he worked alone?

Let $t =$ time it takes Graydon to cut the cheese.

12. When Barb does a job by herself, it takes her 7 hr. If Barb and Wilma work together it takes 6 hr to do the same job. How long does it take Wilma to do the job if she works alone?

Let $t =$ time it takes Wilma to do the job.

13. When each works alone, Burford can mow a lawn in 3 hr less time than Ernie. When they work together, it takes 2 hr. How long does it take each to do the job by himself?

14. It takes Sybil 6 hr longer to paint a room than it takes Joanne. Working together, they can paint the room in 4 hr. How long would it take each working alone to paint the room?

15 A tank can be filled by an inlet pipe in 4 hr. It can be drained by an outlet pipe in 10 hr. If the inlet and outlet pipes are opened simultaneously, how long will it take to fill the empty tank?

16. When Arnie was a small child, he ''helped'' his mother plant onions. She could plant the entire garden in 3 hr. However, Arnie, without his mother realizing it, followed behind her, picking up the onion sets at a rate that would have completely unplanted the garden in 5 hr. Under these circumstances, how long would it have taken to plant the onions?

FOR REVIEW

Solve.

17. If a car used 16 gal of gas for a trip of 264 mi, how much gas will be used for a trip of 1100 mi?

18. A girl casts a shadow 5.5 ft long at the same time that a tree which is 35 ft tall casts a shadow 55 ft long. How tall is the girl?

Write the equation of variation for each of the following.

19. w varies directly as a and inversely as the square root of b.

20. u varies jointly as w and the cube of z and inversely as the square root of x.

Solve.

21. The gravitational attraction A between two masses varies inversely as the square of the distance d between them. If the force of attraction is 20 lb when the masses are 5 ft apart, find the attraction when the masses are 10 ft apart.

22. The resistance R of a wire at a constant temperature varies directly as the length L and inversely as the square of the diameter D. A section of wire having a diameter of 0.01 in and a length of 1 ft has a resistance of 8.2 ohms. Find the approximate resistance of a wire 1 mi in length with diameter 0.05 in.

Exercises 23 and 24 will help you prepare for the next section. Simplify each expression and write as a single fraction.

23. $\dfrac{\frac{1}{2}+1}{\frac{1}{2}-1}$

24. $\dfrac{\frac{3}{4}+\frac{1}{8}}{\frac{1}{2}-\frac{1}{4}}$

ANSWERS: 1. freight train: 65 mph; passenger train: 70 mph 2. 15 mph 3. 135 mph 4. 6 mph 5. walk: 4 mph; ride: 12 mph 6. 200 mph going; 160 mph returning 7. 30 mph 8. 16 miles; 11:40 A.M. 9. $\frac{24}{11}$ days 10. $\frac{60}{7}$ hr 11. 6 hr 12. 42 hr 13. Burford: 3 hr; Ernie: 6 hr 14. Sybil: 12 hr; Joanne: 6 hr 15. $\frac{20}{3}$ hr 16. $\frac{15}{2}$ hr 17. $66\frac{2}{3}$ gal 18. 3.5 ft 19. $w = \frac{ca}{\sqrt{b}}$ 20. $u = \frac{cwz^3}{\sqrt{x}}$ 21. 5 lb 22. 1732 ohms 23. -3 24. $\frac{7}{2}$

6.6 EXERCISES B

Solve.

1. Sam and Carolyn agree to meet in Las Vegas for the weekend. Sam travels 240 mi while Carolyn travels 220 mi. If Sam's rate of travel is 5 mph more than Carolyn's, and they travel the same length of time, at what speed does each travel?

2. John drives 12 mph faster than his mother. If John travels 310 mi in the same time that his mother travels 250 mi, find the speed of each.

3. A plane flies 1160 km with the wind and 840 km against the wind in the same length of time. If the speed of the plane in still air is 250 km/hr, what is the wind speed?

4. Merrilee can fly her plane 100 mph in still air. When the wind is blowing, she flies 345 mi with the wind in the same time that she flies 255 mi against the wind. What is the wind speed?

5. The Bells drive their car a distance of 150 mi at a rate of 30 mph faster than the rate they drive their boat a distance of 50 mi. If the total time of the trip is $5\frac{1}{2}$ hr, how fast do they drive each vehicle?

6. Craig Slack traveled from Newark, New Jersey to a convention in another city 300 mi away. Going, he traveled at a rate of 10 mph more than the rate returning. If the total time of the trip was 11 hr, what was his rate going and his rate returning?

7. A boat takes $1\frac{1}{2}$ times longer to go 72 mi upstream than to return. If the boat cruises at 30 mph in still water, what is the rate of the current?

8. Dick's boat can travel 35 mph in still water. If he leaves at 8:00 A.M., how far downstream can Dick travel, in a river with current of 5 mph, if he must be back to where he starts in 7 hr? At what time must he turn back upstream?

9. If Wylie Smith, Sports Information Director at a leading state university, can prepare a game program in 3 days and his student assistant can do the same job in 8 days, how long will it take them to prepare the program if they work together?

10. Nancy, a file clerk, can file a stack of folders in 75 min. Her assistant can file the same stack in 125 min. How long will it take them to do the job if they start together?

11. Ken can make a machine part in 5 hr. If he and his assistant work together it requires 3 hr to make the part. How long would it take the assistant by herself to make the part?

12. Pipe A can fill a tank in 8 hr. When both pipe A and B are turned on, the tank is filled in 2 hr. How long would be required for pipe B to fill the tank when it is turned on alone?

13. A new machine can produce 1000 units in 6 hr less time than the old machine. When they are both used, the 1000 units are produced in 4 hr. How long does it take each to produce 1000 units working alone?

14. It takes Howard twice as long to stock the freezer in a market as it does Mary. Together they can do the job in 3 hr. How long would it take each working alone?

15. A pond can be filled by an inlet pipe in 10 days and drained by an outlet pipe in 20 days. If the inlet and outlet pipes are opened simultaneously, how long will it take to fill the empty pond?

16. A barrel can be filled by an inlet pipe in 12 min, while it can be drained by an outlet pipe in 16 min. Through an error, both pipes are opened simultaneously. How long will it take to fill the empty barrel?

FOR REVIEW

Solve.

17. A man 6 ft tall casts a shadow 20 ft long. What is the length of the shadow cast at the same time by a building which is 51 ft tall?

18. In an election, the winning candidate won by a 5:3 margin. If he received 1250 votes, how many did the loser receive?

Write the equation of variation for each of the following.

19. y varies directly with x and inversely with the square of z.

20. m varies jointly as the square root of a and the square of b, and inversely as the cube of t.

Solve.

21. The period of vibration of a pendulum varies directly as the square root of its length l. If the period of a pendulum of length 25 cm is $\frac{1}{4}$ sec, find the period of a pendulum of length 100 cm.

22. The time T it takes an elevator to lift a weight w varies jointly as the weight and the distance d to be lifted and inversely as the power p of the motor. If it takes 24 sec for a 3-horsepower motor to lift 500 lb 60 ft, what power is needed to lift 700 lb 100 ft in 35 sec?

Exercises 23 and 24 will help you prepare for the next section. Simplify each expression and write as a single fraction.

23. $\dfrac{\frac{1}{4} - 1}{\frac{1}{4} + 1}$

24. $\dfrac{\frac{2}{5} + \frac{3}{10}}{\frac{1}{6} + \frac{5}{12}}$

6.6 EXERCISES C

Solve.

1. One of the events in the University Days Games involves swimming, running, and riding a bike over a fixed distance. The winner of the event, Mike Ratliff, averaged 2 mph swimming, 10 mph running, and 28 mph riding. The total distance was broken into three parts so that he swam $\frac{1}{20}$ of the total, ran $\frac{1}{4}$ of the total, and rode the remaining portion. If the winning time was 7.5 hr, what was the total distance covered?
[Answer: 100 mi]

2. A large reservoir can be filled by one inlet pipe in 5 days. It takes 8 days to fill it using a second pipe. The reservoir can be drained by an outlet pipe in 10 days. If all three pipes are opened simultaneously, how long will it take to fill the empty reservoir?

3. The Walters traveled to their mountain cabin located 330 miles away. For most of the trip they traveled 65 mph on an interstate highway, but for the final portion they drove on a winding mountain road averaging only 15 mph. If the total time of the trip was 7 hours, how many hours did they drive on the interstate?

4. Cynthia Allen traveled 340 miles west of Fargo. For most of the trip she averaged 60 mph, but for one period of time she was slowed to 20 mph due to a major accident. If the total time of travel was 7 hours, how many miles did she drive at the reduced speed?

6.7 SIMPLIFYING COMPLEX FRACTIONS

A fractional expression that contains at least one other fraction within it is called a **complex fraction.** The following are examples of complex fractions.

$$\frac{\frac{1}{5}}{\frac{2}{3}}, \quad \frac{a}{\frac{3}{4}}, \quad \frac{\frac{x}{y}}{7}, \quad \frac{2+\frac{1}{y}}{3}, \quad \frac{2+\frac{a}{b}}{2-\frac{a}{b}}$$

A complex fraction is **simplified** when all of its component fractions have been eliminated and a simple fraction obtained. There are two basic methods for simplifying a complex fraction. The first involves writing both the numerator and the denominator as single fractions and then dividing.

To Simplify a Complex Fraction (Method 1)

1. Change the numerator and denominator to single fractions.
2. Divide the two fractions.
3. Reduce to lowest terms.

EXAMPLE 1 SIMPLIFYING A COMPLEX FRACTION

Simplify $\dfrac{\frac{x}{y}+2}{2-\frac{x}{y}}$.

$$\frac{x}{y}+2=\frac{x}{y}+\frac{2}{1}=\frac{x}{y}+\frac{2\cdot y}{1\cdot y}=\frac{x+2y}{y}$$ Add the numerator fractions

$$2-\frac{x}{y}=\frac{2}{1}-\frac{x}{y}=\frac{2\cdot y}{1\cdot y}-\frac{x}{y}=\frac{2y-x}{y}$$ Subtract the denominator fractions

PRACTICE EXERCISE 1

Simplify $\dfrac{5+\frac{a}{b}}{\frac{a}{b}-5}$

Thus,

$$\frac{\frac{x}{y}+2}{2-\frac{x}{y}} = \frac{\frac{x+2y}{y}}{\frac{2y-x}{y}} = \frac{x+2y}{y} \div \frac{2y-x}{y} = \frac{x+2y}{y} \cdot \frac{y}{2y-x}$$

$$= \frac{(x+2y)\not{y}}{\not{y}(2y-x)} = \frac{x+2y}{2y-x}.$$

Answer: $\frac{5b+a}{a-5b}$

The second method involves clearing all fractions within the expression by multiplying numerator and denominator by the LCD of all the internal fractions.

To Simplify a Complex Fraction (Method 2)

1. Find the LCD of all fractions within the complex fraction.
2. Multiply numerator and denominator of the complex fraction by the LCD to obtain an equivalent fraction.
3. Reduce to lowest terms.

EXAMPLE 2 SIMPLIFYING A COMPLEX FRACTION

Simplify $\dfrac{1+\dfrac{y}{x-y}}{\dfrac{y}{x+y}-1}$.

$$\frac{\left(1+\frac{y}{x-y}\right)(x-y)(x+y)}{\left(\frac{y}{x+y}-1\right)(x-y)(x+y)}$$ Multiply by LCD, $(x-y)(x+y)$, and distribute

$$= \frac{(1)(x-y)(x+y) + \frac{y}{\cancel{x-y}}\cancel{(x-y)}(x+y)}{\frac{y}{\cancel{(x+y)}}(x-y)\cancel{(x+y)} - (1)(x-y)(x+y)}$$

$$= \frac{(x^2-y^2)+y(x+y)}{y(x-y)-(x^2-y^2)}$$ Use parentheses

$$= \frac{x^2-y^2+yx+y^2}{yx-y^2-x^2+y^2}$$ Watch signs

$$= \frac{x^2+yx}{yx-x^2} = \frac{\cancel{x}(x+y)}{\cancel{x}(y-x)}$$

$$= \frac{x+y}{y-x}$$

PRACTICE EXERCISE 2

Simplify $\dfrac{\dfrac{a}{a-b}-1}{1-\dfrac{a}{a+b}}$.

Answer: $\frac{a+b}{a-b}$

When we evaluate an algebraic expression, the result is often a complex fraction that must be simplified.

EXAMPLE 3 Evaluation Leading to a Complex Fraction

Evaluate $\frac{a-2}{a+3}$ when $a = \frac{x-y}{y}$.

Substitute $\frac{x-y}{y}$ for a.

$$\frac{\frac{x-y}{y}-2}{\frac{x-y}{y}+3} = \frac{\left(\frac{x-y}{y}-2\right)\cdot y}{\left(\frac{x-y}{y}+3\right)\cdot y}$$ Multiply by the LCD, y

$$= \frac{\frac{x-y}{\cancel{y}}\cdot\cancel{y}-2\cdot y}{\frac{x-y}{\cancel{y}}\cdot\cancel{y}+3\cdot y}$$ Distribute

$$= \frac{x-y-2y}{x-y+3y} = \frac{x-3y}{x+2y}$$

Practice Exercise 3

Evaluate $\frac{u-1}{u+1}$ when $u = \frac{a+b}{a}$.

Answer: $\frac{b}{2a+b}$

We often have complex fractions when using negative exponents. After applying the definition to remove the negative exponent, we simplify the result using Method 1 in the next example.

EXAMPLE 4 Physics Problem

The focal length f of a simple convex lens is given by

$$f = (d_1^{-1} + d_2^{-1})^{-1}$$

where d_1 is the distance from an object to the lens and d_2 is the distance from the lens to the image of the object. Remove negative exponents, simplify the complex fraction, and use the result to find f when $d_1 = 20$ inches and $d_2 = 12$ inches.

$$f = (d_1^{-1} + d_2^{-1})^{-1} = \frac{1}{d_1^{-1} + d_2^{-1}}$$ $u^{-1} = \frac{1}{u}$

$$= \frac{1}{\frac{1}{d_1} + \frac{1}{d_2}}$$

$$= \frac{1}{\frac{d_2}{d_1 d_2} + \frac{d_1}{d_1 d_2}}$$ The LCD of $\frac{1}{d_1}$ and $\frac{1}{d_2}$ is $d_1 d_2$.

$$= \frac{1}{\frac{d_2 + d_1}{d_1 d_2}} = 1 \div \frac{d_2 + d_1}{d_1 d_2}$$

$$= 1 \cdot \frac{d_1 d_2}{d_2 + d_1}$$ Multiply by the reciprocal of the divisor

$$= \frac{d_1 d_2}{d_2 + d_1}$$

Practice Exercise 4

Simplify $(x^{-1} - y^{-1})^{-1}$. Evaluate when $x = -2$ and $y = -6$.

To find f, substitute 20 for d_1 and 12 for d_2.

$$f = \frac{(20)(12)}{12 + 20} = \frac{240}{32} = 7.5$$

The focal length is 7.5 inches.

Answer: $\frac{xy}{y - x}$; -3

6.7 EXERCISES A

Simplify.

1. $\dfrac{\frac{x}{2}}{\frac{3}{x+1}}$

2. $\dfrac{\frac{a+1}{3}}{\frac{2a-1}{6}}$

3. $\dfrac{\frac{a}{b}}{\frac{b^2}{a+1}}$

4. $\dfrac{\frac{y+3}{5}}{\frac{(y+3)^2}{10}}$

5. $\dfrac{\frac{1}{a}+1}{\frac{1}{a}-1}$

6. $\dfrac{1-\frac{2}{3y}}{y-\frac{4}{9y}}$

7. $\dfrac{\frac{2}{a}-\frac{2}{b}}{\frac{1}{a^2}-\frac{1}{b^2}}$

8. $\dfrac{1-\frac{x}{y}}{y-\frac{x^2}{y}}$

9. $\dfrac{\frac{1}{a}-\frac{1}{b}}{\frac{a}{b}-\frac{b}{a}}$

10. $\dfrac{x-\frac{4}{x}}{1+\frac{2}{x}}$

11. $\dfrac{a-3+\frac{2}{a}}{a-4+\frac{3}{a}}$

12. $\dfrac{\frac{1}{x+y}-\frac{1}{x-y}}{\frac{-2}{x-y}}$

Evaluate and simplify.

13. $\dfrac{a+1}{a-1}$ for $a = \dfrac{x+y}{y}$

14. $\dfrac{u-1}{u+1}$ for $u = \dfrac{a}{b}$

15. $\dfrac{x^2+y^2}{xy}$ for $x = \dfrac{1}{u}$, $y = \dfrac{1}{v}$

Simplify.

16. $1 - \dfrac{1}{1 - \dfrac{1}{a}}$

17 $a - \dfrac{a}{1 - \dfrac{a}{1 - a}}$

18. $\dfrac{\dfrac{1}{xy} + \dfrac{1}{yz} + \dfrac{1}{xz}}{\dfrac{x + y + z}{xyz}}$

19. $x^{-1} + 1$

20. $u^{-1} + v^{-1}$

21. $a^{-2} + b^{-2}$

22. $(x^{-1} + 1)^{-1}$

23 $\dfrac{2x^{-1} + 2y^{-1}}{(xy)^{-1}}$

24. $\dfrac{u^{-1} + v^{-1}}{u^{-1} - v^{-1}}$

Use the formula $f = \dfrac{d_1 d_2}{d_2 + d_1}$ *to find the focal length f, to the nearest tenth of a centimeter, for the values of* d_1 *and* d_2 *given in Exercises 25–26.*

25 $d_1 = 8.0$ cm and $d_2 = 14.0$ cm

26. $d_1 = 16.0$ cm and $d_2 = 9.0$ cm

27 Calculating an average rate of speed uses a complex fraction. Suppose Marla hikes to the bottom of the Grand Canyon at a rate of 4 mph and returns to the top at a rate of 2 mph. What is her average rate for the hike? At first glance we might assume the average rate is 3 mph, the average of 4 and 2. However, her rate must be closer to 2 mph than to 4 mph since she walks for a longer period of time at this slower rate. Let D represent the distance hiked one direction. From the distance formula distance = (rate)(time), we have

$$\text{time} = \frac{\text{distance}}{\text{rate}}.$$

$$\text{time hiking down} = \frac{\text{distance down}}{\text{rate down}} = \frac{D}{4}$$

$$\text{time hiking up} = \frac{\text{distance up}}{\text{rate up}} = \frac{D}{2}$$

$$\text{average rate} = \frac{\text{total distance}}{\text{total time}} = \frac{2D}{\dfrac{D}{4} + \dfrac{D}{2}}$$

Simplify the complex fraction and find Marla's average rate on her hike. Give answer correct to the nearest tenth.

28. Repeat Exercise 27 assuming that Marla hiked down the canyon at a rate of 5 mph and returned to the top at a rate of 1.5 mph. Give answer correct to the nearest tenth.

FOR REVIEW

Solve.

29. The speed of a boat in still water is 30 km/hr. If the boat travels 72 km upstream in the same time that it takes to travel 108 km downstream, what is the speed of the stream?

30. Winston can prepare a window display in 4 hr and Wanda can do the same job in 3 hr. How long will it take to prepare the display if they work together?

Perform the indicated operations and simplify.

31. $\dfrac{x}{x-5}+\dfrac{2x+1}{5-x}$

32. $\dfrac{12}{z^3+8}+\dfrac{z}{z^2-2z+4}-\dfrac{1}{z+2}$

Exercises 33–36 review topics from Sections 1.7 and 5.1 to help you prepare for the next section. Perform the indicated operations.

33. $\dfrac{15x^3y^2}{5x^2y}$

34. $\dfrac{-22a^3b^4}{11ab^5}$

35. Subtract.
$$\begin{array}{r} 3x^2+2x \\ \underline{3x^2+3x} \end{array}$$

36. Subtract.
$$\begin{array}{r} 2y^3-5x^2 \\ \underline{2y^3-4x^2} \end{array}$$

ANSWERS: 1. $\frac{x(x+1)}{6}$ 2. $\frac{2(a+1)}{2a-1}$ 3. $\frac{a(a+1)}{b^3}$ 4. $\frac{2}{y+3}$ 5. $\frac{1+a}{1-a}$ 6. $\frac{3}{3y+2}$ 7. $\frac{2ab}{b+a}$ 8. $\frac{1}{y+x}$ 9. $\frac{-1}{a+b}$ 10. $x-2$ 11. $\frac{a-2}{a-3}$ 12. $\frac{y}{x+y}$ 13. $\frac{x+2y}{x}$ 14. $\frac{a-b}{a+b}$ 15. $\frac{v^2+u^2}{uv}$ 16. $\frac{-1}{a-1}$ 17. $\frac{-a^2}{1-2a}$ 18. 1 19. $\frac{1+x}{x}$ 20. $\frac{u+v}{uv}$ 21. $\frac{a^2+b^2}{a^2b^2}$ 22. $\frac{x}{1+x}$ 23. $2(x+y)$ 24. $\frac{v+u}{v-u}$ 25. 5.1 cm 26. 5.8 cm 27. 2.7 mph 28. 2.3 mph 29. 6 km/hr 30. $\frac{12}{7}$ hr 31. $\frac{x+1}{5-x}$ 32. $\frac{4}{z^2-2z+4}$ 33. $3xy$ 34. $-\frac{2a^2}{b}$ 35. $-x$ 36. $-x^2$

6.7 EXERCISES B

Simplify.

1. $\dfrac{\dfrac{3}{y}}{\dfrac{5}{y-1}}$

2. $\dfrac{\dfrac{b+3}{6}}{\dfrac{4b-5}{3}}$

3. $\dfrac{\dfrac{x}{2y}}{\dfrac{xy}{4}}$

4. $\dfrac{\dfrac{(a-1)^2}{3}}{\dfrac{a-1}{9}}$

5. $\dfrac{1+\dfrac{1}{x}}{1-\dfrac{1}{x^2}}$

6. $\dfrac{\dfrac{2}{z}+3}{\dfrac{4}{z^2}-9}$

7. $\dfrac{1-\dfrac{a^2}{b^2}}{1-\dfrac{a}{b}}$

8. $\dfrac{\dfrac{2}{x}-\dfrac{5}{y}}{\dfrac{4}{x^2}-\dfrac{25}{y^2}}$

9. $\dfrac{\dfrac{2}{a-2}+1}{\dfrac{2}{a+2}-1}$

10. $\dfrac{1-\dfrac{2}{x}-\dfrac{3}{x^2}}{1+\dfrac{1}{x}}$

11. $\dfrac{1-\dfrac{7}{x}+\dfrac{10}{x^2}}{1+\dfrac{1}{x}-\dfrac{6}{x^2}}$

12. $\dfrac{\dfrac{3}{2u-v}-\dfrac{3}{2u+v}}{\dfrac{-6v}{2u+v}}$

Evaluate and simplify.

13. $\dfrac{1-x}{x}$ for $x=\dfrac{a}{b}$

14. $\dfrac{x+1}{x-1}$ for $x=\dfrac{u}{u-v}$

15. $\dfrac{a+b}{a^2+b^2}$ for $a=\dfrac{1}{x}$, $b=\dfrac{1}{y}$

Simplify.

16. $1-\dfrac{1}{1-\dfrac{1}{x+1}}$

17. $\dfrac{\dfrac{x+1}{x-1}-\dfrac{x-1}{x+1}}{\dfrac{x+1}{x-1}+\dfrac{x-1}{x+1}}$

18. $\dfrac{\dfrac{a}{bc}+\dfrac{b}{ac}+\dfrac{c}{ab}}{\dfrac{1}{b^2c^2}+\dfrac{1}{a^2c^2}+\dfrac{1}{a^2b^2}}$

19. $a^{-1}-1$

20. $x^{-1}-y^{-1}$

21. $u^{-2}-v^{-2}$

22. $(a^{-1}-1)^{-1}$

23. $\dfrac{2a^{-1}-2b^{-1}}{(ab)^{-1}}$

24. $\dfrac{(x^{-1}-y^{-1})^{-1}}{xy}$

Use the formula $f=\dfrac{d_1d_2}{d_2+d_1}$ to find the focal length f, to the nearest tenth of a centimeter, for the values of d_1 and d_2 given in Exercises 25–26.

25. $d_1 = 5.0$ cm and $d_2 = 11.0$ cm

26. $d_1 = 18.0$ cm and $d_2 = 7.0$ cm

27. Finding an average rate of speed results in a complex fraction. Suppose a hiker walks a distance D at a rate of 4 mph and returns distance D at a rate of 3 mph. The average for the trip is

$$\frac{2D}{\dfrac{D}{4}+\dfrac{D}{3}}$$

Simplify the complex fraction to find the average rate walked by the hiker. Give answer correct to the nearest tenth.

28. Repeat Exercise 27 assuming the hiker walked one way at a rate of 5.2 mph and returned at a rate of 3.6 mph. Give answer correct to the nearest tenth.

FOR REVIEW

Solve.

29. A plane flies 1650 km with the wind and 1230 km against the wind in the same length of time. If the speed of the wind is 70 km/hr, what is the speed of the plane in still air?

30. When Simon paints a room by himself, it takes 5 hr. If Simon and Jane work together it takes 2 hr to paint the room. How long does it take Jane to do the job if she works alone?

Perform the indicated operations and simplify.

31. $\dfrac{1}{a^2-4a+4}-\dfrac{3-a}{a^2-4a+4}$

32. $\dfrac{2y-1}{y^2-4y+4}+\dfrac{2y+1}{4-y^2}$

Exercises 33–36 review topics from Sections 1.7 and 5.1 to help you prepare for the next section. Perform the indicated operations.

33. $\dfrac{12a^4b^6}{3a^5b^3}$

34. $\dfrac{-30x^3y^7}{10x^4y^2}$

35. Subtract.
$$\begin{array}{r} 2y^2 + 3y \\ \underline{2y^2 + 7y} \end{array}$$

36. Subtract.
$$\begin{array}{r} 4x^3 - 6x^2 \\ \underline{4x^3 - 3x^2} \end{array}$$

6.7 EXERCISES C

Simplify.

1. $\dfrac{\dfrac{1}{\frac{1}{x} - 1} + x}{\dfrac{1}{\frac{1}{x} + 1} + x}$

2. $\left[\dfrac{1 + \dfrac{a-1}{a+1}}{1 - \dfrac{a-1}{a+1}}\right]^{-2}$

3. $\left[\dfrac{(x^{-1} - y^{-1})^{-1}}{(x^{-1} + y^{-1})^{-1}}\right]^{-1}$ $\left[\text{Answer: } \dfrac{y-x}{y+x}\right]$

4. $\left[1 - \dfrac{1 - \dfrac{1}{a+1}}{1 + \dfrac{1}{a-1}}\right]^{-1}$ $\left[\text{Answer: } \dfrac{a+1}{2}\right]$

6.8 DIVISION OF POLYNOMIALS

STUDENT GUIDEPOSTS

1. Dividing by a Monomial
2. Dividing by a Binomial
3. Dividing by a Polynomial

1 DIVIDING BY A MONOMIAL

We learned how to add, subtract, and multiply polynomials in Chapter 5. Since the quotient of two polynomials is actually a rational expression, division was saved until now. We begin with dividing a polynomial by a monomial (a polynomial with only one term).

To Divide a Polynomial by a Monomial

Divide each term of the polynomial by the monomial and simplify the resulting sum or difference of fractions.

EXAMPLE 1 Dividing by a Monomial

Divide.

(a) $\dfrac{5x^3 + 10x^2 - 20x}{5x} = \dfrac{5x^3}{5x} + \dfrac{10x^2}{5x} - \dfrac{20x}{5x} = x^2 + 2x - 4$

Notice that the first step is the reverse of adding fractions with the same denominators.

PRACTICE EXERCISE 1

Divide.

(a) $\dfrac{12y^4 - 24y^3 + 6y}{6y}$

(b) $\dfrac{3a^3b^4 + 6a^2b^3 - 9ab^5 + 3ab}{3a^2b^2}$

$= \dfrac{3a^3b^4}{3a^2b^2} + \dfrac{6a^2b^3}{3a^2b^2} - \dfrac{9ab^5}{3a^2b^2} + \dfrac{3ab}{3a^2b^2}$

$= ab^2 + 2b - \dfrac{3b^3}{a} + \dfrac{1}{ab}$ Remember the rules of exponents

(b) $\dfrac{4u^4v^5 + 16u^3v^3 - 8u^2v^4 + 12u^5v^5}{4u^4v^4}$

Answers: (a) $2y^3 - 4y^2 + 1$
(b) $v + \dfrac{4}{uv} - \dfrac{2}{u^2} + 3uv$

When dividing by a monomial, practice will allow you to skip the middle step and proceed directly to the final answer.

2 DIVIDING BY A BINOMIAL

Dividing a polynomial by a binomial (a polynomial with two terms) is more complex than dividing by a monomial. The method is shown in the examples that follow.

To Divide a Polynomial by a Binomial

1. Arrange the terms of both in descending order.
2. Divide the first term of the polynomial (the dividend) by the first term of the binomial (the divisor) to obtain the first term of the quotient.
3. Multiply the first (new) term of the quotient by the binomial and subtract the result from the dividend. Bring down the next term to obtain a new polynomial which becomes the new dividend.
4. Divide the first term of the new dividend polynomial by the first term of the binomial and continue this process until the variable in the first term of the remainder dividend is raised to a lower power than the variable in the first term of the divisor.

EXAMPLE 2 Dividing by a Binomial

Divide $x^2 - 15 + 2x$ by $x - 3$.

1. Arrange terms in descending order. $x - 3\overline{)x^2 + 2x - 15}$
2. Divide the first term of the polynomial by the first term of the binomial.

equals

$$\begin{array}{r} x \\ x - 3\overline{)x^2 + 2x - 15} \end{array}$$

divided by

$x^2 \div x = x$

3. Multiply the first term of the quotient by the binomial and subtract the results from the dividend. Bring down the next term to obtain a new dividend.

$$\begin{array}{r} x \\ x - 3\overline{)x^2 + 2x - 15} \\ \underline{x^2 - 3x} \\ 5x - 15 \end{array}$$

$x(x - 3) = x^2 - 3x$

Subtract $x^2 - 3x$ from $x^2 + 2x$ by changing the signs on $x^2 - 3x$ and adding; then bring down -15

PRACTICE EXERCISE 2

Divide $a^2 - 5a - 14$ by $a + 2$.

4. Divide the new dividend polynomial by the binomial using Steps 2 and 3.

$$
\begin{array}{rll}
 & x + 5 & \text{Divide } 5x - 15 \text{ by } x - 3 \\
x - 3 \,\big) & x^2 + 2x - 15 & \\
 & \underline{x^2 - 3x} & \\
 & 5x - 15 & \\
 & \underline{5x - 15} & \text{Multiply } x - 3 \text{ by } 5 \\
 & 0 & \text{No variable in the new dividend; the process terminates}
\end{array}
$$

The quotient is $x + 5$. To check, we multiply the quotient $x + 5$ by the divisor $x - 3$ to obtain the dividend $x^2 + 2x - 15$.

Answer: $a - 7$

EXAMPLE 3 Dividing by a Binomial

Divide $21y + 18y^2 + 40 + 20y^3$ by $7 + 5y$.

First, arrange terms in descending order of y.

$$
\begin{array}{rll}
 & 4y^2 - 2y + 7 & \leftarrow \text{The quotient of } 20y^3 \text{ and } 5y\text{; the quotient of } -10y^2 \text{ and } 5y\text{; the quotient of } 35y \text{ and } 5y \\
5y + 7 \,\big) & 20y^3 + 18y^2 + 21y + 40 & \\
 & \underline{20y^3 + 28y^2} & \leftarrow \text{The quotient } 4y^2 \text{ times the divisor } 5y + 7 \\
 & -10y^2 + 21y & \text{Subtract } 20y^3 + 28y^2 \text{ from } 20y^3 + 18y^2 \text{ and bring down } 21y \\
 & \underline{-10y^2 - 14y} & \leftarrow \text{The quotient } -2y \text{ times } 5y + 7 \\
 & 35y + 40 & \text{Subtract } -10y^2 - 14y \text{ from } -10y^2 + 21y \text{ and bring down } 40 \\
 & \underline{35y + 49} & \leftarrow \text{The quotient } 7 \text{ times } 5y + 7 \\
 & -9 & \text{Subtract; no variable in new dividend}
\end{array}
$$

The answer can be expressed as either $4y^2 - 2y + 7$ with remainder -9, or $4y^2 - 2y + 7 - \frac{9}{5y+7}$. To check this, multiply the quotient $4y^2 - 2y + 7$ by the divisor $5y + 7$ and add the remainder -9. The result is the dividend $20y^3 + 18y^2 + 21y + 40$.

Practice Exercise 3

Divide $20x^2 - 6x + 16x^3 - 10$ by $3 + 4x$.

Answer: $4x^2 + 2x - 3 - \frac{1}{4x+3}$

When terms are missing, either leave space for them or write them with zero coefficients. This is illustrated in the next example.

EXAMPLE 4 Dividing by a Binomial

Divide $a^3 + 8$ by $a + 2$.

Either leave space for the missing terms or write them with zero coefficients.

$$
\begin{array}{rl}
 & a^2 - 2a + 4 \\
a + 2 \,\big) & a^3 \qquad\qquad + 8 \\
 & \underline{a^3 \div 2a^2} \\
 & -2a^2 \\
 & \underline{-2a^2 - 4a} \\
 & 4a + 8 \\
 & \underline{4a + 8} \\
 & 0
\end{array}
\qquad\qquad
\begin{array}{rl}
 & a^2 - 2a + 4 \\
a + 2 \,\big) & a^3 + 0a^2 + 0a + 8 \\
 & \underline{a^3 + 2a^2} \\
 & -2a^2 + 0a \\
 & \underline{-2a^2 - 4a} \\
 & 4a + 8 \\
 & \underline{4a + 8} \\
 & 0
\end{array}
$$

You should recognize that $a + 2$ and $a^2 - 2a + 4$ are the factors of the sum of cubes $a^3 + 8$.

Practice Exercise 4

Divide $y^3 - 64$ by $y - 4$.

Answer: $y^2 + 4y + 16$

To divide polynomials with two variables, we arrange the terms of each in order of descending powers of one variable and divide as before.

EXAMPLE 5 Dividing Polynomials in Two Variables

$(2x^2 + 5xy - 3y^2) \div (2x - y)$

Here the terms are arranged in descending order relative to the variable x.

$$\begin{array}{r}
x + 3y \\
2x - y \,\overline{)\, 2x^2 + 5xy - 3y^2} \\
\underline{2x^2 - xy} \\
6xy - 3y^2 \\
\underline{6xy - 3y^2} \\
0
\end{array}$$

We could also arrange in descending order relative to y and then divide.

$$\begin{array}{r}
3y + x \\
-y + 2x \,\overline{)\, -3y^2 + 5xy + 2x^2} \\
\underline{-3y^2 + 6xy} \\
-xy + 2x^2 \\
\underline{-xy + 2x^2} \\
0
\end{array}$$

In either case, the quotient is $x + 3y$, and $2x^2 + 5xy - 3y^2 = (2x - y)(x + 3y)$.

Practice Exercise 5

Divide $6a^2 + 5ab + b^2$ by $3a + b$.

Answer: $2a + b$

3 DIVIDING BY A POLYNOMIAL

The procedure for dividing a polynomial by a binomial extends to division by other polynomials. This is illustrated in the next example.

EXAMPLE 6 Dividing by a Polynomial

Divide $z^4 - 3z^3 + 2z - 5$ by $z^2 - z + 1$.

$$\begin{array}{r}
z^2 - 2z - 3 \\
z^2 - z + 1 \,\overline{)\, z^4 - 3z^3 + 2z - 5} \\
\underline{z^4 - z^3 + z^2} \\
-2z^3 - z^2 + 2z \\
\underline{-2z^3 + 2z^2 - 2z} \\
-3z^2 + 4z - 5 \\
\underline{-3z^2 + 3z - 3} \\
z - 2
\end{array}$$

Thus, the quotient is $z^2 - 2z - 3$ with remainder of $z - 2$. Alternatively, we could give the answer as

$$z^2 - 2z - 3 + \frac{z - 2}{z^2 - z + 1}.$$

To check this, multiply and add $z - 2$.

Practice Exercise 6

Divide $x^4 + 3x^3 + x^2 - 2x - 5$ by $x^2 + x - 2$.

Answer: $x^2 + 2x + 1 + \frac{x - 3}{x^2 + x - 2}$

6.8 EXERCISES A

Divide.

1. $\dfrac{28a^5 + 44a^4 - 8a^3}{-4a}$

2. $\dfrac{5x^5 - 30x^4 + 25x^3 - 5}{5x^2}$

3. $(-27a^5 + 9a^3 - 81a^2) \div (-3a)$

4. $\dfrac{-72y^3 + 24y^2 + 88y}{-8y}$

5. $\dfrac{15a^4b^3 - 3a^2b^2 + 9a^3b}{-3a^2b^2}$

6. $\dfrac{5x^2y^3z - 10x^3y^2z^5 + 20xyz^4}{-5xy^2z^3}$

7. $a - 2\overline{)3a^2 - 8a + 4}$

8. $(4a - 12 + a^2) \div (a + 6)$

9. $(8y^3 - 1) \div (2y - 1)$

10. $2x - 3\overline{)4x^4 \qquad - 7x^2 + x + 3}$

11. $\dfrac{4y^4 - 7y^2 + y + 1}{2y - 1}$

12. $x + 2\overline{)x^3 + 2x^2 - x - 2}$

13. $\dfrac{9a^3 + 3a^2 + 7a + 1}{3a + 2}$

14. $(5xy + x^2 + 6y^2) \div (x + 2y)$

15. $(a^3 + b^3) \div (a + b)$

16. $\dfrac{x^4 + 3x^3 + 2x^2 - x + 5}{x^2 - x + 1}$

17. $\dfrac{z^4 + z^3 + 4}{z^3 - z + 1}$

18. $\dfrac{2y^4 + 4y^3 - y + 2}{2y^3 + 1}$

19. $3y^2 - 1\overline{)9y^4 + 12y^2 - 5}$

20. $2x^2 + 3\overline{)6x^4 + 2x^3 + 5x^2 - x + 1}$

21 Find a value for p so that the polynomial $y^3 - y^2 + 2y + p$ is divisible by $y - 1$ with remainder zero.

22. Find a value for m so that when the polynomial $2x^3 + x^2 - 4x + m$ is divided by $2x + 1$, the remainder is 5.

23. The area of a rectangle is $6m^2 + 13m + 6$. Find the width if the length is $3m + 2$.

24. If a boat travels a distance of $a^3 + 2a^2 - 40a + 25$ mi in $a - 5$ hr, find its rate of speed.

FOR REVIEW

Simplify.

25. $\dfrac{y - \dfrac{9}{y}}{1 + \dfrac{3}{y}}$

26. $\dfrac{\dfrac{3}{a} + \dfrac{7}{b}}{\dfrac{9b}{a} - \dfrac{49a}{b}}$

27. Evaluate $\dfrac{1 - a}{a - 1}$ for $a = \dfrac{x + y}{x - y}$.

28. Simplify $\dfrac{(u^{-1} + 2v^{-1})^{-1}}{2uv}$.

ANSWERS: 1. $-7a^4 - 11a^3 + 2a^2$ 2. $x^3 - 6x^2 + 5x - \frac{1}{x^2}$ 3. $9a^4 - 3a^2 + 27a$ 4. $9y^2 - 3y - 11$ 5. $-5a^2b + 1 - \frac{3a}{b}$ 6. $\frac{-xy}{z^2} + 2x^2z^2 - \frac{4z}{y^2}$ 7. $3a - 2$ 8. $a - 2$ 9. $4y^2 + 2y + 1$ 10. $2x^3 + 3x^2 + x + 2 + \frac{9}{2x - 3}$ 11. $2y^3 + y^2 - 3y - 1$ 12. $x^2 - 1$ 13. $3a^2 - a + 3 - \frac{5}{3a + 2}$ 14. $x + 3y$ 15. $a^2 - ab + b^2$ 16. $x^2 + 4x + 5$ 17. $z + 1 + \frac{z^2 + 3}{z^3 - z + 1}$ 18. $y + 2 - \frac{2y}{2y^3 + 1}$ 19. $3y^2 + 5$ 20. $3x^2 + x - 2 + \frac{-4x + 7}{2x^2 + 3}$ 21. $p = -2$ 22. $m = 3$ 23. $2m + 3$ 24. $a^2 + 7a - 5$ 25. $y - 3$ 26. $\frac{1}{3b - 7a}$ 27. -1 28. $\frac{1}{2(v + 2u)}$

6.8 EXERCISES B

Divide.

1. $\dfrac{45y^3 - 15y^2 + 20y}{-5y}$

2. $\dfrac{30a^6 - 36a^5 + 24a^4 - 6}{6a^3}$

3. $\dfrac{13z^5 + 39z^4 - 169}{-13z^2}$

4. $(-16x^3 + 28x^2 - 8x) \div (-4x)$

5. $\dfrac{35x^5y^3 - 49x^4y^5 + 28x^2y}{-7x^3y^2}$

6. $\dfrac{49a^3b^2c - 21a^3bc^3 - 28a^2b}{-7a^2bc^2}$

7. $(x^3 - 27) \div (x - 3)$

8. $x - 1\overline{)x^3 \quad + x - 3}$

9. $c + 3\overline{)4c^2 + 7c - 15}$

10. $2x - 5\overline{)12x^3 + 4x^2 - 41x + 140}$

11. $\dfrac{6c^4 - 18c^2 - 8c - 8}{2c - 4}$

12. $3a + 2\overline{)6a^3 + 13a^2 + 24a + 12}$

13. $\dfrac{16x^3 + 8x^2 + x - 3}{4x + 3}$

14. $3a + b\overline{)3a^3 + a^2b - 6a - 2b}$

15. $(x^3 - y^3) \div (x - y)$

16. $\dfrac{y^4 + 5y^3 + y^2 - 13y + 6}{y^2 + 2y - 3}$

17. $\dfrac{a^4 + 3a^2 + 7}{a^3 + a^2 - 2}$

18. $\dfrac{6x^4 + 3x^3 - 1}{3x^3 - 1}$

19. $4z^2 + 1\overline{)12z^4 + 7z^2 + 1}$

20. $3y^2 - 2\overline{)9y^4 + 6y^3 + 12y^2 + 2y + 7}$

21. Find a value for m so that the polynomial $x^3 + x^2 + x + m$ is divisible by $x + 1$ with remainder zero.

22. Find a value for p so that when the polynomial $3x^3 + x^2 + 4x + p$ is divided by $3x - 2$, the remainder is -3.

23. The area of a rectangle is $8u^2 - 10u - 3$. Find the length if the width is $2u - 3$.

24. Linda walks a distance of $2x^3 + 3x^2 - 5x - 3$ km in $2x + 1$ hr. Find her walking rate.

FOR REVIEW

Simplify.

25. $\dfrac{x + \dfrac{5}{x}}{\dfrac{6}{x} - 1}$

26. $\dfrac{\dfrac{5a}{b} - \dfrac{2b}{a}}{\dfrac{8}{a} + \dfrac{11}{b}}$

27. Evaluate $\dfrac{u^2}{u + 1}$ for $u = \dfrac{x}{y}$.

28. Simplify $\dfrac{x^{-1} + y^{-1}}{x^{-1} - y^{-1}}$.

6.8 EXERCISES C

Divide.

1. $a^2 + ab + b^2\overline{)a^5 + 2a^4b + 2a^3b^2 + 2a^2b^3 + ab^4 + b^5}$

2. Determine the value of a so that $x^4 + 2x^2 + a$ is divisible by $x - 1$ with remainder of -1. [Answer: $a = -4$]

3. Evaluate $x^3 - 2x^2 + x - 3$ when $x = 2$. Divide $x^3 - 2x^2 + x - 3$ by $x - 2$ and compare the value of the remainder with the value obtained first.

4. Evaluate $x^4 - 2x^2 + 7$ when $x = -1$. Divide $x^4 - 2x^2 + 7$ by $x + 1$ and compare the value of the remainder with the value obtained first. [Answer: Both are 6.]

6.9 SYNTHETIC DIVISION

Ordinary long division of a polynomial by a binomial of the form $x - a$, presented in the previous section, can be shortened using **synthetic division.** Suppose that we divide $2x^3 - 3x^2 + 3x - 4$ by $x - 2$.

$$\begin{array}{r l}
2x^2 + x + 5 & \text{The quotient is } 2x^2 + x + 5 \\
x - 2\overline{)2x^3 - 3x^2 + 3x - 4} & \\
\underline{2x^3 - 4x^2} & \\
x^2 + 3x & \\
\underline{x^2 - 2x} & \\
5x - 4 & \\
\underline{5x - 10} & \\
6 & \text{The remainder is 6}
\end{array}$$

We could shorten our work by writing only the coefficients since the position of each coefficient tells us what power of x is associated with it.

$$\begin{array}{r l}
2 + 1 + 5 & 2 + 1 + 5 \text{ corresponds to } 2x^2 + x + 5 \\
1 - 2\overline{)2 - 3 + 3 - 4} & \\
\underline{2 - 4} & \\
1 + 3 & \\
\underline{1 - 2} & \\
5 - 4 & \\
\underline{5 - 10} & \\
6 &
\end{array}$$

The first term in each line above (in color) is unnecessary since it is exactly the same as the number directly above it. Hence, we abbreviate further by eliminating these numbers.

$$\begin{array}{r}
2 + 1 + 5 \\
1 - 2\overline{)2 - 3 + 3 - 4} \\
\underline{- 4} \\
1 + 3 \\
\underline{- 2} \\
5 - 4 \\
\underline{- 10} \\
6
\end{array}$$

The 1 in the divisor $1 - 2$ is unnecessary since we are no longer writing the first number in other lines which it was used to obtain. Also, the process of "bringing down" the next term merely wastes space since all subtractions could be performed in one line, as shown in color below.

$$\begin{array}{r}
2 + 1 + 5 \\
-2\overline{)2 - 3 + 3 - 4} \\
\underline{- 4 - 2 - 10} \\
1 + 5 + 6
\end{array}$$

The initial 2 in the quotient $2 + 1 + 5$ is always the same as the initial term in the dividend $2 - 3 + 3 - 4$, and the remaining terms in the quotient $(1 + 5)$ are duplicated in front of the remainder (which is 6) in the bottom line $1 + 5 + 6$. Thus, we can abbreviate further by writing the quotient only in the bottom line by first "bringing down" the 2.

$$\begin{array}{r}
-2\overline{)2 - 3 + 3 - 4} \\
\underline{- 4 - 2 - 10} \\
2 + 1 + 5 + 6
\end{array}$$

Finally, since the division lines are unnecessary, we omit them. Also, since most of us are able to add more quickly than we subtract, by changing the sign of the divisor (from -2 to $+2$) we can obtain the coefficients in the quotient by addition instead of subtraction.

$$\begin{array}{r|rrrrrrr} +2 & 2 & - & 3 & + & 3 & - & 4 \\ & & + & 4 & + & 2 & + & 10 \\ \hline & 2 & + & 1 & + & 5 & + & 6 \end{array}$$

The first 2 is brought down below the line and multiplied by the divisor $+2$, and the product, 4, is placed above the line. Add -3 to 4, obtaining 1, multiply 1 by the divisor $+2$, and place the product, 2, above the line. Add 3 to 2, obtaining 5, multiply 5 by the divisor $+2$, and place the product, 10, above the line. The final addition of -4 and 10 gives 6, the last entry below the line, which is the remainder, while $2 + 1 + 5$ corresponds to the quotient $2x^2 + x + 5$.

We now summarize the steps to follow for synthetic division.

Synthetic Division

To divide a polynomial by $x - a$:

Step 1: Write the coefficients of the polynomial in descending order with a 0 for any missing term.

Step 2: Write $\underline{a|}$ to the left of the coefficients.

Step 3: Bring down the first coefficient, multiply it by a, place the product under the second coefficient, and add it to the second coefficient.

Step 4: Multiply the sum found in Step 3 by a, place the product under the next coefficient, and add.

Step 5: Continue this process until finished.

Step 6: The quotient has coefficients found in the last row with the first term of degree one less than the degree of the polynomial. The last number in the row is the remainder.

Suppose we illustrate these steps by dividing $x^3 + 5x^2 + 1$ by $x + 3$.

$$\begin{array}{rrrr} 1 & 5 & 0 & 1 \end{array}$$

Step 1: Since the x^2-term is missing, write 0.

$$\begin{array}{r|rrrr} -3 & 1 & 5 & 0 & 1 \end{array}$$

Step 2: Since $x - a = x + 3 = x - (-3)$, $a = -3$.

$$\begin{array}{r|rrrr} -3 & 1 & 5 & 0 & 1 \\ & & -3 & & \\ \hline & 1 & 2 & & \end{array}$$

Step 3: Bring down 1, multiply 1 by -3, write the product -3 under 5 and add to obtain 2.

$$\begin{array}{r|rrrr} -3 & 1 & 5 & 0 & 1 \\ & & -3 & -6 & \\ \hline & 1 & 2 & -6 & \end{array}$$

Step 4: Multiply 2 by -3, write the product -6 under 0 and add to obtain -6.

$$\begin{array}{r|rrrr} -3 & 1 & 5 & 0 & 1 \\ & & -3 & -6 & 18 \\ \hline & 1 & 2 & -6 & 19 \end{array}$$

Step 5: Continue the process. The final number 19 is the remainder.

$$x^2 + 2x - 6$$

Step 6: Write the quotient and remainder.

The quotient is $x^2 + 2x - 6$ with a remainder of 19.

Of course all of these steps are performed in one writing so your work will appear as in Step 5.

CAUTION

Remember that in synthetic division the dividing number, a, comes from $x - a$, and a may be positive or negative. For example, to divide by $x - 3$, $a = 3$ and is positive; and to divide by $x + 5 = x - (-5)$, $a = -5$ and is negative. Simply change the sign of the number in the dividing binomial and place it in the position next to the coefficients of the polynomial.

EXAMPLE 1 Using Synthetic Division

Use synthetic division to find $(x^4 - 2x^2 + x + 3) \div (x - 1)$.

Be sure to supply the zero in the position of the missing x^3 term. The polynomial is $x^4 + 0x^3 - 2x^2 + x + 3$, and $a = 1$.

$$\begin{array}{r|l} 1 & 1 + 0 - 2 + 1 + 3 \\ & \; 1 + 1 - 1 + 0 \\ \hline & 1 + 1 - 1 + 0 + 3 \end{array}$$

The quotient is $x^3 + x^2 - x$ and the remainder is 3.

Practice Exercise 1

Use synthetic division to find $(y^4 + y^3 - 10y - 4) \div (y - 2)$.

Answer: $y^3 + 3y^2 + 6y + 2$

CAUTION

When dividing a polynomial by a binomial using synthetic division, arrange the polynomial in descending powers of the variable and insert a zero for each of the terms with zero coefficients.

EXAMPLE 2 Using Synthetic Division

Use synthetic division to find $(3x^5 - 2x^3 + 7 - x + x^2) \div (x + 2)$.

The divisor is $x + 2$ so we have $x + 2 = x - (-2)$. Be sure to supply the coefficient zero and rearrange the terms in descending order. The polynomial is $3x^5 + 0x^4 - 2x^3 + x^2 - x + 7$.

$$\begin{array}{r|l} -2 & 3 + 0 - 2 + 1 - 1 + 7 \\ & \; - 6 + 12 - 20 + 38 - 74 \\ \hline & 3 - 6 + 10 - 19 + 37 - 67 \end{array}$$

The quotient is $3x^4 - 6x^3 + 10x^2 - 19x + 37$, and the remainder is -67.

With practice, the process of synthetic division becomes quite mechanical, and can be completed in a very short time.

Practice Exercise 2

Use synthetic division to find $(5a^4 + 2a^2 + 4a^5 + 4a + 5) \div (a + 1)$.

Answer: The quotient is $4a^4 + a^3 - a^2 + 3a + 1$ with remainder 4.

Remember that synthetic division works only when the divisor binomial is of the form $x - a$. For example, we could not use synthetic division to find $(3x^4 - x^3 + x^2 + 4x - 7) \div (x^2 - 1)$ since the divisor has an x^2-term.

6.9 EXERCISES A

Use synthetic division to find the quotient and remainder in each of the following.

1. $(x^3 - x^2 + 2x - 3) \div (x - 1)$

2. $(2x^3 - 3x + 5) \div (x - 2)$

3. $(x^3 - x^2 - 5x + 2) \div (x + 2)$

4. $(x^3 + 5x^2 + 4x - 6) \div (x + 3)$

5. $(x^4 - 2x^3 + x^2 - x + 5) \div (x + 2)$

6. $(x^4 + 2x^3 - 4x^2 + 2x - 2) \div (x - 1)$

7 $(x^3 - x + x^4 - 3) \div (x + 1)$

8. $(2x - 2x^3 - x^2 + x^4 - 5) \div (x - 2)$

9. $(2x^5 - 3x^4 - 6x^3 + x + 8) \div (x - 3)$

10. $(x^5 - 2x^4 - 10x^3 + 9x^2 - 5x + 4) \div (x - 4)$

11. $(6x^3 - x + 4) \div \left(x - \frac{1}{2}\right)$

12. $(8x^3 + 2x + 3) \div \left(x + \frac{1}{2}\right)$

13 $(3x^5 + x^2 - 5) \div (x + 1)$

14. $(2x^4 - 3x^2 + x + 7) \div (x^2 - 1)$

FOR REVIEW

Divide.

15. $\dfrac{42ab^2c - 60a^2bc + 6ab}{-6a^2b^2}$

16. $2x + 3\overline{)4x^3 + 6x^2 + 10x + 8}$

17. $(x^4 - 1) \div (x - 1)$

18. $\dfrac{x^4 + 3x^2 + 6}{x^2 - x + 3}$

Exercises 19–24 review material covered in Sections 1.2 and 1.6 to help you prepare for the next section. Find each absolute value.

19. $|0|$

20. $|7|$

21. $|-7|$

22. $|x|$ if $x > 0$

23. $|x|$ if $x < 0$

24. $|x|$ if $x = 0$

ANSWERS: 1. $x^2 + 2; -1$ 2. $2x^2 + 4x + 5; 15$ 3. $x^2 - 3x + 1; 0$ 4. $x^2 + 2x - 2; 0$ 5. $x^3 - 4x^2 + 9x - 19; 43$ 6. $x^3 + 3x^2 - x + 1; -1$ 7. $x^3 - 1; -2$ 8. $x^3 - x; -5$ 9. $2x^4 + 3x^3 + 3x^2 + 9x + 28; 92$ 10. $x^4 + 2x^3 - 2x^2 + x - 1; 0$ 11. $6x^2 + 3x + \frac{1}{2}; \frac{17}{4}$ 12. $8x^2 - 4x + 4; 1$ 13. $3x^4 - 3x^3 + 3x^2 - 2x + 2; -7$ 14. cannot be done using synthetic division 15. $-\frac{7c}{a} + \frac{10c}{b} - \frac{1}{ab}$ 16. $2x^2 + 5 - \frac{7}{2x+3}$ 17. $x^3 + x^2 + x + 1$ 18. $x^2 + x + 1 + \frac{-2x+3}{x^2 - x + 3}$ 19. 0 20. 7 21. 7 22. x 23. $-x$ 24. 0

6.9 EXERCISES B

Use synthetic division to find the quotient and remainder in each of the following.

1. $(x^3 - 2x^2 + x - 5) \div (x - 1)$

2. $(2x^3 + 5x - 8) \div (x - 2)$

3. $(x^3 + 2x^2 - 3x - 6) \div (x + 2)$

4. $(x^3 + 8x^2 + 10x - 15) \div (x + 3)$

5. $(x^4 - 3x^3 + 2x^2 + 2x - 3) \div (x + 2)$

6. $(x^4 + 4x^3 - 2x^2 - 5x + 7) \div (x - 1)$

7. $(x^2 - 8x + x^4 - 2) \div (x + 1)$

8. $(9 - 3x^2 + x^3 - x^4 + 2x) \div (x - 2)$

9. $(x^5 - 6x^4 + 4x^3 + x^2 - x + 3) \div (x - 3)$

10. $(2x^5 - 7x^4 - 15x^2 - x + 5) \div (x - 4)$

11. $(4x^3 - 2x^2 + 8x - 2) \div \left(x - \frac{1}{2}\right)$

12. $(8x^3 + 4x - 5) \div \left(x + \frac{1}{2}\right)$

13. $(6x^6 + x^5 - 2) \div (x + 1)$

14. $(3x^5 - 5x^4 + x^2 - 8) \div (x^2 + 1)$

FOR REVIEW

Divide.

15. $\dfrac{35x^2y^2z - 49x^2z^2 + 14yz^2}{7xy^2}$

16. $3x + 2\overline{)6x^3 - 5x^2 - 8}$

17. $(x^5 - 1) \div (x - 1)$

18. $\dfrac{x^4 + 5x^3 + x^2 + 1}{x^2 + x - 5}$

Exercises 19–24 review material from Sections 1.2 and 1.6 to help prepare for the next section. Find each absolute value.

19. $|5 - 5|$

20. $|10|$

21. $|-10|$

22. $|y|$ if $y = 0$

23. $|y|$ if $y < 0$

24. $|y|$ if $y > 0$

6.9 EXERCISES C

Use synthetic division to find the quotient and remainder in each of the following.

1. $(5.6x^3 + 6.2x^2 - 7.2x + 1.35) \div (x - 0.5)$ [Answer: $5.6x^2 + 9x - 2.7; 0$]

2. $(1.05x^4 - 8.04672x^2 + 80.604x - 80.0236) \div (x + 5.08)$

3. Use synthetic division to find the remainder when $4x^4 - 3x^2 + 2x - 1$ is divided by $x + 2$. Evaluate the polynomial when $x = -2$ and compare this result with the remainder. [Answer: Both are 47.]

4. Use synthetic division to show that $x - 1$ is a factor of $x^4 + x^3 - x^2 - 2x + 1$.

CHAPTER 6 REVIEW

KEY WORDS

6.1 An **algebraic fraction** is a fraction that contains a variable in the numerator, the denominator, or both.

A **rational expression** is an algebraic fraction that is the ratio of two polynomials.

Two algebraic fractions are **equivalent** if one can be obtained from the other by multiplying or dividing both numerator and denominator by the same nonzero expression.

A fraction is **reduced to lowest terms** when 1 (or -1) is the only number or expression that divides both numerator and denominator.

A factor is **prime** when its only divisors are itself and 1.

6.2 When one expression is divided by a second, the second expression is called the **divisor.**

6.3 **Like fractions** have the same denominator. **Unlike fractions** have different denominators.

The **least common denominator (LCD)** of two fractions is the smallest expression that has both denominators as factors.

6.4 A **fractional equation** is an equation that contains one or more algebraic fractions or rational expressions.

6.5 The **ratio** of one number a to another number b is the quotient $\frac{a}{b}$.

An equation stating that two ratios are equal is called a **proportion.**

Two triangles are **similar** if the angles of one are equal respectively to the angles of the other.

6.7 A fractional expression that contains at least one other fraction within it is called a **complex fraction.**

KEY CONCEPTS

6.1
1. Fundamental principle of fractions: If the numerator and denominator of an algebraic fraction are multiplied or divided by the same nonzero expression or number, the resulting fraction is equivalent to the original.
2. When reducing fractions, divide out factors only, *never divide out terms.*

6.2
1. When multiplying or dividing fractions, factor numerators and denominators and divide out common factors. Do *not* find the LCD.
2. The reciprocal of $a + b$ is $\frac{1}{a+b}$ and *not* $\frac{1}{a} + \frac{1}{b}$.

6.3
1. When subtracting fractions, use parentheses to avoid sign errors.
2. Before finding the LCD of several fractions, reduce all fractions to lowest terms.
3. Do not divide out terms when adding or subtracting fractions. Only factors can be divided out.
4. Form the cross-product equation only when two fractions are equated.

6.4
1. To solve a fractional equation, multiply both sides by the LCD of all fractions. Check all answers in the original equation and exclude any that result in division by zero.
2. Do not confuse simplifying an algebraic expression with solving a fractional equation.

6.5
1. Means-extremes property: In a proportion, the product of the extremes is equal to the product of the means.
2. The means-extremes property is used to solve a proportion by forming the cross-product equation.
3. y varies directly as x: $y = cx$
4. y varies inversely as x: $y = \frac{c}{x}$
5. z varies jointly as x and y: $z = cxy$

6.7 To simplify a complex fraction, multiply all terms by the LCD of all fractions involved.

6.8 When dividing a polynomial by a binomial, remember to subtract at each intermediate step. *Do not add.*

6.9 When using synthetic division, remember to add at each step. *Do not subtract.*

REVIEW EXERCISES

Part I

6.1

1. Find the values for the variable that must be excluded.

$$\frac{x+2}{x(x-5)}$$

2. Are the given fractions equivalent?

$$\frac{2}{x-y}, \frac{-2}{y-x}$$

Reduce to lowest terms. (Assume values that make the denominator zero are excluded.)

3. $\dfrac{6x^3y^3z^2}{2xy^5z^2}$

4. $\dfrac{8u^3-1}{1-2u}$

6.2 *Multiply.*

5. $\dfrac{x^2+2xy+y^2}{x-2} \cdot \dfrac{x^2-4}{(x+y)^2}$

6. $\dfrac{a^2+2ab+b^2}{a^2-b^2} \cdot \dfrac{b-a}{4b+4a}$

Divide.

7. $\dfrac{x^2-x-6}{x^2-36} \div \dfrac{2x-6}{x^2-4x-12}$

8. $\dfrac{a^2-ab-2b^2}{a^2-3ab+2b^2} \div \dfrac{2a+b}{a-b}$

Perform the indicated operations.

9. $\dfrac{2u^2+uv-v^2}{u^2-uv} \cdot \dfrac{u^2-v^2}{4u^2-v^2} \div \dfrac{u^2+2uv+v^2}{v^2+2uv}$

6.3 *Perform the indicated operations.*

10. $\dfrac{3a}{a+2} - \dfrac{2a}{a+2}$

11. $\dfrac{4x+2}{x-6} + \dfrac{3x-5}{6-x}$

Find the LCD of each pair of fractions.

12. $\dfrac{6}{5x^2}, \dfrac{4}{3x}$

13. $\dfrac{2y}{y^2-5y+6}, \dfrac{y+3}{y^2-6y+9}$

Perform the indicated operations.

14. $\dfrac{b}{a^2+ab+b^2} + \dfrac{a^2-ab}{a^3-b^3}$

15. $\dfrac{7x}{x^2-2x+1} - \dfrac{3}{x-1}$

16. $\dfrac{a^2 - 3ab}{a^2 - b^2} - \dfrac{-2ab - b^2}{2a^2 - ab - b^2}$

17. $\dfrac{-8}{y^2 - y - 6} + \dfrac{y}{y^2 - 6y + 9} - \dfrac{1}{y + 2}$

6.4 *Solve.*

18. $1 - \dfrac{a + 2}{a - 1} = \dfrac{3}{a}$

19. $3y = \dfrac{16}{y - 1} - \dfrac{10}{y - 1}$

20. The reciprocal of 8 more than a number is three times the reciprocal of the number. Find the number.

21. Three times the reciprocal of a number is the same as the number increased by 2. Find the number.

Solve each equation for the indicated variable.

22. $\dfrac{1}{a + b} = c$ for a

23. $x^{-1} + \dfrac{1}{y} = \dfrac{1}{z}$ for y

6.5 *Solve.*

24. If 150 ft of wire weighs 45 lb, what will 270 ft of the same wire weigh?

25. If a car travels 700 mi on 40 gal of gas, how many miles can it travel on 56 gal of gas?

26. The volume V of a given mass of gas varies directly as the temperature T and inversely as the pressure P. If $V = 400$ in^3 when $T = 350°$ and $P = 25$ lb/in^2, find the pressure when $T = 420°$ and $V = 600$ in^3.

27. The intensity I of light varies inversely as the square of the distance d from the source. If the intensity of illumination on a screen 15 ft from a light is 3.5 foot-candles, find the intensity on a screen 5 ft from the light.

6.6 *Solve.*

28. A plane takes $2\frac{1}{2}$ times as long to fly between two cities against the wind as with the wind. If the cities are 500 mi apart and the speed of the plane in still air is 280 mph, find the speed of the wind.

29. The speed of a river is 2 mph. If a boat can travel 104 mi upstream in the same time it can travel 120 mi downstream, what is the speed of the boat in still water?

30. If pipe A can fill a tank in 5 hr and pipe B requires 2 hr to fill the same tank, how long would it take to fill the tank if the pipes are turned on together?

31. If Lizette can do a job in 12 hr less time than Bo, and together they can do the job in 8 hr, how long would it take each to do it alone?

6.7 *Simplify.*

32. $\dfrac{\dfrac{x}{y} - \dfrac{y}{x}}{\dfrac{1}{y} - \dfrac{1}{x}}$

33. $\dfrac{(a^{-1} + 2b^{-1})^{-1}}{\dfrac{ab}{2}}$

6.8 *Divide.*

34. $\dfrac{75z^5 - 50z^4 + 100z^2}{-25z^3}$

35. $\dfrac{64a^3b^3c^3 - 8a^2bc^4 + 24a^2b^2}{8a^2b^2c^2}$

36. $x - 4\overline{)5x^3 - 18x^2 - 14x + 24}$

37. $2x + 1\overline{)4x^3 - 4x^2 - x + 5}$

6.9 *Use synthetic division to find the quotient and remainder in each of the following.*

38. $(x^4 - 3x^3 + 2x^2 - 1) \div (x + 1)$

39. $(x^5 - 5x^3 - 5x + 2) \div (x - 2)$

Part II

40. If $\frac{3}{4}$ in on a map represents 20 mi, how many miles will be represented by 12 in?

41. Are the given fractions equivalent?
$$\frac{x}{a}, \frac{x+7}{a+7}$$

42. Simplify.
$$\frac{x - 1 - \dfrac{6}{x}}{1 + \dfrac{2}{x}}$$

43. Find the LCD of the pair of fractions.
$$\frac{x+3}{x^3 - 8}, \frac{3x}{x^2 + 2x + 4}$$

44. Suppose y varies jointly as x and the square of z and inversely as the square root of w. Find the equation of variation if $y = 10$ when $x = 3$, $z = 2$, and $w = 36$.

45. Find the values for the variable that must be excluded.
$$\frac{a+8}{a^3 + a^2 - 2a}$$

46. Reduce to lowest terms.
$$\frac{b - w}{w - b}$$

47. Add.
$$\frac{6}{x+y} + \frac{3x+y}{x^2 - y^2}$$

48. Perform the indicated operations and simplify.

$$\frac{x^3 - y^3}{x^2 - 3x + xy - 3y} \cdot \frac{x^2 + 2xy - 3x - 6y}{x^3 + x^2y + xy^2} \div \frac{x^2 + xy - 2y^2}{x^3 - xy^2}$$

49. It takes 5 hours to fill the Bonnett's swimming pool and 10 hours to drain it. If the pool is empty, the filler pipe is turned on but the drain is accidentally left open, how many hours will it take to fill the pool?

50. Larry is 6 years older than Amy and the ratio of their ages is $\frac{7}{6}$. How old is each?

51. Solve. $\dfrac{1}{a} + \dfrac{a}{a+1} = \dfrac{a^2}{a^2 + a}$

52. Solve the gravitational force formula

$$F = G\frac{m_1 m_2}{d^2} \quad \text{for } G.$$

53. Perform the indicated operations and simplify.

$$\frac{x}{x^2 + x - 6} + \frac{2}{2 + x - x^2} + \frac{3}{x^2 + 4x + 3}$$

54. Divide. $x - 3\overline{)x^3 - 2x^2 + 5x - 8}$

55. The wind speed is 50 mph. An airplane can travel a distance of 1750 miles with the wind in the same time that it can travel 1250 miles against the wind. What is the speed of the airplane in calm air?

56. The stopping distance d of a car varies directly as the square of the speed v it is traveling. If a car traveling 50 mph needs 150 ft to stop, what is the stopping distance of a car traveling 100 mph?

57. Multiply. $\dfrac{x^2 + x - 2}{x^2 + 6x + 5} \cdot \dfrac{x^2 - 2x - 3}{x^2 - x - 6}$

58. Use synthetic division to find the quotient and remainder.

$$(x^4 - 3x^2 + 2x - 5) \div (x + 2)$$

59. Write $\dfrac{3x}{x - 2}$ as an equivalent fraction with denominator $3x^2 - 12$.

60. Divide. $\dfrac{42x^3y^2 + 36x^4y - 72xy^5}{-6x^3y^4}$

ANSWERS: 1. 0, 5 2. yes 3. $\frac{3x^2}{y^2}$ 4. $-4u^2 - 2u - 1$ 5. $x + 2$ 6. $-\frac{1}{4}$ 7. $\frac{(x+2)^2}{2(x+6)}$ 8. $\frac{a+b}{2a+b}$ 9. $\frac{v}{u}$ 10. $\frac{a}{a+2}$ 11. $\frac{x+7}{x-6}$ 12. $15x^2$ 13. $(y-2)(y-3)^2$ 14. $\frac{a+b}{a^2+ab+b^2}$ 15. $\frac{4x+3}{(x-1)^2}$ 16. $\frac{a-b}{a+b}$ 17. $\frac{15}{(y-3)^2(y+2)}$ 18. $\frac{1}{2}$ 19. 2, −1 20. −12 21. 1 or −3 22. $a = \frac{1-bc}{c}$ 23. $y = \frac{xz}{x-z}$ 24. 81 lb 25. 980 mi 26. 20 lb/in^2 27. 31.5 foot-candles 28. 120 mph 29. 28 mph 30. $\frac{10}{7}$ hr 31. Lizette: 12 hr; Bo: 24 hr 32. $x + y$ 33. $\frac{2}{b+2a}$ 34. $-3z^2 + 2z - \frac{4}{z}$ 35. $8abc - \frac{c^2}{b} + \frac{3}{c^2}$ 36. $5x^2 + 2x - 6$ 37. $2x^2 - 3x + 1 + \frac{4}{2x+1}$ 38. $x^3 - 4x^2 + 6x - 6$; 5 39. $x^4 + 2x^3 - x^2 - 2x - 9$; −16 40. 320 mi 41. no 42. $x - 3$ 43. $x^3 - 8$ 44. $y = \frac{5xz^2}{\sqrt{w}}$ 45. 0, 1, −2 46. −1 47. $\frac{9x-5y}{(x-y)(x+y)}$ 48. $x - y$ 49. 10 hr 50. Larry is 42, Amy is 36. 51. no solution 52. $G = \frac{d^2F}{m_1m_2}$ 53. $\frac{x^2-2x-12}{(x-2)(x+3)(x+1)}$ 54. $x^2 + x + 8 + \frac{16}{x-3}$ 55. 300 mph 56. 600 ft 57. $\frac{x-1}{x+5}$ 58. $x^3 - 2x^2 + x$; −5 59. $\frac{9x(x+2)}{3x^2-12}$ 60. $-\frac{7}{y^2} - \frac{6x}{y^3} + \frac{12y}{x^2}$

CHAPTER 6 TEST

1. Find the values for the variable that must be excluded in $\dfrac{3x}{x^2 - 1}$. **1.** ______

2. Reduce $\dfrac{x^3 - y^3}{x - y}$ to lowest terms. **2.** ______

Perform the indicated operations.

3. $\dfrac{a^2 - a - 2}{a^2 + 3a} \cdot \dfrac{a^2 - 3a}{a^2 - 5a + 6}$ **3.** ______

4. $\dfrac{u^2 - v^2}{u^3 + v^3} \div \dfrac{u - v}{u^2 - uv + v^2}$ **4.** ______

5. $\dfrac{x^3 - y^3}{x^2 - xy + 5x - 5y} \cdot \dfrac{x^2 + 3x - 10}{x^2 + 7x} \div \dfrac{x^3 + x^2y + xy^2}{x^2 + 5x - 14}$ **5.** ______

6. $\dfrac{3}{x + 1} + \dfrac{4 - 2x}{x^2 - 1}$ **6.** ______

7. $\dfrac{3a}{a^2 - 2ab + b^2} - \dfrac{3}{a - b}$ **7.** ______

8. $\dfrac{a}{a^2 + 5a + 6} + \dfrac{2}{4 - a^2} - \dfrac{1}{a + 3}$ **8.** ______

Solve.

9. $a + \dfrac{1}{a + 2} = \dfrac{4}{a + 2}$ **9.** ______

10. Jeff is 2 years older than Sue and the ratio of their ages is $\frac{6}{5}$. How old is each? **10.** ______

CHAPTER 6 TEST

CONTINUED

11. Solve $\dfrac{2}{x + y} = w$ for y.

11. ____________

12. Suppose y varies jointly as x and the square root of z, and inversely as w. Find the equation of variation if $y = 5$ when $x = 3$, $z = 25$, and $w = 6$.

12. ____________

13. An older machine can do a job in 104 minutes; it takes 40 minutes when a newer machine works simultaneously. How long would it take the newer machine working alone to do the job?

13. ____________

14. The speed of a stream is 3 mph. If a canoe can travel 27 mi upstream in the same time it can travel 45 mi downstream, what is the speed of the canoe in still water?

14. ____________

15. Simplify. $\dfrac{x + 2 - \dfrac{15}{x}}{1 + \dfrac{5}{x}}$

15. ____________

Divide.

16. $\dfrac{24a^5 - 16a^3 - 4a^2}{-4a^2}$

16. ____________

17. $2a + 1\overline{)8a^3 + 2a^2 + 5a + 3}$

17. ____________

18. Use synthetic division to find the quotient and remainder.
$(x^4 + 2x^3 - 2x^2 - 4x - 3) \div (x + 1)$

18. ____________

CHAPTER 7

Radicals and Exponents

7.1 ROOTS AND RADICALS

STUDENT GUIDEPOSTS

1. Square Roots
2. Cube Roots
3. Higher Order Roots
4. Approximating Square Roots

1 SQUARE ROOTS

We begin this section by reviewing several concepts presented in Chapter 1. Remember that if a and x are real numbers such that

$$x = a \cdot a = a^2,$$

then a is a square root of x. If a is an integer, then x is a perfect square. Every *positive* real number has two square roots since if a is a square root of x, then

$$x = a^2 \quad \text{and} \quad x = (-a)^2.$$

Thus, both a and $-a$ are square roots of x. We called the positive square root the principal square root and indicated it by using a radical, $\sqrt{}$. For example, the principal square root of 9 is $\sqrt{9}$ or 3. To indicate the negative square root of 9 using a radical, we write $-\sqrt{9} = -3$. For both square roots of 9 we write $\pm\sqrt{9}$ or ± 3, where ± 3 means both 3 and -3.

Even though every positive real number has two square roots, 0 has only one, since $-0 = 0$. Negative real numbers have no real square roots. For example, $\sqrt{-25}$ is not a real number since no real number squared is -25. In general, since $a^2 \geq 0$ for each real number a, *no negative number can have a real square root*.

Notice that

$$\sqrt{7^2} = \sqrt{49} = 7 \quad \text{and} \quad \sqrt{(-7)^2} = \sqrt{49} = 7.$$

Thus, if $x = 7$, a positive number, then $\sqrt{x^2} = x$, but if $x = -7$, a negative number, then $\sqrt{x^2} = -x = -(-7) = 7$. Remember the definition of absolute value from Chapter 1.

$$|x| = \begin{cases} x \text{ if } x \geq 0 \\ -x \text{ if } x < 0 \end{cases}$$

Since $\sqrt{x^2} = x$ or $\sqrt{x^2} = -x$, with absolute value we can define square roots in a single expression.

Square Roots

Let x be any real number. Then
$$\sqrt{x^2} = |x|.$$

EXAMPLE 1 Simplifying Square Roots

Simplify the square roots.

(a) $\sqrt{36} = \sqrt{6^2} = |6| = 6$ $\quad 6^2 = 36$

(b) $\sqrt{(-6)^2} = |-6| = 6$

(c) $\sqrt{-36}$ is not a real number.

(d) $\sqrt{0^2} = |0| = 0$ $\quad 0^2 = 0$

(e) $\sqrt{(2x)^2} = |2x| = |2||x| = 2|x|$
This follows by a property of absolute value: $|ab| = |a||b|$

(f) $\sqrt{(a+1)^2} = |a+1|$

(g) $\sqrt{y^2 - 4y + 4} = \sqrt{(y-2)^2} = |y-2|$

(h) $-\sqrt{u^2} = -|u|$

Practice Exercise 1

Simplify the square roots.

(a) $\sqrt{100}$

(b) $\sqrt{(-10)^2}$

(c) $\sqrt{-100}$

(d) $-\sqrt{0^2}$

(e) $\sqrt{(3y)^2}$

(f) $\sqrt{(x-3)^2}$

(g) $\sqrt{a^2 + 10a + 25}$

(h) $-\sqrt{y^2}$

Answers: (a) 10 (b) 10 (c) not a real number (d) 0 (e) $3|y|$ (f) $|x-3|$ (g) $|a+5|$ (h) $-|y|$

2 CUBE ROOTS

We can also define a cube root of a number. If x and a are real numbers such that
$$x = a^3,$$
then a is a **cube root** of x. If a is an integer, we call x a **perfect cube.** Unlike square roots, there is only one cube root of a number, and negative numbers have cube roots. We denote the cube root by $\sqrt[3]{}$. Thus,
$$\sqrt[3]{8} = \sqrt[3]{2^3} = 2 \quad \text{and} \quad \sqrt[3]{-8} = \sqrt[3]{(-2)^3} = -2.$$

Cube Roots

Let x be any real number. Then
$$\sqrt[3]{x^3} = x.$$

Notice that when finding cube roots, absolute value is not needed since a real number has only one cube root. The cube root of a negative number is negative, while the cube root of a positive number is positive.

EXAMPLE 2 Simplifying Cube Roots

Simplify the cube roots.

(a) $\sqrt[3]{27} = \sqrt[3]{3^3} = 3$ $\quad 3^3 = 27$

(b) $\sqrt[3]{-27} = \sqrt[3]{(-3)^3} = -3$ $\quad (-3)^3 = -27$

(c) $\sqrt[3]{125} = \sqrt[3]{5^3} = 5$ $\quad 5^3 = 125$

Practice Exercise 2

Simplify the cube roots.

(a) $\sqrt[3]{8}$

(b) $\sqrt[3]{-8}$

(c) $\sqrt[3]{216}$

(d) $\sqrt[3]{-64} = \sqrt[3]{(-4)^3} = -4$ $(-4)^3 = -64$

(e) $\sqrt[3]{1} = \sqrt[3]{1^3} = 1$ $1^3 = 1$

(f) $\sqrt[3]{0} = \sqrt[3]{0^3} = 0$ $0^3 = 0$

(g) $\sqrt[3]{(4a)^3} = 4a$

(d) $\sqrt[3]{-216}$

(e) $-\sqrt[3]{1}$

(f) $-\sqrt[3]{0}$

(g) $-\sqrt[3]{(4a)^3}$

Answers: (a) 2 (b) -2 (c) 6 (d) -6 (e) -1 (f) 0 (g) $-4a$

3 HIGHER ORDER ROOTS

We may also define fourth roots, fifth roots, and so on. In general, if x and a are real numbers such that

$$x = a^k, \quad k \text{ is a natural number greater than 1}$$

then a is a **kth root** of x. If a is an integer, x is a **perfect kth power.** The symbol $\sqrt[k]{x}$ indicates the **principal kth root** of x. k is the **index** on the radical. The index must be specified on all radical expressions except square roots, where the index is understood to be 2.

Just as with square roots, when k is an even number, only nonnegative numbers can have kth roots. Likewise, when k is even, principal kth roots are defined using absolute value. For example,

$$\sqrt[4]{2^4} = |2| = 2 \quad \text{and} \quad \sqrt[4]{(-2)^4} = |-2| = 2.$$

However, we saw that negative numbers can have cube roots, and cube roots do not involve absolute value. The same is true for all kth roots where k is an odd number.

Higher Order Roots

Let x be any real number, with k a natural number.

1. **If k is even, $\sqrt[k]{x^k} = |x|$.**
2. **If k is odd, $\sqrt[k]{x^k} = x$.**

EXAMPLE 3 Simplifying Higher Order Roots

Simplify.

(a) $\sqrt[4]{81} = \sqrt[4]{3^4} = |3| = 3$ $3^4 = 81$

(b) $\sqrt[4]{-81}$ is not a real number. Notice that no real number can be multiplied by itself four times and have a negative product like -81.

(c) $\sqrt[4]{(-3)^4} = |-3| = 3$

(d) $\sqrt[5]{243} = \sqrt[5]{3^5} = 3$ $3^5 = 243$

(e) $\sqrt[5]{-243} = \sqrt[5]{(-3)^5} = -3$ $(-3)^5 = -243$

(f) $\sqrt[6]{-64}$ is not a real number.

(g) $-\sqrt[6]{64} = -\sqrt[6]{2^6} = -|2| = -2$ $2^6 = 64$

(h) $\sqrt[5]{(2x)^5} = 2x$

Practice Exercise 3

Simplify.

(a) $\sqrt[4]{625}$

(b) $\sqrt[4]{-625}$

(c) $\sqrt[4]{(-5)^4}$

(d) $\sqrt[5]{5^5}$

(e) $\sqrt[5]{(-5)^5}$

(f) $\sqrt[6]{-1}$

(g) $-\sqrt[6]{1}$

(h) $\sqrt[5]{(3u)^5}$

(i) $\sqrt[4]{(3y)^4} = |3y| = |3||y| = 3|y|$

(j) $\sqrt[6]{(a-3)^6} = |a-3|$

(k) $\sqrt[4]{0} = \sqrt[4]{0^4} = |0| = 0$ $\quad 0^4 = 0$

(i) $\sqrt[4]{(5x)^4}$

(j) $\sqrt[5]{(a-3)^5}$

(k) $\sqrt[5]{0^5}$

Answers: (a) 5 (b) not a real number (c) 5 (d) 5 (e) −5 (f) not a real number (g) −1 (h) $3u$ (i) $5|x|$ (j) $a-3$ (k) 0

When expressions under radicals are nonnegative, taking even roots is much simpler since absolute value need not be used. As a result, *unless we specifically state otherwise,* we will make the following assumption from now on.

Taking Roots from This Point Forward

All variables and algebraic expressions under even-indexed radicals represent nonnegative numbers so that for any index k,

$$\sqrt[k]{x^k} = x.$$

EXAMPLE 4 Simplifying Roots

Simplify. (Variables and expressions under even-indexed radicals are nonnegative.)

(a) $\sqrt{(2x)^2} = 2x$
We do not need to write $2|x|$ as in Example 1(e) since here $|x| = x$.

(b) $\sqrt{(a+1)^2} = a+1$
Now $|a+1| = a+1$ so absolute value is not necessary.

(c) $\sqrt[4]{(3y)^4} = 3y$

(d) $\sqrt[6]{(a-3)^6} = a-3$

(e) $\sqrt{y^2-4y+4} = \sqrt{(y-2)^2} = y-2$

Practice Exercise 4

Simplify.

(a) $\sqrt{(3y)^2}$

(b) $\sqrt{(x-3)^2}$

(c) $\sqrt[4]{(5x)^4}$

(d) $\sqrt[8]{(u+5)^8}$

(e) $\sqrt{a^2+10a+25}$

Answers: (a) $3y$ (b) $x-3$ (c) $5x$ (d) $u+5$ (e) $a+5$

4 APPROXIMATING SQUARE ROOTS

Square roots of perfect squares such as $\sqrt{9} = \sqrt{3^2} = 3$ and $\sqrt{0.25} = \sqrt{(0.5)^2} = 0.5$ are rational numbers. In many applications we have to approximate square roots of numbers such as $\sqrt{2}$ and $\sqrt{0.3}$ that are not perfect squares. The best way to do this is with a calculator that has a square root key $\boxed{\sqrt{\ }}$. For example, to approximate $\sqrt{2}$, we use the following steps on a calculator*:

$$2\boxed{\sqrt{\ }} \longrightarrow \boxed{1.4142136}$$

The display shows 1.4142136 which is a rational number approximation of the irrational number $\sqrt{2}$. In some situations we might use 1.4 for $\sqrt{2}$, correct to the nearest tenth, or 1.41 correct to the nearest hundredth.

Scientific notation and significant digits, introduced in Section 1.8, are sometimes used for approximating roots.

*Calculator steps are given using algebraic logic. If you have a calculator that uses Reverse Polish Notation (RPN), consult your operator's manual.

EXAMPLE 5 Approximating Roots Using a Calculator

(a) Approximate $\frac{-2+\sqrt{53}}{4}$, correct to the nearest hundredth.
The steps to use on a calculator are:

2 [±] [+] 53 [√] [=] [÷] 4 [=] ⟶ [1.3200275]

The display shows 1.3200275, which rounded to the nearest hundredth is 1.32. Note that [±] is the change sign key which made 2 into −2 in the first step.

(b) Approximate $\sqrt{243{,}000{,}000}$, correct to three significant digits.
Since very large or very small numbers do not fit on a calculator display, such numbers must first be expressed in scientific notation. To evaluate $\sqrt{243{,}000{,}000}$ on a scientific calculator we use these steps:

2.43 [EE] 8 [√] ⟶ [1.5588 04]

The display gives the answer in scientific notation, 1.5588×10^4. This can be written as 15,600, correct to three significant digits.

Practice Exercise 5

(a) Approximate $\frac{4-\sqrt{61}}{3}$, correct to the nearest hundredth.

(b) Approximate $\sqrt{0.0000000195}$, correct to three significant digits.

Answers: (a) −1.27
(b) $1.3964 \times 10^{-4} \approx 0.000140$

Many applied problems involve radicals. Exact answers that are irrational numbers must often be approximated to give us a better idea of their value, as in the next example.

EXAMPLE 6 Physics Problem

The distance in miles that can be seen on the surface of the ocean can be approximated by

$$d = 1.4\sqrt{h},$$

where h is the height of the viewer in feet above the surface of the water. From a blimp at an altitude of 3000 ft, is it possible to see a ship at a distance of 90 mi?

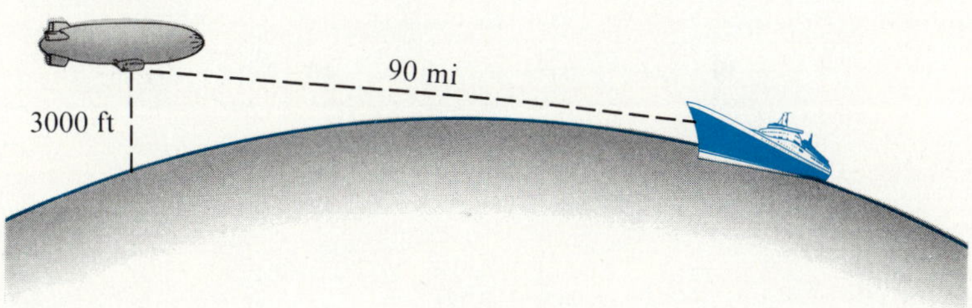

Figure 7.1

Substitute 3000 for h and evaluate d.

$$\begin{aligned} d &= 1.4\sqrt{h} \\ &= 1.4\sqrt{3000} \\ &\approx 76.7 \end{aligned}$$

Since the viewing distance is about 77 miles and since the ship is 90 miles away, the ship is below the horizon and cannot be seen from the blimp. The exact viewing distance is $1.4\sqrt{3000}$ miles, but to most of us, this does not have the same meaning as "about 77 miles."

Practice Exercise 6

What is the approximate viewing distance of an airplane flying at an altitude of 33,000 ft?

Answer: 254 mi

7.1 EXERCISES A

Simplify each radical, writing it as an integer or a rational number.

1. $\sqrt{36}$ **2.** $\sqrt{-16}$ **3.** $\sqrt[3]{27}$ **4.** $\sqrt[3]{-64}$

5. $-\sqrt[3]{8}$ **6.** $-\sqrt{4}$ **7.** $(\sqrt{25})^3$ **8.** $(\sqrt[3]{27})^2$

9. $\sqrt{\frac{1}{4}}$ **10.** $\sqrt[3]{\frac{1}{8}}$ **11.** $\sqrt[4]{\frac{1}{16}}$ **12.** $\sqrt[5]{-32}$

13. $\sqrt[6]{64}$ **14.** $\sqrt[7]{128}$ **15.** $\sqrt[7]{-128}$ **16.** $\sqrt[5]{(-6)^5}$

Simplify each radical in Exercises 17–32. For these exercises do not assume that expressions under even-indexed radicals necessarily represent nonnegative numbers. That is, use absolute value when necessary.

17. $\sqrt{x^2}$ **18.** $\sqrt[3]{x^3}$ **19.** $\sqrt{-x^2}$ **20.** $-\sqrt{x^2}$

21. $-\sqrt[3]{x^3}$ **22.** $\sqrt[3]{-x^3}$ **23.** $\sqrt[4]{y^4}$ **24.** $\sqrt[8]{y^8}$

25. $\sqrt[5]{w^5}$ **26.** $\sqrt[5]{(6xy)^5}$ **27** $\sqrt[4]{(-2x)^4}$ **28.** $\sqrt[5]{(-2x)^5}$

29. $-\sqrt[4]{(-2x)^4}$ **30** $-\sqrt[5]{(-2x)^5}$ **31** $\sqrt{(x+7)^2}$ **32** $\sqrt[3]{(x+7)^3}$

Simplify each radical in Exercises 33–52. For these exercises, return to the assumption that expressions under even-indexed radicals represent nonnegative numbers. That is, absolute value will not be necessary.

33. $\sqrt{u^2}$ **34.** $\sqrt[3]{u^3}$ **35.** $\sqrt{(2a)^2}$ **36.** $\sqrt[3]{(2a)^3}$

37. $\sqrt{49a^2}$ **38.** $\sqrt{(xy)^2}$ **39.** $\sqrt{(x-9)^2}$ **40.** $\sqrt[3]{(x-9)^3}$

41. $\sqrt[3]{(-u)^3}$ **42.** $\sqrt[4]{(7y)^4}$ **43.** $\sqrt[3]{-8x^3}$ **44.** $\sqrt[4]{-x^4}$

45. $\sqrt[7]{x^7}$ **46.** $\sqrt[5]{(-2x)^5}$ **47.** $\sqrt[8]{(6xy)^8}$ **48.** $\sqrt[6]{(-2)^6a^6}$

49 $\sqrt{x^2-14x+49}$ **50.** $\sqrt[3]{(-1)^3x^3}$ **51.** $\sqrt[4]{(-1)^4x^4}$ **52.** $\sqrt[5]{(-7)^5a^5w^5}$

Use a calculator to approximate each number correct to the nearest hundredth.

53. $\sqrt{291}$ **54.** $3+\sqrt{5}$ **55.** $\frac{-4+\sqrt{60}}{3}$ **56.** $\frac{3 \pm \sqrt{122}}{5}$

Use a calculator and scientific notation to approximate each number. Give answers correct to three significant digits.

57. $\sqrt{693{,}000{,}000}$

58. $\sqrt{0.0000000355}$

59 $\dfrac{\sqrt{1{,}250{,}000{,}000}}{0.00623}$

In a retail outlet the cost c is related to the number of items sold n by the equation $c = \sqrt[3]{n}$.

60. Find c when $n = 27$.

61. Find c when $n = 5^3$.

62. Find n when $c = 10$.

In a scientific study a quantity p is related to another variable w by the equation $p = \sqrt[4]{w}$.

63. Find p when $w = 81$.

64. Find p when $w = -81$.

65. Find w when $p = 5$.

The distance in miles that can be seen on the surface of the ocean can be approximated by $d = 1.4\sqrt{h}$, *where h is the height of the viewer in feet above the surface of the water. Use this formula in Exercises 66–67.*

66 From a balloon at an altitude of 1800 ft, is it possible to view a ship at a distance of 50 mi?

67. From a small airplane at an altitude of 4200 ft, is it possible to view an island at a distance of 100 mi?

The time T in seconds required for a pendulum of length L feet to make one swing is given by $T = 2\pi\sqrt{\frac{L}{32}}$. *In Exercises 68–69, find the time, to the nearest tenth of a second, that is needed for a pendulum of the given length to make one swing. Use 3.14 for* π.

68. 10 ft

69. 32.5 ft

FOR REVIEW

The following exercises review material from Chapter 1 and will help you prepare for the next section. Evaluate each pair of radical expressions and compare the results.

70. $\sqrt{4 \cdot 9}$, $\sqrt{4}\,\sqrt{9}$

71. $\dfrac{\sqrt{144}}{\sqrt{36}}$, $\sqrt{\dfrac{144}{36}}$

72. $\dfrac{\sqrt{81}\,\sqrt{49}}{\sqrt{9}}$, $\sqrt{\dfrac{(81)(49)}{9}}$

ANSWERS: 1. 6 2. not a real number 3. 3 4. −4 5. −2 6. −2 7. 125 8. 9 9. $\frac{1}{2}$ 10. $\frac{1}{2}$ 11. $\frac{1}{2}$ 12. −2 13. 2 14. 2 15. −2 16. −6 17. $|x|$ 18. x 19. not a real number unless $x = 0$, then $\sqrt{-0^2} = \sqrt{0} = 0$ 20. $-|x|$ 21. $-x$ 22. $-x$ 23. $|y|$ 24. $|y|$ 25. w 26. $6xy$ 27. $2|x|$ 28. $-2x$ 29. $-2|x|$ 30. $2x$ 31. $|x + 7|$ 32. $x + 7$ 33. u 34. u 35. $2a$ 36. $2a$ 37. $7a$ 38. xy 39. $x - 9$ 40. $x - 9$ 41. $-u$ 42. $7y$ 43. $-2x$ 44. not a real number unless $x = 0$, then $\sqrt[4]{-0^4} = \sqrt[4]{0} = 0$ 45. x 46. $-2x$ 47. $6xy$ 48. $2a$ 49. $x - 7$ 50. $-x$ 51. x 52. $-7aw$ 53. 17.06 54. 5.24 55. 1.25 56. 2.81 and −1.61 57. 26,300 58. 0.000188 59. 5,680,000 60. 3 61. 5 62. 1000 63. 3 64. not a real number 65. 625 66. yes, the viewing distance is about 59.4 mi 67. no, the viewing distance is about 90.7 mi 68. 3.5 sec 69. 6.3 sec 70. 6, 6; equal 71. 2, 2; equal 72. 21, 21; equal

7.1 EXERCISES B

Simplify each radical, writing it as an integer or a rational number.

1. $\sqrt{16}$

2. $\sqrt{-25}$

3. $\sqrt[3]{8}$

4. $\sqrt[3]{-27}$

5. $-\sqrt[3]{64}$ 6. $-\sqrt{9}$ 7. $(\sqrt{36})^3$ 8. $(\sqrt[3]{8})^2$

9. $\sqrt{\frac{1}{9}}$ 10. $\sqrt[3]{\frac{1}{27}}$ 11. $\sqrt[4]{\frac{1}{81}}$ 12. $\sqrt[5]{-243}$

13. $-\sqrt[6]{64}$ 14. $\sqrt[7]{-128}$ 15. $\sqrt[7]{(-3)^7}$ 16. $\sqrt[6]{(-3)^6}$

Simplify each radical in Exercises 17–32. For these exercises do not assume that expressions under even-indexed radicals necessarily represent nonnegative numbers. That is, use absolute value when necessary.

17. $\sqrt{y^2}$ 18. $\sqrt[3]{y^3}$ 19. $\sqrt{-y^2}$ 20. $-\sqrt{y^2}$

21. $-\sqrt[3]{a^3}$ 22. $\sqrt[3]{-a^3}$ 23. $\sqrt[6]{y^6}$ 24. $\sqrt[8]{z^8}$

25. $\sqrt[7]{w^7}$ 26. $\sqrt[3]{(2ab)^3}$ 27. $\sqrt[6]{(-5w)^6}$ 28. $\sqrt[3]{(-5w)^3}$

29. $-\sqrt{(-3x)^2}$ 30. $-\sqrt[3]{(-3x)^3}$ 31. $\sqrt{(y-4)^2}$ 32. $\sqrt[5]{(y-4)^5}$

Simplify each radical in Exercises 33–52. For these exercises, return to the assumption that expressions under even-indexed radicals represent nonnegative numbers. That is, absolute value will not be necessary.

33. $\sqrt{a^2}$ 34. $\sqrt[5]{u^5}$ 35. $\sqrt{(5x)^2}$ 36. $\sqrt[3]{(5x)^3}$

37. $\sqrt{81y^2}$ 38. $\sqrt{(ab)^2}$ 39. $\sqrt{(y+10)^2}$ 40. $\sqrt[3]{(y+10)^3}$

41. $\sqrt[3]{(-x)^3}$ 42. $\sqrt[6]{(8w)^6}$ 43. $\sqrt[3]{-27y^3}$ 44. $\sqrt[6]{-x^6}$

45. $\sqrt[9]{u^9}$ 46. $\sqrt[3]{(-4x)^3}$ 47. $\sqrt[8]{(2wy)^8}$ 48. $\sqrt[4]{(-2)^4a^4}$

49. $\sqrt{x^2+12x+36}$ 50. $\sqrt[5]{(-1)^5w^5}$ 51. $\sqrt[6]{(-1)^6y^6}$ 52. $\sqrt[3]{(-3)^3x^3y^3}$

Use a calculator to approximate each number correct to the nearest hundredth.

53. $\sqrt{483}$ 54. $2-\sqrt{3}$ 55. $\frac{-3+\sqrt{20}}{4}$ 56. $\frac{2\pm\sqrt{141}}{7}$

Use a calculator and scientific notation to approximate each number. Give answers correct to three significant digits.

57. $\sqrt{392{,}000{,}000}$ 58. $\sqrt{0.00000000515}$ 59. $\frac{\sqrt{2{,}340{,}000{,}000}}{0.000442}$

In a retail outlet the cost c is related to the number of items sold n by the equation $c = \sqrt[3]{n}$.

60. Find c when $h = 8$. 61. Find c when $n = 7^3$. 62. Find n when $c = 3$.

In a scientific study a quantity p is related to another variable w by the equation $p = \sqrt[4]{w}$.

63. Find p when $w = 16$. 64. Find p when $w = -16$. 65. Find w when $p = 3$.

The distance in miles that can be seen on the surface of the ocean can be approximated by $d = 1.4\sqrt{h}$, where h is the height of the viewer in feet above the surface of the water. Use this formula in Exercises 66–67.

66. From a blimp at an altitude of 2300 ft, is it possible to view a ship at a distance of 75 mi?

67. From a jet airplane at an altitude of 25,000 ft, is it possible to view the shoreline of a country at a distance of 200 mi?

The time T in seconds required for a pendulum of length L feet to make one swing is given by $T = 2\pi\sqrt{\frac{L}{32}}$. *In Exercises 68–69, find the time, to the nearest tenth of a second, that is required for a pendulum of the given length to make one swing. Use* 3.14 *for* π.

68. 24 ft

69. 7.5 ft

FOR REVIEW

The following exercises review material from Chapter 1 and will help you prepare for the next section. Evaluate each pair of radical expressions and compare the results.

70. $\sqrt{36 \cdot 16}$, $\sqrt{36}\sqrt{16}$

71. $\frac{\sqrt{25}}{\sqrt{100}}$, $\sqrt{\frac{25}{100}}$

72. $\frac{\sqrt{64}}{\sqrt{4}\sqrt{16}}$, $\sqrt{\frac{64}{(4)(16)}}$

7.1 EXERCISES C

For Exercises 1–10 assume that x is a negative number.

1. Is $-x$ positive or negative?

2. True or false: $|x| = -x$?

Simplify.

3. $\sqrt{x^2}$
[Answer: $-x$]

4. $\sqrt[3]{x^3}$

5. $\sqrt{x^2} - \sqrt[3]{x^3}$

6. $\sqrt{x^2} + \sqrt[3]{x^3}$
[Answer: 0]

7. $\sqrt[4]{x^4}$
[Answer: $-x$]

8. $\sqrt[5]{x^5}$

9. $\sqrt{(-x)^2}$

10. $\sqrt[3]{(-x)^3}$

7.2 SIMPLIFYING RADICALS

STUDENT GUIDEPOSTS

1. Simplifying Products
2. Simplifying Quotients
3. Rationalizing a Denominator
4. Rationalizing a Numerator
5. Simplest Form

1 SIMPLIFYING PRODUCTS

In Section 7.1 we used the rule $\sqrt[k]{x^k} = x$ to simplify radicals such as

$$\sqrt{25} = 5, \quad \sqrt[3]{27x^3} = 3x, \quad \text{and} \quad \sqrt[6]{y^6} = y.$$

To simplify radicals involving radicands that are not perfect powers we can use the following rule.

Simplifying Rule for Products

If a and b are any real numbers or expressions, where a and b are nonnegative if k is even, then

$$\sqrt[k]{ab} = \sqrt[k]{a}\sqrt[k]{b}.$$

The kth root of a product is equal to the product of the kth roots.

This rule can be used to simplify radicals when one or more of the factors of the radicand is a perfect kth power. For example,

$$\begin{aligned} \sqrt{50} &= \sqrt{25 \cdot 2} && \text{25 is a perfect square} \\ &= \sqrt{25}\,\sqrt{2} && \sqrt{ab} = \sqrt{a}\,\sqrt{b} \\ &= 5\sqrt{2}. && \text{Simplified form} \end{aligned}$$

In the case of square roots, we look for perfect squares, isolate each one using $\sqrt{ab} = \sqrt{a}\,\sqrt{b}$, and simplify using $\sqrt{x^2} = x$. If you have trouble recognizing perfect square factors, factor the radicand into primes. In the above example, we could have written $50 = 2 \cdot 5 \cdot 5 = 2 \cdot 5^2 = 2 \cdot 25$.

EXAMPLE 1 Simplifying Square Roots

Simplify the square roots.

(a) $$\begin{aligned} \sqrt{75} &= \sqrt{25 \cdot 3} && \text{25 is a perfect square} \\ &= \sqrt{25}\,\sqrt{3} && \sqrt{ab} = \sqrt{a}\,\sqrt{b} \\ &= 5\sqrt{3} && \sqrt{25} = \sqrt{5^2} = 5 \end{aligned}$$

(b) $$\begin{aligned} \sqrt{7^3} &= \sqrt{7^2 \cdot 7} && 7^3 = 7^{2+1} = 7^2 \cdot 7 \\ &= \sqrt{7^2}\,\sqrt{7} && \sqrt{ab} = \sqrt{a}\,\sqrt{b} \\ &= 7\sqrt{7} \end{aligned}$$

(c) $$\begin{aligned} \sqrt{35} &= \sqrt{5 \cdot 7} && \text{Cannot be simplified since neither 5 nor 7 is a perfect square} \\ &= \sqrt{35} \end{aligned}$$

(d) $$\begin{aligned} 5\sqrt{x^6} &= 5\sqrt{(x^3)^2} && x^6 = x^{3 \cdot 2} = (x^3)^2 \\ &= 5x^3 \end{aligned}$$

(e) $$\begin{aligned} 5\sqrt{x^7} &= 5\sqrt{x^6 \cdot x} && x^7 = x^{6+1} = x^6 \cdot x \\ &= 5\sqrt{x^6}\,\sqrt{x} && \sqrt{ab} = \sqrt{a}\,\sqrt{b} \\ &= 5\sqrt{(x^3)^2}\,\sqrt{x} \\ &= 5x^3\sqrt{x} \end{aligned}$$

(f) $$\begin{aligned} \sqrt{45a^3} &= \sqrt{9 \cdot 5 \cdot a^2 \cdot a} && \text{9 and } a^2 \text{ are perfect squares} \\ &= \sqrt{9 \cdot a^2 \cdot 5a} && 5a \text{ will be left under the radical} \\ &= \sqrt{9}\,\sqrt{a^2}\,\sqrt{5a} && \text{Rule extended to } \sqrt{abc} = \sqrt{a}\,\sqrt{b}\,\sqrt{c} \\ &= 3a\sqrt{5a} \end{aligned}$$

Practice Exercise 1

Simplify the square roots.

(a) $\sqrt{88}$

(b) $\sqrt{11^3}$

(c) $\sqrt{42}$

(d) $4\sqrt{y^8}$

(e) $4\sqrt{y^9}$

(f) $\sqrt{108u^3}$

Answers: (a) $2\sqrt{22}$ (b) $11\sqrt{11}$ (c) cannot be simplified (d) $4y^4$ (e) $4y^4\sqrt{y}$ (f) $6u\sqrt{3u}$

To simplify cube roots we look for perfect cubes under the radical, use $\sqrt[3]{ab} = \sqrt[3]{a}\,\sqrt[3]{b}$, and then $\sqrt[3]{x^3} = x$. As with square roots, prime factorization may be used to find perfect cubes.

EXAMPLE 2 Simplifying Cube Roots

Simplify the cube roots.

(a) $$\begin{aligned} \sqrt[3]{54} &= \sqrt[3]{27 \cdot 2} && \text{27 is a perfect cube} \\ &= \sqrt[3]{27} \cdot \sqrt[3]{2} && \sqrt[3]{ab} = \sqrt[3]{a}\,\sqrt[3]{b} \\ &= 3\sqrt[3]{2} && \sqrt[3]{27} = \sqrt[3]{3^3} = 3 \end{aligned}$$

(b) $$\begin{aligned} 7\sqrt[3]{x^5} &= 7\sqrt[3]{x^3 \cdot x^2} && x^5 = x^{3+2} = x^3 \cdot x^2 \\ &= 7\sqrt[3]{x^3}\,\sqrt[3]{x^2} && \sqrt[3]{ab} = \sqrt[3]{a}\,\sqrt[3]{b} \\ &= 7x\sqrt[3]{x^2} && \text{Simplest form} \end{aligned}$$

Practice Exercise 2

Simplify the cube roots.

(a) $\sqrt[3]{88}$

(b) $6\sqrt[3]{y^8}$

(c) $\sqrt[3]{8x^7y^{11}} = \sqrt[3]{8 \cdot x^6 \cdot x \cdot y^9 \cdot y^2}$ — 8, x^6, and y^9 are perfect cubes

$= \sqrt[3]{8 \cdot x^6 \cdot y^9 \cdot xy^2}$ — xy^2 will be left under the radical

$= \sqrt[3]{8}\,\sqrt[3]{x^6}\,\sqrt[3]{y^9}\,\sqrt[3]{xy^2}$ — $\sqrt[3]{abcd} = \sqrt[3]{a}\,\sqrt[3]{b}\,\sqrt[3]{c}\,\sqrt[3]{d}$

$= 2x^2y^3\sqrt[3]{xy^2}$ — $x^6 = x^{2\cdot 3} = (x^2)^3$ and $y^9 = y^{3\cdot 3} = (y^3)^3$

(c) $\sqrt[3]{27a^5b^{13}}$

Answers: (a) $2\sqrt[3]{11}$ (b) $6y^2\sqrt[3]{y^2}$ (c) $3ab^4\sqrt[3]{a^2b}$

In general, to simplify a radical with index k, we look for perfect kth powers, use $\sqrt[k]{ab} = \sqrt[k]{a}\,\sqrt[k]{b}$, and then $\sqrt[k]{x^k} = x$.

EXAMPLE 3 Simplifying Higher Order Roots

Simplify.

(a) $\sqrt[4]{48} = \sqrt[4]{16 \cdot 3}$ — 16 is a perfect fourth power

$= \sqrt[4]{16}\,\sqrt[4]{3}$ — $\sqrt[4]{ab} = \sqrt[4]{a}\,\sqrt[4]{b}$

$= \sqrt[4]{2^4}\,\sqrt[4]{3}$ — $16 = 2^4$

$= 2\sqrt[4]{3}$ — Don't forget the index; $\sqrt{3} \neq \sqrt[4]{3}$

(b) $\sqrt[5]{y^7} = \sqrt[5]{y^5 \cdot y^2}$ — $y^7 = y^{5+2} = y^5 \cdot y^2$

$= \sqrt[5]{y^5}\,\sqrt[5]{y^2}$ — $\sqrt[5]{ab} = \sqrt[5]{a}\,\sqrt[5]{b}$

$= y\sqrt[5]{y^2}$

(c) $y\sqrt[6]{64x^3y^8} = y\sqrt[6]{2^6x^3y^6y^2}$ — $y^8 = y^{6+2} = y^6y^2$

$= y\sqrt[6]{2^6y^6x^3y^2}$ — x^3y^2 will be left under the radical

$= y\sqrt[6]{2^6}\,\sqrt[6]{y^6}\,\sqrt[6]{x^3y^2}$ — $\sqrt[6]{abc} = \sqrt[6]{a}\,\sqrt[6]{b}\,\sqrt[6]{c}$

$= 2y^2\sqrt[6]{x^3y^2}$

PRACTICE EXERCISE 3

Simplify.

(a) $\sqrt[5]{96}$

(b) $\sqrt[4]{u^{11}}$

(c) $xy\sqrt[6]{12x^7y^{14}}$

Answers: (a) $2\sqrt[5]{3}$ (b) $u^2\sqrt[4]{u^3}$ (c) $x^2y^3\sqrt[6]{12xy^2}$

2 SIMPLIFYING QUOTIENTS

When a radicand involves a quotient, the following simplifying rule may be applied. To use it we look for perfect powers in both the numerator and denominator.

Simplifying Rule for Quotients

If a and b are any real numbers ($b \neq 0$) or expressions, where a is nonnegative and b is positive if k is even, then

$$\sqrt[k]{\frac{a}{b}} = \frac{\sqrt[k]{a}}{\sqrt[k]{b}}.$$

The kth root of a quotient is equal to the quotient of the kth roots.

EXAMPLE 4 Simplifying Roots of Quotients

Simplify.

(a) $\sqrt{\frac{49}{25}} = \frac{\sqrt{49}}{\sqrt{25}}$ — $\sqrt{\frac{a}{b}} = \frac{\sqrt{a}}{\sqrt{b}}$

$= \frac{7}{5}$

PRACTICE EXERCISE 4

Simplify.

(a) $\sqrt{\frac{121}{64}}$

(b) $\sqrt[3]{\dfrac{54}{125}} = \dfrac{\sqrt[3]{54}}{\sqrt[3]{125}}$ $\quad \sqrt[3]{\dfrac{a}{b}} = \dfrac{\sqrt[3]{a}}{\sqrt[3]{b}}$

$= \dfrac{\sqrt[3]{27 \cdot 2}}{\sqrt[3]{125}}$ $\quad$ 27 and 125 are perfect cubes

$= \dfrac{\sqrt[3]{3^3}\sqrt[3]{2}}{\sqrt[3]{5^3}}$ $\quad \sqrt[3]{ab} = \sqrt[3]{a}\sqrt[3]{b}$

$= \dfrac{3\sqrt[3]{2}}{5}$

(c) $\sqrt[4]{\dfrac{48x^4}{3y^8}} = \sqrt[4]{\dfrac{16 \cdot \cancel{3} \cdot x^4}{\cancel{3} \cdot y^8}}$ $\quad$ Simplify under radical first

$= \dfrac{\sqrt[4]{2^4x^4}}{\sqrt[4]{(y^2)^4}}$ $\quad \sqrt[4]{\dfrac{a}{b}} = \dfrac{\sqrt[4]{a}}{\sqrt[4]{b}}$, $y^8 = (y^2)^4$

$= \dfrac{\sqrt[4]{2^4}\sqrt[4]{x^4}}{\sqrt[4]{(y^2)^4}}$ $\quad \sqrt[4]{ab} = \sqrt[4]{a}\sqrt[4]{b}$

$= \dfrac{2x}{y^2}$

(b) $\sqrt[3]{\dfrac{56}{125}}$

(c) $\sqrt[4]{\dfrac{162x^5}{2y^{12}}}$

Answers: (a) $\frac{11}{8}$ (b) $\frac{2\sqrt[3]{7}}{5}$ (c) $\dfrac{3x\sqrt[4]{x}}{y^3}$

3 RATIONALIZING A DENOMINATOR

In some problems the denominator is not a perfect power. For example, $\frac{6x}{\sqrt{5}}$ has a square root in the denominator (the denominator is not a rational number). The process of removing radicals from the denominator is called **rationalizing the denominator.** A radical expression is not considered completely simplified until all radicals are removed from the denominator.

To Rationalize a Denominator

1. Simplify the fraction.
2. Multiply both numerator and denominator of the fraction by a radical that makes the radicand in the denominator a perfect power, equal to the index on the radical.

To rationalize the fraction $\frac{6x}{\sqrt{5}}$, multiply both numerator and denominator by $\sqrt{5}$.

$$\frac{6x}{\sqrt{5}} = \frac{6x\sqrt{5}}{\sqrt{5}\sqrt{5}}$$

$$= \frac{6x\sqrt{5}}{5} \qquad \sqrt{a}\sqrt{a} = \sqrt{a^2} = a$$

To see why this works and to find the necessary multipliers for radicals with a higher index, we use $\sqrt[k]{ab} = \sqrt[k]{a}\sqrt[k]{b}$ in reverse, that is,

$$\sqrt[k]{a}\sqrt[k]{b} = \sqrt[k]{ab},$$

to construct the following table.

Radical in denominator	*Multiply and divide by*	*Reason*
$\sqrt{7}$	$\sqrt{7}$	$\sqrt{7}\,\sqrt{7} = \sqrt{7^2} = 7$
$\sqrt[3]{7}$	$\sqrt[3]{7^2}$	$\sqrt[3]{7}\,\sqrt[3]{7^2} = \sqrt[3]{7 \cdot 7^2} = \sqrt[3]{7^3} = 7$
$\sqrt[4]{x}$	$\sqrt[4]{x^3}$	$\sqrt[4]{x}\,\sqrt[4]{x^3} = \sqrt[4]{x \cdot x^3} = \sqrt[4]{x^4} = x$
$\sqrt[4]{x^3}$	$\sqrt[4]{x}$	$\sqrt[4]{x^3}\,\sqrt[4]{x} = \sqrt[4]{x^3 \cdot x} = \sqrt[4]{x^4} = x$
$\sqrt[5]{y^2}$	$\sqrt[5]{y^3}$	$\sqrt[5]{y^2}\,\sqrt[5]{y^3} = \sqrt[5]{y^2y^3} = \sqrt[5]{y^5} = y$

EXAMPLE 5 Rationalizing Denominators

Rationalize the denominator.

(a) $\sqrt{\dfrac{36x^2}{y}} = \dfrac{\sqrt{36x^2}}{\sqrt{y}}$ $\quad \sqrt{\dfrac{a}{b}} = \dfrac{\sqrt{a}}{\sqrt{b}}$

$= \dfrac{\sqrt{36}\,\sqrt{x^2}}{\sqrt{y}}$

$= \dfrac{6x}{\sqrt{y}}$

$= \dfrac{6x\sqrt{y}}{\sqrt{y}\,\sqrt{y}}$ Multiply numerator and denominator by $\sqrt{y}$

$= \dfrac{6x\sqrt{y}}{y}$ $\quad \sqrt{y}\,\sqrt{y} = y$

(b) $\dfrac{\sqrt[3]{x}}{\sqrt[3]{5}} = \dfrac{\sqrt[3]{x}\,\sqrt[3]{5^2}}{\sqrt[3]{5}\,\sqrt[3]{5^2}}$ Multiply numerator and denominator by $\sqrt[3]{5^2}$

$= \dfrac{\sqrt[3]{x \cdot 5^2}}{\sqrt[3]{5^3}}$ $\quad \sqrt[3]{a}\,\sqrt[3]{b} = \sqrt[3]{ab}$

$= \dfrac{\sqrt[3]{25x}}{5}$

(c) $\dfrac{\sqrt[5]{2y}}{\sqrt[5]{4x^2y^3}} = \dfrac{\sqrt[5]{2y}}{\sqrt[5]{2^2x^2y^3}}$

$= \dfrac{\sqrt[5]{2y}\,\sqrt[5]{2^3x^3y^2}}{\sqrt[5]{2^2x^2y^3}\,\sqrt[5]{2^3x^3y^2}}$

$= \dfrac{\sqrt[5]{2y \cdot 2^3x^3y^2}}{\sqrt[5]{2^2x^2y^3 \cdot 2^3x^3y^2}}$

$= \dfrac{\sqrt[5]{2^4x^3y^3}}{\sqrt[5]{2^5x^5y^5}}$

$= \dfrac{\sqrt[5]{16x^3y^3}}{2xy}$

Practice Exercise 5

Rationalize the denominator.

(a) $\sqrt{\dfrac{49u^2}{v}}$

(b) $\dfrac{\sqrt[3]{xy}}{\sqrt[3]{25}}$

(c) $\dfrac{\sqrt[4]{7x^2}}{\sqrt[4]{2x^3y}}$

Answers: (a) $\dfrac{7u\sqrt{v}}{v}$ (b) $\dfrac{\sqrt[3]{5xy}}{5}$ (c) $\dfrac{\sqrt[4]{56x^3y^3}}{2xy}$

We could have rationalized the denominator in Example 5(a) in another way. If we multiply the numerator and denominator of the radicand by y, we will obtain the same result.

$$\sqrt{\frac{36x^2}{y}} = \sqrt{\frac{36x^2 \cdot y}{y \cdot y}} = \frac{\sqrt{36}\,\sqrt{x^2}\,\sqrt{y}}{\sqrt{y^2}} = \frac{6x\sqrt{y}}{y}$$

4 RATIONALIZING A NUMERATOR

Sometimes in more advanced courses it is necessary to change the form of a radical expression by **rationalizing the numerator.** The procedure is similar to rationalizing denominators.

EXAMPLE 6 Rationalizing Numerators

Rationalize the numerator.

(a) $\dfrac{\sqrt{6}}{\sqrt{10}} = \dfrac{\sqrt{6}\sqrt{6}}{\sqrt{10}\sqrt{6}}$ Multiply numerator and denominator by $\sqrt{6}$

$= \dfrac{6}{\sqrt{10 \cdot 6}}$ $\sqrt{a}\sqrt{b} = \sqrt{ab}$

$= \dfrac{6}{\sqrt{2 \cdot 5 \cdot 2 \cdot 3}}$ Factor and look for perfect squares

$= \dfrac{6}{\sqrt{2^2 \cdot 15}} = \dfrac{6}{\sqrt{2^2}\sqrt{15}}$

$= \dfrac{6}{2\sqrt{15}} = \dfrac{3}{\sqrt{15}}$

(b) $\dfrac{\sqrt[4]{x^3}}{\sqrt[4]{y}} = \dfrac{\sqrt[4]{x^3}\sqrt[4]{x}}{\sqrt[4]{y}\sqrt[4]{x}}$ Multiply numerator and denominator by $\sqrt[4]{x}$

$= \dfrac{\sqrt[4]{x^3 \cdot x}}{\sqrt[4]{yx}}$ $\sqrt[4]{a}\sqrt[4]{b} = \sqrt[4]{ab}$

$= \dfrac{\sqrt[4]{x^4}}{\sqrt[4]{yx}} = \dfrac{x}{\sqrt[4]{yx}}$

Practice Exercise 6

Rationalize the numerator.

(a) $\dfrac{\sqrt{x}}{\sqrt{y}}$

(b) $\dfrac{\sqrt[3]{y^2}}{\sqrt[3]{3}}$

Answers: (a) $\dfrac{x}{\sqrt{xy}}$ (b) $\dfrac{y}{\sqrt[3]{3y}}$

5 SIMPLEST FORM

The techniques in this section can be used to change a radical expression into *simplest form.*

Simplest Form of a Radical Expression

A radical expression is written in **simplest form** provided:

1. The radicand contains no factor raised to a power greater than or equal to the index.
2. There are no radicals in a denominator.
3. There are no fractions in a radicand.

The radicals $\sqrt{a^3}$ and $\sqrt[3]{24}$ are not in simplest form since they violate property 1.

$$\sqrt{a^3} = \sqrt{a^2 a} = a\sqrt{a} \quad \text{and} \quad \sqrt[3]{24} = \sqrt[3]{8 \cdot 3} = 2\sqrt[3]{3}$$

The radicals $\frac{1}{\sqrt{2}}$ and $\frac{\sqrt[3]{3}}{\sqrt[3]{x}}$ are not in simplest form since they violate property 2.

$$\frac{1}{\sqrt{2}} = \frac{1 \cdot \sqrt{2}}{\sqrt{2}\sqrt{2}} = \frac{\sqrt{2}}{2} \quad \text{and} \quad \frac{\sqrt[3]{3}}{\sqrt[3]{x}} = \frac{\sqrt[3]{3}\sqrt[3]{x^2}}{\sqrt[3]{x}\sqrt[3]{x^2}} = \frac{\sqrt[3]{3x^2}}{x}$$

Finally, the radicals $\sqrt{\frac{2}{3}}$ and $\sqrt[3]{\frac{x^2}{y}}$ are not in simplest form since they violate property 3.

$$\sqrt{\frac{2}{3}} = \frac{\sqrt{2}}{\sqrt{3}} = \frac{\sqrt{2}\sqrt{3}}{\sqrt{3}\sqrt{3}} = \frac{\sqrt{6}}{3} \quad \text{and} \quad \sqrt[3]{\frac{x^2}{y}} = \frac{\sqrt[3]{x^2}\sqrt[3]{y^2}}{\sqrt[3]{y}\sqrt[3]{y^2}} = \frac{\sqrt[3]{x^2y^2}}{y}$$

We now look at an application that requires simplifying radicals.

EXAMPLE 7 Physics Problem

An object dropped from a height of h ft takes t seconds to reach the ground where $t = \sqrt{\frac{h}{16}}$. How long would it take a rock to hit the water if dropped from a bridge 100 ft above the surface of the water?

Figure 7.2

Substitute 100 for h and simplify.

$$\begin{aligned} t &= \sqrt{\frac{h}{16}} \\ &= \sqrt{\frac{100}{16}} \\ &= \frac{\sqrt{100}}{\sqrt{16}} = \frac{\sqrt{10^2}}{\sqrt{4^2}} = \frac{10}{4} = \frac{5}{2} \end{aligned}$$

It would take $\frac{5}{2}$ or 2.5 seconds for the rock to hit the water.

PRACTICE EXERCISE 7

How long would it take a ball to reach the ground if it is dropped from a window in a building from a height of 320 ft? Give answer to the nearest tenth of a second.

Answer: 4.5 sec

7.2 EXERCISES A

Simplify each of the radicals. Assume all variables and expressions under even-indexed radicals are nonnegative.

1. $\sqrt{125}$ **2.** $6\sqrt{98}$ **3.** $\sqrt{75y^3}$ **4.** $3\sqrt{x^8}$

5. $7\sqrt{50x^3y^5}$ **6.** $\sqrt[3]{24}$ **7.** $\sqrt[3]{27x^4}$ **8.** $5\sqrt[3]{y^8}$

9. $\sqrt[3]{16x^4y^6}$ **10.** $\sqrt[3]{-125x^9y^5}$ **11.** $\sqrt[4]{32}$ **12.** $\sqrt[4]{81x^7}$

13. $4\sqrt[3]{81x^3}$ **14.** $\sqrt[4]{-16x^4}$ **15.** $\sqrt{20x^7y^9}$ **16.** $\sqrt{15xy}$

17. $\sqrt[4]{625x^8y^{12}}$ **18.** $\sqrt[5]{-243x^6y^7}$ **19.** $\sqrt[6]{a^5b^7c^{12}}$ **20.** $\sqrt[3]{-27x^4y^5z^8}$

21. $\sqrt{\frac{45}{16}}$ **22.** $\sqrt{\frac{125}{y^2}}$ **23.** $\sqrt[3]{\frac{125}{8}}$ **24.** $\sqrt[3]{\frac{56x^5}{7y^3}}$

25. $\sqrt[3]{\frac{-27x^4}{8}}$ **26.** $\sqrt[3]{\frac{-27x^4}{8x}}$ **27.** $\sqrt[4]{\frac{16}{81}}$ **28.** $\sqrt[5]{\frac{32x}{y^5}}$

29. $\frac{3x}{y}\sqrt[5]{\frac{x^5y^6}{243}}$ **30.** $\sqrt[3]{\frac{10x^6y^7}{-27z^9}}$ **31.** $\sqrt{\frac{x^2+2x+1}{x^2-2x+1}}$ **32.** $\sqrt[3]{\frac{-16x^8y^{12}}{z^{15}}}$

If the given radical is a denominator, determine the radical which must be multiplied by numerator and denominator to rationalize the denominator.

33. $\sqrt{3x}$ **34.** $\sqrt[3]{5x^2}$ **35.** $\sqrt[5]{x^3}$ **36.** $\sqrt[4]{x^3y}$

Simplify. Rationalize all denominators.

37. $\frac{3}{\sqrt{2}}$ **38.** $\frac{8}{\sqrt{x}}$ **39.** $\sqrt{\frac{25}{y}}$ **40.** $\frac{3}{\sqrt[3]{7}}$

41. $\frac{x^2}{\sqrt[3]{y}}$ **42.** $\frac{x^2}{\sqrt[3]{y^2}}$ **43.** $\frac{7}{\sqrt[4]{y}}$ **44.** $\frac{7}{\sqrt[4]{y^3}}$

45. $\sqrt[3]{\frac{81x^3}{y^2}}$ **46.** $\sqrt[5]{\frac{-32x^7}{y^3}}$ **47.** $\frac{\sqrt[3]{3z^4}}{\sqrt[3]{2x^2y}}$ **48.** $\frac{\sqrt[3]{x^2z^4}}{\sqrt[3]{2y^2}}$

Rationalize the numerator.

49. $\dfrac{\sqrt{5}}{\sqrt{2x}}$

50. $\dfrac{2\sqrt{x}}{\sqrt{6y}}$

51. $\sqrt{\dfrac{2}{5}}$

52. $\dfrac{\sqrt[3]{3}}{\sqrt[3]{x}}$

53. $\dfrac{\sqrt[4]{x^3}}{\sqrt[4]{7}}$

54. $\sqrt[3]{\dfrac{2}{3}}$

55. $\dfrac{\sqrt[3]{4}}{\sqrt[3]{4a}}$

56. $\dfrac{\sqrt{xy}}{6}$

In a manufacturing process, the three quantities p, r, and s are related by the equation $p = \sqrt{\frac{r}{s}}$.

57. Find p when $r = 45$ and $s = 36$.

58. Find p when $r = 1$ and $s = 3$.

59. Find r when $p = 5$ and $s = 1$.

60. Find s when $p = 5$ and $r = 1$.

If an object is dropped from a cliff of height h in feet, the time t in seconds it takes the object to reach the ground is given by $t = \sqrt{\frac{h}{16}}$. *Use this information in Exercises 61–62.*

61. Find the time it would take for a rock to reach the water if it is dropped from a cliff 169 ft above the water level.

62. Find the time it would take for a rock to reach the water if it is dropped from a cliff 200 ft above the water level. Give answer correct to the nearest tenth of a second.

A simple formula for estimating the speed in miles per hour at which a car was traveling prior to an accident is $S = \sqrt{20.5\,l}$, *where l is the length of its skid marks in feet. Use this formula in Exercises 63–64 to estimate the speed of a car with skid marks of the given length.*

63. 50 ft

64. 85 ft

FOR REVIEW

Simplify the radicals in Exercises 65–70. Use absolute value bars when necessary. (That is, do not assume variables are nonnegative for these six exercises.)

65. $\sqrt{36y^2}$ **66.** $\sqrt[3]{64x^3y^3}$ **67.** $\sqrt[5]{-32(x+y)^5}$

68. $\sqrt{a^6}$ **69.** $\sqrt[5]{32a^{20}b^{10}}$ **70.** $\sqrt[4]{(2xy^2)^4}$

ANSWERS: 1. $5\sqrt{5}$ 2. $42\sqrt{2}$ 3. $5y\sqrt{3y}$ 4. $3x^4$ 5. $35xy^2\sqrt{2xy}$ 6. $2\sqrt[3]{3}$ 7. $3x\sqrt[3]{x}$ 8. $5y^2\sqrt[3]{y^2}$ 9. $2xy^2\sqrt[3]{2x}$ 10. $-5x^3y\sqrt[3]{y^2}$ 11. $2\sqrt[4]{2}$ 12. $3x\sqrt[4]{x^3}$ 13. $12x\sqrt[3]{3}$ 14. not a real number 15. $2x^3y^4\sqrt{5xy}$ 16. $\sqrt{15xy}$ 17. $5x^2y^3$ 18. $-3xy\sqrt[5]{xy^2}$ 19. $bc^2\sqrt[6]{a^5b}$ 20. $-3xyz^2\sqrt[3]{xy^2z^2}$ 21. $\frac{3\sqrt{5}}{4}$ 22. $\frac{5\sqrt{5}}{y}$ 23. $\frac{5}{2}$ 24. $\frac{2x\sqrt[3]{x^2}}{y}$ 25. $\frac{-3x\sqrt[3]{x}}{2}$ 26. $\frac{-3x}{2}$ 27. $\frac{2}{3}$ 28. $\frac{2\sqrt[5]{x}}{y}$ 29. $x^2\sqrt[5]{y}$ 30. $-\frac{x^2y^2\sqrt[3]{10y}}{3z^3}$ 31. $\frac{x+1}{x-1}$ 32. $\frac{-2x^2y^4\sqrt[3]{2x^2}}{z^5}$ 33. $\sqrt{3x}$ 34. $\sqrt[3]{5^2x}$ 35. $\sqrt[5]{x^2}$ 36. $\sqrt[4]{xy^3}$ 37. $\frac{3\sqrt{2}}{2}$ 38. $\frac{8\sqrt{x}}{x}$ 39. $\frac{5\sqrt{y}}{y}$ 40. $\frac{3\sqrt[3]{49}}{7}$ 41. $\frac{x^2\sqrt[3]{y^2}}{y}$ 42. $\frac{x^2\sqrt[3]{y}}{y}$ 43. $\frac{7\sqrt[4]{y^3}}{y}$ 44. $\frac{7\sqrt[4]{y}}{y}$ 45. $\frac{3x\sqrt[3]{3y}}{y}$ 46. $\frac{-2x\sqrt[5]{x^2y^2}}{y}$ 47. $\frac{z\sqrt[3]{12xy^2z}}{2xy}$ 48. $\frac{z\sqrt[3]{4x^2yz}}{2y}$ 49. $\frac{5}{\sqrt{10x}}$ 50. $\frac{2x}{\sqrt{6xy}}$ 51. $\frac{2}{\sqrt{10}}$ 52. $\frac{3}{\sqrt[3]{9x}}$ 53. $\frac{x}{\sqrt[4]{7x}}$ 54. $\frac{2}{\sqrt[3]{12}}$ 55. $\frac{1}{\sqrt[3]{a}}$ 56. $\frac{xy}{6\sqrt{xy}}$ 57. $\frac{\sqrt{5}}{2}$ 58. $\frac{\sqrt{3}}{3}$ 59. 25 60. $\frac{1}{25}$ 61. $\frac{13}{4}$ sec 62. 3.5 sec 63. 32 mph 64. 42 mph 65. $6|y|$ 66. $4xy$ 67. $-2(x+y)$ 68. $|a^3|$ 69. $2a^4b^2$ 70. $2|x|y^2$

7.2 EXERCISES B

Simplify each of the radicals. Assume all variables under even indexed radicals are nonnegative.

1. $\sqrt{175}$ **2.** $10\sqrt{147}$ **3.** $\sqrt{18x^3}$ **4.** $8\sqrt{x^{10}}$

5. $9\sqrt{75x^3y^7}$ **6.** $\sqrt[3]{40}$ **7.** $\sqrt[3]{8x^4}$ **8.** $3\sqrt[3]{z^{10}}$

9. $\sqrt[3]{81x^5y^3}$ **10.** $\sqrt[3]{-216x^5y^9}$ **11.** $\sqrt[4]{243}$ **12.** $\sqrt[4]{16x^5}$

13. $5\sqrt[3]{16x^6}$ **14.** $\sqrt[4]{-81x^4}$ **15.** $\sqrt{45x^3y^7}$ **16.** $\sqrt{6xyz}$

17. $\sqrt[4]{81x^{12}y^8}$ **18.** $\sqrt[5]{-32x^8y^6}$ **19.** $\sqrt[6]{a^{10}b^5c^{12}}$ **20.** $\sqrt[3]{-8x^5y^8z^4}$

21. $\sqrt{\frac{20}{49}}$ **22.** $\sqrt{\frac{50}{y^2}}$ **23.** $\sqrt[3]{\frac{64}{125}}$ **24.** $\sqrt[3]{\frac{54x^4}{2y^3}}$

25. $\sqrt[3]{\frac{-8x^5}{27}}$ **26.** $\sqrt[3]{\frac{-8x^5}{27x^2}}$ **27.** $\sqrt[4]{\frac{625}{16}}$ **28.** $\sqrt[5]{\frac{243x^2}{y^5}}$

29. $\frac{2x}{y}\sqrt[5]{\frac{x^6y^5}{32}}$ **30.** $\sqrt[3]{\frac{15x^7y^6}{-8z^{12}}}$ **31.** $\sqrt{\frac{x^2-6x+9}{x^2+6x+9}}$ **32.** $\sqrt[3]{\frac{-81x^9y^{11}}{z^9}}$

If the given radical is a denominator, determine the radical which must be multiplied by numerator and denominator to rationalize the denominator.

33. $\sqrt{5x}$ **34.** $\sqrt[3]{3x^2}$ **35.** $\sqrt[5]{x^2}$ **36.** $\sqrt[4]{xy^3}$

Simplify. Rationalize all denominators.

37. $\frac{5}{\sqrt{3}}$ **38.** $\frac{7}{\sqrt{x}}$ **39.** $\sqrt{\frac{16}{y}}$ **40.** $\frac{7}{\sqrt[3]{5}}$

41. $\frac{2x^3}{\sqrt[3]{y}}$ **42.** $\frac{2x^3}{\sqrt[3]{y^2}}$ **43.** $\frac{8}{\sqrt[4]{y}}$ **44.** $\frac{8}{\sqrt[4]{y^3}}$

45. $\sqrt[3]{\frac{54x^3}{y}}$ 46. $\sqrt[5]{\frac{-243x^6}{y^2}}$ 47. $\frac{\sqrt[3]{4x^4}}{\sqrt[3]{3yz^2}}$ 48. $\frac{\sqrt[3]{x^4z^2}}{\sqrt[3]{3y^2}}$

Rationalize the numerator.

49. $\frac{\sqrt{3}}{\sqrt{5y}}$ 50. $\frac{4\sqrt{y}}{\sqrt{2a}}$ 51. $\sqrt{\frac{3}{7}}$ 52. $\frac{\sqrt[3]{2}}{\sqrt[3]{x}}$

53. $\frac{\sqrt[4]{y}}{\sqrt[4]{5}}$ 54. $\sqrt[3]{\frac{3}{5}}$ 55. $\frac{\sqrt[3]{9}}{\sqrt[3]{18a}}$ 56. $\frac{\sqrt{ab}}{5}$

In an ecological study, the three quantities u, v, and w are related by the equation $u = \sqrt[3]{vw}$.

57. Find u when $v = 4$ and $w = 10$.

58. Find u when $v = 2$ and $w = \frac{1}{5}$.

59. Find v when $u = 4$ and $w = 1$.

60. Find w when $u = 4$ and $v = 2$.

If an object is dropped from a cliff of height h in feet, the time t in seconds it takes the object to reach the ground is given by $t = \sqrt{\frac{h}{16}}$. *Use this information in Exercises 61–62.*

61. Find the time it would take for a rock to reach the water if it is dropped from a cliff 225 ft above the water level.

62. Find the time it would take for a rock to reach the water if it is dropped from a cliff 300 ft above the water level. Give answer correct to the nearest tenth of a second.

A simple formula for estimating the speed in miles per hour at which a car was traveling prior to an accident is $S = \sqrt{20.5\, l}$, *where l is the length of its skid marks in feet. Use this formula in Exercises 63–64 to estimate the speed of a car with skid marks of the given length.*

63. 45 ft

64. 100 ft

FOR REVIEW

Simplify the radicals in Exercises 65–70. Use absolute value bars when necessary. (That is, do not assume variables are nonnegative for these six exercises.)

65. $\sqrt{49a^2}$ 66. $\sqrt[3]{125x^3y^3}$ 67. $\sqrt{-243(x-y)^5}$

68. $\sqrt[4]{a^4}$ 69. $\sqrt[3]{27x^9y^6}$ 70. $\sqrt[6]{(2uv^2)^6}$

7.2 EXERCISES C

Simplify. Rationalize all denominators.

1. $\sqrt{800x^5y^3}$ 2. $(\sqrt[3]{a^3b^6})^2$ 3. $\sqrt{x^2 + 4x^3}$

4. $\sqrt[5]{\frac{2a}{81a^2}}$ $\left[\text{Answer: } \frac{\sqrt[5]{6a^4}}{3a}\right]$

5. $\frac{1}{\sqrt[3]{(x+y)^2}}$

6. $\sqrt[4]{4x\sqrt{16x^6}}$ [Answer: $2x$]

7. Give an example to show that $\sqrt{a+b} \neq \sqrt{a} + \sqrt{b}$.

8. Give an example to show that $\sqrt{a^2+b^2} \neq |a| + |b|$.

7.3 MULTIPLICATION AND DIVISION OF RADICALS

STUDENT GUIDEPOSTS

1 Multiplying Radicals
2 Dividing Radicals

1 MULTIPLYING RADICALS

In the previous section we used the product rule $\sqrt[k]{ab} = \sqrt[k]{a}\sqrt[k]{b}$ to simplify radicals. The emphasis was on writing a radical as a product of radicals. Using the symmetric law of equality, we can change the order and emphasize writing a product of radicals as a single radical.

$$\sqrt[k]{a}\sqrt[k]{b} = \sqrt[k]{ab}$$

After applying this product rule to combine radicals, we can sometimes simplify the resulting radical by removing perfect powers.

EXAMPLE 1 MULTIPLYING RADICALS

Multiply and simplify.

(a) $\sqrt{5}\,\sqrt{45} = \sqrt{5 \cdot 45}$ — $\sqrt{a}\,\sqrt{b} = \sqrt{ab}$

$= \sqrt{5 \cdot 5 \cdot 9}$ — Factor to find perfect squares

$= \sqrt{25}\,\sqrt{9}$ — $\sqrt{ab} = \sqrt{a}\,\sqrt{b}$

$= 5 \cdot 3 = 15$

(b) $\sqrt{2x}\,\sqrt{8x} = \sqrt{2x \cdot 8x}$ — $\sqrt{a}\,\sqrt{b} = \sqrt{ab}$

$= \sqrt{16x^2}$ — 16 and x^2 are perfect squares

$= \sqrt{16}\,\sqrt{x^2}$ — $\sqrt{ab} = \sqrt{a}\,\sqrt{b}$

$= 4x$

(c) $\sqrt{2xy}\,\sqrt{6x^3y^3} = \sqrt{2xy \cdot 6x^3y^3}$ — $\sqrt{a}\,\sqrt{b} = \sqrt{ab}$

$= \sqrt{2^2 \cdot x^4 \cdot y^4 \cdot 3}$ — $x \cdot x^3 = x^{1+3} = x^4$ and $y \cdot y^3 = y^4$

$= \sqrt{2^2}\,\sqrt{(x^2)^2}\,\sqrt{(y^2)^2}\,\sqrt{3}$ — $x^4 = (x^2)^2$ and $y^4 = (y^2)^2$

$= 2x^2y^2\sqrt{3}$

(d) $5\sqrt[3]{3a^2}\,\sqrt[3]{9a} = 5\sqrt[3]{3a^2 \cdot 9a}$ — $\sqrt[3]{a}\,\sqrt[3]{b} = \sqrt[3]{ab}$

$= 5\sqrt[3]{27a^3}$ — 27 and a^3 are perfect cubes

$= 5\sqrt[3]{27}\,\sqrt[3]{a^3}$ — $\sqrt[3]{ab} = \sqrt[3]{a}\,\sqrt[3]{b}$

$= 15a$

(e) $\sqrt[4]{8x^2}\,\sqrt[4]{4x^3} = \sqrt[4]{2^3x^2 \cdot 2^2x^3}$ — $\sqrt[4]{a}\,\sqrt[4]{b} = \sqrt[4]{ab}$

$= \sqrt[4]{2^5x^5}$

$= \sqrt[4]{2^4x^4 \cdot 2x}$ — $2x$ will be left under the radical

$= \sqrt[4]{2^4}\,\sqrt[4]{x^4}\,\sqrt[4]{2x}$ — $\sqrt[4]{abc} = \sqrt[4]{a}\,\sqrt[4]{b}\,\sqrt[4]{c}$

$= 2x\sqrt[4]{2x}$

PRACTICE EXERCISE 1

Multiply and simplify.

(a) $\sqrt{7}\,\sqrt{175}$

(b) $\sqrt{3y}\,\sqrt{27y^3}$

(c) $\sqrt{12x^2y}\,\sqrt{9x^2y^3}$

(d) $4\sqrt[3]{4a^4}\,\sqrt[3]{16a^2}$

(e) $\sqrt[4]{9y^6}\,\sqrt[4]{27y^3}$

Answers: (a) 35 (b) $9y^2$ (c) $6x^2y^2\sqrt{3}$ (d) $16a^2$ (e) $3y^2\sqrt[4]{3y}$

2 DIVIDING RADICALS

When an expression involves the quotient of two radicals, it may be simplified in a similar manner.

$$\frac{\sqrt[k]{a}}{\sqrt[k]{b}} = \sqrt[k]{\frac{a}{b}}$$

Here we are writing a quotient of radicals as a single radical. It can sometimes be simplified by dividing out common factors and removing perfect powers.

EXAMPLE 2 Dividing Radicals

Divide and simplify.

(a) $\dfrac{\sqrt{27a^3}}{\sqrt{3a}} = \sqrt{\dfrac{27a^3}{3a}}$ $\qquad \dfrac{\sqrt{a}}{\sqrt{b}} = \sqrt{\dfrac{a}{b}}$

$= \sqrt{\dfrac{9 \cdot \cancel{3} \cdot a^2 \cdot \cancel{a}}{\cancel{3}\cancel{a}}}$ $\qquad$ Divide out common factors

$= \sqrt{9a^2}$ $\qquad$ 9 and a^2 are perfect squares

$= \sqrt{9}\,\sqrt{a^2} = 3a$

(b) $\dfrac{\sqrt{5x^3y}}{\sqrt{9xy^3}} = \sqrt{\dfrac{5x^3y}{9xy^3}}$ $\qquad \dfrac{\sqrt{a}}{\sqrt{b}} = \sqrt{\dfrac{a}{b}}$

$= \sqrt{\dfrac{5x^2 \cdot \cancel{x} \cdot \cancel{y}}{9\cancel{x} \cdot y^2 \cdot \cancel{y}}}$ $\qquad$ x and y are common factors

$= \sqrt{\dfrac{5x^2}{9y^2}}$

$= \dfrac{\sqrt{x^2}\,\sqrt{5}}{\sqrt{9}\,\sqrt{y^2}}$ $\qquad \sqrt{\dfrac{a}{b}} = \dfrac{\sqrt{a}}{\sqrt{b}}$ and $\sqrt{ab} = \sqrt{a}\,\sqrt{b}$

$= \dfrac{x\sqrt{5}}{3y}$

(c) $\dfrac{\sqrt[3]{16x^7y}}{\sqrt[3]{2xy^4}} = \sqrt[3]{\dfrac{16x^7y}{2xy^4}}$ $\qquad \dfrac{\sqrt[3]{a}}{\sqrt[3]{b}} = \sqrt[3]{\dfrac{a}{b}}$

$= \sqrt[3]{\dfrac{8 \cdot \cancel{2} \cdot x^6 \cdot \cancel{x} \cdot \cancel{y}}{\cancel{2} \cdot \cancel{x} \cdot y^3 \cdot \cancel{y}}}$ $\qquad$ Divide out common factors

$= \sqrt[3]{\dfrac{8x^6}{y^3}}$

$= \dfrac{\sqrt[3]{8}\,\sqrt[3]{(x^2)^3}}{\sqrt[3]{y^3}}$ $\qquad \sqrt[3]{\dfrac{a}{b}} = \dfrac{\sqrt[3]{a}}{\sqrt[3]{b}}$ and $\sqrt[3]{ab} = \sqrt[3]{a}\,\sqrt[3]{b}$

$= \dfrac{2x^2}{y}$

(d) $\dfrac{\sqrt{3x^4}}{\sqrt{15y}} = \sqrt{\dfrac{3x^4}{15y}}$ $\qquad$ Quotient rule

$= \sqrt{\dfrac{x^4}{5y}}$ $\qquad$ Divide out the common factor 3

$= \dfrac{\sqrt{x^4}}{\sqrt{5y}}$

$= \dfrac{x^2}{\sqrt{5y}}$

$= \dfrac{x^2\sqrt{5y}}{\sqrt{5y}\,\sqrt{5y}}$ $\qquad$ Rationalize the denominator

$= \dfrac{x^2\sqrt{5y}}{5y}$

PRACTICE EXERCISE 2

Divide and simplify.

(a) $\dfrac{\sqrt{125b^5}}{\sqrt{5b^3}}$

(b) $\dfrac{\sqrt{8x^4y^3}}{\sqrt{9x^2y^5}}$

(c) $\dfrac{\sqrt[3]{81x^9y^3}}{\sqrt[3]{3x^3y^9}}$

(d) $\dfrac{\sqrt{7x^3}}{\sqrt{14y}}$

Answers: (a) $5b$ (b) $\dfrac{2x\sqrt{2}}{3y}$ (c) $\dfrac{3x^2}{y^2}$ (d) $\dfrac{x\sqrt{2xy}}{2y}$

CAUTION

The multiplication and division rules apply only when the radicals have the same index. For example,

$$\sqrt{2}\,\sqrt[3]{5} \quad \text{and} \quad \frac{\sqrt{3}}{\sqrt[4]{2}}$$

cannot be simplified using the above rules.

EXAMPLE 3 Simplifying Products and Quotients

Perform the indicated operation and simplify.

(a) $\sqrt{4a^2y}\,\sqrt[3]{8a^3y^2} = \sqrt{2^2a^2y}\,\sqrt[3]{2^3a^3y^2}$

$= 2a\sqrt{y} \cdot 2a\sqrt[3]{y^2}$

$= 4a^2\sqrt{y}\,\sqrt[3]{y^2}$

Do not try to use the product rule on $\sqrt{y}\,\sqrt[3]{y^2}$ since the indexes are different.

(b) $\dfrac{\sqrt[4]{16x^4y}}{\sqrt[3]{8x^3y}} = \dfrac{\sqrt[4]{2^4x^4y}}{\sqrt[3]{2^3x^3y}}$

$= \dfrac{2x\sqrt[4]{y}}{2x\sqrt[3]{y}} = \dfrac{\sqrt[4]{y}}{\sqrt[3]{y}}$

We cannot use the quotient rule at this point since the indexes are different. We can, however, rationalize the denominator.

$$\frac{\sqrt[4]{y}\,\sqrt[3]{y^2}}{\sqrt[3]{y}\,\sqrt[3]{y^2}} = \frac{\sqrt[4]{y}\,\sqrt[3]{y^2}}{\sqrt[3]{y^3}} = \frac{\sqrt[4]{y}\,\sqrt[3]{y^2}}{y}.$$

Practice Exercise 3

Perform the indicated operation and simplify.

(a) $\sqrt{9xw^2}\,\sqrt[4]{16a^4x}$

(b) $\dfrac{\sqrt[3]{27a^9}}{\sqrt{9a}}$

Answers: (a) $6aw\sqrt{x}\,\sqrt[4]{x}$ (b) $a^2\sqrt{a}$

Many types of geometry problems, such as the one in the following example, involve radicals.

EXAMPLE 4 Geometry Problem

The radius r of a circle inscribed in a triangle with sides a, b, and c (see Figure 7.3) is given by

$$r = \frac{\sqrt{s-a}\,\sqrt{s-b}\,\sqrt{s-c}}{\sqrt{s}}$$

where $s = \frac{1}{2}(a + b + c)$. Express r as a single radical and use the result to find the radius of a circle inscribed in a triangle with sides 12.0 cm, 14.0 cm, and 20.0 cm.

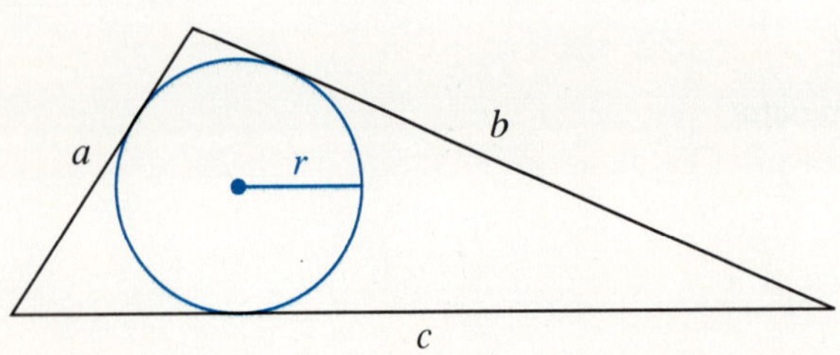

Figure 7.3

Practice Exercise 4

Find the radius of a circle inscribed in a triangle with sides 13.1 ft, 17.3 ft, and 18.4 ft.

Use the product rule to write r as

$$r = \frac{\sqrt{(s-a)(s-b)(s-c)}}{\sqrt{s}}.$$

Now use the quotient rule.

$$r = \sqrt{\frac{(s-a)(s-b)(s-c)}{s}}$$

Next we find the value of s using 12.0 for a, 14.0 for b, and 20.0 for c.

$$s = \frac{1}{2}(a+b+c)$$

$$= \frac{1}{2}(12.0 + 14.0 + 20.0) = \frac{1}{2}(46) = 23$$

Then

$$r = \sqrt{\frac{(23-12)(23-14)(23-20)}{23}}$$

$$= \sqrt{\frac{(11)(9)(3)}{23}} = \sqrt{\frac{297}{23}} \approx 3.5935.$$

The radius of the circle is approximately 3.6 cm.

Answer: 4.4 ft

7.3 EXERCISES A

Multiply and simplify.

1. $\sqrt{18}\sqrt{98}$

2. $2\sqrt{44}\sqrt{33}$

3. $\sqrt{7x}\sqrt{21x}$

4. $\sqrt{5x}\sqrt{5x}$

5 $\sqrt{5x^2y}\sqrt{35xy^2}$

6. $\sqrt{3a^3b^3}\sqrt{12a^3b^2}$

7. $(\sqrt{3xy})^2$

8 $\sqrt{5xy}\sqrt{15x^2y}\sqrt{3xy^2}$

9. $\sqrt{7a^2b}\sqrt{14ab^2}\sqrt{ab^2}$

10. $7\sqrt[3]{6}\sqrt[3]{9}$

11. $\sqrt[3]{2}\sqrt[3]{32}$

12. $\sqrt[4]{4}\sqrt[4]{4}$

13. $\sqrt[3]{3x}\sqrt[3]{9x^2}$

14. $\sqrt[4]{9x^2}\sqrt[4]{9x^2}$

15. $\sqrt[5]{16x^2}\sqrt[5]{4x^3}$

16. $\sqrt[3]{5x^2y^4}\sqrt[3]{50x^3y^2}$

17. $\sqrt[3]{6ab}\sqrt[3]{18a^2b^3}$

18 $\sqrt[4]{9x^2y^5}\sqrt[4]{45x^3y^2}$

19. $\sqrt{(x+y)^3}\sqrt{(x+y)}$

20. $\sqrt{16x^2y^2}\sqrt[3]{16x^3y^3}$

21. $\sqrt[3]{27x^6y^3}\sqrt[4]{48x^8y^5}$

Divide and simplify.

22. $\dfrac{\sqrt{2}}{\sqrt{50}}$

23. $\dfrac{10\sqrt{15}}{\sqrt{75}}$

24. $\dfrac{\sqrt{50x^3}}{\sqrt{2x}}$

25. $\dfrac{\sqrt{75x^2}}{\sqrt{3x}}$

26. $\dfrac{3\sqrt{5x^2y^3}}{\sqrt{4x^4y}}$

27. $\dfrac{\sqrt{98x^3y^5}}{\sqrt{18x^2y^7}}$

28. $\dfrac{\sqrt{25(x+y)^3}}{\sqrt{x+y}}$

29. $\dfrac{\sqrt[3]{81}}{\sqrt[3]{3}}$

30. $\dfrac{\sqrt[3]{6x^2y^3}}{\sqrt[3]{48x^5y^2}}$

31. $\dfrac{\sqrt[4]{48}}{\sqrt[4]{243}}$

32. $\dfrac{\sqrt[4]{a^7b^9}}{\sqrt[4]{16a^3b^2}}$

33. $\dfrac{\sqrt[5]{96x^8}}{\sqrt[5]{3x^2}}$

34. $\dfrac{\sqrt{3x^2}}{\sqrt{3y}}$

35. $\dfrac{\sqrt[3]{27x^3}}{\sqrt[3]{2y}}$

36. $\dfrac{\sqrt[4]{16y^5}}{\sqrt[4]{x^3y}}$

37. $\dfrac{\sqrt[3]{(2x-y)^4(x-2y)^2}}{\sqrt[3]{(2x-y)(x-2y)^5}}$

38. $\dfrac{\sqrt{9x^4y^3}}{\sqrt[3]{8x^6y^9}}$

39. $\dfrac{\sqrt[4]{16x^{12}y^9}}{\sqrt[3]{27x^{12}y^9}}$

During a scientific experiment, Dr. Gordon Johnson found that the three quantities p, q, and r are related by the equation $r = \dfrac{\sqrt{p}}{\sqrt{q}}$.

40. Find r when $p = 98$ and $q = 50$.

41. Find p when $r = 2$ and $q = 9$.

42. Find r when $p = 1$ and $q = 3$.

43. Find r when $p = 25q$.

Use the formula for the radius of a circle inscribed in a triangle obtained in Example 4 to find r for the values of a, b, and c given in Exercises 44–45.

44. $a = 8.0$ ft, $b = 30.0$ ft, $c = 26.0$ ft

45. $a = 11.2$ cm, $b = 16.4$ cm, $c = 21.8$ cm

46 Three circles with radii 3.0 cm, 4.0 cm and 5.0 cm are tangent externally as shown in the figure at right. If their centers are joined to form a triangle, what is the radius of the inscribed circle?

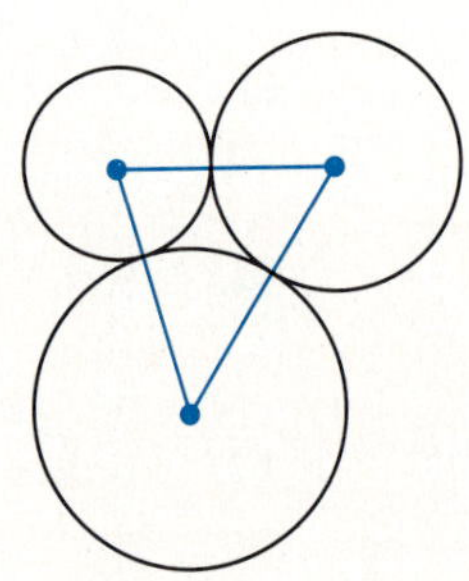

FOR REVIEW

Simplify.

47. $\sqrt{45x^2y^3}$

48. $\sqrt[3]{54a^4b^4}$

49. $\sqrt[4]{16x^9y^4z^3}$

50. $\sqrt{\dfrac{8x^2y^3}{2y}}$ **51.** $\sqrt[3]{\dfrac{27x^2y^2}{x^5y}}$ **52.** $\sqrt{\dfrac{25x^2}{y}}$

Exercises 53–60 review material covered in Sections 1.5 and 5.2 to help you prepare for the next section. Use the distributive law to collect like terms.

53. $7x + 3x$ **54.** $3y - 5y$ **55.** $2x - 5x + 8x$ **56.** $6x - 3y + 2x + y$

Multiply the polynomials.

57. $3x(5 + 2x)$ **58.** $(x - 2y)^2$ **59.** $(x + 5y)(x - 5y)$ **60.** $(3x - 2y)(2x + 5y)$

ANSWERS: 1. 42 2. $44\sqrt{3}$ 3. $7x\sqrt{3}$ 4. $5x$ 5. $5xy\sqrt{7xy}$ 6. $6a^3b^2\sqrt{b}$ 7. $3xy$ 8. $15x^2y^2$ 9. $7a^2b^2\sqrt{2b}$ 10. $21\sqrt[3]{2}$ 11. 4 12. 2 13. $3x$ 14. $3x$ 15. $2x\sqrt[5]{2}$ 16. $5xy^2\sqrt[3]{2x^2}$ 17. $3ab\sqrt[3]{4b}$ 18. $3xy\sqrt[4]{5xy^3}$ 19. $(x + y)^2$ 20. $8x^2y^2\sqrt[3]{2}$ 21. $6x^4y^2\sqrt[4]{3y}$ 22. $\frac{1}{5}$ 23. $2\sqrt{5}$ 24. $5x$ 25. $5\sqrt{x}$ 26. $\frac{3y\sqrt{5}}{2x}$ 27. $\frac{7\sqrt{x}}{3y}$ 28. $5(x + y)$ 29. 3 30. $\frac{\sqrt[3]{y}}{2x}$ 31. $\frac{2}{3}$ 32. $\frac{ab\sqrt[4]{b^3}}{2}$ 33. $2x\sqrt[5]{x}$ 34. $\frac{x\sqrt{y}}{y}$ 35. $\frac{3x\sqrt[3]{4y^2}}{2y}$ 36. $\frac{2y\sqrt[4]{x}}{x}$ 37. $\frac{2x - y}{x - 2y}$ 38. $\frac{3\sqrt{y}}{2y^2}$ 39. $\frac{2\sqrt[4]{y}}{3xy}$ 40. $\frac{7}{5}$ 41. 36 42. $\frac{\sqrt{3}}{3}$ 43. 5 44. 3.0 ft 45. 3.6 cm 46. 2.2 cm 47. $3xy\sqrt{5y}$ 48. $3ab\sqrt[3]{2ab}$ 49. $2x^2y\sqrt[4]{xz^3}$ 50. $2xy$ 51. $\frac{3\sqrt[3]{y}}{x}$ 52. $\frac{5x\sqrt{y}}{y}$ 53. $10x$ 54. $-2y$ 55. $5x$ 56. $8x - 2y$ 57. $15x + 6x^2$ 58. $x^2 - 4xy + 4y^2$ 59. $x^2 - 25y^2$ 60. $6x^2 + 11xy - 10y^2$

7.3 EXERCISES B

Multiply and simplify.

1. $\sqrt{50}\sqrt{18}$ **2.** $5\sqrt{63}\sqrt{21}$ **3.** $\sqrt{3x}\sqrt{21x}$

4. $\sqrt{8x}\sqrt{8x}$ **5.** $\sqrt{3xy^2}\sqrt{15x^2y}$ **6.** $\sqrt{5a^2b^3}\sqrt{45a^3b^3}$

7. $(\sqrt{5xy})^2$ **8.** $\sqrt{4xy}\sqrt{2xy^2}\sqrt{8x^2y}$ **9.** $\sqrt{3ab^2}\sqrt{12a^2b}\sqrt{3a^2b}$

10. $8\sqrt[3]{15}\sqrt[3]{75}$ **11.** $\sqrt[3]{9}\sqrt[3]{81}$ **12.** $\sqrt[4]{25}\sqrt[4]{25}$

13. $\sqrt[3]{5x}\sqrt[3]{25x^2}$ **14.** $\sqrt[4]{4x^2}\sqrt[4]{4x^2}$ **15.** $\sqrt[5]{8x^2}\sqrt[5]{8x^3}$

16. $\sqrt[3]{3x^4y^2}\sqrt[3]{54x^2y^3}$ **17.** $\sqrt[3]{15ab}\sqrt[3]{36a^2b^3}$ **18.** $\sqrt[4]{8x^5y^2}\sqrt[4]{10x^2y^3}$

19. $\sqrt{3(x - y)}\sqrt{3(x - y)^5}$ **20.** $\sqrt[3]{27x^6y^9}\sqrt{27x^3y^3}$ **21.** $\sqrt[3]{24x^5y^7}\sqrt[4]{16x^4y^{12}}$

Divide and simplify.

22. $\dfrac{\sqrt{3}}{\sqrt{75}}$ **23.** $\dfrac{7\sqrt{12}}{\sqrt{147}}$ **24.** $\dfrac{\sqrt{75x^3}}{\sqrt{3x}}$

25. $\dfrac{\sqrt{50x^2}}{\sqrt{2x}}$ **26.** $\dfrac{6\sqrt{7x^3y^2}}{\sqrt{9xy^4}}$ **27.** $\dfrac{\sqrt{50x^5y^3}}{\sqrt{98x^7y^2}}$

28. $\dfrac{\sqrt{36(x - y)^2}}{\sqrt{x - y}}$ **29.** $\dfrac{\sqrt[3]{250}}{\sqrt[3]{2}}$ **30.** $\dfrac{\sqrt[3]{15x^3y^2}}{\sqrt[3]{120x^2y^5}}$

31. $\dfrac{\sqrt[4]{32}}{\sqrt[4]{162}}$

32. $\dfrac{\sqrt[4]{81a^9b^7}}{\sqrt[4]{16a^2b^3}}$

33. $\dfrac{\sqrt[5]{3x^{10}}}{\sqrt[5]{96x^4}}$

34. $\dfrac{\sqrt{5y^2}}{\sqrt{5x}}$

35. $\dfrac{\sqrt[3]{27y^3}}{\sqrt[3]{2x^2}}$

36. $\dfrac{\sqrt[4]{81x^5}}{\sqrt[4]{xy^3}}$

37. $\dfrac{\sqrt[3]{(3x+y)(x-3y)^4}}{\sqrt[3]{(3x+y)^4(x-3y)}}$

38. $\dfrac{\sqrt[3]{27x^9y^4}}{\sqrt{121x^4y^6}}$

39. $\dfrac{\sqrt[3]{16x^4y^5}}{\sqrt[4]{16x^8y^{12}}}$

In a wholesale operation the three quantities c, n, and k are related by the equation $c = \sqrt{n}\,\sqrt{k}$.

40. Find c when $n = 8$ and $k = 18$.

41. Find k when $c = 6$ and $n = 4$.

42. Find c when $n = 1$ and $k = \dfrac{1}{5}$.

43. Find c when $n = \dfrac{25}{k}$.

Use the formula for the radius of a circle inscribed in a triangle obtained in Example 4 to find r for the values of a, b, and c given in Exercises 44–45.

44. $a = 7.0$ ft, $b = 25.0$ ft, $c = 20.0$ ft

45. $a = 14.3$ cm, $b = 17.1$ cm, $c = 19.3$ cm

46. Ben Whitney owns a triangular pasture with sides 75.0 yd, 100.0 yd, and 115.0 yd. He wishes to build a circular exercise track for his horses so that the circuit is as large as possible. What will be the radius of the track? [*Hint:* The largest circle in a triangle is the inscribed circle.]

FOR REVIEW

Simplify.

47. $\sqrt{20x^3y^2}$

48. $\sqrt[3]{40a^5b^5}$

49. $\sqrt[4]{81x^4y^3z^9}$

50. $\sqrt{\dfrac{45x^3y^2}{5x}}$

51. $\sqrt[3]{\dfrac{125x^4y^4}{x^3y^7}}$

52. $\sqrt{\dfrac{36y^2}{x}}$

Exercises 53–60 review material covered in Sections 1.5 and 5.2 to help you prepare for the next section. Use the distributive law to collect like terms.

53. $6y + 11y$

54. $5x - 13x$

55. $8y - 3y + 12y$

56. $4x + 7y + x - 10y$

Multiply the polynomials.

57. $2y(3 - 4y)$

58. $(y + 3x)^2$

59. $(y + 4x)(y - 4x)$

60. $(2x - 3y)(4x + 5y)$

7.3 EXERCISES C

Perform the indicated operations and simplify.

1. $\sqrt{70a^3b^4c}\,\sqrt{42ab^2c^7}$

2. $\dfrac{\sqrt[3]{256xyz^2}}{\sqrt[3]{4xy^2z^4}}$

3. $\sqrt{6x^{-3}}\,\sqrt{2x}$

4. $\dfrac{\sqrt{3x^{-1}y^3}}{\sqrt{15xy^{-1}}}$

$\left[\text{Answer: } \dfrac{y^2\sqrt{5}}{5x}\right]$

5. $\sqrt[3]{5x^{-2}y^5}\,\sqrt[3]{50x^{-1}y^{-1}}$

6. $\dfrac{\sqrt{x^2y^3}\,\sqrt{36xy^2}}{\sqrt{x^{-1}y^{-2}}\,\sqrt{25x^{-2}y^3}}$

$\left[\text{Answer: } \dfrac{6x^3y^2}{5}\right]$

7.4 MORE OPERATIONS ON RADICALS

STUDENT GUIDEPOSTS

1. Like and Unlike Radicals
2. Adding and Subtracting Radicals
3. Summary of Simplifying Techniques
4. Multiplying Binomial Radicals
5. Rationalizing a Binomial Denominator
6. Rationalizing a Binomial Numerator

1 LIKE AND UNLIKE RADICALS

There are no rules for addition and subtraction similar to the rules for multiplication and division. In general,

$$\sqrt{a+b} \neq \sqrt{a} + \sqrt{b},$$

since, for example,

$$5 = \sqrt{25} = \sqrt{9+16} \neq \sqrt{9} + \sqrt{16} = 3 + 4 = 7.$$

Also, in general,

$$\sqrt{a-b} \neq \sqrt{a} - \sqrt{b},$$

since, for example,

$$4 = \sqrt{16} = \sqrt{25-9} \neq \sqrt{25} - \sqrt{9} = 5 - 3 = 2.$$

We may, however, use the distributive laws and the rules for multiplication and division to simplify some expressions involving addition and subtraction. Before we can add or subtract radicals, they must be *like radicals*. Two radicals that have the same index and the same radicand are **like radicals.** If the indexes or the radicands differ, they are **unlike radicals.** The following tables give examples.

Like radicals	*Index*	*Radicand*
$3\sqrt{7}$, $-\sqrt{7}$, $8\sqrt{7}$	2	7
$6\sqrt[3]{x}$, $9\sqrt[3]{x}$, $-3\sqrt[3]{x}$	3	x
$-3\sqrt[4]{2xy^3}$, $5\sqrt[4]{2xy^3}$	4	$2xy^3$

Unlike radicals	*Reason*
$5\sqrt{11}$, $-3\sqrt[3]{11}$	Indexes are different
$4\sqrt{5}$, $3\sqrt{5x}$	Radicands are different
$2\sqrt[3]{xy^2}$, $2\sqrt[5]{x^2y}$	Indexes and radicands are different

2 ADDING AND SUBTRACTING RADICALS

We add or subtract like radicals using the distributive law just as we collect like terms of polynomials. Recall,

$$6\,xy - 2\,xy + 3\,xy = (6 - 2 + 3)\,xy = 7xy.$$

The process is the same for like radicals.

$$6\sqrt{xy} - 2\sqrt{xy} + 3\sqrt{xy} = (6 - 2 + 3)\sqrt{xy} = 7\sqrt{xy}$$

EXAMPLE 1 Adding and Subtracting Like Radicals

Add or subtract and simplify.

(a) $2\sqrt{3} + 7\sqrt{3} = (2 + 7)\sqrt{3} = 9\sqrt{3}$ Distributive law

(b) $\sqrt{x} - 3\sqrt{x} = (1 - 3)\sqrt{x} = -2\sqrt{x}$ Distributive law

(c) $5\sqrt{2x} + 3\sqrt{x}$ cannot be simplified since the radicands are different.

(d) $5\sqrt[3]{y} + 3\sqrt[4]{y}$ cannot be simplified since the indexes are different.

(e) $6\sqrt[4]{xy} - 3\sqrt[4]{xy} + 2\sqrt[3]{xy} = (6 - 3)\sqrt[4]{xy} + 2\sqrt[3]{xy}$ Distributive law

$= 3\sqrt[4]{xy} + 2\sqrt[3]{xy}$ These two terms cannot be combined

Practice Exercise 1

Add or subtract and simplify.

(a) $7\sqrt{5} + 10\sqrt{5}$

(b) $2\sqrt{y} - 6\sqrt{y}$

(c) $-3\sqrt[3]{5y} + 8\sqrt[3]{6y}$

(d) $9\sqrt[4]{11x} - 5\sqrt[3]{11x}$

(e) $6\sqrt[3]{4xy} - 2\sqrt[4]{4xy} + 7\sqrt[3]{4xy}$

Answers: (a) $17\sqrt{5}$ (b) $-4\sqrt{y}$ (c) cannot be simplified (d) cannot be simplified (e) $13\sqrt[3]{4xy} - 2\sqrt[4]{4xy}$

Some unlike radicals can be changed into like radicals by simplifying one or both. For example $\sqrt{12}$ and $\sqrt{75}$ are not like radicals, but if we simplify

$$\sqrt{12} = \sqrt{4 \cdot 3} = 2\sqrt{3} \quad \text{and} \quad \sqrt{75} = \sqrt{25 \cdot 3} = 5\sqrt{3},$$

we obtain like radicals. Thus

$$\sqrt{12} + \sqrt{75} = 2\sqrt{3} + 5\sqrt{3} = 7\sqrt{3}.$$

To Add or Subtract Radicals

1. Simplify each radical as much as possible.
2. Use the distributive law to collect any like radicals.

EXAMPLE 2 Adding and Subtracting Radicals

Add or subtract.

(a) $6\sqrt{125} + 3\sqrt{20} = 6\sqrt{25 \cdot 5} + 3\sqrt{4 \cdot 5}$ Factor to obtain perfect squares

$= 6\sqrt{25}\,\sqrt{5} + 3\sqrt{4}\,\sqrt{5}$ $\sqrt{a \cdot b} = \sqrt{a} \cdot \sqrt{b}$

$= 6 \cdot 5 \cdot \sqrt{5} + 3 \cdot 2 \cdot \sqrt{5}$ $6\sqrt{25 \cdot 5} = 6 \cdot 5\sqrt{5}$ *not* $(6 + 5)\sqrt{5}$

$= 30\sqrt{5} + 6\sqrt{5}$

$= (30 + 6)\sqrt{5}$ Distributive law

$= 36\sqrt{5}$

(b) $-7\sqrt[3]{54} + 3\sqrt[3]{16} = -7\sqrt[3]{27 \cdot 2} + 3\sqrt[3]{8 \cdot 2}$ 27 and 8 are perfect cubes

$= -7\sqrt[3]{27} \cdot \sqrt[3]{2} + 3\sqrt[3]{8}\,\sqrt[3]{2}$ $\sqrt[3]{ab} = \sqrt[3]{a}\,\sqrt[3]{b}$

$= -7 \cdot 3 \cdot \sqrt[3]{2} + 3 \cdot 2 \cdot \sqrt[3]{2}$

$= -21\sqrt[3]{2} + 6\sqrt[3]{2}$

$= (-21 + 6)\sqrt[3]{2}$

$= -15\sqrt[3]{2}$

Practice Exercise 2

Add or subtract.

(a) $4\sqrt{75} + 8\sqrt{12}$

(b) $9\sqrt[3]{40} - 3\sqrt[3]{135}$

(c) $8\sqrt{2x^3} - 2x\sqrt{18x} - 5\sqrt{8x^3}$
$= 8\sqrt{x^2 \cdot 2x} - 2x\sqrt{9 \cdot 2x} - 5\sqrt{4x^2 \cdot 2x}$
$= 8\sqrt{x^2}\,\sqrt{2x} - 2x\sqrt{9}\,\sqrt{2x} - 5\sqrt{4}\,\sqrt{x^2}\,\sqrt{2x}$
$= 8x\sqrt{2x} - 2x \cdot 3\sqrt{2x} - 5 \cdot 2 \cdot x\sqrt{2x}$
$= 8x\sqrt{2x} - 6x\sqrt{2x} - 10x\sqrt{2x}$
$= (8x - 6x - 10x)\sqrt{2x}$
$= -8x\sqrt{2x}$

(d) $4\sqrt[3]{8y^5} - 3\sqrt[3]{27y^5} = 4\sqrt[3]{8 \cdot y^3 \cdot y^2} - 3\sqrt[3]{27 \cdot y^3 \cdot y^2}$
$= 4\sqrt[3]{8}\,\sqrt[3]{y^3}\,\sqrt[3]{y^2} - 3\sqrt[3]{27}\,\sqrt[3]{y^3}\,\sqrt[3]{y^2}$
$= 4 \cdot 2 \cdot y\sqrt[3]{y^2} - 3 \cdot 3 \cdot y\sqrt[3]{y^2}$
$= 8y\sqrt[3]{y^2} - 9y\sqrt[3]{y^2}$
$= -y\sqrt[3]{y^2}$

(c) $-3\sqrt{5y^5} - 2y^2\sqrt{5y} + 8y\sqrt{5y^3}$

(d) $7\sqrt[3]{27x^7} - 5x\sqrt[3]{27x^4}$

Answers: (a) $36\sqrt{3}$ (b) $9\sqrt[3]{5}$ (c) $3y^2\sqrt{5y}$ (d) $6x^2\sqrt[3]{x}$

Sometimes a radical expression can be simplified by first rationalizing a denominator and then adding or subtracting.

EXAMPLE 3 Rationalizing before Adding or Subtracting

Rationalize the denominator and then add or subtract.

(a) $3\sqrt{2x} + \dfrac{5x}{\sqrt{2x}} = 3\sqrt{2x} + \dfrac{5x\sqrt{2x}}{\sqrt{2x}\,\sqrt{2x}}$ Rationalize the denominator
$= 3\sqrt{2x} + \dfrac{5x\sqrt{2x}}{2x}$
$= 3\sqrt{2x} + \dfrac{5\sqrt{2x}}{2}$ Divide out the common factors
$= \left(3 + \dfrac{5}{2}\right)\sqrt{2x}$
$= \dfrac{11}{2}\sqrt{2x} = \dfrac{11\sqrt{2x}}{2}$

(b) $\dfrac{3}{\sqrt[3]{y}} - \dfrac{7\sqrt[3]{y^2}}{y} = \dfrac{3\sqrt[3]{y^2}}{\sqrt[3]{y}\,\sqrt[3]{y^2}} - \dfrac{7\sqrt[3]{y^2}}{y}$ Rationalize the denominator
$= \dfrac{3\sqrt[3]{y^2}}{y} - \dfrac{7\sqrt[3]{y^2}}{y}$
$= \left(\dfrac{3}{y} - \dfrac{7}{y}\right)\sqrt[3]{y^2}$
$= -\dfrac{4}{y}\sqrt[3]{y^2} = \dfrac{-4\sqrt[3]{y^2}}{y}$

Practice Exercise 3

Rationalize the denominator and then add or subtract.

(a) $\dfrac{7y}{\sqrt{5y}} + 6\sqrt{5y}$

(b) $\dfrac{9\sqrt[3]{2y}}{2y} - \dfrac{13}{\sqrt[3]{4y^2}}$

Answers: (a) $\frac{37\sqrt{5y}}{5}$ (b) $-\frac{2\sqrt[3]{2y}}{y}$

3 SUMMARY OF SIMPLIFYING TECHNIQUES

On the next page we present a summary of all the techniques used to simplify radical expressions.

To Simplify a Radical Expression

1. Combine all like radicals by adding and subtracting.
2. When indicated, use the rules for multiplication and division,
$$\sqrt[k]{a}\sqrt[k]{b} = \sqrt[k]{ab} \quad \text{and} \quad \frac{\sqrt[k]{a}}{\sqrt[k]{b}} = \sqrt[k]{\frac{a}{b}}.$$
3. Remove all perfect powers from under the radicals.
4. Rationalize all denominators.

These techniques, as well as the distributive laws, are used in the following multiplication problems.

EXAMPLE 4 MULTIPLYING RADICAL EXPRESSIONS

Multiply.

(a) $\sqrt{7}(\sqrt{35} + \sqrt{7}) = \sqrt{7}\sqrt{35} + \sqrt{7}\sqrt{7}$ Distributive law

$= \sqrt{7 \cdot 35} + \sqrt{7 \cdot 7}$ $\sqrt{a}\sqrt{b} = \sqrt{ab}$

$= \sqrt{7 \cdot 7 \cdot 5} + \sqrt{7 \cdot 7}$ Factor

$= \sqrt{7^2 \cdot 5} + \sqrt{7^2}$ 7^2 is a perfect square

$= \sqrt{7^2}\sqrt{5} + \sqrt{7^2}$

$= 7\sqrt{5} + 7$

(b) $\sqrt{3}(\sqrt{15} - \sqrt{60}) = \sqrt{3}\sqrt{15} - \sqrt{3}\sqrt{60}$ Distributive law

$= \sqrt{3 \cdot 15} - \sqrt{3 \cdot 60}$ $\sqrt{a}\sqrt{b} = \sqrt{ab}$

$= \sqrt{3^2 \cdot 5} - \sqrt{3^2 \cdot 2^2 \cdot 5}$ 3^2 and 2^2 are perfect squares

$= \sqrt{3^2}\sqrt{5} - \sqrt{3^2}\sqrt{2^2}\sqrt{5}$ $\sqrt{ab} = \sqrt{a}\sqrt{b}$

$= 3\sqrt{5} - 3 \cdot 2\sqrt{5}$ Remove perfect squares

$= 3\sqrt{5} - 6\sqrt{5} = -3\sqrt{5}$

(c) $\sqrt[3]{x}(2\sqrt[3]{8x^2} - \sqrt[3]{27x^2})$

$= 2 \cdot \sqrt[3]{x}\sqrt[3]{8x^2} - \sqrt[3]{x}\sqrt[3]{27x^2}$ Distributive law

$= 2\sqrt[3]{x \cdot 8x^2} - \sqrt[3]{x \cdot 27x^2}$ $\sqrt[3]{a}\sqrt[3]{b} = \sqrt[3]{ab}$

$= 2\sqrt[3]{8x^3} - \sqrt[3]{27x^3}$

$= 2\sqrt[3]{8}\sqrt[3]{x^3} - \sqrt[3]{27}\sqrt[3]{x^3}$ $\sqrt[3]{ab} = \sqrt[3]{a}\sqrt[3]{b}$

$= 2 \cdot 2 \cdot x - 3 \cdot x$ Remove perfect cubes

$= 4x - 3x = x$

PRACTICE EXERCISE 4

Multiply.

(a) $\sqrt{11}(\sqrt{22} + \sqrt{11})$

(b) $\sqrt{5}(\sqrt{35} - \sqrt{140})$

(c) $\sqrt[3]{2y}(4\sqrt[3]{32y^2} - \sqrt[3]{108y^2})$

Answers: (a) $11\sqrt{2} + 11$ (b) $-5\sqrt{7}$ (c) $10y$

4 MULTIPLYING BINOMIAL RADICALS

Many radical expressions can be multiplied just as we multiplied binomials in Chapter 5. The FOIL method can be used.

EXAMPLE 5 Multiplying Binomial Radicals

Multiply.

(a) $(\sqrt{3} - \sqrt{2})(\sqrt{3} + 5\sqrt{2})$

$$\begin{aligned} &\overset{F}{=} \sqrt{3}\,\sqrt{3} + \overset{O}{\sqrt{3}(5\sqrt{2})} - \overset{I}{\sqrt{2}\,\sqrt{3}} - \overset{L}{\sqrt{2}(5\sqrt{2})} \quad \text{Use FOIL} \\ &= 3 + 5\sqrt{6} - \sqrt{6} - 5(2) \\ &= 3 - 10 + 5\sqrt{6} - \sqrt{6} \\ &= -7 + 4\sqrt{6} \end{aligned}$$

(b)
$$\begin{aligned} (\sqrt{x} - 5)^2 &= (\sqrt{x} - 5)(\sqrt{x} - 5) \\ &= \sqrt{x}\,\sqrt{x} - 5\sqrt{x} - 5\sqrt{x} + 5 \cdot 5 \quad \text{Use FOIL} \\ &= \sqrt{x^2} - 10\sqrt{x} + 25 \\ &= x - 10\sqrt{x} + 25 \end{aligned}$$

(c) $(\sqrt{5} + \sqrt{7})(\sqrt{5} - \sqrt{7})$

$$\begin{aligned} &= \sqrt{5}\,\sqrt{5} - \sqrt{5}\,\sqrt{7} + \sqrt{5}\,\sqrt{7} - \sqrt{7}\,\sqrt{7} \quad \text{Use FOIL} \\ &= \sqrt{5^2} - \sqrt{35} + \sqrt{35} - \sqrt{7^2} \\ &= 5 - 7 = -2 \end{aligned}$$

Practice Exercise 5

Multiply.

(a) $(\sqrt{5} + \sqrt{11})(\sqrt{5} - 3\sqrt{11})$

(b) $(\sqrt{2y} + 7)^2$

(c) $(\sqrt{13} + \sqrt{3})(\sqrt{13} - \sqrt{3})$

Answers: (a) $-28 - 2\sqrt{55}$ (b) $2y + 14\sqrt{2y} + 49$ (c) 10

5 RATIONALIZING A BINOMIAL DENOMINATOR

Notice that in Example 5(c) the product was a rational number. This occurred because the expressions multiplied fit the formula, $(a + b)(a - b) = a^2 - b^2$, which always gives a rational number.

$$\begin{aligned} (\sqrt{5} + \sqrt{7})(\sqrt{5} - \sqrt{7}) &= (\sqrt{5})^2 - (\sqrt{7})^2 \quad \sqrt{5} = a \text{ and } \sqrt{7} = b \\ &= 5 - 7 = -2 \end{aligned}$$

Square root expressions such as $\sqrt{5} + \sqrt{7}$ and $\sqrt{5} - \sqrt{7}$ are called **conjugates** of each other. Thus $\sqrt{5} - \sqrt{7}$ is the conjugate of $\sqrt{5} + \sqrt{7}$, and $\sqrt{5} + \sqrt{7}$ is the conjugate of $\sqrt{5} - \sqrt{7}$. Since the product of conjugates is a rational number (-2 is our example), we can use this concept to rationalize binomial denominators (denominators with two terms). First consider several examples.

Binomial	*Conjugate*	*Product*
$\sqrt{7} - \sqrt{5}$	$\sqrt{7} + \sqrt{5}$	$(\sqrt{7} - \sqrt{5})(\sqrt{7} + \sqrt{5}) = (\sqrt{7})^2 - (\sqrt{5})^2 = 7 - 5 = 2$
$2\sqrt{3} + \sqrt{2}$	$2\sqrt{3} - \sqrt{2}$	$(2\sqrt{3} + \sqrt{2})(2\sqrt{3} - \sqrt{2}) = (2\sqrt{3})^2 - (\sqrt{2})^2 = 4 \cdot 3 - 2 = 10$
$\sqrt{x} - \sqrt{y}$	$\sqrt{x} + \sqrt{y}$	$(\sqrt{x} - \sqrt{y})(\sqrt{x} + \sqrt{y}) = (\sqrt{x})^2 - (\sqrt{y})^2 = x - y$
$3\sqrt{x} + 2$	$3\sqrt{x} - 2$	$(3\sqrt{x} + 2)(3\sqrt{x} - 2) = (3\sqrt{x})^2 - (2)^2 = 9x - 4$

To Rationalize a Binomial Denominator

1. Simplify the fraction as much as possible.
2. Multiply both numerator and denominator of the fraction by the conjugate of the denominator and simplify.

EXAMPLE 6 Rationalizing a Binomial Denominator

Rationalize the denominator.

(a) $\frac{2}{\sqrt{7}-\sqrt{5}} = \frac{2(\sqrt{7}+\sqrt{5})}{(\sqrt{7}-\sqrt{5})(\sqrt{7}+\sqrt{5})}$ Multiply numerator and denominator by $\sqrt{7}+\sqrt{5}$

$= \frac{2(\sqrt{7}+\sqrt{5})}{(\sqrt{7})^2-(\sqrt{5})^2}$ $(a-b)(a+b)=a^2-b^2$

$= \frac{2(\sqrt{7}+\sqrt{5})}{7-5}$

$= \frac{\cancel{2}(\sqrt{7}+\sqrt{5})}{\cancel{2}} = \sqrt{7}+\sqrt{5}$

(b) $\frac{\sqrt{3}}{5+2\sqrt{3}} = \frac{\sqrt{3}(5-2\sqrt{3})}{(5+2\sqrt{3})(5-2\sqrt{3})}$ Multiply numerator and denominator by $5-2\sqrt{3}$

$= \frac{\sqrt{3}(5-2\sqrt{3})}{(5)^2-(2\sqrt{3})^2}$ $(a+b)(a-b)=a^2-b^2$

$= \frac{\sqrt{3}(5-2\sqrt{3})}{25-4\cdot 3}$ $(2\sqrt{3})^2 = 2^2(\sqrt{3})^2 = 4\cdot 3$

$= \frac{5\sqrt{3}-2\sqrt{3}\,\sqrt{3}}{25-12}$

$= \frac{5\sqrt{3}-6}{13}$ $2\sqrt{3}\,\sqrt{3} = 2(\sqrt{3})^2 = 2\cdot 3 = 6$

(c) $\frac{x-9}{\sqrt{x}-3} = \frac{(x-9)(\sqrt{x}+3)}{(\sqrt{x}-3)(\sqrt{x}+3)}$ Multiply numerator and denominator by $\sqrt{x}+3$

$= \frac{(x-9)(\sqrt{x}+3)}{(\sqrt{x})^2-(3)^2}$ $(a-b)(a+b)=a^2-b^2$

$= \frac{\cancel{(x-9)}(\sqrt{x}+3)}{\cancel{(x-9)}}$ Divide out common factors

$= \sqrt{x}+3$

Practice Exercise 6

Rationalize the denominator.

(a) $\frac{5}{\sqrt{13}+\sqrt{3}}$

(b) $\frac{\sqrt{7}}{3\sqrt{7}-2}$

(c) $\frac{x-25}{\sqrt{x}+5}$

Answers: (a) $\frac{\sqrt{13}-\sqrt{3}}{2}$ (b) $\frac{21+2\sqrt{7}}{59}$ (c) $\sqrt{x}-5$

CAUTION

The technique for rationalizing a binomial denominator works only when the radicals involved are square roots. For example, it will not rationalize the denominator in

$$\frac{1}{\sqrt[3]{x}+\sqrt[3]{y}}$$

if we multiply numerator and denominator by $\sqrt[3]{x}-\sqrt[3]{y}$ or if we try $\sqrt[3]{x^2}-\sqrt[3]{y^2}$.

6 RATIONALIZING A BINOMIAL NUMERATOR

A binomial numerator can also be rationalized using a similar technique. This time, however, we multiply the numerator and denominator by the conjugate of the *numerator*.

$$\frac{\sqrt{2}+\sqrt{5}}{3} = \frac{(\sqrt{2}+\sqrt{5})(\sqrt{2}-\sqrt{5})}{3(\sqrt{2}-\sqrt{5})}$$ Multiply numerator and denominator by $\sqrt{2}-\sqrt{5}$, the conjugate of the numerator

$$= \frac{(\sqrt{2})^2-(\sqrt{5})^2}{3(\sqrt{2}-\sqrt{5})}$$ Use $(a+b)(a-b)=a^2-b^2$

$$= \frac{2-5}{3(\sqrt{2}-\sqrt{5})}$$ Simplify

$$= \frac{-\cancel{3}}{\cancel{3}(\sqrt{2}-\sqrt{5})}$$ Divide out common factor 3 to reduce the fraction

$$= \frac{-1}{\sqrt{2}-\sqrt{5}}$$

CAUTION

Notice that we did not use the distributive law to write $3(\sqrt{2}-\sqrt{5})$ in the form $3\sqrt{2}-3\sqrt{5}$ in the denominators above. It is best to leave expressions in factored form so we can reduce the fraction (if possible) more easily.

Some equations we will study in the next chapter have solutions similar in form to

$$\frac{-2+2\sqrt{5}}{6}.$$

Such expressions can often be reduced to lowest terms by factoring the numerator and dividing out all common factors. For example,

$$\frac{-2+2\sqrt{5}}{6} = \frac{\cancel{2}(-1+\sqrt{5})}{\cancel{2}\cdot 3} = \frac{-1+\sqrt{5}}{3}.$$

7.4 EXERCISES A

Add or subtract.

1. $\sqrt{5}+3\sqrt{5}$

2. $4\sqrt{7}-9\sqrt{7}$

3. $3\sqrt{5x}-4\sqrt{5x}$

4. $6\sqrt[3]{5}-4\sqrt[4]{5}$

5. $6\sqrt[3]{x^2y}-7\sqrt[3]{x^2y}$

6. $3\sqrt[4]{3y^3}+8\sqrt[4]{3x^3}$

7. $\frac{1}{2}\sqrt{32}-\frac{1}{3}\sqrt{162}$

8. $2\sqrt[3]{2}+6\sqrt[3]{16}$

9 $9\sqrt[3]{250}-4\sqrt[3]{128}$

10. $3\sqrt[4]{48}+\sqrt[4]{243}$

11. $\sqrt[4]{625}-\sqrt[3]{125}$

12. $4\sqrt{98}-3\sqrt{72}+2\sqrt{32}$

13. $-6\sqrt{16y^3} + 2\sqrt{9y^3}$

14 $6\sqrt{48y^4} + 4\sqrt{12y^4}$

15. $y\sqrt{x^3} + x\sqrt{xy^2}$

16. $7\sqrt{5xy^2} - 3\sqrt{5xy^2}$

17. $6\sqrt[3]{8x} - 5\sqrt[3]{27x}$

18. $5\sqrt[3]{3x^4} + 7x\sqrt[3]{3x}$

19. $-7\sqrt[3]{40a^5} + 10a\sqrt[3]{135a^2}$

20. $-\sqrt[4]{x^5y^7} + 2x\sqrt[4]{xy^7}$

21 $3\sqrt[5]{a^{10}b^{12}} - ab\sqrt[5]{a^5b^7}$

22. $-8\sqrt{75a^2b^5} + 3b\sqrt{147a^2b^3}$

23. $2\sqrt[3]{27x^4y^6} - 4y\sqrt[3]{125x^4y^3}$

24. $8\sqrt[3]{54} - 4\sqrt[3]{16} - 5\sqrt[3]{250}$

Rationalize the denominator and then add or subtract.

25. $3\sqrt{5} - \dfrac{1}{\sqrt{5}}$

26. $\dfrac{6}{\sqrt{7}} - 2\sqrt{7}$

27. $3\sqrt{12} - \dfrac{5}{\sqrt{3}}$

28. $6\sqrt{5x} - \dfrac{x}{\sqrt{5x}}$

29. $\dfrac{8\sqrt{xy}}{x} + \dfrac{7y}{\sqrt{xy}}$

30 $\dfrac{\sqrt[3]{x^2y^2}}{xy} + \dfrac{1}{\sqrt[3]{xy}}$

An engineer finds that P, s, and t are related by the equation $P = 2\sqrt{s} + \frac{6}{\sqrt{t}}$.

31. Find P when $s = 27$ and $t = 3$.

32. Find s when $P = 3$ and $t = 36$.

Multiply and simplify.

33. $\sqrt{3}(\sqrt{5} + \sqrt{3})$

34. $\sqrt{5}(\sqrt{15} - \sqrt{35})$

35. $\sqrt{3}(2 + \sqrt{6})$

36. $\sqrt{3}\left(\sqrt{3} + \dfrac{2}{\sqrt{3}}\right)$

37. $\sqrt{8}(\sqrt{2} + 3\sqrt{12})$

38. $\sqrt{xy}\left(\dfrac{1}{\sqrt{x}} + \dfrac{1}{\sqrt{y}}\right)$

39. $\sqrt{3x}(\sqrt{3} - \sqrt{x})$

40 $\sqrt[4]{x^3}(\sqrt[4]{xy^3} + \sqrt[4]{xy^4})$

41. $\sqrt[3]{5}(\sqrt[3]{25} - \sqrt[3]{200})$

42. $(\sqrt{5} + \sqrt{2})(\sqrt{5} + \sqrt{2})$

43. $(3\sqrt{3} + \sqrt{7})(3\sqrt{3} - \sqrt{7})$

44 $(2\sqrt{5} - \sqrt{2})(3\sqrt{5} + 4\sqrt{2})$

45. $(\sqrt{12} - \sqrt{5})^2$

46. $(\sqrt{x} + 1)(\sqrt{x} - 1)$

47. $(\sqrt{x} + 1)^2$

48. $(\sqrt{y - 3})^2$

49. $(\sqrt{y} + 5)(\sqrt{y} - 3)$

50. $(2\sqrt{x} + \sqrt{y})(2\sqrt{x} - \sqrt{y})$

Rationalize the denominators.

51. $\dfrac{1}{2 + \sqrt{3}}$

52. $\dfrac{4}{1 - \sqrt{5}}$

53. $\dfrac{\sqrt{7}}{\sqrt{2} - \sqrt{3}}$

54 $\dfrac{\sqrt{2}}{2\sqrt{2} + \sqrt{3}}$

55. $\dfrac{\sqrt{8} - \sqrt{3}}{\sqrt{2} + \sqrt{3}}$

56. $\dfrac{\sqrt{x} + 1}{\sqrt{x} - 1}$

57. $\dfrac{\sqrt{x} + \sqrt{y}}{\sqrt{x} - \sqrt{y}}$

58 $\dfrac{4x - 25}{2\sqrt{x} - 5}$

Rationalize the numerators.

59. $\dfrac{\sqrt{3} + 2}{4}$

60. $\dfrac{2 - \sqrt{2}}{2}$

61. $\dfrac{\sqrt{5} + \sqrt{3}}{\sqrt{3}}$

62. $\dfrac{\sqrt{2} - 1}{\sqrt{2} + 1}$

63. $\dfrac{\sqrt{y} - 1}{\sqrt{y} + 1}$

64. $\dfrac{\sqrt{y} - \sqrt{x}}{\sqrt{y} + \sqrt{x}}$

65. $\dfrac{\sqrt{x} + 2}{x - 4}$

66 $\dfrac{\sqrt{x + h} - \sqrt{x}}{h}$

Reduce to lowest terms.

67. $\dfrac{3 + 6\sqrt{2}}{3}$

68 $\dfrac{5 + \sqrt{125}}{10}$

69. $\dfrac{4 - \sqrt{8}}{2}$

Dr. Gordon Johnson's assistant, Burford, obtained the false results in Exercises 70–72. How would you correct the right side of each equation to keep Burford from being fired?

70. $5\sqrt{2y} = \sqrt{10y}$

71. $(\sqrt{y} + 2)^2 = y + 4$

72. $(4\sqrt{3})^2 = 12$

FOR REVIEW

Simplify.

73. $\sqrt[3]{5xy^3}\,\sqrt[3]{20x^2y}\,\sqrt[3]{5x^2y}$

74. $\dfrac{\sqrt{98x^7y^3}}{\sqrt{50x^5y^2}}$

75. $\dfrac{\sqrt[3]{250x^4}}{\sqrt[3]{2y}}$

The following exercises review material from Section 1.7 and will help you prepare for the next section. Use the rules of exponents to simplify.

76. $(x^2y)^3$

77. $\left(\dfrac{x^3y^5}{xy^2}\right)^4$

78. $\left(\dfrac{x^{-4}y}{x^{-1}y^3}\right)^{-2}$

ANSWERS: 1. $4\sqrt{5}$ 2. $-5\sqrt{7}$ 3. $-\sqrt{5x}$ 4. cannot be simplified 5. $-\sqrt[3]{x^2y}$ 6. cannot be simplified 7. $-\sqrt{2}$ 8. $14\sqrt[3]{2}$ 9. $29\sqrt[3]{2}$ 10. $9\sqrt[4]{3}$ 11. 0 12. $18\sqrt{2}$ 13. $-18y\sqrt{y}$ 14. $32y^2\sqrt{3}$ 15. $2xy\sqrt{x}$ 16. $4y\sqrt{5x}$ 17. $-3\sqrt[3]{x}$ 18. $12x\sqrt[3]{3x}$ 19. $16a\sqrt[3]{5a^2}$ 20. $xy\sqrt[4]{xy^3}$ 21. $2a^2b^2\sqrt[5]{b^2}$ 22. $-19ab^2\sqrt{3b}$ 23. $-14xy^2\sqrt[3]{x}$ 24. $-9\sqrt[3]{2}$ 25. $\frac{14\sqrt{5}}{5}$ 26. $\frac{-8\sqrt{7}}{7}$ 27. $\frac{13\sqrt{3}}{3}$ 28. $\frac{29\sqrt{5x}}{5}$ 29. $\frac{15\sqrt{xy}}{x}$ 30. $\frac{2\sqrt[3]{x^2y^2}}{xy}$ 31. $8\sqrt{3}$ 32. 1 33. $\sqrt{15}+3$ 34. $5\sqrt{3}-5\sqrt{7}$ 35. $2\sqrt{3}+3\sqrt{2}$ 36. 5 37. $4+12\sqrt{6}$ 38. $\sqrt{y}+\sqrt{x}$ 39. $3\sqrt{x}-x\sqrt{3}$ 40. $x\sqrt[4]{y^3}+xy$ 41. -5 42. $7+2\sqrt{10}$ 43. 20 44. $22+5\sqrt{10}$ 45. $17-4\sqrt{15}$ 46. $x-1$ 47. $x+2\sqrt{x}+1$ 48. $y-3$ 49. $y+2\sqrt{y}-15$ 50. $4x-y$ 51. $2-\sqrt{3}$ 52. $-1-\sqrt{5}$ 53. $-\sqrt{14}-\sqrt{21}$ 54. $\frac{4-\sqrt{6}}{5}$ 55. $3\sqrt{6}-7$ 56. $\frac{x+2\sqrt{x}+1}{x-1}$ 57. $\frac{x+2\sqrt{xy}+y}{x-y}$ 58. $2\sqrt{x}+5$ 59. $\frac{-1}{4(\sqrt{3}-2)}$ 60. $\frac{1}{2+\sqrt{2}}$ 61. $\frac{2}{\sqrt{15}-3}$ 62. $\frac{1}{(\sqrt{2}+1)^2}$ 63. $\frac{y-1}{(\sqrt{y}+1)^2}$ 64. $\frac{y-x}{(\sqrt{y}+\sqrt{x})^2}$ 65. $\frac{1}{\sqrt{x}-2}$ 66. $\frac{1}{\sqrt{x+h}+\sqrt{x}}$ 67. $1+2\sqrt{2}$ 68. $\frac{1+\sqrt{5}}{2}$ 69. $2-\sqrt{2}$ 70. $\sqrt{50y}$ 71. $y+4\sqrt{y}+4$ 72. 48 73. $5xy\sqrt[3]{4x^2y^2}$ 74. $\frac{7x\sqrt{y}}{5}$ 75. $\frac{5x\sqrt[3]{xy^2}}{y}$ 76. x^6y^3 77. x^8y^{12} 78. x^6y^4

7.4 EXERCISES B

Add or subtract.

1. $\sqrt{3}+8\sqrt{3}$

2. $8\sqrt{5}-11\sqrt{5}$

3. $4\sqrt{3x}-6\sqrt{3x}$

4. $7\sqrt[3]{11}-3\sqrt[4]{11}$

5. $5\sqrt[3]{xy^2}-9\sqrt[3]{xy^2}$

6. $3\sqrt[4]{7x}+9\sqrt[4]{7y}$

7. $\dfrac{2}{3}\sqrt{45}-\dfrac{3}{4}\sqrt{80}$

8. $8\sqrt[3]{3}+2\sqrt[3]{24}$

9. $7\sqrt[3]{128}-\sqrt[3]{54}$

10. $2\sqrt[4]{162}+\sqrt[4]{32}$

11. $\sqrt[3]{64}-\sqrt[3]{216}$

12. $3\sqrt{147}+2\sqrt{108}-2\sqrt{48}$

13. $-2\sqrt{25y^3}+5\sqrt{49y^3}$

14. $3\sqrt{8y^6}+2\sqrt{18y^6}$

15. $5x\sqrt{y^3}-y\sqrt{x^2y}$

16. $6\sqrt{3x^2y}-2\sqrt{3x^2y}$

17. $8\sqrt[3]{125x}-10\sqrt[3]{64x}$

18. $3x\sqrt[3]{2x}+5\sqrt[3]{2x^4}$

19. $-5\sqrt[3]{24a^5}+3a\sqrt[3]{81a^2}$

20. $-2\sqrt[4]{x^7y^5}+3y\sqrt[4]{x^7y}$

21. $2ab\sqrt[5]{2a^7b^5}-\sqrt[5]{2a^{12}b^{10}}$

22. $-4\sqrt{125a^5b^2}+5a\sqrt{20a^3b^2}$

23. $5\sqrt[3]{8x^6y^4}-3x\sqrt[3]{27x^3y^4}$

24. $2\sqrt[3]{81}+3\sqrt[3]{24}-4\sqrt[3]{375}$

Rationalize the denominator and then add or subtract.

25. $2\sqrt{7}-\dfrac{3}{\sqrt{7}}$

26. $\dfrac{8}{\sqrt{5}}-3\sqrt{5}$

27. $2\sqrt{27}-\dfrac{8}{\sqrt{3}}$

28. $2\sqrt{7x}-\dfrac{3x}{\sqrt{7x}}$

29. $\dfrac{5\sqrt{xy}}{y}+\dfrac{3x}{\sqrt{xy}}$

30. $\dfrac{\sqrt[3]{xy}}{xy}+\dfrac{1}{\sqrt[3]{x^2y^2}}$

An economist finds that C, n, and r are related by the equation $C = 3\sqrt{n} - \frac{10}{\sqrt{r}}$.

31. Find C when $n = 18$ and $r = 2$.

32. Find n when $C = 2$ and $r = 100$.

Multiply and simplify.

33. $\sqrt{5}(\sqrt{3} + \sqrt{5})$

34. $\sqrt{3}(\sqrt{21} - \sqrt{15})$

35. $\sqrt{5}(2 + \sqrt{10})$

36. $\sqrt{2}\left(\sqrt{2} - \frac{3}{\sqrt{2}}\right)$

37. $\sqrt{27}(\sqrt{3} + 2\sqrt{12})$

38. $\sqrt{xy}\left(\frac{1}{\sqrt{y}} - \frac{1}{\sqrt{x}}\right)$

39. $\sqrt{5x}(\sqrt{5} + \sqrt{x})$

40. $\sqrt[4]{y^3}(\sqrt[4]{x^3y} - \sqrt[4]{x^4y})$

41. $\sqrt[3]{3}(\sqrt[3]{9} + \sqrt[3]{72})$

42. $(\sqrt{7} - \sqrt{3})(\sqrt{7} - \sqrt{3})$

43. $(2\sqrt{5} + \sqrt{3})(2\sqrt{5} - \sqrt{3})$

44. $(\sqrt{3} - 4\sqrt{7})(2\sqrt{3} + \sqrt{7})$

45. $(\sqrt{18} - \sqrt{2})^2$

46. $(\sqrt{x} - 2)(\sqrt{x} + 2)$

47. $(\sqrt{x} + 2)^2$

48. $(\sqrt{y - 5})^2$

49. $(\sqrt{x} + 1)(\sqrt{x} - 3)$

50. $(\sqrt{x} - 3\sqrt{y})(\sqrt{x} + 3\sqrt{y})$

Rationalize the denominators.

51. $\frac{3}{1 - \sqrt{2}}$

52. $\frac{2}{2 + \sqrt{3}}$

53. $\frac{\sqrt{2}}{\sqrt{3} + \sqrt{7}}$

54. $\frac{\sqrt{3}}{3\sqrt{3} + \sqrt{2}}$

55. $\frac{\sqrt{8} - 3}{3\sqrt{2} - 2}$

56. $\frac{\sqrt{y} - 1}{\sqrt{y} + 1}$

57. $\frac{\sqrt{x} - \sqrt{y}}{\sqrt{x} + \sqrt{y}}$

58. $\frac{4x - 9}{2\sqrt{x} + 3}$

Rationalize the numerators.

59. $\frac{\sqrt{3} - 2}{4}$

60. $\frac{2 + \sqrt{2}}{2}$

61. $\frac{\sqrt{2} - \sqrt{5}}{\sqrt{3}}$

62. $\frac{\sqrt{3} + 1}{\sqrt{3} - 1}$

63. $\frac{\sqrt{x} + 1}{\sqrt{x} - 1}$

64. $\frac{\sqrt{a} + \sqrt{b}}{a - b}$

65. $\frac{\sqrt{y} - 2}{y - 4}$

66. $\frac{\sqrt{3(x + h)} - \sqrt{3x}}{h}$

Reduce to lowest terms.

67. $\frac{8 - \sqrt{48}}{6}$

68. $\frac{7 - 14\sqrt{3}}{7}$

69. $\frac{3 - \sqrt{18}}{3}$

Dr. Gordon Johnson's assistant, Burford, obtained the false results in Exercises 70–72. How would you correct the right side of each equation to keep Burford from being fired?

70. $3(2\sqrt{xy}) = 6\sqrt{3xy}$

71. $(\sqrt{x} - 5)^2 = x - 25$

72. $(2 + \sqrt{x - 1})^2 = 4 + (x - 1)$

FOR REVIEW

Simplify.

73. $\sqrt[3]{7x^2y}\,\sqrt[3]{28xy^2}\,\sqrt[3]{7xy^2}$

74. $\frac{\sqrt{32x^8y^5}}{\sqrt{162x^7y^3}}$

75. $\frac{\sqrt[3]{54y^4}}{\sqrt[3]{2x}}$

The following exercises review material from Section 1.7 and will help you prepare for the next section. Use the rules of exponents to simplify.

76. $(xy^3)^4$

77. $\left(\frac{x^6y^7}{x^5y}\right)^3$

78. $\left(\frac{x^{-1}y^{-6}}{x^3y^{-2}}\right)^{-3}$

7.4 EXERCISES C

Add or subtract.

1. $3a\sqrt{(a-b)^3} + 2ab\sqrt{a-b} + b\sqrt{(a-b)^3}$
[Answer: $(3a^2 - b^2)\sqrt{a-b}$]

2. $\sqrt[5]{64x^7y^5} - x\sqrt[5]{486x^2y^5}$

Rationalize the denominator and add or subtract.

3. $\sqrt{20} + \sqrt{\frac{1}{5}} - \frac{\sqrt{5}}{5}$

4. $\dfrac{\sqrt{x} + \frac{1}{\sqrt{x}}}{\frac{1}{\sqrt{x}}}$ [*Hint:* Multiply numerator and denominator by $\sqrt{x}$.]

5. $\sqrt{5x^2 + 10x + 5} + \sqrt{20x^2 + 40x + 20}$

6. $5x\sqrt[3]{16xy^3} + 3\sqrt[3]{54x^4y^3} - xy\sqrt[3]{2x}$

Multiply and simplify.

7. $(\sqrt[4]{a} - \sqrt[4]{b^3})(\sqrt[4]{a^3} + \sqrt[4]{b})$

8. $(\sqrt{3+\sqrt{5}} - \sqrt{3-\sqrt{5}})^2$
[Answer: 2]

Rationalize the denominators.

9. $\dfrac{5}{5 - \sqrt{x-5}}$

10. $\dfrac{x - \sqrt{y}}{x - y - \sqrt{y}}$

11. Multiply.
$(\sqrt[3]{a} + \sqrt[3]{b})(\sqrt[3]{a^2} - \sqrt[3]{ab} + \sqrt[3]{b^2})$
[Answer: $a + b$]

12. Rationalize the denominator.
$\dfrac{1}{\sqrt[3]{a} + \sqrt[3]{b}}$
[*Hint*: Consider Exercise 11.]

7.5 RATIONAL EXPONENTS

STUDENT GUIDEPOSTS

1. Defining $a^{1/n}$
2. Defining $a^{m/n}$
3. Simplifying Radicals Using Exponents
4. Rationalizing Denominators
5. Evaluating Expressions with Rational Exponents

In Chapter 1 we discussed the exponential expression a^n, where *base a* is a real number and *exponent n* is an integer. The properties of integer exponents that were established then are reviewed below.

Properties of Exponents

For real numbers a and b and integers m and n,

1. $a^n = \underbrace{a \cdot a \cdot a \ldots a}_{n \text{ factors}}$ $(n \geq 1)$
2. $a^m a^n = a^{m+n}$ (Product rule)
3. $a^m \div a^n = \frac{a^m}{a^n} = a^{m-n}$ $(a \neq 0)$ (Quotient rule)
4. $(a^m)^n = a^{mn}$ (Power rule)
5. $a^0 = 1$ for $a \neq 0$
6. $a^{-n} = \frac{1}{a^n}$ and $\frac{1}{a^{-n}} = a^n$ if $a \neq 0$
7. $(ab)^n = a^n b^n$
8. $\left(\frac{a}{b}\right)^n = \frac{a^n}{b^n}$ $(b \neq 0)$.

1 DEFINING $a^{1/n}$

We now look at exponential expressions in which the exponent is any rational number (not just an integer). We begin by considering exponents of the form $1/n$, where n is a natural number. For our definition of $a^{1/n}$ to be useful, the properties of exponents should still apply when one or more of the exponents is any rational number. For example, consider $a^{1/2}$. If the power rule $(a^m)^n = a^{mn}$ is to apply, then

$$(a^{1/2})^2 = a^{(1/2)2} = a^1 = a.$$

Similarly,

$$(a^{1/3})^3 = a^{(1/3)3} = a^1 = a.$$

As a result, $a^{1/2}$ appears to be a square root of a while $a^{1/3}$ is a cube root of a. In general,

$$a^{1/n} \text{ must be an } n\text{th root of } a$$

since

$$(a^{1/n})^n = a^{(1/n)n} = a^{n/n} = a^1 = a.$$

Definition of $a^{1/n}$

Let a be a real number, with n a natural number. Then

$$a^{1/n} = \sqrt[n]{a}.$$

Since $\sqrt[n]{a}$ is not a real number when $a < 0$ and n is even, the definition of $a^{1/n}$ does not apply when a is negative and n is even.

EXAMPLE 1 Writing $a^{1/n}$ as a Radical

Express each as a radical and simplify.

(a) $3^{1/2} = \sqrt{3}$

(b) $5^{1/3} = \sqrt[3]{5}$

(c) $9^{1/2} = \sqrt{9} = 3$ $\quad 9 = 3^2$

(d) $125^{1/3} = \sqrt[3]{125} = 5$ $\quad 125 = 5^3$

(e) $16^{1/4} = \sqrt[4]{16} = 2$ $\quad 16 = 2^4$

(f) $(243)^{1/5} = \sqrt[5]{243} = 3$ $\quad 243 = 3^5$

Practice Exercise 1

Express each as a radical and simplify.

(a) $11^{1/2}$

(b) $7^{1/3}$

(c) $25^{1/2}$

(d) $64^{1/3}$

(e) $81^{1/4}$

(f) $32^{1/5}$

(g) $x^{1/5} = \sqrt[5]{x}$

(h) $(xy)^{1/3} = \sqrt[3]{xy}$

(g) $y^{1/3}$

(h) $(2ab)^{1/4}$

Answers: (a) $\sqrt{11}$ (b) $\sqrt[3]{7}$ (c) 5 (d) 4 (e) 3 (f) 2 (g) $\sqrt[3]{y}$ (h) $\sqrt[4]{2ab}$

2 DEFINING $a^{m/n}$

We can now extend the definition of exponents to include rational numbers of the form $\frac{m}{n}$. Suppose we consider the radical $\sqrt[3]{a^2}$. This expression can be rewritten using exponents.

$$\sqrt[3]{a^2} = (a^2)^{1/3} = a^{2 \cdot 1/3} = a^{2/3} \qquad (a^m)^n = a^{m \cdot n}$$

Also, $$(\sqrt[3]{a})^2 = (a^{1/3})^2 = a^{1/3 \cdot 2} = a^{2/3}.$$

Thus, $$\sqrt[3]{a^2} = a^{2/3} = (\sqrt[3]{a})^2.$$

This illustrates the following definition.

Definition of $a^{m/n}$

Let a be a real number, with m and n natural numbers. Then

$$a^{m/n} = \sqrt[n]{a^m} = (\sqrt[n]{a})^m$$

provided that $a^{1/n}$ exists. (That is, if a is not negative when n is even.) Also,

$$a^{-m/n} = \frac{1}{a^{m/n}}$$

provided that $a \neq 0$ and $a^{m/n}$ exists.

All of the properties of integer exponents also hold for rational number exponents whenever the individual exponential expressions exist.

EXAMPLE 2 Expressions with Rational-Number Exponents

Simplify and write with only positive exponents.

(a) $2^{1/3} \cdot 2^{1/4} = 2^{1/3+1/4} = 2^{7/12} \qquad a^m a^n = a^{m+n}$

(b) $\dfrac{3^{2/3}}{3^{5/3}} = 3^{2/3-5/3} = 3^{-3/3} = 3^{-1} = \dfrac{1}{3} \qquad \dfrac{a^m}{a^n} = a^{m-n}$

(c) $(5^{1/3})^{3/4} = 5^{(1/3)(3/4)} = 5^{1/4} \qquad (a^m)^n = a^{mn}$

(d) $81^{-3/4} = \dfrac{1}{81^{3/4}} \qquad a^{-m/n} = \dfrac{1}{a^{m/n}}$

$= \dfrac{1}{(\sqrt[4]{81})^3} = \dfrac{1}{3^3} = \dfrac{1}{27} \qquad$ Also using $a^{3/4} = (\sqrt[4]{a})^3$

(e) $(3 \cdot 5)^{2/3} = 3^{2/3} \cdot 5^{2/3} \qquad (ab)^n = a^n b^n$

(f) $\left(\dfrac{8}{27}\right)^{2/3} = \dfrac{8^{2/3}}{27^{2/3}} \qquad \left(\dfrac{a}{b}\right)^n = \dfrac{a^n}{b^n}$

$= \dfrac{(\sqrt[3]{8})^2}{(\sqrt[3]{27})^2} = \dfrac{2^2}{3^2} = \dfrac{4}{9} \qquad$ Also using $a^{2/3} = (\sqrt[3]{a})^2$

Practice Exercise 2

Simplify and write with only positive exponents.

(a) $3^{1/5} \cdot 3^{1/3}$

(b) $\dfrac{2^{1/4}}{2^{5/4}}$

(c) $(7^{2/3})^{3/4}$

(d) $27^{-4/3}$

(e) $(4 \cdot 7)^{3/2}$

(f) $\left(\dfrac{81}{16}\right)^{3/4}$

Answers: (a) $3^{8/15}$ (b) $\frac{1}{2}$ (c) $7^{1/2}$ (d) $\frac{1}{81}$ (e) $8 \cdot 7^{3/2}$ (f) $\frac{27}{8}$

Notice that in Example 2(f) we could have written

$$\left(\frac{8}{27}\right)^{2/3} = \frac{8^{2/3}}{27^{2/3}} = \frac{\sqrt[3]{8^2}}{\sqrt[3]{27^2}} = \frac{\sqrt[3]{64}}{\sqrt[3]{729}} = \frac{4}{9}.$$

However, to square first and then find the cube root of a large number is more difficult. Thus, in general we take the root first and then raise to the power using

$$a^{m/n} = (\sqrt[n]{a})^m.$$

EXAMPLE 3 SIMPLIFYING EXPONENTIAL EXPRESSIONS

Simplify. Assume all variables are positive and write with only positive exponents.

(a) $(8x^3)^{2/3} = (\sqrt[3]{8x^3})^2 = (2x)^2 = 4x^2$

(b) $x^{2/3}x^{1/3} = x^{2/3+1/3} = x^1 = x$

(c) $x^{2/3}x^{-1/3} = x^{2/3-1/3} = x^{1/3}$

(d) $(y^2)^{-1/3} = y^{2\cdot(-1/3)} = y^{-2/3} = \dfrac{1}{y^{2/3}}$

(e) $\dfrac{y^{1/2}}{y^{1/4}} = y^{1/2-1/4} = y^{1/4}$

(f) $(9y^2)^{3/2} = (\sqrt[2]{9y^2})^3 = (3y)^3 = 27y^3$

(g) $(x^6)^{1/8} = x^{6\cdot 1/8} = x^{6/8} = x^{3/4}$

PRACTICE EXERCISE 3

Simplify and write with only positive exponents.

(a) $(27x^6)^{1/3}$

(b) $y^{4/3}y^{2/3}$

(c) $y^{4/3}y^{-1/3}$

(d) $(x^3)^{-2/3}$

(e) $\dfrac{y^{3/5}}{y^{1/5}}$

(f) $(32y^{10})^{2/5}$

(g) $(u^{12})^{3/4}$

Answers: (a) $3x^2$ (b) y^2 (c) y (d) $\frac{1}{x^2}$ (e) $y^{2/5}$ (f) $4y^4$ (g) u^9

3 SIMPLIFYING RADICALS USING EXPONENTS

We sometimes use rational exponents to simplify radical expressions.

EXAMPLE 4 SIMPLIFYING RADICALS USING EXPONENTS

Simplify.

(a)
$$\begin{aligned}\sqrt{25x^8y^6} &= (25x^8y^6)^{1/2} && \sqrt{a} = a^{1/2}\\ &= (5^2)^{1/2}(x^8)^{1/2}(y^6)^{1/2} && (ab)^n = a^nb^n\\ &= 5^{2\cdot 1/2}x^{8\cdot 1/2}y^{6\cdot 1/2} && (a^m)^n = a^{mn}\\ &= 5x^4y^3\end{aligned}$$

(b)
$$\begin{aligned}\sqrt[3]{\frac{8x^{10}}{27y^9}} &= \left(\frac{2^3x^{10}}{3^3y^9}\right)^{1/3} && \sqrt[3]{a} = a^{1/3}\\ &= \frac{(2^3)^{1/3}(x^{10})^{1/3}}{(3^3)^{1/3}(y^9)^{1/3}} && (ab)^n = a^nb^n \text{ and } \left(\frac{a}{b}\right)^n = \frac{a^n}{b^n}\\ &= \frac{2x^{10/3}}{3y^{9/3}} && (a^m)^n = a^{mn}\\ &= \frac{2x^{9/3}x^{1/3}}{3y^3} && a^{10/3} = a^{9/3+1/3} = a^{9/3}\cdot a^{1/3}\\ &= \frac{2x^3\sqrt[3]{x}}{3y^3}\end{aligned}$$

PRACTICE EXERCISE 4

Simplify.

(a) $\sqrt{36x^{10}y^{12}}$

(b) $\sqrt[3]{\dfrac{81y^8}{125x^{15}}}$

Answers: (a) $6x^5y^6$ (b) $\frac{3y^2\sqrt[3]{3y^2}}{5x^5}$

4 RATIONALIZING DENOMINATORS

To rationalize denominators using rational exponents, multiply numerator and denominator by an expression that makes the power in the denominator a whole number.

EXAMPLE 5 Rationalizing Denominators

Rationalize the denominator.

(a) $\dfrac{1}{\sqrt[3]{7^2}} = \dfrac{1}{7^{2/3}} = \dfrac{1 \cdot 7^{1/3}}{7^{2/3} \cdot 7^{1/3}}$

$= \dfrac{7^{1/3}}{7^{2/3+1/3}} = \dfrac{7^{1/3}}{7^1} = \dfrac{7^{1/3}}{7} = \dfrac{\sqrt[3]{7}}{7}$

(b) $\dfrac{3}{\sqrt[4]{x}} = \dfrac{3}{x^{1/4}} = \dfrac{3 \cdot x^{3/4}}{x^{1/4} \cdot x^{3/4}} = \dfrac{3x^{3/4}}{x^{1/4+3/4}} = \dfrac{3x^{3/4}}{x} = \dfrac{3\sqrt[4]{x^3}}{x}$

PRACTICE EXERCISE 5

Rationalize the denominator.

(a) $\dfrac{1}{\sqrt[4]{11^3}}$

(b) $\dfrac{\sqrt[3]{y}}{\sqrt[3]{x}}$

Answers: (a) $\frac{\sqrt[4]{11}}{11}$ (b) $\frac{\sqrt[3]{x^2y}}{x}$

5 EVALUATING EXPRESSIONS WITH RATIONAL EXPONENTS

Some expressions involving rational number exponents are themselves rational numbers and can be simplified, while others represent irrational numbers and cannot be simplified, only approximated. For example,

$$4^{1/2} = 2, \quad 8^{2/3} = 4, \quad \text{and} \quad \left(\frac{8}{27}\right)^{2/3} = \frac{4}{9}$$

all represent rational numbers, whereas

$$5^{1/4}, \quad 2^{13/12}, \quad \text{and} \quad \left(\frac{8}{27}\right)^{3/2}$$

represent irrational numbers and cannot be simplified further. To approximate numbers such as these, we can use a calculator with a $\boxed{y^x}$ key. For example, if we use the following steps on a calculator,

$$5 \boxed{y^x} 4 \boxed{1/x} \boxed{=} \rightarrow \boxed{1.49535},$$

The display shows 1.49535 as an approximation of $5^{1/4}$ or $\sqrt[4]{5}$. Similarly, if we use the steps

$$2 \boxed{y^x} \boxed{(} 13 \boxed{\div} 12 \boxed{)} \boxed{=} \rightarrow \boxed{2.11893},$$

the display shows 2.11893 as an approximation of $2^{13/12}$. Finally, $\left(\frac{8}{27}\right)^{3/2}$ can be approximated as follows:

$$8 \div 27 \boxed{=} \boxed{y^x} \boxed{(} 3 \boxed{\div} 2 \boxed{)} \boxed{=} \rightarrow \boxed{0.161283}.$$

To evaluate expressions with rational number exponents, a calculator with a $\boxed{y^x}$ key is necessary in many applications.

EXAMPLE 6 Investment Problem

The annual rate of return r on an investment of P dollars that is worth A dollars after t years is given by $r = \left(\frac{A}{P}\right)^{1/t} - 1$. Find the approximate rate of return on a painting which was purchased for \$1500 and sold 8 years later for \$9400.

PRACTICE EXERCISE 6

Find the rate of return on a necklace that was purchased for \$2110 and sold 6 years later for \$3950.

We use the given formula with $P = \$1500$, $A = \$9400$, and $t = 8$ years.

$$r = \left(\frac{9400}{1500}\right)^{1/8} - 1$$

Use the following calculator steps.

9400 [÷] 1500 [=] [y^x] 8 [1/x] [=] [−] 1 [=] → 0.257852

The return is approximately 25.8%.

Answer: 11.0%

7.5 EXERCISES A

Simplify by writing each expression as an integer or a rational number, reduced to lowest terms.

1. $36^{1/2}$ **2.** $(-16)^{1/2}$ **3.** $-144^{1/2}$ **4.** $27^{1/3}$

5. $(-64)^{1/3}$ **6.** $-8^{1/3}$ **7.** $(-4)^{1/2}$ **8.** $25^{3/2}$

9. $16^{-1/2}$ **10.** $\left(\frac{9}{16}\right)^{1/2}$ **11.** $\left(\frac{1}{16}\right)^{-1/4}$ **12.** $\frac{1}{25^{-1/2}}$

13. $(2^3)^{4/3}$ **14.** $4^{2/3}4^{1/3}$ **15.** $(4^{-6})^{-1/12}$ **16.** $\frac{81^{3/4}}{81^{1/4}}$

In Exercises 17–28 simplify by writing the expression so that each variable occurs only once and all exponents are positive. All variables represent positive real numbers.

17. $x^{2/3}x^{1/3}$ **18.** $\frac{w^{3/4}}{w^{1/4}}$ **19.** $(x^{1/2})^3$

20. $x^{-1/3}x^{4/3}$ **21.** $(w^3y^6)^{1/3}$ **22.** $(a^{1/3}b)^{1/2}$

23 $\left(\frac{a^3}{b^{-6}}\right)^{1/3}$ **24.** $(8a^3b^{-6})^{1/3}$ **25.** $(3a^{1/2}b^{3/2})^2$

26. $\left(\frac{x^{-1}y^{-1/2}z}{x^{-2/3}y^{1/3}z^{2/3}}\right)^6$ **27** $\left(\frac{96a^3b^{-2}}{3a^{-2}b^8}\right)^{1/5}$ **28.** $\left(\frac{x^{-2/3}}{y^{-1/4}}\right)^{12}(x^{-1/3}y)^{-3}$

Write all radicals in exponential notation, simplify, and give answers in radical notation.

29. $\sqrt[3]{125y^6}$ **30.** $\sqrt[4]{a^8b^{12}}$ **31.** $\sqrt{8x^4y^8}$

32. $\sqrt[3]{8x^4}$

33. $\sqrt[3]{\dfrac{16x^9}{y^{12}}}$

34 $\sqrt[5]{\dfrac{a^{20}}{625b^{15}}}$

Use rational exponents to rationalize the denominator. Give answers in radical form.

35. $\dfrac{1}{\sqrt{5}}$

36 $\dfrac{1}{\sqrt[3]{5^2}}$

37. $\dfrac{3}{\sqrt[4]{3}}$

38. $\dfrac{2}{\sqrt[3]{x}}$

39. $\dfrac{6}{\sqrt[4]{y^3}}$

40. $\dfrac{2}{\sqrt[5]{ab}}$

A chemist learns that C and r are related by the equation $C = 3r^{2/3}$.

41. Find C when $r = 8$.

42. Find C when $r = 0$.

43. Find C when $r = 125$.

44. Find C when $r = -8$.

Use a calculator with a $\boxed{y^x}$ key to approximate each number correct to the nearest tenth.

45. $7^{1/3}$

46. $3^{3/4}$

47. $5^{-2/3}$

48. $(1.08)^{7/3}$

49. $(1.06)^{-5/4}$

50. $\left(\dfrac{3}{5}\right)^{5/3}$

51. The length of an edge of a cube with volume V is given by the formula $e = V^{1/3}$. A storage bin to be constructed in the shape of a cube must hold 50 ft^3 of grain. To the nearest tenth of a foot, what must be the length of an edge of the bin?

52. The radius r of a sphere with volume V is given by $r = \sqrt[3]{\frac{3V}{4\pi}}$. Find the radius (to the nearest tenth of an inch) of a sphere with volume 65.5 in^3. Use 3.14 for π.

53. The annual rate of return r on an investment of P dollars that is worth A dollars after t years is given by $r = \left(\frac{A}{P}\right)^{1/t} - 1$. Find the approximate annual rate of return on a set of rare books purchased for \$1000 and sold 5 years later for \$2400.

Burford worked Exercises 54 and 55. What is wrong with each of them?

54. $-2 = (-2)^1 = (-2)^{2/2} = [(-2)^2]^{1/2} = 4^{1/2} = 2$

55. $3 = 3^1 = 3^{3/3} = (3^3)^{1/3} = 27^{1/3} = \dfrac{1}{3}(27) = 9$

FOR REVIEW

Add or subtract.

56. $-3\sqrt{147} - 5\sqrt{48}$

57. $2\sqrt{9x^3} - 3\sqrt{4x^3}$

58. $\sqrt[3]{8x^4y^6} + y\sqrt[3]{27x^4y^3}$

59. $\sqrt{3x} - \dfrac{x}{\sqrt{3x}}$

Simplify. Rationalize denominators.

60. $\sqrt{5}(\sqrt{125} - \sqrt{45})$

61. $(\sqrt{x} - \sqrt{2})(\sqrt{x} + 2\sqrt{2})$

62. $\dfrac{\sqrt{3}}{\sqrt{3} - \sqrt{2}}$

63. $\dfrac{x - 25}{\sqrt{x} + 5}$

Exercises 64 and 65 will help you prepare for the next section. Find each power.

64. $(1 + \sqrt{y - 2})^2$

65. $(1 - 2\sqrt{x + 1})^2$

ANSWERS: 1. 6 2. not a real number 3. -12 4. 3 5. -4 6. -2 7. not a real number 8. 125 9. $\frac{1}{4}$ 10. $\frac{3}{4}$ 11. 2 12. 5 13. 16 14. 4 15. 2 16. 9 17. x 18. $w^{1/2}$ 19. $x^{3/2}$ 20. x 21. wy^2 22. $a^{1/6}b^{1/2}$ 23. ab^2 24. $\frac{2a}{b^2}$ 25. $9ab^3$ 26. $\frac{z^2}{x^2y^5}$ 27. $\frac{2a}{b^2}$ 28. $\frac{1}{x^7}$ 29. $5y^2$ 30. a^2b^3 31. $2x^2y^4\sqrt{2}$ 32. $2x\sqrt[3]{x}$ 33. $\frac{2x^3\sqrt[3]{2}}{y^4}$ 34. $\frac{a^4\sqrt[5]{5}}{5b^3}$ 35. $\frac{\sqrt{5}}{5}$ 36. $\frac{\sqrt[3]{5}}{5}$ 37. $\sqrt[4]{27}$ 38. $\frac{2\sqrt[3]{x^2}}{x}$ 39. $\frac{6\sqrt[4]{y}}{y}$ 40. $\frac{2\sqrt[5]{a^4b^4}}{ab}$ 41. 12 42. 0 43. 75 44. 12 45. 1.9 46. 2.3 47. 0.3 48. 1.2 49. 0.9 50. 0.4 51. 3.7 ft 52. 2.5 in 53. 19.1% 54. -2 cannot be written as $[(-2)^2]^{1/2}$ which is $4^{1/2} = 2$. 55. $27^{1/3}$ is $\sqrt[3]{27}$ *not* $\frac{1}{3}(27) = 9$ 56. $-41\sqrt{3}$ 57. 0 58. $5xy^2\sqrt[3]{x}$ 59. $\frac{2\sqrt{3x}}{3}$ 60. 10 61. $x + \sqrt{2}x - 4$ 62. $3 + \sqrt{6}$ 63. $\sqrt{x} - 5$ 64. $-1 + 2\sqrt{y - 2} + y$ 65. $5 - 4\sqrt{x + 1} + 4x$

7.5 EXERCISES B

Simplify by writing each expression as an integer or a rational number, reduced to lowest terms.

1. $64^{1/2}$

2. $(-49)^{1/2}$

3. $-9^{1/2}$

4. $64^{1/3}$

5. $(-27)^{1/3}$

6. $-27^{1/3}$

7. $(-9)^{1/2}$

8. $64^{2/3}$

9. $27^{-1/3}$

10. $\left(\dfrac{27}{8}\right)^{2/3}$

11. $\left(\dfrac{1}{81}\right)^{-1/4}$

12. $\dfrac{1}{9^{-1/2}}$

13. $(3^2)^{5/2}$

14. $7^{2/5}7^{3/5}$

15. $(3^{-8})^{-1/4}$

16. $\dfrac{27^{1/3}}{27^{2/3}}$

Simplify by writing the expression so that each variable occurs only once and all exponents are positive. All variables represent positive real numbers.

17. $z^{1/4}z^{3/4}$ **18.** $\dfrac{y^{5/6}}{y^{1/6}}$ **19.** $(y^5)^{3/5}$ **20.** $z^{-1/5}z^{2/5}$

21. $(a^4b^8)^{1/4}$ **22.** $(u^{1/4}v)^{1/3}$ **23.** $\left(\dfrac{w^4}{z^{-8}}\right)^{1/4}$ **24.** $(4u^2v^{-6})^{1/2}$

25. $(5x^{1/3}y^{2/3})^3$ **26.** $\left(\dfrac{a^{-2}b^{-1/4}c^3}{a^{-1/2}b^{3/4}c^0}\right)^4$ **27.** $\left(\dfrac{b^{-1/12}}{8b^{1/3}b^{-1/4}}\right)^2$ **28.** $\left(\dfrac{a^{1/2}}{b^{-3/4}}\right)^8(a^{-1/5}b)^5$

Write all radicals in exponential notation, simplify, and give answers in radical notation.

29. $\sqrt[3]{64x^9}$ **30.** $\sqrt[5]{a^{10}b^{15}}$ **31.** $\sqrt{12x^{10}y^6}$

32. $\sqrt[3]{27x^5}$ **33.** $\sqrt[3]{\dfrac{81y^{12}}{x^{15}}}$ **34.** $\sqrt[5]{\dfrac{a^{25}}{81b^{15}}}$

Use rational exponents to rationalize each denominator. Give answers in radical form.

35. $\dfrac{1}{\sqrt{7}}$ **36.** $\dfrac{1}{\sqrt[3]{7^2}}$ **37.** $\dfrac{5}{\sqrt[4]{5}}$

38. $\dfrac{-6}{\sqrt[3]{x^2}}$ **39.** $\dfrac{9}{\sqrt[5]{y^4}}$ **40.** $\dfrac{3}{\sqrt[4]{ab}}$

A recording company finds that R and s are related by the equation $R = 5s^{3/4}$.

41. Find R when $s = 16$. **42.** Find R when $s = 0$.

43. Find R when $s = 81$. **44.** Find R when $s = -81$.

Use a calculator with a $\boxed{y^x}$ key to approximate each number correct to the nearest tenth.

45. $12^{1/5}$ **46.** $7^{2/3}$ **47.** $9^{-3/4}$ **48.** $(1.09)^{7/3}$ **49.** $(1.07)^{-9/4}$ **50.** $\left(\dfrac{6}{7}\right)^{7/6}$

51. The length of an edge of a cube with volume V is given by the formula $e = V^{1/3}$. What is the length of an edge, to the nearest hundredth of an inch, of a cubical container that holds 1 quart of water? Use the fact that 1 qt = 57.75 in^3.

52. The radius r of a sphere with volume V is given by $r = \sqrt[3]{\frac{3V}{4\pi}}$. Find the radius, to the nearest tenth of a foot, of a sphere with volume 126.4 ft^3.

53. The annual rate of return r on an investment of P dollars that is worth A dollars after t years is given by $r = \left(\frac{A}{P}\right)^{1/t} - 1$. Find the approximate annual rate of return on an antique car purchased for \$6000 and sold 4 years later for \$8500.

Burford worked Exercises 54 and 55. What is wrong with each of them?

54. $-4 = (-4)^1 = (-4)^{2/2} = [(-4)^2]^{1/2} = 16^{1/2} = 4$

55. $2 = 2^1 = 2^{3/3} = (2^3)^{1/3} = 8^{1/3} = \dfrac{1}{3}(8) = \dfrac{8}{3}$

FOR REVIEW

Add or subtract.

56. $-4\sqrt{80} - 3\sqrt{45}$

57. $-7\sqrt{16x^3} + 5\sqrt{49x^3}$

58. $x\sqrt[3]{x^5y^9} - y\sqrt[3]{8x^8y^6}$

59. $\sqrt{5x} + \dfrac{x}{\sqrt{5x}}$

Simplify. Rationalize denominators.

60. $\sqrt{3}(\sqrt{12} + \sqrt{147})$

61. $(2\sqrt{y} + \sqrt{3})(\sqrt{y} - \sqrt{3})$

62. $\dfrac{\sqrt{5}}{\sqrt{5} - 2}$

63. $\dfrac{x - 36}{\sqrt{x} - 6}$

Exercises 64 and 65 will help you prepare for the next section. Find each power.

64. $(2 - \sqrt{x + 1})^2$

65. $(1 + 2\sqrt{x - 3})^2$

7.5 EXERCISES C

Simplify and give answers in radical form.

1. Is $(a + b)^{1/2}$ the same as $a^{1/2} + b^{1/2}$?

2. Is $(a + b)^{1/2}$ the same as $\frac{1}{2}(a + b)$?

3. For what values of x does $(x^2)^{1/2} = x$?

4. For what values of x does $(x^2)^{1/2} = -x$?

Multiply and simplify expressing all answers with positive exponents. All variables represent positive numbers.

5. $a^{1/4}(a^{3/4} + 2a^{1/4})$

6. $2b^{1/3}(b^{-1/3} - b^{1/3})$

7. $(a^{1/2} + b^{1/2})(a^{1/2} - b^{1/2})$

8. $(a^{1/2} - b^{1/2})^2$

Simplify each expression. Assume that m is a positive integer and that x and y are positive real numbers.

9. $x^m x^{m/3}$ [Answer: $x^{4m/3}$]

10. $(x^3)^{m/3}(y^2)^{m/2}$

11. $\dfrac{x^{m/2}}{x^{m/4}}$

12. $\left(\dfrac{x^3}{y}\right)^m \left(\dfrac{y^m}{x^3}\right)^m$

13. $\dfrac{x^{m/3}y^{m/2}}{(xy)^m}$

14. $\left(\dfrac{x^{2m}y^{4m}}{x^m}\right)^{1/m}$

15. Factor $a^{1/2}$ from $a^{3/2} - a^{1/2}$ [Answer: $a^{1/2}(a - 1)$]

16. Factor $x^{1/2} + x^{1/4} - 2$ as if it were a trinomial. [*Hint:* Let $u = x^{1/4}$.] [Answer: $(x^{1/4} - 1)(x^{1/4} + 2)$]

7.6 SOLVING RADICAL EQUATIONS

STUDENT GUIDEPOSTS

1. Radical Equations
2. Solving Radical Equations
3. Literal Equations

1 RADICAL EQUATIONS

An equation containing radicals with variables in the radicand is called a **radical equation.** To solve a radical equation we change it to an equation containing no radicals by using the rule on the following page.

Power Rule

If a and b are algebraic expressions such that $a = b$, then $a^n = b^n$ for any natural number n.

When this rule is applied to an equation involving a variable, the resulting equation may have more solutions than the original. For example, the equation

$$x = 5$$

has only one solution, 5, but if we square both sides,

$$x^2 = 25$$

has two solutions, 5 and -5. Thus, since $a = b$ and $a^n = b^n$ may not be equivalent, we need to check any possible solutions in the original equation after using the power rule. Solutions that do not check in the original equation are called **extraneous roots.**

2 SOLVING RADICAL EQUATIONS

We now present a method for solving radical equations.

To Solve a Radical Equation

1. If only one radical is present, isolate this radical on one side of the equation, simplify, and proceed to step 3.
2. If two radicals are present, isolate one of the radicals on one side of the equation.
3. Raise both sides to a power equal to the index on the radical.
4. Solve the resulting equation. If a radical remains in the equation, isolate it and raise to the power again.
5. Check all possible solutions in the *original* equation since the equations involved may not be equivalent. If all roots are extraneous, the equation has no solution.

CAUTION

It is necessary to isolate the radicals, as stated in step 1 above, since the entire quantity on each side of the equation must be raised to the power. *You cannot square term by term.*

$(\sqrt{4})^2 + (\sqrt{9})^2 = 4 + 9 = 13 \neq (\sqrt{4} + \sqrt{9})^2 = (2 + 3)^2 = 5^2 = 25$

EXAMPLE 1 Solving a Radical Equation

Solve.

$3\sqrt{2y - 5} - \sqrt{y + 23} = 0$

$3\sqrt{2y - 5} = \sqrt{y + 23}$ Isolate the radicals

$(3\sqrt{2y - 5})^2 = (\sqrt{y + 23})^2$ Use the power rule

$9(2y - 5) = y + 23$ Do not forget to square 3

$18y - 45 = y + 23$

$17y = 68$

$y = 4$

Practice Exercise 1

Solve.

$2\sqrt{4y - 3} - \sqrt{2y + 9} = 0$

Check: $3\sqrt{2(4) - 5} - \sqrt{4 + 23} \stackrel{?}{=} 0$

$3\sqrt{3} - \sqrt{27} \stackrel{?}{=} 0$

$3\sqrt{3} - 3\sqrt{3} \stackrel{?}{=} 0$, so the solution is 4.

Answer: $\frac{3}{2}$

To stress the need for checking solutions we consider an equation with no solution.

EXAMPLE 2 Solving a Radical Equation

Solve.

$\sqrt{x^2 + 5} - x + 5 = 0$

$\sqrt{x^2 + 5} = x - 5$ Isolate the radical

$(\sqrt{x^2 + 5})^2 = (x - 5)^2$ Use the power rule

$x^2 + 5 = x^2 - 10x + 25$ $(x - 5)^2$ is *not* $x^2 - 25$; do not forget the middle term $-10x$

$5 = -10x + 25$ Subtract x^2 from both sides

$-20 = -10x$

$2 = x$

Check: $\sqrt{2^2 + 5} - 2 + 5 \stackrel{?}{=} 0$

$\sqrt{9} - 2 + 5 \stackrel{?}{=} 0$

$3 - 2 + 5 \stackrel{?}{=} 0$

$6 \neq 0$, so the equation has no solution.

Practice Exercise 2

Solve.

$\sqrt{x^2 - 9} + x - 1 = 0$

Answer: no solution

EXAMPLE 3 Squaring Twice to Solve an Equation

Solve.

$\sqrt{y + 3} - \sqrt{y - 2} = 1$

$\sqrt{y + 3} = 1 + \sqrt{y - 2}$ Isolate $\sqrt{y + 3}$

$(\sqrt{y + 3})^2 = (1 + \sqrt{y - 2})^2$ Use the power rule

$y + 3 = 1 + 2\sqrt{y - 2} + y - 2$ Don't forget the middle term, $2\sqrt{y - 2}$

$y + 3 = 2\sqrt{y - 2} + y - 1$

$3 = 2\sqrt{y - 2} - 1$ A radical remains

$4 = 2\sqrt{y - 2}$ To isolate $\sqrt{y - 2}$, add 1 to both sides

$2 = \sqrt{y - 2}$ Simplify; divide both sides by 2

$2^2 = (y - 2)^2$ Use power rule again

$4 = y - 2$

$6 = y$

The solution is 6. Check this in the original equation.

Practice Exercise 3

Solve.

$\sqrt{y + 6} - \sqrt{y - 1} = 1$

Answer: 10

Some radical equations result in equations that have to be solved by the factoring technique studied in Chapter 5.

EXAMPLE 4 Solving a Radical Equation that Is Quadratic

Solve.

$$\sqrt{x+15} - x + 5 = 0$$

$\sqrt{x+15} = x - 5$	Isolate the radical
$(\sqrt{x+15})^2 = (x-5)^2$	Square both sides
$x + 15 = x^2 - 10x + 25$	$(a-b)^2 = a^2 - 2ab + b^2$
$0 = x^2 - 11x + 10$	Subtract x and 15
$0 = (x-1)(x-10)$	Factor
$x - 1 = 0$ or $x - 10 = 0$	Zero-product rule
$x = 1$ $\quad$ $x = 10$	

Check:

$\sqrt{1+15} - 1 + 5 \stackrel{?}{=} 0$	$\sqrt{10+15} - 10 + 5 \stackrel{?}{=} 0$
$\sqrt{16} - 1 + 5 \stackrel{?}{=} 0$	$\sqrt{25} - 10 + 5 \stackrel{?}{=} 0$
$4 - 1 + 5 \stackrel{?}{=} 0$	$5 - 10 + 5 \stackrel{?}{=} 0$
$8 \neq 0$	$0 = 0$
1 does not check.	10 does check.

The only solution is 10.

Practice Exercise 4

Solve.

$$\sqrt{x+14} - x + 6 = 0$$

Answer: 11

EXAMPLE 5 A Radical Equation Involving Cube Roots

Solve.

$$\sqrt[3]{z+2} - \sqrt[3]{3z+8} = 0$$

$\sqrt[3]{z+2} = \sqrt[3]{3z+8}$	Isolate radicals
$(\sqrt[3]{z+2})^3 = (\sqrt[3]{3z+8})^3$	Cube both sides
$z + 2 = 3z + 8$	
$-2z = 6$	
$z = -3$	

The solution is -3. Check this.

Practice Exercise 5

Solve.

$$\sqrt[3]{2x-5} - \sqrt[3]{5x+1} = 0$$

Answer: -2

If the radicals in an equation are expressed using rational exponents, the solution technique is the same.

EXAMPLE 6 Solving Equations with Rational Exponents

Solve.

(a) $x^{1/2} = 5$

$(x^{1/2})^2 = 5^2$	Square both sides
$x = 25$	This does check

(b) $(x-2)^{1/3} = 3$

$[(x-2)^{1/3}]^3 = 3^3$	Cube both sides
$x - 2 = 27$	$(a^{1/3})^3 = a$
$x = 29$	This does check

Practice Exercise 6

Solve.

(a) $x^{1/4} = 3$

(b) $(2x-1)^{1/3} = 2$

Answers: (a) 81 (b) $\frac{9}{2}$

3 LITERAL EQUATIONS

In Chapter 2 we solved various literal equations that were linear in the variables. In this section we consider literal equations that involve powers and radicals. We shall assume that all radical expressions are defined.

EXAMPLE 7 Solving Literal Equations

Solve each literal equation.

(a) $r = \sqrt{\dfrac{A}{\pi}}$ for A

		Similar numerical equation
$r = \sqrt{\dfrac{A}{\pi}}$		$5 = \sqrt{\dfrac{A}{3}}$
$r^2 = \dfrac{A}{\pi}$	Square both sides	$5^2 = \dfrac{A}{3}$
$\pi r^2 = \dfrac{A}{\pi} \cdot \pi$	Multiply by π	$3 \cdot 5^2 = \dfrac{A}{3} \cdot 3$
$\pi r^2 = A$		$75 = A$

(b) $a = \sqrt{b - c}$ for c

	Similar numerical equation
$a = \sqrt{b - c}$	$3 = \sqrt{5 - c}$
$a^2 = (\sqrt{b - c})^2$	$(3)^2 = (\sqrt{5 - c})^2$
$a^2 = b - c$	$9 = 5 - c$
$a^2 - b = b - b - c$	$9 - 5 = 5 - 5 - c$
$a^2 - b = -c$	$9 - 5 = -c$
$(-1)(a^2 - b) = (-1)(-c)$	$(-1)(4) = (-1)(-c)$
$b - a^2 = c$	$-4 = c$

Practice Exercise 7

Solve each literal equation.

(a) $u = \sqrt{\dfrac{5}{v}}$ for v

(b) $2m = \sqrt{n - k}$ for n

Answers: (a) $v = \frac{5}{u^2}$
(b) $n = 4m^2 + k$

We conclude this section with an example of an applied problem that involves a radical equation.

EXAMPLE 8 Physics Problem

The time T in seconds it takes for a pendulum to make one swing is related to the length L in feet of the pendulum by

$$T = 2\pi\sqrt{\frac{L}{32}}.$$

Find the length of a pendulum (correct to the nearest hundredth of a foot) if it makes one swing in 4 seconds.

Practice Exercise 8

Find the length of a pendulum (correct to the nearest tenth of a foot) if it makes one swing in 1.5 seconds.

Substitute 4 for T and solve.

$$4 = 2\pi\sqrt{\frac{L}{32}} \qquad T = 4$$

$$\frac{4}{2\pi} = \sqrt{\frac{L}{32}} \qquad \text{Divide by } 2\pi$$

$$\frac{2}{\pi} = \sqrt{\frac{L}{32}} \qquad \text{Reduce fraction}$$

$$\left(\frac{2}{\pi}\right)^2 = \frac{L}{32} \qquad \text{Square both sides}$$

$$32\left(\frac{4}{\pi^2}\right) = L \qquad \text{Multiply by 32}$$

We use the following calculator steps to find L.

32 [×] 4 [÷] [π] [x^2] [=] → 12.969112

The pendulum is approximately 12.97 ft in length.

Answer: 1.8 ft

7.6 EXERCISES A

Solve.

1. $\sqrt{z-3} + 3 = 6$
2. $4\sqrt{2y-1} - 2 = 0$
3. $8\sqrt{3x-1} + 7 = 0$
4. $\sqrt{5x-1} - 2\sqrt{x+1} = 0$
5. $\sqrt{x^2-5} + x - 5 = 0$
[*Hint:* $(5-x)^2 = 25 - 10x + x^2$, *not* $25 - x^2$.]
6. $\sqrt{a^2+9} + a + 1 = 0$
7. $\sqrt{y+6} + \sqrt{y+11} = 5$
[*Hint:* $(5 - \sqrt{y-11})^2 = 25 - 10\sqrt{y-11} + y - 11$, *not* $25 - (y-11)$.]
8. $\sqrt{x+1} + \sqrt{x+7} = -1$
9. $\sqrt{w+2} + \sqrt{w+6} = 4$
10. $\dfrac{8}{\sqrt{a}} = 2$
11. $\dfrac{15}{\sqrt{x}} = \dfrac{20}{\sqrt{x+7}}$
12. $\sqrt{x-1} = x - 1$
13. $\sqrt{y-2} - y + 2 = 0$
14. $\sqrt{2x+1} - \sqrt{x-3} = 2$
15. $\sqrt{3x-2} - \sqrt{2x+5} = -1$

16. $\sqrt[3]{a + 1} = \sqrt[3]{2a + 7}$

17. $\sqrt[3]{x^2 + x} = \sqrt[3]{2 + x^2}$

18. $\sqrt[4]{w + 2} = \sqrt[4]{5w}$

19. $\sqrt[3]{3a + 2} + 4 = 6$

20. $\sqrt{x^2 + x - 3} = x$

21. $\sqrt{a^2 + a + 1} = a$

22. $(x + 1)^{1/2} = 2$

23. $(x - 5)^{1/3} = -3$

24. $(y + 7)^{1/4} = -2$

Solve. Assume all variables are suitably chosen so that all radicals are meaningful.

25. $z = y\sqrt{x}$ for x

26. $x + \sqrt{y} = z$ for y

27. $a = b\sqrt{1 + x}$ for x

28. $a = \sqrt{1 + \dfrac{b}{x}}$ for x

29. $\sqrt{ab + d} = c$ for b

30. $z^{1/3} + x = y$ for z

Solve.

31. Twice the square root of a number is the same as the square root of 9 more than three times the number. Find the number.

32. If the square root of 1 less than a number is subtracted from the square root of 15 more than the number, the result is 2. Find the number.

A manufacturer finds that the cost per day of producing an airplane part is given by $c = 100\sqrt[3]{n} + 1200$ *where* c *is the cost and* n *is the number of parts produced.*

33. What is the cost when no parts (overhead) are produced?

34. Find the cost when 8 parts are produced.

35. Find the cost when 125 parts are produced.

36. How many parts were made on a day when the cost was \$2800?

In an environmental study it is found that A and n are related by the equation $A = \sqrt{n + 16} - \sqrt{n}$.

37. Find A when $n = 0$.

38. Find A when $n = 20$.

39. Find n when $A = 2$.

40. Find n when $A = 0$.

The head trainer at NAU, Mike Nesbitt, has found that he can estimate the number of boxes of tape he needs for a season of competition from the formula $t = 50\sqrt[3]{n - 11}$, *where n is the number of members on the team.*

41. Find the number of boxes needed when the team has 12 members.

42. If at the end of the season Mike had used a total of 200 boxes, how many team members were there?

43 The time T in seconds it takes a pendulum of length L feet to make one swing is given by $T = 2\pi\sqrt{\frac{L}{32}}$. How long is a pendulum (to the nearest hundredth of a foot) if it makes one swing in 1.5 seconds? Use 3.14 for π.

44. The distance d in miles that can be seen on the surface of the ocean is given by $d = 1.4\sqrt{h}$ where h is the height in feet above the surface of the water. How high (to the nearest foot) would the crow's nest on a ship need to be in order to view another ship at a distance of 18.5 miles?

45. The speed at which a car was traveling at the time of an accident can be estimated from the length of the skid mark l as follows: A test car (the car involved if possible) is driven under the same conditions at a test speed S, and skidded to a stop. If L is the length of the test skid, $s = S\sqrt{\frac{l}{L}}$. At the scene of an accident, a policeman determines that $S = 35$ mph, $l = 170$ ft, and $L = 51$ ft. What was the approximate speed of the car involved in the accident?

FOR REVIEW

Simplify by writing the expression so that each variable occurs only once and all exponents are positive.

46. $x^{-1/5}x^{6/5}$

47. $\dfrac{y^{3/4}}{y^{1/2}}$

48. $(27u^3v^{-9})^{1/3}$

Write in exponential notation, simplify, and give answers in radical notation. Rationalize denominators.

49. $\sqrt[3]{64x^3y^{12}}$

50. $\sqrt[4]{\dfrac{16u^5}{v^8}}$

51. $\dfrac{5}{\sqrt[5]{ab}}$

52. Rationalize the denominator.
$$\frac{\sqrt{2}}{\sqrt{2} + \sqrt{11}}$$

53. Rationalize the numerator.
$$\frac{\sqrt{2 + h} - \sqrt{2}}{h}$$

The following exercises will help you prepare for the material in the next section.

54. If $c = \sqrt{a^2 + b^2}$, find c to the nearest tenth if $a = 2.1$ and $b = 4.5$.

55. If $a = \sqrt{c^2 - b^2}$, find a to the nearest tenth if $c = 8.3$ and $b = 6.1$.

Evaluate. Leave answers in radical form but simplify.

56. $\sqrt{(2 - 5)^2 + (-1 - 6)^2}$

57. $\sqrt{(-1 - (-3))^2 + (4 - (-2))^2}$

ANSWERS: 1. 12 2. $\frac{5}{8}$ 3. no solution 4. 5 5. 3 6. no solution 7. -2 8. no solution 9. $\frac{1}{4}$ 10. 16 11. 9 12. 2; 1 13. 2; 3 14. 4; 12 15. 2 16. -6 17. 2 18. $\frac{1}{2}$ 19. 2 20. 3 21. no solution 22. 3 23. -22 24. no solution 25. $x = \frac{z^2}{y^2}$ 26. $y = (z - x)^2$ 27. $x = \frac{a^2}{b^2} - 1$ 28. $x = \frac{b}{a^2 - 1}$ 29. $b = \frac{c^2 - d}{a}$ 30. $z = (y - x)^3$ 31. 9 32. 10 33. \$1200 34. \$1400 35. \$1700 36. 4096 37. 4 38. $6 - 2\sqrt{5}$ 39. 9 40. no solution 41. 50 42. 75 43. 1.83 ft 44. 175 ft 45. 64 mph 46. x 47. $y^{1/4}$ 48. $\frac{3u}{v^3}$ 49. $4xy^4$ 50. $\frac{2u\sqrt[4]{u}}{v^2}$ 51. $\frac{5\sqrt[5]{a^4b^4}}{ab}$ 52. $\frac{\sqrt{22} - 2}{9}$ 53. $\frac{1}{\sqrt{2 + h} + \sqrt{2}}$ 54. 5.0 55. 5.6 56. $\sqrt{58}$ 57. $2\sqrt{10}$

7.6 EXERCISES B

Solve.

1. $\sqrt{4x - 3} - 2 = 3$

2. $5\sqrt{a - 3} + 10 = 0$

3. $3\sqrt{x - 4} - 9 = 0$

4. $\sqrt{2z + 9} = \sqrt{4 + 2z}$

5. $\sqrt{y^2 + 2} - y - 2 = 0$

6. $\sqrt{v^2 - 13} - v + 1 = 0$

7. $\sqrt{z + 20} - \sqrt{z + 4} = 2$

8. $\sqrt{a - 1} + \sqrt{a + 8} = -2$

9. $\sqrt{y + 10} - \sqrt{y - 6} = 2$

10. $\dfrac{9}{\sqrt{x}} = 3$

11. $\dfrac{1}{\sqrt{y}} = \dfrac{2}{\sqrt{y + 12}}$

12. $\sqrt{x + 1} = x + 1$

13. $\sqrt{y - 2} - y + 4 = 0$

14. $\sqrt{2x + 3} - \sqrt{x + 5} = -1$

15. $\sqrt{3x + 1} - \sqrt{2x - 1} = 1$

16. $\sqrt[3]{z + 2} = \sqrt[3]{2z - 23}$

17. $\sqrt[3]{1 - y^2} = \sqrt[3]{y + 1 - y^2}$

18. $\sqrt[4]{3x} = \sqrt[4]{4 - x}$

19. $3\sqrt[3]{3z - 1} - 2 = 4$

20. $\sqrt{y^2 - y + 2} = y$

21. $\sqrt{z^2 + 2z + 2} = z$

22. $(x - 1)^{1/2} = 2$

23. $(x + 6)^{1/3} = -3$

24. $(y + 5)^{1/4} = -3$

Solve. Assume all variables are suitably chosen so that all radicals are meaningful.

25. $z = x\sqrt{y}$ for y

26. $\sqrt{x} + y = z$ for x

27. $c = d\sqrt{2 - x}$ for x

28. $a = \sqrt{1 + \dfrac{b}{x}}$ for b

29. $\sqrt{ab + d} = c$ for d

30. $a^{1/4} - b = c$ for a

31. Three times the square root of a number is the same as the square root of 32 more than the number. Find the number.

32. If the square root of 3 less than a number is added to the square root of 8 less than the number, the result is 5. Find the number.

In a retail outlet the cost per day, c, is related to the number of small appliances sold, n, by $c = 10\sqrt[3]{n} + 200$.

33. What is the cost when no appliances (overhead) are sold?

34. Find the cost when 8 appliances are sold.

35. Find the cost when 64 appliances are sold.

36. How many appliances were sold on a day when the cost was \$230?

A scientist finds that R and s are related by the equation $R = \sqrt{s + 25} + \sqrt{s + 9}$.

37. Find R when $s = 0$.

38. Find R when $s = 16$.

39. Find s when $R = 4$.

40. Find s when $R = 0$.

A manufacturer of puzzles has found that his daily expenses for production can be estimated by $e = 100\sqrt[3]{n + 8}$, *where e represents total expenses and n is the number of puzzles produced.*

41. Find the total expenses when no puzzles are produced.

42. How many puzzles were made on a day when the expenses were \$300?

43. The time T in seconds it takes for a pendulum of length L feet to make one swing is given by $T = 2\pi\sqrt{\frac{L}{32}}$. How long is a pendulum (to the nearest hundredth of a foot) if it makes one swing in 7.2 seconds? Use 3.14 for π.

44. The distance in miles d that can be seen on the surface of the ocean is given by $d = 1.4\sqrt{h}$ where h is the height in feet above the surface of the water. How far would a periscope on a submarine have to be raised in order to view a destroyer located 2.9 miles away? (Give answer to the nearest tenth of a foot.)

45. The speed at which a car was traveling at the time of an accident can be estimated from the length of the skid mark l as follows: A test car (the car involved if possible) is driven under the same conditions at a test speed S, and skidded to a stop. If L is the length of the test skid, $s = S\sqrt{\frac{l}{L}}$. Estimate the speed of an accident vehicle when $S = 30$ mph, $l = 150$ ft, and $L = 24$ ft.

FOR REVIEW

Simplify by writing the expression so that each variable occurs only once and all exponents are positive.

46. $x^{-2/3}x^{5/3}$

47. $\dfrac{y^{5/4}}{y^{3/4}}$

48. $(81u^4v^{-12})^{1/4}$

Write in exponential notation, simplify, and give answers in radical notation. Rationalize denominators.

49. $\sqrt[3]{125x^6y^9}$

50. $\sqrt[4]{\dfrac{625u^9}{v^{16}}}$

51. $\dfrac{3}{\sqrt[5]{a^2b^2}}$

52. Rationalize the denominator.
$$\frac{\sqrt{2} + 1}{\sqrt{2} - 1}$$

53. Rationalize the numerator.
$$\frac{\sqrt{7 + h} - \sqrt{7}}{h}$$

The following exercises will help you prepare for the material in the next section.

54. If $c = \sqrt{a^2 + b^2}$, find c to the nearest tenth if $a = 3.6$ and $b = 7.2$.

55. If $b = \sqrt{c^2 - a^2}$, find b to the nearest tenth if $c = 10.4$ and $a = 7.3$.

Evaluate. Leave answers in radical form but simplify.

56. $\sqrt{(3-6)^2 + (-1-3)^2}$

57. $\sqrt{(-2-(-4))^2 + (7-(-1))^2}$

7.6 EXERCISES C

Solve.

1. $\sqrt{\sqrt{x} + 5} = \sqrt{x} - 1$

[Answer: 16]

2. $\sqrt{x+2} - \dfrac{6}{\sqrt{x+7}} = 0$

3. $\dfrac{x+1}{\sqrt{x^2 + 2x + 1}} = \dfrac{1}{3}$

[Answer: no solution]

4. $\sqrt{x+5} + \sqrt{x-3} = \sqrt{4x}$

5. $\sqrt{\sqrt{x} + \sqrt{x+1}} + 5 = 0$

6. Solve $c^{1/2} = \sqrt{b + \sqrt{a}}$ for a.

[Answer: $a = b^2 - 2bc + c^2$]

Assume that $c = 5\sqrt[3]{n} + 3.6$ *for Exercises 7 and 8.*

7. Find c when n is 2.8.

8. Find n when c is 7.4.

[Answer: approximately 0.44]

7.7 THE PYTHAGOREAN THEOREM AND THE DISTANCE AND MIDPOINT FORMULAS

STUDENT GUIDEPOSTS

1. The Pythagorean Theorem
2. Applications of the Pythagorean Theorem
3. The Distance Formula
4. The Midpoint Formula

1 THE PYTHAGOREAN THEOREM

Many applications involve a property of right triangles called *the Pythagorean theorem*. Remember that the longest side of a right triangle, the side opposite the right angle, is the **hypotenuse** of the triangle and the other two sides are called **legs.** (See Figure 7.4 on the following page.)

The Pythagorean Theorem

In a right triangle with hypotenuse of length c and legs of lengths a and b,

$$c^2 = a^2 + b^2.$$

We can remember the Pythagorean theorem in words as:

In a right triangle, the square of the hypotenuse is equal to the sum of the squares of the legs.

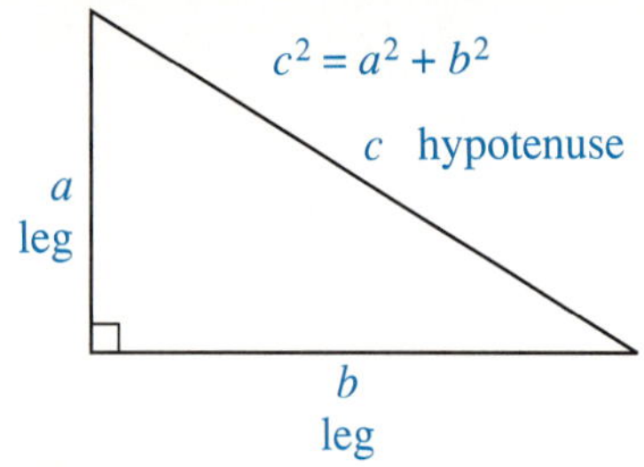

Figure 7.4 The Pythagorean Theorem

We can solve $c^2 = a^2 + b^2$ for c by taking the principle square root of both sides.

$$c = \sqrt{a^2 + b^2}$$

To solve the equation for a, first subtract b^2 from both sides.

$$c^2 = a^2 + b^2$$

$$c^2 - b^2 = a^2 + b^2 - b^2 \quad \text{Subtract } b^2$$

$$c^2 - b^2 = a^2$$

$$\sqrt{c^2 - b^2} = a \quad \text{Take the principal square root of both sides}$$

Similarly, $$\sqrt{c^2 - a^2} = b.$$

We now use these formulas to solve problems involving right triangles.

EXAMPLE 1 USING THE PYTHAGOREAN THEOREM

Find the missing side in each right triangle.

(a) Find the hypotenuse c when $a = 3$ and $b = 4$. See Figure 7.5.

$$c^2 = a^2 + b^2$$
$$c = \sqrt{a^2 + b^2}$$
$$= \sqrt{3^2 + 4^2}$$
$$= \sqrt{9 + 16}$$
$$= \sqrt{25} = 5$$

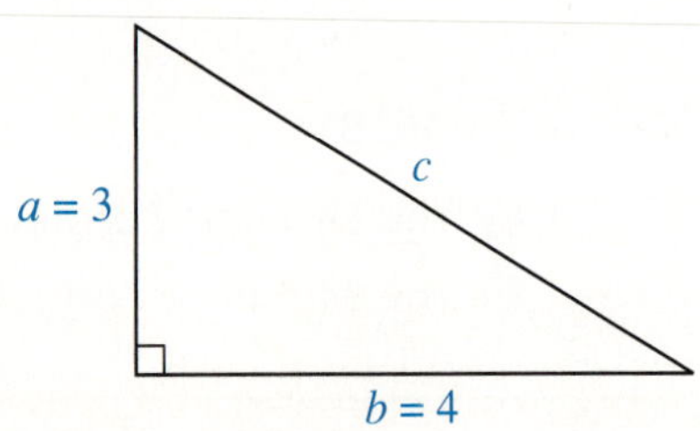

Figure 7.5

(b) Find the leg b when $c = 10$ and $a = 4$. See Figure 7.6.

$$c^2 = a^2 + b^2$$
$$b^2 = c^2 - a^2$$
$$b = \sqrt{c^2 - a^2}$$
$$= \sqrt{10^2 - 4^2}$$
$$= \sqrt{100 - 16}$$
$$= \sqrt{84}$$
$$= \sqrt{4 \cdot 21}$$
$$= 2\sqrt{21}$$

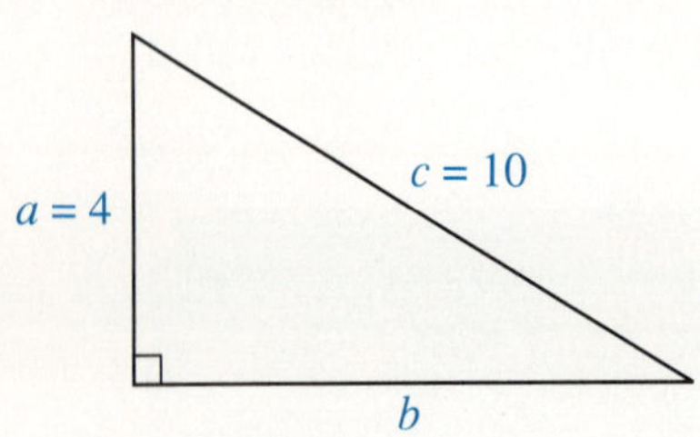

Figure 7.6

PRACTICE EXERCISE 1

Find the missing side in each right triangle.

(a) Find the hypotenuse c when $a = 1$ and $b = 3$.

(b) Find the leg a when $c = 5$ and $b = 2$.

Answers: (a) $\sqrt{10}$ (b) $\sqrt{21}$

CAUTION

Mistakes can be made when calculating square roots in problems involving the Pythagorean theorem. In Example 1(a), $\sqrt{9+16}=\sqrt{25}=5$ *not* $\sqrt{9}+\sqrt{16}=3+4=7$. *Do not take roots term by term in a radicand.*

In Example 1(b) we left the value of b in exact irrational form, $2\sqrt{21}$. In many problems we will approximate such values using a calculator. For example, $2\sqrt{21}\approx 9.2$, and 9.2 gives us a better idea of the length of a side than $2\sqrt{21}$.

2 APPLICATIONS OF THE PYTHAGOREAN THEOREM

In most applied problems involving the Pythagorean theorem, a sketch of the appropriate triangle is helpful.

EXAMPLE 2 Navigation Problem

An airplane flew 20 mi west and then 30 mi south from an airport. To the nearest mile, how many miles was it from the starting point? See Figure 7.7.

$$d^2=(20)^2+(30)^2$$
$$d=\sqrt{(20)^2+(30)^2}$$
$$=\sqrt{400+900}$$
$$=\sqrt{1300}$$
$$=\sqrt{100\cdot 13}$$
$$=10\sqrt{13}$$
$$\approx 36.1$$

The airplane was about 36 mi from the airport.

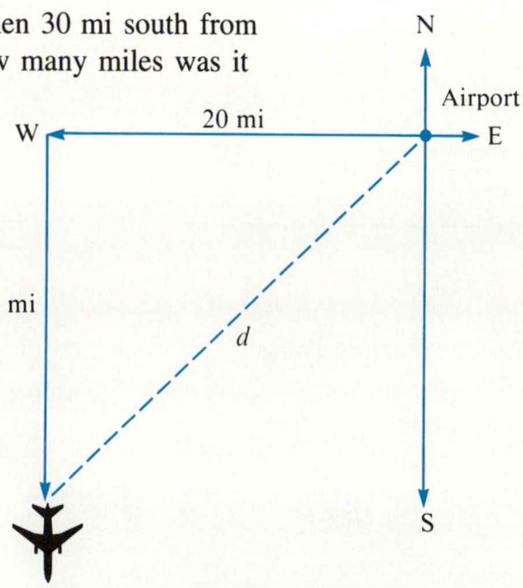

Figure 7.7

Practice Exercise 2

A small boat left an island and sailed 25 miles north and then 10 miles east. To the nearest mile, how far was the boat from the island?

Answer: 27 mi

EXAMPLE 3 Geometry Problem

A baseball diamond is a square 90 feet in length on each side. How far is it directly across the diamond from first base to third base? See Figure 7.8.

$$x^2=90^2+90^2$$
$$x=\sqrt{90^2+90^2}$$
$$=\sqrt{2(90)^2}$$
$$=\sqrt{(90)^2}\sqrt{2}$$
$$=90\sqrt{2}$$
$$\approx 127.3$$

It's about 127.3 ft from first base to third base.

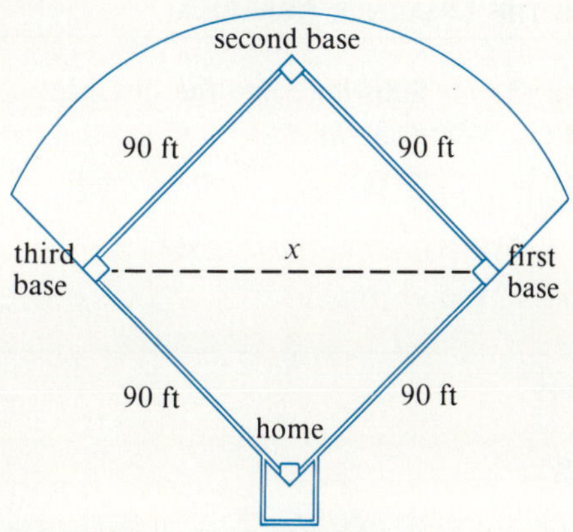

Figure 7.8

Practice Exercise 3

A rectangular lot is to be divided by a fence across the diagonal. Approximately how long will the fence need to be if the lot is 40 m wide and 60 m long?

Answer: approximately 72 m

3 THE DISTANCE FORMULA

Finding the distance between two points in a rectangular coordinate system is a direct application of the Pythagorean theorem. Suppose $P(x_1, y_1)$ and $Q(x_2, y_2)$ are two points as shown in Figure 7.9. Then the numbers $x_2 - x_1$ and $y_2 - y_1$ are the lengths of the legs of the right triangle with vertices P, Q, and R. Applying the Pythagorean theorem, we obtain

$$d^2 = (x_2 - x_1)^2 + (y_2 - y_1)^2,$$

and since d is positive,

$$d = \sqrt{(x_2 - x_1)^2 + (y_2 - y_1)^2}$$

gives the length of the hypotenuse of the triangle—that is, the distance between the points (x_1, y_1) and (x_2, y_2). Since $(x_2 - x_1)$ and $(y_2 - y_1)$ are both squared, $(x_1 - x_2)^2 = (x_2 - x_1)^2$ and $(y_1 - y_2)^2 = (y_2 - y_1)^2$, and it is immaterial which point we label (x_1, y_1) and which point we label (x_2, y_2). Although our particular points are in quadrants I and II, the same results hold regardless of the location of the two points.

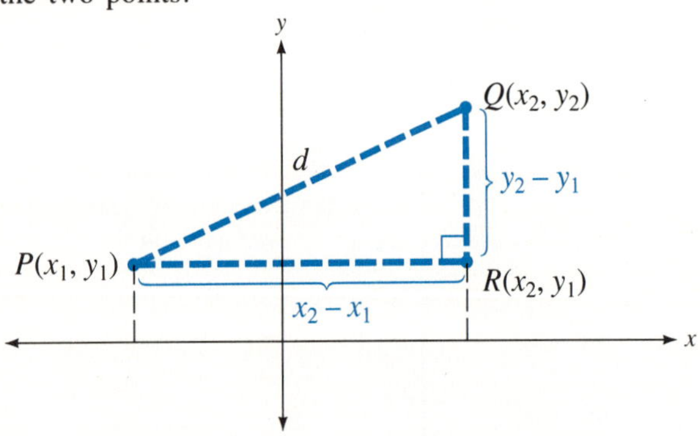

Figure 7.9 Distance between Two Points

These results are summarized in the following definition.

Distance Formula

The distance between two points with coordinates (x_1, y_1) and (x_2, y_2) is given by

$$d = \sqrt{(x_2 - x_1)^2 + (y_2 - y_1)^2}.$$

EXAMPLE 4 Using the Distance Formula

Find the distance between $(-5, 1)$ and $(3, 7)$. Substitute into the distance formula using $(x_1, y_1) = (-5, 1)$ and $(x_2, y_2) = (3, 7)$.

$$\begin{aligned} d &= \sqrt{[3 - (-5)]^2 + (7 - 1)^2} \\ &= \sqrt{8^2 + 6^2} \\ &= \sqrt{64 + 36} = \sqrt{100} = 10 \end{aligned}$$

PRACTICE EXERCISE 4

Find the distance between (2, 5) and (6, 8).

Answer: 5

4 THE MIDPOINT FORMULA

Another formula, the *midpoint formula,* can be established using similar triangles. It shows that the coordinates of a point midway between two given points are found by averaging the corresponding x-coordinates and y-coordinates.

Midpoint Formula

The coordinates $(\bar{x}, \bar{y})$ of the midpoint of a line segment joining (x_1, y_1) and (x_2, y_2) are given by the **midpoint formula**

$$(\bar{x}, \bar{y}) = \left(\frac{x_1 + x_2}{2}, \frac{y_1 + y_2}{2}\right).$$

That is, the x-coordinate of the midpoint is the average of the x-coordinates of the points, and the y-coordinate of the midpoint is the average of the y-coordinates of the points.

CAUTION

In the distance formula we subtract x-coordinates and y-coordinates while in the midpoint formula we add x-coordinates and y-coordinates. Do not confuse these operations when working with these two formulas.

EXAMPLE 5 Using the Midpoint Formula

Find the midpoint between the points $(-2, 1)$ and $(6, -3)$.

Substitute into the midpoint formula.

$$(\bar{x}, \bar{y}) = \left(\frac{x_1 + x_2}{2}, \frac{y_1 + y_2}{2}\right) = \left(\frac{-2 + 6}{2}, \frac{1 + (-3)}{2}\right)$$

$$= \left(\frac{4}{2}, \frac{-2}{2}\right) = (2, -1)$$

The midpoint is $(2, -1)$.

Practice Exercise 5

Find the midpoint between the points $(-3, -7)$ and $(5, 4)$.

Answer: $\left(1, -\frac{3}{2}\right)$

The Pythagorean theorem, the distance formula, and the midpoint formula are frequently used tools in algebra, and should be committed to memory.

7.7 EXERCISES A

Find the unknown leg or hypotenuse in each right triangle with the given sides. Leave answer in radical form, where appropriate.

1. $b = 12,\ c = 13$

2. $a = 7,\ c = 10$

3. $a = \sqrt{2},\ c = 3$

4 $b = 5,\ c = 5\sqrt{3}$

Find the missing side in each right triangle, correct to the nearest tenth of a unit.

5. $a = 1.7$ cm, $b = 3.4$ cm

6. $a = 15.2$ ft, $c = 21.3$ ft

7. $b = 6.5$ in, $c = 17.3$ in

8. $a = 12.4$ m, $c = 10.6$ m

Solve.

9 A field in the shape of a square has a diagonal 2 mi in length. What is the length of a side?

10. A machinist has a rectangular part that measures 15 cm by 20 cm. What is the length of a diagonal?

11. The base of a ladder which is 16 ft long is placed 4 ft from a wall of a building. How high up the wall is the top of the ladder?

12. An airplane left an airport and flew 120 km north and then 200 km west. How far was the plane from the airport?

13. The diagonal of a square measures 16 cm. What is the length of each equal side, to the nearest tenth of a centimeter?

14 The triangle below is an isosceles right triangle. Find a formula for the hypotenuse h in terms of a.

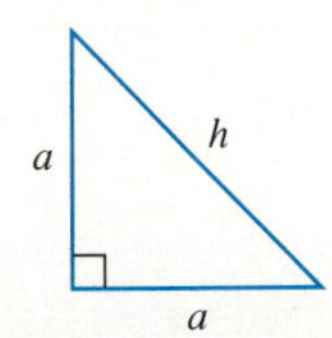

15. Bridges are built with expansion joints to prevent buckling during the heat of summer. Assume, however, that a bridge 1 mile long (5280 ft) has no such joint and expands 2 ft in length during July and that it bulges up in the middle. How high is the bulge? The answer may be surprising!

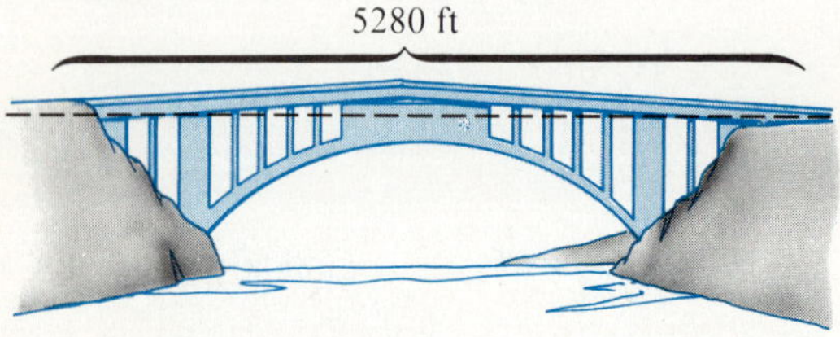

16. Find the area of an isosceles triangle with equal sides of 20 ft and third side 30 ft. [*Hint:* Find the height of the triangle first.]

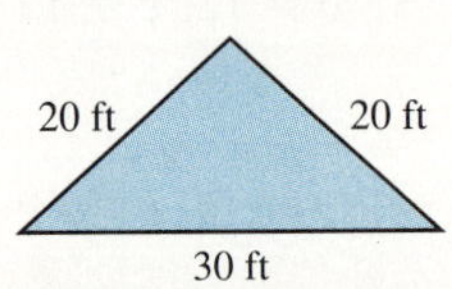

Find the distance between the given points.

17. (3, 6) and (−3, −2)

18. (3, 2) and (−2, −10)

19. $(-2, 1)$ and $(5, -3)$

20. $(2, 2)$ and $(-2, -2)$

The converse of the Pythagorean theorem is also true. That is, if $c^2 = a^2 + b^2$, then a triangle with sides a, b, and c is a right triangle. Find the lengths of the sides of the triangle ABC with the given vertices. Is the triangle a right triangle?

21. $A(-2, 1)$; $B(2, 5)$; $C(1, -2)$

22. $A(1, 2)$; $B(1, -3)$; $C(-2, 1)$

Find the midpoint of the segment between each pair of points.

23. $(2, 3)$ and $(8, 1)$

24. $(-4, 7)$ and $(5, -3)$

25. $(-8, -3)$ and $(-8, 11)$

26. $\left(-5, \frac{1}{2}\right)$ and $\left(3, -\frac{3}{2}\right)$

27 Find the values of a and b if the midpoint of the segment joining $(a, 3)$ and $(-2, b + 1)$ is $(2, 1)$.

28. Find the value of a if the distance between the points $(a, 1)$ and $(-1, 4)$ is 5.

FOR REVIEW

Solve.

29. $\sqrt{y^2 + 9} - y - 1 = 0$

30. $\sqrt{2y + 3} - \sqrt{y - 2} = 2$

The following exercises review material from Chapter 5 and this chapter to help you prepare for the next section. Simplify each radical in Exercises 31–33.

31. $\sqrt{16}$

32. $\sqrt{8}$

33. $\sqrt{108}$

Perform the indicated operations in Exercises 34 and 35.

34. $(2x - y) - (5x - 7y)$

35. $(5 + x\sqrt{2})(5 - x\sqrt{2})$

36. Rationalize the denominator. $\dfrac{7}{3 - \sqrt{2}}$

ANSWERS: 1. 5 2. $\sqrt{51}$ 3. $\sqrt{7}$ 4. $5\sqrt{2}$ 5. 3.8 cm 6. 14.9 ft 7. 16.0 in 8. There is no triangle since c must be longer than a. 9. $\sqrt{2}$ mi $\approx$ 1.4 mi 10. 25 cm 11. $4\sqrt{15}$ ft $\approx$ 15.5 ft 12. $40\sqrt{34}$ km $\approx$ 233 km 13. $8\sqrt{2}$ cm $\approx$ 11.3 cm 14. $h = a\sqrt{2}$ 15. about 72.7 ft 16. about 198.4 ft^2 17. 10 18. 13 19. $\sqrt{65}$ 20. $4\sqrt{2}$ 21. yes 22. no 23. $(5, 2)$ 24. $\left(\frac{1}{2}, 2\right)$ 25. $(-8, 4)$ 26. $\left(-1, -\frac{1}{2}\right)$ 27. $a = 6$; $b = -2$ 28. $a = 3$ or $a = -5$ 29. 4 30. 3; 11 31. 4 32. $2\sqrt{2}$ 33. $6\sqrt{3}$ 34. $-3x + 6y$ 35. $25 - 2x^2$ 36. $3 + \sqrt{2}$

7.7 EXERCISES B

Find the unknown leg or hypotenuse in each right triangle with the given sides. Leave answer in radical form, where appropriate.

1. $a = 6,\ c = 10$ **2.** $b = 2,\ c = 5$ **3.** $b = \sqrt{3},\ c = 3$ **4.** $a = 7,\ c = 7\sqrt{3}$

Find the missing side in each right triangle, correct to the nearest tenth of a unit.

5. $a = 5.2$ cm, $b = 6.9$ cm **6.** $a = 24.3$ ft, $c = 36.7$ ft

7. $b = 2.4$ in, $c = 23.5$ in **8.** $b = 16.5$ m, $c = 11.5$ m

Solve.

9. A machine part in the shape of a square has a diagonal 6 cm in length. What is the length of a side?

10. A field in the shape of a rectangle is 300 yd long and 200 yd wide. What is the length of a diagonal?

11. How long must a ladder be to reach 8 m up the side of a building when the base of the ladder is 2 m from the building?

12. An airplane is 350 mi from the airport. To get to the present position, it flew 200 mi due south and then flew due west. How far west has it gone?

13. A guy wire reaches from the top of a 25-ft pole to the ground 12 ft from the pole. How long is the wire?

14. In a right triangle, one leg is x and the other is $x + 1$. Find a formula for the hypotenuse h in terms of x.

15. The softball diamond in a city park is a square 70 ft on a side. How far is it from home plate to second base?

16. Find the area of an equilateral triangle with sides of length 10 cm.

Find the distance between the given points.

17. $(3, -1)$ and $(-9, 4)$ **18.** $(-2, 5)$ and $(2, 2)$ **19.** $(-3, 2)$ and $(4, 1)$ **20.** $(0, -3)$ and $(7, 2)$

The converse of the Pythagorean theorem is also true. That is, if $c^2 = a^2 + b^2$, then a triangle with sides a, b, and c is a right triangle. Find the lengths of the sides of the triangle ABC with the given vertices. Is the triangle a right triangle?

21. $A(-2, -1);\ B(1, -2);\ C(3, 3)$ **22.** $A(-3, 2);\ B(4, 1);\ C(1, -2)$

Find the midpoint of the segment between each pair of points.

23. $(5, 2)$ and $(3, 12)$ **24.** $(-6, 5)$ and $(8, -3)$ **25.** $(6, -2)$ and $(-5, -2)$ **26.** $\left(3, -\frac{1}{2}\right)$ and $\left(-3, \frac{7}{2}\right)$

27. Find the values of a and b if the midpoint of the segment joining $(a - 1, 5)$ and $(-6, b)$ is $\left(-\frac{3}{2}, 2\right)$.

28. Find the value of a if the distance between the points $(4, -2)$ and $(a, -6)$ is 5.

FOR REVIEW

Solve.

29. $\sqrt{y^2 + 7} - y - 1 = 0$ **30.** $\sqrt{3x + 1} - \sqrt{2x - 1} = 1$

The following exercises will help you prepare for the next section. Simplify each radical in Exercises 31–33.

31. $\sqrt{81}$ **32.** $\sqrt{12}$ **33.** $\sqrt{300}$

Perform the indicated operations in Exercises 34 and 35.

34. $(3x - 2y) - (4x - 10y)$ **35.** $(2 - x\sqrt{3})(2 + x\sqrt{3})$

36. Rationalize the denominator. $\dfrac{2}{\sqrt{11} - 3}$

7.7 EXERCISES C

Exercises 1–4 refer to the regular hexagon shown at right. A calculator may be helpful in some of these problems.

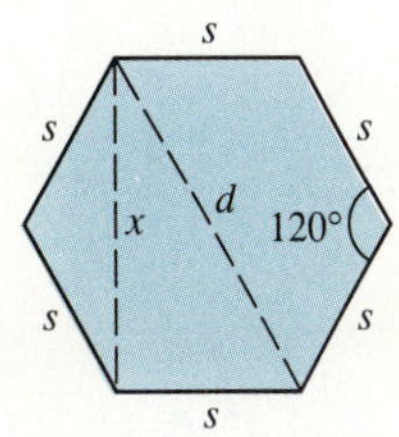

1. Find the distance between parallel sides, x, in terms of the equal sides s. [*Hint:* The side opposite the 30° acute angle in a right triangle is half the hypotenuse.]

2. Find the length of diagonal d in terms of the equal sides s. [Answer: $d = 2s$]

3. What is the value of x when $s = 5$ cm? What is the value of d?

4. What is the area of the hexagon when $s = 5$ cm?

5. Find the equation of the perpendicular bisector of the line segment joining the two points $(-3, 2)$ and $(5, -4)$. [Answer: $4x - 3y - 7 = 0$]

6. A square has area 144 in^2. What is the area of the square formed by joining the midpoints of the sides of this square?

7.8 COMPLEX NUMBERS

STUDENT GUIDEPOSTS

1. Imaginary Numbers
2. Complex Numbers
3. Equality of Complex Numbers
4. Operations on Complex Numbers

1 IMAGINARY NUMBERS

Until now we have discussed only square roots of *nonnegative* numbers. The square root of a negative number, such as -4, is not a real number since no real number a, when squared, can be -4. That is, if a is any real number, $a^2 \geq 0$ so that $a^2 \neq -4$. We now develop a new set of numbers in which square roots of negative numbers are defined. These numbers, imaginary numbers, play an important role in mathematics with many applications in physics and engineering.

Heron of Alexandria (circa 150 B.C. to 250 A.D.) first noticed that the square root of a negative number posed an interesting problem. René Descartes (1596–1650) first used the terms *real* and *imaginary numbers*. Don't let the term *imaginary* mislead you, however, since imaginary numbers are no more ''imaginary'' than real numbers.

By agreeing to follow rules for radicals similar to those we followed with real numbers, only one new numeral is needed to develop the imaginary numbers. We use i as the numeral corresponding to $\sqrt{-1}$.

$$\sqrt{-1} = i \quad \text{so that} \quad i^2 = -1.$$

Remember that the product rule for radicals,

$$\sqrt{a}\sqrt{b} = \sqrt{ab}$$

applies only when $a \geq 0$ and $b \geq 0$. However, if we agree to extend this rule to the case when $a \geq 0$ and $b = -1$, every imaginary number can be written as the product of a real number and the number i. For example,

$$\sqrt{-4} = \sqrt{(4)(-1)} = \sqrt{4}\sqrt{-1} = 2i.$$

EXAMPLE 1 Simplifying Imaginary Numbers

Simplify.

(a) $\sqrt{-9} = \sqrt{9(-1)} = \sqrt{9}\sqrt{-1} = 3i$ $\quad \sqrt{-1} = i$

(b) $\sqrt{-2}\sqrt{-5} = \sqrt{2(-1)}\sqrt{5(-1)} = \sqrt{2}\cdot i \cdot \sqrt{5}\cdot i$

$= \sqrt{2}\sqrt{5}i^2 = \sqrt{10}(-1) = -\sqrt{10}$ $\quad i^2 = -1$

Practice Exercise 1

Simplify.

(a) $\sqrt{-144}$

(b) $\sqrt{-6}\sqrt{-11}$

Answers: (a) $12i$ (b) $-\sqrt{66}$

CAUTION

Do not write $\sqrt{-4}\sqrt{-9} = \sqrt{(-4)(-9)} = \sqrt{36} = 6$ since $\sqrt{-4}\sqrt{-9} = (2i)(3i) = 6i^2 = -6$. The rule $\sqrt{a}\sqrt{b} = \sqrt{ab}$ does not apply when a and b are both negative. We can use this rule when $a \geq 0$ and $b = -1$ to express square roots of negative numbers in terms of the imaginary number i.

EXAMPLE 2 More on Imaginary Numbers

Simplify.

(a) $\dfrac{\sqrt{-30}}{\sqrt{-6}} = \dfrac{\sqrt{30(-1)}}{\sqrt{6(-1)}} = \dfrac{\sqrt{30}\not{i}}{\sqrt{6}\not{i}} = \sqrt{\dfrac{30}{6}} = \sqrt{5}$

(b) $\sqrt{-16} + \sqrt{-49} = \sqrt{16(-1)} + \sqrt{49(-1)}$

$= \sqrt{16}i + \sqrt{49}i = 4i + 7i = 11i$

Practice Exercise 2

Simplify.

(a) $\dfrac{\sqrt{-45}}{\sqrt{-5}}$

(b) $\sqrt{-4} + \sqrt{-121}$

Answers: (a) 3 (b) $13i$

Powers of i cycle through the same four values: i, -1, $-i$, and 1. This is shown in the following table that lists the first sixteen powers of i.

Powers of the Imaginary Number i

$i^1 = i$	$i^5 = i$	$i^9 = i$	$i^{13} = i$
$i^2 = -1$	$i^6 = -1$	$i^{10} = -1$	$i^{14} = -1$
$i^3 = -i$	$i^7 = -i$	$i^{11} = -i$	$i^{15} = -i$
$i^4 = 1$	$i^8 = 1$	$i^{12} = 1$	$i^{16} = 1$

We can use this cyclical pattern to find any higher power of i by writing the expression using factors of $i^4 = 1$.

EXAMPLE 3 Finding Powers of i

Find each power of i.

(a) $i^{21} = i^{20}i = (i^4)^5(i)$ $\quad$ Express 21 as a multiple of 4(20) plus 1 (a number less than 4)

$= (1)^5 i = (1)i = i$

Practice Exercise 3

Find each power of i.

(a) i^{32}

(b) i^{53}

(b) $i^{40} = (i^4)^{10} = (1)^{10} = 1$

(c) $i^{122} = i^{120}i^2 = (i^4)^{30}i^2 = (1)^{30}i^2 = i^2 = -1$

(d) $i^{363} = i^{360}i^3 = (i^4)^{90}i^3 = (1)^{90}i^3 = i^3 = -i$

(c) i^{146}

(d) i^{411}

Answers: (a) 1 (b) i (c) -1 (d) $-i$

2 COMPLEX NUMBERS

Leonhard Euler (1707–83) is credited as being the first to use i for $\sqrt{-1}$. But Carl Friedrich Gauss (1777–1855) first saw that the real numbers and the imaginary numbers could be combined to form a powerful number system, the *complex numbers*. A number of the form

$$a + bi \qquad a \text{ and } b \text{ real numbers}$$

is called a **complex number.** The number a is the **real part** of $a + bi$, and b is the **imaginary part.** If $b = 0$, then the complex number $a + bi = a + 0i = a$ is simply the real number a. If $a = 0$, it is the imaginary number bi. Thus, the complex numbers include both the real numbers and the imaginary numbers.

Figure 7.10 summarizes the relationships among the sets of numbers we have studied. Some examples of each type of number are listed.

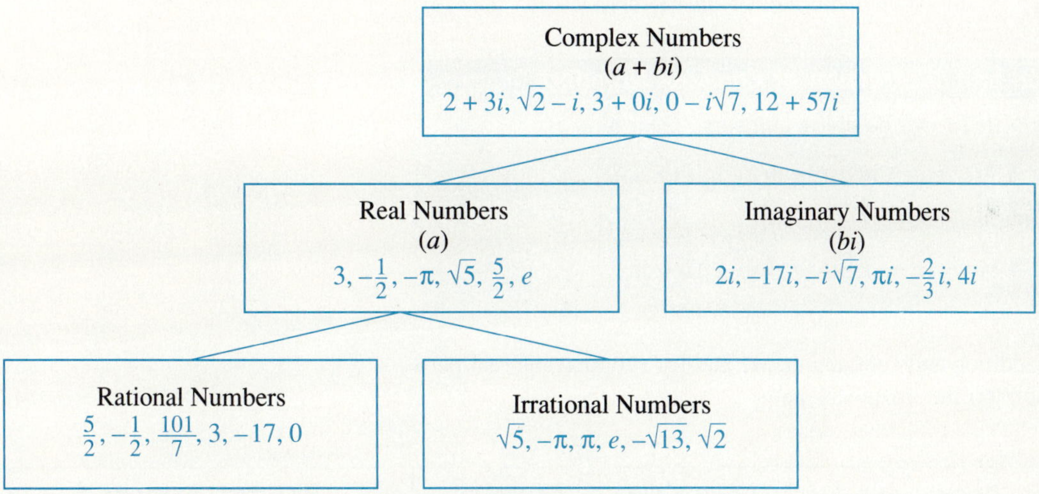

Figure 7.10 Sets of Numbers

3 EQUALITY OF COMPLEX NUMBERS

Before we can do arithmetic on the set of complex numbers, we need to decide when two complex numbers are *equal*.

Equality of Complex Numbers

Two complex numbers $a + bi$ and $c + di$ are **equal,**

$$a + bi = c + di,$$

if and only if their real parts are equal ($a = c$) and their imaginary parts are equal ($b = d$).

EXAMPLE 4 Equality of Complex Numbers

(a) If $a + bi = 3 + 2i$, then $a = 3$ and $b = 2$.

PRACTICE EXERCISE 4

(a) If $x + yi = 5 - 2i$, find x and y.

(b) If $c + di = 5$, then $c = 5$ and $d = 0$ since $5 = 5 + 0i$.

(c) If $u + vi = -\sqrt{3}i$, then $u = 0$ and $v = -\sqrt{3}$ since $-\sqrt{3}i = 0 - \sqrt{3}i$.

(d) If $2x + yi = 3x - 2 + 7i$, then

$$2x = 3x - 2 \quad \text{and} \quad y = 7.$$
$$-x = -2$$
$$x = 2$$

(b) If $a + bi = -8$, find a and b.

(c) If $w + ui = i\sqrt{5}$, find w and u.

(d) If $4a - bi = a + 3 - 9i$, find a and b.

Answers: (a) $x = 5$, $y = -2$ (b) $a = -8$, $b = 0$ (c) $w = 0$, $u = \sqrt{5}$ (d) $a = 1$, $b = 9$

CAUTION

When a radical is the coefficient of i, as in $-\sqrt{3}\,i$ in Example 4(c), notice that i is not under the radical. Since it is a common mistake to put i under the radical, we often write $-\sqrt{3}\,i$ as $-i\sqrt{3}$.

4 OPERATIONS ON COMPLEX NUMBERS

The basic operations on complex numbers are easy to learn since the number i acts in the same way as any other literal number in calculations.

Addition and Subtraction of Complex Numbers

Let $a + bi$ and $c + di$ be two complex numbers. Their **sum** is

$$(a + bi) + (c + di) = (a + c) + (b + d)i,$$

and their **difference** is

$$(a + bi) - (c + di) = (a - c) + (b - d)i.$$

Notice that for addition and subtraction, we add (or subtract) the real parts and then add (or subtract) the imaginary parts.

EXAMPLE 5 Adding and Subtracting Complex Numbers

(a) Add.

$(3 + 2i) + (-4 - i) = 3 - 4 + 2i - i$ Use commutative rule

$= (3 - 4) + (2 - 1)i$ Add real parts and imaginary parts

$= -1 + i$

(b) Subtract.

$(2 - i\sqrt{3}) - (4 + 3i\sqrt{3}) = 2 - 4 - i\sqrt{3} - 3i\sqrt{3}$

$= (2 - 4) - i(\sqrt{3} + 3\sqrt{3})$

$= -2 - 4i\sqrt{3}$

Practice Exercise 5

(a) Add.
$(5 - 3i) + (-2 - i)$

(b) Subtract.
$(8 + i\sqrt{5}) - (-3 - 2i\sqrt{5})$

Answers: (a) $3 - 4i$ (b) $11 + 3i\sqrt{5}$

Multiplication of Complex Numbers

Let $a + bi$ and $c + di$ be two complex numbers. Their **product** is

$$(a + bi)(c + di) = (ac - bd) + (ad + bc)i.$$

Although the definition of multiplication of complex numbers may appear strange at first, it seems more reasonable when we multiply as if the numbers are binomials using the FOIL method.

$$\begin{aligned}(a + bi)(c + di) &= \overset{F}{ac} + \overset{O}{adi} + \overset{I}{bci} + \overset{L}{bdi^2}\\ &= ac + bdi^2 + adi + bci\\ &= ac + bd(-1) + adi + bci \qquad i^2 = -1\\ &= (ac - bd) + (ad + bc)i\end{aligned}$$

In fact, rather than memorizing the definition, it is probably better to use the FOIL method in particular cases, keeping in mind that $i^2 = -1$.

EXAMPLE 6 Multiplying Complex Numbers

Multiply.

(a) $(4 + 2i)(3 - 5i)$

$$\begin{aligned}(4 + 2i)(3 - 5i) &= \overset{F}{4 \cdot 3} - \overset{O}{4 \cdot 5i} + \overset{I}{2 \cdot 3i} - \overset{L}{2 \cdot 5 \cdot i^2}\\ &= 12 - 20i + 6i - 10i^2\\ &= 12 - 14i - 10(-1)\\ &= 12 + 10 - 14i\\ &= 22 - 14i\end{aligned}$$

(b) $$\begin{aligned}(5 + i\sqrt{2})(5 - i\sqrt{2}) &= 5 \cdot 5 - 5i\sqrt{2} + 5i\sqrt{2} - i^2\sqrt{2}\,\sqrt{2}\\ &= 25 - i^2 \cdot 2\\ &= 25 - (-1) \cdot 2\\ &= 25 + 2 = 27\end{aligned}$$

Practice Exercise 6

Multiply.

(a) $(2 - i)(5 + 7i)$

(b) $(8 + i\sqrt{5})(8 - i\sqrt{5})$

Answers: (a) $17 + 9i$ (b) 69

In Example 6(b), a special product was found. The numbers $5 + i\sqrt{2}$ and $5 - i\sqrt{2}$ are *conjugates,* and whenever two conjugates are multiplied, the result is a real number. The **conjugate** of the complex number $a + bi$ is $a - bi$, and the **conjugate** of $a - bi$ is $a + bi$. Thus, $a + bi$ and $a - bi$ are **conjugates** of each other. (Compare this definition to the one for conjugates of irrational numbers in Section 7.4.) One application of conjugates is in division of complex numbers.

Division of Complex Numbers

Let $a + bi$ and $c + di$ be two complex numbers. The **quotient** of $a + bi$ and $c + di$ can be simplified by multiplying numerator and denominator by the conjugate of $c + di$, $c - di$. That is,

$$\frac{(a + bi)}{(c + di)} = \frac{(a + bi)(c - di)}{(c + di)(c - di)} = \frac{(ac + bd) + (bc - ad)i}{c^2 + d^2}$$

As with multiplying complex numbers, when dividing complex numbers it is better to use the method of multiplying numerator and denominator by the conjugate of the denominator rather than memorizing the definition.

EXAMPLE 7 Dividing Complex Numbers

Divide $(3 + 7i)$ by $(2 - 5i)$.

$$\frac{3+7i}{2-5i} = \frac{(3+7i)(2+5i)}{(2-5i)(2+5i)}$$
$$= \frac{6 + 15i + 14i + 35i^2}{4 + 10i - 10i - 25i^2}$$
$$= \frac{6 + 29i - 35}{4 + 25} \quad i^2 = -1$$
$$= \frac{-29 + 29i}{29}$$
$$= \frac{-29}{29} + \frac{29i}{29} = -1 + i$$

Practice Exercise 7

Divide $(2 - 3i)$ by $(1 + i)$.

Answer: $\frac{-1 - 5i}{2}$

EXAMPLE 8 Finding the Reciprocal of a Complex Number

Find the reciprocal of $3 + 2i$ and express it in the form $a + bi$.

The reciprocal of $3 + 2i$ is $\frac{1}{3 + 2i}$. We need to divide the complex number $1 = 1 + 0i$ by the complex number $3 + 2i$. Thus, finding a reciprocal is simply a division problem.

$$\frac{1}{3+2i} = \frac{1(3-2i)}{(3+2i)(3-2i)} = \frac{3-2i}{9 - 6i + 6i - 4i^2}$$
$$= \frac{3-2i}{9+4} = \frac{3-2i}{13} = \frac{3}{13} - \frac{2}{13}i$$

Practice Exercise 8

Find the reciprocal of $4 - 3i$ and express it in the form $a + bi$.

Answer: $\frac{4}{25} + \frac{3}{25}i$

Many applications of complex numbers are found in the study of electricity.

EXAMPLE 9 Electricity Problem

Ohm's law of electricity is $E = IR$ where E is electromotive force in volts, I is current in amperes, and R is resistance in ohms. Find the current if $E = 1 + 4i$ volts and $R = 1 + i$ ohms.

Since $E = IR$, solving for I gives

$$I = \frac{E}{R}.$$

Substitute $1 + 4i$ for E and $1 + i$ for R and divide.

$$I = \frac{1+4i}{1+i} = \frac{(1+4i)(1-i)}{(1+i)(1-i)} = \frac{1 - i + 4i - 4i^2}{1 - i^2}$$
$$= \frac{1 + 3i + 4}{1 + 1} \quad i^2 = -1$$
$$= \frac{5+3i}{2} = \frac{5}{2} + \frac{3}{2}i$$

Thus, the current is $\frac{5}{2} + \frac{3}{2}i$ amperes.

Practice Exercise 9

Use Ohm's law to find I when $E = 1 + 2i$ volts and $R = 3 + i$ ohms.

Answer: $\frac{1}{2} + \frac{1}{2}i$ amperes

7.8 EXERCISES A

Express in terms of the imaginary number i.

1. $\sqrt{-9}$ **2.** $\sqrt{-36}$ **3.** $-\sqrt{-36}$

4. $\sqrt{-8}$ **5.** $-\sqrt{-3}$ **6.** $\sqrt{-\pi^2}$

Simplify and express each answer as a real number or in terms of i.

7. $\sqrt{-4} + \sqrt{-9}$ **8.** $\sqrt{-4} - \sqrt{-9}$ **9.** $\sqrt{-7} + 2\sqrt{-7}$ **10** $\sqrt{-4}\,\sqrt{-9}$

11. $\sqrt{-3}\,\sqrt{-3}$ **12.** $\sqrt{-7}\,\sqrt{-3}$ **13** $\dfrac{\sqrt{-4}}{\sqrt{-9}}$ **14.** $\dfrac{\sqrt{-7}}{\sqrt{-3}}$

Using the definition of equality of complex numbers, determine x and y.

15. $x + yi = 2 - 3i$ **16.** $7x + 3yi = 21$ **17** $(4x + 1) + 7yi = 5 - i$

Perform the indicated operations.

18. $(2 + 3i) + (5 + 8i)$ **19.** $(2 + 3i) - (5 + 8i)$ **20.** $3 + (5 - 2i)$

21. $-6i + (8 + 2i)$ **22.** $(-3 - 2i) - (6 - 7i)$ **23.** $(-5 + 4i) + (5 - 4i)$

24. $7(2 - 3i)$ **25.** $7i(2 - 3i)$ **26.** $(1 - 2i)(3 + 4i)$

27. $(8 + 4i)(-7 + 3i)$ **28.** $(2 + 3i)(2 - 3i)$ **29.** $(2 + 3i)^2$

30. i^{85} **31.** $i^{15} + i^{33}$ **32.** $i^{40} - i^{22}$

Give the conjugate of each of the following complex numbers.

33. $4 + 6i$ **34.** $-4 - 6i$ **35.** 8 **36.** $6i$ **37.** $-7i$

Find the following quotients.

38. $\dfrac{3 + 5i}{1 + i}$ **39.** $\dfrac{7}{4 - 3i}$ **40.** $\dfrac{3i}{5 - 4i}$ **41.** $\dfrac{8 - 7i}{5 + 4i}$

Find the reciprocal of each number and express it in the form $a + bi$.

42. $1 - i$ **43.** $-4 + 8i$ **44.** $-2i$ **45.** 3

Determine whether each statement in Exercises 46–50 is true or false.

46. Every rational number is a complex number.

47. Every complex number is a real number.

48. Every imaginary number is a real number.

49. Every complex number is a rational number.

50. The product of a complex number and its conjugate is always a real number.

In Exercises 51–52, use Ohm's law $E = IR$, given in Example 9, to find the missing variable.

51. $E = 3 + i$, $I = 2 - i$; find R. **52.** $I = 2 + i$, $R = 3 - i$; find E.

FOR REVIEW

Find the missing side in each right triangle, correct to the nearest tenth of a unit.

53. $a = 19.3$ m, $b = 14.5$ m **54.** $b = 12.9$ ft, $c = 34.4$ ft

55. Find the distance between the points $(4, -7)$ and $(-3, -1)$.

56. Find the midpoint of the segment between the points $(-4, 3)$ and $(12, -6)$.

57. Find the length of x in the square shown at right. Give answer correct to the nearest tenth of a centimeter.

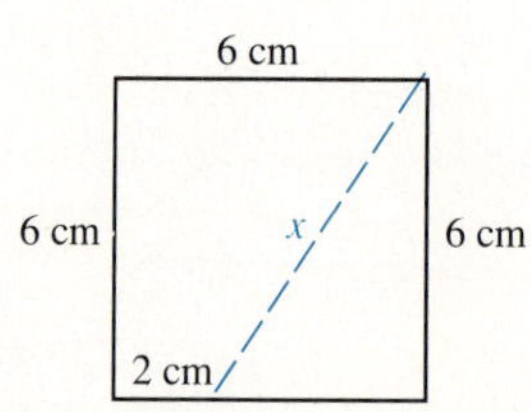

Exercises 58–60 review material from Chapter 5 to help you prepare for the next chapter. Factor each polynomial.

58. $x^2 + 2x - 15$ **59.** $x^2 - 7x$ **60.** $4x^2 + 4x + 1$

ANSWERS: 1. $3i$ 2. $6i$ 3. $-6i$ 4. $2i\sqrt{2}$ 5. $-i\sqrt{3}$ 6. πi 7. $5i$ 8. $-i$ 9. $3i\sqrt{7}$ 10. -6 11. -3 12. $-\sqrt{21}$ 13. $\frac{2}{3}$ 14. $\frac{\sqrt{7}}{\sqrt{3}}$ 15. $x = 2$; $y = -3$ 16. $x = 3$; $y = 0$ 17. $x = 1$; $y = -\frac{1}{7}$ 18. $7 + 11i$ 19. $-3 - 5i$ 20. $8 - 2i$ 21. $8 - 4i$ 22. $-9 + 5i$ 23. 0 24. $14 - 21i$ 25. $21 + 14i$ 26. $11 - 2i$ 27. $-68 - 4i$ 28. 13

29. $-5 + 12i$ 30. i 31. 0 32. 2 33. $4 - 6i$ 34. $-4 + 6i$ 35. 8 36. $-6i$ 37. $7i$ 38. $4 + i$ 39. $\frac{28}{25} + \frac{21}{25}i$ 40. $-\frac{12}{41} + \frac{15}{41}i$ 41. $\frac{12}{41} - \frac{67}{41}i$ 42. $\frac{1}{2} + \frac{1}{2}i$ 43. $-\frac{1}{20} - \frac{1}{10}i$ 44. $\frac{1}{2}i$ 45. $\frac{1}{3}$ 46. true 47. false 48. false 49. false 50. true 51. $1 + i$ ohms 52. $7 + i$ volts 53. 24.1 m 54. 31.9 ft 55. $\sqrt{85}$ 56. $\left(4, -\frac{3}{2}\right)$ 57. 7.2 cm 58. $(x - 3)(x + 5)$ 59. $x(x - 7)$ 60. $(2x + 1)^2$

7.8 EXERCISES B

Express in terms of the imaginary number i.

1. $\sqrt{-16}$ **2.** $\sqrt{-121}$ **3.** $-\sqrt{-121}$ **4.** $\sqrt{-12}$ **5.** $-\sqrt{-5}$ **6.** $-\sqrt{-\pi^2}$

Simplify and express each answer as a real number or in terms of i.

7. $\sqrt{-16} + \sqrt{-25}$ **8.** $\sqrt{-16} - \sqrt{-25}$ **9.** $\sqrt{-7} - 2\sqrt{-7}$ **10.** $\sqrt{-25}\,\sqrt{-16}$

11. $\sqrt{-7}\,\sqrt{-7}$ **12.** $\sqrt{-5}\,\sqrt{-11}$ **13.** $\dfrac{\sqrt{-25}}{\sqrt{-16}}$ **14.** $\dfrac{\sqrt{-5}}{\sqrt{-11}}$

Using the definition of equality of complex numbers, determine x and y.

15. $x + yi = 5 - 15i$ **16.** $2x - 8yi = 10$ **17.** $(2x - 1) + 8yi = 3 - 16i$

Perform the indicated operation.

18. $(2 - 3i) + (3 + 7i)$ **19.** $(2 - 3i) - (3 + 7i)$ **20.** $-7 + (2 - 3i)$

21. $-5i + (3 + i)$ **22.** $(-2 - 11i) - (5 - i)$ **23.** $(-8 + 2i) + (8 - 2i)$

24. $5(3 - 5i)$ **25.** $11i(1 - 5i)$ **26.** $(2 - 3i)(1 + 4i)$

27. $(1 + i)(1 - i)$ **28.** $(2 - 3i)^2$ **29.** $(1 + i\sqrt{2})^2$

30. i^{82} **31.** $i^{12} + i^{49}$ **32.** $i^{19} - i^{25}$

Give the conjugate of each of the following complex numbers.

33. $7 + 3i$ **34.** $-7 - 3i$ **35.** -12 **36.** $11i$ **37.** $-15i$

Find the following quotients.

38. $\dfrac{4 + 3i}{2 + i}$ **39.** $\dfrac{6}{5 - 3i}$ **40.** $\dfrac{2i}{4 - 3i}$ **41.** $\dfrac{6 - 5i}{1 + 3i}$

Find the reciprocal of each number and express it in the form a + bi.

42. $2 - i$ **43.** $-4 - 8i$ **44.** $5i$ **45.** 5

Determine whether each statement in Exercises 46–50 is true or false.

46. Every real number is a complex number.

47. Every imaginary number is a complex number.

48. Every irrational number is a complex number.

49. The sum of a complex number and its conjugate is always a real number.

50. The difference between a complex number and its conjugate is always a real number.

In Exercises 51–52, use Ohm's law $E = IR$, given in Example 9, to find the missing variable.

51. $E = 3 + i$, $I = 2 + i$; find R. **52.** $I = 2 + i$, $R = 4 - i$; find E.

FOR REVIEW

Find the missing side in each right triangle, correct to the nearest tenth of a unit.

53. $a = 27.3$ cm, $b = 8.6$ cm

54. $a = 17.1$ yd, $c = 42.3$ yd

55. Find the distance between the points $(-2, 5)$ and $(3, -5)$.

56. Find the midpoint of the segment between the points $(-9, -3)$ and $(5, 2)$

57. Which triangle at right, ABC or PQR has the greater area?

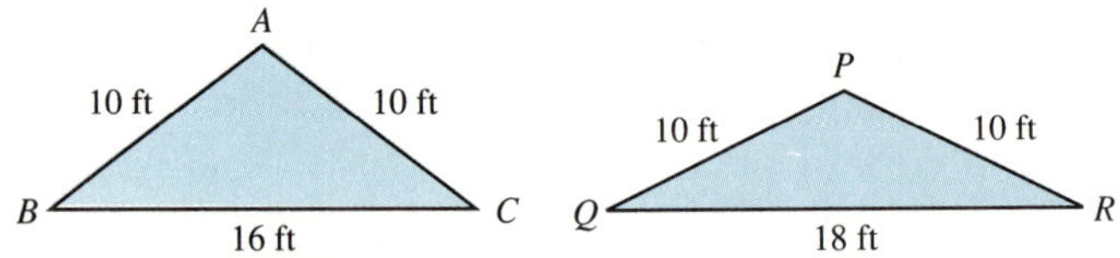

Exercises 58–60 review material from Chapter 5 to help you prepare for the next chapter. Factor each polynomial.

58. $3x^2 + x - 2$

59. $4x^2 - 4x + 1$

60. $16x^2 - 49$

7.8 EXERCISES C

Perform the indicated operations and express the result in the form $a + bi$.

1. $1 + \dfrac{1}{1 + \dfrac{1}{i}}$ $\left[\text{Answer: } \dfrac{3}{2} + \dfrac{1}{2}i\right]$

2. $(2 - i)^3(2 + i)^3$

3. $\dfrac{2}{1 - i} + \dfrac{3}{2 + i}$

CHAPTER 7 REVIEW

KEY WORDS

7.1 A number a is a **cube root** of x if $a^3 = x$. In this case, x is called a **perfect cube.**

A number a is called a ***k*th root** of x if $a^k = x$. In this case x is a **perfect *k*th power.**

7.2 The process of removing radicals from the denominator (or numerator) of a fraction is called **rationalizing the denominator (or numerator).**

7.4 **Like radicals** have the same index and the same radicand.

Unlike radicals have either different indexes or different radicands, or both.

Square root expressions such as $\sqrt{5} + \sqrt{3}$ and $\sqrt{5} - \sqrt{3}$ are called **conjugates** of each other.

7.6 A **radical equation** contains radicals with variables in the radicand.

A solution to a radical equation that has been introduced by raising both sides to a power and that does not check in the original equation is called an **extraneous root.**

7.8 **Imaginary numbers** come from taking square roots of negative real numbers.

A number of the form $a + bi$, where a and b are real and $i = \sqrt{-1}$ is called a **complex number.**

The **real part** of $a + bi$ is a, and the **imaginary part** is b.

Two complex numbers $a + bi$ and $c + di$ are **equal** if and only if $a = c$ and $b = d$.

The complex numbers $a + bi$ and $a - bi$ are **conjugates** of each other.

KEY CONCEPTS

7.1 The **principal *k*th root** of x is denoted by $\sqrt[k]{x}$. If $x \geq 0$, then $\sqrt[k]{x} \geq 0$.

If $x < 0$, then (a) $\sqrt[k]{x} < 0$ when k is odd. (b) $\sqrt[k]{x}$ is not a real number when k is even.

7.2 1. If $\sqrt[k]{a}$ and $\sqrt[k]{b}$ are defined, then $\sqrt[k]{a}\sqrt[k]{b} = \sqrt[k]{ab}$.

2. If $\sqrt[k]{a}$ and $\sqrt[k]{b}$ are defined and $b \neq 0$, then $\dfrac{\sqrt[k]{a}}{\sqrt[k]{b}} = \sqrt[k]{\dfrac{a}{b}}$.

7.4 1. $\sqrt[k]{a+b}$ *is not* $\sqrt[k]{a} + \sqrt[k]{b}$.

2. $\sqrt[k]{a-b}$ *is not* $\sqrt[k]{a} - \sqrt[k]{b}$.

3. To rationalize a binomial denominator, multiply both the numerator and denominator by the conjugate of the denominator.

7.5 1. If a is a real number and n is a natural number, then

$$a^{1/n} = \sqrt[n]{a}.$$

Also, if m is a natural number and $a^{1/n}$ exists, then

$$a^{m/n} = \sqrt[n]{a^m} = (\sqrt[n]{a})^m.$$

2. All rules of integer exponents apply to rational exponents.

7.6 1. To solve a radical equation, isolate a radical on one side of the equation and raise both sides to a power equal to the index on the radical. Check all solutions in the original equation.

2. When solving a literal equation, it is sometimes helpful to solve a similar numerical equation first.

7.7 1. The Pythagorean theorem: In a right triangle with hypotenuse of length c and legs of length a and b,

$$c^2 = a^2 + b^2.$$

2. The distance formula: The distance between two points (x_1, y_1) and (x_2, y_2) is

$$d = \sqrt{(x_2 - x_1)^2 + (y_2 - y_1)^2}.$$

3. The midpoint formula: The coordinates of the midpoint of the segment joining (x_1, y_1) and (x_2, y_2) are

$$(\bar{x}, \bar{y}) = \left(\frac{x_1 + x_2}{2}, \frac{y_1 + y_2}{2}\right).$$

4. Remember that in the distance formula the coordinates are subtracted, and in the midpoint formula, they are added.

7.8 1. $i = \sqrt{-1}$; that is $i^2 = -1$.

2. We can add and subtract complex numbers just like we added and subtracted binomials.

3. Use the FOIL method to multiply complex numbers. Remember that $i^2 = -1$.

4. To divide one complex number by another, multiply numerator and denominator by the conjugate of the denominator.

REVIEW EXERCISES

Part I

7.1 *Simplify each radical in Exercises 1–8. For these exercises do not assume that variables and algebraic expressions under even-indexed radicals necessarily represent nonnegative numbers.*

1. $\sqrt[3]{w^3}$ **2.** $\sqrt{y^2}$ **3.** $-\sqrt[3]{(2x)^3}$ **4.** $\sqrt[4]{(3w)^4}$

5. $\sqrt{(-5x)^2}$ **6.** $-\sqrt[5]{(-4y)^5}$ **7.** $\sqrt{(x+10)^2}$ **8.** $-\sqrt[3]{(y-1)^3}$

For the remainder of these exercises, return to the assumption that all variables and algebraic expressions under even-indexed radicals are positive. Simplify each radical.

9. $\sqrt{(-3)^2}$ **10.** $\sqrt{-16}$ **11.** $\sqrt{81a^2}$ **12.** $\sqrt{9x^2y^2}$

13. $\sqrt[3]{a^3b^3}$ **14.** $\sqrt[3]{-8x^3}$ **15.** $\sqrt[4]{16u^4v^4}$ **16.** $\sqrt[6]{(-3)^6u^6}$

Use a calculator to approximate each number correct to the nearest hundredth.

17. $\sqrt{453}$ **18.** $2 - \sqrt{7}$ **19.** $\dfrac{-3 + \sqrt{29}}{2}$

20. The distance in miles that can be seen on the surface of the ocean can be approximated by $d = 1.4\sqrt{h}$ where h is the height of the viewer in feet above the surface of the water. To the nearest mile, what is the viewing distance from a helicopter at an altitude of 3500 ft?

7.2 *Simplify each radical. Assume all variables under even-indexed radicals are positive.*

21. $\sqrt{50}$ **22.** $2\sqrt{x^4y^6}$ **23.** $4\sqrt{20x^3y^5}$

24. $\sqrt[3]{40}$ **25.** $\sqrt[4]{16x^8y^{12}}$ **26.** $\sqrt[5]{32x^6}$

27. $\sqrt{\frac{45}{y^2}}$ **28.** $\sqrt[3]{\frac{-8x^3y^4}{125}}$ **29.** $\sqrt[4]{\frac{625x^5}{y^8}}$

Rationalize the denominators.

30. $\frac{5}{\sqrt{3}}$ **31.** $\frac{y}{\sqrt[3]{x}}$ **32.** $\frac{\sqrt{x^3}}{\sqrt{2y}}$

Rationalize the numerators.

33. $\frac{\sqrt{2}}{5}$ **34.** $\frac{\sqrt[3]{x^2}}{3}$ **35.** $\frac{\sqrt{2}}{\sqrt{6}}$

36. The formula $S = \sqrt{20.5\,l}$, where l is the length of the skid marks, can be used to approximate the speed of a vehicle prior to an accident. What was the approximate speed of a car that left 75 ft of skid marks before a collision?

7.3 *Multiply and simplify.*

37. $\sqrt{54}\,\sqrt{50}$ **38.** $\sqrt[3]{6ab}\,\sqrt[3]{18a^2b^3}$ **39.** $\sqrt[4]{27x^3y}\,\sqrt[4]{9x^2y^3}$

Divide and simplify.

40. $\frac{\sqrt{3}}{\sqrt{75}}$ **41.** $\frac{\sqrt{48ab^7}}{\sqrt{3a^5}}$ **42.** $\frac{\sqrt[3]{27a^3}}{\sqrt[3]{2a^2b^2}}$

7.4 *Add or subtract. Rationalize all denominators.*

43. $2\sqrt{5x} + 9\sqrt{5x}$ **44.** $5\sqrt{147} - 4\sqrt{48}$ **45.** $-6\sqrt[3]{-54} + 8\sqrt[3]{250}$

46. $5a\sqrt{63a} - 4\sqrt{175a^3}$ **47.** $\sqrt[3]{27x^5y^9} - 3y\sqrt[3]{125x^5y^6}$ **48.** $4\sqrt{3x} + \frac{x}{\sqrt{3x}}$

Multiply and simplify.

49. $\sqrt{5}(\sqrt{15} - \sqrt{3})$ **50.** $\sqrt{5}\left(\sqrt{5} - \dfrac{3}{\sqrt{5}}\right)$ **51.** $\sqrt[3]{xy}(\sqrt[3]{x^2} + \sqrt[3]{y^2})$

52. $(\sqrt{3} - 2)^2$ **53.** $(\sqrt{x} - 3)(\sqrt{x} + 2)$ **54.** $(\sqrt{x} - \sqrt{y})(\sqrt{x} - 2\sqrt{y})$

Rationalize the denominators.

55. $\dfrac{2}{\sqrt{5} - \sqrt{3}}$ **56.** $\dfrac{\sqrt{3}}{\sqrt{3} - 1}$ **57.** $\dfrac{x - 9}{\sqrt{x} - 3}$

7.5 *Simplify, writing the expression so that each variable occurs only once and all exponents are positive.*

58. $x^{-2/3}x^{5/3}$ **59.** $\left(\dfrac{a^{-6}}{b^6}\right)^{1/3}$ **60.** $(y^{-3/4})^{-8}$

61. $\left(\dfrac{x^{2/3}y^{-4/3}}{z^{-5/3}}\right)^3$ **62.** $\left(\dfrac{25x^5y^{-3}}{x^{-5}y^{-9}}\right)^{1/2}$ **63.** $\left(\dfrac{54a^{-6}b^2}{9a^{-3}b^8}\right)^{-2}$

Write in exponential notation, simplify, and give answers in radical notation. Rationalize all denominators.

64. $\sqrt{45x^4y^6}$ **65.** $\sqrt[3]{\dfrac{40x^4}{y^9}}$ **66.** $\dfrac{2x}{\sqrt[3]{3y}}$

Use a calculator with a $\boxed{y^x}$ *key to approximate each number correct to the nearest tenth.*

67. $13^{1/4}$ **68.** $6^{2/5}$ **69.** $(1.08)^{-11/4}$ **70.** $\left(\dfrac{7}{8}\right)^{8/7}$

71. The length of an edge of a cube with volume V is given by $e = \sqrt[3]{V}$. To the nearest tenth of an inch, what is the length of an edge of a cube that holds 400 in^3?

7.6 *Solve.*

72. $\sqrt{x + 5} - \sqrt{x - 4} = 1$ **73.** $\sqrt{y^2 + 4} + y + 2 = 0$ **74.** $(y - 5)^{1/4} = 2$

75. $\sqrt{ab + c} = 5$ for a

76. $z^{1/3}x^{1/3} = y$ for z

77. $x = z\sqrt{1 - y}$ for y

78. The formula $T = 2\pi\sqrt{\frac{L}{32}}$ gives the time of swing for a pendulum of length L. If $T = 5$ sec, find L correct to the nearest tenth of a foot. Use 3.14 for π.

79. Twice the square root of one more than a number is the same as three times the square root of 4 less than the number. Find the number.

7.7 **80.** In a right triangle with leg $b = 50$ and hypotenuse $c = 80$, find a.

81. What is the length of a diagonal in a rectangle with sides 4.5 cm and 7.8 cm? Give answer to the nearest tenth of a centimeter.

82. Find the distance between the points $(-7, 2)$ and $(4, -5)$.

83. Find the midpoint of the segment joining $(-2, -5)$ and $(6, -1)$.

7.8 *Simplify.*

84. $\sqrt{-16} - 4\sqrt{-9}$

85. $\sqrt{-16} \cdot \sqrt{-9}$

86. $\dfrac{\sqrt{-16}}{\sqrt{-9}}$

Perform the indicated operations.

87. $(7 - 4i) + (3 + 6i)$

88. $(7 - 4i) - (3 + 6i)$

89. $2i - (3 - 6i)$

90. $-6i(5 - 2i)$

91. $(4 - 3i)(4 + 3i)$

92. $(4 - 3i)^2$

93. i^{25}

94. $\dfrac{1}{2 + i}$

95. $\dfrac{3 - 2i}{4 + 5i}$

Part II

Perform the indicated operations. Simplify each radical. Assume all variables under even-indexed radicals are positive. Rationalize all denominators.

96. $\sqrt{75}$

97. $\sqrt[3]{54}$

98. $\dfrac{\sqrt{8xy^5}}{\sqrt{2x^2y}}$

99. $3a\sqrt{45a} + 2\sqrt{20a^3}$

100. $\sqrt{3}(\sqrt{3} - \sqrt{6})$

101. $(\sqrt{2} + 1)^2$

102. $\sqrt[3]{25x^2y}\,\sqrt[3]{10xy^4}$

103. $\dfrac{\sqrt{7}}{\sqrt{7} + 1}$

104. $\sqrt[4]{\dfrac{32a^6}{a^2b^5}}$

Use a calculator to approximate each number correct to the nearest tenth.

105. $40^{1/3}$ **106.** $\sqrt[7]{65}$ **107.** $\dfrac{-2-\sqrt{17}}{3}$

Perform the indicated operations.

108. $(2-3i)-(7-4i)$ **109.** $\dfrac{1-4i}{2+i}$ **110.** $(3-5i)(2-i)$

111. Find the midpoint of the segment joining $(4, -3)$ and $(-2, 6)$.

112. If x is any real number, positive, negative, or zero, simplify $\sqrt{(-6x)^2}$.

Solve.

113. $\sqrt{2x-3}-\sqrt{x-2}=1$ **114.** $T=2\pi\sqrt{\dfrac{L}{32}}$ for L.

115. What is the length of a diagonal, to the nearest tenth of an inch, in a square with sides 2.5 inches?

116. Find the distance between the points $(3, -3)$ and $(5, 1)$.

117. Rationalize the numerator.
$$\frac{\sqrt{x}+4}{x-16}$$

118. A researcher finds that M and d are related by the formula
$M=\sqrt{d+2}-\sqrt{d-3}$. Find d when $M=1$.

Simplify, writing the expression so that each variable occurs only once and all exponents are positive. Write answers in radical notation. Rationalize denominators.

119. $\dfrac{(x^{-2/3})^2}{x^{1/3}}$ **120.** $\left(\dfrac{36a^4b^3}{a^2b^4}\right)^{1/2}$

ANSWERS: 1. w 2. $|y|$ 3. $-2x$ 4. $3|w|$ 5. $5|x|$ 6. $4y$ 7. $|x+10|$ 8. $1-y$ 9. 3 10. not a real number 11. $9a$ 12. $3xy$ 13. ab 14. $-2x$ 15. $2uv$ 16. $3u$ 17. 21.28 18. -0.65 19. 1.19 20. 83 mi 21. $5\sqrt{2}$ 22. $2x^2y^3$ 23. $8xy^2\sqrt{5xy}$ 24. $2\sqrt[3]{5}$ 25. $2x^2y^3$ 26. $2x\sqrt[5]{x}$ 27. $\frac{3\sqrt{5}}{y}$ 28. $\frac{-2xy\sqrt[3]{y}}{5}$ 29. $\frac{5x\sqrt[4]{x}}{y^2}$ 30. $\frac{5\sqrt{3}}{3}$ 31. $\frac{y\sqrt[3]{x^2}}{x}$ 32. $\frac{x\sqrt{2xy}}{2y}$ 33. $\frac{2}{5\sqrt{2}}$ 34. $\frac{x}{3\sqrt[3]{x}}$ 35. $\frac{1}{\sqrt{3}}$ 36. 39 mph 37. $30\sqrt{3}$ 38. $3ab\sqrt[3]{4b}$ 39. $3xy\sqrt[4]{3x}$ 40. $\frac{1}{5}$ 41. $\frac{4b^3\sqrt{b}}{a^2}$ 42. $\frac{3\sqrt[3]{4ab}}{2b}$ 43. $11\sqrt{5x}$ 44. $19\sqrt{3}$ 45. $58\sqrt[3]{2}$ 46. $-5a\sqrt{7a}$ 47. $-12xy^3\sqrt[3]{x^2}$ 48. $\frac{13\sqrt{3x}}{3}$ 49. $5\sqrt{3}-\sqrt{15}$ 50. 2 51. $x\sqrt[3]{y}+y\sqrt[3]{x}$ 52. $7-4\sqrt{3}$ 53. $x-\sqrt{x}-6$ 54. $x-3\sqrt{xy}+2y$ 55. $\sqrt{5}+\sqrt{3}$ 56. $\frac{3+\sqrt{3}}{2}$ 57. $\sqrt{x}+3$ 58. x 59. $\frac{1}{a^2b^2}$ 60. y^6 61. $\frac{x^2z^5}{y^4}$ 62. $5x^5y^3$ 63. $\frac{a^6b^{12}}{36}$ 64. $3x^2y^3\sqrt{5}$ 65. $\frac{2x\sqrt[3]{5x}}{y^3}$ 66. $\frac{2x\sqrt[3]{9y^2}}{3y}$ 67. 1.9 68. 2.0 69. 0.8 70. 0.9 71. 7.4 in 72. 20 73. no solution 74. 21 75. $a=\frac{25-c}{b}$ 76. $z=\frac{y^3}{x}$ 77. $y=\frac{z^2-x^2}{z^2}$ 78. 20.3 ft 79. 8 80. $10\sqrt{39}$ 81. 9.0 cm 82. $\sqrt{170}$ 83. $(2, -3)$ 84. $-8i$ 85. -12 86. $\frac{4}{3}$ 87. $10+2i$ 88. $4-10i$ 89. $-3+8i$ 90. $-12-30i$ 91. 25 92. $7-24i$ 93. i 94. $\frac{2-i}{5}$ 95. $\frac{2-23i}{41}$ 96. $5\sqrt{3}$ 97. $3\sqrt[3]{2}$ 98. $\frac{2y^2\sqrt{x}}{x}$ 99. $13a\sqrt{5a}$ 100. $3-3\sqrt{2}$ 101. $3+2\sqrt{2}$ 102. $5xy\sqrt[3]{2y^2}$ 103. $\frac{7-\sqrt{7}}{6}$ 104. $\frac{2a\sqrt[4]{2b^3}}{b^2}$ 105. 3.4 106. 1.8 107. -2.0 108. $-5+i$ 109. $\frac{-2-9i}{5}$ 110. $1-13i$ 111. $(1, \frac{3}{2})$ 112. $6|x|$ 113. 2; 6 114. $L=\frac{8T^2}{\pi^2}$ 115. 3.5 in 116. $2\sqrt{5}$ 117. $\frac{1}{\sqrt{x}-4}$ 118. 7 119. $\frac{\sqrt[3]{x}}{x^2}$ 120. $\frac{6a\sqrt{b}}{b}$

CHAPTER 7 TEST

Assume that algebraic expressions under even-indexed radicals are positive. Evaluate.

1. $\sqrt{(-7)^2}$ 1. ____________

2. $-\sqrt{7^2}$ 2. ____________

3. $\sqrt{16a^2b^2}$ 3. ____________

4. $\sqrt[3]{8x^3}$ 4. ____________

5. $\sqrt[5]{(-2)^5y^5}$ 5. ____________

Simplify.

6. $\sqrt{75}$ 6. ____________

7. $\sqrt[3]{54}$ 7. ____________

8. $\sqrt[4]{16y^5}$ 8. ____________

9. $\sqrt[4]{\dfrac{81x^7}{y^8}}$ 9. ____________

Multiply and simplify.

10. $\sqrt{5}(\sqrt{10} - \sqrt{2})$ 10. ____________

11. $(\sqrt{3} + 1)^2$ 11. ____________

12. $\sqrt[4]{27a^2b^5}\ \sqrt[4]{3a^2b^2}$ 12. ____________

13. $(\sqrt{x} - \sqrt{a})(\sqrt{x} + 3\sqrt{a})$ 13. ____________

Add or subtract.

14. $\sqrt{18} + 2\sqrt{2} - 4\sqrt{8}$ 14. ____________

15. $\sqrt[3]{8x^4y^5} - 2y\sqrt[3]{x^4y^2}$ 15. ____________

16. Express $27^{4/3}$ as a radical and simplify. 16. ____________

17. Evaluate $\dfrac{-9 + \sqrt{20}}{3}$ using a calculator, correct to the nearest tenth. 17. ____________

CHAPTER 7 TEST

CONTINUED

18. Simplify and rationalize the denominator. $\dfrac{\sqrt{50xy^5}}{\sqrt{2x^3}}$ 18. ______

19. Rationalize the denominator. $\dfrac{3}{\sqrt{2} - \sqrt{5}}$ 19. ______

20. Rationalize the numerator. $\dfrac{\sqrt{6} + 3}{-3}$ 20. ______

Solve.

21. $\sqrt{3x + 1} - \sqrt{4x - 3} = -1$ 21. ______

22. $(y - 4)^{1/3} = -2$ 22. ______

23. Given the two points $(2, -1)$ and $(-4, 1)$, 23.(a) ______ 23.(b) ______
    (a) find the distance between them,
    (b) find the midpoint of the line segment joining them.

24. A hiker walked 3 mi south, then 4 mi west. How far was he from the starting point? 24. ______

25. Solve $a^{1/2}b^{1/2} = c$ for a. 25. ______

Perform the indicated operations.

26. $(-2 + 5i) - (-5 + 4i)$ 26. ______

27. $(7 - 2i)(-2 + 4i)$ 27. ______

28. $\dfrac{2 + i}{5 - i}$ 28. ______

29. At Buster's Restaurant, the number n of seats needed in the waiting area can be approximated by $n = 1.5\sqrt{a}$, where a is the number of expected arrivals for dinner. About how many seats are needed when 120 people will dine in an evening? 29. ______

30. If a is any real number, positive, negative, or zero, simplify $\sqrt{(-5a)^2}$. 30. ______

CHAPTER 8

Quadratic Equations and Inequalities

8.1 FACTORING AND TAKING ROOTS

STUDENT GUIDEPOSTS

1. Zero-Product Rule
2. Quadratic Equations in General Form
3. Method of Factoring
4. Number of Solutions
5. Equations of the Form $ax^2 + bx = 0$
6. Method of Taking Roots

1 ZERO-PRODUCT RULE

In Section 5.6 we solved quadratic equations such as $2x^2 - x - 15 = 0$ by factoring and using the zero-product rule, repeated below.

Zero-Product Rule

If a and b are numbers or expressions and

$$\text{if } ab = 0, \quad \text{then} \quad a = 0 \text{ or } b = 0.$$

The zero-product rule tells us that the solutions of an equation, such as $(x - 3)(2x + 5) = 0$, can be found by setting each factor equal to zero and solving both equations for x. This is reviewed in Example 1.

EXAMPLE 1 USING THE ZERO-PRODUCT RULE

Solve $(x - 3)(2x + 5) = 0$.

$$(x - 3)(2x + 5) = 0$$

$$x - 3 = 0 \quad \text{or} \quad 2x + 5 = 0 \qquad \text{Zero-product rule}$$

$$x = 3 \qquad 2x = -5$$

$$x = -\frac{5}{2}$$

The solutions are 3 and $-\frac{5}{2}$. These can be checked easily by substituting the values into the original equation.

PRACTICE EXERCISE 1

Solve $(y + 1)(5y - 4) = 0$.

Answer: $-1, \frac{4}{5}$

Remember that the zero-product rule can be extended to three or more factors. For example, to solve $x(x-3)(2x+3)=0$,

$$x=0 \quad \text{or} \quad x-3=0 \quad \text{or} \quad 2x+3=0 \qquad \text{Zero-product rule}$$
$$x=3 \qquad\qquad 2x=-3$$
$$x=-\frac{3}{2}$$

The solutions are 0, 3, and $-\frac{3}{2}$.

2 QUADRATIC EQUATIONS IN GENERAL FORM

The equation in Example 1 is equivalent to the quadratic equation

$$2x^2-x-15=0.$$

Before solving more equations of this type, we consider the form of a quadratic equation. Any quadratic equation (second-degree equation) can be written in the **general form**

$$ax^2+bx+c=0, \quad a\neq 0$$

where x is the variable and a, b, and c are constants. Since a quadratic equation must have an x^2 term, $a \neq 0$. If we also require $a>0$, it is easier to factor the trinomial. In a given equation, if the coefficient of x^2 is negative, we can always multiply both sides by -1 to make $a>0$.

EXAMPLE 2 Writing Equations in General Form

(a) Write $3x^2=-5x+4$ in general form.
Adding $5x$ and -4 to both sides of the equation, we obtain

$$3x^2+5x-4=0.$$

This is in general form with $a=3$, $b=5$, and $c=-4$ (not 4).

(b) Write $2x^2+9=7x^2-4x$ in general form.
Adding $-7x^2$ and $4x$ to both sides, we have

$$-5x^2+4x+9=0.$$

If we multiply both sides of the equation by (-1), we obtain the general form

$$5x^2-4x-9=0,$$

where $a=5$, $b=-4$, and $c=-9$.

Practice Exercise 2

Write each equation in general form and identify a, b, and c.

(a) $2x^2=5+x$

(b) $4y+y^2=1-2y+y^2$

Answers: (a) $2x^2-x-5=0$; $a=2$, $b=-1$, $c=-5$ (b) not a quadratic equation

3 METHOD OF FACTORING

The method for solving some quadratic equations by factoring and using the zero-product rule is summarized below.

To Solve a Quadratic Equation by Factoring

1. Write the equation in general form by clearing any fractions, dividing by any constant factors, and making the coefficient of x^2 positive by multiplying the equation by -1, if necessary.
2. Factor the left side into a product of two factors.
3. Use the zero-product rule to solve the resulting equation.
4. Check your answers in the original equation.

EXAMPLE 3 SOLVING A QUADRATIC EQUATION BY FACTORING

Solve $\frac{2}{3}x^2 - \frac{1}{3}x = 1$.

$$\frac{2}{3}x^2 - \frac{1}{3}x - 1 = 0 \quad \text{Subtract 1 from both sides}$$

$$2x^2 - x - 3 = 0 \quad \text{Multiply by 3 to clear fractions and obtain general form}$$

$$(2x-3)(x+1) = 0 \quad \text{Factor}$$

$$2x - 3 = 0 \quad \text{or} \quad x + 1 = 0 \quad \text{Zero-product rule}$$

$$2x = 3 \qquad x = -1$$

$$x = \frac{3}{2}$$

Check: $x = \frac{3}{2}$ $\qquad$ $x = -1$

$$\frac{2}{3}\left(\frac{3}{2}\right)^2 - \frac{1}{3}\left(\frac{3}{2}\right) \stackrel{?}{=} 1 \qquad \frac{2}{3}(-1)^2 - \frac{1}{3}(-1) \stackrel{?}{=} 1$$

$$\frac{3}{2} - \frac{1}{2} \stackrel{?}{=} 1 \qquad \frac{2}{3} + \frac{1}{3} \stackrel{?}{=} 1$$

$$1 = 1 \qquad 1 = 1$$

The solutions are $\frac{3}{2}$ and -1.

PRACTICE EXERCISE 3

Solve $\frac{3}{2}y^2 + \frac{5}{2}y = 1$.

Answer: $\frac{1}{3}$, -2

4 NUMBER OF SOLUTIONS

Every quadratic equation has two solutions. Sometimes a solution occurs twice as in the next example. In more advanced courses, we often refer to such a solution as a **root of multiplicity two,** or a **double root.**

EXAMPLE 4 SOLVING AN EQUATION WITH A DOUBLE ROOT

Solve $x^2 = 6x - 9$.

$$x^2 - 6x + 9 = 0 \quad \text{Write in general form}$$

$$(x-3)^2 = 0 \quad \text{Factor}$$

$$x - 3 = 0 \quad \text{or} \quad x - 3 = 0 \quad \text{Zero-product rule}$$

$$x = 3 \qquad x = 3$$

Both solutions are 3, that is, 3 is a root of multiplicity two. We will usually say simply that the solution is 3. Check 3 in the original equation.

PRACTICE EXERCISE 4

Solve $y^2 + 14y = -49$.

Answer: -7

More complicated-looking equations can often be transformed into the general form and solved as in Example 5 on the next page.

EXAMPLE 5 Solving a More Complex Quadratic Equation

Solve $3(y^2 - 6) = y(y + 7) - 3$.

$$3y^2 - 18 = y^2 + 7y - 3 \quad \text{First clear parentheses}$$
$$2y^2 - 18 = 7y - 3$$
$$2y^2 - 7y - 18 = -3$$
$$2y^2 - 7y - 15 = 0 \quad \text{Write in general form}$$
$$(2y + 3)(y - 5) = 0 \quad \text{Factor}$$
$$2y + 3 = 0 \quad \text{or} \quad y - 5 = 0 \quad \text{Zero-product rule}$$
$$2y = -3 \qquad y = 5$$
$$y = -\frac{3}{2}.$$

The solutions are $-\frac{3}{2}$ and 5. Check in the original equation.

Practice Exercise 5

Solve $2(z^2 - 7) = 3(z + 2)$.

Answer: $4, -\frac{5}{2}$

5 EQUATIONS OF THE FORM $ax^2 + bx = 0$

If a quadratic equation in general form has no constant term, then $c = 0$, and the equation has the form $ax^2 + bx = 0$. For example,

$$4x^2 - 7x = 0$$

is a quadratic equation in general form for which $a = 4$, $b = -7$, and $c = 0$. Factoring out an x, we obtain

$$x(4x - 7) = 0,$$

so that $x = 0$ and $x = \frac{7}{4}$. In general, *any quadratic equation for which $c = 0$ will always have 0 as one of the solutions.*

EXAMPLE 6 Solving an Equation in which $c = 0$

Solve $2x^2 = 9x$.

$$2x^2 - 9x = 0 \quad \text{Note that } c = 0$$
$$x(2x - 9) = 0 \quad \text{Factor}$$
$$x = 0 \quad \text{or} \quad 2x - 9 = 0 \quad \text{Zero-product rule}$$
$$2x = 9$$
$$x = \frac{9}{2}$$

The solutions are 0 and $\frac{9}{2}$. Check in the original equation.

Practice Exercise 6

Solve $5y^2 = -3y$.

Answer: $0, -\frac{3}{5}$

CAUTION

Is $2x^2 = 9x$ equivalent to the equation $2x = 9$, which is obtained by dividing both sides by x? The answer is no since $2x = 9$ has only one solution, $x = \frac{9}{2}$, while $2x^2 = 9x$ has two solutions, $\frac{9}{2}$ *and* 0. Never divide both sides of an equation by the variable or by an expression containing the variable.

6 METHOD OF TAKING ROOTS

If a quadratic equation in general form has no x term, then $b = 0$, and the equation takes the form

$$ax^2 + c = 0.$$

Recall that

$$\sqrt{x^2} = |x| = \begin{cases} x \text{ if } x \geq 0 \\ -x \text{ if } x < 0. \end{cases}$$

Thus, taking the square root of both sides of an equation such as

$$x^2 = 9$$

we obtain

$$\sqrt{x^2} = \sqrt{9}$$
$$|x| = 3.$$

But since $|x| = 3$ is equivalent to $x = 3$ or $-x = 3$, then

$$x = 3 \quad \text{or} \quad x = -3.$$

Thus, the equation $x^2 = 9$ has solutions 3 and -3. To solve an equation such as $x^2 - 9 = 0$, our work would appear as follows:

$$x^2 - 9 = 0$$
$$x^2 = 9 \quad \text{Solve the equation for } x^2$$
$$x = \pm\sqrt{9} \quad \text{Take the square root of both sides}$$
$$x = \pm 3. \quad x = +3 \text{ (or 3) and } x = -3$$

The method for solving a quadratic equation when $b = 0$, shown above, is called the **method of taking roots.** We summarize this method below.

To Solve the Quadratic Equation $ax^2 + c = 0$ by Taking Roots

1. Subtract c from both sides, then divide both sides by a to obtain
$$x^2 = -\frac{c}{a}.$$
2. Take the square root of both sides to obtain
$$x = \pm\sqrt{-\frac{c}{a}}.$$

The equation $x^2 = 9$, solved above by taking roots, could be written as $x^2 - 9 = 0$ and solved by factoring, since $x^2 - 9 = (x - 3)(x + 3) = 0$ also gives 3 and -3 as the solutions. However, not all equations of this type can be factored using integers, as shown in Example 7 (b) below.

EXAMPLE 7 Solving Quadratic Equations by Taking Roots

(a) Solve $3x^2 - 48 = 0$.

$$3x^2 = 48 \quad \text{Add 48 to both sides}$$
$$x^2 = 16 \quad \text{Divide by 3}$$
$$x = \pm\sqrt{16} = \pm 4 \quad \text{Take square root of both sides}$$

The solutions are 4 and -4. Check.

Practice Exercise 7

(a) Solve $2y^2 - 50 = 0$.

(b) Solve $6y^2 - 72 = 0$.

$$6y^2 = 72 \quad \text{Add 72}$$
$$y^2 = 12 \quad \text{Divide by 6}$$
$$y = \pm\sqrt{12} = \pm 2\sqrt{3} \quad \text{Take square root of both sides}$$

The solutions are $2\sqrt{3}$ and $-2\sqrt{3}$. Check.

(b) Solve $5z^2 - 40 = 0$.

Answers: (a) 5, −5 (b) $\pm 2\sqrt{2}$

Sometimes solutions to a quadratic equation are not real numbers but are complex or imaginary numbers. One reason we introduced complex numbers in Chapter 7 was to provide a number system in which *all* quadratic equations have solutions.

EXAMPLE 8 Quadratic Equations with Complex Solutions

(a) Solve $3x^2 + 12 = 0$.

$$3x^2 = -12 \quad \text{Subtract 12 from both sides}$$
$$x^2 = -4 \quad \text{Divide both sides by 3}$$
$$x = \pm\sqrt{-4} \quad \text{Take square root of both sides}$$
$$x = \pm\sqrt{4(-1)} = \pm\sqrt{4}\sqrt{-1} = \pm 2i$$

Check: $x = 2i$ | $x = -2i$

$$3(2i)^2 + 12 \stackrel{?}{=} 0 \qquad 3(-2i)^2 + 12 \stackrel{?}{=} 0$$
$$3(-4) + 12 \stackrel{?}{=} 0 \qquad 3(-4) + 12 \stackrel{?}{=} 0$$
$$-12 + 12 = 0 \qquad -12 + 12 \stackrel{?}{=} 0$$
$$0 = 0 \qquad 0 = 0$$

The solutions are $\pm 2i$, that is, $2i$ and $-2i$.

(b) Solve $(y - 2)^2 + 3 = 0$.

Although this equation is not exactly in the form for taking roots, we can still use that method and solve for $y - 2$ first.

$$(y-2)^2 = -3 \quad \text{Subtract 3 from both sides}$$
$$y - 2 = \pm\sqrt{-3} \quad \text{Take square root of both sides}$$
$$y - 2 = \pm i\sqrt{3} \quad \sqrt{-3} = \sqrt{(-1)(3)} = \sqrt{-1}\sqrt{3} = i\sqrt{3}$$
$$y = 2 \pm i\sqrt{3} \quad \text{Add 2 to both sides}$$

The solutions are $2 + i\sqrt{3}$ and $2 - i\sqrt{3}$, which do check in the original equation.

Practice Exercise 8

(a) Solve $9x^2 + 4 = 0$.

(b) Solve $(y + 1)^2 + 16 = 0$

Answers: (a) $\pm\frac{2}{3}i$ (b) $-1 \pm 4i$

Many applied problems involve quadratic equations.

EXAMPLE 9 Interest Problem

The amount of money A which will result if a principal P is invested at r percent interest compounded annually for two years is given by $A = P(1 + r)^2$. If \$2000 grows to \$2599.20 in 2 years using this formula, what is the rate of interest?

The formula above is a quadratic equation that can be solved by tak-

Practice Exercise 9

Use $A = P(1 + r)^2$ to find the rate of interest used for \$5000 to grow to \$5832 in 2 years.

ing roots. In this case we first solve for $1 + r$ and then for r.

$A = P(1 + r)^2$	Formula
$2599.20 = 2000(1 + r)^2$	$A = \$2599.20$ and $P = \$2000$
$1.2996 = (1 + r)^2$	Divide by 2000
$\pm\sqrt{1.2996} = 1 + r$	Take the square root
$1.14 = 1 + r$	Use only the positive square root obtained with a calculator
$0.14 = r$	

The interest rate is 14%. Had we used the $1 + r = -1.14$, we would have $r = -2.14$, which has no meaning in this problem.

Answer: 8%

8.1 EXERCISES A

Solve using the zero-product rule.

1. $(x - 1)(x + 2) = 0$

2. $(4y - 1)(4y + 1) = 0$

3. $x(x + 8) = 0$

4. $\frac{1}{3}y\left(y - \frac{2}{3}\right) = 0$

5. $(8.2x - 24.6)(7.2x - 14.4) = 0$

6. $(3x + 2)(5x - 1) = 0$

7. $(y - 1) - (2y + 3) = 0$

8 $(x - 1)(x + 1)(x - 2) = 0$

9. $3z(z - 1)(z + 7) = 0$

Simplify each equation, express it in general form, and identify the constants a, b, and c.

10. $2x^2 + x = 5$

11. $z^2 = 3z$

12. $2(x + x^2) = 3x - 1$

13. $z^2 + 1 = 0$

14. $3z = 3(z^2 + 1)$

15. $(x - 2)(x + 2) = 5$

Solve the following quadratic equations. All of these equations have real number solutions.

16. $y^2 - y - 2 = 0$

17. $2x^2 - 2x - 12 = 0$

18. $5z^2 - 15z = 0$

19. $4y^2 - 20y + 25 = 0$

20. $z(8 - 2z) = 6$

21 $5x^2 - 4x = 3x^2 + 9x + 7$

22. $(3y - 1)(2y + 1) = 3(2y + 1)$

23. $9(x^2 + 1) = 24x - 7$

24. $5 + z^2 - 2z = z(z - 2)$

25. $5y(y + 1) = 5y^2 + 5y$

26. $x^2 = 16$

27. $3z^2 - 10 = -10$

28. $x^2 = \frac{1}{4}$

29. $(x - 2)^2 = 25$

30. $4y^2 = 72$

31. $2z^2 = 5 + z^2$

32. $y^2 + 2y = 2y + 4$

33. $x^3 + 3x^2 + 2x = 0$

Solve the following quadratic equations. All of these equations have complex number solutions.

34. $x^2 + 9 = 0$

35. $4y^2 + 1 = 0$

36. $25z^2 + 75 = 0$

37. $(x + 1)^2 = -144$

38. $(2x - 1)^2 + 4 = 0$

39. $(z - 8)^2 + 7 = 0$

Solve the following quadratic equations. Solutions to these are either real or complex numbers.

40. $6x = -12x^2$

41. $-4y^2 + 64 = 0$

42. $4z^2 + 64 = 0$

43. $(x - 4)^2 + 5 = 0$

44. $(x - 4)^2 - 5 = 0$

45. $6z^3 - z^2 - 12z = 0$

Solve.

46. The length of a rectangle is 5 cm more than the width. Find the length and width if the area is 84 cm^2.

Let w = the width
w + 5 = the length

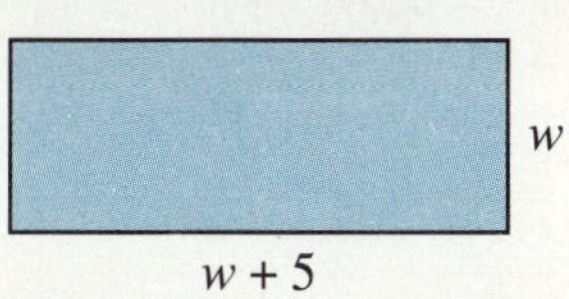

47. If you are given the investment formula $A = P(1 + r)^2$ and $A = \$3630$ when $P = \$3000$, find r.

48. The pressure P in pounds per square foot of a wind blowing V miles per hour can be approximated by $P = \frac{3V^2}{1000}$. Suppose the pressure of the wind against the side of a building is 12.25 lb/ft^2. What is the approximate velocity of the wind?

49 During the winter, the demand equation for heating oil is $D = \frac{500}{P}$ while the supply equation is $S = p - 95$, where p is the price in cents per gallon, D is the number of gallons demanded (in thousands), and S is the number of gallons supplied (also in thousands). Find the price per gallon when the supply equals the demand.

Determine what is wrong with each of the "solutions" in Exercises 50–51.

50.
$$\begin{aligned} x^2 + 5x &= 0 \\ x(x+5) &= 0 \\ \frac{x(x+5)}{x} &= \frac{0}{x} \\ x + 5 &= 0 \\ x &= -5 \end{aligned}$$

51.
$$\begin{aligned} x^2 - x &= 5 \\ x(x-1) &= 5 \\ x = 5 \quad \text{or} \quad x - 1 &= 5 \\ x &= 6 \end{aligned}$$

FOR REVIEW

The following exercises review material that we covered in Chapters 5 and 7. They will help you prepare for the next section. Find the products in Exercises 52–53.

52. $(x + 7)^2$

53. $\left(x - \frac{1}{2}\right)^2$

Factor the trinomials in Exercises 54–55.

54. $x^2 - 6x + 9$

55. $x^2 + x + \frac{1}{4}$

Simplify the square roots in Exercises 56–58.

56. $\sqrt{12}$

57. $\sqrt{\frac{81}{16}}$

58. $\sqrt{\frac{10}{9}}$

Evaluate each expression in Exercises 59–64 if $a = 3$, $b = -2$, and $c = -5$.

59. $-b$

60. $4ac$

61. $b^2 - 4ac$

62. $\sqrt{b^2 - 4ac}$

63. $-b + \sqrt{b^2 - 4ac}$

64. $-b - \sqrt{b^2 - 4ac}$

ANSWERS: 1. 1, -2 2. $\frac{1}{4}, -\frac{1}{4}$ 3. 0, -8 4. 0, $\frac{2}{3}$ 5. 3, 2 6. $\frac{1}{5}, -\frac{2}{3}$ 7. -4 8. 1, -1, 2 9. 0, 1, -7 10. $2x^2 + x - 5 = 0$; $a = 2$, $b = 1$, $c = -5$ 11. $z^2 - 3z = 0$; $a = 1$, $b = -3$, $c = 0$ 12. $2x^2 - x + 1 = 0$; $a = 2$, $b = -1$, $c = 1$ 13. $z^2 + 1 = 0$; $a = 1$, $b = 0$, $c = 1$ 14. $z^2 - z + 1 = 0$; $a = 1$, $b = -1$, $c = 1$ 15. $x^2 - 9$; $a = 1$, $b = 0$, $c = -9$ 16. 2, -1 17. 3, -2 18. 0, 3 19. $\frac{5}{2}$ 20. 1, 3 21. 7, $-\frac{1}{2}$ 22. $\frac{4}{3}, -\frac{1}{2}$ 23. $\frac{4}{3}$ 24. no solution 25. Any number is a solution. 26. ± 4 27. 0 28. $\pm\frac{1}{2}$ 29. -3, 7 30. $\pm 3\sqrt{2}$ 31. $\pm\sqrt{5}$ 32. ± 2 33. 0, -1, -2 34. $\pm 3i$ 35. $\pm\frac{1}{2}i$ 36. $\pm i\sqrt{3}$ 37. $-1 \pm 12i$ 38. $\frac{1}{2} \pm i$ 39. $8 \pm i\sqrt{7}$ 40. 0, $-\frac{1}{2}$ 41. ± 4 42. $\pm 4i$ 43. $4 \pm i\sqrt{5}$ 44. $4 \pm \sqrt{5}$ 45. 0, $-\frac{4}{3}, \frac{3}{2}$ 46. 12 cm, 7 cm 47. 10% 48. 64 mph 49. 100¢/gal 50. The solution $x = 0$ has been lost by incorrectly dividing both sides by x. 51. Do not use the zero-product rule on a product equal to 5. 52. $x^2 + 14x + 49$ 53. $x^2 - x + \frac{1}{4}$ 54. $(x - 3)^2$ 55. $\left(x + \frac{1}{2}\right)^2$ 56. $2\sqrt{3}$ 57. $\frac{9}{4}$ 58. $\frac{\sqrt{10}}{3}$ 59. 2 60. -60 61. 64 62. 8 63. 10 64. -6

8.1 EXERCISES B

Solve using the zero-product rule.

1. $(y - 2)(y + 5) = 0$ **2.** $(6z + 5)(6z - 5) = 0$ **3.** $(y - 7)y = 0$

4. $\frac{3}{4}z(z + 7) = 0$ **5.** $(1.1y - 3.3)(2.2y + 6.6) = 0$ **6.** $(4x - 3)(7x + 1) = 0$

7. $(z + 3) + (z - 3) = 0$ **8.** $(x - 2)(x + 5)(x - 3) = 0$ **9.** $7z(z - 2)(z + 10) = 0$

Simplify each equation, express it in general form, and identify the constants a, b, and c.

10. $y - y^2 = 5$ **11.** $x^2 = -7x - 12$ **12.** $2y^2 + 18y = 4$

13. $x^2 = 5$ **14.** $5y^2 = 8(y^2 - y)$ **15.** $(x + 3)(x - 3) = 4x$

Solve the following quadratic equations. All of these equations have real-number solutions.

16. $x^2 - 5x + 6 = 0$ **17.** $2z^2 - 8z - 24 = 0$ **18.** $4y^2 - 8y = 0$

19. $4x^2 + 12x + 9 = 0$ **20.** $\frac{1}{2}z^2 - 3z + 4 = 0$ **21.** $4y^2 + 4 = 7y^2 + 11y + 14$

22. $(x + 1)(x - 5) = -5x + 7$ **23.** $2z(z + 1) = 1 + 3z$ **24.** $3y^2 - 6y + 2 = 3y(y - 2)$

25. $2x^2 + 6x = 2x(x + 3)$ **26.** $(2z - 3)(z + 1) = 4(2z - 3)$ **27.** $y^2 - 81 = 0$

28. $3z^2 - 27 = 0$ **29.** $(y + 3)^2 = 81$ **30.** $2x^2 - 150 = 0$

31. $7y^2 - \frac{1}{7} = 0$ **32.** $3 - x^2 = x^2 - 3$ **33.** $3z^2 - 8z = 147 - 8z$

Solve the following quadratic equations. All of these equations have complex-number solutions.

34. $x^2 + 49 = 0$ **35.** $9y^2 + 1 = 0$ **36.** $10z^2 + 50 = 0$

37. $(x + 2)^2 = -100$ **38.** $(2x - 1)^2 + 1 = 0$ **39.** $(z + 6)^2 + 6 = 0$

Solve the following quadratic equations. Solutions to these equations are either real or complex numbers.

40. $y^2 = 6 + y$ **41.** $-3z^2 + 6 = 0$ **42.** $3z^2 + 6 = 0$

43. $(x + 5)^2 + 8 = 0$ **44.** $(x + 5)^2 - 8 = 0$ **45.** $y^3 + 6y^2 + 8y = 0$

Solve.

46. The area of a triangle is 30 in^2. Find the base and height if the base is 4 in more than the height.

47. If you are given the investment formula $A = P(1 + r)^2$, and $A = \$9331.20$ when $P = \$8000$, find r.

48. The pressure P in pounds per square foot of a wind blowing V miles per hour can be approximated by $P = \frac{3V^2}{1000}$. Suppose the pressure of the wind against the side of a bridge is 1.88 lb/ft^2. What is the approximate velocity of the wind?

49. The demand equation for gasoline is $D = \frac{1100}{P}$ and the supply equation is $S = p - 100$, where p is the price in cents per gallon, D is the number of gallons demanded (in ten thousands), and S is the number of gallons supplied (also in ten thousands). Find the price per gallon when the supply equals the demand.

Tell what is wrong with each of the "solutions" in Exercises 50–51.

50.
$$\begin{aligned} x^2 + 6x &= 0 \\ x(x+6) &= 0 \\ \frac{x(x+6)}{x} &= \frac{0}{x} \\ x + 6 &= 0 \\ x &= -6 \end{aligned}$$

51.
$$\begin{aligned} x^2 + x &= 5 \\ x(x+1) &= 5 \\ x = 5 \quad \text{or} \quad x + 1 &= 5 \\ x &= 4 \end{aligned}$$

FOR REVIEW

The following exercises review material that we covered in Chapters 5 and 7. They will help you prepare for the next section. Find the products in Exercises 52–53.

52. $(x - 10)^2$

53. $\left(x + \frac{1}{2}\right)^2$

Factor the trinomials in Exercises 54–55.

54. $x^2 + 6x + 9$

55. $x^2 - x + \frac{1}{4}$

Simplify the square roots in Exercises 56–58.

56. $\sqrt{20}$

57. $\sqrt{\frac{121}{25}}$

58. $\sqrt{\frac{3}{16}}$

Evaluate each expression in Exercises 59–64 if $a = 2$, $b = -1$, and $c = -4$.

59. $-b$

60. $4ac$

61. $b^2 - 4ac$

62. $\sqrt{b^2 - 4ac}$

63. $-b + \sqrt{b^2 - 4ac}$

64. $-b - \sqrt{b^2 - 4ac}$

8.1 EXERCISES C

Solve for x. Assume that a and b are positive constants.

1. $x^2 - 9a^2 = 0$

2. $x^2 - (b + a)^2 = 0$
[Answer: $\pm(b + a)$]

3. $x^2 - bx - 2b^2 = 0$

4. $(20x^2 + 23x - 7)(21x^2 + 5x - 6) = 0$

$\left[\text{Answer: } -\frac{7}{5}, \frac{1}{4}, -\frac{2}{3}, \frac{3}{7}\right]$

5. $(x - 1)(x + 2) + (x - 3)(x + 2) = 0$

The distance h in feet that a free-falling object falls without being given an initial velocity is given by the formula $h = 16t^2$. Use this formula in Exercises 6 and 7.

6. How far will an object fall in 2.75 sec?

7. How long would it take an object to fall from a bridge to the bottom of a gorge if the bridge is 650 ft above the level of the river? Give answer to the nearest tenth of a second.

8.2 COMPLETING THE SQUARE AND THE QUADRATIC FORMULA

STUDENT GUIDEPOSTS

1. Completing the Square
2. The Quadratic Formula
3. Choosing the Method to Use

1 COMPLETING THE SQUARE

Factoring and taking roots are the easiest and best ways to solve many quadratic equations, but some quadratic equations cannot be solved by these techniques. For such equations, we usually use the *quadratic formula,* which is derived by a process called *completing the square*.

Suppose we want to solve an equation of the form

$$(x + 5)^2 = 4.$$

By the method of taking roots from Section 8.1, we take the square root of both sides obtaining

$$x + 5 = \pm\sqrt{4}$$
$$x + 5 = \pm 2$$
$$x = -5 \pm 2. \quad \text{Subtract 5 from both sides}$$

Thus, $x = -5 + 2 = -3$ or $x = -5 - 2 = -7$.

If a general quadratic equation

$$ax^2 + bx + c = 0$$

is changed into an equation of the form

$$(x - A)^2 = B,$$

where A and B are real numbers, the solutions are

$$x = A \pm \sqrt{B}. \quad \text{One is } A + \sqrt{B}\text{, the other is } A - \sqrt{B}$$

To see how this transformation can be made, we begin by considering the square of the binomial $x + d$.

$$(x + d)^2 = x^2 + 2dx + d^2$$

Notice that in the expression $x^2 + 2dx + d^2$, the constant term, d^2, is the square of one half the coefficient of x, $2d$. Had we been given

$$x^2 + 2dx,$$

we could **complete the square** by adding the square of one half the coefficient of x,

$$\left[\frac{1}{2}(2d)\right]^2 = d^2.$$

Several examples are given in the following table.

To complete square on:	*Add (one-half the coefficient of x)²:*	*To obtain the perfect square:*
$x^2 + 6x$	$9\left[9 = \left(\frac{1}{2} \cdot 6\right)^2\right]$	$x^2 + 6x + 9 = (x + 3)^2$
$x^2 - 10x$	$25\left[25 = \left(\frac{1}{2} \cdot (-10)\right)^2\right]$	$x^2 - 10x + 25 = (x - 5)^2$
$x^2 - x$	$\frac{1}{4}\left[\frac{1}{4} = \left(\frac{1}{2} \cdot (-1)\right)^2\right]$	$x^2 - x + \frac{1}{4} = \left(x - \frac{1}{2}\right)^2$

We now use the process of completing the square to solve an equation.

EXAMPLE 1 Solving an Equation by Completing the Square

Solve $x^2 - 6x + 8 = 0$ by completing the square.

First we isolate the constant term on the right side of the equation, leaving space as indicated.

$$x^2 - 6x \qquad = -8$$

Then we complete the square on the left side by adding $\left[\frac{1}{2} \cdot (-6)\right]^2 = (-3)^2 = 9$. But if 9 is added on the left side, to keep both sides equal, 9 must be added to the right side.

$$x^2 - 6x + 9 = -8 + 9$$

$$(x - 3)^2 = 1 \qquad \text{Check by squaring the left side}$$

To complete the solution, we take the square root of both sides of this equation.

$x - 3 = \pm\sqrt{1}$ — $\pm\sqrt{1}$ represents two numbers, $\sqrt{1}$ and $-\sqrt{1}$

$x - 3 = \pm 1$ — $\sqrt{1} = 1$

$x = 3 \pm 1$ — Solve for x by adding 3 to both sides

$x = 3 + 1$ or $x = 3 - 1$ — $x = 3 \pm 1$ means $x = 3 + 1$ or $x = 3 - 1$

$x = 4 \qquad x = 2$

Thus, the solutions are 2 and 4. We can also solve this equation by the factoring method: $(x - 2)(x - 4) = 0$, which gives us the same solutions of 2 and 4. Notice that the factoring method is much easier and would be preferred in this case.

Practice Exercise 1

Solve $y^2 - 4y - 5 = 0$ by completing the square.

First write the equation in the form

$$y^2 - 4y \qquad = 5.$$

Then add $\left[\frac{1}{2}(-4)\right]^2 = [-2]^2 = 4$ to both sides.

Answer: 5, −1

If the coefficient of x^2 in a quadratic equation is not 1, we *cannot* complete the square by taking one-half the coefficient of x and squaring it. For example, $(2x + 3)^2 = 4x^2 + 12x + 9$ but $\left(\frac{1}{2} \cdot 12\right)^2 = 6^2 \neq 9$. In such a case, we need to divide both sides of the equation by a number that will change the coefficient of x^2 to 1, as in the following example.

EXAMPLE 2 Solving an Equation by Completing the Square

Solve $2x^2 + 2x - 3 = 0$ by completing the square.

$$2x^2 + 2x = 3 \quad \text{Isolate the constant}$$

$$x^2 + x = \frac{3}{2} \quad \text{Divide by 2 to make the coefficient of } x^2 \text{ equal to 1}$$

$$x^2 + x + \frac{1}{4} = \frac{3}{2} + \frac{1}{4} \quad \left(\frac{1}{2}\cdot 1\right)^2 = \frac{1}{4}$$

$$\left(x + \frac{1}{2}\right)^2 = \frac{7}{4} \quad \frac{3}{2} + \frac{1}{4} = \frac{6}{4} + \frac{1}{4} = \frac{7}{4}$$

$$x + \frac{1}{2} = \pm\sqrt{\frac{7}{4}} = \pm\frac{\sqrt{7}}{\sqrt{4}} \quad \sqrt{\frac{a}{b}} = \frac{\sqrt{a}}{\sqrt{b}}$$

$$= \frac{\pm\sqrt{7}}{2}$$

$$x = -\frac{1}{2} \pm \frac{\sqrt{7}}{2} = \frac{-1 \pm \sqrt{7}}{2}$$

The solutions are $\frac{-1+\sqrt{7}}{2}$ and $\frac{-1-\sqrt{7}}{2}$.

Practice Exercise 2

Solve $3x^2 - x - 5 = 0$ by completing the square.

First isolate the constant.

$$3x^2 - x = 5$$

Then divide each term by 3.

$$x^2 - \frac{1}{3}x = \frac{5}{3}$$

Answer: $\frac{1 \pm \sqrt{61}}{6}$

To Solve a Quadratic Equation by Completing the Square

1. Simplify the equation as much as possible and write it with both variable terms on the left side and the constant term on the right side.
2. If the coefficient of x^2 is not 1, divide each term by that coefficient.
3. Add the square of one-half the coefficient of x to both sides.
4. The left side is now the perfect square of a binomial.
5. Use the method of taking roots to complete the solution.

2 THE QUADRATIC FORMULA

Rather than continuing to solve particular quadratic equations by completing the square, we can solve a general quadratic equation using this method, and in the process, derive the *quadratic formula*.

$$ax^2 + bx + c = 0 \quad a \neq 0$$

$$ax^2 + bx = -c \quad \text{Isolate the constant}$$

$$\frac{\cancel{a}x^2}{\cancel{a}} + \frac{b}{a}x = -\frac{c}{a} \quad \text{Divide by } a \text{ to make the coefficient of } x^2 \text{ equal to 1}$$

$$x^2 + \frac{b}{a}x + \left(\frac{b}{2a}\right)^2 = -\frac{c}{a} + \left(\frac{b}{2a}\right)^2 \quad \left(\frac{1}{2}\cdot\frac{b}{a}\right)^2 = \left(\frac{b}{2a}\right)^2$$

$$\left(x + \frac{b}{2a}\right)^2 = -\frac{c}{a} + \frac{b^2}{4a^2}$$

$$= -\frac{4ac}{4a^2} + \frac{b^2}{4a^2} \quad 4a^2 \text{ is LCD}$$

$$= \frac{b^2 - 4ac}{4a^2} \quad \text{Subtract}$$

$$x + \frac{b}{2a} = \pm\sqrt{\frac{b^2 - 4ac}{4a^2}} = \frac{\pm\sqrt{b^2 - 4ac}}{\sqrt{4a^2}} \qquad \sqrt{\frac{a}{b}} = \frac{\sqrt{a}}{\sqrt{b}}$$

$$= \frac{\pm\sqrt{b^2 - 4ac}}{2a}$$

$$x = -\frac{b}{2a} \pm \frac{\sqrt{b^2 - 4ac}}{2a} \qquad \text{Subtract } \frac{b}{2a}$$

$$\boldsymbol{x = \frac{-b \pm \sqrt{b^2 - 4ac}}{2a}}$$

$2a$ is the denominator of the *entire expression*

This last formula is called the **quadratic formula** and it must be memorized. To solve a quadratic equation, identify the constants a, b, and c and substitute them into the quadratic formula.

To Solve a Quadratic Equation Using the Quadratic Formula

1. Write the equation in general form, $ax^2 + bx + c = 0$. Be sure that all common factors have been divided out and all fractions have been removed.
2. Identify the constants a, b, and c.
3. Substitute the values of a, b, and c into the quadratic formula,
$$x = \frac{-b \pm \sqrt{b^2 - 4ac}}{2a}.$$
4. Simplify the numerical expression to obtain the solutions.

EXAMPLE 3 SOLVING AN EQUATION BY THE QUADRATIC FORMULA

Solve $x^2 - 6x + 8 = 0$ using the quadratic formula.

The equation is already in general form with $a = 1$, $b = -6$ (not 6), and $c = 8$.

$$x = \frac{-b \pm \sqrt{b^2 - 4ac}}{2a}$$

$$= \frac{-(-6) \pm \sqrt{(-6)^2 - 4(1)(8)}}{2(1)} \qquad \text{Multiply } 4(1)(8)\text{, then subtract the product from } (-6)^2$$

$$= \frac{6 \pm \sqrt{36 - 32}}{2}$$

$$= \frac{6 \pm \sqrt{4}}{2} = \frac{6 \pm 2}{2}$$

$$x = \frac{6 + 2}{2} = \frac{8}{2} = 4 \quad \text{or} \quad x = \frac{6 - 2}{2} = \frac{4}{2} = 2$$

The solutions are 4 and 2.

PRACTICE EXERCISE 3

Solve $x^2 + 2x - 15 = 0$ using the quadratic formula.

Answer: 3, −5

It would have been easier to solve the equation in Example 3 by factoring, since $x^2 - 6x + 8 = (x - 2)(x - 4)$. By using the zero-product rule we obtain solutions 2 and 4, the same as were obtained using the quadratic formula.

EXAMPLE 4 Solving an Equation with Irrational Solutions

Solve $6x^2 - 10 = -8x$ using the quadratic formula.

First write the equation in general form and simplify.

$$6x^2 + 8x - 10 = 0 \quad \text{Add } 8x \text{ to both sides}$$
$$2(3x^2 + 4x - 5) = 0 \quad \text{Factor out 2}$$
$$3x^2 + 4x - 5 = 0 \quad \text{Divide by 2 to simplify}$$

Thus $a = 3$, $b = 4$, and $c = -5$. $c = -5$, not 5

$$x = \frac{-b \pm \sqrt{b^2 - 4ac}}{2a} = \frac{-4 \pm \sqrt{(4)^2 - 4(3)(-5)}}{2(3)}$$
$$= \frac{-4 \pm \sqrt{16 + 60}}{6} \quad \text{Watch the signs}$$
$$= \frac{-4 \pm \sqrt{76}}{6} = \frac{-4 \pm \sqrt{4 \cdot 19}}{6} \quad \text{4 is a perfect square}$$
$$= \frac{-4 \pm \sqrt{4}\sqrt{19}}{6} \quad \sqrt{ab} = \sqrt{a}\sqrt{b}$$
$$= \frac{-4 \pm 2\sqrt{19}}{6} \quad \sqrt{4} = 2$$
$$= \frac{\cancel{2}(-2 \pm \sqrt{19})}{\cancel{2} \cdot 3} \quad \text{Factor and divide out 2}$$
$$= \frac{-2 \pm \sqrt{19}}{3} \quad \text{This is simplified}$$

The solutions are $\frac{-2 + \sqrt{19}}{3}$ and $\frac{-2 - \sqrt{19}}{3}$.

Practice Exercise 4

Solve $5x^2 + 2x - 1 = 0$ using the quadratic formula.

Answer: $\frac{-1 \pm \sqrt{6}}{5}$

3 CHOOSING THE METHOD TO USE

In general, to solve a quadratic equation, the methods of factoring and taking roots are the easiest and quickest to apply. If these two techniques are not appropriate, it is better to go directly to the quadratic formula rather than to use the method of completing the square.

We now solve a quadratic equation with complex solutions using the quadratic formula.

EXAMPLE 5 A Quadratic Equation with Complex Solutions

Solve $x^2 - 2x + 3 = 0$.

Use the quadratic formula with $a = 1$, $b = -2$, and $c = 3$.

$$x = \frac{-(-2) \pm \sqrt{(-2)^2 - 4(1)(3)}}{2(1)}$$
$$= \frac{2 \pm \sqrt{4 - 12}}{2} = \frac{2 \pm \sqrt{-8}}{2} = \frac{2 \pm 2i\sqrt{2}}{2}$$
$$= \frac{\cancel{2}(1 \pm i\sqrt{2})}{\cancel{2}} = 1 \pm i\sqrt{2}$$

Practice Exercise 5

Solve $x^2 - 6x + 10 = 0$.

Answer: $3 \pm i$

Many applied problems can be described by quadratic equations and solved using the quadratic formula. We often approximate solutions using a calculator to make the values more meaningful.

EXAMPLE 6 Geometry Problem

If the hypotenuse of a right triangle is 7 inches long and one leg is 3 inches longer than the other, find the measure of each leg.

The triangle is shown in Figure 8.1. Recall that the Pythagorean theorem states that the sum of the squares of the legs of a right triangle is equal to the square of the hypotenuse.

Let x = the measure of one leg,
$x + 3$ = the measure of the other leg.
Use the Pythagorean theorem.

$$\begin{aligned} x^2 + (x+3)^2 &= 7^2 \\ x^2 + x^2 + 6x + 9 &= 49 \\ 2x^2 + 6x + 9 &= 49 \\ 2x^2 + 6x - 40 &= 0 \\ x^2 + 3x - 20 &= 0 \end{aligned}$$

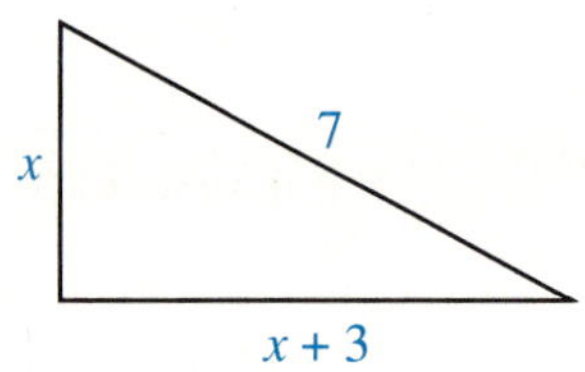

Figure 8.1

Since we cannot factor, we use the quadratic formula.

$$x = \frac{-b \pm \sqrt{b^2 - 4ac}}{2a}$$

$$x = \frac{-3 \pm \sqrt{(3)^2 - 4(1)(-20)}}{2(1)} \qquad a = 1,\ b = 3,\ c = -20$$

$$= \frac{-3 \pm \sqrt{9 + 80}}{2} = \frac{-3 \pm \sqrt{89}}{2}$$

We obtain $\frac{-3+\sqrt{89}}{2}$ and $\frac{-3-\sqrt{89}}{2}$ as solutions to the quadratic equation. But since $\frac{-3-\sqrt{89}}{2}$ is negative, it is discarded so that

$$x = \frac{-3+\sqrt{89}}{2} = \frac{\sqrt{89}-3}{2} \quad \text{and} \quad x + 3 = \frac{\sqrt{89}-3}{2} + 3 = \frac{\sqrt{89}-3}{2} + \frac{6}{2} = \frac{\sqrt{89}+3}{2}.$$

Although the legs are exactly $\frac{\sqrt{89}-3}{2}$ inches and $\frac{\sqrt{89}+3}{2}$ inches, in a practical situation, we would probably give approximations to these lengths. Since $\sqrt{89} \approx 9.4$, the legs are approximately $\frac{9.4-3}{2} = 3.2$ inches and $\frac{9.4+3}{2} = 6.2$ inches.

Practice Exercise 6

The hypotenuse of a right triangle is 11 km long and one leg is 4 km shorter than the other. Find the length of each leg to the nearest tenth of a kilometer.

Answer: 9.5 km, 5.5 km

EXAMPLE 7 Physics Problem

If an object is propelled upward from an initial height of s feet with initial velocity of v ft/sec, ignoring air resistance, the height h of the object t seconds after it is released is given by

$$h = -16t^2 + vt + s.$$

A child throws a rock upward from the edge of a building 30 ft high with initial velocity 32 ft/sec. At what time, to the nearest tenth of a second,

Practice Exercise 7

A rocket is fired from an underground silo with initial velocity 450 ft/sec. If the silo is 50 ft below ground level, at what time, to the nearest tenth of a second, will the rocket be half a mile high? [*Hint:* $s = -50$ and 1 mile = 5280 ft.]

will the rock be 20 ft above ground level? See Figure 8.2.

Substitute 30 for s, 32 for v, and 20 for h into $h = -16t^2 + vt + s$.

$$20 = -16t^2 + 32t + 30$$
$$0 = -16t^2 + 32t + 10$$
$$0 = 8t^2 - 16t - 5$$

Use the quadratic formula with $a = 8$, $b = -16$, and $c = -5$.

$$t = \frac{-b \pm \sqrt{b^2 - 4ac}}{2a}$$
$$= \frac{-(-16) \pm \sqrt{(-16)^2 - 4(8)(-5)}}{2(8)}$$
$$= \frac{16 \pm \sqrt{256 + 160}}{16} = \frac{16 \pm \sqrt{416}}{16}$$
$$= \frac{16 \pm 4\sqrt{26}}{16} = \frac{4 \pm \sqrt{26}}{4}$$

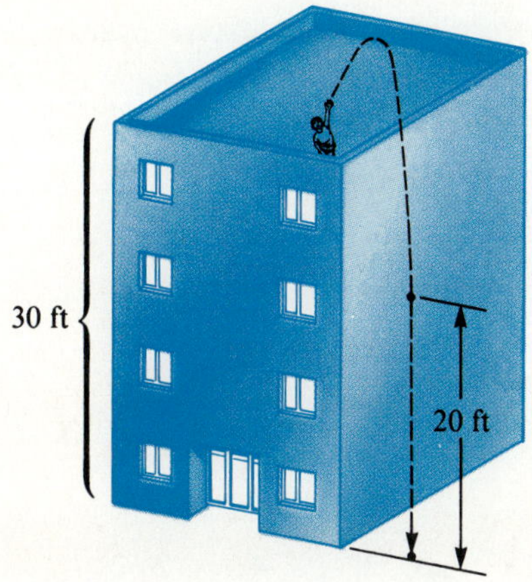

Figure 8.2

If we approximate $\sqrt{26}$ with a calculator, $\sqrt{26} \approx 5.1$. Then

$$t = \frac{4 + 5.1}{4} \approx 2.3 \quad \text{or} \quad t = \frac{4 - 5.1}{4} \approx -0.3.$$

Since t represents time, we discard the solution -0.3. Thus, the rock will be 20 ft from the ground about 2.3 seconds after it is tossed.

Answer: 8.6 sec (on the way up) and 19.5 sec (on the way down)

8.2 EXERCISES A

Solve.

1. $(z - 5)^2 = 16$ **2.** $(y + 8)^2 - 3 = 0$ **3.** $(2x + 3)^2 = 9$ **4.** $(5z + 1)^2 - 81 = 0$

5. $(2y - 3)^2 + 11 = 0$ **6.** $(3x - 7)^2 = 8$ **7** $(3z - 1)^2 = \frac{4}{9}$ **8.** $(3x + 2)^2 + 6 = 0$

What must be added to complete the square in each expression?

9. $y^2 - 6y$ **10.** $x^2 + x$ **11** $z^2 + \frac{8}{5}z$

Solve by completing the square.

12. $x^2 + 6x + 5 = 0$ **13.** $z^2 - 2z - 8 = 0$

14 $2u^2 + 3u - 1 = 0$

15. $x^2 + 2x + 3 = 0$

Solve by using the quadratic formula.

16. $v^2 - 8v + 15 = 0$

17. $y^2 - 25 = 0$

18. $3u^2 + 10u = -3u - 4$

19. $4x(x + 3) = 9x$

20. $5u^2 - u = 1$

21. $6x^2 - 9x - 12 = 0$

22. $8u^2 = 4(u^2 + 4)$

23 $\frac{1}{3}x^2 - \frac{4}{9}x - \frac{2}{3} = 0$

24. $(2u - 1)(u - 2) = 2(u + 4) + 3$

Solve. (First try factoring or taking roots, then use the quadratic formula if necessary.)

25. $x^2 + 6x + 8 = 0$

26. $3x^2 + 4x = 5$

27. $z^2 - 6 = 1 - 6z$

28. $3u^2 - 3u + 2 = 1 + 2u$

29. $x^2 + 2x + 2 = 0$

30. $10w^2 = 12w + 6$

31 $2x^2 - 3x + 4 = 0$

32 $(2z - 1)(z - 2) - 11 = 2(z + 4) - 8$

33. $4w^2 + 9w + 3 = 0$

Solve.

34 The hypotenuse of a right triangle is 11 m, and one leg is 2 m longer than the other. Rounded to the nearest tenth, find the length of each leg.

35. The sum of the squares of a number and 2 more than the number is 6. Find the number.

36. Ivan Danielson, the water commissioner in southwestern Colorado, uses a special gauge to measure stream velocity. When this gauge, an L-shaped tube, is placed in a stream in such a way that the water flows into one end, the height h (in feet) of the water in the tube is related to the velocity v (in miles per hour) of the stream by the formula $v^2 = 64h$. What is the approximate velocity of West Deer Creek if Mr. Danielson has taken a reading of $h = 0.4$ ft?

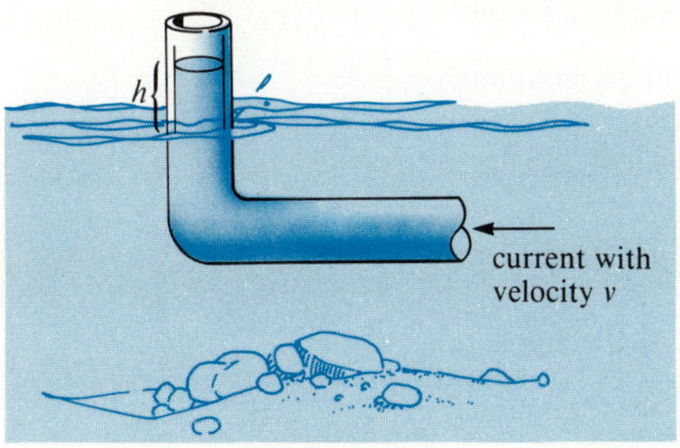

37 A baseball diamond is a square 90 ft in length on each side. In an attempt to throw a runner out, the second baseman throws the ball to the catcher who is 6 ft up the line towards third base. How far does the ball travel?

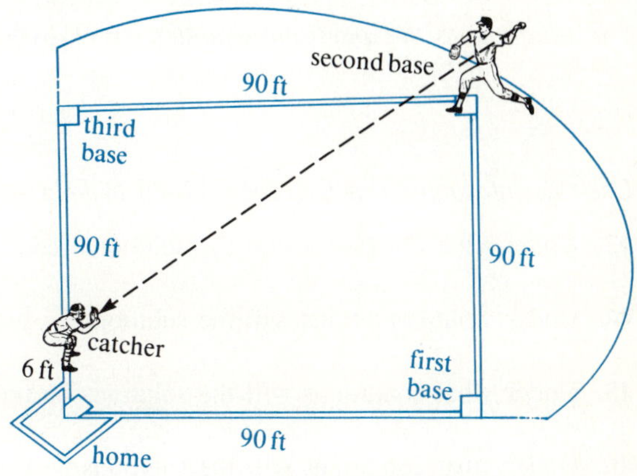

38. If an object is shot upward that has initial velocity v_0 ft/sec, in t seconds its height above ground level is given by $h = v_0t - 16t^2$. If a man fires a gun with muzzle velocity (initial velocity) of 2500 ft/sec at a helicopter 4936 ft directly above him, how long will it take for the bullet to hit the helicopter?

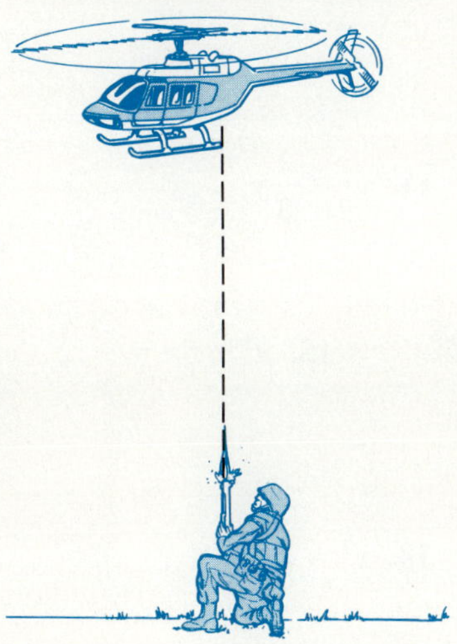

39. The square of a number is 5 more than four times the number. Find the number.

FOR REVIEW

Solve each equation in Exercises 40–42.

40. $(x - 5)(x + 12) = 0$ **41.** $z(z - 5)(z + 8) = 0$ **42.** $(y - 3) - (2y + 7) = 0$

The solutions to the quadratic equation $ax^2 + bx + c = 0$ *are*

$$x = \frac{-b \pm \sqrt{b^2 - 4ac}}{2a}.$$

Use this information in Exercises 43–46 to help you prepare for the next section.

43. Under what conditions will the solutions reduce to one real-number solution, $-\frac{b}{2a}$?

44. Under what conditions will the solutions be two different real numbers?

45. Under what conditions will the solutions be two different complex numbers?

46. Under what conditions will the solutions be two different imaginary numbers?

ANSWERS: 1. 9, 1 2. $-8 \pm \sqrt{3}$ 3. $-3, 0$ 4. $-2, \frac{8}{5}$ 5. $\frac{3 \pm i\sqrt{11}}{2}$ 6. $\frac{7 \pm 2\sqrt{2}}{3}$ 7. $\frac{5}{9}, \frac{1}{9}$ 8. $\frac{-2 \pm i\sqrt{6}}{3}$ 9. 9 10. $\frac{1}{4}$ 11. $\frac{16}{25}$ 12. $-1, -5$ 13. $4, -2$ 14. $\frac{-3 \pm \sqrt{17}}{4}$ 15. $1 \pm i\sqrt{2}$ 16. 3, 5 17. $5, -5$ 18. $-\frac{1}{3}, -4$ 19. $0, -\frac{3}{4}$ 20. $\frac{1 \pm \sqrt{21}}{10}$ 21. $\frac{3 \pm \sqrt{41}}{4}$ 22. ± 2 23. $\frac{2 \pm \sqrt{22}}{3}$ 24. $\frac{9}{2}, -1$ 25. $-2, -4$ 26. $\frac{-2 \pm \sqrt{19}}{3}$ 27. $1, -7$ 28. $\frac{5 \pm \sqrt{13}}{6}$ 29. $-1 \pm i$ 30. $\frac{3 \pm 2\sqrt{6}}{5}$ 31. $\frac{3 \pm i\sqrt{23}}{4}$ 32. $\frac{9}{2}, -1$ 33. $\frac{-9 \pm \sqrt{33}}{8}$ 34. 6.7 m, 8.7 m 35. $-1 \pm \sqrt{2}$ 36. 5 mph 37. 123.1 ft 38. 2 sec 39. 5 or -1 40. $5, -12$ 41. $0, 5, -8$ 42. -10 43. when $b^2 - 4ac = 0$ 44. when $b^2 - 4ac > 0$ 45. when $b^2 - 4ac < 0$ 46. when $b = 0$ and $-4ac < 0$

8.2 EXERCISES B

Solve.

1. $(x - 7)^2 = 49$ **2.** $(z + 2)^2 - 5 = 0$ **3.** $(3y - 1)^2 = 4$ **4.** $(3x + 5)^2 - 7 = 0$

5. $(4z + 7)^2 + 8 = 0$ **6.** $(2y + 1)^2 = \frac{1}{4}$ **7.** $(2x - 5)^2 = 1$ **8.** $(3y + 1)^2 + 10 = 0$

What must be added to complete the square in each expression?

9. $x^2 - 10x$ **10.** $z^2 + 3z$ **11.** $y^2 - \frac{1}{3}y$

Solve by completing the square.

12. $y^2 - 10y + 21 = 0$ **13.** $2x^2 = 7 - 3x$ **14.** $3v^2 + 12v = 135$ **15.** $n^2 + 5n = -12$

Solve by using the quadratic formula.

16. $x^2 + 5x - 14 = 0$ **17.** $z^2 - 16 = 0$ **18.** $x^2 = 4x + 21$

19. $4v^2 = 20v - 25$

20. $y(y + 1) + 2y = 4$

21. $4v^2 + 1 = 2v(v + 2)$

22. $(3u + 1)(u - 2) = 3(u + 1)$

23. $\frac{1}{2}y^2 - \frac{5}{4}y + \frac{1}{4} = 0$

24. $10v^2 = 5(v^2 + 25)$

Solve. (First try factoring or taking roots, then use the quadratic formula if necessary.)

25. $x^2 - 6x + 8 = 0$

26. $2x^2 + 2x = 3$

27. $y^2 - 9 = 3y$

28. $5x^2 + 3x = 2$

29. $y^2 + 4y + 5 = 0$

30. $3x(x - 2) = 12 - 6x$

31. $\frac{2}{3}v^2 - \frac{1}{3}v - 1 = 0$

32. $-3z^2 + 2z - 4 = 0$

33. $(y + 1)^2 = 5 + 3y$

Solve.

34. The hypotenuse of a right triangle is 9 in, and one leg is 3 in longer than the other. Rounded to the nearest tenth, find the length of each leg.

35. The number of square yards in the area of a square is 3.75 less than the number of yards in its perimeter. Find the length of each side.

36. Find the length of a side of an equilateral triangle if the height is 12 m.

37. Two boats leave an island at the same time, one traveling due north and the other traveling due west. If both are traveling at a rate of 30 mph, approximately how far apart will the two be in 2 hr?

38. A child with a slingshot fires a rock upward from the edge of a building 20 ft high with an initial velocity of 70 ft/sec. Ignoring air resistance, when will the rock be 30 ft high? Give answer correct to the nearest tenth of a second. Use the formula in Example 7.

39. If the square of Waldo's age is decreased by 300, the result is twenty times his age. How old is Waldo?

FOR REVIEW

Solve each equation in Exercises 40–42.

40. $(x + 3)(2x - 7) = 0$

41. $y(y + 2)(y - 10) = 0$

42. $(z - 3) + (z - 5) = 0$

The solutions to the quadratic equation with $a = 1$, $x^2 + bx + c = 0$, *are*

$$x = \frac{-b \pm \sqrt{b^2 - 4c}}{2}.$$

Use this information in Exercises 43–46 to help you prepare for the next section.

43. Under what conditions will the solutions reduce to one real-number solution, $-\frac{b}{2}$?

44. Under what conditions will the solutions be two different real numbers?

45. Under what conditions will the solutions be two different complex numbers?

46. Under what conditions will the solutions be two different imaginary numbers?

8.2 EXERCISES C

Solve by completing the square. Assume that b and c are constant real numbers.

1. $4x^2 + 7x - 2 = 0$

2. $x^2 + bx + c = 0$

Solve by using the quadratic formula.

3. $x^2 + \sqrt{2}x - 3 = 0$

4. $\sqrt{3}x^2 + 3x - \frac{1}{2} = 0$

[Answer: $\frac{-3\sqrt{3} \pm \sqrt{27 + 6\sqrt{3}}}{6}$]

8.3 USING THE DISCRIMINANT

STUDENT GUIDEPOSTS

1 The Discriminant

2 Finding an Equation Given the Solutions

In the following example we use the quadratic formula to solve three different quadratic equations with three different types of solutions. (The equations in the first two parts would normally be solved by the method of factoring, but we are using the formula to illustrate a point.)

EXAMPLE 1 Solving Equations Using the Quadratic Formula

Solve each equation using the quadratic formula.

(a) $x^2 - 2x - 3 = 0$

$$x = \frac{-b \pm \sqrt{b^2 - 4ac}}{2a}$$

$$= \frac{-(-2) \pm \sqrt{(-2)^2 - 4(1)(-3)}}{2(1)} \qquad a = 1,\ b = -2,\ c = -3$$

$$= \frac{2 \pm \sqrt{4 + 12}}{2} = \frac{2 \pm \sqrt{16}}{2}$$

$$= \frac{2 \pm 4}{2} = 1 \pm 2$$

In this case we obtain exactly two real solutions, $x = 3$ and $x = -1$.

(b) $x^2 - 2x + 1 = 0$

$$x = \frac{-(-2) \pm \sqrt{(-2)^2 - 4(1)(1)}}{2(1)} \qquad a = 1,\ b = -2,\ c = 1$$

$$= \frac{2 \pm \sqrt{4 - 4}}{2} = \frac{2 \pm \sqrt{0}}{2}$$

$$= \frac{2}{2} = 1$$

In this case we obtain exactly one real solution, $x = 1$, a double root.

PRACTICE EXERCISE 1

Solve each equation using the quadratic formula.

(a) $x^2 + 7x - 8 = 0$

(b) $x^2 - 6x + 9 = 0$

(c) $x^2 - 2x + 2 = 0$

$$x = \frac{-(-2) \pm \sqrt{(-2)^2 - 4(1)(2)}}{2(1)} \qquad a = 1,\ b = -2,\ c = 2$$

$$= \frac{2 \pm \sqrt{4 - 8}}{2}$$

$$= \frac{2 \pm \sqrt{-4}}{2}$$

$$= \frac{2 \pm 2i}{2}$$

$$= \frac{2(1 \pm i)}{2} = 1 \pm i$$

In this case we obtain two complex solutions which are conjugates, $x = 1 + i$ and $x = 1 - i$.

(c) $x^2 - 2x + 5 = 0$

Answers: (a) 1, −8 (b) 3 (c) $1 \pm 2i$

1 THE DISCRIMINANT

If we examine the solutions given by the quadratic formula in each part of Example 1,

$$x = \frac{-b \pm \sqrt{b^2 - 4ac}}{2a},$$

we see that the kind of solution is determined by the number under the radical,

$$b^2 - 4ac.$$

This number is called the **discriminant** of the quadratic equation $ax^2 + bx + c = 0$.

If $b^2 - 4ac > 0$, then there are two real solutions,

$$x_1 = \frac{-b + \sqrt{b^2 - 4ac}}{2a} \quad \text{and} \quad x_2 = \frac{-b - \sqrt{b^2 - 4ac}}{2a}$$

In Example 1(a), $x^2 - 2x - 3 = 0$ has two real solutions and the discriminant is 16, which is greater than zero. In (b), the discriminant of $x^2 - 2x + 1 = 0$ is zero, and there is exactly one real solution (a double root). In (c), the discriminant of $x^2 - 2x + 2 = 0$ is -4 which is less than zero, and there are two complex solutions that are conjugates of each other. These results are generalized in the following.

Types of Solutions to a Quadratic Equation

The quadratic equation $ax^2 + bx + c = 0$ has one of the following.

1. Two real-number solutions if $b^2 - 4ac > 0$.
2. Exactly one real-number solution if $b^2 - 4ac = 0$ (sometimes considered to be two real but equal solutions, that is, a double root).
3. Two nonreal complex-number solutions if $b^2 - 4ac < 0$. (The solutions are conjugates of each other.)

EXAMPLE 2 Finding Types of Solutions to an Equation

Tell the nature of the solutions to the following quadratic equations by calculating the discriminants.

(a) $x^2 - 6x + 7 = 0$ $\quad a = 1, b = -6, c = 7$

$$\begin{aligned} b^2 - 4ac &= (-6)^2 - 4(1)(7) \\ &= 36 - 28 \\ &= 8 > 0 \end{aligned}$$

There are two real-number solutions.

(b) $2x^2 + 9 = 0$ $\quad a = 2, b = 0, c = 9$

$$\begin{aligned} b^2 - 4ac &= (0)^2 - 4(2)(9) \\ &= 0 - 72 \\ &= -72 < 0 \end{aligned}$$

There are two nonreal complex-number solutions.

(c) $-x^2 + 10x - 25 = 0$ $\quad a = -1, b = 10, c = -25$

$$\begin{aligned} b^2 - 4ac &= (10)^2 - 4(-1)(-25) \\ &= 100 - 100 = 0 \end{aligned}$$

There is only one real-number solution, a double root.

Practice Exercise 2

Use the discriminant to tell the nature of the solutions.

(a) $4x^2 - 20x + 25 = 0$

(b) $5x^2 + 45 = 0$

(c) $7x^2 - 20x - 3 = 0$

Answers: (a) one real solution (b) two complex solutions (c) two real solutions

2 FINDING AN EQUATION GIVEN THE SOLUTIONS

We now turn to the problem of finding a quadratic equation when we are given its solutions. To see how this is done, solve the following equation by factoring.

$$\begin{aligned} x^2 - 2x - 3 &= 0 \\ (x + 1)(x - 3) &= 0 \\ x + 1 = 0 \quad &\text{or} \quad x - 3 = 0 \\ x = -1 \quad & \quad\quad\; x = 3 \end{aligned}$$

We need only reverse the steps in the problem above to construct the quadratic equation from the solutions. Suppose that we are given two solutions, -1 and 3.

$$\begin{aligned} x = -1 \quad &\text{or} \quad x = 3 \\ x - (-1) = 0 \quad &\text{or} \quad x - 3 = 0 \\ x + 1 = 0 \quad &\text{or} \quad x - 3 = 0 \\ (x + 1)(x - 3) &= 0 \\ x^2 - 2x - 3 &= 0 \end{aligned}$$

Finding a Quadratic Equation with Given Solutions

If x_1 and x_2 are two real numbers or two conjugate complex numbers, the equation

$$(x - x_1)(x - x_2) = 0$$

is a quadratic equation with real coefficients having solutions x_1 and x_2.

EXAMPLE 3 Finding Equations Given the Solutions.

Find the quadratic equation that has the given solutions.

(a) $x_1 = -2$ and $x_2 = 5$

$$(x - x_1)(x - x_2) = 0$$

$$(x - (-2))(x - 5) = 0 \quad \text{Substitute the given solutions}$$

$$(x + 2)(x - 5) = 0$$

$$x^2 - 5x + 2x - 10 = 0 \quad \text{Multiply the binomial factors}$$

$$x^2 - 3x - 10 = 0$$

(b) $x_1 = 1 + 3i$ and $x_2 = 1 - 3i$

Notice that x_1 and x_2 are conjugates of each other.

$$(x - x_1)(x - x_2) = 0$$

$$[x - (1 + 3i)][x - (1 - 3i)] = 0 \quad \text{Substitute the solutions}$$

$$x^2 - (1 - 3i)x - (1 + 3i)x + (1 + 3i)(1 - 3i) = 0 \quad \text{Multiply the binomial factors}$$

$$x^2 - 2x + 10 = 0 \quad \text{Simplify}$$

Practice Exercise 3

Find the quadratic equation that has the given solutions.

(a) $x_1 = 3$ and $x_2 = -8$

(b) $x_1 = 2 - i$ and $x_2 = 2 + i$

Answers: (a) $x^2 + 5x - 24 = 0$ (b) $x^2 - 4x + 5 = 0$

8.3 EXERCISES A

Use the discriminant to find the nature of the solutions.

1. $x^2 - 3x + 2 = 0$

2. $x^2 + 4x + 5 = 0$

3. $3x^2 - 2x + 5 = 0$

4. $x^2 + 6x + 9 = 0$

5. $-2x^2 - 3x + 1 = 0$

6. $5x^2 + 7 = 0$

7. $-3x^2 + 2x - 4 = 0$

8. $4x^2 - 4x + 1 = 0$

9. $7x^2 - 3 = 0$

Find the quadratic equation that has the following solutions.

10. 2, 3

11. −2, 3

12. $\sqrt{3}, -\sqrt{3}$

13. $2i, -2i$

14. $2 + i, 2 - i$ **15.** $0, 4$ **16.** $\frac{2}{3}, -1$ **17.** $2 + 3i, 2 - 3i$

FOR REVIEW

What must be added to complete the square in each of the following?

18. $x^2 + 12x$

19. $z^2 + \frac{3}{4}z$

20. $y^2 + \frac{2}{3}y$

Solve.

21. $3(x^2 - 1) = 4x$

22. $z^2 - 5z + 2 = 2 - z$

23. $3(y^2 + 2y) - 8 = 0$

24. The pressure P in lb/ft^2 of a wind blowing V mph can be approximated by the equation

$$P = \frac{3V^2}{1000}.$$

What is the approximate velocity of a wind which generates a pressure of 8.4 lb/ft^2 against the side of Jean Benning's mobile home?

25. Solve the equation $3x^2 - 2x + 7 = 0$. It has complex solutions.

The following exercises review material from Chapters 6 and 7 to help you prepare for the next section. Solve each equation in Exercises 26 and 27.

26. $\frac{1}{x-2} + \frac{x}{x+2} = \frac{x^2}{x^2-4}$

27. $\sqrt{x^2 + 1} - x = 1$

28. Anthony Burgess can paint a shed in 6 hours and his father can paint the same shed in 3 hours. How long would it take them to paint the shed if they worked together?

ANSWERS: 1. two real solutions 2. two complex solutions 3. two complex solutions 4. one real solution 5. two real solutions 6. two complex solutions 7. two complex solutions 8. one real solution 9. two real solutions 10. $x^2 - 5x + 6 = 0$ 11. $x^2 - x - 6 = 0$ 12. $x^2 - 3 = 0$ 13. $x^2 + 4 = 0$ 14. $x^2 - 4x + 5 = 0$ 15. $x^2 - 4x = 0$ 16. $3x^2 + x - 2 = 0$ 17. $x^2 - 4x + 13 = 0$ 18. 36 19. $\frac{9}{64}$ 20. $\frac{1}{9}$ 21. $\frac{2 \pm \sqrt{13}}{3}$ 22. 0, 4 23. $\frac{-3 \pm \sqrt{33}}{3}$ 24. 52.9 mph 25. $\frac{1 \pm 2i\sqrt{5}}{3}$ 26. no solution 27. 0 28. 2 hr

8.3 EXERCISES B

Use the discriminant to determine the nature of the solutions.

1. $x^2 - 6x + 8 = 0$
2. $x^2 + 6x + 10 = 0$
3. $x^2 - 6x + 9 = 0$
4. $2x^2 - 3x + 7 = 0$
5. $-3x^2 - 4x + 2 = 0$
6. $4x^2 + 9 = 0$
7. $-4x^2 + x - 3 = 0$
8. $25x^2 + 10x + 1 = 0$
9. $4x^2 - 17 = 0$

Find the quadratic equation that has the following solutions.

10. $3, 5$
11. $-3, 5$
12. $5i, -5i$
13. $\sqrt{5}, -\sqrt{5}$
14. $5 - i, 5 + i$
15. $\frac{1}{4}, -1$
16. $-8, 0$
17. $3 - 4i, 3 + 4i$

FOR REVIEW

What must be added to complete the square in each of the following?

18. $z^2 + 3z$
19. $y^2 + \frac{1}{2}y$
20. $x^2 - x$

Solve.

21. $2(y^2 - 2y) = 3$
22. $9x^2 + 12x + 4 = 0$
23. $4(z^2 - 1) = 7z$
24. If an object is thrown upward with initial velocity v_0 ft/sec, in t sec its height above ground level is given by $h = v_0t - 16t^2$. If the initial velocity is 128 ft/sec, how long will it take for an object thrown upward to reach a height of 240 ft?
25. Solve the equation
$$4x^2 + 81 = 0.$$
It has complex solutions.

The following exercises review material from Chapters 6 and 7 to help you prepare for the next section. Solve each equation in Exercises 26 and 27.

26. $\dfrac{1}{x+1} + \dfrac{2}{x-4} = \dfrac{1}{2}$
27. $\sqrt{x^2 + 5} - 1 = x$
28. John Babbitt can repair his Blazer in 4 hours, and when his sister Jane helps him, the job takes 3 hours. How long would it take Jane to do the job by herself?

8.3 EXERCISES C

1. Find the discriminant and use it to tell the nature of the solutions. $x^2 + \sqrt{2}x + 3 = 0$
2. Find the quadratic equation that has the following solutions. $\sqrt{2}, 3\sqrt{2}$
[Answer: $x^2 - 4\sqrt{2}x + 6 = 0$]
3. Verify that the product of the solutions of $ax^2 + bx + c = 0$ is $\frac{c}{a}$.
4. Verify that the sum of the solutions of $ax^2 + bx + c = 0$ is $-\frac{b}{a}$.

5. Find a quadratic equation that has $-\frac{7}{2}$ as the sum of its solutions and $\frac{13}{2}$ as the product of its solutions.

6. One solution of $mx^2 - 14x - 5 = 0$ is 5. Find m and then find the other solution.
[Answer: $m = 3$; $-\frac{1}{3}$]

The discriminant can also be used for factoring a trinomial with integer coefficients. It can be shown that $ax^2 + bx + c$, with a, b, and c integers, is factorable (with integer coefficient factors) whenever the discriminant satisfies $b^2 - 4ac \geq 0$ and $b^2 - 4ac$ is a perfect square. Use this in Exercises 7–12 to tell whether the given polynomial is factorable. If it is, find the factors.

7. $3x^2 - 26x - 77$ [Answer: $b^2 - 4ac = (40)^2$; $(3x + 7)(x - 11)$]

8. $4x^2 - 28x + 21$

9. $49x^2 + 84x + 36$

10. $2x^2 - 3$ [Answer: $b^2 - 4ac = 24$; cannot factor]

11. $11x^2 - 99$

12. $63x^2 + 19x - 182$

8.4 EQUATIONS REDUCIBLE TO QUADRATIC EQUATIONS

STUDENT GUIDEPOSTS

1. Equations Quadratic in Form
2. The *u*-Substitution Method
3. Solving Fractional Equations
4. Solving Radical Equations
5. Literal Equations

1 EQUATIONS QUADRATIC IN FORM

Some equations that are not quadratic can be solved using the techniques of Sections 8.1 and 8.2 by making an appropriate substitution. For example,

$$x^4 - 13x^2 + 36 = 0$$

is not a quadratic equation (it contains x^4), but if we substitute u for x^2, we obtain the equation

$$u^2 - 13u + 36 = 0,$$

which is quadratic in u. Equations like this are **quadratic in form.**

EXAMPLE 1 Solving an Equation Quadratic in Form

Solve.

$$x^4 - 13x^2 + 36 = 0$$

$$(x^2)^2 - 13x^2 + 36 = 0 \qquad x^4 = (x^2)^2$$

$$u^2 - 13u + 36 = 0 \qquad \text{Let } u = x^2\text{, then } u^2 = (x^2)^2 = x^4$$

$$(u - 4)(u - 9) = 0 \qquad \text{Use the zero-product rule}$$

$$u - 4 = 0 \quad \text{or} \quad u - 9 = 0$$

$$u = 4 \qquad\qquad u = 9$$

Since $u = x^2$,

$$x^2 = 4 \qquad x^2 = 9 \qquad \text{Solve for the original variable}$$

$$x = \pm 2 \qquad x = \pm 3.$$

PRACTICE EXERCISE 1

Solve $x^6 - 9x^3 + 8 = 0$.
Let $u = x^3$ and substitute.

Check: $x^4 - 13x^2 + 36 = 0$

$$(2)^4 - 13(2)^2 + 36 \stackrel{?}{=} 0$$
$$16 - 52 + 36 \stackrel{?}{=} 0$$
$$0 = 0$$

2 checks and -2 will check also.

$$(3)^4 - 13(3)^2 + 36 \stackrel{?}{=} 0$$
$$81 - 117 + 36 = 0$$

3 checks and -3 will check also.

The solutions are 2, -2, 3, and -3.

Answer: 1, 2

CAUTION

Notice in Example 1 that the solutions to the original equation are not 4 and 9, which are the values of the variable u and not of x. Always remember to find the solutions for the original variable by back-substituting.

2 THE *u*-SUBSTITUTION METHOD

The technique for solving an equation quadratic in form, illustrated in Example 1, is sometimes called the ***u*-substitution method.**

EXAMPLE 2 Using the *u*-Substitution Method

Solve.

$$y^{2/3} - 5y^{1/3} - 6 = 0$$
$$(y^{1/3})^2 - 5y^{1/3} - 6 = 0 \qquad \text{Let } u = y^{1/3}\text{; then } u^2 = y^{2/3}$$
$$u^2 - 5u - 6 = 0$$
$$(u + 1)(u - 6) = 0$$
$$u + 1 = 0 \quad \text{or} \quad u - 6 = 0$$
$$u = -1 \qquad u = 6$$

Since $u = y^{1/3}$,

$$y^{1/3} = -1 \qquad y^{1/3} = 6 \qquad \text{Back-substitute}$$
$$(y^{1/3})^3 = (-1)^3 \qquad (y^{1/3})^3 = 6^3 \qquad \text{Cube both sides}$$
$$y = -1 \qquad y = 216.$$

Check:

$$(-1)^{2/3} - 5(-1)^{1/3} - 6 \stackrel{?}{=} 0$$
$$(-1^{1/3})^2 - 5(-1) - 6 \stackrel{?}{=} 0$$
$$(-1)^2 + 5 - 6 \stackrel{?}{=} 0$$
$$1 + 5 - 6 = 0$$

$$(216)^{2/3} - 5(216)^{1/3} - 6 \stackrel{?}{=} 0$$
$$(216^{1/3})^2 - 5(6) - 6 \stackrel{?}{=} 0$$
$$6^2 - 30 - 6 \stackrel{?}{=} 0$$
$$36 - 30 - 6 = 0$$

The solutions are -1 and 216.

PRACTICE EXERCISE 2

Solve $(a + 5) - 5(a + 5)^{1/2} + 4 = 0$.
Let $u = (a + 5)^{1/2}$.
Then $u^2 = (a + 5)$

Answer: 11, -4

3 SOLVING FRACTIONAL EQUATIONS

In Chapters 6 and 7 when a fractional or radical equation resulted in a quadratic equation, the solution was obtained by using the method of factoring. Now that we have studied other solution techniques, we can solve a greater variety of problems. First, we review the solution method for fractional equations.

To Solve a Fractional Equation

1. Find the LCD of all fractions in the equation.
2. Multiply each term on both sides of the equation by the LCD to clear fractions.
3. Solve the resulting equation which is free of fractions.
4. Check to make sure that the solutions are not values that make a denominator zero in one of the fractions in the original equation.

EXAMPLE 3 Solving Fractional Equations

(a) Solve. $\dfrac{3}{x+1}+\dfrac{5}{x-1}=1$

$$(x+1)(x-1)\left[\frac{3}{x+1}+\frac{5}{x-1}\right]=1(x+1)(x-1) \quad \text{Multiply by LCD}$$

$$\frac{3(x+1)(x-1)}{x+1}+\frac{5(x+1)(x-1)}{x-1}=x^2-1 \quad \text{Distribute and divide out common factors}$$

$$3(x-1)+5(x+1)=x^2-1$$

$$3x-3+5x+5=x^2-1$$

$$x^2-8x-3=0 \quad \text{Write in general form}$$

$$x=\frac{-(-8)\pm\sqrt{(-8)^2-4(1)(-3)}}{2(1)} \quad \text{Use the quadratic formula}$$

$$=\frac{8\pm\sqrt{64+12}}{2}$$

$$=\frac{8\pm\sqrt{76}}{2}=\frac{8\pm2\sqrt{19}}{2}=4\pm\sqrt{19}$$

The solutions are $4+\sqrt{19}$ and $4-\sqrt{19}$. Since the only values that make a denominator zero in the original equation are 1 and -1, these two solutions will check.

(b) $1+\dfrac{3}{x-5}=\dfrac{1}{x}$

$$x(x-5)\left[1+\frac{3}{x-5}\right]=x(x-5)\left[\frac{1}{x}\right] \quad \text{Multiply by LCD}$$

$$x(x-5)+3x=x-5 \quad \text{Distribute}$$

$$x^2-5x+3x=x-5$$

$$x^2-3x+5=0 \quad \text{Write in general form}$$

$$x=\frac{-(-3)\pm\sqrt{(-3)^2-4(1)(5)}}{2(1)} \quad \text{Use the quadratic formula}$$

$$=\frac{3\pm\sqrt{9-20}}{2}=\frac{3\pm\sqrt{-11}}{2}=\frac{3\pm i\sqrt{11}}{2}$$

This time we obtain complex solutions $\frac{3+i\sqrt{11}}{2}$ and $\frac{3-i\sqrt{11}}{2}$. Since 0 and 5 are the only numbers that make a denominator zero in the original equation, these two solutions will check.

PRACTICE EXERCISE 3

(a) Solve $\dfrac{1}{y+2}-\dfrac{3}{y-2}=4$

(b) $\dfrac{2}{x}-\dfrac{1}{x-3}=\dfrac{4x}{x-3}$

Answers: (a) $\frac{-1\pm\sqrt{33}}{4}$ (b) $\frac{1\pm i\sqrt{95}}{8}$

Some applied problems, when translated to a mathematical model, result in quadratic equations. Since most quadratic equations have two distinct solutions, and since it is possible that only one of these is an appropriate solution to the given problem, care must be taken to *check all possible solutions in the words of the original problem*. Remember when solving applied problems, write complete and detailed definitions of the variables.

Many work problems similar to those we studied in Section 6.6 translate into quadratic equations, as in the next example.

EXAMPLE 4 Work Problem

When Laura and Will work together, they can complete a lab test in 2 hr less time than Laura takes to do it alone. If Will can do the experiment in 6 hr working alone, find the time Laura needs if she works alone and the time needed when they work together. Report the answers to the nearest tenth of an hour.

Let x = time Laura needs working alone,

$x - 2$ = time working together,

6 = time Will needs working alone,

$\frac{1}{x}$ = amount done by Laura in one hour,

$\frac{1}{x-2}$ = amount done working together in one hour,

$\frac{1}{6}$ = amount done by Will in one hour.

(amount by Laura) + (amount by Will) = (amount together)

$$\frac{1}{x} + \frac{1}{6} = \frac{1}{x-2}$$

$$6x(x-2)\left[\frac{1}{x} + \frac{1}{6}\right] = \frac{6x(x-2)}{x-2} \qquad \text{The LCD is } 6x(x-2)$$

$$\frac{6\cancel{x}(x-2)}{\cancel{x}} + \frac{\cancel{6}x(x-2)}{\cancel{6}} = \frac{6x\cancel{(x-2)}}{\cancel{x-2}} \qquad \text{Distribute and divide out common factors}$$

$$6(x-2) + x(x-2) = 6x$$

$$6x - 12 + x^2 - 2x = 6x$$

$$x^2 - 2x - 12 = 0 \qquad \text{General form}$$

$$x = \frac{-(-2) \pm \sqrt{(-2)^2 - 4(1)(-12)}}{2} \qquad \text{Quadratic formula}$$

$$= \frac{2 \pm \sqrt{4 + 48}}{2}$$

$$= \frac{2 \pm \sqrt{52}}{2}$$

$$= \frac{2 \pm 2\sqrt{13}}{2}$$

$$= 1 \pm \sqrt{13}$$

Since $1 - \sqrt{13}$ is negative it must be discarded as a value for time. The solution is $1 + \sqrt{13}$ or approximately 4.6 hr for Laura. Thus,

$$x = 4.6 \text{ hr} \qquad \text{Laura's time}$$

$$x - 2 = 2.6 \text{ hr.} \qquad \text{Time together}$$

Practice Exercise 4

Julie and her father can complete a task in 4 hr working together. If it takes Julie 3 hr longer than her father to do the same task working alone, to the nearest tenth of an hour, how long does it take each alone?

Answer: Julie: 9.8 hr; her father: 6.8 hr

Some motion problems, involving the formula $d = rt$, translate into fractional equations that reduce to quadratic equations.

EXAMPLE 5 Motion Problem

A trucker transported a heavy load of steel girders from Los Angeles to Albuquerque, a distance of 825 miles, and returned to Los Angeles empty. His rate of speed going was 11 mph slower than his rate returning. If the total time of travel was 27.5 hours, what was his rate of speed with the full load?

Let x = the rate returning to Los Angeles,
$x - 11$ = the rate traveling with a full load.

We use the distance formula $d = rt$, solved for $t = \frac{d}{r}$, to express the two times. Since the distance going and the distance returning are both 825, we have

$$\frac{825}{x} = \text{time spent returning to Los Angeles,}$$

$$\frac{825}{x-11} = \text{time spent traveling with a full load.}$$

Since the total time of the trip was 27.5 hours, we have the following equation.

$$\frac{825}{x} + \frac{825}{x-11} = 27.5 \quad \text{The LCD is } x(x-11)$$

$$x(x-11)\left[\frac{825}{x} + \frac{825}{x-11}\right] = 27.5\,x(x-11) \quad \text{Multiply both sides by the LCD}$$

$$x(x-11)\frac{825}{x} + x(x-11)\frac{825}{x-11} = 27.5\,x(x-11) \quad \text{Use distributive law}$$

$$825(x-11) + 825x = 27.5x^2 - 302.5x$$

$$825x - 9075 + 825x = 27.5x^2 - 302.5x$$

$$1650x - 9075 = 27.5x^2 - 302.5x$$

$$16{,}500x - 90{,}750 = 275x^2 - 3025x \quad \text{Clear decimals}$$

$$0 = 275x^2 - 19{,}525x + 90{,}750$$

$$0 = 275(x^2 - 71x + 330) \quad \text{Factor out 275}$$

$$0 = x^2 - 71x + 330 \quad \text{Divide out 275}$$

$$0 = (x-5)(x-66)$$

$$x - 5 = 0 \quad \text{or} \quad x - 66 = 0$$

$$x = 5 \qquad\qquad x = 66$$

We can discard 5 as a solution since $x - 11 = 5 - 11 = -6$, which cannot be a rate. Thus, the rate returning to Los Angeles was 66 mph, making the rate traveling with a full load 55 mph ($x - 11 = 66 - 11 = 55$).

PRACTICE EXERCISE 5

A boat sailed downstream 270 mi and then returned to the dock. The rate of travel downstream was 3 mph faster than the rate upstream. If the total time of the trip was 19 hr, what was the rate downstream?

Answer: 30 mph

4 SOLVING RADICAL EQUATIONS

We now solve equations involving radicals that reduce to quadratic equations. Before we restate the method for solving these equations, recall the power rule, that is needed in their solution.

If $a = b$, then $a^n = b^n$.

To Solve a Radical Equation

1. If only one radical is present, isolate this radical on one side of the equation, simplify, and proceed to step 3.
2. If two radicals are present, isolate one of the radicals on one side of the equation.
3. Raise both sides to a power equal to the index on the radical.
4. Solve the resulting equation. If a radical remains in the equation, isolate it and raise to the power again.
5. Check all possible solutions in the *original* equation since the equations involved may not be equivalent. If all extraneous roots are discarded, the equation has no solution.

EXAMPLE 6 Solving a Radical Equation

Solve $\sqrt{3x+1} - \sqrt{x+9} = 2$.

$$\sqrt{3x+1} = 2 + \sqrt{x+9} \quad \text{Isolate one radical}$$
$$(\sqrt{3x+1})^2 = (2 + \sqrt{x+9})^2 \quad \text{Square both sides}$$
$$3x + 1 = 4 + 4\sqrt{x+9} + x + 9 \quad \text{Use } (a+b)^2 = a^2 + 2ab + b^2\text{: } 4\sqrt{x+9} \text{ is } 2ab$$
$$2x - 12 = 4\sqrt{x+9} \quad \text{Isolate the radical}$$
$$x - 6 = 2\sqrt{x+9} \quad \text{Factor out 2 and divide by 2}$$
$$(x-6)^2 = (2\sqrt{x+9})^2 \quad \text{Square both sides}$$
$$x^2 - 12x + 36 = 4(x+9) \quad \text{Square 2 on the right}$$
$$x^2 - 12x + 36 = 4x + 36$$
$$x^2 - 16x = 0$$
$$x(x - 16) = 0$$
$$x = 0 \quad \text{or} \quad x - 16 = 0$$
$$x = 16$$

Check: $\sqrt{3(0)+1} - \sqrt{0+9} \stackrel{?}{=} 2$ $\qquad$ $\sqrt{3(16)+1} - \sqrt{16+9} \stackrel{?}{=} 2$

$\sqrt{1} - \sqrt{9} \stackrel{?}{=} 2$ $\qquad$ $\sqrt{49} - \sqrt{25} \stackrel{?}{=} 2$

$1 - 3 \neq 2$ $\qquad$ $7 - 5 = 2$

The only solution is 16.

Practice Exercise 6

Solve $\sqrt{2y+5} - \sqrt{y+6} = 1$.

Answer: 10

CAUTION

Is $(2 + \sqrt{x+9})^2$ equal to $4 + (x + 9)$? The answer is no. Remember that $(a+b)^2 = a^2 + \mathbf{2ab} + b^2$, so that

$$(2 + \sqrt{x+9})^2 = 2^2 + \mathbf{2(2)\sqrt{x+9}} + (\sqrt{x+9})^2 \quad a = 2 \text{ and } b = \sqrt{x+9}$$
$$= 4 + 4\sqrt{x+9} + x + 9.$$

Also, remember that you *must check* possible solutions in the *original* equation.

Sometimes we can recognize that a radical equation has no solution before going through all the steps used to obtain "solutions" that will not check. For example, an equation such as

$$\sqrt{4x-3}+\sqrt{2x+2}=-1$$

can have no solution since the left side must be nonnegative but the right side is -1. Remember that $\sqrt{4x-3}$ and $\sqrt{2x+2}$ represent the principal (nonnegative) square roots of numbers, so their sum is also nonnegative and cannot equal -1. If we did not see this at first, solving would give possible solutions of 1 and 7, which do not check.

5 LITERAL EQUATIONS

Many literal equations require the methods of this chapter for solution.

EXAMPLE 7 Solving a Literal Equation

Solve $V = \pi r^2 h$ (volume of a cylinder) for r.

We can use the method of taking roots on this formula.

$$V = \pi r^2 h \quad \text{Volume of a cylinder}$$

$$\frac{V}{\pi h} = r^2 \quad \text{Divide by } \pi h$$

$$\pm\sqrt{\frac{V}{\pi h}} = r \quad \text{Square root of both sides}$$

$$\sqrt{\frac{V}{\pi h}} = r \quad \text{Use positive radical since } r \text{ is the radius of the cylinder}$$

Practice Exercise 7

Solve $d = \frac{1}{2}gt^2$ for t, where t is positive.

Answer: $t = \sqrt{\frac{2d}{g}}$

8.4 EXERCISES A

Solve, using the u-substitution method.

1. $x^4 - 5x^2 + 4 = 0$

2. $(z+3)^2 - 6(z+3) + 8 = 0$

3. $y - 5\sqrt{y} + 4 = 0$

4. $x - 6 - 3\sqrt{x-6} + 2 = 0$

5. $2(y-5)^2 + 5(y-5) = 3$

6. $x^4 - 16 = 0$

7. $z^4 + 20 = 9z^2$

8. $y - 11\sqrt{y} + 28 = 0$

9 $x^{-2} - x^{-1} - 12 = 0$
[*Hint:* Let $u = x^{-1}$.]

10. $(z-7)^4 - 13(z-7)^2 + 42 = 0$

11. $y^6 - 26y^3 - 27 = 0$

12. $x^4 - x^2 - 12 = 0$

Solve.

13. $\dfrac{3y}{2} = \dfrac{y^2+1}{y}$

14. $\dfrac{1}{x} = \dfrac{-5}{x^2+6}$

15. $\dfrac{2v}{v+2} - 1 = \dfrac{3}{v}$

16. $\dfrac{x^2}{2x+4} = 1$

17. $\dfrac{2x-1}{x+1} = \dfrac{x-3}{x+2}$

18. $\dfrac{2}{x+3} + \dfrac{3}{x-3} = 1$

19. $\dfrac{2}{x+3} - \dfrac{1}{x-2} = 1$

20. $\dfrac{-2}{v^2-2v} = \dfrac{1}{v} + \dfrac{3v}{v-2}$

21. $\dfrac{1}{2} + \dfrac{1}{u+5} = \dfrac{1}{u-1}$

22. $\sqrt{x^2+3} - 2 = 0$

23. $\sqrt{x} = 2 - x$

24. $z\sqrt{2} = \sqrt{6-4z}$

25. $\sqrt{x+4} - x + 8 = 0$

26. $\sqrt{5z-1} + z - 3 = 0$

27. $\sqrt{y^2-2y+10} - \sqrt{3y+4} = 0$

28. $\sqrt{x^2-3x} + \sqrt{3x+7} = 0$

29. $\sqrt{2z+5} - \sqrt{z+2} = 1$

30. $\sqrt{3y+3} + \sqrt{y-1} = 4$

31. $(x-5)^{1/2} = x^{1/2} + 1$

32. $\sqrt[3]{z^2} = 2$

33. If the reciprocal of the square of a number is the same as the reciprocal of 5 more than the number, find the number. Use your calculator to give answers correct to three decimal places.

34. A number divided by 6 less than the number is equal to -2 divided by the number. Find the number. Give answers as decimals correct to three decimal places.

35 Maria can do a job in 3 hr less time than Wanda. If they work together they can do the job in 4 hr. How long will it take each working alone to do the job? Report the answer to the nearest tenth of an hour.

36. If Les and Chip work together they can paint a room in 2 hr less time than Les can do the job working alone. If it takes Chip 8 hr to paint the room working alone, how long does it take Les by himself and how long does it take them working together? Report the answer to the nearest tenth of an hour.

37. When Alene and Zelda work together, they can build a room on the house in 3 days less time than Alene can do the job by herself. If it takes Zelda 10 days to build the room working alone, how long does it take Alene by herself and how long does it take them working together? Report the answer to the nearest tenth of a day.

38. The amount of money A which will result if a principal P is invested at r percent interest compounded annually for two years is given by $A = P(1 + r)^2$. If \$1000 grows to \$1254.40 in 2 years using this formula, what is the rate of interest?

The distance d in miles to the horizon from an object h miles above the surface of the earth can be approximated by $d = \sqrt{h(h + 8000)}$. *Use this formula in Exercises 39–40.*

39. What is the approximate distance to the horizon from an aircraft cruising at an altitude of 6.5 miles?

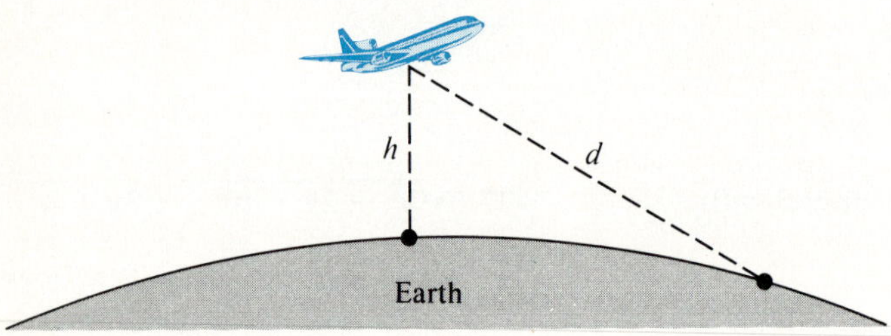

40. What is the approximate altitude of an airplane if the horizon is 130 miles away?

41. On a vacation, Shirley Cicero traveled 900 mi to her destination and returned home. Going, her average rate was 10 mph faster than the rate returning. If the total time of travel going and returning was 33 hr, what was the rate going?

42 An airplane traveled a distance of 2000 miles and returned to the airport. If the rate going was 25 mph faster than the rate returning, and if the total time of travel was 9.5 hr, to the nearest mile per hour, how fast did the airplane travel on the return trip?

If a car is traveling at a rate of v mph, the shortest distance in feet, d, needed to bring the car to a full stop under ideal conditions (including the reaction time of the driver) can be approximated by the formula $d = 0.05v^2 + 1.05v$.

43. If a child chases a ball into the street 60 ft in front of a car traveling 25 mph, can the driver avoid a tragedy?

44. If a car is traveling 50 mph, what is the shortest distance needed to bring the car to a full stop?

Solve for the indicated variable.

45. $A = \pi r^2$ for r $(r > 0)$

46. $\sqrt{c^2 - a^2} = b$ for c $(c > 0)$

47. $\dfrac{u^2}{1 + v^2} = 2$ for v

48 $\dfrac{1}{x} + \dfrac{n}{x + 1} = 1$ for x

FOR REVIEW

Use the discriminant to tell the nature of the solutions to the following quadratic equations (two real, one real, or two complex).

49. $x^2 - 3x - 4 = 0$

50. $9x^2 + 6x + 1 = 0$

51. $2x^2 - 2x + 5 = 0$

Find the quadratic equation that has the following solutions.

52. $3, -11$

53. $\dfrac{1}{3}, -\dfrac{1}{2}$

54. $7i, -7i$

The following exercises review material from Chapter 2 to help you prepare for the next section. Solve each compound inequality.

55. $3x - 2 < 1 - (1 - x)$ and $4x + 1 > 2x - (x + 1)$

56. $4x + 3 > 5x - (4 - x)$ or $6(x - 5) > 2x + 6$

ANSWERS: 1. $\pm 1, \pm 2$ 2. $1, -1$ 3. $1, 16$ 4. $7, 10$ 5. $2, \frac{11}{2}$ 6. $\pm 2, \pm 2i$ 7. $\pm 2, \pm\sqrt{5}$ 8. $16, 49$ 9. $\frac{1}{4}, -\frac{1}{3}$ 10. $7 \pm \sqrt{6}, 7 \pm \sqrt{7}$ 11. $3, -1$ 12. $\pm i\sqrt{3}, \pm 2$ 13. $\pm\sqrt{2}$ 14. $-2, -3$ 15. $6, -1$ 16. $1 \pm \sqrt{5}$ 17. $\frac{-5 \pm \sqrt{21}}{2}$ 18. $\frac{5 \pm \sqrt{73}}{2}$ 19. $\pm i$ 20. $-\frac{1}{3}$ 21. $-2 \pm \sqrt{21}$ 22. $1, -1$ 23. 1 24. 1 25. 12 26. 1 27. $2, 3$ 28. no solution 29. $2, -2$ 30. 2 31. no solution 32. $\pm 2\sqrt{2}$ 33. $2.791, -1.791$ 34. $2.606, -4.606$ 35. Maria: 6.8 hr; Wanda: 9.8 hr 36. Les: 5.1 hr; together: 3.1 hr 37. Alene: 7.2 days; together: 4.2 days 38. 12% 39. 228 mi 40. 2.1 mi 41. 60 mph 42. 409 mph 43. Yes; he can stop in 57.5 ft. 44. 177.5 ft 45. $r = \sqrt{\frac{A}{\pi}}$ 46. $c = \sqrt{a^2 + b^2}$ 47. $v = \pm\sqrt{\frac{u^2 - 2}{2}}$ 48. $x = \frac{n \pm \sqrt{n^2 + 4}}{2}$ 49. two real solutions 50. one real solution 51. two complex solutions 52. $x^2 + 8x - 33 = 0$ 53. $6x^2 + x - 1 = 0$ 54. $x^2 + 49 = 0$

8.4 EXERCISES B

Solve using the u-substitution method.

1. $y^4 - 29y^2 + 100 = 0$

2. $(x - 1)^2 - 2(x - 1) - 15 = 0$

3. $z + 3\sqrt{z} - 10 = 0$

4. $y + 1 - 8\sqrt{y + 1} + 7 = 0$

5. $3(z + 4)^2 - 7(z + 4) = 6$

6. $y^4 - 81 = 0$

7. $x^4 + 24 = 10x^2$

8. $z^{2/3} + 2z^{1/3} = 15$

9. $y^{-2} + 3y^{-1} - 40 = 0$

10. $(x - 1)^4 - 8(x - 1)^2 + 15 = 0$

11. $z^6 - 3z^3 - 10 = 0$

12. $y^4 + 6y^2 + 5 = 0$

Solve.

13. $\dfrac{2y^2 + 3}{y} = \dfrac{5y}{2}$

14. $\dfrac{3}{x^2 - 4} = \dfrac{1}{x}$

15. $\dfrac{3u}{u - 2} - 2 = \dfrac{1}{u}$

16. $\dfrac{1}{x^2} = \dfrac{1}{3x - 1}$

17. $\dfrac{2x + 1}{x - 1} = \dfrac{x + 2}{x - 3}$

18. $\dfrac{3}{x - 4} + \dfrac{2}{x + 4} = 1$

19. $\dfrac{2}{x + 2} - \dfrac{1}{x - 1} = 1$

20. $\dfrac{1}{3} - \dfrac{1}{v + 2} = \dfrac{1}{v + 3}$

21. $\dfrac{3}{x^2 + 3x} + \dfrac{4}{x} = \dfrac{2x}{x + 3}$

22. $\sqrt{y^2 - 7} - 3 = 0$

23. $6 - y = \sqrt{y}$

24. $x\sqrt{3} = \sqrt{9x + 30}$

25. $\sqrt{2y + 11} - y + 2 = 0$

26. $\sqrt{3x + 1} - 2x + 6 = 0$

27. $\sqrt{z^2 + 6z} + \sqrt{2z + 21} = 0$

28. $\sqrt{10y - 3} - \sqrt{3y^2 - 7y + 7} = 0$

29. $\sqrt{3x + 1} - \sqrt{x - 4} = 3$

30. $\sqrt{5z + 4} - \sqrt{3z - 2} = 2$

31. $(y - 3)^{1/2} = y^{1/2} + 1$

32. $\sqrt[3]{y^2 + 1} = 3$

33. The reciprocal of 1 more than the square of a number is equal to the reciprocal of 3 times the number. Find the number. Give answers correct to three decimal places.

34. The ratio of 2 more than a number and 3 less than the number is equal to 4 times the number. Find the number. Give answers correct to three decimal places.

35. José needs 4 days more to build a garage than his father needs. If working together they can do the job in 6 days, how long does it take each working alone? Report the answer to the nearest tenth of a day.

36. If Heidi and Heather work together they can prepare a display in 3 hr less time than Heidi can do the job working alone. If Heather can prepare the display in 5 hr, how long would it take Heidi to do the job working alone and how long does it take them working together? Report the answer to the nearest tenth of an hour.

37. Art can complete a lab test in 2 hr less time than Wes. If they work together they can do the test in 3 hr. How long will it take each working alone to do the test? Report the answer to the nearest tenth of an hour.

38. The amount of money A which will result if a principal P is invested at r percent interest compounded annually for two years is given by $A = P(1 + r)^2$. What is the rate of interest if \$5000 grows to \$6612.50 in 2 years?

The distance d in miles to the horizon from an object h miles above the surface of the earth can be approximated by $d = \sqrt{h(h + 8000)}$. Use this formula in Exercises 39–40.

39. If an airplane is flying at an altitude of 5 miles, approximately how far is it from the airplane to the horizon?

40. What is the approximate altitude of an aircraft if the horizon is 155 miles away?

41. A man out for exercise walks a distance of 12 miles at a rate of 8 mph slower than the rate he rides a bicycle for a distance of 24 miles. If the total time of the trip is 5 hours, how fast does he walk? How fast does he ride?

42. A woman flew from Phoenix to Denver (a distance of 800 miles) at a rate 40 mph faster than on the return trip. If the total time of the trip was 8 hours, to the nearest tenth, what was her rate going to Denver, and what was her rate returning to Phoenix?

If a car is traveling at a rate of v mph, the shortest distance in feet, d, needed to bring the car to a full stop under ideal conditions (including the reaction time of the driver) can be approximated by the formula $d = 0.05v^2 + 1.05v$.

43. If a dog runs into the street 40 ft in front of a car traveling 18 mph, can the driver avoid hitting the dog?

44. If a car is traveling 60 mph, what is the shortest distance needed to bring the car to a full stop?

Solve for the indicated variable.

45. $A = 4\pi r^2$ for r $(r > 0)$

46. $\sqrt{c^2 - b^2} = a$ for b $(b > 0)$

47. $\dfrac{4v}{u^2 + v^2} = 3$ for u

48. $\dfrac{n}{x} - \dfrac{n}{x - 1} = 1$ for x

FOR REVIEW

Use the discriminant to tell the nature of the solutions to the following quadratic equations (two real, one real, or two complex).

49. $x^2 - 5x + 6 = 0$

50. $36x^2 + 12x + 1 = 0$

51. $3x^2 + 2x + 3 = 0$

Find the quadratic equation which has the following solutions.

52. $5, -7$

53. $\dfrac{2}{3}, -\dfrac{1}{5}$

54. $-9i, 9i$

The following exercises review material from Chapter 2 to help you prepare for the next section. Solve each compound inequality.

55. $4(1 - 2x) > 3 - (x - 1)$ and $(x - 4) - (3x + 2) < 5$

56. $6x - (3 - x) < -(2x - 6)$ or $x - (3x + 1) > -3x + 5$

8.4 EXERCISES C

Solve.

1. $\left(\dfrac{x^2 - 3}{x}\right)^2 - 5\left(\dfrac{x^2 - 3}{x}\right) + 6 = 0$

$\left[\text{Answer: } -1, 3, \dfrac{3 \pm \sqrt{21}}{2}\right]$

2. $\sqrt[3]{x + 1} + \sqrt[6]{x + 1} - 2 = 0$

3. $\dfrac{x + 3}{x - 5} + \dfrac{x}{x - 3} = 1$

4. $\dfrac{1}{x + \sqrt{2}} + \dfrac{1}{x - \sqrt{2}} = \dfrac{x^2}{x^2 - 2}$

[*Hint:* $x^2 - 2 = (x - \sqrt{2})(x + \sqrt{2})$]

5. $\sqrt{2x+3} + \sqrt{7-x} = \sqrt{8x+1}$
[Answer: 3]

6. $(x^6 - 2)^{1/3} = x$

7. $\dfrac{1}{x^2 - x - 2} = \dfrac{3}{x^2 - 2x - 3} + \dfrac{1}{x^2 - 5x + 6}$
[Answer: $\frac{2}{3}$]

8. $\dfrac{1}{\sqrt{x+1}} + \dfrac{3}{2} = 2$

9. If a train is traveling at v mph, the shortest distance in feet, d, needed to bring the train to a full stop under ideal conditions can be approximated by $d = 0.05v^2 + 1.2v$. A car, stalled on the railroad tracks, is seen by the engineer at a distance of 450 ft. If the crew applies the brakes immediately and stops the train within inches of the car, what was the velocity of the train prior to braking? Give answer correct to the nearest tenth.

8.5 SOLVING QUADRATIC AND RATIONAL INEQUALITIES

STUDENT GUIDEPOSTS

1 Solving Quadratic Inequalities
2 Solving Rational Inequalities

1 SOLVING QUADRATIC INEQUALITIES

Inequalities such as

$$x^2 + x - 6 < 0 \quad \text{and} \quad 2x^2 - 3x + 7 \geq 0$$

are called **quadratic inequalities.** In both instances, if the inequality symbol were replaced by an equal sign, a quadratic equation would result. There are many techniques for solving these inequalities, but the method we present is perhaps the simplest. We begin by replacing the inequality symbol with an equal sign, then solve and graph the solution as in Figure 8.3.

$$x^2 + x - 6 = 0$$

$$(x-2)(x+3) = 0 \qquad \text{Factor}$$

$$x - 2 = 0 \quad \text{or} \quad x + 3 = 0 \qquad \text{Zero-product rule}$$

$$x = 2 \qquad\qquad x = -3$$

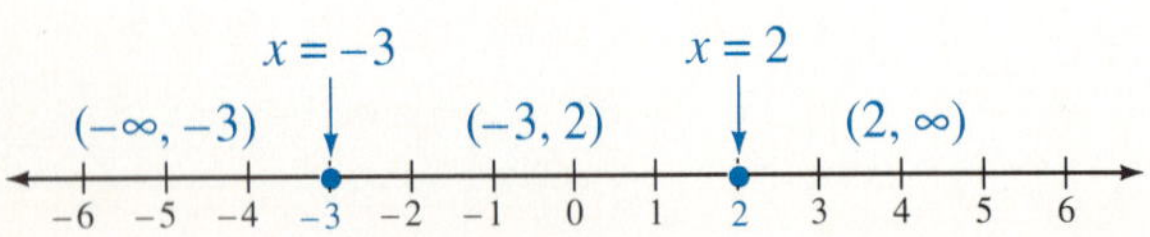

Figure 8.3

The solutions to the equation separate the number line into three intervals:

$$(-\infty, -3), \quad (-3, 2), \quad \text{and} \quad (2, \infty).$$

We know that -3 and 2 are *not* solutions to $x^2 + x - 6 < 0$ since they *are* solutions to the equation $x^2 + x - 6 = 0$. Thus, solutions to the inequality must be found in one or more of the three intervals.

It can be shown that all numbers in a given interval produce the same results when substituted into the inequality; that is, either they all solve the inequality or else none of them solves the inequality. To get a feeling for this, place the value of the quadratic expression directly above the number on the number line used to obtain it, as in Figure 8.4. For example, if $x = 1$, $x^2 + x - 6 = (1)^2 + (1) - 6 = -4$. We place -4 directly above 1 on the number line. The other values, 14, 6, 0, -4, -6, -6, -4, 0, 6, and 14 are obtained the same way.

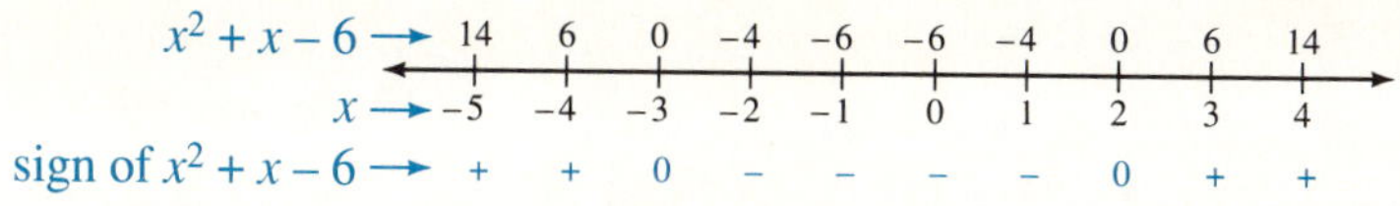

Figure 8.4

Notice that:

$$\begin{aligned}&\text{for } x \text{ in } (-\infty, -3), && x^2 + x - 6 > 0 \text{ (is +)},\\ &\textbf{for } \boldsymbol{x} \textbf{ in } \mathbf{(-3, 2)}, && \boldsymbol{x^2 + x - 6 < 0} \text{ (is −)},\\ &\text{for } x \text{ in } (2, \infty), && x^2 + x - 6 > 0 \text{ (is +)}.\end{aligned}$$

Thus, the solution to $x^2 + x - 6 < 0$ is

$$(-3, 2) \quad \text{or} \quad -3 < x < 2.$$

On the other hand, the inequality $x^2 + x - 6 > 0$ has solution

$$(-\infty, -3) \text{ or } (2, \infty), \quad \textit{or} \quad x < -3 \text{ or } 2 < x.$$

Similarly, the solution to $x^2 + x - 6 \le 0$ is

$$[-3, 2] \quad \text{or} \quad -3 \le x \le 2.$$

Notice that equality holds this time since -3 and 2 solve the equation $x^2 + x - 6 = 0$, hence the inequality $x^2 + x - 6 \le 0$. Also, the solution to $x^2 + x - 6 \ge 0$ is

$$(-\infty, -3] \text{ or } [2, \infty), \quad \textit{or} \quad x \le -3 \text{ or } 2 \le x.$$

In general, to solve a quadratic inequality, we solve the corresponding equation, the solutions to which are the **critical points** of the inequality. The critical points separate the number line into intervals. Since all numbers in an interval either solve or do not solve the given inequality, it is necessary to select only one **test point** in each interval. We illustrate this procedure in the next examples.

EXAMPLE 1 A Quadratic Inequality (Two Critical Points)

Solve $2x^2 + 5x - 3 \ge 0$.

First we find the critical points by solving $2x^2 + 5x - 3 = 0$.

$$(2x - 1)(x + 3) = 0 \qquad \text{Factor}$$

$$2x - 1 = 0 \quad \text{or} \quad x + 3 = 0 \qquad \text{Zero-product rule}$$

$$2x = 1 \qquad\qquad x = -3$$

$$x = \frac{1}{2}$$

The critical points separate the number line into the three intervals shown in Figure 8.5.

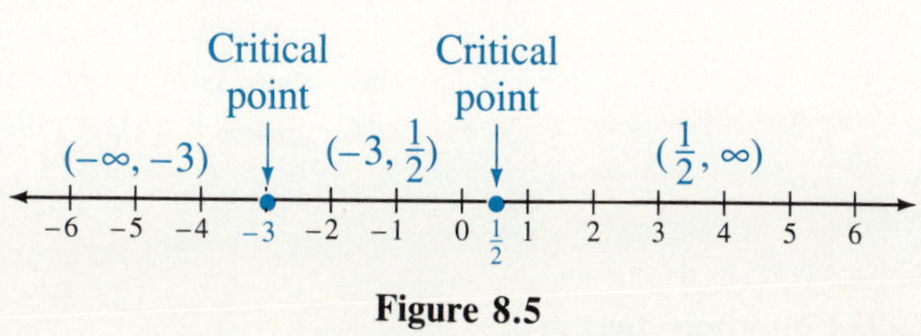

Figure 8.5

The numbers $\frac{1}{2}$ and -3 are solutions to $2x^2 + 5x - 3 \ge 0$ since they *are* solutions to $2x^2 + 5x - 3 = 0$. We find other solutions by choosing test points, such as -4, 0, and 1, from each interval.

Practice Exercise 1

(a) Solve $x^2 - x - 6 \le 0$.

First solve $x^2 - x - 6 = 0$ to obtain the critical points. Use -3, 0, and 4 as test points.

Let $x = -4$: $2x^2 + 5x - 3 = 2(-4)^2 + 5(-4) - 3 = 32 - 20 - 3 = 9 > 0$.

Let $x = \;\; 0$: $2x^2 + 5x - 3 = 2(0)^2 + 5(0) - 3 = 0 + 0 - 3 = -3 < 0$.

Let $x = \;\; 1$: $2x^2 + 5x - 3 = 2(1)^2 + 5(1) - 3 = 2 + 5 - 3 = 4 > 0$.

(b) Using your calculations in the first part, what is the solution to $x^2 - x - 6 \geq 0$?

Then, for any x in $(-\infty, -3)$, $2x^2 + 5x - 3 > 0$. Similarly, if x is in $\left(-3, \frac{1}{2}\right)$, $2x^2 + 5x - 3 < 0$, and if x is in $\left(\frac{1}{2}, \infty\right)$, $2x^2 + 5x - 3 > 0$. We summarize this by writing the appropriate inequality above each interval on a number line, as in Figure 8.6.

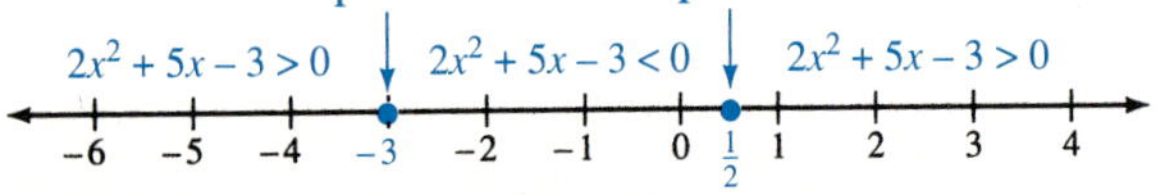

Figure 8.6

The solution to the inequality $2x^2 + 5x - 3 \geq 0$ is

$$(-\infty, -3] \text{ or } \left[\frac{1}{2}, \infty\right), \quad \text{or} \quad x \leq -3 \text{ or } \frac{1}{2} \leq x.$$

The graph of the solution is shown in Figure 8.7. (Remember from Chapter 2 that the square brackets mean that the critical points are part of the solution. Parentheses are used when they are not part of the solution.)

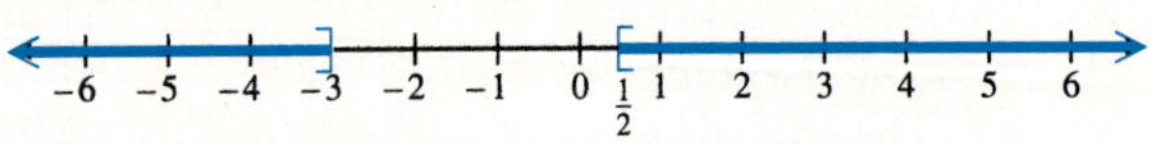

Figure 8.7 $(-\infty, -3]$ or $\left[\frac{1}{2}, \infty\right)$

In view of the results above, if we had been asked to solve $2x^2 + 5x - 3 \leq 0$, the solution would have been

$$\left[-3, \frac{1}{2}\right] \quad \text{or} \quad -3 \leq x \leq \frac{1}{2},$$

as shown in Figure 8.8.

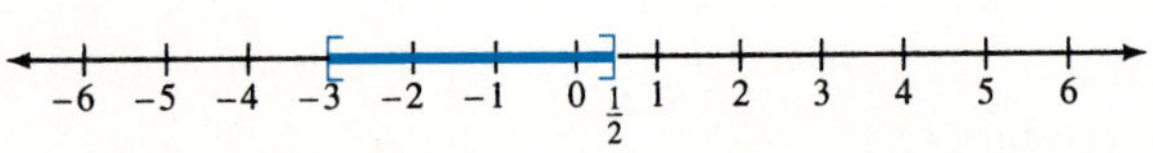

Figure 8.8 $\left[-3, \frac{1}{2}\right]$

Answers: (a) $[-2, 3]$ or $-2 \leq x \leq 3$ (b) $(-\infty, -2]$ or $[3, \infty)$, *or* $x \leq -2$ or $x \geq 3$.

EXAMPLE 2 A QUADRATIC INEQUALITY (ONE CRITICAL POINT)

Solve $x^2 - 4x + 4 > 0$.

We first solve $x^2 - 4x + 4 = 0$.

$$(x - 2)(x - 2) = 0$$

$$x - 2 = 0 \quad \text{or} \quad x - 2 = 0$$

$$x = 2 \qquad\qquad x = 2$$

With only one critical point, the number line is divided into only two intervals,

$$(-\infty, 2) \quad \text{and} \quad (2, \infty).$$

PRACTICE EXERCISE 2

(a) Solve $x^2 + 6x + 9 < 0$.

First solve $x^2 + 6x + 9 = 0$ and show there is only one critical point. Use -4 and 0 as test points.

Then evaluate at test points 0 and 3.

Let $x = 0$: $0^2 - 4(0) + 4 = 4 > 0$.

Let $x = 3$: $3^2 - 4(3) + 4 = 9 - 12 + 4 = 1 > 0$.

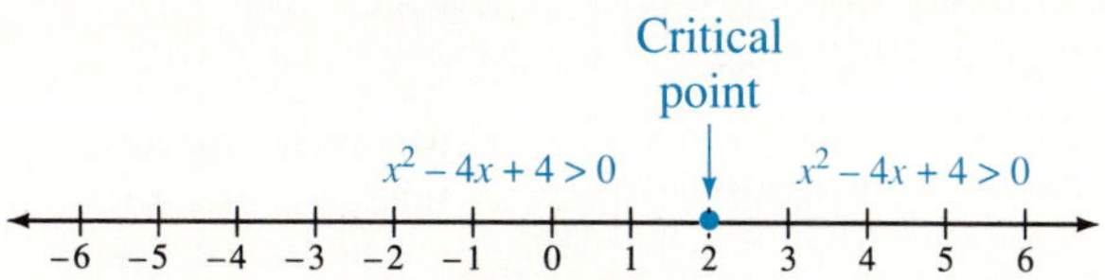

Figure 8.9

The results are shown in Figure 8.9. The solution to $x^2 - 4x + 4 > 0$ is

$$(-\infty, 2) \text{ or } (2, \infty) \quad \textit{or} \quad x < 2 \text{ or } x > 2,$$

and it is graphed in Figure 8.10.

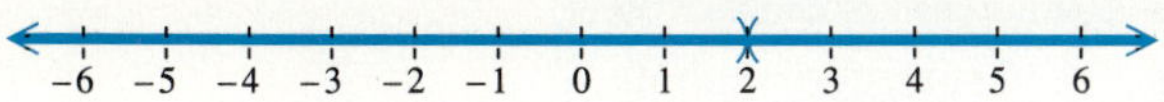

Figure 8.10 $(-\infty, 2)$ or $(2, \infty)$

In view of the results above, we can see that the solution to the inequality

$$x^2 - 4x + 4 \geq 0$$

is the entire collection of real numbers, or $(-\infty, \infty)$. Also, the solution to

$$x^2 - 4x + 4 \leq 0$$

is $x = 2$, and the inequality

$$x^2 - 4x + 4 < 0$$

has no solution.

(b) Solve $x^2 + 6x + 9 > 0$.

(c) Solve $x^2 + 6x + 9 \leq 0$.

(d) Solve $x^2 + 6x + 9 \geq 0$.

Answers: (a) no solution (b) $(-\infty, -3)$ or $(-3, \infty)$, *or* $x < -3$ *or* $x > -3$ (c) $x = -3$ (d) Every real number is a solution, or $(-\infty, \infty)$.

EXAMPLE 3 A Quadratic Inequality (No Critical Points)

Solve $x^2 + x + 2 \geq 0$.

We first solve $x^2 + x + 2 = 0$.

$$x = \frac{-1 \pm \sqrt{(1)^2 - 4(1)(2)}}{2(1)} \qquad \text{This time we cannot factor}$$

$$= \frac{-1 \pm \sqrt{-7}}{2} = \frac{-1 \pm i\sqrt{7}}{2}$$

Since only real (not complex) numbers are identified with points on a number line, there are no points on the line associated with the solutions $(-1 \pm i\sqrt{7})/2$. As a result, the solutions in this case do not divide the number line into separate parts, so $x^2 + x + 2$ has the same sign regardless of the choice of value for the variable x. Since

$$0^2 + 0 + 2 = 2 > 0,$$

every real number is a solution to the given inequality, as graphed in Figure 8.11 on the next page. In interval notation this is $(-\infty, \infty)$.

Practice Exercise 3

(a) Solve $x^2 + 3x + 5 \leq 0$.

Show that the solutions to $x^2 + 3x + 5 = 0$ are not real. Use 0 as a test point.

(b) Solve $x^2 + 3x + 5 < 0$.

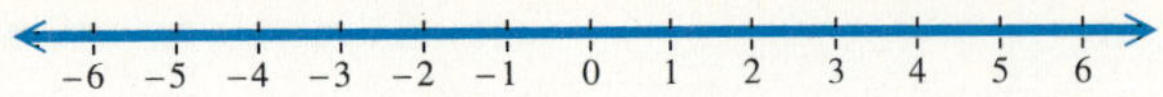

Figure 8.11 $(-\infty, \infty)$

In view of the above, the solution to

$$x^2 + x + 2 > 0$$

is also the entire set of real numbers, or $(-\infty, \infty)$, and the inequalities

$$x^2 + x + 2 < 0 \quad \text{and} \quad x^2 + x + 2 \le 0$$

have no solution.

(c) Solve $x^2 + 3x + 5 \ge 0$.

(d) Solve $x^2 + 3x + 5 > 0$.

Answers: (a) no solution (b) no solution (c) Every real number is a solution, or $(-\infty, \infty)$. (d) Every real number is a solution, or $(-\infty, \infty)$.

We now summarize the method for solving a quadratic inequality.

Method for Solving a Quadratic Inequality

1. Replace the inequality symbol with an equal sign and solve the resulting quadratic equation, obtaining the critical points.
2. If there are two real critical points, they separate the number line into three distinct intervals. Evaluate the quadratic expression at an arbitrary test point chosen from each interval and, using these values, specify the solution.
3. If there is only one real critical point, it separates the number line into two distinct intervals. Evaluate at an arbitrary test point in each interval and specify the solution.
4. If there are no real critical points (only complex solutions to the equation), either every number is a solution or else there is no solution, as determined by trying a single test point.

Sometimes the critical points are irrational numbers and we use rational approximations of them to determine possible test points.

EXAMPLE 4 Quadratic Inequality (Irrational Critical Points)

Solve $(x + 3)(x - 1) \le -2$.

Simplify $x^2 + 2x - 3 \le -2$.

$$x^2 + 2x - 1 \le 0$$

We need to solve the following equation.

$$x^2 + 2x - 1 = 0 \qquad \text{Writing in general form}$$

$$x = \frac{-2 \pm \sqrt{2^2 - 4(1)(-1)}}{2(1)}$$

$$= \frac{-2 \pm \sqrt{8}}{2} = \frac{-2 \pm 2\sqrt{2}}{2}$$

$$= \frac{\cancel{2}(-1 \pm \sqrt{2})}{\cancel{2}} = -1 \pm \sqrt{2}$$

Since $\sqrt{2} \approx 1.4$, $-1 \pm \sqrt{2} \approx -1 \pm 1.4$. Then the critical points are, approximately, 0.4 and -2.4. Knowing this, we choose test points -3, 0, and 1.

Practice Exercise 4

Solve $(x - 7)(x + 1) \ge -13$.

Let $x = -3$: $(-3)^2 + 2(-3) - 1 = 9 - 6 - 1 = 2 > 0$

Let $x = 0$: $0^2 + 2(0) - 1 = -1 < 0$

Let $x = 1$: $1^2 + 2(1) - 1 = 2 > 0$

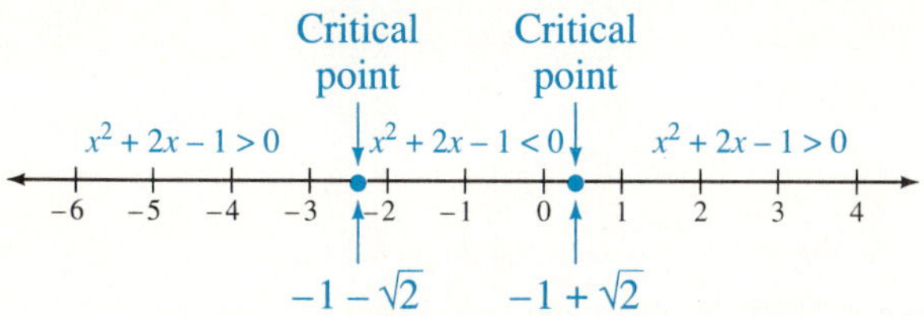

Figure 8.12

The solution, $[-1 - \sqrt{2}, -1 + \sqrt{2}]$ or $-1 - \sqrt{2} \le x \le -1 + \sqrt{2}$, is shown in Figure 8.13.

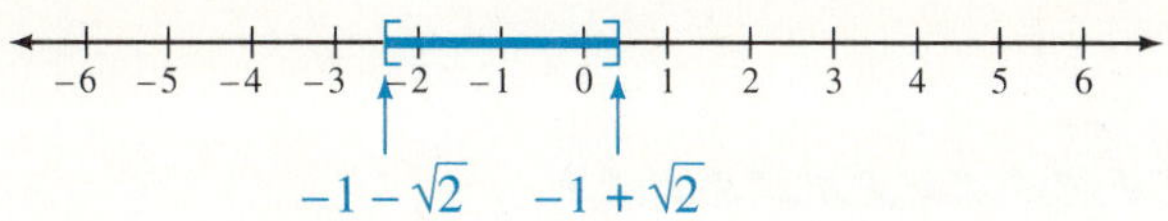

Figure 8.13 $[-1 - \sqrt{2}, -1 + \sqrt{2}]$

Answer: $(-\infty, 3 - \sqrt{3}]$ or $[3 + \sqrt{3}, \infty)$, *or* $x \le 3 - \sqrt{3}$ or $x \ge 3 + \sqrt{3}$

2 SOLVING RATIONAL INEQUALITIES

Inequalities such as $\frac{x-1}{x+2} \ge 0$, $\frac{3y-7}{y} \le 0$, $\frac{z^2 - z - 2}{z+3} < 0$ are called **rational inequalities.** By adjusting the definition of a critical point to include any value of the variable that makes the numerator zero *or* the denominator zero, rational inequalities can be solved using a method similar to that for quadratic inequalities. The critical points of a rational inequality divide the number line into various intervals, and a test point is chosen from each. The solution to the inequality is determined by the sign of the rational expression at each test point.

EXAMPLE 5 Solving a Rational Inequality

Solve $\frac{x-1}{x+2} \ge 0$ and graph the solution.

To find the critical points, set both the numerator and the denominator equal to zero and solve each equation. Here the critical points are 1 and −2. These separate the number line into three intervals:

$$(-\infty, -2), \quad (-2, 1), \quad (1, \infty).$$

In this case, 1 *is* a solution but −2 *is not*. In general, critical points that result from setting the denominator equal to zero will never be solutions to the inequality since division by zero is not defined.

We use test points −3, 0, and 2.

$$\text{Let } x = -3: \quad \frac{-3 - 1}{-3 + 2} = \frac{-4}{-1} = 4 > 0$$

$$\text{Let } x = 0: \quad \frac{0 - 1}{0 + 2} = -\frac{1}{2} < 0$$

$$\text{Let } x = 2: \quad \frac{2 - 1}{2 + 2} = \frac{1}{4} > 0$$

These results are shown in Figure 8.14 on the next page.

Practice Exercise 5

Solve.

$$\frac{x-1}{x+2} < 0$$

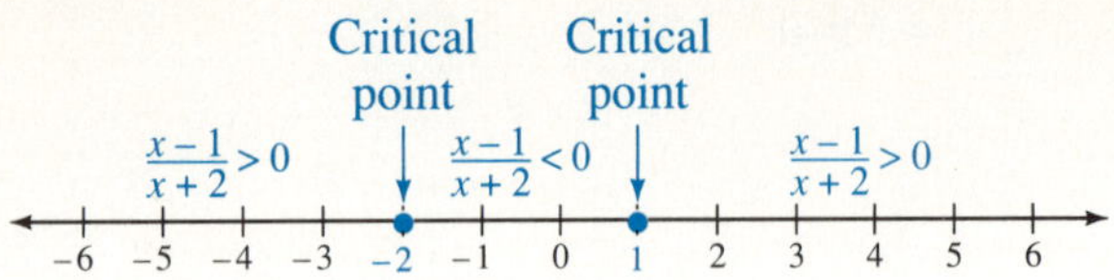

Figure 8.14

Thus, the solution to the given inequality is

$$(-\infty, -2) \text{ or } [1, \infty), \quad or \quad x < -2 \text{ or } 1 \le x.$$

Notice that 1 is included in the solution but -2 is not. The graph of the solution is given in Figure 8.15.

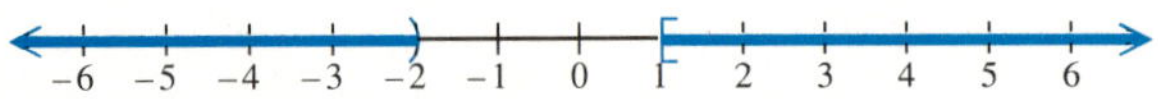

Figure 8.15

Answer: $(-2, 1)$ or $-2 < x < 1$

CAUTION

To solve rational inequalities such as the one in Example 5, can we multiply both sides by the denominator to clear the fraction? The answer is no. For example,

$$\frac{x-1}{x+2} \ge 0$$

$$(x+2)\frac{x-1}{x+2} \ge 0(x+2) \qquad \text{THIS IS WRONG}$$

$$x - 1 \ge 0$$

$$x \ge 1.$$

Remember that inequalities change signs when multiplied by a negative expression, but the inequality symbol remains the same when the expression is positive. Since $x + 2$ can be either positive or negative, we cannot simply multiply both sides by it.

If the right side of a rational inequality is not zero, we must change it to an equivalent inequality with a zero on the right.

EXAMPLE 6 Solving a Rational Inequality

Solve and graph $\frac{2x+1}{x-1} > 3$.

To solve an inequality of this form, we need to change the inequality to an equivalent inequality having a fractional expression greater than zero. In this case, subtract 3 from both sides and simplify on the left to obtain

$$\frac{4-x}{x-1} > 0.$$

The critical points are 4 and 1, so we can use 0, 2, and 5 as test points.

PRACTICE EXERCISE 6

Solve.

$$\frac{x-3}{x+1} \le 2$$

$$\text{Let } x = 0: \frac{4 - 0}{0 - 1} = \frac{4}{-1} = -4 < 0$$

$$\text{Let } x = 2: \frac{4 - 2}{2 - 1} = \frac{2}{1} = 2 > 0$$

$$\text{Let } x = 5: \frac{4 - 5}{5 - 1} = \frac{-1}{4} = -\frac{1}{4} < 0$$

We summarize these results in Figure 8.16.

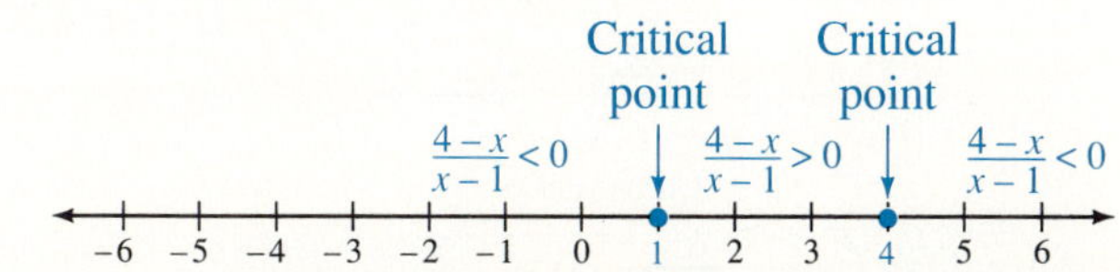

Figure 8.16

Thus, the solution is

$$(1, 4) \quad \text{or} \quad 1 < x < 4,$$

which is graphed in Figure 8.17.

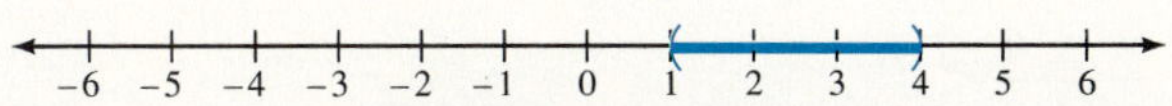

Figure 8.17 (1,4)

Answer: $(-\infty, -5]$ or $(-1, \infty)$, *or* $x \leq -5$ or $x > -1$

EXAMPLE 7 Solving a More Complex Rational Inequality

Solve $\dfrac{z^2 - z - 2}{z + 3} \leq 0$ and graph the solution.

Solve $z^2 - z - 2 = 0$ and $z + 3 = 0$ to obtain critical points -1, 2, and -3, and notice that -1 and 2 are solutions to the given inequality while -3 is not. This time the number line is separated into four intervals:

$$(-\infty, -3), \quad (-3, -1), \quad (-1, 2), \quad (2, \infty).$$

Evaluate at test points -4, -2, 0, and 3.

$$\text{Let } z = -4: \frac{(-4)^2 - (-4) - 2}{(-4) + 3} = -18 < 0$$

$$\text{Let } z = -2: \frac{(-2)^2 - (-2) - 2}{(-2) + 3} = 4 > 0$$

$$\text{Let } z = 0: \frac{(0)^2 - (0) - 2}{(0) + 3} = -\frac{2}{3} < 0$$

$$\text{Let } z = 3: \frac{(3)^2 - (3) - 3}{(3) + 3} = \frac{2}{3} > 0$$

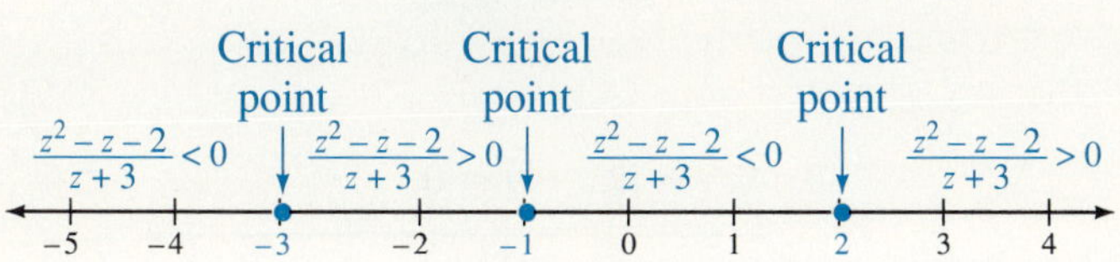

Figure 8.18

Practice Exercise 7

Solve.

$$\frac{x^2 - x - 12}{x - 2} \geq 0$$

The results, shown in Figure 8.18, indicate that the solution to $\frac{z^2 - z - 2}{z + 3} \leq 0$ is

$$(-\infty, -3) \text{ or } [-1, 2] \quad \textit{or} \quad z < -3 \text{ or } -1 \leq z \leq 2,$$

which is graphed in Figure 8.19.

Figure 8.19 $(-\infty, -3)$ or $[-1, 2]$

Answer: $[-3, 2)$ or $[4, \infty)$, *or* $-3 \leq x < 2$ or $x \geq 4$

Looking at Figure 8.18, we can also see that the solution to $\frac{z^2 - z - 2}{z + 3} > 0$ is

$$(-3, -1) \text{ or } (2, \infty) \quad \textit{or} \quad -3 < z < -1 \text{ or } 2 < z.$$

Notice that this time the numerator critical points, -1 and 2, are not included in the solution.

Other types of inequalities, such as

$$(x - 1)(x - 2)(x - 5) \geq 0 \quad \text{and} \quad \frac{y^2 - 4}{y^2 + 9} < 0$$

can be solved using exactly the same critical-test point technique.

8.5 EXERCISES A

Solve each inequality and give the answer using inequalities and interval notation. Graph the solution.

1. $x^2 - 5x + 6 > 0$

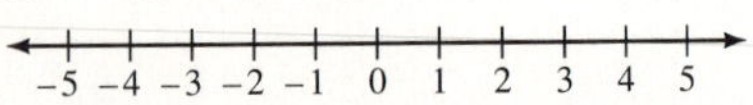

2. $x^2 - x < 6$

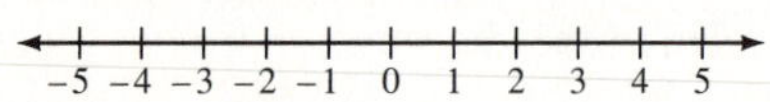

3. $2x^2 \leq 1 - x$

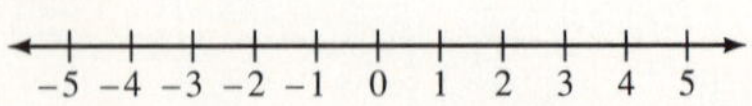

4. $2x^2 + x \geq 15$

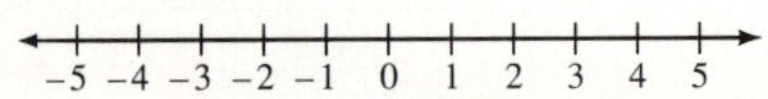

5. $-x^2 + 2x + 2 < 0$

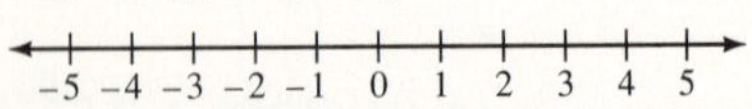

6. $x^2 + 2x + 2 > 0$

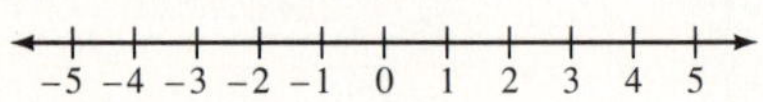

7. $4x^2 - 12x + 9 \leq 0$

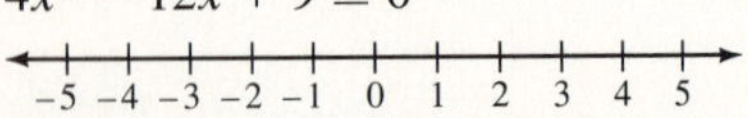

8. $4x^2 - 12x + 9 < 0$

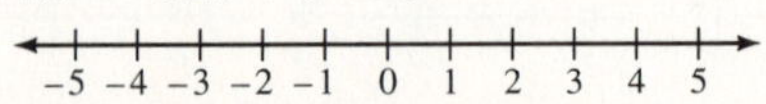

9 $3x^2 - 2x \le 2$

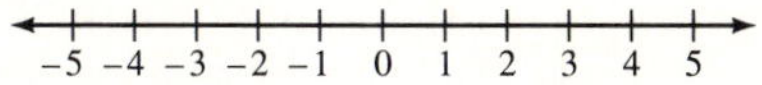

10. $(x - 2)(x + 2) < 3x$

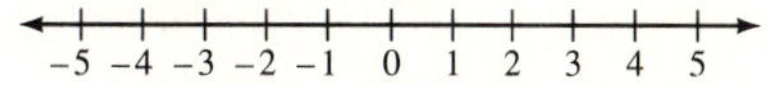

11. $x(x - 5) > 0$

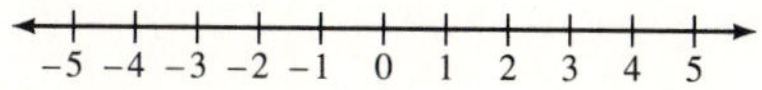

12. $x^2 - 9 \le 0$

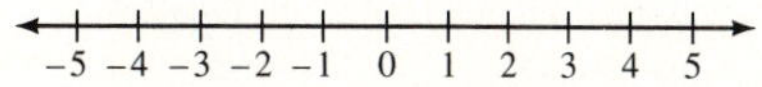

13 $x^2 + 9 \le 0$

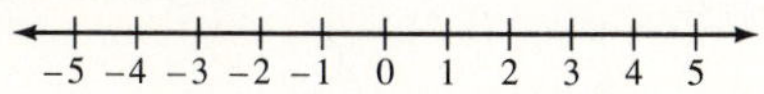

14. $x(x - 3) < x + 5$

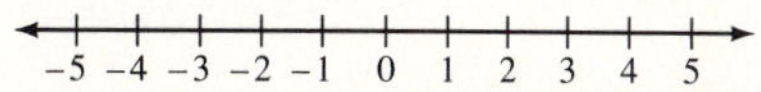

15. $\dfrac{x - 3}{x + 1} < 0$

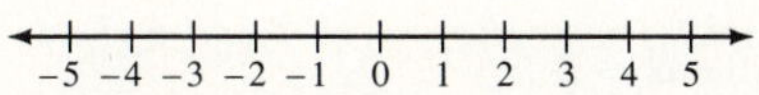

16. $\dfrac{x - 2}{x + 3} \ge 0$

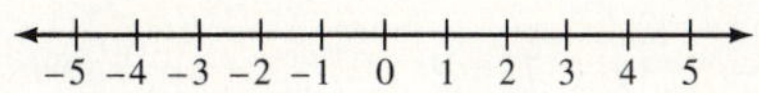

17 $\dfrac{2x + 1}{x - 1} < 3$

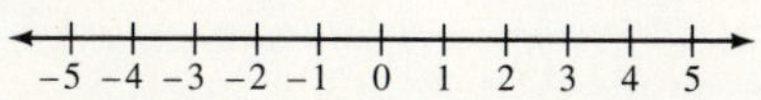

18. $\dfrac{2x - 1}{x + 3} \ge 1$

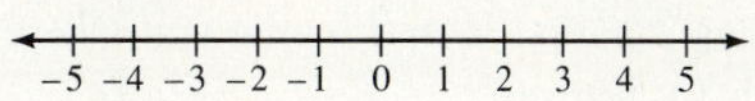

19. $\dfrac{3x - 2}{x - 3} > 2$

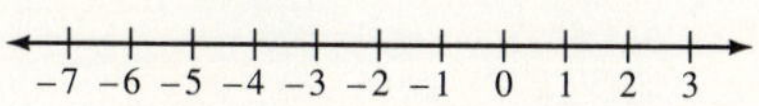

20. $\dfrac{3}{x - 1} \le 1$

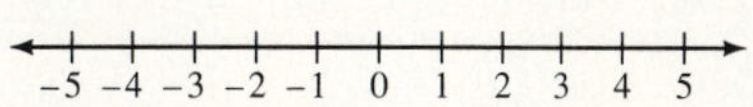

21. $\dfrac{x}{x + 1} < 2$

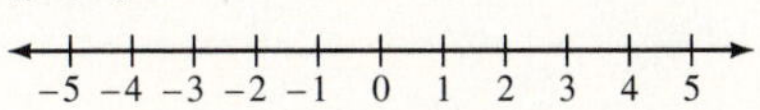

22 $\dfrac{(x - 3)(x + 2)}{x - 1} > 0$

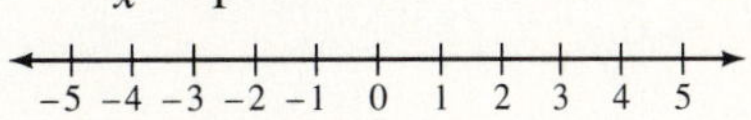

23. $\dfrac{x^2 - 25}{x} \le 0$

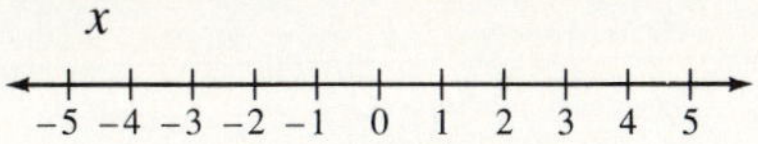

24. $(x - 1)(x + 4)(x - 5) \ge 0$

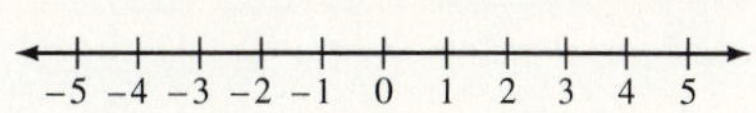

25. Burford worked this problem. What is wrong with his "solution"?

$$\frac{x-3}{x+5} \le 0$$

$$(x+5)\frac{x-3}{x+5} \le (x+5)(0)$$

$$x - 3 \le 0$$

$$x \le 3$$

FOR REVIEW

Solve each equation or applied problem in Exercises 26–32.

26. $6x^{-2} + x^{-1} = 2$

27. $(y-8)^2 + 5(y-8) - 6 = 0$

28. $\frac{1}{x^2-1} = \frac{1}{x-1}$

29. $\frac{14}{y^2+8y-33} = \frac{y}{y-3} - \frac{1}{y+11}$

30. $\sqrt{7y+23} - \sqrt{3y+7} = 2$

31. $\sqrt{x+8} - x + 4 = 0$

32. When each works alone, Ernie can mow a lawn in 3 hours less time than Burford. When they work together, it takes 2 hours. How long does it take each to do the job by himself?

The following exercises, reviewing material from Section 8.2, will help you prepare for the next chapter. In Exercises 33 and 34, what must be added to complete the square?

33. $x^2 + 4x$

34. $x^2 - x$

Solve the quadratic equation in Exercises 35 and 36 by completing the square.

35. $x^2 + 4x - 21 = 0$

36. $x^2 - 2x - 2 = 0$

37. What is the smallest possible value y can assume if x is a real number and $y = (x-1)^2 + 4$?

38. What is the largest possible value y can assume if x is a real number and $y = -(x+1)^2 + 2$?

ANSWERS:

1. $x < 2$ or $3 < x$; $(-\infty, 2)$ or $(3, \infty)$

2. $-2 < x < 3$; $(-2, 3)$

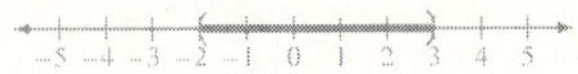

3. $-1 \le x \le \frac{1}{2}$; $\left[-1, \frac{1}{2}\right]$

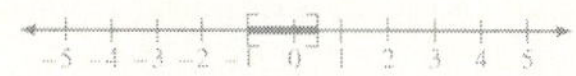

4. $x \le -3$ or $x \ge \frac{5}{2}$; $(-\infty, -3]$ or $\left[\frac{5}{2}, \infty\right)$

5. $x > 1 + \sqrt{3}$ or $x < 1 - \sqrt{3}$; $(-\infty, 1 - \sqrt{3})$ or $(1 + \sqrt{3}, \infty)$

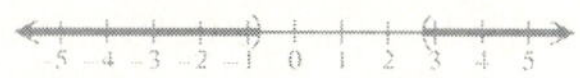

6. every real number is a solution; $(-\infty, \infty)$

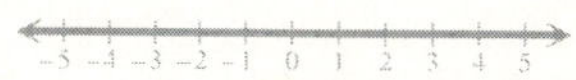

7. $x = \frac{3}{2}$; $\left[\frac{3}{2}, \frac{3}{2}\right]$

8. no solution

9. $\frac{1-\sqrt{7}}{3} \le x \le \frac{1+\sqrt{7}}{3}$; $\left[\frac{1-\sqrt{7}}{3}, \frac{1+\sqrt{7}}{3}\right]$

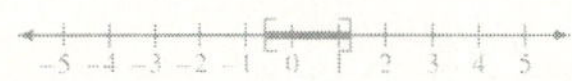

10. $-1 < x < 4$; $(-1, 4)$

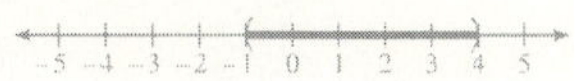

11. $x < 0$ or $5 < x$; $(-\infty, 0)$ or $(5, \infty)$

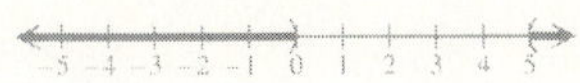

12. $-3 \le x \le 3$; $[-3, 3]$

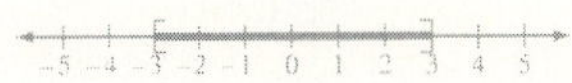

13. no solution

14. $-1 < x < 5$; $(-1, 5)$

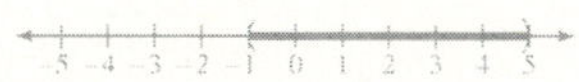

15. $-1 < x < 3$; $(-1, 3)$

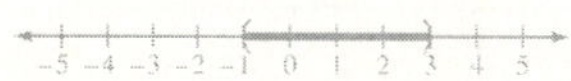

16. $x < -3$ or $2 \le x$; $(-\infty, -3)$ or $[2, \infty)$

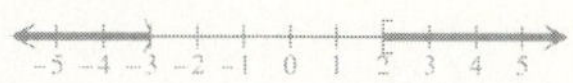

17. $x < 1$ or $x > 4$; $(-\infty, 1)$ or $(4, \infty)$

18. $x < -3$ or $4 \le x$; $(-\infty, -3)$ or $[4, \infty)$

19. $x < -4$ or $3 < x$; $(-\infty, -4)$ or $(3, \infty)$

20. $x < 1$ or $4 \le x$; $(-\infty, 1)$ or $[4, \infty)$

21. $x < -2$ or $-1 < x$; $(-\infty, -2)$ or $(-1, \infty)$

22. $-2 < x < 1$ or $x > 3$; $(-2, 1)$ or $(3, \infty)$

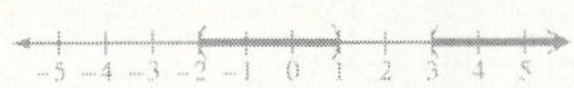

23. $x \le -5$ or $0 < x \le 5$; $(-\infty, -5]$ or $(0, 5]$

24. $-4 \le x \le 1$ or $5 \le x$; $[-4, 1]$ or $[5, \infty)$

25. It is incorrect to multiply both sides by $x + 5$ since $x + 5$ can be positive or negative. 26. $2, -\frac{3}{2}$ 27. 2, 9 28. 0 29. 1 30. $-1, -2$ 31. 8 32. Ernie: 3 hr; Burford: 6 hr 33. 4 since $x^2 + 4x + 4 = (x + 2)^2$ 34. $\frac{1}{4}$ since $x^2 - x + \frac{1}{4} = \left(x - \frac{1}{2}\right)^2$ 35. 3, -7 36. $1 \pm \sqrt{3}$ 37. 4 when $x = 1$ 38. 2 when $x = -1$

8.5 EXERCISES B

Solve each inequality and give the answer using inequalities and interval notation. Graph the solution.

1. $x^2 - 5x + 6 \le 0$

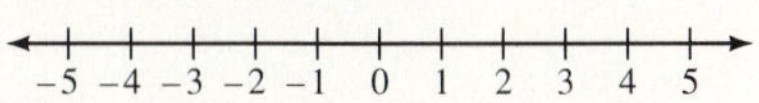

2. $x^2 - x \ge 6$

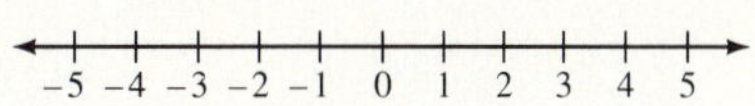

3. $2x^2 > 1 - x$

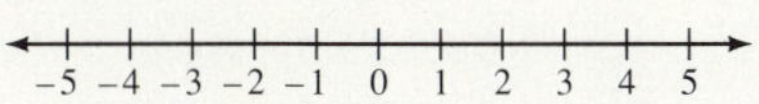

4. $2x^2 + x < 15$

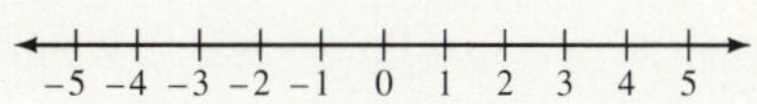

5. $-x^2 + 2x + 2 \ge 0$

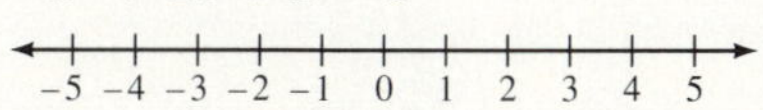

6. $x^2 + 2x + 2 \le 0$

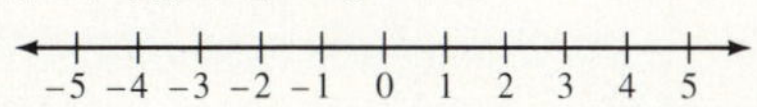

7. $4x^2 - 12x + 9 \geq 0$

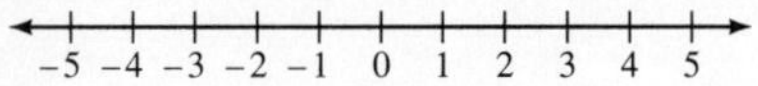

8. $4x^2 - 12x + 9 > 0$

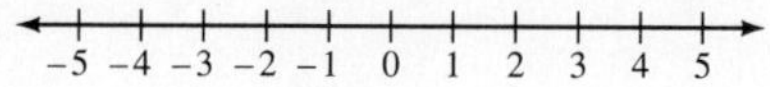

9. $3x^2 - 2x > 2$

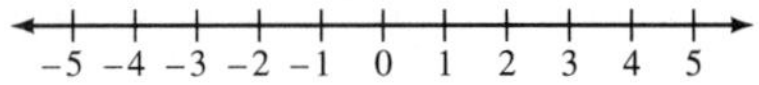

10. $(x - 2)(x + 2) \geq 3x$

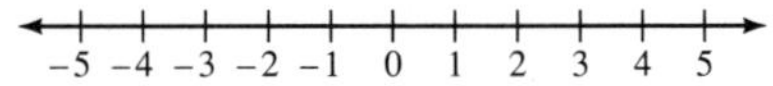

11. $x(x - 5) \leq 0$

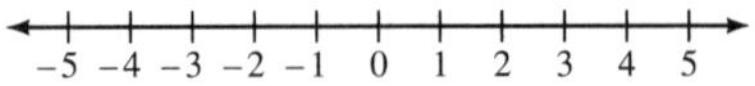

12. $x^2 - 9 > 0$

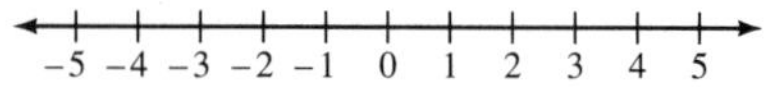

13. $x^2 + 9 > 0$

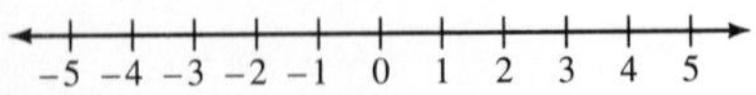

14. $x(x - 3) \geq x + 5$

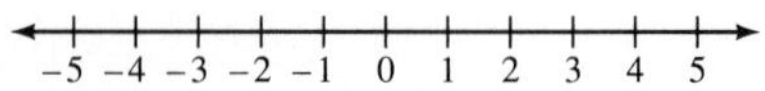

15. $\dfrac{x - 3}{x + 1} \geq 0$

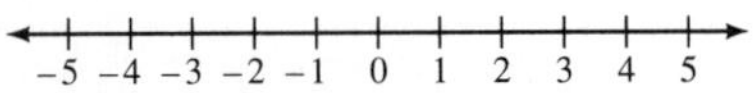

16. $\dfrac{x - 2}{x + 3} < 0$

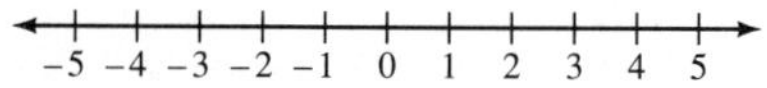

17. $\dfrac{2x + 1}{x - 1} \geq 3$

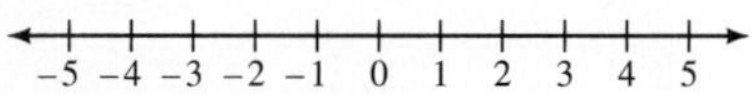

18. $\dfrac{2x - 1}{x + 3} < 1$

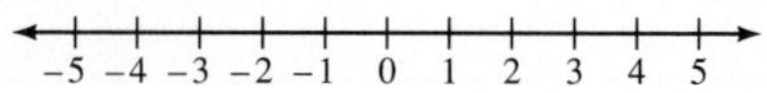

19. $\dfrac{3x - 2}{x - 3} \leq 2$

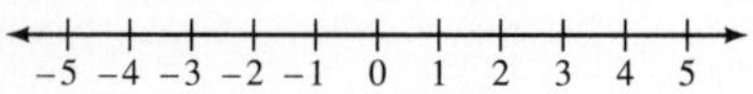

20. $\dfrac{3}{x - 1} > 1$

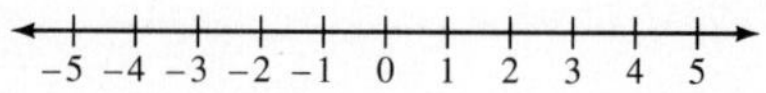

21. $\dfrac{x}{x + 1} \geq 2$

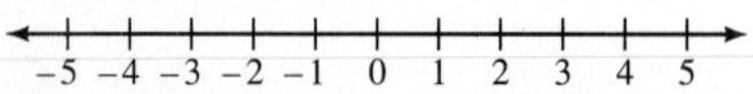

22. $\dfrac{(x - 3)(x + 2)}{x - 1} \leq 0$

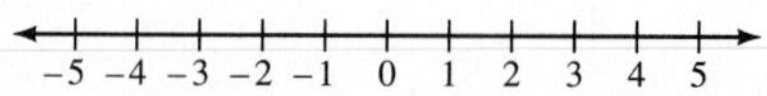

23. $\dfrac{x^2 - 25}{x} > 0$

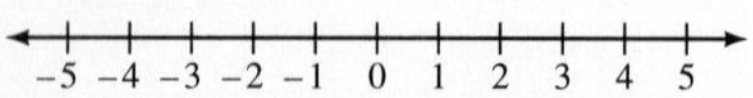

24. $(x - 1)(x + 4)(x - 5) < 0$

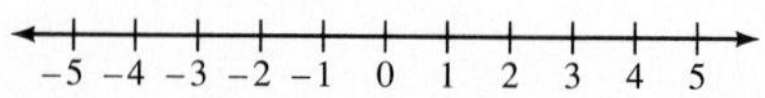

25. Burford worked this problem. What is wrong with his "solution"?

$$\frac{x^2 - 4}{x} > 0$$

$$x\left(\frac{x^2 - 4}{x}\right) > x(0)$$

$$x^2 - 4 > 0$$

$$x^2 > 4$$

$$x > 2 \text{ or } x > -2$$

FOR REVIEW

Solve each equation or applied problem in Exercises 26–32.

26. $x^4 - 5x^2 = -6$

27. $(y + 1)^2 - 2(y + 1) - 35 = 0$

28. $\frac{x}{3} - \frac{6}{x} = 1$

29. $\frac{1}{x + 2} - \frac{2}{x + 1} = 5$

30. $\sqrt{2x - 1} - \sqrt{x - 1} = 1$

31. $x - 3 - 4\sqrt{x - 3} + 3 = 0$

32. The distance d in miles from an object h miles above the surface of the earth to the horizon can be approximated by $d = \sqrt{h(h + 8000)}$. If an airplane is flying at an altitude of 4.2 mi, approximately how far is it from the airplane to the horizon?

The following exercises, reviewing material from Section 8.2, will help you prepare for the next chapter. In Exercises 33 and 34, what must be added to complete the square?

33. $x^2 + 8x$

34. $x^2 + x$

Solve the quadratic equation in Exercises 35 and 36 by completing the square.

35. $x^2 - 2x - 15 = 0$

36. $x^2 - x - 5 = 0$

37. What is the smallest possible value y can assume if x is a real number and $y = (x + 2)^2 + 5$?

38. What is the largest possible value y can assume if x is a real number and $y = -(x + 2)^2 + 3$?

8.5 EXERCISES C

Solve each inequality and give the answer using inequalities and interval notation. Graph the solution.

1. $\frac{x^2 - 4}{x^2 + 9} < 0$

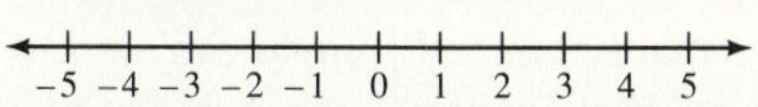

2. $\frac{x^2 - 3}{2x} < 1$

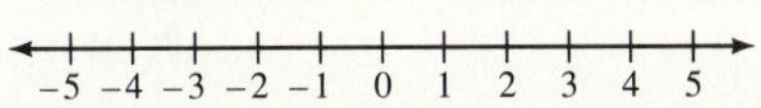

3. $\frac{2}{x - 2} \geq \frac{1}{x + 1}$

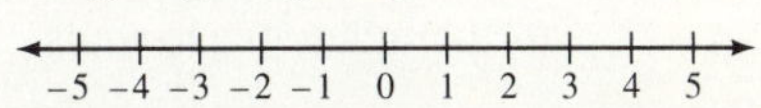

[Answer: $-4 \leq x < -1$ or $x > 2$; $[-4, -1)$ or $(2, \infty)$]

CHAPTER 8 REVIEW

KEY WORDS

8.1 The **general form** of a quadratic equation is $ax^2 + bx + c = 0$, $a > 0$.

When a solution to a quadratic equation occurs twice, it is a **root of multiplicity two** or a **double root.**

8.3 The **discriminant** of the quadratic equation $ax^2 + bx + c = 0$ is the number $b^2 - 4ac$.

8.5 An inequality of the form $ax^2 + bx + c < 0$ or $ax^2 + bx + c \geq 0$ is a **quadratic inequality.**

The **critical points** are solutions to the quadratic equation that is formed by changing the inequality symbol to an equal sign.

A **test point** is a number in one of the intervals formed by critical points that is used to determine the sign of the quadratic expression in that interval.

KEY CONCEPTS

8.1

1. Zero-product rule: If $ab = 0$, then $a = 0$ or $b = 0$. This rule only applies when one side of the equation is zero, and the other side is a product.
2. Do not use the zero-product rule on a zero-sum or zero-difference equation. For example, $(2x + 1) - (x + 5) = 0$ is *not* a zero-product equation. To solve it, clear parentheses and combine like terms. *Do not* set $2x + 1$ and $x + 5$ equal to zero!
3. The method of taking roots applies only when $b = 0$; that is, to a quadratic equation of the form $ax^2 + c = 0$.
4. Every quadratic equation in which $c = 0$, that is, of the form $ax^2 + bx = 0$, has 0 as one solution.

8.2

1. To complete the square on $x^2 + dx$, add $\left(\frac{1}{2}d\right)^2$.
2. The method of completing the square by adding the square of half the coefficient of the x term applies only when the coefficient of x^2 is $+1$.
3. Quadratic formula: The solutions of $ax^2 + bx + c = 0$ are given by
$$x = \frac{-b \pm \sqrt{b^2 - 4ac}}{2a}.$$
4. When solving a quadratic equation, always try factoring or taking roots (when applicable) first. If these methods fail, go directly to the quadratic formula.

8.3

1. The discriminant of a quadratic equation is used to determine the nature of the solutions.
If $b^2 - 4ac > 0$, the quadratic equation has two real solutions.
If $b^2 - 4ac = 0$, the quadratic equation has exactly one real solution, a double root.
If $b^2 - 4ac < 0$, the quadratic equation has two complex solutions.
2. If x_1 and x_2 are two real numbers or two conjugate complex numbers, then
$$(x - x_1)(x - x_2) = 0$$
is the quadratic equation with real coefficients having solutions x_1 and x_2.

8.4

1. When solving an equation quadratic in form using a u-substitution, be sure to give solutions for the original variable and not for the substituted variable u.
2. Solving radical equations depends on the power rule:
$$\text{If } a = b \text{ then } a^n = b^n.$$
3. When solving an equation involving square roots, *do not* square term by term; rather, square both sides and remember the middle term if one side is a binomial.
4. Be sure to check your answers in the original equation.

8.5

1. To solve $ax^2 + bx + c > 0$ or $ax^2 + bx + c < 0$, first solve $ax^2 + bx + c = 0$, obtaining the critical points. Then use test points, finding the value of $ax^2 + bx + c$ in the intervals formed by the critical points, to solve the inequality.
2. Rational inequalities are solved in the same way as quadratic inequalities using the critical-test point technique.

REVIEW EXERCISES

Part I

8.1, 8.2 *Solve.*

1. $y^2 - 2y - 48 = 0$
2. $3x(x + 5) = -5x - 25$
3. $5x^2 - 125 = 0$
4. $y^2 + 100 = 20y$
5. $y^2 - \frac{15}{2}y + 14 = 0$
6. $(x - 5)(3x + 7) = -30$

7. $x^2 - 4x = -7$

8. $x^2 + 12 = 0$

9. $3x^2 = 15x$

10. The amount of money A that will result if a principal P is invested at r percent interest compounded annually for 2 years is given by $A = P(1 + r)^2$. If \$1000 grows to \$1188.10 in 2 years using this formula, what is the rate of interest?

11. The hypotenuse of a right triangle is 16 meters, and one leg is 3 meters longer than the other. Rounded to the nearest tenth, find the length of each leg.

12. If an object is thrown downward with initial velocity of v_0 ft/sec, in t seconds its distance traveled is given by $h = v_0 t + 16t^2$. If the initial velocity is 64 ft/sec, how long will it take for an object thrown from a balloon 960 feet high to reach the ground?

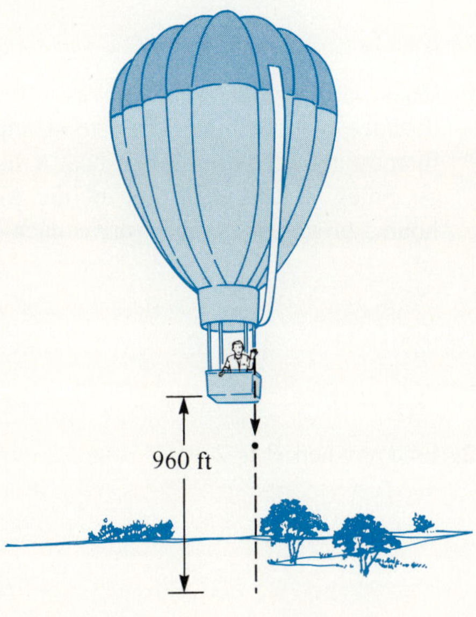

13. A radio tower was bent over one third of the distance from the bottom during a heavy windstorm. If the top of the tower now rests on the ground 50 feet from the base of the tower, to the nearest foot, how tall was the tower?

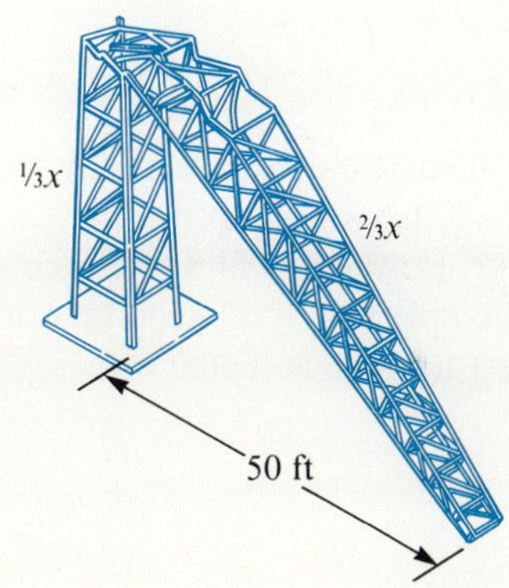

8.3 *Use the discriminant to tell the nature of the solutions.*

14. $9x^2 + 6x + 1 = 0$

15. $2x^2 - 3x = 7$

16. $2x^2 = 3x - 7$

Give the quadratic equation that has the following solutions.

17. 4, 1

18. $1 + i$, $1 - i$

19. $-1, \frac{3}{2}$

8.4 *Solve.*

20. $x^4 - 11x^2 + 28 = 0$

21. $y + 4 - 5\sqrt{y + 4} + 6 = 0$

22. $(y + 2)^2 - (y + 2) = 6$

23. $\dfrac{3}{x + 2} + \dfrac{2}{x - 2} = 2$

24. $\dfrac{1}{3} - \dfrac{2}{y - 4} = \dfrac{1}{y + 1}$

25. $\dfrac{u + 2}{u - 3} - \dfrac{u}{u + 4} = -1$

26. $\sqrt{3x^2 - 5} - 4 = 0$

27. $\sqrt{4x + 1} - x + 5 = 0$

28. $\sqrt{2y^2 - 5y + 4} - \sqrt{y^2 + 6y - 20} = 0$

29. Fred can complete a lab test in 2 hr less time than Walt. If they work together they can do the test in 3 hr. How long will it take each working alone to complete the test? Report the answer to the nearest tenth of an hour?

30. On a vacation, the Hagoods drive their car a distance of 420 miles at a rate 40 mph faster than the rate they drive their boat a distance of 50 miles. If the total time of the trip is $9\frac{1}{2}$ hours, how fast do they drive each vehicle?

In a physics experiment the equation $P = \sqrt{3y + 1} - \sqrt{y - 1}$ *is used.*

31. Find P when y is 9. Give both the exact answer and an approximate answer correct to the nearest hundredth.

32. Find y when P is 2.

Solve for the indicated variable.

33. $E = mc^2$ for c

34. $\dfrac{n}{x + 1} + \dfrac{n}{x - 1} = -1$ for x

8.5 *Solve. Give answer using inequalities and interval notation. Graph the solution.*

35. $9x^2 - 6x + 1 > 0$

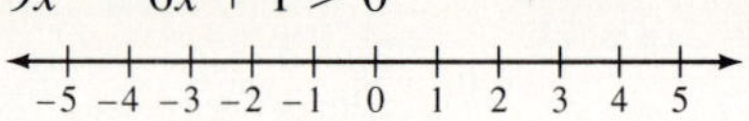

36. $3x^2 + 4x - 4 \leq 0$

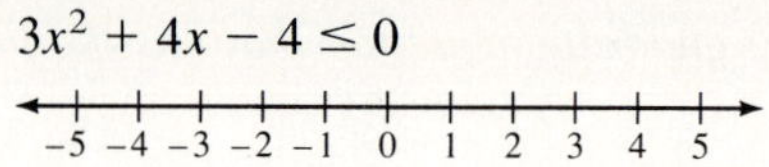

37. $\dfrac{x-5}{x+3} \le 0$

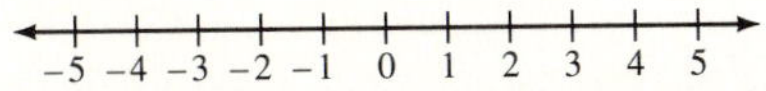

38. $\dfrac{(x+1)(x-4)}{x+2} > 0$

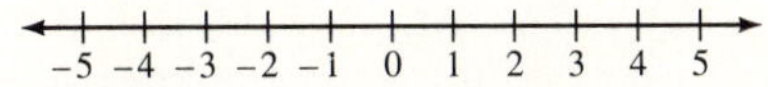

Part II

Solve.

39. $v^4 - 6v^2 + 8 = 0$

40. $y^2 = 4y - 2$

41. $7x^2 + 21 = 0$

42. $\sqrt{6z+4} - \sqrt{2z+5} = 3$

43. $\dfrac{x}{x-1} - \dfrac{1}{x+1} = \dfrac{2}{x^2-1}$

44. $(z+1)^2 - 5(z+1) + 6 = 0$

45. $x^2 + 2x + 2 = 0$

46. $x^2 - 7x = 0$

47. Twice the product of two positive consecutive odd integers is 390. Find the integers.

48. Ten less than the square of Ezra's age is the same as 8 more than three times his age. How old is Ezra?

49. Five times the number of square centimeters in the area of a square is the same as the number of centimeters in the perimeter increased by 3. Find the length of a side correct to the nearest hundredth of a centimeter.

50. A polygon having n sides has d diagonals where $d = \frac{n^2 - 3n}{2}$. If a polygon has 14 diagonals, how many sides does it have?

51. It takes Sybil 6 hours longer to paint a room than it takes Joanna. Working together, they can paint the room in 4 hours. How long would it take each working alone to paint the room?

52. Jeff walked a distance of 15 miles at a rate of 4 mph slower than the rate he jogged 10 miles. If the total time of the trip was 6 hours, to the nearest tenth, how fast did he jog?

53. Use the discriminant to tell the nature of the solutions to $2x^2 - 5x + 7 = 0$.

54. Solve by completing the square.

$$3x^2 - 4x + 1 = 0$$

55. Give the quadratic equation that has -2 and $\frac{1}{2}$ as solutions.

56. Solve $W = gt^2 + 5$ for t.

Solve. Give answer using inequalities and interval notation. Graph the solution.

57. $x^2 - 7x + 10 \le 0$

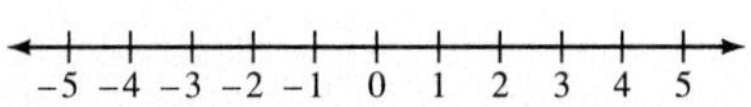

58. $\dfrac{x + 8}{x + 4} \le 3$

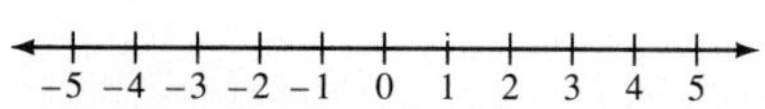

ANSWERS: 1. 8; −6 2. $-\frac{5}{3}$; −5 3. 5; −5 4. 10 5. $\frac{7}{2}$; 4 6. $\frac{4 \pm \sqrt{31}}{3}$ 7. $2 \pm i\sqrt{3}$ 8. $\pm 2i\sqrt{3}$ 9. 0; 5 10. 9% 11. 9.7 m; 12.7 m 12. 6 sec 13. 87 ft 14. one real 15. two real 16. two complex 17. $x^2 - 5x + 4 = 0$ 18. $x^2 - 2x + 2 = 0$ 19. $2x^2 - x - 3 = 0$ 20. ±2; $\pm\sqrt{7}$ 21. 0; 5 22. 1; −4 23. $\frac{5 \pm \sqrt{73}}{4}$ 24. $6 \pm \sqrt{34}$ 25. $-5 \pm \sqrt{29}$ 26. $\pm\sqrt{7}$ 27. 12 28. 3; 8 29. Fred: 5.2 hr; Walt: 7.2 hr 30. car: 60 mph; boat: 20 mph 31. $2\sqrt{7} - 2\sqrt{2}$; 2.46 32. 1; 5 33. $c = \pm\sqrt{\frac{E}{m}}$ 34. $-n \pm \sqrt{n^2 + 1}$
35. $x < \frac{1}{3}$ or $x > \frac{1}{3}$; $\left(-\infty, \frac{1}{3}\right)$ or $\left(\frac{1}{3}, \infty\right)$ 
36. $-2 \le x \le \frac{2}{3}$; $\left[-2, \frac{2}{3}\right]$ 37. $-3 < x \le 5$; $(-3, 5]$
38. $-2 < x < -1$ or $x > 4$; $(-2, -1)$ or $(4, \infty)$ 39. ±2; $\pm\sqrt{2}$ 40. $2 \pm \sqrt{2}$
41. $\pm i\sqrt{3}$ 42. 10 43. no solution 44. 1; 2 45. $-1 \pm i$ 46. 0; 7 47. 13; 15 48. 6 yr 49. 1.27 cm 50. 7 sides 51. Sybil: 12 hr; Joanna: 6 hr 52. 7.2 mph 53. two complex 54. $\frac{1}{3}$; 1 55. $2x^2 + 3x - 2 = 0$
56. $t = \pm\sqrt{\frac{W - 5}{g}}$ 57. $2 \le x \le 5$; $[2, 5]$
58. $x < -4$ or $x \ge -2$; $(-\infty, -4)$ or $[-2, \infty)$

CHAPTER 8 TEST

Solve.

1. $2x^2 = 5 - 9x$ | **1.** ______

2. $y^2 + 2y - 7 = 0$ | **2.** ______

3. $x^2 + 2x + 3 = 0$ | **3.** ______

4. $(y + 3)^2 - 4(y + 3) - 12 = 0$ | **4.** ______

5. $\dfrac{x^2}{4 - 3x} = 1$ | **5.** ______

6. The base of a triangle is 2 cm more than the height. Find the base and height if the area is 24 cm^2. | **6.** ______

7. Three times the reciprocal of the square of a number is the same as twice the reciprocal of the number. Find the number. | **7.** ______

8. In an experiment the equation $P = \sqrt{x + 1} - \sqrt{2x}$ is used. Find x when P is -1. | **8.** ______

9. Solve $a + b = 9c^2$ for c. | **9.** ______

10. It takes a 5-inch pipe 9 days longer to fill a reservoir than a 7-inch pipe. If the two pipes are turned on together, they can fill the reservoir in 20 days. How long would it take the 7-inch pipe to fill the reservoir alone? | **10.** ______

CHAPTER 8 TEST

CONTINUED

11. The amount of money A that will result if a principal P is invested at r percent interest compounded annually for two years is given by $A = P(1 + r)^2$. If \$10,000 grows to \$11,556.25 in 2 years, what is the rate of interest?

11. ____________

12. When an object is launched upward with initial velocity of 96 ft/sec from an initial height of 112 ft, ignoring air resistance, the height h of the object after t seconds is given by $h = -16t^2 + 96t + 112$. Use the discriminant of the appropriate quadratic equation to tell the number of times that the object is at a height of 256 ft. At what time(s) is the object actually at that height?

12. ____________

13. Use the discriminant to find the nature of the solutions of $4x^2 - 5x + 8 = 0$.

13. ____________

14. Give the quadratic equation that has solutions $\frac{3}{5}$, -2.

14. ____________

15. Solve $-x^2 + 4x + 5 < 0$. Graph the solution.

15. −5 −4 −3 −2 −1 0 1 2 3 4 5

16. Solve $\dfrac{y-1}{y-4} \leq 0$. Graph the solution.

16. −5 −4 −3 −2 −1 0 1 2 3 4 5

CHAPTER 9

More on Functions

9.1 QUADRATIC FUNCTIONS

STUDENT GUIDEPOSTS

1 Quadratic Functions	3 Graphing Quadratic Functions
2 Parabolas	4 Maximum and Minimum Problems

When a baseball player hits a pop fly, the path taken by the ball is a curve like the one shown in Figure 9.1. In this section we study *quadratic functions* whose graphs are curves similar to this.

Figure 9.1

1 QUADRATIC FUNCTIONS

In Chapter 3 we discussed linear functions defined by the equation $f(x) = mx + b$. We now consider another class of functions that have many applications in mathematics. Any function that can be described by an equation of the form

$$f(x) = ax^2 + bx + c \qquad (a, b, c \text{ constants}, a \neq 0)$$

is a **quadratic function.** If $a = 0$, we would lose the squared term and have a linear function. Like linear functions, quadratic functions have the set of real numbers for their domain.

2 PARABOLAS

The graph of a quadratic function is a ∪-shaped curve called a **parabola.** Consider the two special quadratic functions $f(x) = x^2$ $(a = 1, b = 0, c = 0)$ and $g(x) = -x^2$ $(a = -1, b = 0\ c = 0)$. Remember that $f(x)$ and $g(x)$ are names for y. The graphs of these two functions, obtained by making a table of values, are given in Figures 9.2 and 9.3, respectively. Notice that for $f(x) = x^2$, $a = 1 > 0$

and the parabola opens upward; for $g(x) = -x^2$, $a = -1 < 0$ and the parabola opens downward.

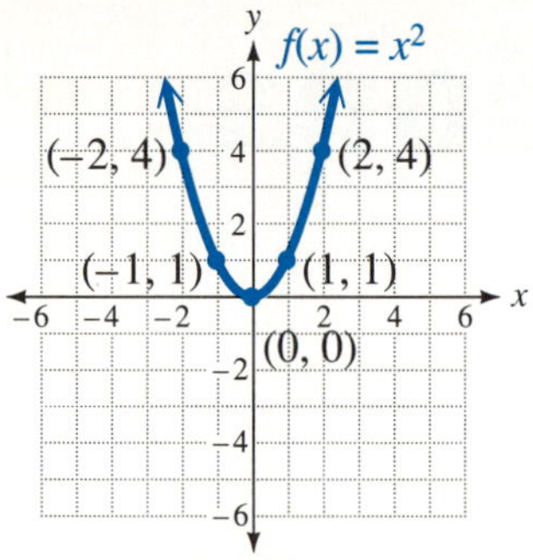

Figure 9.2

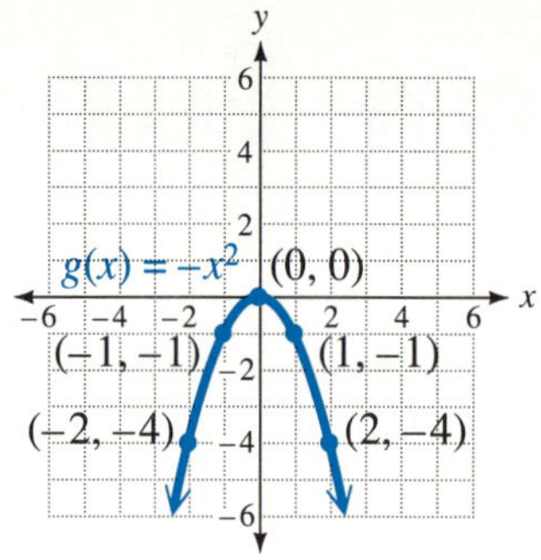

Figure 9.3

When a parabola opens upward, it has a low point or minimum and when it opens downward, it has a high point or maximum. In both cases, this point is called the **vertex** of the parabola. In our two example functions the vertex of each is at the origin (0, 0).

If we memorize the graphs of $f(x) = x^2$ and $g(x) = -x^2$, we can use them to find the graphs of any quadratic function of the form

$$f(x) = x^2 + c \quad \text{or} \quad f(x) = -x^2 + c.$$

Study the graphs in Figure 9.4. Notice that the graph of $f(x) = x^2 + 2$ can be obtained by "sliding" the graph of $f(x) = x^2$ up 2 units. The graph of $f(x) = x^2 - 1$ is obtained by sliding the graph of $f(x) = x^2$ down 1 unit since $c = -1$. Similarly, we can graph $f(x) = -x^2 + 3$ by sliding the graph of $g(x) = -x^2$ up 3 (+3) units, and graph $f(x) = -x^2 - 2$ by sliding the graph of $g(x) = -x^2$ down 2 (−2) units. By using this information, we can quickly sketch the graphs of a multitude of quadratic functions by simply moving or sliding a memorized graph.

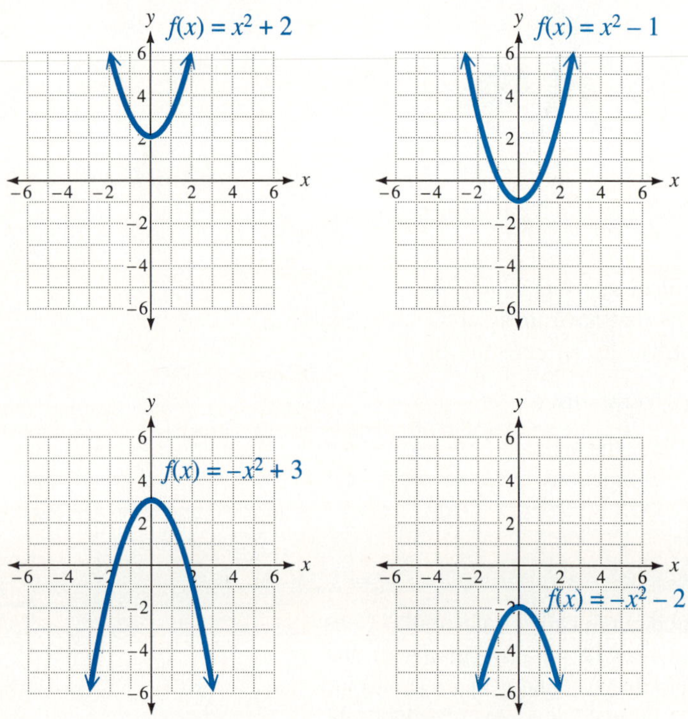

Figure 9.4 Quadratic Functions

We now consider graphing general quadratic functions. Suppose we begin by making a table of values and plot the graph of $f(x) = x^2 - 2x - 3$ in Figure 9.5. Notice that the graph crosses the x-axis when $f(x)$ (or y) is 0 and x is -1 and 3. The points $(-1, 0)$ and $(3, 0)$ are the **x-intercepts.** The parabola opens up (notice that $a = 1 > 0$) and has a low point at the vertex $(1, -4)$. Also, there are pairs of points at the same distance on either side of the line $x = 1$, the **line of symmetry** of the parabola.

x	$f(x)$
0	-3
1	-4
-1	0
2	-3
-2	5
3	0
4	5

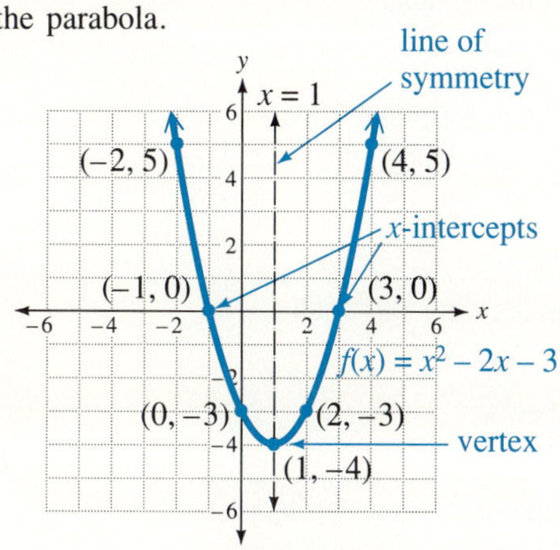

Figure 9.5

3 GRAPHING QUADRATIC FUNCTIONS

Knowing that the graph of a general quadratic function $f(x) = ax^2 + bx + c$ is a parabola, we can minimize the number of points to plot by determining whether the graph opens up or down and finding the x-intercepts, the vertex, and the line of symmetry. In view of the preceding discussion we can conclude, as shown in Figure 9.6, that the graph of a parabola $f(x) = ax^2 + bx + c$

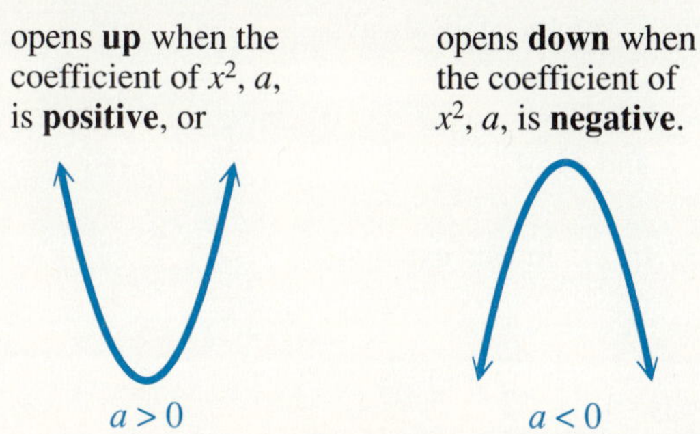

Figure 9.6

To find the x-intercepts, we set $f(x)$ equal to zero and solve the resulting quadratic equation. In our example, $f(x) = x^2 - 2x - 3$, so we solve

$$x^2 - 2x - 3 = 0 \quad f(x) = 0$$
$$(x + 1)(x - 3) = 0$$
$$x + 1 = 0 \quad \text{or} \quad x - 3 = 0$$
$$x = -1 \quad \text{or} \quad x = 3.$$

The intercepts are $(-1, 0)$ and $(3, 0)$.

One way to find the vertex is to complete the square in $f(x) = x^2 - 2x - 3$.

$$f(x) = x^2 - 2x \qquad -3 \quad \text{Rewrite leaving space as indicated}$$

To complete the square on $x^2 - 2x$, we add the square of half the coefficient of x, $\left[\frac{1}{2}(-2)\right]^2 = \left[-1\right]^2 = 1$. But to maintain equality with $f(x)$, we must also subtract this value as shown below.

$$f(x) = x^2 - 2x + 1 - 3 - 1$$ Add 1 and subtract 1, the square of half the coefficient of x

$$= (x^2 - 2x + 1) - 4$$ Group the terms in the trinomial

$$= (x - 1)^2 - 4$$ Factor the trinomial

Since the graph opens up, the smallest value that $f(x)$ can assume occurs when $(x - 1)^2$ is as small as possible. This is when $(x - 1)^2$ is 0, or when $x = 1$. When $x = 1$,

$$f(1) = (1 - 1)^2 - 4 = 0 - 4 = -4.$$

Thus, the vertex has coordinates $(1, -4)$, which agrees with our findings in Figure 9.5. Also, the line of symmetry has equation

$$\boldsymbol{x = x\text{-coordinate of the vertex}}$$

which in our example is $x = 1$.

Rather than complete the square to find the vertex each time we have a quadratic function, we can complete the square on a general function $f(x) = ax^2 + bx + c$, obtaining the x-coordinate of the vertex as $-\frac{b}{2a}$. In our example, since $b = -2$ and $a = 1$,

$$-\frac{b}{2a} = -\frac{(-2)}{2(1)} = \frac{2}{2} = 1.$$ Substitute -2 for b and 1 for a

To find the y-coordinate, we substitute the value found for the x-coordinate into the formula for $f(x)$. Again for our particular example,

$$f(1) = (1)^2 - 2(1) - 3 = 1 - 2 - 3 = -4.$$ Substitute 1 for x

Thus, the vertex is $\left(-\frac{b}{2a}, f\left(-\frac{b}{2a}\right)\right) = (1, -4)$. The line of symmetry is always the vertical line

$$\boldsymbol{x = -\frac{b}{2a}}.$$ Which is $x = 1$ when $b = -2$ and $a = 1$

One other point that is easy to find is the y-intercept $(0, c)$. In our example $c = -3$, so the y-intercept is $(0, -3)$.

We summarize these remarks in the following.

To Graph the Quadratic Function $f(x) = ax^2 + bx + c$

1. Find whether the graph opens up (if $a > 0$) or down (if $a < 0$).
2. Find the vertex and line of symmetry using the fact that the x-coordinate of the vertex is $-\frac{b}{2a}$ and the line of symmetry is $x = -\frac{b}{2a}$. The y-coordinate of the vertex is found by evaluating $f\left(-\frac{b}{2a}\right)$.
3. If the vertex is located above the x-axis and the graph opens down, or if the vertex is located below the x-axis and the graph opens up, find the x-intercepts by solving the equation
$$ax^2 + bx + c = 0.$$
4. Plot the vertex and the x-intercepts (or two other points on either side of the line of symmetry if there are no x-intercepts) and sketch the parabola.

EXAMPLE 1 Graphing a Quadratic Function

Use the x-intercepts, the vertex, and the line of symmetry to graph $g(x) = -2x^2 + 12x - 10$.

Since the coefficient of x^2 is negative ($a = -2$), the parabola opens down. With $a = -2$ and $b = 12$,

$$-\frac{b}{2a} = -\frac{12}{2(-2)} = -\frac{12}{-4} = 3.$$

Thus, the x-coordinate of the vertex is 3, and the y-coordinate is

$$\begin{aligned} g(3) &= -2(3)^2 + 12(3) - 10 \\ &= -2(9) + 36 - 10 \\ &= -18 + 36 - 10 = 8 \end{aligned}$$

With the vertex at (3, 8) and the graph opening down, we know that it will also cross the x-axis. Thus we look for the x-intercepts.

$$\begin{aligned} -2x^2 + 12x - 10 &= 0 && g(x) = 0 \text{ gives intercepts} \\ -2(x^2 - 6x + 5) &= 0 && \text{Factor out } -2\text{, change all signs} \\ x^2 - 6x + 5 &= 0 && \text{Divide by } -2 \\ (x - 1)(x - 5) &= 0 && \text{Factor} \end{aligned}$$

$$x - 1 = 0 \quad \text{or} \quad x - 5 = 0$$
$$x = 1 \qquad\qquad x = 5$$

The x-intercepts are (1, 0) and (5, 0). Finally, the line of symmetry, $x = -\frac{b}{2a}$, is $x = 3$, and the graph is given in Figure 9.7.

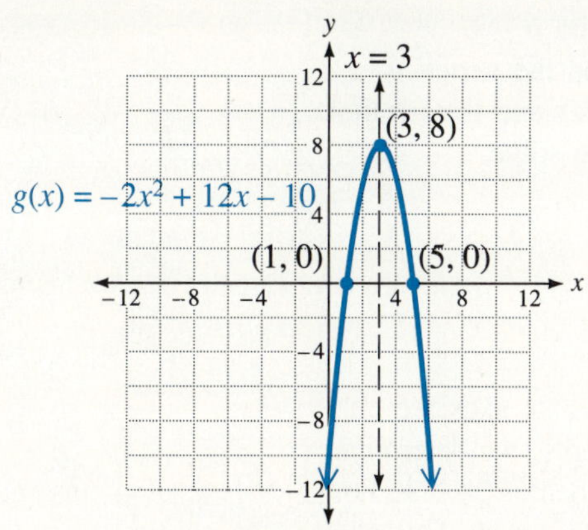

Figure 9.7

PRACTICE EXERCISE 1

Use the x-intercepts, the vertex, and the line of symmetry to graph $f(x) = x^2 - 2x - 8$. Since $a = 1 > 0$, the graph opens __________.
The x-coordinate of the vertex is given by $-\frac{b}{2a}$ = __________.

The y-coordinate of the vertex is given by $f\left(-\frac{b}{2a}\right)$ = __________. To find the x-intercepts, set $f(x) = 0$ and solve.

$$x^2 - 2x - 8 = 0$$

Finally, the line of symmetry is

$x = -\frac{b}{2a}$ = __________.

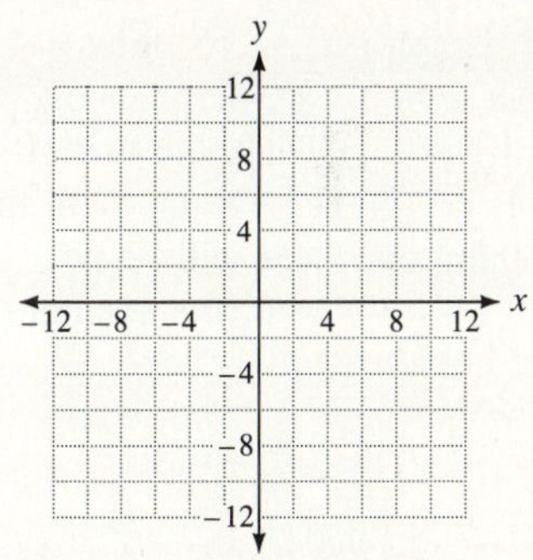

Answer: The graph is a parabola that opens up, has vertex (1, −9), has x-intercepts (−2, 0) and (4, 0), and line of symmetry $x = 1$.

EXAMPLE 2 Graphing a Quadratic Function

(a) Graph $f(x) = x^2 + 5x + 4$.
Since $a = 1 > 0$, the parabola opens up. With $a = 1$ and $b = 5$, $-\frac{b}{2a} = -\frac{5}{2(1)} = -\frac{5}{2}$, so that the line of symmetry is $x = -\frac{5}{2}$. Also,

$$\begin{aligned} f\left(-\frac{5}{2}\right) &= \left(-\frac{5}{2}\right)^2 + 5\left(-\frac{5}{2}\right) + 4 \\ &= \frac{25}{4} - \frac{25}{2} + 4 = -\frac{9}{4}, \end{aligned}$$

PRACTICE EXERCISE 2

(a) Graph $g(x) = -x^2 - 5x - 4$.

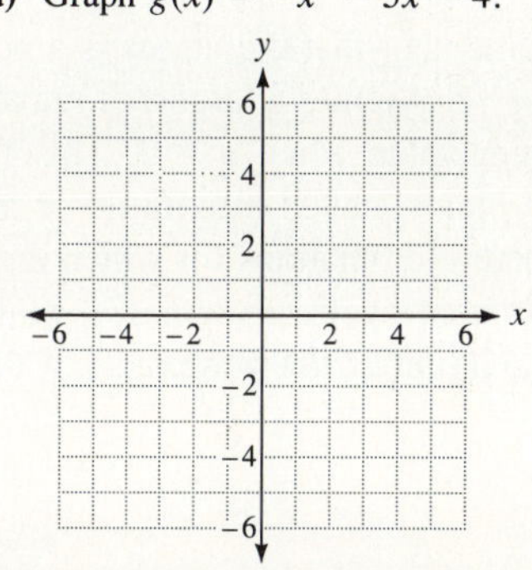

so that the vertex is $\left(-\frac{5}{2}, -\frac{9}{4}\right)$. With the vertex located below the x-axis, and the parabola opening up, we look for the x-intercepts.

$$x^2 + 5x + 4 = 0$$
$$(x + 4)(x + 1) = 0$$
$$x + 4 = 0 \quad \text{or} \quad x + 1 = 0$$
$$x = -4 \qquad\qquad x = -1$$

Thus, the x-intercepts are $(-4, 0)$ and $(-1, 0)$, and the graph is given in Figure 9.8.

(b) Graph $g(x) = x^2 - 2x + 3$.

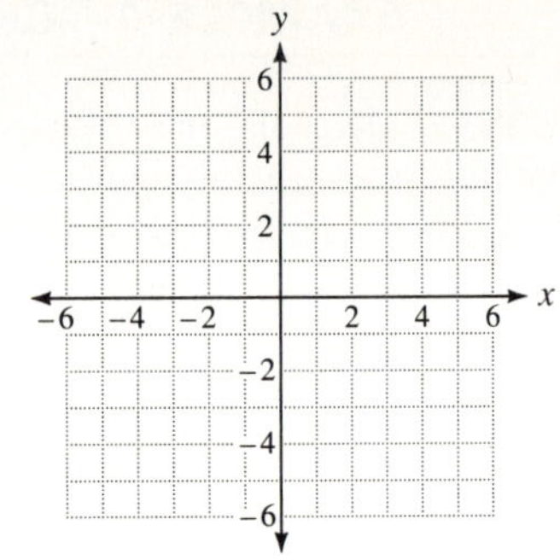

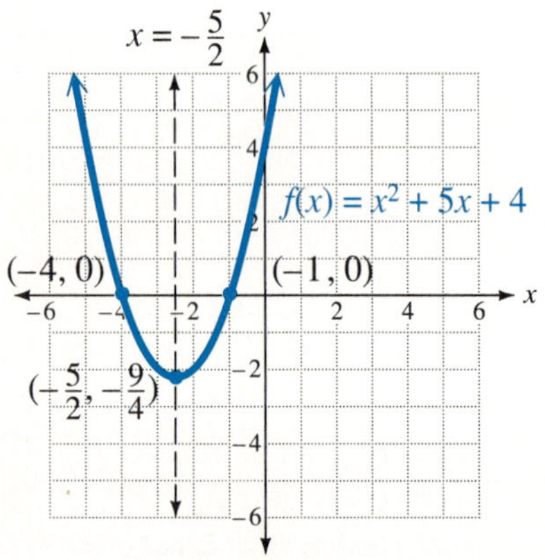

Figure 9.8

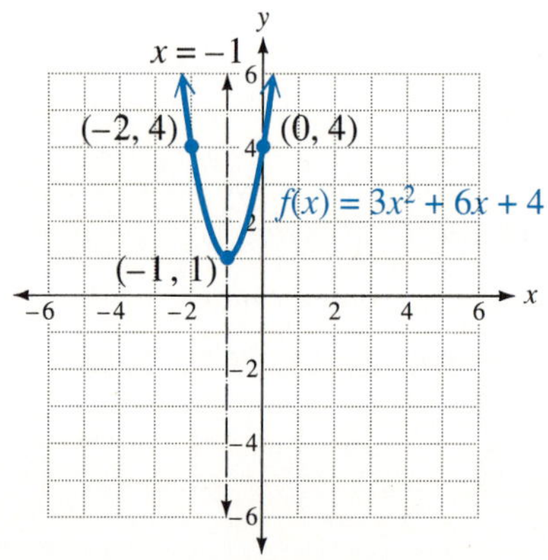

Figure 9.9

(b) Graph $f(x) = 3x^2 + 6x + 4$.
Since $a = 3 > 0$, the parabola opens up. With $a = 3$ and $b = 6$, $-\frac{b}{2a} = -\frac{6}{2(3)} = -1$. Thus, the line of symmetry is $x = -1$ and the vertex is $(-1, f(-1)) = (-1, 1)$. With the vertex above the x-axis and the parabola opening up, there will be no x-intercepts. If we were to solve $3x^2 + 6x + 4 = 0$, the solutions we obtain would be

$$\frac{-3 \pm i\sqrt{3}}{3},$$

which are not real numbers. (Remember that the coordinate axes are only used to plot real numbers.) Since there are no x-intercepts, we determine two additional points on either side of the line of symmetry $x = -1$. For example, when $x = 0$, $f(0) = 4$, and when $x = -2$, $f(-2) = 4$. This gives us the points $(0, 4)$ and $(-2, 4)$. The graph is given in Figure 9.9.

Answers: (a) The graph is a parabola that opens down, has vertex $\left(-\frac{5}{2}, \frac{9}{4}\right)$, has x-intercepts $(-4, 0)$ and $(-1, 0)$, and line of symmetry $x = -\frac{5}{2}$. (b) The graph is a parabola that opens up, has vertex $(1, 2)$, has no x-intercepts, has line of symmetry $x = 1$, and passes through $(0, 3)$ and $(2, 3)$.

4 MAXIMUM AND MINIMUM PROBLEMS

Remember that the vertex is a point at which a quadratic function $f(x) = ax^2 + bx + c$ assumes a largest or maximum value when $a < 0$ and a smallest or minimum value when $a > 0$. This fact can be used to solve a variety of applied problems, called **maximum** or **minimum problems,** that can be translated into quadratic functions for which the maximum or minimum value is desired. If the parabola opens up it has a minimum at the vertex; if it opens down it has a maximum at the vertex.

EXAMPLE 3 Maximum Problem

A man has 36 yd of fencing material. What are the dimensions of the largest rectangular pen that he can enclose? What is the maximum area?

Let l = the length of the desired pen,

w = the width of the desired pen.

Make a sketch as in Figure 9.10. We know that

$$2l + 2w = 36. \quad \text{The perimeter} = 36 \text{ yd}$$

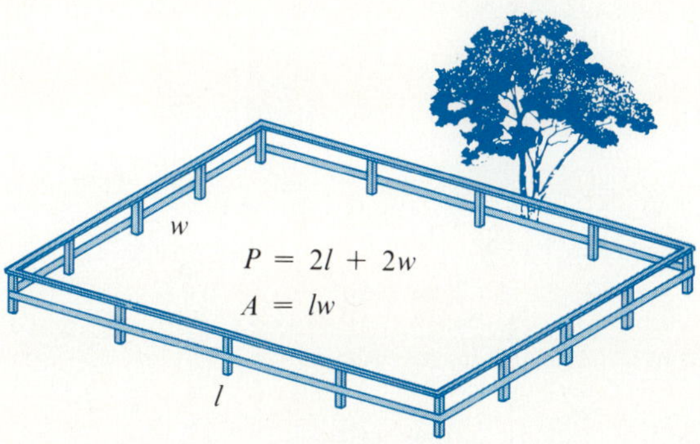

Figure 9.10

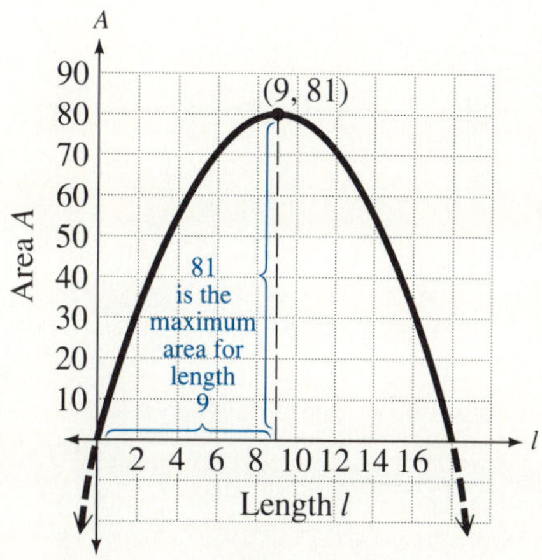

Figure 9.11

The quantity we wish to maximize is the area, A, of the pen.

$$A = lw$$

By solving $2l + 2w = 36$ for w,

$$w = 18 - l,$$

and substituting, we obtain

$$A = lw = l(18 - l) = -l^2 + 18l.$$

Thus, quadratic function $A = -l^2 + 18l$ gives the area of the pen as a function of the length l. We can graph this function as in Figure 9.11, where the area A is shown along the y-axis and the length l along the x-axis. The maximum point of the graph (the vertex) will give the maximum area (the y-coordinate) determined by a particular length (the x-coordinate). With $a = -1$, and $b = 18$, $-\frac{b}{2a} = -\frac{18}{2(-1)} = 9$. Thus, the y-coordinate of the vertex is

$$-(9)^2 + 18(9) = -81 + 162 = 81. \quad \text{Substitute 9 for } x$$

Since $a = -1 < 0$, the parabola $A = -l^2 + 18l$ opens down. Then 81 is the maximum value of A, occurring when $l = 9$. Since, when $l = 9$, $w = 18 - l = 9$, the dimensions that give the maximum area of 81 yd^2 are 9 yd by 9 yd.

Practice Exercise 3

What is the minimum product of two numbers whose difference is 4? What are the numbers?

Let x = one number,

y = second number.

We know that $x - y = 4$ or that $x = y + 4$. We must minimize the product $P = xy = (y + 4)y$.

Answer: The minimum product is -4 and the numbers are 2 and -2.

9.1 EXERCISES A

In Exercises 1–4, sketch the graph of each quadratic function by sliding the graph of $f(x) = x^2$ or $g(x) = -x^2$ up or down.

1. $h(x) = x^2 + 1$

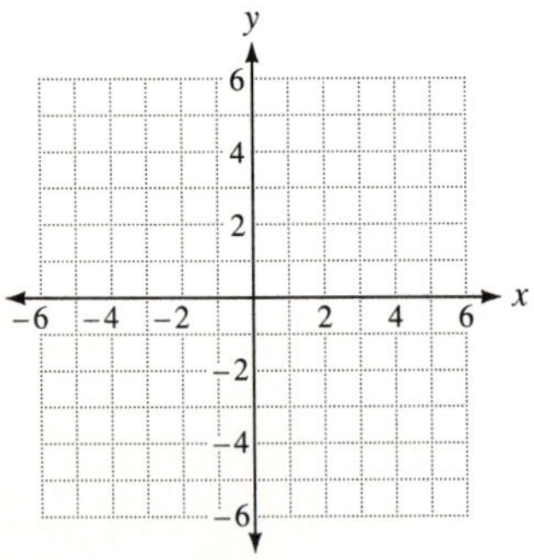

2. $h(x) = x^2 - 2$

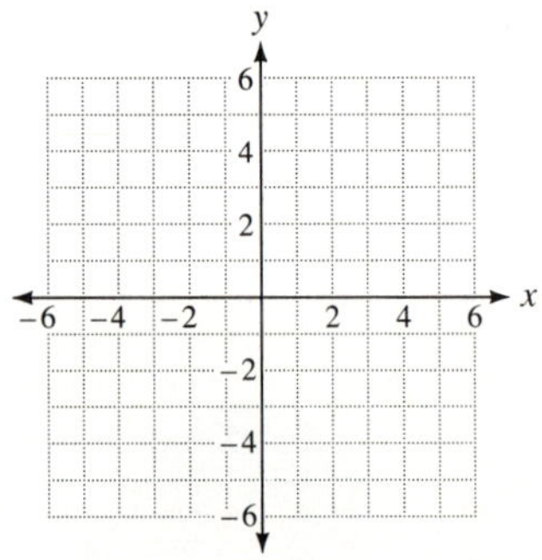

3. $h(x) = -x^2 + 1$

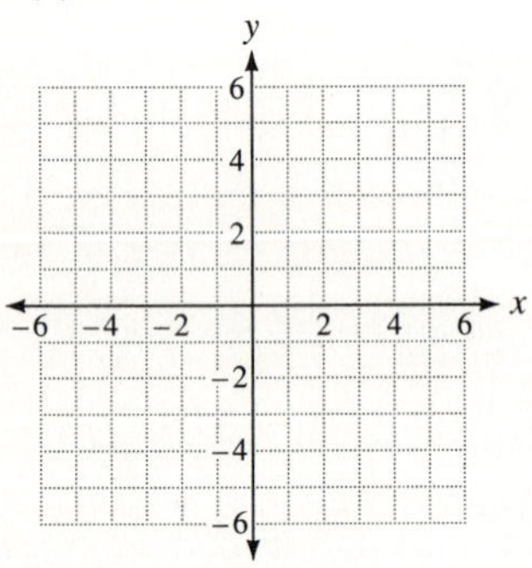

4. $h(x) = -x^2 - 3$

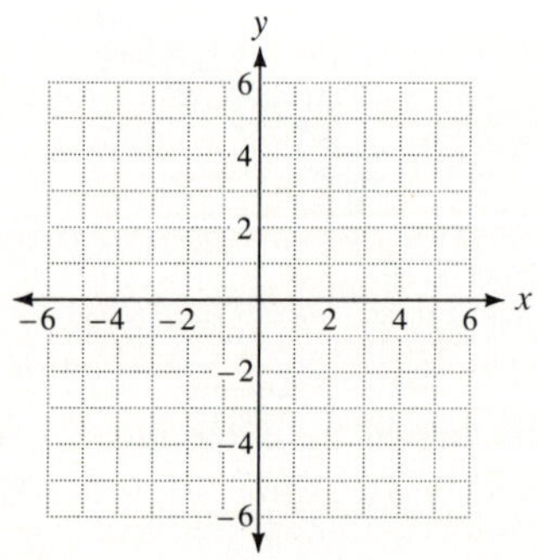

Without graphing, give the x-intercepts (if they exist) and the vertex, and tell whether the graph of each quadratic function opens up or down.

5. $f(x) = x^2 - 4x + 3$

6. $g(x) = x^2 + 4x + 3$

7. $f(x) = x^2 + 2x + 2$

8. $g(x) = x^2 + 8x$

9. $f(x) = -x^2 + 5x - 6$

10. $g(x) = 3x^2 - 6x + 3$

11. $f(x) = -2x^2 + 4x - 3$

12. $g(x) = 3x^2 + 5x + 1$

Find the x-intercepts (if they exist), the vertex, and graph of the function.

13. $f(x) = x^2 + 2x - 3$

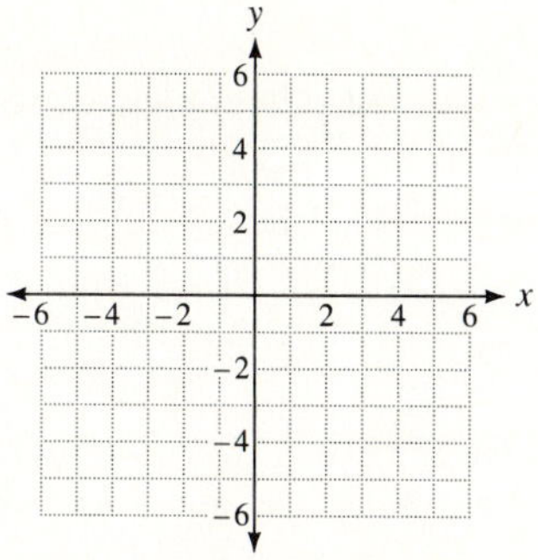

14. $g(x) = -x^2 - 2x + 3$

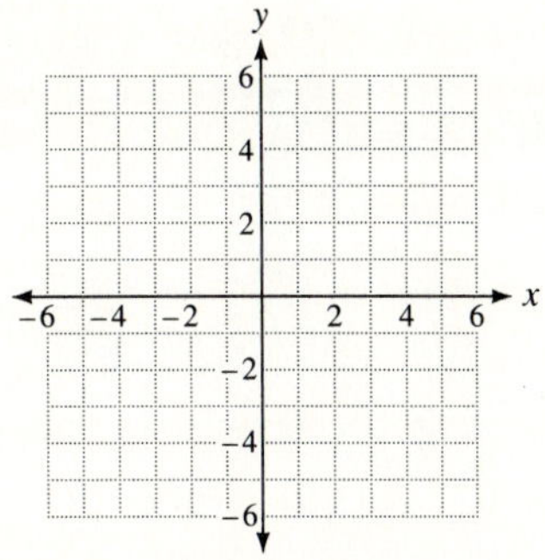

15. $f(x) = x^2 - 4x$

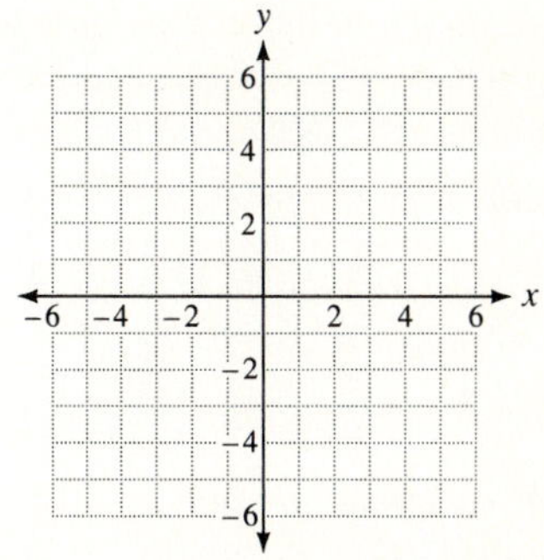

16. $g(x) = -x^2 + 4x$

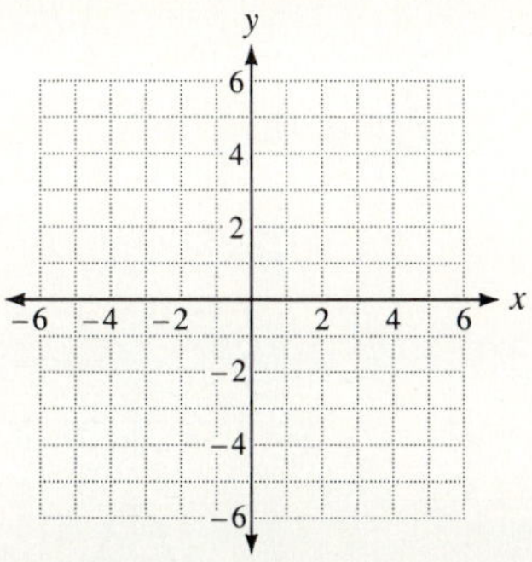

17. $f(x) = -4x^2 - 8x - 4$

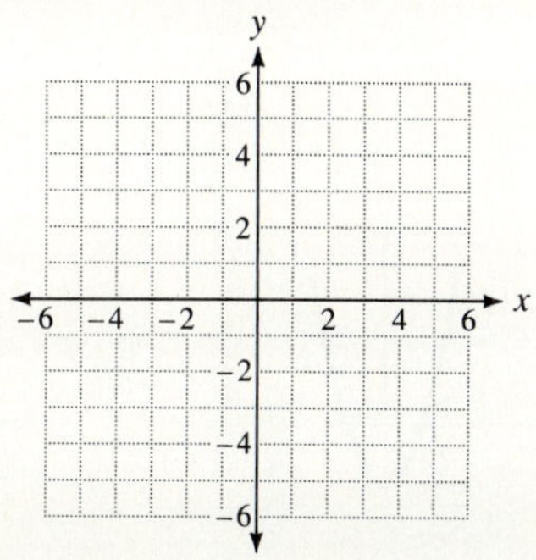

18. $f(x) = 2x^2 - 8x + 11$

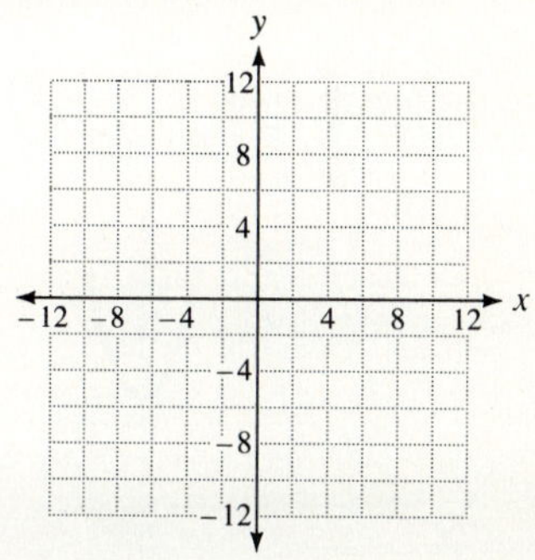

Solve.

19. A man has 40 m of fencing material. What are the dimensions of the largest rectangular pen which he can enclose? What is the maximum area?

20. What is the minimum product of two numbers whose difference is 8? What are the numbers?

21 Over a period of years, by studying the daily sales at Perko's Delicatessen, Perko has discovered that his profit can be approximated by $P = -x^2 + 80x - 1500$, where x represents the number of sandwiches sold. How many sandwiches should he sell daily to maximize his profit? What is the profit under these ideal conditions?

22. A ball is thrown upward with initial velocity of 32 ft/sec. If its height after x sec is given by $h = -16x^2 + 32x$, find the maximum height that it will attain. At what time will the ball hit the ground?

FOR REVIEW

The following exercises review material from Chapter 3 to help you prepare for the next section. Specify the domain of each function defined by the equation given in Exercises 23–25.

23. $y = 2x + 8$ **24.** $y = x^2 + 5$ **25.** $y = \dfrac{1}{x + 9}$

For each function in Exercises 26–27, evaluate: **(a)** $g(0)$ **(b)** $g(-1)$ **(c)** $g(2)$ **(d)** $g(x + h)$ **(e)** $g(x + h) - g(x)$.

26. $g(x) = 5x - 1$ **27.** $g(x) = x^2 - 1$

ANSWERS: 1.

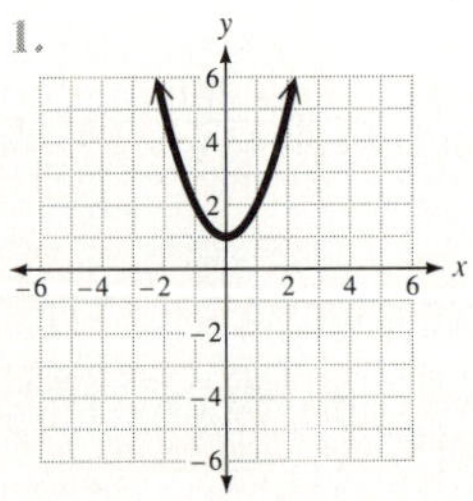

2.

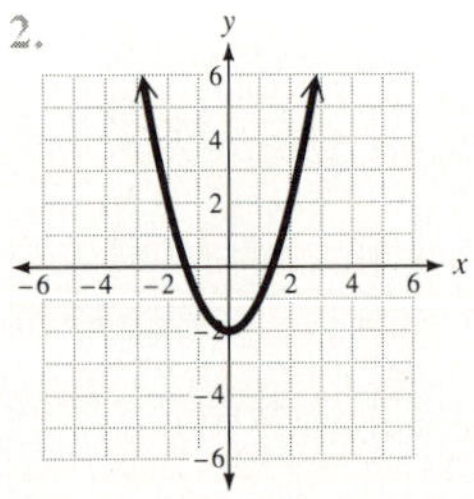

3.

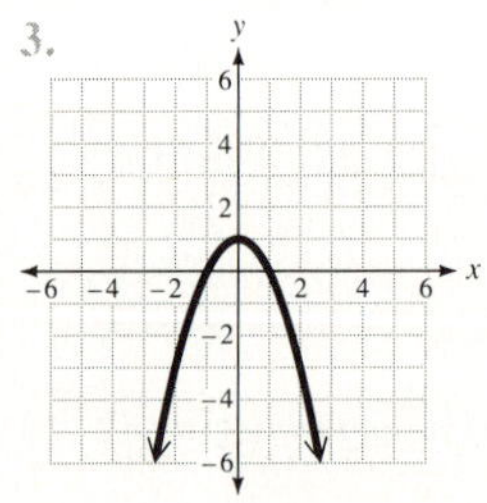

4.

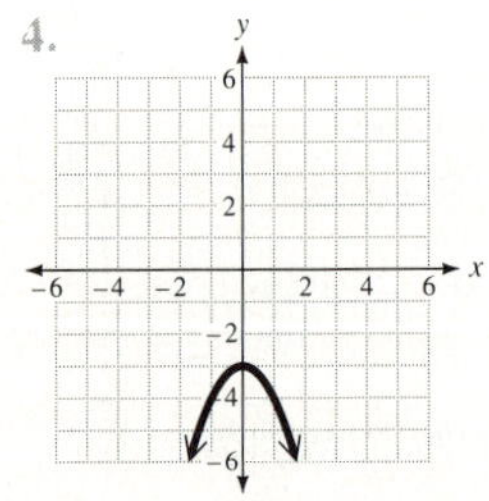

5. (1, 0) and (3, 0); (2, −1); up 6. (−1, 0) and (−3, 0); (−2, −1); up 7. no x-intercepts; (−1, 1); up 8. (0, 0) and (−8, 0); (−4, −16); up 9. (2, 0) and (3, 0); $\left(\frac{5}{2}, \frac{1}{4}\right)$; down 10. (1, 0) is the only x-intercept; (1, 0); up 11. no x-intercepts; (1, −1); down 12. $\left(\frac{-5+\sqrt{13}}{6}, 0\right)$ and $\left(\frac{-5-\sqrt{13}}{6}, 0\right)$; $\left(-\frac{5}{6}, -\frac{13}{12}\right)$; up

13.

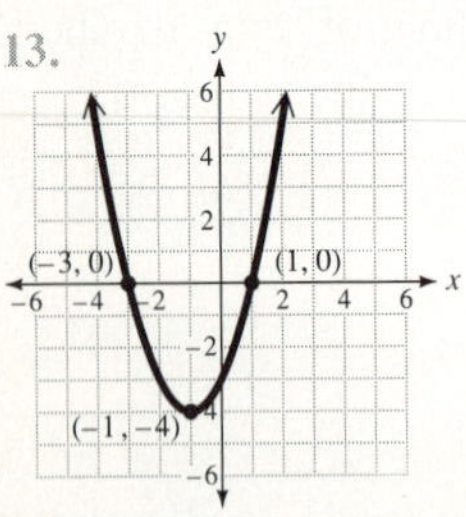

14.

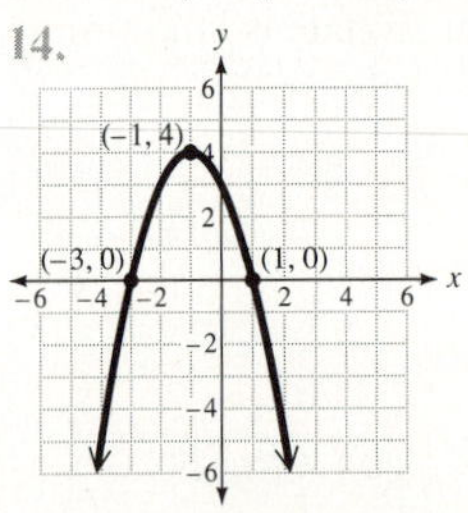

15.

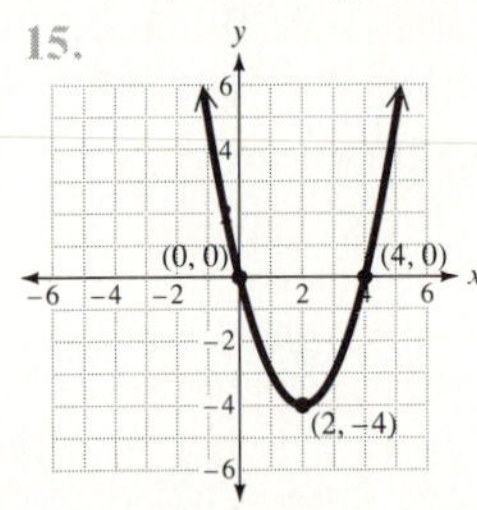

16.

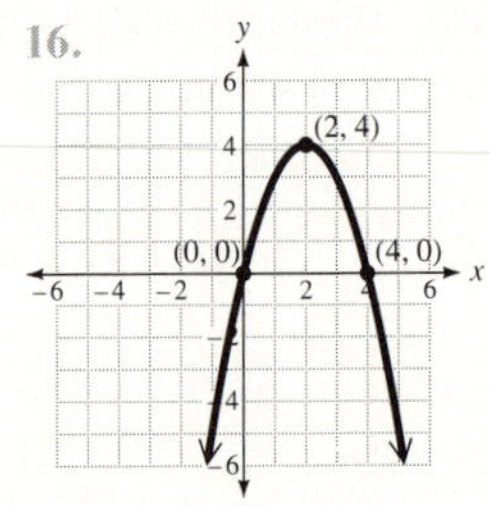

17.

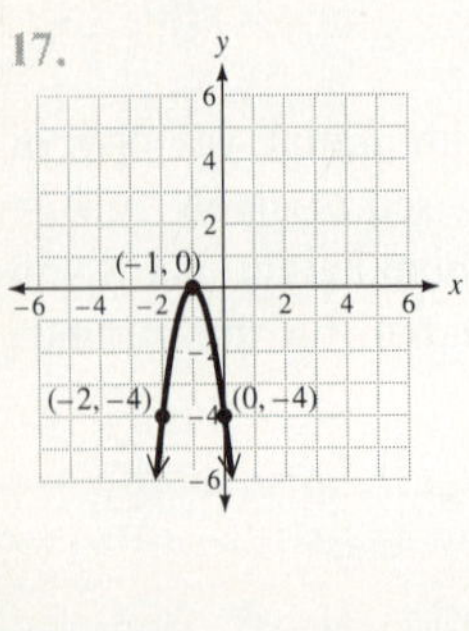

18.

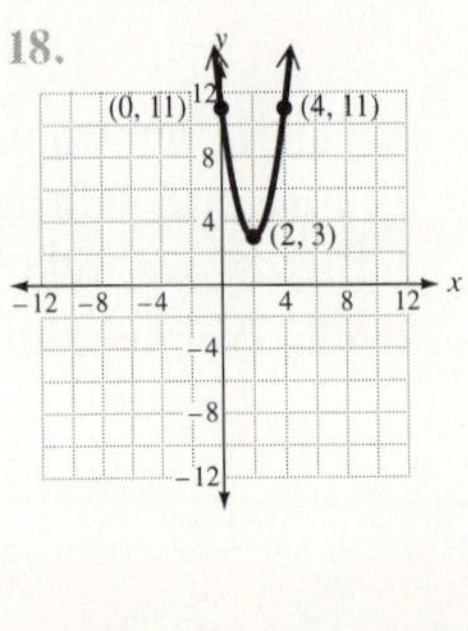

19. 10 m by 10 m; 100 m² 20. −16; 4 and −4 21. 40 sandwiches; $100 22. 16 ft; 2 sec after it is thrown upward 23. all real numbers 24. all real numbers 25. all real numbers except −9 26. (a) −1 (b) −6 (c) 9 (d) $5x + 5h - 1$ (e) $5h$ 27. (a) −1 (b) 0 (c) 3 (d) $x^2 + 2xh + h^2 - 1$ (e) $2xh + h^2$

9.1 EXERCISES B

In Exercises 1–4, sketch the graph of each quadratic function by sliding the graph of $f(x) = x^2$ or $g(x) = -x^2$ up or down.

1. $h(x) = x^2 + 3$

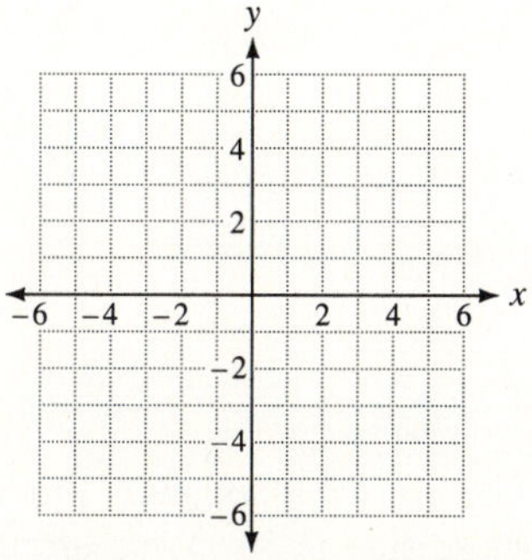

2. $h(x) = x^2 - 3$

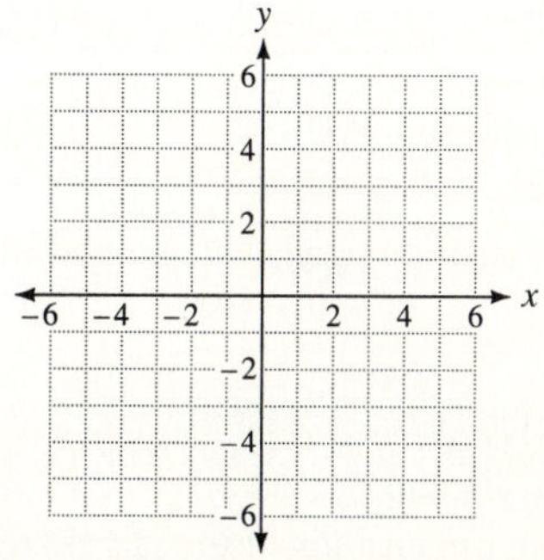

3. $h(x) = -x^2 + 2$

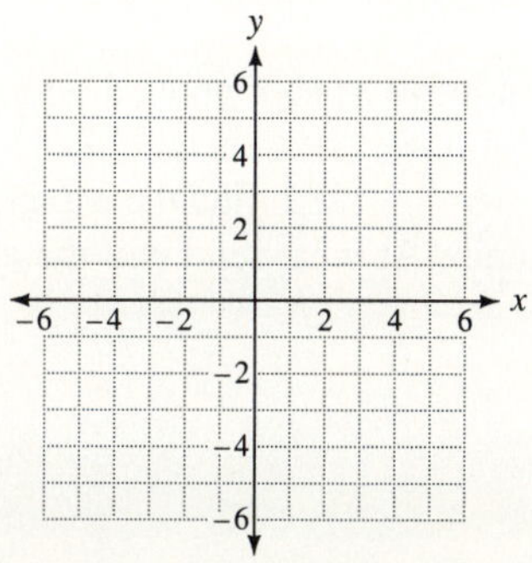

4. $h(x) = -x^2 - 1$

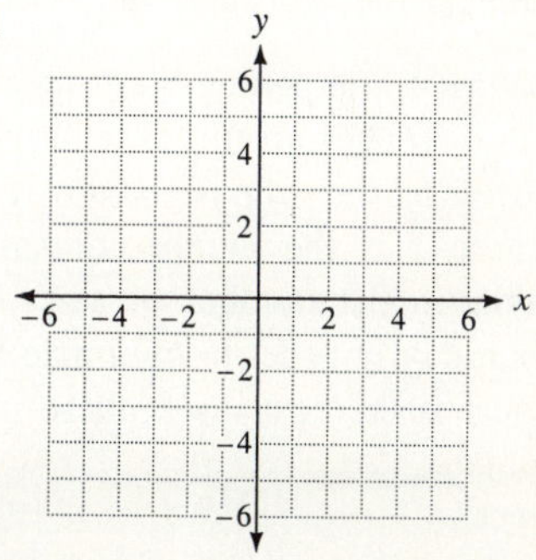

Without graphing, give the x-intercepts (if they exist) and the vertex, and tell whether the graph of each quadratic function opens up or down.

5. $f(x) = -x^2 + 4x - 3$

6. $g(x) = -x^2 - 4x - 3$

7. $h(x) = -x^2 - 2x - 2$

8. $f(x) = -x^2 - 8x$

9. $g(x) = x^2 - 5x + 6$

10. $h(x) = -3x^2 + 6x - 3$

11. $f(x) = 2x^2 - 4x + 3$

12. $g(x) = -3x^2 - 5x - 1$

Find the x-intercepts (if they exist), the vertex, and graph of the function.

13. $f(x) = x^2 - 2x - 3$

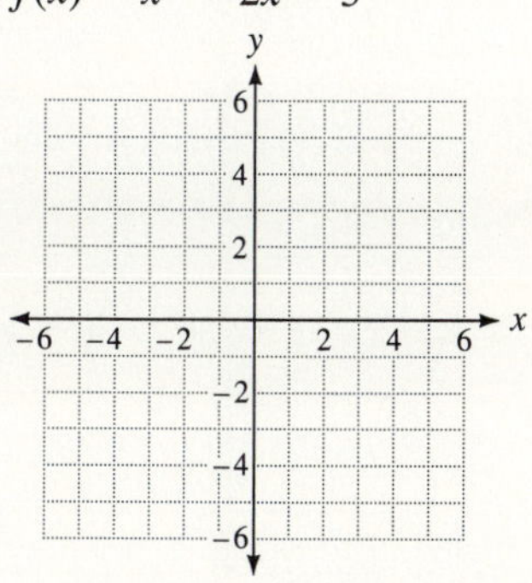

14. $g(x) = -x^2 + 2x + 3$

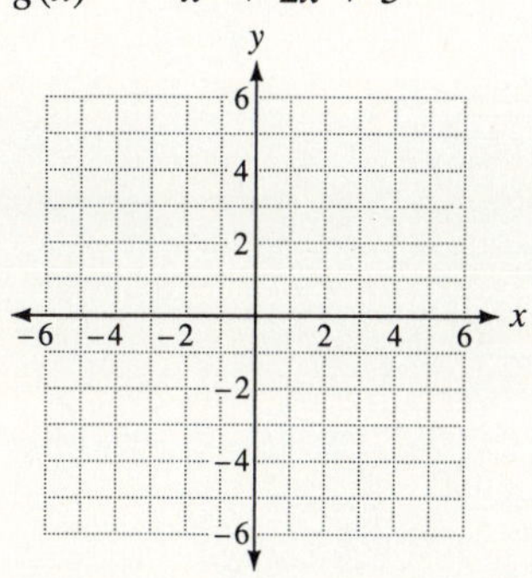

15. $f(x) = x^2 + 4x$

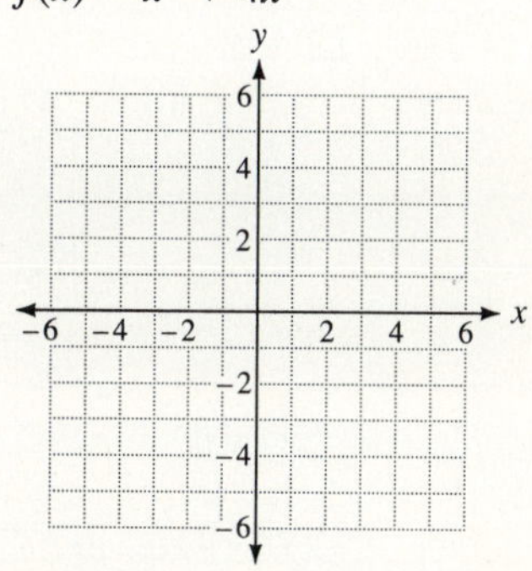

16. $g(x) = -x^2 - 4x$

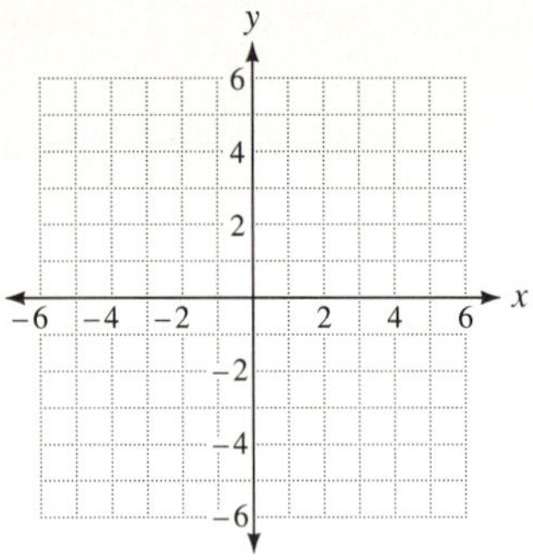

17. $f(x) = -4x^2 + 8x - 4$

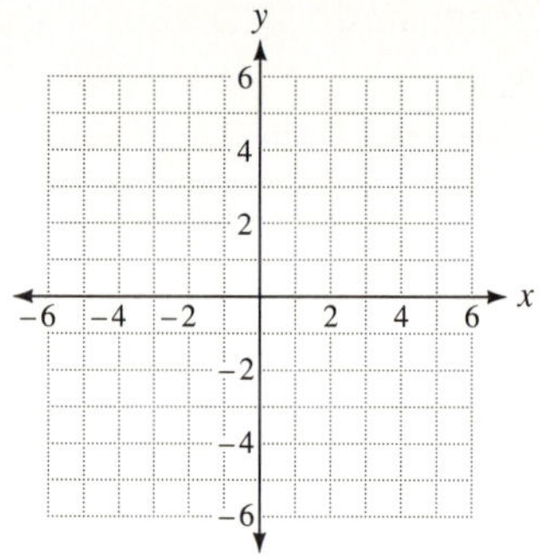

18. $g(x) = 4x^2 - 8x + 4$

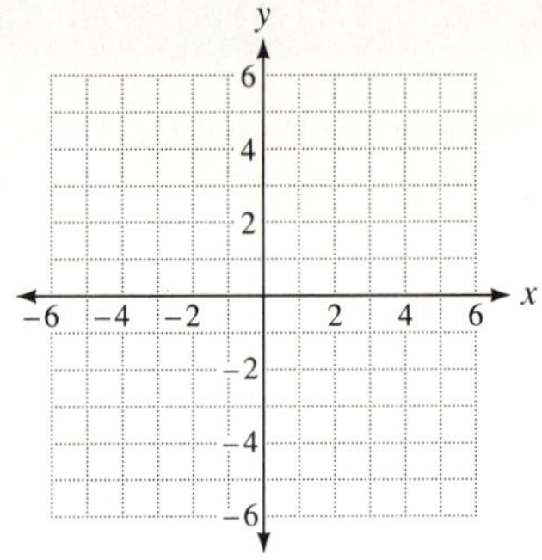

Solve.

19. A woman has 60 yd of fencing material. What are the dimensions of the largest rectangular pen which she can enclose? What is the maximum area?

20. What is the maximum product of two numbers whose sum is 50? What are the numbers?

21. Larsen's Sharpening Shop specializes in sharpening saws. Terry Larsen has determined that the cost of operating the shop is given by $c = x^2 - 40x + 430$, where x is the number of saws sharpened daily. What is the number of saws Terry must sharpen each day in order to minimize the operational cost? What is the cost under these ideal conditions?

22. An arrow is shot upward with an initial velocity of 128 ft/sec. If its height after x sec is given by $h = -16x^2 + 128x$, find the maximum height that it will attain. At what time will the arrow hit the ground?

FOR REVIEW

The following exercises review material from Chapter 3 to help you prepare for the next section. Specify the domain of each function defined by the equation given in Exercises 23–25.

23. $y = -3x + 12$

24. $y = -3x^2 + x$

25. $y = \dfrac{2}{x - 5}$

For each function in Exercises 26–27, evaluate: **(a)** $g(0)$ **(b)** $g(-1)$ **(c)** $g(2)$ **(d)** $g(x + h)$ **(e)** $g(x + h) - g(x)$.

26. $g(x) = 2x + 3$

27. $g(x) = -x^2 + 2$

9.1 EXERCISES C

Find the vertex, the x-intercepts, and graph of the function.

1. $f(x) = 2x^2 - 8x + 5$

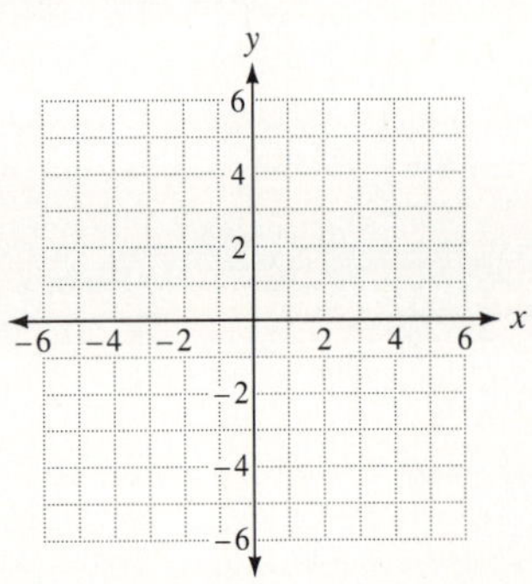

2. $f(x) = -2x^2 + 5x + 7$

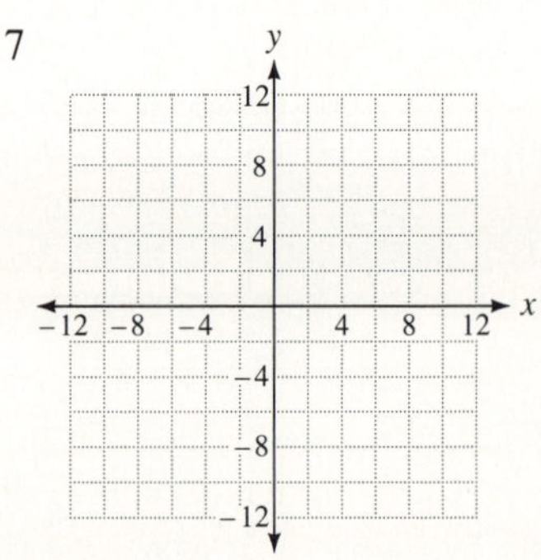

3. Suppose that (1, 0), (−2, −3), and (4, 39) are three ordered pairs in the quadratic function with equation $y = f(x) = ax^2 + bx + c$. Find the values of a, b, and c by solving a system of three linear equations. Use these values to find the value of y when x is 2. [Answer: $a = 2$, $b = 3$, $c = -5$; $y = 9$ when $x = 2$]

4. When a department store opens, the number of people in the store increases from zero to a maximum number then down to zero again at closing time. Suppose the number N of people in a store is given as a function of time t by $N = -15t^2 + 180t$, where $t = 0$ corresponds to the time the store opens at 10:00 A.M. At what time will the store have the maximum number of customers? What is the maximum number of customers? At what times does the store close?

5. Chuck Little plans to build a rectangular pen adjacent to a stream. What is the maximum area he can enclose with 100 ft of fencing if no fence is required next to the stream?

9.2 PROPERTIES OF FUNCTIONS AND SPECIAL FUNCTIONS

STUDENT GUIDEPOSTS

1. The Domain of a Function
2. Algebra of Functions
3. Composition of Functions
4. Polynomial Functions
5. Square Root Functions
6. Absolute Value Functions

1 THE DOMAIN OF A FUNCTION

Remember from Section 3.5 that the domain of a function given by an equation $y = f(x)$ is the largest set of real numbers for which the expression in x is defined. For example,

$$f(x) = 2x + 1 \quad \text{and} \quad g(x) = \frac{2}{x-3}$$

are functions with domains the set of all real numbers and the set of all real numbers except 3, respectively. Since division by zero is undefined, 3 must be excluded from the domain of g. When a radical is used in the definition of a function, such as

$$h(x) = \sqrt{x+2},$$

the domain of the function includes only those real numbers for which the radicand is nonnegative. Thus, $x + 2$ must be nonnegative so that $x + 2 \geq 0$ or $x \geq -2$. The domain of h is then all real numbers $x \geq -2$.

EXAMPLE 1 Finding Domains

Give the domain of each function.

(a) $f(x) = x^2 + 2x - 3$
Since $x^2 + 2x - 3$ is defined for every real number x, the domain of f is the set of all real numbers.

(b) $g(x) = \sqrt{x - 5}$
Since $x - 5$ must be nonnegative,

$$\begin{aligned} x - 5 &\geq 0 \\ x &\geq 5. \end{aligned}$$

The domain of g is all real numbers $x \geq 5$.

(c) $h(x) = \dfrac{1}{\sqrt{2 - x}}$
Since $2 - x$ must be nonnegative,

$$\begin{aligned} 2 - x &\geq 0. \\ -x &\geq -2 \\ x &\leq 2 \quad \text{Reverse the inequality} \end{aligned}$$

However in this case, x cannot be 2 since the denominator $\sqrt{2 - x}$ would be zero. Thus, the domain of h is all real numbers $x < 2$.

Practice Exercise 1

Give the domain of each function.

(a) $f(x) = -x^2 + 8$

(b) $g(x) = \sqrt{x + 7}$

(c) $h(x) = \dfrac{5}{\sqrt{x - 8}}$

Answers: (a) all real numbers
(b) all real numbers $x \geq -7$
(c) all real numbers $x > 8$

2 ALGEBRA OF FUNCTIONS

When two functions are given, new functions can be formed by algebraically combining them as defined below.

Algebra of Functions

Let f and g be two functions.

$(f + g)(x) = f(x) + g(x)$	Sum of f and g
$(f - g)(x) = f(x) - g(x)$	Difference of f and g
$(fg)(x) = f(x) \cdot g(x)$	Product of f and g
$\left(\dfrac{f}{g}\right)(x) = \dfrac{f(x)}{g(x)},\ g(x) \neq 0$	Quotient of f and g

EXAMPLE 2 Finding Algebraic Combinations of Functions

Let $f(x) = 2x + 1$ and $g(x) = x - 5$.

(a) Find $(f + g)(x) = f(x) + g(x) = (2x + 1) + (x - 5)$ $f(x) = 2x + 1$ and $g(x) = x - 5$

$$= 3x - 4$$

(b) Find $(f + g)(-3)$. Since $(f + g)(x) = 3x - 4$ from (a), substitute -3 for x.

$$\begin{aligned} (f + g)(-3) &= 3(-3) - 4 \\ &= -9 - 4 = -13 \end{aligned}$$

Practice Exercise 2

Let $f(x) = x + 5$ and $g(x) = x^2 - 1$.

(a) Find $(f + g)(x)$.

(b) Find $(f + g)(-2)$.

(c) Find $(f-g)(-2)$. Since $(f-g)(x) = f(x) - g(x)$ $= (2x+1) - (x-5) = x + 6$,

$$(f-g)(-2) = (-2) + 6 = 4.$$

(d) Find $(fg)(0)$. Since $(fg)(x) = f(x) \cdot g(x)$ $= (2x+1)(x-5) = 2x^2 - 9x - 5$,

$$(fg)(0) = 2(0)^2 - 9(0) - 5$$
$$= 0 - 0 - 5 = -5.$$

(e) Find $\left(\frac{f}{g}\right)(5)$. Since $\left(\frac{f}{g}\right)(x) = \frac{f(x)}{g(x)} = \frac{2x+1}{x-5}$, we must substitute 5 for x to find $\left(\frac{f}{g}\right)(5)$.

$$\left(\frac{f}{g}\right)(5) = \frac{2(5)+1}{(5)-5} = \frac{11}{0}, \text{ which is not defined.}$$

Notice that the quotient function is not defined at any value of x that makes the denominator 0.

(c) Find $(f-g)(-3)$.

(d) Find $(fg)(1)$.

(e) Find $\left(\frac{f}{g}\right)(-5)$.

Answers: (a) $(f+g)(x) = x^2 + x + 4$ (b) $(f+g)(-2) = 6$ (c) $(f-g)(-3) = -6$ (d) $(fg)(1) = 0$ (e) $\left(\frac{f}{g}\right)(-5) = 0$

3 COMPOSITION OF FUNCTIONS

Suppose that $f(x) = 3x + 2$ and $g(x) = 1 - x$. Then

$$g(3) = 1 - (3) = -2$$
$$\text{and} \quad f(-2) = 3(-2) + 2 = -6 + 2 = -4.$$

If these results are combined, we have actually found

$$f[g(3)] = f[-2] \quad \text{Since } g(3) = -2$$
$$= -4. \quad \text{Since } f(-2) = -4$$

Combining two functions in this way is defined below.

Composition of Functions

Let f and g be two functions. The expression $f[g(x)]$ is the **composition of** f **with** g, and $g[f(x)]$ is the **composition of** g **with** f.

EXAMPLE 3 Composition of Functions

Let $f(x) = 5x + 2$ and $g(x) = x^2 - 1$.

(a) Find $f[g(2)]$.
We first find $g(2)$.

$$g(2) = (2)^2 - 1 = 4 - 1 = 3$$

Now we find $f[g(2)]$.

$$f[g(2)] = f[3] \quad g(2) = 3$$
$$= 5(3) + 2$$
$$= 15 + 2 = 17$$

(b) Find $g[f(0)]$.
First find $f(0)$.

$$f(0) = 5(0) + 2 = 2$$

Practice Exercise 3

Let $f(x) = 2x^2 + 3$ and $g(x) = x - 4$.

(a) Find $f[g(4)]$.

(b) Find $g[f(0)]$.

Now we find $g[f(0)]$.

$$\begin{aligned} g[f(0)] &= g[2] && f(0) = 2 \\ &= (2)^2 - 1 \\ &= 4 - 1 = 3 \end{aligned}$$

Answers: (a) 3 (b) −1

4 POLYNOMIAL FUNCTIONS

Linear and quadratic functions studied in Sections 3.5 and 9.1 are actually special cases of *polynomial functions*. A function of the form

$$f(x) = a_nx^n + a_{n-1}x^{n-1} + \cdots + a_1x + a_0$$

where $a_n, a_{n-1}, \ldots, a_1, a_0$ are real numbers, with $a_n \neq 0$, is a **polynomial function** of degree n. Notice that if $n = 1$,

$$f(x) = a_1x + a_0$$

is a linear function, and if $n = 2$

$$f(x) = a_2x^2 + a_1x + a_0$$

is a quadratic function. The domain of every polynomial function is the entire set of real numbers.

Although graphing polynomial functions is a topic usually considered in more advanced courses, we can obtain a rough graph by plotting points.

EXAMPLE 4 GRAPHING A POLYNOMIAL FUNCTION

Graph the function $f(x) = x^3 - 2x^2 - 5x + 6$.

Construct a table of values, plot the corresponding points, and join them with a smooth curve. The graph, which is typical of many third-degree polynomials, is given in Figure 9.12.

x	$f(x)$
−2	0
−1	8
0	6
1	0
2	−4
3	0
4	18

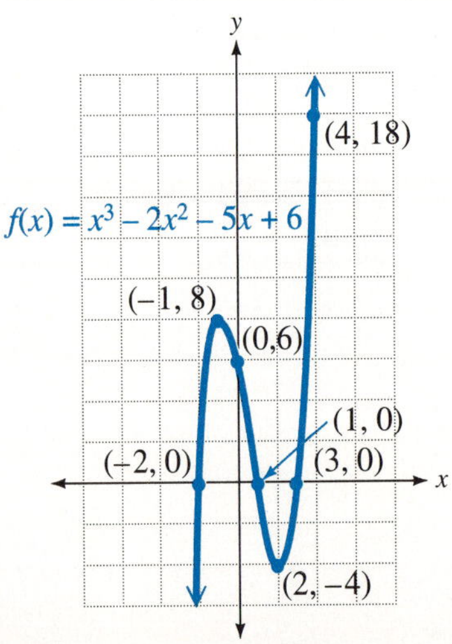

Figure 9.12 Third-Degree Polynomial Function

PRACTICE EXERCISE 4

Graph $f(x) = -x^4 + 5x^2 - 4$.
Complete the table below.

x	$f(x)$
−3	
−2	
−1	
0	
1	
2	
3	

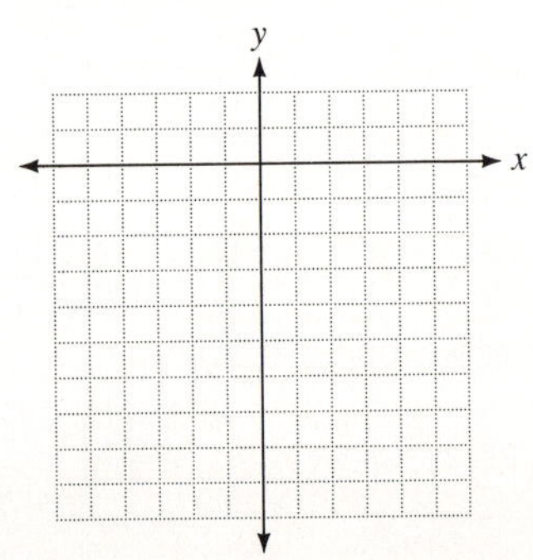

Answer: The graph passes through the points (−3, −40), (−2, 0), (−1, 0), (0, −4), (1, 0), (2, 0), and (3, −40).

5 SQUARE ROOT FUNCTIONS

Another special function is the square root function. Remember that the radical denotes the principal (nonnegative) square root of a number. For example, $\sqrt{4}$ represents 2, not -2 nor ± 2. The function

$$f(x) = \sqrt{x}, \quad \text{for } x \geq 0$$

is called the **square root function.** Notice that the domain of the square root function is the set of all nonnegative real numbers. The graph of this function, given in Figure 9.13, is actually half of a parabola opening to the right with vertex (0, 0). We will consider relations like this in more detail in Chapter 11.

x	$f(x)$
0	0
1	1
4	2
9	3

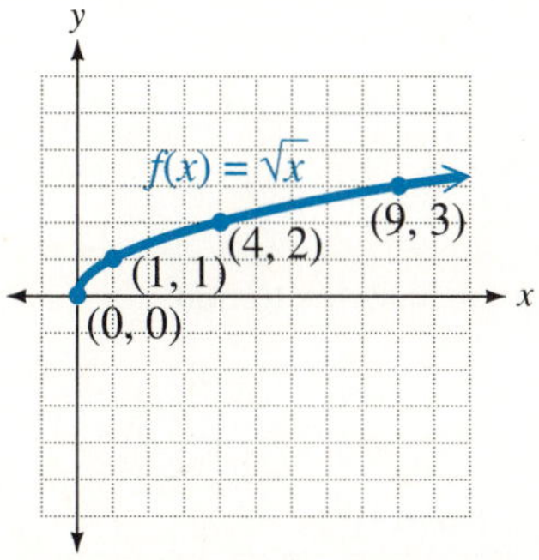

Figure 9.13 Square Root Function

EXAMPLE 5 Graphing a Square Root Function

Graph the function $f(x) = \sqrt{x + 3}$.

Since $x + 3$ must be nonnegative, the domain of this function is all real numbers $x \geq -3$. Its graph, given in Figure 9.14, has the same shape as the square root function, but shifted to the left 3 units.

x	$f(x)$
−3	0
−2	1
1	2
6	3

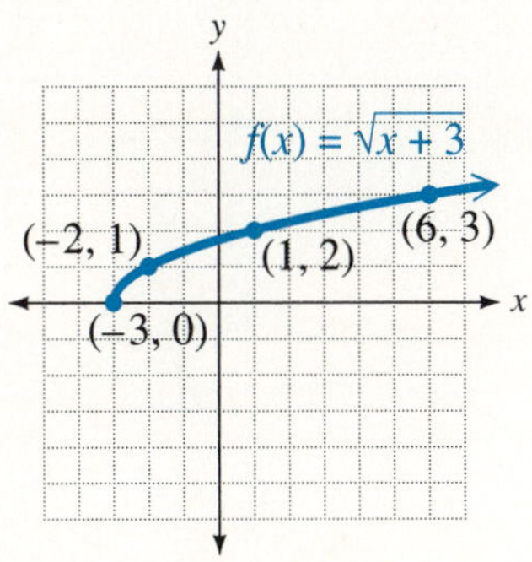

Figure 9.14

Practice Exercise 5

Graph $f(x) = \sqrt{x - 1}$.

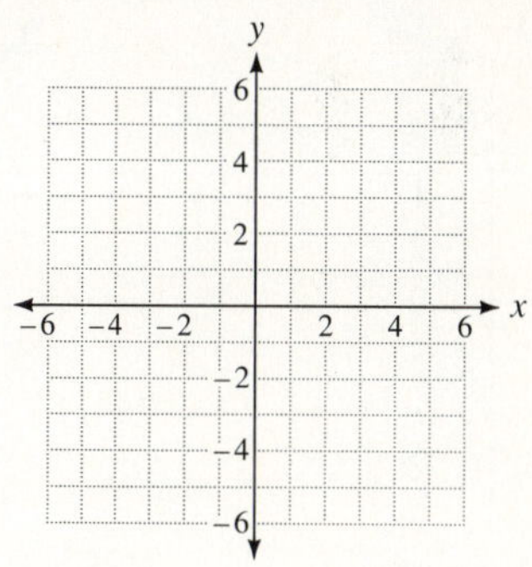

Answer: The graph has the same shape as the square root function, but shifted right 1 unit.

6 ABSOLUTE VALUE FUNCTIONS

Recall the definition of absolute value, given in Chapter 1.

$$|x| = \begin{cases} x \text{ if } x \geq 0 \\ -x \text{ if } x < 0 \end{cases}$$

We use this to define the **absolute value function**

$$f(x) = |x|.$$

In Section 3.1 we graphed the equation $y = |x|$ which gives us the graph of the absolute value function, a V-shaped curve shown in Figure 9.15.

x	$f(x)$
0	0
1	1
-1	1
2	2
-2	2
3	3
-3	3

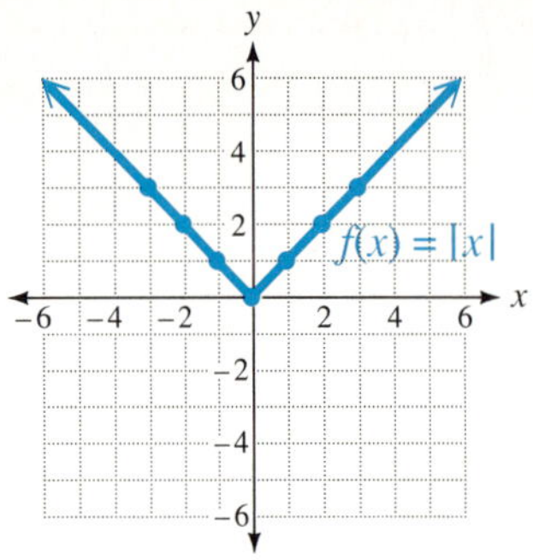

Figure 9.15 Absolute Value Function

Variations of the absolute value function have graphs similar in shape to that of $f(x) = |x|$, as shown in the next example.

EXAMPLE 6 Graphing an Absolute Value Function

Graph the function $f(x) = |x + 1|$.

Construct a table of values like the one below. The graph of $f(x) = |x + 1|$ is given in Figure 9.16. Notice that the graph could be obtained by shifting the graph of the absolute value function to the left 1 unit.

x	$f(x)$
0	1
1	2
-1	0
2	3
-2	1
-3	2
-4	3

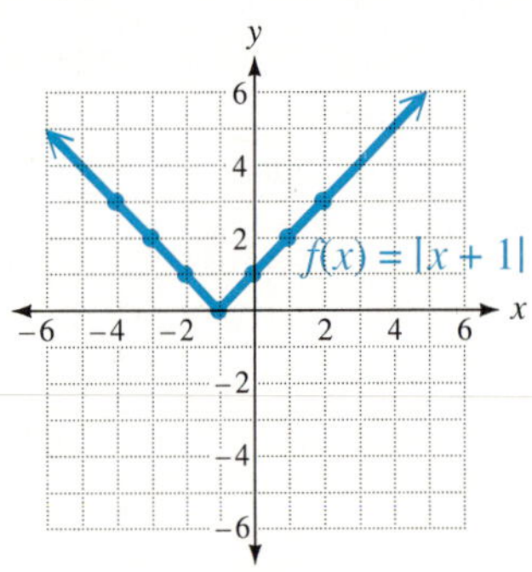

Figure 9.16

Practice Exercise 6

Graph $f(x) = |x - 3|$.

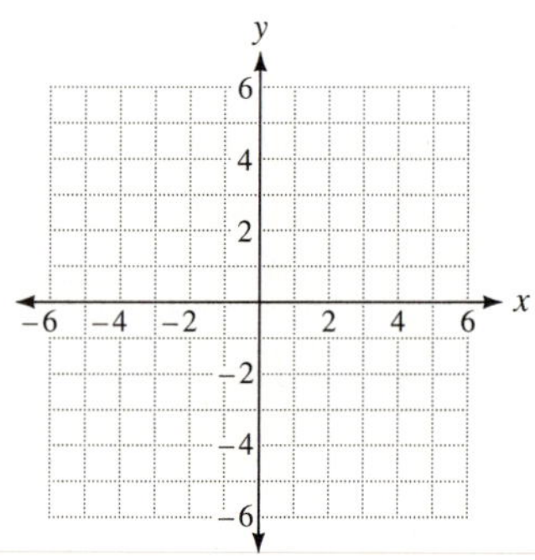

Answer: The graph is similar to $f(x) = |x|$ but shifted right 3 units.

9.2 EXERCISES A

Find the domain of the function defined by each equation.

1. $y = \sqrt{4 - x}$

2. $y = \dfrac{5}{x + 5}$

3. $y = \dfrac{x}{x - 1}$

4. $y = \dfrac{2}{\sqrt{x + 2}}$

5. $y = \sqrt{x^2 + 10}$

6. $f(x) = \sqrt{x^2 - 1}$

7. $f(x) = \dfrac{1}{\sqrt{x^2 - 1}}$

8 $f(x) = \sqrt{x^2 - 8x + 12}$

Assume that $f(x) = 4 + 3x$ *and* $g(x) = x^2 - 3$. *Find each function value.*

9. $(f + g)(-1)$ **10.** $(f + g)(-4)$ **11.** $(f - g)(2)$ **12.** $(f - g)(3)$

13. $(fg)(-2)$ **14.** $(fg)(-1)$ **15.** $\left(\frac{f}{g}\right)(-3)$ **16.** $\left(\frac{f}{g}\right)(-4)$

17 $f[g(-1)]$ **18.** $f[g(-4)]$ **19.** $g[f(2)]$ **20.** $g[f(-2)]$

Make a table of values and sketch the graph of each function.

21. $f(x) = x^3 - 4x$

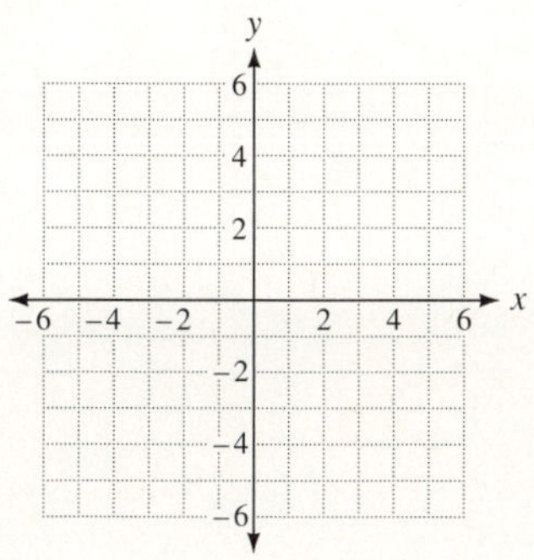

22. $f(x) = -x^3 + x^2 + 9x - 9$

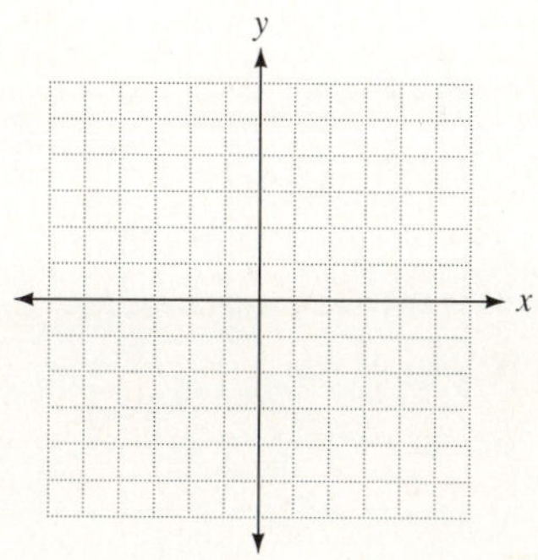

23. $f(x) = x^4 - 20x^2 + 64$

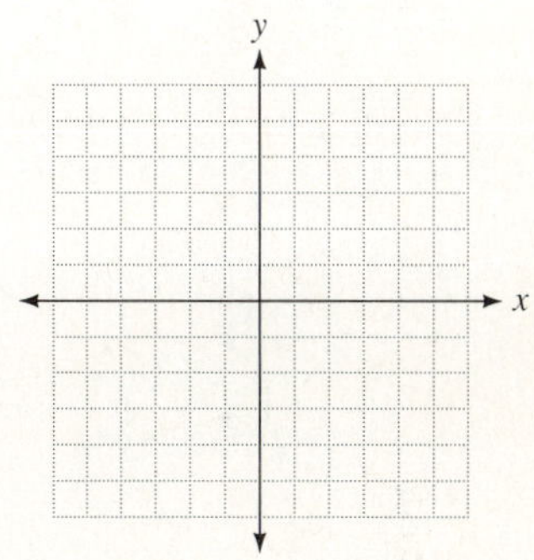

24. $f(x) = \sqrt{x + 2}$

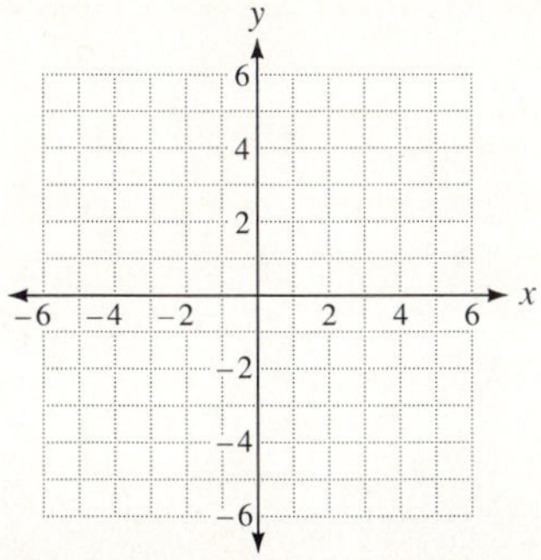

25. $f(x) = \sqrt{x - 3}$

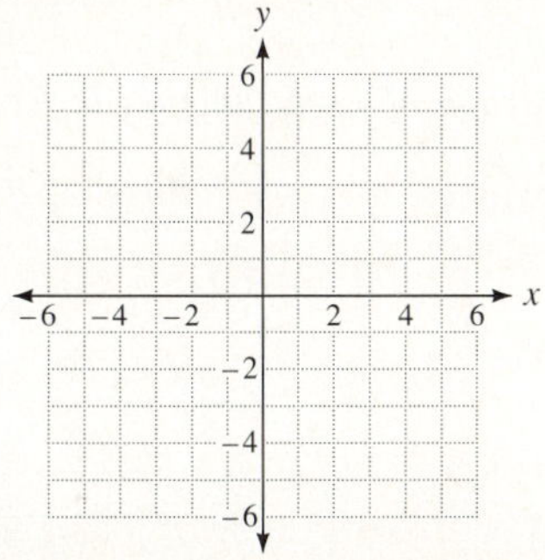

26. $f(x) = -\sqrt{x}$

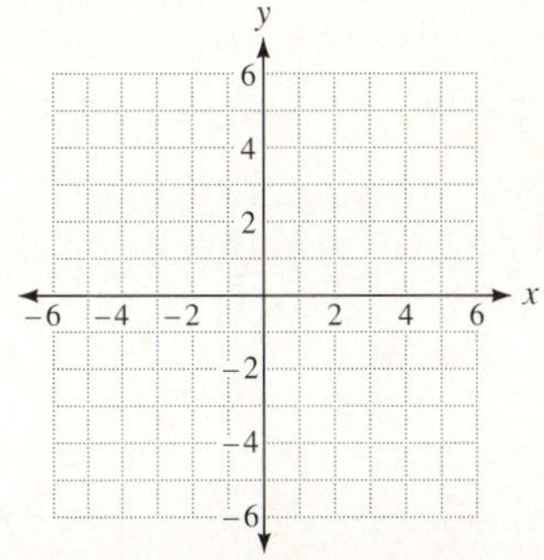

27. $f(x) = |x - 1|$

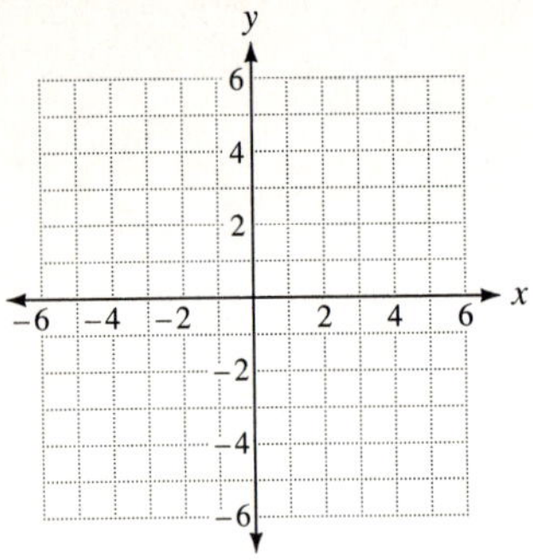

28. $f(x) = |x + 2|$

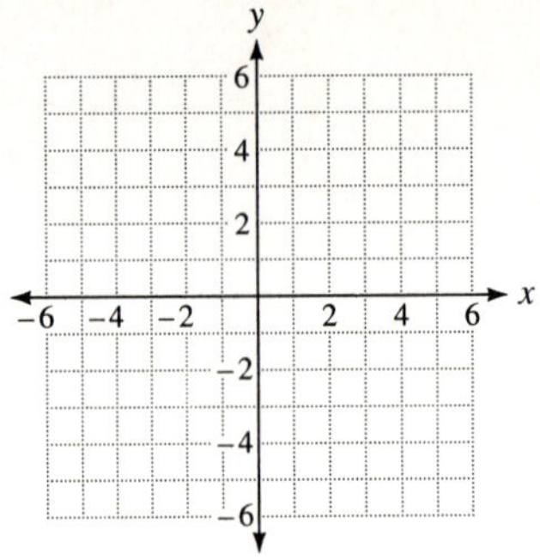

29. $f(x) = |x| + 1$

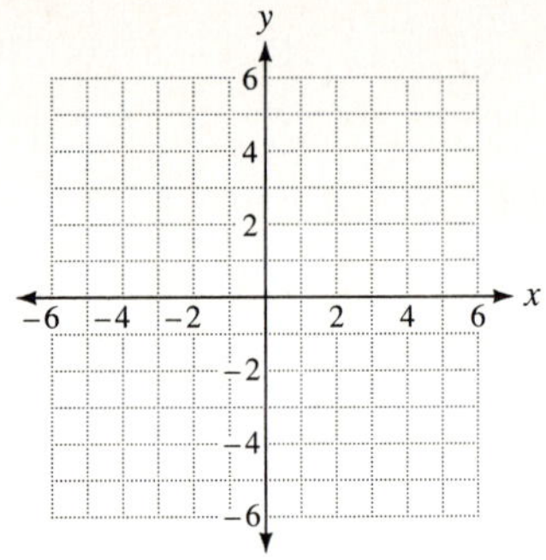

30. $f(x) = |x| - 1$

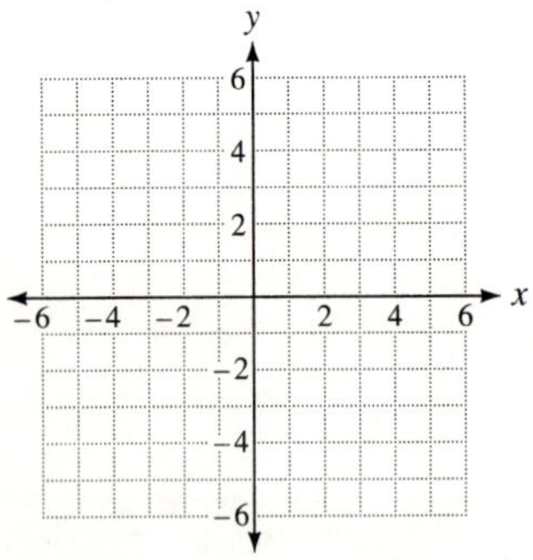

31. $f(x) = -|x|$

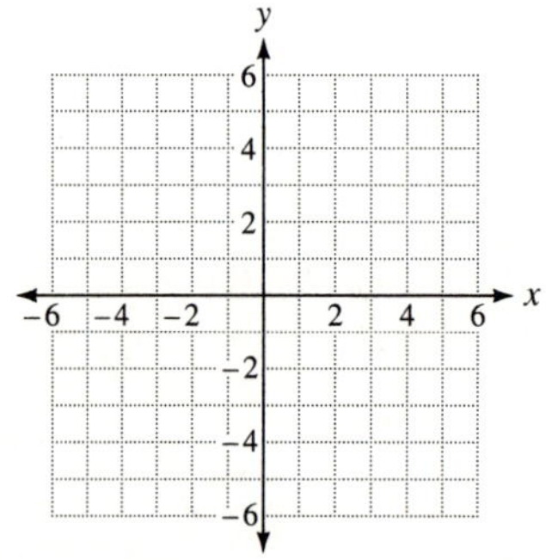

32. $f(x) = -|x + 2|$

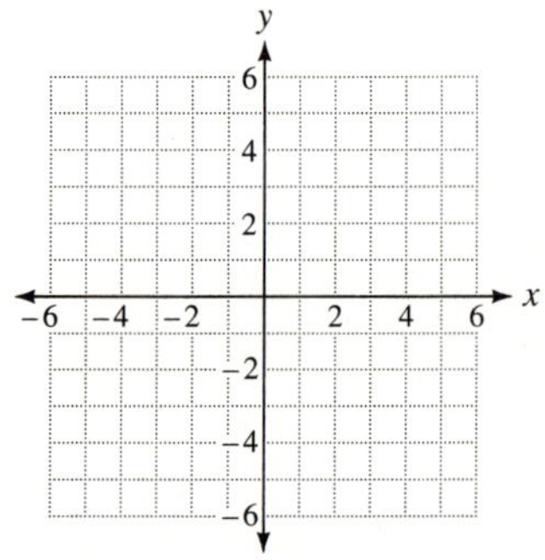

FOR REVIEW

Without graphing, tell the x-intercepts, the vertex, the line of symmetry, and whether the graph of each quadratic function opens up or down in Exercises 33–34.

33. $f(x) = x^2 - 2x - 15$

34. $g(x) = -x^2 + 2x + 15$

Find the vertex, the x-intercepts, and the graph of each function in Exercises 35–36.

35. $f(x) = x^2 - 2x - 3$

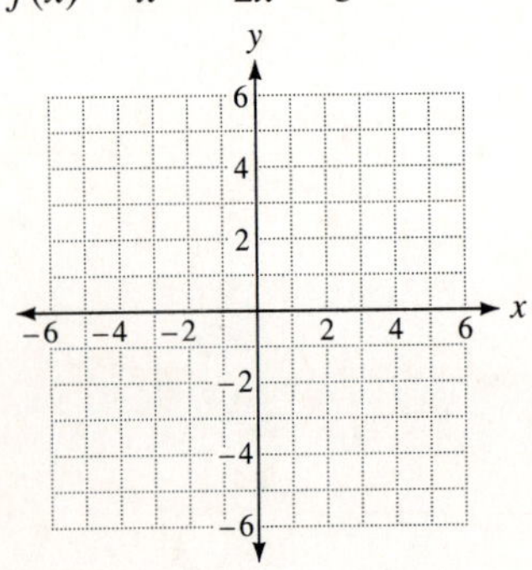

36. $g(x) = -x^2 - 2x + 3$

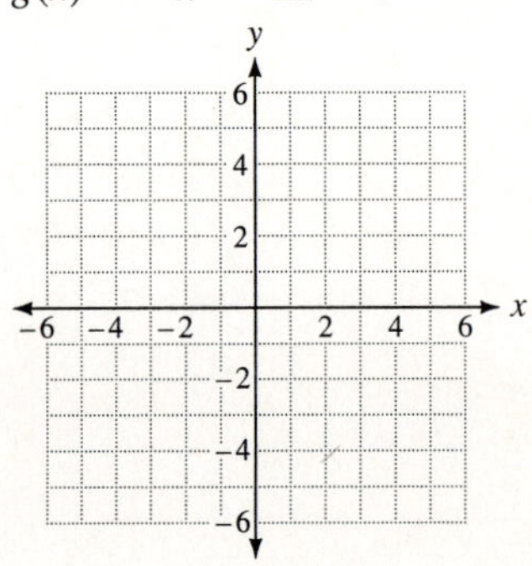

37. The length and width of a rectangle have a sum of 100 feet. What width will produce a maximum area?

The following exercise reviews material from Chapter 3 to help you prepare for the next section.

38. Consider the relation $r = \{(1, 3), (2, 1), (3, 1), (4, 2)\}$.

(a) Graph the relation r.

(b) Is the relation r a function? Explain.

(c) Give the relation s formed by interchanging the x and y coordinates of each ordered pair in r.

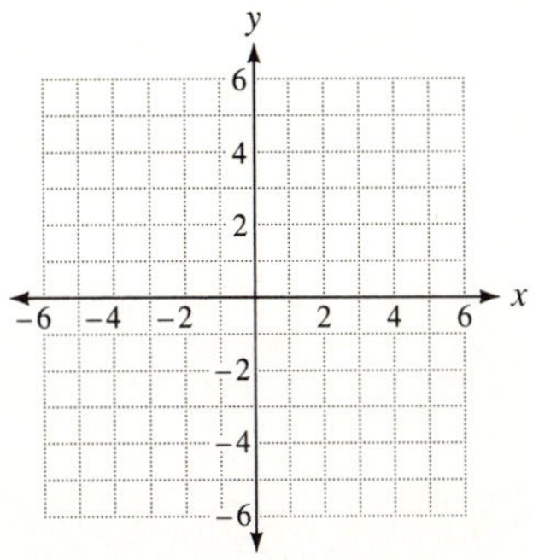

(d) Graph the new relation s.

(e) Is the new relation s a function? Explain.

(f) Could you eliminate one ordered pair from r so that the relation s formed by interchanging the coordinates is a function?

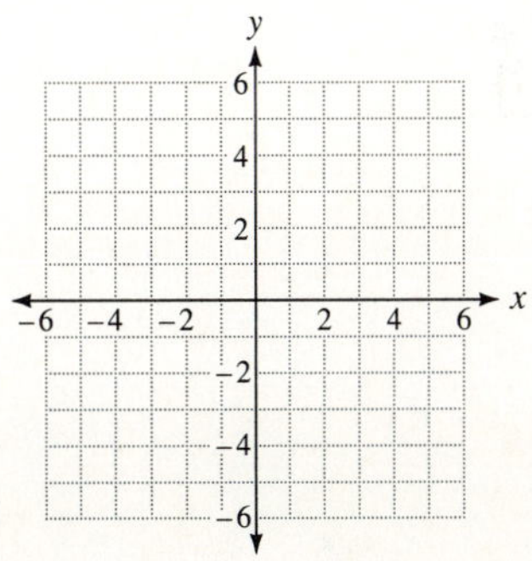

ANSWERS: **1.** all real numbers $x \le 4$ **2.** all real numbers except -5 **3.** all real numbers except 1 **4.** all real numbers $x > -2$ **5.** all real numbers **6.** all real numbers $x \le -1$ or $x \ge 1$ **7.** all real numbers $x < -1$ or $x > 1$ **8.** all real numbers $x \le 2$ or $x \ge 6$ **9.** -1 **10.** 5 **11.** 9 **12.** 7 **13.** -2 **14.** -2 **15.** $-\frac{5}{6}$ **16.** $-\frac{8}{13}$ **17.** -2 **18.** 43 **19.** 97 **20.** 1

21.

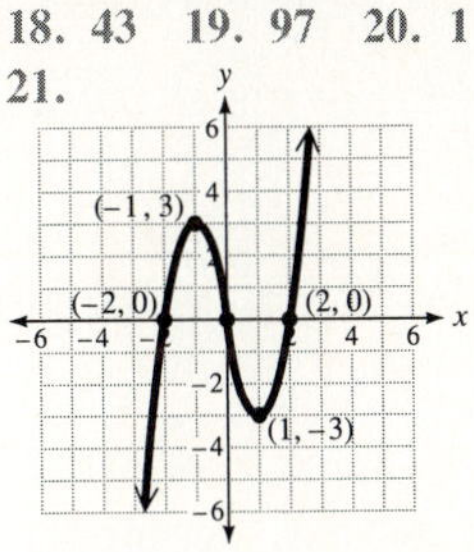

22.

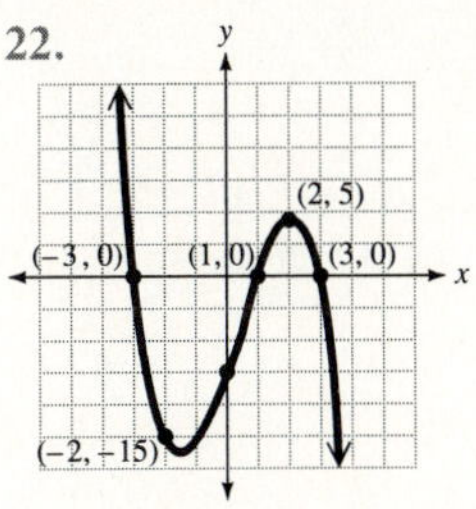

23.

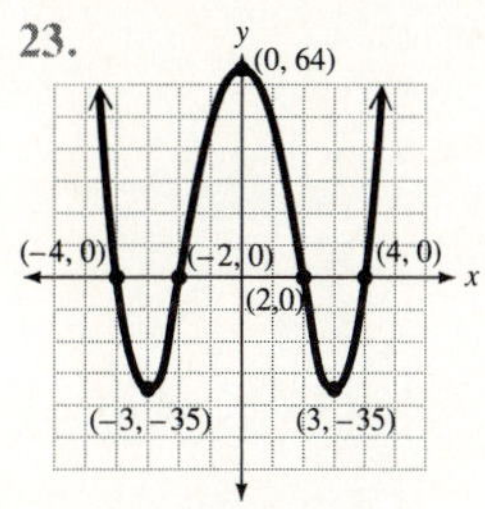

24.

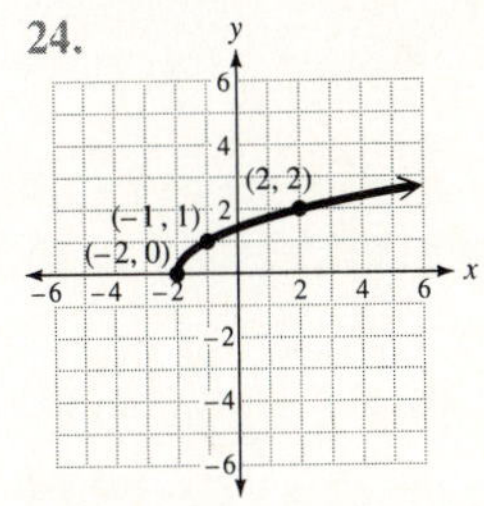

25.

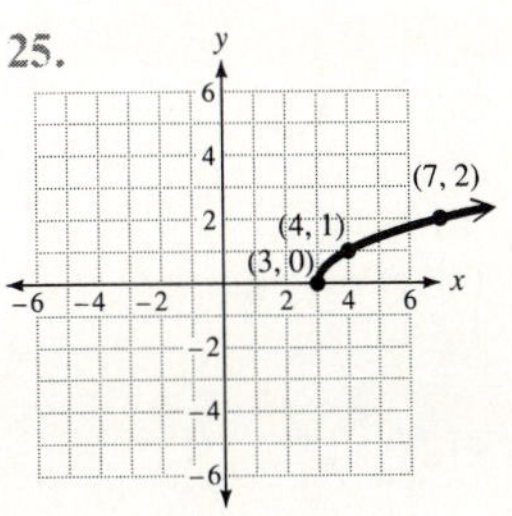

26.

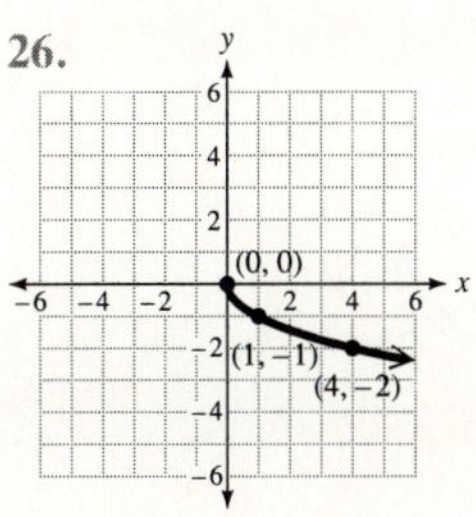

27.

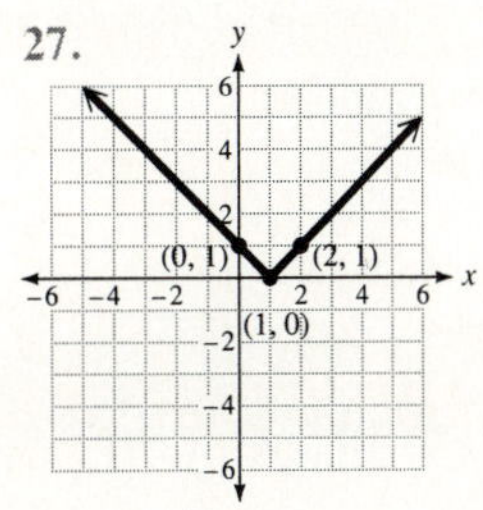

28.

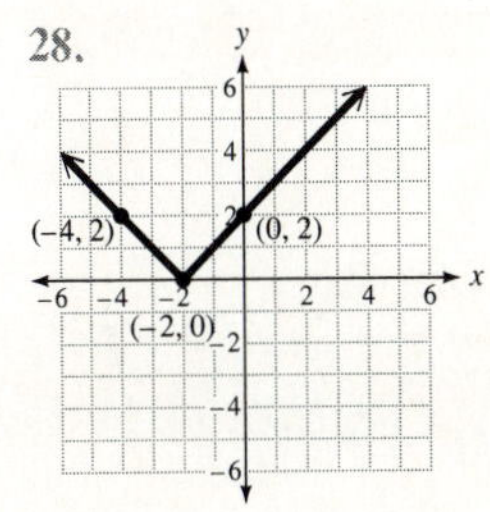

29. 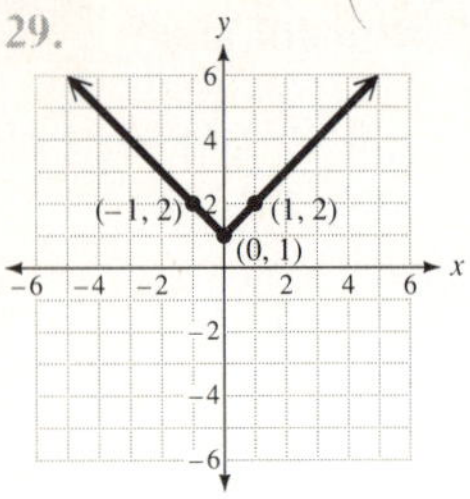

30. 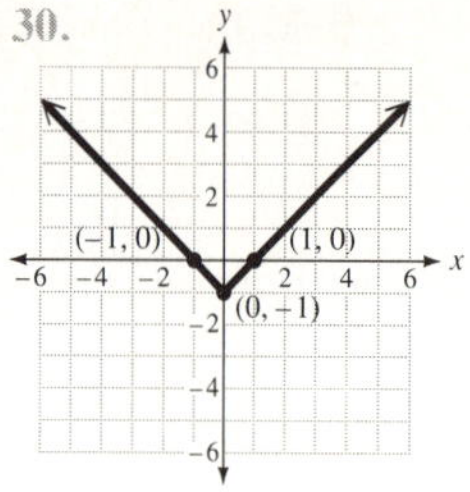

31. 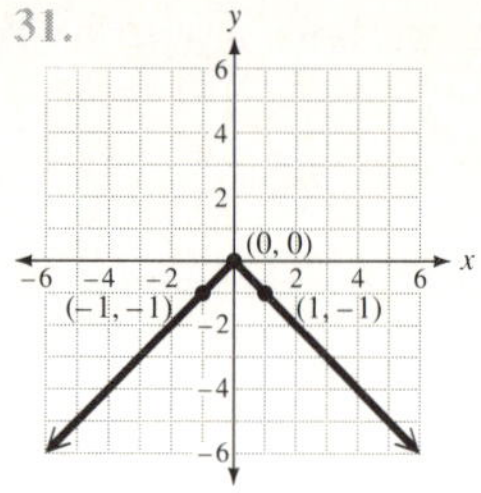

32. 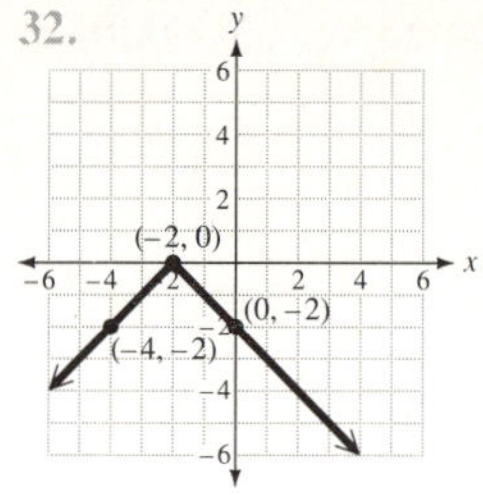

33. (5, 0) and (−3, 0); (1, −16); $x = 1$; up 34. (5, 0) and (−3, 0); (1, 16); $x = 1$; down

35. 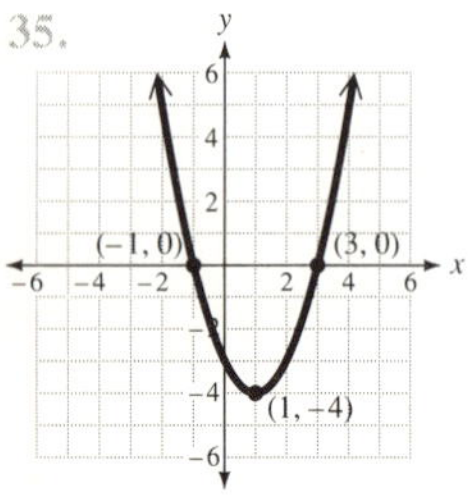

36. 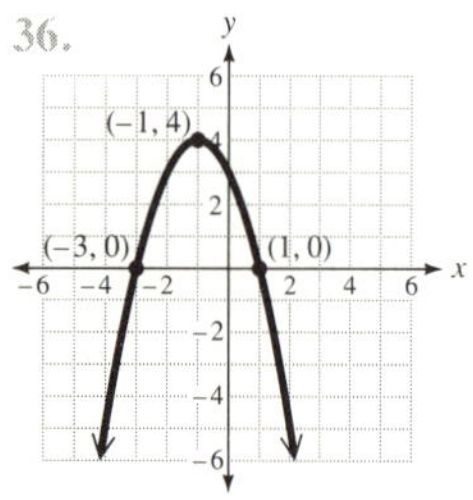

37. 50 ft 38. (a) 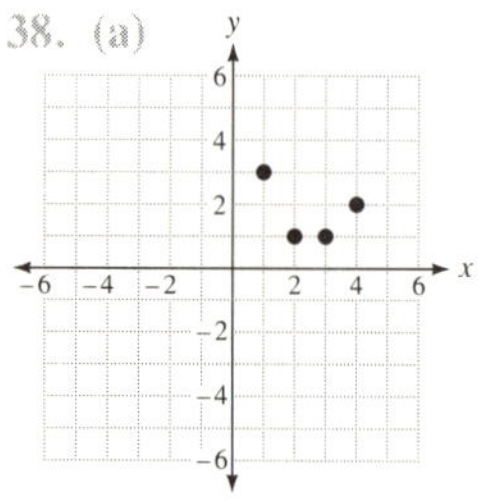

(b) yes (c) $s = \{(3, 1), (1, 2), (1, 3), (2, 4)\}$

(d) 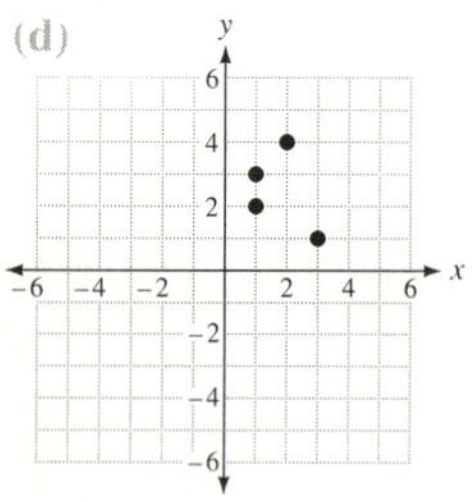

(e) no (f) Yes; if either (2, 1) or (3, 1) were omitted from r then s would be a function.

9.2 EXERCISES B

Find the domain of the function defined by each equation.

1. $y = \sqrt{x - 2}$
2. $y = \dfrac{x}{x + 5}$
3. $y = \dfrac{2}{x - 3}$
4. $y = \dfrac{7}{\sqrt{x - 3}}$
5. $y = \dfrac{1}{\sqrt{x^2 + 8}}$
6. $g(x) = \sqrt{x^2 - 4}$
7. $g(x) = \dfrac{1}{\sqrt{x^2 - 4}}$
8. $g(x) = \sqrt{x^2 - 5x - 14}$

Assume that $f(x) = x^2 + 2$ and $g(x) = 1 - 3x$. Find each function value.

9. $(f + g)(-1)$
10. $(f + g)(-4)$
11. $(f - g)(2)$
12. $(f - g)(3)$
13. $(fg)(-2)$
14. $(fg)(-1)$
15. $\left(\dfrac{f}{g}\right)(-3)$
16. $\left(\dfrac{f}{g}\right)(-4)$
17. $f[g(-1)]$
18. $f[g(-4)]$
19. $g[f(2)]$
20. $g[f(-2)]$

Make a table of values and sketch the graph of each function.

21. $f(x) = x^3 - 9x$

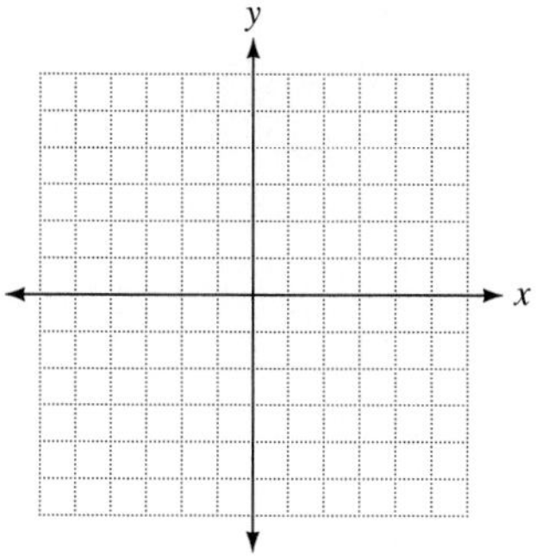

22. $f(x) = -x^3 + 4x^2 + 4x - 16$

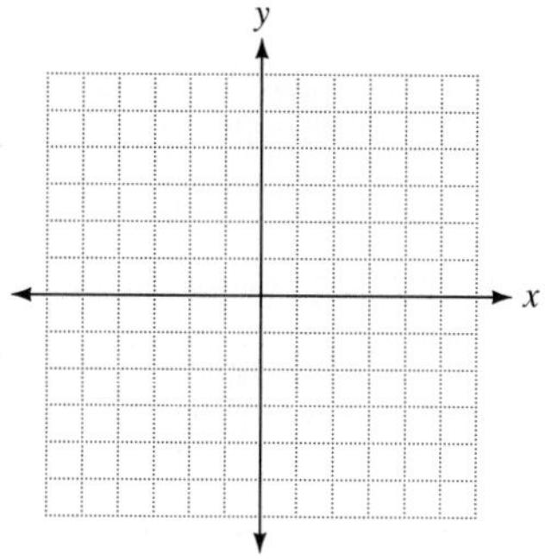

23. $f(x) = x^4 - 10x^2 + 9$

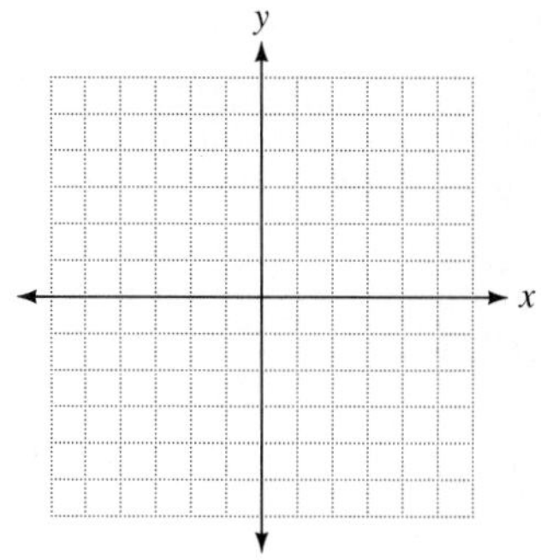

24. $f(x) = \sqrt{x + 4}$

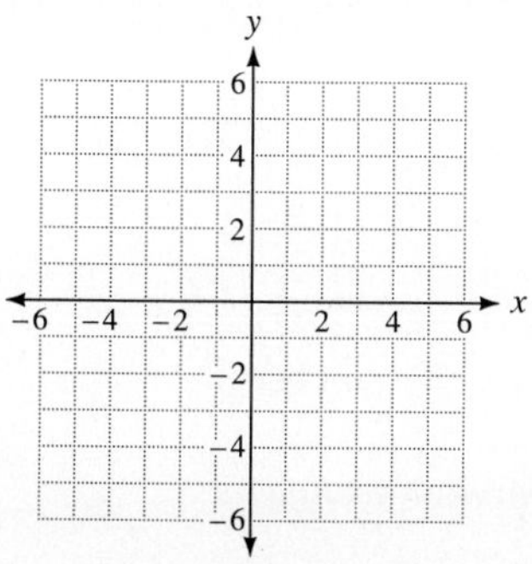

25. $f(x) = \sqrt{x - 2}$

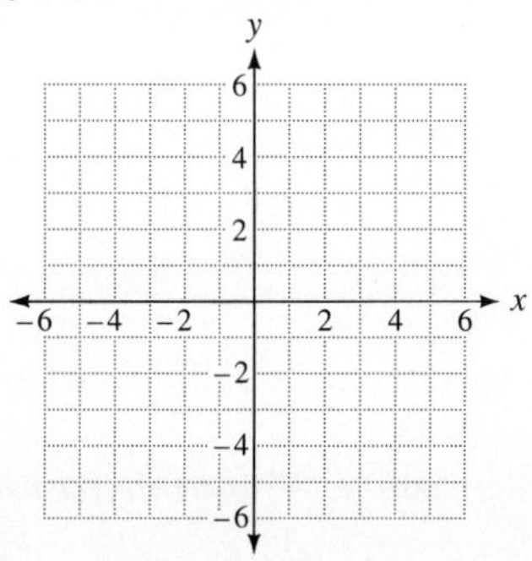

26. $f(x) = -\sqrt{x + 4}$

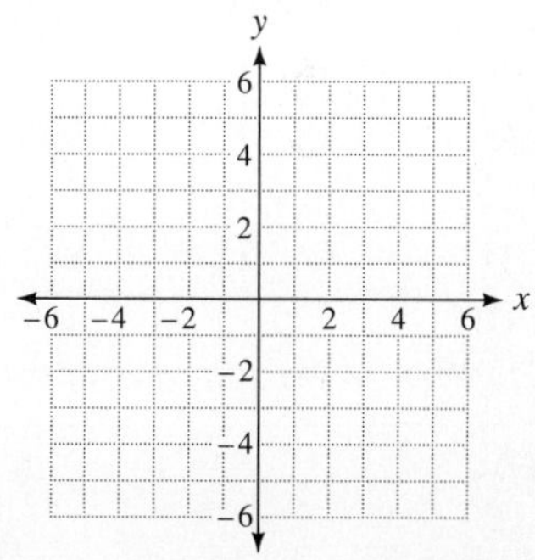

27. $f(x) = |x - 2|$

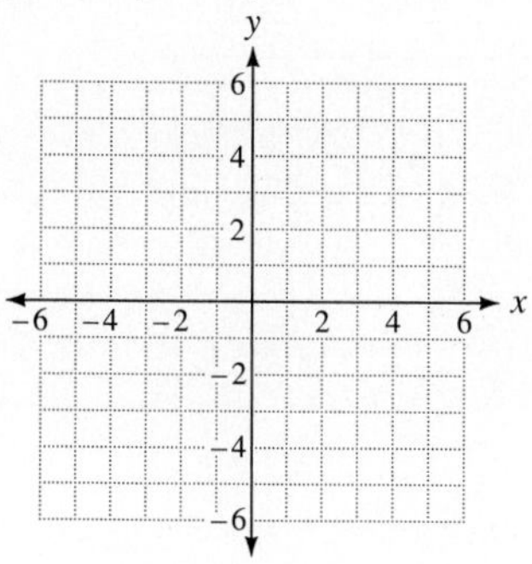

28. $f(x) = |x + 3|$

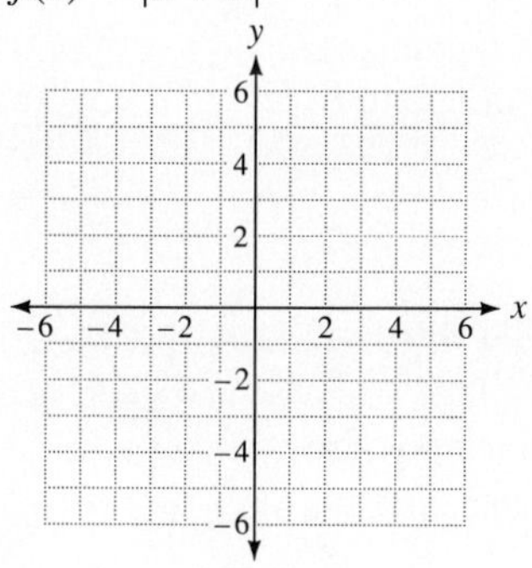

29. $f(x) = |x| + 2$

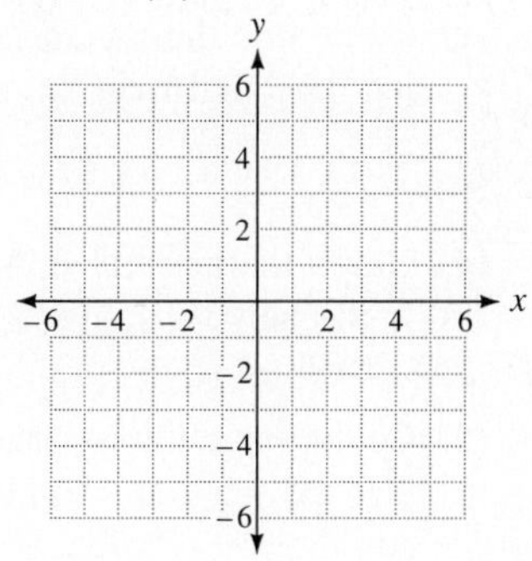

30. $f(x) = |x| - 2$

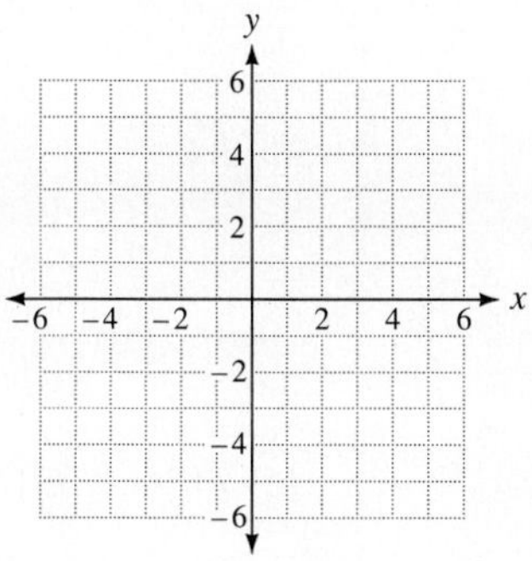

31. $f(x) = 2|x|$

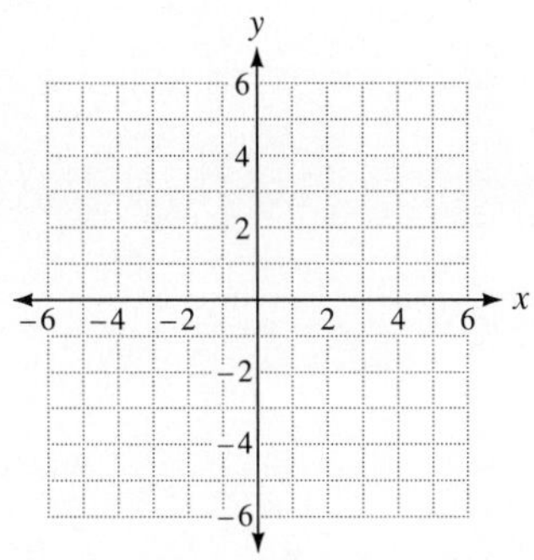

32. $f(x) = -|x - 1|$

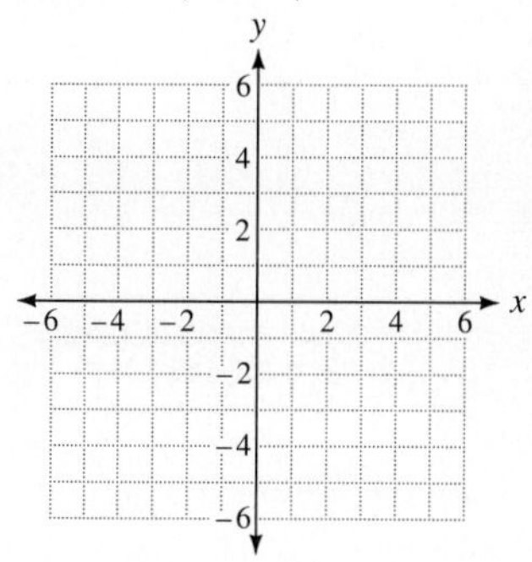

FOR REVIEW

Without graphing, tell the x-intercepts, the vertex, the line of symmetry, and whether the graph of each quadratic function opens up or down in Exercises 33–34.

33. $g(x) = 4x^2 - 25$

34. $f(x) = -3x^2 + 2x - 1$

Find the vertex, the x-intercepts, and the graph of each function in Exercises 35–36.

35. $f(x) = x^2 - x + 5$

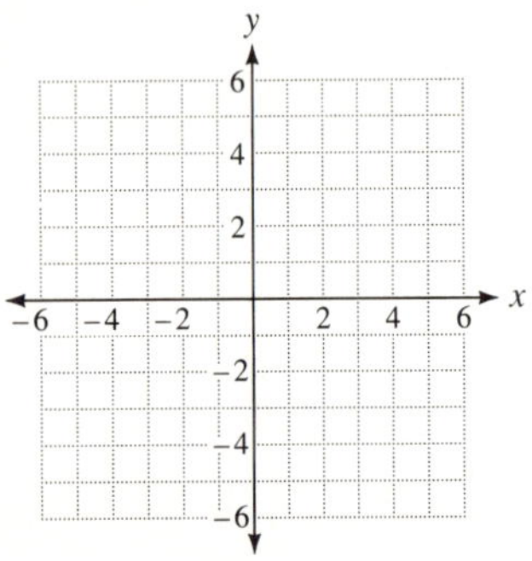

36. $g(x) = -x^2 + 2x + 2$

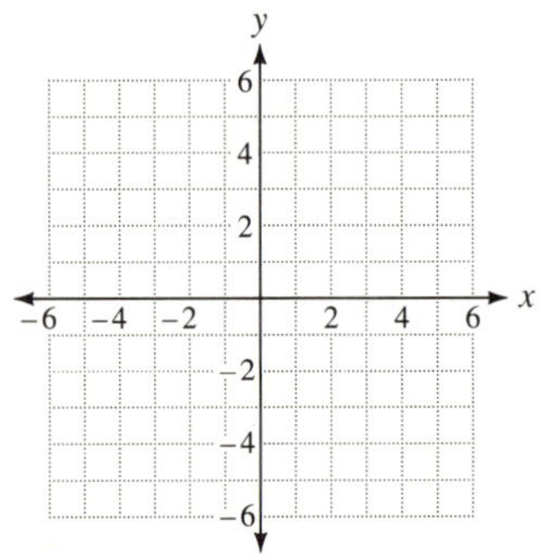

37. What is the minimum product of two numbers whose difference is 10? What are the numbers?

The following exercise reviews material from Chapter 3 to help you prepare for the next section.

38. Consider the relation $r = \{(1, 2), (2, 2), (3, 4), (4, 3)\}$.

(a) Graph the relation r.

(b) Is the relation r a function? Explain.

(c) Determine the relation s formed by interchanging the x and y coordinates of each ordered pair in r.

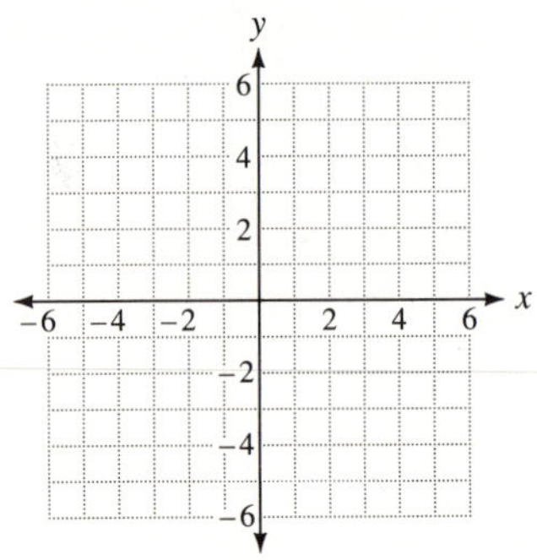

(d) Graph the new relation s.

(e) Is the new relation s a function? Explain.

(f) Could you eliminate one ordered pair from r so that the relation s formed by interchanging the coordinates is a function?

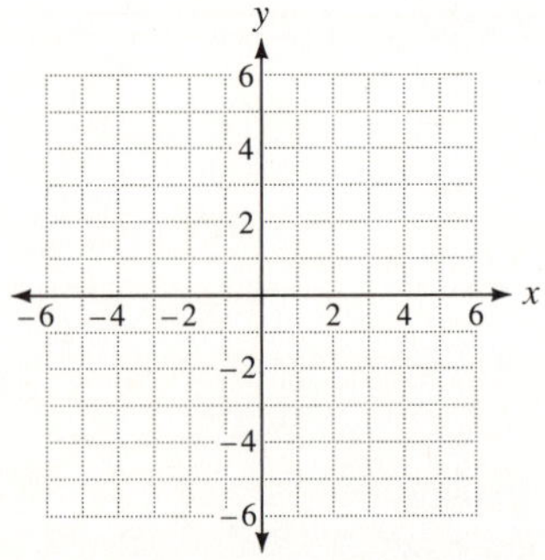

9.2 EXERCISES C

1. Assume that $f(x) = 5x + 4$ and $g(x) = x^2 - x + 1$. Find **(a)** $f[g(x)]$, **(b)** $g[f(x)]$, and **(c)** $f[g(x + h)]$.

For each function in Exercises 2–4, evaluate **(a)** $f(x+h)$ **(b)** $f(x+h)-f(x)$ **(c)** $\frac{f(x+h)-f(x)}{h}$. *(Assume that* $h \neq 0$.*) Expressions like these are needed in calculus.*

2. $f(x) = x + 1$

3. $f(x) = x^2 - 1$

4. $f(x) = 5$

5. A tray is to be made from a rectangular piece of aluminum measuring 12 inches by 12 inches by cutting equal-sized squares from each corner and bending up the sides. Assume that the measure of the side of each square is x. Express the volume V of the tray as a function of x. Give the domain of this function. Graph the function on its domain and use the graph to estimate the size of the square to be cut from each corner to maximize the volume. What is the maximum volume?

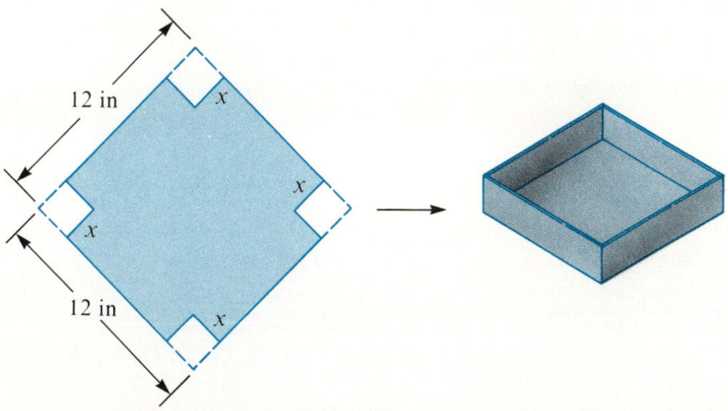

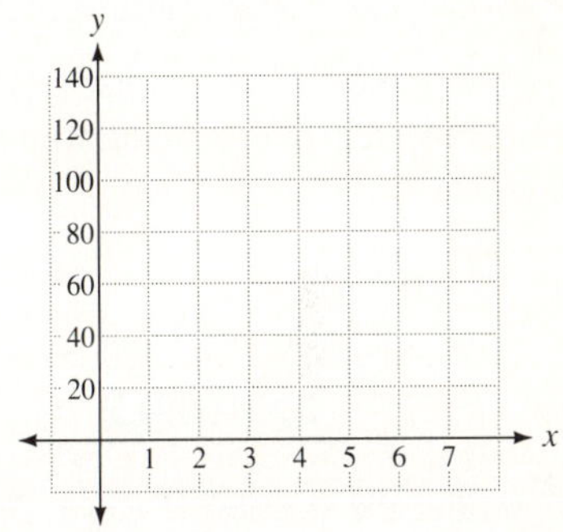

9.3 INVERSE FUNCTIONS

STUDENT GUIDEPOSTS

1. Inverse Relations
2. One-to-One Functions
3. Horizontal Line Test
4. Graphical Interpretation of Inverses
5. Finding Inverse Functions

1 INVERSE RELATIONS

Suppose we interchange the x and y coordinates of each ordered pair in a relation to obtain a new relation. The resulting relation and the original are called *inverses* of each other. For example, consider the relation

$$\{(1, 2), (2, 3), (3, 2)\}.$$

Then the inverse relation is

$$\{(2, 1), (3, 2), (2, 3)\}.$$

Using an arrow diagram, the given relation is

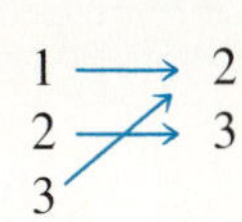

while the inverse relation is

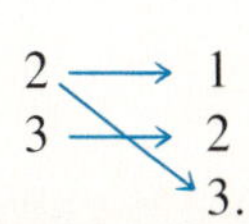

The inverse could have been found by simply reversing the direction of each arrow.

Inverse of a Relation

If R is a relation, the **inverse** of R, denoted by R^{-1}, is given by

$$R^{-1} = \{(y, x) | (x, y) \in R\}.$$

The domain of R^{-1} is equal to the range of R, and the range of R^{-1} is equal to the domain of R.

CAUTION

The symbol for an inverse relation, R^{-1}, is read "R inverse." This symbol should not be confused with $\frac{1}{R} = R^{-1}$. We can tell them apart only by context.

EXAMPLE 1 Finding Inverse Relations

Consider the relation $R = \{(0, 2), (1, 1), (2, 3), (3, 2)\}$. Find R^{-1}. Is R a function? Is R^{-1} a function?

To find R^{-1}, we interchange the coordinates of each pair in R. Thus,

$$R^{-1} = \{(2, 0), (1, 1), (3, 2), (2, 3)\}.$$

R is a function since no two ordered pairs have the same x-coordinate and different y-coordinates. The same is not true for R^{-1} since (2, 0) and (2, 3) are in R^{-1} with the same x-coordinate, 2, but different y-coordinates, 0 and 3.

Practice Exercise 1

Consider the relation $S = \{(0, 1), (0, 4), (2, 3), (3, 4)\}$. Find S^{-1}. Is S a function? Is S^{-1} a function?

Answer: $S^{-1} = \{(1, 0), (4, 0), (3, 2), (4, 3)\}$; S is not a function. S^{-1} is not a function.

2 ONE-TO-ONE FUNCTIONS

It is clear from Example 1 that even though a relation is a function, its inverse need not be a function. For a relation to be a function, each element in the domain must correspond to exactly one element in the range. In order for the inverse of a function to be a function (not just a relation), the reverse must be true. Functions with this reverse property are called *one-to-one*.

One-to-One Functions

A function is a **one-to-one function** is each element in the range of the function corresponds to exactly one element in the domain.

EXAMPLE 2 Finding One-to-One Functions

Tell whether each function is one-to-one.

(a) $f = \{(1, 2), (2, -1), (3, 0), (4, 3)\}$
The elements in the range of f are 2, -1, 0, and 3. Since 2 corresponds only to 1, -1 corresponds only to 2, 0 corresponds only to 3, 3 corresponds only to 4, f is one-to-one.

Practice Exercise 2

Tell whether each function is one-to-one.

(a) $h = \{(1, 5), (2, 5), (3, 4)\}$

(b) $g = \{(1, 2), (2, 3), (3, 2)\}$
Since 2 in the range of g corresponds to both 1 *and* 3 in the domain, g is not one-to-one.

(b) $f = \{(2, 3), (3, 4), (4, 5)\}$
Answers: (a) h is not one-to-one. (b) f is one-to-one.

3 HORIZONTAL LINE TEST

It is easy to tell if a function is one-to-one by looking at its graph. Remember that for a relation to be a function, its graph must pass the vertical line test. Like this, for a function to be one-to-one, its graph must pass the **horizontal line test.**

Horizontal Line Test

If a horizontal line can be passed through the graph of a function in such a way that the line intersects two or more points on the graph, the function *is not* one-to-one.

The function $g = \{(1, 2), (2, 3), (3, 2)\}$ in Example 2(b) has its graph given in Figure 9.17(a). The horizontal line l through (1, 2) and (3, 2) shows that the function is not one-to-one.

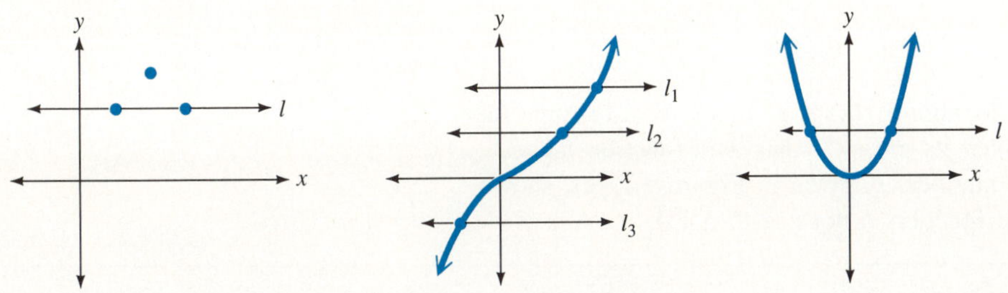

(a) Not a one-to-one function (b) One-to-one function (c) Not a one-to-one function

Figure 9.17 Horizontal Line Test

The function graphed in Figure 9.17(b) is one-to-one since no matter what horizontal line is passed through the graph, it will intersect the curve in no more than one point. On the other hand, the function graphed in Figure 9.17(c) is not one-to-one since the line l intersects the graph in two points.

4 GRAPHICAL INTERPRETATION OF INVERSES

One way to tell if two functions f and g are inverses is to look at their graphs. The function g is the inverse of function f provided the graph of g is symmetric to the graph of f with respect to the line $y = x$. Two points that are symmetric with respect to the line $y = x$ have coordinates of the form (a, b) and (b, a), as in Figure 9.18.

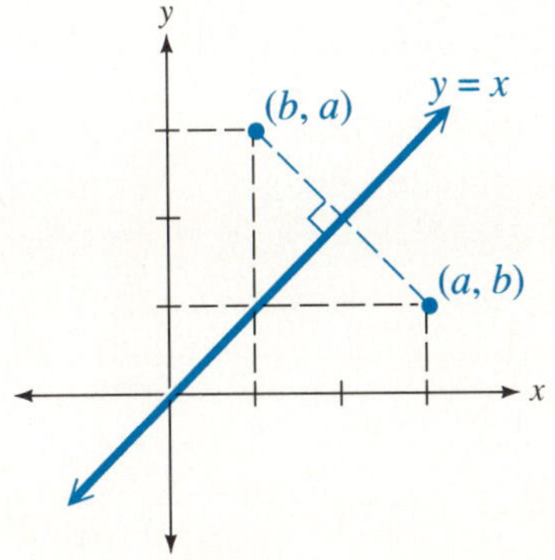

Figure 9.18 Symmetry with Respect to $y = x$

EXAMPLE 3 Graphing Inverse Functions

Graph $f(x) = x^2$ $(x \geq 0)$ and $g(x) = \sqrt{x}$, and consider symmetry with respect to $y = x$.

A careful check of the graphs of these functions in Figure 9.19 shows symmetry with respect to $y = x$. Thus, f and g are inverses.

x	$f(x)$
0	0
$\frac{1}{2}$	$\frac{1}{4}$
1	1
2	4

x	$g(x)$
0	0
$\frac{1}{4}$	$\frac{1}{2}$
1	1
4	2

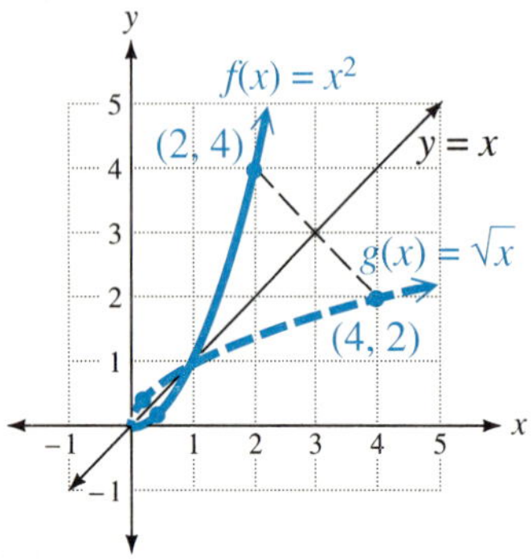

Figure 9.19 Inverse Functions

The reason for restricting the domain of $f(x) = x^2$ to $x \geq 0$ is to force the function to be one-to-one, so that its inverse exists as a function. (The graph of $f(x) = x^2$ for all real numbers x, given in Figure 9.17(c), shows that without this restriction the function is not one-to-one.)

Practice Exercise 3

Consider the graphs of $f(x) = 3x$ and $g(x) = \frac{1}{3}x$ shown below. Are f and g inverse functions?

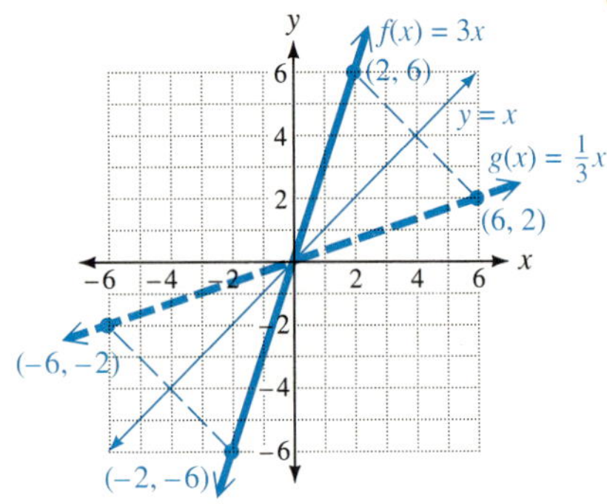

Answer: yes

5 FINDING INVERSE FUNCTIONS

Notice that in Example 3, the points plotted for the inverse function pairs involve an interchange of the x and y coordinates. For example, (2, 4) is on $f(x) = x^2$ while (4, 2) is on $g(x) = \sqrt{x}$. This observation leads to a method for finding the equation of the inverse of a given function.

Consider the function $f(x) = 3x$, and write it as $y = 3x$. If we interchange x and y and solve for y, we obtain $y = \frac{1}{3}x$.

$$y = 3x$$

$$x = 3y \quad \text{Interchange } x \text{ and } y$$

$$\frac{1}{3}x = y \quad \text{Solve for } y$$

Also, if we write $y = x^2$, assume that $x \geq 0$ (to make the function one-to-one) as in Example 3, and follow the same procedure, we obtain $y = \sqrt{x}$.

$$y = x^2$$

$$x = y^2 \quad \text{Interchange } x \text{ and } y$$

$$\sqrt{x} = y \quad \text{Take square root of both sides; remember } x \text{ and } y \text{ are nonnegative}$$

These examples illustrate the method for finding the equation of the inverse of a given one-to-one function.

Finding the Inverse of a Function

To find the inverse of a one-to-one function f defined by $y = f(x)$, interchange x and y, and then solve this new equation for y. The inverse of f is denoted by $y = f^{-1}(x)$.

EXAMPLE 4 Finding the Inverse of a Function

Find the inverse of each function.

(a) $f(x) = 3x - 5$.

$y = 3x - 5$ Replace $f(x)$ with y

$x = 3y - 5$ Interchange x and y

$x - 5 = 3y$ Solve for y

$\frac{x-5}{3} = y$ The defining relation for the inverse of f

Thus, $$f^{-1}(x) = \frac{1}{3}x - \frac{5}{3}.$$

(b) $g(x) = x^3 + 2$.

$y = x^3 + 2$ $y = g(x)$

$x = y^3 + 2$ Interchange x and y

$x - 2 = y^3$ Solve for y

$\sqrt[3]{x-2} = y$ Take cube root of both sides

Thus, $g^{-1}(x) = \sqrt[3]{x-2}$.

Practice Exercise 4

Find the inverse of each function.

(a) $f(x) = \frac{1}{2}x + \frac{3}{2}$

(b) $g(x) = 4 - 8x^3$

Answers: (a) $f^{-1}(x) = 2x - 3$ (b) $g^{-1}(x) = \frac{1}{2}\sqrt[3]{4-x}$

9.3 EXERCISES A

Find the inverse of each relation. Is the original relation a function? Is the inverse relation also a function?

1. $\{(0, 1), (1, 2)\}$

2. $\{(-1, 1), (0, 1), (1, 1)\}$

3. $\{(-3, 1), (-2, 2), (-1, 0), (1, 5), (2, 4)\}$

4. $\{(0, 1), (0, 2), (1, 5), (2, 7)\}$

5. $\{(1, 5), (1, -5), (2, 5), (2, -5)\}$

6. $\{(2, -1), (3, -1), (4, -1), (5, -1), (6, -1)\}$

Tell whether each graph is the graph of a one-to-one function.

7.

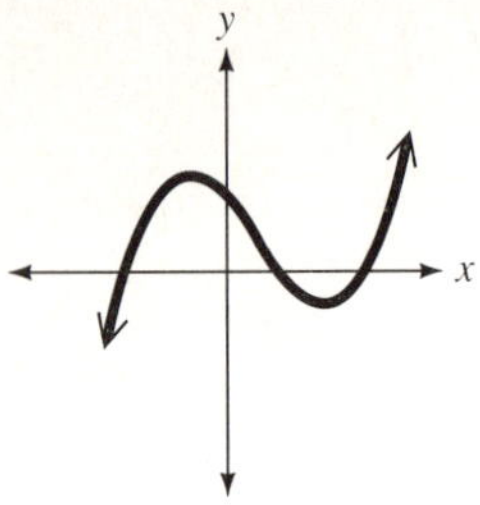

8.

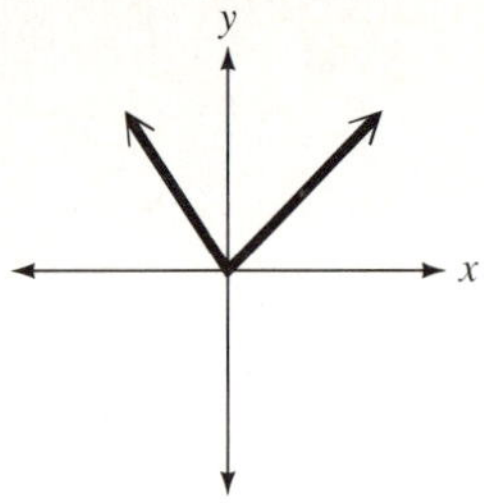

9.

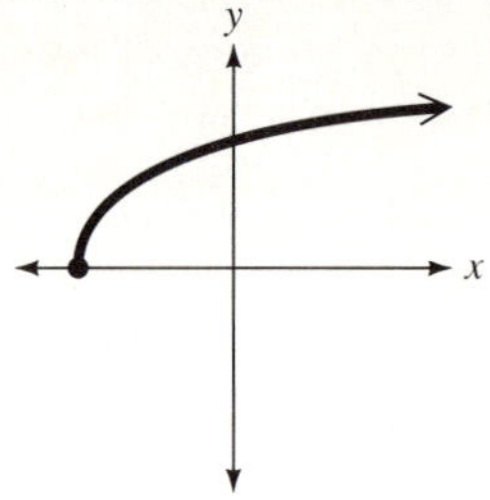

10.

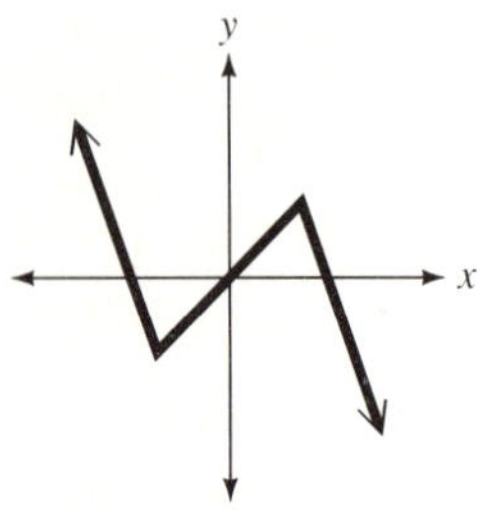

11.

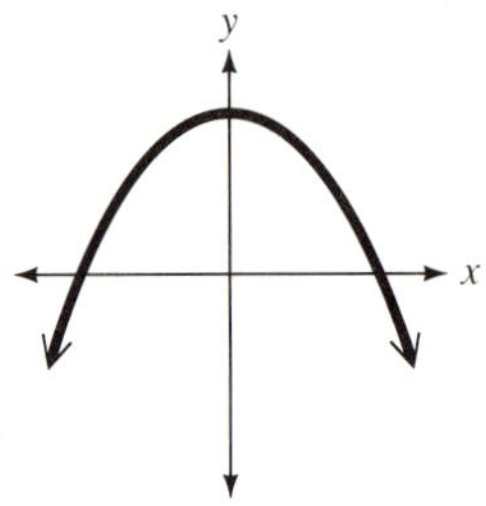

12.

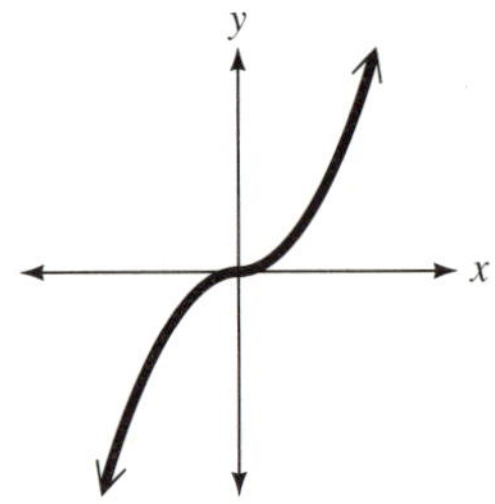

Graph each pair of functions in the same coordinate system. Are the functions inverses of each other?

13. $f(x) = 2x$ and $g(x) = -2x$

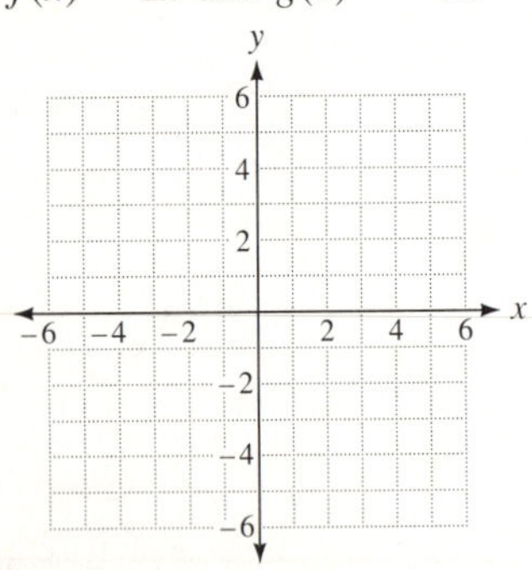

14. $f(x) = \sqrt{x - 2}$ and $g(x) = x^2 + 2,\ x \geq 0$

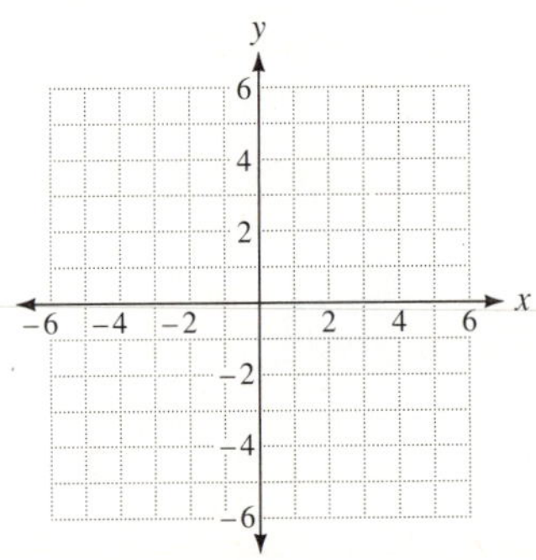

Without graphing, find the inverse of each function.

15. $f(x) = 2x + 3$

16 $f(x) = 2x^3$

17. $f(x) = \frac{1}{3}x + 2$

18. $f(x) = -x$

19. $f(x) = 2$

20 $g(x) = x^2 + 1,\ x \geq 0$

In Exercises 21–22, the graph of a function is given. Sketch the graph of its inverse on the same grid.

21.

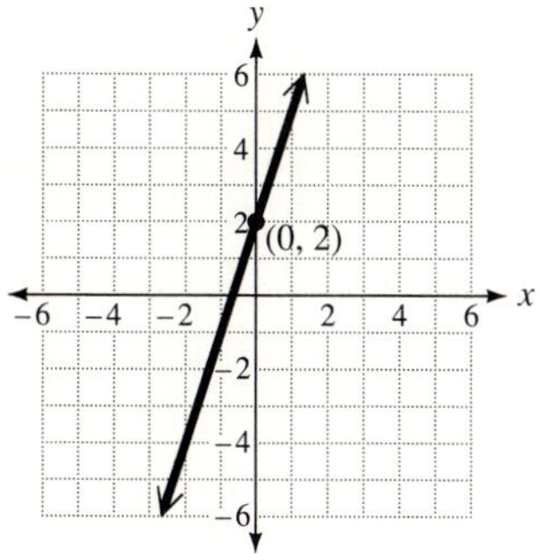

22.

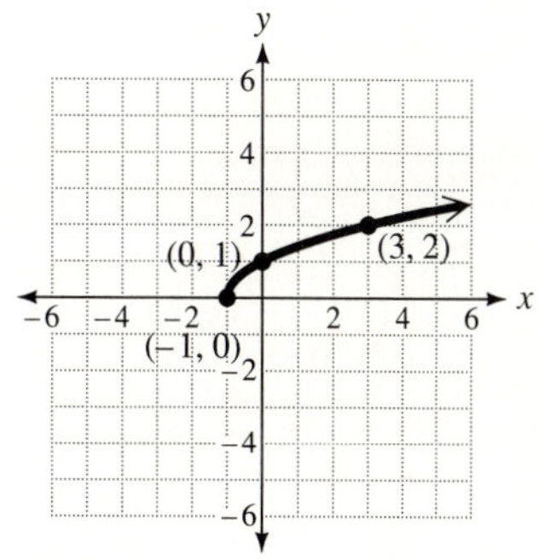

23. Consider the function $f(x) = 2x + 1$. Find the equation for the inverse of f. Use these two functions to find the following.

(a) $f(3)$ **(b)** $f^{-1}(3)$ **(c)** $f[f^{-1}(3)]$ **(d)** $f^{-1}[f(3)]$

24. Consider the function $g(x) = 8x^3$. Find the equation for the inverse of g. Use these two functions to find the following.

(a) $g(-2)$ **(b)** $g^{-1}(-2)$ **(c)** $g[g^{-1}(-2)]$ **(d)** $g^{-1}[g(-2)]$

FOR REVIEW

Sketch the graph of each function.

25. $f(x) = |x - 3|$

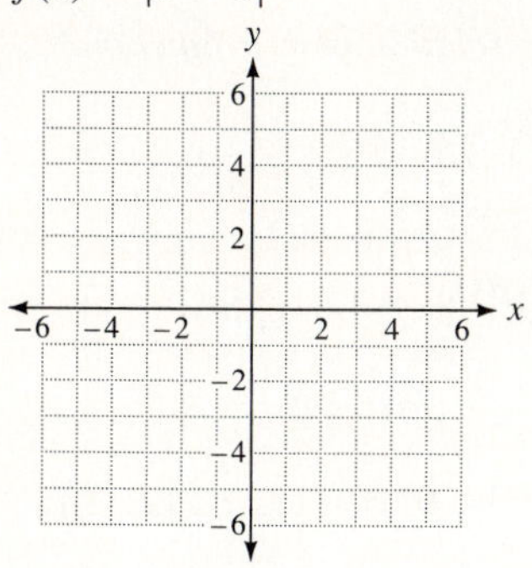

26. $f(x) = |x| - 3$

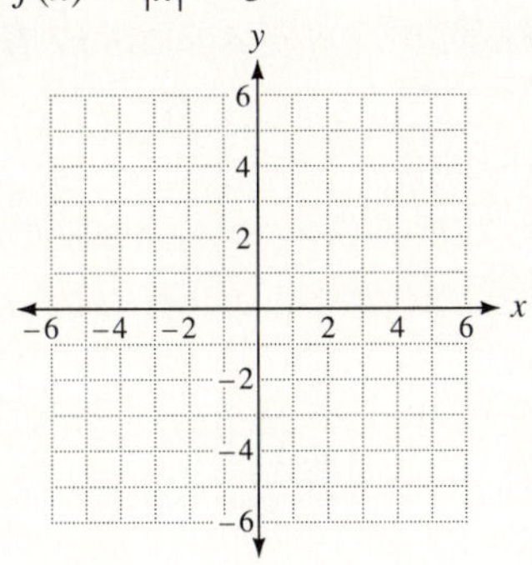

27. $f(x) = \sqrt{x + 4}$

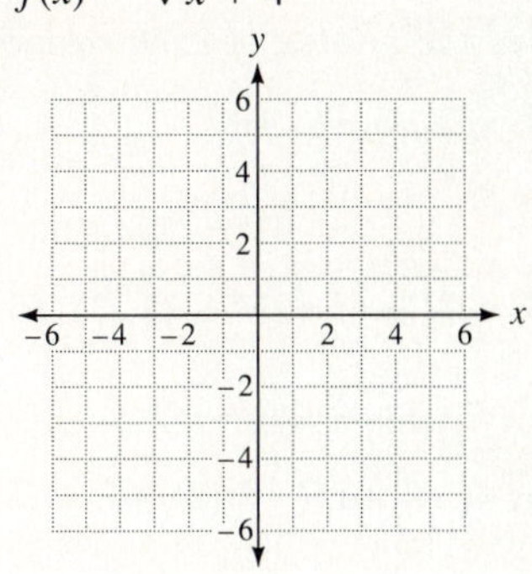

Exercises 28–30 will help you prepare for the material in Chapter 10. Give the value of x in each equation.

28. $x^2 = 25$ **29.** $2^x = 4$ **30.** $3^x = \frac{1}{9}$

ANSWERS: 1. $\{(1, 0), (2, 1)\}$; yes; yes 2. $\{(1, -1), (1, 0), (1, 1)\}$; yes; no 3. $\{(1, -3), (2, -2), (0, -1), (5, 1), (4, 2)\}$; yes; yes 4. $\{(1, 0), (2, 0), (5, 1), (7, 2)\}$; no; yes 5. $\{(5, 1), (-5, 1), (5, 2), (-5, 2)\}$; no; no 6. $\{(-1, 2), (-1, 3), (-1, 4), (-1, 5), (-1, 6)\}$; yes; no 7. not one-to-one 8. not one-to-one 9. is one-to-one 10. not one-to-one 11. not one-to-one 12. is one-to-one

13. no;

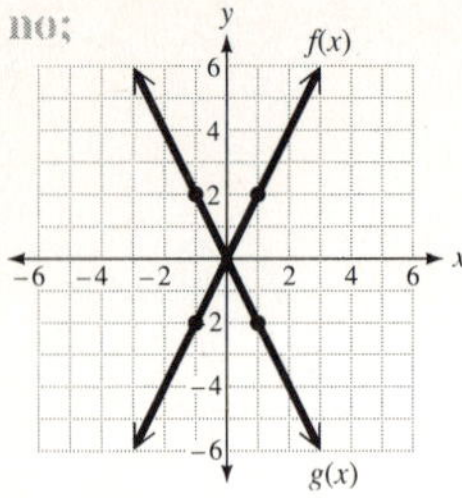

14. yes;

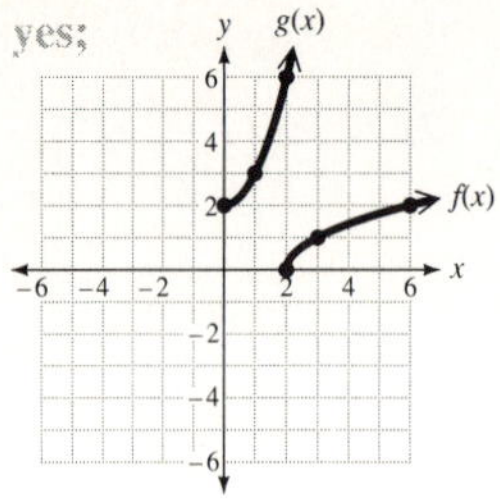

15. $f^{-1}(x) = -\frac{1}{2}x - \frac{3}{2}$ 16. $f^{-1}(x) = \sqrt[3]{\frac{x}{2}}$ 17. $f^{-1}(x) = 3x - 6$ 18. $f^{-1}(x) = -x$ 19. $f^{-1}(x)$ does not exist since $f(x) = 2$ is not one-to-one 20. $g^{-1}(x) = \sqrt{x - 1}$

21.

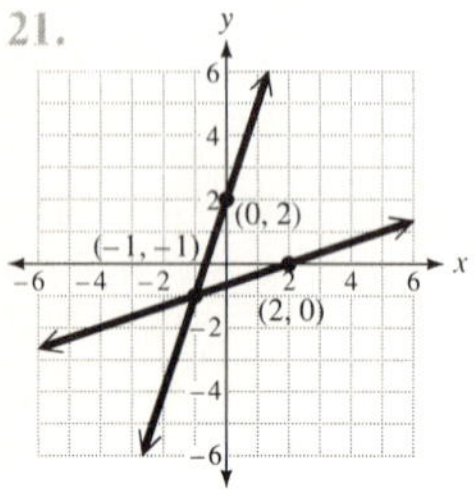

22.

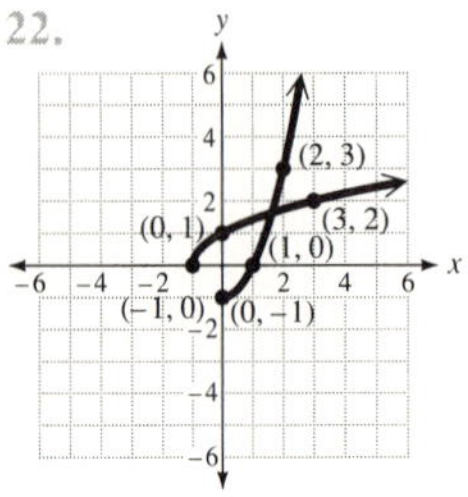

23. $f^{-1}(x) = \frac{1}{2}x - \frac{1}{2}$; (a) 7 (b) 1 (c) 3 (d) 3

24. $g^{-1}(x) = \frac{\sqrt[3]{x}}{2}$; (a) -64 (b) $\frac{-\sqrt[3]{2}}{2}$ (c) -2 (d) -2

25.

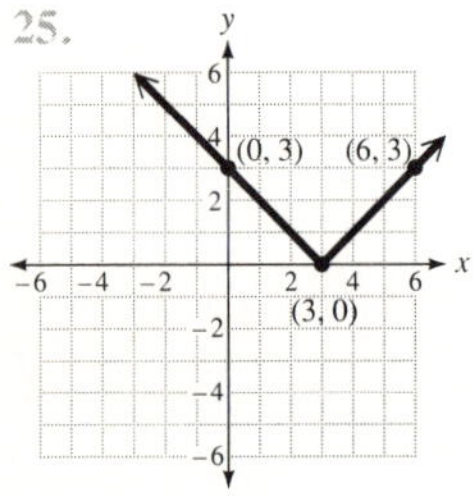

26.

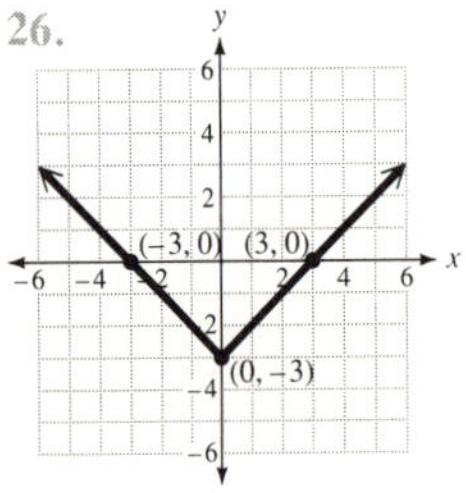

27.

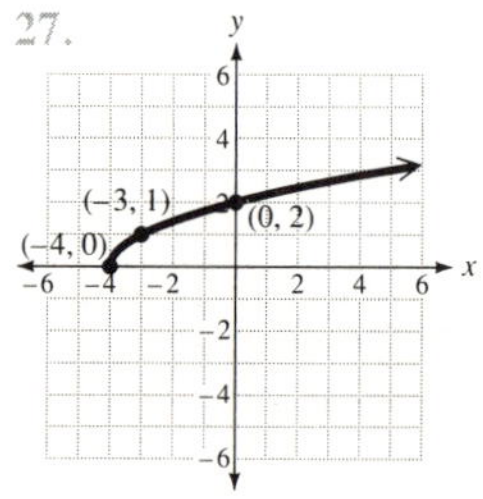

28. 5 or -5 29. 2 30. -2

9.3 EXERCISES B

Find the inverse of each relation. Is the original relation a function? Is the inverse relation also a function?

1. $\{(2, 1), (5, 4)\}$
2. $\{(5, 0), (10, 0), (20, 0)\}$
3. $\{(3, -1), (2, 5), (1, 7), (0, 8), (-1, 4)\}$
4. $\{(-1, 0), (-1, 1), (-1, 2), (-1, 3)\}$
5. $\{(-3, 0), (-3, 2), (0, -2), (2, -2)\}$
6. $\{(-2, -2), (-1, -1), (0, 0), (1, 1), (2, 2)\}$

Tell whether each graph is the graph of a one-to-one function.

7.

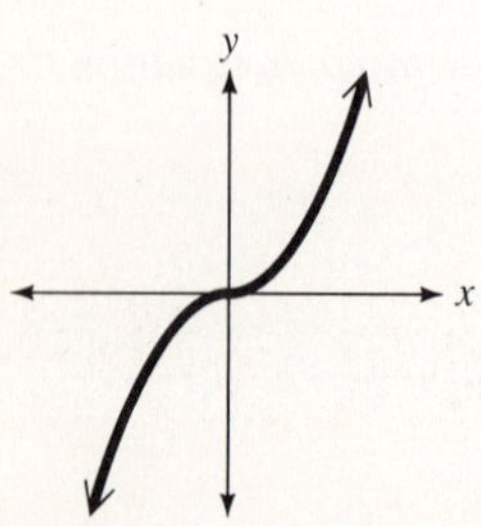

8.

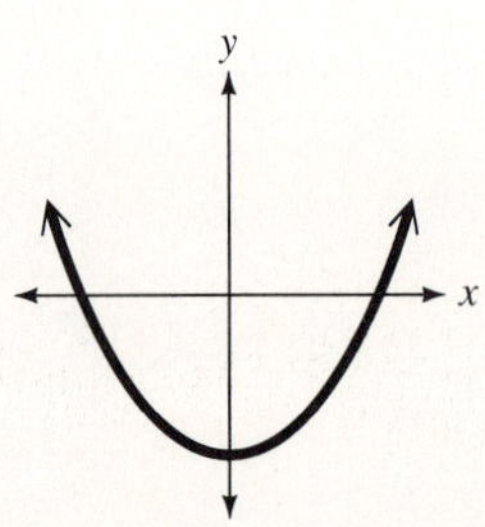

9.

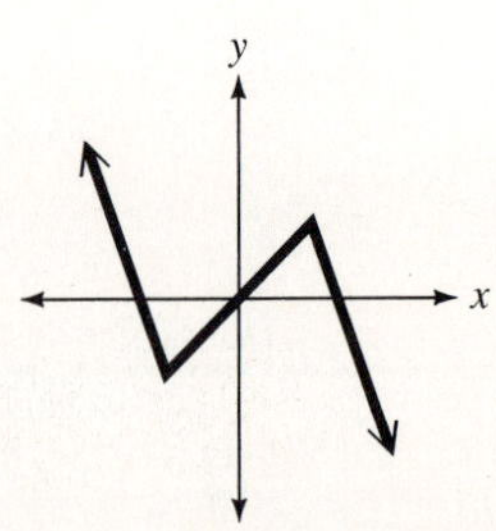

10.

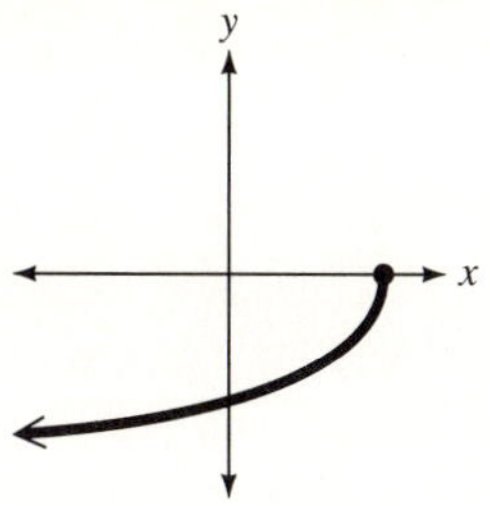

11.

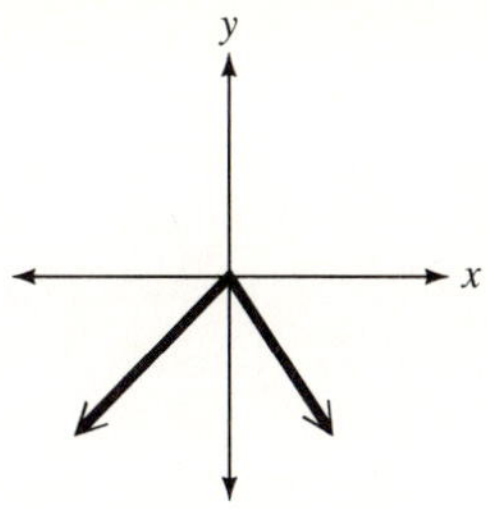

12.

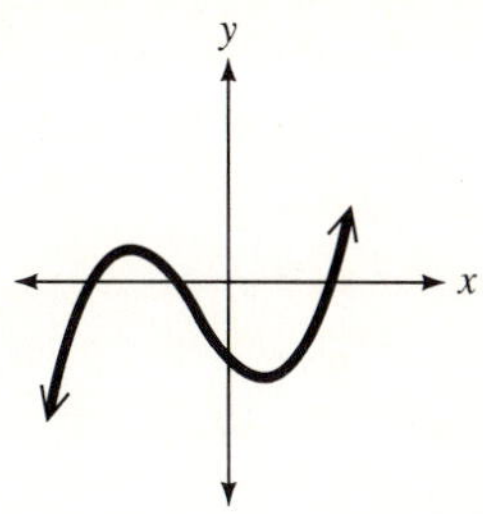

Graph each pair of functions in the same coordinate system. Are the functions inverses of each other?

13. $f(x) = -3x$ and $g(x) = 3x$

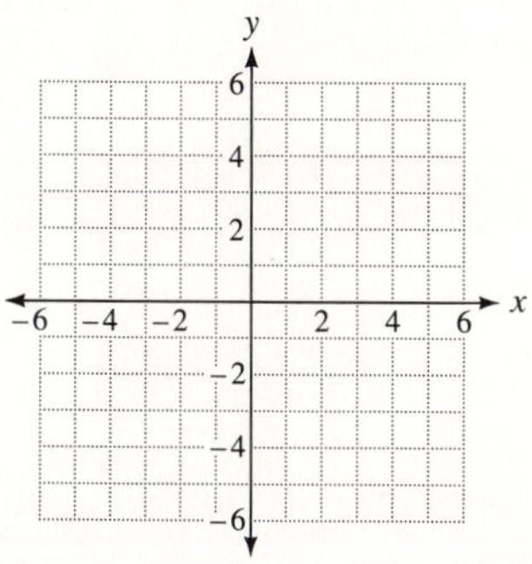

14. $f(x) = \sqrt{x+4}$ and $g(x) = x^2 - 4,\ x \geq 0$

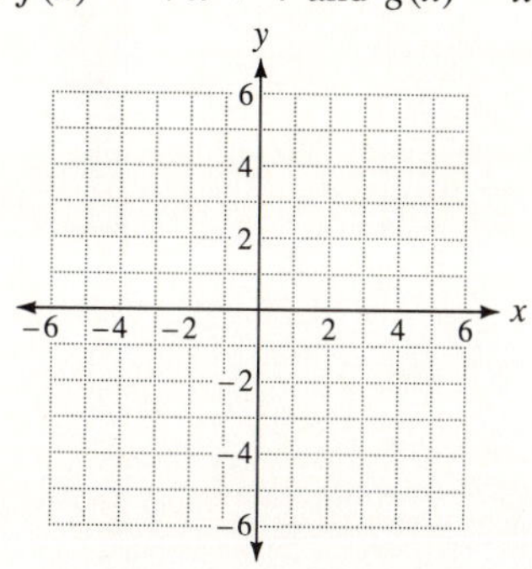

Without graphing, find the inverse of each function.

15. $g(x) = \dfrac{1}{2}x - 1$

16. $g(x) = -x^3$

17. $g(x) = 5x - 6$

18. $g(x) = x$

19. $f(x) = -3$

20. $f(x) = -x^2 + 1,\ x \geq 0$

In Exercises 21–22, the graph of a function is given. Sketch the graph of its inverse on the same grid.

21.

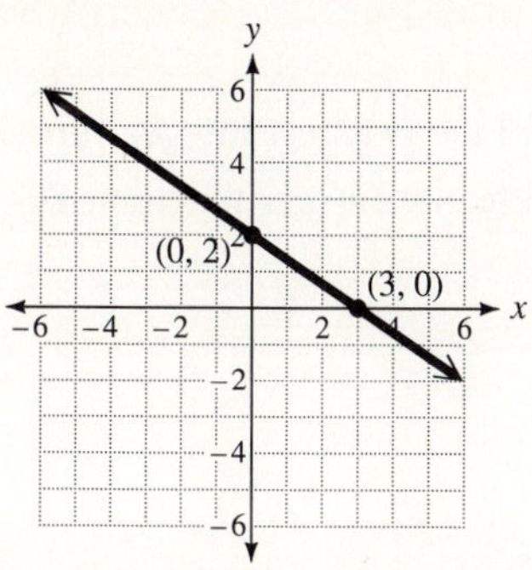

22.

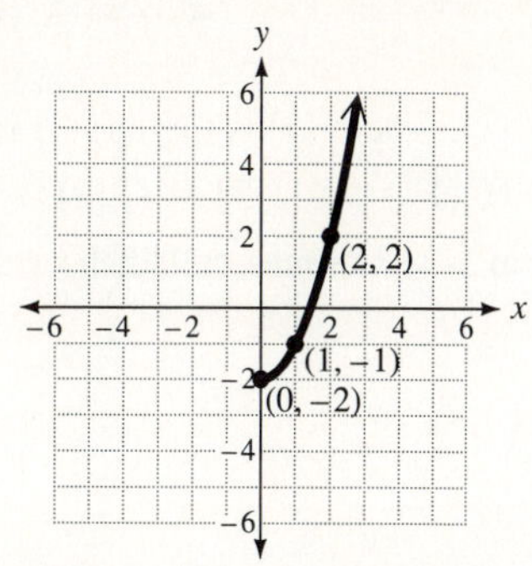

23. Consider the function $g(x) = \frac{1}{3}x - 1$. Find the equation for the inverse of g. Use these two functions to find the following.

(a) $g(-2)$ **(b)** $g^{-1}(-2)$ **(c)** $g[g^{-1}(-2)]$ **(d)** $g^{-1}[g(-2)]$

24. Consider the function $f(x) = -8x^3$. Find the equation for the inverse of f. Use these two functions to find the following.

(a) $f(-1)$ **(b)** $f^{-1}(-1)$ **(c)** $f[f^{-1}(-1)]$ **(d)** $f^{-1}[f(-1)]$

FOR REVIEW

Sketch the graph of each function.

25. $f(x) = |x - 4|$

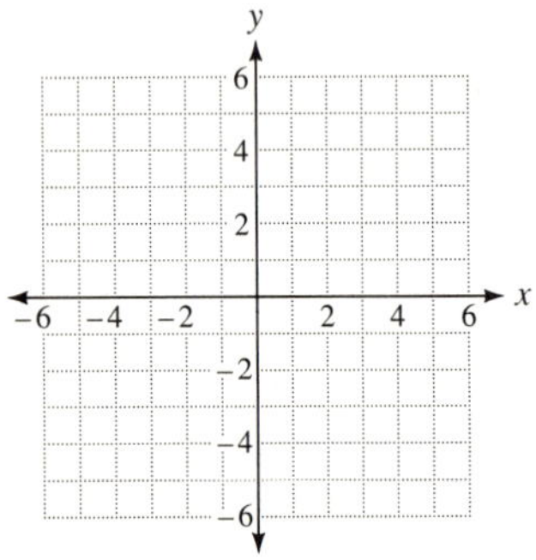

26. $f(x) = x^3 - 16x$

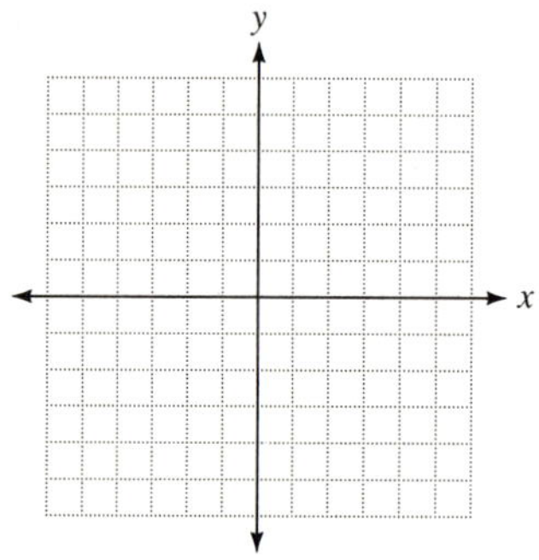

27. $f(x) = -\sqrt{x + 4}$

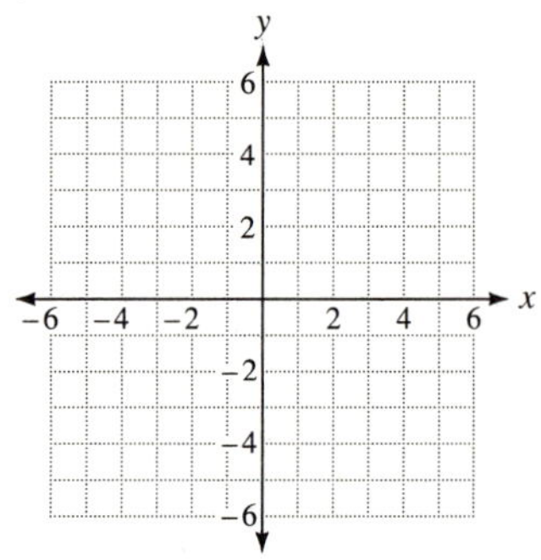

Exercises 28–30 will help you prepare for the material in Chapter 10. Find the value of x in each equation.

28. $x^2 = 9$

29. $3^x = 27$

30. $2^x = \frac{1}{8}$

9.3 EXERCISES C

*A function f is called **even** if $f(x) = f(-x)$ for every x in the domain of f, and **odd** if $f(x) = -f(-x)$ for every x in the domain of f. Tell whether each function is even, odd, or neither.*

1. $f(x) = x$

2. $f(x) = x^2$

3. $f(x) = x^3$

4. $f(x) = 2x^3 - x$

5. $f(x) = x^2 - x$

6. $f(x) = 5$

7. The function $f(x) = [x]$, where $[x]$ is the largest integer n such that $n \leq x$, is called the **greatest integer function.** For example, $\left[\frac{1}{2}\right] = 0$, $[1] = 1$, $[1.5] = 1$, $[2] = 2$, etc. Sketch the graph of the greatest integer function. Do you see why this function is sometimes called the *stairstep function?*

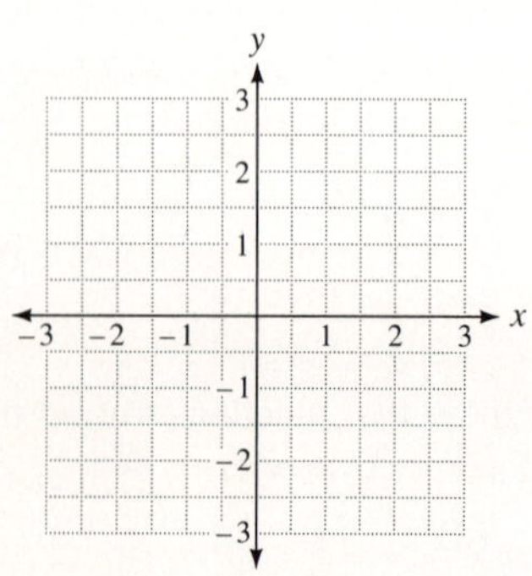

CHAPTER 9 REVIEW

KEY WORDS

9.1 A function defined by the equation $f(x) = ax^2 + bx + c$, $a \neq 0$, is a **quadratic function.**

The graph of a quadratic function is a **parabola.**

The high point or low point on a parabola is its **vertex.**

The line through the vertex of a parabola about which the curve is symmetric is called the **line of symmetry** of the parabola.

9.2 If f and g are two functions, the expression $f[g(x)]$ is the **composition of f with g**, and $g[f(x)]$ is the **composition of g with f.**

A **polynomial function** of degree n is a function defined by $f(x) = a_nx^n + a_{n-1}x^{n-1} + \ldots + a_1x + a_0$, $a_n \neq 0$.

The **square root function** is defined by $f(x) = \sqrt{x}$, $x \geq 0$.

The **absolute value function** is defined by $f(x) = |x|$.

9.3 The **inverse** of a relation R is $R^{-1} = \{(y, x) | (x, y) \in R\}$.

A function is **one-to-one** if each element in the range corresponds to exactly one element in the domain.

KEY CONCEPTS

9.1 Consider the quadratic function $f(x) = ax^2 + bx + c$, $a \neq 0$, whose graph is a parabola.

1. If $a > 0$ the parabola opens up; if $a < 0$ the parabola opens down.
2. The x-coordinate of the vertex is $-\frac{b}{2a}$, and the y-coordinate is $f\left(-\frac{b}{2a}\right)$.
3. The line of symmetry is the vertical line $x = -\frac{b}{2a}$.
4. If the parabola opens up and the vertex is above the x-axis, there are no x-intercepts. Likewise, if the parabola opens down and the vertex is below the x-axis there are no x-intercepts.
5. When there are x-intercepts, they can be found by solving the quadratic equation $ax^2 + bx + c = 0$.
6. The y-intercept is easy to find and has coordinates $(0, c)$.
7. Some problems that require finding a maximum or minimum value of a quadratic expression can be solved by finding the vertex of the parabola.

9.2

1. When a function is defined by an equation, the domain is the largest set of real numbers for which the expression in x is defined. This means that values of x that make a denominator zero or a radicand negative must be excluded.
2. If f and g are two functions,
   (a) $(f + g)(x) = f(x) + g(x)$
   (b) $(f - g)(x) = f(x) - g(x)$
   (c) $(fg)(x) = f(x) \cdot g(x)$
   (d) $\left(\frac{f}{g}\right)(x) = \frac{f(x)}{g(x)}$, $g(x) \neq 0$.
3. The graph of every square root function is half of a parabola.
4. The graph of every absolute value function is a V-shaped curve. Do not confuse the graph of a quadratic function (a U-shaped parabola) with the graph of an absolute value function.

9.3

1. To find the inverse of a relation given as a set of ordered pairs, form the set of ordered pairs obtained by interchanging the x- and y-coordinates.
2. The horizontal line test is used to tell whether a function is one-to-one in the same way the vertical line test is used to tell whether a relation is a function.
3. If two functions are inverses, their graphs are symmetric with respect to the line $y = x$.
4. To find the inverse of a function f defined by the equation $y = f(x)$, interchange x and y and solve the resulting equation for y to obtain the equation of the inverse, $y = f^{-1}(x)$.

REVIEW EXERCISES

Part I

9.1 *Sketch the graph of each function by sliding the graph of $f(x) = x^2$ or $g(x) = -x^2$ up or down.*

1. $h(x) = x^2 - 5$

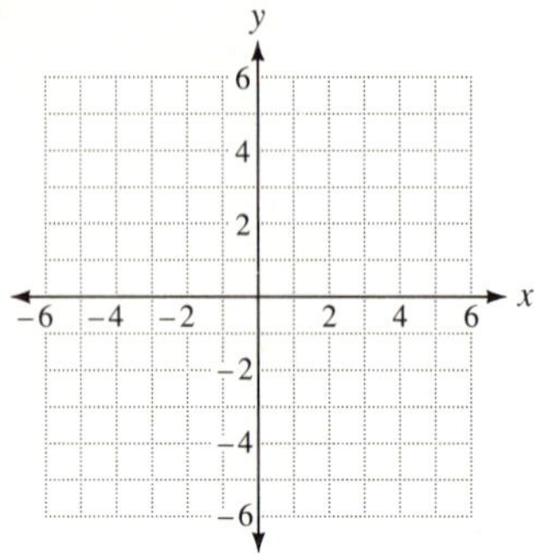

2. $h(x) = -x^2 + 4$

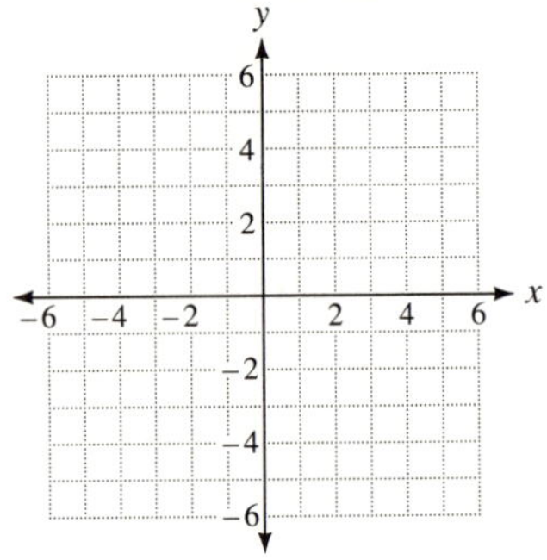

Without graphing, give the x-intercepts (if they exist) and the vertex, and tell whether the graph of each quadratic function opens up or down.

3. $f(x) = x^2 + 7x - 30$

4. $f(x) = -x^2 + 2x - 5$

Find the x-intercepts (if they exist), the vertex, and graph of the function.

5. $f(x) = x^2 - 2x - 3$

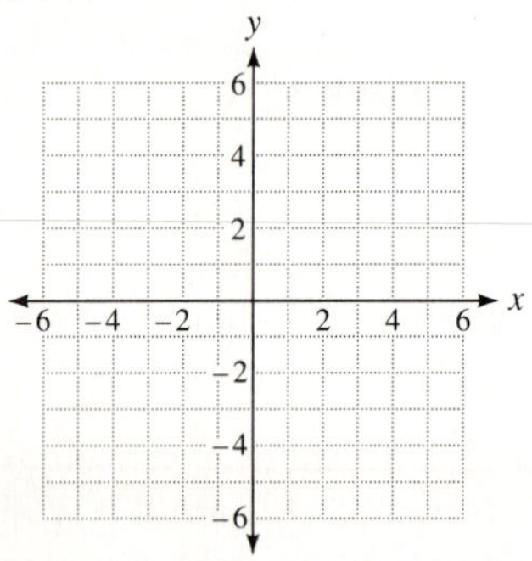

6. $g(x) = -x^2 + 4x - 5$

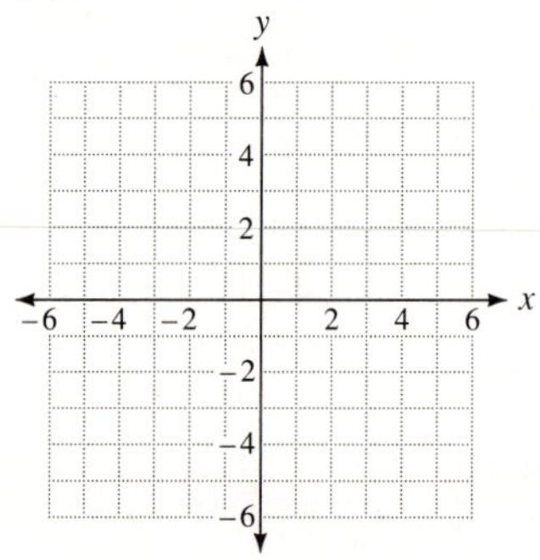

7. Jack Pritchard is building a rectangular room with a fixed perimeter of 72 ft. What dimensions would give the maximum area? What is the maximum area?

9.2 *Find the domain of the function defined by each equation.*

8. $y = \dfrac{3}{x - 8}$

9. $y = \sqrt{2x - 1}$

10. $f(x) = \dfrac{7}{\sqrt{2x - 1}}$

Assume that $f(x) = 2 - x$ and $g(x) = x^2 - 2$. Find each function value.

11. $(f + g)(3)$

12. $(f - g)(-2)$

13. $(fg)(1)$

14. $\left(\frac{f}{g}\right)(-1)$

15. $f[g(2)]$

16. $g[f(-3)]$

Make a table of values and sketch the graph of each function.

17. $f(x) = x^3 + 1$

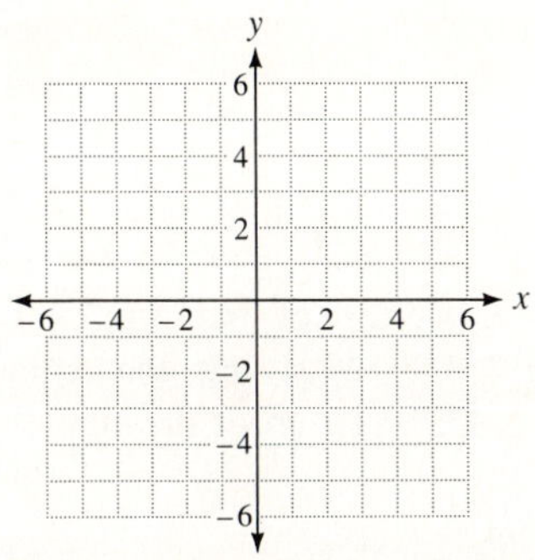

18. $f(x) = -\sqrt{x + 6}$

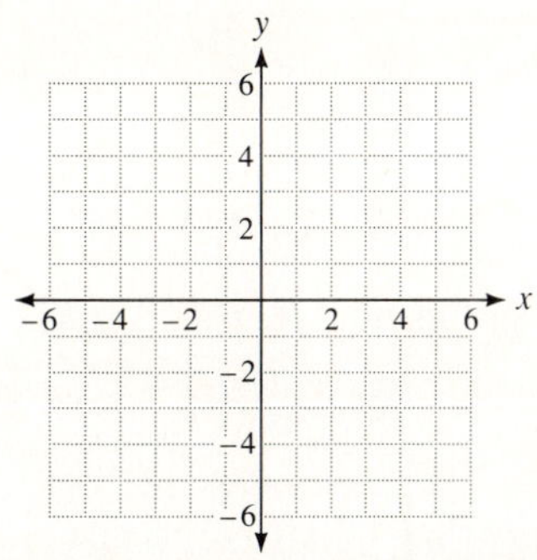

19. $f(x) = |x - 5|$

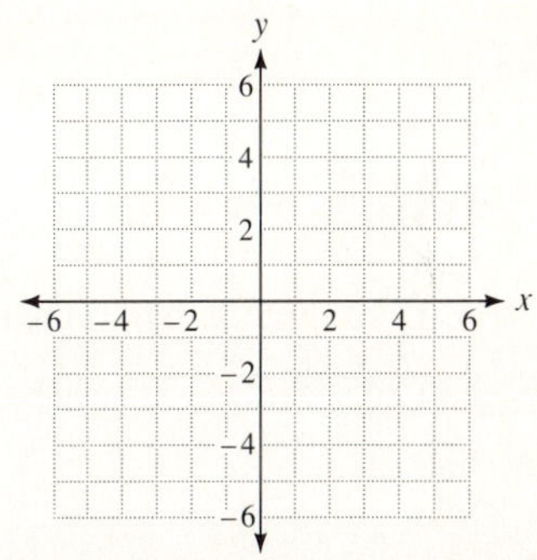

9.3 **20.** Find the inverse of the relation $\{(-2, 5), (0, 3), (1, 3)\}$. Is the original relation a function? Is the inverse a function?

21. Is this the graph of a one-to-one function?

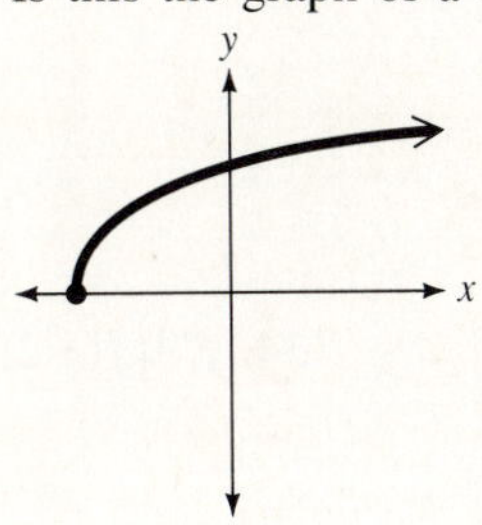

22. Graph the functions $f(x) = \sqrt{x - 3}$ and $g(x) = x^2 + 3$, $x \geq 0$. Are these inverses of each other?

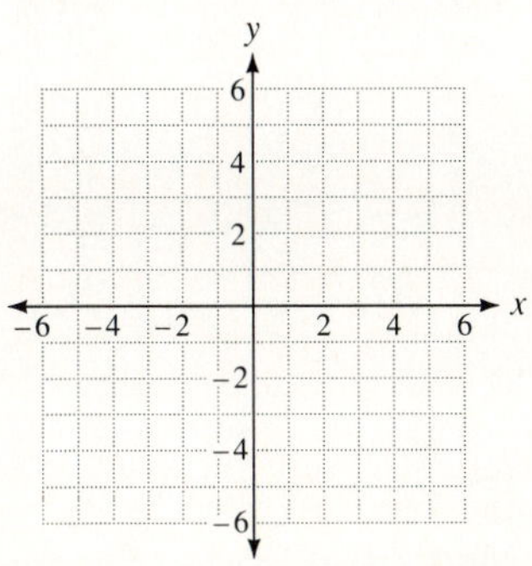

23. Consider the function $f(x) = 4x - 1$. Find the equation for the inverse of f. Use these to find the following.

(a) $f(-2)$
(b) $f^{-1}(-2)$
(c) $f[f^{-1}(-2)]$
(d) $f^{-1}[f(-2)]$

Part II

Graph each function and tell whether it is one-to-one.

24. $f(x) = 2x^2 - 5$

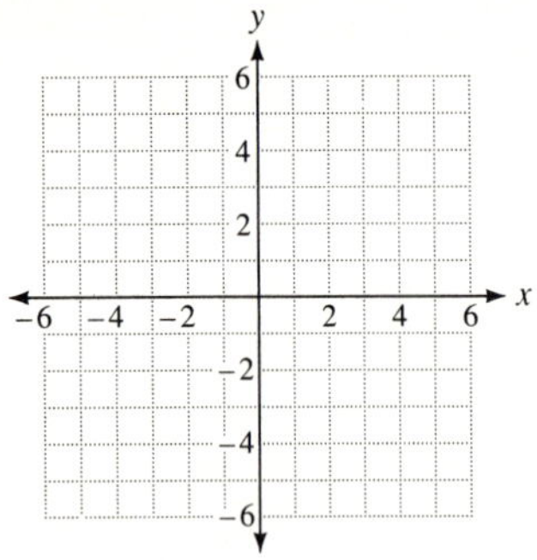

25. $f(x) = |x| - 6$

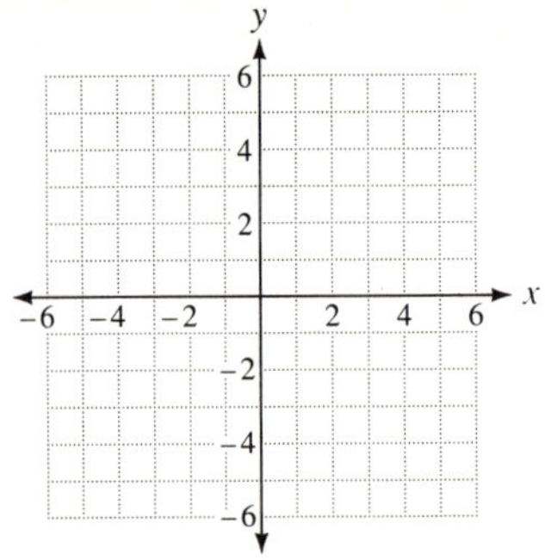

26. $f(x) = \sqrt{x + 5}$

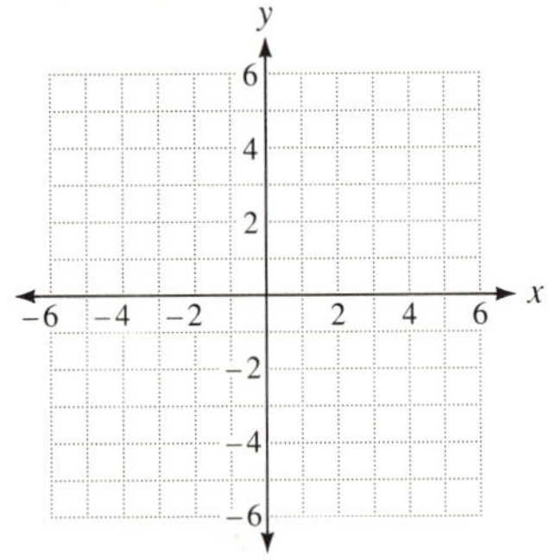

27. What is the domain of the function $f(x) = \frac{6}{\sqrt{6 - x}}$?

28. What is the vertex of the graph of $g(x) = -x^2 + 2x + 35$? What are the x-intercepts? Does the graph open up or down?

If $f(x) = 1 + 2x$ *and* $g(x) = \frac{1}{3}x - \frac{2}{3}$, *find each of the following.*

29. $f^{-1}(x)$

30. $g^{-1}(x)$

31. $f^{-1}(-1)$

32. $g^{-1}(-1)$

33. $f[g(-1)]$

34. $f^{-1}[f(-2)]$

35. $g[g^{-1}(7)]$

36. $(f + g)(3)$

37. $\left(\frac{f}{g}\right)(0)$

38. $(fg)(-1)$

39. $(f - g)(-2)$

40. $f^{-1}[g^{-1}(-1)]$

41. What is the minimum product of two numbers whose difference is 30? What are the two numbers?

ANSWERS: 1.

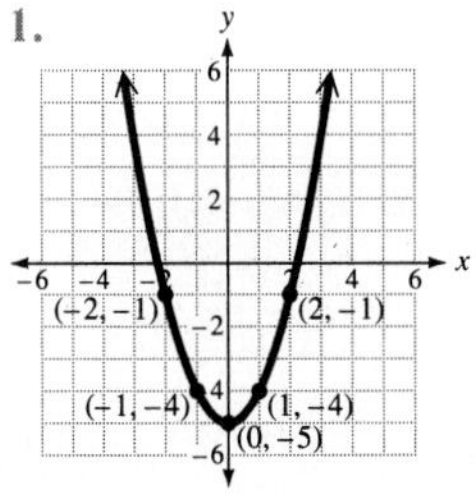

2.

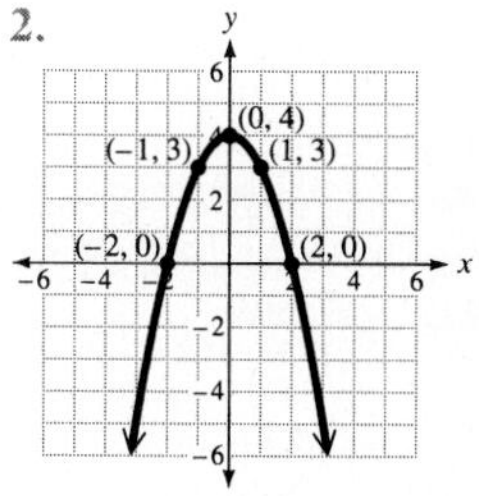

3. (3, 0) and (−10, 0); $\left(-\frac{7}{2}, -\frac{169}{4}\right)$; up
4. no x-intercepts; (1, −4); down

5.

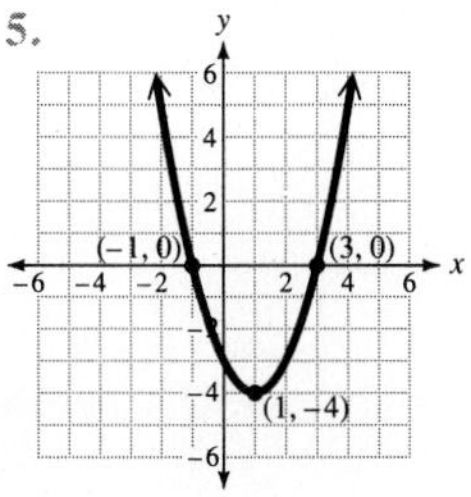

6.

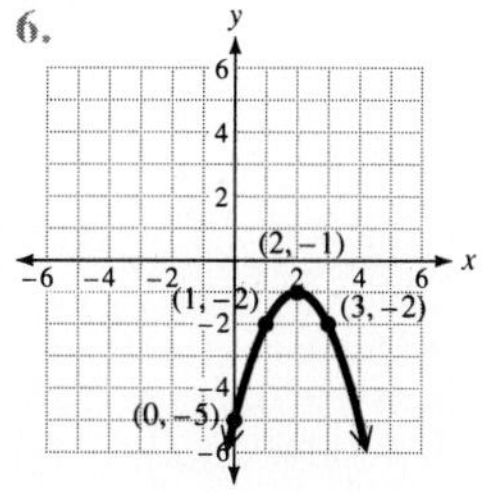

no x-intercepts 7. 18 ft by 18 ft; 324 ft^2 8. all real numbers except 8 9. all real numbers $x \geq \frac{1}{2}$ 10. all real numbers $x > \frac{1}{2}$ 11. 6 12. 2 13. −1 14. −3 15. 0 16. 23

17.

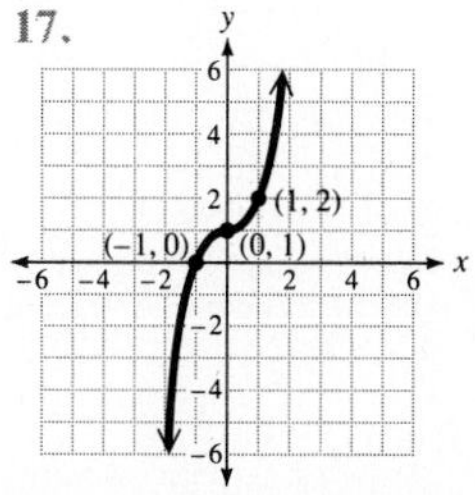

18.

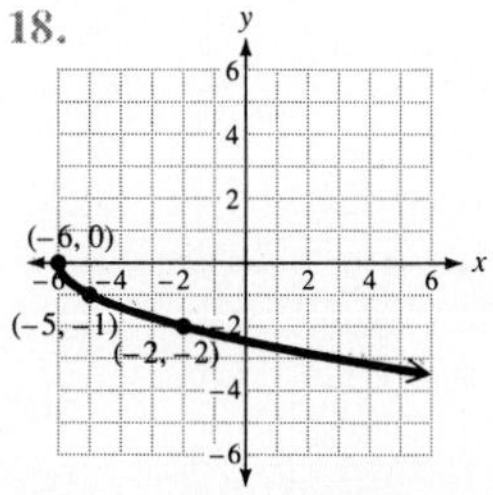

19.

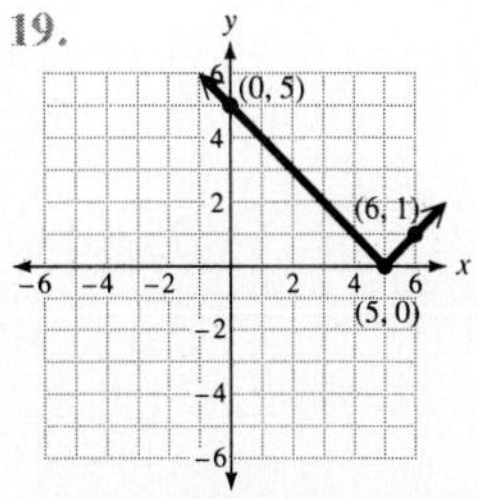

20. {(5, −2), (3, 0), (3, 1)}; yes; no 21. yes

22. yes;

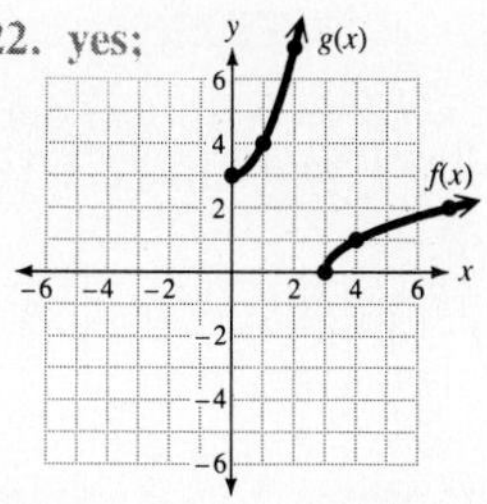

23. $f^{-1}(x) = \frac{1}{4}x + \frac{1}{4}$;
(a) −9 (b) $-\frac{1}{4}$
(c) −2 (d) −2

24.

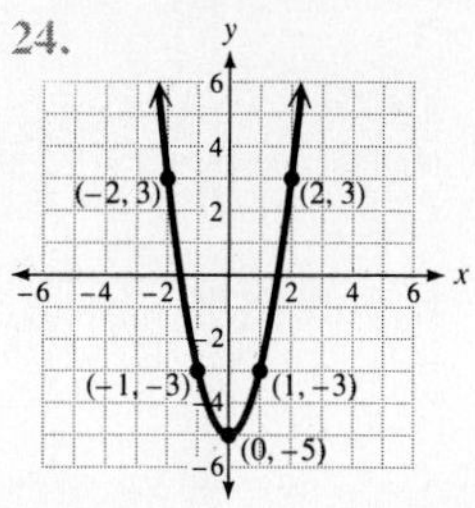

not one-to-one

25.

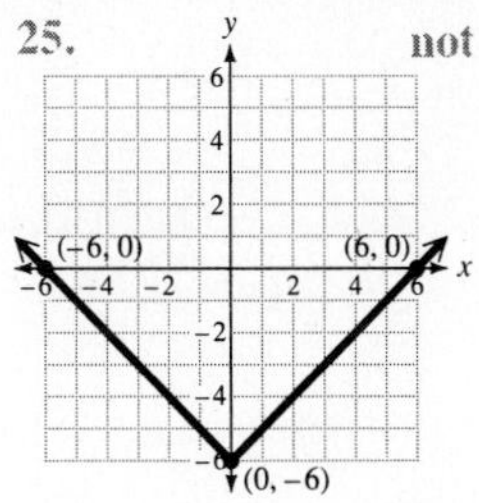

not one-to-one

26.

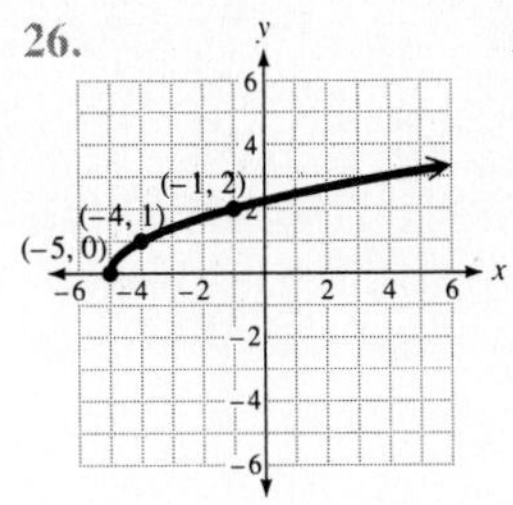

is one-to-one 27. all real numbers $x < 6$ 28. (1, 36); (−5, 0) and (7, 0); down 29. $f^{-1}(x) = \frac{1}{2}x - \frac{1}{2}$ 30. $g^{-1}(x) = 3x + 2$ 31. −1 32. −1 33. −1 34. −2 35. 7 36. $\frac{22}{3}$ 37. $-\frac{3}{2}$ 38. 1 39. $-\frac{5}{3}$ 40. −1 41. −225; 15 and −15

CHAPTER 9 TEST

1. How would you obtain the graph of $h(x) = x^2 + 3$ from the graph of $f(x) = x^2$? — 1. ______

2. Find the x-intercepts, the vertex, and graph of the function $f(x) = x^2 - 4x + 3$. — 2.

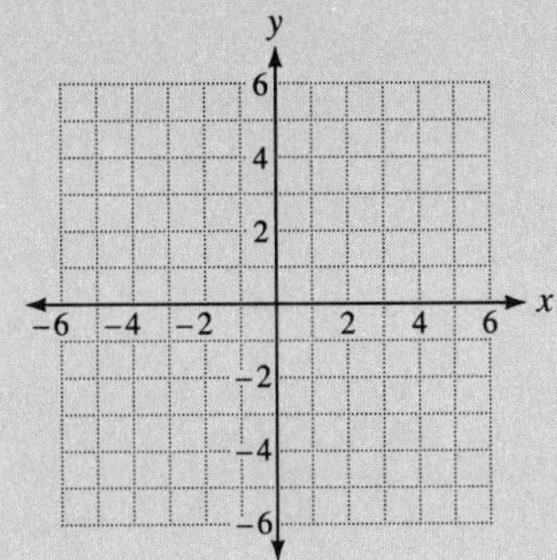

3. What is the maximum product of two numbers whose sum is 24? — 3. ______

4. Find the domain of the function $f(x) = \dfrac{1}{\sqrt{x + 10}}$. — 4. ______

Assume that $f(x) = x - 7$ and $g(x) = -x^2 + 3$. Find each function value in problems 5–10.

5. $(f + g)(-3)$ — 5. ______

6. $\left(\dfrac{f}{g}\right)(2)$ — 6. ______

7. $(f - g)(-1)$ — 7. ______

8. $(fg)(-2)$ — 8. ______

9. $f[g(-3)]$ — 9. ______

10. $g[f(5)]$ — 10. ______

11. Is this the graph of a one-to-one function? — 11. ______

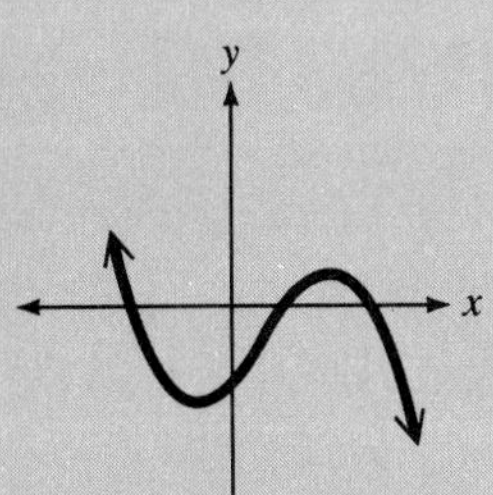

CHAPTER 9 TEST

CONTINUED

12. Give the equation for the inverse of the function $f(x) = \frac{1}{4}x - \frac{3}{4}$ and use it to find **(a)** $f^{-1}(2)$, and **(b)** $f[f^{-1}(2)]$.

12. ______________________
12. (a) ______________________
12. (b) ______________________

Graph the following functions.

13. $f(x) = -\sqrt{x + 2}$

13.

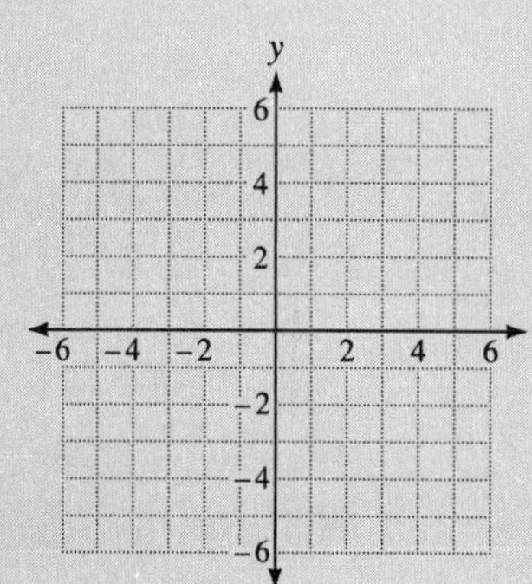

14. $f(x) = |x| + 2$

14.

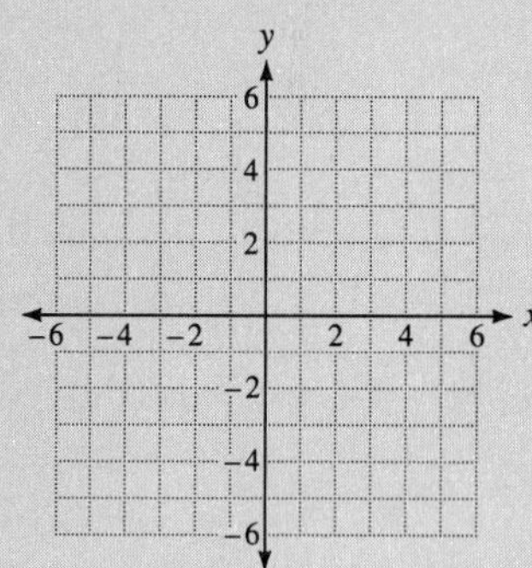

15. $f(x) = |x + 2|$

15.

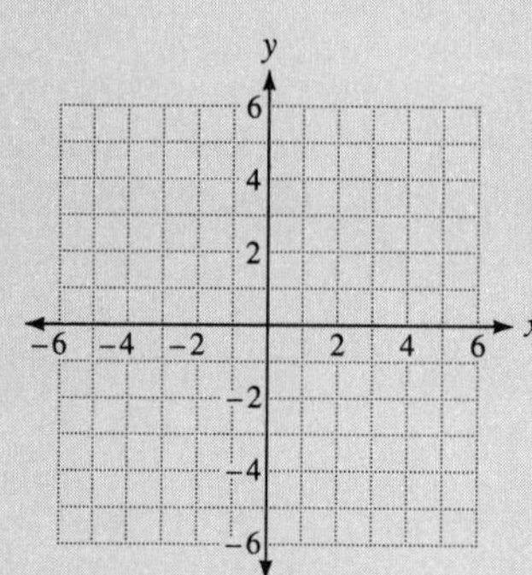

CHAPTER 10

Exponential and Logarithmic Functions

10.1 BASIC CONCEPTS

STUDENT GUIDEPOSTS

1. Logarithms
2. Exponential and Logarithmic Forms
3. Exponential and Logarithmic Properties
4. Calculating Exponentials

Historically, logarithms were developed to help carry out complicated numerical calculations. With the advent of computers and calculators, however, computing with logarithms is no longer a necessary skill. Nevertheless, logarithmic and exponential equations and functions remain important, with numerous applications in mathematics.

Calculators are invaluable tools for working with exponentials and logarithms; hence, our presentation is geared toward their use. If you prefer to use a table of logarithms, the appendix at the back of the text provides detailed instructions. In all calculator computations, we have stored the intermediate results with no rounding until the final step. There can be small differences in answers you get from a logarithmic table and from various calculators. But answers should be close enough to verify the accuracy of your work.

1 LOGARITHMS

In previous chapters we solved equations such as

$$3x + 2 = 7 \quad \text{and} \quad 2x^2 - 5x - 3 = 0.$$

In these equations, the variable does not appear in an exponent. We now consider equations such as

$$3^x = 9$$

in which the variable does occur in an exponent. The equation-solving rules we have already learned are of little help for these equations. At present, we must discover the solution by inspection. In this case, the problem is not too difficult, and we would probably recognize that when 3 is raised to the second power, the result is 9. Thus, the solution to the equation $3^x = 9$ is

$$x = 2.$$

Alternatively, we might give the solution in words.

x is the exponent on 3 which gives the number 9

The word *logarithm* can be used instead of *exponent*. The solution might be written

x is the logarithm on 3 which gives the number 9.

Since 3 is called the *base* in the expression 3^x, the solution can also be written

x is the logarithm to the base 3 of 9.

This final statement is usually symbolized by

$$x = \log_3 9.$$

2 EXPONENTIAL AND LOGARITHMIC FORMS

We showed above that

$$3^x = 9 \quad \text{and} \quad x = \log_3 9$$

are two forms of the same equation, with $3^x = 9$ called the **exponential form** and $x = \log_3 9$ called the **logarithmic form.** That is, they are equivalent equations, with the logarithmic form being "solved" for the variable x.

Definition of Logarithm

Let a, x, and y be real numbers, $a > 0$, $a \neq 1$, that satisfy the exponential equation

$$a^x = y.$$

Then x is the **logarithm to the base a of y,** and the equivalent logarithmic equation is

$$x = \log_a y.$$

Notice in the definition, that we restrict a to values greater than 0 since if $a < 0$, then a^x would not be defined for numbers such as $x = \frac{1}{2}$. Also, if $a = 1$ we have $a^x = a^y$ for all x and y. Remember also that the word *logarithm* means *exponent*.

EXAMPLE 1 CONVERTING FORMS

Give an equivalent equation for each of the following.

Equation	*Equivalent equation*	
$2^x = 32$	$x = \log_2 32$	$x = 5$ in this case
$7^x = c$	$x = \log_7 c$	
$x = \log_3 8$	$3^x = 8$	
$x = \log_5 125$	$5^x = 125$	$x = 3$ in this case

PRACTICE EXERCISE 1

Give an equivalent equation for each of the following.

(a) $8^x = 64$

(b) $b^x = 5$

(c) $x = \log_4 37$

(d) $x = \log_9 81$

Answers: (a) $x = \log_8 64$ (b) $x = \log_b 5$ (c) $4^x = 37$ (d) $9^x = 81$

To convert an equation from exponential to logarithmic form, or vice versa, it may help to remember that in both cases the base is written below the level of the logarithm (exponent), as indicated in the diagram on the next page.

exponent

$a^x = y$ $\qquad$ $x = \log_a y$

base

Since solutions to simple exponential equations can often be found by direct inspection, solutions to some logarithmic equations are more easily found by converting to exponential form.

EXAMPLE 2 Finding the Value of a Logarithm

Give the numerical value of x in $x = \log_4 16$.

We first convert $x = \log_4 16$ to exponential form: $4^x = 16$. At this point, it might be clear that x must be 2 since $4^2 = 16$. By writing 16 as a power of 4, $16 = 4^2$, we have

$$4^x = 4^2.$$

Since for any a ($a \neq 0$ and $a \neq 1$), $a^x = a^y$ implies that $x = y$, we conclude that $x = 2$.

Practice Exercise 2

Give the numerical value of x in $x = \log_3 81$.

Answer: 4

EXAMPLE 3 Solving a Logarithmic Equation

Find the numerical value of x in $\log_x \frac{1}{8} = -3$.

Convert to exponential form.

$$x^{-3} = \frac{1}{8}$$

$$x^{-3} = 2^{-3} \qquad \frac{1}{8} = \frac{1}{2^3} = 2^{-3}$$

It is now clear that x is 2 since the exponents in both sides of the equation are -3, forcing the bases in both sides to be equal. That is, in general, if $a^x = b^x$, then $a = b$ ($x \neq 0$).

Practice Exercise 3

Find the numerical value of x in $\log_x \frac{1}{16} = -2$.

Answer: 4

3 EXPONENTIAL AND LOGARITHMIC PROPERTIES

Notice in Examples 2 and 3, we worked toward writing an equation in which each side was an exponential form, and then equated either exponents or bases. These properties of exponentials, summarized below, will become even more apparent when we discuss exponential functions in Section 10.2.

Properties of Exponentials

Let x and y be real numbers, a and b positive real numbers with $a \neq 1$ and $b \neq 1$.

1. **$a^x = a^y$ if and only if $x = y$.**
2. **$a^x = b^x$ if and only if $a = b$.**

We now develop two important properties of logarithms. Since

$$a^0 = 1 \quad \text{(for } a \text{ any number except 0)},$$

rewriting this equation in logarithmic form, we have

$$\log_a 1 = 0.$$

Likewise, since $a^1 = a$ (for a any number),

converting to logarithmic form, we have

$$\log_a a = 1.$$

These two results are summarized next.

Properties of Logarithms

If a is any base ($a > 0$, $a \neq 1$),

1. $\log_a a = 1$ $(a^1 = a)$
2. $\log_a 1 = 0.$ $(a^0 = 1)$

For example, $\log_5 5 = 1$ (since $5^1 = 5$), $\log_5 1 = 0$ (since $5^0 = 1$), and $\log_{10} 10 = 1$ (since $10^1 = 10$).

4 CALCULATING EXPONENTIALS

A calculator with a [y^x] key can be helpful for calculating exponentials. For example, if we wanted to approximate 2^π, we would use this sequence of keys:

2 [y^x] [π] [=] ⟶ [8.82498]

The display would show 8.82498.

EXAMPLE 4 Calculating Exponentials

Use a calculator to approximate each number correct to four decimal places.

(a) $2^{\sqrt{2}}$

2 [y^x] 2 [$\sqrt{\ }$] [=] ⟶ [2.66514]

Thus, $2^{\sqrt{2}} \approx 2.6651$.

(b) $(1.04)^{60}$

1.04 [y^x] 60 [=] ⟶ [10.5196]

Thus, $(1.04)^{60} \approx 10.5196$.

Practice Exercise 4

Calculate each number correct to three decimal places.

(a) $\pi^{\sqrt{3}}$

(b) $(1.085)^{40}$

Answers: (a) 7.263 (b) 26.133

Many applied problems, such as the one in the next example, involve exponentials.

EXAMPLE 5 Biology Problem

The number of bacteria N in a culture can be approximated by $N = (1000)2^t$ where t is time in hours. How many bacteria are present in 12.5 hours?

Substitute 12.5 for t and find the value of N using a calculator.

$$N = (1000)2^{12.5} = 5{,}792{,}620$$

1000 [×] 2 [y^x] 12.5 [=] ⟶ [5792620]

There are about 5,792,620 bacteria after 12.5 hours.

Practice Exercise 5

Use the formula $N = (1000)2^t$ to find the number of bacteria present after 20 hours.

Answer: 1.0486×10^9

10.1 EXERCISES A

Convert each equation to logarithmic form.

1. $2^3 = 8$ **2.** $5^x = 100$ **3.** $m = n^p$

4. $8^0 = 1$ **5.** $7^{1/2} = c$ **6.** $10^{-2} = 0.01$

Convert each equation to exponential form.

7. $\log_a 3 = x$ **8.** $\log_{10} 100 = 2$ **9.** $\log_8 4 = \frac{2}{3}$

10. $\log_a 1 = 0$ **11.** $u = \log_v 7$ **12.** $\log_b \left(\frac{1}{b}\right) = -1$

Give the numerical value of x in each of the following exponential equations.

13. $2^x = 8$ **14.** $3^x = 81$ **15.** $5^x = \frac{1}{5}$

16. $10^x = 0.0001$ **17.** $x^{-1} = \frac{1}{4}$ **18.** $x^{1/2} = 3$

Find the numerical value of x by first converting to the equivalent exponential equation.

19. $x = \log_2 16$ **20.** $x = \log_8 2$ **21.** $\log_x \frac{1}{4} = -1$

22. $\log_x 11 = 1$ **23.** $\log_4 x = \frac{1}{2}$ **24.** $\log_2 x = 5$

25. If we know that $5^x = 5^{3.7}$, then what is x? **26.** If we know that $7^{2.3} = y^{2.3}$, then what is y?

Evaluate each logarithm.

27. $\log_4 4$ **28.** $\log_7 7$ **29.** $\log_3 1$ **30.** $\log_8 1$

**Use a calculator to approximate each number, correct to three decimal places.*

31. $5^{\sqrt{2}}$ **32.** 5^{π} **33.** $(1.08)^{50}$

34. Change $y = \log_3 (-9)$ to exponential form. Is there a value of y that makes this true?

35. In a scientific study, it has been shown that the variables w and v satisfy the equation $w = \log_{100} v$. **(a)** What is the value of v when w is $\frac{1}{2}$? **(b)** What is the value of w when v is 10,000?

*Since a calculator may be helpful in working many of the problems throughout this chapter, individual problems will not be designated as calculator exercises.

36. An electrical engineer uses the formula $g = 10 \log_{10} \frac{a}{b}$ to calculate the gain on an amplifier. What is the value of g when a is 200 and b is 20?

37. A banker uses the formula $A = P(1.06)^n$ to calculate interest compounded annually. What is the value of A when P is \$1000 and n is 20 years?

38 The number of bacteria N in a culture is given in terms of time t, in hours, by $N = (2000)2^t$. How many bacteria are present initially when $t = 0$? How many are present in 7 hours?

39. The atmospheric pressure P on an object, in pounds per square inch, can be approximated by $P = 14.7(2.7)^{-0.2x}$ where x is the height of the object in miles above sea level. What is the pressure on an object 2 miles high?

FOR REVIEW

Exercises 40–42 review material from Chapter 3 to help you prepare for the next section. Graph each function.

40. $y = 2x$

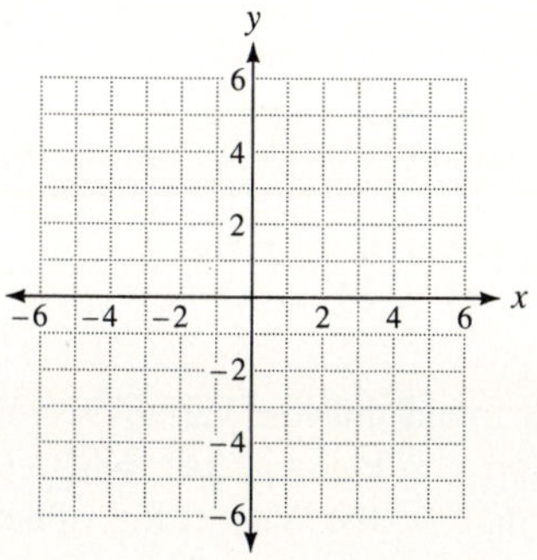

41. $y = -2x$

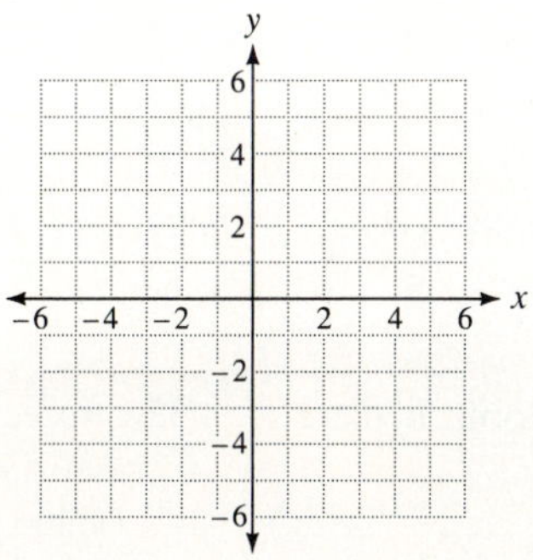

42. $y = x^2 + 1$

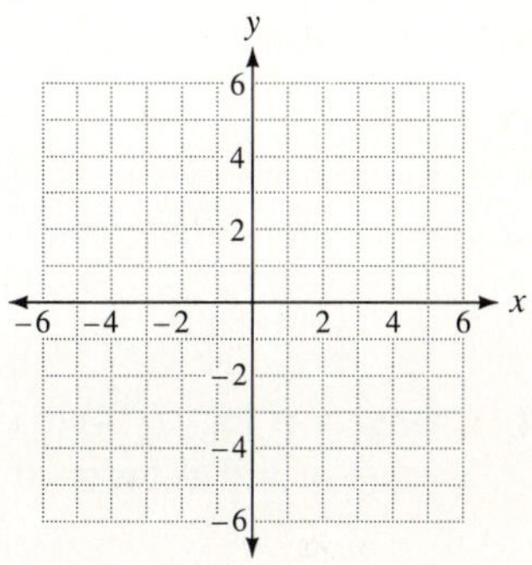

ANSWERS: 1. $\log_2 8 = 3$ 2. $\log_5 100 = x$ 3. $\log_n m = p$ 4. $\log_8 1 = 0$ 5. $\log_7 c = \frac{1}{2}$ 6. $\log_{10}(0.01) = -2$ 7. $a^x = 3$ 8. $10^2 = 100$ 9. $8^{2/3} = 4$ 10. $a^0 = 1$ 11. $v^u = 7$ 12. $b^{-1} = \frac{1}{b}$ 13. $x = 3$ 14. $x = 4$ 15. $x = -1$ 16. $x = -4$ 17. $x = 4$ 18. $x = 9$ 19. $x = 4$ 20. $x = \frac{1}{3}$ 21. $x = 4$ 22. $x = 11$ 23. $x = 2$ 24. $x = 32$ 25. 3.7 26. 7 27. 1 28. 1 29. 0 30. 0 31. 9.739 32. 156.993 33. 46.902 34. $3^y = -9$; no 35. (a) 10 (b) 2 36. 10 37. \$3207.14 38. 2000; 256,000 39. 9.88 lb/in²

40.

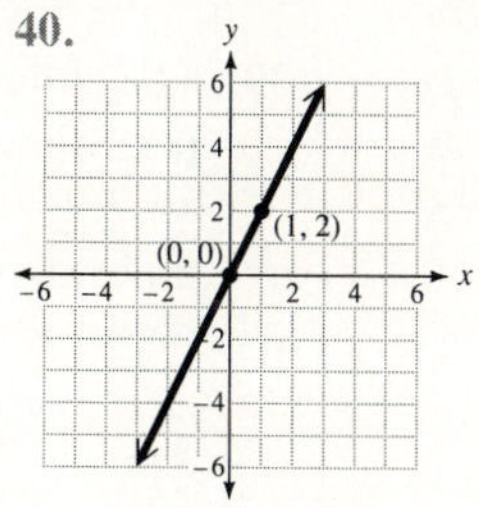

41.

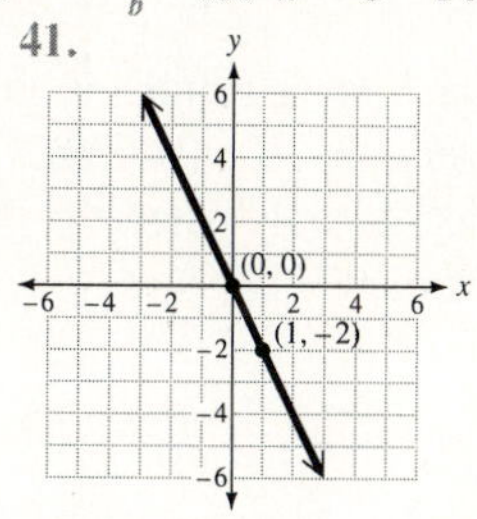

42.

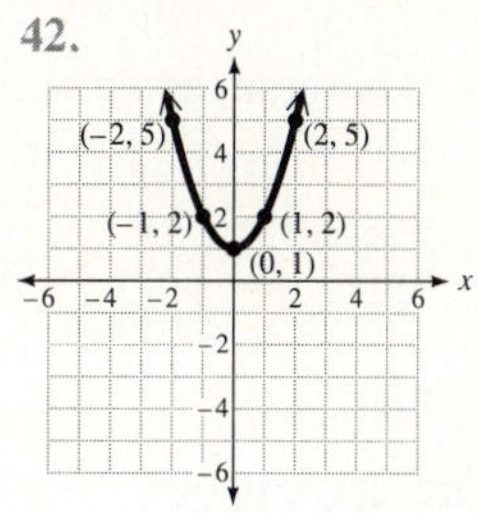

10.1 EXERCISES B

Convert each equation to logarithmic form.

1. $2^4 = 16$

2. $3^x = 200$

3. $u = v^w$

4. $9^0 = 1$

5. $11^{1/2} = a$

6. $10^{-3} = 0.001$

Convert each equation to exponential form.

7. $\log_a 5 = x$

8. $\log_{10} 1000 = 3$

9. $\log_{27} 9 = \dfrac{2}{3}$

10. $\log_b 1 = 0$

11. $w = \log_v 11$

12. $\log_a \left(\frac{1}{a}\right) = -1$

Give the numerical value of x in each of the following exponential equations.

13. $2^x = 4$

14. $3^x = 27$

15. $7^x = \frac{1}{7}$

16. $10^x = 100$

17. $x^5 = 243$

18. $x^{1/3} = 2$

Find the numerical value of x by first converting to the equivalent exponential equation.

19. $x = \log_2 8$

20. $x = \log_{16} 2$

21. $\log_x \frac{1}{3} = -1$

22. $\log_x 15 = 1$

23. $\log_9 x = \frac{1}{2}$

24. $\log_2 x = 6$

25. If we know that $9^x = 9^{4.1}$, then what is x?

26. If we know that $12^{-3.9} = y^{-3.9}$, then what is y?

Evaluate each logarithm.

27. $\log_8 8$

28. $\log_\pi \pi$

29. $\log_7 1$

30. $\log_{\sqrt{2}} 1$

Use a calculator to approximate each number, correct to three decimal places.

31. $4^{\sqrt{3}}$

32. $\pi^{\sqrt{2}}$

33. $(1.20)^{25}$

34. Change $y = \log_4 (-4)$ to exponential form. Is there a value of y that makes this true?

35. An economist has found that the variables c and p satisfy the equation $c = \log_{25} p$. **(a)** What is the value of p when c is $\frac{1}{2}$? **(b)** What is the value of c when p is $\frac{1}{25}$?

36. An electrical engineer uses the formula $g = 10 \log_{10} \frac{a}{b}$ to calculate the gain on an amplifier. What is the value of g when a is 300 and b is 30?

37. A banker uses the formula $A = P(1.06)^n$ to calculate interest compounded annually. What is the value of A when P is \$1000 and n is 30 years?

38. The number of bacteria N in a culture is given in terms of time t, in hours, by $N = (4000)2^t$. How many bacteria are present initially when $t = 0$? How many are present in 7 hours?

39. The atmospheric pressure P on an object, in pounds per square inch, can be approximated by $P = 14.7(2.7)^{-0.2x}$ where x is the height of the object in miles above sea level. What is the pressure on an object 5 miles high?

FOR REVIEW

Exercises 40–42 review material from Chapter 3 to help you prepare for the next section. Graph each function.

40. $y = 3x$

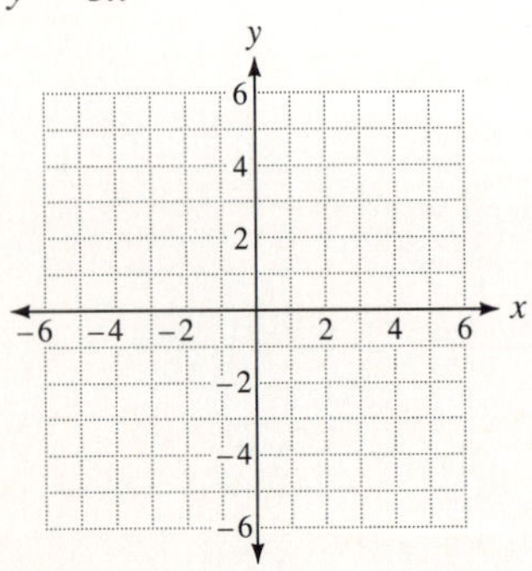

41. $y = -3x$

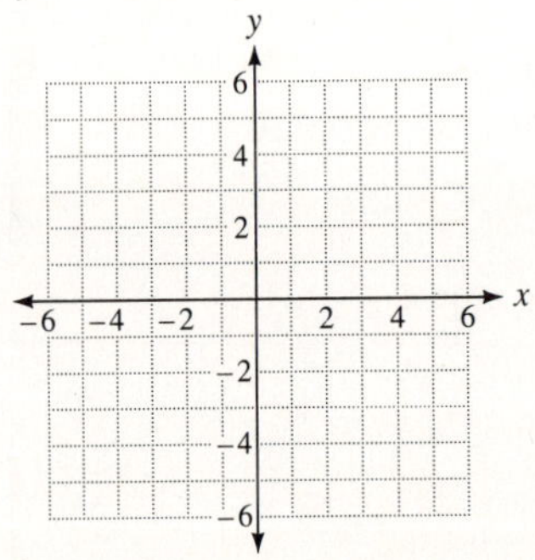

42. $y = x^2 - 1$

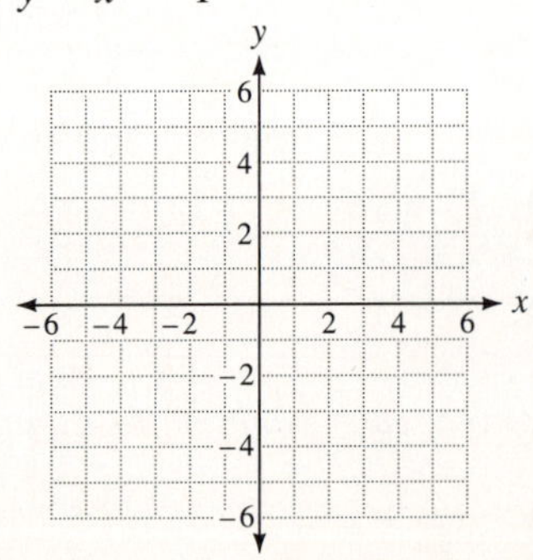

10.1 EXERCISES C

Give the numerical value of x.

1. $\log_3 (x^2 + 8x) = 2$

2. $\log_{\sqrt{2}} x = -2$

3. $\log_3 \sqrt[3]{9} = x$

4. $\log_x \sqrt{7} = \frac{1}{2}$

5. $\log_x \sqrt[3]{5} = \frac{1}{3}$

6. $|\log_5 x^2| = 2$

[Answer: $\pm 5, \pm \frac{1}{5}$]

10.2 LOGARITHMIC AND EXPONENTIAL FUNCTIONS

STUDENT GUIDEPOSTS

1. Exponential Functions
2. Logarithmic Functions

1 EXPONENTIAL FUNCTIONS

In Chapters 3 and 9 we considered several functions, including constant, linear, and quadratic functions, and their graphs. Two other important kinds of functions are exponential and logarithmic functions. A function defined by the equation

$$y = a^x, \quad a > 0, a \neq 1$$

is an **exponential function,** and the number a is called the **base.** It is not difficult to calculate the value of y when x is a rational number p/q since $a^{p/q}$ is defined to be the qth root of a raised to the pth power. Although we have not defined a^x for irrational x such as $x = \sqrt{2}$ or $x = \pi$, we shall assume that $a^{\sqrt{2}}$ can be approximated as closely as we wish by using rational approximations of $\sqrt{2}$. For example, since $\sqrt{2} \approx 1.414$,

$$a^{\sqrt{2}} \approx a^{1.414} = a^{1414/1000},$$

which does have meaning for us. The base a to other irrational powers can be approximated in a similar way, or by using a calculator with a $\boxed{y^x}$ key.

We now graph $y = 2^x$ and show that our assumption about irrational powers is reasonable. We select values of x and calculate the corresponding values of y to obtain the table of values in Figure 10.1. We can use 1.41 as an approximation for $\sqrt{2}$, and from the graph, we see that 2.7 is an approximation for $2^{\sqrt{2}}$. Similarly, using 3.14 as an approximation for π, we can estimate 2^{π} to be about 8.8. A calculator gives 2.66514 for $2^{\sqrt{2}}$ and 8.82498 for 2^{π}. These values are of course more accurate than the values we are able to read from the graph.

To compare exponential functions, we graph $y = 2^x$, $y = \left(\frac{1}{2}\right)^x$, $y = 3^x$, $y = \left(\frac{1}{3}\right)^x$, $y = 10^x$, and $y = \left(\frac{1}{10}\right)^x$ in the same coordinate system. First we construct a table of values. The scale for our graphs in Figure 10.2 does not allow us to plot all of the points listed in the table, but the values are included there for purposes of comparison.

x	y
0	1
−1	$\frac{1}{2}$
1	2
−2	$\frac{1}{4}$
2	4
−3	$\frac{1}{8}$
3	8
−4	$\frac{1}{16}$

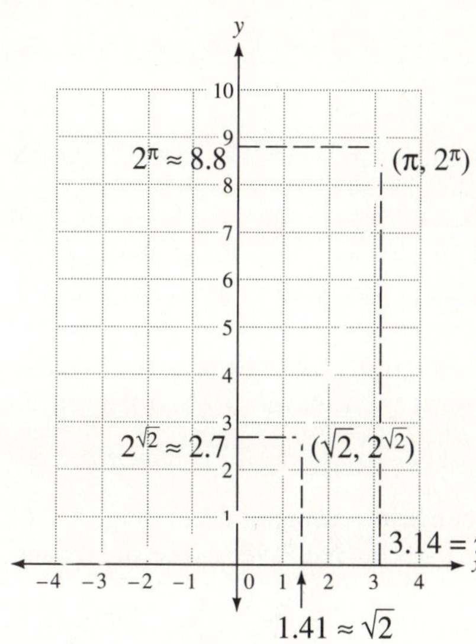

Figure 10.1

x	$y = 2^x$	$y = \left(\frac{1}{2}\right)^x$	$y = 3^x$	$y = \left(\frac{1}{3}\right)^x$	$y = 10^x$	$y = \left(\frac{1}{10}\right)^x$
0	1	1	1	1	1	1
-1	$\frac{1}{2}$	2	$\frac{1}{3}$	3	$\frac{1}{10}$	10
1	2	$\frac{1}{2}$	3	$\frac{1}{3}$	10	$\frac{1}{10}$
-2	$\frac{1}{4}$	4	$\frac{1}{9}$	9	$\frac{1}{100}$	100
2	4	$\frac{1}{4}$	9	$\frac{1}{9}$	100	$\frac{1}{100}$
-3	$\frac{1}{8}$	8	$\frac{1}{27}$	27	$\frac{1}{1000}$	1000
3	8	$\frac{1}{8}$	27	$\frac{1}{27}$	1000	$\frac{1}{1000}$
-4	$\frac{1}{16}$	16	$\frac{1}{81}$	81	$\frac{1}{10,000}$	10,000
4	16	$\frac{1}{16}$	81	$\frac{1}{81}$	10,000	$\frac{1}{10,000}$

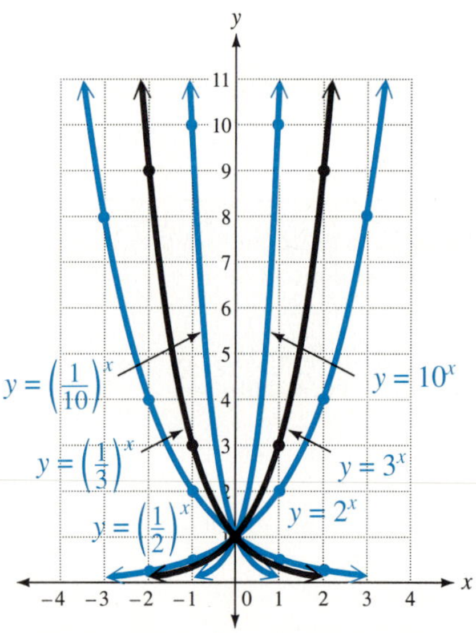

Figure 10.2 Exponential Functions

❷ LOGARITHMIC FUNCTIONS

Several observations can be made about the functions above. In all cases the graph passes through the point (0, 1), since $a^0 = 1$ for any value of a. For every base a greater than 1, the graph rises from left to right, and toward the left it gets closer and closer to (but never crosses) the x-axis. For every base a less than 1, the graph decreases from left to right, and toward the right it gets closer and closer to (but never crosses) the x-axis. Since each exponential function

$$y = a^x \qquad a > 0 \text{ and } a \neq 1$$

is either increasing or decreasing, it is one-to-one and has an inverse. To find the equation of that inverse, we interchange x and y,

$$x = a^y \qquad x \text{ and } y \text{ have been interchanged}$$

and solve for y. To solve this equation for y we simply convert it to logarithmic form.

$$y = \log_a x.$$

A **logarithmic function** is a function defined by an equation of the form

$$y = \log_a x.$$

To graph a logarithmic function we can graph its equivalent exponential form, keeping in mind that the roles of x and y have been interchanged. Suppose we graph $y = \log_2 x$ by plotting $x = 2^y$.

$x = 2^y$	y
1	0
$\frac{1}{2}$	-1
2	1
$\frac{1}{4}$	-2
4	2
$\frac{1}{8}$	-3
8	3

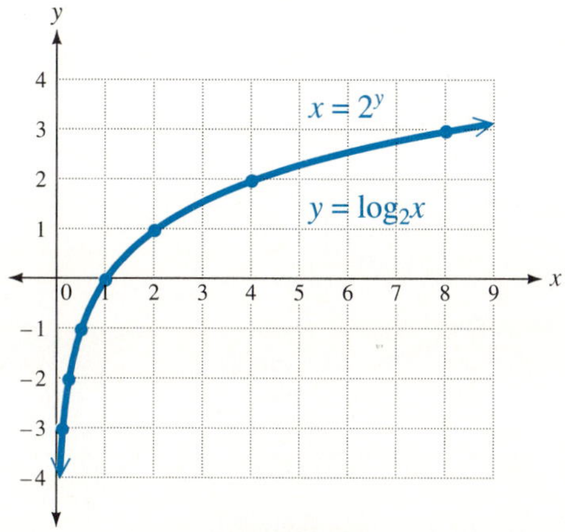

Figure 10.3

To construct the table of values in Figure 10.3, we find x for a given value of y, then plot the points, and join them with a smooth curve.

If we restrict ourselves to $a > 1$, the graph of each exponential function $y = a^x$ has the same basic shape, shown in Figure 10.4, and passes through the point (0, 1). The graph of every logarithmic function $y = \log_a x$ has the same basic shape, shown in Figure 10.5, and passes through (1, 0). With this in mind, we can quickly graph an exponential or logarithmic function by plotting a few points.

We can also observe, by plotting the graphs of

$$y = \log_a x \quad \text{and} \quad y = a^x$$

in the same coordinate system, as in Figure 10.6, that each graph is the reflection of the other across the line $y = x$. This illustrates the fact that these functions are inverses of each other. If the graph of either is known, the graph of the other can easily be found by reflection in the line $y = x$.

There are many applications of logarithmic and exponential functions. The chemistry problem on the next page is one example.

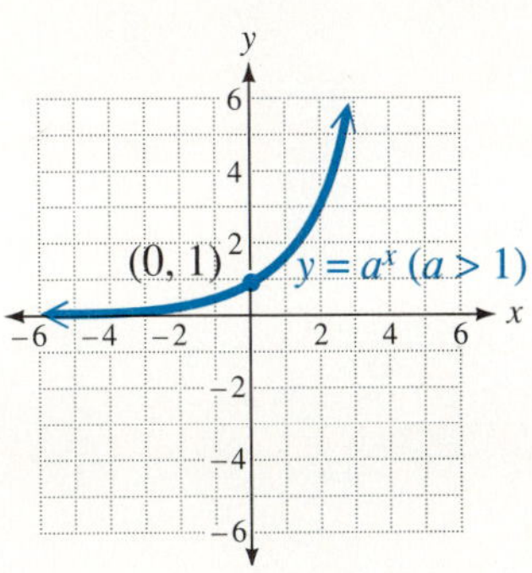

Figure 10.4
General Exponential Function

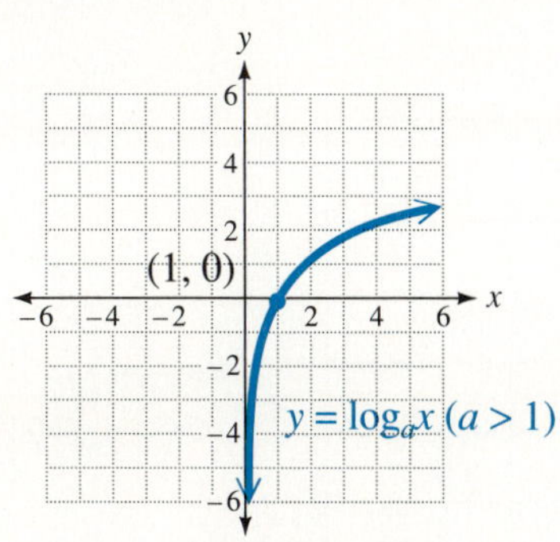

Figure 10.5
General Logarithmic Function

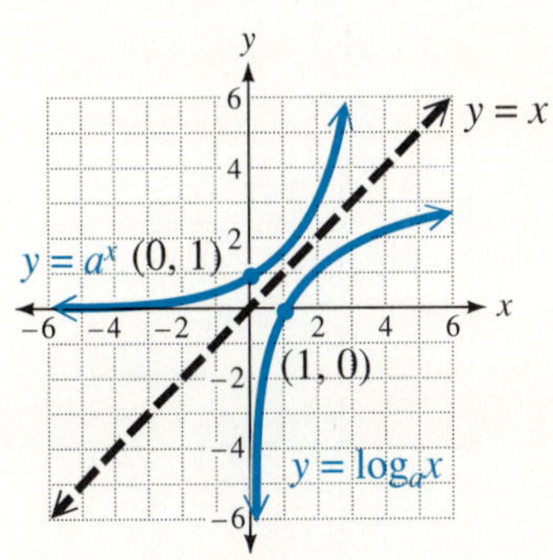

Figure 10.6
Inverse Functions

EXAMPLE 1 CHEMISTRY PROBLEM

A radioactive isotope has a half-life of 1200 years. This means that if 1000 grams of the isotope are allowed to decay, the amount A that remains after t years is given by the exponential function $A = (1000)2^{-t/1200}$. Find the amount remaining at the end of 2400 years and sketch the graph of the function for $t \geq 0$.

To find the amount remaining after 2400 years, substitute 2400 for t.

$$\begin{aligned} A &= (1000)2^{-2400/1200} \\ &= (1000)2^{-2} \\ &= (1000)\frac{1}{2^2} \\ &= (1000)\frac{1}{4} \\ &= 250 \end{aligned}$$

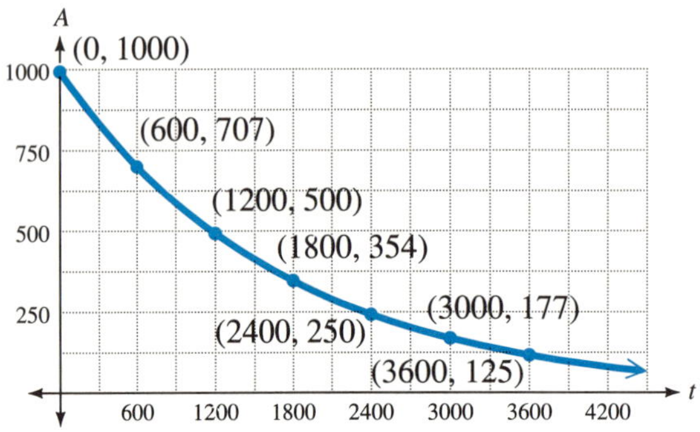

Figure 10.7 $A = (1000)2^{-t/1200}$

Thus 250 grams remain after 2400 years. The graph of the function for $t \geq 0$ is given in Figure 10.7.

PRACTICE EXERCISE 1

Refer to Example 1 and find the amount of radioactive isotope that would remain after 5000 years.

Answer: About 55.7 grams

10.2 EXERCISES A

1. Given an exponential function $y = a^x$, a is called the ______________.
2. The graph of the logarithmic function $y = \log_a x$ can be found by plotting the points that satisfy the exponential equation ______________.
3. The graph of $y = \log_3 x$ can be found by reflecting the graph of $y = 3^x$ across the line with equation ______________.
4. The graph of $y = 5^x$ and $y =$ ______________ are reflections of each other across the line $y = x$.
5. Complete the following table and sketch the graphs of the functions $y = 2^x$, $y = \left(\frac{1}{3}\right)^x$, $y = 3^x$, $y = 4^x$ in the given coordinate system.

x	$y = 2^x$	$y = \left(\frac{1}{3}\right)^x$	$y = 3^x$	$y = 4^x$
0				
−1				
1				
−2				
2				
−3				
3				

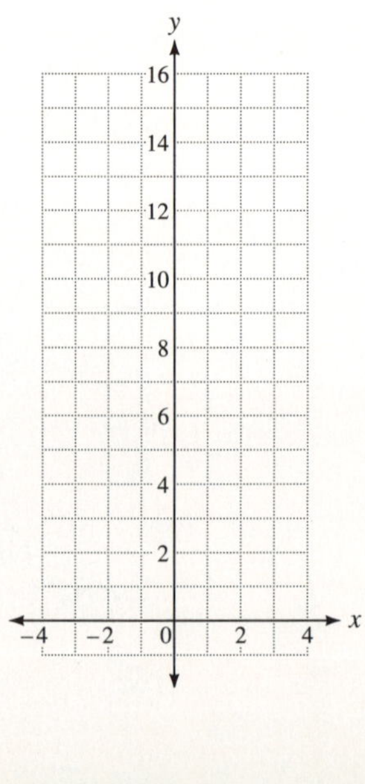

6. Sketch the graph of $y = \log_3 x$ by plotting $x = 3^y$.

x	y
	0
	1
	−1
	2
	−2

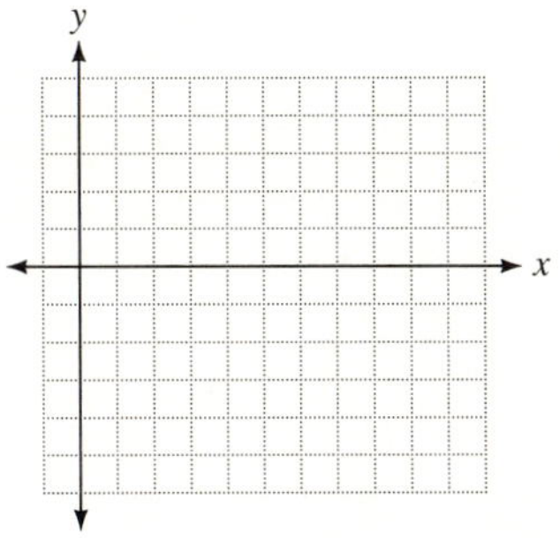

7. Having plotted $y = 3^x$ in Exercise 5, could you have used this information to sketch the graph required in Exercise 6? Explain.

8. The half-life of an antibiotic in the bloodstream is about 8 hours. This means that when 64 milligrams of the drug are absorbed in your bloodstream, the number D of milligrams remaining after t hours is given by the function $D = (64)2^{-t/8}$. Graph this function for $t \geq 0$ and find the number of milligrams of the antibiotic remaining in the bloodstream after **(a)** 0 hours, **(b)** 8 hours, **(c)** 16 hours, and **(d)** 24 hours.

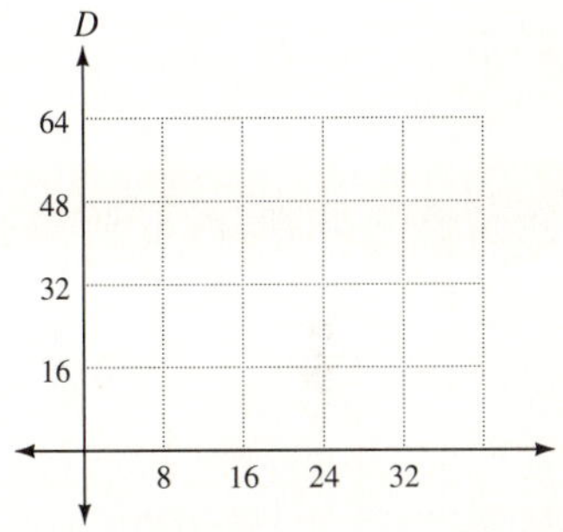

9. A bacteria culture containing 100 cells doubles in number each hour. The number N of bacteria present after t hours is given by the exponential function $N = (100)2^t$. Graph this function for $t \geq 0$ and find the number of bacteria present after **(a)** 0 hours, **(b)** 2 hours, **(c)** 4 hours, and **(d)** 6 hours.

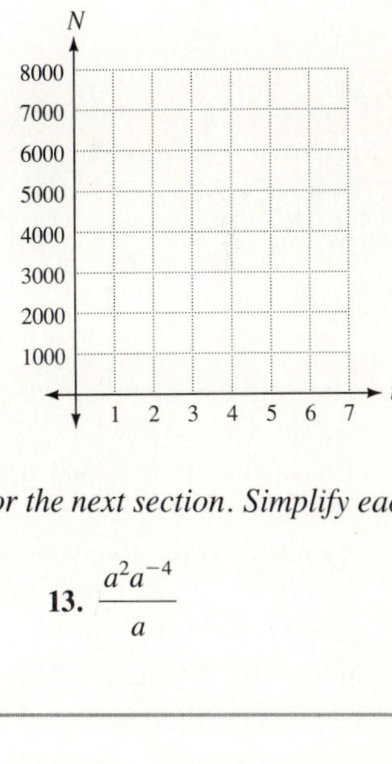

FOR REVIEW

Exercises 10–13 review properties of exponents from Section 1.7 to help you prepare for the next section. Simplify each expression.

10. a^3a^5 **11.** $\dfrac{a^8}{a^2}$ **12.** $(a^5)^3$ **13.** $\dfrac{a^2a^{-4}}{a}$

ANSWERS:

1. base
2. $x = a^y$
3. $y = x$
4. $\log_5 x$

5.

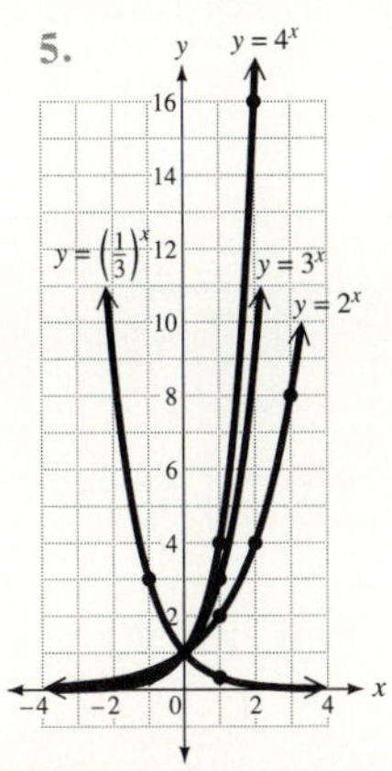

6.

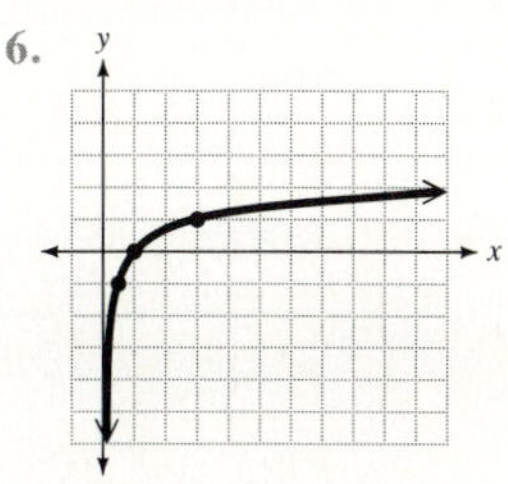

7. yes; by reflecting the graph across the line $y = x$

8.

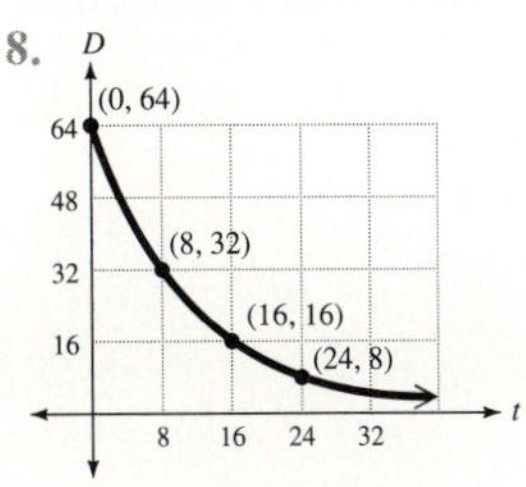

(a) 64 mg (b) 32 mg (c) 16 mg (d) 8 mg

9.

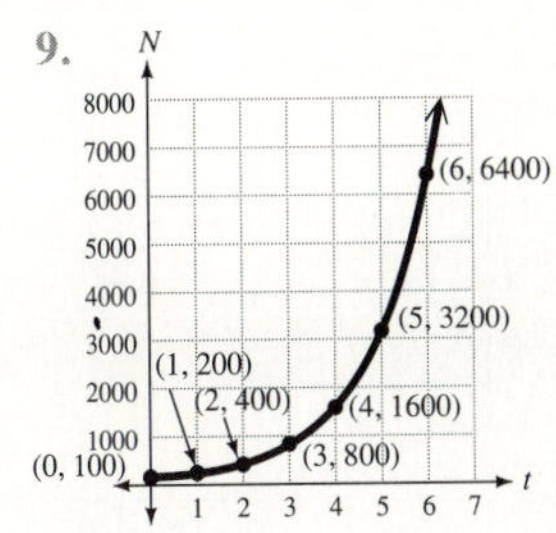

(a) 100 (b) 400 (c) 1600 (d) 6400

10. a^8 11. a^6 12. a^{15} 13. $\frac{1}{a^3}$

10.2 EXERCISES B

1. A function described by the equation $y = a^x$, $a > 0$, is called a(n) ________________ function.

2. The graph of the logarithmic function, ________________, can be found by plotting the points that satisfy the exponential equation $x = a^y$.

3. The graph of $y = \log_5 x$ can be found by reflecting the graph of ________________ across the line with equation $y = x$.

4. The graph of ________________ and $y = \log_{12} x$ are reflections of each other across the line $y = x$.

5. Sketch the graphs of the functions $y = 5^x$, $y = 5^{-x}$, $y = 2^{-x}$, $y = 4^x$ in the same coordinate system.

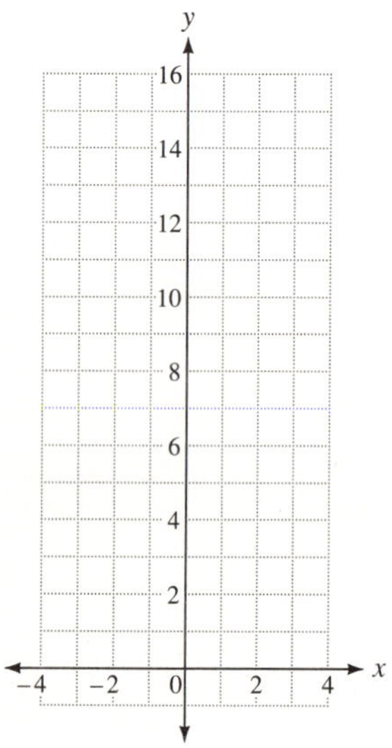

6. Sketch the graph of $y = \log_4 x$ by plotting $x = 4^y$.

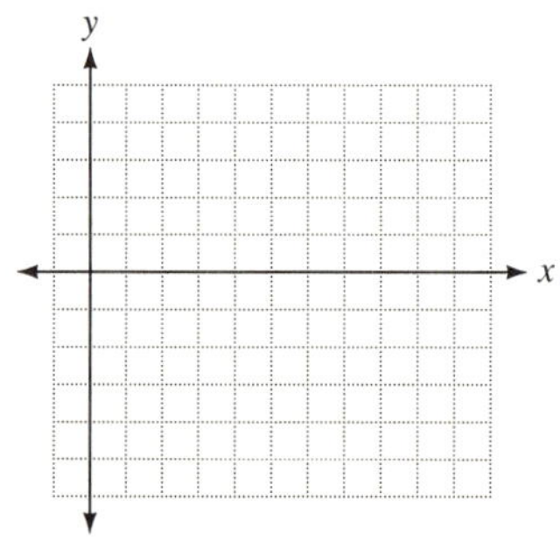

7. Having plotted $y = 4^x$ in Exercise 5, could you have used this information to sketch the graph required in Exercise 6? Explain.

8. The half-life of radioactive strontium-90, used for fuel in nuclear reactors, is approximately 28 years. This means that if 500 grams of strontium-90 are allowed to decay, the number A of grams remaining after t years is given by the exponential equation $A = (500)2^{-t/28}$. Graph this function for $t \geq 0$ and find the number of grams of strontium-90 remaining after **(a)** 0 years, **(b)** 28 years, **(c)** 56 years, and **(d)** 140 years.

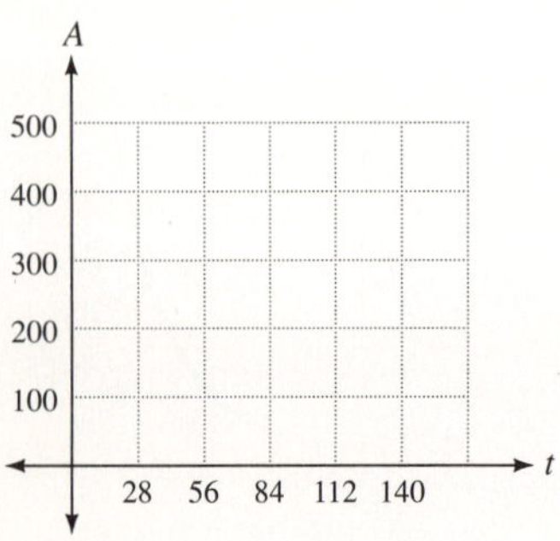

9. The value of a rare coin purchased for \$100 triples each year. This means that the value V after t years is given by the exponential function $V = (100)3^t$. Graph this function for $t \geq 0$ and determine the value of the coin after **(a)** 0 years, **(b)** 1 year, **(c)** 2 years, and **(d)** 5 years.

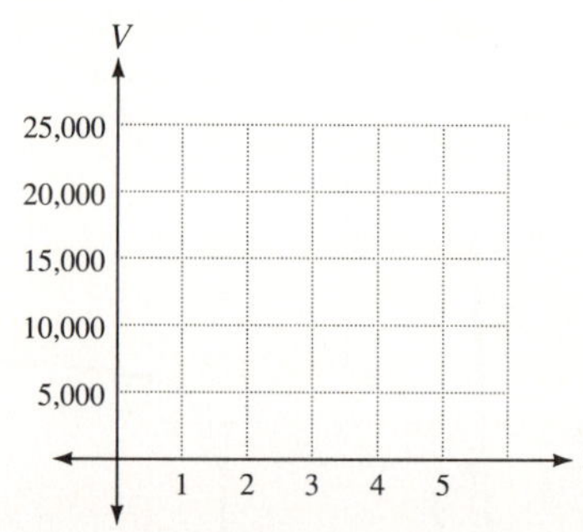

FOR REVIEW

Exercises 10–13 review properties of exponents from Section 1.7 to help you prepare for the next section. Simplify each expression.

10. a^4a^{10} **11.** $\frac{a^7}{a^3}$ **12.** $(a^6)^4$ **13.** $\frac{a^3a^{-2}}{a^{-1}}$

10.2 EXERCISES C

Graph the functions.

1. $y = 2^{x+1}$

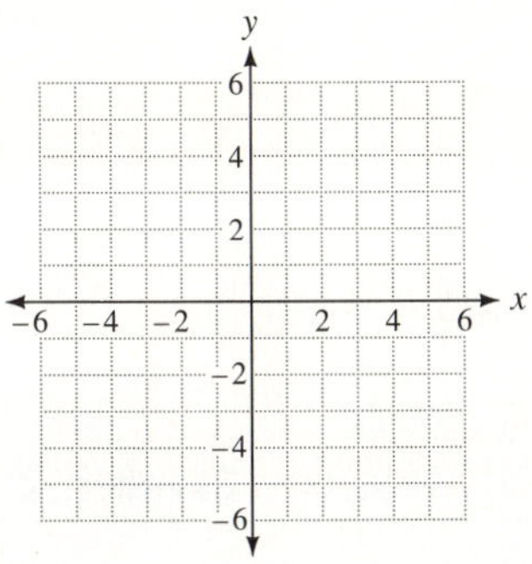

2. $y = \log_2 (x + 1)$

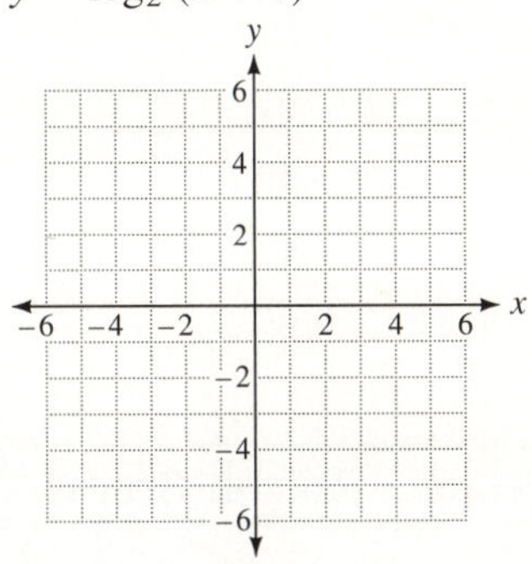

3. $y = 3^{2x}$

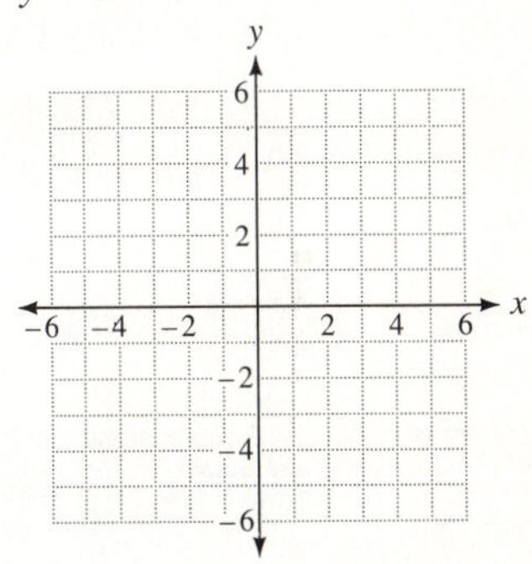

10.3 PROPERTIES OF LOGARITHMS

STUDENT GUIDEPOSTS

1. The Product Rule
2. The Quotient Rule
3. The Power Rule
4. Using a Combination of Rules

1 THE PRODUCT RULE

Since logarithms are exponents, the basic rules of exponents can be used to develop several useful properties of logarithms.

Suppose that a, x, and y are numbers, and a is a suitable base for a logarithm ($a > 0$, $a \neq 1$). The product rule for exponents is

$$a^x \cdot a^y = a^{x+y}.$$

Set $\quad u = a^x \quad \text{and} \quad v = a^y$

and write each in logarithmic form.

$$x = \log_a u \quad \text{and} \quad y = \log_a v$$

Then $\quad u \cdot v = a^x \cdot a^y = a^{x+y}.$

Write $u \cdot v = a^{x+y}$ in logarithmic form.

$$\log_a (u \cdot v) = x + y$$

Substitute $\log_a u$ for x and $\log_a v$ for y.

$$\log_a (u \cdot v) = \log_a u + \log_a v$$

This gives us the **product rule** for logarithms.

Product Rule

For any base a ($a > 0$, $a \neq 1$) and positive numbers u and v,

$$\log_a (u \cdot v) = \log_a u + \log_a v.$$

(The log of a product is the sum of the logs.)

EXAMPLE 1 Using the Product Rule

Express $\log_2 (4 \cdot 8)$ as a sum of logarithms.

$$\log_2 (4 \cdot 8) = \log_2 4 + \log_2 8 \qquad \text{Product rule}$$

In this case, we can verify the equation directly.

$$\log_2 (4 \cdot 8) = \log_2 32 = 5 \qquad 2^5 = 32$$
$$\log_2 4 = 2 \qquad 2^2 = 4$$
$$\log_2 8 = 3 \qquad 2^3 = 8$$

Hence, $\log_2 (4 \cdot 8) = 5 = 2 + 3 = \log_2 4 + \log_2 8.$

Practice Exercise 1

Express $\log_2 (16 \cdot 32)$ as a sum of logarithms.

Answer: $\log_2 16 + \log_2 32$

EXAMPLE 2 Using the Product Rule in Reverse

Express $\log_a 3 + \log_a w$ as a single logarithm.

$$\log_a 3 + \log_a w = \log_a (3 \cdot w) \qquad \text{Product rule in reverse}$$

Practice Exercise 2

Express $\log_a 17 + \log_a m$ as a single logarithm.

Answer: $\log_a (17 \cdot m)$

CAUTION

Do not conclude that the sum of two logs is the log of a sum.

$$\log_a u + \log_a v \neq \log_a (u + v)$$

For example,

$$\log_a 6 = \log_a (2 \cdot 3) = \log_a 2 + \log_a 3 \neq \log_a (2 + 3) = \log_a 5.$$

Also note that $\log_a (u \cdot v) \neq \log_a u \cdot \log_a v$.

2 THE QUOTIENT RULE

If a is a base for a logarithm, and x and y are numbers, the quotient rule for exponents is

$$\frac{a^x}{a^y} = a^{x-y}.$$

As before, we set

$$u = a^x \quad \text{and} \quad v = a^y$$

and express these in logarithmic form.

$$x = \log_a u \quad \text{and} \quad y = \log_a v$$

Then

$$\frac{u}{v} = \frac{a^x}{a^y} = a^{x-y}.$$

Writing $\frac{u}{v} = a^{x-y}$ in logarithmic form, we obtain

$$\log_a \frac{u}{v} = x - y.$$

Substituting $\log_a u$ for x and $\log_a v$ for y,

$$\log_a \frac{u}{v} = \log_a u - \log_a v.$$

This is the **quotient rule** for logarithms.

Quotient Rule

For any base a ($a > 0$, $a \neq 1$) and positive numbers u and v,

$$\log_a \frac{u}{v} = \log_a u - \log_a v.$$

(The log of a quotient is the difference of the logs.)

EXAMPLE 3 Using the Quotient Rule

Express $\log_2 \frac{32}{8}$ as a difference of logarithms.

$$\log_2 \frac{32}{8} = \log_2 32 - \log_2 8 \quad \text{Quotient rule}$$

In this case, we can verify the equation directly.

$$\log_2 \frac{32}{8} = \log_2 4 = 2 \qquad 2^2 = 4$$

$$\log_2 32 = 5 \qquad 2^5 = 32$$

$$\log_2 8 = 3 \qquad 2^3 = 8$$

Hence, $\log_2 \frac{32}{8} = 2 = 5 - 3 = \log_2 32 - \log_2 8.$

Practice Exercise 3

Express $\log_4 \frac{4}{64}$ as a difference of logarithms.

Answer: $\log_4 4 - \log_4 64$

EXAMPLE 4 Using the Quotient Rule in Reverse

Express $\log_a 3 - \log_a w$ as a single logarithm.

$$\log_a 3 - \log_a w = \log_a \frac{3}{w} \quad \text{Quotient rule in reverse}$$

Practice Exercise 4

Express $\log_a 17 - \log_a m$ as a single logarithm.

Answer: $\log_a \frac{17}{m}$

CAUTION

Do not conclude that the difference of two logs is the log of the difference.

$$\log_a u - \log_a v \neq \log_a (u - v)$$

For example,

$$\log_a 3 = \log_a \frac{6}{2} = \log_a 6 - \log_a 2 \neq \log_a (6 - 2) = \log_a 4.$$

Note also that $\log_a \frac{u}{v} \neq \frac{\log_a u}{\log_a v}$.

3 THE POWER RULE

Assume that

$$u = a^x$$

and raise both sides of this equation to the cth power.

$$u^c = (a^x)^c$$

By the power rule for exponents,

$$u^c = (a^x)^c = a^{x \cdot c}.$$

Converting $u^c = a^{x \cdot c}$ to logarithmic form, we obtain

$$\log_a u^c = x \cdot c.$$

But since $u = a^x$, $x = \log_a u$ so that

$$\log_a u^c = x \cdot c = (\log_a u) \cdot c = c \log_a u.$$

Thus, we have the **power rule** for logarithms.

Power Rule

For any base a ($a > 0$, $a \neq 1$), any positive number u and any number c,

$$\log_a u^c = c \log_a u.$$

(The log of a number to a power is the power times the log of the number.)

EXAMPLE 5 Using the Product Rule

(a) Express $\log_2 4^3$ as a product.

$$\log_2 4^3 = 3 \log_2 4 \quad \text{Power rule}$$

We can verify the above equation directly.

$$\log_2 4^3 = \log_2 64 = 6 \quad 2^6 = 64$$

$$3 \log_2 4 = 3 \cdot 2 = 6 \quad 2^2 = 4$$

(b) Express $\log_2 \sqrt[3]{64}$ as a product.

First we write $\sqrt[3]{64}$ as $64^{1/3}$. Then

$$\log_2 \sqrt[3]{64} = \log_2 64^{1/3} = \frac{1}{3} \log_2 64. \quad \text{Power rule}$$

The truth of this equation is clear since

$$\log_2 \sqrt[3]{64} = \log_2 4 = 2 \quad \sqrt[3]{64} = 4 \text{ and } 2^2 = 4$$

$$\text{and } \frac{1}{3} \log_2 64 = \frac{1}{3} \cdot 6 = 2. \quad 2^6 = 64$$

Practice Exercise 5

(a) Express $\log_7 8^2$ as a product.

(b) Express $\log_7 27^{1/3}$ as a product.

Answers: (a) $2 \log_7 8$ (b) $\frac{1}{3} \log_7 27$

CAUTION

Notice the difference between $\log_a u^c$ and $(\log_a u)^c$. For example,

$$\log_a u^2 = 2 \log_a u \quad \text{but} \quad (\log_a u)^2 = (\log_a u)(\log_a u).$$

4 USING A COMBINATION OF RULES

Logarithms of expressions that involve more complicated products, quotients, or powers can be simplified using a combination of the rules.

EXAMPLE 6 Using a Combination of Rules

Express $\log_a \frac{x\sqrt{y}}{z^3}$ in terms of logarithms of x, y, and z.

$$\log_a \frac{x\sqrt{y}}{z^3} = \log_a \frac{x \cdot y^{1/2}}{z^3} \quad \text{Convert radical to fractional exponent}$$

$$= \log_a (x \cdot y^{1/2}) - \log_a z^3 \quad \text{Quotient rule}$$

$$= \log_a x + \log_a y^{1/2} - \log_a z^3 \quad \text{Product rule}$$

$$= \log_a x + \frac{1}{2} \log_a y - 3 \log_a z \quad \text{Power rule}$$

Practice Exercise 6

Express $\log_a \frac{u^3}{v^2\sqrt{w}}$ in terms of logarithms of u, v, and w.

Answer: $3 \log_a u - 2 \log_a v - \frac{1}{2} \log_a w$

Notice in Example 6 that the quotient and product rules are used first with the power rule last. When a combination of logarithms of this nature is simplified to a single logarithm using the rules in reverse, the power rule is used first followed by the quotient and product rules.

EXAMPLE 7 Using a Combination of Rules in Reverse

Express $\frac{1}{3} \log_a x - 5 \log_a y + \log_a z$ as a single logarithm.

$$\frac{1}{3} \log_a x - 5 \log_a y + \log_a z$$

$$= \log_a x^{1/3} - \log_a y^5 + \log_a z \quad \text{Always use power rule first}$$

$$= \log_a \frac{x^{1/3}}{y^5} + \log_a z \quad \text{Quotient rule}$$

$$= \log_a \left(\frac{x^{1/3}}{y^5} \cdot z\right) \quad \text{Product rule}$$

$$= \log_a \frac{z\sqrt[3]{x}}{y^5} \quad \text{Simplify}$$

Practice Exercise 7

Express $-4 \log_a u + \frac{1}{2} \log_a v - 8 \log_a w$ as a single logarithm.

Answer: $\log_a \frac{\sqrt{v}}{u^4 w^8}$

EXAMPLE 8 Using the Rules to Evaluate Logarithms

Given that $\log_a 2 = 0.3010$ and $\log_a 3 = 0.4771$, find the following logarithms.

(a) $\log_a 6 = \log_a (2 \cdot 3) = \log_a 2 + \log_a 3$
$= 0.3010 + 0.4771$
$= 0.7781$

(b) $\log_a 9 = \log_a 3^2 = 2 \log_a 3 = 2(0.4771) = 0.9542$

(c) $\log_a \frac{3}{2} = \log_a 3 - \log_a 2 = 0.4771 - 0.3010 = 0.1761$

Practice Exercise 8

Given that $\log_a 2 = 0.3010$ and $\log_a 5 = 0.6990$, find the following logarithms.

(a) $\log_a 10$

(b) $\log_a 50$

(c) $\log_a \frac{25}{4}$

Answers: (a) 1.0000 (b) 1.6990 (c) 0.7960

10.3 EXERCISES A

Express as a sum of logarithms.

1. $\log_3 (8 \cdot 80)$ **2.** $\log_5 (20 \cdot u)$ **3.** $\log_c (a \cdot b)$

Express as a difference of logarithms.

4. $\log_4 \frac{7}{30}$ **5.** $\log_2 \frac{12}{v}$ **6.** $\log_c \frac{a}{b}$

Express as a product.

7. $\log_2 x^5$ **8.** $\log_3 \sqrt{y}$ **9.** $\log_a b^{-3}$

Express as a single logarithm.

10. $\log_4 3 + \log_4 8$ **11.** $\log_b x + \log_b y$ **12.** $\log_2 5 - \log_2 9$ **13.** $\log_c x - \log_c y$

Express in terms of logarithms of x, y, and z.

14. $\log_a \frac{x^3y}{z}$ **15** $\log_a xy^2\sqrt{z}$ **16.** $\log_a \sqrt[3]{\frac{xy^2}{z}}$

Express as a single logarithm.

17 $\frac{1}{2} \log_a y - \log_a x + 2 \log_a z$ **18.** $3 \log_a x - \log_a y - \log_a z$ **19.** $\frac{1}{2}(\log_a x - \log_a y + 3 \log_a z)$

Given that $\log_a 3 = 0.4771$ *and* $\log_a 5 = 0.6990$, *find the following logarithms.*

20. $\log_a 15$ **21.** $\log_a \frac{5}{3}$ **22.** $\log_a 5^2$ **23.** $\log_a 75$

24. $\log_a \frac{25}{3}$ **25** $\log_a 0.2$ **26.** $\log_a \sqrt[3]{5}$ **27.** $\log_a \sqrt{15}$

Tell whether the following are true or false.

28. $\frac{\log_a u}{\log_a v} = \log_a u - \log_a v$ **29.** $\log_a a = 1$ **30.** $\log_a 1 = 0$

31. $\log_a u^c = (\log_a c)(\log_a u)$ **32.** $\log_a (u + v) = \log_a u + \log_a v$ **33.** $\log_a (u \cdot v) = (\log_a u)(\log_a v)$

Solve.

34. Dr. Gordon Johnson, a well-known meteorologist, uses the formula $W = \log_2 (m + n)$. Calculate the value of W when m is 10 and n is 6.

35. When Dr. Johnson's assistant, Burford, was asked to find the value of W (Exercise 34) when m and n are both 8, he performed the following calculation:

$$W = \log_2 (m + n) = \log_2 (8 + 8)$$
$$= \log_2 8 + \log_2 8 = 3 + 3 = 6.$$

Why was Burford fired? What is the actual value of W?

36. If $Y = \log_2 (a - b)$, calculate the value of Y when $a = 8$ and $b = 4$, and show that $Y \neq \log_2 a - \log_2 b$.

37. If $V = \log_3 \frac{a}{b}$, calculate the value of V when $a = 27$ and $b = 3$, and show that $V \neq \frac{\log_3 a}{\log_3 b}$.

FOR REVIEW

38. Convert $x^a = y$ to logarithmic form.

39. Convert $\log_b m = p$ to exponential form.

Give the numerical value of x in the following.

40. $2^x = 128$

41. $x = \log_3 81$

42. $\log_x \dfrac{1}{8} = -3$

43. If you have the graph of $y = 5^x$, how could you use it to obtain the graph of $y = \log_5 x$?

44. The number n of deer in a newly opened game preserve is given by $n = 75 \log_4 (3t + 1)$, where t is time in years after the opening. Find the number of deer in the preserve when t is 5.

ANSWERS: 1. $\log_3 8 + \log_3 80$ 2. $\log_5 20 + \log_5 u$ 3. $\log_c a + \log_c b$ 4. $\log_4 7 - \log_4 30$ 5. $\log_2 12 - \log_2 v$ 6. $\log_c a - \log_c b$ 7. $5 \log_2 x$ 8. $\frac{1}{2} \log_3 y$ 9. $-3 \log_a b$ 10. $\log_4 (3 \cdot 8)$ 11. $\log_b xy$ 12. $\log_2 \frac{5}{9}$ 13. $\log_c \frac{x}{y}$ 14. $3 \log_a x + \log_a y - \log_a z$ 15. $\log_a x + 2 \log_a y + \frac{1}{2} \log_a z$ 16. $\frac{1}{3} (\log_a x + 2 \log_a y - \log_a z)$ 17. $\log_a \frac{z^2\sqrt{y}}{x}$ 18. $\log_a \frac{x^3}{yz}$ 19. $\log_a \sqrt{\frac{xz^3}{y}}$ 20. 1.1761 21. 0.2219 22. 1.3980 23. 1.8751 24. 0.9209 25. −0.6990 26. 0.2330 27. 0.5881 28. false 29. true (convert to exponential form) 30. true (convert to exponential form) 31. false 32. false 33. false 34. $W = \log_2 16 = 4$ 35. He was fired because he assumed that the logarithm of a sum is the sum of the logarithms, which is *not true*. The actual value of W when m and n are both 8 is 4. 36. $Y = 2 \neq 3 - 2 = 1$ 37. $V = 2 \neq \frac{3}{1} = 3$ 38. $a = \log_x y$ 39. $b^p = m$ 40. $x = 7$ 41. $x = 4$ 42. $x = 2$ 43. Reflect it across the line $y = x$ since the functions are inverses. 44. 150

10.3 EXERCISES B

Express as a sum of logarithms.

1. $\log_4 (5 \cdot 6)$

2. $\log_8 (a \cdot 9)$

3. $\log_b (u \cdot v)$

Express as a difference of logarithms.

4. $\log_5 \frac{8}{9}$ **5.** $\log_3 \frac{u}{7}$ **6.** $\log_b \frac{x}{y}$

Express as a product.

7. $\log_3 x^4$ **8.** $\log_4 \sqrt{z}$ **9.** $\log_a u^{-2}$

Express as a single logarithm.

10. $\log_3 4 + \log_3 9$ **11.** $\log_a w + \log_a y$ **12.** $\log_2 8 - \log_2 3$ **13.** $\log_a w - \log_a u$

Express in terms of logarithms of x, y, and z.

14. $\log_a \frac{x^2y}{z^2}$ **15.** $\log_a xy^3\sqrt{z}$ **16.** $\log_a \sqrt[3]{\frac{x}{y^2z}}$

Express as a single logarithm.

17. $\frac{1}{2} \log_a x - \log_a y + 3 \log_a z$ **18.** $2 \log_a z - \log_a y - \log_a x$ **19.** $\frac{1}{2} (\log_a y + \log_a x - 5 \log_a z)$

Given that $\log_a 3 = 0.4771$ *and* $\log_a 7 = 0.8451$, *find the following logarithms.*

20. $\log_a 21$ **21.** $\log_a \frac{7}{3}$ **22.** $\log_a 7^2$ **23.** $\log_a 147$

24. $\log_a \frac{49}{3}$ **25.** $\log_a \frac{15}{35}$ **26.** $\log_a \sqrt[3]{49}$ **27.** $\log_a \sqrt{63}$

Tell whether the following are true or false.

28. $\log_a 0 = 1$ **29.** $\log_a (x + y) = (\log_a x)(\log_a y)$ **30.** $\log_a 1 = a$

31. $x - y = \log_a x - \log_a y$ **32.** $\log_a x^c = (\log_a x)^c$ **33.** $c \log_a x = \frac{c}{\log_a x}$

Solve.

34. An environmental chemist might use the formula $H = \log_2 (a - p)$. Calculate the value of H when a is 10 and p is 6.

35. When Margo used the formula in Exercise 34 to find H when a is 8 and p is 4, she did the following calculation:

$$H = \log_2 (8 - 4) = \log_2 8 - \log_2 4 = 3 - 2 = 1.$$

What did she do wrong? What is the actual value of H?

36. If $U = \log_2 (a + b)$, calculate the value of U when $a = 8$ and $b = 8$, and show that $U \neq \log_2 a + \log_2 b$.

37. If $W = \log_3 ab$, calculate the value of W when $a = 1$ and $b = 9$, and show that $W \neq (\log_3 a)(\log_3 b)$.

FOR REVIEW

38. Convert $m^w = t$ to logarithmic form.

39. Convert $\log_p k = d$ to exponential form.

Find the numerical value of x in the following.

40. $2^x = 256$

41. $x = \log_5 625$

42. $\log_x \frac{1}{25} = -2$

43. What exponential equation would you graph to obtain the graph of $y = \log_3 x$?

44. The number n of items in stock in a distribution center is given by $n = 25 \log_3 (2t + 1)$, where t is time in days following the opening of the center. Find the number of items in stock when t is 4.

10.3 EXERCISES C

1. Express $\log_3 (a^2 - ab + b^2) + \log_3 (a + b)$ as a single logarithm and simplify.

2. Express $\log_a \sqrt[3]{\frac{x^2\sqrt{y}}{(x + y)^3}}$ as a sum or difference of logarithms.

3. Find the numerical value of x if $\frac{1}{3} \log_a 8 - \frac{5}{2} \log_a 4 - \log_a \frac{1}{16} = \log_a x$.

[Answer: 1]

4. Find the value of $\log_{10} [\log_4 (\log_2 16)]$.

[Answer: 0]

10.4 COMMON AND NATURAL LOGARITHMS

STUDENT GUIDEPOSTS

1. Common Logarithms
2. Base Conversion Formulas
3. Antilogarithms
4. Natural Logarithms
5. Natural Logarithmic Function

1 COMMON LOGARITHMS

Since $1^y = x$ can be true only if $x = 1$ ($1^y = 1$ for every y), $y = \log_1 x$ has very little importance. Thus, 1 is not used as a base for logarithms. However, every positive number except 1 can be used. Since our number system is based on 10, logarithms to the base 10 were commonly used for computation. As a result, base 10 logarithms are called **common logarithms.** For convenience, we omit the base number 10 in common logarithmic expressions. For example,

$$\log_{10} x \quad \text{can be written as} \quad \log x.$$

Thus, when you see a logarithm written without the base, the base is understood to be 10.

The fundamental property that makes common logarithms useful is the fact that every positive number can be expressed as a power of 10. Consider the list of integer powers of 10 on the following page and their logarithmic form.

Power of 10		*Logarithmic form*		
10^5	$= 100{,}000$	$\log_{10} 100{,}000$	$= \log 100{,}000$	$= 5$
10^4	$= 10{,}000$	$\log_{10} 10{,}000$	$= \log 10{,}000$	$= 4$
10^3	$= 1000$	$\log_{10} 1000$	$= \log 1000$	$= 3$
10^2	$= 100$	$\log_{10} 100$	$= \log 100$	$= 2$
10^1	$= 10$	$\log_{10} 10$	$= \log 10$	$= 1$
10^0	$= 1$	$\log_{10} 1$	$= \log 1$	$= 0$
10^{-1}	$= 0.1$	$\log_{10} 0.1$	$= \log 0.1$	$= -1$
10^{-2}	$= 0.01$	$\log_{10} 0.01$	$= \log 0.01$	$= -2$
10^{-3}	$= 0.001$	$\log_{10} 0.001$	$= \log 0.001$	$= -3$
10^{-4}	$= 0.0001$	$\log_{10} 0.0001$	$= \log 0.0001$	$= -4$
10^{-5}	$= 0.00001$	$\log_{10} 0.00001$	$= \log 0.00001$	$= -5$

Several observations can be made from the list. The logarithm of 1 is 0, the logarithm of a number greater than 1 is positive, and the logarithm of a number less than 1 is negative. On the assumption that numbers between two integer powers of 10 have logarithms between the logarithms of the integer powers of 10 (that is, between the two integers), the log of a number between 1 and 10 must be a number between 0 and 1. For example,

$$10^0 = 1 < \mathbf{5} < 10 = 10^1$$

so that

$$0 = \log 1 < \mathbf{\log 5} < \log 10 = 1.$$

Thus, the logarithm of 5 must be a positive decimal between 0 and 1. Similarly, the logarithm of a number between 10 and 100 must be a number between 1 and 2 and so forth. Prior to the availability of calculators, tables like that at the back of this book were used to approximate logarithms. The appendix of this text provides instructions and examples for these tables.

As an example, suppose we find the value of log 726. This means we are looking for the exponent (logarithms are exponents) on 10 that would give 726. That is, if $a = \log 726$, a is an exponent, and in fact, $10^a = 726$. With $10^2 = 100$, and $10^3 = 1000$, we would anticipate that since

$$10^2 < 10^a < 10^3$$
$$\mathbf{100 < 726 < 1000}$$

a is a number between 2 and 3. We find this number in Example 1(a).

EXAMPLE 1 Using a Calculator to Find Common Logarithms

Use a calculator to find the following common logarithms.

(a) log 726

Enter 726 and press the key labeled LOG, as in the sequence below.

726 LOG → 2.8609366

The display shows log 726. Remember that this means 10 raised to the power 2.8609366 is approximately 726.

(b) log 0.00459

Use the following sequence.

0.00459 LOG → −2.3381873

Correct to four decimal places, $\log 0.00459 \approx -2.3382$.

Practice Exercise 1

Use a calculator to find the following common logarithms, correct to four decimal places.

(a) log 249

(b) log 0.000641

(c) $\log(-4562)$

The key [+/−] changes the sign of 4562 in the steps below.

4562 [+/−] [LOG] → [ERROR]

In this case, the display shows *error* since logarithms of negative numbers are not defined.

(c) $\log(-0.0543)$

Answers: (a) 2.3962 (b) −3.1931 (c) none

2 BASE CONVERSION FORMULAS

Now that we know how to find common logarithms using a calculator, we might wonder how to find logarithms to other bases. To answer this we will explore the relationship among logarithms to different bases. For example, consider $\log_a x$ and $\log_b x$. If we let

$$u = \log_b x,$$

then

$$b^u = x.$$

Take the logarithm to the base a of both sides.

$$\log_a b^u = \log_a x \qquad \text{If } w = z \text{ then } \log_a w = \log_a z$$

$$u \log_a b = \log_a x \qquad \text{Power rule}$$

$$(\log_b x)(\log_a b) = \log_a x \qquad \text{Substitute } \log_b x \text{ for } u$$

$$\log_b x = \frac{\log_a x}{\log_a b} \qquad \textit{Not } \log_a \frac{x}{b} = \log_a x - \log_a b$$

A special case of this formula results if $x = a$.

$$\log_b x = \frac{\log_a x}{\log_a b}$$

$$\log_b a = \frac{\log_a a}{\log_a b} \qquad \text{Substitute } a \text{ for } x$$

$$\log_b a = \frac{1}{\log_a b} \qquad \text{Log}_a\, a = 1$$

We have just established the following conversion formulas.

Base Conversion Formulas

If a and b are any two bases, and $x > 0$, then

$$\log_b x = \frac{\log_a x}{\log_a b} \quad \text{and} \quad \log_b a = \frac{1}{\log_a b}.$$

EXAMPLE 2 Using Base Conversion Formulas

(a) Express $\log_{100} x$ in terms of $\log_{10} x$ and simplify.

$$\log_{100} x = \frac{\log_{10} x}{\log_{10} 100} = \frac{\log x}{\log 10^2} = \frac{\log x}{2} = \frac{1}{2} \log x$$

PRACTICE EXERCISE 2

(a) Express $\log_{16} 4$ in terms of $\log_2 4$ and simplify.

(b) Express $\log_5 9$ in terms of base 9 logarithms.

$$\log_5 9 = \frac{1}{\log_9 5}$$

(b) Express $\log_3 7$ in terms of base 7 logarithms.

Answers: **(a)** $\frac{1}{2}$ **(b)** $\frac{1}{\log_7 3}$

We can now find the logarithm of a number to any base using the first base conversion formula, common logarithms, and a calculator.

EXAMPLE 3 Finding Logarithms to Other Bases

Find $\log_7 428$ correct to four decimal places.

Using the first base conversion formula with $x = 428$, $b = 7$, and $a = 10$, we have the following.

$$\log_7 428 = \frac{\log 428}{\log 7} \approx 3.1138$$

The calculator steps are:

428 [LOG] [÷] 7 [LOG] [=] → [3.1137734].

Practice Exercise 3

Find $\log_5 743$, correct to four decimal places.

Answer: 4.1075

CAUTION

In the formula $\log_b x = \frac{\log_a x}{\log_a b}$, we divide the two logarithms $\log_a x$ and $\log_a b$; we don't subtract them. Do not confuse this with $\log_a \frac{x}{b}$, which is $\log_a x - \log_a b$. Notice in Example 3 that we divided log 428 by log 7 to obtain 3.1138; we did not subtract.

In Example 3 we found that $\log_7 428 \approx 3.1138$. Remembering that a logarithm is an exponent, this means that $7^{3.1138} \approx 428$. We can use the [y^x] key on a calculator to verify this result.

7 [y^x] 3.1138 [=] → [428.022]

Notice that due to round-off errors, the display is not exactly 428, but rather an approximation of it.

3 ANTILOGARITHMS

If we are given a number, we now know how to find its common logarithm. Suppose we reverse this procedure and find a number when its logarithm is given. The number that corresponds to a given logarithm is called its **antilogarithm (antilog).** In the equation

$$x = \log n$$

x is the logarithm of n and n is the antilogarithm of x. In exponential form, $n = 10^x$. To find an antilogarithm using a calculator, we use either the [10^x] key or the [INV] key followed by the [LOG] key. Both sequences are given below where we find the antilogarithm of 2.8609366.

2.8609366 [10^x] → [725.99997]

2.8609366 [INV] [LOG] → [725.99997]

In Example 1(a), we found log 726 to be 2.8609366. Notice that the antilogarithm of 2.8609366 found above does not give the exact value 726, but rather a close approximation to it due to round-off differences in the calculator.

EXAMPLE 4 Finding Antilogarithms

Use a calculator to find the following antilogs correct to three significant digits.

(a) $\log n = 0.6278$

We need to find n where $n = 10^{0.6278}$.

0.6278 [10^x] → 4.2442406

or 0.6278 [INV] [LOG] → 4.2442406

Thus, n is approximately 4.24.

(b) $\log n = -4.3205$

4.3205 [+/−] [10^x] → 0.0000478

or 4.3205 [+/−] [INV] [LOG] → 0.0000478

Thus, n is approximately 0.0000478.

Practice Exercise 4

Use a calculator to find each antilog, correct to three significant digits.

(a) $\log n = 1.6275$

(b) $\log n = -2.0058$

Answers: (a) 42.4 (b) 0.00987

4 NATURAL LOGARITHMS

Logarithm bases other than 10 occur frequently in science and mathematics. One of the most widely used base numbers is e. The number e is an irrational number, approximately equal to 2.71828, and originates in a natural way in calculus. For this reason, logarithms to the base e are called **natural logarithms.** The standard notation for the natural logarithm of a number x is

$$\ln x \text{ instead of } \log_e x.$$

The [ln x] key on a calculator is used to find natural logarithms the same way the [LOG] key is used to find common logarithms.

EXAMPLE 5 Finding Natural Logarithms

Use a calculator to find the following natural logarithms, correct to four decimal places.

(a) ln 329

The steps to follow are given below.

329 [ln x] → 5.7960578

Thus, $\ln x$ is approximately 5.7961, correct to four decimal places.

(b) ln 0.00541

0.00541 [ln x] → −5.2195062

Thus, $\ln 0.00541 \approx -5.2195$.

Practice Exercise 5

Find the following natural logarithms, correct to four decimal places.

(a) ln 1598

(b) ln 0.000641

Answers: (a) 7.3765 (b) −7.3525

To find antilogarithms of natural logarithms, we use either the $\boxed{e^x}$ key or the $\boxed{\text{INV}}$ and $\boxed{\ln x}$ keys, much as we did for common logarithms.

EXAMPLE 6 Finding Natural Antilogarithms

Use a calculator to find the following antilogs correct to three significant digits.

(a) $\ln x = 3.2541$

We need to find x where $x = e^{3.2541}$

$$3.2541 \ \boxed{e^x} \rightarrow \boxed{25.896297}$$

$$\text{or} \quad 3.2541 \ \boxed{\text{INV}} \ \boxed{\ln x} \rightarrow \boxed{25.896297}$$

Thus, x is approximately 25.9.

(b) $\ln x = -1.4738$

$$1.4738 \ \boxed{+/-} \ \boxed{e^x} \rightarrow \boxed{0.2290534}$$

$$\text{or} \quad 1.4738 \ \boxed{+/-} \ \boxed{\text{INV}} \ \boxed{\ln x} \rightarrow \boxed{0.2290534}$$

Thus, x is approximately 0.229.

Practice Exercise 6

Use a calculator to find each antilog, correct to three significant digits.

(a) $\ln x = 2.6117$

(b) $\ln x = -3.7542$

Answers: (a) 13.6 (b) 0.0234

5 NATURAL LOGARITHMIC FUNCTION

In Section 10.2 we graphed exponential functions $y = a^x$ and their inverse logarithmic functions $y = \log_a x$. The **natural logarithmic function** $y = \ln x$ and its inverse $y = e^x$ are important functions in mathematics. So important is $y = e^x$ that it is recognized above all other exponential functions and sometimes called *the* **exponential function.** Representative values of each function are given in the tables with the graphs shown in Figure 10.8.

x	$\ln x$
0.5	−0.7
1	0
2	0.7
3	1.1
4	1.4
5	1.6

x	e^x
−3	0.05
−2	0.1
−1	0.4
0	1
1	2.7
2	7.4

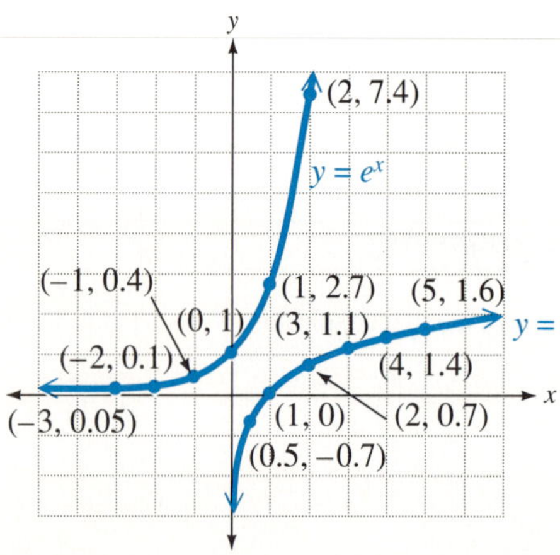

Figure 10.8 Natural Logarithmic and Exponential Functions

Variations of the exponential function appear in numerous applications. Two of these variations are graphed in Figure 10.9.

x	e^{-x}	$e^{-x/2}$
0	1.00	1.00
−1	2.72	1.65
1	0.37	0.61
−2	7.40	2.72
2	0.14	0.37
−3	20.1	4.49
3	0.05	0.22
−4	54.7	7.40
4	0.02	0.14

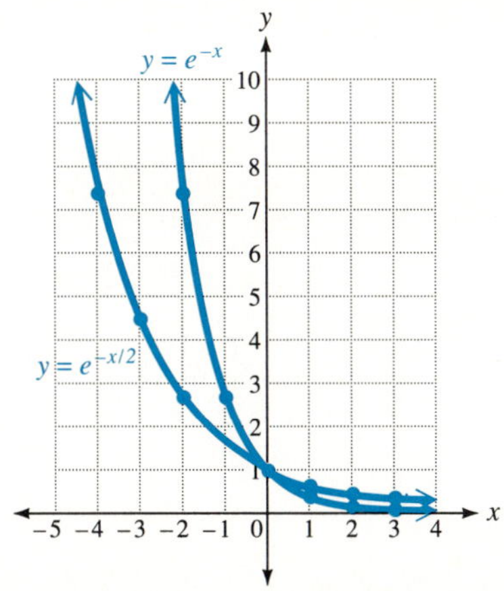

Figure 10.9 Variations of Exponential Function

10.4 EXERCISES A

Tell whether the statement is true *or* false. *If the statement is false, explain why.*

1. Common logarithms use 10 as the base number.

2. Instead of writing $\log_{10} x$ we usually write $\log x$.

3. In the equation $y = \log m$, y is called the antilogarithm of m.

Find each common logarithm, correct to four decimal places.

4. $\log 253$ **5.** $\log 0.159$ **6.** $\log 17.6$

7. $\log (-43)$ **8.** $\log 1{,}265{,}000$ **9.** $\log 0.00391$

Express the given logarithms in terms of the indicated base, and simplify.

10. $\log_{1000} x$; 10 **11.** $\log_{32} 8$; 2 **12.** $\log_7 12$; 12

13. $\log_{100} 10^6$; 10 **14.** $\log_{36} x$; 6 **15.** $\log_6 x$; 36

Evaluate each logarithm using common logarithms.

16. $\log_{100} 10$ **17.** $\log_7 12$ **18.** $\log_5 30$

19. $\log_7 1.32$ **20.** $\log_2 0.0352$ **21.** $\log_{4.5} 275$

Find each common antilogarithm, correct to three significant digits.

22. $\log n = 2.0359$ **23.** $\log n = 0.6298$ **24.** $\log n = -1.7645$ **25.** $\log n = 3.6415$

Find each natural logarithm, correct to four decimal places.

26. $\ln 627$ **27.** $\ln 0.654$ **28.** $\ln 18.5$ **29.** $\ln 23{,}400$

Find each natural antilogarithm, correct to three significant digits.

30. $\ln x = 1.4532$ **31.** $\ln x = 0.6428$ **32.** $\ln x = 4.6565$ **33.** $\ln x = -3.4115$

34 The demand equation for x units of a product relative to the price per unit p is given by $p = 200 - 0.4\, e^{0.004x}$. Find the price if the demand is 100 units.

35. The atmospheric pressure P, measured in pounds per square inch, at an altitude x, measured in miles above sea level, can be approximated by $P = 14.7\, e^{-0.2x}$. What is the atmospheric pressure on the outside of a jet airplane cruising at an altitude of 6 miles?

36. In chemistry, the **pH (hydrogen potential)** of a substance is defined by

$$pH = -\log\, [H^+],$$

where $[H^+]$ is the concentration of hydrogen ions in the solution. The pH scale varies from 0 to 14, with the pH of distilled water equal to 7. A substance is an acid if its pH is less than 7, and a base if its pH is greater than 7. Find the pH of a soft drink with $[H^+] = 3.45 \times 10^{-7}$.

37 Stars are categorized by brightness on a scale measured in magnitudes with the faintest star visible to the naked eye assigned a magnitude of 6. Telescopes are designed to view stars beyond a magnitude of 6. The limiting magnitude L of a telescope depends on the diameter d of its lens and is given by $L = 8.8 + 2.2 \ln d$. What is the limiting magnitude of a telescope with a lens 15 inches in diameter?

FOR REVIEW

Express as a single logarithm.

38. $\frac{1}{3}(\log_a x - 2 \log_a y)$ **39.** $3 \log_a x - \log_a y - 4 \log_a z$

Express in terms of logarithms of x, y, and z.

40. $\log_a \frac{x^3\sqrt{z}}{y^5}$ **41.** $\log_a \sqrt{xyz}$

Exercises 42–51 review some of the properties of logarithms.

42. $\log 2 =$ ______ **43.** $\log 3 =$ ______ **44.** $\log 5 =$ ______ **45.** $\log 6 =$ ______

46. Does $\log (2 + 3) = \log 2 + \log 3$?

47. Does $\log (2 \cdot 3) = (\log 2)(\log 3)$?

48. Does $\log (2 \cdot 3) = \log 2 + \log 3$?

49. Does $\log \frac{6}{3} = \frac{\log 6}{\log 3}$?

50. Does $\log \frac{6}{3} = \log 6 - \log 3$?

51. Does $\log_2 5 = \log 5 - \log 2$?

The following exercises will help you prepare for the next section. Tell whether the statement in Exercises 52–54 is true *or* false.

52. If $x = y$ then $\log_a x = \log_a y$.

53. If $2^x = 2^y$ then $x = 2y$.

54. If $\log_a x = \log_a y$ then $x = y$.

Solve each equation.

55. $\frac{x + 10}{x + 1} = 10$

56. $x(x + 1) = 2$

ANSWERS: 1. true 2. true 3. false, y is the logarithm of m 4. 2.4031 5. −0.7986 6. 1.2455 7. Logarithms of negative numbers are not defined. 8. 6.1021 9. −2.4078 10. $\frac{1}{3} \log x$ 11. $\frac{3}{5}$ 12. $\frac{1}{\log_{12} 7}$ 13. 3 14. $\frac{1}{2} \log_6 x$ 15. $2 \log_{36} x$ 16. $\frac{1}{2}$ 17. 1.2770 18. 2.1133 19. 0.1427 20. −4.8283 21. 3.7344 22. 109 23. 4.26 24. 0.0172 25. 4380 26. 6.4409 27. −0.4246 28. 2.9178 29. 10.0605 30. 4.28 31. 1.90 32. 105 33. 0.0330 34. \$199.40 35. approximately 4.43 lb/in^2 36. approximately 6.46 37. approximately 14.8 38. $\log_a \sqrt[3]{\frac{x}{y^2}}$ 39. $\log_a \frac{x^3}{yz^4}$ 40. $3 \log_a x + \frac{1}{2} \log_a z - 5 \log_a y$ 41. $\frac{1}{2} (\log_a x + \log_a y + \log_a z)$ 42. 0.3010 43. 0.4771 44. 0.6990 45. 0.7782 46. no 47. no 48. yes 49. no 50. yes 51. no 52. true 53. false 54. true 55. 0 56. −2, 1

10.4 EXERCISES B

Tell whether the statement is true *or* false. *If the statement is false, explain why.*

1. Natural logarithms use 2 as the base number.

2. The natural logarithm of x is written $\ln x$.

3. If n is the common antilogarithm of m, then $n = 10^m$.

Find each common logarithm, correct to four decimal places.

4. $\log 478$ **5.** $\log 0.987$ **6.** $\log 38.4$

7. $\log (-67)$ **8.** $\log 3{,}415{,}000$ **9.** $\log 0.00628$

Express the given logarithms in terms of the indicated base, and simplify.

10. $\log_{100} x$; 10 **11.** $\log_{64} 8$; 2 **12.** $\log_5 8$; 8

13. $\log_{1000} 10^6$; 10 **14.** $\log_{81} x$; 3 **15.** $\log_3 x$; 81

Evaluate each logarithm using common logarithms.

16. $\log_{1000} 10$ **17.** $\log_{12} 7$ **18.** $\log_8 24$

19. $\log_4 62.4$ **20.** $\log_3 0.000915$ **21.** $\log_{1.6} 0.134$

Find each common antilogarithm, correct to three significant digits.

22. $\log n = 4.1378$ **23.** $\log n = 0.4134$ **24.** $\log n = -2.6643$ **25.** $\log n = 5.1167$

Find each natural logarithm, correct to four decimal places.

26. $\ln 431$ **27.** $\ln 0.238$ **28.** $\ln 27.6$ **29.** $\ln 0.00395$

Find each natural antilogarithm, correct to three significant digits.

30. $\ln x = 2.3597$ **31.** $\ln x = 0.7415$ **32.** $\ln x = 5.2138$ **33.** $\ln x = -2.0015$

34. The demand equation for x units of a product relative to the price per unit p is given by $p = 200 - 0.4\ e^{0.004x}$. Find the price if the demand is 150 units.

35. The atmospheric pressure P, measured in pounds per square inch, at an altitude x, measured in miles above sea level, can be approximated by $P = 14.7\ e^{-0.2x}$. What is the atmospheric pressure on the outside of an airplane cruising at an altitude of 4 miles?

36. In chemistry, the **pH (hydrogen potential)** of a substance is defined by

$$\text{pH} = -\log\ [\text{H}^+],$$

where $[\text{H}^+]$ is the concentration of hydrogen ions in the solution. The pH scale varies from 0 to 14, with the pH of distilled water equal to 7. A substance is an acid if its pH is less than 7, and a base if its pH is greater than 7. Find the pH of a spa with $[\text{H}^+] = 2.15 \times 10^{-8}$.

37. Stars are categorized by brightness on a scale measured in magnitudes with the faintest star visible to the naked eye assigned a magnitude of 6. Telescopes are designed to view stars beyond a magnitude of 6. The limiting magnitude L of a telescope depends on the diameter d of its lens and is given by $L = 8.8 + 2.2 \ln d$. What is the limiting magnitude of a telescope with a lens 10 inches in diameter?

Express as a single logarithm.

38. $\frac{1}{3}(\log_a x + 2 \log_a y)$ **39.** $3 \log_a x + \log_a y - 4 \log_a z$

Express in terms of logarithms of x, y, and z.

40. $\log_a \frac{xy^3}{\sqrt{z}}$ **41.** $\log_a \sqrt[3]{xyz}$

Exercises 42–51 review some of the properties of logarithms.

42. $\log 2 =$ ______ **43.** $\log 4 =$ ______ **44.** $\log 6 =$ ______ **45.** $\log 8 =$ ______

46. Does $\log(2 + 4) = \log 2 + \log 4$? **47.** Does $\log(2.4) = (\log 2)(\log 4)$?

48. Does $\log(2 \cdot 4) = \log 2 + \log 4$? **49.** Does $\log \frac{8}{4} = \frac{\log 8}{\log 4}$?

50. Does $\log \frac{8}{2} = \log 8 - \log 2$? **51.** Does $\log_4 8 = \log 8 - \log 4$?

The following exercises will help you prepare for the next section. Tell whether the statement in Exercises 52–54 is true or false.

52. If $x \neq y$ then $\log_a x \neq \log_a y$.

53. If $2^x = 2^y$ then $x = y$.

54. If $\log_a x = \log_a y$ then $x = ay$.

Solve each equation.

55. $\dfrac{x+2}{x-2} = 4$

56. $x(x-3) = 4$

10.4 EXERCISES C

Sketch the graph of each function.

1. $y = e^{x+1}$

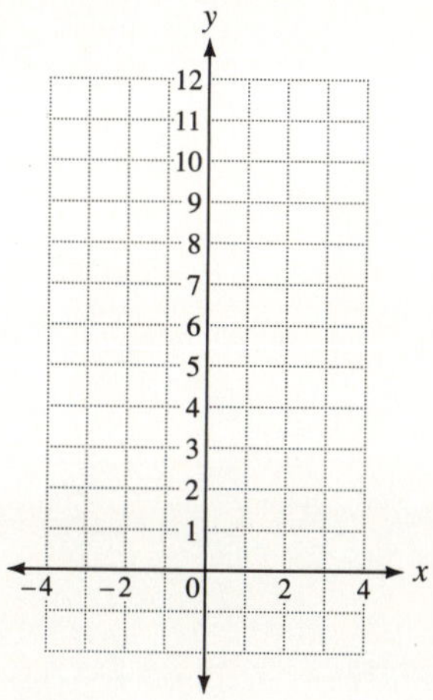

2. $y = e^{x-1}$

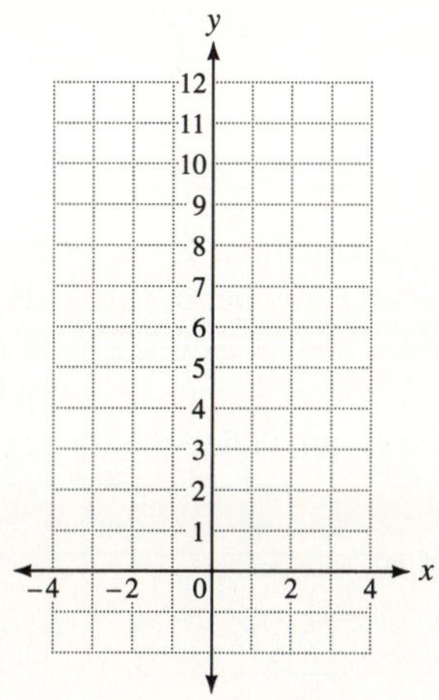

3. $y = e^{3x}$

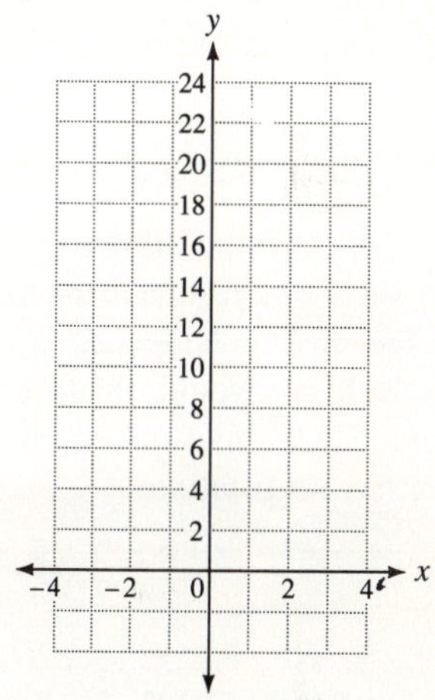

4. $y = e^{x^2}$

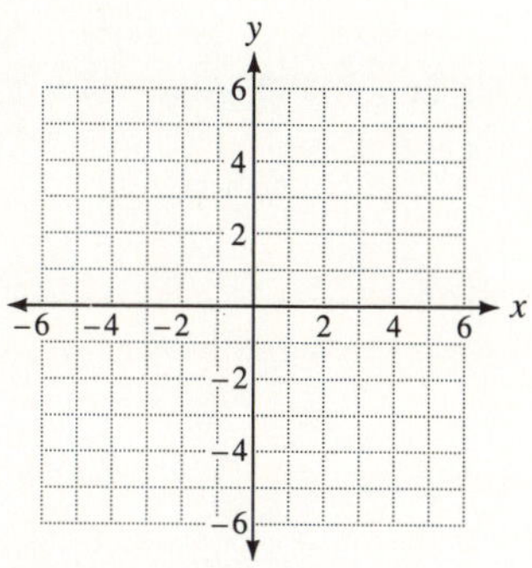

5. $y = \ln(x+1)$

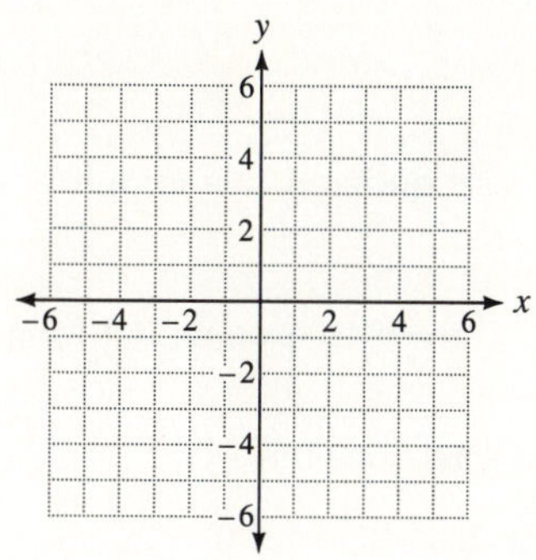

6. $y = \ln x^2$

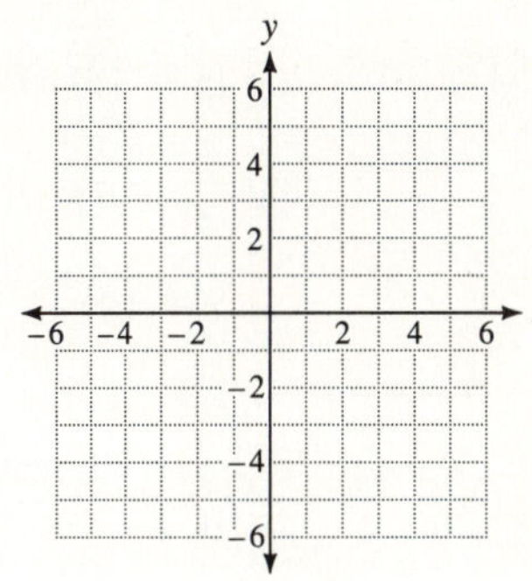

10.5 LOGARITHMIC AND EXPONENTIAL EQUATIONS

STUDENT GUIDEPOSTS

1 Logarithmic Equations
2 Exponential Equations

1 LOGARITHMIC EQUATIONS

Equations that contain logarithms of expressions involving the variable are called **logarithmic equations.** Solving these equations requires a thorough knowledge of the properties of logarithms.

EXAMPLE 1 Solving a Logarithmic Equation

Solve $\log_5 x = \log_5 (3x - 8)$.

If the logarithms of two numbers using the same base are equal, then the numbers are equal. Thus,

$$x = 3x - 8. \quad \text{If } \log_a n = \log_a m, \text{ then } n = m$$
$$-2x = -8$$
$$x = 4$$

Check: $\log_5 4 \stackrel{?}{=} \log_5 (3(4) - 8)$

$\log_5 4 \stackrel{?}{=} \log_5 (12 - 8)$

$\log_5 4 = \log_5 4$

The solution is 4.

Practice Exercise 1

Solve $\log_3 2x = \log_3 (x - 4)$.
Be sure to check!

Answer: no solution

CAUTION

Always check possible solutions in the original equation. Remember that any value that results in the logarithm of a negative number or zero must be excluded. This happened in Practice Exercise 1 above. Solving as in Example 1, we would obtain $x = -4$, but $\log 2(-4) = \log (-8)$ is not defined. Since this solution must be discarded, we concluded that the equation has no solution.

EXAMPLE 2 Solving a Logarithmic Equation

Solve.

$\log (x + 10) - \log (x + 1) = 1$

$$\log \frac{x + 10}{x + 1} = 1 \quad \text{Use quotient rule on left side}$$

$$\frac{x + 10}{x + 1} = 10^1 = 10 \quad \text{Convert to exponential form}$$

$$x + 10 = 10(x + 1) \quad \text{Clear fraction}$$
$$x + 10 = 10x + 10$$
$$0 = 9x$$
$$0 = x$$

Check: $\log (0 + 10) - \log (0 + 1) \stackrel{?}{=} 1$

$\log 10 - \log 1 \stackrel{?}{=} 1$

$1 - 0 \stackrel{?}{=} 1$

$1 = 1$

The solution is 0.

Practice Exercise 2

Solve.

$\log (x + 8) - \log (x - 1) = 1$

Answer: 2

To Solve a Logarithmic Equation

1. Obtain a single logarithmic expression using the same base on one side of the equation, or write each side as a logarithm using the same base.
2. Convert the result to an exponential equation or find the antilog of both sides and solve the result.
3. Check all possible solutions in the original equation. (Negative numbers and zero do not have logarithms.)

EXAMPLE 3 Solving a Logarithmic Equation

Solve.

$$\log x + \log (x + 1) = \log 2$$

$\log x(x + 1) = \log 2$ — Product rule

$x(x + 1) = 2$ — If $\log n = \log m$ then $n = m$

$x^2 + x - 2 = 0$ — Write the quadratic equation in general form

$(x + 2)(x - 1) = 0$ — Factor

$x + 2 = 0$ or $x - 1 = 0$ — Use the zero-product rule

$x = -2$ $\qquad x = 1$

When we try to check -2, we obtain

$$\log (-2) + \log (-2 + 1) \stackrel{?}{=} \log 2$$

which means that -2 is not a solution since $\log (-2)$ and $\log (-1)$ are not defined. The only solution is 1. (It does check.)

Practice Exercise 3

Solve.

$$\log (x - 2) + \log (x + 2) = \log 5$$

Answer: 3

2 EXPONENTIAL EQUATIONS

Equations with variable exponents are called **exponential equations.** Such equations can often be solved by taking the logarithm (using an appropriate base) of both sides and then applying the power rule.

EXAMPLE 4 Solving an Exponential Equation

Solve the exponential equation for x.

$$10^{2x+1} = 100$$

$\log 10^{2x+1} = \log 100$ — Take common log of both sides

$(2x + 1) \log 10 = \log 100$ — Power rule

$(2x + 1)(1) = 2$ — $\log 10 = 1$ and $\log 100 = 2$

$2x + 1 = 2$

$2x = 1$

$x = \frac{1}{2}$

The solution is $\frac{1}{2}$. Check in the original equation.

Practice Exercise 4

Solve the exponential equation for x.

$$5^{3x-2} = 25$$

Answer: $\frac{4}{3}$

When both sides of an exponential equation can be expressed as a power having the same base, it is much faster to equate the corresponding exponents (set them equal to each other) and solve. Suppose we work Example 4 using this method.

$$10^{2x+1} = 100$$
$$10^{2x+1} = 10^2 \qquad \text{Express as powers with same base}$$
$$2x + 1 = 2 \qquad \text{Equate exponents, since } 10^n = 10^m \text{ means } n = m$$
$$2x = 1$$
$$x = \frac{1}{2}$$

In such cases we can eliminate a few steps.

To Solve an Exponential Equation

1. Try to express each side as a power using the same base, and equate the resulting exponents.
2. If Step 1 fails, take the logarithm of each side and use the power rule to eliminate the variable exponents.
3. Solve the resulting equation and check in the original.

EXAMPLE 5 Solving an Exponential Equation

Solve $3^{x+1} = \dfrac{1}{27}$.

Since $\dfrac{1}{27} = \dfrac{1}{3^3} = 3^{-3}$, we need to solve the equation below.

$$3^{x+1} = 3^{-3} \qquad \text{Express as powers with same base}$$
$$x + 1 = -3 \qquad \text{Equate exponents}$$
$$x = -4 \qquad \text{This does check}$$

The solution is -4.

Practice Exercise 5

Solve $7^{2x-1} = \dfrac{1}{49}$.

Answer: $-\frac{1}{2}$

EXAMPLE 6 Solving an Exponential Equation

Solve $5^{2x} = 8$.

Since 8 is not easily expressible as a power with base 5, we take the common logarithm of both sides.

$$\log 5^{2x} = \log 8$$
$$2x \log 5 = \log 8 \qquad \text{Power rule}$$
$$x = \frac{\log 8}{2 \log 5} \qquad \text{Divide by 2 log 5}$$

This is the exact form of the answer. If we need a decimal approximation, such as in an applied problem, we would use a calculator. The steps are given below.

8 [LOG] [÷] 5 [LOG] [÷] 2 [=] → [0.6460148]

Thus, $x \approx 0.646$, correct to three decimal places.

Practice Exercise 6

Solve $9^{3x} = 4$.

Answer: $\frac{\log 4}{3 \log 9} \approx 0.210$

Many applied problems use formulas that involve natural logarithmic or exponential equations.

EXAMPLE 7 Physics Problem

When a beam of light passes through water, as shown in Figure 10.10, the intensity of the light I is diminished by the number of feet x from the surface according to the formula $I = I_0 e^{-kx}$, where I_0 is the intensity of the light at the surface and k is the coefficient of extinction depending on the nature of the water. If the coefficient of extinction of Lake Mary is 0.06, at what depth (to the nearest tenth of a foot) will the intensity of a light be reduced to 50% of that at the surface?

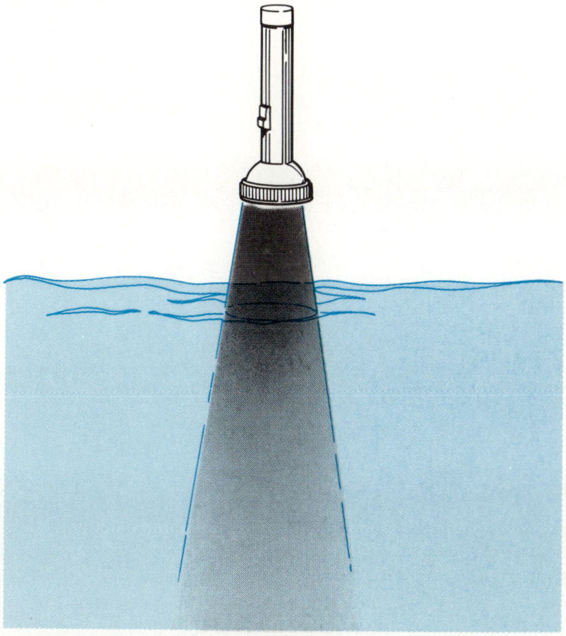

Figure 10.10

We are asked to find the depth at which $I = 0.5I_0$ (50% of the original intensity) when $k = 0.06$. Substitute these values in the formula.

$$0.5I_0 = I_0 e^{-0.06x}$$

$$0.5 = e^{-0.06x} \quad \text{Divide both sides by } I_0$$

$$\ln 0.5 = \ln e^{-0.06x} \quad \text{Take the natural log of both sides}$$

$$\ln 0.5 = -0.06x \quad \ln e^u = u$$

$$\frac{\ln 0.5}{-0.06} = x \quad \text{Divide both sides by } -0.06$$

Using a calculator we find $x \approx 11.552453$.

At a depth of about 11.6 ft, the light has been diminished by 50%.

Practice Exercise 7

Refer to Example 7. At what depth will the intensity of light be 25% of the original intensity?

Answer: 23.1 ft

10.5 EXERCISES A

Solve the following for x. When appropriate, give answer using three significant digits.

1. $\log(3x + 1) = \log(2x + 5)$ **2.** $\log_3(x + 6) = 2\log_3 x$ **3.** $\log(9x + 1) - \log x = 1$

4. $\log(x-3) + \log x = 1$

5 $\log_2(x+1) + \log_2(x-1) = 3$

6. $\log(x^2+1) - \log(x-2) = 1$

7. $\log_5 x + \log_5(x-1) = \log_5 12$

8. $\log x - \log(7x+6) = -1$

9. $\ln x + \ln(x+2) = \ln 3$

10. $10^x = 1000$

11. $3^x = 16$

12. $10^{3x+5} = 100$

13. $2^{x+7} = 8$

14. $5^{x^2-1} = 125$

15. $2^{3x} = 7$

16 $3^{x^2}9^x = \dfrac{1}{3}$

17. $6^{x+1} = 27$

18 $2^{2x}3^{x+1} = 5$

19. If the coefficient of extinction of Crystal Lake is 0.008, use the formula in Example 7 and find the depth (to the nearest tenth of a foot) at which light will be reduced by 50% from that at the surface.

20 Repeat Exercise 19 for an 80% reduction of the light intensity at the surface.

FOR REVIEW

Express each logarithm in terms of the indicated base and simplify.

21. $\log_8 128$; 2

22. $\log_6 4$; 4

Evaluate each logarithm.

23. $\log_2 18$

24. $\log 38.6$

25. $\ln 0.233$

ANSWERS: 1. 4 2. 3 3. 1 4. 5 5. 3 6. 3, 7 7. 4 8. 2 9. 1 10. 3 11. 2.52 12. -1 13. -4 14. ± 2 15. 0.936 16. -1 17. 0.839 18. 0.206 19. 86.6 ft 20. 201.2 ft 21. $\frac{7}{3}$ 22. $\frac{1}{\log_4 6}$ 23. 4.1699 24. 1.5866 25. -1.4567

10.5 EXERCISES B

Solve the following for x. When appropriate, give answer using three significant digits.

1. $\log(x+5) = \log(1-5x)$

2. $2\log(x+1) = \log 4x$

3. $\log(3x+1) - \log(4-x) = 1$

4. $\log(6x+2) + \log 4x = 1$

5. $\log(x+1) - \log(x^2+19) = -1$

6. $\log_3(2x+1) + \log_3(x-1) = 3$

7. $\log_8 (x + 2) + \log_8 (x - 3) = \log_8 6$

8. $\log (2x - 1) - \log (9x + 1) = -1$

9. $\ln (x + 1) + \ln (x - 1) = \ln 3$

10. $100^x = 10$

11. $8^x = 22$

12. $1000^{x+2} = 10$

13. $3^{2x+4} = 27$

14. $2^{2x^2+3} = 32$

15. $8^{3x}2^{x^2} = 2^{10}$

16. $5^{x+2} = 18$

17. $3^{2x} = 4^{x+1}$

18. $4^{x-5}3^{2x} = 10$

19. If the coefficient of extinction of Lake Pleasant is 0.09, use the formula in Example 7 and find the depth (to the nearest tenth of a foot) at which light will be reduced by 50% from that at the surface.

20. Repeat Exercise 19 for an 80% reduction of the light intensity at the surface.

FOR REVIEW

Express each logarithm in terms of the indicated base and simplify.

21. $\log_{32} 64$; 2

22. $\log_5 x$; 25

Evaluate each logarithm.

23. $\log_3 41$

24. $\log_8 0.0159$

25. $\log 0.0045$

26. $\ln 62.9$

10.5 EXERCISES C

Solve for x.

1. $(\log_5 x)^2 = 2 \log_5 x$

2. $7^{2x}3^x = 10$

3. $2^{x^2}3^{x^2} = 6^{5x-6}$
[Answer: 2, 3]

Solve for n.

4. $3170 = 1000(1.08)^n$

5. $397 = 1000(1.08)^{-n}$

6. $S = A(1 + i)^n$

7. $A = S(1 + i)^{-n}$

8. $y = 2.5e^{-3n}$

9. $y = 800e^{2n}$

10. $y = ce^{n/3}$
[Answer: $n = 3(\ln y - \ln c)$]

11. $y = ce^{-n/2}$

12. $y = 500(1 + x)^{n/2}$
$\left[\text{Answer: } n = \dfrac{2 \log y - 2 \log 500}{\log (1 + x)}\right]$

10.6 MORE APPLICATION PROBLEMS

Many applied problems in science and business use formulas that involve logarithmic or exponential expressions. The compound interest formula is one of the most common in this category. The sum S of an amount A invested for n compounding periods at a rate of interest i per period is given by the formula

$$S = A(1 + i)^n.$$

Solving this formula for A gives us the initial or present value necessary for accumulating with interest the future sum S.

$$A = S(1 + i)^{-n}$$

EXAMPLE 1 Compound Interest Problem

If \$1000 is invested at an annual rate of 8% compounded quarterly, what is the value of the account at the end of 10 years?

The formula we use is

$$S = A(1 + i)^n,$$

where $A = \$1000$, the quarterly interest rate i is $\frac{0.08}{4} = 0.02$, and the number of compounding periods n is $4(10) = 40$ (4 per year for 10 years).

$$\begin{aligned} S &= 1000(1 + 0.02)^{40} \\ &= 1000(1.02)^{40} \end{aligned}$$

We can use a calculator with a $\boxed{y^x}$ key to evaluate S as follows:

$$1.02 \ \boxed{y^x} \ 40 \ \boxed{\times} \ 1000 \ \boxed{=} \rightarrow \boxed{2208.04}$$

In 10 years the investment will be worth \$2208.04.

Practice Exercise 1

Repeat Example 1. However, this time assume compounding is semiannual (twice a year).

Answer: \$2191.12

EXAMPLE 2 Compound Interest Problem

The Barkers wish to have \$50,000 available for their newborn child's college education. How much should they deposit today in a savings account that pays 9% interest compounded semiannually, to accumulate the needed amount by the end of 18 years?

The formula to use is

$$A = S(1 + i)^{-n}$$

where $S = \$50{,}000$, the semiannual interest rate i is $\frac{0.09}{2} = 0.045$, and the number of compounding periods n is $2(18) = 36$ (2 per year for 18 years).

$$\begin{aligned} A &= 50{,}000(1 + 0.045)^{-36} \\ &= 50{,}000(1.045)^{-36} \end{aligned}$$

$$1.045 \ \boxed{y^x} \ 36 \ \boxed{+/-} \ \boxed{\times} \ 50{,}000 \rightarrow \boxed{10251.41}$$

The Barkers should deposit \$10,251.41 to have the necessary \$50,000 in 18 years.

Practice Exercise 2

Repeat Example 2. This time assume compounding is quarterly.

Answer: \$10,074.21

Many scientific applications can be described by an equation of the form

$$y = c \log_a \frac{x}{x_0}.$$

For the measurement of sound and earthquakes, the base a is 10, while for cell-generation problems, the base a is 2.

EXAMPLE 3 Biology Problem

The equation that describes the number of cell fissions that occur in a given period of time is the number y given by

$$y = \log_2 \frac{x}{x_0},$$

where x_0 is the initial cell count and x is the final cell count. Find the number of cell fissions in a given time if the initial cell count is 25 and the final cell count is 630,000.

Since $x_0 = 25$ and $x = 630{,}000$, we evaluate the following expression.

$$\begin{aligned} y &= \log_2 \frac{630{,}000}{25} \\ &= \log_2 25{,}200 && \text{Simplify fraction} \\ &= \frac{\log 25{,}200}{\log 2} \approx 14.6 \end{aligned}$$

25,200 [LOG] [÷] 2 [LOG] [=] → 14.621136

There are approximately 14.6 fissions during that period.

Practice Exercise 3

Find the number of cell fissions if the initial cell count is 150 and the final cell count is 325,000.

Answer: approximately 11.1 fissions

EXAMPLE 4 Physics Problem

The loudness of sound in decibels is given by the formula

$$y = 10 \log \frac{x}{x_0}$$

where x_0 is the weakest sound detected by the observer and x is the intensity of a given sound. Find the value of y if x is 51,000 times x_0.

$$\begin{aligned} y &= 10 \log \frac{51{,}000x_0}{x_0} && x = 51{,}000x_0 \\ &= 10 \log 51{,}000 \\ &\approx 47.1 \end{aligned}$$

The loudness of sound is about 47.1 decibels.

Practice Exercise 4

What is the loudness of a sound 200,000 times more intense than the weakest sound detected by an observer?

Answer: 53.0 decibels

The growth of a population of organisms that is not limited by predators, food supply, or living space, can be approximated exponentially by

$$y = Ie^{kt},$$

where y is the number of organisms after time t starting with an initial population I. The value of k is the percent of growth per unit of time t.

EXAMPLE 5 Population Growth Problem

The population of Upstate, Maine, varies according to the equation

$$y = 320e^{0.12t}$$

where t is measured in years. Find the time it will take for the population to double.

Practice Exercise 5

How long will it take for the population of Upstate, Maine to triple?

Letting $t = 0$, we get $y = 320e^0 = 320(1) = 320$. Thus, when $t = 0$, the population is 320. We need to find t when $y = 2(320) = 640$.

$$640 = 320e^{0.12t}$$

$$\frac{640}{320} = e^{0.12t}$$

$$2 = e^{0.12t}$$

$$\ln 2 = \ln e^{0.12t}$$

$$\ln 2 = 0.12t \qquad \text{Power rule}$$

$$t = \frac{\ln 2}{0.12} \approx 5.78$$

The population will double in about 5.78 years.

Answer: about 9.2 yr

10.6 EXERCISES A

Solve.

1. If \$1250 is invested at an annual rate of 12% compounded quarterly, what will be the value of the account at the end of 10 years?

2. What amount must Cathy invest at 8% interest compounded semiannually, to accumulate \$12,000 at the end of 15 years?

3. Margot invested \$685 in an account 25 years ago. What is the value of the account if the annual interest is 8% compounded **(a)** semiannually? **(b)** quarterly?

4. What amount must Tim invest at 8% interest compounded quarterly, to accumulate \$4000 at the end of 12 years?

5. Use the equation $y = \log_2 \frac{x}{x_0}$ to find the number of cell fissions that occur during the time of an experiment if there are 125 cells initially and 9525 cells finally.

6. Use the equation $y = \log_2 \frac{x}{x_0}$ to find the final number of cells in an experiment if there were 420 cells initially and the number of fissions was 12.

7. Use the sound equation $y = 10 \log \frac{x}{x_0}$ to determine the loudness of a sound in decibels if the given sound has intensity 63,000 times the weakest sound detected.

8. Use the equation $y = 10 \log \frac{x}{x_0}$ to find $\frac{x}{x_0}$ if the loudness of the sound is 90 decibels.

9. The magnitude y on the Richter scale of the intensity x of an earthquake is given by $y = \log \frac{x}{x_0}$ where x_0 is the minimum intensity used for comparison. If an earthquake has intensity 50,100 times the minimum intensity, what is its Richter scale magnitude?

10. Use the equation in Exercise 9 to find the ratio $\frac{x}{x_0}$ if the magnitude of an earthquake is 7.0 on the Richter scale.

11. The population of Warm, Florida, is increasing according to the equation $y = 5000e^{0.02t}$, where t is in years. Find the population 12 years after the initial count.

12. How long will it take for the population in Exercise 11 to double?

13. Use the equation $y = 250e^{-0.5t}$ to find the half-life (the time for one-half of a substance to decay) of a radioactive substance where y is the amount of the substance remaining and t is in years.

14. Use the equation $y = 1200e^{-0.08t}$ to determine when one-fourth of a given radioactive substance will be gone.

15. Given that $\text{pH} = -\log [H_3O^+]$ and that $[H_3O^+] = 10^{-8.2}$, find the pH of the solution.

16. If the pH of an acid solution is 3.2, find $[H_3O^+]$ using $\text{pH} = -\log [H_3O^+]$.

FOR REVIEW

Solve for x.

17. $\log x + \log (x - 2) = \log 3$

18. $2^{3x+1} = 16$

The following exercises review material from Chapters 3, 8, and 9 to help you prepare for the next chapter. Graph each function.

19. $x + 3y - 6 = 0$

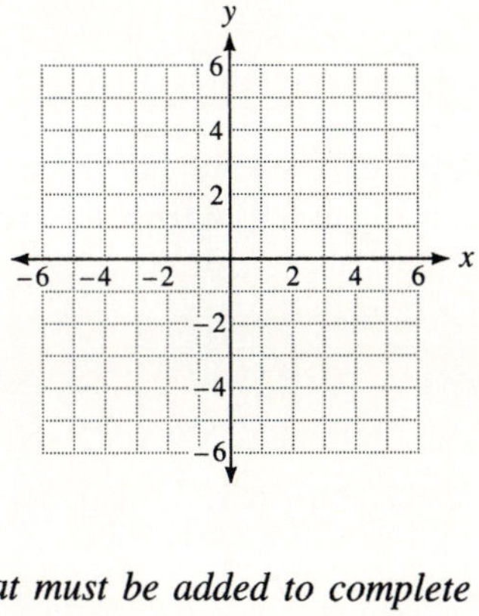

20. $y = x^2 - 4x - 5$

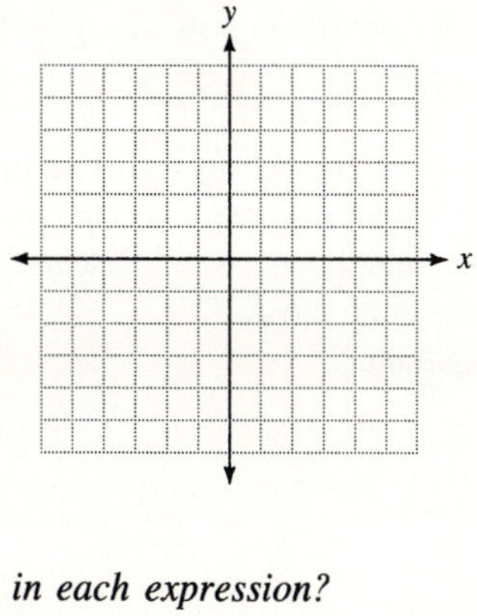

21. $y = \sqrt{x - 5}$

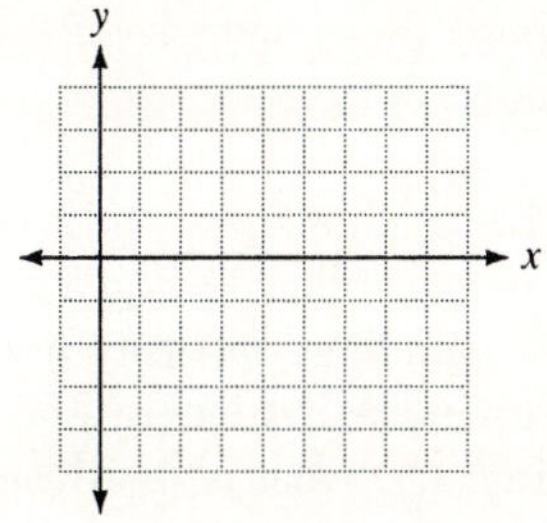

What must be added to complete the square in each expression?

22. $x^2 - 8x$

23. $y^2 + 10y$

24. $x^2 - x$

ANSWERS: 1. \$4077.55 2. \$3699.83 3. (a) \$4868.08 (b) \$4962.58 4. \$1546.15 5. 6.25 fissions 6. 1,720,320 7. 48.0 decibels 8. 1.0×10^9 9. 4.7 10. 10,000,000 11. 6356 12. 34.7 years 13. 1.39 years 14. 3.6 years 15. 8.2 16. $10^{-3.2}$ 17. 3 18. 1 19.

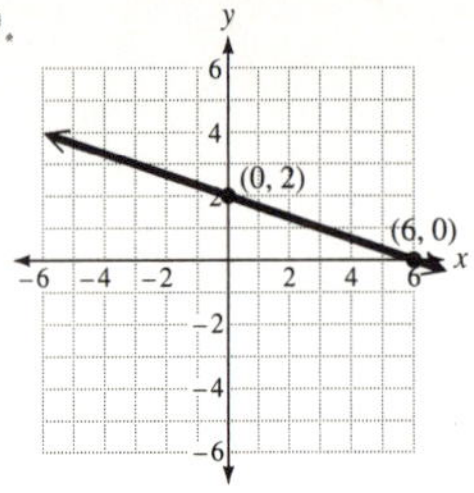

20.

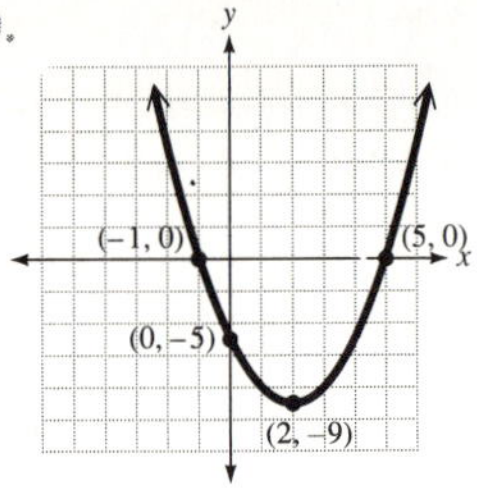

21.

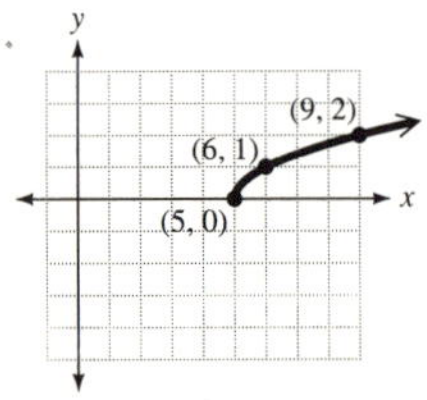

22. 16 23. 25 24. $\frac{1}{4}$

10.6 EXERCISES B

Solve.

1. Ms. Rutter invests \$5000 at 16% interest compounded semiannually. How much will she have in her account at the end of 9 years?

2. Larry wants to have \$9000 in his account in three years so he can buy a new car. How much must he invest now at 12% compounded quarterly to accumulate the \$9000?

3. Walter invested \$4360 in an account 15 years ago. What is the value of the account if interest is 12% compounded **(a)** quarterly? **(b)** monthly?

4. Susan must have \$750 in two years to pay a debt. How much must she invest today to have this amount in her account at the end of that time if interest is 12% compounded monthly?

5. Use the equation $y = \log_2 \frac{x}{x_0}$ to find the number of cell fissions that occur in a given period of time if there are 30 cells initially and 437,000 cells finally.

6. Use the equation $y = \log_2 \frac{x}{x_0}$ to find the number of cell fissions that occur if at the end of the experiment there are 320 times as many cells as there were at the beginning.

7. Use the sound equation $y = 10 \log \frac{x}{x_0}$ to determine the loudness of a sound in decibels if the given sound has intensity 210,000 times the weakest sound detected.

8. Use the equation $y = 10 \log \frac{x}{x_0}$ to find $\frac{x}{x_0}$ if the loudness of the sound is 36 decibels.

9. The magnitude y on the Richter scale of the intensity x of an earthquake is given by $y = \log \frac{x}{x_0}$ where x_0 is the minimum intensity used for comparison. If an earthquake has intensity 75,000 times the minimum intensity, what is its Richter scale magnitude?

10. Use the equation in Exercise 9 to find the ratio $\frac{x}{x_0}$ if the magnitude of an earthquake is 6.0 on the Richter scale.

11. The population of a state is decreasing according to the equation $y = 3{,}200{,}000e^{-0.03t}$. Find the population after 6 years.

12. How long will it take for the population in Exercise 11 to be one-half the original amount?

13. Use the equation $y = 1000e^{-0.4t}$ to find the half-life (the time for one-half of a substance to decay) of a radioactive substance where y is the amount of the substance remaining and t is in years.

14. Use the equation $y = 500e^{-0.05t}$ to determine when one-fourth of a given radioactive substance will be gone.

15. Given that $\text{pH} = -\log\,[H_3O^+]$ and that $[H_3O^+] = 10^{-7.3}$, find the pH of the solution.

16. If the pH of a solution is 4.5, find $[H_3O^+]$ using $\text{pH} = -\log\,[H_3O^+]$.

FOR REVIEW

Solve for x.

17. $\log_2 (x + 5) - \log_2 (x - 7) = 2$

18. $2^x 3^x = 11$

The following exercises review material from Chapters 3, 8, and 9 to help you prepare for the next chapter. Graph each function.

19. $2x - 3y = 0$

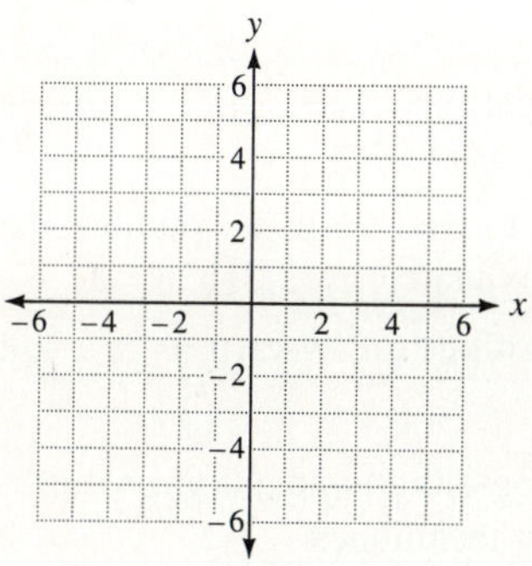

20. $y = -x^2 + 4x + 5$

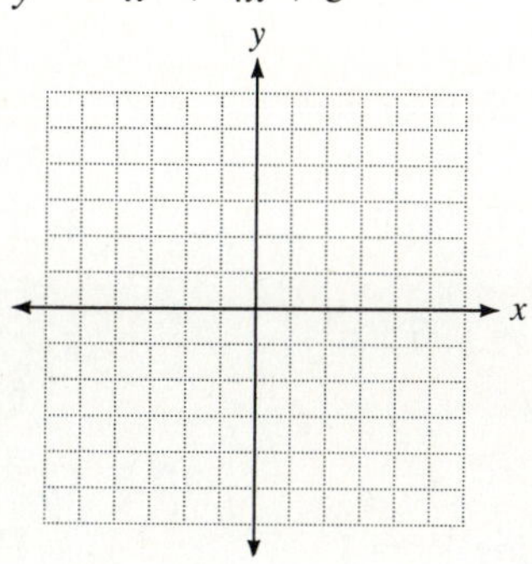

21. $y = -\sqrt{x - 5}$

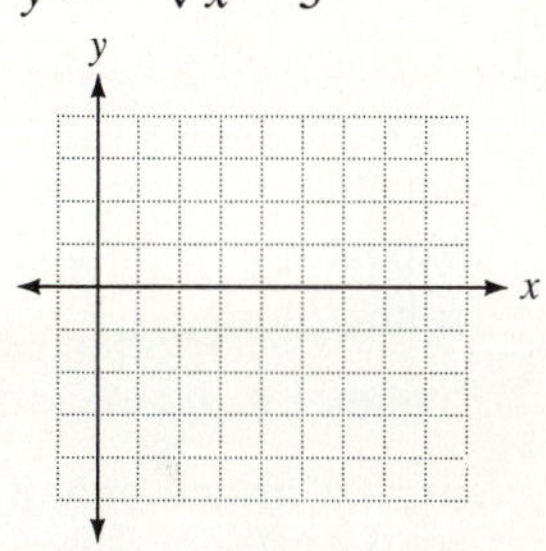

What must be added to complete the square in each expression?

22. $x^2 - 4x$

23. $y^2 + 16y$

24. $x^2 + x$

10.6 EXERCISES C

The formula used for population growth, $y = Ie^{kt}$, is called the Malthusian model after Thomas Malthus who, in the 18th century, made predictions about growing populations. Remember that k is the percent of growth per unit of time. Use this information in the following exercises.

1. The population of the United States was 237 million in 1985. If the annual growth rate is approximately 0.9%, what will be the population in the year 2000?

2. When the interstate freeway bypassed a small town in New Mexico, the population began to decline from 12,000 residents at an annual rate of 4.1%. How many years will it take for the population to reach 8000?

3. The population of Alpine, Oregon was 800 in 1970 and 1150 in 1990. Assuming the same growth rate find the value of k (correct to three decimal places) and use it to predict the population of Alpine in the year 2010.

4. The population of the southwestern United States is expected to double in 30 years. What will be the approximate annual growth rate of this region?

CHAPTER 10 REVIEW

KEY WORDS

10.1 A **logarithm** is an exponent. In fact, if $a^x = y$ where $a > 0$, $a \neq 1$, then x is the **logarithm to the base *a* of *y*.**

10.2 A function defined by the equation $y = a^x$, $a > 0$, $a \neq 1$, is an **exponential function** with **base *a*.**

A function defined by the equation $y = \log_a x$, $a > 0$, $a \neq 1$, is a **logarithmic function.**

10.4 Logarithms using 10 as a base are **common logarithms.**

A number that corresponds to a given logarithm is its **antilogarithm (antilog).**

Logarithms using e as a base are **natural logarithms.**

The function defined by the equation $y = \ln x$ is the **natural logarithmic function.**

10.5 An equation that contains logarithms of expressions involving the variable is a **logarithmic equation.**

An equation that contains variable exponents is an **exponential equation.**

KEY CONCEPTS

10.1 **1.** If a, x, and y are numbers, $a > 0$ $a \neq 1$, then the equation $y = a^x$ is equivalent to the equation $x = \log_a y$, with the first being the exponential form and the second the logarithmic form.

2. Suppose $a, b > 0$, $a \neq 1$, and $b \neq 1$. Then

$a^x = a^y$ if and only if $x = y$, and
$a^x = b^x$ if and only if $a = b$.

3. If a is any base ($a > 0$, $a \neq 1$), $\log_a a = 1$ and $\log_a 1 = 0$.

10.2 **1.** The logarithmic function $y = \log_a x$ and the exponential function $y = a^x$ are inverses. The graph of each is the reflection of the graph of the other across the line $y = x$.

2. To graph a logarithmic function $y = \log_a x$, it is easiest to plot a few points determined by the equivalent exponential equation $x = a^y$.

10.3 **1.** Product rule: $\log_a (uv) = \log_a u + \log_a v$ [which is not $\log_a (u + v)$]

2. Quotient rule: $\log_a \frac{u}{v} = \log_a u - \log_a v$ [which is not $\log_a (u - v)$]

3. Power rule: $\log_a u^c = c \log_a u$ [which is not $\log_a cu$]

10.4 **1.** Common logarithms are written as $\log x$ rather than $\log_{10} x$.

2. If a and b are any two bases, and $x > 0$, then

$$\log_b x = \frac{\log_a x}{\log_a b} \text{ and } \log_b a = \frac{1}{\log_a b}.$$

3. If $x = \log n$, x is the logarithm of n and n is the antilogarithm (antilog) of x. In exponential form, $n = 10^x$.

4. The number e is an irrational number that can be approximated by 2.71828.

5. Natural logarithms are written as $\ln x$ rather than $\log_e x$.

10.5 **1.** To solve a logarithmic equation use one of the following techniques.

(a) Obtain a single logarithmic expression using the same base on one side of the equation and convert the result to an exponential equation.

(b) Write each side as a logarithm using the same base and find the antilog of both sides.

2. To solve an exponential equation: try to express each side as a power using the same base and equate exponents; if this fails, take the log of both sides and use the power rule to eliminate the variable exponents.

3. When solving logarithmic and exponential equations, always check your answers in the original equation. Remember that the logarithm of a negative number or zero is not defined so that "solutions" that require finding such logarithms must be discarded.

REVIEW EXERCISES

Part I

10.1 **1.** Another term for the word *logarithm* is ______________.

2. The exponential form of $y = \log_a x$ is ______________.

3. The logarithmic form of $u = b^v$ is ______________.

4. Every positive number except ______________ can be used as a base for logarithms.

Convert to logarithmic form.

5. $2^x = 50$

6. $x^3 = 27$

7. $u^v = q$

Convert to exponential form.

8. $\log_3 a = b$

9. $\log_a a = 1$

10. $\log_c w = m$

Give the numerical value of x.

11. $x^4 = 81$

12. $x^{-1} = \frac{1}{7}$

13. $x^{1/2} = 4$

14. $x = \log_2 8$

15. $\log_x 125 = 3$

16. $\log_3 x = -2$

17. For any base a, $\log_a a =$ ______________.

18. For any base a, $\log_a 1 =$ ______________.

10.2 **19.** Sketch the graph of $y = 5^x$ and use it to sketch the graph of $y = \log_5 x$.

x	$y = 5^x$
0	
1	
−1	
2	
−2	

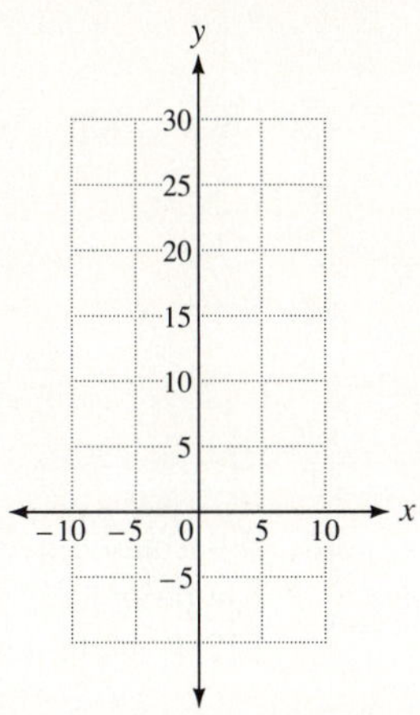

10.3 **20.** If a is any base and u and v are positive numbers, $\log_a (u \cdot v) =$ ______________.

21. If a is any base and u and v are positive numbers, $\log_a \left(\frac{u}{v}\right) =$ ______________.

22. If a is any base, u is a positive number, and c is any number, $\log_a u^c =$ ______________.

23. The rule given in Exercise 20 is called the ________________ rule.

24. The rule given in Exercise 21 is called the ________________ rule.

25. The rule given in Exercise 22 is called the ________________ rule.

26. Express in terms of logarithms of x, y, and z.
$\log_a \frac{x^3\sqrt{z}}{y^4}$

27. Express as a single logarithm.
$\frac{1}{3}\log_a x + 5\log_a y$

Decide whether the following are true or false. If false, change the right side to make it true.

28. $\log_a \frac{u}{v} = \frac{\log_a u}{\log_a v}$

29. $\log_a a = a$

30. $\log_a (u \cdot v) = (\log_a u)(\log_a v)$

31. $\log_a u^c = u \log_a c$

10.4 *Find each logarithm.*

32. $\log 4135$

33. $\log 0.000789$

34. $\log_3 6.29$

35. $\ln 125.6$

36. $\ln 0.0625$

37. $\log_8 0.00788$

Find each antilogarithm, correct to three significant digits.

38. $\log x = 2.2547$

39. $\log x = -1.6554$

40. $\ln x = 1.7983$

10.5 *Solve the following equations.*

41. $\log (x + 3) + \log x = 1$

42. $\log x - \log (3x + 7) = \log \frac{1}{10}$

43. $3^{5x-1} = 81$

44. $5^{2x} = 8$

10.6 *Solve each applied problem.*

45. If \$875 is invested at an annual rate of 4% compounded quarterly, what will be the value of the account at the end of 20 years?

46. What amount must Paul invest at 12% interest compounded semiannually, to accumulate \$2750 at the end of 10 years?

47. Use $y = \log_2 \frac{x}{x_0}$ to find the number of cell fissions that occur during an experiment if there were 540 cells initially and the final count was 8960 cells.

48. An earthquake measured 5.2 on the Richter scale. Use $y = \log \frac{x}{x_0}$ to determine the ratio $\frac{x}{x_0}$.

Part II

49. Express as a single logarithm.
$\frac{1}{2}(\log_a x - \log_a z - 3 \log_a y)$

50. Express in terms of logarithms of x, y, and z.
$\log_a \sqrt[5]{\frac{xy}{z^2}}$

Find each logarithm.

51. $\log 0.00378$

52. $\ln 17.2$

53. $\log_5 2.59$

Solve the following equations.

54. $\ln x + \ln (x + 4) = \ln 5$

55. $2^x 2^{x^2} = 4$

Solve.

56. If \$8000 is invested at an annual rate of 14% compounded semiannually, what is the value of the account at the end of 5 years?

57. Use the sound equation $y = 10 \log \frac{x}{x_0}$ to find the loudness of a sound in decibels if the given sound has intensity 85,000 times the weakest sound detected.

Find each antilogarithm, correct to three significant digits.

58. $\log x = 2.4137$

59. $\log x = -0.3742$

60. $\ln x = -2.4558$

61. Convert to logarithmic form.
$8^{-3} = x$

62. Convert to exponential form.
$\ln 6 = x$

Solve.

63. The population of a state is increasing according to the equation $y = 2{,}200{,}000e^{0.04t}$. Find the time it will take for the population to double (t is in years).

64. Annette wants to have \$20,000 in her account in four years to have a down payment for a piece of property. How much must she invest now at 10% compounded quarterly to accumulate the \$20,000?

Answer true *or* false *in the following. If false, explain why.*

65. The exponential form of $y = \log_a x$ is $a^x = y$.

66. The logarithmic form of $u = b^v$ is $v = \log_b u$.

67. If a is any base and u and v are positive numbers, then $\log_a (u + v) = (\log_a u)(\log_a v)$.

68. If a is any base and u and v are positive numbers, then $\log_a \frac{u}{v} = \log_a u - \log_a v$.

69. If a is any base and u is a positive number, then $\log_a \sqrt{u} = 2 \log_a u$.

70. Base 10 logarithms are called common logarithms.

71. In the equation $x = \log n$, n is called the antilogarithm of x.

72. The graph of $y = \log_a x$ can be obtained from the graph of $y = a^x$ by reflection in the line $y = x$.

73. If a and b are bases, then $\dfrac{1}{\log_b a} = \log_a b$.

74. If a and b are bases and $x > 0$, then $\dfrac{\log_a x}{\log_a b} = \log_x a$.

75. If $y = \ln x$ then $e^y = x$.

ANSWERS: 1. exponent 2. $x = a^y$ 3. $v = \log_b u$ 4. 1 5. $x = \log_2 50$ 6. $3 = \log_x 27$ 7. $v = \log_u q$ 8. $3^b = a$ 9. $a^1 = a$ 10. $c^m = w$ 11. 3 12. 7 13. 16 14. 3 15. 5 16. $\frac{1}{9}$ 17. 1 18. 0 19. (Reflect the graph of $y = 5^x$ across the line $y = x$ to obtain the graph of $y = \log_5 x$.)

x	$y = 5^x$
0	1
1	5
-1	$\frac{1}{5}$
2	25
-2	$\frac{1}{25}$

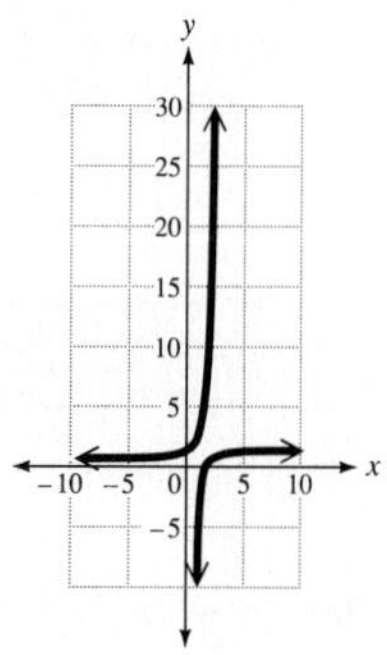

20. $\log_a u + \log_a v$ 21. $\log_a u - \log_a v$ 22. $c \log_a u$ 23. product 24. quotient 25. power 26. $3 \log_a x + \frac{1}{2} \log_a z - 4 \log_a y$ 27. $\log_a y^5 \sqrt[3]{x}$ 28. false; $\log_a u - \log_a v$ 29. false; 1 30. false; $\log_a u + \log_a v$ 31. false; $c \log_a u$ 32. 3.6165 33. -3.1029 34. 1.6739 35. 4.8331 36. -2.7726 37. -2.3292 38. 180 39. 0.0221 40. 6.04 41. 2 42. 1 43. 1 44. 0.646 45. \$1939.63 46. \$857.46 47. 4.1 fissions 48. $10^{5.2}$ 49. $\log_a \sqrt{\frac{x}{zy^3}}$ 50. $\frac{1}{5}(\log_a x + \log_a y - 2 \log_a z)$ 51. -2.4225 52. 2.8449 53. 0.5913 54. 1 55. 1, -2 56. \$15,737.20 57. 49.3 decibels 58. 259 59. 0.422 60. 0.0858 61. $-3 = \log_8 x$ 62. $e^x = 6$ 63. 17.3 years 64. \$6312.42 65. false; $x = a^y$ 66. true 67. false; $\log_a (u + v)$ cannot be simplified 68. true 69. false; $\log_a \sqrt{u} = \frac{1}{2} \log_a u$ 70. true 71. true 72. true 73. true 74. false; $\frac{\log_a x}{\log_a b} = \log_b x$ 75. true

CHAPTER 10 TEST

1. Convert $\log_b 5 = x$ to exponential form. 1. ______

2. Convert $u^v = 7$ to logarithmic form. 2. ______

Give the numerical value of x.

3. $x^{-3} = \dfrac{1}{64}$ 3. ______

4. $\log_5 \dfrac{1}{5} = x$ 4. ______

5. $\log_9 x = \dfrac{1}{2}$ 5. ______

6. $\log_x \sqrt{11} = \dfrac{1}{2}$ 6. ______

7. Express $\log_a \dfrac{x^2y^3}{z}$ as logarithms of x, y, and z. 7. ______

8. Express $2 \log_a z - \dfrac{1}{3} \log_a y$ as a single logarithm. 8. ______

Decide whether the following are true or false.

9. $\log_a (u + v) = \log_a u + \log_a v$ 9. ______

10. $\log_a u^{1/2} = \dfrac{\log_a u}{2}$ 10. ______

11. For any base a, $\log_a a =$ ____. 11. ______

12. Find log 0.612. 12. ______

13. Find ln 6.35. 13. ______

CHAPTER 10 TEST

CONTINUED

14. Find the antilogarithm, correct to three significant digits.
$\log x = 3.2923$

14. ______

15. Find the antilogarithm, correct to three significant digits.
$\ln x = -0.4351$

15. ______

Solve.

16. $2^{x^2} = 16$

16. ______

17. $\log x - \log (x + 1) = -1$

17. ______

18. Use the sound equation $y = 10 \log \frac{x}{x_0}$ to find the loudness of a sound in decibels if the given sound has intensity 120,000 times the weakest sound detected.

18. ______

19. The loan officer at the First Local Bank uses the formula $A = P(1.02)^n$. Find the value of A when P is \$5620 and n is 20.

19. ______

20. The population of a city is increasing according to $y = 120{,}000e^{0.15t}$. In how many years will the population be 10 times the original count?

20. ______

21. Graph $y = \log_2 x$.

21.

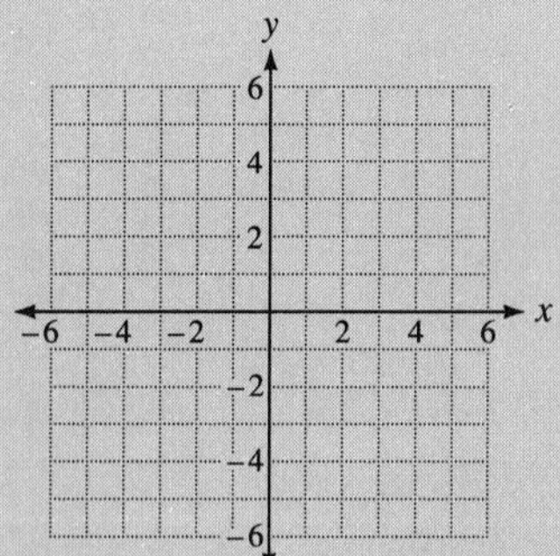

CHAPTER 11

Conic Sections

11.1 THE PARABOLA

STUDENT GUIDEPOSTS

1. Second-Degree Equations in Two Variables
2. Conic Sections
3. Parabolas Opening Up or Down
4. Parabolas Opening Left or Right
5. Standard Forms of a Parabola

The study of conics, including parabolas, circles, ellipses, and hyperbolas, dates back to the early Greeks who used these curves to solve various construction problems. However, applications of conics continue to grow today. The reflecting properties of parabolic surfaces are used in telescopes, signal lights, radar units, television antennae, and navigational systems. Planets and satellites travel in elliptical orbits, and atomic particles and comets travel in hyperbolic paths.

1 SECOND-DEGREE EQUATIONS IN TWO VARIABLES

In Chapter 3 we studied linear equations in two variables of the form $ax + by + c = 0$, with a and b not both zero, whose graphs are straight lines. Linear equations are sometimes called first-degree equations since the variables x and y appear only to the first power. An equation of the form

$$Ax^2 + Bxy + Cy^2 + Dx + Ey + F = 0,$$

where at least one of A, B, or C is not zero, is called a **second-degree equation in two variables** x and y.

2 CONIC SECTIONS

Many second degree equations have graphs that can be obtained by intersecting a cone and a plane. These curves are called **conic sections,** or **conics.** The four primary conic sections are parabolas, circles, ellipses, and hyperbolas. See Figure 11.1 on the following page.

3 PARABOLAS OPENING UP OR DOWN

Since we are already familiar with parabolas from our work in Chapter 9, it is appropriate to begin our study of the conic sections with the parabola. Geometrically, a **parabola** is the set of all points in a plane that are equidistant from a point (called the **focus**) and a straight line not containing the point. Although equations of parabolas can be derived from this definition, such a development is usually deferred to more advanced courses.

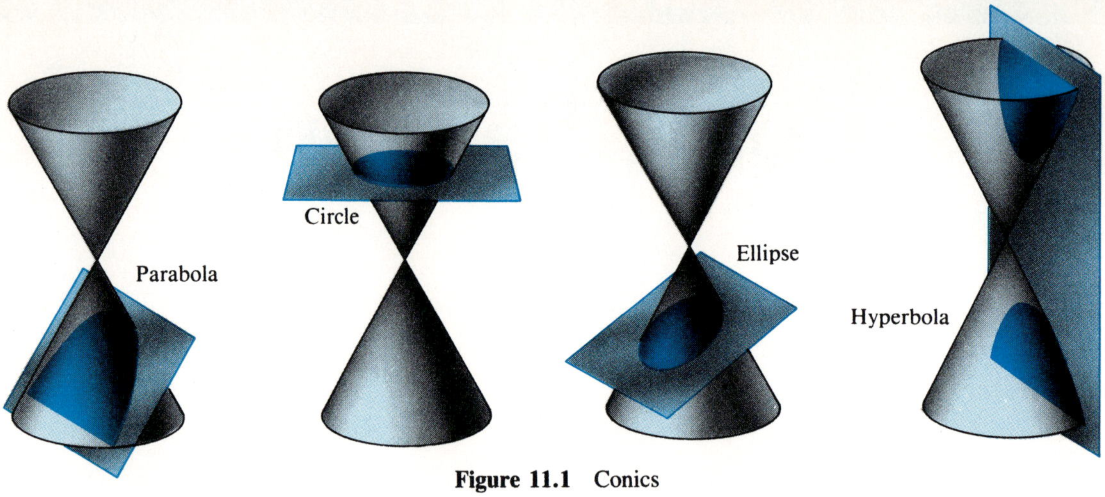

Figure 11.1 Conics

Recall from Section 9.1 that the graph of a quadratic function

$$y = ax^2 + bx + c \qquad a \neq 0$$

is a parabola that opens up if $a > 0$ and down if $a < 0$. Notice that such equations are special types of second-degree equations in the two variables x and y. If we complete the square in the variable x, the equation can be written in the form

$$y = a(x - h)^2 + k. \qquad a \neq 0$$

For example, consider the equation $y = 2x^2 - 4x - 1$. Suppose we complete the square on x. We first factor a 2 from the terms involving x and rewrite the equation leaving space as shown.

$$y = 2(x^2 - 2x \qquad) - 1$$

To complete the square on x inside the parentheses, we must add the square of half the coefficient of x, $\left[\frac{1}{2} \cdot (-2)\right]^2 = \left[-1\right]^2 = 1$. But if 1 is added inside the parentheses, since it is multiplied by 2, to maintain equality we must subtract 2 as shown.

$$y = 2(x^2 - 2x + 1) - 1 - 2 \qquad 2(+1) = 2 \text{ so subtract 2 to maintain equality}$$

Thus, we have $y = 2(x - 1)^2 - 3$. $\quad x^2 - 2x + 1 = (x - 1)^2$

Since $a = 2 > 0$, the parabola opens up and has vertex at the lowest point on the curve where the value of y is as small as possible. Since $2(x - 1)^2 \geq 0$, the smallest value y can assume occurs when $2(x - 1)^2 = 0$, that is, when $x = 1$. If x is 1, then $y = 2(1 - 1)^2 - 3 = 0 - 3 = -3$. Thus, the vertex is at $(1, -3)$ and the line of symmetry is $x = 1$. Since the vertex is below the x-axis and the parabola opens up, the curve crosses the x-axis at x-intercepts found by solving

$$2x^2 - 4x - 1 = 0.$$

Using the quadratic formula, we can obtain the solutions

$$x = \frac{2 \pm \sqrt{6}}{2},$$

which are approximately 2.2 and -0.2. The graph is given in Figure 11.2.

4 PARABOLAS OPENING LEFT OR RIGHT

The graph of a second-degree equation of the form

$$x = ay^2 + by + c$$

is also a parabola, but this relation was not considered in Section 9.1 since it is

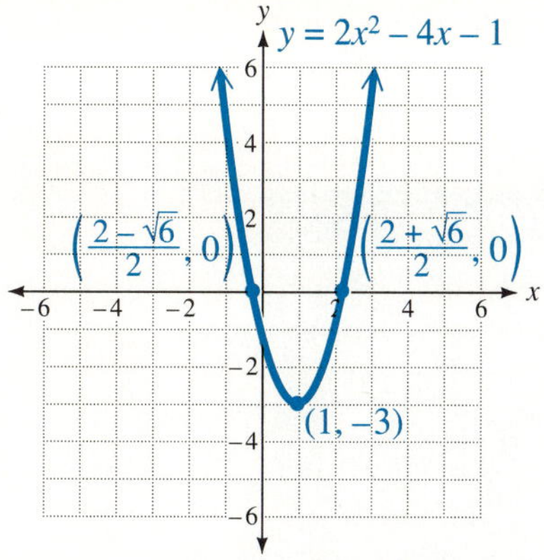

Figure 11.2 Parabola Opening Up

not a function. The graph of $y = ax^2 + bx + c$ will open up or down, and the graph of $x = ay^2 + by + c$ will open left or right.

EXAMPLE 1 Graphing a Parabola Opening Right

Graph $x = y^2 - 2y - 1$.

To make a table of values, we substitute values of y into the equation and simplify to obtain the corresponding values of x. The points in the table are plotted in Figure 11.3.

x	y
−1	0
−2	1
2	−1
−1	2
7	−2
2	3
7	4

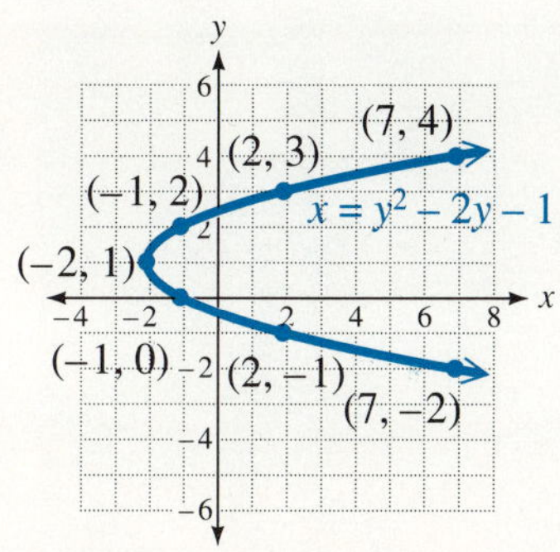

Figure 11.3 Parabola Opening Right

Practice Exercise 1

Graph $x = -y^2 + 2y + 1$.

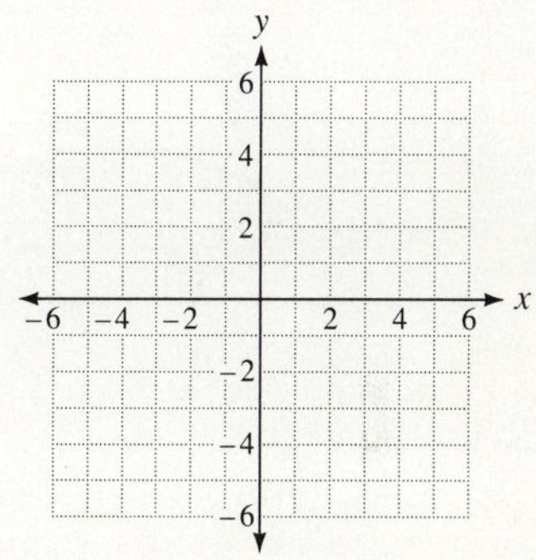

Answer: The graph is a parabola opening left with vertex (2, 1) and passing through (1, 2) and (1, 0).

5 STANDARD FORMS OF A PARABOLA

Notice that $x = y^2 - 2y - 1$ is not a function since for a given value of x, 2 for example, there are two values of y, 3 and −1. Thus, this relation does not pass the vertical line test. Rather than simply plotting points, we can identify the graph more directly by completing the square in y.

$$x = (y^2 - 2y \qquad) - 1 \qquad \text{Rewrite leaving space}$$

Add the square of half the coefficient of y, $\left[\frac{1}{2} \cdot (-2)\right]^2 = [-1]^2 = 1$, inside the parentheses and balance this out by subtracting 1 outside the parentheses.

$$\begin{aligned} x &= (y^2 - 2y + 1) - 1 - 1 \\ &= (y - 1)^2 - 2 \end{aligned}$$

We have now written the equation in the form

$$x = a(y - k)^2 + h,$$

where $a = 1 > 0$ and $(h, k) = (-2, 1)$. In general, whenever $a > 0$, the parabola opens to the right and the vertex is at (h, k). This agrees with what we found in Example 1. Also, if $a < 0$, the parabola will open to the left. We summarize these remarks in the following.

Standard Forms of the Equation of a Parabola

The second-degree equation $y = ax^2 + bx + c$ can be written in the **standard form** $y = a(x - h)^2 + k$ by completing the square on x. The graph has one of the following forms.

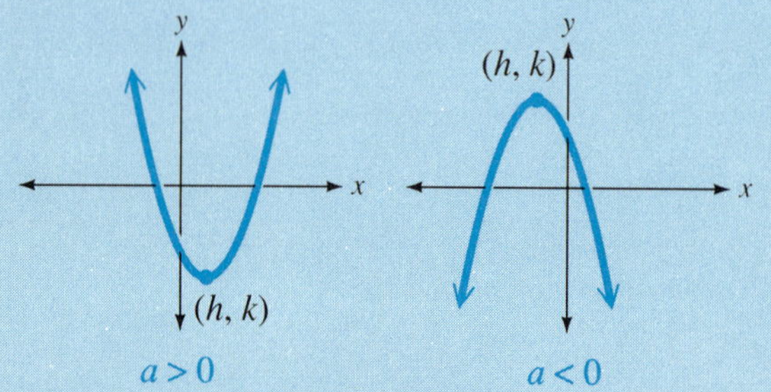

Figure 11.4

The second-degree equation $x = ay^2 + by + c$ can be written in the **standard form** $x = a(y - k)^2 + h$ by completing the square on y. The graph has one of the following forms.

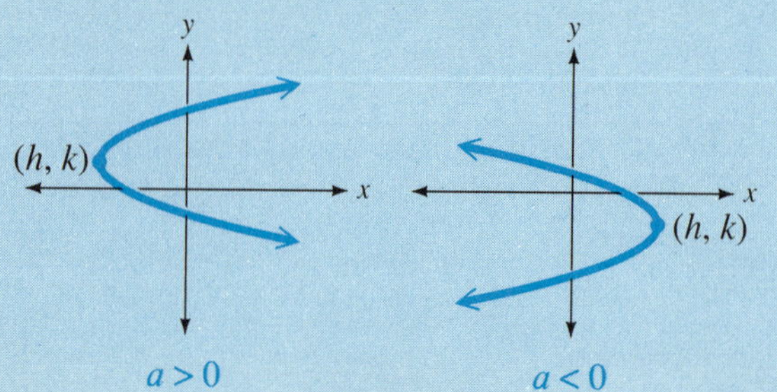

Figure 11.5

EXAMPLE 2 Using Standard Form to Graph a Parabola

Graph $x = -y^2 - 4y - 7$ by writing the equation in standard form.

We complete the square on y.

$$x = -(y^2 + 4y \qquad) - 7 \qquad \text{Factor out } -1 \text{ and leave space}$$
$$= -(y^2 + 4y + 4) - 7 + 4$$

The square of half the coefficient of y is $\left(\frac{1}{2} \cdot 4\right)^2 = (2)^2 = 4$. Since adding 4 inside the parentheses is the same as adding -4 to the right side (it is multiplied by $a = -1$), we add 4 outside the parentheses to maintain equality. Thus

$$x = -(y + 2)^2 - 3.$$

The vertex is $(-3, -2)$ since $y + 2 = y - k$ means that $k = -2$. Also, $a = -1 < 0$ so the parabola opens to the left. There are no y-intercepts (which would normally be found by solving $-y^2 - 4y - 7 = 0$). To find two additional points on the graph, let $x = -4$ and solve for y.

$$-4 = -y^2 - 4y - 7$$

Practice Exercise 2

Graph $y = -x^2 + 4x - 1$ by writing the equation in standard form.

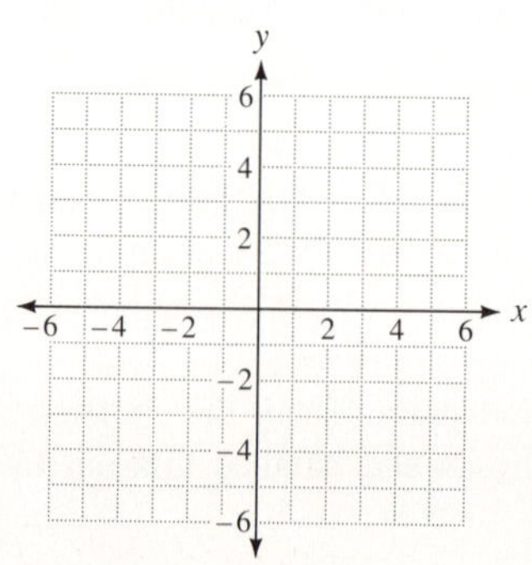

$$
\begin{aligned}
y^2 + 4y + 3 &= 0 \\
(y + 3)(y + 1) &= 0 \\
y + 3 = 0 \quad &\text{or} \quad y + 1 = 0 \\
y = -3 \quad & \quad\quad y = -1
\end{aligned}
$$

Thus, $(-4, -3)$ and $(-4, -1)$ are two points on the parabola along with the x-intercept $(-7, 0)$. The graph is given in Figure 11.6.

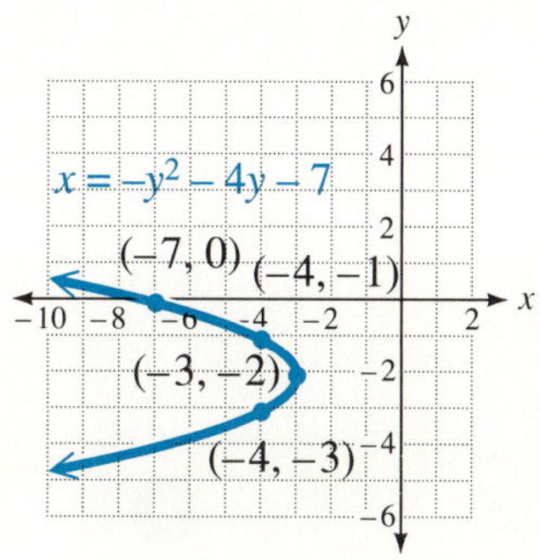

Figure 11.6

Answer: $y = -(x - 2)^2 + 3$, the parabola opens down with vertex $(2, 3)$ and passes through $(0, -1)$ and $(4, -1)$.

The next example illustrates one of the many applications of parabolas.

EXAMPLE 3 Architecture Problem

The entry to a shopping mall is a parabolic arch 20 ft high and 16 ft wide at the base. Find the equation for the parabola if the vertex is placed at the origin of a coordinate system. See Figure 11.7. What is the vertical clearance 4 ft from the center of the entryway?

Since the vertex is $(h, k) = (0, 0)$, and since the parabola opens down, the equation has the form

$$
\begin{aligned}
y &= a(x - h)^2 + k \\
&= a(x - 0)^2 + 0 = ax^2.
\end{aligned}
$$

One point on the parabola is $(8, -20)$. Substitute 8 for x and -20 for y to find the value of a.

$$
\begin{aligned}
y &= ax^2 \\
-20 &= a(8)^2 \\
-\frac{20}{64} &= a \\
-\frac{5}{16} &= a
\end{aligned}
$$

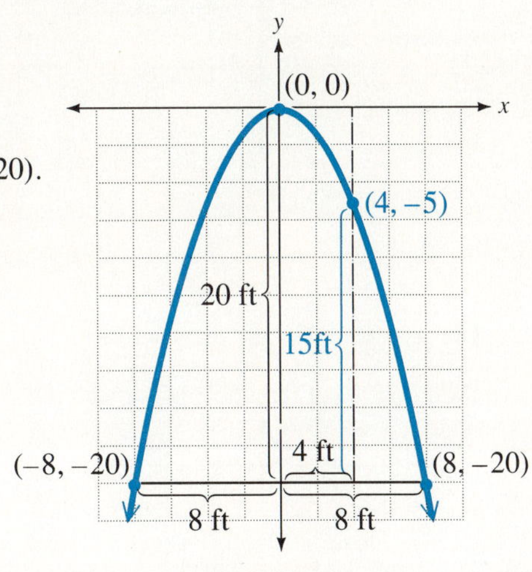

Figure 11.7

Thus, the equation of the parabola is $y = -\frac{5}{16}x^2$. To find the vertical clearance 4 ft from the center, we need to find y when x is 4.

$$
y = -\frac{5}{16}x^2 = -\frac{5}{16}(4)^2 = -5
$$

Hence, the vertical clearance 4 ft from the center of the arch is 5 ft lower than the total height 20 ft, or 15 ft.

Practice Exercise 3

Can a delivery truck with a load 9 ft wide and 13.5 ft tall drive through the arch in Example 3?

Answer: Yes; the vertical clearance 4.5 ft from the center of the arch is about 13.7 ft.

11.1 EXERCISES A

For each parabola, **(a)** *give the vertex, and* **(b)** *state whether the parabola opens up, down, left, or right.*

1. $y = 3(x - 1)^2 + 5$

2. $y = -2(x + 1)^2 - 3$

3. $x = -2(y - 3)^2 + 4$

4. $x = 5(y + 4)^2 - 7$

5. $y = x^2 - 5$

6. $x = -y^2 + 6$

7. $y = x^2 - 2x + 9$

8. $x = -y^2 + 6y - 15$

9. $x = -2y^2 - 4y - 7$

Graph each parabola by writing the equation in standard form.

10. $x = y^2$

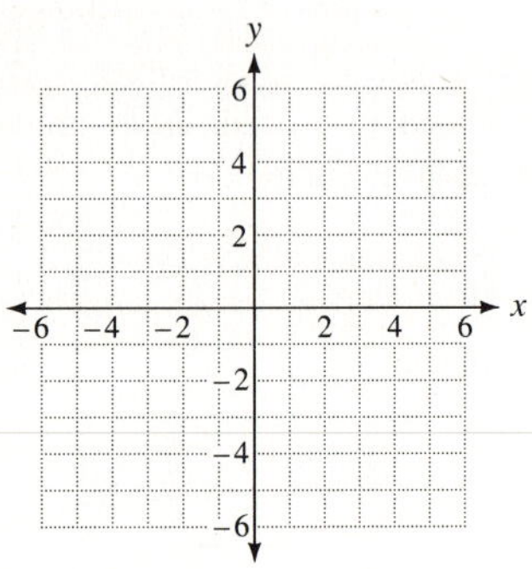

11. $y = -x^2$

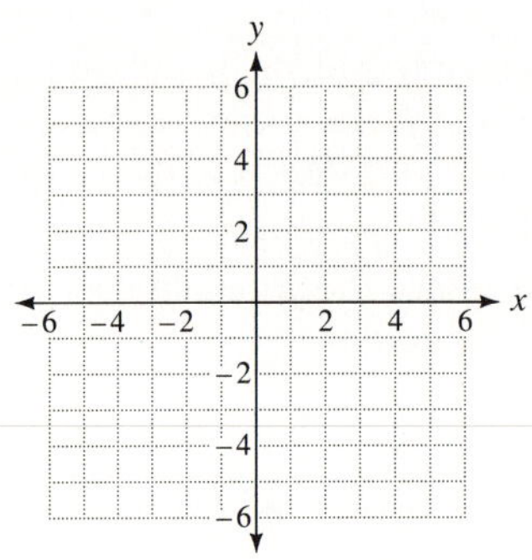

12. $x = -y^2$

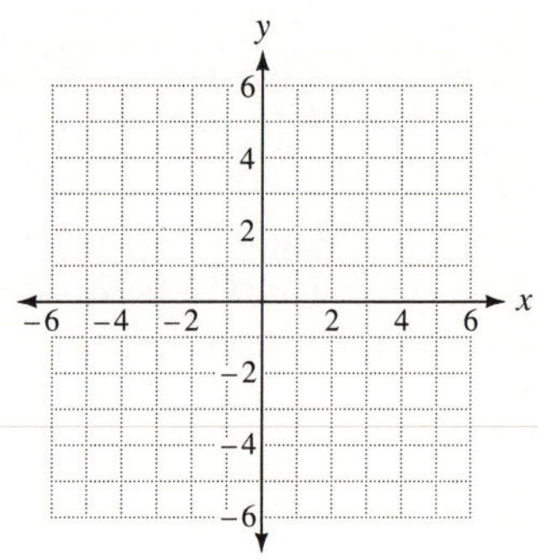

13. $y = x^2 - 2x + 3$

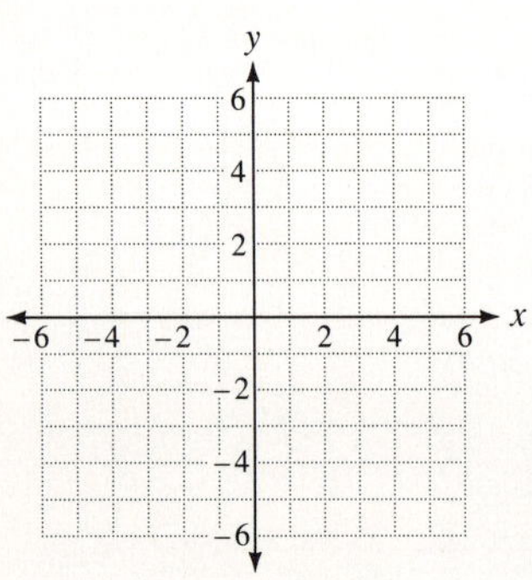

14. $x = y^2 - 2$

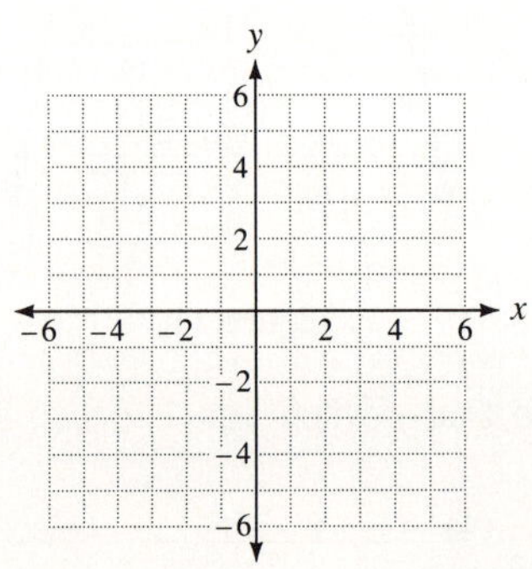

15. $x = y^2 - 2y - 3$

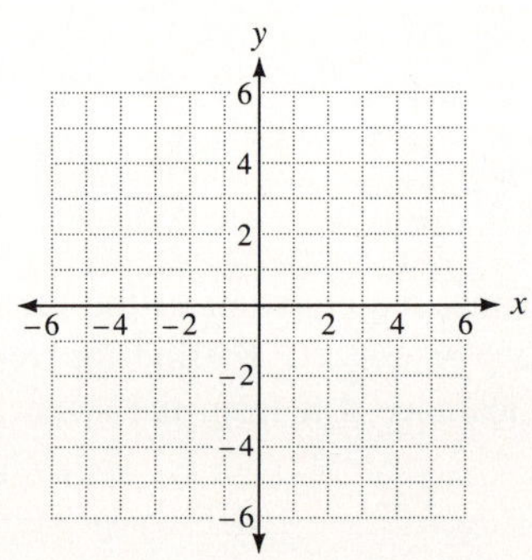

16. $4x = -y^2$

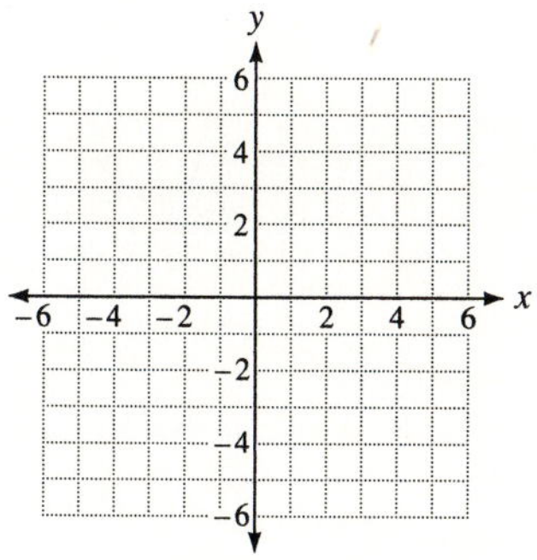

17. $2y = -x^2 + 2x + 1$

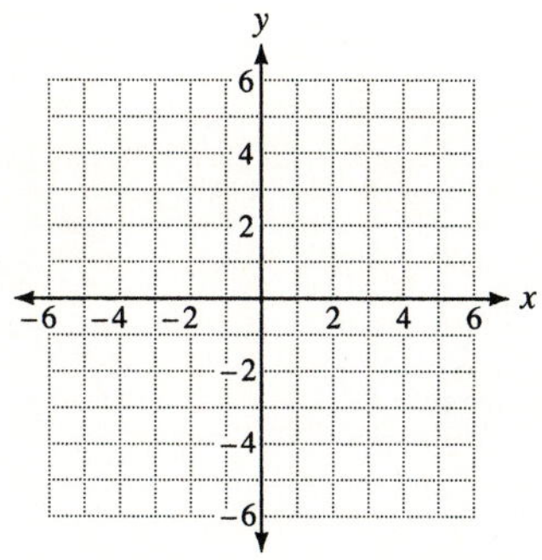

18. $8x = y^2 - 16$

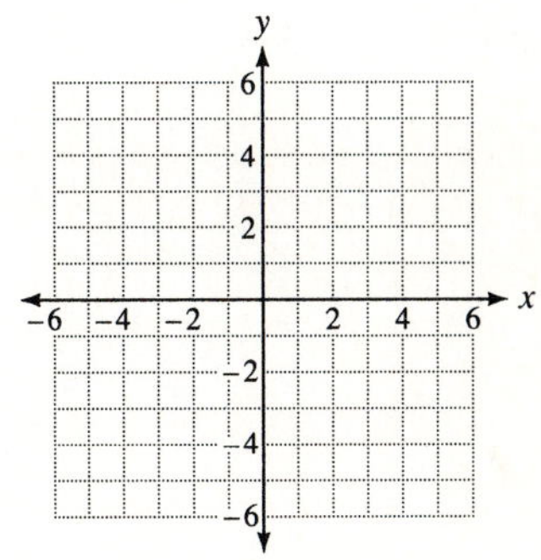

19. The cross-section of a ditch is a parabola. If the surface of the water in a full ditch is 10 ft wide and the ditch is 6 ft deep at the center, how deep is the water 2 ft from the edge of the ditch?

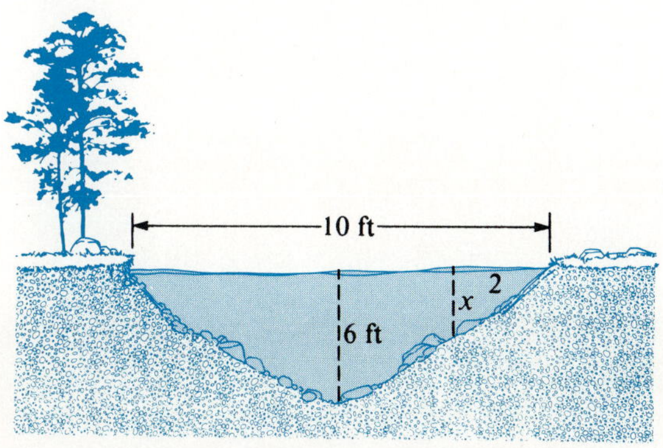

20. A footbridge over a creek is in the shape of a parabola. If the span of the bridge is 32 ft and the arch rises 4 ft above the bank of the creek, find the equation of the parabola assuming that its vertex is at the origin and it opens down.

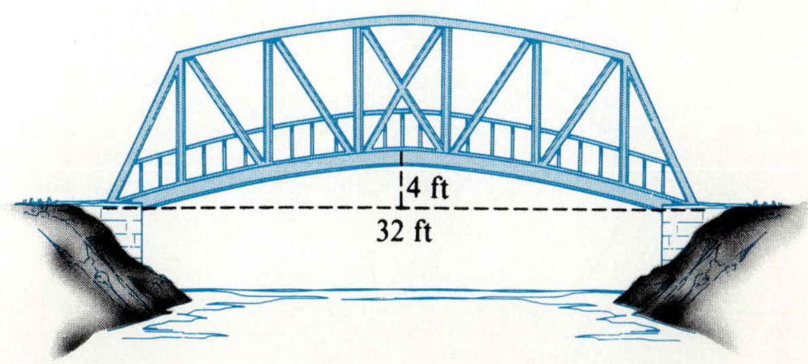

FOR REVIEW

Exercises 21–24 review topics covered in Chapter 10.

21. If $x = 439$, find **(a)** $\log x$, and **(b)** $\ln x$, correct to four decimal places.

22. If $x = -2.4351$, find **(a)** the common antilogarithm of x, and **(b)** the natural antilogarithm of x, correct to three significant digits.

23. Solve. $\log (x - 3) + \log x = 1$

24. Express $\log_a \frac{x^2\sqrt{y}}{z}$ using sums, differences, or multiples of logarithms of a single variable.

Exercises 25 and 26 review the distance formula from Section 7.7 to help you prepare for the next section.

25. Find the distance between the two points $(3, -7)$ and $(-1, 4)$.

26. The point (x, y) is 2 units from $(-1, 3)$. Find an equation involving x and y that describes this relationship.

ANSWERS: 1. (a) $(1, 5)$ (b) up 2. (a) $(-1, -3)$ (b) down 3. (a) $(4, 3)$ (b) left 4. (a) $(-7, -4)$ (b) right 5. (a) $(0, -5)$ (b) up 6. (a) $(6, 0)$ (b) left 7. (a) $(1, 8)$ (b) up 8. (a) $(-6, 3)$ (b) left 9. (a) $(-5, -1)$ (b) left

10.

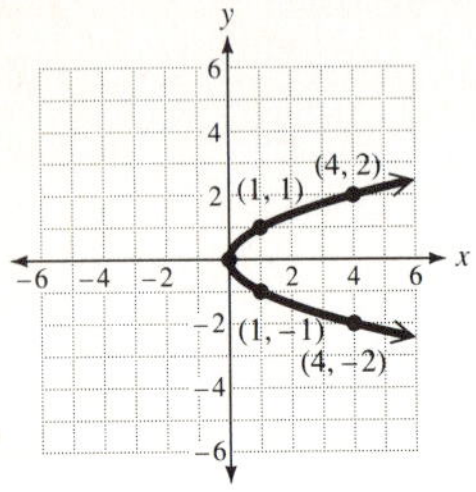

11.

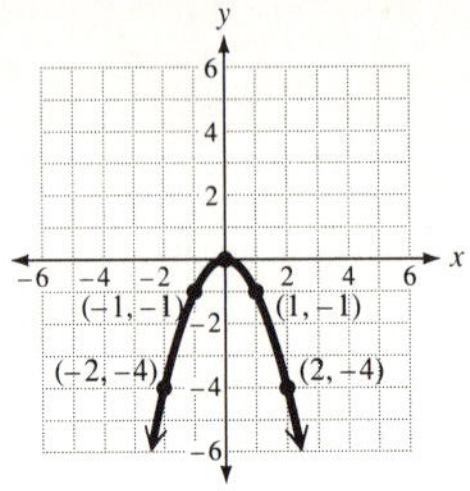

12.

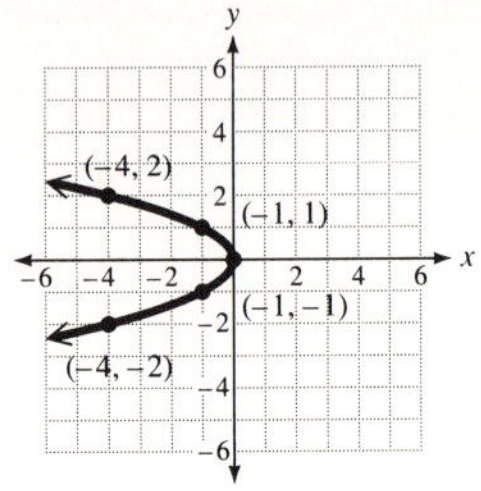

13.

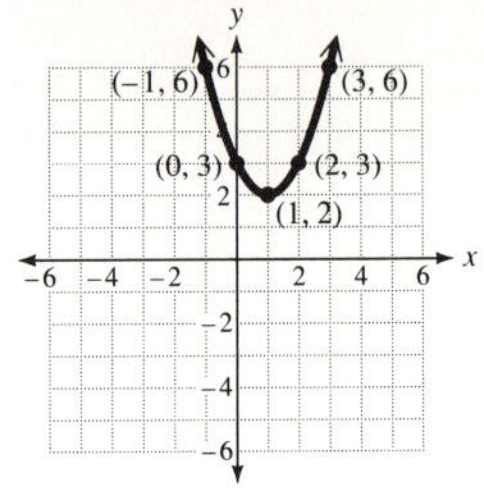

14.

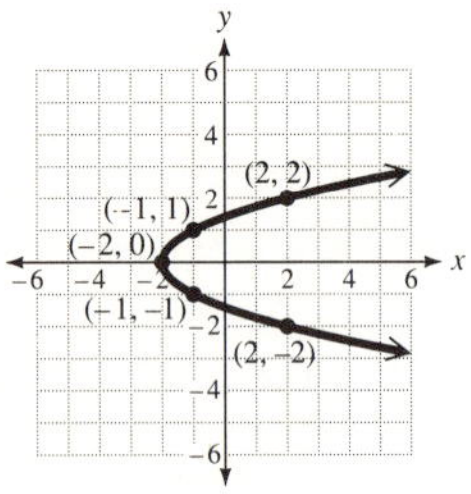

15.

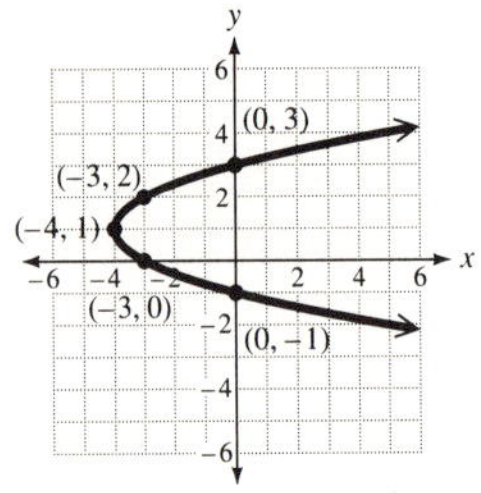

16.

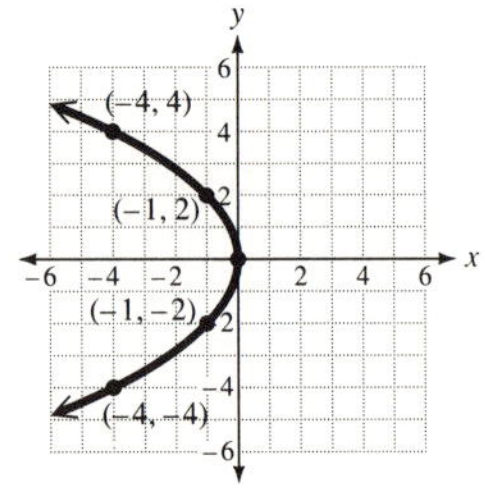

17.

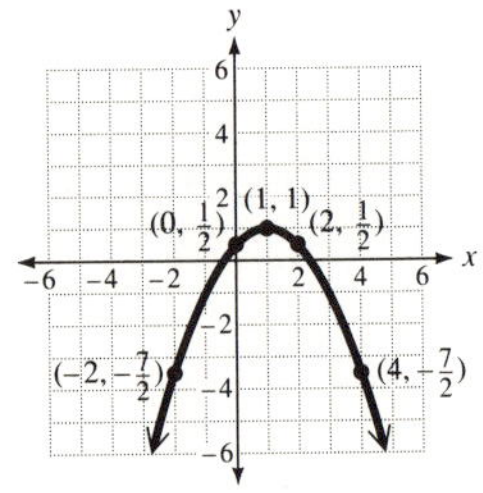

18.

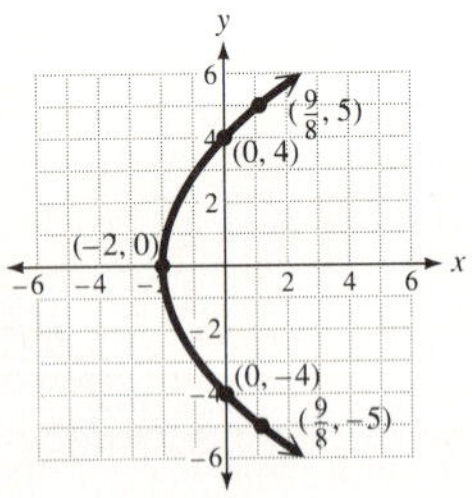

19. 3.84 ft 20. $y = -\frac{1}{64}x^2$ 21. (a) 2.6425 (b) 6.0845 22. (a) 0.00367 (b) 0.0876 23. 5 24. $2 \log_a x + \frac{1}{2} \log_a y - \log_a z$ 25. $\sqrt{137}$ 26. $(x + 1)^2 + (y - 3)^2 = 4$

11.1 EXERCISES B

For each parabola, ***(a)*** *give the vertex, and* ***(b)*** *state whether the parabola opens up, down, left, or right.*

1. $y = 2(x + 1)^2 - 5$

2. $y = -4(x - 2)^2 + 6$

3. $x = -5(y + 3)^2 - 5$

4. $x = 3(y - 4)^2 + 5$

5. $x = y^2 + 2$

6. $y = -x^2 - 3$

7. $x = y^2 + 10y + 24$

8. $y = 3x^2 - 12x + 20$

9. $2y = -x^2 - 2x - 7$

Graph each parabola by writing the equation in standard form.

10. $x = 2y^2$

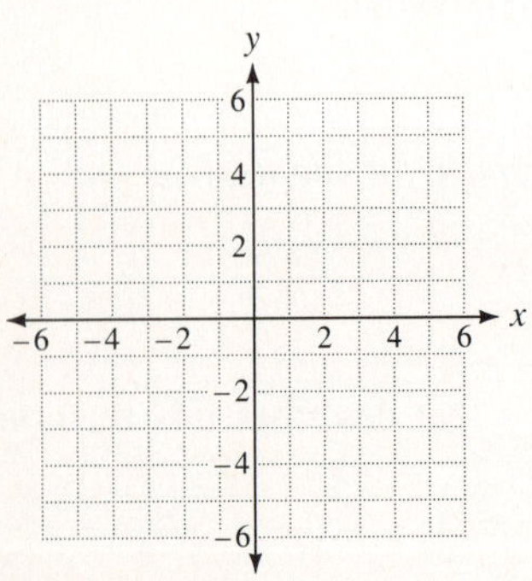

11. $y = -\dfrac{1}{2}x^2$

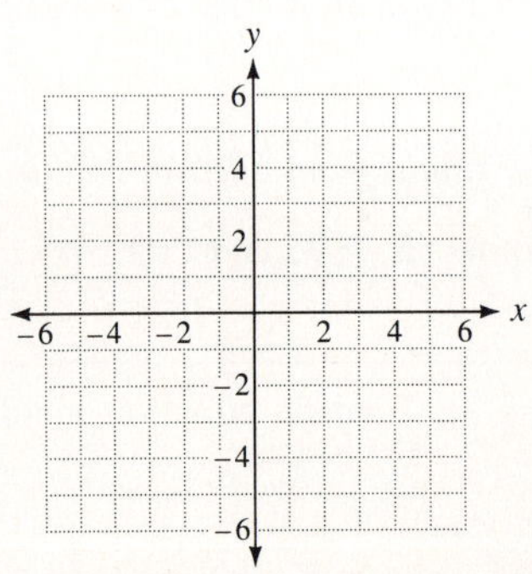

12. $x = -y^2 + 2$

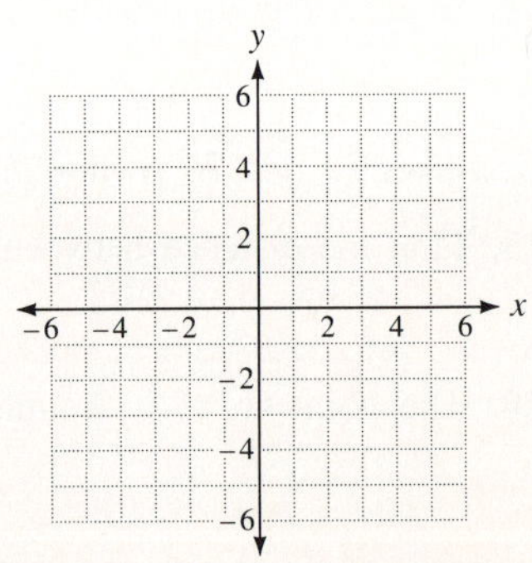

13. $y = x^2 - 4x + 3$

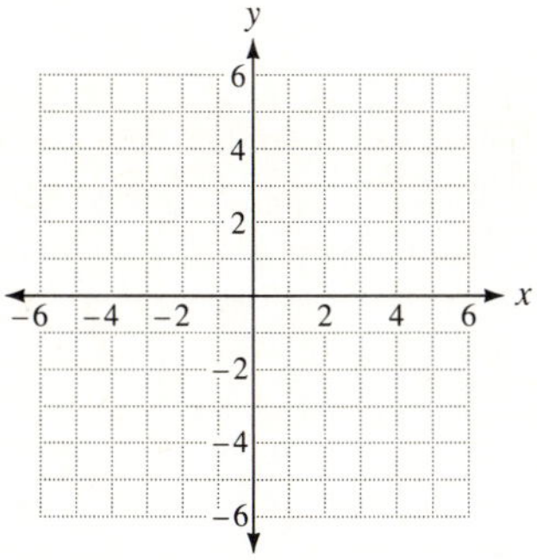

14. $x = y^2 + 1$

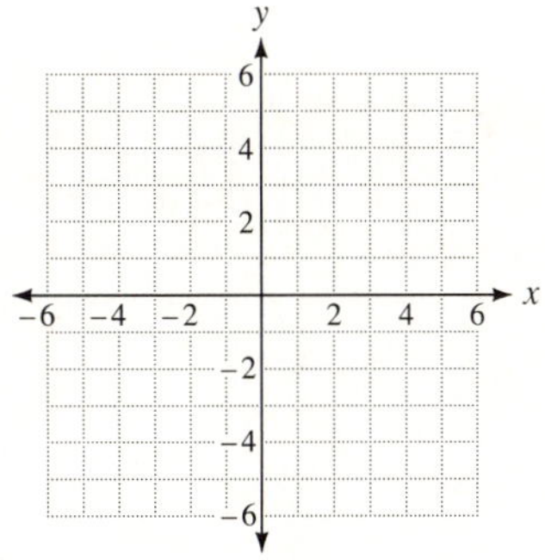

15. $x = -y^2 - 2y + 3$

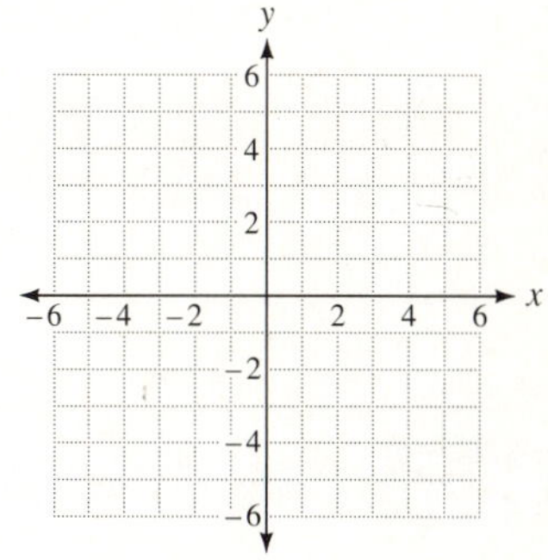

16. $2y = x^2$

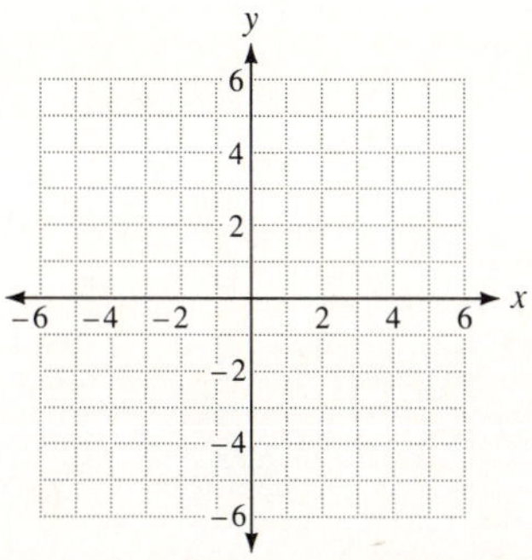

17. $4x = -y^2 - 2y + 3$

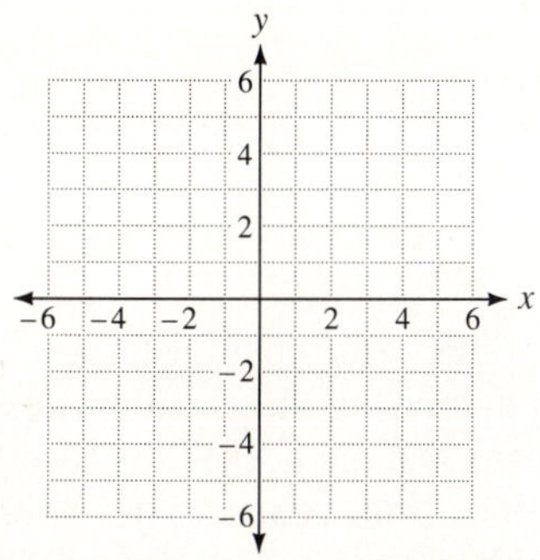

18. $8y = -x^2 - 6x + 7$

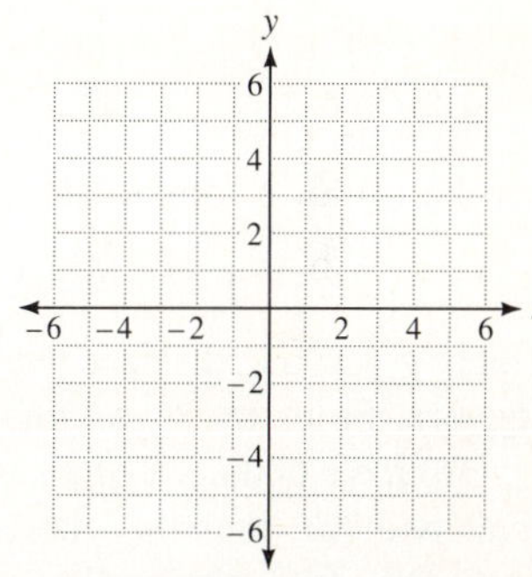

19. A tunnel is in the shape of a parabola. The maximum height is 30 ft and it is 24 ft wide at the base. Place the vertex at the origin and find the equation of the parabola. What is the vertical clearance 4 ft from the edge of the tunnel?

20. A television antenna is formed by rotating a parabola around its axis of symmetry. If the diameter of the antenna is 6 meters and it is $\frac{3}{4}$ meter deep, find the equation of the parabola by placing the vertex at the origin and assuming that it opens to the right.

FOR REVIEW

Exercises 21–24 review topics covered in Chapter 10.

21. If $x = 0.385$, find **(a)** $\log x$, and **(b)** $\ln x$, correct to four decimal places.

22. If $x = -1.5496$, find **(a)** the common antilogarithm of x, and **(b)** the natural antilogarithm of x, correct to three significant digits.

23. Solve. $2^{3x^2} = 16^x$

24. Express $\log_a \frac{x^3y^4}{\sqrt{z}}$ using sums, differences, or multiples of logarithms of a single variable.

Exercises 25 and 26 review the distance formula from Section 7.7 to help you prepare for the next section.

25. Find the distance between the two points $(-2, 3)$ and $(4, -1)$.

26. The point (x, y) is 3 units from $(2, -1)$. Find an equation involving x and y that describes this relationship.

11.1 EXERCISES C

Write each equation in standard form and identify the vertex. Does the parabola open up, down, left, or right?

1. $4y^2 + 4y + 12x + 9 = 0$

2. $36x^2 - 24x - 180y + 49 = 0$

11.2 THE CIRCLE

STUDENT GUIDEPOSTS

1. Standard Form of a Circle
2. General Form of a Circle

To study the next conic section, the circle, we begin with its definition and then use the distance formula developed in Section 7.7,

$$d = \sqrt{(x_2 - x_1)^2 + (y_2 - y_1)^2},$$

which gives the distance between two points with coordinates (x_1, y_1) and (x_2, y_2). From these we find the standard form of the equation of a circle.

A **circle** is the set of all points in a plane that are a fixed distance r from a given point (h, k) in that plane. The given point is the **center** of the circle and the fixed distance is the **radius** of the circle.

1 STANDARD FORM OF A CIRCLE

Let (x, y) be an arbitrary point on the circle with radius r centered at the point (h, k), as in Figure 11.8. Since every point (x, y) is r units from (h, k), by the distance formula we have

$$r = \sqrt{(x - h)^2 + (y - k)^2}.$$

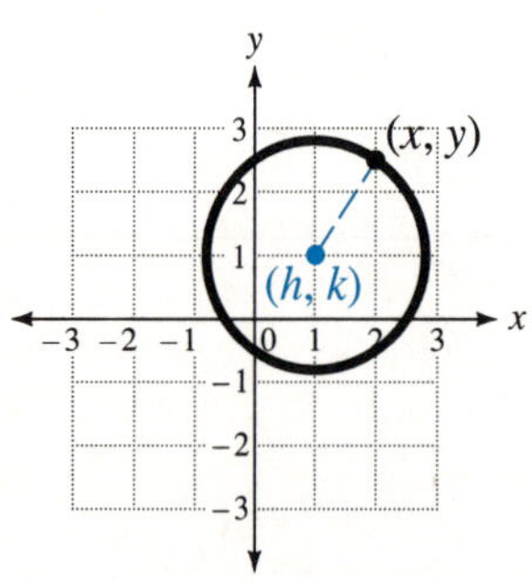

Figure 11.8 Circle

Squaring both sides yields the standard form of the equation of a circle.

Standard Form of the Equation of a Circle

The **standard form** of the equation of a circle with center (h, k) and radius r is

$$(x - h)^2 + (y - k)^2 = r^2.$$

Circles centered at the origin have a simplified standard form. Consider the equation

$$x^2 + y^2 = 4.$$

We can write this equation in the form

$$(x - 0)^2 + (y - 0)^2 = 2^2$$

to see that the center is indeed (0, 0) and the radius is 2. See the graph in Figure 11.9.

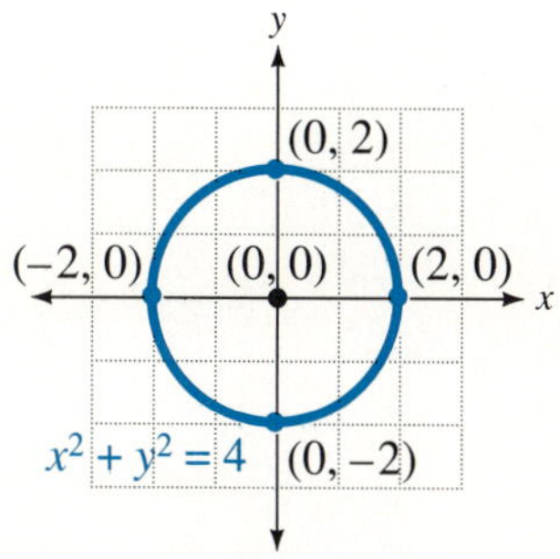

Figure 11.9

EXAMPLE 1 Graphing a Circle

Graph $(x - 1)^2 + (y + 2)^2 = 9$.

We rewrite the equation in the form $(x - h)^2 + (y - k)^2 = r^2$.

$$(x - 1)^2 + [y - (-2)]^2 = 3^2$$

Thus, the center is $(1, -2)$, and the radius is 3. See Figure 11.10.

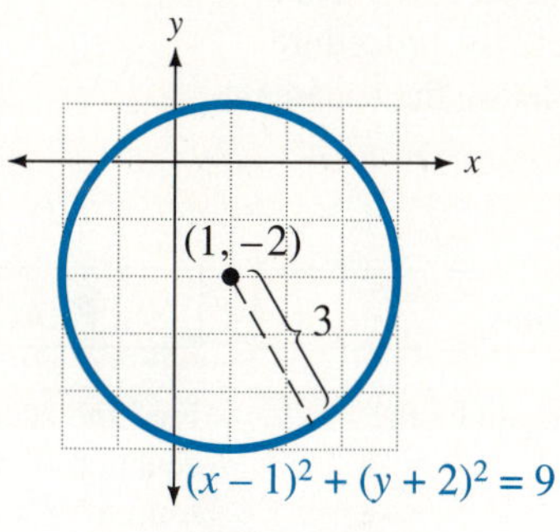

Figure 11.10

Practice Exercise 1

Graph $(x + 1)^2 + (y - 2)^2 = 4$.

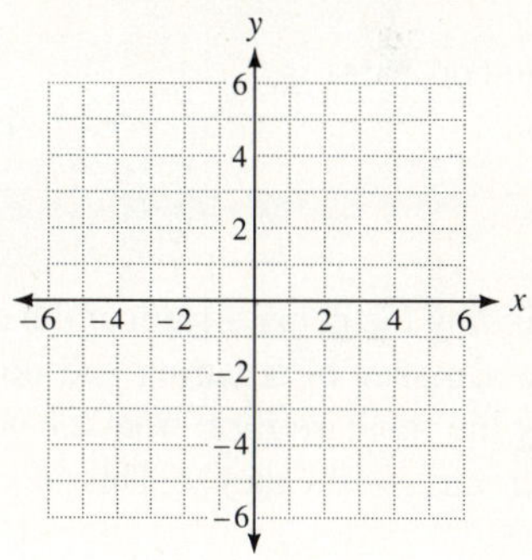

Answer: The graph is a circle centered at $(-1, 2)$ with radius 2.

If we are given the center and radius of a circle, we can find the standard form of its equation by substituting h, k, and r into $(x - h)^2 + (y - k)^2 = r^2$.

EXAMPLE 2 Finding the Standard Form

Find the standard form of the equation of the circle centered at $(2, -1)$ with radius 3.

Substitute $h = 2$, $k = -1$, and $r = 3$ in the standard form.

$$(x - h)^2 + (y - k)^2 = r^2$$

$$(x - 2)^2 + (y - (-1))^2 = 3^2 \quad \text{Watch the sign on } k$$

$$(x - 2)^2 + (y + 1)^2 = 9$$

Practice Exercise 2

Find the standard form of the equation of the circle centered at $(-4, 3)$ with radius $\sqrt{5}$.

Answer: $(x + 4)^2 + (y - 3)^2 = 5$

2 GENERAL FORM OF A CIRCLE

When the standard form of the equation of a circle is expanded, the resulting equation, called the **general form of the equation of a circle,** is actually a second-degree equation of the form

$$x^2 + y^2 + Dx + Ey + F = 0.$$

EXAMPLE 3 Finding the General Form

Find the general form of the equation of the circle centered at $(2, -1)$ with radius 3.

From Example 2, the standard form of this equation is

$$(x - 2)^2 + (y + 1)^2 = 9.$$

Expanding, we have

$$x^2 - 4x + 4 + y^2 + 2y + 1 = 9.$$

The general form is

$$x^2 + y^2 - 4x + 2y - 4 = 0.$$

Practice Exercise 3

Find the general form of the equation of the circle centered at $(-4, 3)$ with radius $\sqrt{5}$.

Answer: $x^2 + y^2 + 8x - 6y + 20 = 0$

Finding the general form of the equation of a circle from the standard form is simply a matter of squaring and collecting like terms. The reverse procedure, finding the standard form from the general form, requires completing the squares in both variables x and y. This is shown in the next example.

EXAMPLE 4 Finding the Standard Form

Find the standard form of the equation of the circle with general form

$$x^2 + y^2 - 6x + 10y + 29 = 0.$$

Give the center and radius, and graph the equation.

Start by writing the equation with the variable terms on the left side and the constant term on the right, leaving space as shown.

$$x^2 - 6x \qquad + y^2 + 10y \qquad = -29$$

Next, to complete the square in x, add $\left[\frac{1}{2}(-6)\right]^2 = 9$, and to complete the square in y, add $\left[\frac{1}{2}(10)\right]^2 = 25$ to the left side. Then, to maintain equality add 9 and 25 to the right side.

Practice Exercise 4

Find the standard form of the equation of the circle with general form $x^2 + y^2 + 4x - 6y + 9 = 0$.

What is the center of the circle?

What is the radius?

$$x^2 - 6x + 9 + y^2 + 10y + 25 = -29 + 9 + 25$$
$$(x^2 - 6x + 9) + (y^2 + 10y + 25) = 5$$
$$(x - 3)^2 + (y + 5)^2 = 5 \quad \text{Factor}$$

This is the standard form of the equation of the circle with center $(3, -5)$. Notice that since $r^2 = 5$, the radius is $\sqrt{5}$. See Figure 11.11.

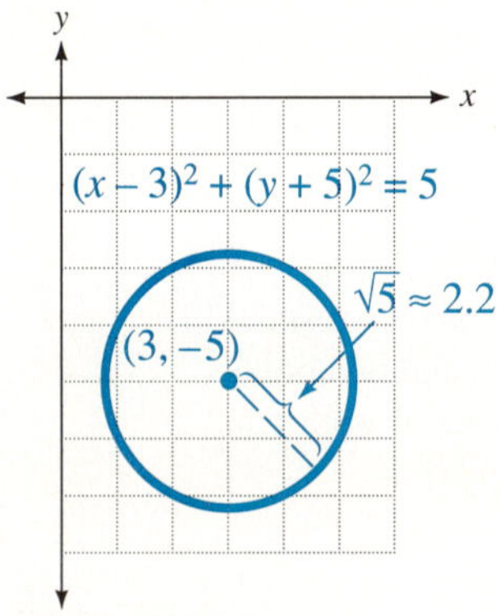

Figure 11.11

Graph the circle.

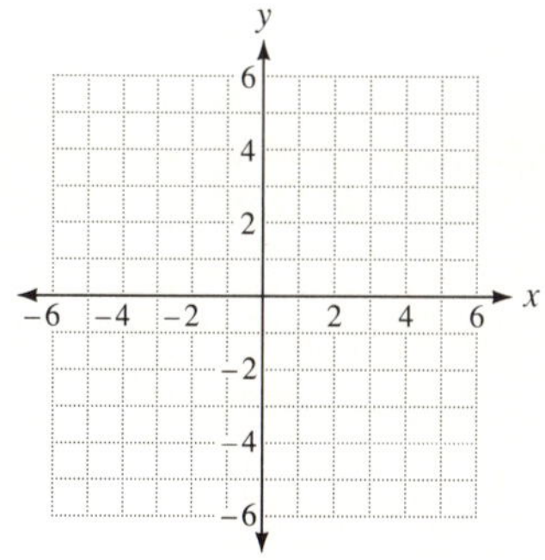

Answer: $(x + 2)^2 + (y - 3)^2 = 4$; $(-2, 3)$; $r = 2$

The circle is used in many applications. The following example illustrates its use in construction.

EXAMPLE 5 Engineering Problem

A railroad tunnel for a small-scale train is designed in the shape of a semicircle with diameter 6 meters. Give the equation for this semicircle and find the vertical clearance 1 meter from the centerline of the tunnel.

Suppose we set a coordinate system with the origin (0, 0) at the centerline of the tunnel as in Figure 11.12.

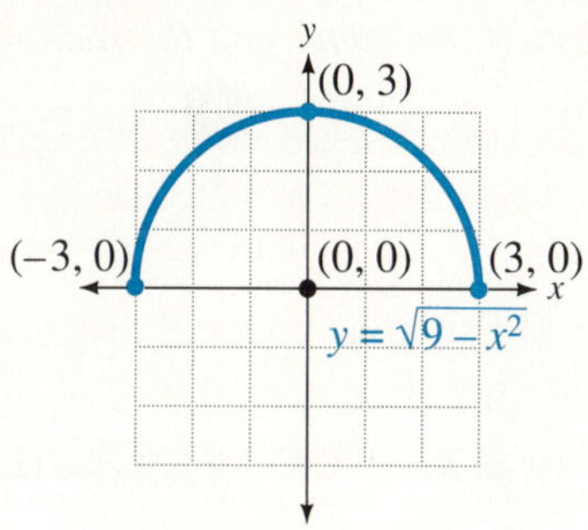

Figure 11.12 Railroad Tunnel

The equation of the circle, the top half of which is our semicircle is

$$(x - 0)^2 + (y - 0)^2 = 3^2 \quad r = 3 \text{ since the diameter is } 6$$
$$x^2 + y^2 = 9$$

If we solve this equation for y,

$$y^2 = 9 - x^2$$
$$y = \pm\sqrt{9 - x^2},$$

Practice Exercise 5

Can a boxcar 3 meters wide and 3 meters high pass through the tunnel if the tracks are centered in the tunnel?

the top half of the circle is described with the positive value for y,

$$y = \sqrt{9 - x^2}.$$

To find the vertical clearance 1 meter from the center line, let $x = 1$ in this equation.

$$y = \sqrt{9 - (1)^2} = \sqrt{9 - 1} = \sqrt{8} \approx 2.83$$

The clearance is approximately 2.83 meters.

Answer: No; the vertical clearance 1.5 meters from the center is only about 2.6 meters.

The equation of the bottom half of the circle in Example 5 is $y = -\sqrt{9 - x^2}$. In Section 9.2 we graphed square root functions of the forms $y = \sqrt{x - 1}$ and $y = -\sqrt{x - 1}$ which were the upper and lower halves, respectively, of the parabola $y^2 = x - 1$ or $x = y^2 + 1$. Although circles and parabolas opening left and right are not functions, the upper and lower halves of each can be described as functions using variations of the square root function.

11.2 EXERCISES A

Find the standard and general forms of the equation of a circle centered at the given point and with the given radius.

1. $(-3, 2)$, $r = 1$

2. $(5, -3)$, $r = 2$

3. $(-1, -1)$, $r = 3$

4. $\left(-\frac{1}{2}, 0\right)$, $r = 1$

5. $(0, 0)$, $r = 1$

6. $(-2, -4)$, $r = 6$

Find the standard form of the equation of each circle and identify the center and the radius.

7. $x^2 + y^2 - 2x + 4y - 20 = 0$

8. $x^2 + y^2 + 6x = 0$

9 $4x^2 + 4y^2 + 40x - 4y + 37 = 0$

Graph each equation.

10. $x^2 + y^2 = 25$

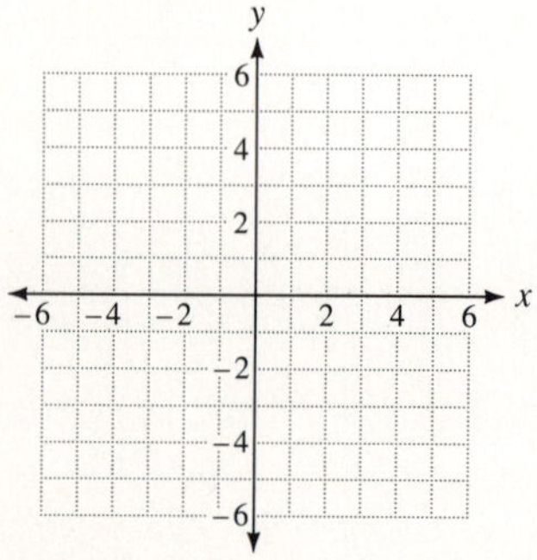

11. $(x - 2)^2 + (y + 3)^2 = 4$

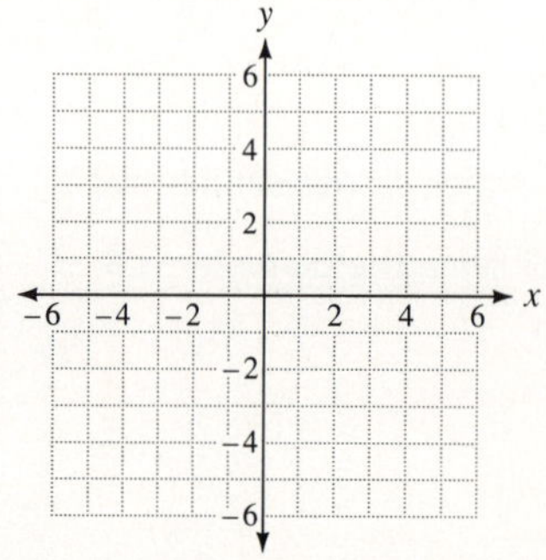

12. $(x - 1)^2 + (y - 2)^2 = 9$

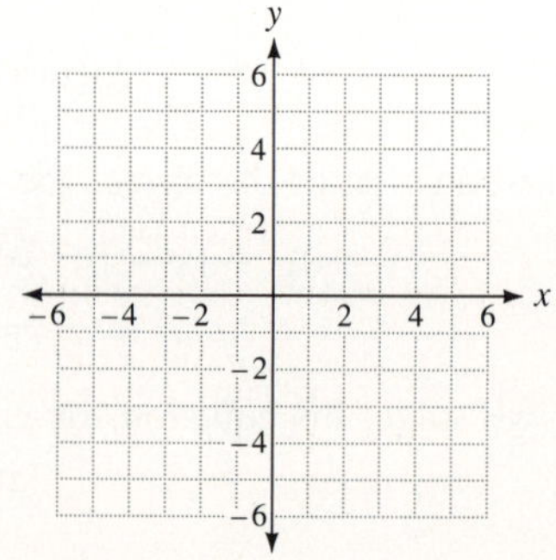

13. $x^2 + (y - 3)^2 = 4$

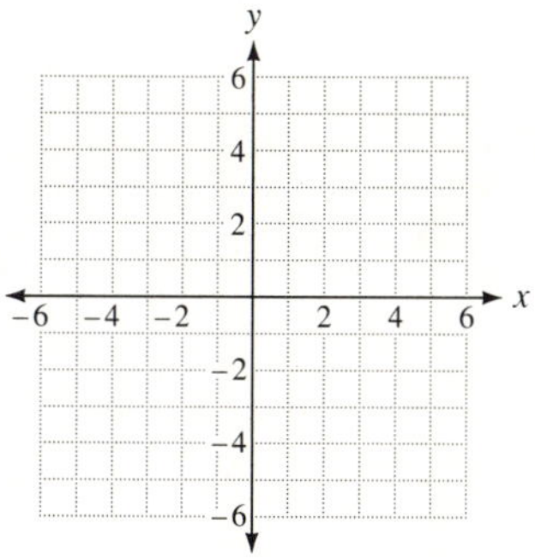

14. $y = \sqrt{25 - x^2}$

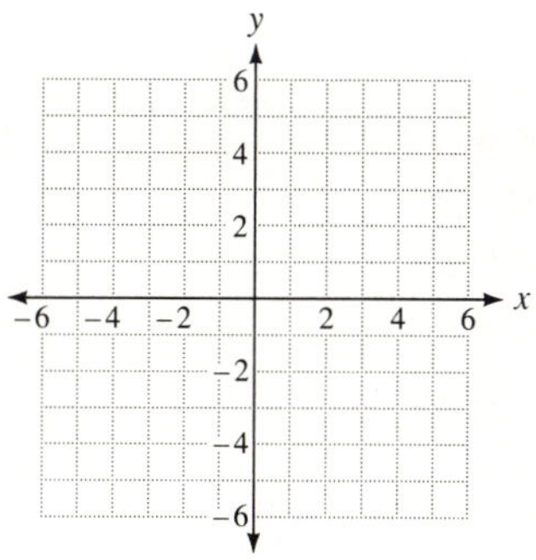

15. $y = -\sqrt{16 - x^2}$

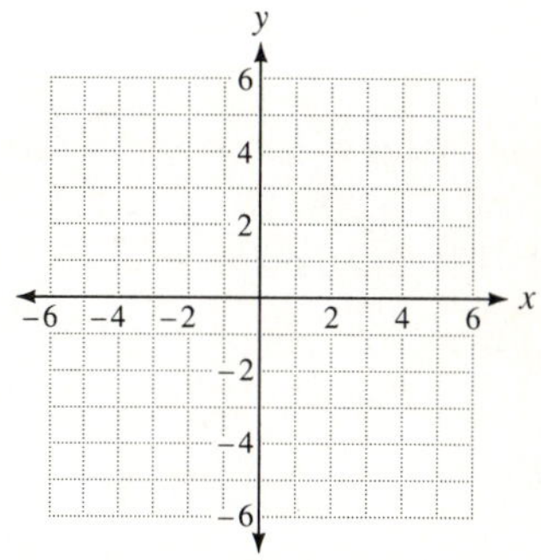

16. A highway tunnel in the shape of a semicircle has a diameter measuring 30 ft. Find the equation of this semicircle using a coordinate system with origin at the center line of the roadway. What is the vertical clearance 5 ft from the edge of the tunnel?

17. A canal with cross-section a semicircle is to have a depth of 3 ft at a distance of 4 ft from the center of the waterline in a full canal. Find the equation of the semicircle. What is the maximum depth of the water in the canal?

FOR REVIEW

Graph each equation.

18. $y^2 = 4x$

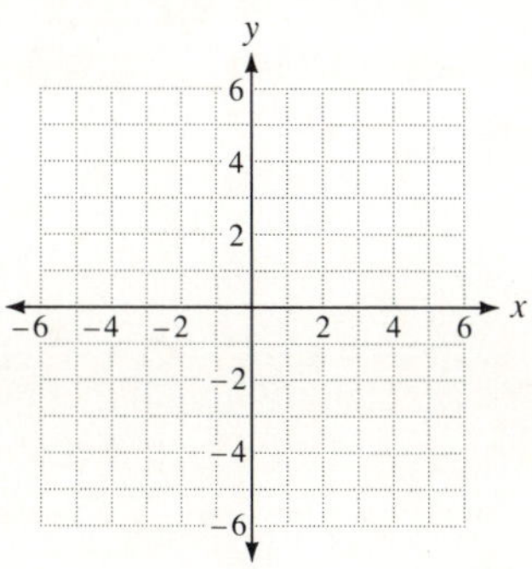

19. $x^2 - 4x + 8y + 4 = 0$

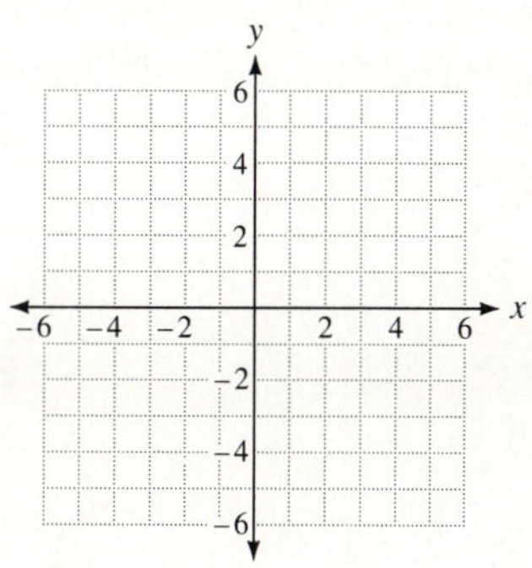

20. $y = -\sqrt{x - 3}$

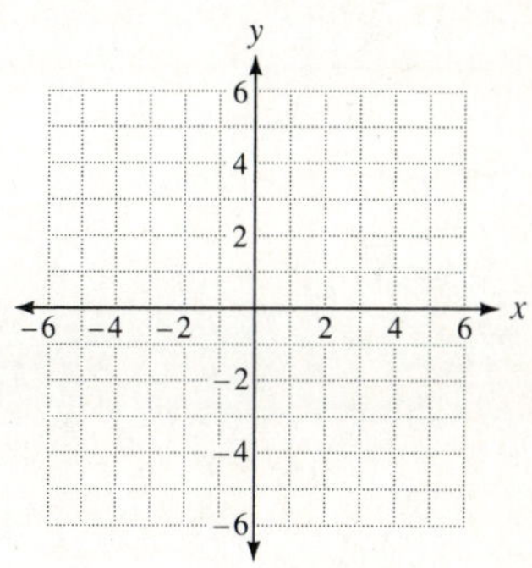

Exercises 21–23 will help you prepare for the next section.

21. For the equation $\frac{x^2}{4} + \frac{y^2}{9} = 1$, **(a)** find the value(s) of y when $x = 0$, and **(b)** find the value(s) of x when $y = 0$.

22. For the equation $\frac{x^2}{4} - \frac{y^2}{9} = 1$, **(a)** find the value(s) of y when $x = 0$, and **(b)** find the value(s) of x when $y = 0$.

23. Solve the equation $\frac{x^2}{4} - \frac{y^2}{9} = 0$ for y. Discuss the graphs of the two solutions.

ANSWERS: 1. $(x + 3)^2 + (y - 2)^2 = 1^2$; $x^2 + y^2 + 6x - 4y + 12 = 0$ 2. $(x - 5)^2 + (y + 3)^2 = 2^2$; $x^2 + y^2 - 10x + 6y + 30 = 0$ 3. $(x + 1)^2 + (y + 1)^2 = 3^2$; $x^2 + y^2 + 2x + 2y - 7 = 0$ 4. $\left(x + \frac{1}{2}\right)^2 + y^2 = 1^2$; $x^2 + y^2 + x - \frac{3}{4} = 0$ or $4x^2 + 4y^2 + 4x - 3 = 0$ 5. $x^2 + y^2 = 1^2$; $x^2 + y^2 - 1 = 0$ 6. $(x + 2)^2 + (y + 4)^2 = 6^2$; $x^2 + y^2 + 4x + 8y - 16 = 0$ 7. $(x - 1)^2 + (y + 2)^2 = 5^2$; $(1, -2)$; $r = 5$ 8. $(x + 3)^2 + y^2 = 3^2$; $(-3, 0)$; $r = 3$ 9. $(x + 5)^2 + \left(y - \frac{1}{2}\right)^2 = 4^2$; $\left(-5, \frac{1}{2}\right)$; $r = 4$

10.

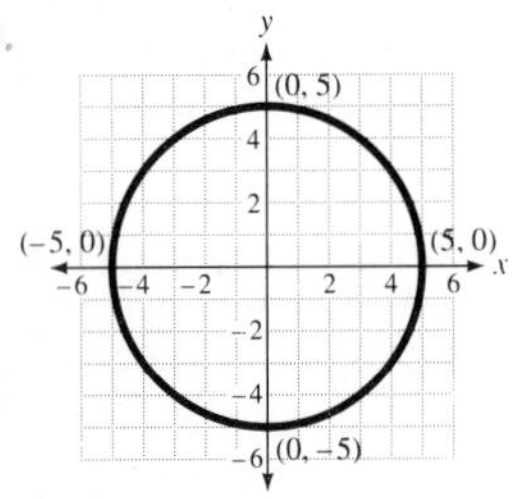

11.

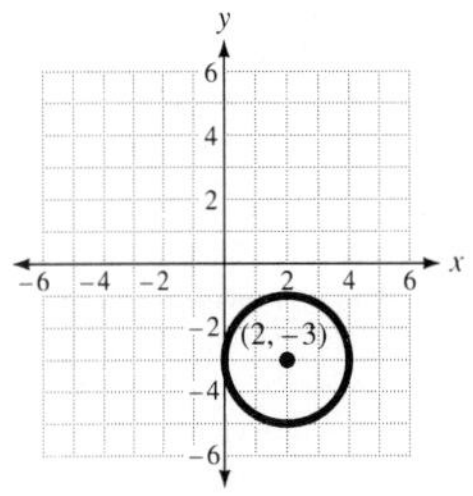

12.

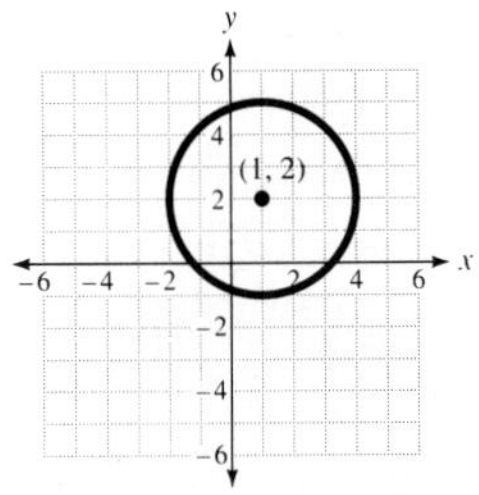

13.

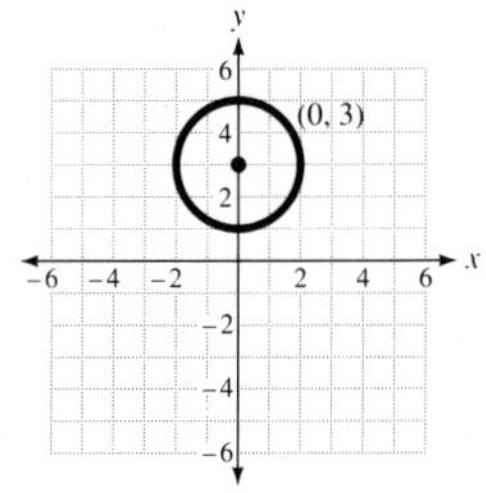

14.

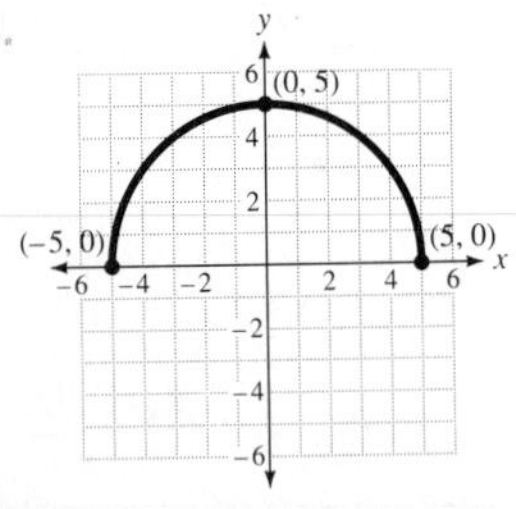

15.

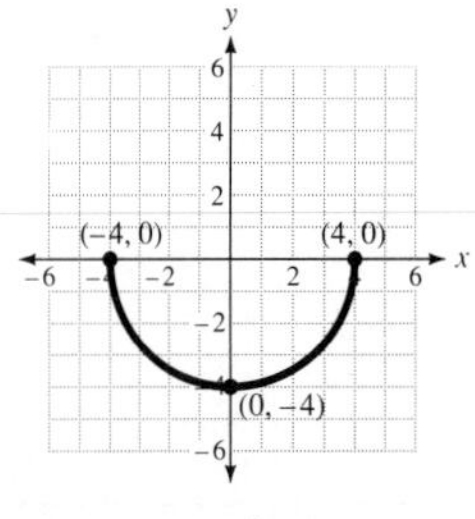

16. $y = \sqrt{225 - x^2}$; approximately 11.2 ft
17. $y = -\sqrt{25 - x^2}$; 5 ft

18.

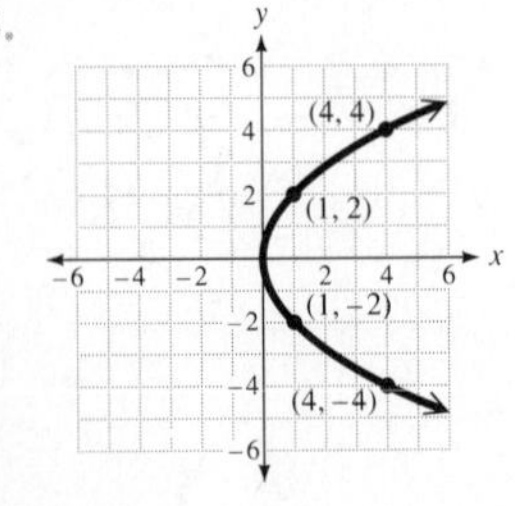

19.

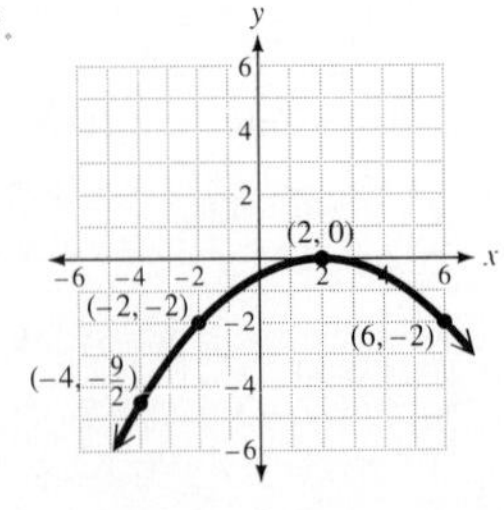

20.

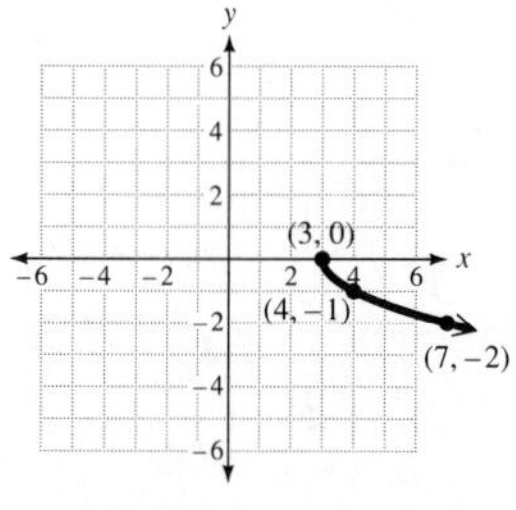

21. (a) ± 3 (b) ± 2 22. (a) The values are imaginary, $\pm 3i$ (b) ± 2 23. $y = \pm\frac{3}{2}x$; the graphs are intersecting lines passing through the origin (0, 0) with slopes $\frac{3}{2}$ and $-\frac{3}{2}$.

11.2 EXERCISES B

Find the standard and general forms of the equation of a circle centered at the given point and with the given radius.

1. $(2, -1)$, $r = 3$

2. $(1, 4)$, $r = 5$

3. $(-3, -2)$, $r = 6$

4. $(0, 4)$, $r = 2$

5. $\left(2, \frac{1}{2}\right)$, $r = 5$

6. $(0, 0)$, $r = 7$

Find the standard form of the equation of each circle and identify the center and the radius.

7. $x^2 + y^2 - 10x + 4y + 20 = 0$

8. $x^2 + y^2 + 8y = 0$

9. $4x^2 + 4y^2 + 4x - 8y + 1 = 0$

Graph each equation.

10. $x^2 + y^2 = 9$

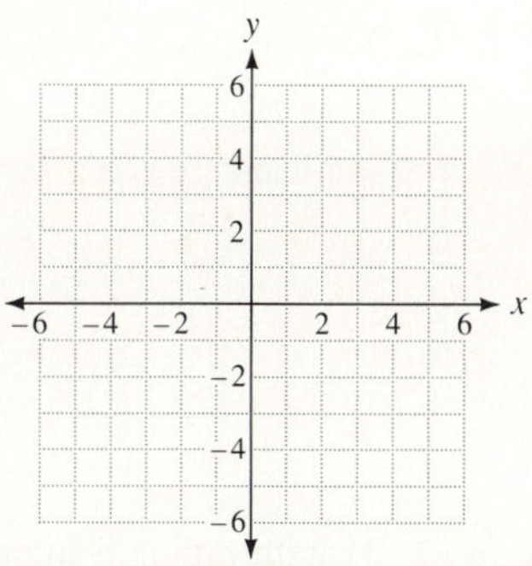

11. $(x - 1)^2 + (y + 1)^2 = 4$

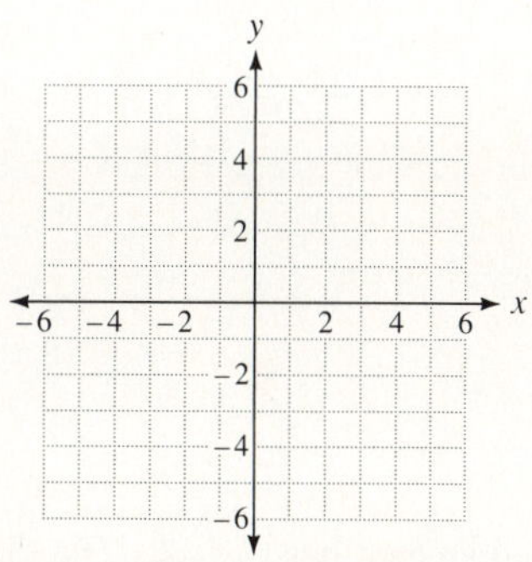

12. $(x + 2)^2 + (y + 2)^2 = 4$

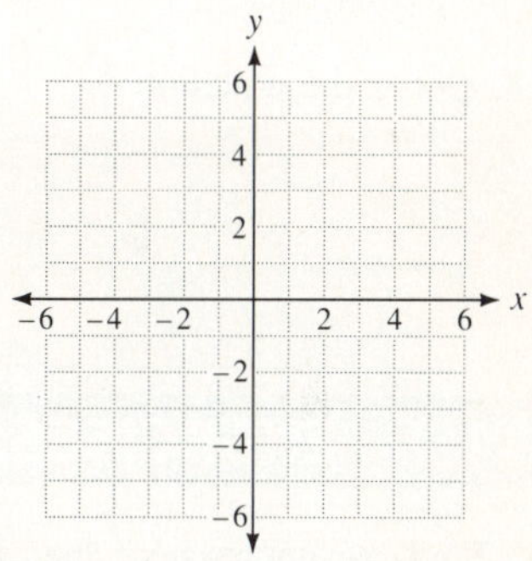

13. $(x - 3)^2 + y^2 = 9$

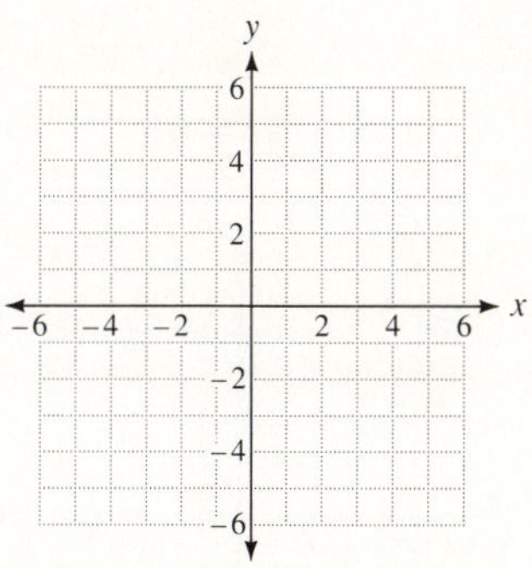

14. $y = -\sqrt{25 - x^2}$

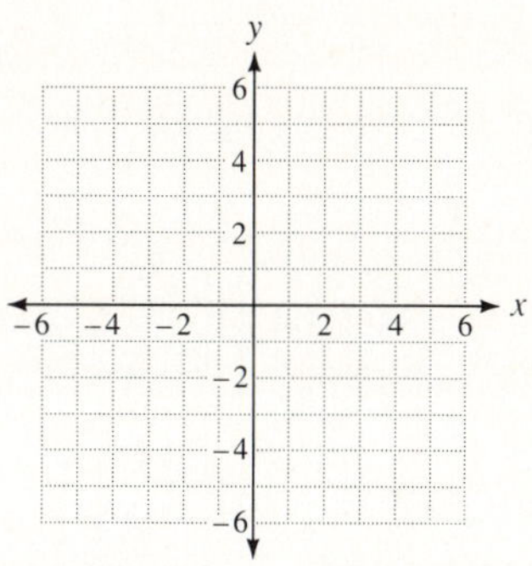

15. $y = \sqrt{16 - x^2}$

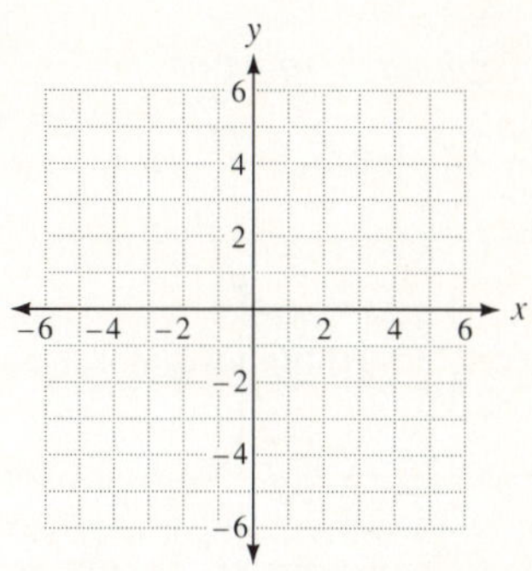

16. A canal with cross-section a semicircle is 8 ft deep at the center. Find the equation of this semicircle by placing a coordinate system with the origin at the center of the water level in a full canal. What is the depth of the canal 2 ft from the edge of the canal?

17. A railroad tunnel is in the shape of a semicircle. If the vertical clearance of the tunnel is 12 ft at a distance of 5 ft from the centerline, what is the equation of the semicircle? What is the maximum width of the tunnel at ground level?

FOR REVIEW

Graph each equation.

18. $x = 4y^2$

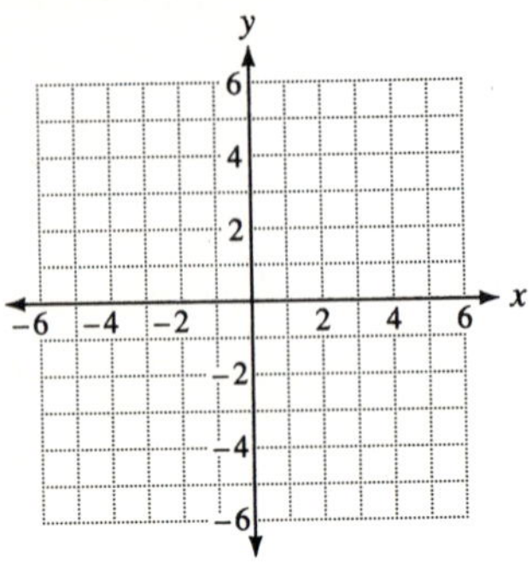

19. $y^2 - 4y + 8x + 4 = 0$

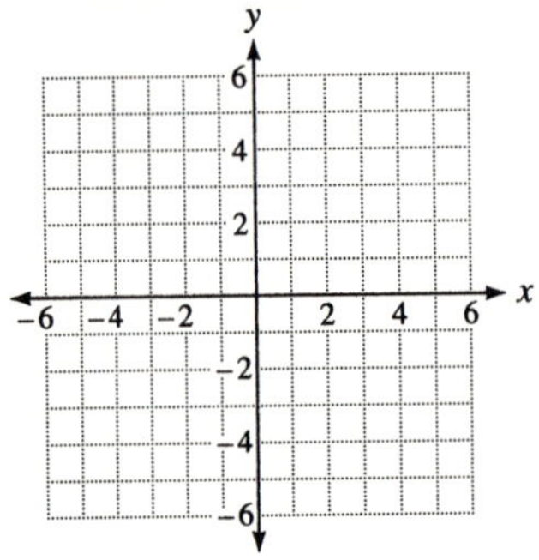

20. $y = \sqrt{x + 5}$

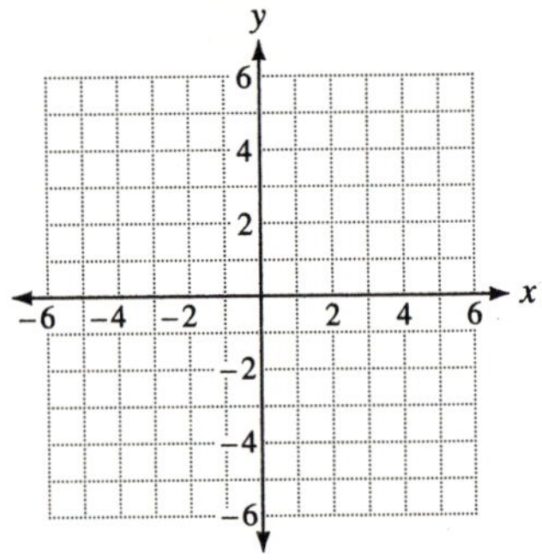

Exercises 21–23 will help you prepare for the next section.

21. For the equation $\frac{x^2}{16} + \frac{y^2}{4} = 1$, **(a)** find the value(s) of y when $x = 0$, and **(b)** find the value(s) of x when $y = 0$.

22. For the equation $\frac{x^2}{16} - \frac{y^2}{4} = 1$, **(a)** find the value(s) of y when $x = 0$, and **(b)** find the value(s) of x when $y = 0$.

23. Solve the equation $\frac{x^2}{16} - \frac{y^2}{4} = 0$ for y. Discuss the graphs of the two solutions.

11.2 EXERCISES C

1. Find the equation of the circle whose diameter has endpoints $(-2, -3)$ and $(6, -3)$.
[Answer: $(x - 2)^2 + (y + 3)^2 = 16$]

2. The circle centered at $(2, 3)$ with radius 5 intersects the x-axis at what two points?

Graph each equation.

3. $3x^2 + 3y^2 = 27$

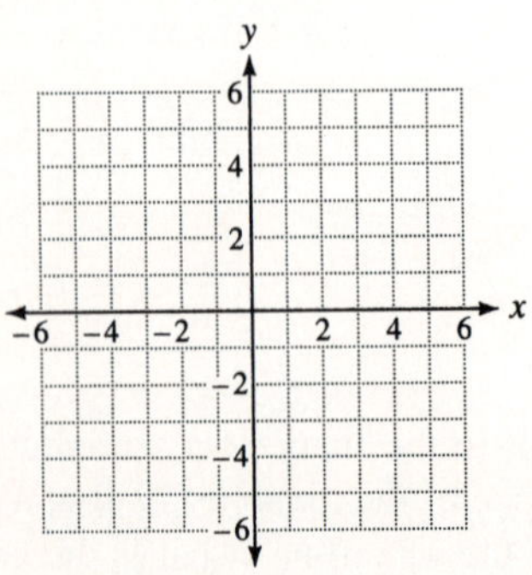

4. $x^2 + y^2 - 4x = 0$

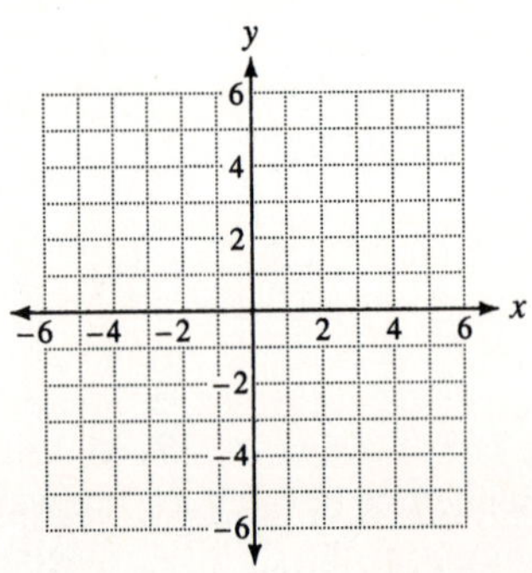

5. $x^2 + y^2 + 6y = 0$

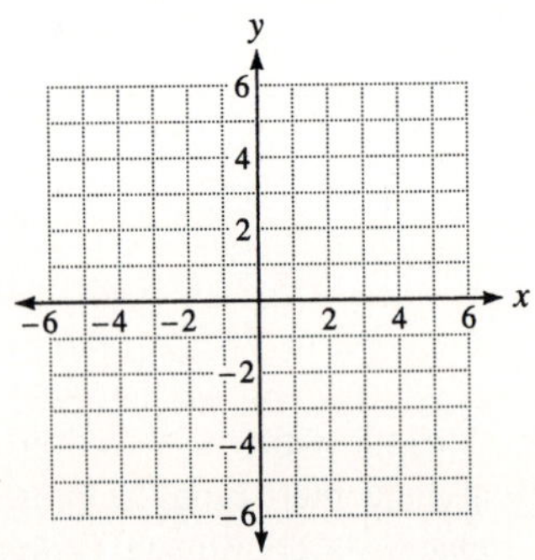

11.3 THE ELLIPSE AND THE HYPERBOLA

STUDENT GUIDEPOSTS

1. Standard Form of an Ellipse
2. Standard Forms of a Hyperbola
3. Asymptotes
4. Alternate Form of a Hyperbola
5. Summary of Conics

A circle is the set of all points in a plane whose distance from *one* fixed point (the center) is a constant. An **ellipse** is the set of all points in a plane the sum of whose distances from *two* fixed points (called **foci,** plural of **focus**) is a constant. If a string is attached to two pins and drawn taut using a pencil, the oval-shaped curve traced out by moving the pencil around the pins while keeping the string taut is an ellipse. See Figure 11.13.

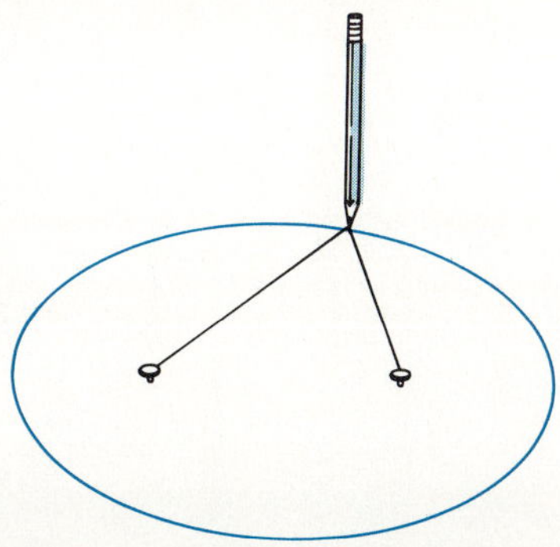

Figure 11.13 Construction of Ellipse

1 STANDARD FORM OF AN ELLIPSE

The orbit through which the earth travels around the sun and the orbits of communication satellites around the earth can all be approximated by ellipses. To simplify the discussion, we will consider only ellipses whose graphs are centered at the origin of a coordinate system.

Standard Form of the Equation of an Ellipse

The **standard form** of the equation of an ellipse centered at the origin with x-intercepts $(a, 0)$ and $(-a, 0)$ and y-intercepts $(0, b)$ and $(0, -b)$ is

$$\frac{x^2}{a^2} + \frac{y^2}{b^2} = 1. \quad a > 0, b > 0, a \neq b$$

If $a = b$ in $\frac{x^2}{a^2} + \frac{y^2}{b^2} = 1$, the equation can be written as $\frac{x^2}{a^2} + \frac{y^2}{a^2} = 1$ or as $x^2 + y^2 = a^2$. This is the equation of a circle centered at the origin, not an ellipse.

EXAMPLE 1 FINDING THE EQUATION OF AN ELLIPSE

Find the standard form of the equation of the ellipse centered at the origin with x-intercepts $(3, 0)$ and $(-3, 0)$ and y-intercepts $(0, 2)$ and $(0, -2)$.

Substitute $a = 3$ and $b = 2$ in the standard form.

PRACTICE EXERCISE 1

Find the standard form of the equation of the ellipse centered at the origin with intercepts $(6, 0)$, $(-6, 0)$, $(0, 1)$ and $(0, -1)$.

$$\frac{x^2}{3^2} + \frac{y^2}{2^2} = 1$$

$$\frac{x^2}{9} + \frac{y^2}{4} = 1$$

Answer: $\frac{x^2}{36} + \frac{y^2}{1} = 1$

Graphing an ellipse when its equation is given in standard form is simply a matter of plotting the four intercepts.

EXAMPLE 2 Graphing an Ellipse

Graph the ellipse with equation $\frac{x^2}{25} + \frac{y^2}{4} = 1.$

We write the equation as $\frac{x^2}{5^2} + \frac{y^2}{2^2} = 1.$

Thus, the x-intercepts are (5, 0) and (−5, 0), and the y-intercepts (0, 2) and (0, −2). The graph is given in Figure 11.14.

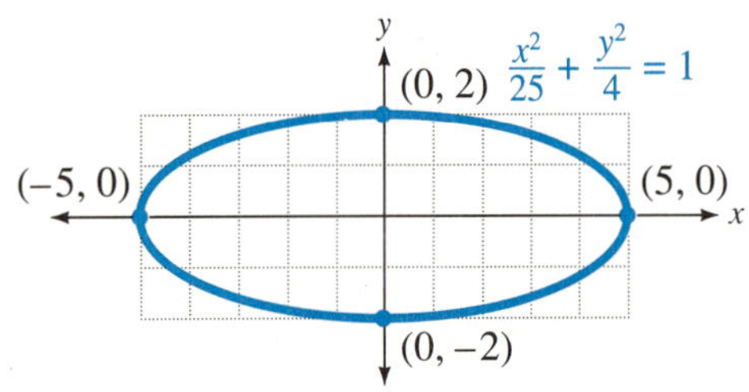

Figure 11.14

Practice Exercise 2

Graph $\frac{x^2}{4} + \frac{y^2}{25} = 1.$

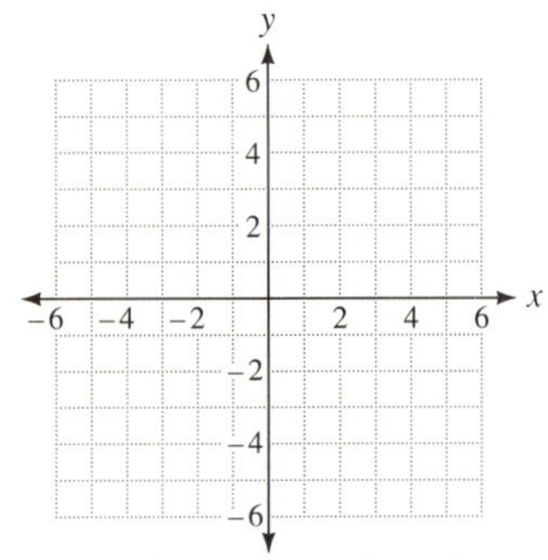

Answer: The graph is the ellipse with intercepts (2, 0), (−2, 0), (0, 5), and (0, −5).

A *hyperbola* is defined much like an ellipse. A **hyperbola** is the set of all points in a plane the *difference* of whose distances from two fixed points (called foci) is constant. Many modern buildings have cross-sectional structures in the shape of a hyperbola.

2 STANDARD FORMS OF A HYPERBOLA

As we did with ellipses, we will concentrate on hyperbolas centered at the origin of a coordinate system.

Standard Forms of the Equation of a Hyperbola

The **standard form** of the equation of a hyperbola centered at the origin with x-intercepts $(a, 0)$ and $(-a, 0)$ and opening left and right is

$$\frac{x^2}{a^2} - \frac{y^2}{b^2} = 1. \quad a > 0 \text{ and } b > 0$$

The **standard form** of the equation of a hyperbola centered at the origin with y-intercepts $(0, b)$ and $(0, -b)$ and opening up and down is

$$\frac{y^2}{b^2} - \frac{x^2}{a^2} = 1. \quad a > 0 \text{ and } b > 0$$

3 ASYMPTOTES

To plot an ellipse centered at the origin, the four intercepts are sufficient. For a hyperbola centered at the origin, however, there are only two intercepts, so additional information is needed. Associated with every hyperbola are two intersecting lines with the property that points on the hyperbola get closer to the lines as they get farther from the center. Any graph that approaches a line in this way is said to be **asymptotic** to the line, and the line is an **asymptote** of the graph.

Asymptotes of a Hyperbola

The asymptotes of either of the hyperbolas

$$\frac{x^2}{a^2} - \frac{y^2}{b^2} = 1 \quad \text{or} \quad \frac{y^2}{b^2} - \frac{x^2}{a^2} = 1$$

are the lines containing the diagonals of the rectangle with sides parallel to the axes, passing through the points $(a, 0)$, $(-a, 0)$, $(0, b)$, and $(0, -b)$.

Figure 11.15 shows the graphs of the two kinds of hyperbolas centered at the origin. To graph any hyperbola, plot the points $(a, 0)$, $(-a, 0)$, $(0, b)$, and $(0, -b)$, construct the rectangle through these points with sides parallel to the axes, and draw the diagonals of the rectangle, extending them to form the asymptotes. The hyperbola is the pair of curves that pass through the intercepts and approach the asymptotes. Notice that the hyperbola opens left and right when the coefficient of the x^2-term is positive, as in Figure 11.15(a), and opens up and down when the coefficient of the y^2-term is positive, as in Figure 11.15(b).

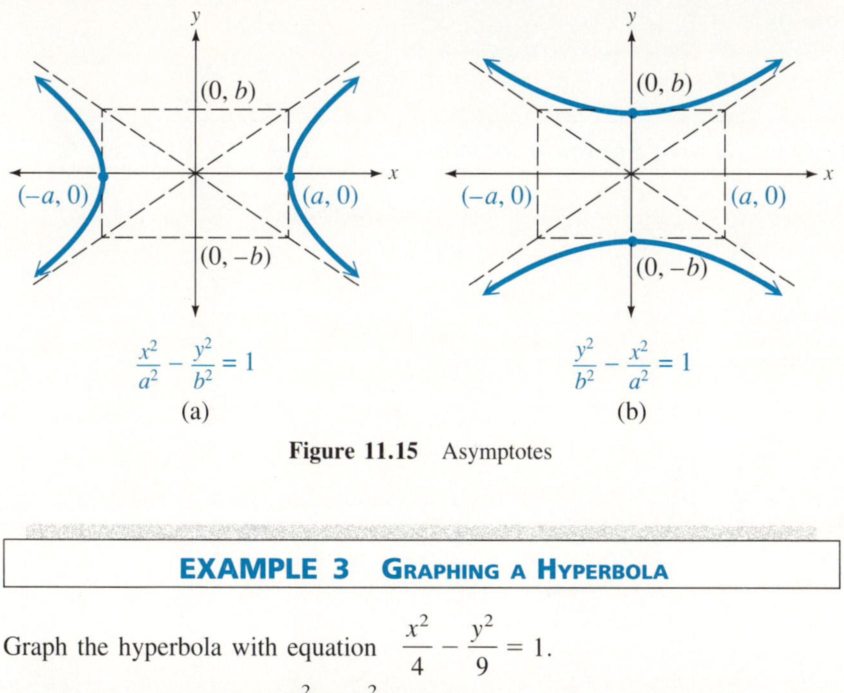

Figure 11.15 Asymptotes

EXAMPLE 3 GRAPHING A HYPERBOLA

Graph the hyperbola with equation $\dfrac{x^2}{4} - \dfrac{y^2}{9} = 1$.

This can be written $\dfrac{x^2}{2^2} - \dfrac{y^2}{3^2} = 1$.

Since $a = 2$ and $b = 3$, the rectangle passes through the points $(2, 0)$, $(-2, 0)$, $(0, 3)$, and $(0, -3)$. We construct the rectangle, extend the diagonals to get the asymptotes, plot the intercepts, and sketch the hyperbola (it opens right and left since the x^2 term is positive) as in Figure 11.16.

PRACTICE EXERCISE 3

Graph $\dfrac{y^2}{9} - \dfrac{x^2}{4} = 1$.

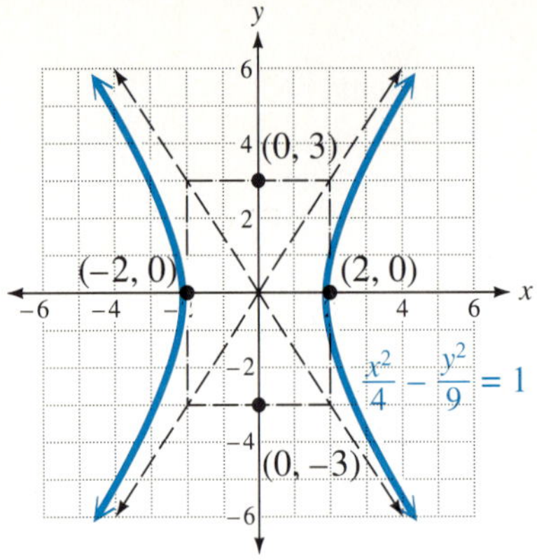

Figure 11.16

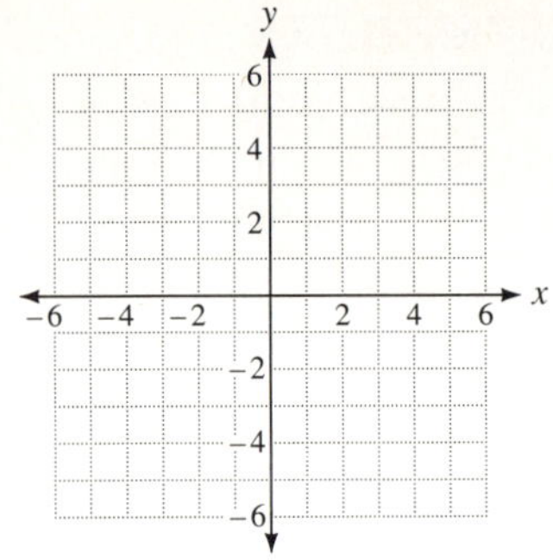

Answer: The hyperbola opens up and down with y-intercepts (0, 3) and (0, −3) and same asymptotes as in Figure 11.16.

4 ALTERNATE FORM OF A HYPERBOLA

The hyperbolas we have studied thus far open up and down or right and left with intercepts on the y-axis or x-axis, respectively. Other hyperbolas have the axes as asymptotes, and these are considered now.

Alternate Form of the Equation of a Hyperbola

Hyperbolas having the x-axis and y-axis as asymptotes have equations

$$xy = c, \quad c \text{ a constant.}$$

EXAMPLE 4 Graphing a Hyperbola with Equation $xy = c$

Graph $xy = 6$.

First we solve for y,

$$y = \frac{6}{x},$$

and then make a table of values. Notice that we cannot use 0 for x. The graph is given in Figure 11.17.

x	y
6	1
3	2
2	3
1	6
−1	−6
−2	−3
−3	−2
−6	−1

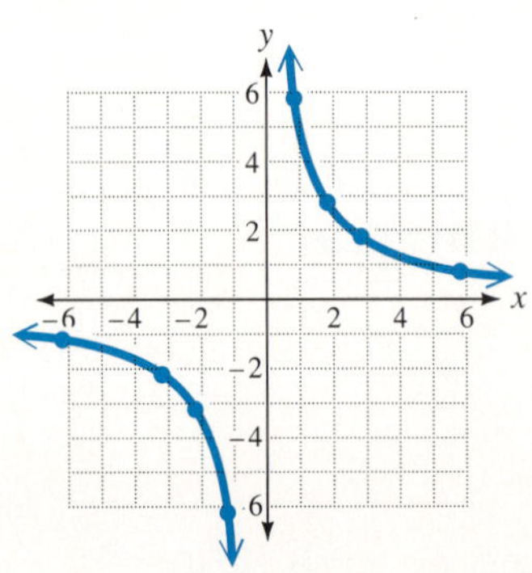

Figure 11.17

Practice Exercise 4

Graph $xy = -6$.

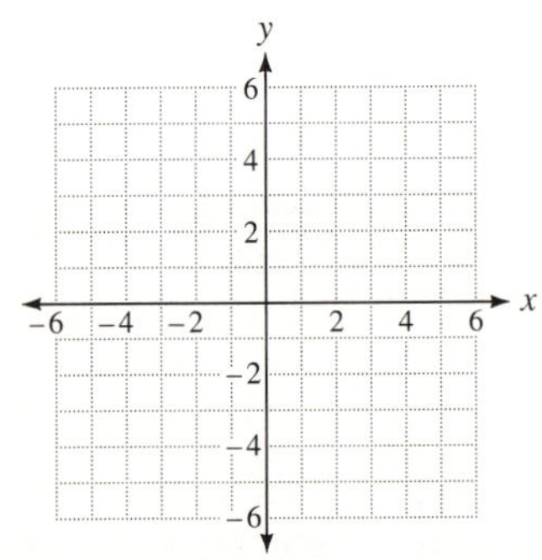

Answer: The graph is much like that in Figure 11.17 but in quadrants II and IV instead of I and III.

The two parts of the hyperbola in Example 4 are in quadrants I and III. This happens when $c > 0$ in $xy = c$. When $c < 0$, the two parts are in quadrants II and IV as in Practice Exercise 4. Hyperbolas of the form $xy = c$ have the same shape as hyperbolas with standard-form equations but they are rotated 45°.

5 SUMMARY OF CONICS

We conclude this section with a summary of the equations and graphs of the conic sections studied in this chapter.

Summary of the Conic Sections

Equation	*Description of graph*	*What to look for*
$y = a(x - h)^2 + k$	Parabola: opens up if $a > 0$, opens down if $a < 0$, vertex at (h, k)	x is squared but y is not
$x = a(y - k)^2 + h$	Parabola: opens right if $a > 0$, opens left if $a < 0$, vertex at (h, k)	y is squared but x is not
$(x - h)^2 + (y - k)^2 = r^2$	Circle: center at (h, k) with radius r	x^2 and y^2 have the same coefficient
$\frac{x^2}{a^2} + \frac{y^2}{b^2} = 1$	Ellipse: center at $(0, 0)$, x-intercepts $(\pm a, 0)$ and y-intercepts $(0, \pm b)$	x^2 and y^2 have different coefficients with the same sign
$\frac{x^2}{a^2} - \frac{y^2}{b^2} = 1$	Hyperbola: center at $(0, 0)$, x-intercepts $(\pm a, 0)$, opens left and right	x^2 has positive coefficient and y^2 has negative coefficient
$\frac{y^2}{b^2} - \frac{x^2}{a^2} = 1$	Hyperbola: center at $(0, 0)$, y-intercepts $(0, \pm b)$, opens up and down	y^2 has positive coefficient and x^2 has negative coefficient
$xy = c$	Hyperbola: in quadrants I and III if $c > 0$, and in II and IV if $c < 0$	xy-term

EXAMPLE 5 IDENTIFYING CONICS

Identify the graph of each equation.

(a) $y^2 = x + 5$

Since y is squared and x is not, the graph is a parabola. In fact, since $x = y^2 - 5$, $a > 0$ and the parabola opens to the right.

(b) $2x^2 = 5 - 2y^2$

We can write the equation in the form $2x^2 + 2y^2 = 5$ or $x^2 + y^2 = \frac{5}{2}$. Since x and y are both squared and have the same coefficient, the graph is a circle centered at the origin with radius $\sqrt{\frac{5}{2}} = \frac{\sqrt{10}}{2}$.

PRACTICE EXERCISE 5

Identify the graph of each equation.

(a) $9x^2 + 4y^2 = 36$
[*Hint:* Divide both sides by 36.]

(b) $\frac{x}{5} = \frac{1}{y}$

Answers: (a) ellipse
(b) hyperbola

11.3 EXERCISES A

Give the standard form of the equation of the ellipse centered at the origin with the given x-intercepts and y-intercepts.

1. (2, 0), (−2, 0) and (0, 3), (0, −3)

2. (9, 0), (−9, 0) and (0, 11), (0, −11)

3. (7, 0), (−7, 0) and (0, 1), (0, −1)

4. (10, 0), (−10, 0) and (0, 2), (0, −2)

Give the standard form of the equation of the hyperbola centered at the origin with the given a, b, and direction of opening.

5. $a = 2$, $b = 3$, up and down

6. $a = 9$, $b = 11$, left and right

7. $a = 7$, $b = 1$, left and right

8. $a = 10$, $b = 2$, up and down

Graph.

9. $\frac{x^2}{16} + \frac{y^2}{4} = 1$

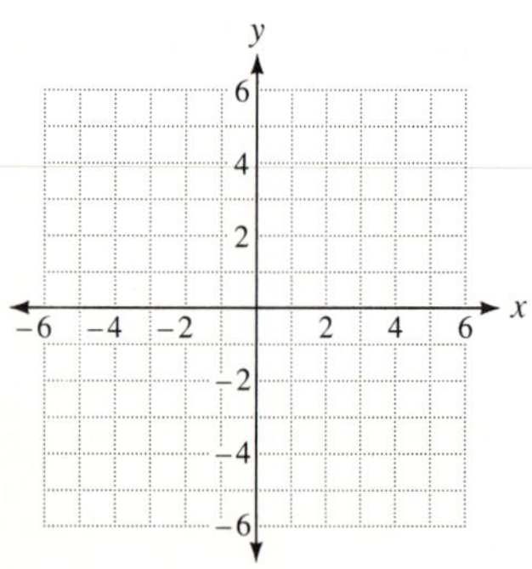

10. $\frac{x^2}{1} + \frac{y^2}{9} = 1$

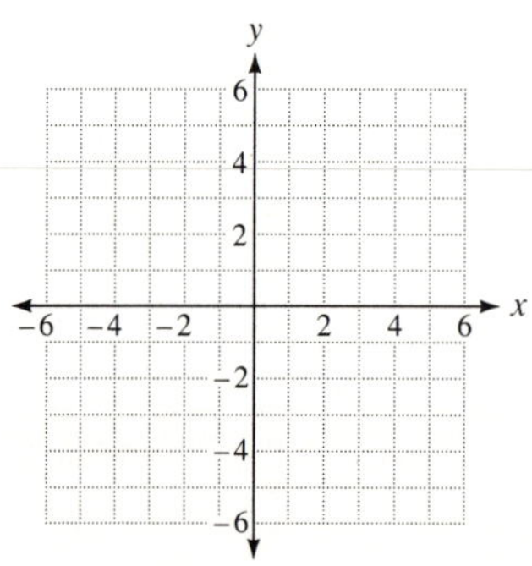

11. $\frac{x^2}{36} + \frac{y^2}{16} = 1$

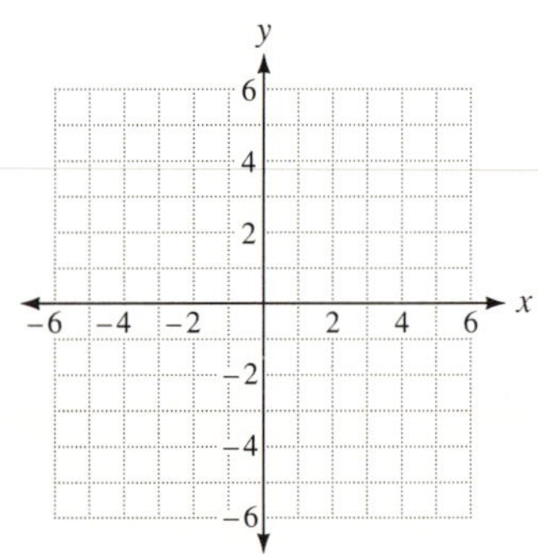

12. $25x^2 + 9y^2 = 225$

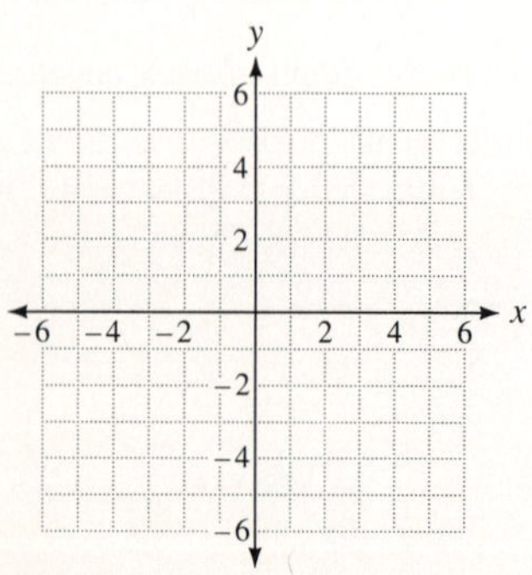

13. $\frac{x^2}{4} - \frac{y^2}{25} = 1$

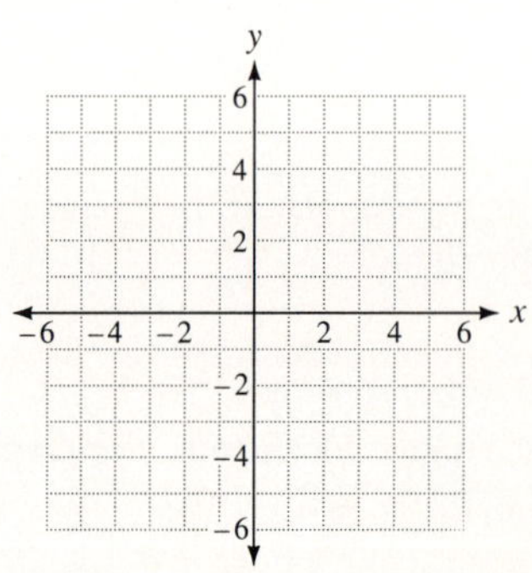

14. $\frac{x^2}{25} - \frac{y^2}{4} = 1$

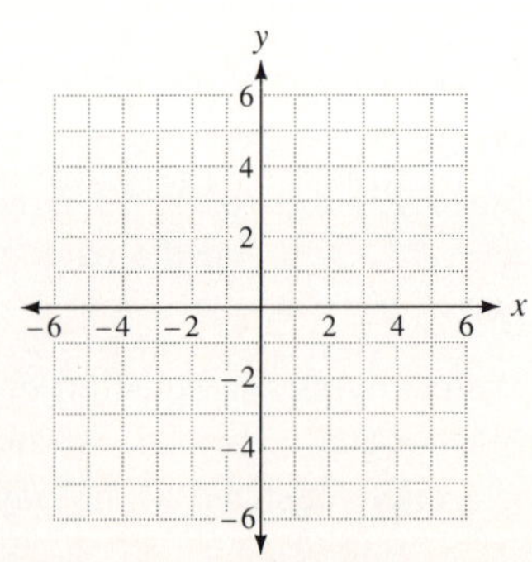

15. $\dfrac{y^2}{4} - \dfrac{x^2}{25} = 1$

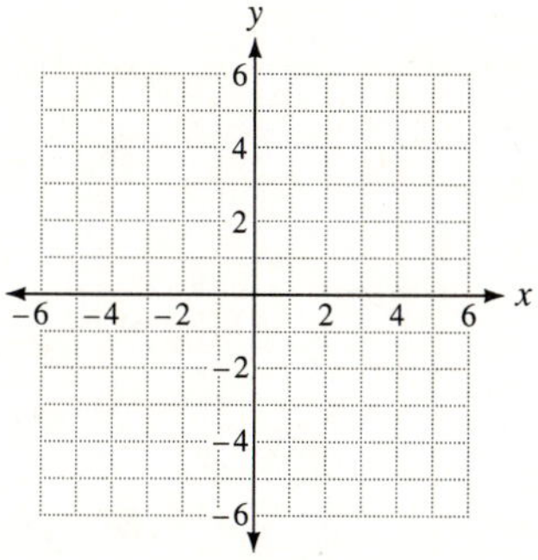

16. $x^2 - y^2 = 4$

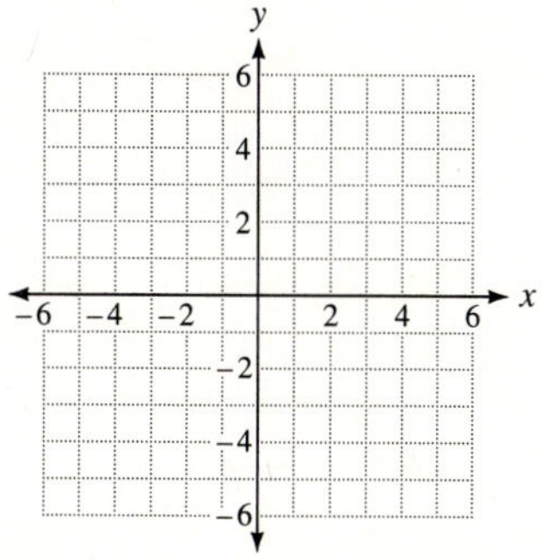

17. $y^2 - 4x^2 = 16$

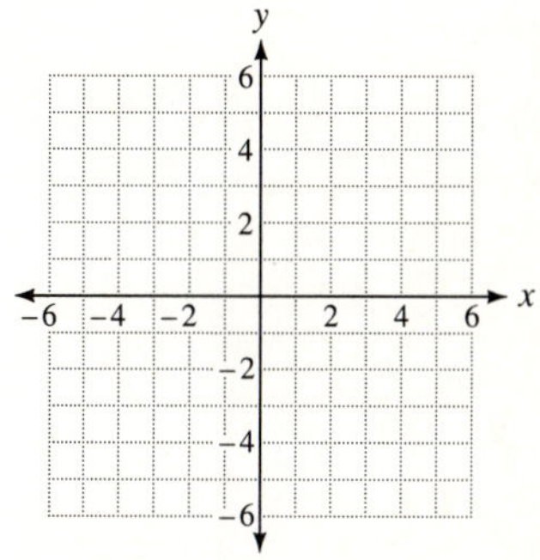

18. $xy = 4$

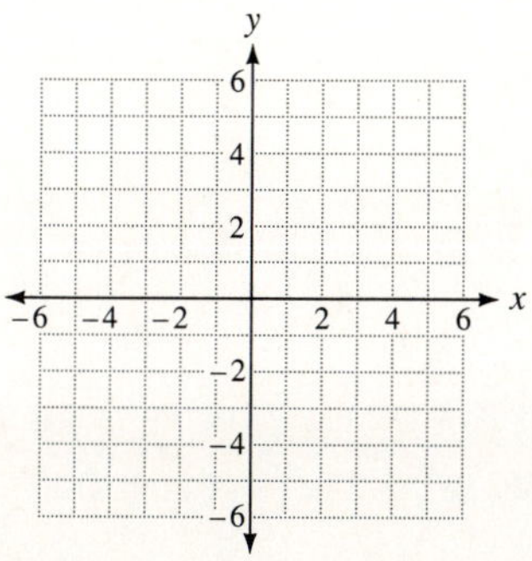

19. $xy = -4$

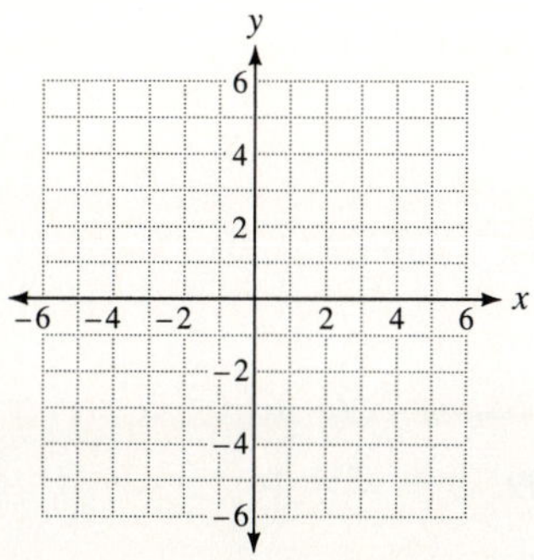

20. $xy = 12$

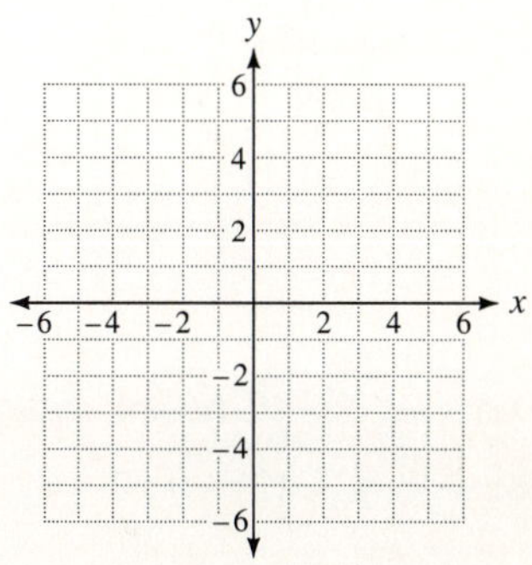

Identify each of the following as the equation of a parabola, circle, ellipse, or hyperbola.

21. $x^2 + y^2 = 25$

22. $x^2 = 2y$

23. $\dfrac{x^2}{4} - \dfrac{y^2}{49} = 1$

24. $(x - 1)^2 + (y + 3)^2 = 16$

25. $\dfrac{x^2}{36} + \dfrac{y^2}{16} = 1$

26. $y^2 - x^2 = 16$

27. $y = \dfrac{1}{x}$

28. $y^2 - 2x + y - 5 = 0$

29. $x^2 + y^2 - 2x + 2y - 2 = 0$

30. A point is sometimes called a *degenerate circle* or *degenerate ellipse* since a plane can intersect a cone in a single point. Make a sketch to show how this can happen.

31. What is the graph of $x^2 + y^2 = 0$. Compare this with Exercise 30.

FOR REVIEW

Find the standard and general forms of the equation of a circle centered at the given point and with the given radius.

32. $(2, 2)$, $r = 2$ **33.** $(0, 0)$, $r = 8$ **34.** $(-1, 5)$, $r = 7$

Graph.

35. $x = y^2 + 4y$

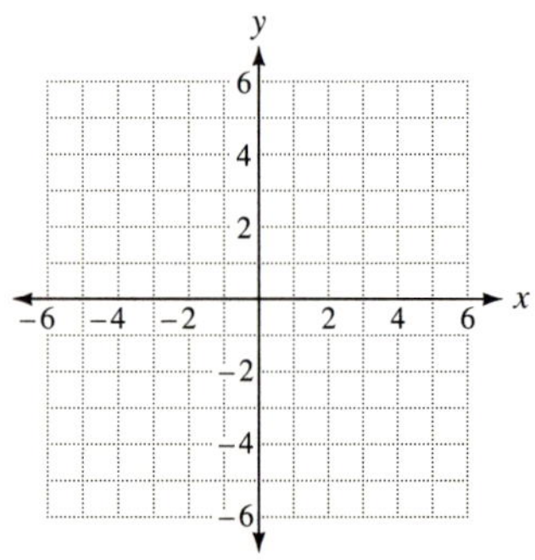

36. $(x - 1)^2 + (y + 1)^2 = 4$

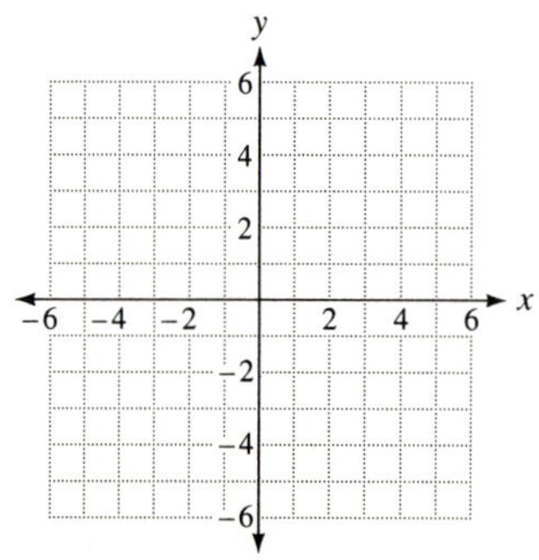

37. $y = -\sqrt{36 - x^2}$

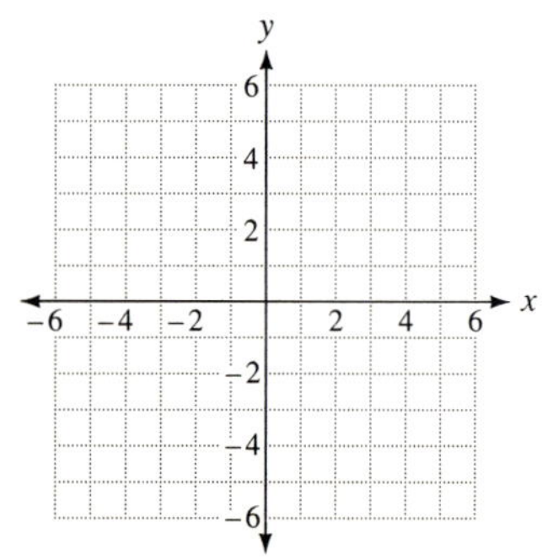

Exercises 38–40 review material from Chapter 8 to help you prepare for the next section. Solve each equation.

38. $x^2 - x - 12 = 0$ **39.** $9 + y^2 = 14$ **40.** $x^4 - 25x^2 + 144 = 0$

ANSWERS: 1. $\frac{x^2}{4} + \frac{y^2}{9} = 1$ 2. $\frac{x^2}{81} + \frac{y^2}{121} = 1$ 3. $\frac{x^2}{49} + \frac{y^2}{1} = 1$ 4. $\frac{x^2}{100} + \frac{y^2}{4} = 1$ 5. $\frac{y^2}{9} - \frac{x^2}{4} = 1$ 6. $\frac{x^2}{81} - \frac{y^2}{121} = 1$ 7. $\frac{x^2}{49} - \frac{y^2}{1} = 1$ 8. $\frac{y^2}{4} - \frac{x^2}{100} = 1$

9.

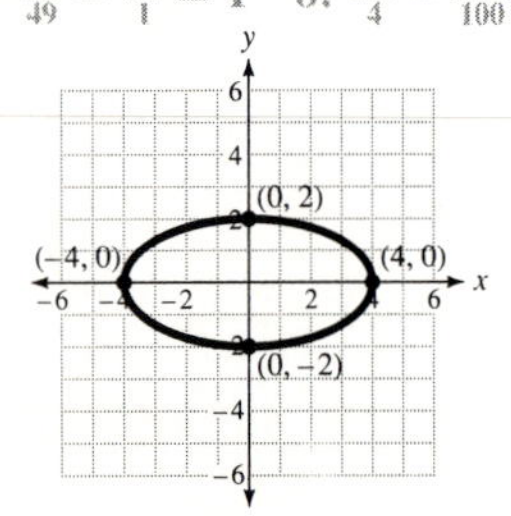

10.

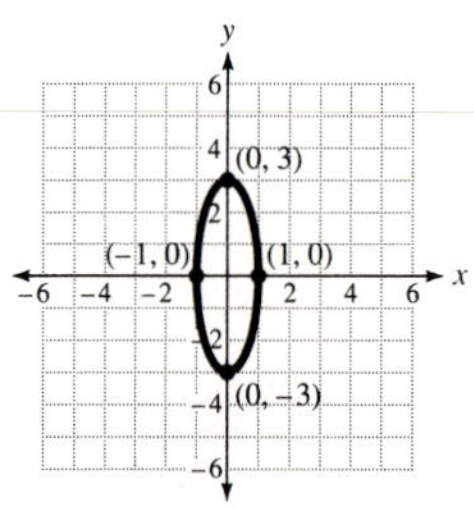

11.

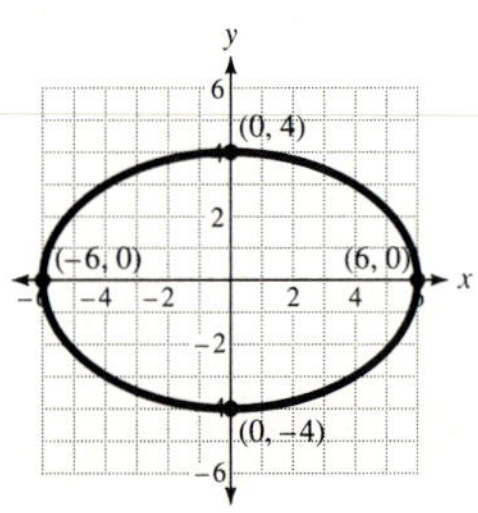

12.

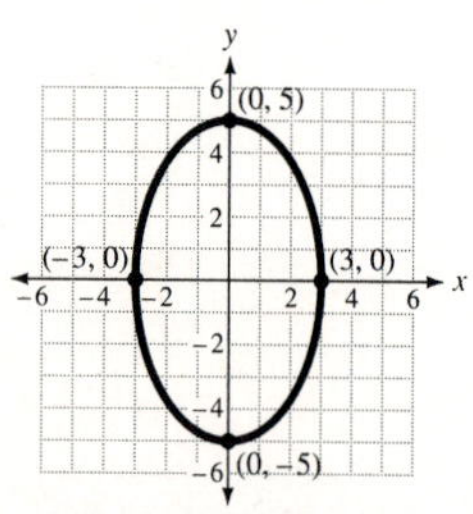

13.

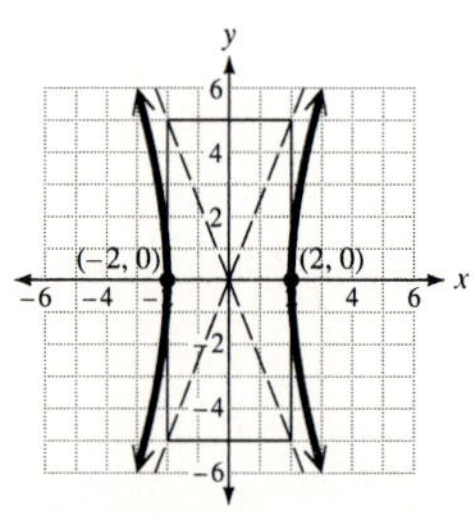

14.

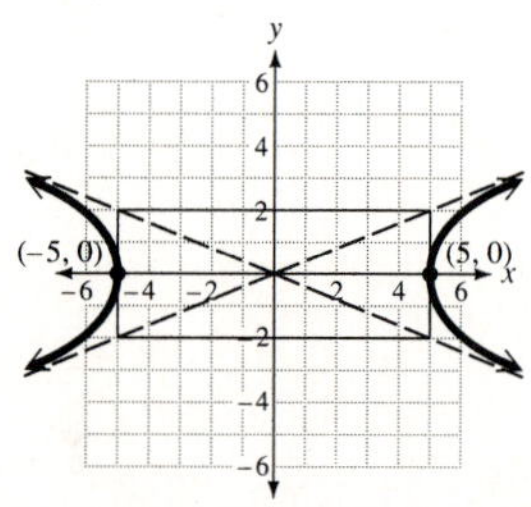

15.

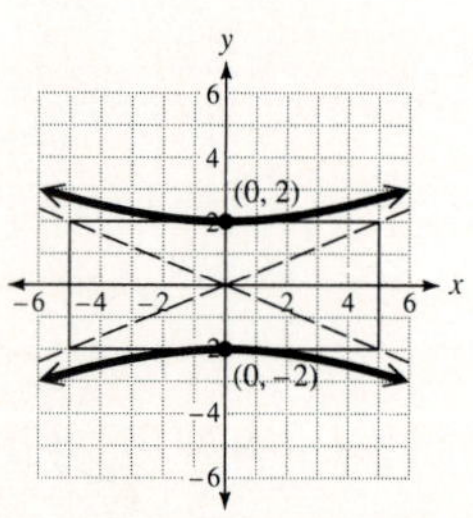

16.

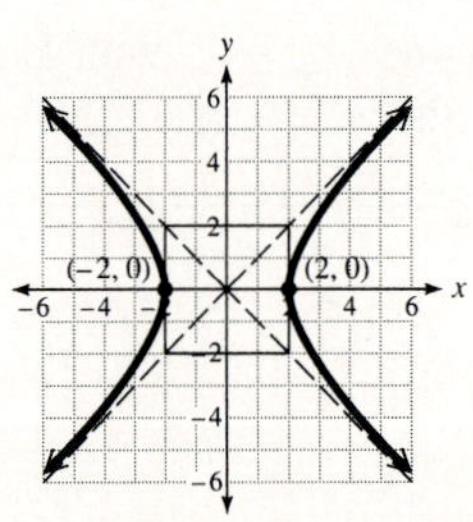

17.

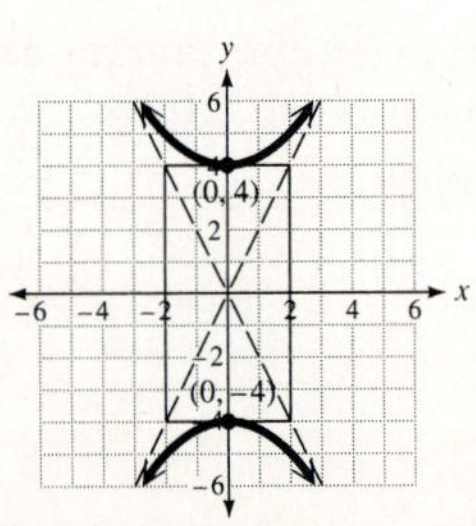

18.
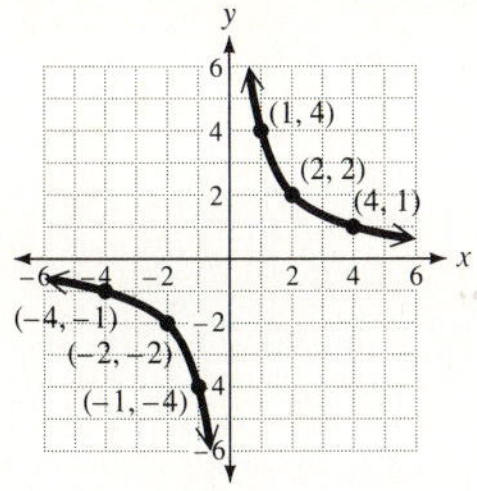

19.
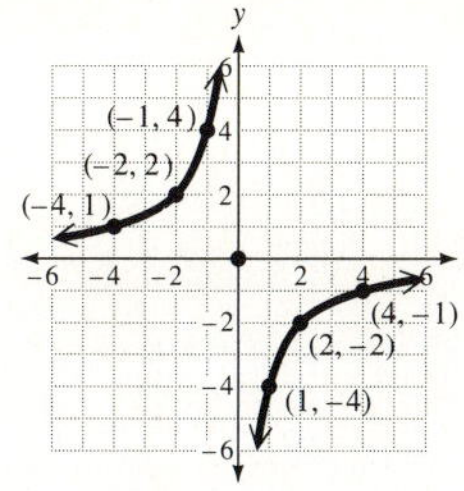

20.
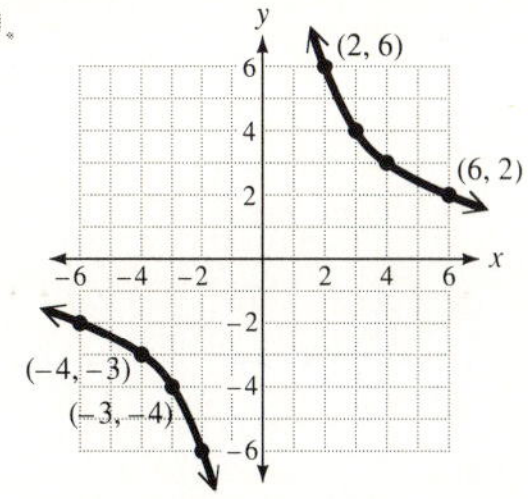

21. circle 22. parabola 23. hyperbola 24. circle 25. ellipse 26. hyperbola 27. hyperbola 28. parabola 29. circle 30. Have the plane pass through the vertex of the cone and only through this point. 31. The graph is the point (0, 0). Since the equation has the appearance of the equation of a circle, this could be called a point circle. 32. $(x-2)^2+(y-2)^2=2^2$; $x^2+y^2-4x-4y+4=0$ 33. $x^2+y^2=8^2$; $x^2+y^2-64=0$ 34. $(x+1)^2+(y-5)^2=7^2$; $x^2+y^2+2x-10y-23=0$

35.
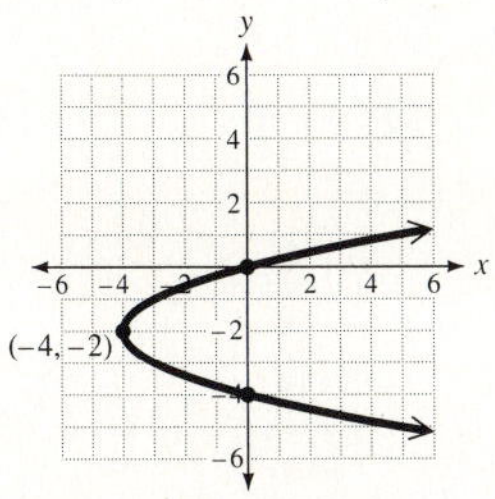

36.
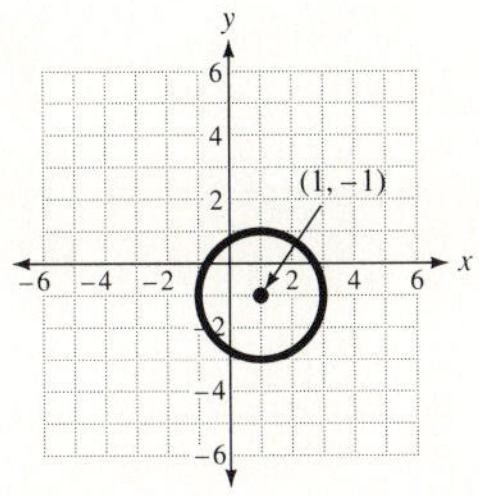

37.
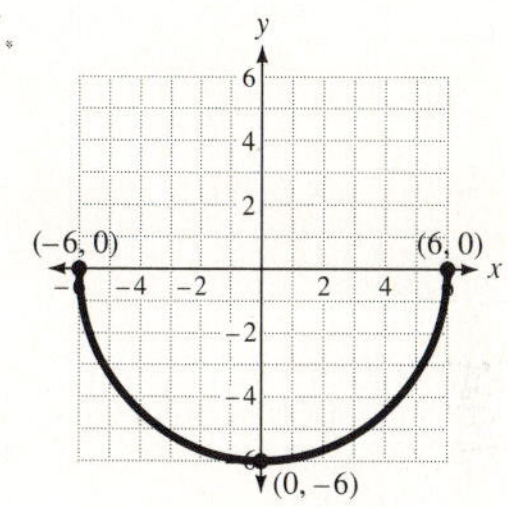

38. −3, 4 39. $\pm\sqrt{5}$ 40. ±3, ±4

11.3 EXERCISES B

Give the standard form of the equation of the ellipse centered at the origin with the given x-intercepts and y-intercepts.

1. (6, 0), (−6, 0) and (0, 4), (0, −4)

2. (12, 0), (−12, 0) and (0, 9), (0, −9)

3. (5, 0), (−5, 0) and (0, 8), (0, −8)

4. (1, 0), (−1, 0) and (0, 7), (0, −7)

Give the standard form of the equation of the hyperbola centered at the origin with the given a, b, and direction of opening.

5. $a=6$, $b=4$, up and down

6. $a=12$, $b=9$, up and down

7. $a=5$, $b=8$, left and right

8. $a=1$, $b=7$, left and right

Graph.

9. $\dfrac{x^2}{25}+\dfrac{y^2}{9}=1$

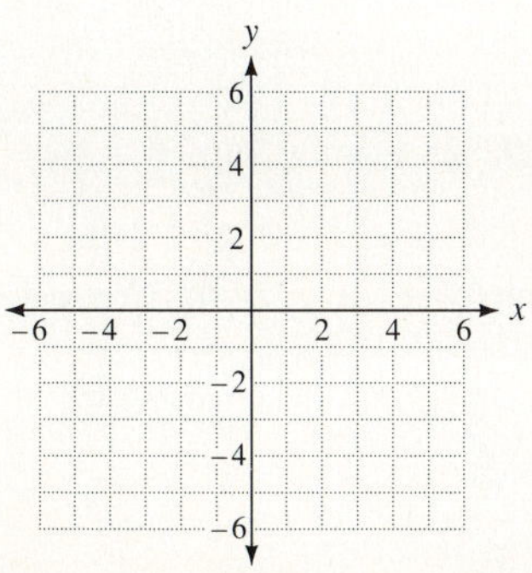

10. $\dfrac{x^2}{4}+\dfrac{y^2}{9}=1$

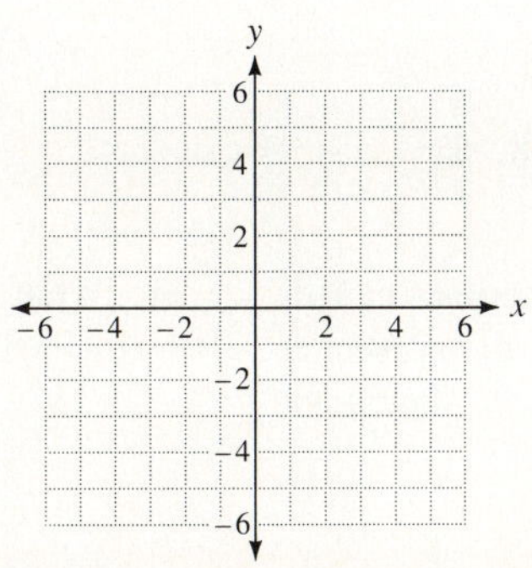

11. $\dfrac{x^2}{16}+\dfrac{y^2}{25}=1$

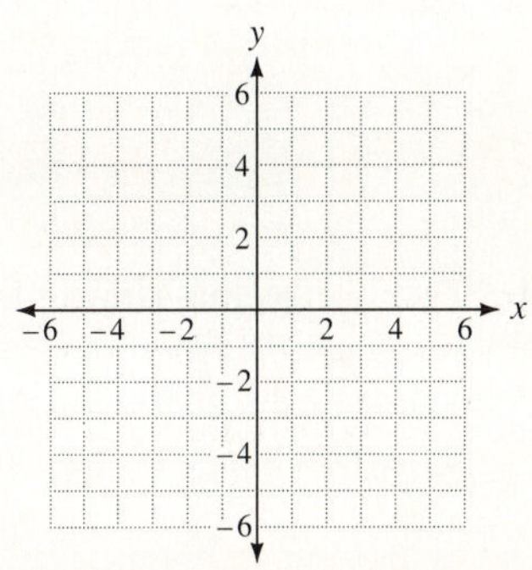

12. $x^2 + 25y^2 = 25$

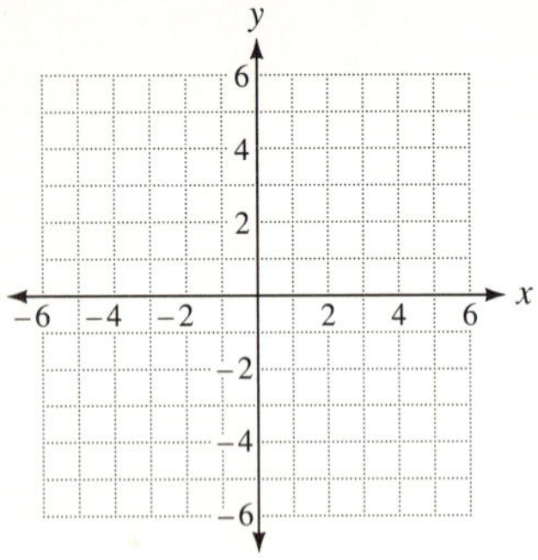

13. $\dfrac{x^2}{16} - \dfrac{y^2}{9} = 1$

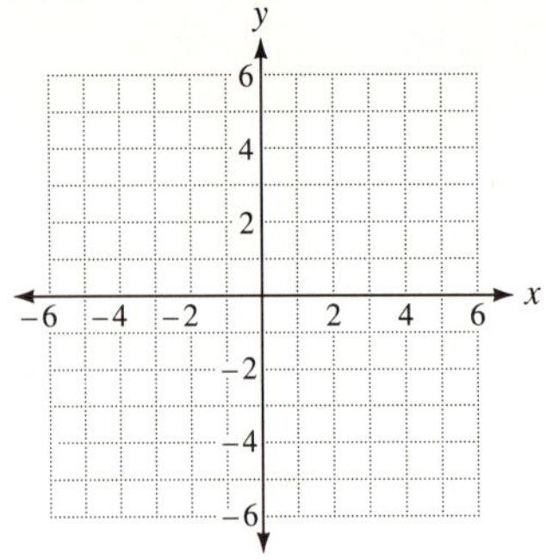

14. $\dfrac{y^2}{16} - \dfrac{x^2}{9} = 1$

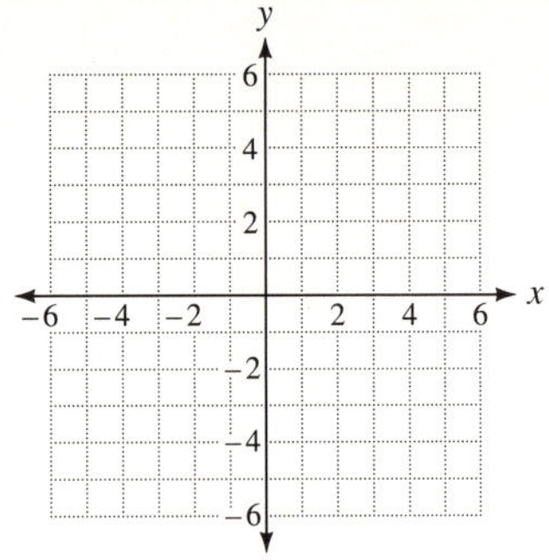

15. $x^2 - y^2 = 1$

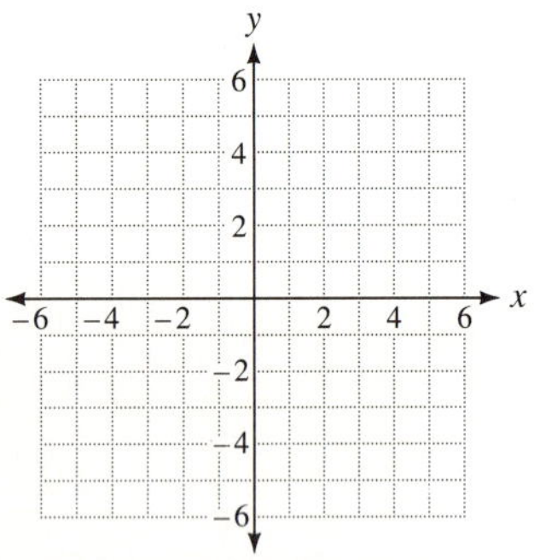

16. $y^2 - x^2 = 1$

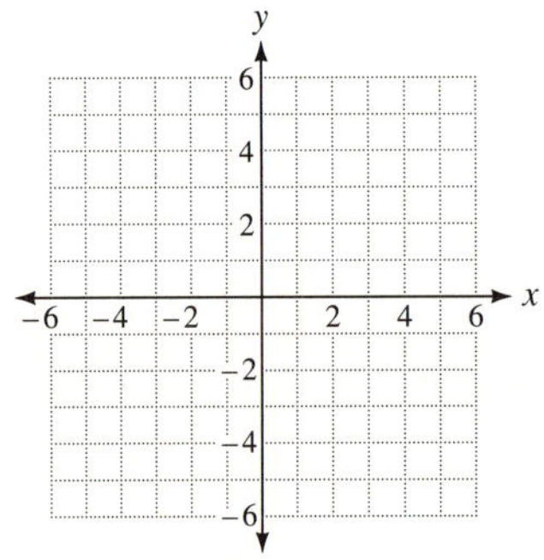

17. $9x^2 - 4y^2 = 36$

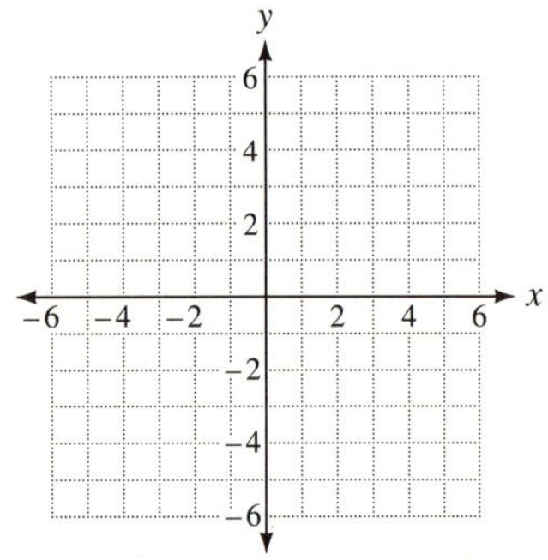

18. $xy = 8$

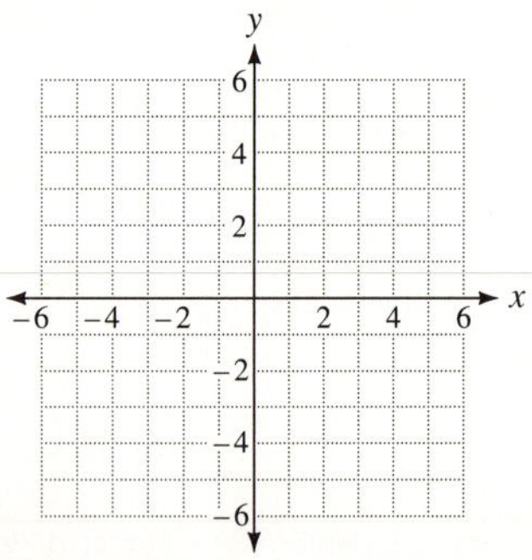

19. $xy = -8$

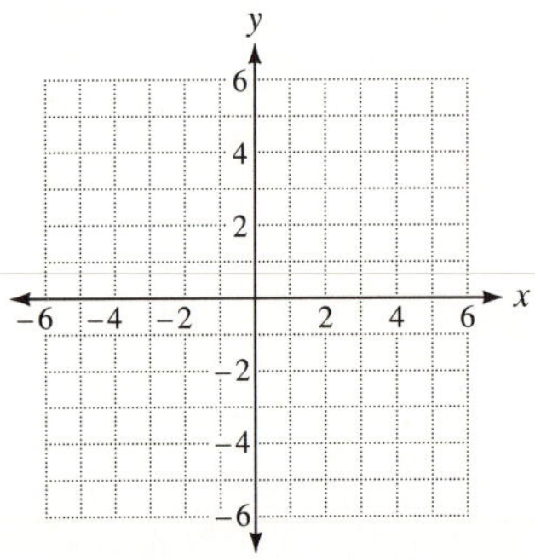

20. $xy = -12$

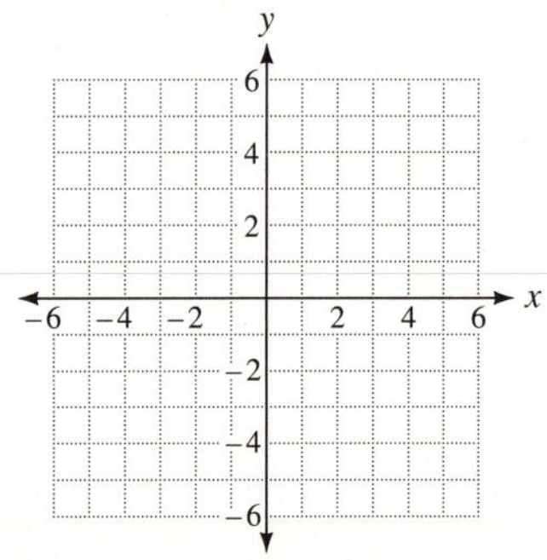

Identify each of the following as the equation of a parabola, circle, ellipse, or hyperbola.

21. $(x - 1)^2 + (y + 5)^2 = 49$

22. $x^2 + y^2 = 49$

23. $\dfrac{x^2}{49} - \dfrac{y^2}{4} = 1$

24. $\dfrac{x^2}{49} + \dfrac{y^2}{4} = 1$

25. $x = y^2 - y + 1$

26. $y^2 - x^2 = 36$

27. $x = \dfrac{7}{y}$

28. $8x^2 + 3y^2 = 24$

29. $x^2 + y^2 - 2x - 8 = 0$

30. Two intersecting straight lines are sometimes called a *degenerate hyperbola* since a plane can intersect a cone in two intersecting lines. Make a sketch to show how this can happen.

31. What is the graph of $x^2 - y^2 = 0$? Compare this with Exercise 30.

FOR REVIEW

Find the standard and general forms of the equation of a circle centered at the given point and with the given radius.

32. $(3, 4)$, $r = 1$ **33.** $(0, 0)$, $r = 11$ **34.** $(-2, 3)$, $r = \sqrt{5}$

Graph.

35. $x = -y^2 + 4$

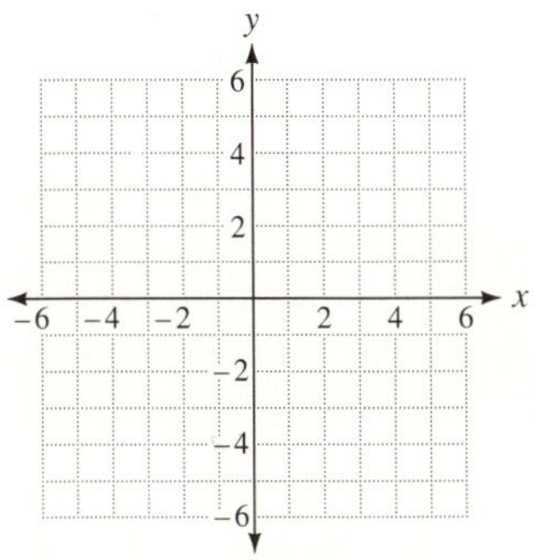

36. $x^2 + y^2 = 1$

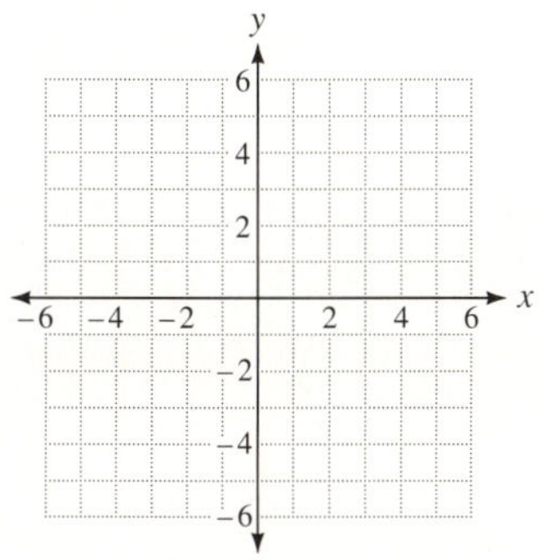

37. $y = -\sqrt{9 - x^2}$

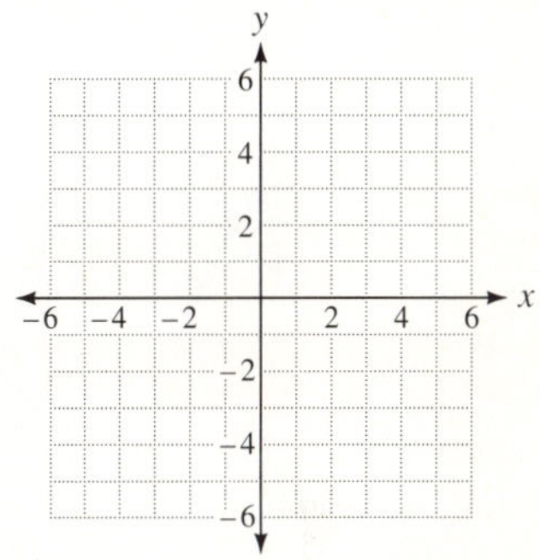

Exercises 38–40 review material from Chapter 8 to help you prepare for the next section. Solve each equation.

38. $2x^2 + x - 15 = 0$ **39.** $y^2 + 6 = 13$ **40.** $w^4 - 11w^2 + 18 = 0$

11.3 EXERCISES C

1. Give the intercepts of the ellipse with equation $4x^2 + 25y^2 = 100$.

2. Give the intercepts of the hyperbola with equation $9x^2 - 4y^2 = 72$.

3. Find the equation of the ellipse passing through $\left(2, \frac{2\sqrt{5}}{3}\right)$ with x-intercepts $(3, 0)$ and $(-3, 0)$.
$\left[\text{Answer: } \frac{x^2}{9} + \frac{y^2}{4} = 1\right]$

4. Find the equation of the hyperbola passing through $(2\sqrt{2}, 1)$ and $(-2\sqrt{2}, 1)$ with x-intercepts $(2, 0)$ and $(-2, 0)$.

5. A straight line is sometimes called a *degenerate parabola* since a plane can intersect a cone in a straight line. Make a sketch to show how this can happen.

11.4 NONLINEAR SYSTEMS OF EQUATIONS

STUDENT GUIDEPOSTS

1. Nonlinear Systems
2. Solving Nonlinear Systems
3. The Elimination Method

1 NONLINEAR SYSTEMS

When we studied systems of two linear (first-degree) equations in Chapter 4, the graphs of the equations gave useful information about the solutions. The same is true for **nonlinear systems** that involve second-degree equations. A solution to the nonlinear system

$$x^2 + y^2 = 25$$
$$x + y = 1$$

is an ordered pair of numbers that satisfies both equations. Graphically, a solution is a point of intersection of the two graphs. In the system above, the graph of the first equation is a circle, and the graph of the second is a line. In general, the graphs of a circle and a line can be related in one of three different ways, as shown in Figure 11.18.

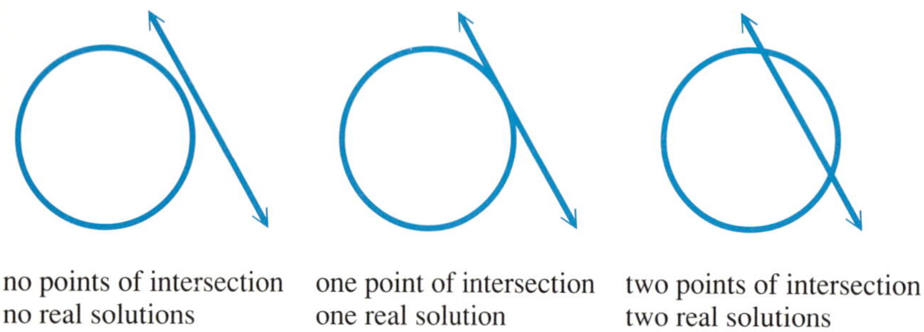

Figure 11.18 Intersecting a Circle and a Line

If we graph both equations in the same coordinate system, as in Figure 11.19, the points of intersection appear to have coordinates $(-3, 4)$ and $(4, -3)$. By substitution, it can be shown that these pairs of numbers satisfy both equations, hence are solutions to the system.

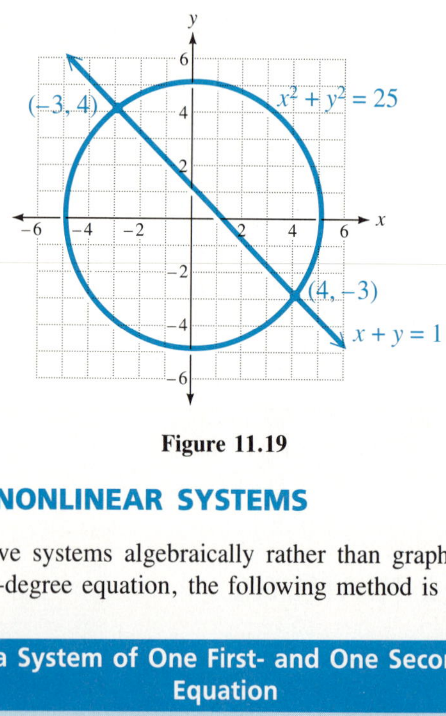

Figure 11.19

2 SOLVING NONLINEAR SYSTEMS

It is better to solve systems algebraically rather than graphically. When one equation is a first-degree equation, the following method is easy to apply.

To Solve a System of One First- and One Second-Degree Equation

1. Solve the first-degree equation for either variable.
2. Substitute in the second-degree equation and solve the resulting single-variable equation.
3. Substitute each of these values in the first-degree equation to find the corresponding values of the second variable.
4. Check your answer by substitution in both equations.

EXAMPLE 1 Solving a Nonlinear System

Solve $x^2 + y^2 = 25$
$x + y = 1.$

Solve $x + y = 1$ for y.

$$y = 1 - x$$
$$x^2 + (1 - x)^2 = 25 \quad \text{Substitute } 1 - x \text{ for } y$$
$$x^2 + 1 - 2x + x^2 = 25$$
$$2x^2 - 2x - 24 = 0$$
$$x^2 - x - 12 = 0 \quad \text{Divide by 2}$$
$$(x - 4)(x + 3) = 0$$
$$x - 4 = 0 \quad \text{or} \quad x + 3 = 0$$
$$x = 4 \qquad x = -3$$
$$y = 1 - x \qquad y = 1 - x$$
$$= 1 - 4 = -3 \qquad = 1 - (-3) = 4$$

Check: $(4, -3)$ $(-3, 4)$

$$4^2 + (-3)^2 \stackrel{?}{=} 25 \qquad 4 + (-3) \stackrel{?}{=} 1 \qquad (-3)^2 + 4^2 \stackrel{?}{=} 25 \qquad (-3) + 4 \stackrel{?}{=} 1$$
$$16 + 9 \stackrel{?}{=} 25 \qquad 1 = 1 \qquad 9 + 16 \stackrel{?}{=} 25 \qquad 1 = 1$$
$$25 = 25 \qquad 25 = 25$$

The solutions are $(4, -3)$ and $(-3, 4)$, which were obtained previously by graphing.

PRACTICE EXERCISE 1

Solve $2x^2 + y^2 = 9$
$y - 2x = 3.$

Answer: $(0, 3)$, $(-2, -1)$

3 THE ELIMINATION METHOD

Many systems of second-degree equations can be solved using an **elimination method** similar to the one used for systems of linear equations in Chapter 4. Such systems have zero, one, two, three, or four solutions.

The Elimination Method

1. If the equations can be added or subtracted to eliminate a variable, add or subtract and proceed to (3). Otherwise,
2. Multiply one or both equations by a constant, then add or subtract them to eliminate a variable.
3. Solve the resulting single-variable equation. Substitute each value into one of the original equations and solve. The resulting pair(s) of numbers are solution(s) to the system.
4. Check your answer by substitution in both equations.

EXAMPLE 2 Using the Elimination Method

(a) Solve $x^2 + y^2 = 14$
$x^2 - y^2 = 4.$

Add to eliminate y.

$$2x^2 = 18$$
$$x^2 = 9$$
$$x = \pm 3$$

PRACTICE EXERCISE 2

(a) Solve $x^2 + 2y^2 = 8$
$x^2 - 2y^2 = 0.$

Substitute 3 for x in the first equation.

$$(3)^2 + y^2 = 14$$
$$9 + y^2 = 14$$
$$y^2 = 5$$
$$y = \pm\sqrt{5}$$

Two of the solutions are $(3, \sqrt{5})$ and $(3, -\sqrt{5})$. When we substitute -3 for x in the first equation we obtain $y = \pm\sqrt{5}$. Thus, there are four solutions: $(3, \sqrt{5})$, $(3, -\sqrt{5})$, $(-3, \sqrt{5})$, and $(-3, -\sqrt{5})$. Figure 11.20 shows the four points of intersection of the circle and the hyperbola.

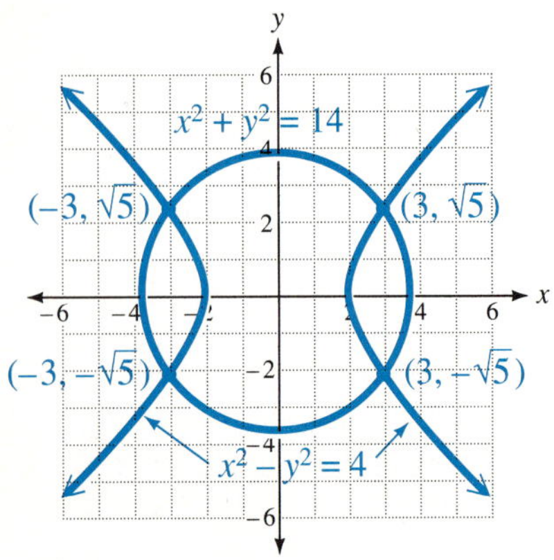

Figure 11.20

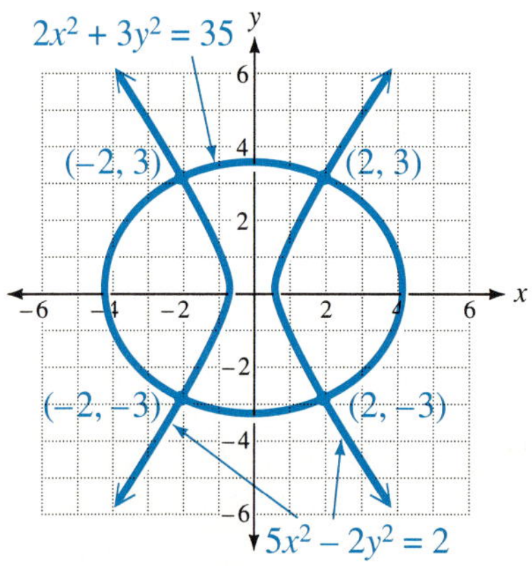

Figure 11.21

(b) Solve $5x^2 + y^2 = 10$
$2x^2 + 3y^2 = 17$.

(b) Solve

$$2x^2 + 3y^2 = 35$$
$$5x^2 - 2y^2 = 2.$$

If we multiply the first equation by 2 and the second equation by 3, we obtain the following system.

$$4x^2 + 6y^2 = 70$$
$$15x^2 - 6y^2 = 6$$

Solving this system by the elimination method, we obtain four solutions: (2, 3), (2, −3), (−2, 3), and (−2, −3). Figure 11.21 shows the four points of intersection of the ellipse and the hyperbola.

Answers: (a) $(2, \sqrt{2})$, $(2, -\sqrt{2})$, $(-2, \sqrt{2})$, $(-2, -\sqrt{2})$
(b) $(1, \sqrt{5})$, $(1, -\sqrt{5})$, $(-1, \sqrt{5})$, $(-1, -\sqrt{5})$

When one of the equations in a system is of the form $xy = c$, it is best to solve this equation for one of the variables and substitute in the other equation.

EXAMPLE 3 Solving a System with One Equation $xy = c$

Solve

$$x^2 + y^2 = 25$$
$$xy = 12.$$

$$x = \frac{12}{y} \quad \text{Solve } xy = 12 \text{ for } x$$

$$\left(\frac{12}{y}\right)^2 + y^2 = 25 \quad \text{Substitute in the first equation}$$

$$\frac{144}{y^2} + y^2 = 25$$

$$144 + y^4 = 25y^2 \quad \text{Multiply by } y^2$$

PRACTICE EXERCISE 3

Solve $x^2 - y^2 = -5$
$xy = 6$.

$$\begin{aligned} y^4 - 25y^2 + 144 &= 0 \\ u^2 - 25u + 144 &= 0 \qquad \text{Let } u = y^2 \\ (u-9)(u-16) &= 0 \\ u - 9 = 0 \quad &\text{or} \quad u - 16 = 0 \\ u = 9 \quad & \quad u = 16 \\ y^2 = 9 \quad & \quad y^2 = 16 \qquad y^2 = u \\ y = \pm 3 \quad & \quad y = \pm 4 \end{aligned}$$

When $y = 3$, $x = \frac{12}{3} = 4$; when $y = -3$, $x = \frac{12}{-3} = -4$;
when $y = 4$, $x = \frac{12}{4} = 3$; when $y = -4$, $x = \frac{12}{-4} = -3$.
Thus, the solutions are (4, 3), (−4, −3), (3, 4), and (−3, −4). Check.

Answer: (2, 3), (−2, −3)

Remember that the equation $xy = 12$ is the graph of a hyperbola with asymptotes the x- and y-axes. In Example 3 we found that this hyperbola intersects the circle $x^2 + y^2 = 25$ in the four points (4, 3), (−4, −3), (3, 4), and (−3, −4).

In the next example we solve one of the many applied problems involving nonlinear systems of equations.

EXAMPLE 4 Business Problem

Outdoor Industries manufactures and sells gas barbeque grills. If x represents the number of grills made and sold each week, the total weekly cost is given by

$$C = 50x + 1000,$$

and the total weekly revenue produced is given by

$$R = 100x - 0.2x^2.$$

Find the break-even value of x, the value of x for which the weekly cost and revenue are equal.

We need to find the value of x when $C = R$, that is, when

$$\begin{aligned} 50x + 1000 &= 100x - 0.2x^2. \\ 0.2x^2 - 50x + 1000 &= 0 \qquad \text{Collect like terms} \\ 2x^2 - 500x + 10{,}000 &= 0 \qquad \text{To clear the decimal, multiply by 10} \\ x^2 - 250x + 5000 &= 0 \qquad \text{Factor out 2 and divide both sides by 2.} \end{aligned}$$

Since the left side cannot be factored, we use the quadratic formula.

$$\begin{aligned} x &= \frac{-b \pm \sqrt{b^2 - 4ac}}{2a} = \frac{-(-250) \pm \sqrt{(-250)^2 - 4(1)(5000)}}{2(1)} \\ &= \frac{250 \pm \sqrt{62{,}500 - 20{,}000}}{2} = \frac{250 \pm \sqrt{42{,}500}}{2} \end{aligned}$$

Using a calculator, we find that $\sqrt{42{,}500}$ is approximately 206.2. Thus,

$$x \approx \frac{250 \pm 206.2}{2} = 228.1 \text{ and } 21.9.$$

Since x is the number of grills manufactured and sold weekly we use 228 and 22 for the approximate break-even points. That is, if either 228 or 22 grills are made and sold each week, the cost of production and the revenue realized will be about the same.

Practice Exercise 4

The area of a rectangular pasture must be 30,000 yd^2, and the length must be three times the width. Find the dimensions of the pasture.

Answer: 300 yd by 100 yd

11.4 EXERCISES A

Solve each system.

1. $x^2 + y^2 = 10$
$x - y = 2$

2. $x + y = 3$
$x^2 - y^2 = 3$

3. $xy = 20$
$x + y = 9$

4 $5x^2 + xy - y^2 = -1$
$y - 2x = 1$

5. $xy - x^2 = 5$
$x - 2y = 3$

6. $x^2 + y^2 = 25$
$x^2 - y^2 = 25$

7. $x^2 + 3y^2 = 37$
$2x^2 - y^2 = 46$

8. $x^2 + y^2 = 5$
$xy = 2$

9 $3xy - y^2 = -13$
$2xy + y^2 = -2$
[*Hint:* Eliminate the xy term.]

10. $x^2 + y^2 = 4$
$xy = 10$

11. $x^2 + 2xy + y^2 = 9$
$x^2 - 2xy - y^2 = 9$

12. $x^2 + y^2 = 169$
$x + y = 17$

Write as a nonlinear system and solve.

13 The sum of the squares of two numbers is 100. The difference of the numbers is 2. Find the numbers.

14. The product of two numbers is 12 and the difference of their squares is 7. Find the numbers.

15. The area of a rectangular field must be 4050 m^2 and the length must be twice the width. Find the dimensions.

16. At what points do the ellipse $\frac{x^2}{1} + \frac{y^2}{4} = 1$ and the hyperbola $x^2 - y^2 = 1$ intersect?

17. Larson Manufacturing makes and sells beehives. If x represents the number of hives made and sold each week, the total weekly cost and revenue produced are $C = 10x + 80$ and $R = 25x - 0.2x^2$. Find the break-even value of x (when $C = R$).

FOR REVIEW

Graph.

18. $\frac{x^2}{9} + \frac{y^2}{25} = 1$

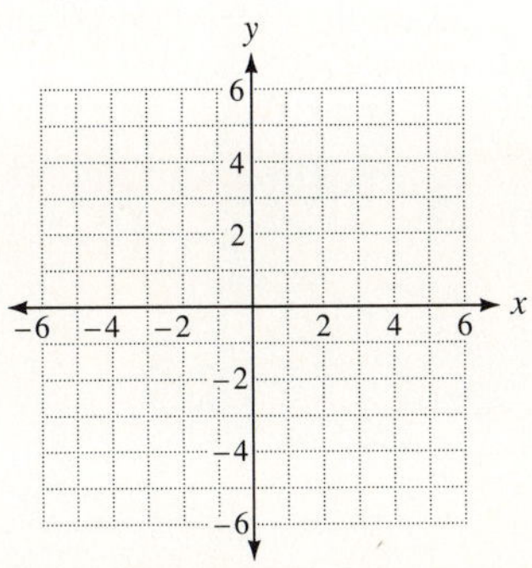

19. $\frac{y^2}{25} - \frac{x^2}{9} = 1$

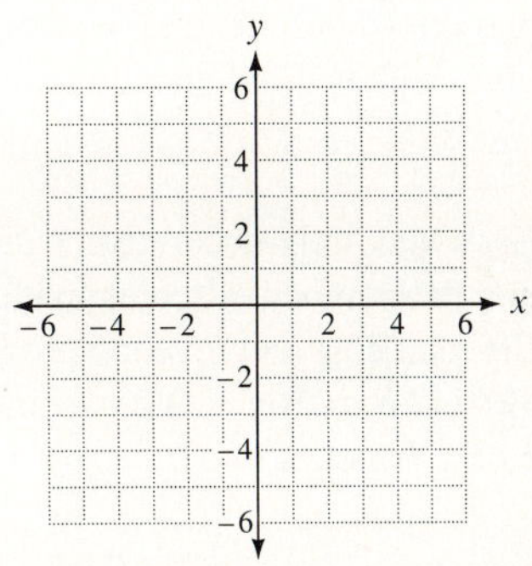

ANSWERS: **1.** (3, 1), (−1, −3) **2.** (2, 1) **3.** (5, 4), (4, 5) **4.** (0, 1), (1, 3) **5.** no real solutions **6.** (5, 0), (−5, 0) **7.** (5, 2), (5, −2), (−5, 2), (−5, −2) **8.** (2, 1), (−2, −1), (1, 2), (−1, −2) **9.** $\left(-\frac{3}{2}, 2\right), \left(\frac{3}{2}, -2\right)$ **10.** no real solutions **11.** (3, 0), (3, −6), (−3, 0), (−3, 6) **12.** (5, 12), (12, 5) **13.** (8, 6); (−6, −8) **14.** (4, 3); (−4, −3) **15.** 45 m by 90 m **16.** (1, 0), (−1, 0) **17.** approximately 6 and 69 **18.**

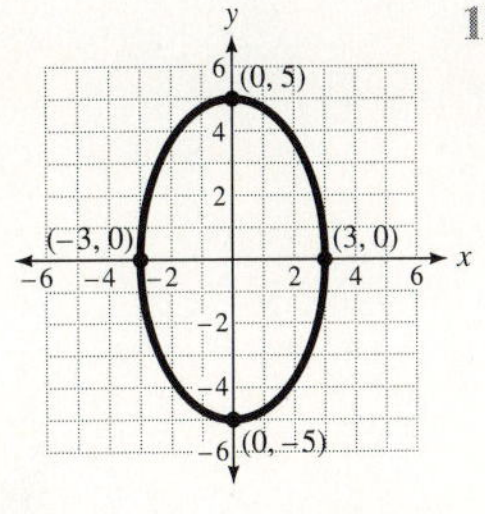

19.

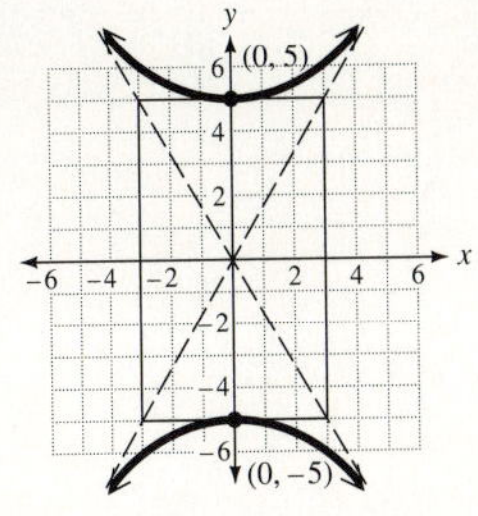

11.4 EXERCISES B

Solve each system.

1. $x^2 + y^2 = 8$
   $x + y = 4$

2. $x^2 - 2y^2 = 7$
   $2x + y = 7$

3. $x - y = -4$
   $x^2 - y^2 = -16$

4. $xy = 12$
   $2x - y = -2$

5. $x^2 - xy + y^2 = 1$
   $3x - y = 2$

6. $y^2 - xy = 3$
   $2x + y = 7$

7. $x^2 - 3y^2 = -8$
   $x^2 + 4y^2 = 13$

8. $y^2 - x^2 = 0$
   $x^2 + y^2 = 8$

9. $x^2 + 2y^2 = 1$
   $xy = -2$

10. $x^2 - 3xy = 18$
    $x^2 - 2xy = 15$

11. $2x^2 - 3xy + y^2 = 6$
    $x^2 + 3xy - y^2 = 69$

12. $x^2 + y^2 = 200$
    $x - 2y = -10$

Write as a nonlinear system and solve.

13. The sum of the squares of two numbers is 106. If the sum of the numbers is 4, find the numbers.

14. The product of two real numbers is -20 and the difference of their squares is 9. Find the numbers.

15. The area of a large room is 2800 ft^2 and the length is 30 ft more than the width. Find the dimensions.

16. At what points do the circle $x^2 + y^2 = 16$ and the parabola $y^2 = -6x$ intersect?

17. Tim Chandler makes and sells decorative patio tables. If x represents the number of tables made and sold each week, the total cost and revenue produced are $C = 30x + 200$ and $R = 50x - 0.2x^2$. Find the break-even value of x (when $C = R$).

FOR REVIEW

Graph.

18. $\dfrac{x^2}{36} + \dfrac{y^2}{9} = 1$

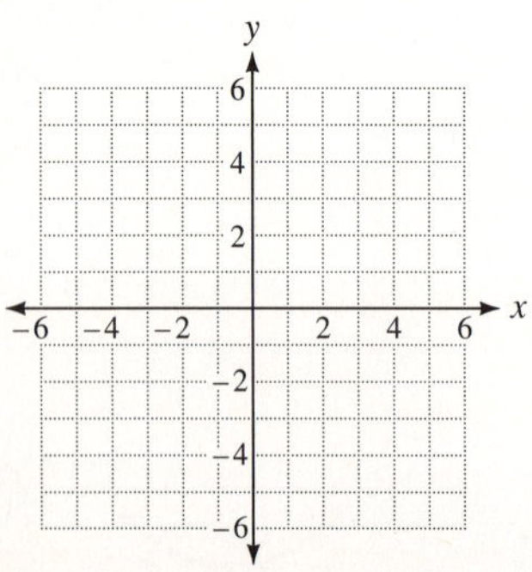

19. $\dfrac{x^2}{4} - \dfrac{y^2}{1} = 1$

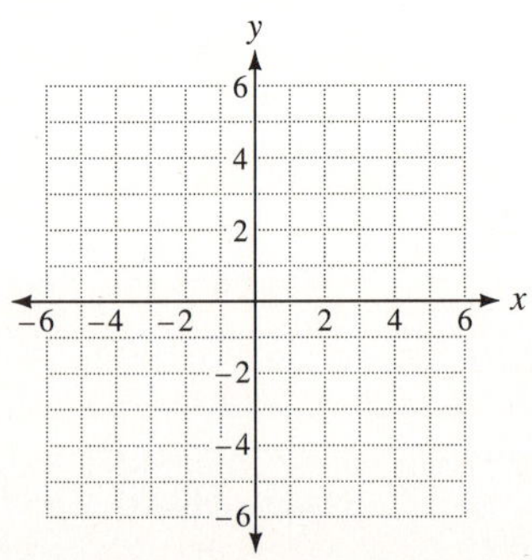

11.4 EXERCISES C

Write as a nonlinear system and solve.

1. The area of a rectangle is 21 cm^2, and its diagonal is $\sqrt{58}$ cm long. Find the dimensions of the rectangle. [Answer: 7 cm by 3 cm]

2. The perimeter of a rectangular field is 13.8 mi, and its area is 11 mi^2. Find the dimensions of the field.

Solve.

3. $x^2 + xy + y^2 = 9$
   $3x^2 - xy = 0$

4. $x^2 + xy + y^2 = 6$
   $x^2 - y^2 = 0$ [Answer: $(\sqrt{2}, \sqrt{2})$, $(-\sqrt{2}, -\sqrt{2})$, $(\sqrt{6}, -\sqrt{6})$, $(-\sqrt{6}, \sqrt{6})$]

CHAPTER 11 REVIEW

KEY WORDS

11.1 A **second-degree equation in two variables** x and y is an equation of the form $Ax^2 + Bxy + Cy^2 + Dx + Ey + F = 0$.

Conic sections (or **conics**) are curves that can be formed by intersecting a plane and a cone.

A **parabola** is the set of all points in a plane equidistant from a point (the **focus**) and a straight line not containing the point.

11.2 A **circle** is a set of all points in a plane that are a fixed distance r, called the **radius,** from a given point (h, k) in the plane, called the **center** of the circle.

11.3 An **ellipse** is the set of all points in a plane the sum of whose distances from two fixed points (called **foci**) is constant.

A **hyperbola** is the set of all points in a plane the difference of whose distances from two fixed points (called **foci**) is constant.

The lines approached by a hyperbola are called **asymptotes.**

11.4 A system of equations in which at least one equation is not linear is called a **nonlinear system.**

KEY CONCEPTS

11.1 Parabola:

1. $y = ax^2 + bx + c$ opens up or down;
   $y = a(x - h)^2 + k$ Standard form
   vertex (h, k), opens up if $a > 0$, opens down if $a < 0$.
2. $x = ay^2 + by + c$ opens left or right;
   $x = a(y - k)^2 + h$ Standard form
   vertex (h, k), opens right if $a > 0$, opens left if $a < 0$.

11.2

1. Circle with center (h, k) and radius r:
   $(x - h)^2 + (y - k)^2 = r^2$ Standard form
2. When the standard form is expanded, the general form of a circle is obtained.
   $x^2 + y^2 + Dx + Ey + F = 0$ General form

11.3

1. Ellipse with center (0, 0): $\frac{x^2}{a^2} + \frac{y^2}{b^2} = 1$
2. Hyperbola with center (0, 0):
   $\frac{x^2}{a^2} - \frac{y^2}{b^2} = 1$ opens left and right;
   $\frac{y^2}{b^2} - \frac{x^2}{a^2} = 1$ opens up and down.
3. Hyperbola with the axes as asymptotes:
   $xy = c$
   If $c > 0$, curves are in quadrants I and III.
   If $c < 0$, curves are in quadrants II and IV.

11.4

1. A system of one linear and one quadratic equation may have zero, one, or two real solutions. To solve, use the substitution method.
2. A system of two second-degree equations may have zero, one, two, three, or four real solutions. To solve, use the elimination method or the substitution method.

REVIEW EXERCISES

Part I

11.1 *Graph.*

1. $y = x^2 - 2x - 2$

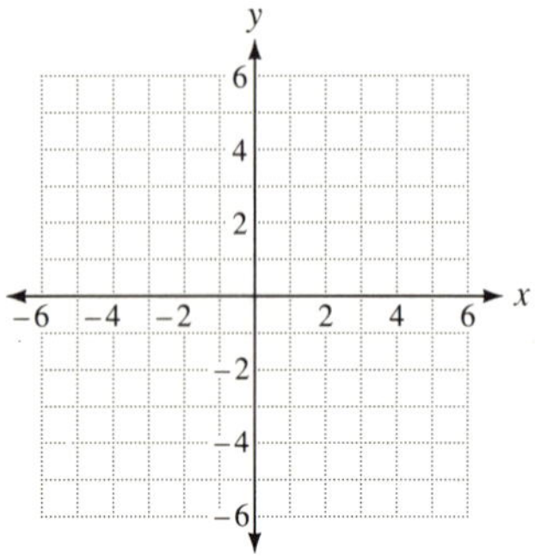

2. $x = -y^2 + 2y + 3$

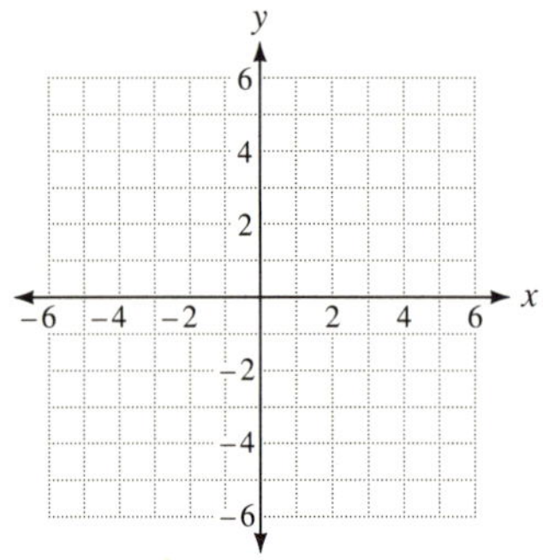

3. $8x = y^2 - 4y - 4$

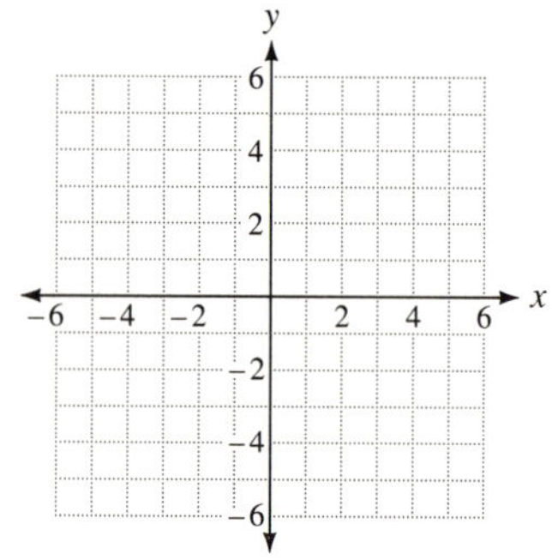

11.2 *Find the standard and general forms of the equation of a circle centered at the given point with the given radius.*

4. $(-5, 2)$, $r = 4$

5. $(2, 6)$, $r = 3$

6. $\left(\frac{1}{2}, \frac{1}{2}\right)$, $r = 2$

7. $(0, 5)$, $r = 1$

Find the standard form of the equation of each circle and give the center and radius.

8. $x^2 + y^2 - 4x + 16y + 19 = 0$

9. $4x^2 + 4y^2 + 4x - 24y + 33 = 0$

Graph.

10. $(x + 2)^2 + (y - 2)^2 = 16$

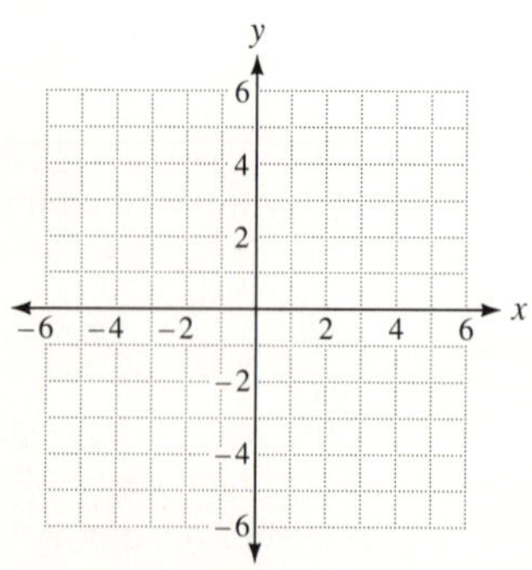

11. $y = \sqrt{9 - x^2}$

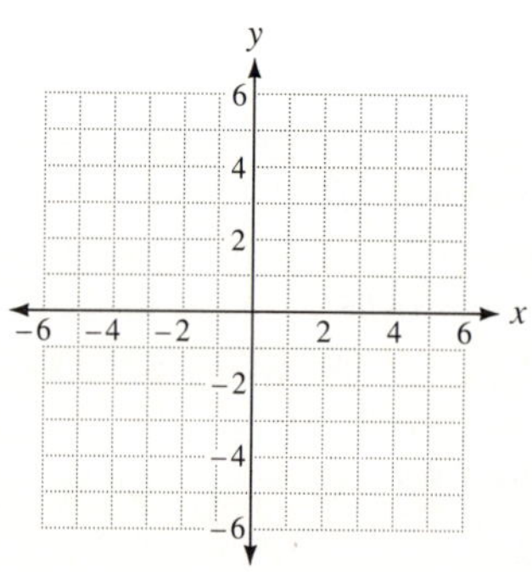

12. $y = -\sqrt{4 - x^2}$

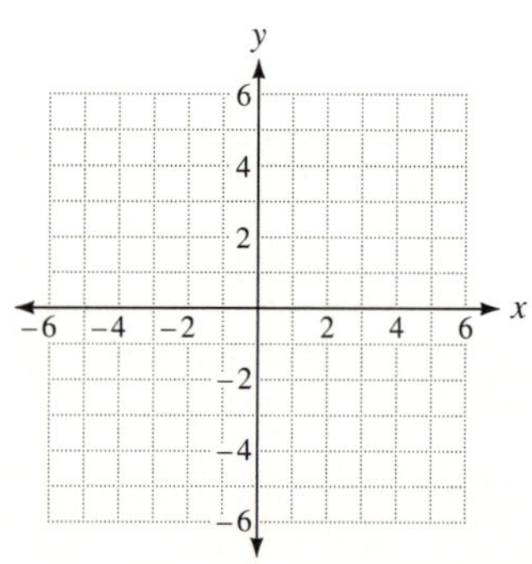

13. A tunnel in the shape of a semicircle must have a vertical clearance of 8 ft at a distance of 6 ft on either side of the center line. Find the equation of the semicircle by placing the origin at the centerline, and give the maximum height of the tunnel.

11.3 *Give the standard form of each equation.*

14. Ellipse: x-intercepts $(1, 0)$, $(-1, 0)$
y-intercepts $(0, 5)$, $(0, -5)$

15. Ellipse: x-intercepts $(4, 0)$, $(-4, 0)$
y-intercepts $(0, 3)$, $(0, -3)$

16. Hyperbola: $a = 5$, $b = 3$, opens up and down

17. Hyperbola: $a = 2$, $b = 7$, opens left and right

Graph.

18. $\dfrac{x^2}{9} + \dfrac{y^2}{4} = 1$

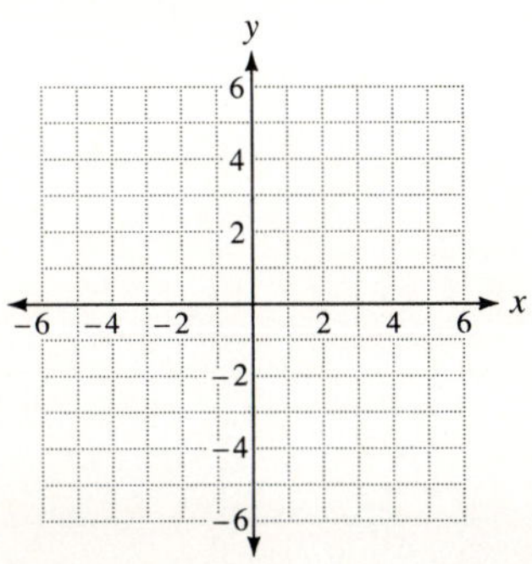

19. $\dfrac{x^2}{16} - \dfrac{y^2}{4} = 1$

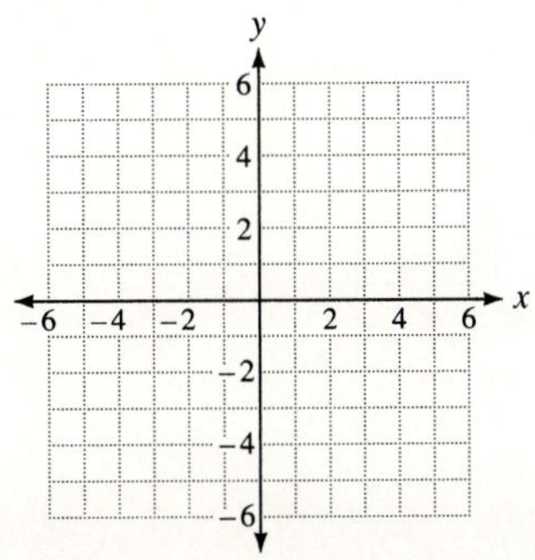

20. $\dfrac{y^2}{4} - \dfrac{x^2}{16} = 1$

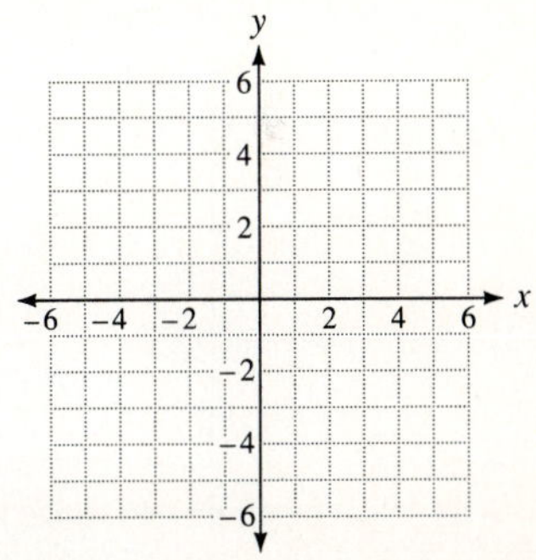

Identify each as the equation of a parabola, circle, ellipse, or hyperbola.

21. $x^2 + y^2 = 49$

22. $x^2 - 8y = 0$

23. $(x + 4)^2 + (y - 1)^2 = 20$

24. $\dfrac{x^2}{4} + \dfrac{y^2}{9} = 1$

25. $\dfrac{x^2}{4} - \dfrac{y^2}{9} = 1$

26. $x^2 = y - 2$

27. $y^2 + 2x - 4y = 0$

28. $y^2 - 9x^2 = 9$

29. $16x^2 + 25y^2 = 400$

11.4 *Solve the following systems.*

30. $2x^2 + y^2 = 3$
$x - y = 2$

31. $x^2 + y^2 = 20$
$xy = 8$

32. $3x^2 + y^2 = 3$
$x^2 - 2y^2 = 1$

Write as nonlinear systems and solve.

33. At what points do the parabola $y^2 = 3x$ and the hyperbola $\frac{x^2}{4} - \frac{y^2}{4} = 1$ intersect?

34. A triangular reflecting surface is in the shape of a right triangle with hypotenuse 5 ft. If one leg must be one foot longer than the other, what are the lengths of the legs?

Part II

Graph.

35. $4y = x^2 + 2x + 13$

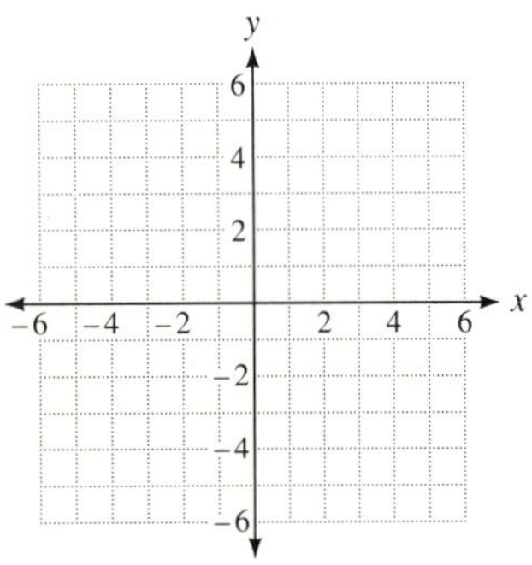

36. $(x - 1)^2 + (y - 3)^2 = 9$

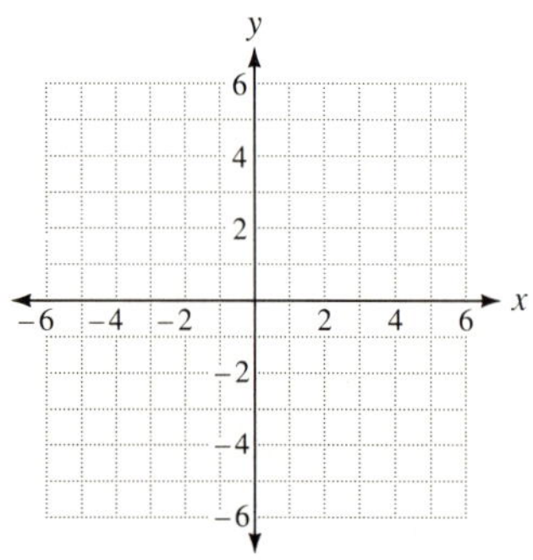

37. $xy = 3$

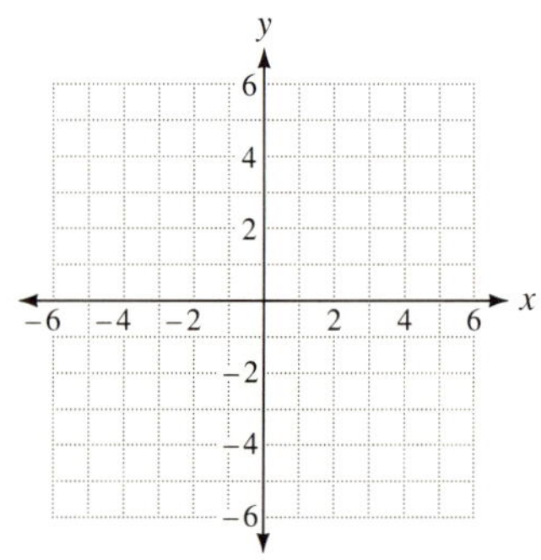

Identify each as the equation of a parabola, circle, ellipse, or hyperbola.

38. $x^2 - y^2 = 4$

39. $x^2 + y^2 - 8x + 2y - 3 = 0$

40. $y = \dfrac{16}{x}$

41. $x = -y^2 + 4y$

42. $4x^2 - 8y^2 = 32$

43. $x^2 + y^2 = 4$

44. Solve.

$$xy + 2x^2 = -8$$
$$x + y = 5$$

45. Write in standard form and give the center and radius of the circle with equation $x^2 + y^2 + 8x + 2y - 3 = 0$.

ANSWERS: 1.

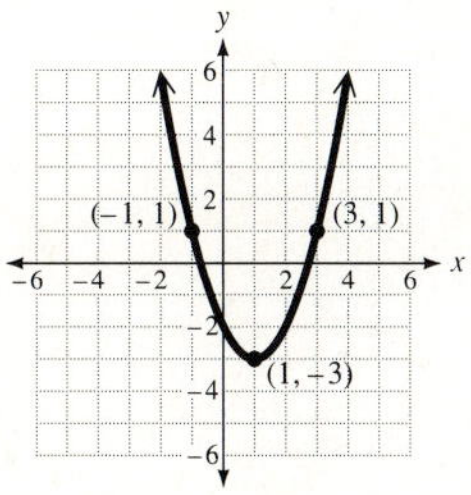

2.

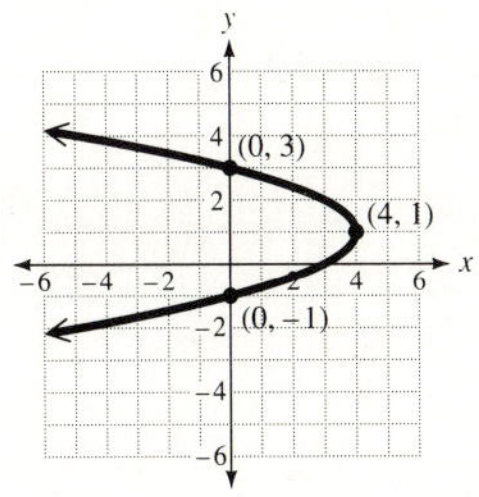

3.

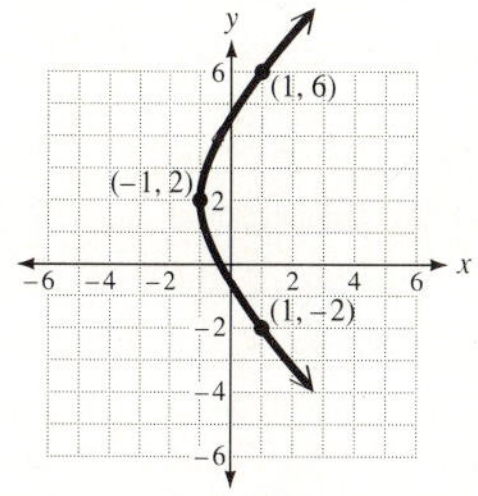

4. $(x+5)^2+(y-2)^2=4^2$, $x^2+y^2+10x-4y+13=0$ 5. $(x-2)^2+(y-6)^2=3^2$, $x^2+y^2-4x-12y+31=0$ 6. $\left(x-\frac{1}{2}\right)^2+\left(y-\frac{1}{2}\right)^2=2^2$, $2x^2+2y^2-2x-2y-7=0$ 7. $x^2+(y-5)^2=1^2$, $x^2+y^2-10y+24=0$ 8. $(x-2)^2+(y+8)^2=7^2$; $(2,-8)$; $r=7$ 9. $\left(x+\frac{1}{2}\right)^2+(y-3)^2=1^2$; $\left(-\frac{1}{2},3\right)$; $r=1$

10.

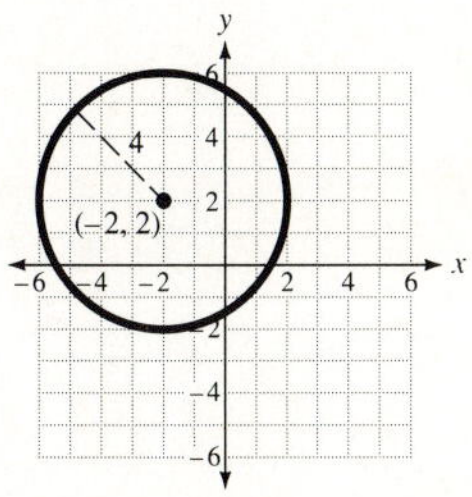

11.

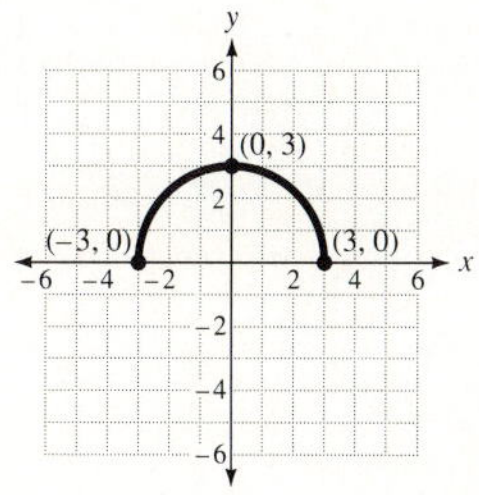

12.

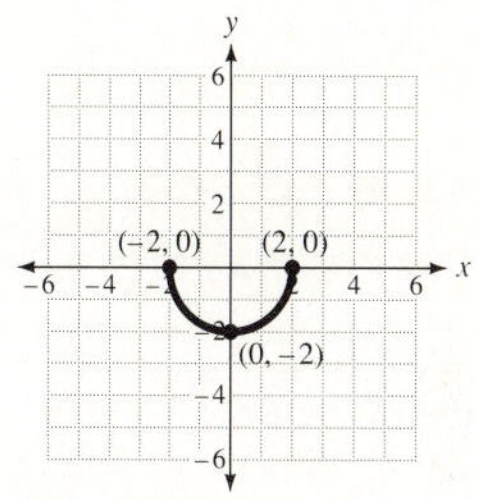

13. $y=\sqrt{100-x^2}$; 10 ft 14. $\frac{x^2}{1}+\frac{y^2}{25}=1$ 15. $\frac{x^2}{16}+\frac{y^2}{9}=1$ 16. $\frac{y^2}{9}-\frac{x^2}{25}=1$ 17. $\frac{x^2}{4}-\frac{y^2}{49}=1$

18.

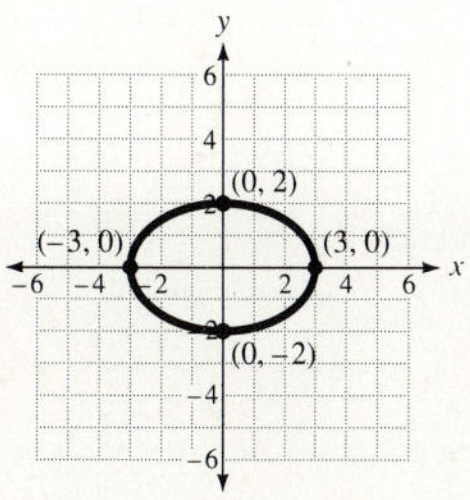

19.

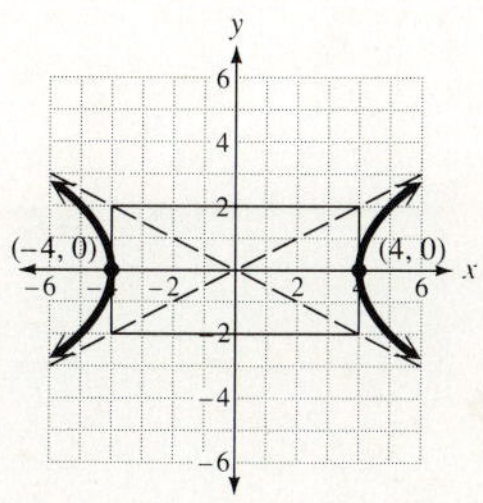

20.

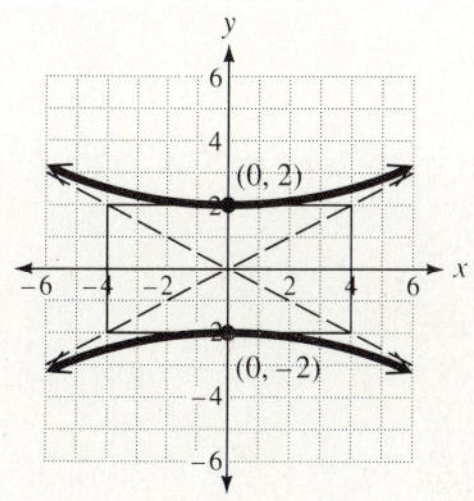

21. circle 22. parabola 23. circle 24. ellipse 25. hyperbola 26. parabola 27. parabola 28. hyperbola 29. ellipse 30. $\left(\frac{1}{3},-\frac{5}{3}\right)$, $(1,-1)$ 31. $(2,4)$, $(-2,-4)$, $(4,2)$, $(-4,-2)$ 32. $(1,0)$, $(-1,0)$ 33. $(4,2\sqrt{3})$, $(4,-2\sqrt{3})$ 34. 3 ft; 4 ft

35.

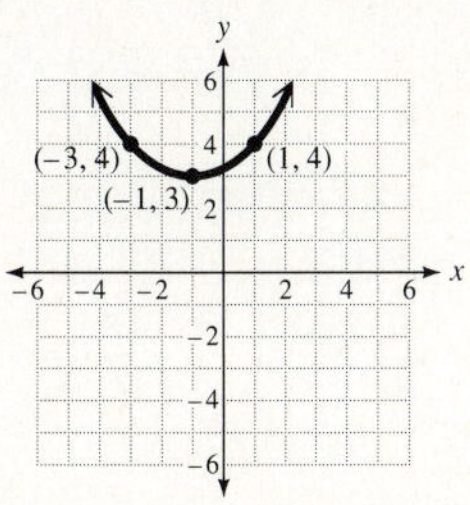

36.

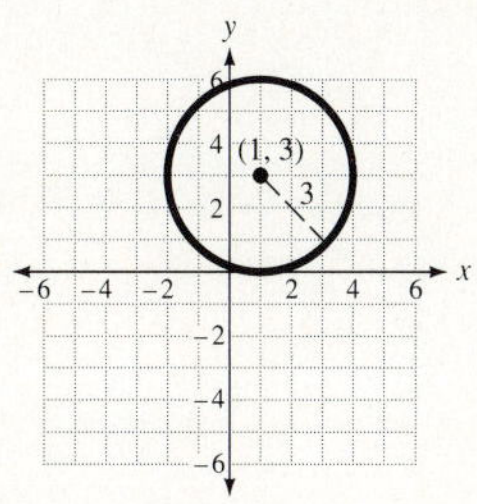

37.

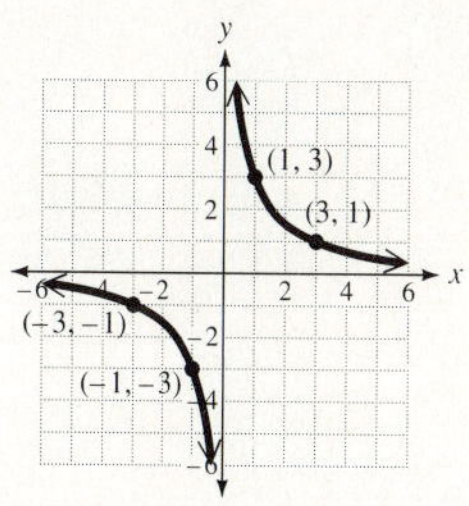

38. hyperbola 39. circle 40. hyperbola 41. parabola 42. hyperbola 43. circle 44. no real solutions 45. $(x+4)^2+(y+1)^2=20$; $(-4,-1)$; $r=2\sqrt{5}$

CHAPTER 11 TEST

Graph.

1. $y = -x^2 + 2$

1. 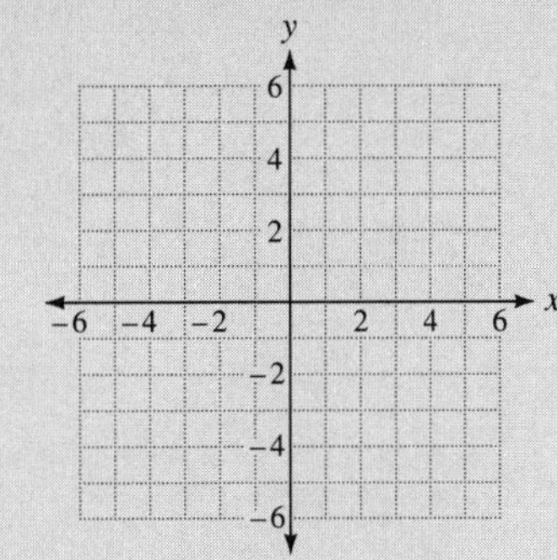

2. $x = y^2 + 4y + 3$

2. 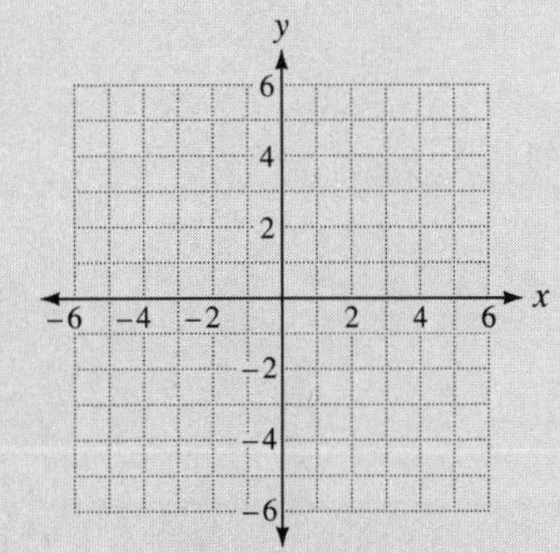

3. $x^2 + y^2 = 16$

3. 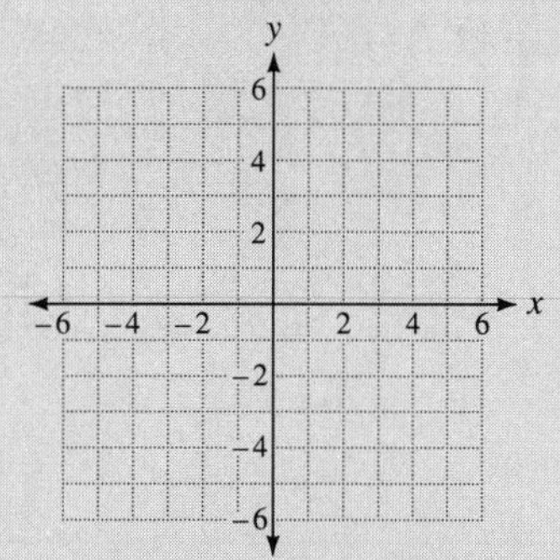

4. $(x - 2)^2 + (y + 1)^2 = 9$

4. 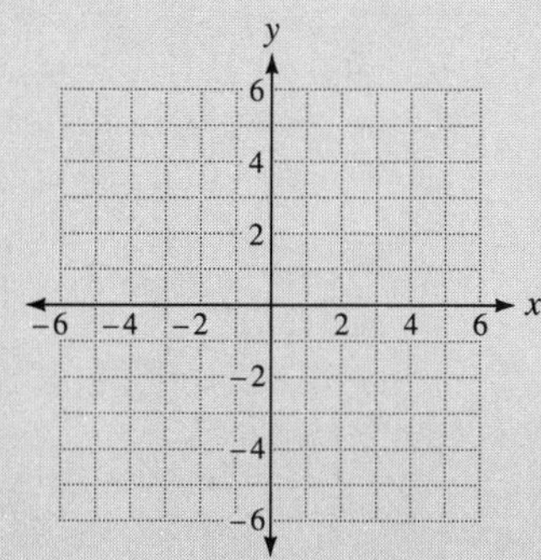

5. Find the standard and general forms of the equation of a circle centered at $(-2, -1)$ with radius $r = 5$.

5. ______________________

6. Find the standard form of the equation of the circle with general equation $x^2 + y^2 + 14x - 12y + 60 = 0$. What are the center and radius of the circle?

6. ____________________

Graph.

7. $\dfrac{x^2}{16} + \dfrac{y^2}{4} = 1$

7.

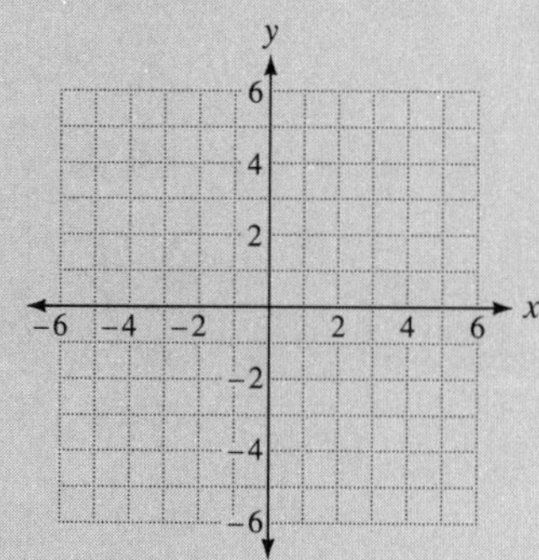

8. $\dfrac{y^2}{4} - \dfrac{x^2}{9} = 1$

8.

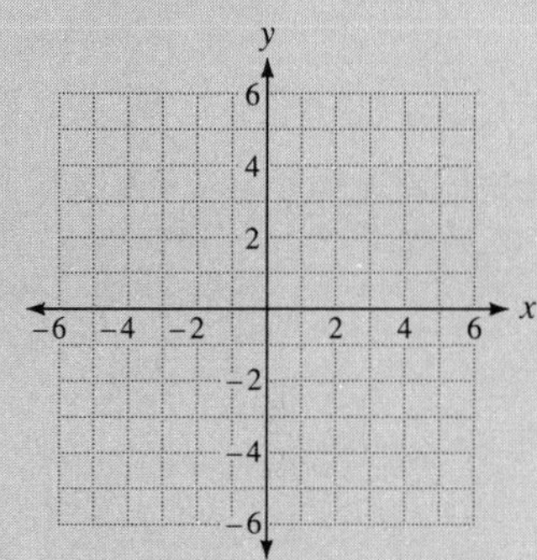

Identify each as the equation of a parabola, circle, ellipse, or hyperbola.

9. $\dfrac{x^2}{4} - \dfrac{y^2}{9} = 1$ 9. ____________________

10. $x^2 + (y - 1)^2 = 9$ 10. ____________________

11. $y = x^2 - 3x + 5$ 11. ____________________

12. $4x^2 + y^2 = 4$ 12. ____________________

Solve.

13. $3x^2 - y^2 = -13$
$2x + y = -1$

13. ____________________

14. At what points do the circle $x^2 + y^2 = 9$ and the parabola $x^2 = 8y$ intersect?

14. ____________________

CHAPTER 12

Geometry

12.1 INDUCTIVE AND DEDUCTIVE REASONING

STUDENT GUIDEPOSTS

1 Inductive Reasoning

2 Deductive Reasoning

1 INDUCTIVE REASONING

To help us understand geometry, we begin by considering the two basic ways that we reason, or think. The first can be illustrated by studying the following list of numbers:

$$4, \ 11, \ 18, \ 25.$$

What is the next number in this list? Most of us will try to discover what 4, 11, 18, and 25 have in common, and shortly realize that $4 + 7 = 11$, $11 + 7 = 18$, $18 + 7 = 25$, so it would seem that each number after 4 is obtained from the one in front of it by adding 7. As a result, we would probably conclude that the next number should be 32 because $25 + 7 = 32$. Our reasoning process involved considering several *specific* observations, and based on these, formulating the *general* conclusion that the list would continue in the same pattern if we always added 7 to one number to obtain the next.

Inductive Reasoning

We use **inductive reasoning** when we reach a general conclusion based on a limited collection of specific observations.

Natural and social scientists frequently use inductive reasoning. When a laboratory experiment is performed several times with the same result, the physicist might form a general conclusion based on this experimentation. Or a sociologist might collect information from a limited number of people and attempt to draw a general conclusion about the total population. Also, when a new medicine is tested on a sample of several hundred people, the test results might lead a scientist to draw conclusions about the drug's effectiveness. In all these cases, a general conclusion is drawn from specific observations.

Does inductive reasoning always lead to the same conclusion? The answer is no, not always. If you return to our earlier example of four numbers, in which we concluded that the next number must be 32, you will see that another conclusion is also possible. The next number in the series can be 1, followed by 8, 15, 22, etc. This sequence gives the dates of the Fridays in the year 1991, starting with Friday, January 4, 1991!

As you can see, although inductive reasoning is widely used, there are no guarantees that the conclusion drawn is always correct or that it is the only possible conclusion. The primary flaw with inductive reasoning is that we cannot be sure of what will generally be true based on a limited number of cases.

EXAMPLE 1 Using Inductive Reasoning

Use inductive reasoning to determine the next element in each list. Remember that there might be more than one answer.

(a) 2, 4, 8, 16, 32

We might conclude that the next number is 64 because each number after the first is twice the one before it.

(b) o, t, t, f, f, s, s, e

This one is a bit more difficult. This is a list of the first letters in the words *one, two, three, four,* etc. Thus, the next letter would be *n*, the first letter in the word *nine*.

Practice Exercise 1

Use inductive reasoning to determine the next element in each list.

(a) 1, 4, 9, 16, 25

(b)

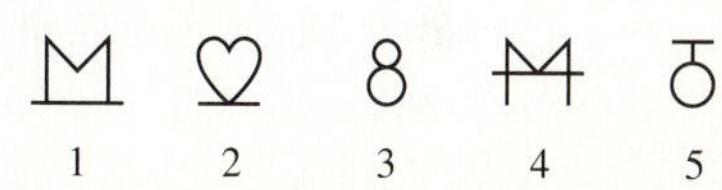

Answers: (a) One answer is 36 since each number is the square of the numbers 1, 2, 3, 4, and 5. That is, $1 = 1^2$, $4 = 2^2$, $9 = 3^2$, $16 = 4^2$, $25 = 5^2$, and $36 = 6^2$. (b) The next element is  since each element is one of the numbers 1, 2, 3, 4, and 5 placed next to its mirror image.

EXAMPLE 2 Application of Inductive Reasoning

To determine the preferred candidate in the upcoming student body presidential election, a student decides to interview the first 50 students who pass by the entrance to the student union building. He discovers that 42 prefer Abby Bayona and 8 prefer Shawn Herman. Based on these 50 specific observations, he concludes that Abby Bayona will win the election. Is this conclusion necessarily correct? Can you think of situations in which the results of his poll might not accurately reflect the feelings of the total student body?

Obviously the interviewer's conclusion is not necessarily correct. For example, the limited sample of 50 students interviewed might have included sorority sisters of Abby Bayona who happened to be attending a meeting in the union building.

Practice Exercise 2

Can you suggest possible changes in the data that might occur if only male students were included in the survey?

Answer: Male students might tend to give Shawn Herman more votes.

2 DEDUCTIVE REASONING

Although inductive reasoning might not always lead to the same conclusion, it is still an important process, one that is used even by mathematicians. For example, consider the following observations:

$$1 + 1 = 2, \quad 1 + 3 = 4, \quad 3 + 3 = 6, \quad 3 + 5 = 8,$$
$$5 + 1 = 6, \quad 7 + 3 = 10.$$

After examining these equations, a mathematician might conclude that the sum of two odd integers is always an even integer. Certainly these six observations do not prove that this statement is true. What if we listed several hundred such observations with the same results? Would that have proved the statement? The mathematician would say you cannot prove a general statement by giving any number of specific cases, unless, of course, the number of specific cases is limited. The mathematician usually requires a different kind of proof, one that is based on *deductive reasoning*.

Deductive Reasoning

We use **deductive reasoning** when we reach a specific conclusion based on a collection of generally accepted assumptions.

When we reason deductively, we start with one or more **premises,** which are previously known facts, and attempt to arrive at a conclusion that necessarily follows if the premises are accepted. An example of deductive reasoning is given below.

Premise: Terri is a happy person.
Premise: Every happy person is pleasant.
Conclusion: Terri is pleasant.

Because Terri is a happy person, and every happy person is pleasant, it follows logically that Terri, as one of the happy persons, must also be pleasant.

The next example gives practice in recognizing deductive and inductive reasoning.

EXAMPLE 3 Comparing Inductive and Deductive Reasoning

Determine if each conclusion follows logically from the premises, and state whether the reasoning is inductive or deductive.

(a) *Premise:* My coat is tan.
Premise: Bob's coat is tan.
Premise: Di's coat is tan.
Conclusion: All coats are tan.

Because we are reasoning from three specific examples and drawing a general conclusion, the process involves inductive reasoning. It is obvious that the conclusion does not logically follow from the premises.

(b) *Premise:* All athletes are in good physical condition.
Premise: Shelly is an athlete.
Conclusion: Shelly is in good physical condition.

Practice Exercise 3

Determine if each conclusion follows logically from the premises, and state whether the reasoning is inductive or deductive.

(a) *Premise:* This year is leap year.
Conclusion: Next year will not be leap year.

(b) *Premise:* The Boston Celtics won the NBA Championship in the years 1960–66.
Premise: The Boston Celtics won the NBA Championship in the years 1968–69.
Conclusion: The Celtics won the NBA Championship every year during the 1960s.

In this case, if we accept the premises, then the conclusion logically follows. This is an example of deductive reasoning. Notice that the conclusion may be true or false; we do not actually know whether Shelly is in good condition. However, this is beside the point because the conclusion logically follows from the premises. Certainly, if the premises are true and a conclusion follows, then the conclusion is also true. In mathematics we assume premises are true so that deduced conclusions are also true statements.

Answers: (a) Using deductive reasoning, the conclusion follows logically from the given premise and the accepted but unstated premise that leap year occurs every four years. (b) The conclusion does not necessarily follow from the premises. It was reached by inductive reasoning. It is a false statement because in 1967 the Philadelphia 76ers won the NBA Championship.

We now give an illustration of a practical application of deductive reasoning. Assume the following warranty is given with a new radio.

> This radio is warranted for one year from the date of purchase against defects in materials or workmanship. During this period any such defects will be repaired or the radio will be replaced at the company's option without charge. This warranty is void in the case of misuse or negligence.

Assume the premises stated in this warranty and answer the following questions.

(a) Bill purchased a radio and when he got home and opened the box, he discovered that the case was cracked. What could Bill conclude?

Because Bill had just purchased the radio, one year had not gone by. He had not been negligent, nor had he misused the radio. He concluded that the company would either repair or replace the radio.

(b) Beth purchased a radio and fourteen months later the digital display burned out. What could Beth conclude?

Because fourteen months exceeds the time period of the warranty, Beth concluded that the company would not be required by the warranty to repair or replace the radio.

(c) Jamie purchased a radio and two months later left it outside after listening to it while sunbathing. That night it rained, and the next day the radio would not play. What could Jamie conclude?

Although the radio became defective during the warranty period, the defect was due to Jamie's negligence. The company would not be required by the warranty to repair or replace it.

12.1 EXERCISES A

Use inductive reasoning in Exercises 1–6 to determine the next element in each list.

1. 3, 8, 13, 18, 23

2. 14, 12, 10, 8, 6

3. 1, 3, 9, 27, 81

4 1, 2, 4, 8, 16

5. 1, 1, 2, 3, 5, 8, 13, 21

6. S, M, T, W, T, F

In Exercises 7–11, determine if each conclusion follows logically from the premises, and state whether the reasoning is inductive or deductive.

7. *Premise:* If you are a mathematics major, then you can compute discounts on sale items.
Premise: Becky is a mathematics major.
Conclusion: Becky can compute discounts on sale items.

8. *Premise:* If you are a mathematics major, then you can compute discounts on sale items.
Premise: Becky can compute discounts on sale items.
Conclusion: Becky is a mathematics major.

9. *Premise:* It rained on Tuesday.
Premise: It rained on Wednesday.
Conclusion: It will rain on Thursday.

10. *Premise:* If I buy a car, then it will be a Buick.
Premise: If I receive a check from my father, then I will buy a car.
Premise: I received a check from my father.
Conclusion: I will buy a Buick.

11. *Premise:* If it is a frog, then it is green.
Premise: If it hops, then it is a frog.
Conclusion: If it hops, then it is green.

12. Suppose we add the following premise to those given in Exercise 11. What conclusion can you reach?
Premise: It hops.

The puzzles in Exercises 13–17 are classics, and solving them requires a certain amount of deductive reasoning. Some of these puzzles are quite challenging. Do not be discouraged if you have trouble finding solutions immediately. We hope the puzzles will get you thinking, and also provide a bit of entertainment.

13. Mary has two current coins in her purse. Together they total 55¢. One is not a nickel. What are the two coins? (*Hint:* We do *not* say that neither coin is a nickel.)

14 If you take 5 apples from 8 apples what have you got?

15. We know there are 12 one-cent stamps in a dozen, but how many two-cent stamps are in a dozen?

16. What arithmetic symbol can be placed between 2 and 3 to form a number greater than 2 and less than 3?

17. A boat will carry 200 pounds maximum. How can a man weighing 200 pounds and his two daughters, each of whom weighs 100 pounds, use the boat to reach an island?

ANSWERS: 1. 28 2. 4 3. 243 4. 32 5. 34 6. S 7. follows logically; deductive reasoning 8. does not follow logically; many people can compute discounts on sale items but they are not necessarily mathematics majors 9. does not follow logically; this is inductive reasoning 10. follows logically; deductive reasoning 11. follows logically; deductive reasoning 12. Conclusion: It is green. 13. a nickel and a 50¢ piece (We are told that *one* is not a nickel, not *both* are not nickels.) 14. You have 5 apples. (We are not asked what was left.) 15. 12 16. A decimal point, since 2.3 is greater than 2 and less than 3. 17. The two daughters cross, and one returns with the boat and stays while the man crosses. The other daughter returns with the boat and picks up the remaining daughter.

12.1 EXERCISES B

Use inductive reasoning in Exercises 1–6 to determine the next element in each list.

1. 2, 8, 14, 20, 26

2. 24, 19, 14, 9

3. 1, 5, 25, 125

4. 40, 20, 10, 5

5. 1, 3, 4, 7, 11, 18, 29

6. J, F, M, A, M

In Exercises 7–11, determine if each conclusion follows logically from the premises, and state whether the reasoning is inductive or deductive.

7. *Premise:* If you are a home buyer, then you make payments.
Premise: Doug is a home buyer.
Conclusion: Doug makes payments.

8. *Premise:* If you are a home buyer, then you make payments.
Premise: Doug makes payments.
Conclusion: Doug is a home buyer.

9. *Premise:* Last year I won money in Las Vegas.
Premise: The year before I won money in Las Vegas.
Conclusion: I will win in Las Vegas this year.

10. *Premise:* If you are going to be an engineer, then you will study mathematics.
Premise: If you study mathematics, then you will get a good job.
Premise: Roy is going to be an engineer.
Conclusion: Roy will get a good job.

11. *Premise:* If you are an ogg, then you are an arg.
Premise: If you are a pon, then you are an ogg.
Conclusion: If you are a pon, then you are an arg.

12. Suppose we add the following premise to those given in Exercise 11. What conclusion can you reach?
Premise: You are a pon.

The puzzles in Exercises 13–17 are classics, and solving them requires a certain amount of deductive reasoning. Some of these puzzles are quite challenging. Do not be discouraged if you have trouble finding solutions immediately. We hope the puzzles will get you thinking, and also provide a bit of entertainment.

13. A younger dog and an older dog are in the back yard. The younger dog is the older dog's daughter, but the older dog is not the younger dog's mother. Explain.

14. A rancher had 20 cattle. All but 12 died. How many did he have left?

15. A museum fired an archaeologist who claimed she found a coin dated 300 B.C. Why?

16. The number of marbles in a jar doubles every minute and is full in 10 minutes. When was the jar half full?

17. How many times can you subtract 5 from 25?

12.1 EXERCISES C

1. A judge wishing to convict a defendant puts two pieces of paper in a hat. He tells the jury that if the defendant draws the piece marked "guilty" he will be convicted, but if he draws the piece marked "innocent" he will be set free. The hitch is that the judge wrote "guilty" on both pieces of paper. But when the crafty defendant showed the jury one piece of paper, the judge was forced to let him go free. How did the defendant outwit the judge?

2. You have 3 sacks, each containing 3 coins. Two of the sacks contain real coins and each coin weighs 1 pound. The third contains counterfeit coins and each weighs 1 pound 1 ounce. A scale is available, but it can be used one time, and one time only, to obtain a particular measure of weight. How might you use the scale to determine which sack contains the counterfeit coins? (*Note:* You cannot add or subtract coins to a total since any change of reading on the scale up or down will cause it to zero out.)

12.2 LOGICAL SYSTEMS

STUDENT GUIDEPOSTS

1. Logical Statements
2. Conditional Statements
3. Biconditional Statements
4. Negating Statements

1 LOGICAL STATEMENTS

In the previous section we discussed the differences between inductive and deductive reasoning, and noted that in mathematics, we are primarily interested in the latter. Recall that deductive reasoning involves arriving at particular conclusions based on accepted premises or statements. To understand better how this process relates to mathematics in general, and more specifically to geometry, we need to recognize some of the ways we make statements when we reason deductively, or think logically. In mathematics we must be precise in the manner in which we state facts; it must be clear what is true and what is false. In a sense, we are forced to make *logical statements* according to the following definition.

Logical Statements

A statement is called a **logical statement** if the statement has **truth value,** that is, it is understood to be either true or false, not both.

EXAMPLE 1 LOGICAL STATEMENTS

Determine whether the following are logical statements.

(a) Texas is one of the fifty United States.
This is a logical statement since it has truth value, it is true. There is no confusion as to whether it is true or false.

(b) $2 + 3 = 7$
This is a logical statement, it has truth value, it is false.

PRACTICE EXERCISE 1

Are the following logical statements?

(a) Christmas day is in December.

(b) $5 - 3 = 1$

(c) A bird in the hand is worth two in the bush.
This is not a logical statement, its truth value is open to question. Statements such as this are vague and difficult to interpret. Such imprecise statements are avoided in mathematics.

(d) John is a good person.
Descriptive words such as "good" are open to interpretation. As a result, a simple statement such as this does not really have truth value. However, within a certain context, we might be able to agree that this statement is true or false, and, in this context, it might be considered to be a logical statement.

(e) All dogs are animals.
This is a logical statement, it is true.

(f) Some cars are painted white.
This is a logical statement, it is true.

(c) Still waters run deep.

(d) It is hot today.

(e) All men are mortal.

(f) Some people live in Arkansas.

Answers: (a) yes (b) yes (c) no (d) within a particular context, yes (e) yes (f) yes

2 CONDITIONAL STATEMENTS

The two statements in Example 1(e) and (f) are of special importance. The words "all" and "some" are used often in statements in mathematics, and their meaning must be well understood. For a statement of the form

All A are B,

to be true, every object in the category described by A must also be an object in category B. One way to say this is "If something satisfies A, then it also satisfies B." In this form, we are making a **conditional statement,** a statement that involves using the words "If" something, "then" something. Most of the important statements in mathematics have this form. For example, we say:

If a number is divisible by 6, *then* it is divisible by 2.
If a number is rational, *then* it can be written as a fraction.
If x satisfies the equation $2x - 6 = 0$, *then* $x = 3$.

The **converse** of a conditional "If A, then B," is the conditional "If B, then A." If a given conditional is true, its converse can be either true or false. For example, the converse of the first statement above, "If a number is divisible by 2, then it is divisible by 6" is false, although the original conditional is true since, for example, 4 is divisible by 2 but not divisible by 6. However, the converse of the second statement, "If a number can be written as a fraction, then it is rational" is true and the original statement is also true. One of the greatest fallacies of logic involves "assuming the converse." That is, we often assume that if a given conditional is true, then its converse is also true.

EXAMPLE 2 Converse of a Statement

Give the converse of each statement. Is the given statement true or false? Is the converse true or false?

(a) If $x = 5$, then $x + 1 = 6$.
The converse is: If $x + 1 = 6$, then $x = 5$.
The statement is true, and the converse is true.

Practice Exercise 2

Give the converse of each statement. Is the statement true or false? Is the converse true or false?

(a) If $2x = 10$, then $x = 5$.

(b) If I plan to be an architect, then I will study mathematics.

(b) If a person is the U.S. President, then he/she is a U.S. citizen.
The converse is: If a person is a U.S. citizen, then he/she is the President of the U.S.
The statement is true, but the converse is false.

Answers: (a) If $2x = 10$, then $x = 5$; true; true (b) If I study mathematics, then I plan to be an architect; true; false

3 BICONDITIONAL STATEMENTS

A statement in mathematics that combines a conditional and its converse is called a **biconditional.** For example, we might say:

If x is even, then x is divisible by 2,
and if x is divisible by 2, then x is even.

Such statements are often simplified using the phrase "if and only if" which is abbreviated "iff." Thus, we might state:

x is even if and only if x is divisible by 2.

or

x is even iff x is divisible by 2.

A biconditional will be true only when the conditional and its converse that make it up are both true or both false.

A statement of the form "Some A are B" claims that there is at least one object in category A that is also in category B. For example, the statement

Some prime numbers *are* even.

means that there is at least one even prime. This statement is of course true, since 2 is a prime number that is even. However, the statement does not claim that *all* primes are also even, which of course is not a true statement.

4 NEGATING STATEMENTS

Sometimes in mathematics we need to *negate* a statement. The **negation** of a given statement is formed by adding the word "not" either specifically, or in an implied manner. For example, the negation of the statement

My car is a Buick.

is the statement

My car is not a Buick.

In this case, we simply added the word *not,* which negated the verb in the sentence. Sometimes a negation can be formed in a more implied manner by changing the words of the sentence without using *not*. For example, to negate

I am guilty.

we might state

I am innocent.

Negating statements that involve the word *all,* that is, a conditional statement,

All A are B. *or* If A then B.

is a bit more challenging. Consider, for example, the statement:

All dogs have fleas. *or* If an animal is a dog, then it has fleas.

One way to negate this statement is to say

Not all dogs have fleas. *or* It is not true that, if an animal is a dog, then it has fleas.

But what does this really mean? Since the statement "All dogs have fleas" is clearly false, we would like its negation to be true. Our first impression might be that the negation is the same as

All dogs do not have fleas. *or* If an animal is a dog, then it does not have fleas.

However, since this statement is also clearly false, it could not serve as the required negation. After reflecting a bit, we might try.

Some dogs do not have fleas. *or* This animal is a dog, and it does not have fleas.

Since this statement is true, the opposite truth value of the original, and since it seems to capture the feeling of "making negative what was given originally," we agree that this is a way to negate statements of this type.

Negating Statements

The negation of "All *A* are *B*" is

"Some *A* are not *B*."

Using the alternative form involving conditionals, the negation of "If *A*, then *B*" is

"*A* and not *B*."

EXAMPLE 3 Negation of a Statement

Give the negation of each statement.

(a) The Celtics won the championship.
The negation is: The Celtics did not win the championship.

(b) All politicians are honest.
The negation is: Some politicians are not honest.
Or we would probably say: Some politicians are dishonest.

(c) If x is the weight of the person, then $x \geq 0$.
The negation is: x is the weight of the person and $x < 0$.

Practice Exercise 3

Give the negation of each statement.

(a) Today is Friday.

(b) All cats eat mice.

(c) If x is positive, then $2x > 0$.

Answers: (a) Today is not Friday. (b) Some cats do not eat mice. (c) x is positive and $2x \leq 0$.

In the previous section we considered conclusions that were obtained deductively from a set of premises. If a conclusion necessarily follows when a set of premises are assumed, we say that the conclusion is **valid.** Obtaining valid conclusions plays a major role in mathematics. We should note that there is a difference between "valid" and "true." A conclusion can be valid, but not necessarily a true statement. For example, if we accept as premises:

Premise: All athletes are healthy.
Premise: Jerry is an athlete.

then a valid conclusion is:

Conclusion: Jerry is healthy.

Whether the statement "Jerry is healthy" is true or false, of course, is really not known. However, this does not deter us from the fact that the statement is still a valid conclusion *if* we accept the two premises. Generally, however, we can say that if the accepted premises are also true statements, then any valid conclusion will also be true. In mathematics, when we reason deductively, we do assume that our premises are true statements, making our conclusions both valid and true.

To gain more practice at obtaining valid conclusions, we now consider a special logical form called a **syllogism.** The classic example of a syllogism is:

Premise: All men are mortal.
Premise: Socrates is a man.
Conclusion: Socrates is mortal.

Most of us would agree that the conclusion in this syllogism is valid. One visual method for showing this involves using **Euler diagrams,** named for Leonard Euler (1707–83), a famous Swiss mathematician. We begin by drawing a diagram using circles that represents the first premise. Since "men" is included in the set of "mortals," we draw the circle representing "men" inside the circle representing "mortals."

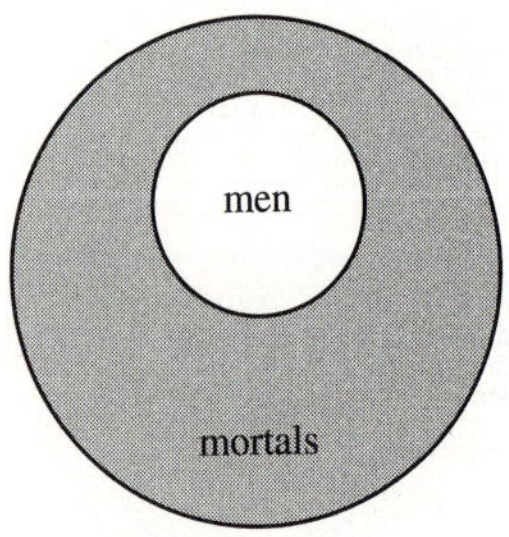

Figure 12.1

The second premise, "Socrates is a man" would seem to indicate that Socrates should be placed inside the circle representing "men." We do this by placing an "x" in this circle.

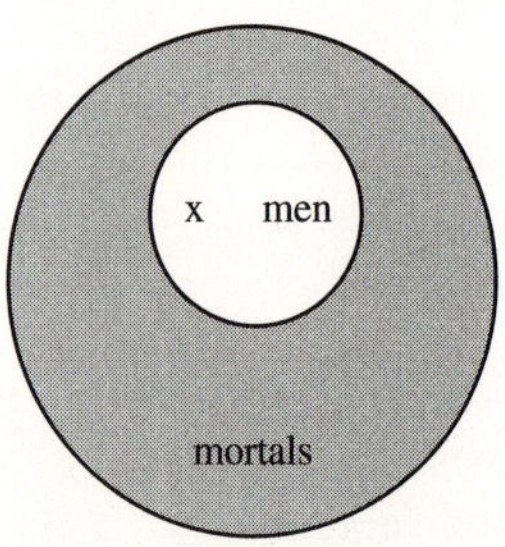

Figure 12.2

Then Figure 12.2, the combination of the two premises, would seem to indicate that the conclusion is indeed valid since the "x" representing Socrates is also included in the circle representing "mortals." We have verified that the conclusion is valid using an Euler diagram. Euler diagrams are especially useful when premises involve the word "some" as shown in the next example.

EXAMPLE 4 Using Euler Diagrams

Use an Euler diagram to determine if the given conclusion is valid.

Premise: Some students study hard.
Premise: I am a student.
Conclusion: I study hard.

The first premise can be sketched using an Euler diagram as in Figure 12.3. Notice that we simply intersect the circles this time; *some* does not mean *all,* so the circle for "students" cannot be placed inside the circle for "persons who study hard."

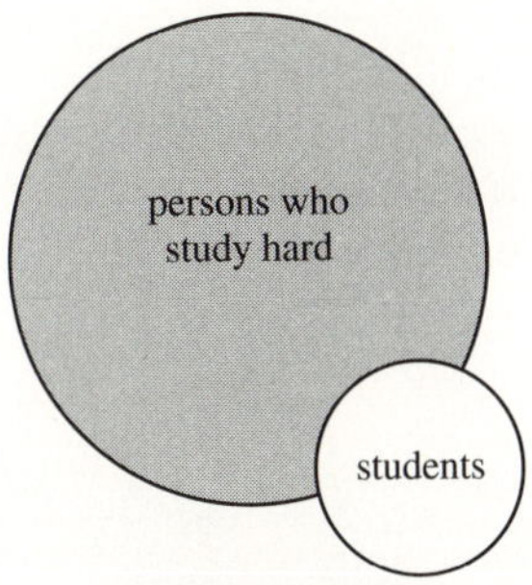

Figure 12.3

Then "I" can be placed inside the circle for students in one of two ways as shown in Figure 12.4. Since one of these ways does not force the conclusion to necessarily follow, we must conclude that "I study hard" is not a valid conclusion.

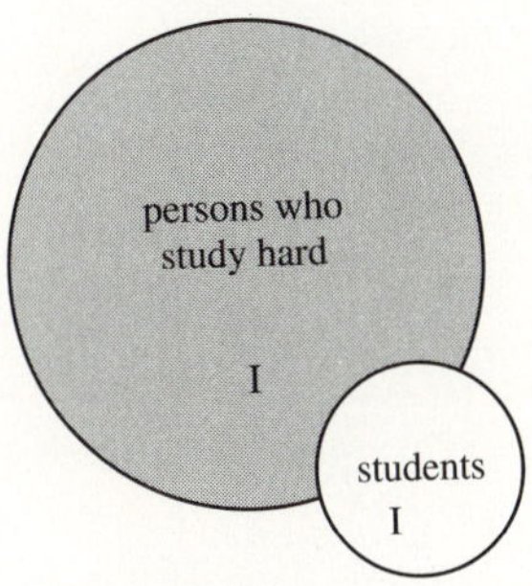

Figure 12.4

Practice Exercise 4

Use an Euler diagram to determine if the given conclusion is valid.
Premise: Some athletes eat oatmeal.
Premise: Sue is an athlete.
Conclusion: Sue eats oatmeal.

Answer: The conclusion is not valid.

Euler diagrams provide one technique for determining whether a conclusion is valid or not. Another technique that is useful for certain types of situations involves using an array or grid to keep track of the given premises, which can help us discover a valid conclusion that might not be stated originally. We illustrate this technique below. Assume the following premises:

(a) Art, Bev, Cheryl, Dot, and Ed all attend the same college where one is a freshman, one is a sophomore, one is a junior, one is a senior, and one is a graduate student.

(b) Art, Bev, and Cheryl have not yet completed their undergraduate work.

(c) Bev is one year ahead of Ed.

(d) Art is not a freshman.

(e) Ed is not a freshman.

(f) Art is in a higher class than Ed.

We must try to conclude each student's class in college. To keep track of the information given we can use the following grid.

	Fr	So	Jr	Sr	Gr
Art					
Bev					
Cheryl					
Dot					
Ed					

We agree to fill the positions in the grid with an ''x'' or an ''o'' to indicate that the statement is true or false, respectively. Premises (b), (d), and (e), allow us to fill the grid as follows.

	Fr	So	Jr	Sr	Gr
Art	o				o
Bev					o
Cheryl					o
Dot					
Ed	o				

Premise (f) allows us to place two more ''o'' entries in Ed's row as shown below, followed by an ''x'' given that Dot is the graduate student. Once we place the ''x'' we can fill out the rest of the row with ''o'' entries.

	Fr	So	Jr	Sr	Gr
Art	o				o
Bev					o
Cheryl					o
Dot	o	o	o	o	x
Ed	o			o	o

Premises (c) and (f) tell us that Bev is a year ahead of Ed, and that Art is in a higher class than Ed, so we must complete the bottom row as follows.

	Fr	So	Jr	Sr	Gr
Art	o				o
Bev					o
Cheryl					o
Dot	o	o	o	o	x
Ed	o	x	o	o	o

We can then place the "o's" in the remaining slots in the sophomore column. Also, premise (c) allows us to place an "x" as follows.

	Fr	So	Jr	Sr	Gr
Art	o	o			o
Bev		o	x		o
Cheryl		o			o
Dot	o	o	o	o	x
Ed	o	x	o	o	o

We can then complete the remaining positions in the row for Bev and the column for juniors to obtain:

	Fr	So	Jr	Sr	Gr
Art	o	o	o		o
Bev	o	o	x	o	o
Cheryl		o	o		o
Dot	o	o	o	o	x
Ed	o	x	o	o	o

Thus, we are forced to place an "x" in the first row showing Art to be the senior and an "x" in the third row showing Cheryl to be the freshman. The completed grid is given below.

	Fr	So	Jr	Sr	Gr
Art	o	o	o	x	o
Bev	o	o	x	o	o
Cheryl	x	o	o	o	o
Dot	o	o	o	o	x
Ed	o	x	o	o	o

Thus, Art is the senior, Bev is the junior, Cheryl is the freshman, Dot is the graduate student, and Ed is the sophomore.

This brief introduction to the study of logical systems will help you appreciate the review of some of the basic concepts of geometry presented in the following sections.

12.2 EXERCISES A

In Exercises 1–6, determine if each is a logical statement.

1. Mickey Mouse is a cartoon character.

2. Joe Montana is a quarterback in the NFL.

3. $(4 + 2) + 7 = 4 + (2 + 7)$

4. There is no sense crying over spilt milk.

5. All cows eat hay.

6. Some books have over 500 pages.

In Exercises 7–12, give the converse of each statement.

7. If Mary walks to work, then she will be healthy.

8. If x is odd, then x is not a multiple of 2.

9. If it is a chicken, then it lays eggs.

10. All boys like girls.

11. All teachers are not wealthy.

12. A person is wise, if he saves his money.

In Exercises 13–16, write each statement in the form of a biconditional using "iff."

13. If Joe loves Lucy, then he will be happy, and if Joe is happy, then he loves Lucy.

14. If $x = 5$, then $x + 2 = 7$, and if $x + 2 = 7$, then $x = 5$.

15. If we win the game, then we will have played hard, and conversely.

16. All cats meow, and all animals that meow are cats.

In Exercises 17–22, give the negation of each statement.

17. This computer has 40k of memory.

18. Larry is an excellent student.

19. All frogs are green.

20. All rockets do not hit their target.

21. If Cindy is smart, she will stay in school.

22. If x is negative, then x is not a natural number.

In Exercises 23–28, use an Euler diagram to determine if the given conclusion to each syllogism is valid.

23. *Premise:* All home buyers make payments.
Premise: Barry is a home buyer.
Conclusion: Barry makes payments.

24. *Premise:* All home buyers make payments.
Premise: Barry makes payments.
Conclusion: Barry is a home buyer.

25. *Premise:* Some cows give milk.
Premise: Bossy is a cow.
Conclusion: Bossy gives milk.

26. *Premise:* Some cows give milk.
Premise: Bossy gives milk.
Conclusion: Bossy is a cow.

27. *Premise:* All cars have four wheels.
Premise: Some vehicles with four wheels are red.
Conclusion: Some cars are red.

28. *Premise:* All cars have four wheels.
Premise: Some cars are red.
Conclusion: Some red vehicles have four wheels.

29. If a conditional statement is true, is its converse always true?

30. Assume the following premises, and use a grid to determine which person drives which type of car.
 (a) Ann, Beth, and Cal drive a compact, midsize, and luxury car, not necessarily in that order.
 (b) Ann earns more money than Cal.
 (c) Beth lives in New York.
 (d) The person with the highest income drives the luxury car.
 (e) Cal is the eldest of the three.
 (f) The youngest lives in California and drives the midsize car.

FOR REVIEW

31. Use inductive reasoning to determine the next element in the list.

$$-2, 4, -6, 8, -10$$

32. Determine if the conclusion follows from the premises, and state whether the reasoning is inductive or deductive.
Premise: Last year I paid my taxes on time.
Premise: The year before I paid my taxes on time.
Conclusion: This year I will pay my taxes on time.

ANSWERS: 1. is a logical statement 2. is a logical statement 3. is a logical statement 4. not a logical statement 5. is a logical statement 6. is a logical statement 7. If Mary is healthy, then she walks to work. 8. If x is not a multiple of 2, then x is odd. 9. If it lays eggs, then it is a chicken. 10. All girls like boys. 11. All people who are not wealthy are teachers. 12. If a person is wise, then he saves money. 13. Joe loves Lucy iff he is happy. 14. $x = 5$ iff $x + 2 = 7$. 15. We will win the game iff we play hard. 16. An animal is a cat iff it meows. 17. This computer does not have 40k of memory. 18. Larry is not an excellent student. 19. Some frogs are not green. 20. Some rockets hit their target. 21. Cindy is smart and she will not stay in school. 22. x is negative and x is a natural number. 23. valid 24. not valid 25. not valid 26. not valid 27. not valid 28. valid 29. no 30. Ann drives the midsize car, Beth drives the luxury car, and Cal drives the compact car. 31. 12 32. does not follow logically (is not a valid conclusion); inductive reasoning

12.2 EXERCISES B

In Exercises 1–6, determine if each is a logical statement.

1. John Wayne often starred as a cowboy in the movies.

2. President Kennedy was assassinated in Dallas.

3. $3(7 + 8) = 21 + 24$

4. A watched pot never boils.

5. All oceans have beaches.

6. Some airlines give discount fares.

In Exercises 7–12, give the converse of each statement.

7. If Johnny drinks his orange juice, he will not get sick.

8. If x is a natural number, then $x \neq 0$.

9. If I play blackjack, then I will win.

10. All whales live in the sea.

11. All lawyers are not expensive.

12. If she is retired, then she has lots of spare time.

In Exercises 13–16, write each statement in the form of a biconditional using "iff."

13. If Donna wins the lottery, then she will take a cruise around the world, and if Donna takes a cruise around the world, then she won the lottery.

14. If $x = -6$, then $x - 1 = -7$, and if $x - 1 = -7$, then $x = -6$.

15. If we buy a video recorder, then we will tape the Super Bowl Game, and conversely.

16. All camels have humps, and all animals with humps are camels.

In Exercises 17–22, give the negation of each statement.

17. Ted has a $50 savings bond.

18. My motorhome has a gasoline generator for electricity.

19. All clouds are white.

20. All cheerleaders are not gymnasts.

21. If Terry has the money, he will hire a tutor.

22. If x is a fraction, then x is not an integer.

In Exercises 23–28, use an Euler diagram to determine if the given conclusion to each syllogism is valid.

23. *Premise:* All roommates are noisy.
Premise: Stephanie is my roommate.
Conclusion: Stephanie is noisy.

24. *Premise:* All roommates are noisy.
Premise: Stephanie is noisy.
Conclusion: Stephanie is my roommate.

25. *Premise:* Some mammals live in a tree.
Premise: Bonzo is a mammal.
Conclusion: Bonzo lives in a tree.

26. *Premise:* Some mammals live in a tree.
Premise: Bonzo lives in a tree.
Conclusion: Bonzo is a mammal.

27. *Premise:* All birds can fly.
Premise: Some things that fly have feathers.
Conclusion: Some things with feathers are birds.

28. *Premise:* All birds can fly.
Premise: Some things with feathers are birds.
Conclusion: Some things with feathers can fly.

29. If a conditional statement is false, is its converse always false?

30. The following puzzle has been attributed to Ernest Dudeney (1857–1930), called by some the greatest puzzle writer of all time. Assume the following premises, and use two grids to determine the name of the engineer, the brakeman, and the fireman.

(a) Smith, Jones, and Rodriquez are the engineer, brakeman, and fireman on a train, not necessarily in that order.
(b) Riding on the train are three passengers with the same last names as the crewmembers, identified as Passenger Smith, Passenger Jones, and Passenger Rodriquez in the following statements.
(c) The brakeman lives in Denver.
(d) Passenger Rodriquez lives in San Francisco.
(e) Passenger Jones long ago forgot all the algebra that he learned in high school.
(f) The passenger with the same name as the brakeman lives in New York.
(g) The brakeman and one of the passengers, a professor of mathematical physics, attend the same church.
(h) Smith beat the fireman in a game of tennis at a court near their homes.

[*Hint:* For one grid use the names of the crewman and the roles they have as crewmembers. For the second grid, use the names of the passengers and the names of the three cities in the premises.]

FOR REVIEW

31. Use inductive reasoning to determine the next element in the list.

$$-1, \frac{1}{2}, -\frac{1}{3}, \frac{1}{4}, -\frac{1}{5}$$

32. Determine if the conclusion follows from the premises, and state whether the reasoning is inductive or deductive.
Premise: If I own a ranch, then I run cattle.
Premise: If I run cattle, then I eat lots of beef.
Premise: I do own a ranch.
Conclusion: I eat lots of beef.

12.2 EXERCISES C

Determine how to negate a statement of the form "Some A are B" and use your technique to negate the statements given in Exercises 1–4.

1. Some apples are red.

2. Some equations have infinitely many solutions.

3. Some equations have no solution.

4. Some evil people will do anything to gain an advantage.

12.3 LINES AND ANGLES

STUDENT GUIDEPOSTS

1. Segments, Lines, and Rays
2. Coinciding, Intersecting, and Parallel Lines
3. Angles
4. Measuring Angles
5. Classifying Angles
6. Perpendicular Lines
7. Parallel Lines Cut by a Transversal

1 SEGMENTS, LINES, AND RAYS

A **point** is a precise location in space often symbolized by a dot and labeled with a capital letter, such as the point A shown at the right. •A

Figure 12.5 displays four geometric figures. An arrowhead indicates that the curve continues in that direction without end.

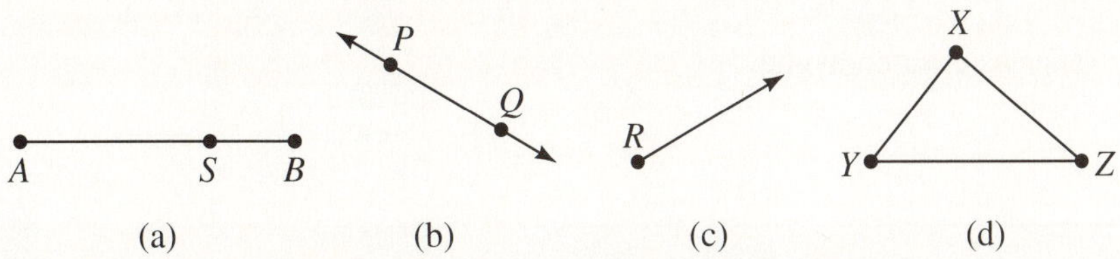

Figure 12.5 Geometric Figures

A **segment** is a geometric figure made up of two **endpoints** and all points between them. The segment with endpoints A and B in Figure 12.5(a) can be denoted by either $\overline{AB}$ or $\overline{BA}$. The point S is said to be **on the segment** $\overline{AB}$. Figure 12.5(b) represents a (straight) **line,** which consists of the points on the segment $\overline{PQ}$ together with all points beyond P and Q. We denote this line by $\overleftrightarrow{PQ}$ or $\overleftrightarrow{QP}$.

Any two distinct points determine exactly one line. The two distinct points E and F determine the unique line $\overleftrightarrow{EF}$ shown in Figure 12.6. For convenience, we often use a small letter such as l to represent a line.

Figure 12.6

A **ray** is that part of a line lying on one side of a point P on the line, and includes the point P. A ray is shown in color in Figure 12.7. By selecting another point Q on the ray, we have a way to denote the ray by $\overrightarrow{PQ}$. Unlike the notations used for segments and lines, the rays $\overrightarrow{PQ}$ and $\overrightarrow{QP}$ are different. A ray has only one endpoint, which is always written first. Taken together, the rays $\overrightarrow{PQ}$ and $\overrightarrow{QP}$ make up the line $\overleftrightarrow{PQ}$. Another example of a ray is shown in Figure 12.5(c).

Figure 12.7 Ray

EXAMPLE 1 SEGMENTS, LINES, AND RAYS

Use Figure 12.8 to answer each question. Deductive reasoning is used with premises consisting of the definitions given above.

Figure 12.8

(a) Is point C on $\overline{AB}$? No

(b) Is point C on $\overleftrightarrow{AB}$? Yes

(c) Is point C on $\overrightarrow{BA}$? No

(d) What are the endpoints of $\overline{CA}$? A and C

(e) What are the endpoints of $\overrightarrow{CA}$? Only C

(f) What are the endpoints of $\overleftrightarrow{CA}$? There are no endpoints.

(g) Are $\overline{AB}$ and $\overline{BA}$ the same? Yes

(h) Are $\overrightarrow{AB}$ and $\overrightarrow{BA}$ the same? No

PRACTICE EXERCISE 1

Draw two points X and Y and place point Z on $\overleftrightarrow{XY}$ but not on $\overline{XY}$. Use deductive reasoning and this figure to answer the following questions.

(a) Is Z on $\overrightarrow{YX}$?

(b) Is Z on $\overline{YX}$?

(c) Is Y on $\overrightarrow{ZX}$?

(d) What are the endpoints of $\overline{XY}$?

(e) What are the endpoints of $\overrightarrow{XY}$?

(f) What are the endpoints of $\overleftrightarrow{ZX}$?

(g) Are $\overline{YZ}$ and $\overline{ZY}$ the same?

(h) Are $\overrightarrow{ZX}$ and $\overrightarrow{XZ}$ the same?

Answers: (a) yes (b) no (c) yes (d) X and Y (e) only X (f) There are no endpoints. (g) yes (h) no

2 COINCIDING, INTERSECTING, AND PARALLEL LINES

A **plane** is a surface with the property that any two points in it can be joined by a line, all points of which are also contained in the surface. Intuitively, a plane can be thought of as any flat surface such as a blackboard or a desktop. If two lines in a plane represent the same line, they are **coinciding lines.** When two lines share exactly one common point, they are **intersecting lines.** When two distinct lines in a plane do not intersect, the lines are called **parallel lines.** These three situations are illustrated in Figure 12.9.

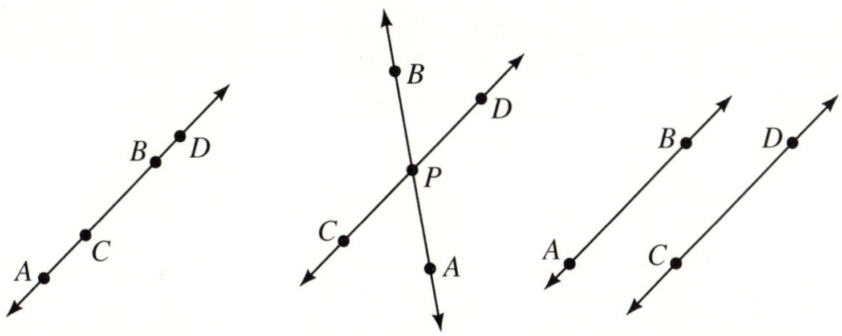

(a) $\overleftrightarrow{AB}$ and $\overleftrightarrow{CD}$ coincide

(b) $\overleftrightarrow{AB}$ and $\overleftrightarrow{CD}$ intersect at point P

(c) $\overleftrightarrow{AB}$ and $\overleftrightarrow{CD}$ are parallel

Figure 12.9

Two segments, two rays, or a segment and a ray can also be called **parallel** if they are on parallel lines. Thus, in Figure 12.9(c) $\overline{AB}$ is parallel to $\overline{CD}$, $\overrightarrow{AB}$ is parallel to $\overrightarrow{DC}$, and $\overline{AB}$ is parallel to $\overrightarrow{CD}$.

3 ANGLES

An **angle** is a geometric figure consisting of two rays that share a common endpoint, called the **vertex** of the angle. In Figure 12.10, the two rays $\overrightarrow{AC}$ and $\overrightarrow{AB}$ form the angle denoted by $\angle BAC$, $\angle CAB$, or simply $\angle A$. The rays $\overrightarrow{AC}$ and $\overrightarrow{AB}$ are the **sides** of $\angle A$.

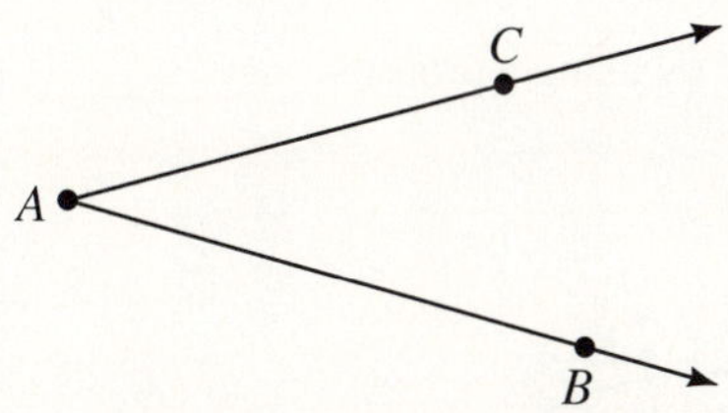

Figure 12.10 $\angle CAB$

We are familiar with measuring the length of segments by some suitable unit of measure such as inch, centimeter, foot, or meter. In order to measure an angle, we need a measuring unit. The most common unit is the degree (°). An angle with measure 0° is formed by two coinciding rays such as $\overrightarrow{AB}$ and $\overrightarrow{AC}$ in Figure 12.11. As the ray $\overrightarrow{AB}$ rotates in a counterclockwise direction from ray $\overrightarrow{AC}$, the two rays form larger and larger angles, as shown in Figure 12.12.

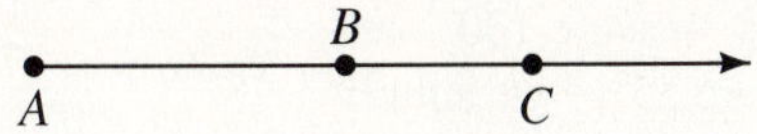

$\angle BAC$ has measure 0°

Figure 12.11

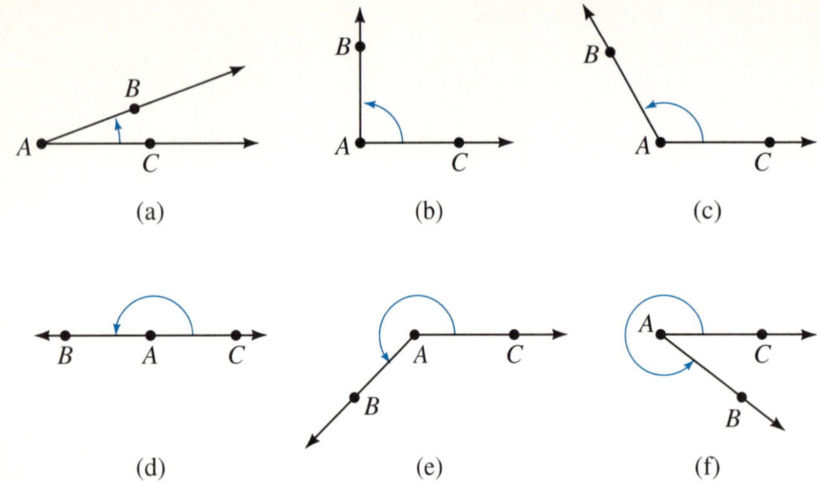

Figure 12.12 Angles

If $\overrightarrow{AB}$ is allowed to rotate completely around until it coincides with ray $\overrightarrow{AC}$ again, the resulting angle is said to measure 360°. Thus, an angle of measure 1° is formed by making $\frac{1}{360}$ of a complete rotation, as in Figure 12.13

Figure 12.13 Angle Measuring 1°

4 MEASURING ANGLES

The approximate measure of an angle in degrees can be found by using a protractor (see Figure 12.14).

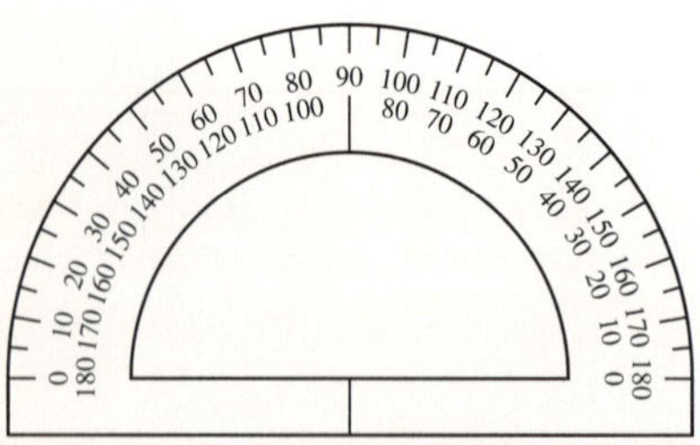

Figure 12.14 Protractor

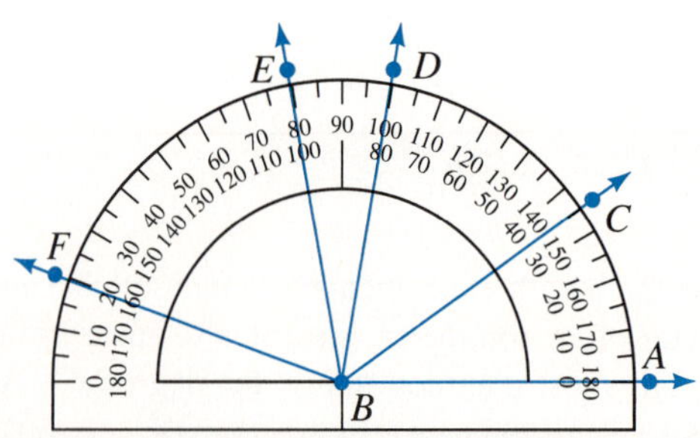

Figure 12.15 Measuring Angles

Figure 12.15 shows how a protractor is used to measure various angles. $\angle ABC$ has measure 35°, $\angle ABD$ has measure 80°, $\angle ABE$ has measure 100°, and $\angle ABF$ has measure 160°. Rather than say "$\angle ABC$ has measure 35°," we simply write $\angle ABC = 35°$. Also, when two angles have the same measure, we simply say that the angles are **equal.**

5 CLASSIFYING ANGLES

When two rays forming an angle are in opposite directions, such as in Figure 12.12(d), the resulting angle is a **straight angle** and has measure 180°. When a ray $\overrightarrow{AB}$ is rotated through one-fourth of a complete revolution from $\overrightarrow{AC}$, as in Figure 12.12(b), the resulting angle is a **right angle** and has measure 90°. Right angles are often marked with a square corner at the vertex, such as $\llcorner$. Angles measuring between 0° and 90°, such as the angle in Figure 12.12(a), are **acute angles.** Angles measuring between 90° and 180°, such as the angle in Figure 12.12(c), are **obtuse angles.**

Two angles with the same measure are **equal** or **congruent.** Two angles whose measures total 90° are **complementary,** and two angles whose measures total 180° are **supplementary.**

When two lines intersect, four angles are formed. The intersecting lines $\overleftrightarrow{AB}$ and $\overleftrightarrow{CD}$ in Figure 12.16 form the four angles $\angle APD$, $\angle DPB$, $\angle BPC$, and $\angle CPA$. $\angle APC$ and $\angle DPB$ are **vertical angles** to each other. The same is true for $\angle APD$ and $\angle BPC$.

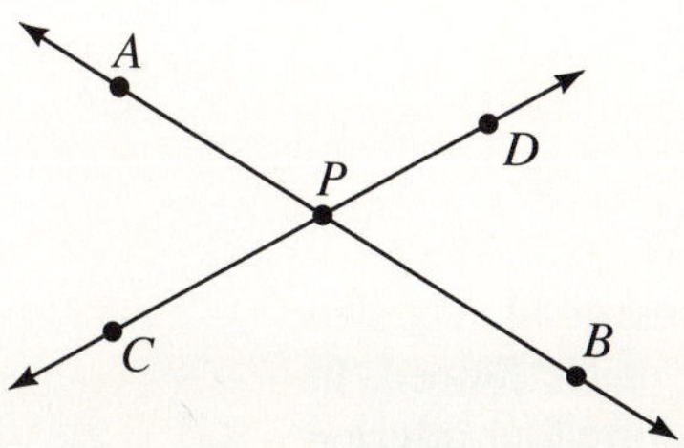

Figure 12.16 Vertical Angles

6 PERPENDICULAR LINES

Vertical angles have the same measure. Thus, in Figure 12.16, $\angle APC$ and $\angle DPB$ are equal, as are $\angle APD$ and $\angle BPC$. When the four angles formed by two intersecting lines all have measure 90°, that is, when they are all right angles, the lines are **perpendicular.**

EXAMPLE 2 Angles and Lines

The following statements refer to Figure 12.17. The premises are the definitions, and we use deductive reasoning to make each statement.

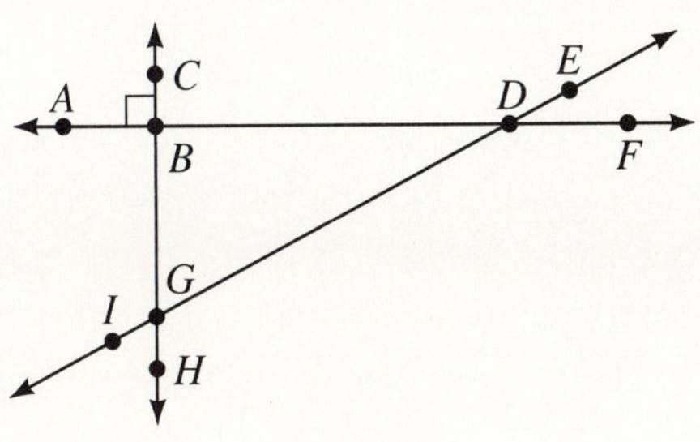

Figure 12.17

(a) $\overleftrightarrow{AB}$ and $\overleftrightarrow{DF}$ are coinciding lines.

Practice Exercise 2

Use the following figure and answer true or false in (a)–(k).

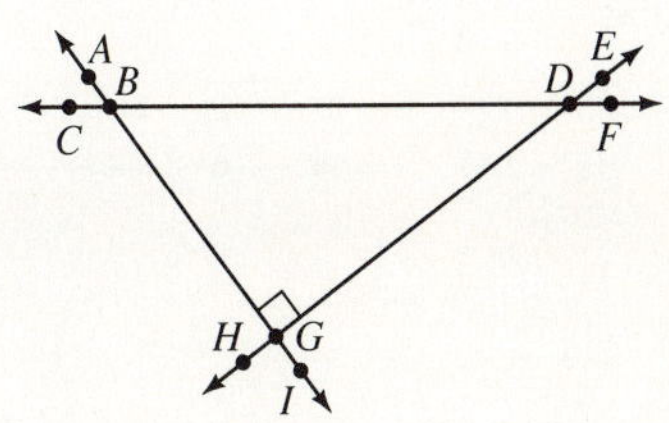

(a) $\overleftrightarrow{HD}$ and $\overleftrightarrow{GE}$ are coinciding lines.

(b) $\overleftrightarrow{AF}$ and $\overleftrightarrow{IE}$ are intersecting lines.

(c) $\overrightarrow{GI}$ and $\overrightarrow{GH}$ are the sides of $\angle IGH$.

(d) $\angle IGD$ is a straight angle.

(e) $\angle ABC$ is a right angle.

(f) $\angle IGB$ and $\angle BGD$ are supplementary.

(g) $\angle EDF$ is an acute angle.

(h) $\angle EDB$ is an obtuse angle.

(i) $\angle IGH$ and $\angle BGD$ are vertical angles.

(j) $\angle EDB$ and $\angle FDG$ are equal (they are vertical angles).

(k) $\overleftrightarrow{AF}$ and $\overleftrightarrow{CH}$ are perpendicular.

(b) $\overleftrightarrow{AI}$ and $\overleftrightarrow{HE}$ are parallel lines.

(c) $\overrightarrow{AB}$ and $\overrightarrow{DB}$ are the sides of $\angle ABD$.

(d) $\angle BGI$ is a straight angle.

(e) $\angle BGD$ is a right angle.

(f) $\angle BDE$ and $\angle EDF$ are complementary.

(g) $\angle ABD$ is an acute angle.

(h) $\angle BDE$ is an obtuse angle.

(i) $\angle ABC$ and $\angle CBG$ are vertical angles.

(j) $\angle DGI$ and $\angle IGH$ are equal.

(k) $\overleftrightarrow{CB}$ and $\overleftrightarrow{HG}$ are intersecting lines.

Answers: (a) true (b) false (c) false (d) true (e) true (f) false (g) false (h) true (i) false (j) true (k) true

7 PARALLEL LINES CUT BY A TRANSVERSAL

A line that intersects each of two parallel lines is called a **transversal.** The two lines are said to be "cut" by the transversal and several angles are formed. In Figure 12.18, transversal $\overleftrightarrow{AD}$ cuts parallel lines $\overleftrightarrow{GF}$ and $\overleftrightarrow{HE}$ to form four **interior angles,** $\angle GBC$, $\angle FBC$, $\angle BCH$, and $\angle BCE$, and four **exterior angles,** $\angle ABG$, $\angle ABF$, $\angle HCD$, and $\angle ECD$. Since $\angle GBC$ and $\angle BCE$ are on alternate sides of the transversal, they are called **alternate interior angles.** Similarly, $\angle FBC$ and $\angle BCH$ are alternate interior angles. In Figure 12.18, $\angle ABF$ and $\angle BCE$ are **corresponding angles.** Three other pairs of corresponding angles are $\angle ABG$ and $\angle BCH$, $\angle GBC$ and $\angle HCD$, and $\angle FBC$ and $\angle ECD$.

Alternate interior angles always have the same measure. Thus, referring to Figure 12.18 again, $\angle FBC$ and $\angle BCH$ have the same measure, as do $\angle GBC$ and $\angle BCE$. Also, corresponding angles always have the same measure so that, for example, $\angle ABF$ has the same measure as $\angle BCF$ in Figure 12.18. (You can draw several figures and use inductive reasoning to conclude that these angles are probably equal, but deductive reasoning is needed to actually prove the statements.)

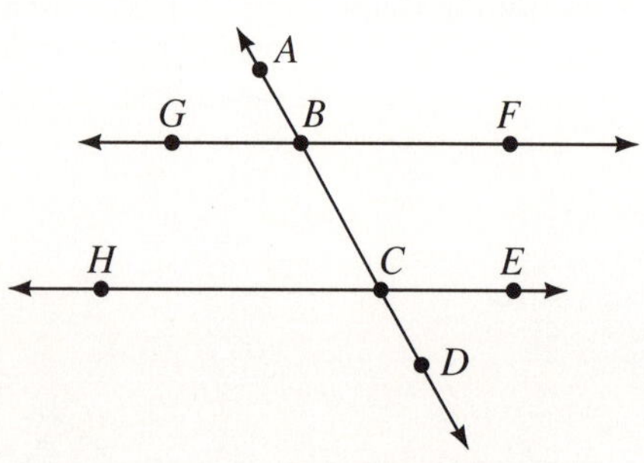

Figure 12.18

EXAMPLE 3 Transversal Cutting Parallel Lines

The following statements refer to Figure 12.19, in which $\overleftrightarrow{AB}$ is parallel to $\overleftrightarrow{DE}$. We use deductive reasoning to make the following statements.

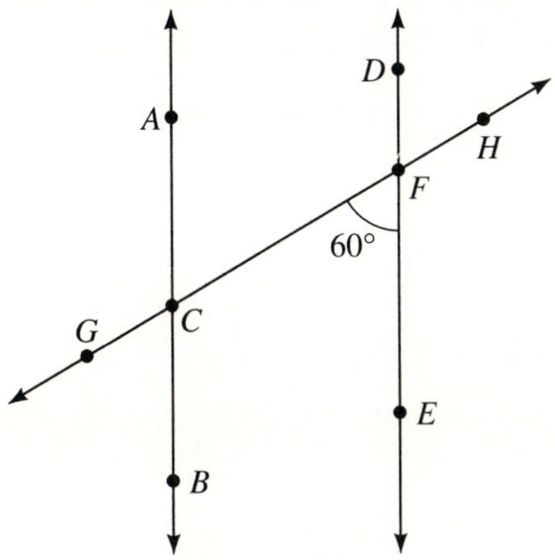

Figure 12.19

(a) $\overleftrightarrow{GH}$ is a transversal.

(b) $\angle DFC$ is an interior angle.

(c) $\angle ACG$ is an exterior angle.

(d) $\angle EFC$ and $\angle ACF$ are alternate interior angles.

(e) $\angle DFC$ has the same measure as $\angle FCB$.

(f) $\angle HFE$ has the same measure as $\angle FCB$.

(g) $\angle DFC$ has measure 120°.

(h) $\angle ACG$ has measure 120°.

(i) $\angle GCB$ has measure 60°.

Practice Exercise 3

Use the following figure, in which $\overleftrightarrow{PS}$ is parallel to $\overleftrightarrow{WU}$, and answer true or false in (a)–(i).

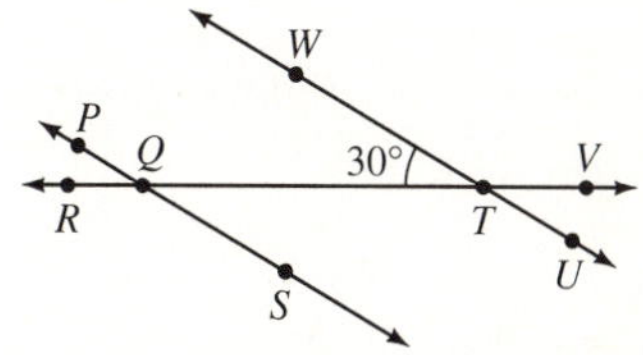

(a) $\overleftrightarrow{QT}$ is a transversal.

(b) $\angle VTU$ is an interior angle.

(c) $\angle RQS$ is an exterior angle.

(d) $\angle WTV$ and $\angle QTU$ are alternate interior angles.

(e) $\angle UTQ$ and $\angle PQT$ have the same measure.

(f) $\angle VTU$ and $\angle PQR$ have the same measure.

(g) $\angle WTV$ has measure 150°.

(h) $\angle RQS$ has measure 30°.

(i) $\angle PQT$ has measure 150°.

Answers: (a) true (b) false (c) true (d) false (e) true (f) true (g) true (h) false (i) true

12.3 EXERCISES A

Use the figure below and deductive reasoning to answer each question in Exercises 1–16.

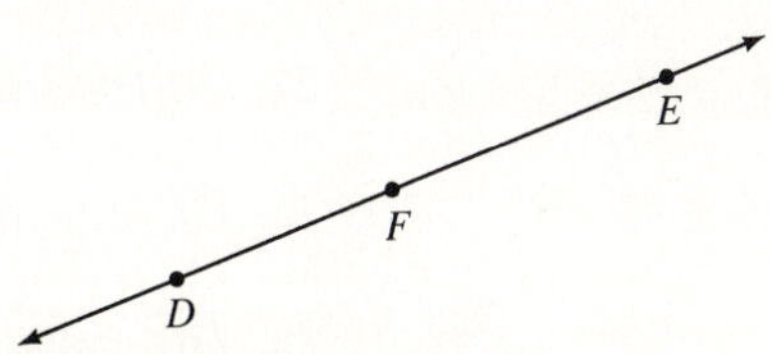

1. Is point E on $\overline{DF}$?

2. Is point E on $\overrightarrow{FD}$?

3. Is point E on $\overleftrightarrow{DF}$?

4. Is point E on $\overrightarrow{DF}$?

5. What are the endpoints of $\overrightarrow{FE}$?

6. What are the endpoints of $\overleftrightarrow{FE}$?

7. What are the endpoints of $\overline{FE}$?

8. Are $\overline{DE}$ and $\overline{ED}$ the same?

9. Are $\overrightarrow{DE}$ and $\overrightarrow{ED}$ the same?

10. Are $\overleftrightarrow{DE}$ and $\overleftrightarrow{ED}$ the same?

11. Do $\overleftrightarrow{DE}$ and $\overleftrightarrow{DF}$ coincide?

12. What is the measure of $\angle DFE$?

13. What is the measure of $\angle EDF$?

14. What is the vertex of $\angle EFD$?

15. Is $\angle EFD$ a right angle?

16. Are $\angle EDF$ and $\angle DFE$ supplementary?

Use the figure with the protractor to determine the measures of the given angles.

17. $\angle ABC$

18. $\angle ABE$

19. $\angle ABD$

20. $\angle ABF$

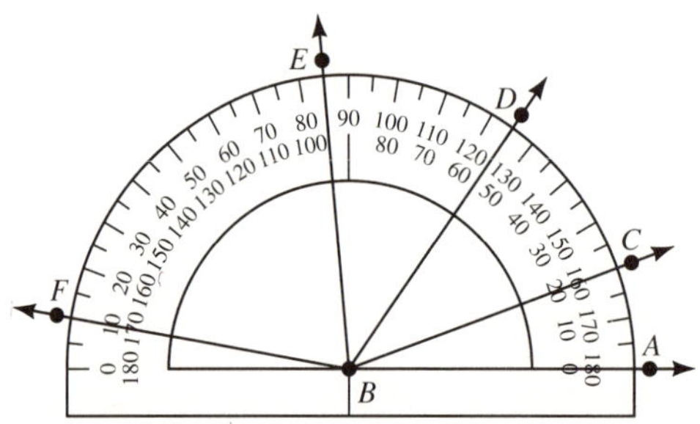

Use the figure below and deductive reasoning to answer true *or* false *in Exercises 21–34.*

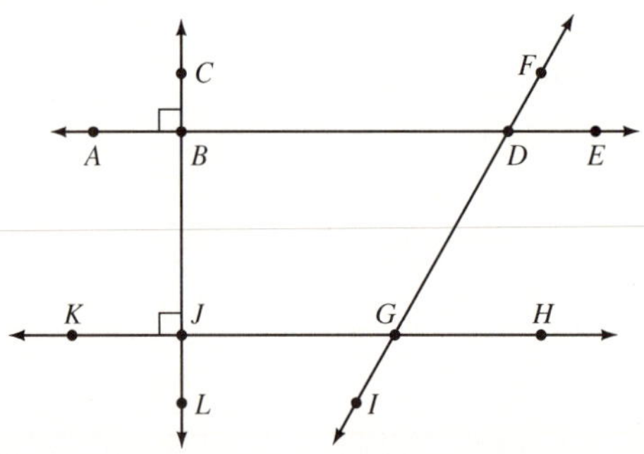

21. $\angle DBJ$ is a right angle.

22. $\angle BDG$ is an acute angle.

23. $\angle DGH$ and $\angle JGI$ are obtuse angles.

24. $\angle FDB$ and $\angle EDG$ have the same measure.

25. $\overleftrightarrow{AE}$ is parallel to $\overleftrightarrow{KH}$.

26. $\overleftrightarrow{CB}$ is perpendicular to $\overleftrightarrow{GH}$.

27. The measure of $\angle GJL$ is 90°.

28. $\overrightarrow{GD}$ and $\overrightarrow{JG}$ are the sides of $\angle JGD$.

29. $\angle IGJ$ and $\angle JGF$ are supplementary.

30. $\overline{BE}$ is parallel to $\overrightarrow{JK}$.

31. $\overleftrightarrow{IF}$ is a transversal that cuts parallel lines $\overleftrightarrow{AE}$ and $\overleftrightarrow{KH}$.

32. $\angle FDE$ and $\angle BDG$ are corresponding angles.

33. $\angle EDG$ and $\angle DGJ$ are alternate interior angles.

34. $\angle BDG$ and $\angle DGH$ have the same measure.

FOR REVIEW

In Exercises 35–36, determine if each conclusion is valid, and state whether the reasoning is inductive or deductive.

35. *Premise:* If the measure of $\angle A$ is 70°, then the measure of $\angle B$ is 20°.
Premise: The measure of $\angle A$ is 70°.
Conclusion: The measure of $\angle B$ is 20°.

36. *Premise:* Point C is on $\overrightarrow{AB}$.
Premise: Point D is on $\overrightarrow{AB}$.
Premise: Point E is on $\overrightarrow{AB}$.
Conclusion: Point F is on $\overrightarrow{AB}$.

ANSWERS: 1. no 2. no 3. yes 4. yes 5. only the point F 6. There are no endpoints. 7. F and E 8. yes 9. no 10. yes 11. yes 12. 180° 13. 0° 14. F 15. no 16. yes 17. 20° 18. 95° 19. 55° 20. 170° 21. true 22. true 23. false 24. true 25. true 26. true 27. true 28. false 29. true 30. true 31. true 32. false 33. true 34. true 35. follows logically; deductive reasoning 36. does not follow logically; inductive reasoning

12.3 EXERCISES B

Use the figure below and deductive reasoning to answer each question in Exercises 1–16.

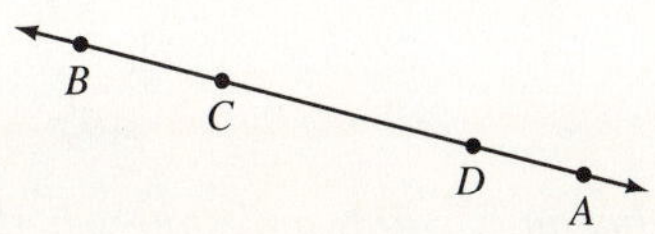

1. Is point C on $\overline{AB}$?

2. Is point B on $\overrightarrow{DC}$?

3. Is point A on $\overleftrightarrow{BD}$?

4. Is A on $\overrightarrow{DC}$?

5. What are the endpoints of $\overrightarrow{DC}$?

6. What are the endpoints of $\overleftrightarrow{DC}$?

7. What are the endpoints of $\overline{DC}$?

8. Are $\overline{BD}$ and $\overline{DB}$ the same?

9. Are $\overrightarrow{BD}$ and $\overrightarrow{DB}$ the same?

10. Are $\overleftrightarrow{BD}$ and $\overleftrightarrow{DB}$ the same?

11. Do $\overleftrightarrow{BA}$ and $\overleftrightarrow{DC}$ coincide?

12. What is the measure of $\angle DBC$?

13. What is the measure of $\angle DCB$?

14. What is the vertex of $\angle DCB$?

15. Is $\angle ADC$ a straight angle?

16. Are $\angle BCD$ and $\angle CBD$ complementary?

Use the figure with the protractor to determine the measures of the given angles.

17. $\angle PQR$

18. $\angle PQT$

19. $\angle PQS$

20. $\angle PQU$

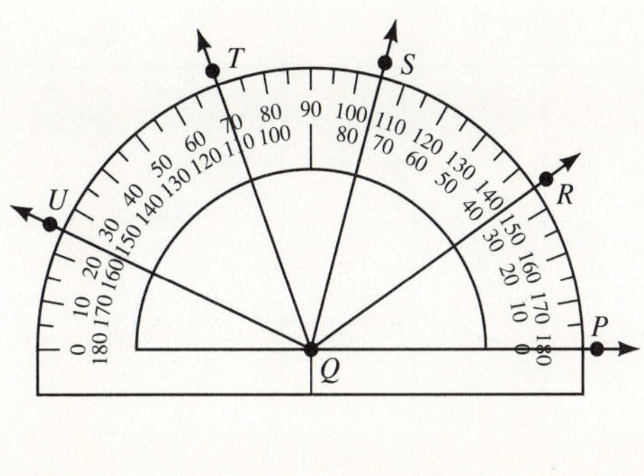

Use the figure below and deductive reasoning to answer true *or* false *in Exercises 21–34.*

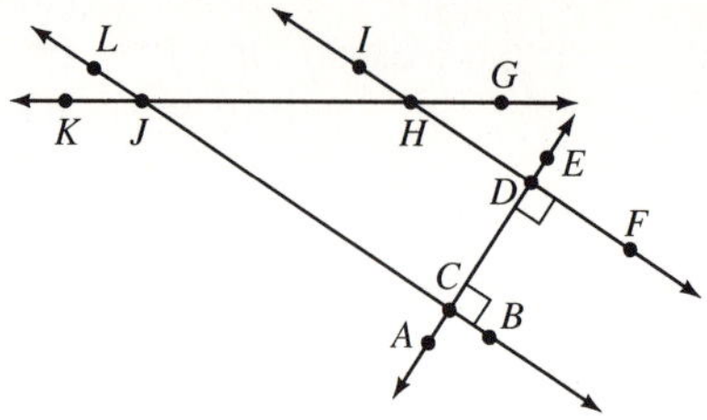

21. $\angle HDC$ is a straight angle.

22. $\angle JHD$ is an obtuse angle.

23. $\angle IHJ$ and $\angle GHD$ are vertical angles.

24. $\angle LJH$ and $\angle KJC$ have the same measure.

25. $\overleftrightarrow{IF}$ is parallel to $\overleftrightarrow{LJ}$.

26. $\overleftrightarrow{AC}$ is perpendicular to $\overleftrightarrow{IH}$.

27. The measure of $\angle JCA$ is 180°.

28. $\overrightarrow{HG}$ and $\overrightarrow{HF}$ are the sides of $\angle GHF$.

29. $\angle JHD$ and $\angle DHG$ are complementary.

30. $\overleftrightarrow{CL}$ is perpendicular to $\overleftrightarrow{IF}$.

31. $\overleftrightarrow{KG}$ is a transversal that cuts parallel lines $\overleftrightarrow{IF}$ and $\overleftrightarrow{LB}$.

32. $\angle GHD$ and $\angle HJC$ are corresponding angles.

33. $\angle DHJ$ and $\angle HJC$ are alternate interior angles.

34. $\angle IHJ$ and $\angle LJK$ have the same measure.

FOR REVIEW

In Exercises 35–36, determine if each conclusion is valid, and state whether the reasoning is inductive or deductive.

35. *Premise:* $\angle A = \angle B$
Premise: $\angle A = \angle C$
Premise: $\angle A = \angle D$
Conclusion: $\angle A = \angle E$

36. *Premise:* If $\overleftrightarrow{AB}$ is perpendicular to $\overleftrightarrow{CD}$, then $\overleftrightarrow{AB}$ is parallel to $\overleftrightarrow{EF}$.
Premise: $\overleftrightarrow{AB}$ is perpendicular to $\overleftrightarrow{CD}$.
Conclusion: $\overleftrightarrow{AB}$ is parallel to $\overleftrightarrow{EF}$.

12.3 EXERCISES C

Use the figure below and deductive reasoning to answer the questions in Exercises 1–8. Lines $\overleftrightarrow{BG}$ *and* $\overleftrightarrow{MI}$ *are parallel.*

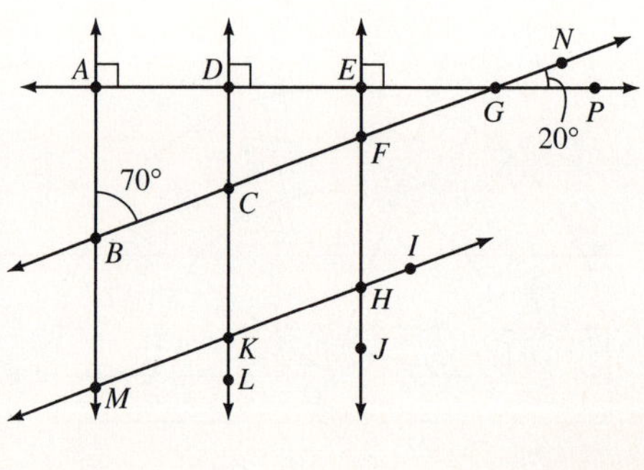

1. What is the measure of $\angle BCK$?

2. What is the measure of $\angle EFG$?

3. What is the measure of $\angle FHI$?

4. What is the measure of $\angle HKL$?

5. What is the measure of $\angle EGF$?

6. What is the sum of the measures of $\angle GEF$, $\angle EFG$, and $\angle FGE$?

7. What is the sum of the measures of $\angle GDC$, $\angle DCG$, and $\angle CGD$?

8. What is the sum of the measures of $\angle CBM$, $\angle BMK$, $\angle MKC$, and $\angle KCB$?

12.4 TRIANGLES

STUDENT GUIDEPOSTS

1. Triangles
2. Classifying Triangles
3. Congruent Triangles
4. Similar Triangles
5. 30°–60° Right Triangles
6. 45°–45° Right Triangles

1 TRIANGLES

In this section we present an introduction to the study of triangles and their properties. Consider three noncollinear points A, B, and C and the corresponding segments $\overline{AB}$, $\overline{BC}$, and $\overline{CA}$, as shown in Figure 12.20. The resulting geometric figure is a **triangle** that has sides $\overline{AB}$, $\overline{BC}$, and $\overline{CA}$, has **vertices** (plural of **vertex**) A, B, and C, and is denoted by $\triangle ABC$ ($\triangle$ represents the word "triangle").

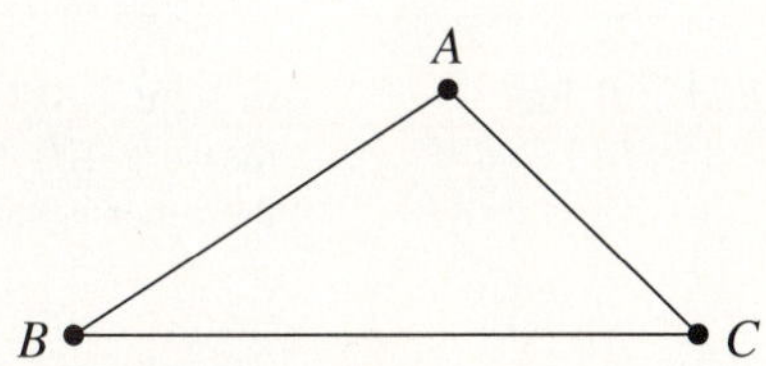

Figure 12.20 Triangle

2 CLASSIFYING TRIANGLES

Triangles are often classified by their angles, as shown in Figure 12.21. They can also be classified by their sides, as in Figure 12.22. We can also describe triangles with a combination of these terms, as in Figure 12.23.

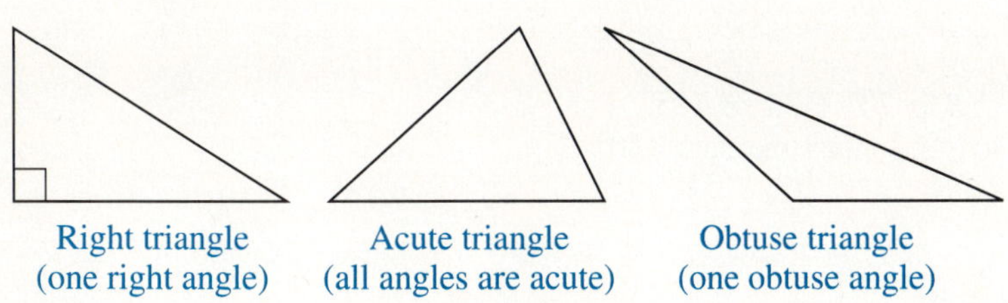

Figure 12.21

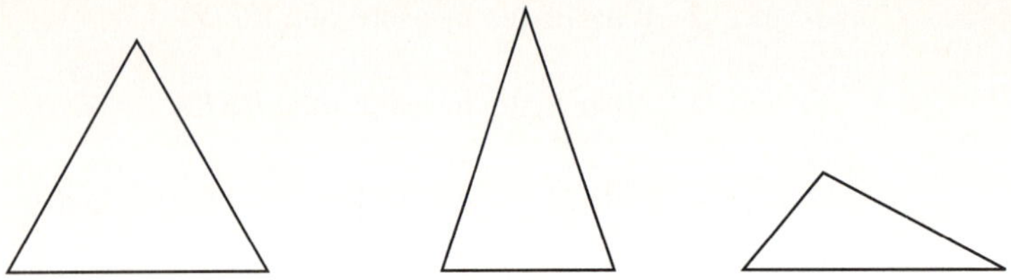

Equilateral triangle (all sides the same length)

Isosceles triangle (two or more sides the same length)

Scalene triangle (all sides of different length)

Figure 12.22

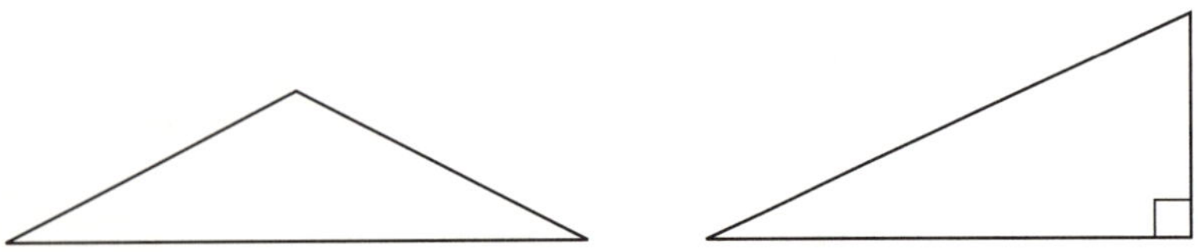

Obtuse isosceles triangle

Right scalene triangle

Figure 12.23

We now consider several properties of triangles.

Sum of Angles

The sum of the measures of the angles in any triangle is 180°.

EXAMPLE 1 Angles of a Triangle

Consider $\triangle ABC$ in Figure 12.24. If $\angle A$ has measure 35° and $\angle B$ has measure 70°, find the measure of $\angle C$.

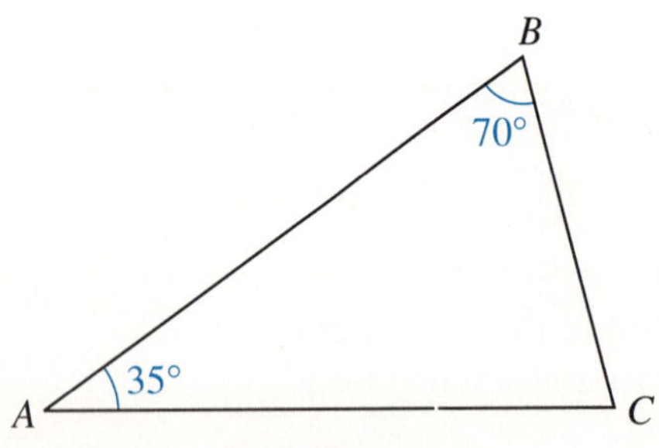

Figure 12.24

The sum of the measures of $\angle A$ and $\angle B$ is $70° + 35° = 105°$. For the sum of the three measures to equal 180°, $\angle C$ must measure $180° - 105° = 75°$.

Practice Exercise 1

In right $\triangle ABC$, one acute angle has measure 42°. What is the measure of the other acute angle?

Answer: 48°

❸ CONGRUENT TRIANGLES

Two triangles are **congruent** if they can be made to coincide by placing one on top of the other, either directly or by flipping one of them over. In Figure 12.25, $\triangle ABC$ and $\triangle DEF$ are congruent, since $\triangle ABC$ could be placed on top of $\triangle DEF$. $\triangle ABC$ and $\triangle GHI$ are also congruent, since if $\triangle GHI$ were flipped over it could be made to coincide with $\triangle ABC$. On the other hand, $\triangle ABC$ and $\triangle JKL$ are not congruent, since they cannot be made to coincide either directly or by flipping one of them over.

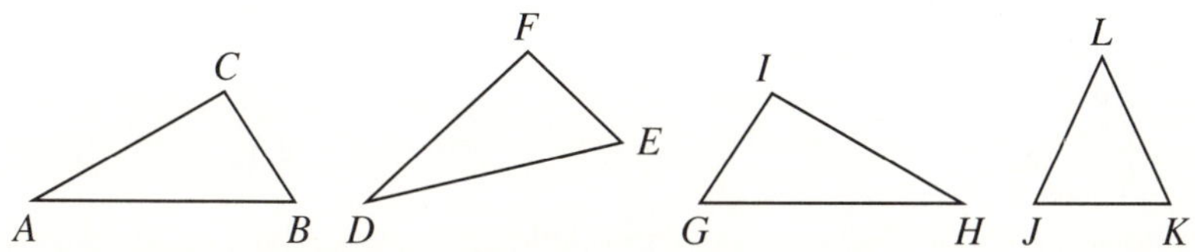

Figure 12.25

More precisely, two triangles will be congruent if the measures of the three sides of one triangle are equal to the measures of the three sides of the other, and the measures of the three angles of one are equal to the measures of the three angles of the other. Fortunately, in practical situations we do not need to establish equality relative to all six parts to show that two triangles are congruent. Three ways to show congruence are summarized as follows.

Congruent Triangles

Two triangles are congruent when one of the following can be shown.

1. Each of the three sides of one triangle is equal in length to a side of the other triangle (congruence by side-side-side, or SSS).
2. Two sides of one triangle are equal in length to two sides of the other triangle, and the angle formed by these sides in one triangle is equal in measure to the angle formed by the corresponding sides in the other triangle (congruence by side-angle-side, or SAS).
3. Two angles of one triangle are equal in measure to two angles of the other triangle, and the side between these two angles in one triangle is equal in length to the corresponding side in the other triangle (congruence by angle-side-angle, or ASA).

EXAMPLE 2 Congruent Triangles

Use deductive reasoning to state why the triangles in Figure 12.26 on the following page are congruent.

Practice Exercise 2

State why the given triangles on the following page are congruent.

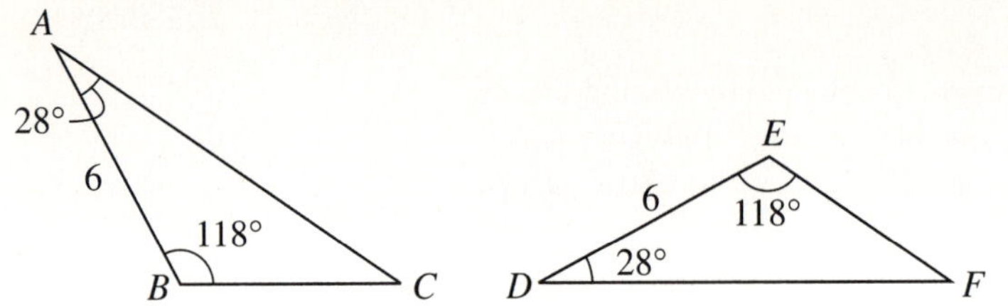

Figure 12.26

Since the measures of $\overline{AB}$ and $\overline{DE}$ are both 6, and $\overline{AB}$ and $\overline{DE}$ are located between the corresponding equal angles, $\angle A$ and $\angle D$, and $\angle B$ and $\angle E$, $\triangle ABC$ is congruent to $\triangle DEF$ by ASA.

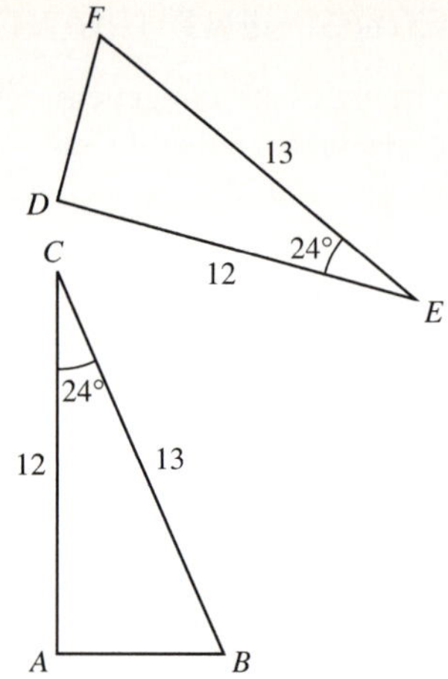

Answer: The triangles are congruent by SAS.

4 SIMILAR TRIANGLES

In Section 6.5 we introduced the notion of similar triangles. Although congruent triangles have the same size and shape, triangles that have the same shape but may differ in size are called **similar.** Congruent triangles are always similar, but similar triangles need not be congruent. For example, the triangles in Figure 12.27 are similar but clearly not congruent.

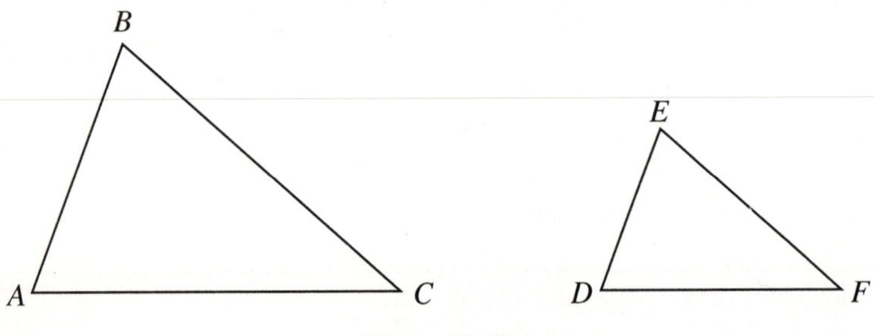

Figure 12.27

To determine whether two triangles are similar, we need only compare the angles.

Similar Triangles

Two triangles are similar when the angles of one are equal in measure to the angles of the other.

Since the sum of the angles of any triangle equals 180°, we know that whenever two angles of one triangle are equal to two angles of a second, the remaining angles are also equal, making the triangles similar.

EXAMPLE 3 Similar Triangles

Use deductive reasoning to state why the triangles in Figure 12.28 are similar.

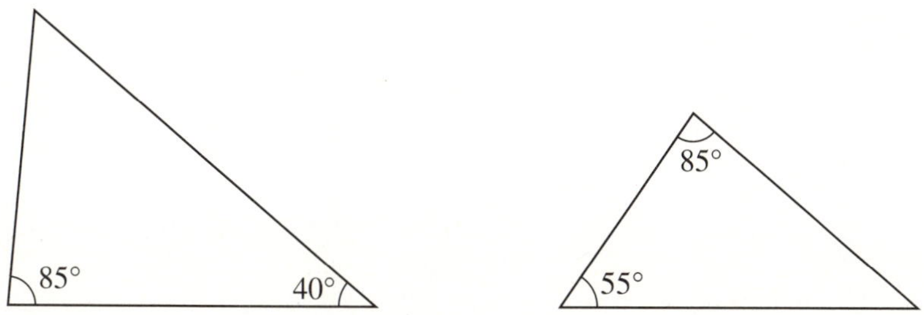

Figure 12.28

Since $85° + 40° = 125°$ and $180° - 125° = 55°$, the angles of the triangles are equal. Thus, the triangles are similar.

Practice Exercise 3

State why the triangles are similar.

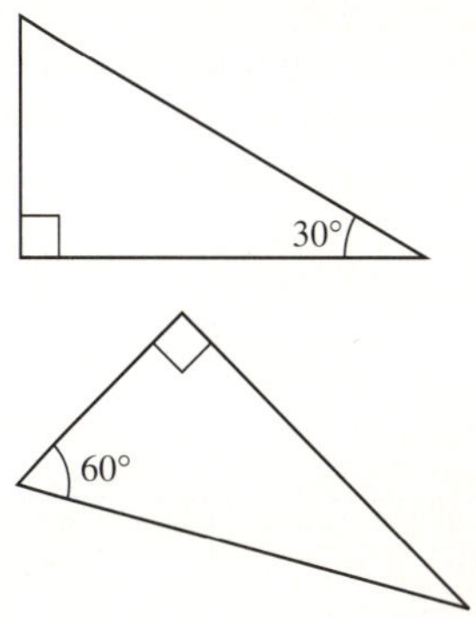

Answer: Both triangles are right triangles. Since $90° + 30° = 120°$ and $180° - 120° = 60°$, the angles of the triangles are equal.

5 30°–60° RIGHT TRIANGLES

We will now look at two special right triangles which have several applications in mathematics. A right triangle in which the two acute angles have measure 30° and 60° is called a **30°–60° right triangle.** The sides in such a triangle satisfy the following property.

> **Property of 30°–60° Right Triangle**
>
> In a 30°–60° right triangle, the length of the leg opposite the 30° angle is one-half the length of the hypotenuse.

Consider the right triangle in Figure 12.29. Since the hypotenuse c is 2, a must be 1 (one-half of c). We can find b using one of the equivalent forms of the Pythagorean theorem.

$$\begin{aligned} b^2 &= c^2 - a^2 \\ &= 2^2 - 1^2 \\ &= 4 - 1 \\ &= 3 \\ b &= \sqrt{3} \quad \text{Take the principal square root of both sides} \end{aligned}$$

In general, if a is the side opposite the 30° angle in a right triangle, $c = 2a$ and $b = \sqrt{3}a$.

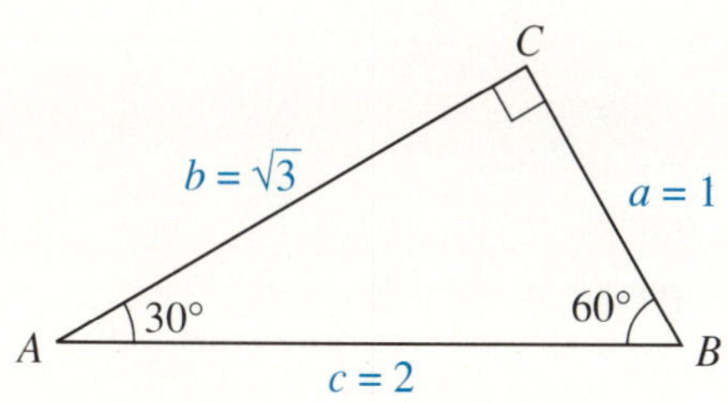

Figure 12.29 30°–60° Right Triangle

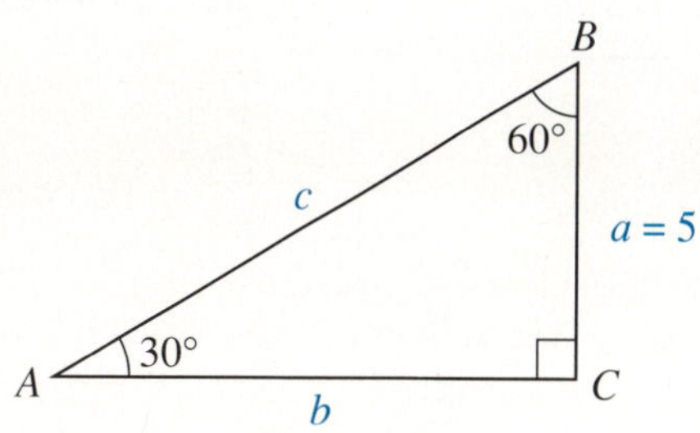

Figure 12.30

EXAMPLE 4 Solving a 30°–60° Right Triangle

In right $\triangle ABC$, $\angle C$ is the right angle, $\angle A$ has measure 30°, and $a = 5$. Find c and b.

First sketch the triangle as in Figure 12.30. Since $a = 5$ is one-half the hypotenuse, c must be 10. Also $b = \sqrt{3}a = \sqrt{3}(5) = 5\sqrt{3}$. We reach the same result using the Pythagorean theorem.

$$\begin{aligned} b^2 &= c^2 - a^2 \\ &= 10^2 - 5^2 \\ &= 100 - 25 = 75 \\ b &= \sqrt{75} = \sqrt{25 \cdot 3} = 5\sqrt{3} \end{aligned}$$

Practice Exercise 4

In right $\triangle ABC$, $\angle C$ is the right angle, $\angle B$ has measure 60°, and $c = 6$. Find a and b.

Answer: $a = 3$, $b = 3\sqrt{3}$

6 45°–45° RIGHT TRIANGLES

A right triangle with both acute angles measuring 45° is called a **45°–45° right triangle.** Since a 45°–45° right triangle is isosceles, the legs have equal length, and the Pythagorean theorem can be used to find the length of the hypotenuse. In Figure 12.31, since a and b are both 1, and

$$\begin{aligned} c^2 &= a^2 + b^2 \\ &= 1^2 + 1^2 = 1 + 1 = 2, \end{aligned}$$

then $c = \sqrt{2}$.

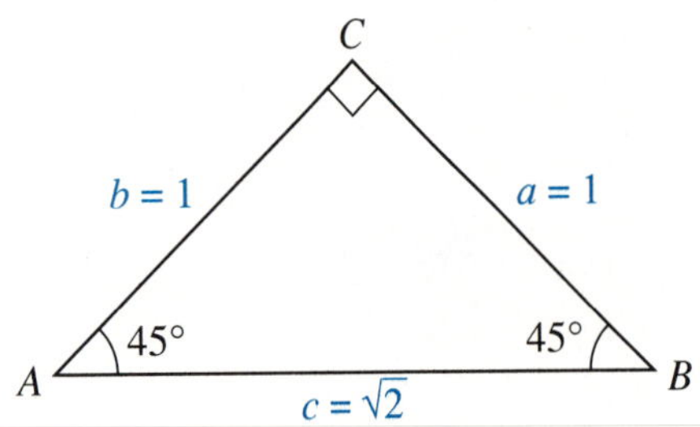

Figure 12.31 45°–45° Right Triangle

In general, if the legs are of length a then the hypotenuse is $\sqrt{2}a$.

In Section 12.6 we will discuss the perimeter and area of a triangle.

12.4 EXERCISES A

1. $\triangle ABC$ is given.

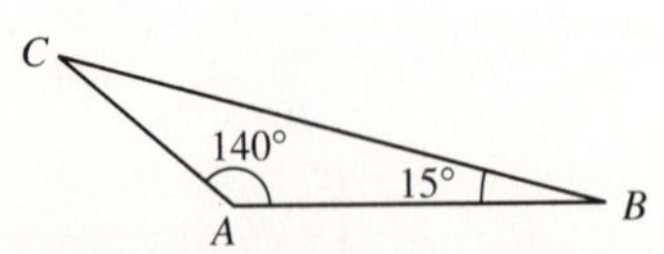

(a) What is the measure of $\angle C$?

(b) Classify the triangle by its angles.

(c) Classify the triangle by its sides.

2. $\triangle DEF$ is given.

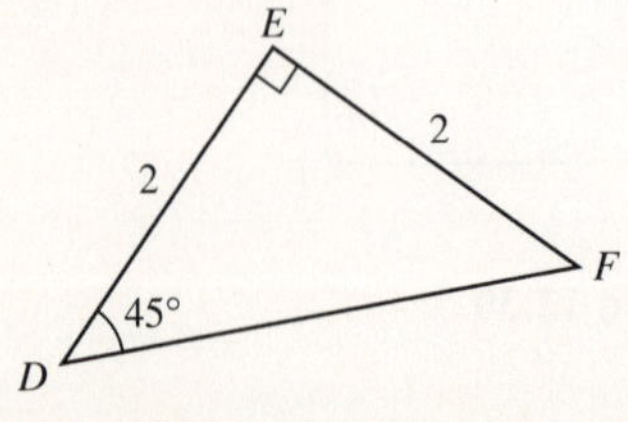

(a) What is the measure of $\angle F$?

(b) Classify the triangle by its angles.

(c) Classify the triangle by its sides.

Use deductive reasoning to decide whether the given triangles are congruent. If so, state why.

3.

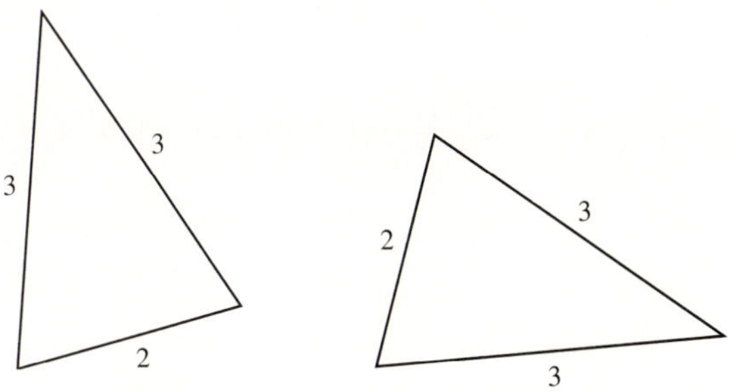

4.

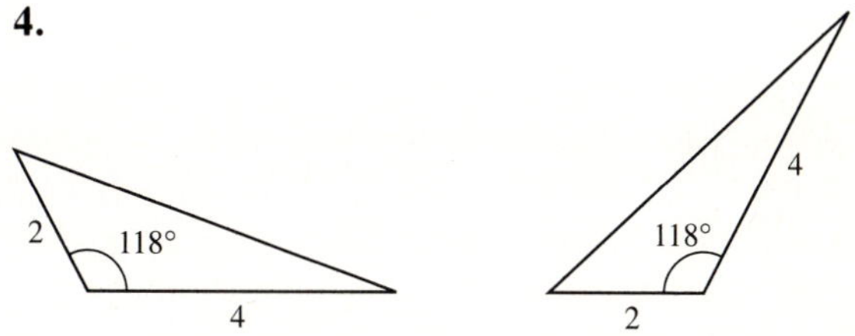

5.

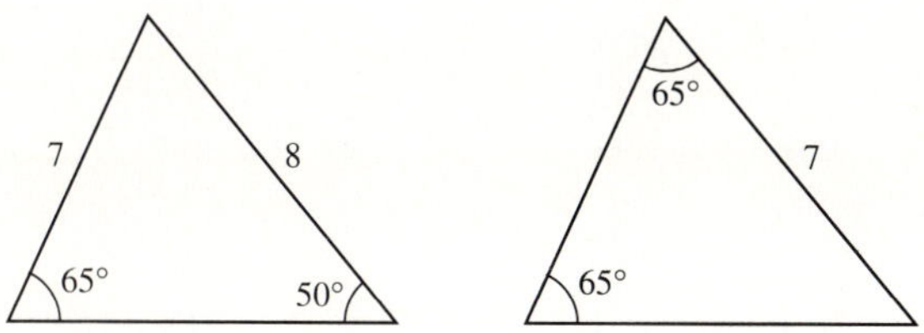

6.

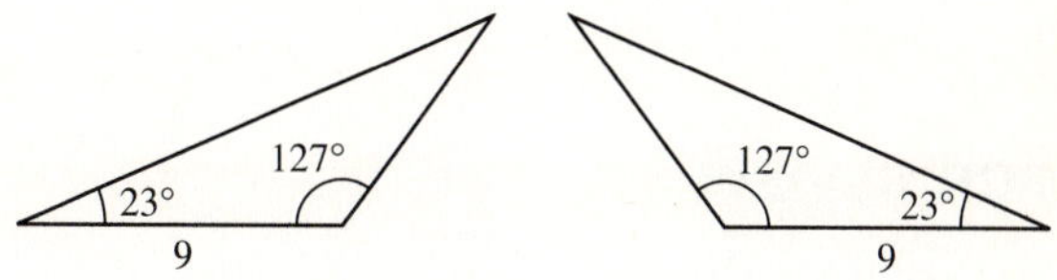

Use deductive reasoning to decide whether the given triangles are similar.

7.

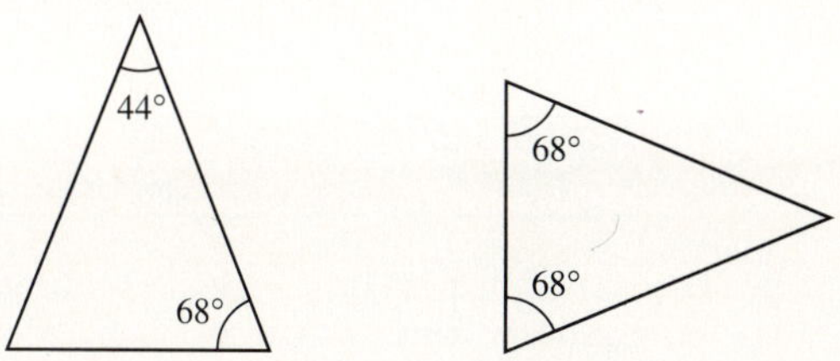

8.

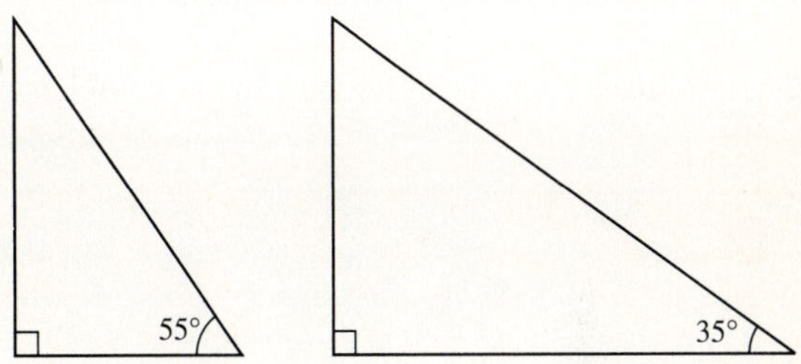

9

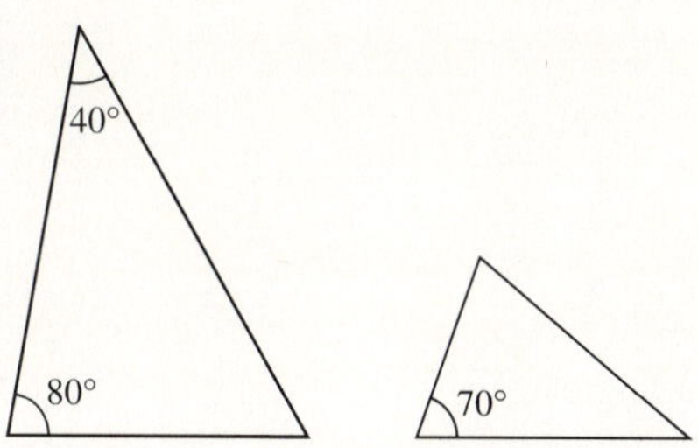

10.

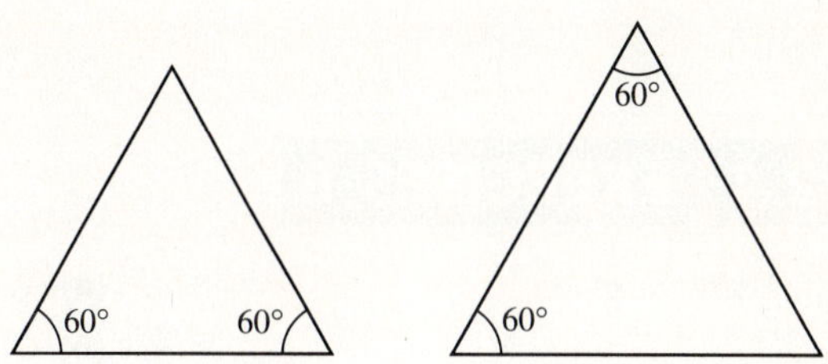

Use the properties of 30°–60° or 45°–45° right triangles to find the missing legs or hypotenuse.

11.

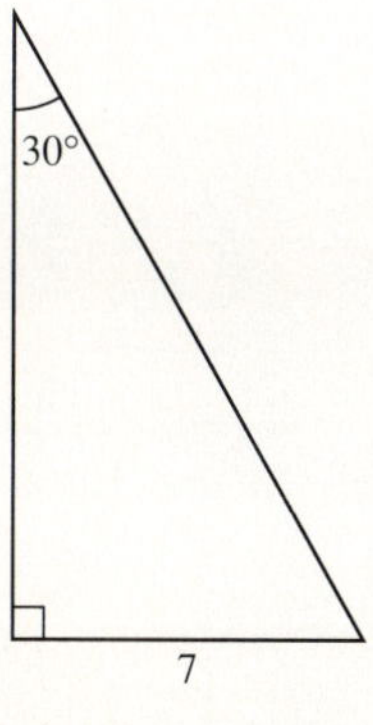

12.

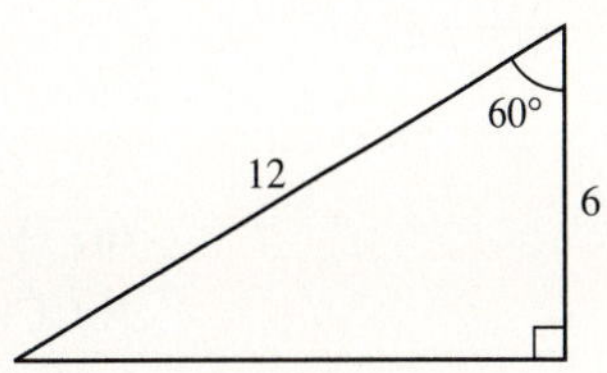

13. 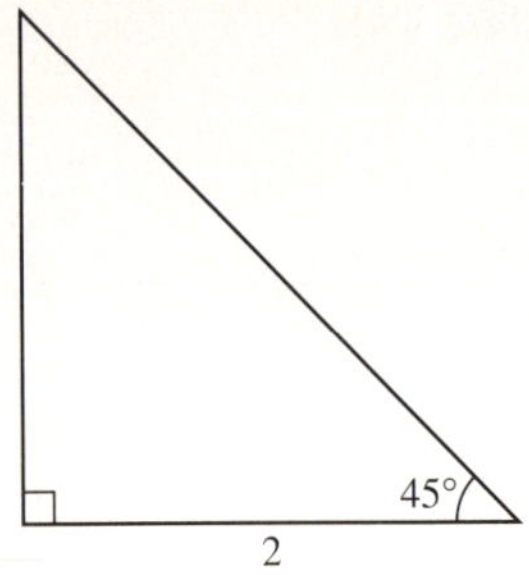

14. 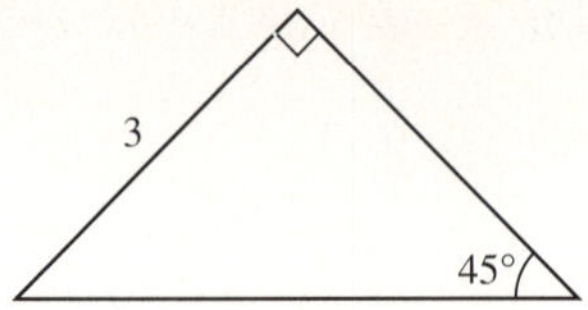

15. Is it possible for an isosceles triangle to be a right triangle?

16. Are all isosceles triangles equilateral?

FOR REVIEW

Answer true *or* false *in Exercises 17–20.*

17. A straight angle has measure 180°.

18. Two angles whose measures total 180° are complementary.

19. Vertical angles have the same measure.

20. Alternate interior angles are always complementary.

ANSWERS: 1. (a) 25° (b) obtuse triangle (c) scalene triangle 2. (a) 45° (b) right triangle (c) isosceles triangle 3. congruent by SSS 4. congruent by SAS 5. The triangles are similar but not congruent. 6. congruent by ASA 7. They are similar. 8. They are similar. 9. They are not similar. 10. They are similar. 11. hypotenuse 14; leg $7\sqrt{3}$ 12. 6; $6\sqrt{3}$ 13. hypotenuse $2\sqrt{2}$; leg 2 14. hypotenuse $3\sqrt{2}$; leg 3 15. yes 16. no 17. true 18. false (supplementary) 19. true 20. false (equal)

12.4 EXERCISES B

1. $\triangle ABC$ is given.

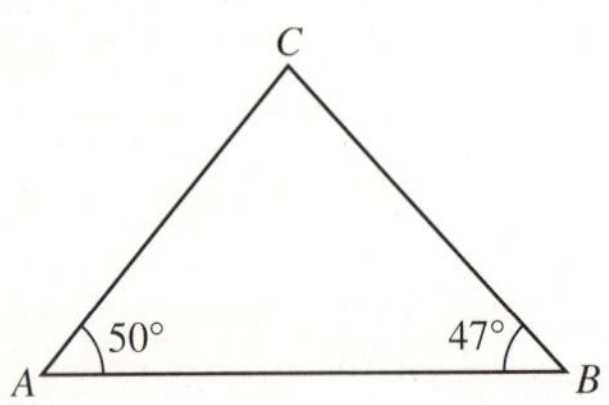

(a) What is the measure of $\angle C$?
(b) Classify the triangle by its angles.
(c) Classify the triangle by its sides.

2. $\triangle DEF$ is given.

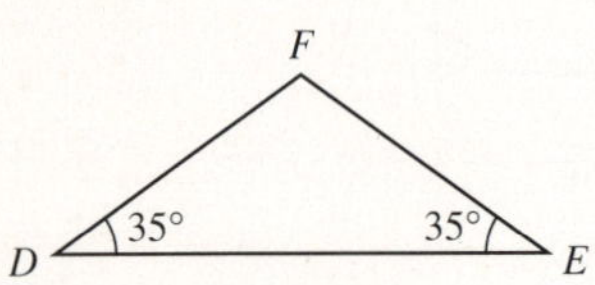

(a) What is the measure of $\angle F$?
(b) Classify the triangle by its angles.
(c) Classify the triangle by its sides.

Use deductive reasoning to decide whether the given triangles are congruent. If so, state why.

3.

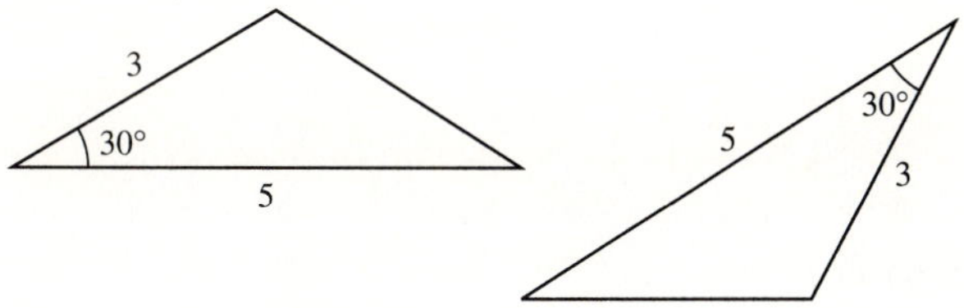

4.

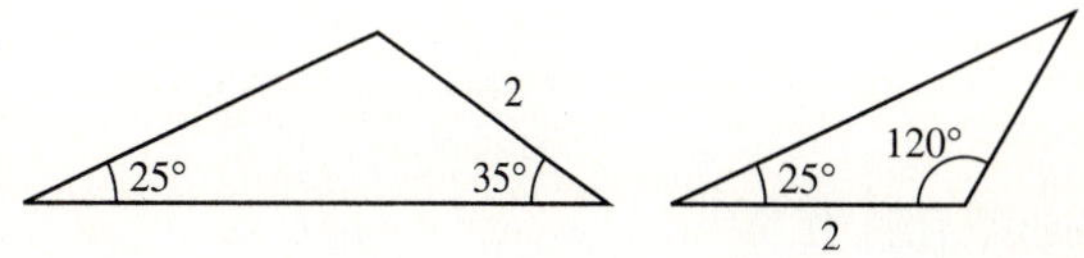

5.

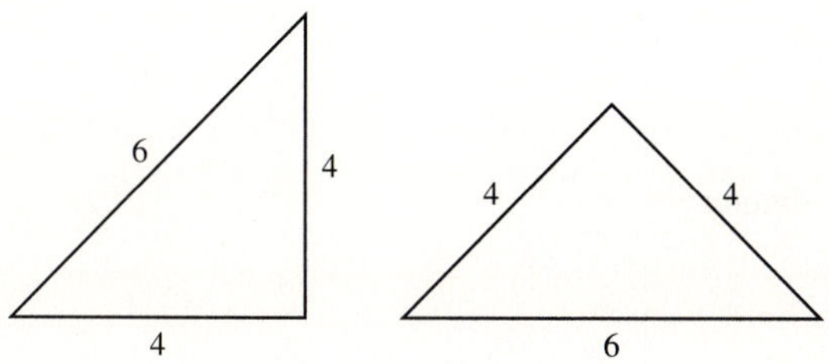

6.

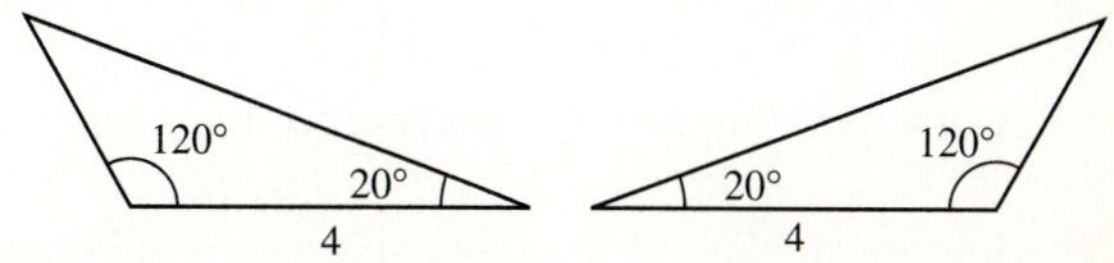

Use deductive reasoning to decide whether the given triangles are similar.

7.

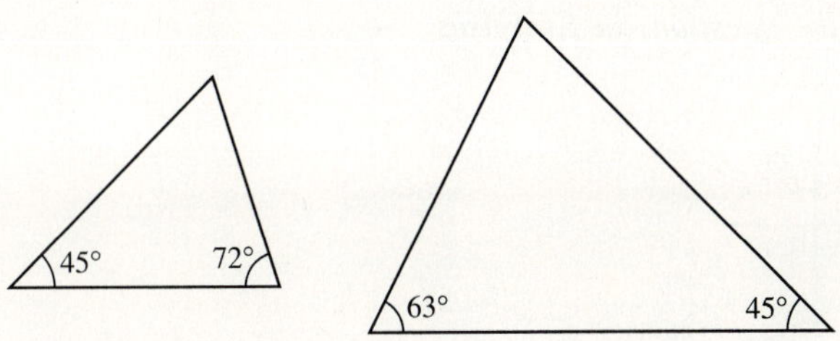

8.

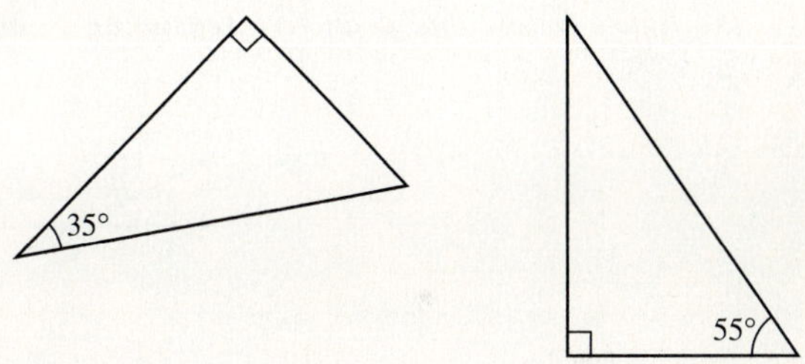

9.

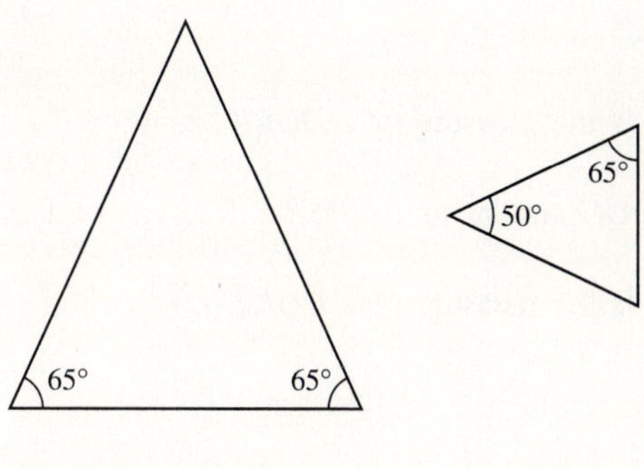

10.

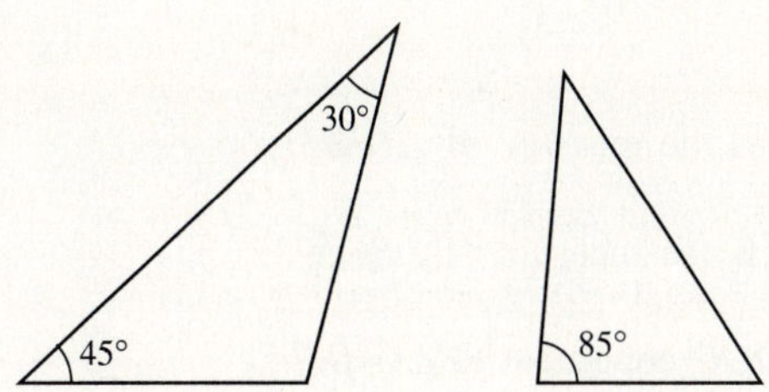

Use the properties of 30°–60° or 45°–45° right triangles to find the missing legs or hypotenuse.

11.

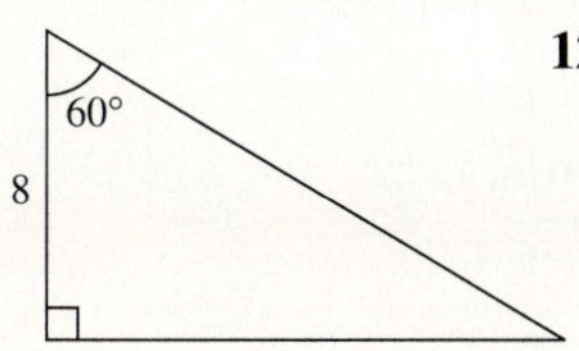

12.

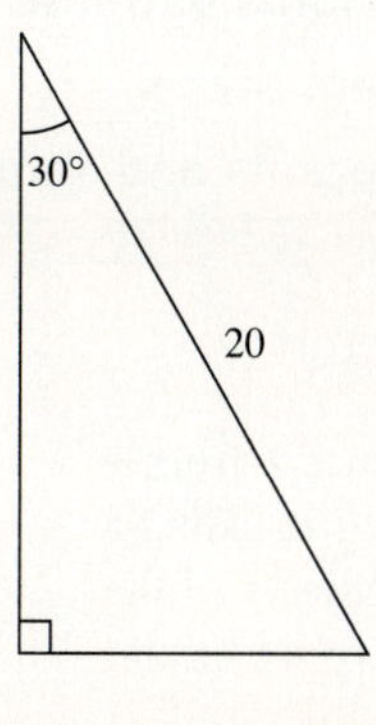

13.

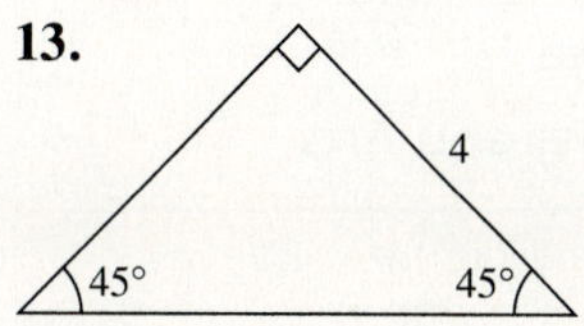

14.

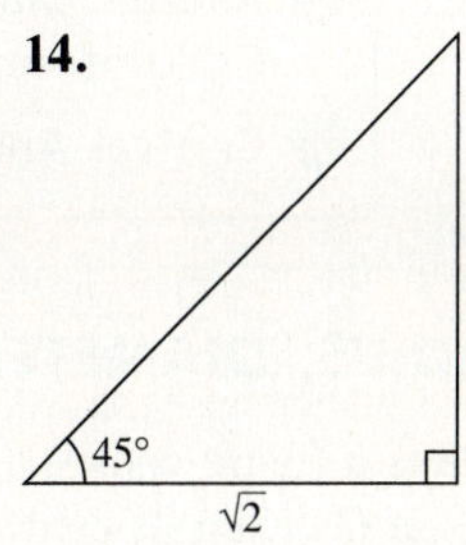

15. Is it possible for an equilateral triangle to be a right triangle?

16. Are all equilateral triangles isosceles?

FOR REVIEW

Answer true *or* false *in Exercises 17–20. If the answer is false, tell why.*

17. An acute angle has measure 90°.

18. Two angles whose measures total 360° are complementary.

19. Corresponding angles always have the same measure.

20. Interior angles on the same side of a transversal are supplementary.

12.4 EXERCISES C

Use the figure below and deductive reasoning to answer the questions in Exercises 1–6.

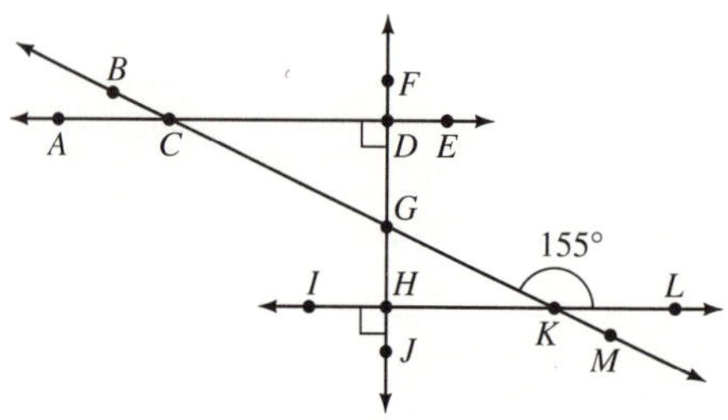

1. What is the measure of $\angle GKH$?

2. What is the measure of $\angle DGC$?

3. What is the measure of $\angle BCA$?

4. Is $\triangle DGC$ similar to $\triangle GHK$?

5. Is $\triangle DGC$ congruent to $\triangle GHK$?

6. What is the measure of $\angle DGK$?

12.5 CIRCLES

STUDENT GUIDEPOSTS

1. Radius, Diameter, and Circumference
2. Central Angles and Arcs
3. Chords and Inscribed Angles
4. Semicircles
5. Secants and Tangents

1 RADIUS, DIAMETER, AND CIRCUMFERENCE

A **circle** is a figure consisting of all points located the same distance r from a fixed point O called its **center.** In Figure 12.32 the segment $\overline{OA}$ or r is called a **radius** of the circle. The segment $\overline{DE}$, with length d, is called a **diameter** of the circle. The distance around the circle is called the **circumference** C of the circle.

Since the diameter is twice the radius, we can write

$$d = 2r \quad \text{and} \quad r = \frac{d}{2}.$$

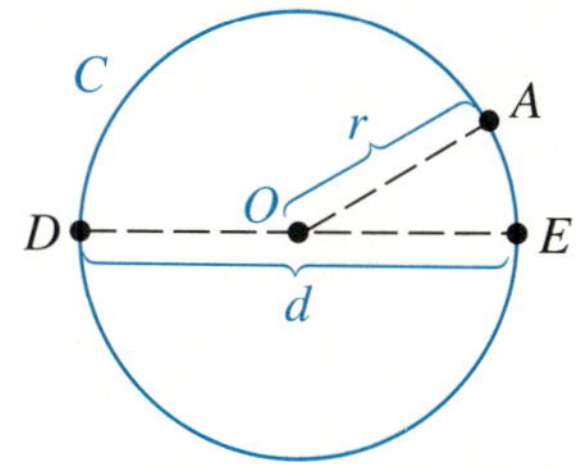

Figure 12.32 Circle

EXAMPLE 1 Diameter of a Circle

Find the diameter of the circle with radius 2.3 cm, shown in Figure 12.33.

$$d = 2r = 2(2.3\text{ cm})$$
$$= 4.6\text{ cm}$$

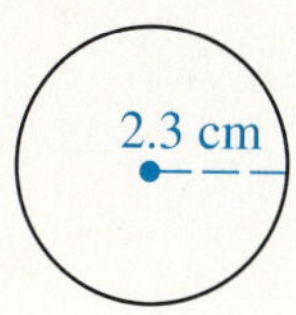

Figure 12.33

Practice Exercise 1

Find the radius of a circle with diameter 22.4 meters.

Answer: 11.2 m

Suppose that we measure the diameter d and the circumference C of the circles formed by the tops of several cans. For example, assume that for one can C is 6.28 inches and d is 2.00 inches. Then

$$\frac{C}{d} = \frac{6.28\text{ in}}{2.00\text{ in}} = 3.14.$$

Measurements from other cans will give us approximately the same result. We can then use inductive reasoning to assume that the ratio of the circumference of any circle to its diameter is a constant approximately equal to 3.14. The number $\frac{22}{7}$ is also used as an approximation of this number, which is called π (the Greek letter *pi*). That is, $\frac{C}{d} = \pi$ or $C = \pi d$ for any circle.

Circumference of a Circle

The circumference of a circle with diameter d and radius r is

$$C = \pi d = 2\pi r.$$

EXAMPLE 2 FINDING CIRCUMFERENCE

Find the circumference of a circle with diameter 32 mm. Use 3.14 for π to give an approximation of the actual value.

$$C = \pi d = \pi(32 \text{ mm}) = 32\pi \text{ mm}$$

The approximate value of C is

$$C \approx 3.14(32 \text{ mm})$$
$$\approx 100.5 \text{ mm.} \quad \text{Rounded to the nearest tenth}$$

PRACTICE EXERCISE 2

Find the circumference of a circle with radius 63 inches. Use $\frac{22}{7}$ for π.

Answer: 396 in

2 CENTRAL ANGLES AND ARCS

A **central angle** of a circle is an angle with its vertex at the center of the circle. In Figure 12.34, $\angle AOB$ is a central angle that intercepts **minor arc** $\widehat{ACB}$, the shorter part of the circle between A and B, and **major arc** $\widehat{ADB}$, shown in color.

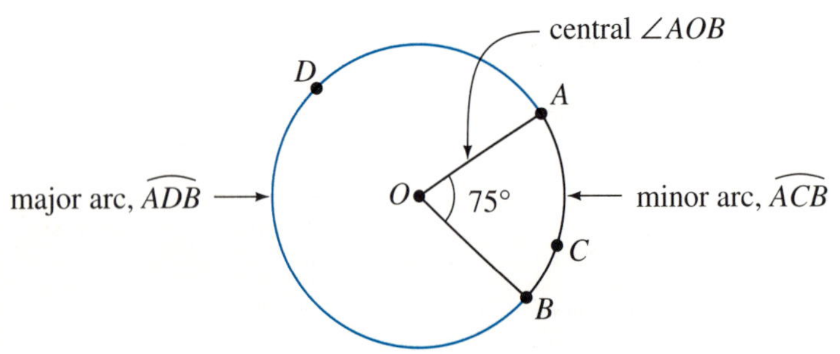

Figure 12.34

The **measure of an arc** is given by the measure of the central angle that determines it. Since the measure of central $\angle AOB$ in Figure 12.34 is 75°, the measure of minor arc $\widehat{ACB}$ is also 75°. Then the measure of major arc $\widehat{ADB}$ is 285°, since $360° - 75° = 285°$.

3 CHORDS AND INSCRIBED ANGLES

A segment whose endpoints are on a circle is called a **chord** of the circle. Any chord that contains the center is also a diameter of the circle. Three distinct points A, B, and C on a circle determine **inscribed angle** $\angle ABC$, with sides determined by the chords $\overline{AB}$ and $\overline{BC}$ as shown in Figure 12.35. The inscribed angle $\angle ABC$ determines **intercepted arc** $\widehat{ADC}$ (shown in color).

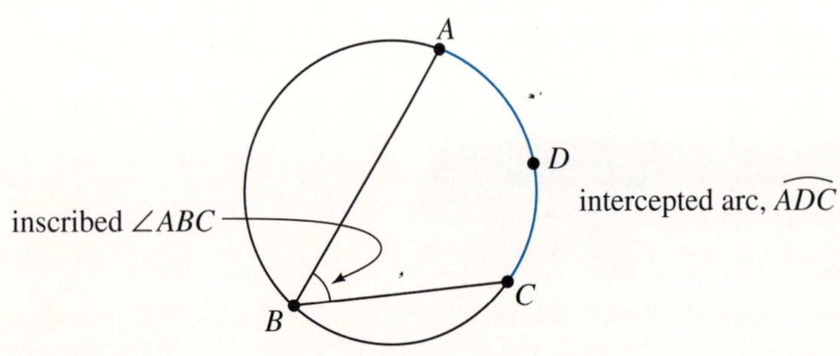

Figure 12.35

Consider the central $\angle AOC$ measuring 60° in Figure 12.36, with inscribed $\angle ABC$. The measure of $\angle ABC$ is 30°, one-half the measure of its intercepted arc, $\overset{\frown}{ADC}$, which has measure 60° (the measure of $\angle AOC$). This is true in general.

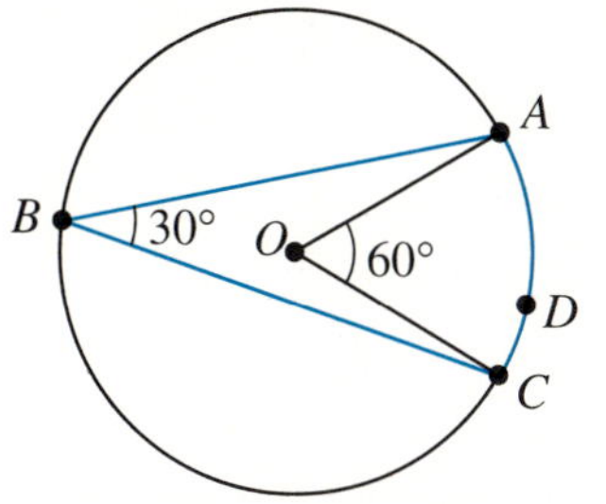

Figure 12.36

Measure of Inscribed Angle

The measure of any inscribed angle is one-half the measure of its intercepted arc.

EXAMPLE 3 MEASURING ARCS

Use Figure 12.37 and deductive reasoning to answer the following questions.

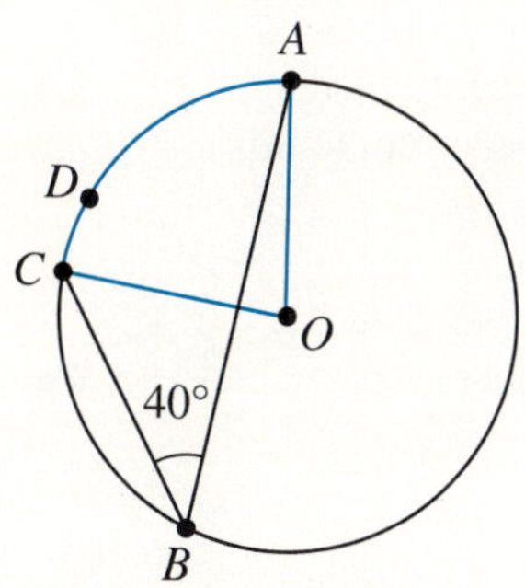

Figure 12.37

(a) What is the measure of $\overset{\frown}{ADC}$?

The measure of $\overset{\frown}{ADC}$ is the same as the measure of central $\angle AOC$. Since the measure of inscribed $\angle ABC$ is one-half the measure of $\angle AOC$, the measure of $\angle AOC$ is 80°. Thus, the measure of $\overset{\frown}{ADC}$ is 80°.

(b) What is the measure of $\overset{\frown}{ABC}$?

The measure of $\overset{\frown}{ABC}$ is 360° less the measure of $\overset{\frown}{ADC}$. Thus, the measure of $\overset{\frown}{ABC}$ is 360° − 80° = 280°.

PRACTICE EXERCISE 3

Use the figure below to answer the following questions.

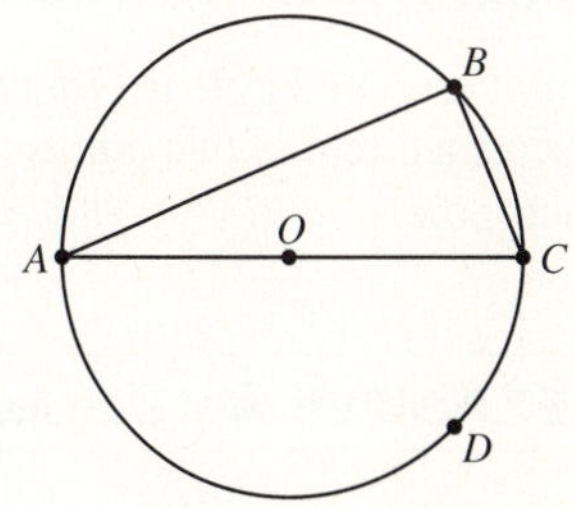

(a) What is the measure of $\overset{\frown}{ADC}$?

(b) What is the measure of inscribed $\angle ABC$?

Answers: (a) 180° (b) 90°

4 SEMICIRCLES

A chord that passes through the center of a circle is a diameter, and an arc whose endpoints are the endpoints of a diameter is called a **semicircle.**

Suppose we inscribe an angle in a semicircle as shown in Figure 12.38.

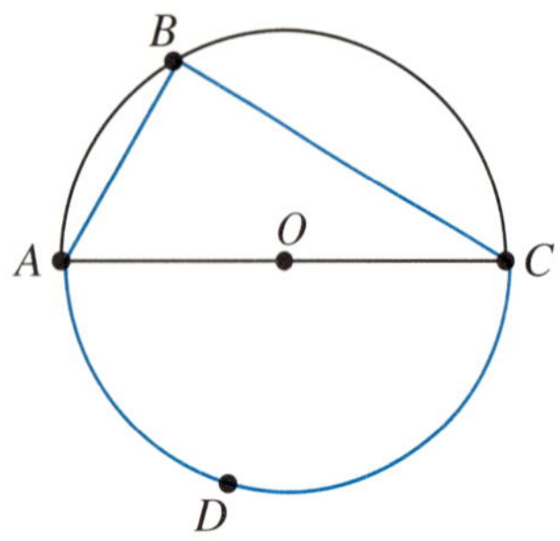

Figure 12.38

We can use deductive reasoning to prove that any angle inscribed in a semicircle is a right angle. First we note that since $\widehat{ADC}$ is one-half of the circle, it has measure 180°. Therefore,

$$\begin{aligned} \angle ABC &= \frac{1}{2}\widehat{ADC} \\ &= \frac{1}{2}(180°) \\ &= 90° \end{aligned}$$

Thus, $\angle ABC$ is a right angle.

5 SECANTS AND TANGENTS

A line can intersect a circle in two points or exactly one point. A **secant** is a line that intersects a circle in two points. A **tangent** is a line that intersects a circle in exactly one point.

EXAMPLE 4 Determining Secants and Tangents

Refer to Figure 12.39 to answer each question.

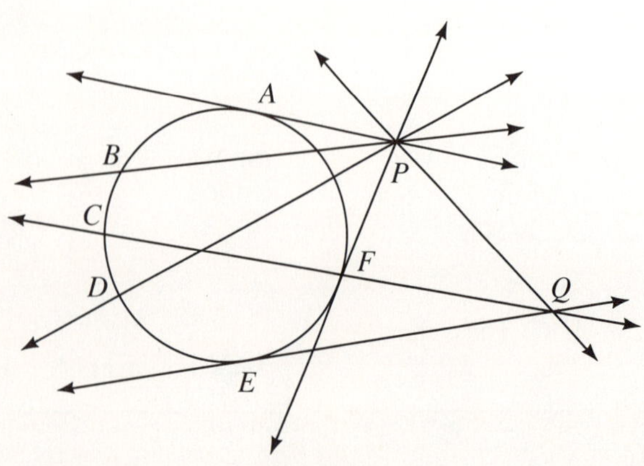

Figure 12.39

Practice Exercise 4

Refer to the figure to answer each question.

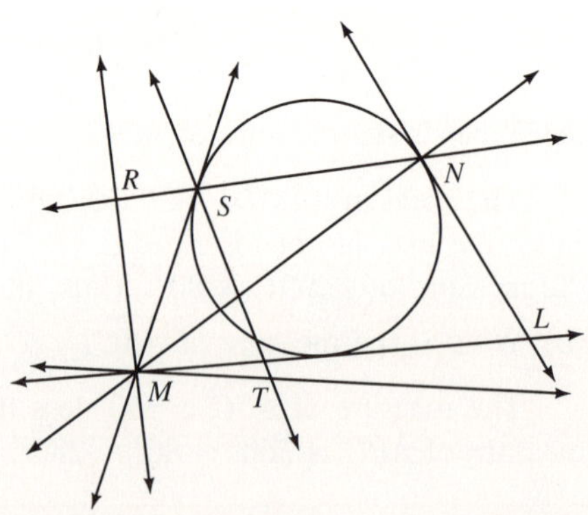

(a) Which lines are secants? Since $\overleftrightarrow{PB}$, $\overleftrightarrow{PD}$, and $\overleftrightarrow{QC}$ intersect the circle in two points, they are secants.

(b) Which lines are tangents? Since $\overleftrightarrow{PA}$, $\overleftrightarrow{PF}$, and $\overleftrightarrow{QE}$ intersect the circle in exactly one point, they are tangents.

(c) Are there lines in Figure 12.39 that are neither secants nor tangents? Yes. $\overleftrightarrow{PQ}$ does not intersect the circle at all.

(a) Which lines are secants?

(b) Which lines are tangents?

(c) Which lines are neither tangents nor secants?

Answers: (a) $\overleftrightarrow{MN}$, $\overleftrightarrow{SN}$, and $\overleftrightarrow{ST}$
(b) $\overleftrightarrow{LN}$, $\overleftrightarrow{MS}$, and $\overleftrightarrow{LM}$
(c) $\overleftrightarrow{MT}$ and $\overleftrightarrow{MR}$

12.5 EXERCISES A

Find the diameter of the circle with the given radius.

1. $r = 11$ in **2.** $r = 5.85$ cm **3.** $r = \frac{3}{4}$ ft

Find the radius of the circle with the given diameter.

4. $d = 24$ m **5.** $d = \frac{5}{2}$ in **6.** $d = 7.24$ mm

Find the circumference of the circle. Leave the answer using π.

7. $r = 7$ yd **8.** $r = \frac{7}{2}$ ft **9** $r = 7.7$ mi

Find the circumference of the circle. Use 3.14 for π.

10. $r = 6.8$ km **11.** $d = 92$ in **12** $r = 0.05$ cm

Use the figure and deductive reasoning to answer the questions in Exercises 13–16.

13. What is the measure of $\widehat{ADC}$?

14. What is the measure of $\widehat{ABC}$?

15. What is the measure of inscribed $\angle ABC$?

16. What is the measure of the arc intercepted by $\angle ABC$?

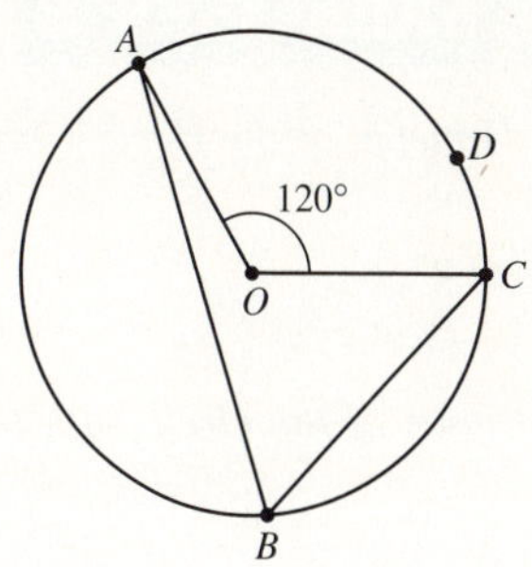

Use the figure and deductive reasoning to answer the questions in Exercises 17–20.

17. What is the measure of central $\angle AOC$?

18. What is the measure of $\widehat{ADC}$?

19. What is the measure of $\widehat{ABC}$?

20. What is the measure of the arc intercepted by $\angle ABC$?

21. If a circle passes through all three vertices of right $\triangle ABC$ with $\angle C = 90°$, what is the measure of $\widehat{ACB}$?

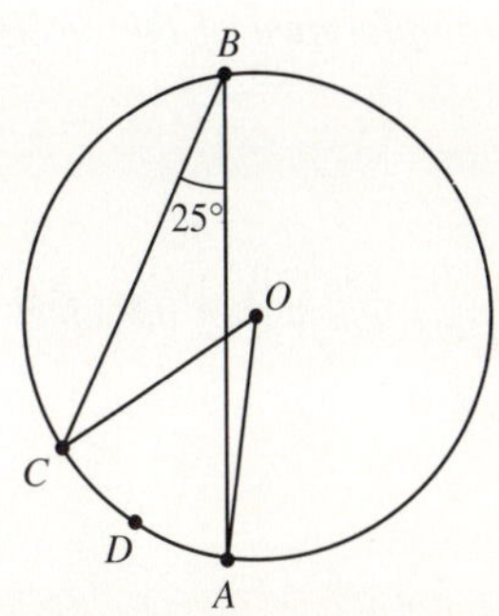

Use the figure and deductive reasoning in Exercises 22–24.

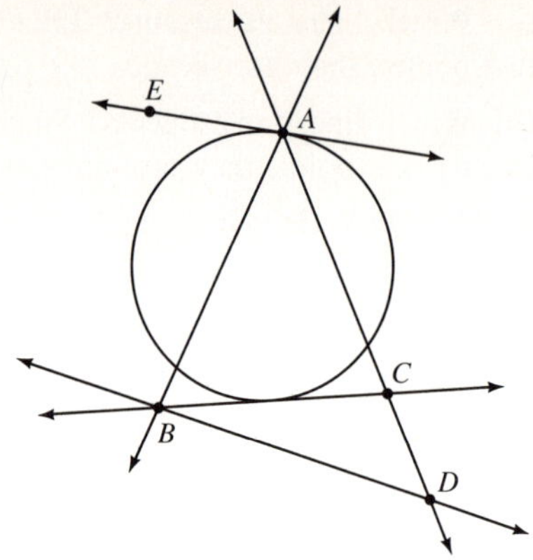

22. Find the lines that are secants.

23. Find the lines that are tangents.

24. Find the lines that are neither secants nor tangents.

FOR REVIEW

25. In a triangle two angles have measures 82° and 54°. What is the measure of the third angle?

26. In a right triangle one acute angle has measure 16°. What is the measure of the other acute angle?

27. In an isosceles triangle one angle has measure 108°. What are the measures of the other angles?

28. What are the measures of the angles in an equilateral triangle?

ANSWERS: 1. 22 in 2. 11.7 cm 3. $\frac{3}{2}$ ft 4. 12 m 5. $\frac{5}{4}$ in 6. 3.62 mm 7. 14π yd 8. 7π ft 9. 15.4π mi 10. 42.7 km 11. 288.9 in 12. 0.314 cm 13. 120° 14. 240° 15. 60° 16. 120° 17. 50° 18. 50° 19. 310° 20. 50° 21. 180° 22. $\overleftrightarrow{AB}$ and $\overleftrightarrow{AC}$ 23. $\overleftrightarrow{AE}$ and $\overleftrightarrow{BC}$ 24. $\overleftrightarrow{BD}$ 25. 44° 26. 74° 27. 36° and 36° 28. 60°, 60°, and 60°

12.5 EXERCISES B

Find the diameter of the circle with the given radius.

1. $r = 28$ cm **2.** $r = 2.75$ ft **3.** $r = \frac{7}{6}$ in

Find the radius of the circle with the given diameter.

4. $d = 120$ mm **5.** $d = \frac{2}{3}$ in **6.** $d = 28.6$ m

Find the circumference of the circle. Leave the answer using π.

7. $r = 14$ m **8.** $r = \frac{14}{11}$ cm **9.** $r = 5.67$ yd

Find the circumference of the circle. Use 3.14 for π.

10. $d = 32.8$ ft **11.** $r = 420$ mm **12.** $d = 1.64$ in

Use the figure and deductive reasoning to answer the questions in Exercises 13–16.

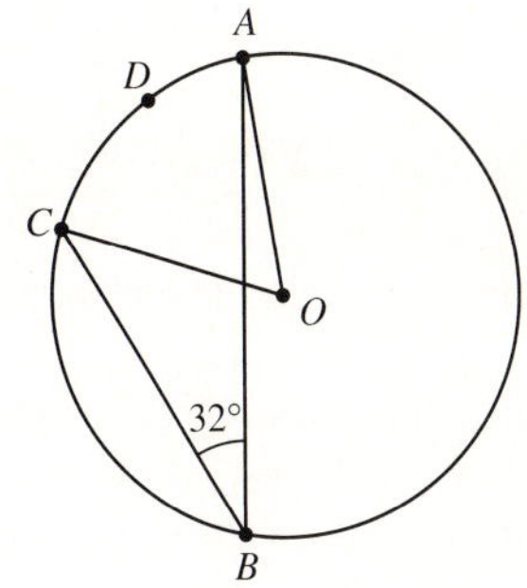

13. What is the measure of $\widehat{ADC}$?

14. What is the measure of $\widehat{ABC}$?

15. What is the measure of $\angle AOC$?

16. What is the measure of the arc intercepted by $\angle ABC$?

Use the figure and deductive reasoning to answer the questions in Exercises 17–20.

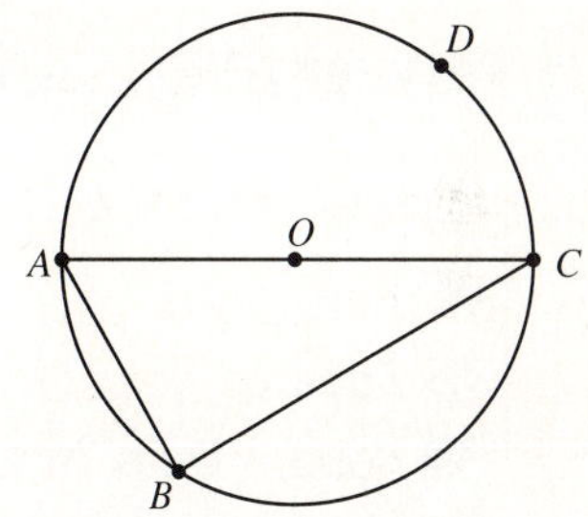

17. What is the measure of central $\angle AOC$?

18. What is the measure of $\widehat{ABC}$?

19. What is the measure of inscribed $\angle ABC$?

20. What is the measure of the arc intercepted by $\angle ABC$?

21. If A, B, and C are points on a circle and $\angle ACB = 90°$, what can be said about $\overline{AB}$ in relation to the circle?

Use the figure and deductive reasoning in Exercises 22–24.

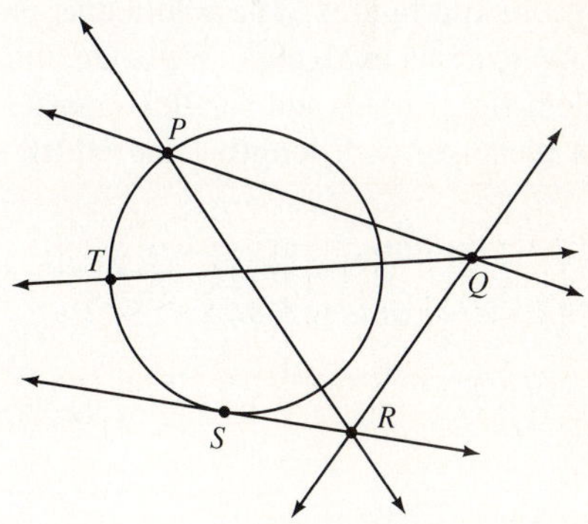

22. Find the lines that are secants.

23. Find the lines that are tangents.

24. Find the lines that are neither secants nor tangents.

FOR REVIEW

25. Two of the angles of a triangle have measures 105° and 55°. What is the measure of the third angle?

26. In a right triangle one acute angle has measure 47°. What is the measure of the other acute angle?

27. If one of the angles of an isosceles triangle has measure 60°, what can be said about the triangle?

28. What is the measure of the equal angles in an isosceles right triangle?

12.5 EXERCISES C

In Exercises 1–3, assume that the earth is 93,000,000 *miles from the sun and that the orbit of the earth is a circle. Use* 3.14 *for* π.

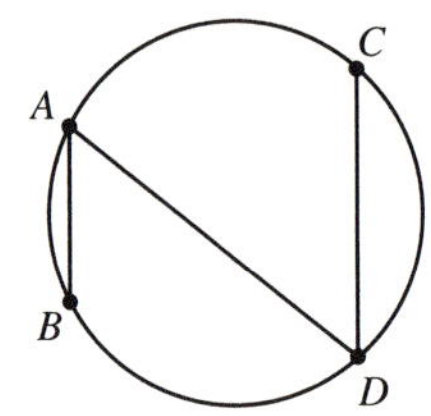

1. What distance does the earth travel in one year?

2. What distance does the earth travel in one day?

3. What distance does the earth travel in one minute?

4. In the figure, $\overline{AB} \parallel \overline{CD}$. Use deductive reasoning to prove that $\widehat{AC} = \widehat{BD}$.

12.6 PERIMETER AND AREA

STUDENT GUIDEPOSTS

1. Rectangles
2. Squares
3. Square Units of Measure
4. Parallelograms
5. Triangles
6. Trapezoids
7. Circles

1 RECTANGLES

In this section we review formulas for finding the perimeters and areas of six geometric figures. These formulas provide problem-solving tools for numerous applications in algebra. We begin with the **rectangle,** a four-sided figure having four right angles and parallel opposite sides of equal length. Figure 12.40 shows a rectangle with **length** l and **width** w.

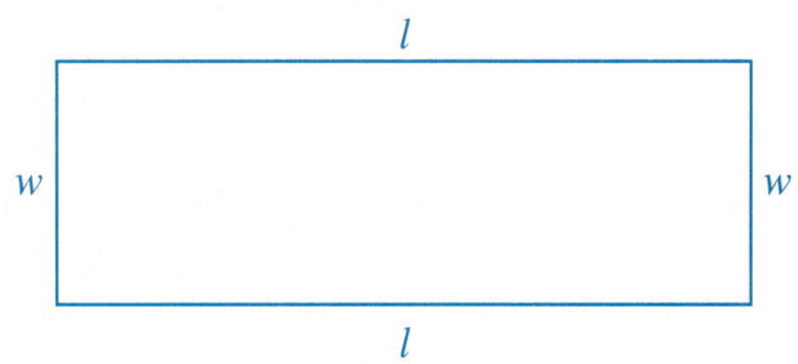

Figure 12.40 Rectangle

The **perimeter** of a geometric figure is the distance around the figure. Thus, the perimeter of a rectangle is

$$
\begin{aligned}
l + w + l + w &= l + l + w + w \\
&= 2l + 2w.
\end{aligned}
$$

Perimeter of a Rectangle

The perimeter P of a rectangle is equal to twice the length plus twice the width.

$$P = 2l + 2w$$

The perimeter of the rectangle in Figure 12.41 is given by

$$\begin{aligned} P &= 2l + 2w \\ &= 2(5\text{ cm}) + 2(3\text{ cm}) \\ &= 10\text{ cm} + 6\text{ cm} = 16\text{ cm}. \end{aligned}$$

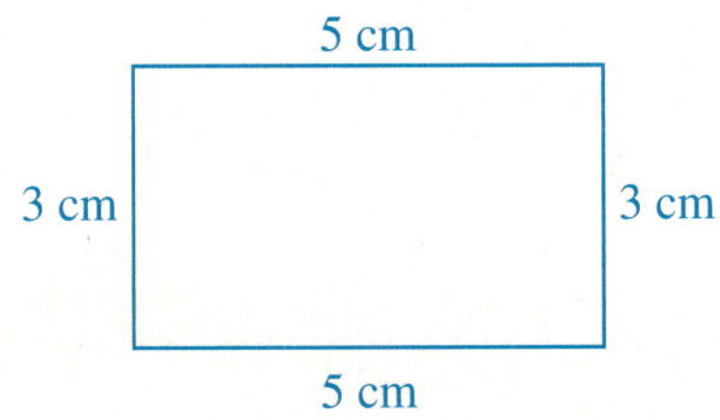

Figure 12.41

EXAMPLE 1 Perimeter of a Rectangle

Find the length of a fence needed to enclose a rectangular pasture that is 72 m wide and 90 m long.

The perimeter of the pasture is the length of the fence needed.

$$\begin{aligned} P &= 2l + 2w \\ &= 2(90\text{ m}) + 2(72\text{ m}) \\ &= 180\text{ m} + 144\text{ m} \\ &= 324\text{ m} \end{aligned}$$

Thus, 324 m of fence are needed.

Practice Exercise 1

A farmer needs to fence a rectangular pasture that is 520 ft long and 470 ft wide. How long must the fence be?

Answer: 1980 ft

2 SQUARES

A special type of rectangle, called a **square,** is shown in Figure 12.42. All its sides have the same length. If s is the length of a side, then the perimeter of a square is

$$\begin{aligned} P &= 2l + 2w \\ &= 2s + 2s \\ &= 4s. \end{aligned}$$

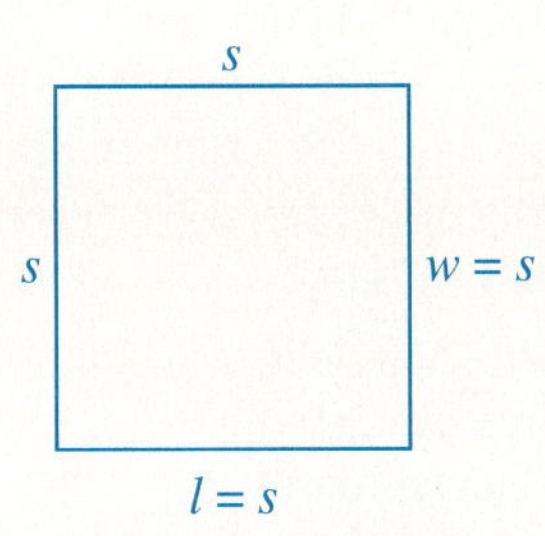

Figure 12.42 Square

Perimeter of a Square

The perimeter of a square is equal to four times the length of a side.

$$P = 4s$$

EXAMPLE 2 Perimeter of a Square

Find the perimeter of a square whose sides are 15.3 m. See Figure 12.43.

$$\begin{aligned} P &= 4s \\ &= 4(\mathbf{15.3\ m}) \\ &= 61.2\text{ m} \end{aligned}$$

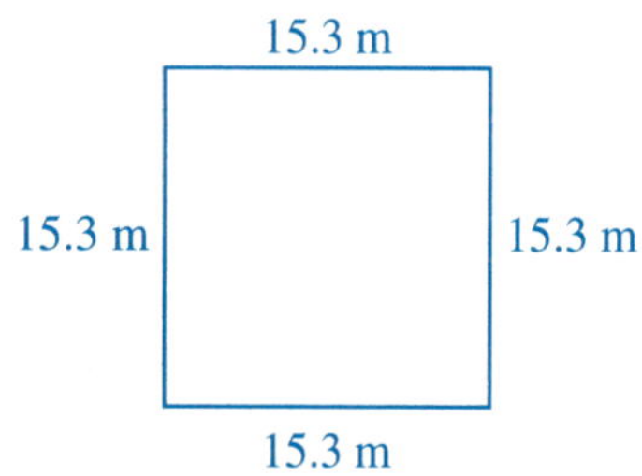

Figure 12.43

Practice Exercise 2

A square has side $3\frac{3}{4}$ ft. Find the perimeter.

Answer: 15 ft

3 SQUARE UNITS OF MEASURE

Two special squares are shown in Figure 12.44. The area of a square that has side 1 cm is one **square centimeter** and is written 1 cm^2. Likewise, if the side is 1 in, the area is one **square inch,** 1 in^2. Similarly, if some other unit of length is used, the area is 1 square unit, or 1 unit^2.

Area = A = 1 square centimeter (sq cm) = 1 cm^2

Area = A = 1 square inch (sq in) = 1 in^2

Figure 12.44

Square units of measure are used for the areas of rectangles. The rectangle in Figure 12.45 has area 10 cm^2 since there are 10 of the 1-cm^2 squares contained in it. In general, this is what is meant by *area*. Notice that 10 cm^2 can be found by multiplying 5 cm by 2 cm.

$$10\text{ cm}^2 = 5\text{ cm} \cdot 2\text{ cm}$$

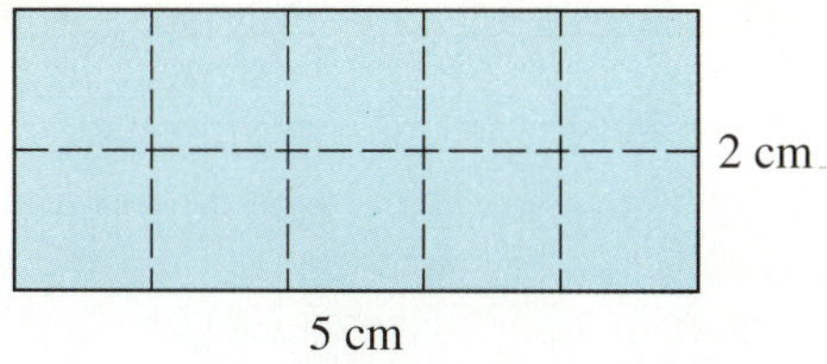

Figure 12.45

Area of a Rectangle

The area A of a rectangle is equal to the length times the width.

$$A = lw$$

EXAMPLE 3 Area of a Rectangle

Find the area of the rectangle with width 6 m and length 12 m, shown in Figure 12.46. (Sometimes we say "the rectangle that is 6 m by 12 m.")

$$\begin{aligned} A &= lw \\ &= (12 \text{ m})(6 \text{ m}) \\ &= (12)(6) \text{ m}^2 = 72 \text{ m}^2 \end{aligned}$$

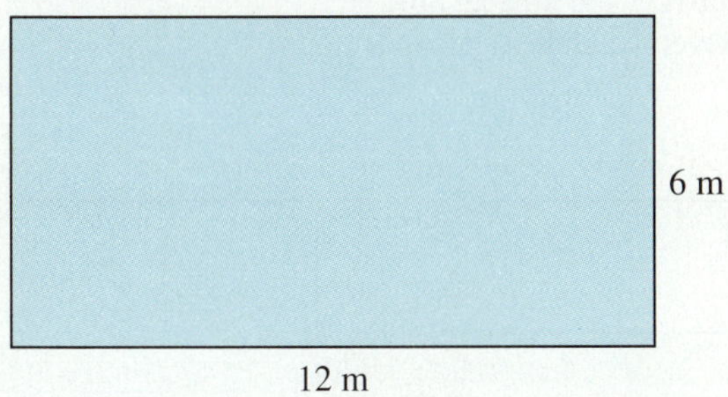

Figure 12.46

The area is 72 square meters.

Practice Exercise 3

Find the area of a rectangle that is 17 cm by 20 cm.

Answer: 340 cm^2

Since all sides of a square have the same measure s, the area of a square is

$$lw = ss = s^2.$$

Area of a Square

The area of a square is equal to the square of a side.

$$A = s^2$$

EXAMPLE 4 AREA OF A COMPOSITE REGION

Find the area of the region in Figure 12.47.

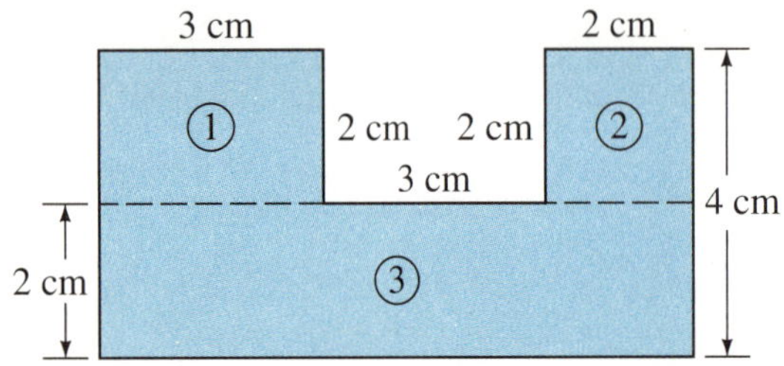

Figure 12.47

The region can be divided into two rectangles and a square.

① is 2 cm by 3 cm   Area ① = 3 cm × 2 cm = 6 cm^2

② is 2 cm by 2 cm   Area ② = 2 cm × 2 cm = 4 cm^2

③ is 2 cm by 8 cm   Area ③ = 8 cm × 2 cm = 16 cm^2

The total area is the sum of the three areas.

$$A = 6 \text{ cm}^2 + 4 \text{ cm}^2 + 16 \text{ cm}^2 = 26 \text{ cm}^2$$

PRACTICE EXERCISE 4

Find the area of the region shown.

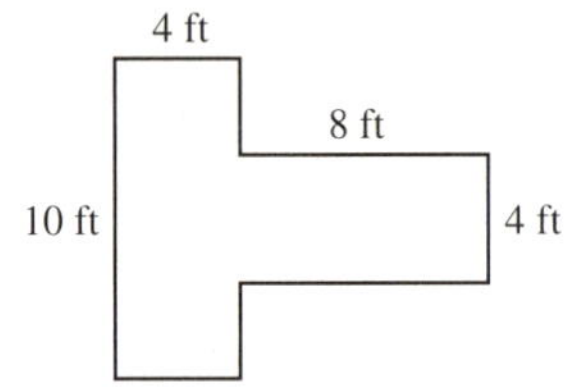

Answer: 72 ft^2

4 PARALLELOGRAMS

A four-sided figure whose opposite sides are parallel is a **parallelogram.** A rectangle is a special case of this type of figure. Several parallelograms are shown in Figure 12.48.

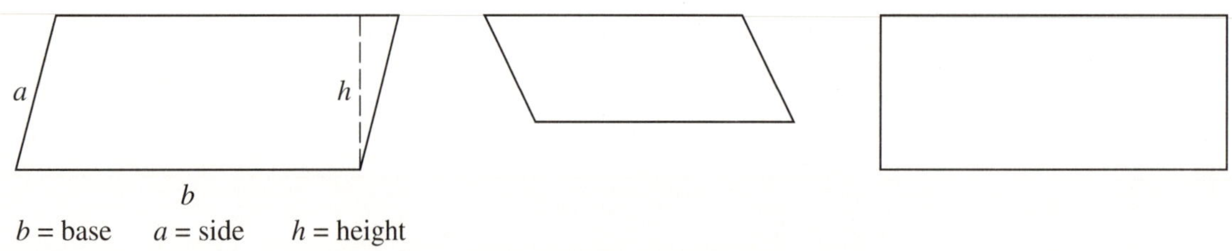

Figure 12.48 Parallelograms

Perimeter of a Parallelogram

The perimeter of a parallelogram with side a and base b is given by

$$P = 2a + 2b.$$

EXAMPLE 5 PERIMETER OF A PARALLELOGRAM

Find the perimeter of the parallelogram shown in Figure 12.49.

$$P = 2(10.5 \text{ m}) + 2(6.5 \text{ m})$$
$$= 34 \text{ m}$$

PRACTICE EXERCISE 5

Find the perimeter of a parallelogram with base $7\frac{1}{2}$ ft and side $4\frac{1}{2}$ ft.

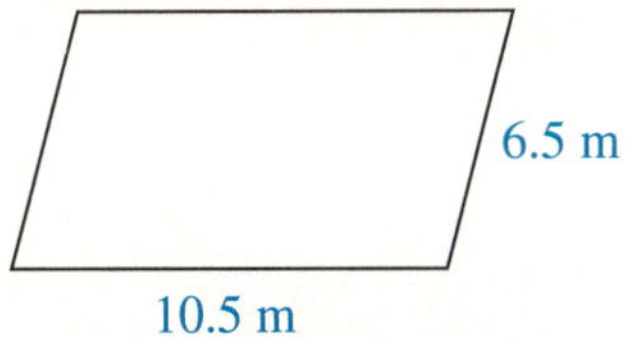

Figure 12.49

Answer: 24 ft

To discover how to find the area of a parallelogram, cut off one end and put it on the other end. This forms a rectangle (Figure 12.50) with length $l = b$ and width $w = h$. Thus, the area of the parallelogram (which is the area of the rectangle) is equal to the base times the height.

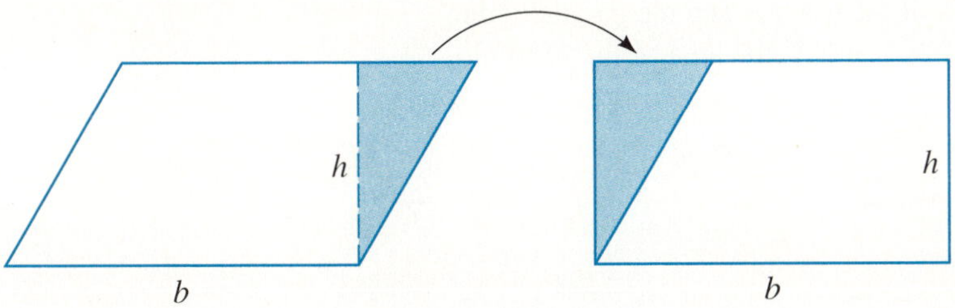

Figure 12.50

Area of a Parallelogram

The area of a parallelogram with base b and height h is given by

$$A = bh.$$

EXAMPLE 6 Area of a Parallelogram

Find the area of the parallelogram shown in Figure 12.51 with height 15.2 cm and base 20.5 cm.

$$\begin{aligned} A &= bh \\ &= \mathbf{(20.5\ cm)(15.2\ cm)} \\ &= (20.5)(15.2)\ \text{cm}^2 \\ &= 311.6\ \text{cm}^2 \end{aligned}$$

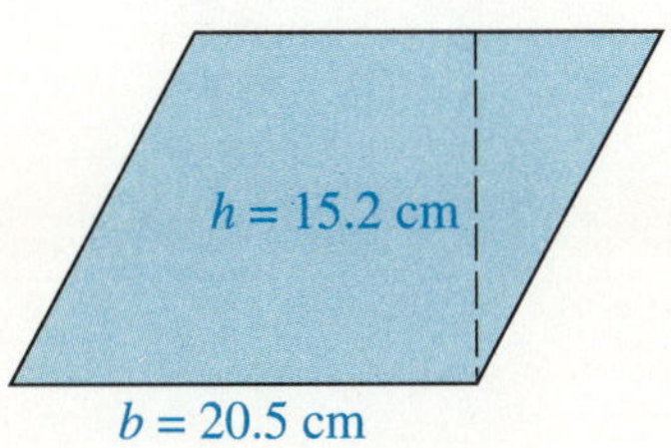

Figure 12.51

Practice Exercise 6

Find the area of a parallelogram with base 10.8 mm and height 12.5 mm.

Answer: 135 mm^2

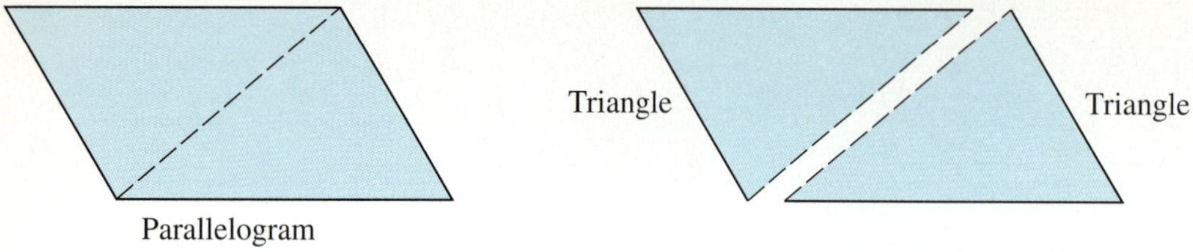

Figure 12.52

5 TRIANGLES

If the parallelogram in Figure 12.52 is cut along the dotted line, the area of each of the two triangles formed is one-half the area of the parallelogram. Thus, a triangle can be made into a parallelogram by adding to it a triangle of the same size. This is shown in Figure 12.53, where triangle B is the same size and shape as the original triangle A. Since the heights of the triangle and the parallelogram are the same, the area of the triangle must be one-half the area of the parallelogram.

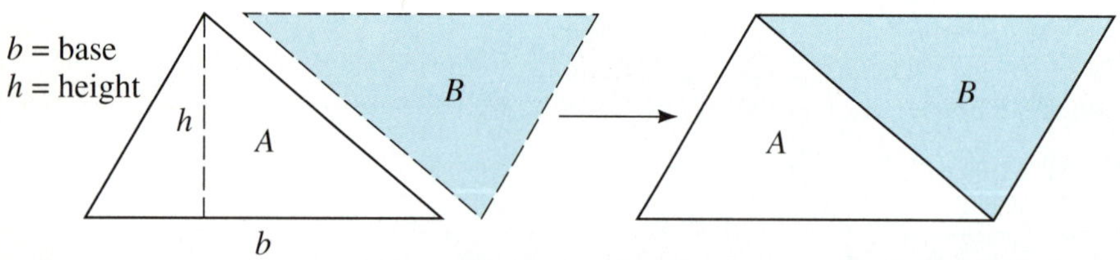

Figure 12.53 Triangles

Area of a Triangle

The area of a triangle with base b and height h is

$$A = \frac{1}{2}bh.$$

EXAMPLE 7 Application of Area

A wall to be painted is in the shape shown in Figure 12.54.

(a) How many square feet must be painted?

First find the area of the rectangle.

$$A = lw = 10 \text{ ft} \cdot 8 \text{ ft} = 80 \text{ ft}^2$$

The base of the triangle is 10 ft and the height is 6 ft.

$$A = \frac{1}{2}bh = \frac{1}{2}(10 \text{ ft})(6 \text{ ft}) = 30 \text{ ft}^2$$

The total area to be covered is

$$80 \text{ ft}^2 + 30 \text{ ft}^2 = 110 \text{ ft}^2.$$

Practice Exercise 7

The wall shown in the figure is to be painted.

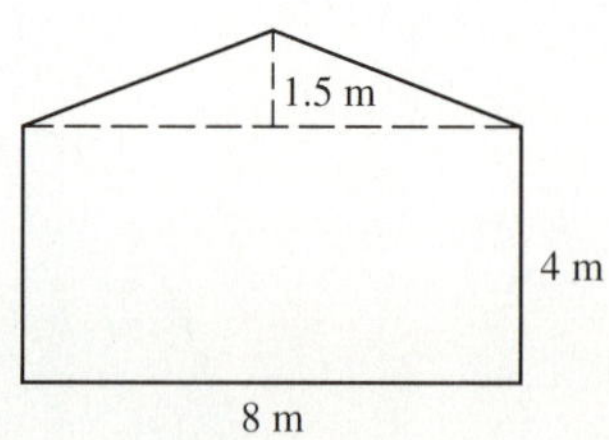

(a) What is the area to be painted?

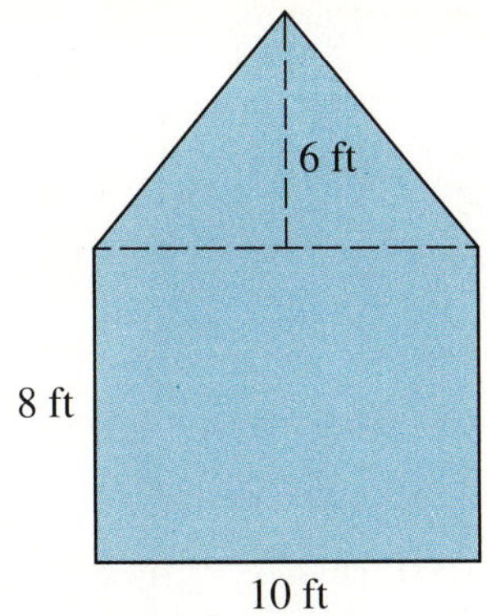

Figure 12.54

(b) How much paint will it take if one gallon covers 220 ft^2?

Multiply 110 ft^2 by $\frac{1 \text{ gal}}{220 \text{ ft}^2}$.

$$\frac{110 \cancel{\text{ft}^2}}{1} \cdot \frac{1 \text{ gal}}{220 \cancel{\text{ft}^2}} = \frac{\cancel{110} \text{ gal}}{2(\cancel{110})} = \frac{1}{2} \text{ gal}$$

It will take $\frac{1}{2}$ gallon of paint to cover the area.

(b) If one liter of paint covers 19 m^2, how many liters of paint are required?

Answers: (a) 38 m^2 (b) 2 L

6 TRAPEZOIDS

A **trapezoid** is a four-sided figure with at least two parallel sides. See Figure 12.55. The parallel lines are called the *bases* and h is the *height.*

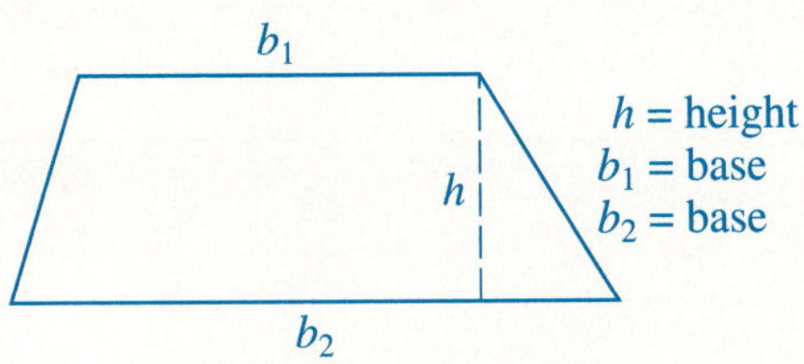

Figure 12.55 Trapezoid

If we take a trapezoid the same shape and size as the given one, turn it over and put the two trapezoids together, as in Figure 12.56, it forms a parallelogram. The area of the parallelogram is equal to the base times the height. The base is $b_1 + b_2$ and the height is h. The area of the trapezoid is one-half their product.

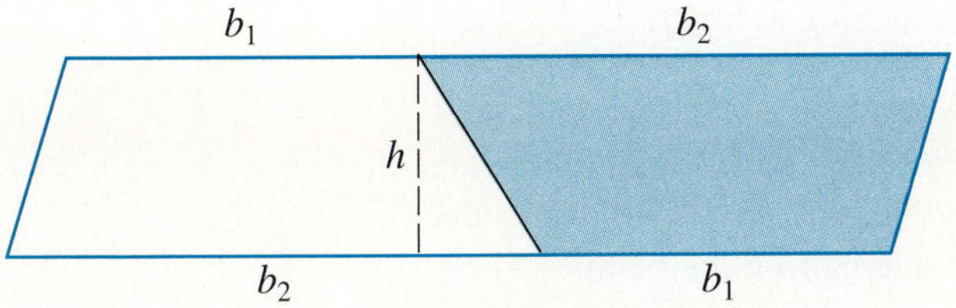

Figure 12.56

Area of a Trapezoid

The area A of a trapezoid with height h and bases b_1 and b_2 is

$$A = \frac{1}{2}(b_1 + b_2)h.$$

EXAMPLE 8 Area of a Trapezoid

Find the area of the trapezoid shown in Figure 12.57.

$$\begin{aligned} A &= \frac{1}{2}(b_1 + b_2)h \\ &= \frac{1}{2}(24 \text{ mm} + 16 \text{ mm})(12 \text{ mm}) \\ &= \frac{1}{2}(40 \text{ mm})(12 \text{ mm}) \\ &= \frac{1}{2}(40)(12) \text{ mm}^2 = 240 \text{ mm}^2 \end{aligned}$$

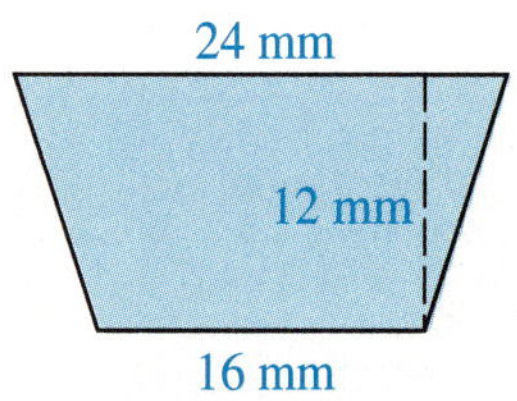

Figure 12.57

Practice Exercise 8

Find the area of the trapezoid with bases 6 inches and 14 inches and height 9 inches.

Answer: 90 in^2

7 CIRCLES

In Section 12.5 we discussed circles. We calculated circumferences but did not determine the area of a circle.

Area of a Circle

The area A of a circle is given by

$$A = \pi r^2.$$

We can get some idea of why this is true by looking at Figure 12.58. Suppose we cut the circle in Figure 12.58(a) along all the radii (plural of radius). When we put the top half of the circle above the bottom half as in Figure 12.58(b) and then slide them together as in Figure 12.58(c), the figure formed is approximately a parallelogram with height r. Its base is πr since $2\pi r$ is the circumference and half the circle is used on top and half on the bottom. Thus,

$$\begin{aligned} A &= bh \\ &= (\pi r)(r) = \pi r^2. \end{aligned}$$

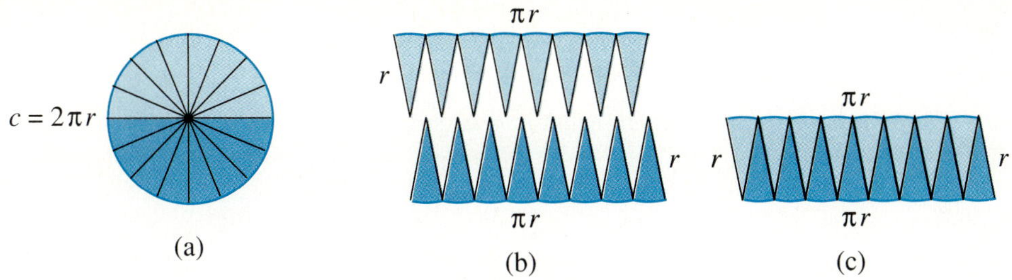

Figure 12.58

EXAMPLE 9 Application of Area of a Circle

In the machine part shown in Figure 12.59, each circular hole has radius 3 cm. Find the area of metal that is left. Use 3.14 for π.

First calculate the area of the trapezoid.

$$A = \frac{1}{2}(b_1 + b_2)h$$
$$= \frac{1}{2}(\mathbf{16\ cm} + \mathbf{24\ cm})(\mathbf{14\ cm})$$
$$= \frac{1}{2}(40\text{ cm})(14\text{ cm})$$
$$= \frac{1}{2}(40)(14)\text{ cm}^2$$
$$= (20)(14)\text{ cm}^2 = 280\text{ cm}^2$$

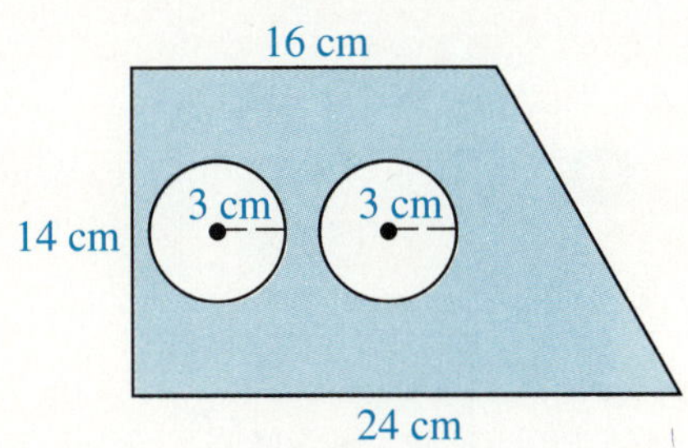

Figure 12.59

Now find the area of each circle.

$$A = \pi r^2 \approx 3.14(\mathbf{3\ cm})^2$$
$$= (3.14)(9)\text{ cm}^2 = 28.26\text{ cm}^2$$

Thus, the two circles have a combined area of

$$2(28.26\text{ cm}^2) = 56.52\text{ cm}^2.$$

The area of metal is the area of the trapezoid minus the area of the circles.

$$\begin{array}{r} 280.00 \\ -\ \ 56.52 \\ \hline 223.48 \end{array}$$

The area left is 223.48 cm^2.

Practice Exercise 9

A triangle with base 10 inches and height 5 inches has three holes of radius 1 inch drilled through it. What is the area left in the triangle after the holes are drilled? Use 3.14 for π.

Answer: 15.58 in^2

12.6 EXERCISES A

Find the perimeter of the rectangle (or square) with the given width and length.

1. 6 ft by 11 ft

2. 2.5 m by 3.6 m

3. 102 mm by 30.5 mm

4. 6.8 mi by $9\frac{1}{2}$ mi

5. 7.2 cm by 7.2 cm

6. $22\frac{1}{2}$ mi by $22\frac{1}{2}$ mi

Find the area of the rectangle (or square) with the given width and length.

7. 6 ft by 11 ft

8. 2.5 m by 3.6 m

9. 102 mm by 30.5 mm

10. 6.8 mi by $9\frac{1}{2}$ mi

11. 7.2 cm by 7.2 cm

12. $22\frac{1}{2}$ mi by $22\frac{1}{2}$ mi

Solve.

13. Find the perimeter and area of the given figure.

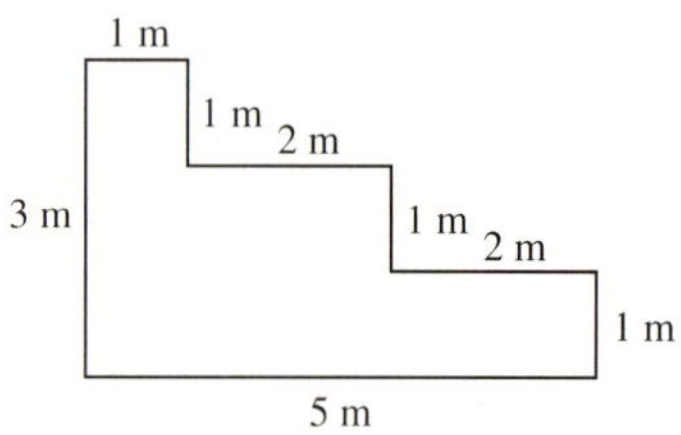

14. Fencing costs \$3.20 per yard. If a pasture 62 yd by 205 yd is to be fenced, how much will it cost?

15. A rose garden is 7.2 m by 10.8 m. If a square with side 2.5 m is used in the middle of the garden for a fountain, what area remains for roses?

16 The area shown is to be carpeted. How much will it cost if carpet costs \$16.50 per square yard?

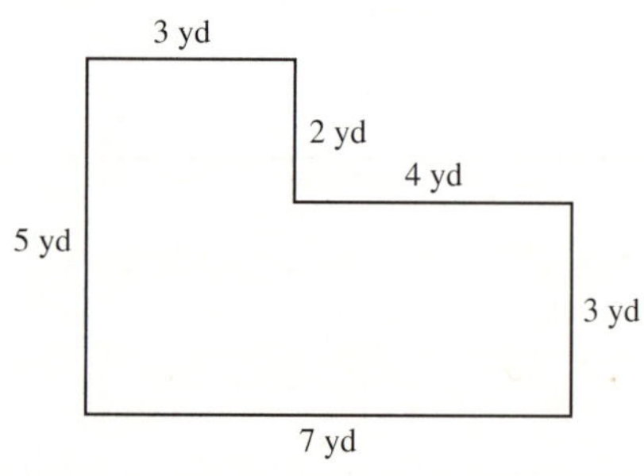

Find the area of each figure.

17.

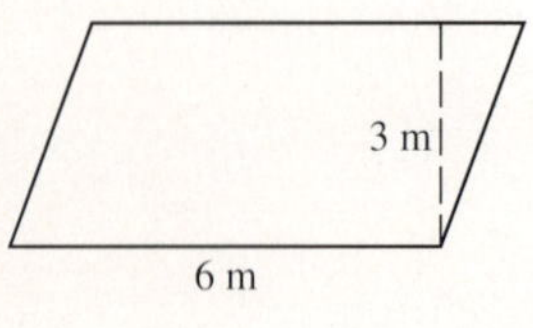

18.

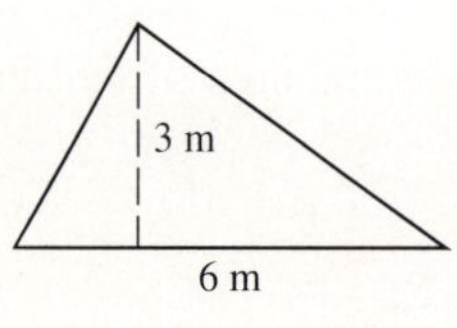

19.

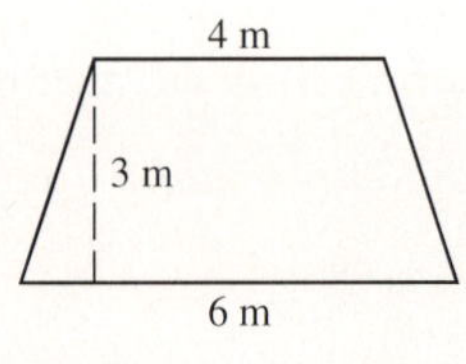

20.

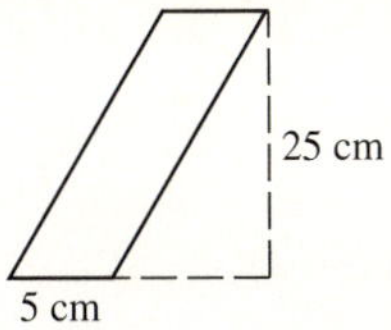

21.

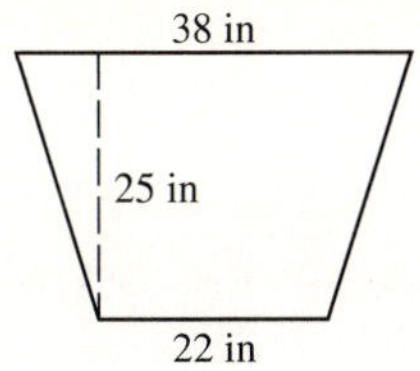

22.

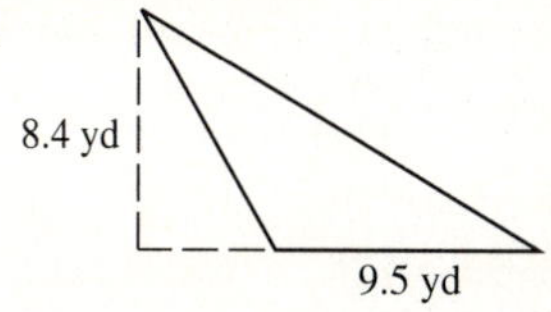

23.

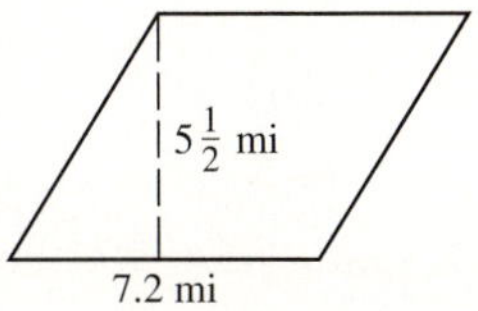

24.

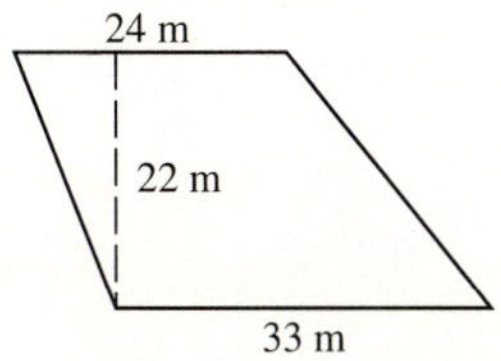

25.

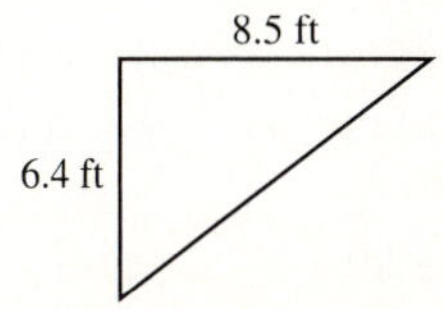

26.

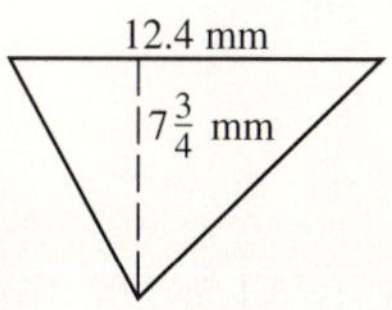

27.

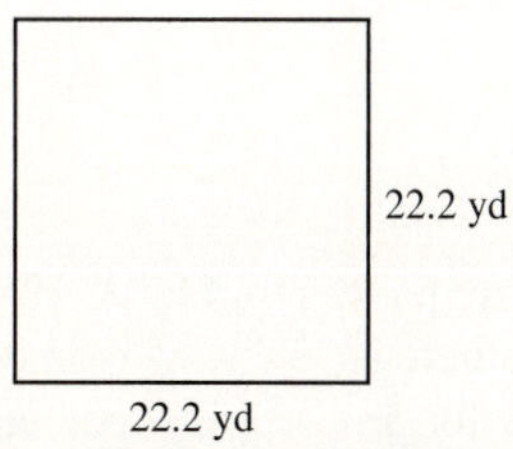

28.

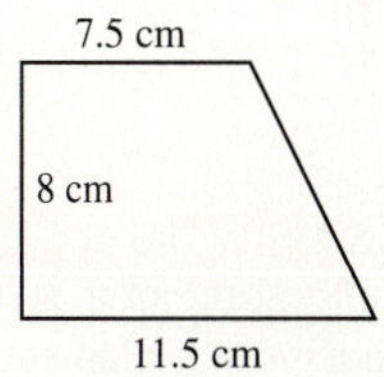

Find the perimeter of each figure.

29.

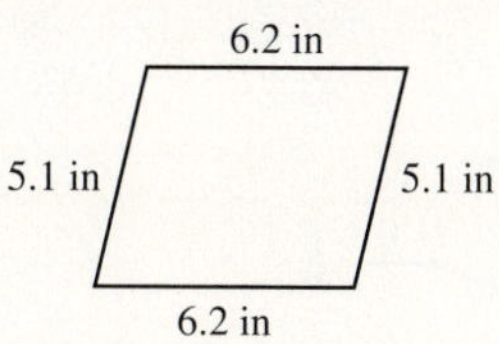

30.

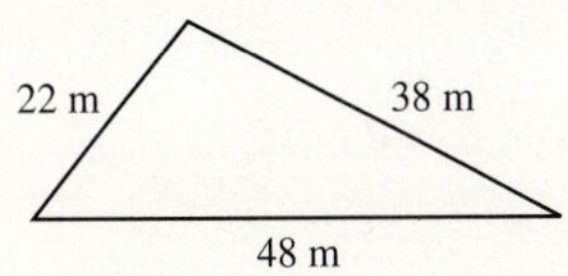

31.

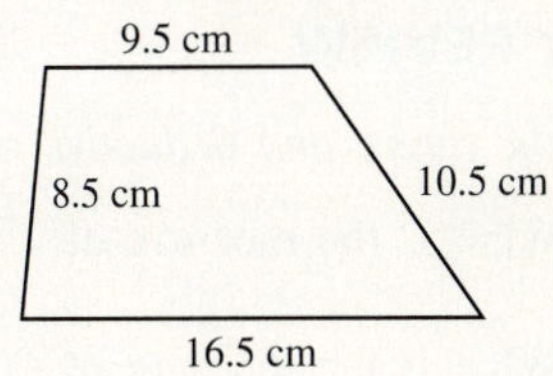

32 Find the area of the figure.

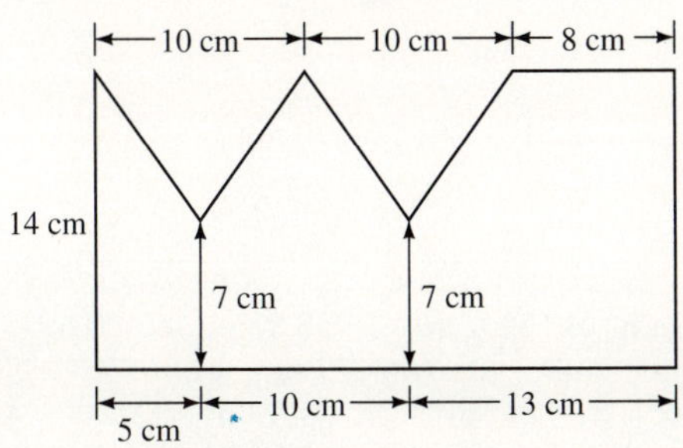

33. The recreation area sketched in the figure is to be carpeted. How much will it cost if carpet is \$14.50 per square yard?

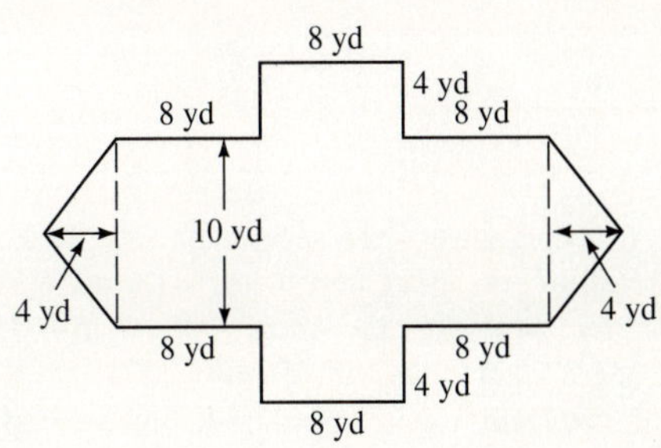

34. For a play, 10 walls shaped like the one shown in the figure must be painted on both sides. How much will it cost if a gallon of paint covers 160 ft^2 and costs $12.50 per gallon?

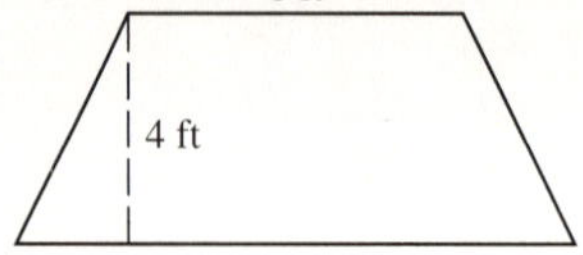

Find the area of the circle. Leave the answer using π.

35. $r = 7$ yd

36. $r = \frac{7}{2}$ ft

37. $r = 7.7$ mi

Find the area of the circle. Use 3.14 for π.

38. $r = 6.8$ km

39. $d = 92$ in

40. $r = 0.05$ cm

Solve.

41. How much more area is there for water to pass through in a $\frac{3}{4}$-inch diameter water hose than there is in a $\frac{1}{2}$-inch diameter water hose? Use 3.14 for π.

42. A 12-inch diameter pizza costs $4.50. A 16-inch diameter pizza costs $7.50. Which pizza costs less per square inch?

FOR REVIEW

Use the figure and deductive reasoning in Exercises 43–45.

43. What is the measure of $\angle AOB$?

44. What is the measure of $\overset{\frown}{ADB}$?

45. What is the measure of $\angle ACB$?

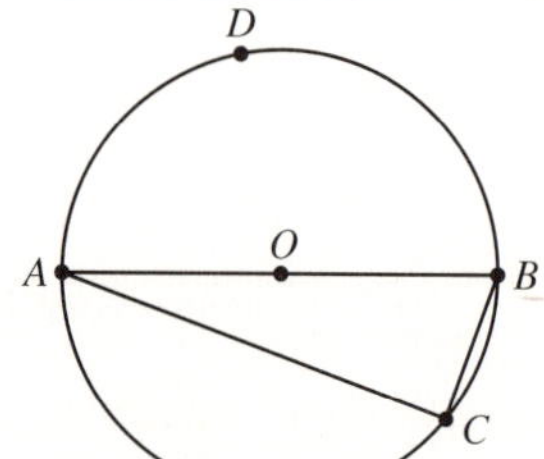

ANSWERS: (Some answers are rounded.) 1. 34 ft 2. 12.2 m 3. 265 mm 4. 32.6 mi 5. 28.8 cm 6. 90 mi 7. 66 ft^2 8. 9.0 m^2 9. 3111 mm^2 10. 64.6 mi^2 11. 51.84 cm^2 12. 506.25 mi^2 13. 16 m; 9 m^2 14. $1708.80 15. 71.51 m^2 16. $445.50 17. 18 m^2 18. 9 m^2 19. 15 m^2 20. 125 cm^2 21. 750 in^2 22. 39.9 yd^2 23. 39.6 mi^2 24. 627 m^2 25. 27.2 ft^2 26. 48.05 mm^2 27. 492.8 yd^2 (to the nearest tenth) 28. 76 cm^2 29. 22.6 in 30. 108 m 31. 45 cm 32. 322 cm^2 33. $4988 34. $50 35. 49π yd^2 36. $\frac{49}{4}\pi$ ft^2 37. 59.29π mi^2 38. 145.2 km^2 39. 6644.2 in^2 40. 0.00785 cm^2 41. 0.246 in^2 42. 12-in: $0.0398 per in^2; 16-in: $0.0373 per in^2; 16-in pizza costs less. 43. 180° 44. 180° 45. 90°

12.6 EXERCISES B

Find the perimeter of the rectangle (or square) with the given width and length.

1. 10 ft by 17 ft

2. 8.2 m by 9.5 m

3. 2500 mm by 45.5 mm

4. 2.2 mi by $6\frac{1}{2}$ mi

5. 40.2 cm by 40.2 cm

6. $10\frac{1}{2}$ mi by $10\frac{1}{2}$ mi

Find the area of the rectangle (or square) with the given width and length.

7. 10 ft by 17 ft

8. 8.2 m by 9.5 m

9. 2500 mm by 45.5 mm

10. 2.2 mi by $6\frac{1}{2}$ mi

11. 40.2 cm by 40.2 cm

12. $10\frac{1}{2}$ mi by $10\frac{1}{2}$ mi

Solve.

13. The play area shown needs to have a new fence around it. How long will the fence need to be?

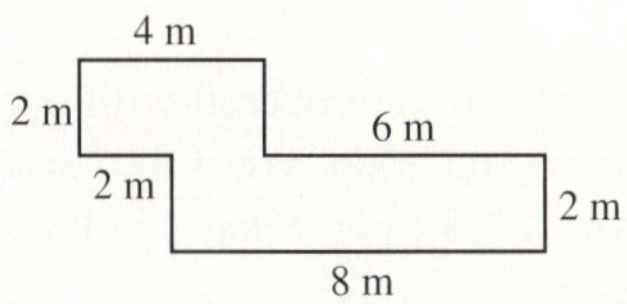

14. A rancher needs to fence an area which is 2500 yd by 4200 yd. If fencing costs \$2.60 per yard, how much will it cost?

15. A flower garden is in the shape of a square which is 22.5 m on a side. If a walk 1.5 m wide is put across the garden parallel to one of the sides, how much area is left for flowers?

16. The area shown is carpeted with carpeting costing \$24.50 per square yard. How much did it cost?

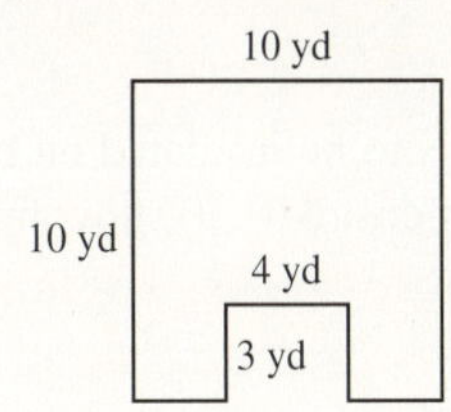

Find the area of each figure.

17.

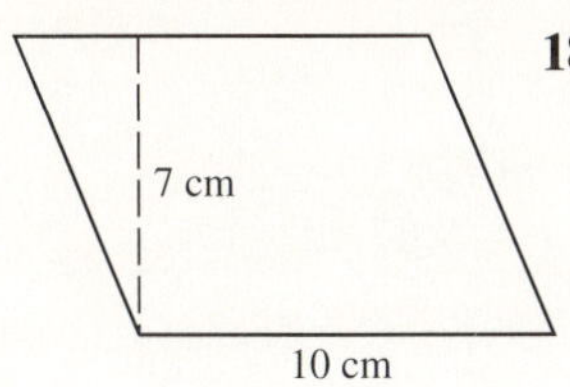

18.

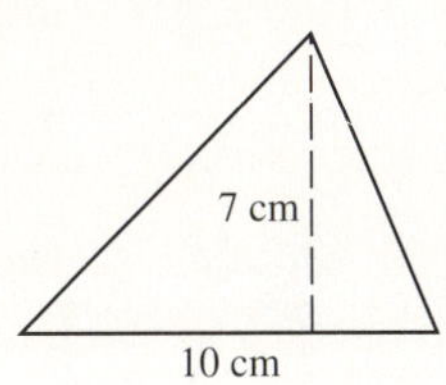

19.

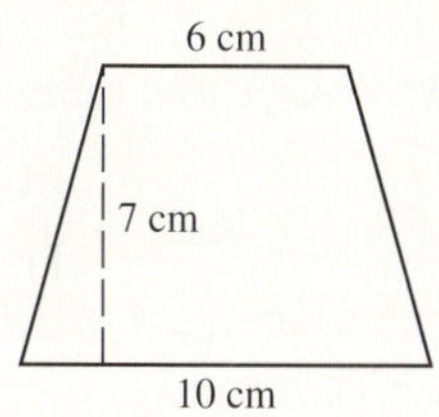

20.

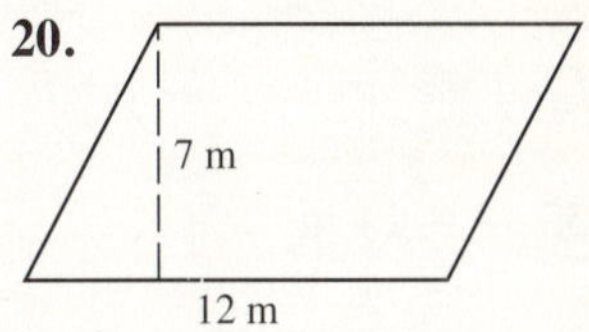

21.

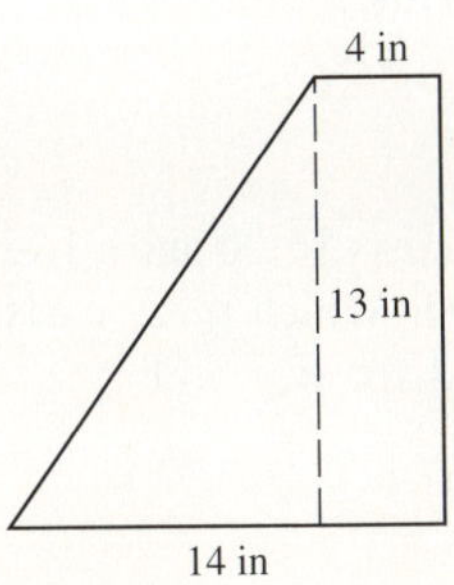

22.

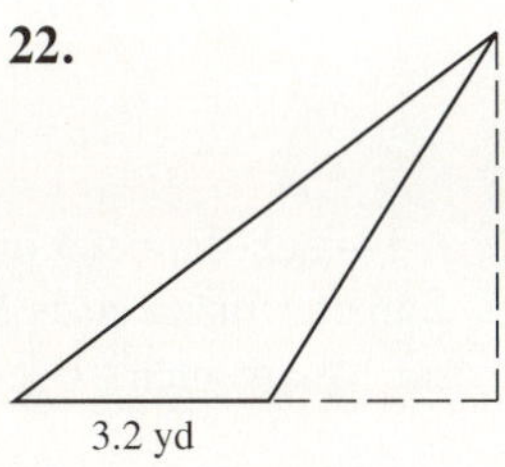

23.

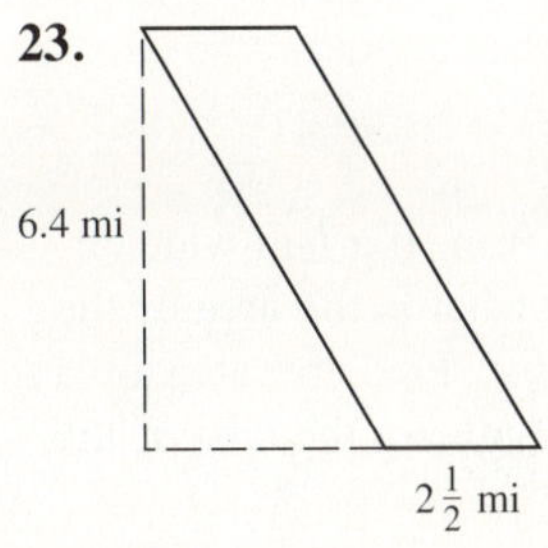

24.

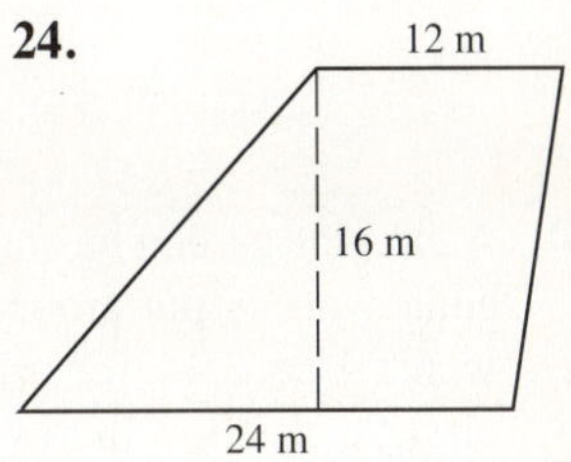

25.

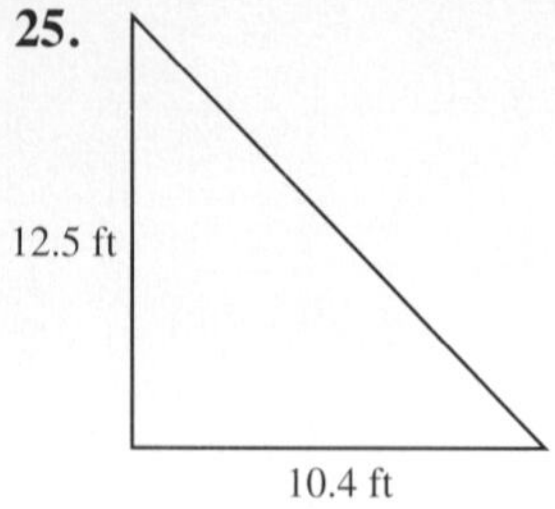

26.

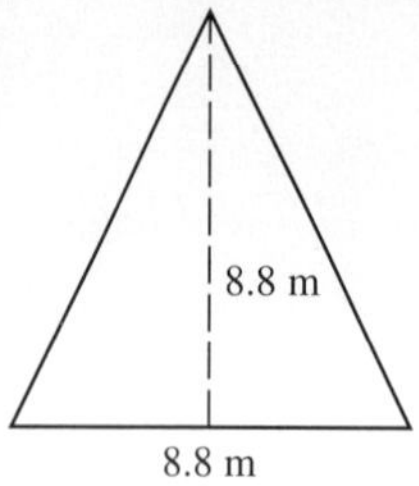

27.

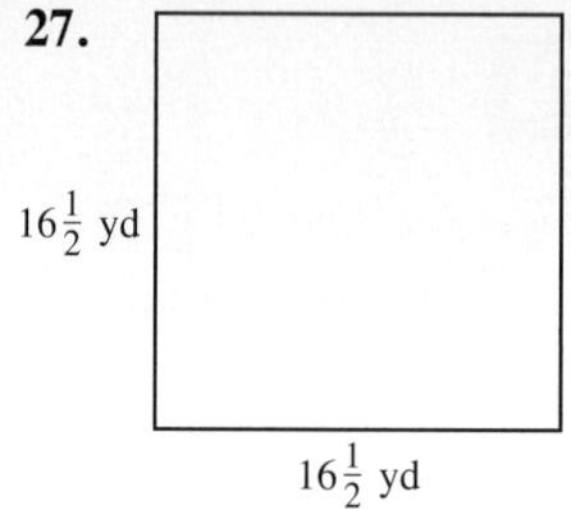

28.

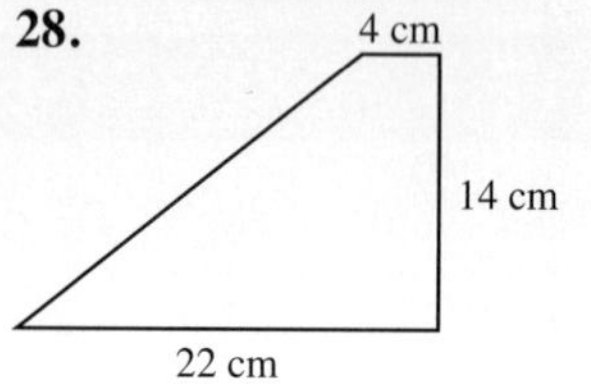

Find the perimeter of each figure.

29.

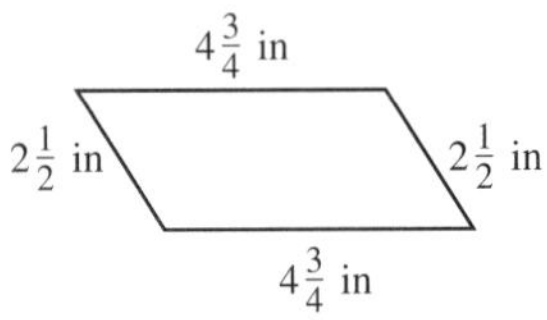

30.

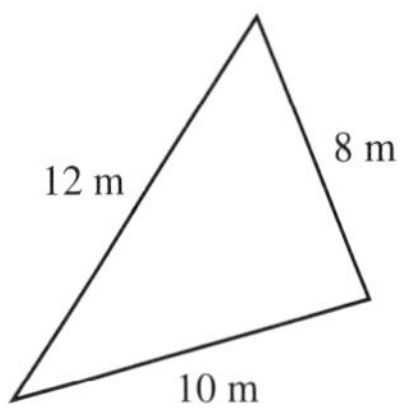

31.

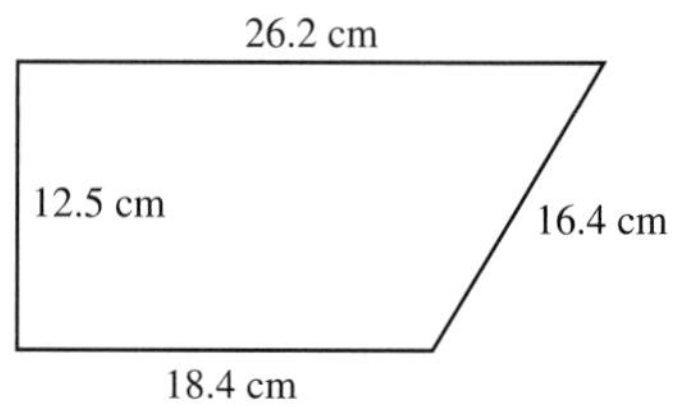

32. Find the area of the figure.

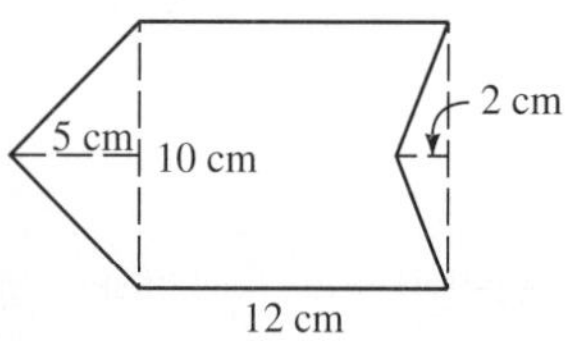

33. The patio area shown is to be covered with outdoor carpeting. What will be the total cost if the carpeting can be installed for \$22.50 per square meter?

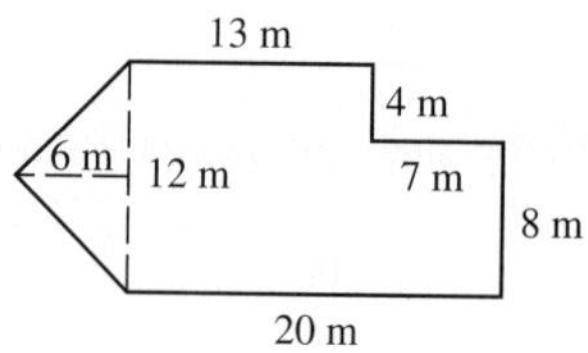

34. The sheet of metal shown is to be insulated on both sides. If 6 m^2 of insulation cost \$10.20, how much will it cost to do the job?

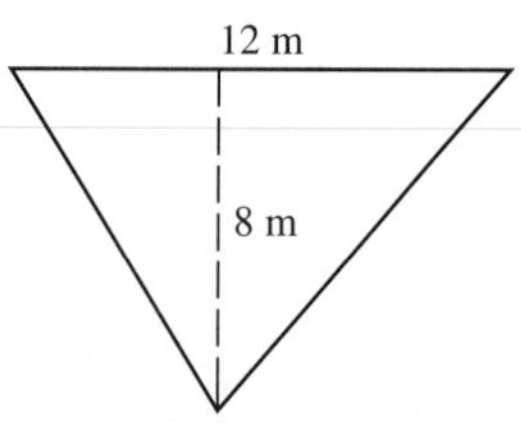

Find the area of the circle. Leave the answer using π.

35. $r = 14$ m

36. $r = \dfrac{14}{11}$ cm

37. $r = 5.67$ yd

Find the area of the circle. Use 3.14 for π.

38. $d = 32.8$ ft

39. $r = 420$ mm

40. $d = 1.64$ in

Solve.

41. A circular garden has radius 9 m. If a 1-m-wide circular walk is put around it, what is the area of the walk? Use 3.14 for π. (*Hint:* Find the area of a circle with 10-m radius and subtract the area of the garden.)

42. A 14-inch-diameter pizza costs \$6.00 and a 16-inch-diameter pizza costs \$8.00. Which pizza costs less per square inch? Use 3.14 for π.

FOR REVIEW

Use the figure and deductive reasoning in Exercises 43–45.

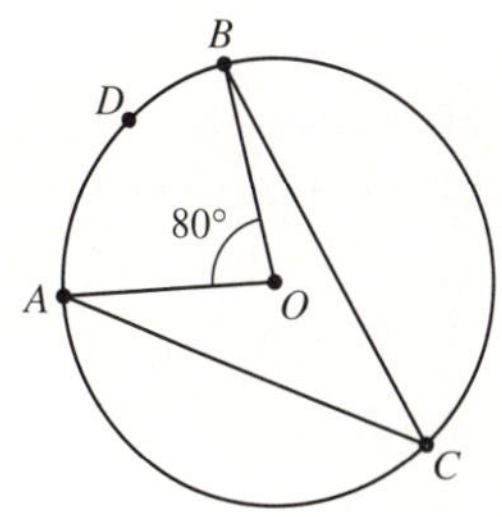

43. What is the measure of $\widehat{ADB}$?

44. What is the measure of $\widehat{ACB}$?

45. What is the measure of $\angle ACB$?

12.6 EXERCISES C

Solve.

1. A patio is in the shape of a trapezoid with bases 8.1 yd and 6.7 yd and height 5.8 yd. Assuming no waste, to the nearest cent how much will it cost to cover the patio with outdoor carpeting that costs \$2.15 a square foot?
[Answer: \$830.50]

2. A sheet of metal is in the shape of a triangle with base 1500 cm and height 800 cm, and has a piece removed from the center of it in the shape of a parallelogram with base 125 cm and height 85 cm. How much will it cost to cover the metal on both sides with rustproofing paint that costs \$0.45 per square meter?

3. Find the area (to the nearest hundredth) of the metal in the machine part shown, which has three circular holes drilled in it. Use 3.14 for π.

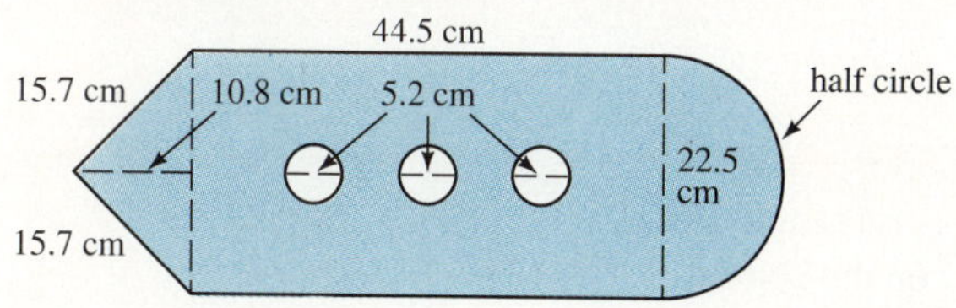

[Answer: 1257.77 cm^2]

4. Find the exterior perimeter (to the nearest tenth) of the machine part in Exercise 3.

5. The shaded portion of the circle at right is called a **sector** of the circle, and determined by central $\angle AOB$. The area of a sector is given by the formula $A = \frac{\pi}{360}r^2a$ where a is the measure of central $\angle AOB$. Use this formula to find the area (to the nearest hundredth) of the sector determined by the given central $\angle AOB$ in a circle of radius 5 cm. Use 3.14 for π.
(a) $\angle AOB = 180°$
(b) $\angle AOB = 90°$
(c) $\angle AOB = 35°$

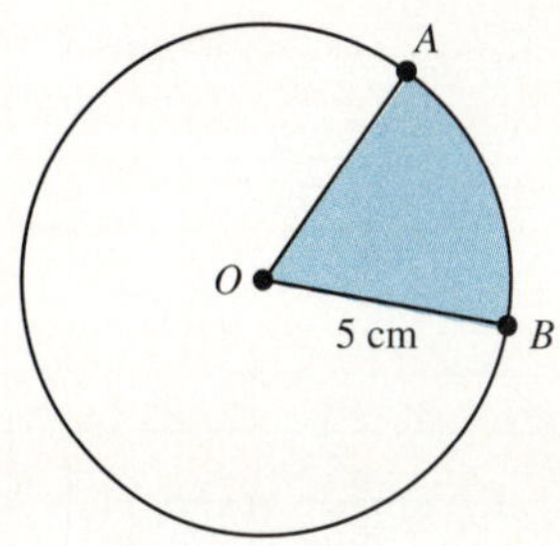

12.7 VOLUME AND SURFACE AREA

STUDENT GUIDEPOSTS

1. Volume
2. Rectangular Solids
3. Cubes
4. Surface Area
5. Cylinders
6. Spheres

1 VOLUME

To measure length we use units such as 1 cm and 1 inch. When we talk about area we use units such as 1 cm^2 and 1 in^2. To find volume we will use cubic units such as 1 cm^3 and 1 in^3.

The **volume** of a cube that is 1 cm by 1 cm by 1 cm is one **cubic centimeter** and is written 1 cm^3. Similarly, a cube that is 1 in by 1 in by 1 in is one **cubic inch,** 1 in^3. See Figure 12.60. If some other unit were used for measuring, the volume of the cube would be 1 cubic unit, or 1 $unit^3$.

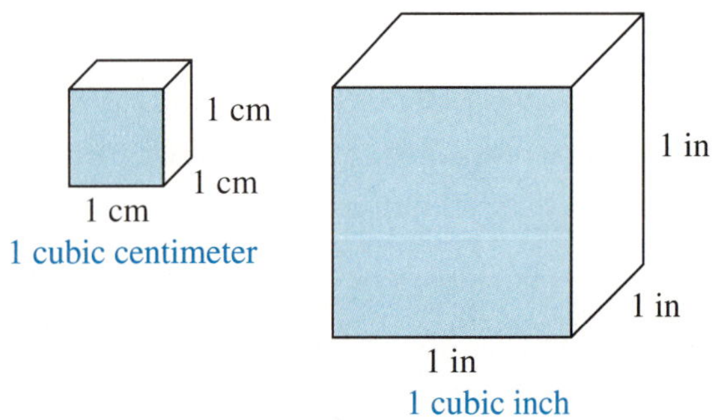

Figure 12.60

2 RECTANGULAR SOLIDS

The cubic unit of measure is used for the volume of solids. The **rectangular solid** in Figure 12.61(a) has volume 20 cm^3 since there are 10 cm^3 on the bottom layer and 10 cm^3 on the top layer. In Figure 12.61(b) there is one more layer (10 cm^3) than in Figure 12.61(a), so the volume is 30 cm^3. These volumes can be found without a diagram by multiplying the area of the base by the height.

$$20 \text{ cm}^3 = \underbrace{5 \text{ cm} \cdot 2 \text{ cm}}_{\text{area of base}} \cdot \underbrace{2 \text{ cm}}_{\text{height}}$$

$$30 \text{ cm}^3 = \underbrace{5 \text{ cm} \cdot 2 \text{ cm}}_{\text{area of base}} \cdot \underbrace{3 \text{ cm}}_{\text{height}}$$

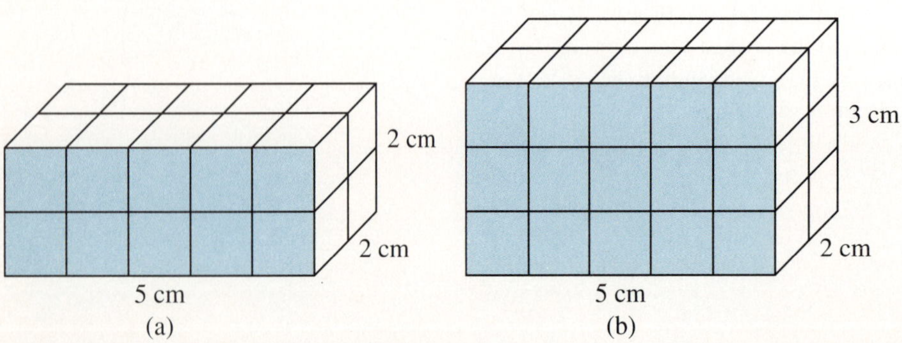

Figure 12.61

Volume of a Rectangular Solid

The volume V of a rectangular solid is equal to the area of the base lw times the height h.

$$V = lwh$$

EXAMPLE 1 VOLUME OF A RECTANGULAR SOLID

A highway construction truck has a bed that is 5.5 yd long, 2.5 yd wide, and 2 yd deep. How many cubic yards of gravel will it hold?

Find the volume of the truck bed.

$$\begin{aligned} V &= lwh \\ &= (5.5 \text{ yd})(2.5 \text{ yd})(2 \text{ yd}) \\ &= (5.5)(2.5)(2) \text{ yd}^3 \\ &= 27.5 \text{ yd}^3 \end{aligned}$$

The truck will hold 27.5 yd^3 of gravel.

PRACTICE EXERCISE 1

A trailer for hauling cotton is 18 ft by 8 ft by 10 ft. What volume of cotton will it hold?

Answer: 1440 ft^3

3 CUBES

A **cube** is a rectangular solid for which $l = w = h$. We talk about *edges* of the cube, and call the length of each edge e.

Volume of a Cube

The volume V of a cube is equal to the cube of the edge.

$$V = e^3 \qquad (e^3 \text{ is } e \text{ times } e \text{ times } e)$$

EXAMPLE 2 VOLUME OF A CUBE

Find the volume of the cube in Figure 12.62 which has an edge of 1.5 ft.

$$\begin{aligned} V &= (1.5 \text{ ft})^3 \\ &= (1.5 \text{ ft})(1.5 \text{ ft})(1.5 \text{ ft}) \\ &= 3.375 \text{ ft}^3 \end{aligned}$$

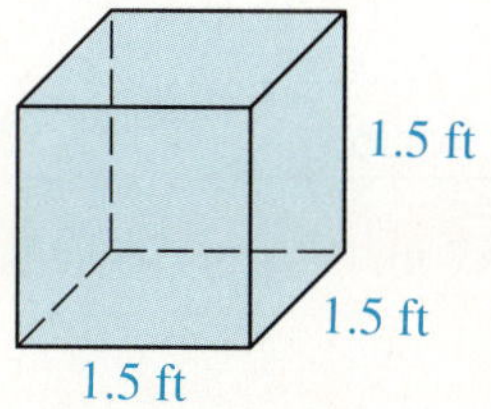

Figure 12.62

PRACTICE EXERCISE 2

Find the volume of a cube which has an edge of $3\frac{1}{2}$ inches.

Answer: $42\frac{7}{8}$ in^3

4 SURFACE AREA

Another type of problem involving a rectangular solid is finding the area of the surface. If we can do this, we can find the amount of material needed to make a rectangular tank or the amount of paint needed to paint a building. The **surface area** of an object is the area of all its surfaces. Consider the solid in Figure 12.63, which was used to count cubic centimeters earlier. Now, we count square centimeters on its surface. There are (5 cm)(3 cm) = 15 cm^2 on the front face. But the back is just like the front, so there are

$$2(15\text{ cm}^2) = 30\text{ cm}^2 \qquad 2lh$$

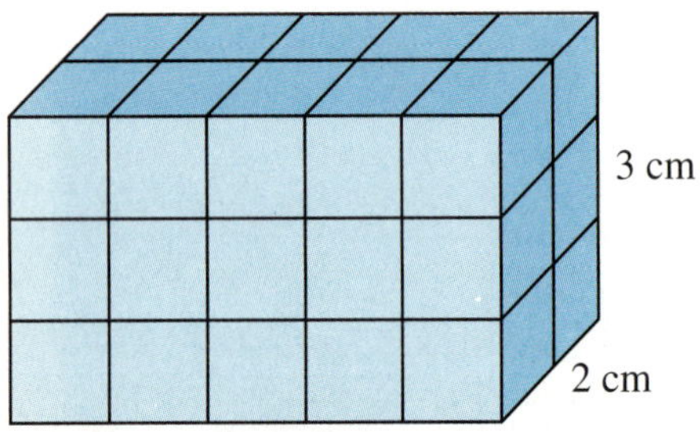

Figure 12.63

on the front and back together. On the two ends there are

$$2(2\text{ cm})(3\text{ cm}) = 12\text{ cm}^2. \qquad 2wh$$

On the bottom there are

$$(5\text{ cm})(2\text{ cm}) = 10\text{ cm}^2,$$

with the same number on the top. The total area of the top and bottom is

$$2(10\text{ cm}^2) = 20\text{ cm}^2. \qquad 2lw$$

The total surface area is the sum of all surfaces.

Area of front and back:	30 cm^2	$2lh$
Area of two ends:	12 cm^2	$2wh$
Area of bottom and top:	20 cm^2	$2lw$
Total surface area:	62 cm^2	

Surface Area of a Rectangular Solid

The total surface area A of a rectangular solid is

$$A = 2lh + 2wh + 2lw.$$

EXAMPLE 3 SURFACE AREA OF A RECTANGULAR SOLID

Find the total surface area of a rectangular solid 6 m by 3 m by 7 m.

$$\begin{aligned} A &= 2lh + 2wh + 2lw \\ &= 2(6\text{ m})(7\text{ m}) + 2(3\text{ m})(7\text{ m}) + 2(6\text{ m})(3\text{ m}) \\ &= 2(6)(7)\text{m}^2 + 2(3)(7)\text{m}^2 + 2(6)(3)\text{m}^2 \\ &= 84\text{ m}^2 + 42\text{ m}^2 + 36\text{ m}^2 = 162\text{ m}^2 \end{aligned}$$

PRACTICE EXERCISE 3

Find the surface area of a rectangular solid 12 ft by 16 ft by 20 ft.

Answer: 1504 ft^2

EXAMPLE 4 Application of Surface Area

The four vertical sides of a storage building are to be painted. One gallon of paint will cover 150 ft^2 of area. How much paint will be used if the building is 16 ft by 12 ft by 8 ft high?

The area of the front and back is twice the length times the height.

$$2(16 \text{ ft})(8 \text{ ft}) = 256 \text{ ft}^2$$

The area of the two ends is twice the width times the height.

$$2(12 \text{ ft})(8 \text{ ft}) = 192 \text{ ft}^2$$

The total area is the sum of these two areas.

Area of front and back:	256 ft^2
Area of two ends:	192 ft^2
Total area:	448 ft^2

Since the paint covers 150 ft^2 per gal, the unit fraction is $\frac{1 \text{ gal}}{150 \text{ ft}^2}$.

$$\frac{448 \cancel{\text{ft}^2}}{1} \times \frac{1 \text{ gal}}{150 \cancel{\text{ft}^2}} = \frac{448}{150} \text{ gal}$$

$$\approx 2.99 \text{ gal} \quad \text{To the nearest hundredth}$$

Thus, it will take about 3 gal to paint the building.

Practice Exercise 4

The outside walls of a workshop are to be covered with siding that costs \$1.50 per square foot. How much will it cost if the building is 16 ft wide, 18 ft long, and 8 ft high?

Answer: \$816

Surface Area of a Cube

The surface area of a cube with edge e is given by

$$A = 6e^2.$$

EXAMPLE 5 Surface Area of a Cube

Find the surface area of a cube with edge 14 cm.

$$A = 6e^2$$
$$= 6(14 \text{ cm})^2$$
$$= 6(196) \text{ cm}^2$$
$$= 1176 \text{ cm}^2$$

Practice Exercise 5

Find the surface area of a cube with edge 200 mm.

Answer: 240,000 mm^2

5 CYLINDERS

A **cylinder** (see Figure 12.64) is a solid with circular ends of the same radius. For example, a can of beans is in the shape of a cylinder.

Remember that the volume of a rectangular solid is the area of the base times the height. The same is true for a cylinder. Since the base is a circle with area πr^2, we have the following formula.

Volume of a Cylinder

The volume V of a cylinder is equal to the area of the base, πr^2, times the height, h.

$$V = \pi r^2 h$$

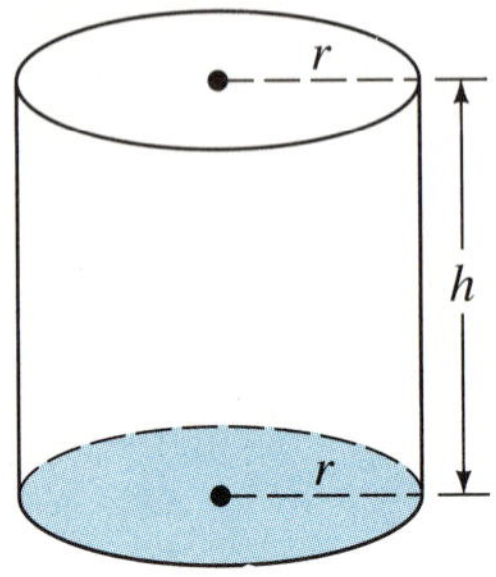

Figure 12.64 Cylinder

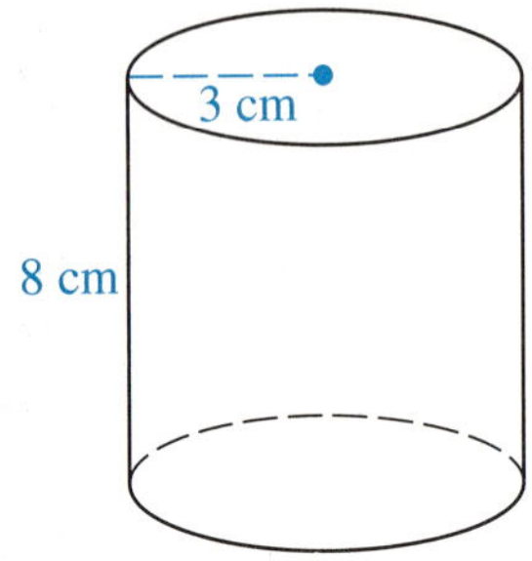

Figure 12.65

EXAMPLE 6 Volume of a Cylinder

Find the volume of the cylinder in Figure 12.65, which is 8 cm high and has a base of radius 3 cm. Use 3.14 for π.

Since $r = 3$ cm and $h = 8$ cm,

$$\begin{aligned} V &= \pi r^2 h \\ &\approx 3.14(\mathbf{3\ cm})^2(\mathbf{8\ cm}) \\ &= (3.14)(9)(8)\ \text{cm}^3 = 226.08\ \text{cm}^3. \end{aligned}$$

Practice Exercise 6

Find the volume of a cylinder with radius of the base 6 cm and height 20 cm. Use 3.14 for π.

Answer: 2260.8 cm^3

EXAMPLE 7 Application of Volume

A can of blueberry pie filling has diameter 3 in and height 4.5 in. How many cans of filling are needed to fill a 9-in diameter pie pan 1 in deep?

First find the volume of pie filling in one can.

$$\begin{aligned} V &= \pi r^2 h \\ &\approx 3.14(1.5\ \text{in})^2(4.5\ \text{in}) \\ &= (3.14)(2.25)(4.5)\ \text{in}^3 \\ &\approx 31.8\ \text{in}^3 \qquad \text{Volume in one can to nearest tenth} \end{aligned}$$

Now find the volume of pie filling in a 9-in diameter (4.5-in radius) pie pan.

$$\begin{aligned} V &= \pi r^2 h \\ &\approx 3.14(4.5\ \text{in})^2(\mathbf{1\ in}) \qquad \text{Filling is 1 in deep} \\ &= (3.14)(20.25)(1)\ \text{in}^3 \\ &\approx 63.6\ \text{in}^3 \qquad \text{Volume in pie pan to nearest tenth} \end{aligned}$$

Since one can has 31.8 in^3 of filling, multiply by the unit fraction $\frac{1\ \text{can}}{31.8\ \text{in}^3}$.

Practice Exercise 7

Water is stored in a cylindrical tank with diameter of the base 12 m and height 10 m. How many water trucks having cylindrical tanks with diameter 1.6 m and length 5 m can be filled from the storage tank?

$$\frac{63.6\ \cancel{\text{in}^3}}{1} \times \frac{1\ \text{can}}{31.8\ \cancel{\text{in}^3}} = \frac{63.6}{31.8}\ \text{can} = 2\ \text{cans}$$

It will take 2 cans of filling to make the pie.

Answer: 112.5

To see how to find the surface area of a cylinder, consider a can without top or bottom. See Figure 12.66(a). If it is cut along the seam and pressed flat, as in Figure 12.66(b), a rectangle is formed. The length of the rectangle is the circumference of the circle and the width is the height. Thus, the surface area of the side of the can is $2\pi rh$.

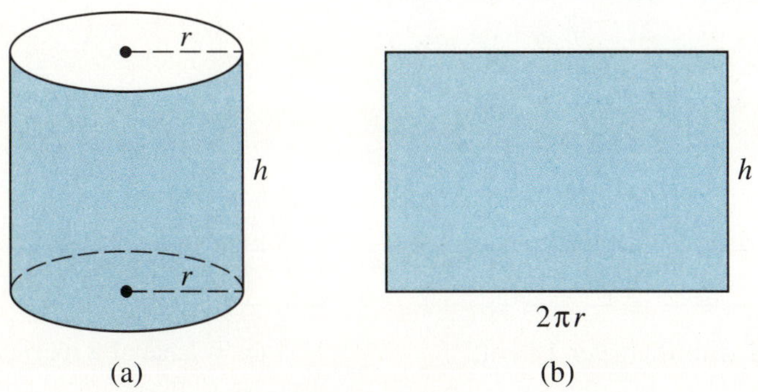

Figure 12.66

Since the area of the top of the can is πr^2, and the area of the bottom is also πr^2, we have the following:

Surface Area of a Cylinder

The surface area A of a cylinder is given by

$$A = 2\pi rh + 2\pi r^2.$$

EXAMPLE 8 Surface Area of a Cylinder

Find the surface area of a cylinder with radius 7 ft and height 9 ft.

$$\begin{aligned} A &= 2\pi rh + 2\pi r^2 \\ &\approx 2(3.14)(7\text{ ft})(9\text{ ft}) + 2(3.14)(7\text{ ft})^2 \\ &= 2(3.14)(7)(9)\text{ ft}^2 + 2(3.14)(49)\text{ ft}^2 \\ &= 703.36\text{ ft}^2 \end{aligned}$$

Practice Exercise 8

Find the surface area of a cylinder with radius 16 ft and height 4 ft.

Answer: 2009.6 ft^2

6 SPHERES

A **sphere** (Figure 12.67) is a solid in the shape of a ball. The radius of a sphere is the distance from its center to the surface. The diameter is twice the radius. If a sphere is cut through the center, the cut surface is a circle.

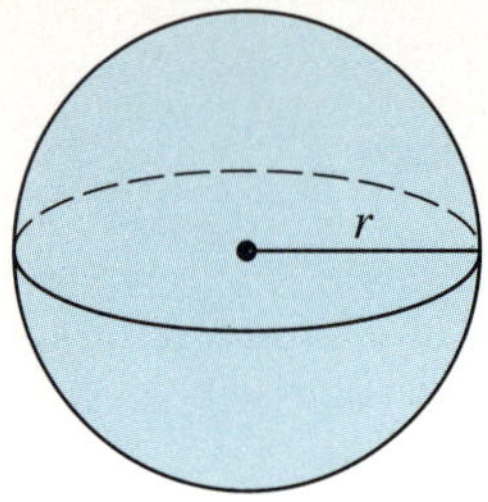

Figure 12.67 Sphere

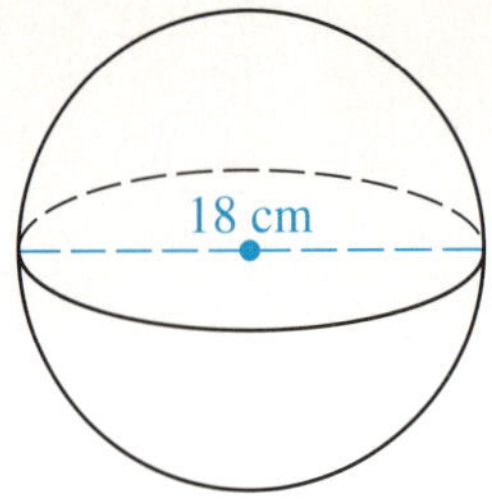

Figure 12.68

Volume of a Sphere

The volume V of a sphere is given by

$$V = \frac{4}{3}\pi r^3.$$

EXAMPLE 9 Volume of a Sphere

Find the volume of the sphere in Figure 12.68 with diameter 18 cm.

If the diameter is 18 cm, then the radius r is $\frac{18}{2}$ cm or 9 cm.

$$\begin{aligned} V &= \frac{4}{3}\pi r^3 \approx \frac{4}{3}(3.14)(9\text{ cm})^3 \\ &= \frac{4}{3}(3.14)729\text{ cm}^3 \\ &= 3052.08\text{ cm}^3 \end{aligned}$$

Practice Exercise 9

Find the volume of a sphere with diameter 10 ft.

Answer: 523.3 ft^3

Surface Area of a Sphere

The surface area A of a sphere is given by

$$A = 4\pi r^2.$$

EXAMPLE 10 Application of Surface Area

A spherical tank with radius 3.2 m is to be covered with insulation. If the insulation costs $2.50 per square meter, how much will the material cost?

$$\begin{aligned} A &= 4\pi r^2 \\ &\approx 4(3.14)(3.2\text{ m})^2 \\ &= 4(3.14)(10.24)\text{ m}^2 \\ &\approx 128.61\text{ m}^2 \end{aligned}$$

To the nearest hundredth

Since the insulation costs $2.50 per square meter, the cost is

$$\frac{128.61\text{ m}^2}{1} \times \frac{\$2.50}{1\text{ m}^2} \approx \$321.53.$$

To the nearest cent

Practice Exercise 10

A spherical tank with diameter 6.4 m is to be prepared for painting. If the labor costs are $4.50 per square meter, how much will the job cost?

Answer: $578.76

12.7 EXERCISES A

Find the volume of the rectangular solid with the given dimensions.

1. 10 m by 5 m by 6 m

2. 2.3 cm by 1.2 cm by 4.5 cm

3 $9\frac{1}{2}$ in by 3.4 in by $8\frac{3}{4}$ in

Find the volume of the cube with the given edge length.

4. 8 m

5. 10.2 cm

6. $1\frac{3}{5}$ in

Find the total surface area of each rectangular solid.

7. 6 ft by 2 ft by 5 ft

8. $1\frac{1}{2}$ in by $\frac{1}{2}$ in by 4 in

9. 10.2 yd by $8\frac{1}{2}$ yd by 4.5 yd

10 15.5 cm by 6.4 cm by 20.2 cm

Find the surface area of the cube with the given edge length.

11. 5 ft

12. 21 mm

13. $1\frac{1}{2}$ yd

Solve.

14. Frank's truck is to be used to carry topsoil. If the truck bed is 2 yd by 1.5 yd by 0.8 yd, how many cubic yards of topsoil will it hold?

15 A tank is in the shape of a cube. How many grams of water will it hold if the edge of the cube is 22 cm? Note that 1 cm^3 of water weighs 1 g.

16 How many square centimeters of glass did it take to make the bottom and sides of the tank in Exercise 15?

17. Elizabeth owns a rectangular building that is 30 ft by 12 ft by 8 ft high. How much will it cost to put siding on the building if the siding costs $0.75 per square foot?

Find the volume of each cylinder.

18. $r = 6$ cm, $h = 5$ cm

19 $d = 24.4$ in, $h = 30$ in

20. $d = 11.4$ m, $h = 4.4$ m

Find the surface area of each cylinder.

21. $r = 6$ cm, $h = 5$ cm

22 $d = 24.4$ in, $h = 30$ in

23. $d = 11.4$ m, $h = 4.4$ m

Find the volume of each sphere.

24. $r = 12$ mm

25 $r = \frac{3}{4}$ in

26. $d = 5.6$ yd

Find the surface area of each sphere.

27. $r = 12$ mm

28. $r = \frac{3}{4}$ in

29 $d = 5.6$ yd

Solve.

30 A can of cherry pie filling has diameter 8 cm and height 12 cm. How many cans (to the nearest tenth) are needed to fill a 20-cm diameter pie pan 3 cm deep?

31. A cylindrical storage tank has radius 3.8 ft and height 9.8 ft. How many gallons (to the nearest tenth) of paint are needed to paint the tank (including top and bottom) if one gallon covers 150 ft^2?

32. A spherical tank has radius 8.6 m. If a rust-preventing material costs $1.50 per square meter, what is the cost to rustproof the tank?

33. A spherical tank with radius 32 cm is filled with gasoline. How many liters are in the tank? Note that 1 $cm^3 = 0.001$ L.

FOR REVIEW

Find the area of metal left on each of the machine parts shown, which have circular holes drilled in them. Use 3.14 for π.

34.

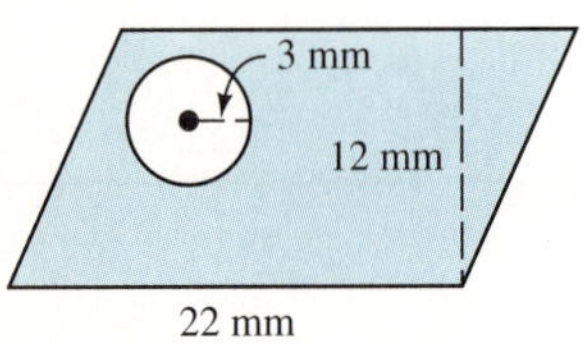

35

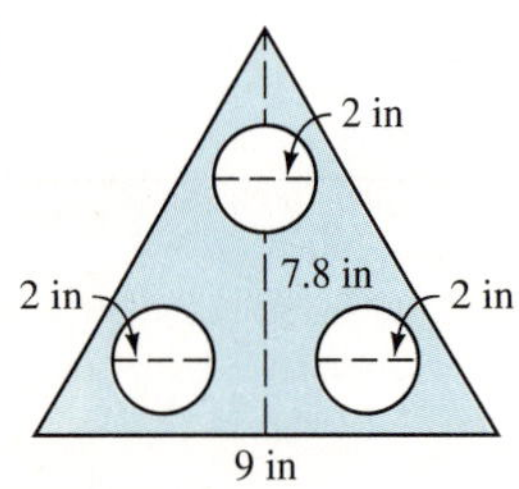

36.

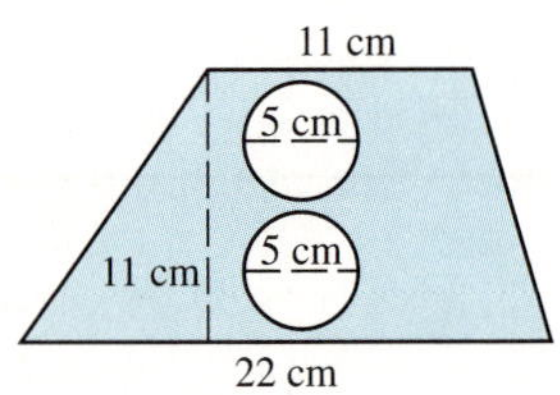

ANSWERS: 1. 300 m^3 2. 12.42 cm^3 3. 282.625 in^3 4. 512 m^3 5. 1061.208 cm^3 6. 4.096 in^3 7. 104 ft^2 8. $17\frac{1}{2}$ in^2 9. 341.7 yd^2 10. 1083.16 cm^2 11. 150 ft^2 12. 2646 mm^2 13. $13\frac{1}{2}$ yd^2 14. 2.4 yd^3 15. 10,648 g 16. 2420 cm^2 17. $504 18. 565.2 cm^3 19. 14,020.7 in^3 (nearest tenth) 20. 448.9 m^3 (nearest tenth) 21. 414.48 cm^2 22. 3233.2 in^2 (nearest tenth) 23. 361.5 m^2 (nearest tenth) 24. 7234.56 mm^3 25. 1.766 in^3 (nearest thousandth) 26. 91.91 yd^3 (nearest hundredth) 27. 1808.64 mm^2 28. 7.065 in^2 29. 98.47 yd^2 (nearest hundredth) 30. 1.6 cans 31. 2.2 gal 32. $1393.41 33. 137.19 L (nearest hundredth) 34. 235.74 mm^2 35. 25.68 in^2 36. 142.25 cm^2

12.7 EXERCISES B

Find the volume of the rectangular solid with the given dimensions.

1. 6 m by 12 m by 15 m

2. 9.2 cm by 1.5 cm by 7.5 cm

3. $6\frac{1}{2}$ in by 5.5 in by $3\frac{1}{5}$ in

Find the volume of the cube with the given edge length.

4. 12 m

5. 0.1 cm

6. $2\frac{1}{6}$ in

Find the total surface area of each rectangular solid.

7. 3 ft by 5 ft by 12 ft

8. $4\frac{1}{2}$ in by $\frac{1}{4}$ in by 6 in

9. 8.6 yd by $4\frac{1}{5}$ yd by 7.5 yd

10. 12.2 cm by 100 cm by 6.5 cm

Find the surface area of the cube with the given edge length.

11. 17 ft

12. 0.01 mm

13. $4\frac{1}{6}$ yd

Solve.

14. A large earth mover has a bed which is 5 yd by 4 yd by 10 yd. How many cubic yards of dirt will it hold?

15. A water tank is 22.5 cm by 10 cm by 5.2 cm. How many grams of water will it hold? Note that 1 cm^3 of water weighs 1 g.

16. A box which is a cube with edge 8.5 dm is to be covered with foil. How much foil is required?

17. A rectangular tank 4 m by 6 m by 5 m (5 m is the height) needs to be insulated on the sides and top. If insulation costs \$2.15 per square meter, how much will it cost?

Find the volume of each cylinder.

18. $r = 9$ cm, $h = 15$ cm

19. $d = 6.8$ m, $h = 10.5$ m

20. $d = 20.4$ in, $h = 8.5$ in

Find the surface area of each cylinder.

21. $r = 9$ cm, $h = 15$ cm

22. $d = 6.8$ m, $h = 10.5$ m

23. $d = 20.4$ in, $h = 8.5$ in

Find the volume of each sphere.

24. $r = 30$ mm

25. $r = \frac{2}{5}$ in

26. $d = 9.8$ yd

Find the surface area of each sphere.

27. $r = 30$ mm

28. $r = \frac{2}{5}$ in

29. $d = 9.8$ yd

Solve.

30. Mercury is stored in a cylinder of radius 16 cm and height 20 cm. How many cylindrical tubes with radius 0.5 cm and height 100 cm can be filled from a full supply?

31. A cylindrical tank is to be made from sheet metal. If the tank is to have radius 2.2 ft and height 12 ft, what will the metal cost if it is $16.50 per square foot?

32. A spherical tank with diameter 36 ft is to be insulated. If insulation costs $0.50 per square foot, how much will the job cost?

33. How many liters of chemical can be stored in a sphere with radius 650 mm? Note that 1 mm^3 = 0.000001 L.

FOR REVIEW

Find the area of metal left on each of the machine parts shown, which have circular holes drilled in them. Use 3.14 for π.

34.

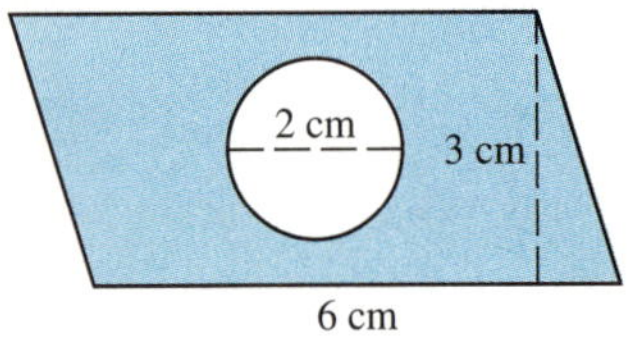

35.

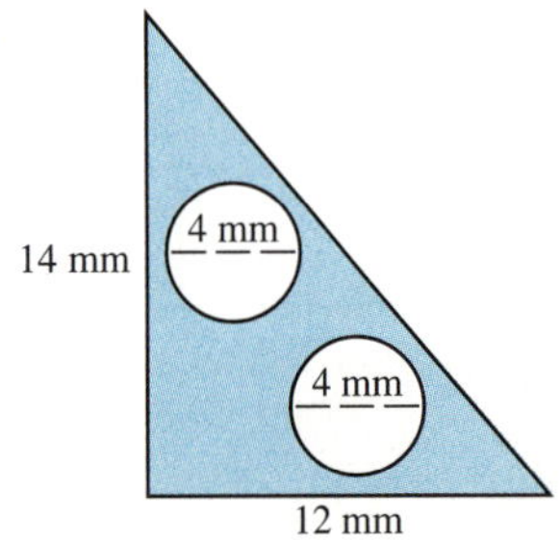

36.

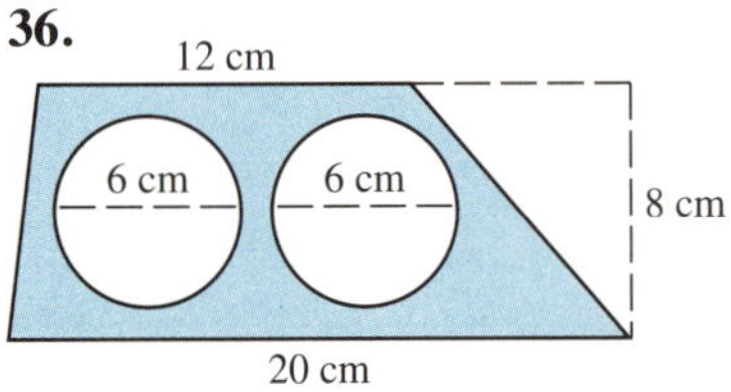

12.7 EXERCISES C

1. Find the volume of the wood remaining when the rectangular block shown at right has a hole bored through it. Use 3.14 for π and give the answer rounded to the nearest hundredth.
[Answer: 0.46 ft^3]

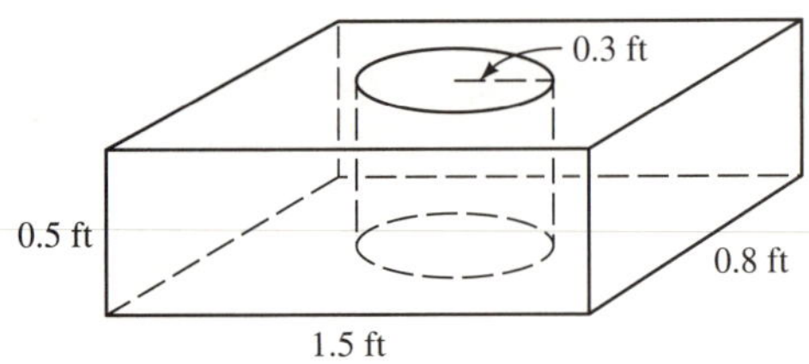

2. Find the total surface area of the wood remaining in the block in Exercise 1.

CHAPTER 12 REVIEW

KEY WORDS

12.1 We use **inductive reasoning** when we reach a general conclusion based on a limited collection of specific observations.

We use **deductive reasoning** when we reach a specific conclusion based on a collection of generally accepted assumptions.

12.2 A statement is a **logical statement** if the statement has **truth value,** that is, it is understood to be either true or false, not both.

A statement of the form "If A, then B" is a **conditional (statement).** Sometimes a conditional is given in the form "All A are B."

The **converse** of the conditional "If A, then B" is the statement "If B, then A."

A **biconditional** is the combination of a conditional and its converse, written A if and only if B, or simply, A iff B.

The **negation** of a statement is formed by adding the word "not," either specifically or implied. The negation of a statement has the opposite truth value from the statement itself.

A conclusion based on a set of premises is **valid** if it necessarily follows from the premises.

A **syllogism** is a special logical form consisting of several premises and a conclusion, the validity of which must be verified.

12.3 A **segment** is a geometric figure made up of two endpoints and all points between them.

A **line** consists of all the points on a segment together with all the points beyond the endpoints of the segment.

A **ray** is that part of a line lying on one side of point P on the line, together with the point P.

A **plane** is a surface with the property that any two points in it can be joined by a line, all points of which are also contained in the surface.

Parallel lines are distinct lines in a plane that do not intersect.

An **angle** is a geometric figure consisting of two rays that share a common endpoint called the **vertex** of the angle.

A **straight** angle has measure 180°.

A **right** angle has measure 90°.

An **acute** angle has measure between 0° and 90°.

An **obtuse** angle has measure between 90° and 180°.

Two angles are **complementary** if their measures total 90°.

Two angles are **supplementary** if their measures total 180°.

Vertical angles are non-adjacent angles formed by intersecting lines.

When the four angles formed by two intersecting lines all have measure 90°, the lines are **perpendicular.**

A **transversal** is a line that intersects each of two parallel lines.

12.4 A **triangle** is a three-sided geometric figure.

Two triangles are **congruent** if they can be made to coincide by placing one on top of the other.

Two triangles are **similar** if the angles of one are equal to the angles of the other.

12.5 A **circle** is a figure with all points the same distance from a point called the **center** of the circle.

The **radius** of a circle is the distance from the center to any point on the circle.

The **diameter** of a circle is twice the radius.

The **circumference** of a circle is the distance around the circle.

A **secant** is a line that intersects a circle in two points.

A **tangent** is a line that intersects a circle in exactly one point.

12.6 A **rectangle** is a four-sided figure with opposite sides parallel and adjacent sides perpendicular.

The **perimeter** of a figure is the distance around the figure.

A **square** is a rectangle with all sides equal.

A **square unit** is the area of a square that measures one unit on each side.

A **parallelogram** is a four-sided figure whose opposite sides are parallel and equal.

A **trapezoid** is a four-sided figure with two parallel sides.

12.7 A **cube** is a rectangular solid with length, width, and height the same.

A **cubic unit** is the volume of a cube that has measure one unit on each edge.

The **surface area** of an object is the area of all of its surfaces.

A **cylinder** is a solid with circular ends of the same radius.

A **sphere** is a solid in the shape of a ball.

KEY CONCEPTS

12.2 **1.** A conditional and its converse do not always have the same truth value. One of the great fallacies of logic is assuming that the converse is true.

2. A statement of the form "Some A are B" only claims that there is at least one thing that satisfies A and also satisfies B. It does not claim that "all" things satisfying A also satisfy B.

3. The negation of "All *A* are *B*" is "Some *A* are not *B*." Equivalently, the negation of "If *A*, then *B*" is "*A* and not *B*."
4. Euler diagrams are used to test the validity of a conclusion in a syllogism.
5. Sometimes an array or grid can be used to analyze given premises to help discover a valid conclusion.

12.3 1. Angles are measured in degrees (°).
2. Vertical angles are equal in measure.
3. Alternate interior angles are equal in measure.
4. Corresponding angles are equal in measure.

12.4 1. Similar triangles have all corresponding angles equal.
2. Triangles can be shown to be congruent by SSS, SAS, or ASA.
3. The sum of the measures of the angles of a triangle is 180°.
4. In the 30°–60° triangle below,

$$c = 2a \text{ and } b = \sqrt{3}a.$$

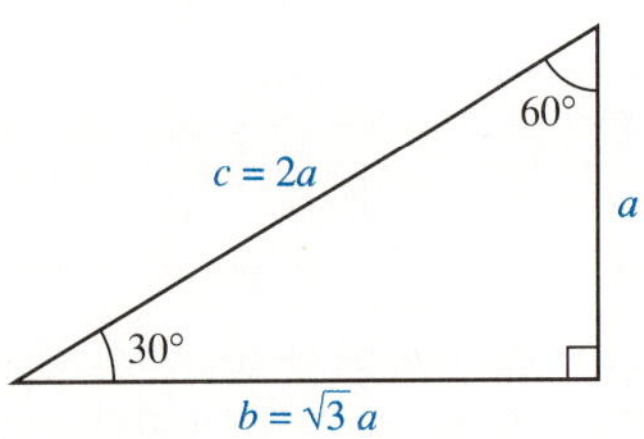

5. In the 45°–45° triangle below,

$$c = \sqrt{2}a \text{ and } b = a.$$

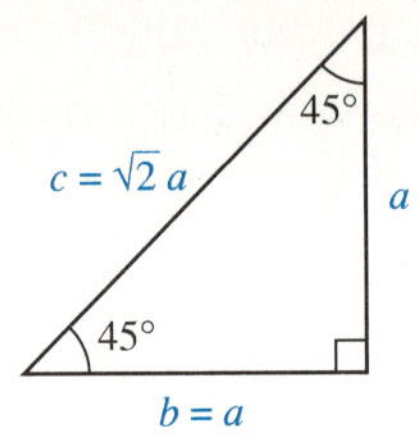

12.5 1. For a circle: $d = 2r$ and $C = \pi d = 2\pi r$.
2. The measure of an arc is given by the measure of the central angle that determines it.
3. The measure of any inscribed angle is one-half the measure of its intercepted arc.
4. An angle inscribed in a semicircle is a right angle.

12.6 Rectangle: $P = 2l + 2w$, $A = lw$
Square: $P = 4s$, $A = s^2$
Parallelogram: $A = bh$
Triangle: $A = \frac{1}{2}bh$
Trapezoid: $A = \frac{1}{2}(b_1 + b_2)h$
Circle: $d = 2r$, $C = \pi d = 2\pi r$, $A = \pi r^2$

12.7 Rectangular solid: $V = lwh$,
$A = 2lh + 2wh + 2lw$
Cube: $V = e^3$, $A = 6e^2$
Cylinder: $V = \pi r^2 h$, $A = 2\pi rh + 2\pi r^2$
Sphere: $V = \frac{4}{3}\pi r^3$, $A = 4\pi r^2$

REVIEW EXERCISES

Part I

12.1 *Use inductive reasoning in Exercises 1–3 to determine the next element in each list.*

1. 2, 7, 12, 17, 22
2. A, C, E, G, I
3. $1, \frac{1}{2}, \frac{1}{4}, \frac{1}{8}, \frac{1}{16}$

In Exercises 4–7, determine if each conclusion follows logically from the premises, and state whether the reasoning is inductive or deductive.

4. *Premise:* If I wash my car, then it will rain.
Premise: I washed my car.
Conclusion: It will rain.

5. *Premise:* If I wash my car, then it will rain.
Premise: It is raining.
Conclusion: I washed my car.

6. *Premise:* Bob Begay owns a VCR.
Premise: Bob Jones owns a VCR.
Premise: Bob Santos owns a VCR.
Conclusion: Everyone named Bob owns a VCR.

7. *Premise:* If it is Sunday, then tomorrow is Monday.
Premise: If tomorrow is Monday, then it is a holiday.
Conclusion: If it is Sunday, tomorrow is a holiday.

8. Explain why
Conclusion: Tomorrow is a holiday.
does not follow logically from the premises in Exercise 7.

12.2 9. Is the following a logical statement?

$(a + 3) + 11 = 11 + (a + 3)$

10. Give the converse of the following statement.

If Jack is lucky, then he will win the game.

11. Write the following as a biconditional.

If $x > 0$, then $x + 1 > 1$, and conversely.

12. Give the negation of the following statement.

All referees are impartial.

13. Use an Euler diagram to test the validity of the conclusion given in the following syllogism.
Premise: All chefs are good cooks.
Premise: Some people named Alphonse are chefs.
Conclusion: Some people named Alphonse are good cooks.

12.3, 12.4 *Use the figure below and deductive reasoning to answer* true *or* false *in Exercises 14–23. If the answer is false, tell why.*

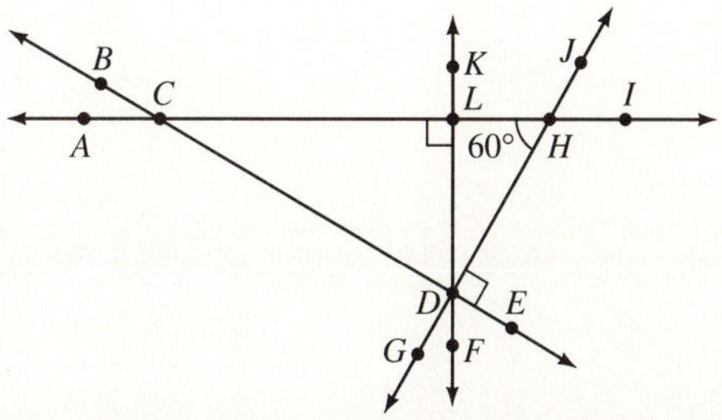

14. $\angle CDH$ is a straight angle.

15. $\angle LDB$ has measure 60°.

16. $\overleftrightarrow{FK}$ is perpendicular to $\overleftrightarrow{AC}$.

17. The measure of $\angle CDE$ is 180°.

18. $\angle LCD$ and $\angle LHD$ are supplementary.

19. $\angle BCA$ has measure 30°.

20. $\overrightarrow{CD}$ is parallel to $\overline{HI}$.

21. $\triangle LHD$ is similar to $\triangle CLD$.

22. $\triangle CDH$ is congruent to $\triangle CLD$.

23. $\triangle CDH$ is a right triangle.

24. In a 30°–60° right triangle, a is the shorter leg and $a = 10$. Find b and c.

25. In a 45°–45° right triangle, $a = 7$. Find b and c.

12.5 26. Find the diameter of the circle with radius 2.6 cm.

27. Find the radius of the circle with diameter $\frac{4}{5}$ yd.

28. Find the approximate circumference of the circle with diameter 9.2 ft. Use 3.14 for π.

Refer to the figure to the right and use deductive reasoning to answer each question in Exercises 29–33.

29. What is $\angle AOB$ called with respect to the circle?

30. What is $\angle ACB$ called with respect to the circle?

31. What is the measure of $\widehat{AB}$?

32. What is the measure of $\widehat{ACB}$?

33. What is the measure of $\angle ACB$?

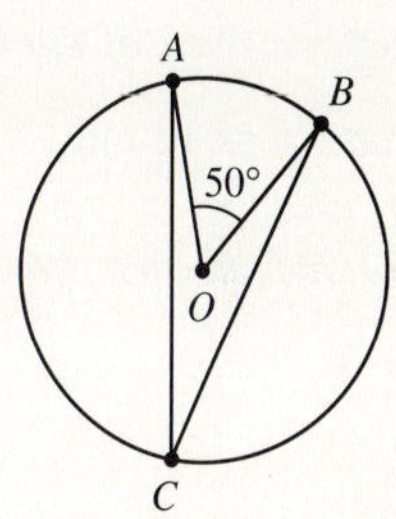

12.6 *Find the perimeter and area of the rectangle or square.*

34. 9 cm by 20 cm

35. $6\frac{1}{2}$ ft by 4.2 ft

36. 3.2 mi by 3.2 mi

Find the area of each figure.

37.

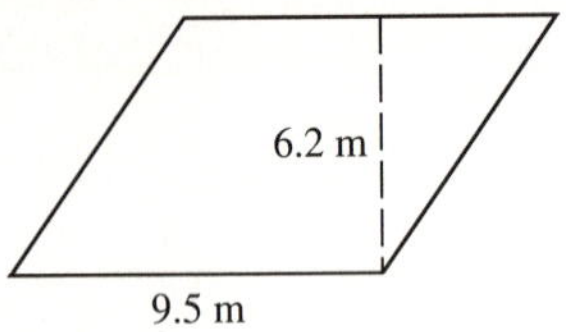

38.

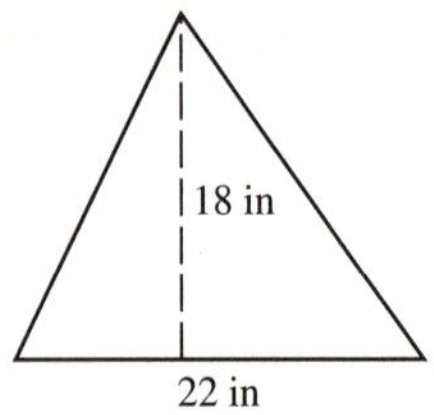

39.

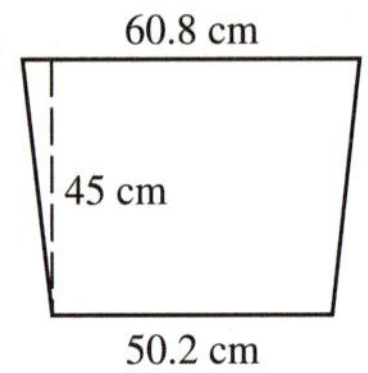

Find the perimeter of each figure.

40.

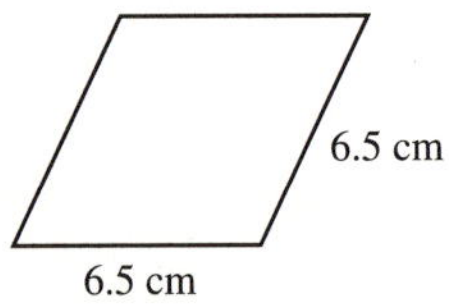

41.

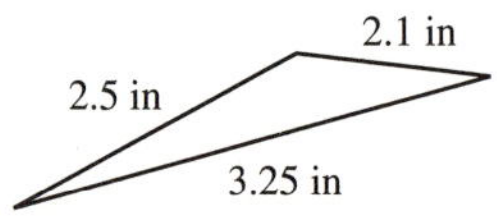

42.

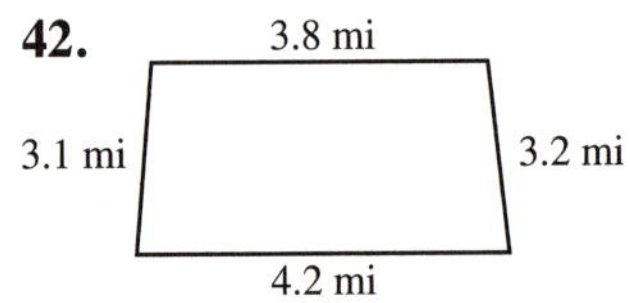

43. What will it cost to carpet the area shown if carpet costs $22.50 per square yard?

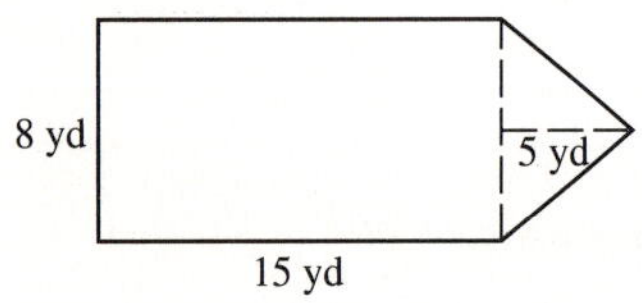

12.7 *Find the volume of each rectangular solid.*

44. 2.2 ft by 1.5 ft by 6.4 ft

45. 52 cm by 22 cm by 8 cm

Find the volume of each cube with the given edge length.

46. 16 m

47. 8.2 in

Find the surface area of each rectangular solid.

48. 2.2 ft by 1.5 ft by 6.4 ft

49. 52 cm by 22 cm by 8 cm

Find the surface area of each cube with the given edge length.

50. 16 m

51. 8.2 m

Find the volume of each cylinder. Use 3.14 for π.

52. $r = 8$ cm, $h = 12$ cm

53. $d = 2.2$ in, $h = 1.5$ in

Find the surface area of each cylinder.

54. $r = 8$ cm, $h = 12$ cm

55. $d = 2.2$ in, $h = 1.5$ in

Find the volume of each sphere.

56. $r = 10$ ft

57. $d = 8.8$ m

Find the surface area of each sphere.

58. $r = 10$ ft

59. $d = 8.8$ m

Part II

Find the circumference and area of each circle. Use 3.14 *for* π*, and round answers to the nearest tenth.*

60. $r = 6.8$ km

61. $d = 22$ in

62. $r = 8\frac{1}{2}$ ft

63. A 12-in-diameter pizza costs \$5.00 and a 16-in-diameter pizza costs \$8.00. Which pizza costs less per square inch?

64. A spherical tank with radius 12 ft is to be insulated. If insulation costs \$0.50 per square foot, how much will it cost to insulate the tank?

65. How many grams of water will a tank 10 cm by 8 cm by 2.5 cm hold? [*Hint:* Water weighs 1 gm per cubic centimeter.]

66. How many square centimeters of glass did it take to make the sides and bottom of the tank in Exercise 65?

67. The area shown is to be carpeted. How much will it cost if carpet is \$19.50 per square yard?

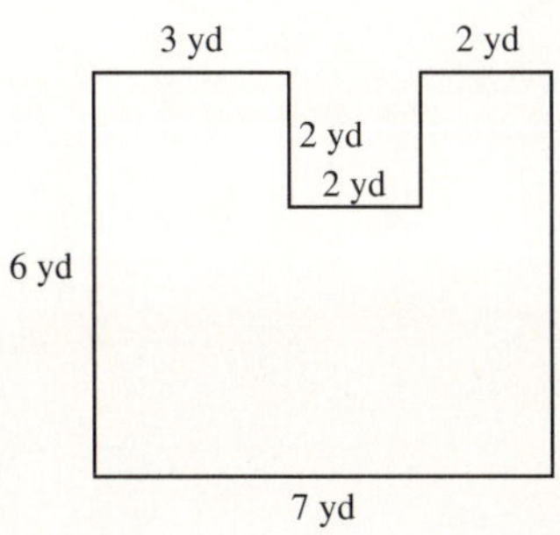

68. Are the given triangles congruent? Explain.

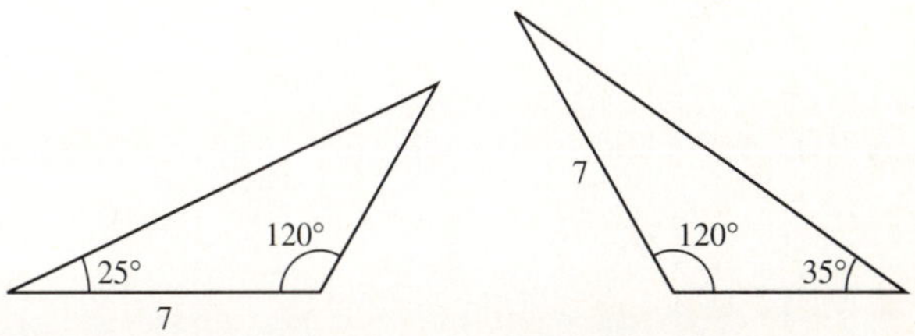

Use the figure to work Exercises 69–71.

69. Find the lines that are secants.

70. Find the lines that are tangents.

71. Find the lines that are neither secants nor tangents.

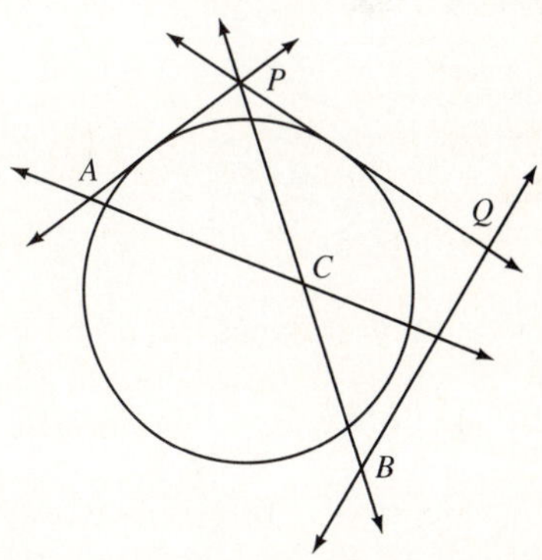

72. If 3 cats kill 3 mice in 3 minutes, how long will it take 100 cats to kill 100 mice?

73. There are 5 white socks and 5 black socks in a drawer. How many socks must you remove without looking, to be assured of having a pair that match?

74. If a is the shorter of the two legs of a 30°–60° right triangle and $a = 12$ cm, find b and c.

75. If the hypotenuse of a 45°–45° right triangle is $7\sqrt{2}$ ft, find the length of each leg.

76. Give the converse of the following statement.

All thin people are healthy.

77. Give the negation of the following statement.

If x is very large, then y is very small.

78. If a conditional is true, does the biconditional formed by combining the conditional with its converse also have to be true?

ANSWERS: 1. 27 2. K 3. $\frac{1}{32}$ 4. follows logically; deductive reasoning 5. does not follow logically; the premise did not say that if it rains, then I washed my car. 6. does not follow logically; inductive reasoning 7. follows logically; deductive reasoning 8. We cannot conclude that tomorrow is a holiday, only that if it is Sunday, then tomorrow is a holiday. 9. yes 10. If Jack wins the game, then he is lucky. 11. $x > 0$ iff $x + 1 > 1$ 12. Some referees are partial 13. The conclusion is valid. 14. false (right angle) 15. true 16. true 17. true 18. false (complementary) 19. true 20. false ($\overleftrightarrow{CD}$ and $\overleftrightarrow{HI}$ intersect at C) 21. true 22. false (only one side equal) 23. true 24. $b = 10\sqrt{3}$; $c = 20$ 25. $b = 7$; $c = 7\sqrt{2}$ 26. 5.2 cm 27. $\frac{2}{5}$ yd 28. 28.9 ft 29. central angle 30. inscribed angle 31. 50° 32. 310° 33. 25° 34. 58 cm; 180 cm^2 35. 21.4 ft; 27.3 ft^2 36. 12.8 mi; 10.24 mi^2 37. 58.9 m^2 38. 198 in^2 39. 2497.5 cm^2 40. 26 m 41. 7.85 in 42. 14.3 mi 43. $3150 44. 21.12 ft^3 45. 9152 cm^3 46. 4096 m^3 47. 551.368 in^3 48. 53.96 ft^2 49. 3472 cm^2 50. 1536 m^2 51. 403.44 in^2 52. 2411.52 cm^3 53. 5.7 in^3 54. 1004.8 cm^2 55. 17.96 in^2 56. 4186.7 ft^3 57. 356.6 m^3 58. 1256 ft^2 59. 243.2 m^2 60. 42.7 km; 145.2 km^2 61. 69.1 in; 379.9 in^2 62. 53.4 ft; 226.9 ft^2 63. 16-in costs less (12-in: $0.044 per square inch; 16-in: $0.040 per square inch) 64. $904.32 65. 200 g 66. 170 cm^2 67. $741 68. yes; ASA 69. $\overleftrightarrow{AC}$ and $\overleftrightarrow{PB}$ 70. $\overleftrightarrow{PQ}$ and $\overleftrightarrow{PA}$ 71. $\overleftrightarrow{BQ}$ 72. 3 minutes 73. 3 74. $b = 12\sqrt{3}$ cm; $c = 24$ cm 75. 7 ft 76. All healthy people are thin. 77. x is very large and y is not very small. 78. no

CHAPTER 12 TEST

1. Use inductive reasoning to give the next element in the following list: 125, 25, 5, 1. — 1. ______

2. Does the conclusion below follow logically from the premises? What type of reasoning is being used? — 2. ______

   Premise: Michael Jordan eats oatmeal.
   Premise: Larry Bird eats oatmeal.
   Premise: Magic Johnson eats oatmeal.
   Conclusion: All great basketball players eat oatmeal.

3. Give the converse of the statement: If x is a multiple of 4, then x is a multiple of 2. — 3. ______

4. Give the negation of the statement: All turkeys live in coops. — 4. ______

Use the figure below and deductive reasoning to answer true *or* false *in problems 5–8.*

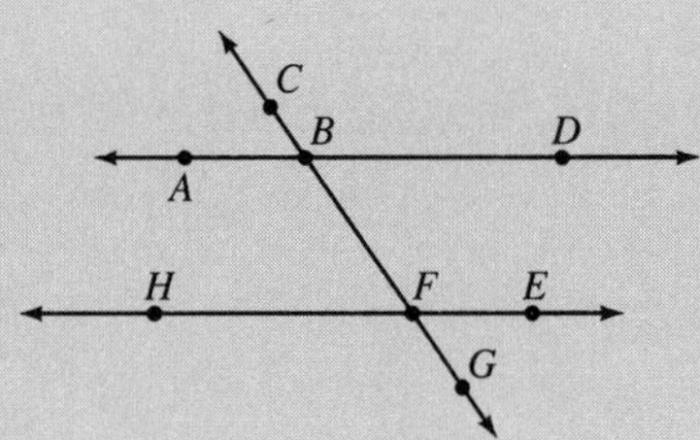

5. $\angle ABD$ is a straight angle. — 5. ______

6. $\angle BFE$ and $\angle BFH$ are complementary. — 6. ______

7. $\overleftrightarrow{AB}$ is parallel to $\overleftrightarrow{FG}$. — 7. ______

8. $\angle EFG$ and $\angle ABC$ are equal. — 8. ______

9. Are the given triangles congruent? — 9. ______

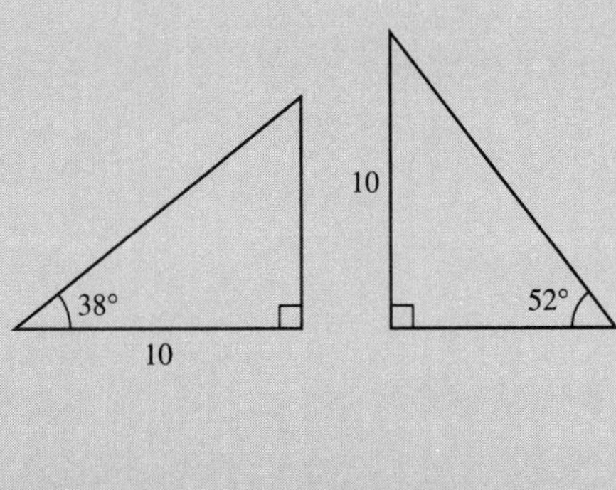

CHAPTER 12 TEST

CONTINUED

Use the figure below and deductive reasoning to answer problems 10–12.

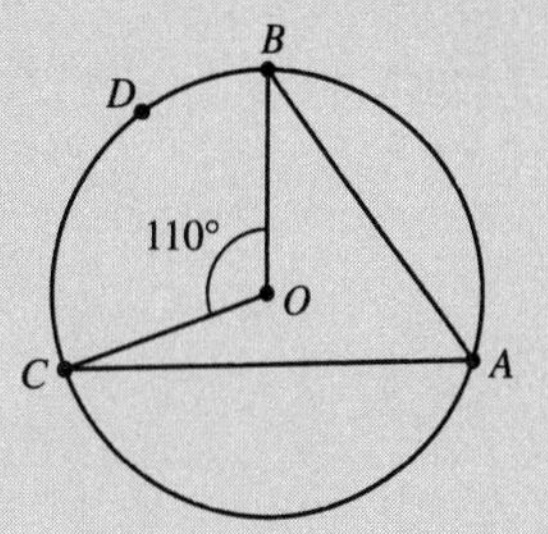

10. What is the measure of $\overset{\frown}{BDC}$? 10. ____________

11. What is the measure of $\overset{\frown}{BAC}$? 11. ____________

12. What is the measure of $\angle BAC$? 12. ____________

Problems 13–22 refer to the figure below. Answer true *or* false.

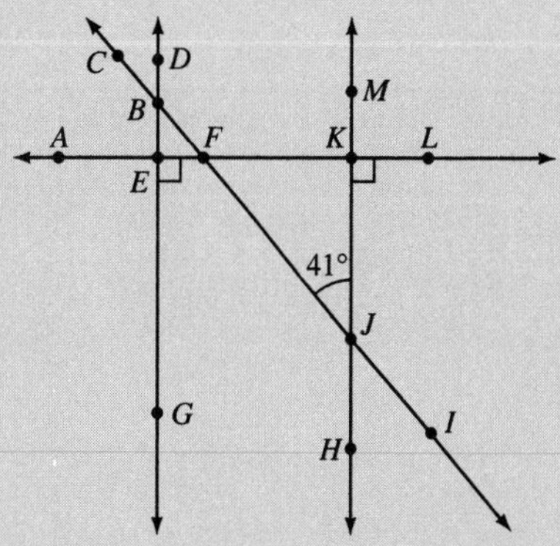

13. $\angle FBE$ is an obtuse angle. 13. ____________

14. $\angle MKF$ is a right angle. 14. ____________

15. $\overline{EG}$ is parallel to $\overrightarrow{HM}$. 15. ____________

16. $\overleftrightarrow{AE}$ is perpendicular to $\overline{JH}$. 16. ____________

17. $\angle IJH$ has measure 41°. 17. ____________

18. $\angle FBE$ has measure 41°. 18. ____________

19. $\angle BFE$ and $\angle BFK$ are complementary. 19. ____________

20. $\triangle KJF$ is a right triangle. 20. ____________

21. $\angle CBD$ has measure 41°. 21. ____________

22. $\triangle BFE$ is similar to $\triangle JKF$. 22. ____________

23. In a right triangle, $\angle A = 30°$ and $a = 20$. Find b and c. 23. ______

24. In a right triangle, $\angle B = 45°$ and the hypotenuse is 8. Find the length of each leg. 24. ______

25. The area shown is to be carpeted. How much will it cost if carpet is $22.50 per square yard? 25. ______

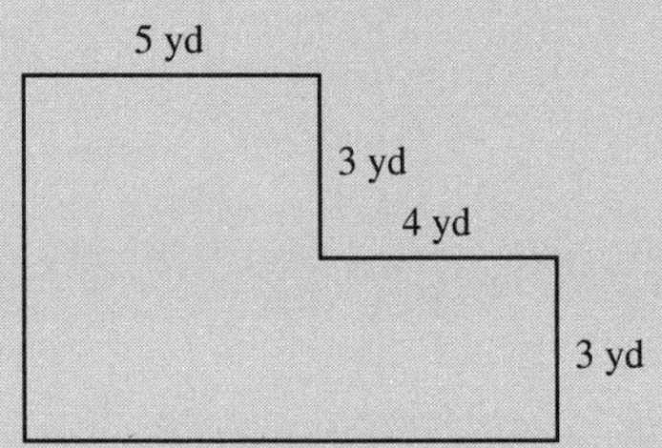

26. Find the circumference of a circle with radius 2.6 ft. Use 3.14 for π. 26. ______

27. Find the area of a circle with diameter 9.2 m. Use 3.14 for π. 27. ______

28. Find the **(a)** volume and **(b)** surface area of a cube with edge 3.6 ft. 28. (a) ______

(b) ______

29. A 14-in-diameter pizza costs $7.50 and a 16-in pizza costs $8.50. Which costs less per square inch? 29. ______

30. Find the area of the given figure. 30. ______

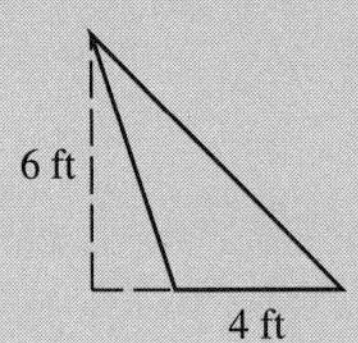

31. Find the perimeter of the given figure. 31. ______

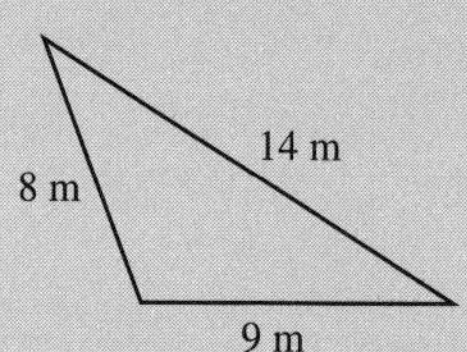

32. Find the volume of a cylinder with radius 8 cm and height 10 cm. 32. ______

33. Find the surface area of a sphere with radius 20 inches. 33. ______

Solve.

34. Gravel costs $1.50 per cubic meter. How much would a truckload cost if the bed of the truck is 4.2 m by 5.0 m by 2.5 m? 34. ______

FINAL REVIEW EXERCISES

CHAPTER 0

1. Reduce $\frac{242}{220}$ to lowest terms.

2. Bill received 28 of the 42 votes cast. What fractional part of the votes did he receive? Reduce the fraction to lowest terms.

Perform the indicated operations.

3. $\dfrac{5}{11} \cdot \dfrac{33}{35}$

4. $\dfrac{20}{39} \div \dfrac{6}{13}$

5. $\dfrac{2}{15} + \dfrac{8}{3}$

6. $\dfrac{11}{14} - \dfrac{3}{4}$

7. $29.705 + 3.01$

8. $17.6 - 9.72$

9. $(0.0049)(52.8)$

10. $2.86 \div 0.052$

11. Change $\frac{43}{18}$ to a mixed number.

12. Change $8\frac{3}{13}$ to an improper fraction.

13. There are 475 light bulbs in a shipment. If $\frac{1}{25}$ of them are defective, how many are defective?

14. Harry bought $2\frac{3}{8}$ pounds of one candy and $3\frac{11}{16}$ pounds of another. What total weight of candy did he buy?

15. Convert $\frac{3}{8}$ to a percent.

16. Convert 2.72 to a fraction.

Exercises 17–20 refer to the bar graph showing the number of students in each class who applied for a scholarship.

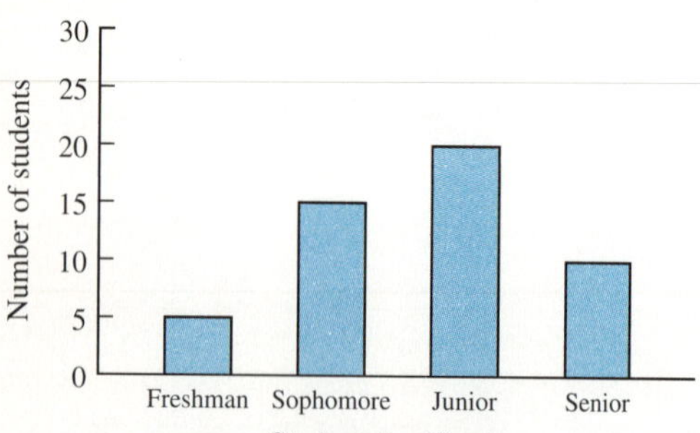

17. How many students applied?

18. What is the ratio of juniors to sophomores who applied?

19. What percent of those applying were freshmen?

20. What is the ratio of freshmen and sophomores to juniors and seniors?

Exercises 21–25 refer to the broken-line graph showing the number of vehicles passing through an intersection (in hundreds) during an 8-hour interval.

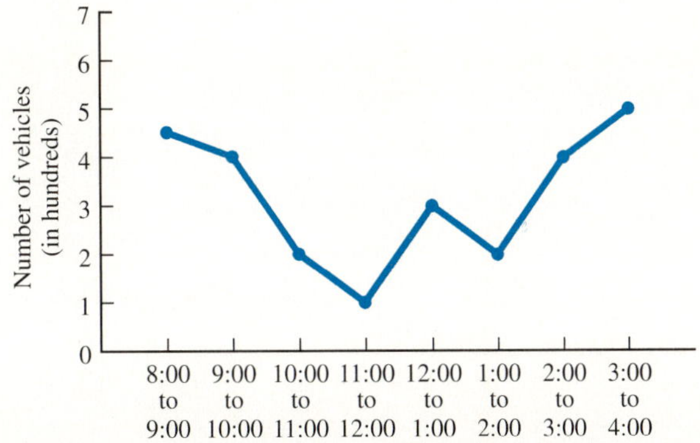

21. How many vehicles passed through the intersection from 8:00 to 9:00?

22. How many vehicles passed through the intersection from 11:00 to 12:00?

23. How many vehicles passed through the intersection from 8:00 to 12:00?

24. How many vehicles passed through the intersection during the period from 8:00 to 4:00?

25. What percent (to the nearest tenth) of the total number of vehicles passed through the intersection from 8:00 to 12:00?

Find the mean and median. Give answers to nearest tenth.

26. 9, 10, 17, 6

27. 6.2, 8.5, 4.7, 3.3, 9.4

28. A truck was driven 826 miles on 203 gallons of fuel. To the nearest tenth, what was the average miles per gallon?

29. Find the mode. 42, 35, 42, 50, 35, 70, 60, 35

Complete.

30. 42 in = _____ ft

31. 5400 cm = _____ km

32. 2.5 g = _____ mg

33. 5 lb = _____ g

34. $100\dfrac{\text{km}}{\text{hr}} = \_\_\_\_\_ \dfrac{\text{mi}}{\text{hr}}$

35. 60°C = _____°F

36. If sugar costs 36¢ per lb, how much will the sugar cost for a cake which requires 24 oz of sugar?

37. If a car's gasoline tank holds 25 gal, how many liters does it hold?

Perform the indicated operations and simplify.

38. 21 hr 15 min
+ 6 hr 48 min

39. 13 yd 1 ft 7 in
− 2 yd 2 ft 4 in

40. 25.2 cm
+ 7.9 cm

41. 421.6 ml
− 382.7 ml

42. $4 \times$ (2 gal 2 qt 1 pt)

43. 62 km $\div$ 8 km

CHAPTER 1

44. Evaluate $x^2 + 4(x - y)$ for $x = 5$ and $y = 3$.

45. If $A = P(1 + r)^t$, find A when P is \$2000, the interest is 8% compounded annually, and t is 2 years.

46. Find the reciprocal of $-\frac{3}{4}$.

47. Find $|-8|$.

Place the correct symbol, =, <, or >, between the pairs of numbers.

48. $-8 \quad -10$

49. $\dfrac{8}{31} \quad \dfrac{4}{17}$

Perform the indicated operations.

50. $-4 - (-16)$

51. $14 + (-24)$

52. $\left(-\dfrac{5}{9}\right) \cdot \left(-\dfrac{3}{20}\right)$

53. $(7.5) \div (-2.5)$

Evaluate.

54. $2 \cdot 4 + (8 - 12) - 2(3 - 1)$

55. $-8 - [-(-8)]$

Factor.

56. $2y + 12$

57. $-14x - 21$

Collect like terms.

58. $-8a + 3b - 4a - 5 - b$

59. $7y - (2y - 5) + 3$

60. Evaluate $2u^2 - v^2$ when $u = -2$ and $v = 4$.

61. Write $-5x^4$ without exponents.

62. Evaluate. $-\sqrt{144}$

63. Evaluate. $\sqrt{\dfrac{12}{75}}$

Simplify and write without negative exponents.

64. $3x^2x^3$

65. $\dfrac{y^6}{y^2}$

66. $(-3a^3)^2$

67. a^{-2}

68. Write 36,200,000 in scientific notation.

69. Write 7.2×10^{-4} using standard notation.

Evaluate using scientific notation.

70. (0.0000021)(3,000,000)

71. $\dfrac{0.000055}{11{,}000{,}000}$

CHAPTER 2

Solve.

72. $3 - 3(x - 1) = 4x - 1$

73. $z(z + 3) = z^2 - 9$

74. A wire 36 cm long is to be cut into two pieces in such a way that one piece is 8 cm longer than the other. How long is each piece?

75. Two cars leave the same city at 9:00 A.M., one heading north and the other heading south. If one is traveling 50 mph and the other 35 mph, at what time will they be 340 mi apart?

76. After a 6% raise, Paul's new salary is \$9752. What was his former salary?

77. The sum of two consecutive odd integers is 96. Find the two integers.

78. Find the measure of each angle of a triangle if the second is four times the first and the third is 5° more than twice the first.

79. After receiving a 20% discount on the selling price, Jeff Altman paid \$33.60, including 5% sales tax, for a pair of shoes. What was the price of the shoes before the discount?

Solve and graph. Give answer in interval notation when appropriate.

80. $3z - 5 \geq 4(z - 1)$

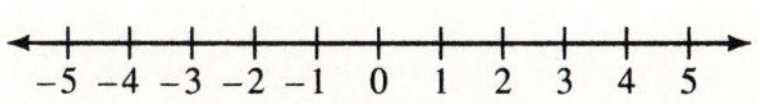

81. $y - 2 < -4$ or $y - 2 \geq 4$

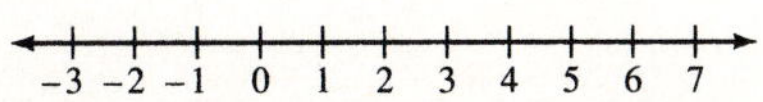

82. $-5 \leq 2x - 1 < 1$

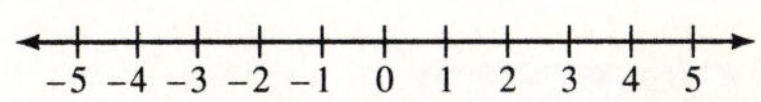

83. $|4 - 3x| = 7$

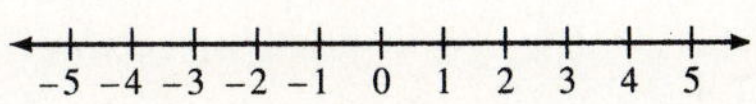

84. $|2y + 1| > 5$

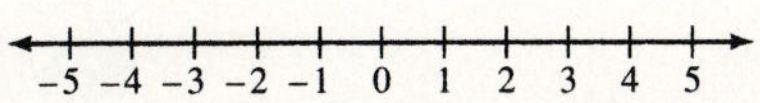

85. $|2z - 3| \leq 7$

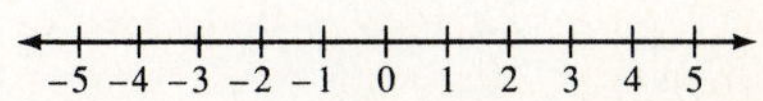

86. Solve $P = q + 50r$ for r.

87. Solve $|x - 1| = |2x - 3|$.

CHAPTER 3

88. Give the x-intercept, y-intercept, slope, and graph of the line with equation $2x - 5y + 10 = 0$.

x-intercept _______
y-intercept _______
slope _______

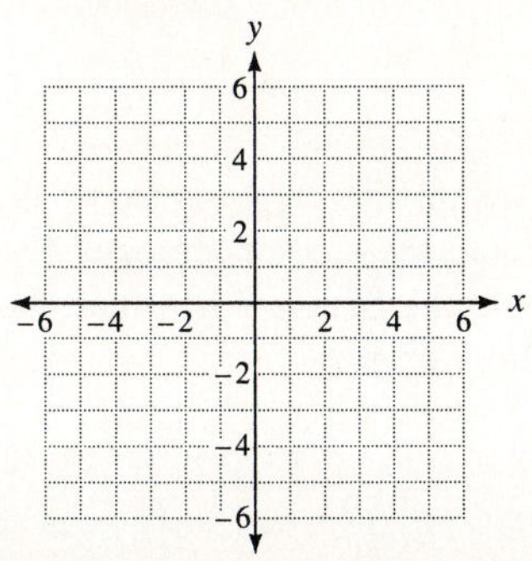

89. Graph using the slope method.

$$y = \frac{2}{3}x + 1$$

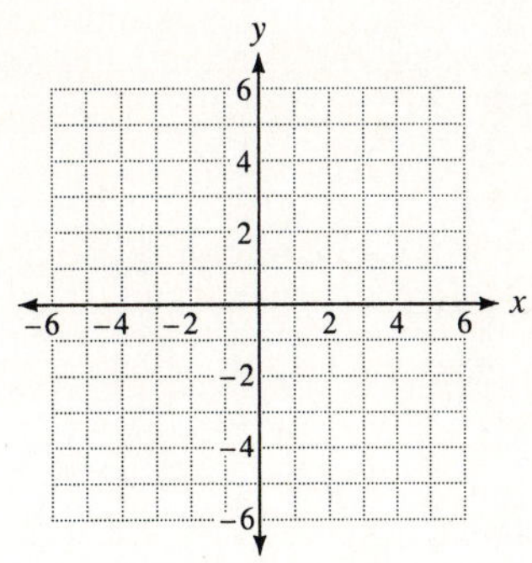

90. Find the general form of the equation of the line passing through the points $(4, -2)$ and $(-1, 3)$.

91. Write the equation, in general form, of the line containing the point $(2, 5)$ and perpendicular to $4x + 3y - 8 = 0$.

92. Is the following a function?

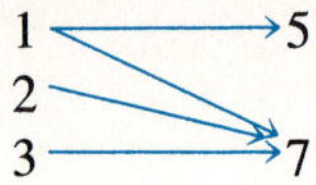

93. Is the following the graph of a function?

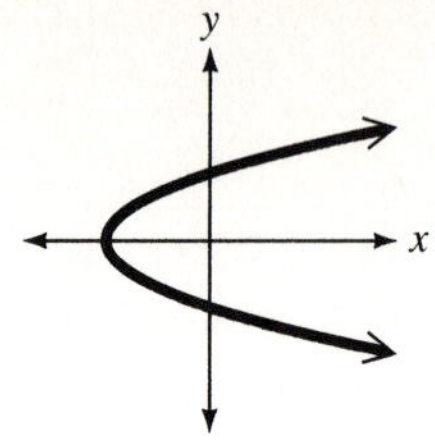

94. Is the relation $\{(1, 3), (2, 3), (3, 3)\}$ a function?

95. Specify the domain of the function defined by

$$y = \frac{1}{x + 5}.$$

96. Given the function $f(x) = x^2 + 3$, find the following.
(a) $f(-2)$
(b) $f(1)$
(c) $f(-a)$

97. Steve Caparella earns an annual salary of \$15,000 plus a 20% commission on all of his sales. Express his annual earnings as a function, f, of his total sales, x. What are his annual earnings when his yearly sales total is \$520,000?

CHAPTER 4

98. Tell whether $(3, -2)$ is a solution to the system.

$$\begin{aligned} x - 5y - 13 &= 0 \\ 2x + 3y &= 0 \end{aligned}$$

99. Without solving, tell the number of solutions to the system.

$$\begin{aligned} 4x - 3y &= 2 \\ -8x + 6y &= -2 \end{aligned}$$

Solve.

100. $$\begin{aligned} 3x + 5y &= -2 \\ x - 3y &= 4 \end{aligned}$$

101. $$\begin{aligned} x - 5y &= 2 \\ -3x + 15y &= -6 \end{aligned}$$

102. Tom Reese wishes to mix two kinds of candy, one worth 90¢/lb and the other worth \$1.50/lb. How many pounds of each must there be in 60 lb of the mix if it is to sell for \$1.30/lb?

103. A collection of dimes and quarters is worth \$3.75. If there are 27 coins in the collection, how many of each are there?

104. $$\begin{aligned} x - 3y + 2z &= 6 \\ 2x + 4y - z &= 2 \\ 4x - 2y + 3z &= 14 \end{aligned}$$

105. $$\begin{aligned} x - y - z &= 2 \\ 3x - y - 3z &= -4 \\ x + y - z &= 4 \end{aligned}$$

106. $$\begin{aligned} x + y - z &= 4 \\ 2x - y + 3z &= -3 \\ x + 2y - z &= 6 \end{aligned}$$

107. The average of a student's three scores is 79. The sum of the first and the second scores is 57 more than the third, and the first is 7 less than the second. Find the three scores.

Evaluate the determinants.

108. $$\begin{vmatrix} 3 & 5 \\ -1 & 7 \end{vmatrix}$$

109. $$\begin{vmatrix} 1 & -2 & 3 \\ 0 & 5 & -1 \\ 2 & -1 & 4 \end{vmatrix}$$

Solve for x only using Cramer's rule. [Hint: Use Exercises 108 and 109.]

110. $$\begin{aligned} 3x + 5y &= 1 \\ -x + 7y &= -9 \end{aligned}$$

111. $$\begin{aligned} x - 2y + 3z &= 7 \\ 5y - z &= -15 \\ 2x - y + 4z &= 5 \end{aligned}$$

Graph in a Cartesian coordinate system.

112. $4x + 2y > 8$

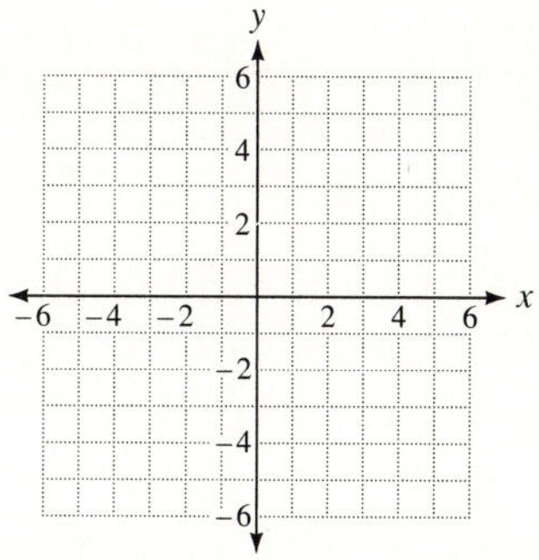

113. $5x - 2y \le 10$
$x + 2 > 0$

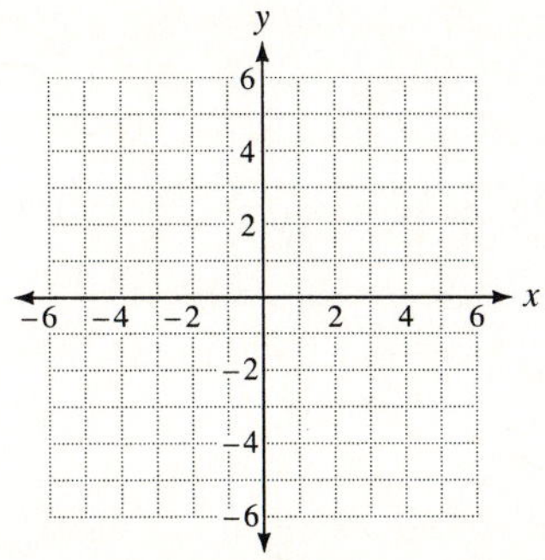

114. $x \ge 0$
$y \ge 0$
$5x + 3y \le 15$

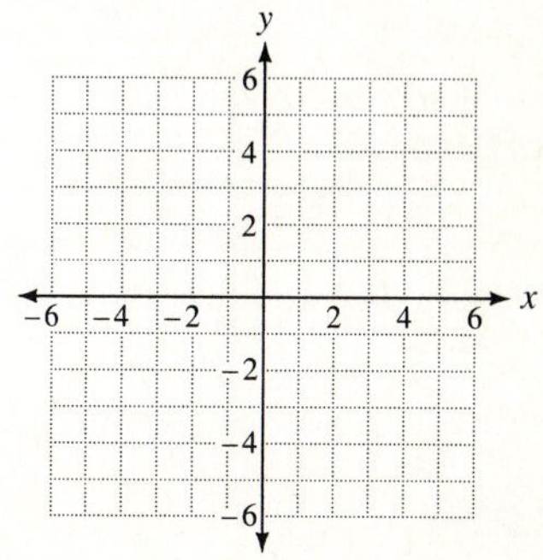

CHAPTER 5

In Exercises 115–116, collect like terms and write in descending powers of x.

115. $3x - 5x^3 + x^2 - x^7 + 2$

116. $3x^2y^2 - 2x^4y + x^2y^2 + 3x^4y - xy$

117. Evaluate $2a^2b - 3ab^2$ when $a = -1$ and $b = 3$.

118. Add $3y^2 - 14y + y^3 - 5$ and $7 - y + 2y^3 - y^2$.

119. Subtract $-2xy + 3x^2y^2 - 9$ from $5x^2y^2 - 2xy + 4$.

120. Perform the indicated operations.
$(3ab - 2a^2b^2) - (a^2b^2 - 5) + (1 - 2ab)$

Multiply.

121. $-2xy^2(3x - 4y)$

122. $(2x + y)(x - 4y)$

123. $(a - 3b)^2$

124. $(3u - v)(3u + v)$

Factor.

125. $3x^2 - 12$

126. $a^2 + 10a + 25$

127. $4y^3 - 32$

128. $3x^2 - 8x + 4$

129. $3x^3 + 3$

130. $-6u^2 + 10uv + 4v^2$

Solve.

131. $2x(x - 3) = 0$

132. $x^2 - 1 = 3(x + 1)$

133. Twice the product of two consecutive positive odd integers is 126. Find the integers.

134. The length of a rectangle is 7 cm more than the width and the area is 44 cm^2. Find its dimensions.

The current I in amperes in an electrical circuit varies according to time t in seconds by the equation $I = 4t^2 - 72t + 180$. *Use this in Exercises 135–36.*

135. What is the initial amperage in the circuit?

136. In how many seconds will the amperage be 0 amperes?

CHAPTER 6

137. What values of the variable must be excluded?

$$\frac{3x + 1}{x^2 - 2x}$$

138. Reduce to lowest terms.

$$\frac{8x^3 + 1}{2x + 1}$$

Perform the indicated operations.

139. $\dfrac{x-y}{4x+4y}\cdot\dfrac{x^2-y^2}{x^2-2xy+y^2}$

140. $\dfrac{a^2-9}{a^2-8a+16}\div\dfrac{3a-9}{a-4}$

141. $\dfrac{4}{x+y}+\dfrac{2x+y}{x^2-y^2}$

142. $\dfrac{2}{a+b}-\dfrac{2a}{a^2+2ab+b^2}$

143. $\dfrac{4y}{y^2-1}-\dfrac{4}{y-1}+\dfrac{3y}{1-y}$

144. $\dfrac{x^2-9}{x^2-x-6}\div\dfrac{2x^2+x-15}{x^2+7x+10}\cdot\dfrac{2x^2-x-10}{x^2-25}$

Solve.

145. $\dfrac{1}{y}=\dfrac{y-2}{24}$

146. $\dfrac{a}{a+3}-\dfrac{18}{a^2-9}=1$

147. The speed of a river is 5 mph. If a boat travels 75 mi downstream in the same time that it takes to travel 45 mi upstream, find the speed of the boat in still water.

148. A small pipe can fill a reservoir in 5 days and a large pipe can fill it in 3 days. How long will it take to fill it if both pipes are turned on together?

149. If a car travels 675 mi on 30 gal of gas, how many miles can it travel on 46 gal of gas?

150. Suppose y varies jointly as x and the square root of z and inversely as the square of w. Find the equation of variation if $y = 2$, when $x = 3$, $z = 4$, and $w = 3$.

151. Solve $\dfrac{1}{a-b}=c$ for a.

152. Simplify. $\dfrac{\dfrac{a}{b}-\dfrac{b}{a}}{\dfrac{1}{a}+\dfrac{1}{b}}$

153. Divide. $\dfrac{9z^6+27z^5-3z^4}{-3z^4}$

154. Divide. Use synthetic division to check your work. $(x^4-x^3+x^2-3x+2)\div(x-1)$

CHAPTER 7

Simplify each radical in Exercises 155–158. For these exercises, do not assume that variables and algebraic expressions under even-indexed radicals necessarily represent nonnegative numbers.

155. $\sqrt{(-5x)^2}$

156. $\sqrt[3]{(-5x)^3}$

157. $-\sqrt{w^2}$

158. $\sqrt{(a+b)^2}$

Evaluate, assuming all algebraic expressions under even-indexed radicals are positive.

159. $\sqrt{(-11)^2}$

160. $\sqrt[3]{27a^3b^6}$

161. $\sqrt[4]{(-2)^4a^8}$

Simplify each radical and assume all variables under even-indexed radicals are positive.

162. $2\sqrt{20a^3b^2}$

163. $\sqrt{\dfrac{18}{z^2}}$

164. $\sqrt[3]{\dfrac{-125x^3y}{8y^4}}$

Multiply or divide. Rationalize all denominators.

165. $\sqrt{6xy}\sqrt{3x^3y^2}$

166. $\dfrac{\sqrt[3]{16a^3}}{\sqrt[3]{2b}}$

Add or subtract and simplify.

167. $\sqrt[3]{27a^5b^9} - 3b\sqrt[3]{125a^5b^6}$

168. $2x\sqrt{63x} + 5\sqrt{28x^3}$

Multiply and simplify.

169. $\sqrt{3}(\sqrt{12} - \sqrt{3})$

170. $(\sqrt{2} + 1)^2$

171. $(\sqrt{x} - 5)(\sqrt{x} + 5)$

172. $(\sqrt{2} - \sqrt{5})(2\sqrt{2} + \sqrt{5})$

173. Rationalize the denominator.

$$\frac{\sqrt{7}}{\sqrt{7} - 2}$$

174. Simplify.

$$\left(\frac{3a^{-1}b^{-1/2}}{a^3b}\right)^{-2}$$

Rationalize the numerators.

175. $\dfrac{\sqrt{14}}{2}$

176. $\dfrac{\sqrt[3]{x}}{7}$

177. $\dfrac{\sqrt{x} - 2}{3}$

Solve.

178. $\sqrt{y + 11} - 1 = \sqrt{y + 4}$

179. $\sqrt{2x + 1} - \sqrt{x + 6} = 0$

180. $x^{1/3}y^{1/3} = a$ for y

181. A hiker walked 9 mi east, then 12 mi north. How far was she then from the starting point?

182. Find the distance between the points $(-2, 6)$ and $(-3, -1)$.

183. Find the midpoint of the segment joining $(-4, -1)$ and $(6, 7)$.

Perform the indicated operations.

184. $(2 + 3i) + (8 - i)$

185. $(3 + i)(2 - 4i)$

186. $\dfrac{2 - i}{3 + 2i}$

187. i^{89}

188. $\sqrt{-25} - \sqrt{-4}$

189. $\dfrac{\sqrt{-9}}{\sqrt{-4}}$

CHAPTER 8

Solve.

190. $y^2 - 6y - 7 = 0$

191. $4x^2 = 8x$

192. $z^2 - 2z - 1 = 0$

193. $y^4 - 3y^2 + 2 = 0$

194. $(x - 3)^2 - 3(x - 3) - 4 = 0$

195. $(z - 3)(z + 3) = -13$

196. $x^2 + 2x + 4 = 0$

197. $\dfrac{1}{x - 1} + \dfrac{2}{x} = 1$

198. $\sqrt{x^2 + 2} - \sqrt{x + 1} = 0$

199. $\sqrt{3y + 1} - \sqrt{y - 4} = 3$

200. Cindy can complete a lab test in 3 hr less time than Burford. If they work together, they can do the test in 4 hr. To the nearest tenth of an hour, how long will it take each working alone to complete the test?

201. A boat sails downstream 96 miles and returns to the dock in a total time of 7 hours. Find the speed of the boat downstream, to the nearest tenth, if the speed downstream is 1 mph faster than the speed upstream.

202. Twice the product of two positive consecutive even integers is 240. Find the integers.

203. The number of square inches in the area of a square is the same as 5 more than the number of inches in its perimeter. Find the length of a side.

204. Solve by completing the square.
$x^2 - 5x + 2 = 0$

205. Solve for x. $\dfrac{n}{x-2} + \dfrac{n}{x+2} = 1$

206. Use the discriminant to determine the nature of the solutions of $2x^2 - 3x + 7 = 0$

207. Find a quadratic equation that has 3, -4 as solutions.

208. Solve $2x^2 - 3x - 2 < 0$ and give the answer using interval notation.

209. Solve $\dfrac{x+2}{x-7} \geq 0$ and give the answer using interval notation.

CHAPTER 9

210. Sketch the graph of $h(x) = -x^2 + 6$ by sliding the graph of $f(x) = -x^2$ up or down.

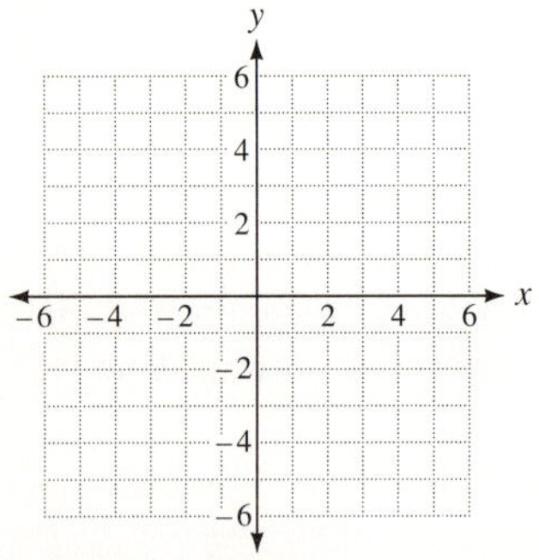

211. Find the x-intercepts, the vertex, and the graph of $f(x) = x^2 - 6x + 5$.

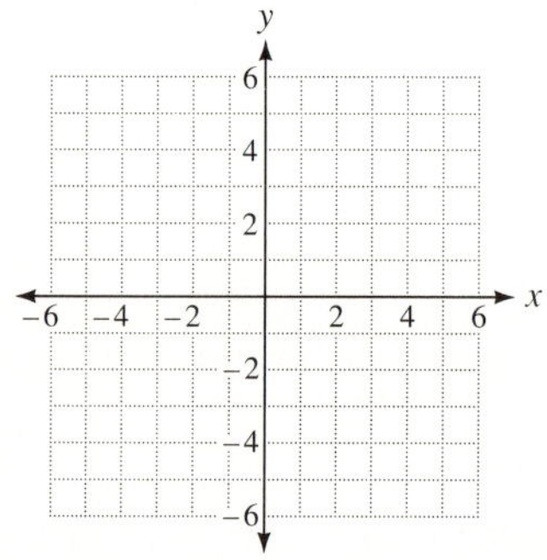

212. Without graphing, determine whether the graph of the quadratic function opens up or down and find the vertex and x-intercepts (if they exist).
$f(x) = -x^2 + 3x - 7$

213. What is the maximum area of a rectangular pasture enclosed by 240 yd of fencing? What are the dimensions of the pasture?

Find the domain of each function.

214. $f(x) = \sqrt{5 - x}$

215. $g(x) = \dfrac{9}{\sqrt{5 - x}}$

216. Assume that $f(x) = 1 - x$ and $g(x) = x^2 + 3$. Find each function value.

(a) $(f + g)(2)$ **(b)** $(f - g)(2)$ **(c)** $(fg)(2)$

(d) $\left(\dfrac{f}{g}\right)(2)$ **(e)** $f[g(2)]$ **(f)** $g[f(2)]$

Make a table of values, sketch the graph of each function, and tell whether the function is one-to-one.

217. $f(x) = x^4 - 6$

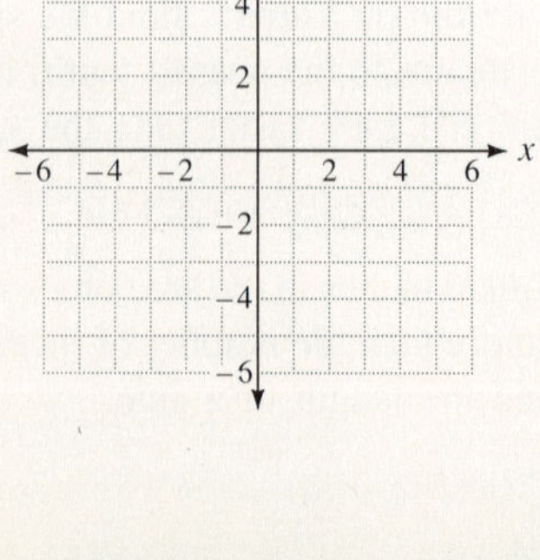

218. $g(x) = |x| - 4$

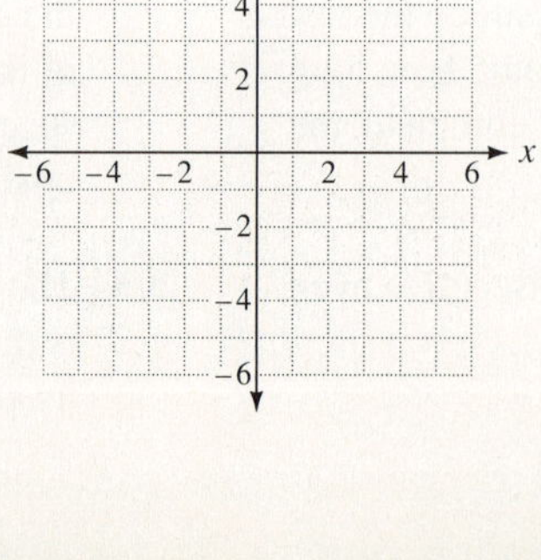

219. $h(x) = -\sqrt{x + 3}$

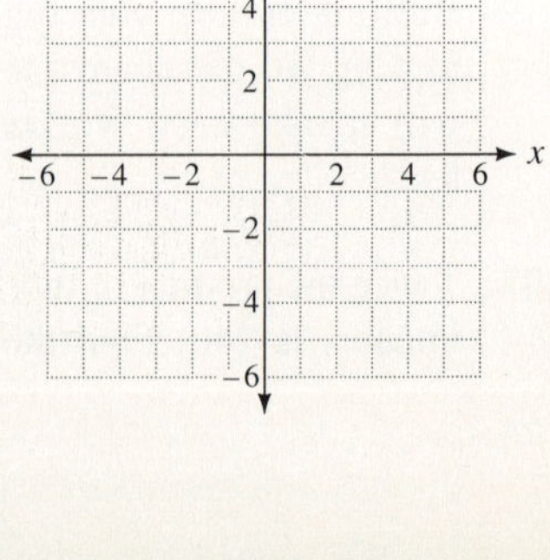

220. Consider the function $f(x) = \frac{1}{5}x - \frac{2}{5}$. Find the equation for the inverse of f and use it to find the following.

(a) $f(-1)$ **(b)** $f^{-1}(-1)$ **(c)** $f[f^{-1}(-1)]$ **(d)** $f^{-1}[f(-1)]$

CHAPTER 10

221. Convert $7^x = 3$ to logarithmic form.

222. Convert $\log_a 25 = 6$ to exponential form.

223. Express $\log_a \frac{x^2}{y\sqrt{z}}$ in terms of logarithms of x, y, and z.

224. Express $-3 \log_a x + \frac{1}{2} \log_a y$ as a single logarithm.

Write true *or* false *for each of the following. If the statement is false, change the right side to make it true.*

225. $\log_a \frac{u}{v} = \frac{\log_a u}{\log_a v}$

226. $\log_a a = 1$

Given that $\log_a 3 = 0.4771$ *and* $\log_a 5 = 0.6990$, *find the following logarithms.*

227. $\log_a 9$

228. $\log_a \frac{3}{5}$

229. $\log_a 15$

Find each logarithm.

230. $\log 0.000328$

231. $\ln 298.3$

232. $\log_5 491$

Find each antilogarithm, correct to three significant digits.

233. $\log x = 2.6255$

234. $\log x = -0.4135$

235. $\ln x = 1.3759$

Solve.

236. $\log x - \log (x + 9) = -1$

237. $3^{2x+6} = 27$

238. Use the sound equation.

$$y = 10 \log \frac{x}{x_0}$$

to find the loudness of a sound in decibels if the given sound has intensity 75,000 times the weakest sound detected.

239. If \$1000 is invested at an annual rate of 12% compounded quarterly, use $S = A(1 + i)^n$ to find the value of the account at the end of 6 years.

240. The population of a county is increasing according to $y = 100{,}000\, e^{0.13t}$. In how many years will the population double?

241. Graph $y = 2^x$ and $y = \log_2 x$ in the same coordinate system. What is the relationship between these two functions?

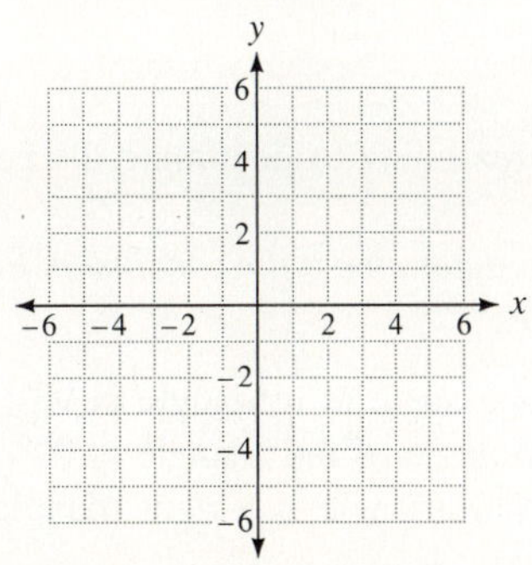

CHAPTER 11

242. Find the standard and general forms of the equation of a circle with center $(-4, 2)$ and radius 6.

243. Find the standard form of the equation of the following circle, and give the center and radius. $x^2 + y^2 - 2x + 8y + 13 = 0$

244. Give the standard form of the equation of the ellipse with intercepts $(2, 0)$, $(-2, 0)$, $(0, 5)$, and $(0, -5)$.

245. Give the standard form of the equation of the hyperbola that opens left and right and has $a = 3$ and $b = 9$.

Graph.

246. $y = -x^2 + 2x - 2$

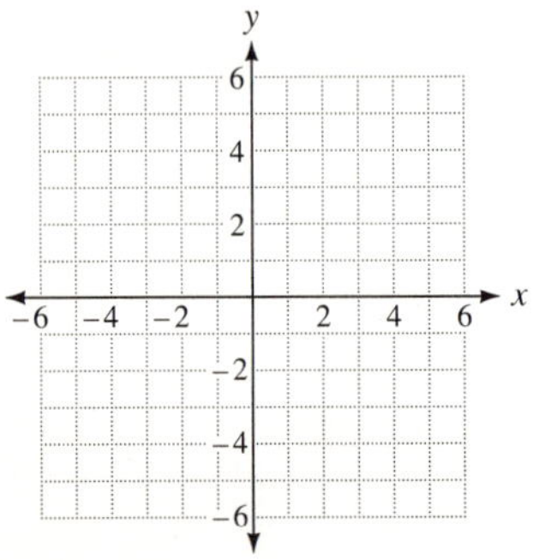

247. $x = y^2 - 4$

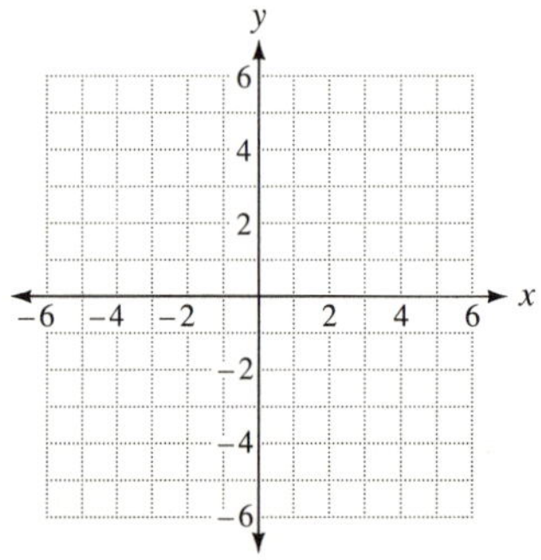

248. $(x + 1)^2 + (y - 3)^2 = 4$

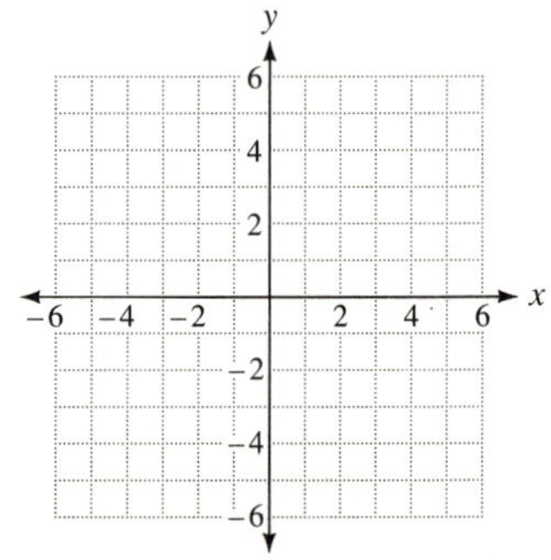

249. $\dfrac{x^2}{16} + \dfrac{y^2}{9} = 1$

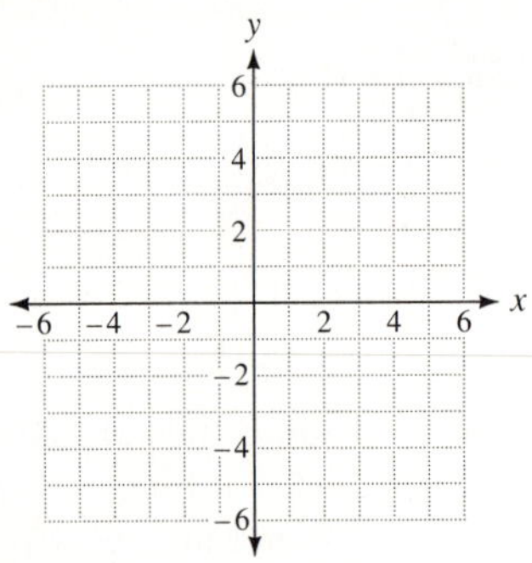

250. $\dfrac{x^2}{4} - \dfrac{y^2}{9} = 1$

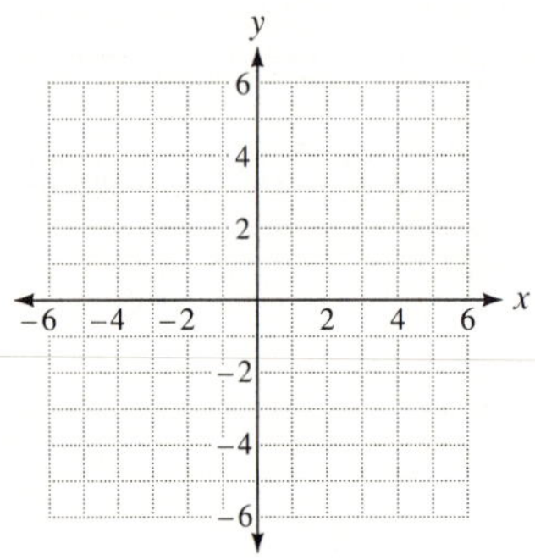

251. $xy = 6$

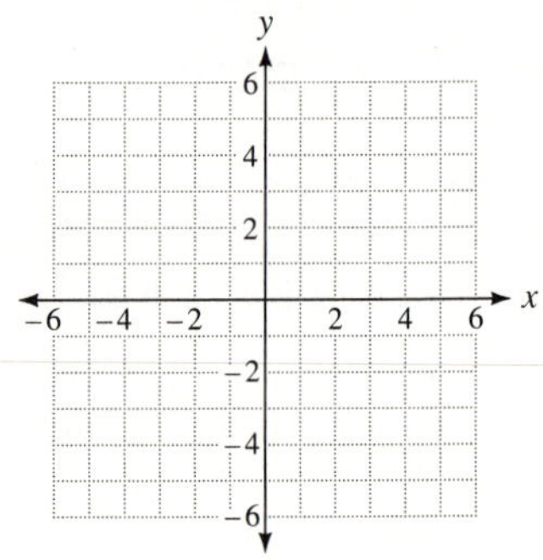

Solve the following systems.

252. $x^2 + 3y^2 = 19$
$x - y = 3$

253. $2x^2 + y^2 = 6$
$x^2 - y^2 = -3$

CHAPTER 12

254. Use inductive reasoning to determine the next element in the list. 1, 3, 9, 27, 81

In Exercises 255–56, determine if the conclusion follows logically from the premises, and state whether the reasoning is inductive or deductive.

255. *Premise:* If one angle in a triangle is 90°, then the triangle is a right triangle.
Premise: One angle of the triangle is 90°.
Conclusion: The triangle is a right triangle.

256. *Premise:* The answer to problem one is correct.
Premise: The answer to problem two is correct.
Premise: The answer to problem three is correct.
Conclusion: The answer to problem four is correct.

257. Is the following a logical statement?

$$3(a + b) = 3a + 3b$$

258. Give the converse of the following statement.

If it rains, then I will wear a raincoat.

259. Give the negation of the following statement.

All bears are brown.

260. Use an Euler diagram to test the validity of the conclusion given in the following syllogism.

Premise: All dogs have three legs.
Premise: Fang is a dog.
Conclusion: Fang has three legs.

Use the figure and deductive reasoning to answer true or false in Exercises 261–66. If the answer is false, tell why.

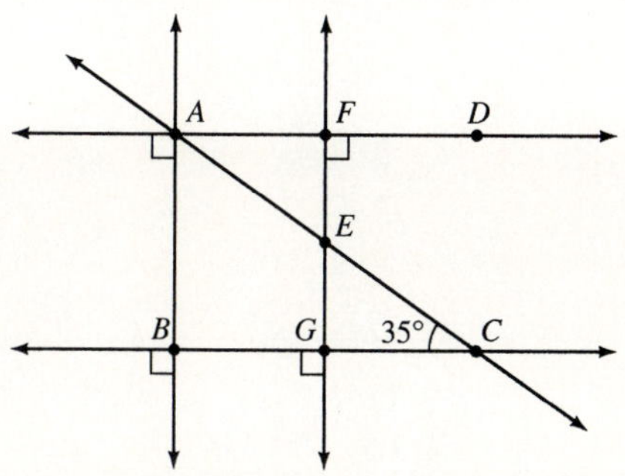

261. ΔABC is a right triangle.

262. $\angle CAD = 35°$.

263. $\overleftrightarrow{AD}$ is parallel to $\overleftrightarrow{AB}$.

264. $\overleftrightarrow{AD}$ is perpendicular to $\overleftrightarrow{FG}$.

265. ΔABC is similar to ΔEGC.

266. If AF = CG, then ΔAFE is congruent to ΔCGE.

Use the figure and deductive reasoning in Exercises 267–68.

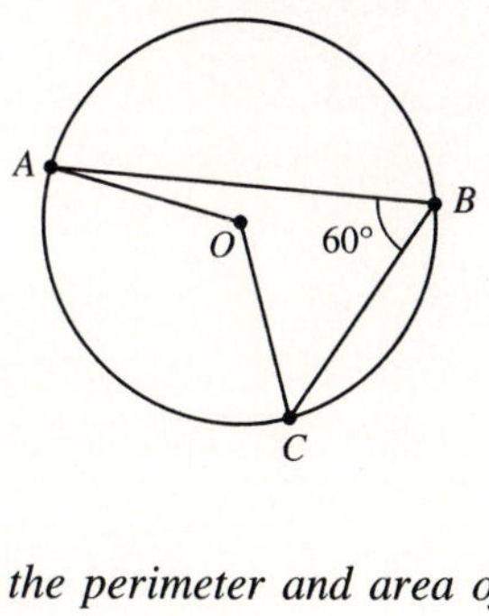

267. What is the measure of $\angle AOC$?

268. What is the measure of $\widehat{ABC}$?

Find the perimeter and area of the rectangle or square.

269. 5 ft by 9 ft

270. $12\frac{1}{2}$ m by 6.2 m

271. The area shown is to be carpeted. How much will it cost if carpet is \$22.50 per square yard?

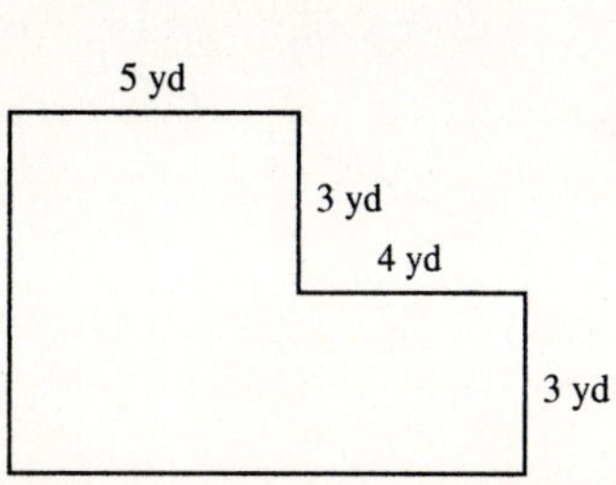

Find the perimeter and area of each figure.

272.

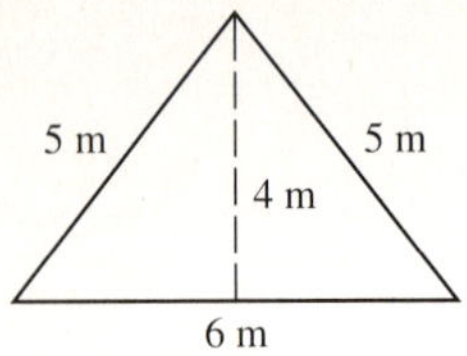

273.

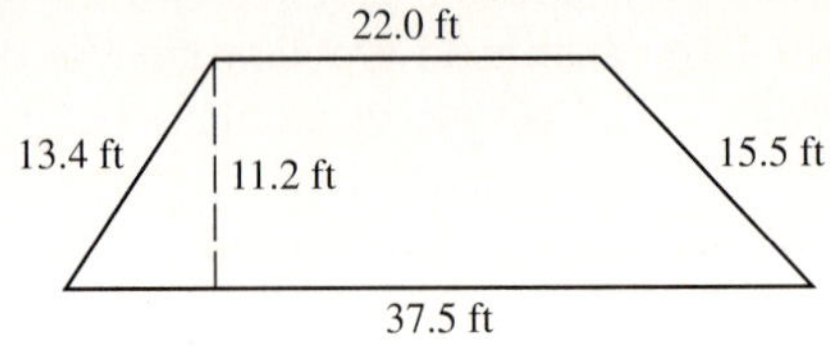

274. In a 30°–60° right triangle, a is the side opposite the 30° angle. If $a = 22$, find b and c.

275. In a 45°–45° right triangle, $b = 45$ cm. Find a and c.

Find the circumference and area of each circle. Use 3.14 for π and give answers to nearest tenth.

276. $r = 2.6$ ft

277. $d = 9.2$ m

278. A 14-in diameter pizza costs \$7.50 and a 16-in pizza costs \$8.50. Which costs less per square inch?

Find the volume and surface area of each solid.

279. Rectangular solid: 5 m by 2 m by 8 m

280. Cube: 3.6 ft edge

281. Cylinder: $d = 5.2$ dm, $h = 7.8$ dm

282. Sphere: $r = 15$ in

283. A spherical tank has radius 42 cm. How many liters of gasoline will it hold (to the nearest liter)?

ANSWERS: 1. $\frac{11}{10}$ 2. $\frac{2}{3}$ 3. $\frac{3}{7}$ 4. $\frac{10}{9}$ 5. $\frac{14}{5}$ 6. $\frac{1}{28}$ 7. 32.715 8. 7.88 9. 0.25872 10. 55 11. $2\frac{7}{18}$ 12. $\frac{107}{13}$ 13. 19 14. $6\frac{1}{16}$ 15. 37.5% 16. $2\frac{18}{25}$ 17. 50 18. $\frac{4}{3}$ 19. 10% 20. $\frac{2}{3}$ 21. 450 22. 100 23. 1150 24. 2550 25. 45.1% 26. 10.5, 9.5 27. 6.4, 6.2 28. 4.1 mpg 29. 35 30. 3.5 31. 0.054 32. 2500 33. 2270 34. 62.1 35. 140 36. 54¢ 37. 94.6 38. 28 hr 3 min 39. 10 yd 2 ft 3 in 40. 33.1 cm 41. 38.9 mi 42. 10 gal 2 qt 43. 7.75 44. 33 45. \$2332.80 46. $-\frac{4}{3}$ 47. 8 48. > 49. > 50. 12 51. −10 52. $\frac{1}{12}$ 53. −3 54. 0 55. −16 56. $2(y + 6)$ 57. $-7(2x + 3)$ 58. $-12a + 2b - 5$ 59. $5y + 8$ 60. −8 61. $-5xxxx$ 62. −12 63. $\frac{2}{5}$ 64. $3x^5$ 65. y^4 66. $9a^6$ 67. $\frac{1}{a^2}$ 68. 3.62×10^7 69. 0.00072 70. $6.3 \times 10^0 = 6.3$ 71. 5×10^{-12} 72. 1 73. −3 74. 14 cm, 22 cm 75. 1:00 P.M. 76. \$9200 77. 47, 49 78. 25°, 100°, 55° 79. \$40.00

80. $(-\infty, -1]$ 

81. $(-\infty, -2)$ or $[6, \infty)$ 

82. $[-2, 1)$ 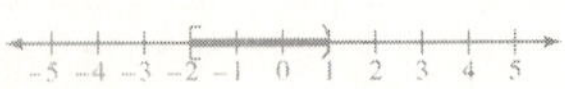

83. $-1, \frac{11}{3}$ 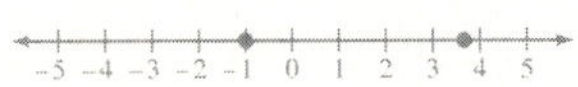

84. $(-\infty, -3)$ or $(2, \infty)$ 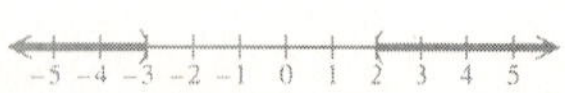

85. $[-2, 5]$ 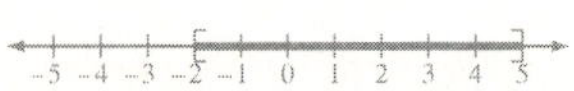

86. $r = \frac{P - q}{50}$ 87. $2, \frac{4}{3}$

88. x-intercept = $(-5, 0)$, y-intercept = $(0, 2)$, slope = $\frac{2}{5}$

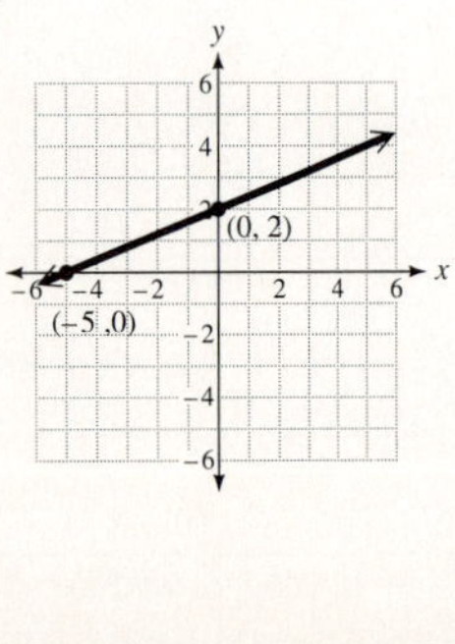

89.

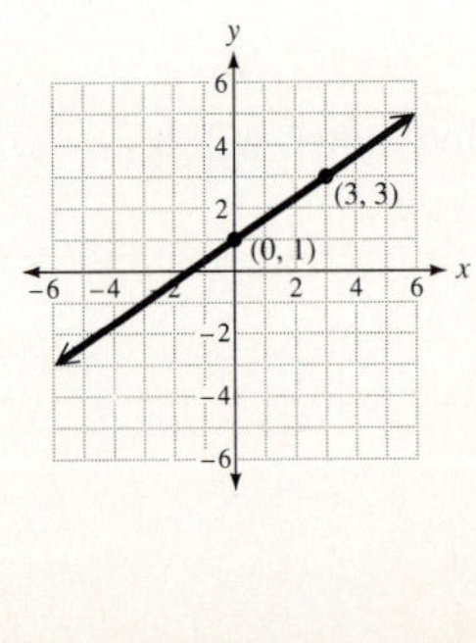

90. $x + y - 2 = 0$
91. $3x - 4y + 14 = 0$
92. no 93. no 94. yes
95. all real numbers except −5
96. (a) 7 (b) 4 (c) $a^2 + 3$
97. $f(x) = (0.20)x + 15{,}000$; $f(520{,}000) = \$119{,}000$ 98. yes
99. no solutions; lines are parallel

100. (1, −1) **101.** infinitely many solutions of the form $\left(x, \frac{1}{5}x - \frac{2}{5}\right)$ for x any real number **102.** 20 lb of 90¢ candy; 40 lbs of \$1.50 candy **103.** 7 quarters, 20 dimes **104.** infinitely many solutions; the system is dependent **105.** no solution; the system is inconsistent **106.** (1, 2, −1) **107.** 70, 77, 90 **108.** 26 **109.** −7 **110.** $A_x = 52$, $A = 26$, $x = 2$ **111.** $A_x = -7$, $A = -7$, $x = 1$

112.

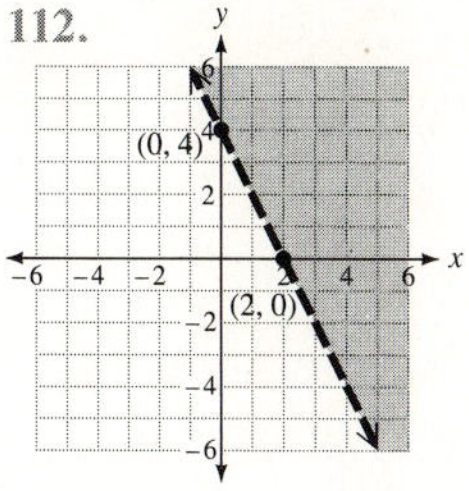

113.

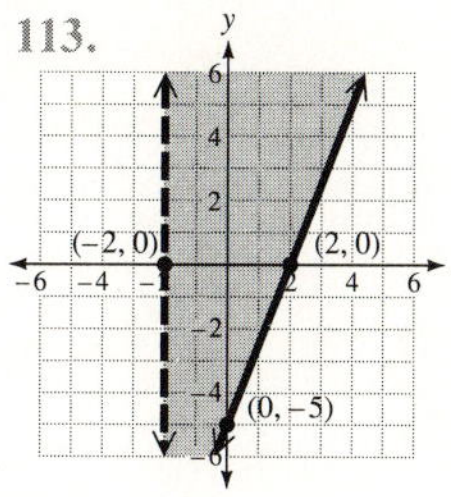

114.

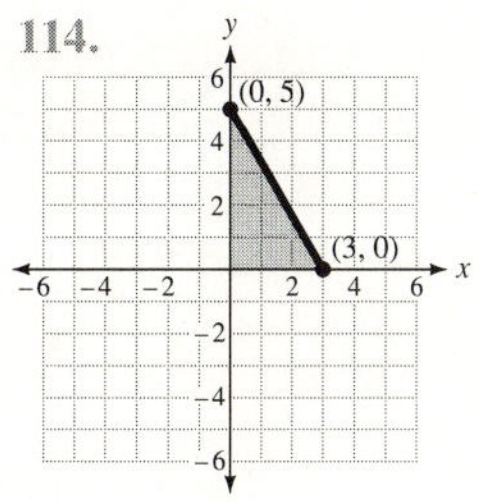

115. $-x^7 - 5x^3 + x^2 + 3x + 2$ **116.** $x^4y + 4x^2y^2 - xy$ **117.** 33 **118.** $3y^3 + 2y^2 - 15y + 2$ **119.** $2x^2y^2 + 13$ **120.** $-3a^2b^2 + ab + 6$ **121.** $-6x^2y^2 + 8xy^3$ **122.** $2x^2 - 7xy - 4y^2$ **123.** $a^2 - 6ab + 9b^2$ **124.** $9u^2 - v^2$ **125.** $3(x - 2)(x + 2)$

126. $(a + 5)(a + 5)$ **127.** $4(y - 2)(y^2 + 2y + 4)$ **128.** $(3x - 2)(x - 2)$ **129.** $3(x + 1)(x^2 - x + 1)$ **130.** $-2(u - 2v)(3u + v)$ **131.** 0, 3 **132.** −1, 4 **133.** 7 and 9 **134.** 4 cm by 11 cm **135.** 180 amperes **136.** first at 3 sec and again at 15 sec **137.** 0, 2 **138.** $4x^2 - 2x + 1$ **139.** $\frac{1}{4}$ **140.** $\frac{a+3}{3(a-4)}$ **141.** $\frac{3(2x-y)}{(x-y)(x+y)}$ **142.** $\frac{2b}{(a+b)(a+b)}$ **143.** $\frac{-3y^2-3y-4}{(y-1)(y+1)}$ **144.** $\frac{x+2}{x-5}$ **145.** 6, −4 **146.** no solution **147.** 20 mph **148.** $\frac{15}{8}$ days **149.** 1035 mi **150.** $y = \frac{3x\sqrt{z}}{w^2}$ **151.** $a = \frac{1}{c} + b$ **152.** $a - b$ **153.** $-3z^2 - 9z + 1$ **154.** $x^3 + x - 2$; 0 **155.** $5|x|$ **156.** $-5x$ **157.** $-|w|$ **158.** $|a + b|$ **159.** 11 **160.** $3ab^2$ **161.** $2a^2$ **162.** $4ab\sqrt{5a}$ **163.** $\frac{3\sqrt{2}}{z}$ **164.** $\frac{-5x}{2y}$ **165.** $3x^2y\sqrt{2y}$ **166.** $\frac{2a\sqrt[3]{b^2}}{b}$ **167.** $-12ab^3\sqrt[3]{a^2}$ **168.** $16x\sqrt{7x}$ **169.** 3 **170.** $3 + 2\sqrt{2}$ **171.** $x - 25$ **172.** $-1 - \sqrt{10}$ **173.** $\frac{7+2\sqrt{7}}{3}$ **174.** $\frac{a^8b^3}{9}$ **175.** $\frac{7}{\sqrt{14}}$ **176.** $\frac{x}{7\sqrt[3]{x^2}}$ **177.** $\frac{x-4}{3(\sqrt{x}+2)}$ **178.** 5 **179.** 5 **180.** $\frac{a^3}{x}$ **181.** 15 mi **182.** $5\sqrt{2}$ **183.** (1, 3) **184.** $10 + 2i$ **185.** $10 - 10i$ **186.** $\frac{4-7i}{13}$ **187.** i **188.** $3i$ **189.** $\frac{3}{2}$ **190.** −1, 7 **191.** 0, 2 **192.** $1 \pm \sqrt{2}$ **193.** $\pm\sqrt{2}, \pm 1$ **194.** 7, 2 **195.** $\pm 2i$ **196.** $-1 \pm i\sqrt{3}$ **197.** $2 \pm \sqrt{2}$ **198.** $\frac{1 \pm i\sqrt{3}}{2}$ **199.** 5, 8 **200.** Cindy takes 6.8 hr; Burford takes 9.8 hr **201.** 27.9 mph **202.** 10, 12 **203.** 5 in **204.** $\frac{5 \pm \sqrt{17}}{2}$ **205.** $n \pm \sqrt{n^2 + 4}$ **206.** two complex **207.** $x^2 + x - 12 = 0$ **208.** $\left(-\frac{1}{2}, 2\right)$ **209.** $(-\infty, 2]$ or $(7, \infty)$ **210.**

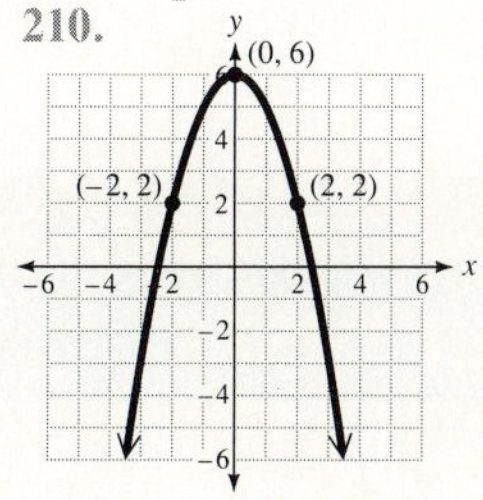

211. x-intercepts: (1, 0) and (5, 0) vertex: (3, −4)

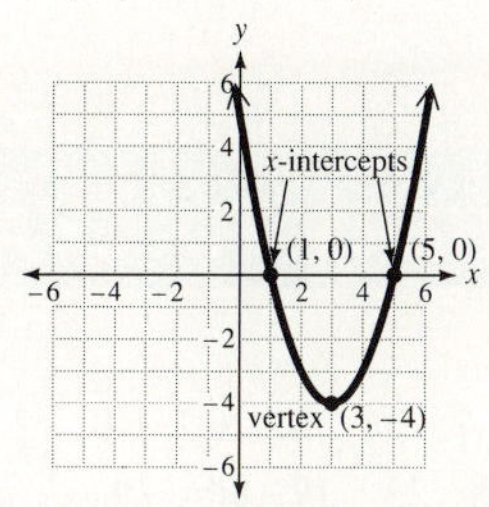

212. down; the vertex is $\left(\frac{3}{2}, -\frac{19}{4}\right)$; there are no x-intercepts **213.** 3600 yd²; 60 yd by 60 yd **214.** all real numbers $x \le 5$ **215.** all real numbers $x < 5$ **216.** (a) 6 (b) −8 (c) −7 (d) $-\frac{1}{7}$ (e) −6 (f) 4

217. not one-to-one

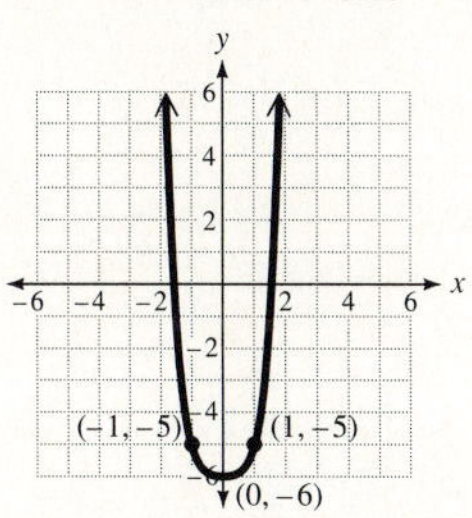

218. not one-to-one

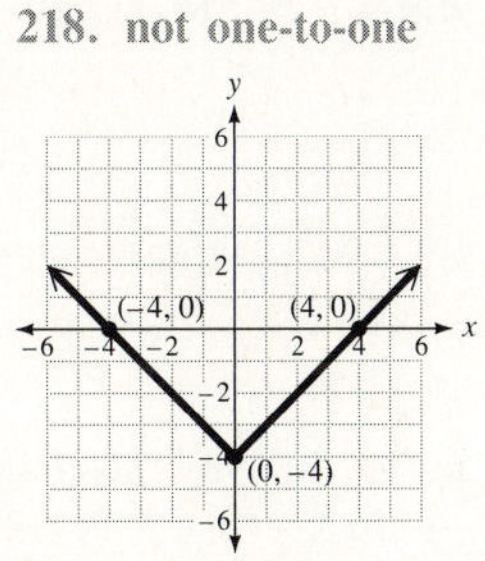

219. is one-to-one

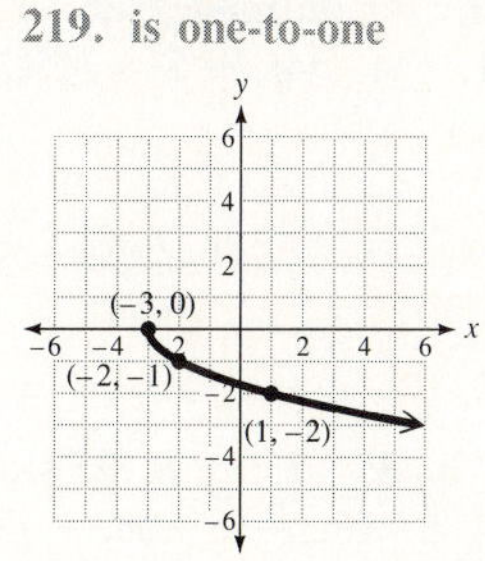

220. $f^{-1}(x) = 5x + 2$; (a) $-\frac{3}{5}$ (b) −3 (c) −1 (d) −1 **221.** $\log_7 3 = x$ **222.** $a^6 = 25$ **223.** $2 \log_a x - \log_a y - \frac{1}{2}\log_a z$ **224.** $\log_a \frac{\sqrt{y}}{x^3}$ **225.** false; $\log_a u - \log_a v$ **226.** true **227.** 0.9542 **228.** −0.2219 **229.** 1.1761 **230.** −3.4841

231. 5.6981 **232.** 3.8501 **233.** 422 **234.** 0.386 **235.** 3.96 **236.** 1 **237.** $-\frac{3}{2}$ **238.** 49 decibels **239.** \$2032.79 **240.** about 5.3 years **241.** The functions are inverses.

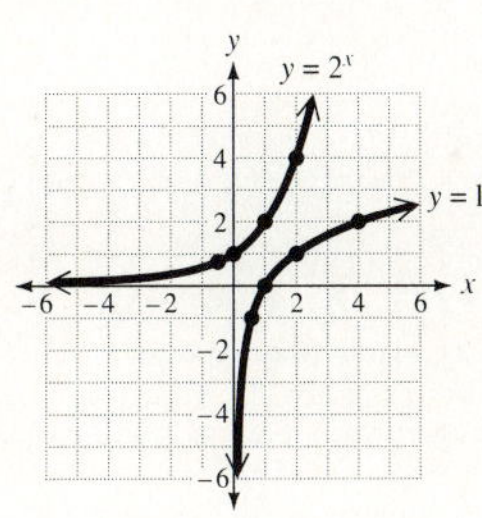

242. $(x + 4)^2 + (y - 2)^2 = 6^2$; $x^2 + y^2 + 8x - 4y - 16 = 0$

243. $(x - 1)^2 + (y + 4)^2 = 2^2$; $(1, -4)$; $r = 2$ **244.** $\frac{x^2}{4} + \frac{y^2}{25} = 1$ **245.** $\frac{x^2}{9} - \frac{y^2}{81} = 1$

246.
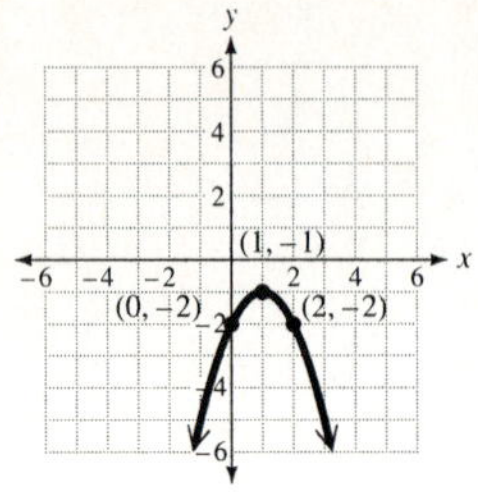

247.
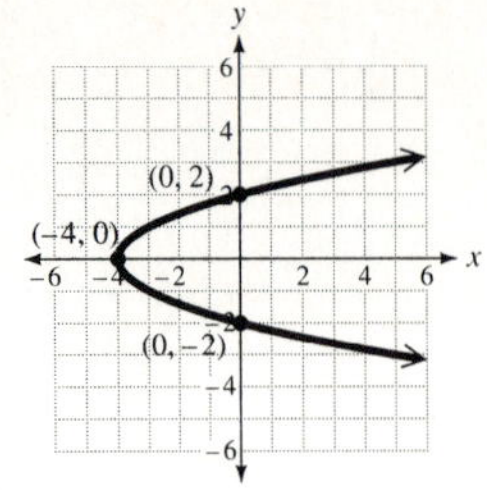

248.
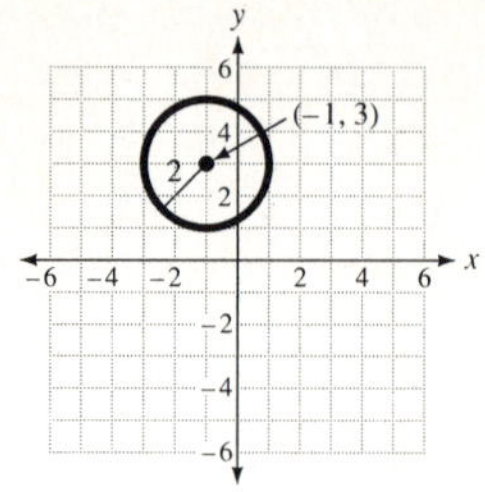

249.
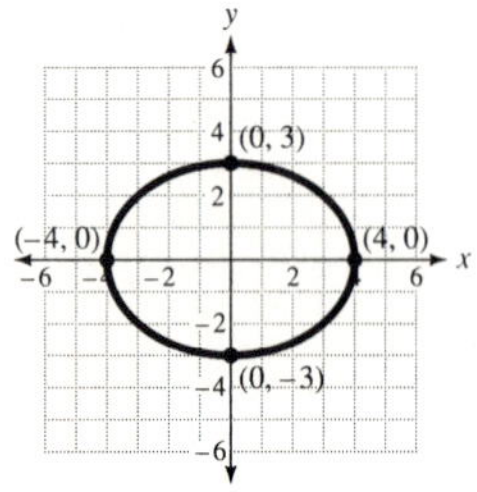

250.
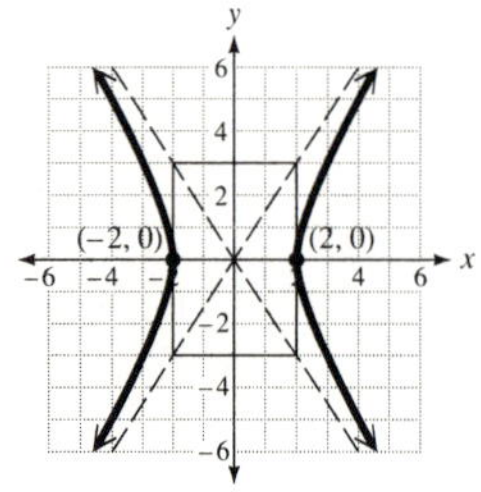

251.
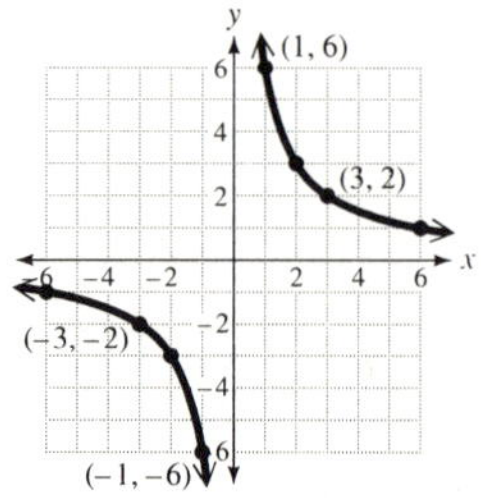

252. $(4, 1), \left(\frac{1}{2}, -\frac{5}{2}\right)$ **253.** (1, 2), (1, −2), (−1, 2), (−1, −2) **254.** 243 **255.** follows logically; deductive reasoning **256.** does not follow logically; inductive reasoning **257.** yes **258.** If I wear a raincoat, then it is raining. **259.** Some bears are not brown. **260.** valid **261.** true **262.** true **263.** false (perpendicular) **264.** true **265.** true **266.** true **267.** 120° **268.** 240° **269.** 28 ft, 45 ft^2 **270.** 37.4 m, 77.5 m^2 **271.** \$945 **272.** 16 m, 12 m^2 **273.** 88.4 ft, 333.2 ft^2 **274.** $b = 22\sqrt{3}$, $c = 44$ **275.** $a = 45$ cm, $c = 45\sqrt{2}$ cm **276.** 16.3 ft, 21.2 ft^2 **277.** 28.9 m, 66.4 m^2 **278.** 14 in: \$0.049 per square inch, 16 in: \$0.042 per square inch, 16 in costs less **279.** 80 m^3, 132 m^2 **280.** 46.7 ft^3, 77.8 ft^2 **281.** 165.6 dm^3, 169.8 dm^2 **282.** 14,130 in, 2826 in^2 **283.** 310 L

ANSWERS TO CHAPTER TESTS

CHAPTER 0

1. 24 **2.** $\frac{11}{15}$ **3.** $\frac{3}{11}$ **4.** $\frac{9}{4}$ **5.** $\frac{29}{18}$ **6.** $\frac{1}{24}$ **7.** $5\frac{3}{5}$ **8.** 20% **9.** $\frac{63}{20}$ **10.** \$1.08 **11.** 432 **12.** $4\frac{7}{10}$ liters **13.** 40 **14.** 90 **15.** $\frac{2}{1}$ or 2 to 1 **16.** 26.6% **17.** 60% **18.** 25 **19.** 80 **20.** $\frac{7}{16}$ or 7 to 16 **21.** 25% **22.** 20 **23.** 18 **24.** 8 **25.** (a) 254.5 pounds (b) 255 pounds (c) no mode **26.** 2.5 **27.** 0.028 **28.** 14.3 **29.** 149 **30.** 8 yd 2 ft 7 in **31.** 23 yd 2 ft 2 in **32.** 4 **33.** 16.3 L **34.** no; 59°F = 15°C

CHAPTER 1

1. a^3 **2.** 1 **3.** 75 **4.** 225 **5.** 125 **6.** 35 **7.** 5 **8.** \$594.05 **9.** false **10.** 13 **11.** true **12.** $<$ **13.** 141° **14.** −22 **15.** 8 **16.** $-\frac{1}{8}$ **17.** 3 **18.** 5 **19.** −18 **20.** −19 **21.** −6 **22.** $-3(y-5)$ **23.** $5-b-2y$ **24.** 3 **25.** 15 **26.** $-12y+6$ **27.** 11 **28.** $-\frac{5}{4}$ **29.** true **30.** false **31.** $2y^7$ **32.** a^4 **33.** $-8x^6$ **34.** $\frac{1}{z^3}$ **35.** $\frac{y^7}{x^8}$ **36.** $-\frac{1}{64}$ **37.** 3×10^{-2}

CHAPTER 2

1. −5 **2.** −6 **3.** 0 **4.** −7 **5.** −60 **6.** 14 in; 19 in **7.** 15°; 45°; 120° **8.** 1:00 P.M. **9.** 350 **10.** \$15,600 **11.** 15% **12.** $a = \frac{W-c}{30}$ **13.** $w = \frac{uv+2}{u+v}$ **14.** $x \le 2$

15. $y < -2$ or $y > 4$
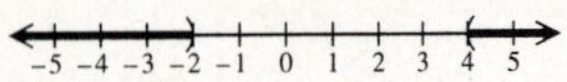

16. $1 < z < 5$
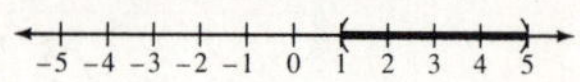

17. $-1, \frac{11}{3}$ **18.** [−2, 7] **19.** $-6, \frac{4}{3}$ **20.** $x > -6$

CHAPTER 3

1. $y = -\frac{3}{2}x + 6$ **2.** (4, 0) **3.** (0, 6) **4.** $-\frac{3}{2}$ **5.**

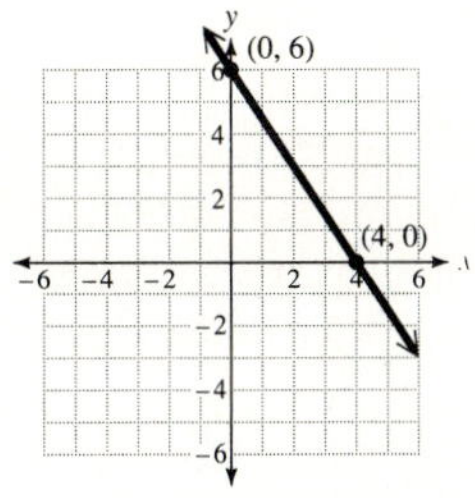

6. (10, 70); \$70 **7.** -2; $\frac{1}{2}$
8. parallel **9.** $3x + y + 2 = 0$
10. $2x + y - 5 = 0$ **11.** $y = 8$ **12.** yes **13.** yes
14. no **15.** -3 **16.** $3a$ **17.** linear
18.

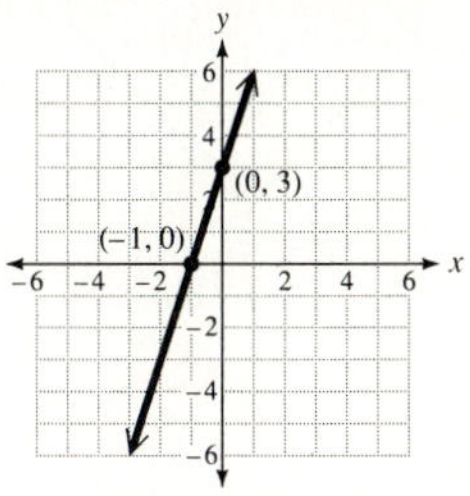

19. all real numbers except -3
20. $g(x) = 20x$

CHAPTER 4

1. no **2.** infinitely many **3.** $(5, -1)$ **4.** $(x, 2x - 3)$ x any real number **5.** 6; 11 **6.** 18 lb of \$2.00/lb nuts and 30 lb of \$2.80/lb nuts **7.** no solution **8.** $(1, 0, -2)$ **9.** $x + y + z = 10{,}000$; $0.08x + 0.07y + 0.10z = 810$; $0.08x + 0.07y - 0.04z = 390$ **10.** 16 **11.** -14 **12.** $x = 1, A_x = 11, A = 11$ **13.** $x = 3, A_x = -9, A = -3$
14.

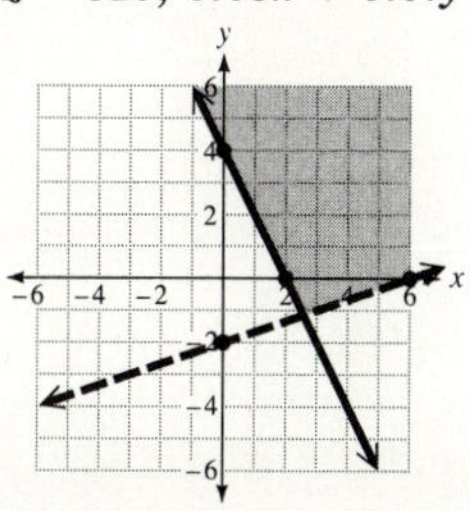

15.

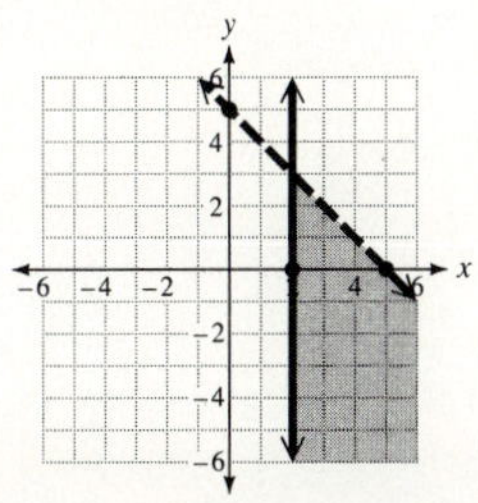

CHAPTER 5

1. $-5x^2y$ **2.** 4 **3.** trinomial **4.** 53 **5.** $3x^3 - 4$ **6.** $-a^3b^2 + 4b^4$ **7.** $5x^2y^3$ **8.** $8ab + 1$ **9.** $-12u^3v + 8u^2v^2$ **10.** $a^2 - 9b^2$ **11.** $z^2 + 10z + 25$ **12.** $z^2 - 10z + 25$ **13.** $12x^2 + 7xy - 10y^2$ **14.** $a^4 - b^4$ **15.** $-11xy^2(xy - 6 + 4x)$ **16.** $(3u + v)(2u - v)$ **17.** $2(x - 2)(x + 2)$ **18.** $(a + 2b)(a^2 - 2ab + 4b^2)$ **19.** $(x - 3y)^2$ **20.** $(b - y)(x - 2)$ **21.** $(x + y - 2)(x + y - 3)$ **22.** 7, -1 **23.** -5, $\frac{1}{2}$ **24.** Mack is 15; Betty is 12 **25.** 30

CHAPTER 6

1. 1, -1 **2.** $x^2 + xy + y^2$ **3.** $\frac{a+1}{a+3}$ **4.** 1 **5.** $\frac{(x-2)^2}{x^2}$ **6.** $\frac{1}{x-1}$ **7.** $\frac{3b}{(a-b)^2}$ **8.** $\frac{-2(2a+1)}{(a+2)(a+3)(a-2)}$ **9.** 1, -3 **10.** Jeff is 12; Sue is 10 **11.** $y = \frac{2-wx}{w}$ **12.** $y = \frac{2x\sqrt{z}}{w}$ **13.** 65 minutes **14.** 12 mph **15.** $x - 3$ **16.** $-6a^3 + 4a + 1$ **17.** $4a^2 - a + 3$ **18.** $x^3 + x^2 - 3x - 1$; -2

CHAPTER 7

1. 7 **2.** -7 **3.** $4ab$ **4.** $2x$ **5.** $-2y$ **6.** $5\sqrt{3}$ **7.** $3\sqrt[3]{2}$ **8.** $2y\sqrt[4]{y}$ **9.** $\frac{3x\sqrt[4]{x^3}}{y^2}$ **10.** $5\sqrt{2} - \sqrt{10}$ **11.** $4 + 2\sqrt{3}$ **12.** $3ab\sqrt[4]{b^3}$ **13.** $x + 2\sqrt{ax} - 3a$ **14.** $-3\sqrt{2}$ **15.** 0 **16.** 81 **17.** -1.5 **18.** $\frac{5y^2\sqrt{y}}{x}$ **19.** $-\sqrt{2} - \sqrt{5}$ **20.** $\frac{1}{\sqrt{6}-3}$ **21.** 21 **22.** -4 **23.** (a) $2\sqrt{10}$ (b) $(-1, 0)$ **24.** 5 mi **25.** $a = \frac{c^2}{b}$ **26.** $3 + i$ **27.** $-6 + 32i$ **28.** $\frac{9+7i}{26}$ **29.** about 16 seats **30.** $5|a|$

CHAPTER 8

1. $\frac{1}{2}$, -5 **2.** $-1 \pm 2\sqrt{2}$ **3.** $-1 \pm i\sqrt{2}$ **4.** 3, -5 **5.** 1, -4 **6.** 8 cm, 6 cm **7.** $\frac{3}{2}$ **8.** 8 **9.** $c = \frac{\pm\sqrt{a+b}}{3}$ **10.** 36 days **11.** 7.5% **12.** one time; 3 sec after it is launched **13.** two complex **14.** $5x^2 + 7x - 6 = 0$ **15.** $x < -1$ or $x > 5$; $(-\infty, -1)$ or $(5, \infty)$

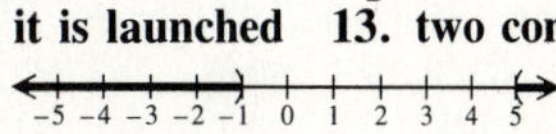

16. $1 \le x < 4$; $[1, 4)$

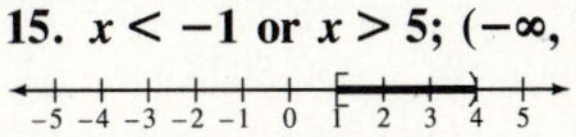

CHAPTER 9

1. Slide the graph of $f(x) = x$ up 3 units.

2.

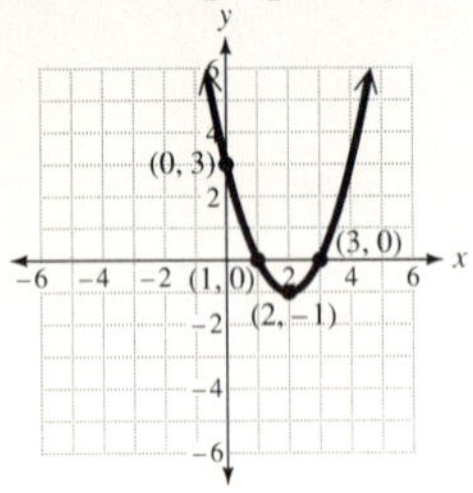

3. 144 **4.** all real numbers $x > -10$ **5.** -16 **6.** 5 **7.** -10 **8.** 9 **9.** -13 **10.** -1 **11.** no **12.** $f^{-1}(x) = 4x + 3$; (a) 11 (b) 2

13.

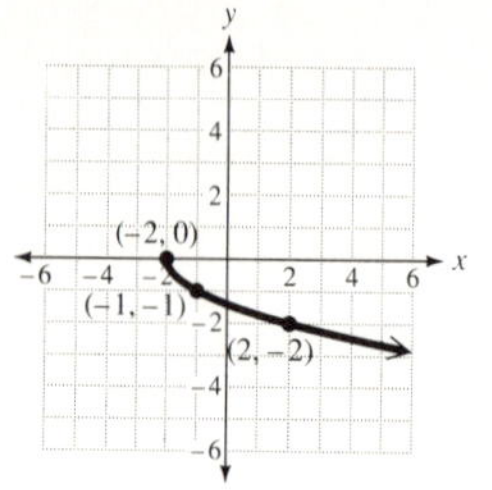

14.

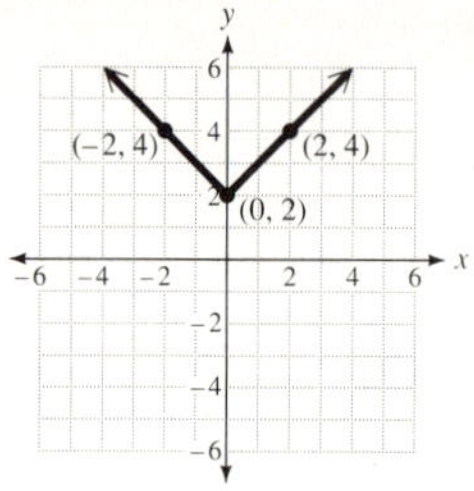

15.

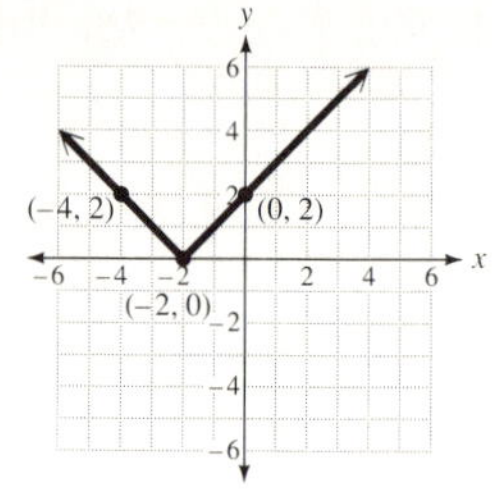

CHAPTER 10

1. $b^x = 5$ **2.** $\log_u 7 = v$ **3.** 4 **4.** -1 **5.** 3 **6.** 11 **7.** $2 \log_a x + 3 \log_a y - \log_a z$ **8.** $\log_a \frac{z^2}{y^{1/3}}$ **9.** false **10.** true **11.** 1 **12.** -0.2132 **13.** 1.8485 **14.** 1960 **15.** 0.647 **16.** 2, -2 **17.** $\frac{1}{9}$ **18.** 50.8 decibels **19.** \$8351.04 **20.** 15.4 years **21.**

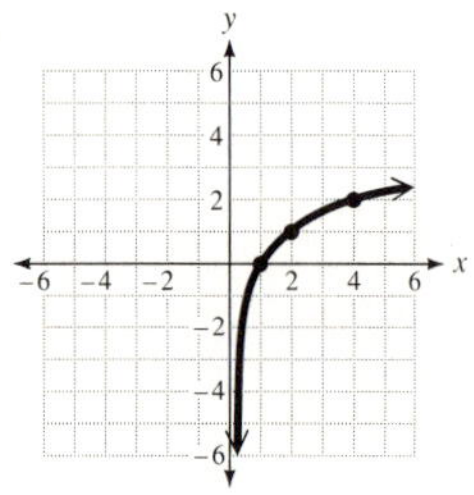

CHAPTER 11

1.

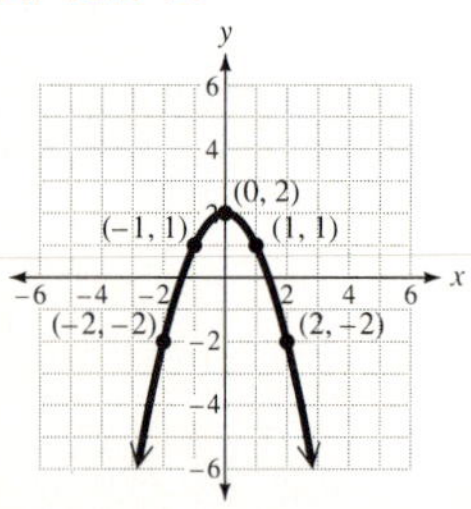

2.

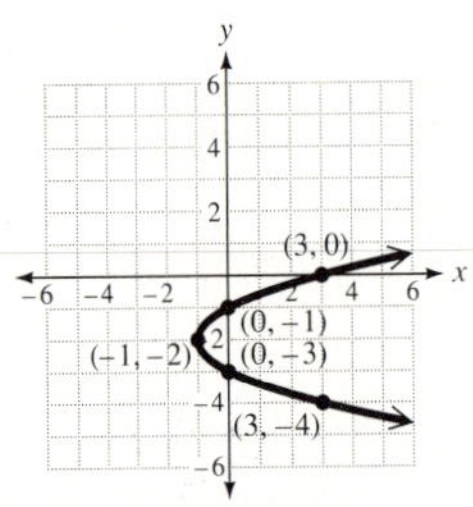

3.

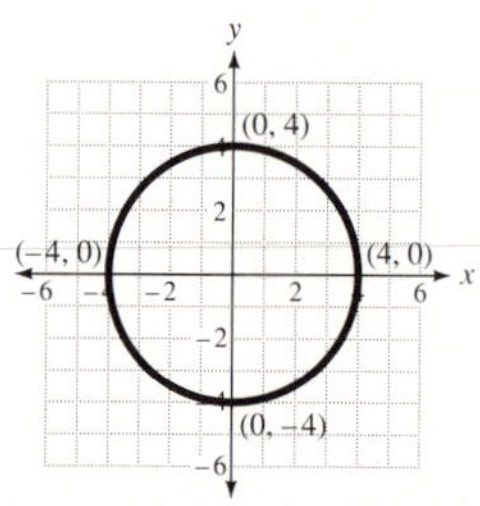

4.

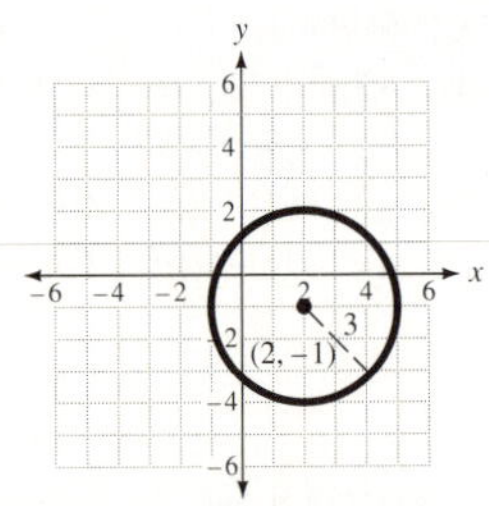

5. $(x + 2)^2 + (y + 1)^2 = 5^2$; $x^2 + y^2 + 4x + 2y - 20 = 0$ **6.** $(x + 7)^2 + (y - 6)^2 = 5^2$; $(-7, 6)$; $r = 5$

7.

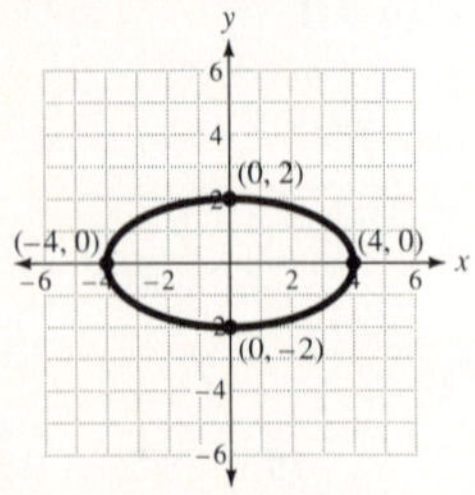

8.

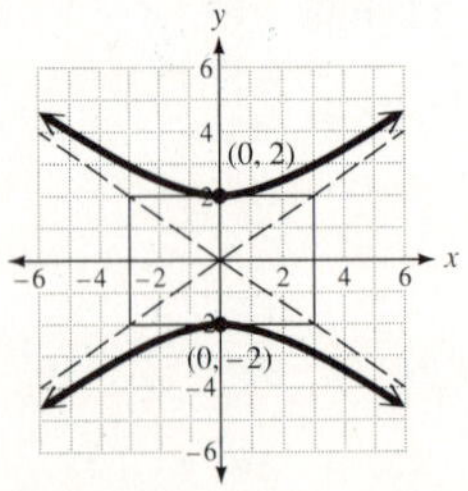

9. hyperbola **10.** circle **11.** parabola **12.** ellipse **13.** $(2, -5)$, $(-6, 11)$ **14.** $(2\sqrt{2}, 1)$, $(-2\sqrt{2}, 1)$

CHAPTER 12

1. $\frac{1}{5}$ **2.** no; inductive reasoning **3.** If x is a multiple of 2, then x is a multiple of 4. **4.** Some turkeys do not live in coops. **5.** true **6.** false (supplementary) **7.** false ($\overleftrightarrow{AB}$ and $\overleftrightarrow{FG}$ intersect at B) **8.** true **9.** yes **10.** 110° **11.** 250° **12.** 55° **13.** false **14.** true **15.** true **16.** true **17.** true **18.** true **19.** false **20.** true **21.** true **22.** true **23.** $b = 20\sqrt{3}$, $c = 40$ **24.** $4\sqrt{2}$ **25.** \$945 **26.** 16.3 ft **27.** 66.4 m^2 **28.** (a) 46.7 ft^3 (b) 77.8 ft^2 **29.** 16-in (14-in: \$0.049 per in^2; 16-in: \$0.042 per in^2) **30.** 12 ft^2 **31.** 31 m **32.** 2009.6 cm^3 **33.** 5024 in^2 **34.** \$78.75

APPENDIX A

Logarithmic Tables and Interpolation

STUDENT GUIDEPOSTS

1. Characteristic and Mantissa
2. Finding Common Logarithms
3. Standard Form of a Logarithm
4. Finding Antilogarithms
5. Linear Interpolation
6. Logarithmic Computation

1 CHARACTERISTIC AND MANTISSA

In the past few years, scientific calculators have all but replaced logarithm tables. However, when a calculator is unavailable, a table of common logarithms can be used. A table such as the one at the end of this appendix is limited in accuracy to three significant digits (four if interpolation is used).

If n is any positive number, n can be written in scientific notation as

$$n = m \times 10^c,$$

where $1 \le m < 10$ and c is an integer. Using the product rule,

$$\log n = \log (m \times 10^c) = \log m + \log 10^c = \log m + c.$$

Log m, the **mantissa** of log n, is a decimal greater than or equal to 0 and less than 1. The **characteristic** c of log n is an integer. Since any positive number can be written in scientific notation, the table of common logarithms, with logarithms of numbers between 1.00 and 9.99 in increments of 0.01, gives us logarithms of any number with three significant digits. The left-hand column in the table shows numbers 1.0 through 9.9 while the numbers 0 through 9 head each of the other columns. To find log 1.23, for example, we look down the left column to 1.2, read across that row to the column headed 3, and find the decimal 0.0899.

n	0	1	2	3	4	5	6	7	8	9
1.0	.0000	.0043	.0086	.0128	.0170	.0212	.0253	.0294	.0334	.0374
1.1	.0414	.0453	.0492	.0531	.0569	.0607	.0645	.0682	.0719	.0755
1.2	.0792	.0828	.0864	**.0899**	.0934	.0969	.1004	.1038	.1072	.1106
1.3	.1139	.1173	.1206	.1239	.1271	.1303	.1335	.1367	.1399	.1430
1.4	.1461	.1492	.1523	.1553	.1584	.1614	.1644	.1673	.1703	.1732

2 FINDING COMMON LOGARITHMS

We now summarize the method for finding a common logarithm using the table.

To Find the Common Logarithm of a Given Number

1. Write the number in scientific notation.
2. Use the table of common logarithms to find the logarithm of the number between 1 and 10 (the mantissa of the logarithm of the given number).
3. The logarithm of the original number is the characteristic (the exponent on 10) plus the mantissa.

EXAMPLE 1 Finding Common Logarithms

(a) Find log 45,600.

$$\begin{aligned} \log 45{,}600 &= \log (4.56 \times 10^4) && \text{Scientific notation} \\ &= \log 4.56 + \log 10^4 && \log uv = \log u + \log v \\ &= \log 4.56 + 4 && \log 10^4 = 4 \\ &= 0.6590 + 4 && \text{From table of common logarithms} \\ &= 4.6590 \end{aligned}$$

(b) Find log 4.56.

$$\begin{aligned} \log 4.56 &= \log (4.56 \times 10^0) \\ &= \log 4.56 + \log 10^0 \\ &= 0.6590 + 0 \\ &= 0.6590 \end{aligned}$$

(c) Find log 0.00456.

$$\begin{aligned} \log 0.00456 &= \log (4.56 \times 10^{-3}) \\ &= \log 4.56 + \log 10^{-3} \\ &= 0.6590 + (-3) \\ &= 0.6590 - 3 && \textit{Not } -3.6590 \end{aligned}$$

PRACTICE EXERCISE 1

(a) Find log 283,000.

(b) Find log 2.83.

(c) Find log 0.0000283.

Answers: (a) 5.4518 (b) 0.4518 (c) 0.4518 − 5

3 STANDARD FORM OF A LOGARITHM

In Example 1(c), notice that we left the characteristic separate from the mantissa in what is called **standard form.** If we subtract, we obtain

$$0.6590 - 3 = -2.3410.$$

Although −2.3410 is a correct form for the logarithm of 0.00456, it does not directly display the characteristic (−3) and positive mantissa (0.6590). When tables are used, it is wise to leave the logarithm in standard form. (If we use a calculator to find logarithms, the result is displayed in the latter form. You should be able to convert from one form of a logarithm to the other. Changing from

$$0.6590 - 3 \quad \text{to} \quad -2.3410$$

is simply a matter of subtracting. Converting from

$$-2.3410 \quad \text{to} \quad 0.6590 - 3$$

is a bit more challenging. In this case we add and subtract 3 (any integer greater than the absolute value of -2.3410 will work, and 3 is the smallest such integer).

$$\begin{array}{r} 3.0000 - 3 \\ -2.3410 \quad\;\; \\ \hline 0.6590 - 3 \end{array}$$

Suppose we add and subtract 10.

$$\begin{array}{r} 10.0000 - 10 \\ -2.3410 \quad\;\;\; \\ \hline 7.6590 - 10 \end{array}$$

We clearly obtain the same result since

$$\begin{aligned} 7.6590 - 10 &= 7 + 0.6590 - 10 \\ &= 0.6590 + 7 - 10 = 0.6590 - 3. \end{aligned}$$

4 FINDING ANTILOGARITHMS

Up to now we have illustrated how to find a logarithm of a given number. Now we show how to find a number when its logarithm is given, that is, how to find an **antilogarithm (antilog).** To use the table, the given logarithm must be in standard form showing the positive mantissa and the integer characteristic.

EXAMPLE 2 Finding Common Antilogarithms

(a) Find the antilog of 2.7832. (That is, find $10^{2.7832}$.)
We need to find n in the equation

$$\begin{aligned} \log n &= 2.7832 \\ &= 0.7832 + 2. \end{aligned}$$

The mantissa is 0.7832, and the characteristic is 2. In the table we find 0.7832 in the row headed by 6.0 and in the column headed by 7.

n	0	1	2	3	4	5	6	7	8	9
5.5	.7404	.7412	.7419	.7427	.7435	.7443	.7451	.7459	.7466	.7474
5.6	.7482	.7490	.7497	.7505	.7513	.7520	.7528	.7536	.7543	.7551
5.7	.7559	.7566	.7574	.7582	.7589	.7597	.7604	.7612	.7619	.7627
5.8	.7634	.7642	.7649	.7657	.7664	.7672	.7679	.7686	.7694	.7701
5.9	.7709	.7716	.7723	.7731	.7738	.7745	.7752	.7760	.7767	.7774
6.0	.7782	.7789	.7796	.7803	.7810	.7818	.7825	.7832	.7839	.7846
6.1	.7853	.7860	.7868	.7875	.7882	.7889	.7896	.7903	.7910	.7917
6.2	.7924	.7931	.7938	.7945	.7952	.7959	.7966	.7973	.7980	.7987
6.3	.7993	.8000	.8007	.8014	.8021	.8028	.8035	.8041	.8048	.8055
6.4	.8062	.8069	.8075	.8082	.8089	.8096	.8102	.8109	.8116	.8122

$$\begin{aligned} n &= 6.07 \times 10^2 \\ &= 607 \end{aligned}$$

Thus, $\quad 10^{2.7832} = 607.$

(b) Find the antilog of -2.2321.
We must first express

$$\log n = -2.2321$$

in standard form by adding and subtracting 3.

Practice Exercise 2

(a) Find the antilog of 5.8075.

(b) Find the antilog of -1.5935

$$
\begin{array}{r}
3.0000 - 3 \\
-2.2321 \\
\hline
0.7679 - 3
\end{array}
$$

Thus, the mantissa is .7679 and the characteristic is -3. From the table,

$$
\begin{aligned}
n &= 5.86 \times 10^{-3} \\
&= 0.00586.
\end{aligned}
$$

Answers: (a) 642,000 (b) 0.0255

5 LINEAR INTERPOLATION

To find the common logarithm of a number with four significant digits, such as 5138, we can use a process called **linear interpolation.** The word *linear* appears because the process uses a straight line to approximate the portion of the graph of $y = \log x$ between two values in the table. Consider Figure A.1.

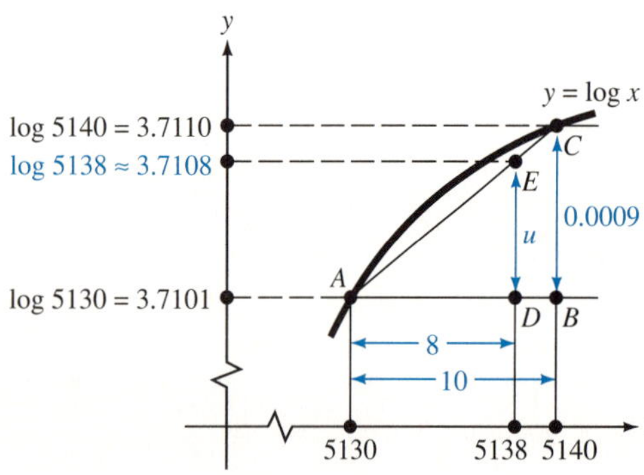

Figure A.1 Interpolation

Noting that triangles ABC and ADE are similar, we conclude that

$$\frac{8}{10} = \frac{u}{0.0009}.$$

Thus,
$$u = \frac{8}{10}(0.0009) \approx 0.0007.$$

To obtain the approximation for log 5138, add 0.0007 to log 5130.

$$\log 5138 \approx \log 5130 + 0.0007 = 3.7101 + 0.0007 = 3.7108.$$

EXAMPLE 3 Using Linear Interpolation (Logarithms)

Use linear interpolation to find log 0.006412. The characteristic is -3 so we concentrate on finding the mantissa for log 6.412.

		x	$\log x$		
0.010	0.002	6.410	0.8069	u	0.0006
		6.412	?		
		6.420	0.8075		

$$\frac{0.002}{0.010} = \frac{2}{10} = \frac{u}{0.0006}$$

Practice Exercise 3

Use linear interpolation to find log 75,620.

$$u = \frac{2}{10}(0.0006) \approx 0.0001$$

$$\begin{aligned} \log 6.412 &\approx \log 6.41 + 0.0001 \\ &= 0.8069 + 0.0001 \\ &= 0.8070 \end{aligned}$$

$$\log 0.006412 = 0.8070 - 3 = -2.1930$$

Answer: 4.8786

If a calculation requires finding the antilog of a logarithm whose mantissa is not in the table, linear interpolation can be used to approximate the four-digit answer.

EXAMPLE 4 Using Linear Interpolation (Antilogs)

Find the antilog of 2.4705.

The characteristic 2 tells us where to place the decimal point in the final answer. Thus we can concentrate on 0.4705. In the table, 0.4705 is between log 2.95 = 0.4698 and log 2.96 = 0.4713.

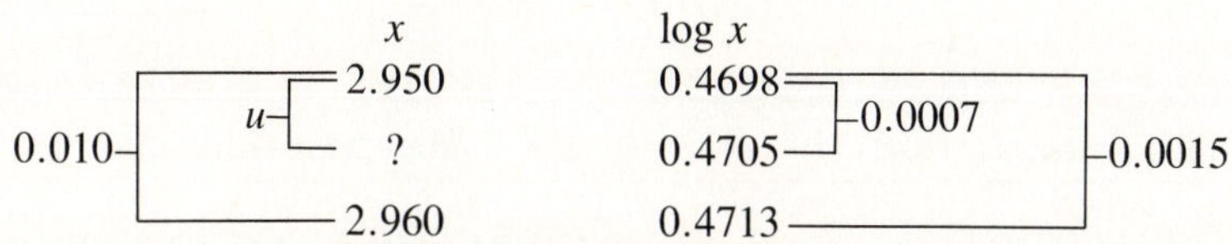

$$\frac{u}{0.010} = \frac{0.0007}{0.0015} = \frac{7}{15}$$

$$u = \frac{7}{15}(0.010) \approx 0.005$$

$$\log 2.955 \approx 0.4705$$

Thus, the antilog of 2.4705 is approximately

$$2.955 \times 10^2 = 295.5.$$

Practice Exercise 4

Find the antilog of -1.6329.

Answer: 0.02329

6 LOGARITHMIC COMPUTATION

In the past, difficult and time-consuming calculations such as

$$\frac{(0.0325) \cdot (42.3)^2}{\sqrt[3]{1.07}}$$

were often done using logarithms. Products, quotients, and powers of decimals can be found by converting to logarithms and adding, subtracting, and multiplying the results. Since, for example, adding two decimals is easier than multiplying them, logarithmic calculation was a time-saving device. Today, however, the calculator has made logarithmic computation nearly obsolete. Nevertheless, in order to understand better the rules of logarithms and to learn more about logarithms in general, we briefly illustrate the techniques of logarithmic computation in the following examples.

EXAMPLE 5 Logarithmic Computation

Use logarithms to find N to three significant digits if $N = \frac{(1.06)^{20}}{1.35}$.

$$N = \frac{(1.06)^{20}}{1.35}$$

$$\log N = \log \frac{(1.06)^{20}}{1.35} \quad \text{If } x = y\text{, then } \log x = \log y$$

$$= \log (1.06)^{20} - \log 1.35 \quad \text{Quotient rule}$$

$$= 20 \log 1.06 - \log 1.35 \quad \text{Power rule}$$

$$= 20(0.0253) - 0.1303$$

$$= 0.3757 \quad \text{Remember that this is } \log N$$

N is the antilog of 0.3757.

$$N = 2.38 \times 10^0 \quad \text{To three significant digits}$$

$$= 2.38$$

Practice Exercise 5

Use logarithms to find N to three significant digits if $N = \frac{(1.15)^{42}}{(1.02)^{15}}$.

$$N = \frac{(1.15)^{42}}{(1.02)^{15}}$$

$$\log N = \log \frac{(1.15)^{42}}{(1.02)^{15}}$$

$$= \log (1.15)^{42} - \log (1.02)^{15}$$

$$= 42 \log 1.15 - 15 \log 1.02$$

Answer: 263

EXAMPLE 6 Logarithmic Computation

Use logarithms to find N to three significant digits if $N = \sqrt[5]{21.5}\,(0.0349)^3$.

$$N = (21.5)^{1/5}(0.0349)^3 \quad \text{Convert to fractional exponent}$$

$$\log N = \log [(21.5)^{1/5}(0.0349)^3] \quad \text{Take log of both sides}$$

$$= \log (21.5)^{1/5} + \log (0.0349)^3 \quad \text{Product rule}$$

$$= \frac{1}{5} \log 21.5 + 3 \log 0.0349 \quad \text{Power rule}$$

$$= \frac{1}{5}(1.3324) + 3(0.5428 - 2) \quad \text{Watch the characteristics}$$

$$= 0.2665 + 1.6284 - 6$$

$$= 1.8949 - 6$$

$$= 0.8949 - 5 \quad \text{This is } \log N\text{, not } N$$

$$N = 7.85 \times 10^{-5} \quad \text{Find the antilog: } 10^{0.8949} = 7.85$$

$$= 0.0000785$$

Practice Exercise 6

Use logarithms to find N to three significant digits if $N = \sqrt[5]{3160}\,(0.00871)^6$.

Answer: 2.45×10^{-11}

These examples show the usefulness of logarithms for power calculations when a calculator with a $\boxed{y^x}$ key is not available. When performing logarithmic calculations, write the problem in detail, as above. If this is not done, we may find ourselves multiplying or dividing when we should be adding or subtracting, and we might not recognize that in the last step, an antilog must be found.

EXAMPLE 7 LOGARITHMIC COMPUTATION

Use logarithms to find N correct to three significant digits if

$$N = \frac{\sqrt[3]{0.0943}}{251}.$$

$$N = \frac{(0.0943)^{1/3}}{251}$$

$$\begin{aligned}\log N &= \log \frac{(0.0943)^{1/3}}{251}\\ &= \log (0.0943)^{1/3} - \log 251\\ &= \frac{1}{3}\log (0.0943) - \log 251\\ &= \frac{1}{3}(0.9745 - 2) - 2.3997\end{aligned}$$

If we divide $0.9745 - 2$ by 3, we no longer have an integer characteristic, since 2 is not evenly divisible by 3. Thus, we need to convert $0.9745 - 2$ to $1.9745 - 3$ and then divide by 3 to obtain

$$\log N = (0.6582 - 1) - (2.3997).$$

Next, we change the form of the characteristic in $0.6582 - 1$ to $9.6582 - 10$ in order to subtract 2.3997 and keep the mantissa positive.

$$\begin{array}{r} 9.6582 - 10 \\ -2.3997 \quad\;\; \\ \hline 7.2585 - 10 \end{array}$$

Then $$\begin{aligned}\log N &= 7.2585 - 10\\ &= 0.2585 - 3\\ N &= 1.81 \times 10^{-3} \qquad \text{Find the antilog: } 10^{0.2585} = 1.81\\ &= 0.00181.\end{aligned}$$

PRACTICE EXERCISE 7

Use logarithms to find N correct to three significant digits if

$$N = \frac{(216)^{1/2}}{\sqrt[5]{0.00375}}.$$

Answer: 44.9

The two difficulties we met in the preceding example are not as bothersome when a calculator is used to find logarithms. In such cases, the nonnegative mantissas necessary for using the table of common logarithms need not be kept.

EXERCISES A

1. The common logarithm of a number can bė expressed in two parts: an integer, called the

(a) ____________________ and a decimal part between 0 and 1, called the

(b) ____________________.

2. Given that $\log 73.2 = 1.8645$, the characteristic is **(a)** ____________________, and the mantissa is

(b) ____________________.

3. Given that $\log 0.0732 = 0.8645 - 2$, the characteristic is **(a)** ______________, and the mantissa is **(b)** ______________.

4. Given that $\log 7.32 = 0.8645$, the characteristic is **(a)** ______________, and the mantissa is **(b)** ______________.

5. When using the table of common logarithms to find the antilog of 3.9974, we find the number whose mantissa is **(a)** ______________ and multiply it by 10 to the **(b)** ______________ power.

Using the table of common logarithms, find the following.

6. log 0.00759

7. log 7590

8. log 0.376

9. log 5

10. log 0.5

11. log 500

Convert each logarithm to standard form, which displays the positive mantissa and the characteristic.

12. -5.1258

13. -2.0013

14. -0.3176

Using the table of common logarithms, find the antilogarithm of each number.

15. $0.6571 - 1$

16. 4.9754

17. $0.9090 - 3$

18. -1.4318

19. -0.9508

20. $9.8401 - 10$

Use the table and linear interpolation to approximate the logarithm of each number, correct to four decimal places.

21. 3.278

22. 6399

23. 437.6

24. 12,420

25. 0.003972

26. 0.05979

Use the table and linear interpolation to approximate the antilogarithm of each number, correct to four significant digits.

27. 1.6974

28. 2.2390

29. $0.5409 - 2$

30. $0.9738 - 1$

31. -3.1155

32. -2.3031

Use logarithmic computation to find N, correct to three significant digits.

33. $N = \dfrac{(1.07)^{30}}{5.13}$

$$\log N = \log \frac{(1.07)^{30}}{5.13}$$
$$= \log (1.07)^{30} - \log 5.13$$
$$= 30 \log 1.07 - \log 5.13$$
$$= 30(\qquad) - (\qquad)$$

34. $N = \sqrt[7]{32.8}\,(0.491)^3$

$$\log N = \log [\sqrt[7]{32.8}\,(0.491)^3]$$
$$= \log (32.8)^{1/7} + \log (0.491)^3$$
$$= \frac{1}{7} \log (32.8) + 3 \log (0.491)$$
$$=$$

35. $N = (1.04)^{25}(3.78)$

36. $N = [(23.5)(1.08)]^{1/5}$

37. $N = \dfrac{(30{,}700)^{1/3}}{(22.5)^3}$

38. $N = [\sqrt{1.03}\,(1.08)^3]^{2/3}$

39. $N = \dfrac{(0.00351)^{1/4}}{(4.12)^2}$

40. $N = \dfrac{(3.78)^2(0.135)^{1/2}}{7280}$

41. $N = \dfrac{(6810)(1.21)^{40}}{225}$

ANSWERS: **1. (a) characteristic (b) mantissa 2. (a) 1 (b) 0.8645 3. (a) −2 (b) 0.8645 4. (a) 0 (b) 0.8645 5. (a) 0.9974 (b) 3rd 6. 0.8802 − 3 or −2.1198 7. 3.8802 8. 0.5752 − 1 or −0.4248 9. 0.6990 10. 0.6990 − 1 or −0.3010 11. 2.6990 12. 0.8742 − 6 13. 0.9987 − 3 14. 0.6824 − 1 15. 0.454 16. 94,500 17. 0.00811 18. 0.0370 19. 0.112 20. 0.692 21. 0.5156 22. 3.8061 23. 2.6411 24. 4.0941 25. 0.5990 − 3 or −2.4010 26. 0.7766 − 2 or −1.2234 27. 49.82 28. 173.4 29. 0.03475 30. 0.9415 31. 0.0007665 32. 0.004976 33. 1.48 34. 0.195 35. 10.1 36. 1.91 37. 0.00275 38. 1.18 39. 0.0143 40. 0.000721 41. 62,000**

EXERCISES B

1. When a logarithm is written so that the characteristic is separate from the mantissa, the logarithm is in ________________ form.

2. Given that log 25.7 = 1.4099, the characteristic is **(a)** ________________ and the mantissa is **(b)** ________________.

3. Given that log 0.0257 = 0.4099 − 2, the characteristic is **(a)** ________________ and the mantissa is **(b)** ________________.

4. Given that log 2.57 = 0.4099, the characteristic is **(a)** ________________ and the mantissa is **(b)** ________________.

5. When using the table of common logarithms to find the antilog of 4.0691, we find the number whose mantissa is **(a)** ________________ and multiply it by 10 to the **(b)** ________________ power.

Using the table of common logarithms, find the following.

6. log 0.00325

7. log 3250

8. log 0.211

9. log 9

10. log 0.9

11. log 900

Convert each logarithm to standard form, which displays the positive mantissa and the characteristic.

12. −6.2215

13. −3.0045

14. −0.2987

Using the table of common logarithms, find the antilogarithm of each number.

15. 0.0607 − 1

16. 5.2201

17. −2.3382

18. −1.3655

19. −0.3002

20. 9.2405 − 10

Use the table and linear interpolation to approximate the logarithm of each number, correct to four decimal places.

21. 7429

22. 6.157

23. 29,340

24. 249.7

25. 0.03995

26. 0.0004256

Use the table and linear interpolation to approximate the antilogarithm of each number, correct to four significant digits.

27. 2.4675

28. 1.2390

29. 0.9286 − 1

30. 3.6754

31. −3.1595

32. −2.8118

Use logarithmic computation to find N, correct to three significant digits.

33. $N = \dfrac{(1.12)^{25}}{62.5}$

34. $N = \sqrt[6]{526}\,(0.028)^4$

35. $N = (1.07)^{42}(5.69)$

36. $N = [(52.6)(1.52)]^{1/5}$

37. $N = \dfrac{(426{,}000)^{1/4}}{(0.0911)^2}$

38. $N = [\sqrt[3]{5.6}\,(7.22)^3]^{1/6}$

39. $N = \dfrac{(291{,}000)^{2/5}}{(0.0861)^3}$

40. $N = \dfrac{(7.99)^3(37.2)^{3/2}}{0.222}$

41. $N = \dfrac{(4090)(1.13)^5}{6630}$

EXERCISES C

Use logarithms to find N, correct to three significant digits.

1. $N = \dfrac{\sqrt[9]{(1.25)^5(0.00367)^2}}{\sqrt[4]{(0.0169)^3}}$

[Answer: 6.95]

2. $N = \sqrt[8]{[(1.08)^3(0.0569)^2]^7}$

TABLE OF COMMON LOGARITHMS

n	0	1	2	3	4	5	6	7	8	9
1.0	.0000	.0043	.0086	.0128	.0170	.0212	.0253	.0294	.0334	.0374
1.1	.0414	.0453	.0492	.0531	.0569	.0607	.0645	.0682	.0719	.0755
1.2	.0792	.0828	.0864	.0899	.0934	.0969	.1004	.1038	.1072	.1106
1.3	.1139	.1173	.1206	.1239	.1271	.1303	.1335	.1367	.1399	.1430
1.4	.1461	.1492	.1523	.1553	.1584	.1614	.1644	.1673	.1703	.1732
1.5	.1761	.1790	.1818	.1847	.1875	.1903	.1931	.1959	.1987	.2014
1.6	.2041	.2068	.2095	.2122	.2148	.2175	.2201	.2227	.2253	.2279
1.7	.2304	.2330	.2355	.2380	.2405	.2430	.2455	.2480	.2504	.2529
1.8	.2553	.2577	.2601	.2625	.2648	.2672	.2695	.2718	.2742	.2765
1.9	.2788	.2810	.2833	.2856	.2878	.2900	.2923	.2945	.2967	.2989
2.0	.3010	.3032	.3054	.3075	.3096	.3118	.3139	.3160	.3181	.3201
2.1	.3222	.3243	.3263	.3284	.3304	.3324	.3345	.3365	.3385	.3404
2.2	.3424	.3444	.3464	.3483	.3502	.3522	.3541	.3560	.3579	.3598
2.3	.3617	.3636	.3655	.3674	.3692	.3711	.3729	.3747	.3766	.3784
2.4	.3802	.3820	.3838	.3856	.3874	.3892	.3909	.3927	.3945	.3962
2.5	.3979	.3997	.4014	.4031	.4048	.4065	.4082	.4099	.4116	.4133
2.6	.4150	.4166	.4183	.4200	.4216	.4232	.4249	.4265	.4281	.4298
2.7	.4314	.4330	.4346	.4362	.4378	.4393	.4409	.4425	.4440	.4456
2.8	.4472	.4487	.4502	.4518	.4533	.4548	.4564	.4579	.4594	.4609
2.9	.4624	.4639	.4654	.4669	.4683	.4698	.4713	.4728	.4742	.4757
3.0	.4771	.4786	.4800	.4814	.4829	.4843	.4857	.4871	.4886	.4900
3.1	.4914	.4928	.4942	.4955	.4969	.4983	.4997	.5011	.5024	.5038
3.2	.5051	.5065	.5079	.5092	.5105	.5119	.5132	.5145	.5159	.5172
3.3	.5185	.5198	.5211	.5224	.5237	.5250	.5263	.5276	.5289	.5302
3.4	.5315	.5328	.5340	.5353	.5366	.5378	.5391	.5403	.5416	.5428
3.5	.5441	.5453	.5465	.5478	.5490	.5502	.5514	.5527	.5539	.5551
3.6	.5563	.5575	.5587	.5599	.5611	.5623	.5635	.5647	.5658	.5670
3.7	.5682	.5694	.5705	.5717	.5729	.5740	.5752	.5763	.5775	.5786
3.8	.5798	.5809	.5821	.5832	.5843	.5855	.5866	.5877	.5888	.5899
3.9	.5911	.5922	.5933	.5944	.5955	.5966	.5977	.5988	.5999	.6010
4.0	.6021	.6031	.6042	.6053	.6064	.6075	.6085	.6096	.6107	.6117
4.1	.6128	.6138	.6149	.6160	.6170	.6180	.6191	.6201	.6212	.6222
4.2	.6232	.6243	.6253	.6263	.6274	.6284	.6294	.6304	.6314	.6325
4.3	.6335	.6345	.6355	.6365	.6375	.6385	.6395	.6405	.6415	.6425
4.4	.6435	.6444	.6454	.6464	.6474	.6484	.6493	.6503	.6513	.6522
4.5	.6532	.6542	.6551	.6561	.6571	.6580	.6590	.6599	.6609	.6618
4.6	.6628	.6637	.6646	.6656	.6665	.6675	.6684	.6693	.6702	.6712
4.7	.6721	.6730	.6739	.6749	.6758	.6767	.6776	.6785	.6794	.6803
4.8	.6812	.6821	.6830	.6839	.6848	.6857	.6866	.6875	.6884	.6893
4.9	.6902	.6911	.6920	.6928	.6937	.6946	.6955	.6964	.6972	.6981
5.0	.6990	.6998	.7007	.7016	.7024	.7033	.7042	.7050	.7059	.7067
5.1	.7076	.7084	.7093	.7101	.7110	.7118	.7126	.7135	.7143	.7152
5.2	.7160	.7168	.7177	.7185	.7193	.7202	.7210	.7218	.7226	.7235
5.3	.7243	.7251	.7259	.7267	.7275	.7284	.7292	.7300	.7308	.7316
5.4	.7324	.7332	.7340	.7348	.7356	.7364	.7372	.7380	.7388	.7396
n	0	1	2	3	4	5	6	7	8	9

TABLE OF COMMON LOGARITHMS

CONTINUED

n	0	1	2	3	4	5	6	7	8	9
5.5	.7404	.7412	.7419	.7427	.7435	.7443	.7451	.7459	.7466	.7474
5.6	.7482	.7490	.7497	.7505	.7513	.7520	.7528	.7536	.7543	.7551
5.7	.7559	.7566	.7574	.7582	.7589	.7597	.7604	.7612	.7619	.7627
5.8	.7634	.7642	.7649	.7657	.7664	.7672	.7679	.7686	.7694	.7701
5.9	.7709	.7716	.7723	.7731	.7738	.7745	.7752	.7760	.7767	.7774
6.0	.7782	.7789	.7796	.7803	.7810	.7818	.7825	.7832	.7839	.7846
6.1	.7853	.7860	.7868	.7875	.7882	.7889	.7896	.7903	.7910	.7917
6.2	.7924	.7931	.7938	.7945	.7952	.7959	.7966	.7973	.7980	.7987
6.3	.7993	.8000	.8007	.8014	.8021	.8028	.8035	.8041	.8048	.8055
6.4	.8062	.8069	.8075	.8082	.8089	.8096	.8102	.8109	.8116	.8122
6.5	.8129	.8136	.8142	.8149	.8156	.8162	.8169	.8176	.8182	.8189
6.6	.8195	.8202	.8209	.8215	.8222	.8228	.8235	.8241	.8248	.8254
6.7	.8261	.8267	.8274	.8280	.8287	.8293	.8299	.8306	.8312	.8319
6.8	.8325	.8331	.8338	.8344	.8351	.8357	.8363	.8370	.8376	.8382
6.9	.8388	.8395	.8401	.8407	.8414	.8420	.8426	.8432	.8439	.8445
7.0	.8451	.8457	.8463	.8470	.8476	.8482	.8488	.8494	.8500	.8506
7.1	.8513	.8519	.8525	.8531	.8537	.8543	.8549	.8555	.8561	.8567
7.2	.8573	.8579	.8585	.8591	.8597	.8603	.8609	.8615	.8621	.8627
7.3	.8633	.8639	.8645	.8651	.8657	.8663	.8669	.8675	.8681	.8686
7.4	.8692	.8698	.8704	.8710	.8716	.8722	.8727	.8733	.8739	.8745
7.5	.8751	.8756	.8762	.8768	.8774	.8779	.8785	.8791	.8797	.8802
7.6	.8808	.8814	.8820	.8825	.8831	.8837	.8842	.8848	.8854	.8859
7.7	.8865	.8871	.8876	.8882	.8887	.8893	.8899	.8904	.8910	.8915
7.8	.8921	.8927	.8932	.8938	.8943	.8949	.8954	.8960	.8965	.8971
7.9	.8976	.8982	.8987	.8993	.8998	.9004	.9009	.9015	.9020	.9025
8.0	.9031	.9036	.9042	.9047	.9053	.9058	.9063	.9069	.9074	.9079
8.1	.9085	.9090	.9096	.9101	.9106	.9112	.9117	.9122	.9128	.9133
8.2	.9138	.9143	.9149	.9154	.9159	.9165	.9170	.9175	.9180	.9186
8.3	.9191	.9196	.9201	.9206	.9212	.9217	.9222	.9227	.9232	.9238
8.4	.9243	.9248	.9253	.9258	.9263	.9269	.9274	.9279	.9284	.9289
8.5	.9294	.9299	.9304	.9309	.9315	.9320	.9325	.9330	.9335	.9340
8.6	.9345	.9350	.9355	.9360	.9365	.9370	.9375	.9380	.9385	.9390
8.7	.9395	.9400	.9405	.9410	.9415	.9420	.9425	.9430	.9435	.9440
8.8	.9445	.9450	.9455	.9460	.9465	.9469	.9474	.9479	.9484	.9489
8.9	.9494	.9499	.9504	.9509	.9513	.9518	.9523	.9528	.9533	.9538
9.0	.9542	.9547	.9552	.9557	.9562	.9566	.9571	.9576	.9581	.9586
9.1	.9590	.9595	.9600	.9605	.9609	.9614	.9619	.9624	.9628	.9633
9.2	.9638	.9643	.9647	.9652	.9657	.9661	.9666	.9671	.9675	.9680
9.3	.9685	.9689	.9694	.9699	.9703	.9708	.9713	.9717	.9722	.9727
9.4	.9731	.9736	.9741	.9745	.9750	.9754	.9759	.9763	.9768	.9773
9.5	.9777	.9782	.9786	.9791	.9795	.9800	.9805	.9809	.9814	.9818
9.6	.9823	.9827	.9832	.9836	.9841	.9845	.9850	.9854	.9859	.9863
9.7	.9868	.9872	.9877	.9881	.9886	.9890	.9894	.9899	.9903	.9908
9.8	.9912	.9917	.9921	.9926	.9930	.9934	.9939	.9943	.9948	.9952
9.9	.9956	.9961	.9965	.9969	.9974	.9978	.9983	.9987	.9991	.9996
n	0	1	2	3	4	5	6	7	8	9

APPENDIX B

Introduction to Probability

STUDENT GUIDEPOSTS

1. Language of Probability
2. Definition of Probability
3. Probability of A or B
4. Complementary Events
5. Experimental Probability

1 LANGUAGE OF PROBABILITY

Probability originated with problems related to games of chance, but today it has grown to include applications in such areas as genetics, insurance, physics, social science, and medicine.

Determining the likelihood of a particular event is the substance of probability. To make this precise, we need a variety of terms. We use the term **experiment** to mean the performing of some activity such as flipping a coin, drawing a card, or rolling dice. If an experiment is performed, the collection of all possible **outcomes** or results is called the **sample space** of the experiment. It is assumed that any particular outcome is just as likely to occur as any other, that is, the outcomes are **equally likely.** Any subset of the sample space is called an **event.** Elements in a particular event are called **favorable outcomes** or **successes,** while outcomes not in the event are called **unfavorable outcomes** or **failures.**

The experiment of flipping two fair coins has sample space

$$S = \{hh, ht, th, tt\},$$

where hh corresponds to the outcome of obtaining a head on each coin, and ht corresponds to the outcome of obtaining a head on the first coin and a tail on the second. Note that ht and th are different. This is clear if one coin is thought to be a dime, and the other a quarter, for then certainly there are two ways to obtain one head and one tail. The event

$$E = \{tt\}$$

is the event of obtaining two tails, while

$$F = \{ht, th\}$$

is the event of obtaining one head and one tail.

The experiment of rolling two unloaded dice has a sample space consisting of thirty-six pairs of numbers, each number varying from 1 through 6. This is because we have insisted that each outcome be as likely as any other. We cannot use {2, 3, 4, 5, 6, 7, 8, 9, 10, 11, 12} to represent the sample space since the total of 7 can be obtained in six ways

$$(1, 6), \quad (2, 5), \quad (3, 4), \quad (4, 3), \quad (5, 2), \quad (6, 1)$$

whereas the total of 2 is less likely since it can only be obtained in one way, (1, 1).

Observe that (2, 5) and (5, 2) are different outcomes. This is perhaps easier to understand if we think of the dice as being of different colors, one red and the other white. A 2 with the red die and a 5 with the white is clearly a different outcome than a 5 with the red and a 2 with the white.

We show the sample space for this experiment as follows.

$$\begin{array}{cccccc}
(1, 1) & (1, 2) & \mathbf{(1, 3)} & (1, 4) & (1, 5) & (1, 6) \\
(2, 1) & \mathbf{(2, 2)} & (2, 3) & (2, 4) & (2, 5) & \mathbf{(2, 6)} \\
\mathbf{(3, 1)} & (3, 2) & (3, 3) & (3, 4) & \mathbf{(3, 5)} & (3, 6) \\
(4, 1) & (4, 2) & (4, 3) & \mathbf{(4, 4)} & (4, 5) & (4, 6) \\
(5, 1) & (5, 2) & \mathbf{(5, 3)} & (5, 4) & (5, 5) & (5, 6) \\
(6, 1) & \mathbf{(6, 2)} & (6, 3) & (6, 4) & (6, 5) & (6, 6)
\end{array}$$

F (diagonal through (3, 1), (2, 2), (1, 3))

E (diagonal through (6, 2), (5, 3), (4, 4), (3, 5), (2, 6))

The event E of obtaining a total of 8 is

$$E = \{(6, 2), (5, 3), (4, 4), (3, 5), (2, 6)\}$$

while the event F of obtaining a total of 4 is

$$F = \{(3, 1), (2, 2), (1, 3)\}.$$

The pairs that give a particular total are all found along a diagonal from lower left to upper right.

2 DEFINITION OF PROBABILITY

We are now able to give a definition of probability along with several important properties. We use $n(S)$, read "n of S," to represent the number of elements in set S, and $n(E)$ for the number of elements in E.

Probability

Let S be the sample space of an experiment, and E an event (E is a subset of S). The **probability that E will occur,** or simply the **probability of E,** denoted by $P(E)$, is given by

$$P(E) = \frac{n(E)}{n(S)} = \frac{\text{the number of favorable outcomes}}{\text{the total number of outcomes}}.$$

Since E is a subset of S, $0 \le n(E) \le n(S)$, so that dividing through by $n(S)$, we obtain

$$\frac{0}{n(S)} \le \frac{n(E)}{n(S)} \le \frac{n(S)}{n(S)}$$

or

$$\mathbf{0 \le P(E) \le 1.}$$

Hence the probability of an event is always a number between 0 and 1, inclusive. If $P(E) = 0$, we say that E is **impossible** (E cannot occur). If $P(E) = 1$, we say that E is **certain** (E must occur). If E and F are two events such that $P(E) < P(F)$, we say that E is **less likely to occur** than F, or that F is **more likely to occur** than E. If $P(E) = P(F)$, the events are called **equally likely.**

EXAMPLE 1 Probabilities Involving Coins

The sample space for the experiment of flipping two fair coins is $S = \{hh, ht, th, tt\}$.

(a) If E is the event of obtaining two heads, what is $P(E)$?

Since $E = \{hh\}$,

$$P(E) = \frac{n(E)}{n(S)} = \frac{1}{4}.$$

Remember that *hh* is one outcome

(b) If F is the event of obtaining at least one head, find $P(F)$.

$$F = \{hh, ht, th\}.$$

Thus,
$$P(F) = \frac{n(F)}{n(S)} = \frac{3}{4}.$$

Practice Exercise 1

Use the sample space in Example 1.

(a) What is the probability of two tails?

(b) What is the probability of no tails?

Answers: (a) $\frac{1}{4}$ (b) $\frac{1}{4}$

EXAMPLE 2 Probabilities Involving Dice

Consider the sample space for the experiment of rolling two dice.

(a) Determine the probability of rolling a total of 8.

The sample space was given earlier in this section.

$$E = \{(6, 2), (5, 3), (4, 4), (3, 5), (2, 6)\}$$

Thus $P(E) = \dfrac{n(E)}{n(S)} = \dfrac{5}{36}$.

(b) Determine the probability of rolling a total of 3.

From the array of outcomes given in this section, the number of ways of obtaining a three is 2. Thus,

$$P(3) = \frac{2}{36} = \frac{1}{18}.$$

Practice Exercise 2

Use the sample space of rolling two dice.

(a) What is the probability of rolling a total of 10?

(b) What is the probability of rolling a total of 5?

Answers: (a) $\frac{1}{12}$ (b) $\frac{1}{9}$

HINT

The probabilities in Example 2 do not mean that if the dice are thrown 36 times exactly 5 sums of 8 and 2 sums of 3 will be obtained. It does mean that if the dice are thrown hundreds of times it is likely that about 5/36 of the tosses will result in a sum of 8, and about 1/18 of the tosses will give a sum of 3.

3 PROBABILITY OF *A* OR *B*

When two or more events are joined by the word *or,* we say that the result is the **disjunction** of the events. If A is the event of rolling a 5 with one die, and B is the event of rolling a 3 with one die, then A *or* B is the event of rolling a 3 or a 5 with one die. In this case, it is clear that

$$\begin{aligned} P(A \text{ or } B) &= P(A) + P(B) \\ \frac{2}{6} &= \frac{1}{6} + \frac{1}{6}. \end{aligned}$$

The **conjunction** of two or more events is formed by joining them with the word *and*. In the above, it is impossible for both A and B to occur on one trial. When this happens, we say that A and B are **mutually exclusive,** and observe that the probability that both A and B occur is 0, that is,

$$P(A \text{ and } B) = 0.$$

Thus, the event (A *and* B) is impossible.

Suppose that A is the event of drawing an ace and B is the event of drawing a spade in the experiment of drawing one card from a well-shuffled deck of 52 cards. Then (A *or* B) is the event of drawing an ace or a spade on one draw of a card. Since there are 13 spades and 4 aces, we are tempted to add to obtain 17 as the number of outcomes favorable to (A *or* B). However, the ace of spades has then been counted twice (both as a spade and as an ace) so that 17 is 1 too many for the number of favorable outcomes. Thus, $P(A \text{ or } B) = 16/52$. This same result can be obtained by the following formula.

$$\begin{aligned} P(A \text{ or } B) &= P(A) + P(B) - P(A \text{ and } B) \\ &= \frac{4}{52} + \frac{13}{52} - \frac{1}{52} \qquad P(A \text{ and } B) \text{ is } \tfrac{1}{52} \\ &= \frac{16}{52} \end{aligned}$$

When A and B are mutually exclusive $P(A \text{ and } B) = 0$ so that the formula still applies. The above examples are specific cases of the next theorem.

Probability of A or B

Let A and B be events. Then

$$P(A \text{ or } B) = P(A) + P(B) - P(A \text{ and } B).$$

If A and B are mutually exclusive, $P(A \text{ and } B) = 0$ and the formula reduces to

$$P(A \text{ or } B) = P(A) + P(B).$$

EXAMPLE 3 Probabilities of Two Events

If a card is drawn from a well-shuffled deck of 52 cards, find the probability of drawing the following cards.

(a) An ace or a king

$$\begin{aligned} P(\text{ace } \textit{or} \text{ king}) &= P(\text{ace}) + P(\text{king}) - P(\text{ace } \textit{and} \text{ king}) \\ &= \frac{4}{52} + \frac{4}{52} - \frac{0}{52} = \frac{8}{52} = \frac{2}{13} \end{aligned}$$

(b) A face card or a heart

$$\begin{aligned} P(\text{face } \textit{or} \text{ heart}) &= P(\text{face}) + P(\text{heart}) - P(\text{face } \textit{and} \text{ heart}) \\ &= \frac{12}{52} + \frac{13}{52} - \frac{3}{52} \qquad \text{3 heart face cards} \\ &= \frac{22}{52} = \frac{11}{26} \end{aligned}$$

(c) An ace and a jack

$P(\text{ace } \textit{and} \text{ jack}) = 0$ since the two events are mutually exclusive.

Practice Exercise 3

Use the sample space in Example 3 to find each probability.

(a) a jack or a queen

(b) a ten or a black card

(c) a five and a seven

Answers: (a) $\frac{2}{13}$ (b) $\frac{7}{13}$ (c) 0

4 COMPLEMENTARY EVENTS

If E is an event and F is the event described as not E, then E and F are **complementary events.** Since either E or F must be true, $P(E \text{ or } F) = 1$. Also, $P(E \text{ and } F) = 0$. Thus,

$$P(E \text{ or } F) = P(E) + P(F) = 1$$
$$\text{or} \quad P(F) = 1 - P(E).$$

Complementary Events

If E is an event and F is the event that E will not occur, then E and F are complementary and

$$P(F) = 1 - P(E).$$

EXAMPLE 4 Application from Medicine

During an outbreak of both measles and mumps, health officials warned parents to have children vaccinated. If they did not, the probability of a child contracting measles was 0.012, the probability of contracting the mumps was 0.031, and the probability of contracting both was 0.005.

(a) Find the probability of a child who was not vaccinated contracting the measles or the mumps.

$$\begin{aligned} P(\text{Me or Mu}) &= P(\text{Me}) + P(\text{Mu}) - P(\text{Me and Mu}) \\ &= 0.012 + 0.031 - 0.005 \\ &= 0.038 \end{aligned}$$

(b) Find the probability of a child not contracting either of the diseases.

Let F be the event of contracting neither of the diseases.

$$\begin{aligned} P(F) &= 1 - P(\text{Me or Mu}) \\ &= 1 - 0.038 = 0.962 \end{aligned}$$

Practice Exercise 4

Use the data in Example 4.

(a) What is the probability of a child not contracting both measles and mumps?

(b) Find the probability of a child not contracting the measles and not contracting the mumps.

Answers: (a) 0.995 (b) 0.962

5 EXPERIMENTAL PROBABILITY

How did the health officials in Example 4 determine the probabilities that are given? They most likely used experimental data. During similar outbreaks of measles and mumps in previous years, they could have selected representative groups of children, counted them, counted the number that contracted measles, and counted the number that contracted mumps. This data could then have been used to determine empirically the given probabilities. For example, if 485 total children were in a group and 6 of them contracted measles, the probability of measles was determined by

$$P(\text{Me}) = \frac{6}{485} \approx 0.0123711 \approx 0.012.$$

This type of probability is used extensively in all the sciences, in engineering, and in business. Two particular areas that are visible to most people are weather forecasting and life insurance. Also, many experiments are carried out to determine various characteristics of our society. For example, an architectural firm might study the heights of people who are likely to use a building in order to improve the design.

EXERCISES A

In Exercises 1–3 answer yes *if the number could be a probability and* no *if it could not.*

1. 0

2. $-\frac{1}{2}$

3. 0.00001

Let $S = \{1, 2, 3, 4, 5, 6\}$ *be the sample space of rolling one unloaded die. Determine the probability of each event in Exercises 4–6.*

4. Rolling a 5

5. Rolling a number greater than 0

6. Rolling a number $n \geq 4$

Let S be the sample space for the experiment of rolling two unloaded dice. Find the probability of each event in Exercises 7–9.

7. Rolling a total of 7

8. Rolling a total of 7 or 11

9. Rolling a total greater than 3

In the game of craps, on the first roll a player wins with a total of 7 or 11 and loses with a total of 2, 3, or 12. Any other total becomes his point, and he must continue to roll until he rolls his point and wins, or rolls a 7 and loses. Find the probability of each event in Exercises 10–11.

10. Winning on the first roll

11. Rolling a total other than a winning or losing total on the first roll

A roulette wheel contains 38 slots numbered 00, 0, 1, 2, 3, 4, . . . , 35, 36. Eighteen of the slots numbered 1 through 36 are colored black and the rest are colored red. The 00 and 0 slots are colored green and are called house numbers. The wheel is spun and a ball is rolled around the rim in the opposite direction. Eventually the ball falls into a slot. Find the probability of each event in Exercises 12–15.

12. Ball falls into the number 3 slot

13. Ball falls into a red slot

14. Ball falls into a blue slot

15. Ball falls into an odd-numbered slot

A card is drawn from a well-shuffled deck of 52 cards. In Exercises 16–22 find the probability of each event.

16. Drawing an ace

17. Drawing a black card

18. Drawing the queen of hearts

19. Drawing a 7 or a queen

20. Drawing a 7 or a diamond

21. Drawing a face card or a heart

22. Drawing a king and a heart

23. If the probability of winning a game is 4/7, what is the probability of losing if there are no ties?

24. A child picks up a telephone and dials seven digits at random. What is the probability that he dials his own number?

25. It has been determined that the number of possible genetic combinations that a child of one couple can have is 2^{48}. Assuming that Mr. and Mrs. Levin have one child, what is the probability that a second child will have the same genetic makeup?

ANSWERS: 1. yes 2. no 3. yes 4. $\frac{1}{6}$ 5. 1 6. $\frac{1}{2}$ 7. $\frac{1}{6}$ 8. $\frac{2}{9}$ 9. $\frac{11}{12}$ 10. $\frac{2}{9}$ 11. $\frac{2}{3}$ 12. $\frac{1}{38}$ 13. $\frac{9}{19}$ 14. 0 15. $\frac{9}{19}$ 16. $\frac{1}{13}$ 17. $\frac{1}{2}$ 18. $\frac{1}{52}$ 19. $\frac{2}{13}$ 20. $\frac{4}{13}$ 21. $\frac{11}{26}$ 22. $\frac{1}{52}$ 23. $\frac{3}{7}$ 24. $\frac{1}{10^7} = \frac{1}{10{,}000{,}000}$ 25. $\frac{1}{2^{48}}$

EXERCISES B

In Exercises 1–3 answer yes *if the number could be a probability and* no *if it could not.*

1. $\dfrac{2}{3}$

2. 1

3. $\dfrac{5}{4}$

Let $S = \{1, 2, 3, 4, 5, 6\}$ be the sample space of rolling one unloaded die. Determine the probability of each event in Exercises 4–6.

4. Rolling a number less than 3

5. Rolling a multiple of 3

6. Rolling a number $n < 1$

Let S be the sample space for the experiment of rolling two unloaded dice. Find the probability of each event in Exercises 7–9.

7. Rolling a total of 11

8. Rolling a total of 5 or 2

9. Rolling a total less than or equal to 11

In the game of craps, on the first roll a player wins with a total of 7 or 11 and loses with a total of 2, 3, or 12. Any other total becomes his point, and he must continue to roll until he rolls his point and wins, or rolls a 7 and loses. Find the probability of the event in Exercise 10 and answer the questions in Exercise 11.

10. Losing on the first roll

11. Of all the points 4, 5, 6, 8, 9, 10, which two are the most likely to be rolled? Least likely to be rolled?

A roulette wheel contains 38 slots numbered 00, 0, 1, 2, 3, 4, . . . , 35, 36. Eighteen of the slots numbered 1 through 36 are colored black and the rest are colored red. The 00 and 0 slots are colored green and are called house numbers. The wheel is spun and a ball is rolled around the rim in the opposite direction. Eventually the ball falls into a slot. Find the probability of each event in Exercises 12–15.

12. Ball falls into a black slot

13. Ball falls into a green slot

14. Ball falls into a black or red slot

15. Ball falls into the number 0 slot

A card is drawn from a well-shuffled deck of 52 cards. In Exercises 16–22 find the probability of each event.

16. Drawing a heart

17. Drawing a face card

18. Drawing a joker

19. Drawing a 7 or a black card

20. Drawing a face card or a black card

21. Drawing a face card or a 3

22. Drawing a face card and a heart

23. If the probability of a horse losing a race is 9/11, what is the probability that the horse will win?

24. Claude has flipped a fair coin fifty times and it has fallen heads each time. He conjectures that on the fifty-first flip the likelihood of a tail will certainly increase. Do you agree? What is the probability of a tail on the fifty-first flip?

25. A couple plans to have two children.
(a) Give the sample space representing these children according to sex. [Hint: Use gg to represent two girls.]
(b) What is the probability of having two boys?
(c) What is the probability of having two children of the same sex?
(d) What is the probability of one girl and one boy?

EXERCISES C

1. If a 5-card hand is dealt from a standard 52-card deck, what is the probability of getting a *royal flush* which is ace-king-queen-jack-ten all of the same suit?

$\left[\text{Answer: } \frac{1}{649{,}740} \approx 0.0000015\right]$

2. If a 5-card hand is dealt from a standard 52-card deck, what is the probability of getting *four of a kind* which is a hand such as two-two-two-two-six or ace-ace-ace-ace-ten?

3. If the given number of dice are thrown, what is the probability of them all coming up four?
(a) 1 **(b)** 2 **(c)** 3 **(d)** 4 **(e)** n

4. If n dice are thrown, what is the probability of them all coming up the same number?

SOLUTIONS TO SELECTED EXERCISES

All exercises are selected from the A set of exercises.

0.2

8. $4\frac{2}{5} \cdot \frac{3}{11} = \frac{22}{5} \cdot \frac{3}{11} = \frac{2 \cdot 11}{5} \cdot \frac{3}{11}$

$= \frac{2 \cdot \cancel{11} \cdot 3}{5 \cdot \cancel{11}} = \frac{6}{5}$

14. $3\frac{1}{5} \div \frac{2}{5} = \frac{16}{5} \div \frac{2}{5} = \frac{16}{5} \cdot \frac{5}{2} =$

$\frac{\cancel{2} \cdot 2 \cdot 2 \cdot 2 \cdot \cancel{5}}{\cancel{5} \cdot \cancel{2}} =$

$\frac{2 \cdot 2 \cdot 2}{1} = \frac{8}{1} = 8$

43. $\frac{5}{6}$ of 72 means $\frac{5}{6} \cdot 72$.

$\frac{5}{6} \cdot 72 = \frac{5}{6} \cdot \frac{72}{1} = \frac{5}{2 \cdot 3} \cdot \frac{2 \cdot 2 \cdot 2 \cdot 3 \cdot 3}{1} =$

$\frac{5 \cdot \cancel{2} \cdot 2 \cdot 2 \cdot \cancel{3} \cdot 3}{\cancel{2} \cdot \cancel{3}} = \frac{5 \cdot 2 \cdot 2 \cdot 3}{1} = 60$ girls

$72 - 60 = 12$ boys

46. Divide $12\frac{1}{2}$ by $\frac{3}{4}$.

$12\frac{1}{2} \div \frac{3}{4} = \frac{25}{2} \div \frac{3}{4} = \frac{25}{2} \cdot \frac{4}{3} = \frac{5 \cdot 5}{2} \cdot \frac{2 \cdot 2}{3} =$

$\frac{5 \cdot 5 \cdot \cancel{2} \cdot 2}{\cancel{2} \cdot 3} = \frac{50}{3} = 16\frac{2}{3}$ gallons

0.3

41. $7\frac{3}{4} + 9\frac{1}{8} = \frac{31}{4} + \frac{73}{8} = \frac{31 \cdot 2}{2 \cdot 2 \cdot 2} + \frac{73}{2 \cdot 2 \cdot 2}$

$= \frac{62 + 73}{2 \cdot 2 \cdot 2} = \frac{135}{8}$ **Total distance**

$\frac{135}{8} \cdot \frac{1}{2} = \frac{135}{16} = 8\frac{7}{16}$ mi **One half distance**

0.4

37. $\frac{15}{20} = \frac{3 \cdot \cancel{5}}{4 \cdot \cancel{5}} = \frac{3}{4}$ **Fraction of shots**

$\frac{3}{4} = 0.75 = 75\%$ **Percent of shots**

39. Find 55% of 5000.
$(0.55)(5000) = 2750$ votes

0.5

10. The ratio of math majors to the total number of students is

$$\frac{300}{1300} = \frac{3}{13}.$$

13. The ratio of science majors to the total number of students is

$$\frac{400}{1300} = \frac{4}{13}.$$

Change $\frac{4}{13}$ to a percent to obtain 30.8%.

49. The number of employees who earn \$9 or less an hour is the total of the numbers who earn \$3 to \$5, \$5 to \$7, and \$7 to \$9. This total is

$$5 + 10 + 18 = 33.$$

52. The number of employees who earn more than \$9 an hour is 18. Thus, the ratio of these employees to the total number of employees is $\frac{18}{51} = \frac{6}{17}$. Changing $\frac{6}{17}$ to a percent gives 35.3%.

0.6

5. To find the mean, add the seven numbers and divide the result by 7.

$$\frac{7.2 + 6.1 + 8.6 + 9.2 + 5.5 + 7.2 + 8.8}{7} = \frac{52.6}{7} \approx 7.5$$

11. To find the median, first arrange the seven numbers by size.

5.5, 6.1, 7.2, 7.2, 8.6, 8.8, 9.2

The median is the middle number, 7.2, which is the fourth number from the left.

23. The right column in the table is:

$$\begin{aligned} (16{,}200)(8) &= 129{,}600 \\ (20{,}800)(10) &= 208{,}000 \\ (28{,}600)(7) &= 200{,}200 \\ (32{,}000)(2) &= 64{,}000 \\ (46{,}000)(1) &= \underline{46{,}000} \\ & 647{,}800. \end{aligned}$$

The total is 647,800. To find the average salary divide 647,800 by

$$8 + 10 + 7 + 2 + 1 = 28.$$

$$\frac{647{,}800}{28} = 23{,}135.714$$

Thus, to the nearest cent, the average salary is \$23,135.71.

Since there are 28 salaries, when ranked by size, the median is the average of the middle two, both of which are \$20,800. Thus, the median salary is \$20,800.

The mode of the salaries is the one that occurs most frequently. Since \$20,800 occurs 10 times, more frequently than any other salary, the mode is \$20,800.

0.7

14. Use two unit fractions, $\frac{1 \text{ min}}{60 \text{ sec}}$ and $\frac{1 \text{ hr}}{60 \text{ min}}$

$$5400 \text{ sec} = \frac{5400 \cancel{\text{sec}}}{1} \times \frac{1 \cancel{\text{min}}}{60 \cancel{\text{sec}}} \times \frac{1 \text{ hr}}{60 \cancel{\text{min}}}$$

$$= 5400 \times \frac{1}{60} \times \frac{1}{60} \text{ hr} = 1.5 \text{ hr}$$

23. $88\frac{\text{ft}}{\text{sec}} = \frac{88 \cancel{\text{ft}}}{1 \cancel{\text{sec}}} \times \frac{60 \cancel{\text{sec}}}{1 \cancel{\text{min}}} \times \frac{60 \cancel{\text{min}}}{1 \text{ hr}} \times \frac{1 \text{ mi}}{5280 \cancel{\text{ft}}}$

$$= \frac{88 \times 60 \times 60}{1 \times 5280} \frac{\text{mi}}{\text{hr}} = 60\frac{\text{mi}}{\text{hr}}$$

31. Since the price is given in cents per foot, change 3 mi to feet.

$$3 \text{ mi} = \frac{3 \cancel{\text{mi}}}{1} \times \frac{5280 \text{ ft}}{1 \cancel{\text{mi}}}$$

$$= 15{,}840 \text{ ft}$$

$$\frac{15{,}840 \cancel{\text{ft}}}{1} \times \frac{20 \text{ cents}}{1 \cancel{\text{ft}}} = 316800¢ = \$3168.00$$

41.

$$\begin{array}{rrr} 6 \text{ yd} & 2 \text{ ft} & 11 \text{ in} \\ + \; 7 \text{ yd} & 1 \text{ ft} & 9 \text{ in} \\ \hline 13 \text{ yd} & 3 \text{ ft} & 20 \text{ in} \end{array} = 13 \text{ yd} + 3 \text{ ft} + 1 \text{ ft} + 8 \text{ in}$$

$$= 13 \text{ yd} + 1 \text{ yd} + 1 \text{ ft} + 8 \text{ in}$$

$$= 14 \text{ yd } 1 \text{ ft } 8 \text{ in}$$

45. 5 da 3 hr 8 min = 4 da + 1 da + 2 hr + 1 hr + 8 min

= 4 da + 24 hr + 2 hr + 60 min + 8 min

= 4 da 26 hr 68 min

$$\begin{array}{rrr} 4 \text{ da} & 26 \text{ hr} & 68 \text{ min} \\ -\; 4 \text{ da} & 6 \text{ hr} & 40 \text{ min} \\ \hline 0 \text{ da} & 20 \text{ hr} & 28 \text{ min} \end{array} = 20 \text{ hr } 28 \text{ min}$$

48. 5 × (4 gal 3 qt) = 20 gal 15 qt

= 20 gal + 12 qt + 3 qt

= 20 gal + 3 gal + 3 qt

= 23 gal 3 qt

54. (7 hr) ÷ 3 = 2 hr with remainder 1 hr

1 hr = 60 min

60 min + 33 min = 93 min

93 min ÷ 3 = 31 min

Thus, (7 hr 33 min) ÷ 3 = 2 hr 31 min.

59. To make 7 loaves it would take 7 times the amount necessary for one loaf.

(1 lb 12 oz) × 7 = 7 lb + 84 oz

= 7 lb + 80 oz + 4 oz

= 7 lb + 5 lb + 4 oz

= 12 lb 4 oz

0.8

17. 1000 g (kg) 100 g (hg) 10 g (dag)

1 2

0.00721 kg = 000 721 dag

= 0.721 dag

23. 100 g (hg) 10 g (dag) 1 g (g) 0.1 g (dg)

3 2 1

4728 dg = 4.728 hg

= 4.728 hg

32. $20\frac{\text{cg}}{\text{da}} = \frac{20 \cancel{\text{cg}}}{1 \cancel{\text{da}}} \times \frac{7 \cancel{\text{da}}}{1 \text{ wk}} \times \frac{1 \text{ g}}{100 \cancel{\text{cg}}}$

$$= 1.4\frac{\text{g}}{\text{wk}}$$

41. 600 g = 0.6 kg

$$\frac{0.6 \cancel{\text{kg}}}{1} \times \frac{6 \text{ dollars}}{1 \cancel{\text{kg}}} = \$3.60$$

57. $40 \text{ L} = \frac{40 \cancel{\text{L}}}{1} \times \frac{1.06 \cancel{\text{qt}}}{1 \cancel{\text{L}}} \times \frac{1 \text{ gal}}{4 \cancel{\text{qt}}}$

$$= \frac{(40)(1.06)}{4} \text{ gal}$$

$$= 10.6 \text{ gal}$$

62. $F = \frac{9}{5}C + 32$

$$= \frac{9}{5}(40) + 32$$

$$= 72 + 32 + 104$$

Thus, 40°C = 104°F.

65. $C = \frac{5}{9}(F - 32)$

$$= \frac{5}{9}(200 - 32)$$

$$= \frac{5}{9}(168) \approx 93$$

Thus, 200°F ≈ 93°C.

1.1

51. $x - [a(b + 1) - c] = 12 - [2(3 + 1) - 5] =$

$12 - [2 \cdot 4 - 5] = 12 - [8 - 5] = 12 - 3 = 9$

60. $A = 2\pi rh + 2\pi r^2 = 2(3.14)(2)(10) + 2(3.14)(2)^2 =$

$125.60 + 25.12 = 150.72 \text{ cm}^2$

1.2

26. Given $\frac{17}{43}$ and $\frac{28}{79}$. The cross product is $17 \cdot 79 = 1343$, and the second cross product is $43 \cdot 28 = 1204$. Since $1343 > 1204$, we have $\frac{17}{43} > \frac{28}{79}$.

1.3

51. The difference in temperature was from $+35°$ to $-13°$. That is, we must find $35 - (-13) = 35 + 13 = 48°$. Thus, the total change in temperature was a decrease of $48°$.

1.4

44. Since 20 penalty points must be deducted from the total 3 times, then $3(-20) = -60$ represents the deduction. Thus, $423 + (-60) = 363$ points is her total after losing the penalty points.

50. $3 - \dfrac{2 + (-4)}{1 - 5} = 3 - \dfrac{-2}{-4}$

$= 3 - \dfrac{1}{2} = \dfrac{6}{2} - \dfrac{1}{2} = \dfrac{5}{2}$

63. When $a = -1$, $b = 2$, and $c = -4$,

$2[3(a + 1) + 2(4 + c)]$

$= 2[3((-1) + 1) + 2(4 + (-4))]$

$= 2[3(0) + 2(0)] = 2[0 + 0] = 2[0] = 0.$

75. -3000 represents the amount owed. $+1200$ represents the amount repaid, and -650 is the amount again borrowed. We must compute $(-3000) + (1200) + (-650) = -2450$. Thus, the Brisebois family owes \$2450, so that -2450 represents the status of their account.

1.5

37. $\dfrac{1}{2}x - \dfrac{2}{3}y - \dfrac{5}{2}x - \dfrac{1}{3}y$

$= \dfrac{1}{2}x - \dfrac{5}{2}x - \dfrac{2}{3}y - \dfrac{1}{3}y$

$= \left(\dfrac{1}{2} - \dfrac{5}{2}\right)x - \left(\dfrac{2}{3} + \dfrac{1}{3}\right)y$

$= -2x - y$

43. $2x - (-x + 1) + 3$

$= 2x + x - 1 + 3$

$= 3x + 2$

47. $a - [3a - (1 - 2a)] = a - [3a - 1 + 2a] =$

$a - [5a - 1] = a - 5a + 1 = -4a + 1$

55. When $x = -3$, $-2x^2 = -2(-3)^2 = -2 \cdot 9 = -18$. Note that we square -3 before multiplying the result by -2.

56. When $x = -3$, $(-2x)^2 = ((-2)(-3))^2 = (6)^2 = 36$. Note that we first multiply -2 by -3, obtaining 6, then square the result. Compare Exercise 55 with Exercise 56.

68. When $x = -3$ and $y = -1$, $|x - y| =$ $|(-3) - (-1)| = |(-3) + 1| = |-2| = 2$. Notice that we first simplify inside the absolute value bars.

1.6

17. Always reduce fractions when they appear under a radical. Thus,

$$\sqrt{\frac{8}{18}} = \sqrt{\frac{\not{2} \cdot 4}{\not{2} \cdot 9}} = \sqrt{\frac{4}{9}} = \frac{2}{3}.$$

29. Since $3^2 = 9 < 15 < 16 = 4^2$,
we have $\sqrt{3^2} = 3 < \sqrt{15} < 4 = \sqrt{4^2}$.

Thus, $\sqrt{15}$ is between 3 and 4.

1.7

27. $\left(\dfrac{2y}{x^3}\right)^{-2} = \dfrac{(2y)^{-2}}{(x^3)^{-2}}$ $\quad \left(\dfrac{a}{b}\right)^n = \dfrac{a^n}{b^n}$

$= \dfrac{2^{-2}y^{-2}}{(x^3)^{-2}}$ $\quad (ab)^n = a^nb^n$

$= \dfrac{2^{-2}y^{-2}}{x^{-6}}$ $\quad (a^m)^n = a^{mn}$

$= \dfrac{x^6}{2^2y^2}$

$= \dfrac{x^6}{4y^2}$

30. $\dfrac{a^{-2}b^2}{a^4b^{-3}} = a^{-2-4}b^{2-(-3)}$ $\quad \dfrac{a^m}{a^n} = a^{m-n}$

$= a^{-6}b^5$

$= \dfrac{1}{a^6} \cdot b^5$ $\quad a^{-n} = \dfrac{1}{a^n}$

$= \dfrac{b^5}{a^6}$

51. When $a = -2$, $a^{-2} = (-2)^{-2} = \dfrac{1}{(-2)^2} = \dfrac{1}{4}$.

52. When $a = -2$, $-2a = -2(-2) = 4$.

53. When $a = -2$, $-a^2 = -(-2)^2 = -(4) = -4$. Note the difference between Exercises 51, 52, and 53.

54. $a^{-2} + b^{-2} = (-2)^{-2} + (3)^{-2}$ Substitute −2 for *a* and 3 for *b*

$= \dfrac{1}{(-2)^2} + \dfrac{1}{3^2}$ $\quad a^{-n} = \dfrac{1}{a^n}$

$= \dfrac{1}{4} + \dfrac{1}{9}$

$= \dfrac{13}{36}$

55. $(a + b)^{-2} = ((-2) + 3)^{-2}$ Substitute −2 for *a* and 3 for *b*

$= (1)^{-2}$

$= \dfrac{1}{1^2} = 1$

Note the difference between Exercises 54 and 55.

59. $a^{-1}b^{-1} = (-2)^{-1}(3)^{-1}$

$= \left(\dfrac{1}{-2}\right)\left(\dfrac{1}{3}\right)$ $\quad a^{-1} = \dfrac{1}{a^1} = \dfrac{1}{a}$

$= -\dfrac{1}{6}$

60. $(ab)^{-1} = ((-2)(3))^{-1}$

$= (-6)^{-1}$

$= \dfrac{1}{(-6)^1}$

$= -\dfrac{1}{6}$

1.8

19. $(0.0000022)(300) = (2.2 \times 10^{-6})(3.0 \times 10^2) = (2.2)(3.0) \times (10^{-6})(10^2) = 6.6 \times 10^{-4}$

2.1

28. Change $2\frac{1}{2}$ to $\frac{5}{2}$ and subtract $\frac{5}{2}$ from both sides of the equation.

$x + 2\dfrac{1}{2} = 5$

$x + \dfrac{5}{2} = 5$

$x + \dfrac{5}{2} - \dfrac{5}{2} = 5 - \dfrac{5}{2}$

$x + 0 = \dfrac{10}{2} - \dfrac{5}{2}$

$x = \dfrac{5}{2} = 2\dfrac{1}{2}$

2.2

15. Since $-\frac{1}{8}$ is the divisor of z we must multiply both sides by $-\frac{1}{8}$.

$\dfrac{z}{-\frac{1}{8}} = 16$

$\left(-\dfrac{1}{8}\right)\dfrac{z}{-\frac{1}{8}} = 16\left(-\dfrac{1}{8}\right)$

$\dfrac{(-\frac{1}{8})}{(-\frac{1}{8})}z = -\dfrac{16}{8}$ $\quad \dfrac{-\frac{1}{8}}{-\frac{1}{8}} = 1$

$1 \cdot z = -2$

$z = -2$

16. First multiply both sides by 3.

$\dfrac{2x}{3} = 4$

$(3)\dfrac{2x}{3} = 4(3)$

$\dfrac{3}{3} \cdot 2x = 12$

$2x = 12$

$\dfrac{2x}{2} = \dfrac{12}{2}$ Now divide by 2

$x = 6$

22. First change $2\frac{1}{2}$ to $\frac{5}{2}$.

$2\dfrac{1}{2}x = 10$

$\dfrac{5}{2}x = 10$

$\dfrac{2}{5} \cdot \dfrac{5}{2}x = 10 \cdot \dfrac{2}{5}$ The product $\frac{2}{5} \cdot \frac{5}{2}$ is 1

$1 \cdot x = \dfrac{10}{1} \cdot \dfrac{2}{5}$

$x = 4$ $\quad 10 \div 5 = 2$ and $2 \cdot 2 = 4$

2.3

11. Collect like terms on the left side.

$-\dfrac{1}{7}y + y = 18$

$\left(-\dfrac{1}{7} + 1\right)y = 18$

$\left(-\dfrac{1}{7} + \dfrac{7}{7}\right)y = 18$

$\dfrac{6}{7}y = 18$

$$\frac{7}{6}\cdot\frac{6}{7}y = 18\cdot\frac{7}{6}$$ **$\frac{7}{6}$ is the reciprocal of $\frac{6}{7}$**

$$1\cdot y = \frac{18}{1}\cdot\frac{7}{6}$$

$$y = 21$$ **$18 \div 6 = 3$ and $3 \cdot 7 = 21$**

14. $-7 + 21y + 23 = 16$

$21y + 16 = 16$ **Collect like terms**

$21y = 0$ **Subtract 16 from both sides**

$y = 0$ **Divide both sides by 21**

17. $6y - 4y + 1 = 12 + 2y - 11$

$2y + 1 = 1 + 2y$ **Collect like terms**

Since this is an identity, every real number is a solution.

19. Clear fractions by multiplying through by 2 and collect like terms on both sides.

$$3y + \frac{5}{2}y + \frac{3}{2} = \frac{1}{2}y + \frac{5}{2}y$$

$$2\left(3y + \frac{5}{2}y + \frac{3}{2}\right) = 2\left(\frac{1}{2}y + \frac{5}{2}y\right)$$

$6y + 5y + 3 = y + 5y$ **Multiply each term by 2**

$11y + 3 = 6y$ **Collect like terms**

$11y = 6y - 3$ **Subtract 3 from both sides**

$11y - 6y = -3$ **Subtract 6*y* from both sides**

$5y = -3$

$y = -\frac{3}{5}$ **Divide both sides by 5**

29. $5(2 - 3x) = 15 - (x + 7)$

$10 - 15x = 15 - x - 7$ **Clear parentheses**

$10 - 15x = 8 - x$ **Collect like terms**

$10 = 8 + 14x$ **Add 15*x* to both sides**

$2 = 14x$ **Subtract 8 from both sides**

$\frac{2}{14} = x$ **Divide both sides by 14**

$\frac{1}{7} = x$

36. The fractions could be cleared by multiplying both sides by the LCD of 2 and 3, which is 6. However, in this exercise the fractions are cleared when we use the distributive law.

$$\frac{1}{3}(6x - 9) = \frac{1}{2}(8x - 4)$$

$$\frac{1}{3}\cdot 6x - \frac{1}{3}\cdot 9 = \frac{1}{2}\cdot 8x - \frac{1}{2}\cdot 4$$

$$2x - 3 = 4x - 2$$

$$2x = 4x - 2 + 3$$

$$2x - 4x = 1$$

$$-2x = 1$$

$$\frac{-2x}{-2} = \frac{1}{-2}$$

$$x = -\frac{1}{2}$$

37. $5(x + 1) - 4x = x - 5$

$5x + 5 - 4x = x - 5$ **Clear parentheses**

$x + 5 = x - 5$ **Collect like terms**

$5 = -5$ **Subtract *x* from both sides**

Since we obtain a contradiction, the equation has no solution.

38. Clear all parentheses first.

$3(2 - 4x) = 4(2x - 1) - 2(1 + x)$

$6 - 12x = 8x - 4 - 2 - 2x$

$6 - 12x = 8x - 2x - 4 - 2$ **Collect like terms**

$-12x = 6x - 6 - 6$

$-12x - 6x = -12$

$-18x = -12$

$$\frac{-18x}{-18} = \frac{-12}{-18}$$

$$x = \frac{2}{3}$$

41. $-4(-x + 1) + 3(2x + 3) - 7x = -10$

$4x - 4 + 6x + 9 - 7x = -10$ **Clear parentheses**

$3x + 5 = -10$ **Collect like terms**

$3x = -15$ **Subtract 5 from both sides**

$x = -5$ **Divide both sides by 3**

2.4

32. Let x = the length of the pole,

$\frac{1}{6}x$ = the length of the pole in the sand,

25 = the length of the pole in the water,

$\frac{2}{3}x$ = the length of the pole in the air.

The equation to solve is

$$x = \frac{1}{6}x + 25 + \frac{2}{3}x.$$

$6x = 6\left[\frac{1}{6}x + 25 + \frac{2}{3}x\right]$ **Multiply by 6 to clear fractions**

$6x = x + 150 + 4x$

$6x = 5x + 150$

$x = 150$ The length of the pole is 150 ft.

33. Let x = the score Becky must make on the fourth test. To average four scores, we add them and divide by 4. Thus, we need to solve

$$\frac{96 + 78 + 91 + x}{4} = 90.$$

$$\not{4}\left[\frac{96 + 78 + 91 + x}{\not{4}}\right] = 4 \cdot 90 \quad \textbf{Multiply both sides by 4}$$

$$96 + 78 + 91 + x = 360$$

$$265 + x = 360$$

$$x = 95$$

Becky must get at least a 95 on the fourth test.

34. Let x = the number of miles Charlene drove. We need to solve

$$2(12.50) + (0.10)x = 73.$$

$$25 + 0.10x = 73$$

$$0.10x = 48$$

$$x = 480$$

Charlene drove 480 miles.

41. Let x = rise in the water level. Since the volume of water is a rectangular solid of length 10 ft, width 8 ft, and height x ft, we know that the volume is

$$(10)(8)x = 80x \text{ ft}^3.$$

But the volume is also equal to the volume of the sphere, which is about 33.5 ft^3. Thus, we need to solve

$$80x = 33.5.$$

$$x = \frac{33.5}{80} \approx 0.4 \text{ ft}$$

2.5

8. Let x = amount of fuel added. The increase in fuel is $1182 - 985 = 197$ gal. The question asked is: 197 is what % of 985? This translates to:

$$197 = (x)(985).$$

$$\frac{197}{985} = x$$

$$0.2 = x$$

The percent increase is 20%.

9. Let x = the population in 1980. The increase in population in 1990 was 35% of x, or $0.35x$. Thus, the population in 1990 was the population in 1980 plus the increase. Then we need to solve

$$x + 0.35x = 567.$$

$$(1 + 0.35)x = 567$$

$$1.35x = 567$$

$$x = \frac{567}{1.35}$$

$$x = 420$$

The population in 1980 was 420.

11. Let x = the amount to invest. Over a five-year period, 11% simple interest is 55%. The equation to solve is

$$x + 0.55x = 11{,}470.$$

$$1.55x = 11{,}470$$

$$x = 7400$$

Paul must invest \$7400.

12. Let x = amount invested in a savings account,
$10{,}000 - x$ = amount invested in a mutual fund,
$0.12x$ = earnings in the savings account,
$0.15(10{,}000 - x)$ = earnings in the mutual fund.
We need to solve

$$0.12x + 0.15(10{,}000 - x) = 1275.$$

$$0.12x + 1500 - 0.15x = 1275$$

$$-0.03x = -225$$

$$x = 7500$$

$$10{,}000 - x = 2500$$

Lucky Lucia invested \$7500 in the savings account and \$2500 in the mutual fund.

15. Let x = the number of miles Jeff drove. If y is the bill less sales tax, then $y + 0.04y = 199.68$. Thus $y(1.04) = 199.68$ so that $y = 192$. We must solve

$$(4)(13) + (0.08)x = 192.$$

$$0.08x = 140$$

$$x = 1750$$

Jeff drove a total of 1750 miles.

18. Let t = the time the woman walks,
$7 - t$ = the time she rides,
$4.5t$ = the distance she walks,
$2.5(7 - t)$ = the distance she rides.
Since the distance down and the distance back out are equal, we need to solve

$$4.5t = 2.5(7 - t).$$

$$4.5t = 17.5 - 2.5t$$

$$7t = 17.5$$

$$t = 2.5$$

She walks 2.5 hours, and the total length of the trip is $2(4.5)(2.5) = 22.5$ miles.

19. Let t = the time of travel,
$55t$ = the distance that one travels,
$50t$ = the distance that the second travels.
When they meet, the total of the two distances traveled is 315, so the equation to solve is

$$55t + 50t = 315.$$

$$105t = 315$$

$$t = 3$$

Since they traveled for 3 hr and left at 9:30 A.M., they will meet at 12:30 P.M.

22. Let x = rate for one press,
$x - 15$ = rate for second press,
$150x$ = number flyers printed by one,
$150(x - 15)$ = number flyers printed by other.
Then we need to solve

$$150x + 150(x - 15) = 14{,}850.$$
$$150x + 150x - 2250 = 14{,}850$$
$$300x = 12{,}600$$
$$x = \frac{12{,}600}{300} = 42$$
$$x - 15 = 42 - 15 = 27$$

One press prints 42 fliers/min and the other 27 fliers/min.

23. Let t = time for cutter to overtake the private boat,
$35t$ = distance cutter travels.
Since the private boat has traveled 10 nautical miles before the cutter starts, the distance the private boat travels can be written as

$$10 + 20t.$$

The two distances are equal so we solve

$$35t = 10 + 20t.$$
$$15t = 10$$
$$t = \frac{10}{15} = \frac{2}{3}$$

It takes $\frac{2}{3}$ hr or 40 minutes for the cutter to overtake the private boat.

Supplementary Applied Problems

5. Let x = the length of the pole. We can obtain an equation to use by setting the length of the pole equal to the sum of the lengths.

(total length) = **(length in sand)** + **(length in water)** + **(length in air)**

$$x = \frac{1}{4}x + 9 + \frac{3}{8}x$$

$$x = \frac{1}{4}x + 9 + \frac{3}{8}x$$
$$8x = 2x + 72 + 3x \quad \text{Multiply by the LCD 8}$$
$$8x = 5x + 72$$
$$3x = 72$$
$$x = \frac{72}{3} = 24$$

The pole is 24 ft long.

10. Let x = measure of first angle of the triangle,
$x + 30$ = measure of second angle of the triangle,
$4(x + 30)$ = measure of third angle of the triangle (if the second is $\frac{1}{4}$ the third, then the third is 4 times the second).

$$x + (x + 30) + 4(x + 30) = 180 \quad \text{Sum of angles is 180}$$
$$x + x + 30 + 4x + 120 = 180$$
$$6x + 150 = 180$$
$$6x = 30$$
$$x = 5$$
$$x + 30 = 35$$
$$4(x + 30) = 140$$

The first angle is 5°, the second is 35°, and the third is 140°.

13. Let x = width of rectangle,
$2x$ = length of rectangle,
$x + 8$ = width increased by 8,
$2x - 8$ = length decreased by 8.
For a square the length and width are equal.

$$x + 8 = 2x - 8$$
$$x = 2x - 16$$
$$-x = -16$$
$$x = 16$$
$$2x = 32$$

Thus, the width is 16 ft and the length is 32 ft.

19. x = population in 1980.
$0.20x$ = amount of decrease.

(population in 1980) − (amt. of decrease) = (pop. in 1990)

$$x - 0.20x = 740$$
$$(1 - 0.20)x = 740$$
$$0.8x = 740$$
$$x = \frac{740}{0.8} = 925$$

There were 925 people in Deserted in 1980.

20. x = amount invested at 14% interest,
$10{,}000 - x$ = amount invested at 11% interest (If x dollars are invested at 14%, the amount left to invest at 11% is $10{,}000 - x$),
$0.14x$ = amount of interest earned on 14% investment,
$0.11(10{,}000 - x)$ = amount of interest earned on 11% investment.

$$0.14x + 0.11(10{,}000 - x) = 1310$$
$$14x + 11(10{,}000 - x) = 131{,}000 \quad \text{Multiply by 100 to clear decimals}$$
$$14x + 110{,}000 - 11x = 131{,}000$$
$$3x = 21{,}000$$
$$x = 7000$$
$$10{,}000 - x = 3000$$

Thus, \$7000 was invested at 14% and \$3000 at 11%.

24. x = score Becky must make to have 90 average. The average of her 4 scores must be 90.

$$\frac{96 + 78 + 91 + x}{4} = 90$$

$$\frac{265 + x}{4} = 90$$

$$4\left(\frac{265 + x}{4}\right) = 90(4)$$

$$265 + x = 360$$

$$x = 95$$

Becky must make at least 95 on her fourth test to get an A.

2.6

14. $$A = \frac{1}{2}(b_1 + b_2)h$$

$$2A = (b_1 + b_2)h \quad \text{Multiply both sides by 2}$$

$$2A = b_1h + b_2h \quad \text{Remove parentheses}$$

$$2A - b_2h = b_1h \quad \text{Subtract } b_2h$$

$$\frac{2A - b_2h}{h} = b_1 \quad \text{Divide by } h$$

26. $2(x - b) + 5x = 4c$

$$2x - 2b + 5x = 4c$$

$$7x - 2b = 4c$$

$$7x = 4c + 2b$$

$$x = \frac{4c + 2b}{7}$$

2.7

11. $\frac{x + 3}{-2} < 17$

$$(-2)\left[\frac{x + 3}{-2}\right] > (-2)(17) \quad \text{Multiply both sides by } -2 \text{ and reverse inequality}$$

$$x + 3 > -34 \quad \text{Do not reverse inequality when subtracting 3}$$

$$x > -37$$

The solution is $(-37, \infty)$.

16. $5(3 - x) - 10 \le 25$

$$15 - 5x - 10 \le 25$$

$$-5x + 5 \le 25$$

$$-5x \le 20$$

$$x \ge -4 \quad \text{Divide both sides by } -5 \text{ and reverse inequality}$$

The solution is $[-4, \infty)$.

28. Let x = the unknown number,

$\frac{1}{3}x$ = one-third the number.

If $\frac{1}{3}x$ is no more than 15, we solve

$$\frac{1}{3}x \le 15.$$

$$x \le 45 \quad \text{Multiply by 3}$$

The solution is $(-\infty, 45]$.

29. Let s = Walt's score on the fourth test. The average of his four scores is

$$\frac{86 + 83 + 97 + s}{4} = \frac{266 + s}{4}.$$

Since the average must be 90 or better, we need to solve

$$\frac{266 + s}{4} \ge 90.$$

$$266 + s \ge (4)(90) \quad \text{Clear fraction}$$

$$266 + s \ge 360$$

$$s \ge 94$$

Walt's score must be 94 or better.

30. Since S must exceed (be greater than) C, and $S = 150n$ while $C = 125n + 350$, we need to solve

$$150n > 125n + 350.$$

$$25n > 350 \quad \text{Subtract } 125n \text{ from both sides}$$

$$n > \frac{350}{25} \quad \text{Divide both sides by 25}$$

$$n > 14$$

To make a profit, the owner must sell 15 or more sets.

2.8

15. Let s = Jake's score on the fourth test. We need to solve

$$80 \le \frac{69 + 82 + 79 + s}{4} \le 90.$$

$$320 \le 69 + 82 + 79 + s \le 360$$

$$320 \le 230 + x \le 360$$

$$90 \le x \le 130$$

But the maximum percent is 100%. Thus, Jake must score between 90% and 100% (he can't score more than 100%), inclusive.

2.9

14. We need to solve

$$\frac{1}{2}z + 1 = \frac{3}{4}z - 1 \quad \text{and} \quad \frac{1}{2}z + 1 = -\left(\frac{3}{4}z - 1\right).$$

$$4\left(\frac{1}{2}z + 1\right) = 4\left(\frac{3}{4}z - 1\right) \qquad 4\left(\frac{1}{2}z + 1\right) = -4\left(\frac{3}{4}z - 1\right)$$

$$2z + 4 = 3z - 4 \qquad 2z + 4 = -3z + 4$$

$$8 = z \qquad 5z = 0$$

$$z = 0$$

The two solutions are 8 and 0.

19. Let x = the desired number,
$|x + 1|$ = absolute value of the sum of the number and 1.

We need to solve $|x + 1| = |x|$. Thus, we get

$$x + 1 = x \quad \text{and} \quad x + 1 = -x.$$
$$1 = 0 \qquad\qquad 1 = -2x$$
$$\text{no solution} \qquad -\frac{1}{2} = x$$

The number is $-\frac{1}{2}$.

35. $\left|\dfrac{x+1}{2}\right| > 3$ This translates into the following.

$$\frac{x+1}{2} > 3 \quad \text{or} \quad \frac{x+1}{2} < -3$$
$$x + 1 > 6 \quad \text{or} \quad x + 1 < -6$$
$$x > 5 \quad \text{or} \quad x < -7$$

The solution is $(-\infty, -7)$ or $(5, \infty)$.

41. Let x = the desired number.
We need to solve $|2x + 3| \geq 7$, which translates into the following.

$$2x + 3 \leq -7 \quad \text{or} \quad 2x + 3 \geq 7$$
$$2x \leq -10 \quad \text{or} \quad 2x \geq 4$$
$$x \leq -5 \quad \text{or} \quad x \geq 2$$

3.1

39. To complete (1,   ), substitute 1 for x in $y = 200x + 30$ and solve for y.

$$y = 200(1) + 30 = 230$$

Similarly, when $x = 5$, $y = 200(5) + 30 = 1030$, and when $x = 10$, $y = 200(10) + 30 = 2030$.
(a) The cost of producing 1 part is \$230, taken from the ordered pair (1,230). **(b)** The cost of producing 5 parts is \$1030. **(c)** The cost of producing 10 parts is \$2030.

3.3

17. The slope of l_1 is

$$m_1 = \frac{2 - (-1)}{0 - 1} = \frac{3}{-1} = -3.$$

The slope of l_2 is

$$m_2 = \frac{-1 - 0}{0 - 3} = \frac{-1}{-3} = \frac{1}{3}.$$

Since -3 and $\frac{1}{3}$ are negative reciprocals of each other, l_1 and l_2 are perpendicular.

3.4

13. First write $-3x - y + 4 = 0$ in slope-intercept form.

$$-y = 3x - 4$$
$$y = -3x + 4$$

The slope of this line is -3, so a line perpendicular to it has slope $\frac{1}{3}$. Use the point-slope form.

$$y - 4 = \frac{1}{3}(x - (-1))$$
$$3y - 12 = x + 1 \quad \textbf{Multiply both sides by 3}$$
$$-x + 3y - 13 = 0$$
$$x - 3y + 13 = 0 \quad \textbf{General form}$$

16. We are given x-intercept (3, 0) and slope $\frac{1}{2}$. Use the x-intercept like any other point in the point-slope form.

$$y - y_1 = m(x - x_1)$$
$$y - 0 = \frac{1}{2}(x - 3)$$
$$2y = x - 3 \quad \textbf{Multiply by 2}$$
$$x - 2y - 3 = 0 \quad \textbf{General form}$$

20. Since lines with undefined slope are vertical lines (parallel to the y-axis), the desired line must be parallel to the y-axis passing through (2, 3). Since all x-coordinates of points on this line are 2, the equation is

$$x = 2 \text{ or } x - 2 = 0.$$

22. Basically we are given two points, (10, 2800) and (25, 6125). Using the formula for m and the point-slope form,

$$y - 2800 = \frac{6125 - 2800}{25 - 10}(x - 10)$$
$$y - 2800 = \frac{3325}{15}(x - 10)$$
$$y - 2800 = \frac{665}{3}(x - 10).$$

Multiply each side of the equation by 3 to obtain

$$3y - 8400 = 665(x - 10)$$
$$3y - 8400 = 665x - 6650$$
$$665x - 3y + 1750 = 0. \quad \textbf{General form}$$

Use this equation to find y when $x = 31$.

3.6

23. Let x = his total sales,
$0.12x$ = his commission.
Thus, his annual earnings are found by adding \$15,000 to his commission.

$$f(x) = 15{,}000 + 0.12x$$

When x is 100,000, we find

$$f(100{,}000) = 15{,}000 + (0.12)(100{,}000)$$
$$= 15{,}000 + 12{,}000$$
$$= 27{,}000.$$

His earnings are \$27,000.

24. We need to evaluate $f(x)$ when $x = 2$.

$$\begin{aligned} f(x) &= -16x^2 + 128x \\ f(2) &= -16(2)^2 + 128(2) \\ &= -16(4) + 256 \\ &= -64 + 256 = 192 \end{aligned}$$

The height of the object is 192 ft in 2 sec.

4.1

20. We graph each line using a scale which will allow us to read the intersection. The solution appears to be $(\frac{1}{2}, -1)$, which does check in both equations.

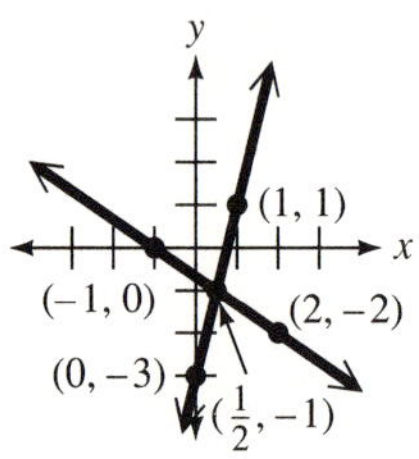

4.2

5. $5x - 5y = 10$
$x - y = 2$

Solve the second equation for y,

$$y = x - 2,$$

and substitute into the first.

$$\begin{aligned} 5x - 5(x - 2) &= 10 \\ 5x - 5x + 10 &= 10 \\ 10 &= 10 \end{aligned}$$

Since we obtain an identity, the system has infinitely many solutions. Since $y = x - 2$, any point on the common line of solutions can be written in the form $(x, x - 2)$, for any real number x.

10. $5x - 7y = 3$
$-10x + 14y = -1$

Multiply the second equation by $\frac{1}{2}$ and add.

$$\begin{aligned} 5x - 7y &= 3 \\ -5x + 7y &= -\tfrac{1}{2} \\ \hline 0 &= \tfrac{5}{2} \end{aligned}$$

Since this is a contradiction, there is no solution.

19. $\frac{3}{4}x - \frac{2}{3}y = 1$

$\frac{3}{8}x - \frac{1}{6}y = 1$

First, clear all fractions by multiplying the first equation by 12 and the second by 24.

$$\begin{aligned} 9x - 8y &= 12 \\ 9x - 4y &= 24 \\ \hline -4y &= -12 \quad \text{Subtract to eliminate } x \\ y &= 3 \end{aligned}$$

Substitute 3 for y in $9x - 8y = 12$.

$$\begin{aligned} 9x - 8(3) &= 12 \\ 9x - 24 &= 12 \\ 9x &= 36 \\ x &= 4 \end{aligned}$$

The solution is $(4, 3)$.

20. $0.6x - 0.7y = 1.3$
$-1.2x + 1.4y = 7.3$

First, multiply both equations by 10 to clear the decimals.

$$\begin{aligned} 6x - 7y &= 13 \\ -12x + 14y &= 73 \\ 12x - 14y &= 26 \\ \hline 0 &= 99 \quad \text{Multiply first equation by 2 and add} \end{aligned}$$

Since we obtain a contradiction, the system has no solution.

21. $\frac{4}{x} - \frac{2}{y} = -2$

$\frac{2}{x} + \frac{1}{y} = 3$

The system can be written in the form

$$4\left(\frac{1}{x}\right) - 2\left(\frac{1}{y}\right) = -2$$

$$2\left(\frac{1}{x}\right) + \left(\frac{1}{y}\right) = 3.$$

Substitute u for $\frac{1}{x}$ and v for $\frac{1}{y}$ to obtain the following system.

$$\begin{aligned} 4u - 2v &= -2 \\ 2u + v &= 3 \end{aligned}$$

The first equation can be simplified by multiplying both sides by $\frac{1}{2}$.

$$\begin{aligned} 2u - v &= -1 \\ 2u + v &= 3 \\ \hline 4u &= 2 \quad \text{Add to eliminate } v \\ u &= \frac{2}{4} \\ u &= \frac{1}{2} \end{aligned}$$

Substitute $\frac{1}{2}$ for u in $2u + v = 3$.

$$\begin{aligned} 2\left(\frac{1}{2}\right) + v &= 3 \\ 1 + v &= 3 \\ v &= 2 \end{aligned}$$

Since $u = \frac{1}{2}$, $\frac{1}{x} = \frac{1}{2}$ so $x = 2$. Also, since $v = 2$, $\frac{1}{y} = 2$ so $y = \frac{1}{2}$. The solution is $\left(2, \frac{1}{2}\right)$.

4.3

2. Let x = Jim's age,
y = Pete's age,
$x + 3$ = Jim's age in 3 years,
$y + 3$ = Pete's age in 3 years.

$y = x - 7$ **Pete is 7 years younger than Jim**

$x + 3 + y + 3 = 33$ **Sum of ages in 3 years is 33**

$x + y = 27$ **Simplify**

$x + x - 7 = 27$ **Substitute $y = x - 7$**

$$2x - 7 = 27$$

$$2x = 34$$

$$x = 17$$

Jim is 17 and Pete is $y = x - 7 = 17 - 7 = 10$.

4. Let x = number of hours at 30 mph,
y = number of hours at 40 mph,
$30x$ = distance traveled at 30 mph,
$40y$ = distance traveled at 40 mph.
The first equation is

$$30x + 40y = 230.$$

By traveling 10 mph *faster* throughout, she would have traveled at 40 mph for x hr and 50 mph for y hr. The second equation is

$$40x + 50y = 300.$$

First divide through both equations by 10 to obtain the following system.

$$3x + 4y = 23$$
$$4x + 5y = 30$$

$12x + 16y = 92$ **Multiply first by 4**

$-12x - 15y = -90$ **Multiply second by −3**

$y = 2$ **Add**

Substitute 2 for y in $3x + 4y = 23$.

$$3x + 4(2) = 23$$

$$3x + 8 = 23$$

$$3x = 15$$

$$x = 5$$

She traveled for 5 hr at 30 mph and 2 hr at 40 mph.

5. Let x = cost of one book in dollars,
y = cost of one pen in dollars,
$4x$ = value of 4 books,
$6y$ = value of 6 pens,
$3x$ = value of 3 books,
$9y$ = value of 9 pens.

$4x + 6y = 9$ **Cost of 4 books and 6 pens**

$3x + 9y = 9$ **Cost of 3 books and 9 pens**

Divide the second equation by 3 to simplify, and solve for x.

$x + 3y = 3$ **Divide by 3**

$x = -3y + 3$ **Solve for x**

$4x + 6y = 9$ **First equation**

$4(-3y + 3) + 6y = 9$ **Substitute $x = -3y + 3$**

$$-12y + 12 + 6y = 9$$

$$-6y + 12 = 9$$

$$-6y = -3$$

$$y = \frac{1}{2} = 0.5$$

$$x = -3(0.5) + 3$$

$$= -1.5 + 3 = 1.5$$

One book costs \$1.50, and one pen costs \$0.50.

8. Let x = number of adults,
y = number of children.
The first equation is

$$x + y = 450.$$

Changing to cents, we obtain the second equation.

$$200x + 75y = 60{,}000$$

Solve the first equation for x,

$$x = 450 - y,$$

and substitute into the second.

$$200(450 - y) + 75y = 60{,}000$$

$$90{,}000 - 200y + 75y = 60{,}000$$

$$-125y = -30{,}000$$

$$y = 240$$

Then $x = 450 - y = 450 - 240 = 210$. Thus, there were 210 adults and 240 children at the play.

14. Let x = amount invested at 10%,
y = amount invested at 9%.
Since \$500 more was invested at 10% than 9%, one equation is

$$x = 500 + y.$$

$0.10x$ = interest earned on 10% account,

$0.09y$ = interest earned on 9% account.

Then the total interest earned, $0.10x + 0.09y$, was \$183 giving us the second equation,

$$0.10x + 0.09y = 183,$$

which can be simplified by multiplying both sides by 100.

$$10x + 9y = 18{,}300$$

Since the first equation is already solved for x, substitute this expression into the second.

$$10(500 + y) + 9y = 18{,}300$$
$$5000 + 10y + 9y = 18{,}300$$
$$19y = 13{,}300$$
$$y = 700$$

Then $x = 500 + y = 500 + 700 = 1200$. Thus, \$1200 was invested at 10%, and \$700 at 9%.

16. Let x = number of liters of 50% acid solution needed,
y = number of liters of 25% acid solution needed,
$0.50x$ = amount of acid in x liters of 50% acid,
$0.25y$ = amount of acid in y liters of 25% acid,
$0.40(10)$ = amount of acid in 10 liters of 40% acid.

$x + y = 10$ **Total amount of solution**

$0.50x + 0.25y = (0.40)(10)$ **Amount of acid**

$50x + 25y = 40(10)$ **Multiply by 100 to clear decimals**

$2x + y = 16$ **Divide by 25**

Now solve the system by subtracting to eliminate y.

$$\begin{array}{r} x + y = 10 \\ \underline{2x + y = 16} \\ -x \quad = -6 \end{array}$$
$$x = 6$$
$$6 + y = 10$$
$$y = 4$$

Thus, 6L of 50% and 4L of 25% acid are required.

18. Let x = dollars charged per day,
y = dollars charged per mile,
$3x$ = dollars charged for 3 days,
$316y$ = dollars charged for 316 miles,
$5x$ = dollars charged for 5 days,
$242y$ = dollars charged for 242 miles.

$3x + 316y = 143.12$ **Mr. Green's total bill**

$5x + 242y = 147.44$ **Mr. Brown's total bill**

Multiply the first equation by 5 and the second by -3 and add.

$$\begin{array}{rcr} 15x + 1580y & = & 715.60 \\ -15x - 726y & = & -442.32 \\ \hline 854y & = & 273.28 \end{array}$$
$$y = \frac{273.28}{854} = 0.32$$
$$5x + 242(0.32) = 147.44$$
$$5x + 77.44 = 147.44$$
$$5x = 70$$
$$x = 14$$

The charge is \$14 per day and \$0.32 per mile.

19. Let x = rate of houseboat in still water,
y = rate of current in river,
$x + y$ = rate traveling downstream,
$x - y$ = rate traveling upstream.

Using the formula $d = rt$, we have:

$30 = (x + y)3$ **Distance downstream**

$30 = (x - y)5.$ **Distance upstream**

Dividing the first equation by 3 and the second by 5 we have the following system.

$$x + y = 10$$
$$x - y = 6$$

Add to eliminate y.

$$2x = 16$$
$$x = 8$$

Substitute 8 for x in $x + y = 10$.

$$8 + y = 10$$
$$y = 2$$

The houseboat traveled 8 mph, and the river current was 2 mph.

4.4

5. (A) $3x + y - z = 4$
(B) $y - 2z = 5$
(C) $2x + z = -1$

If we use (A) and (B) and eliminate y, the result can be paired with (C) immediately.

$$\begin{array}{rr} (A) - (B) & 3x + z = -1 \\ (C) & \underline{2x + z = -1} \\ & x \quad = 0 \end{array}$$ **Subtract to eliminate *z***

Substitute 0 for x in $3x + z = -1$.

$$0 + z = -1$$
$$z = -1$$

Substitute -1 for z in (B) to obtain y.

$$y - 2(-1) = 5$$
$$y + 2 = 5$$
$$y = 3$$

The solution is $(0, 3, -1)$.

12. (A) $0.3x - 2.1y + 1.2z = -3$
(B) $5.7x + 4.2y - 4.8z = -3$
(C) $1.1x - 6.9y + 3.6z = -19$

To eliminate decimals, multiply each equation by 10.

$$\begin{array}{rl} 10(A) & 3x - 21y + 12z = -30 \\ 10(B) & 57x + 42y - 48z = -30 \\ 10(C) & 11x - 69y + 36z = -190 \end{array}$$

$$\begin{array}{rl} 40(A) & 12x - 84y + 48z = -120 \\ 10(B) & \underline{57x + 42y - 48z = -\ 30} \\ (D) & 69x - 42y \qquad = -150 \end{array}$$

$$\begin{array}{rl} -30(A) & -9x + 63y - 36z = 90 \\ 10(C) & \underline{11x - 69y + 36z = -190} \\ (E) & 2x - 6y \qquad = -100 \end{array}$$

$$\begin{array}{rl} \text{(D)} & 69x - 42y = -150 \\ -7\text{(E)} & -14x + 42y = 700 \\ \hline & 55x = 550 \quad \textbf{(D) – 7(E)} \\ & x = 10 \end{array}$$

$$\begin{array}{rl} \text{(E)} & 2(10) - 6y = -100 \\ & -6y = -120 \\ & y = 20 \end{array}$$

$$\begin{array}{rl} 10\text{(A)} & 3(10) - 21(20) + 12z = -30 \\ & 30 - 420 + 12z = -30 \\ & 12z = 360 \\ & z = 30 \end{array}$$

The solution is (10, 20, 30).

18. Let x = amount invested in mutual fund,
y = amount invested in certificates,
z = amount invested in business.

The system to solve is:

$$\begin{aligned} x + y + z &= 10{,}000 \\ 0.08x + 0.07y + 0.10z &= 840 \\ 0.08x + 0.07y - 0.02z &= 360. \end{aligned}$$

Clearing the decimals and subtracting the third from the second gives

$$12z = 48{,}000$$
$$z = 4000.$$

Substituting 4000 for z in the first two equations will give a system in x and y which, when solved, gives $x = 2000$ and $y = 4000$.

21. Let x = price in cents per pound of oranges,
y = price in cents per pound of apples,
z = price in cents per pound of limes.

The average of three numbers is found by adding the numbers and dividing by 3.

$\frac{x + y + z}{3} = 50$ **Average price is 50 cents per pound**

$x + y + z = 150$ **Multiply by three**

$6x + 5y + 2z = 600$ **Cost of 6 lb of oranges, 5 lb of apples, and 2 lb of limes**

$y = z - 25$ **Cost of limes minus 25 is cost of apples**

The system to solve is the following.

$$\begin{array}{rl} \text{(A)} & x + y + z = 150 \\ \text{(B)} & 6x + 5y + 2z = 600 \\ \text{(C)} & y - z = -25 \end{array}$$

Multiply (A) by -6 and add to (B) to get

$$\begin{array}{rl} -6\text{(A)} & -6x - 6y - 6z = -900 \\ \text{(B)} & 6x + 5y + 2z = 600 \\ \hline \text{(D)} & -y - 4z = -300. \end{array}$$

$$\begin{array}{rl} \text{(C)} & y - z = -25 \\ \text{(D)} & -y - 4z = -300 \\ \hline & -5z = -325 \quad \textbf{(C) + (D)} \\ & z = 65 \end{array}$$

$$\begin{array}{rl} \text{(C)} & y - 65 = -25 \\ & y = 40 \end{array}$$

$$\begin{array}{rl} \text{(A)} & x + 40 + 65 = 150 \\ & x = 45 \end{array}$$

Oranges are 45¢/lb, apples are 40¢/lb, and limes are 65¢/lb.

4.5

14. $3x + 2y = 7$, $-x + 4y = 0$

$$A = \begin{vmatrix} 3 & 2 \\ -1 & 4 \end{vmatrix} = (3)(4) - (-1)(2) = 12 + 2 = 14$$

$$A_x = \begin{vmatrix} 7 & 2 \\ 0 & 4 \end{vmatrix} = (7)(4) - (0)(2) = 28 \qquad A_y = \begin{vmatrix} 3 & 7 \\ -1 & 0 \end{vmatrix} = (3)(0) - (-1)(7) = 7$$

$$x = \frac{A_x}{A} = \frac{28}{14} = 2 \qquad y = \frac{A_y}{A} = \frac{7}{14} = \frac{1}{2}$$

18.
$$\begin{array}{ll} 4x - 3y + 8z = 12 & 4x - 3y + 8z = 12 \\ 2x - \frac{3}{2}y + 4z = 11 & 4x - 3y + 8z = 22 \quad \textbf{Remove fractions} \\ x \qquad - 5z = -10 & x \qquad - 5z = -10 \end{array}$$

$$A = \begin{vmatrix} 4 & -3 & 8 \\ 4 & -3 & 8 \\ 1 & 0 & -5 \end{vmatrix} = 4(15) - 4(15) + 1(0) = 0$$

$$A_x = \begin{vmatrix} 12 & -3 & 8 \\ 22 & -3 & 8 \\ -10 & 0 & -5 \end{vmatrix} = 12(15) - 22(15) + (-10)(0) = -150$$

We do not need to calculate A_y and A_z. Since $A = 0$ and $A_x = -150 \neq 0$, there are no solutions to the system.

5.1

20.
$$\begin{aligned} &\frac{1}{2}x^3 - \frac{1}{3}x - \frac{1}{4}x^3 - \frac{1}{9}x + 1 \\ &= \frac{1}{2}x^3 - \frac{1}{4}x^3 - \frac{1}{3}x - \frac{1}{9}x + 1 \\ &= \left(\frac{1}{2} - \frac{1}{4}\right)x^3 + \left(-\frac{1}{3} - \frac{1}{9}\right)x + 1 \\ &= \frac{1}{4}x^3 - \frac{4}{9}x + 1 \end{aligned}$$

25.
$$\begin{aligned} &3x^4y^8 - 2xy^9 - 4x^4y^8 + 5xy^9 - 4x^{10} \\ &= 5xy^9 - 2xy^9 + 3x^4y^8 - 4x^4y^8 - 4x^{10} \\ &= (5 - 2)xy^9 + (3 - 4)x^4y^8 - 4x^{10} \\ &= 3xy^9 - x^4y^8 - 4x^{10} \end{aligned}$$

38.
$$\begin{aligned} &(6a^3b^3 + 3a^2b - 4ab) - (-8a^3b^3 + 5a^2b + 10ab^2) \\ &= 6a^3b^3 + 3a^2b - 4ab + 8a^3b^3 - 5a^2b - 10ab^2 \\ &= 6a^3b^3 + 8a^3b^3 + 3a^2b - 5a^2b - 10ab^2 - 4ab \\ &= (6 + 8)a^3b^3 + (3 - 5)a^2b - 10ab^2 - 4ab \\ &= 14a^3b^3 - 2a^2b - 10ab^2 - 4ab \end{aligned}$$

41. $\left(\frac{2}{3}z^2 - \frac{1}{5}z + 2\right) + \left(\frac{1}{2}z^2 + z - \frac{3}{4}\right)$

$$= \frac{2}{3}z^2 + \frac{1}{2}z^2 - \frac{1}{5}z + z + 2 - \frac{3}{4}$$
$$= \left(\frac{2}{3} + \frac{1}{2}\right)z^2 + \left(-\frac{1}{5} + 1\right)z + \left(2 - \frac{3}{4}\right)$$
$$= \frac{7}{6}z^2 + \frac{4}{5}z + \frac{5}{4}$$

43. $(7.3x^2y^2 - 8.7xy + 14.2) + (-6.7x^2y^2 + 2.3xy + 5.9)$

$$= 7.3x^2y^2 - 6.7x^2y^2 - 8.7xy + 2.3xy + 14.2 + 5.9$$
$$= 0.6x^2y^2 - 6.4xy + 20.1$$

44. $(6x^3y^3 + 5x^2y^2) + (11x^2y^2 - 4xy) - (2x^3y^3 - 2x^2y^2 + 7xy)$

$$= 6x^3y^3 + 5x^2y^2 + 11x^2y^2 - 4xy - 2x^3y^3 + 2x^2y^2 - 7xy$$
$$= 6x^3y^3 - 2x^3y^3 + 5x^2y^2 + 11x^2y^2 + 2x^2y^2 - 4xy - 7xy$$
$$= (6 - 2)x^3y^3 + (5 + 11 + 2)x^2y^2 + (-4 - 7)xy$$
$$= 4x^3y^3 + 18x^2y^2 - 11xy$$

46. $5a^2 - 4b(3a - [2b - b(a - 1) - (ab + b)])$

$$= 5a^2 - 4b(3a - [2b - ba + b - ab - b])$$
$$= 5a^2 - 4b(3a - [2b - 2ab])$$
$$= 5a^2 - 4b(3a - 2b + 2ab)$$
$$= 5a^2 - 12ab + 8b^2 - 8ab^2$$

51. Substitute 5 for u, -2 for v, and 3 for w in $3uvw - 4u^2v$.

$$3(5)(-2)(3) - 4(5)^2(-2) = (15)(-6) - 4(25)(-2)$$
$$= -90 - 4(-50)$$
$$= -90 + 200 = 110$$

53. The total cost of manufacturing and wholesale is $M(u) + W(u)$.

$$M(u) + W(u) = (3u^2 - 4u + 4) + (u^2 - u + 3)$$
$$= 3u^2 - 4u + 4 + u^2 - u + 3$$
$$= 3u^2 + u^2 - 4u - u + 4 + 3$$
$$= (3 + 1)u^2 + (-4 - 1)u + (4 + 3)$$
$$= 4u^2 - 5u + 7$$

56. $P(u) = I(u) - (M(u) + W(u))$

$$= (5u^2 - 2u) - (4u^2 - 5u + 7)$$
$$= 5u^2 - 2u - 4u^2 + 5u - 7$$
$$= u^2 + 3u - 7$$

5.2

28. First we multiply $(a - 5b)$ and $(2a + b)$, and then multiply the result by $2a - 3b$.

$$\begin{array}{rrr} a & - 5b & \\ \underline{2a} & \underline{+ \ b} & \\ \underline{2a^2} & \underline{- 10ab} & \\ & ab & - 5b^2 \\ \hline 2a^2 & - 9ab & - 5b^2 \end{array} \qquad \begin{array}{rrrr} 2a^2 & - 9ab & - 5b^2 & \\ \underline{2a} & \underline{- 3b} & & \\ \underline{4a^3} & \underline{- 18a^2b} & \underline{- 10ab^2} & \\ & -6a^2b & + 27ab^2 & + 15b^3 \\ \hline 4a^3 & - 24a^2b & + 17ab^2 & + 15b^3 \end{array}$$

30. $(a - b)[3a^2 - (a + b)(a - 2b)]$

$$= (a - b)[3a^2 - (a^2 - ab - 2b^2)]$$
$$= (a - b)[3a^2 - a^2 + ab + 2b^2]$$
$$= (a - b)(2a^2 + ab + 2b^2)$$
$$= 2a^3 + a^2b + 2ab^2 - 2a^2b - ab^2 - 2b^3$$
$$= 2a^3 + a^2b - 2a^2b + 2ab^2 - ab^2 - 2b^3$$
$$= 2a^3 - a^2b + ab^2 - 2b^3$$

34. $F(t) = N(t)S(t)$

$$= (t^2 - 2t - 5)(2t^2 + t - 6)$$
$$= 2t^4 - 3t^3 - 18t^2 + 7t + 30$$

$$\begin{array}{rrrrr} & & t^2 & - 2t & - 5 \\ & & \underline{2t^2} & \underline{+ \ t} & \underline{- 6} \\ \underline{2t^4} & \underline{- 4t^3} & \underline{- 10t^2} & & \\ & t^3 & - 2t^2 & - 5t & \\ & & -6t^2 & + 12t & + 30 \\ \hline 2t^4 & - 3t^3 & - 18t^2 & + 7t & + 30 \end{array}$$

48. $(0.5x + 0.1y)^2$

$$= (0.5x)^2 + 2(0.5x)(0.1y) + (0.1y)^2$$
$$= 0.25x^2 + 0.1xy + 0.01y^2$$

56. $[a - (b + c)][a + (b + c)]$

$$= a^2 - (b + c)^2$$
$$= a^2 - (b^2 + 2bc + c^2)$$
$$= a^2 - b^2 - 2bc - c^2$$

62. $S = A(1 + i)^2 = 1000(1 + 0.12)^2$

$$= 1000(1.12)^2$$
$$= 1000(1.2544)$$
$$= \$1254.40$$

$$S = A + 2Ai + Ai^2$$
$$= 1000 + 2(1000)(0.12) + (1000)(0.12)^2$$
$$= 1000 + (2000)(0.12) + (1000)(0.0144)$$
$$= 1000 + 240 + 14.4 = \$1254.40$$

5.3

13. $64xy^5z^2 - 128x^2y^3z^3 - 144xy^2z^4$

$$= (16xy^2z^2)4y^3 - (16xy^2z^2)8xyz - (16xy^2z^2)9z^2$$
$$= 16xy^2z^2(4y^3 - 8xyz - 9z^2)$$

14. Notice that $a + b$ is a common factor of the polynomial.

$$3a(a + b) - 2b(a + b) = (3a - 2b)(a + b)$$

17. $5x^2 + 2xy + 10x + 4y = (5x^2 + 2xy) + (10x + 4y)$

$$= x(5x + 2y) + 2(5x + 2y)$$
$$= (x + 2)(5x + 2y)$$

20. $2a^2b + 6ab - 3a - 9$

$$= 2ab(a + 3) - 3(a + 3)$$
$$= (2ab - 3)(a + 3) \quad \textbf{Factor out } (a + 3)$$

24. $7x^3y^3 + 21x^2y^2 - 10x^3y^2 - 30x^2y$

$= x^2y(7xy^2 + 21y - 10xy - 30)$

$= x^2y[(7xy^2 + 21y) - (10xy + 30)]$

$= x^2y[7y(xy + 3) - 10(xy + 3)]$

$= x^2y(7y - 10)(xy + 3)$

25. $E = 25c^2e^3 - 10c^3e^2$

$= 5c^2e^2(5e - 2c)$ **Factor out $5c^2e^2$, the GCF**

5.4

24. $-4w^2 - 34w - 70 = -2(2w^2 + 17w + 35)$
Factor $2w^2 + 17w + 35$ with $a = 2$, $b = 17$, and $c = 35$.

Factors of a	*Factors of c*
2, 1	35, 1 and 1, 35
	7, 5 and 5, 7

$-2(2w^2 + 17w + 35)$

$= -2(2w + \_\_)(w + \_\_)$

$\stackrel{?}{=} -2(2w + 5)(w + 7)$ **Does not work**

$\stackrel{?}{=} -2(2w + 7)(w + 5)$ **This works**

$-2(2w^2 + 17w + 35) = -2(2w + 7)(w + 5)$

31. $6x^2 + 23xy + 20y^2$

Factors of a	*Factors of c*
1, 6	20, 1 and 1, 20
2, 3	10, 2 and 2, 10
	5, 4 and 4, 5

$6x^2 + 23xy + 20y^2$

$= (\_\_x + \_\_y)(\_\_x + \_\_y)$

$\stackrel{?}{=} (x + 10y)(6x + 2y)$ **Does not work**

$\stackrel{?}{=} (2x + 5y)(3x + 4y)$ **This works**

$6x^2 + 23xy + 20y^2 = (2x + 5y)(3x + 4y)$

37. $(x + 2)^2 - 3(x + 2) + 2$
Substitute u for $x + 2$ and factor.

$$u^2 - 3u + 2 = (u - 2)(u - 1)$$

Then substitute $x + 2$ for u.

$$(x + 2)^2 - 3(x + 2) + 2 = (x + 2 - 2)(x + 2 - 1) = x(x + 1)$$

40. $u^6 - 7u^3 + 12$
Substitute x for u^3 and factor.

$$x^2 - 7x + 12 = (x - 3)(x - 4)$$

Then substitute u^3 for x.

$$u^6 - 7u^3 + 12 = (u^3 - 3)(u^3 - 4)$$

45. $h = -16t^2 + 656t - 640$

First factor the right side.

$h = -16(t^2 - 41t + 40)$ **Factor out −16**

$= -16(t - 40)(t - 1)$ **Factor trinomial**

(a) When $t = 0.5$,

$h = -16(0.5 - 40)(0.5 - 1)$

$= -16(-39.5)(-0.5) = -316$

After 0.5 sec, the missile is 316 ft below the surface of the water.

(b) When $t = 1$,

$h = -16(1 - 40)(1 - 1)$

$= -16(-39)(0) = 0.$

After 1 sec, the missile is at the water level.

(c) When $t = 20$,

$h = -16(20 - 40)(20 - 1)$

$= -16(-20)(19) = 6080.$

After 20 sec, the missile is 6080 ft high.

5.5

14. $(x - y)^2 - 9 = u^2 - 9$ **Let $u = x - y$**

$= u^2 - 3^2$ **$u = a$, $3 = b$, in $a^2 - b^2 = (a + b)(a - b)$**

$= (u + 3)(u - 3)$

$= [(x - y) + 3][(x - y) - 3]$ **$u = x - y$**

$= (x - y + 3)(x - y - 3)$

23. $7u^3 + 56v^3 = 7(u^3 + 8v^3)$

$= 7(u^3 + (2v)^3)$

Substitute u for a and $2v$ for b in $(a^3 + b^3) = (a + b)(a^2 - ab + b^2)$.

$$7(u^3 + (2v)^3) = 7(u + 2v)(u^2 - 2uv + 4v^2)$$

37. $(x + y)^2 + 2(x + y) + 1$
First substitute u for $(x + y)$ and factor.

$$u^2 + 2u + 1 = (u + 1)^2$$

Then substitute $(x + y)$ for u.

$$(x + y)^2 + 2(x + y) + 1 = (x + y + 1)^2$$

39. $3uv + u - 3v^2 - v = (3uv + u) - (3v^2 + v)$

$= u(3v + 1) - v(3v + 1)$

$= (u - v)(3v + 1)$

44. $-18u^2 - 69uv - 60v^2$

$= -3(6u^2 + 23uv + 20v^2)$

$= -3(2u + 5v)(3u + 4v)$

Always factor out the GCF first.

48. $(u + v)^2 + 3(u + v)(x + y) + 2(x + y)^2$

$= w^2 + 3wz + 2z^2$ **Let $w = u + v$ and $z = x + y$**

$= (w + z)(w + 2z)$

$= [(u + v) + (x + y)][(u + v) + 2(x + y)]$ **$w = u + v$ and $z = x + y$**

$= (u + v + x + y)(u + v + 2x + 2y)$

5.6

16. $3u^2 = 5u$ **Do *not* divide by u**

$$3u^2 - 5u = 0$$
$$u(3u - 5) = 0$$
$$u = 0 \quad \text{or} \quad 3u - 5 = 0$$
$$3u = 5$$
$$u = \frac{5}{3}$$

26. $y^3 - 2y^2 + y = 0$

$$y(y^2 - 2y + 1) = 0$$
$$y(y - 1)(y - 1) = 0$$
$$y = 0 \quad \text{or} \quad y - 1 = 0 \quad \text{or} \quad y - 1 = 0$$
$$y = 1 \qquad\qquad y = 1$$

34. Let x = length of a side,
x^2 = area of square,
$4x$ = perimeter of square.
The equation to solve is

$$x^2 = 12 + 4x.$$
$$x^2 - 4x - 12 = 0$$
$$(x + 2)(x - 6) = 0$$
$$x = -2, 6 \quad \textbf{Zero-product rule}$$

We need to discard -2 since the length of a side cannot be negative. Thus, the side is 6 in in length.

37. Let n = the first even integer,
$n + 2$ = the next even integer.
The equation to solve is

$$n^2 + (n + 2)^2 = 100.$$
$$n^2 + n^2 + 4n + 4 = 100$$
$$2n^2 + 4n - 96 = 0$$
$$n^2 + 2n - 48 = 0 \quad \textbf{Divide out 2}$$
$$(n + 8)(n - 6) = 0$$
$$n = -8, 6 \quad \textbf{Zero-product rule}$$

Since the integers were to be positive, we discard -8 and obtain 6 and 8.

40. Let w = width of the box,
$w + 7$ = length of the box.

$$V = 1728 = w(w + 7)(12)$$
$$144 = w(w + 7) \quad \textbf{Divide by 12}$$
$$144 = w^2 + 7w$$
$$0 = w^2 + 7w - 144$$
$$0 = (w + 16)(w - 9)$$
$$w + 16 = 0 \quad \text{or} \quad w - 9 = 0$$
$$w = -16 \qquad\qquad w = 9$$

The width is 9 cm and the length is $9 + 7 = 16$ cm.

41. Let x = height of triangle,
$x - 8$ = base of triangle.
The area of a triangle is given by $A = \frac{1}{2}bh$, so we need to solve

$$42 = \frac{1}{2}(x - 8)(x).$$
$$84 = x^2 - 8x \quad \textbf{Multiply by 2}$$
$$x^2 - 8x - 84 = 0$$
$$(x + 6)(x - 14) = 0$$
$$x = -6, 14 \quad \textbf{Zero-product rule}$$

We discard -6, so the height is 14 cm and the base is 6 cm.

49. We find n when $P(n)$ is 30.

$$30 = 3n^2 - 2n - 10$$
$$0 = 3n^2 - 2n - 40$$
$$0 = (3n + 10)(n - 4)$$
$$n = -\frac{10}{3}, 4$$

Since the number of suits sold cannot be negative (nor a fraction), 4 suits were sold.

6.1

31. $$\frac{xz + xw + yz + yw}{z^2 + zw} = \frac{x(z + w) + y(z + w)}{z(z + w)}$$
$$= \frac{(x + y)\cancel{(z + w)}}{z\cancel{(z + w)}} = \frac{x + y}{z}$$

34. $$\frac{9a^2 - 15a + 25}{27a^3 + 125} = \frac{9a^2 - 15a + 25}{(3a)^3 + 5^3}$$
$$= \frac{\cancel{9a^2 - 15a + 25}}{(3a + 5)\cancel{(9a^2 - 15a + 25)}}$$
$$= \frac{1}{3a + 5}$$

44. We evaluate $P = \dfrac{A}{1 + 0.07\,t}$ when $A = 5000$ and $t = 6$. The steps needed to evaluate

$$\frac{5000}{1 + 0.07(6)}$$

on a calculator follow.

6 [×] 0.07 [+] 1 [=] [1/x] [×] [5000] [=] → 3521.1268

To the nearest cent, $P = \$3521.13$.

6.2

10. $$\frac{4 - a^2}{c^2 + 2cd + d^2} \cdot \frac{c^2 - d^2}{2 - a - a^2} \cdot \frac{c + d - ca - da}{2a - a^2}$$
$$= \frac{(2 - a)(2 + a)(c - d)(c + d)}{(c + d)(c + d)(2 + a)(1 - a)} \cdot \frac{1(c + d) - a(c + d)}{a(2 - a)}$$

$$= \frac{\cancel{(2-a)}\cancel{(2+a)}(c-d)\cancel{(c+d)}\cancel{(1-a)}\cancel{(c+d)}}{\cancel{(c+d)}\cancel{(c+d)}\cancel{(2+a)}\cancel{(1-a)}a\cancel{(2-a)}}$$

$$= \frac{c-d}{a}$$

18. 0.27 is actually equal to $\frac{27}{100}$, so its reciprocal is $\frac{100}{27}$.

21. $$\frac{(u^2v^3)^2}{3uv} \div \frac{(uv^2)^3}{12u^3v}$$

$$= \frac{u^4v^6}{3uv} \cdot \frac{12u^3v}{u^3v^6} = \frac{\overset{u^3}{\cancel{u^4}}\cancel{v^6} \cdot \overset{4}{\cancel{12}}\cancel{u^3}\cancel{v}}{\cancel{3}\cancel{uv}\cancel{u^3}\cancel{v^6}} = 4u^3$$

28. $$\frac{3s^2-5st-2t^2}{t^2+st-2s^2} \cdot \frac{t^2+4st-5s^2}{12s^2+7st+t^2} \div \frac{5s^2-9st-2t^2}{8s^2-2st-t^2}$$

$$= \frac{(3s+t)(s-2t)(t+5s)(t-s)}{(t+2s)(t-s)(3s+t)(4s+t)} \cdot \frac{8s^2-2st-t^2}{5s^2-9st-2t^2}$$

$$= \frac{\cancel{(3s+t)}\cancel{(s-2t)}\cancel{(t+5s)}\cancel{(t-s)}\cancel{(4s+t)}(2s-t)}{(t+2s)\cancel{(t-s)}\cancel{(3s+t)}\cancel{(4s+t)}\cancel{(5s+t)}\cancel{(s-2t)}}$$

$$= \frac{2s-t}{2s+t}$$

6.3

5. $$\frac{x}{8x-6} - \frac{5}{2(3-4x)} = \frac{x}{2(4x-3)} - \frac{5}{2(3-4x)}$$

$$= \frac{x}{2(4x-3)} - \frac{(-1)5}{(-1)(2)(3-4x)}$$

$$= \frac{x}{2(4x-3)} - \frac{-5}{2(4x-3)}$$

$$= \frac{x}{2(4x-3)} + \frac{5}{2(4x-3)}$$

With practice, you can skip from the first step to this one

$$= \frac{x+5}{2(4x-3)}$$

15. $$\frac{z+2}{z^2+4z+4} \quad \text{and} \quad \frac{z-1}{z^2+z-2}$$

$$\frac{\cancel{z+2}}{(z+2)\cancel{(z+2)}} \quad \text{and} \quad \frac{\cancel{z-1}}{\cancel{(z-1)}(z+2)}$$

$$\frac{1}{z+2} \quad \text{and} \quad \frac{1}{z+2}$$

Once the fractions have been reduced to lowest terms, the LCD is clearly $z + 2$.

17. $$\frac{x+3}{5x^4-15x^3-50x^2} \quad \text{and} \quad \frac{x-7}{10x^3-100x^2+250x}$$

$$\frac{x+3}{5x^2(x^2-3x-10)} \quad \text{and} \quad \frac{x-7}{10x(x^2-10x+25)}$$

$$\frac{x+3}{5xx(x-5)(x+2)} \quad \text{and} \quad \frac{x-7}{2 \cdot 5x(x-5)(x-5)}$$

Since the fractions cannot be reduced, we concentrate on the denominators as shown. The LCD must consist of one 2, one 5, two x's, two $(x-5)$'s, and one $(x+2)$. Thus,

$$\begin{aligned} \text{LCD} &= 2 \cdot 5 \cdot x \cdot x(x-5)(x-5)(x+2) \\ &= 10x^2(x-5)^2(x+2). \end{aligned}$$

33. $$a - b - \frac{-ab^2}{a^2+ab} = a - b - \frac{-\overset{1}{\cancel{a}}b^2}{\cancel{a}(a+b)}$$

$$= \frac{(a-b)}{1} + \frac{b^2}{a+b}$$

The LCD = a + b

$$= \frac{(a-b)(a+b)}{(a+b)} + \frac{b^2}{a+b}$$

$$= \frac{a^2-b^2}{a+b} + \frac{b^2}{a+b}$$

$$= \frac{a^2-b^2+b^2}{a+b}$$

$$= \frac{a^2}{a+b}$$

35. $$\frac{2a-1}{a^2-4a+4} + \frac{2a+3}{4-a^2}$$

$$= \frac{2a-1}{(a-2)(a-2)} + \frac{2a+3}{(2-a)(2+a)}$$

$$= \frac{2a-1}{(a-2)(a-2)} + \frac{(-1)(2a+3)}{(a-2)(a+2)}$$

The LCD = $(a-2)^2(a+2)$

$$= \frac{(2a-1)(a+2)}{(a-2)^2(a+2)} - \frac{(2a+3)(a-2)}{(a-2)^2(a+2)}$$

$$= \frac{(2a^2+3a-2)-(2a^2-a-6)}{(a-2)^2(a+2)}$$

Use parentheses

$$= \frac{2a^2+3a-2-2a^2+a+6}{(a-2)^2(a+2)}$$

$$= \frac{4a+4}{(a-2)^2(a+2)} = \frac{4(a+1)}{(a-2)^2(a+2)}$$

44. $$\frac{x+y}{x^2+2xy+y^2} - \frac{x}{x^2-xy-2y^2} + \frac{y}{x^2-4xy-5y^2}$$

$$= \frac{\cancel{(x+y)}}{\cancel{(x+y)}(x+y)} - \frac{x}{(x+y)(x-2y)}$$

$$+ \frac{y}{(x+y)(x-5y)}$$

$$= \frac{(x-2y)(x-5y)}{(x+y)(x-2y)(x-5y)} - \frac{x(x-5y)}{(x+y)(x-2y)(x-5y)}$$

$$+ \frac{y(x-2y)}{(x+y)(x-5y)(x-2y)}$$

$$= \frac{(x^2-7xy+10y^2)-(x^2-5xy)+(xy-2y^2)}{(x+y)(x-2y)(x-5y)}$$

$$= \frac{x^2-7xy+10y^2-x^2+5xy+xy-2y^2}{(x+y)(x-2y)(x-5y)}$$

$$= \frac{8y^2-xy}{(x+y)(x-2y)(x-5y)} = \frac{y(8y-x)}{(x+y)(x-2y)(x-5y)}$$

6.4

7. $\frac{x}{x+4} = \frac{4}{x-4} + \frac{x^2+16}{x^2-16}$

$\frac{x}{x+4} = \frac{4}{x-4} + \frac{x^2+16}{(x+4)(x-4)}$ **The LCD = (x − 4)(x + 4)**

$$(x-4)\cancel{(x+4)}\left[\frac{x}{\cancel{x+4}}\right]$$

$$= (x-4)(x+4)\left[\frac{4}{x-4} + \frac{x^2+16}{(x-4)(x+4)}\right]$$

$$x(x-4) = \cancel{(x-4)}(x+4)\left[\frac{4}{\cancel{x-4}}\right]$$

$$+ \cancel{(x-4)(x+4)}\left[\frac{x^2+16}{\cancel{(x-4)(x+4)}}\right]$$

$$x^2 - 4x = 4x + 16 + x^2 + 16$$

$$-4x = 4x + 32$$

$$-8x = 32$$

$$x = -4$$

But since -4 gives zero in the denominator of $\frac{x}{x+4}$, it cannot be a solution. Thus, there are no solutions.

14. $\frac{-1}{a^2-3a} = \frac{1}{a} + \frac{a}{a-3}$

$$\frac{-1}{a(a-3)} = \frac{1}{a} + \frac{a}{a-3}$$

$$a(a-3)\left[\frac{-1}{a(a-3)}\right] = a(a-3)\left[\frac{1}{a} + \frac{a}{a-3}\right]$$

$$-1 = (a-3) + a^2$$

$$-1 = a - 3 + a^2$$

$$0 = a^2 + a - 2$$

$$0 = (a+2)(a-1)$$

$a = -2, 1$ **Zero-product rule**

Both -2 and 1 will check.

20. Let x = Lorraine's age,
$\frac{3}{4}x$ = Harriet's age.
The equation to solve is

$$x - \frac{3}{4}x = 10.$$

$4x - 3x = 40$ **Multiply through by 4**

$$x = 40$$

Lorraine is 40 and Harriet is 30.

23. Let x = the score to be made on the fourth test. To average the four scores, add them and divide by 4.

$$\frac{75 + 80 + 82 + x}{4} = 83$$

$$\cancel{4}\left[\frac{75 + 80 + 82 + x}{\cancel{4}}\right] = 4[83]$$

$$237 + x = 332$$

$$x = 95$$

She needs a 95 to average 83 on the four tests.

30. $a^{-1} = b + c^{-1}$

$$\frac{1}{a} = b + \frac{1}{c}$$

$$\cancel{a}c\left[\frac{1}{\cancel{a}}\right] = ac[b] + a\cancel{c}\left[\frac{1}{\cancel{c}}\right]$$

$$c = abc + a$$

$$c - abc = a$$

$$c(1 - ab) = a$$

$$c = \frac{a}{1 - ab}$$

6.5

2. Let x = number of votes for loser. The equation to solve is

$$\frac{5}{3} = \frac{820}{x}.$$

$5x = 3(820)$ **Cross-product equation**

$$x = \frac{3(820)}{5} = 3(164) = 492$$

The loser received 492 votes.

7. Let x = the length of one piece,
$65 - x$ = the length of the second piece.

$$[x + (65 - x) = 65.]$$

With the ratio of the two pieces 8:5, we solve

$$\frac{x}{65 - x} = \frac{8}{5}.$$

$$5x = 8(65 - x)$$

$$5x = 520 - 8x$$

$$13x = 520$$

$$x = 40$$

$$65 - x = 65 - 40 = 25$$

The two pieces have length 40 ft and 25 ft.

13. $a = \frac{cu}{vw}$ is the equation describing the variation. Substitute 4 for a, 3 for u, 3 for v, and 4 for w to find the constant of variation, c.

$$4 = \frac{c(3)}{(3)(4)}$$

$$4 = \frac{3c}{12}$$

$$48 = 3c$$

$$16 = c$$

The equation of variation is $a = \frac{16u}{vw}$.

17. $A = \frac{c}{d^2}$ is the equation describing the variation.
Substitute 64 for A and 9 for d to find the constant of variation, c.

$$64 = \frac{c}{9^2}$$
$$(64)(81) = c$$
$$5184 = c$$

Next, substitute 24 for d in $A = \frac{5184}{d^2}$ to find the desired attraction.

$$A = \frac{5184}{(24)^2} = \frac{5184}{576} = 9$$

The attraction is 9 lb.

20. $w = \frac{c}{d^2}$
Substitute 150 for w and 4000 for d to find c.

$$150 = \frac{c}{(4000)^2}, \quad \text{so} \quad (4000)^2(150) = c$$

Then substitute 5000 for d (1000 miles above the surface is 5000 miles from the center) and substitute $150(4000)^2$ for c.

$$w = \frac{(150)(4000)^2}{(5000)^2} = \frac{(150)(4)^2(1000)^2}{5^2(1000)^2} = 96$$

The object weighs 96 lb 1000 mi above the surface.

6.6

6. Let x = her rate going to Denver,
$x + 40$ = her rate returning to Phoenix.
Since the distance going and the distance returning are both 800, using $d = rt$, or $t = \frac{d}{r}$, we have

$$\frac{800}{x} = \text{time going to Denver,}$$
$$\frac{800}{x+40} = \text{time returning to Phoenix.}$$

Since the total time of the trip is 9 hours, we solve $\frac{800}{x} + \frac{800}{x+40} = 9$.

$$x(x+40)\left[\frac{800}{x}\right] + x(x+40)\left[\frac{800}{(x+40)}\right] = 9x(x+40)$$
$$800x + 32{,}000 + 800x = 9x^2 + 360x$$
$$9x^2 - 1240x - 32{,}000 = 0$$
$$(x - 160)(9x + 200) = 0$$

$x = 160$ or $x = -\frac{200}{9}$, which is meaningless

Her rate going was 160 mph and her rate returning was 200 mph ($x + 40 = 160 + 40 = 200$).

8. Let x = number of hours they travel upstream,
$4 - x$ = number of hours they return downstream.

Since they go upstream the same distance that they return downstream, we equate distance using $d = rt$.
$6x$ = distance traveled upstream
$12(4 - x)$ = distance traveled downstream

$$6x = 12(4 - x)$$
$$6x = 48 - 12x$$
$$18x = 48$$
$$x = \frac{48}{18} = \frac{8}{3} = 2\frac{2}{3} \text{ hr}$$

Thus, they must turn back in $2\frac{2}{3}$ hr, or in 2 hr 40 min, which is at 11:40 A.M. At that time they will have traveled $\left(2\frac{2}{3}\right)(6) = \left(\frac{8}{3}\right)(6) = 16$ mi upstream.

15. Let t = time to fill tank when both pipes are open,
$\frac{1}{4}$ = amount filled by inlet in 1 hr,
$\frac{1}{10}$ = amount drained by outlet in 1 hr,
$\frac{1}{t}$ = amount filled in 1 hr when both are open.
We need to solve

$$\frac{1}{t} = \frac{1}{4} - \frac{1}{10}.$$
$$(20t)\frac{1}{t} = 20t\left(\frac{1}{4} - \frac{1}{10}\right)$$
$$20 = 5t - 2t$$
$$20 = 3t$$
$$\frac{20}{3} = t$$

It takes $\frac{20}{3}$ hr or $6\frac{2}{3}$ hr to fill the tank.

6.7

12.
$$\frac{\frac{1}{x+y} - \frac{1}{x-y}}{\frac{-2}{x-y}} = \frac{\left[\frac{1}{x+y} - \frac{1}{x-y}\right](x+y)(x-y)}{\left[\frac{-2}{x-y}\right](x+y)(x-y)}$$
$$= \frac{(x-y) - (x+y)}{-2(x+y)} = \frac{x - y - x - y}{-2(x+y)}$$
$$= \frac{-2y}{-2(x+y)} = \frac{y}{x+y}$$

13.
$$\frac{a+1}{a-1} = \frac{\frac{x+y}{y} + 1}{\frac{x+y}{y} - 1} = \frac{\left[\frac{x+y}{y} + 1\right]y}{\left[\frac{x+y}{y} - 1\right]y}$$
$$= \frac{x + y + y}{x + y - y} = \frac{x + 2y}{x}$$

17. $$a - \frac{a}{1-\frac{a}{1-a}} = a - \frac{a}{\frac{1-a}{1-a} - \frac{a}{1-a}}$$
$$= a - \frac{a}{\frac{1-2a}{1-a}} = a - \left[\frac{a}{1}\cdot\frac{1-a}{1-2a}\right]$$
$$= a - \frac{a(1-a)}{1-2a} = \frac{a(1-2a)}{1-2a} - \frac{a-a^2}{1-2a}$$
$$= \frac{a-2a^2-a+a^2}{1-2a} = \frac{-a^2}{1-2a}$$

23. $$\frac{2x^{-1}+2y^{-1}}{(xy)^{-1}} = \frac{\frac{2}{x}+\frac{2}{y}}{\frac{1}{xy}} = \frac{\left[\frac{2}{x}+\frac{2}{y}\right]xy}{\left[\frac{1}{xy}\right]xy}$$
$$= \frac{2y+2x}{1} = 2(x+y)$$

25. Evaluate

$$f = \frac{d_1 d_2}{d_2 + d_1}$$

when $d_1 = 8.0$ and $d_2 = 14.0$.

$$f = \frac{(8.0)(14.0)}{14.0 + 8.0} = \frac{112}{22} \approx 5.1$$

The focal length is about 5.1 cm.

27. We need to simplify

$$\frac{2D}{\frac{D}{4}+\frac{D}{2}}.$$

First add the fractions in the denominator; they have LCD 4.

$$\frac{2D}{\frac{D}{4}+\frac{2D}{4}} = \frac{2D}{\frac{3D}{4}}$$
$$= \frac{4}{3D}(2D) \quad \text{The } D\text{'s divide out}$$
$$= \frac{8}{3} \approx 2.7$$

Marla's average rate of speed was about 2.7 mph.

6.8

3. $$\frac{-27a^5 + 9a^3 - 81a^2}{-3a}$$
$$= \frac{-27a^5}{-3a} + \frac{9a^3}{-3a} + \frac{-81a^2}{-3a}$$
$$= 9a^4 - 3a^2 + 27a$$

16.
$$\begin{array}{r} x^2 + 4x + 5 \\ x^2 - x + 1\overline{)\,x^4 + 3x^3 + 2x^2 - x + 5} \\ \underline{x^4 - x^3 + x^2} \\ 4x^3 + x^2 - x \\ \underline{4x^3 - 4x^2 + 4x} \\ 5x^2 - 5x + 5 \\ \underline{5x^2 - 5x + 5} \\ 0 \end{array}$$

Thus, the quotient is $x^2 + 4x + 5$.

20.
$$\begin{array}{r} 3x^2 + x - 2 \\ 2x^2 + 3\overline{)\,6x^4 + 2x^3 + 5x^2 - x + 1} \\ \underline{6x^4 \qquad + 9x^2} \\ 2x^3 - 4x^2 \\ \underline{2x^3 \qquad + 3x} \\ -4x^2 - 4x \\ \underline{-4x^2 \qquad - 6} \\ -4x + 7 \end{array}$$

The quotient is $3x^2 + x - 2$ with remainder $-4x + 7$, or $3x^2 + x - 2 + \frac{-4x+7}{2x^2+3}$.

21. Divide $y^3 - y^2 + 2y + p$ by $y - 1$.

$$\begin{array}{r} y^2 + 2 \\ y - 1\overline{)\,y^3 - y^2 + 2y + p} \\ \underline{y^3 - y^2} \\ 0 + 2y + p \\ \underline{2y - 2} \\ p + 2 \end{array}$$

For $p + 2$ to be 0, we solve

$$p + 2 = 0$$
$$p = -2.$$

6.9

7. First write the dividend in descending powers of x.

$$(x^4 + x^3 - x - 3) \div (x + 1)$$

Write coefficients (don't forget $0x^2$) and begin.

$$\begin{array}{r|r} -1 & 1 + 1 + 0 - 1 - 3 \\ & \underline{-1 + 0 + 0 + 1} \\ & 1 + 0 + 0 - 1 - 2 \\ & (x^3 + 0x^2 + 0x - 1) \end{array}$$

We obtain $x^3 - 1$ with a remainder of -2.

13.
$$\begin{array}{r|r} -1 & 3 + 0 + 0 + 1 + 0 - 5 \\ & \underline{-3 + 3 - 3 + 2 - 2} \\ & 3 - 3 + 3 - 2 + 2 - 7 \end{array}$$

We obtain $3x^4 - 3x^3 + 3x^2 - 2x + 2$ with a remainder of -7.

7.1

27. $\sqrt[4]{(-2x)^4} = |-2x| = |-2||x| = 2|x|$

30. $-\sqrt[5]{(-2x)^5} = -(-2x) = 2x$

31. $\sqrt{(x+7)^2} = |x+7|$
Remember that even-indexed radicals require absolute value.

32. $\sqrt[3]{(x+7)^3} = x+7$
Odd-indexed radicals don't require absolute value.

49. $\sqrt{x^2-14x+49} = \sqrt{(x-7)^2}$ **Factor the radicand**
$= x - 7$ **We assume $x - 7$ is non-negative so $|x - 7| = x - 7$**

59. To evaluate $\frac{\sqrt{1{,}250{,}000{,}000}}{0.00623}$, first write 1,250,000,000 in scientific notation as 1.25×10^9. The steps on your calculator are:

1.25 [EE] 9 [√] [÷] 0.00623 [=] → [5.675 06]

The display gives 5.675×10^6 which to three significant digits is $5.68 \times 10^6 = 5{,}680{,}000$.

66. Find d when $h = 1800$ using $d = 1.4\sqrt{h}$.

$d = 1.4\sqrt{1800} = 59.39697$ **Using a calculator**

The viewing distance is about 59.4 mi, which means that an object 50 mi away could be seen.

7.2

20.
$$\sqrt[3]{-27x^4y^5z^8} = \sqrt[3]{(-3)^3x^3\cdot x\cdot y^3\cdot y^2\cdot z^6\cdot z^2}$$
$$= \sqrt[3]{(-3)^3x^3y^3z^6xy^2z^2}$$
$$= -3xyz^2\sqrt[3]{xy^2z^2}$$

29.
$$\frac{3x}{y}\sqrt[5]{\frac{x^5y^6}{243}} = \frac{3x}{y}\sqrt[5]{\frac{x^5y^5y}{3^5}}$$
$$= \frac{3x}{y}\cdot\frac{xy}{3}\sqrt[5]{y}$$
$$= \frac{\cancel{3}x^2\cancel{y}}{\cancel{3y}}\sqrt[5]{y} = x^2\sqrt[5]{y}$$

34. If $\sqrt[3]{5x^2}$ is in a denominator, we need to multiply numerator and denominator by $\sqrt[3]{5^2x}$, for then
$$\sqrt[3]{5x^2}\,\sqrt[3]{5^2x} = \sqrt[3]{5^3x^3}$$
$$= 5x,$$
and the radical in the denominator will be removed.

43.
$$\frac{7}{\sqrt[4]{y}} = \frac{7\sqrt[4]{y^3}}{\sqrt[4]{y}\sqrt[4]{y^3}}$$
$$= \frac{7\sqrt[4]{y^3}}{\sqrt[4]{y\cdot y^3}} = \frac{7\sqrt[4]{y^3}}{y}$$

48.
$$\frac{\sqrt[3]{x^2z^4}}{\sqrt[3]{2y^2}} = \frac{\sqrt[3]{x^2z^3z}}{\sqrt[3]{2y^2}}$$
$$= \frac{z\sqrt[3]{x^2z}}{\sqrt[3]{2y^2}}$$
$$= \frac{z\sqrt[3]{x^2z}\,\sqrt[3]{2^2y}}{\sqrt[3]{2y^2}\,\sqrt[3]{2^2y}}$$
$$= \frac{z\sqrt[3]{4x^2yz}}{\sqrt[3]{2^3y^3}} = \frac{z\sqrt[3]{4x^2yz}}{2y}$$

52.
$$\frac{\sqrt[3]{3}}{\sqrt[3]{x}} = \frac{\sqrt[3]{3}\,\sqrt[3]{3^2}}{\sqrt[3]{x}\,\sqrt[3]{3^2}}$$
$$= \frac{\sqrt[3]{3^3}}{\sqrt[3]{3^2x}} = \frac{3}{\sqrt[3]{9x}}$$

7.3

5.
$$\sqrt{5x^2y}\sqrt{35xy^2}$$
$$= \sqrt{5x^2y\cdot 5\cdot 7\cdot xy^2}$$
$$= \sqrt{5^2x^2y^2\cdot 7xy}$$
$$= 5xy\sqrt{7xy}$$

8.
$$\sqrt{5xy}\,\sqrt{15x^2y}\,\sqrt{3xy^2} = \sqrt{5xy\cdot 15x^2y\cdot 3xy^2}$$
$$= \sqrt{5^2\cdot 3^2\cdot x^4\cdot y^4}$$
$$= 5\cdot 3\cdot x^2\cdot y^2 = 15x^2y^2$$

18.
$$\sqrt[4]{9x^2y^5}\,\sqrt[4]{45x^3y^2} = \sqrt[4]{9x^2y^5\cdot 45x^3y^2}$$
$$= \sqrt[4]{3^4x^4y^4\cdot 5xy^3}$$
$$= 3xy\sqrt[4]{5xy^3}$$

27.
$$\frac{\sqrt{98x^3y^5}}{\sqrt{18x^2y^7}} = \sqrt{\frac{98x^3y^5}{18x^2y^7}}$$
$$= \sqrt{\frac{\cancel{2}\cdot 49\cdot \cancel{x^2}\cdot x\cdot \cancel{y^5}}{\cancel{2}\cdot 9\cdot \cancel{x^2}\cdot y^2\cdot \cancel{y^5}}}$$
$$= \frac{\sqrt{49x}}{\sqrt{9y^2}} = \frac{7\sqrt{x}}{3y}$$

46. Since the circles have radii 3.0, 4.0, and 5.0 cm, the sides of the triangle are
$$3.0 + 4.0 = 7.0 = a$$
$$4.0 + 5.0 = 9.0 = b$$
$$5.0 + 3.0 = 8.0 = c$$

We use the formula
$$r = \sqrt{\frac{(s-a)(s-b)(s-c)}{s}}$$
where $s = \frac{7.0 + 9.0 + 8.0}{2} = \frac{24.0}{2} = 12.0.$

Thus, $r = \sqrt{\frac{(12.0-7.0)(12.0-9.0)(12.0-8.0)}{12.0}}$
$$= \sqrt{\frac{(5.0)(3.0)(4.0)}{12.0}}$$
$$= \sqrt{5.0} \approx 2.236068.$$

The radius of the inscribed circle is about 2.2 cm.

7.4

9.
$$9\sqrt[3]{250} - 4\sqrt[3]{128} = 9\sqrt[3]{125\cdot 2} - 4\sqrt[3]{64\cdot 2}$$
$$= 9\sqrt[3]{125}\,\sqrt[3]{2} - 4\sqrt[3]{64}\,\sqrt[3]{2}$$
$$= 9\cdot 5\sqrt[3]{2} - 4\cdot 4\sqrt[3]{2}$$
$$= 45\sqrt[3]{2} - 16\sqrt[3]{2} = 29\sqrt[3]{2}$$

14. $6\sqrt{48y^4} + 4\sqrt{12y^4} = 6\sqrt{16y^4 \cdot 3} + 4\sqrt{4y^4 \cdot 3}$
$= 6\sqrt{16y^4}\sqrt{3} + 4\sqrt{4y^4}\sqrt{3}$
$= 6(4y^2)\sqrt{3} + 4(2y^2)\sqrt{3}$
$= 24y^2\sqrt{3} + 8y^2\sqrt{3} = 32y^2\sqrt{3}$

21. $3\sqrt[5]{a^{10}b^{12}} - ab\sqrt[5]{a^5b^7}$
$= 3\sqrt[5]{a^{10}b^{10}b^2} - ab\sqrt[5]{a^5b^5b^2}$
$= 3\sqrt[5]{(a^2)^5(b^2)^5}\sqrt[5]{b^2} - ab\sqrt[5]{a^5b^5}\sqrt[5]{b^2}$
$= 3a^2b^2\sqrt[5]{b^2} - ab(ab)\sqrt[5]{b^2}$
$= 3a^2b^2\sqrt[5]{b^2} - a^2b^2\sqrt[5]{b^2} = 2a^2b^2\sqrt[5]{b^2}$

30. $\dfrac{\sqrt[3]{x^2y^2}}{xy} + \dfrac{1}{\sqrt[3]{xy}} = \dfrac{\sqrt[3]{x^2y^2}}{xy} + \dfrac{\sqrt[3]{x^2y^2}}{\sqrt[3]{xy}\,\sqrt[3]{x^2y^2}}$
$= \dfrac{\sqrt[3]{x^2y^2}}{xy} + \dfrac{\sqrt[3]{x^2y^2}}{xy} = \dfrac{2\sqrt[3]{x^2y^2}}{xy}$

40. $\sqrt[4]{x^3}(\sqrt[4]{xy^3} + \sqrt[4]{xy^4}) = \sqrt[4]{x^3}\,\sqrt[4]{xy^3} + \sqrt[4]{x^3}\,\sqrt[4]{xy^4}$
$= \sqrt[4]{x^3xy^3} + \sqrt[4]{x^3xy^4}$
$= \sqrt[4]{x^4y^3} + \sqrt[4]{x^4y^4}$
$= x\sqrt[4]{y^3} + xy$

44. $(2\sqrt{5} - \sqrt{2})(3\sqrt{5} + 4\sqrt{2}) = 2\sqrt{5} \cdot 3\sqrt{5} +$
$2\sqrt{5} \cdot 4\sqrt{2} - \sqrt{2} \cdot 3\sqrt{5} - \sqrt{2} \cdot 4\sqrt{2}$
$= 2 \cdot 3 \cdot 5 + 2 \cdot 4\sqrt{10} - 3\sqrt{10} - 4 \cdot 2$
$= 30 + 8\sqrt{10} - 3\sqrt{10} - 8$
$= 22 + 5\sqrt{10}$

54. $\dfrac{\sqrt{2}}{2\sqrt{2} + \sqrt{3}} = \dfrac{\sqrt{2}(2\sqrt{2} - \sqrt{3})}{(2\sqrt{2} + \sqrt{3})(2\sqrt{2} - \sqrt{3})}$
$= \dfrac{\sqrt{2}(2\sqrt{2} - \sqrt{3})}{4(2) - 3}$
$= \dfrac{2 \cdot 2 - \sqrt{6}}{8 - 3} = \dfrac{4 - \sqrt{6}}{5}$

58. $\dfrac{4x - 25}{2\sqrt{x} - 5} = \dfrac{(4x - 25)(2\sqrt{x} + 5)}{(2\sqrt{x} - 5)(2\sqrt{x} + 5)}$
$= \dfrac{\cancel{(4x - 25)}(2\sqrt{x} + 5)}{\cancel{4x - 25}} = 2\sqrt{x} + 5$

66. $\dfrac{\sqrt{x + h} - \sqrt{x}}{h}$
$= \dfrac{(\sqrt{x + h} - \sqrt{x})(\sqrt{x + h} + \sqrt{x})}{h(\sqrt{x + h} + \sqrt{x})}$
$= \dfrac{(x + h) - x}{h(\sqrt{x + h} + \sqrt{x})}$
$= \dfrac{\cancel{h}}{\cancel{h}(\sqrt{x + h} + \sqrt{x})} = \dfrac{1}{\sqrt{x + h} + \sqrt{x}}$

68. $\dfrac{5 + \sqrt{125}}{10} = \dfrac{5 + \sqrt{25 \cdot 5}}{10}$
$= \dfrac{5 + 5\sqrt{5}}{10}$
$= \dfrac{\cancel{5}(1 + \sqrt{5})}{\cancel{5} \cdot 2} = \dfrac{1 + \sqrt{5}}{2}$

7.5

23. $\left(\dfrac{a^3}{b^{-6}}\right)^{1/3} = \dfrac{a^{(3)(1/3)}}{b^{(-6)(1/3)}} = \dfrac{(a^3)^{1/3}}{(b^{-6})^{1/3}} = \dfrac{a^{3(1/3)}}{b^{(-6)1/3}}$
$= \dfrac{a^1}{b^{-2}} = ab^2$

27. $\left(\dfrac{96a^3b^{-2}}{3a^{-2}b^8}\right)^{1/5} = \left(\dfrac{\cancel{3} \cdot 32a^{3-(-2)}}{\cancel{3}b^{8-(-2)}}\right)^{1/5}$
$= \left(\dfrac{32a^5}{b^{10}}\right)^{1/5}$
$= \dfrac{(32)^{1/5}(a^5)^{1/5}}{(b^{10})^{1/5}} = \dfrac{\sqrt[5]{32a^5}}{b^2} = \dfrac{2a}{b^2}$

34. $\sqrt[5]{\dfrac{a^{20}}{625b^{15}}} = \left(\dfrac{a^{20}}{625b^{15}}\right)^{1/5}$
$= \dfrac{(a^{20})^{1/5}}{(5^4)^{1/5}(b^{15})^{1/5}}$
$= \dfrac{a^4}{(5^4)^{1/5}b^3}$
$= \dfrac{a^4}{b^3\sqrt[5]{5^4}}$
$= \dfrac{a^4\sqrt[5]{5}}{b^3\sqrt[5]{5^4}\,\sqrt[5]{5}} = \dfrac{a^4\sqrt[5]{5}}{5b^3}$

36. $\dfrac{1}{\sqrt[3]{5^2}} = \dfrac{1}{5^{2/3}} = \dfrac{5^{1/3}}{5^{2/3}5^{1/3}} = \dfrac{5^{1/3}}{5} = \dfrac{\sqrt[3]{5}}{5}$

7.6

3. $8\sqrt{3x - 1} + 7 = 0$
Since $8\sqrt{3x - 1}$ is nonnegative for any x and 7 is positive, there is no value of x for which their sum is 0. Thus, there is no solution.

5. $\sqrt{x^2 - 5} + x - 5 = 0$
$\sqrt{x^2 - 5} = 5 - x$
$(\sqrt{x^2 - 5})^2 = (5 - x)^2$
$x^2 - 5 = 25 - 10x + x^2$
$-5 = 25 - 10x$
$-30 = -10x$
$3 = x$

The solution is 3 since it checks in the original equation.

7. $\sqrt{y + 6} + \sqrt{y + 11} = 5$
$\sqrt{y + 6} = 5 - \sqrt{y + 11}$
$(\sqrt{y + 6})^2 = (5 - \sqrt{y + 11})^2$
$y + 6 = 25 - 10\sqrt{y + 11} + y + 11$
$y + 6 = 36 - 10\sqrt{y + 11} + y$
$6 = 36 - 10\sqrt{y + 11}$
$-30 = -10\sqrt{y + 11}$
$3 = \sqrt{y + 11}$
$9 = y + 11$
$-2 = y$ **This checks**

10. $\dfrac{8}{\sqrt{a}} = 2$

$\dfrac{8\sqrt{a}}{\sqrt{a}} = 2\sqrt{a}$ Multiply by $\sqrt{a}$

$8 = 2\sqrt{a}$

$4 = \sqrt{a}$

$16 = a$ Square both sides

This does check.

15. $\sqrt{3x-2} - \sqrt{2x+5} = -1$

$\sqrt{3x-2} = \sqrt{2x+5} - 1$

$(\sqrt{3x-2})^2 = (\sqrt{2x+5} - 1)^2$

$3x - 2 = 2x + 5 - 2\sqrt{2x+5} + 1$

$x - 8 = -2\sqrt{2x+5}$

$(x-8)^2 = (-2\sqrt{2x+5})^2$

$x^2 - 16x + 64 = 4(2x+5)$

$x^2 - 16x + 64 = 8x + 20$

$x^2 - 24x + 44 = 0$

$(x-2)(x-22) = 0$

$x - 2 = 0$ or $x - 22 = 0$

$x = 2$ $\quad x = 22$

2 checks but 22 does not.

18. $\sqrt[4]{w+2} = \sqrt[4]{5w}$

$(\sqrt[4]{w+2})^4 = (\sqrt[4]{5w})^4$

$w + 2 = 5w$

$2 = 4w$

$\dfrac{1}{2} = w$ This does check

28. $a = \sqrt{1 + \dfrac{b}{x}}$ for x

$a^2 = \left(\sqrt{1 + \dfrac{b}{x}}\right)^2$

$a^2 = 1 + \dfrac{b}{x}$

$a^2 - 1 = \dfrac{b}{x}$

$x(a^2 - 1) = b$

$x = \dfrac{b}{a^2 - 1}$

30. $z^{1/3} + x = y$ for z

$z^{1/3} = y - x$

$(z^{1/3})^3 = (y-x)^3$

$z = (y-x)^3$

31. Let x = the number,

$2\sqrt{x}$ = twice the square root of the number,

$\sqrt{3x+9}$ = the square root of 9 more than 3 times the number.

$2\sqrt{x} = \sqrt{3x+9}$

$(2\sqrt{x})^2 = (\sqrt{3x+9})^2$

$4x = 3x + 9$

$x = 9$ This does check

34. $c = 100\sqrt[3]{n} + 1200$

$= 100\sqrt[3]{8} + 1200 = 100(2) + 1200$

$= 200 + 1200 = 1400$

The cost is \$1400.

43. We need to solve

$$1.5 = 2\pi\sqrt{\frac{L}{32}}.$$

$(1.5)^2 = 4\pi^2\left(\dfrac{L}{32}\right)$ Square both sides

$\dfrac{(1.5)^2}{4\pi^2} = \dfrac{L}{32}$ Divide both sides by $4\pi^2$

$\dfrac{32(1.5)^2}{4\pi^2} = L$ Multiply both sides by 32

Using a calculator and 3.14 for π we obtain $L \approx$ 1.8256319. To the nearest hundredth, the pendulum is 1.83 ft long.

7.7

4. Substitute 5 for b and $5\sqrt{3}$ for c into

$a^2 + b^2 = c^2.$

$a^2 + (5)^2 = (5\sqrt{3})^2$

$a^2 + 25 = (25)(3)$

$a^2 + 25 = 75$

$a^2 = 50$

$a = \sqrt{50}$ Use positive root only

$a = \sqrt{25 \cdot 2} = 5\sqrt{2}$

9. The diagonal of a square is the hypotenuse of the right triangle formed with the sides. If x is the length of a side,

$x^2 + x^2 = (2)^2.$

$2x^2 = 4$

$x^2 = 2$

$x = \sqrt{2}$

Thus, each side is $\sqrt{2}$ mi or about 1.4 mi.

14. By the Pythagorean theorem we have:

$a^2 + a^2 = h^2.$

$2a^2 = h^2$

$\sqrt{2a^2} = h$

$a\sqrt{2} = h$

27. The midpoint of the segment joining $(a, 3)$ and $(-2, b + 1)$ is

$$(\bar{x}, \bar{y}) = \left(\frac{a + (-2)}{2}, \frac{3 + (b + 1)}{2}\right) = \left(\frac{a - 2}{2}, \frac{b + 4}{2}\right).$$

Since $(\bar{x}, \bar{y}) = (2, 1)$, $\bar{x} = 2$ and $\bar{y} = 1$ so that

$$\frac{a - 2}{2} = 2 \quad \text{and} \quad \frac{b + 4}{2} = 1.$$

$$a - 2 = 4 \qquad b + 4 = 2$$

$$a = 6 \qquad b = -2$$

Thus, $a = 6$ and $b = -2$.

7.8

10. $\sqrt{-4}\sqrt{-9} = \sqrt{(-1)(4)}\sqrt{(-1)(9)} = \sqrt{-1}\sqrt{4}\sqrt{-1}\sqrt{9} = i(2)i(3) = 6i^2 = -6$

13. $\dfrac{\sqrt{-4}}{\sqrt{-9}} = \dfrac{\sqrt{(-1)(4)}}{\sqrt{(-1)(9)}} = \dfrac{\sqrt{-1}\sqrt{4}}{\sqrt{-1}\sqrt{9}} = \dfrac{i\sqrt{4}}{i\sqrt{9}} = \dfrac{2}{3}$

17. $(4x + 1) + 7yi = 5 - i$

$4x + 1 = 5$ and $7y = -1$

$4x = 4 \qquad y = -\dfrac{1}{7}$

$x = 1$

41. $\dfrac{8 - 7i}{5 + 4i} = \dfrac{(8 - 7i)(5 - 4i)}{(5 + 4i)(5 - 4i)} = \dfrac{40 - 67i + 28i^2}{25 - 16i^2} = \dfrac{40 - 67i - 28}{25 + 16}$ **$i^2 = -1$** $= \dfrac{12 - 67i}{41} = \dfrac{12}{41} - \dfrac{67}{41}i$

8.1

8. $(x - 1)(x + 1)(x - 2) = 0$

The zero-product rule applies to more than 2 factors so we solve

$x - 1 = 0$ or $x + 1 = 0$ or $x - 2 = 0$

$x = 1$ or $x = -1$ or $x = 2$

The three solutions are 1, −1, 2.

21.

$5x^2 - 4x = 3x^2 + 9x + 7$

$2x^2 - 4x = 9x + 7$ **Subtract $3x^2$**

$2x^2 - 13x = 7$ **Subtract $9x$**

$2x^2 - 13x - 7 = 0$ **Subtract 7**

$(2x + 1)(x - 7) = 0$

$2x + 1 = 0$ or $x - 7 = 0$

$2x = -1 \qquad x = 7$

$x = -\dfrac{1}{2}$

The solutions are $-\frac{1}{2}$ and 7.

25. $5y(y + 1) = 5y^2 + 5y$

$5y^2 + 5y = 5y^2 + 5y$ **Clear parentheses**

Since we obtain an identity, any real number is a solution.

33.

$x^3 + 3x^2 + 2x = 0$

$x(x^2 + 3x + 2) = 0$ **Factor x**

$x(x + 1)(x + 2) = 0$

$x = 0$ or $x + 1 = 0$ or $x + 2 = 0$

$x = -1 \qquad x = -2$

The solutions are 0, −1, and −2.

38. $(2x - 1)^2 + 4 = 0$

$(2x - 1)^2 = -4$

$2x - 1 = \pm\sqrt{-4}$ **Take square root of both sides**

$2x - 1 = \pm 2i$ **$\sqrt{-4} = \sqrt{(4)(-1)} = \sqrt{4}\sqrt{-1} = 2i$**

$2x = 1 \pm 2i$

$x = \dfrac{1 \pm 2i}{2} = \dfrac{1}{2} \pm i$

46. Let w = the width of the rectangle,
$w + 5$ = the length of the rectangle.
Use the formula $A = lw$ and substitute.

$(w + 5)w = 84$

$w^2 + 5w = 84$

$w^2 + 5w - 84 = 0$

$(w - 7)(w + 12) = 0$

$w - 7 = 0$ or $w + 12 = 0$

$w = 7 \qquad w = -12$

Since the width cannot be negative, we discard −12. Thus, the width is 7 cm and the length is 12 cm $(w + 5 = 7 + 5 = 12)$.

49. Since the supply must equal the demand, $S = D$, which gives us

$$p - 95 = \frac{500}{p}.$$

$p(p - 95) = 500$ **Multiply both sides by p**

$p^2 - 95p = 500$ **Distribute**

$p^2 - 95p - 500 = 0$ **General form**

$(p + 5)(p - 100) = 0$ **Factor**

$p + 5 = 0$ and $p - 100 = 0$ **Zero-product rule**

$p = -5$ $\quad p = 100$

Since -5 cannot be a price per gallon, we have 100¢/gal as the desired price.

8.2

7. $(3z - 1)^2 = \frac{4}{9}$

$$3z - 1 = \pm\sqrt{\frac{4}{9}} = \pm\frac{2}{3}$$

$$3z = 1 \pm \frac{2}{3}$$

$$3z = 1 + \frac{2}{3} \quad \text{or} \quad 3z = 1 - \frac{2}{3}$$

$$3z = \frac{5}{3} \quad \text{or} \quad 3z = \frac{1}{3}$$

$$z = \frac{5}{9} \quad \text{or} \quad z = \frac{1}{9}$$

The solutions are $\frac{5}{9}$ and $\frac{1}{9}$.

11. We must add $\left(\frac{1}{2} \cdot \frac{8}{5}\right)^2$ or $\left(\frac{8}{10}\right)^2 = \left(\frac{4}{5}\right)^2 = \frac{16}{25}$ to complete the square.

14. $2u^2 + 3u - 1 = 0$

$$u^2 + \frac{3}{2}u - \frac{1}{2} = 0 \quad \textbf{Divide through by 2}$$

$$u^2 + \frac{3}{2}u = \frac{1}{2}$$

$$u^2 + \frac{3}{2}u + \frac{9}{16} = \frac{1}{2} + \frac{9}{16} \quad \textbf{Add } \tfrac{9}{16} = \left[\tfrac{1}{2}\left(\tfrac{3}{2}\right)\right]^2 \textbf{ to both sides}$$

$$\left(u + \frac{3}{4}\right)^2 = \frac{17}{16}$$

$$u + \frac{3}{4} = \pm\sqrt{\frac{17}{16}} = \pm\frac{\sqrt{17}}{4}$$

$$u = -\frac{3}{4} \pm \frac{\sqrt{17}}{4}$$

$$u = \frac{-3 \pm \sqrt{17}}{4}$$

23. $\frac{1}{3}x^2 - \frac{4}{9}x - \frac{2}{3} = 0$

$$9\left(\frac{1}{3}x^2 - \frac{4}{9}x - \frac{2}{3}\right) = 9 \cdot 0 \quad \textbf{Multiply by 9 to clear fractions}$$

$$3x^2 - 4x - 6 = 0$$

$$x = \frac{-b \pm \sqrt{b^2 - 4ac}}{2a}$$

$$= \frac{-(-4) \pm \sqrt{(-4)^2 - 4(3)(-6)}}{2(3)}$$

$$= \frac{4 \pm \sqrt{16 + 72}}{6} = \frac{4 \pm \sqrt{88}}{6}$$

$$= \frac{4 \pm 2\sqrt{22}}{6} = \frac{\not{2}(2 \pm \sqrt{22})}{\not{2} \cdot 3} = \frac{2 \pm \sqrt{22}}{3}$$

31. $2x^2 - 3x + 4 = 0$

$a = 2,\ b = -3,\ c = 4$

$$x = \frac{-b \pm \sqrt{b^2 - 4ac}}{2a}$$

$$= \frac{-(-3) \pm \sqrt{(-3)^2 - 4(2)(4)}}{2(2)}$$

$$= \frac{3 \pm \sqrt{9 - 32}}{4} = \frac{3 \pm \sqrt{-23}}{4}$$

$$= \frac{3 \pm i\sqrt{23}}{4} \quad \sqrt{-23} = \sqrt{23(-1)} = \sqrt{23}\sqrt{-1} = i\sqrt{23}$$

32. $(2z - 1)(z - 2) - 11 = 2(z + 4) - 8$

$$2z^2 - 4z - z + 2 - 11 = 2z + 8 - 8$$

$$2z^2 - 5z - 9 = 2z$$

$$2z^2 - 7z - 9 = 0$$

$$(2z - 9)(z + 1) = 0$$

$$2z - 9 = 0 \quad \text{or} \quad z + 1 = 0$$

$$2z = 9 \qquad z = -1$$

$$z = \frac{9}{2}$$

The solutions are $\frac{9}{2}$ and -1.

34. Let x = the length of one leg,
$x + 2$ = the length of the second leg.
Use the Pythagorean theorem to obtain the desired equation.

$$x^2 + (x + 2)^2 = 11^2$$

$$x^2 + x^2 + 4x + 4 = 121$$

$$2x^2 + 4x - 117 = 0$$

$$x = \frac{-b \pm \sqrt{b^2 - 4ac}}{2a} = \frac{-4 \pm \sqrt{16 - 4(2)(-117)}}{4}$$

$$= \frac{-4 \pm \sqrt{952}}{4} = \frac{-4 \pm 2\sqrt{238}}{4}$$

Since the length of a leg cannot be negative, we discard $\frac{-4 - 2\sqrt{238}}{4}$. Use 15.4 for an approximation of $\sqrt{238}$.

$$\frac{-4 + 2(15.4)}{4} = 6.7$$

One leg is about 6.7 m and the other is 8.7 m.

37. Since the catcher is 6 ft down the line towards third base, he is actually 84 ft from third base. Let x = distance from the catcher to second base. Then x is the hypotenuse of a right triangle with legs 90 ft and 84 ft. Use the Pythagorean theorem.

$$90^2 + 84^2 = x^2$$

$$15{,}156 = x^2$$

$$\sqrt{15{,}156} = x \quad \textbf{Use positive root only}$$

$$123.10971 \approx x$$

The ball must travel about 123.1 ft.

8.3

14.
$$[x-(2+i)][x-(2-i)]=0$$
$$[x-2-i][x-2+i]=0$$
$$x^2-2x+ix-2x+4-2i-ix+2i-i^2=0$$
$$x^2-4x+4+1=0 \quad \textbf{$i^2=-1$}$$
$$x^2-4x+5=0$$

17. With $x_1 = 2+3i$ and $x_2 = 2-3i$, we have:
$$(x-x_1)(x-x_2)=0$$
$$[x-(2+3i)][x-(2-3i)]=0$$
$$x^2-2x-3ix-2x+3ix+(2+3i)(2-3i)=0$$
$$x^2-4x+(4-9i^2)=0$$
$$x^2-4x+(4+9)=0$$
$$x^2-4x+13=0.$$

8.4

9. $x^{-2}-x^{-1}-12=0$ **Let $u=x^{-1}$; then $u^2=x^{-2}$**
$$u^2-u-12=0$$
$$(u+3)(u-4)=0$$
$$u+3=0 \quad \text{or} \quad u-4=0$$
$$u=-3 \qquad u=4$$

To complete the solution, we substitute $x^{-1}=\frac{1}{x}$ for u and solve for x.
$$\frac{1}{x}=-3 \quad \text{or} \quad \frac{1}{x}=4$$
$$x=-\frac{1}{3} \qquad x=\frac{1}{4}$$

10. $(z-7)^4-13(z-7)^2+42=0$
Substitute u for $(z-7)^2$.
$$u^2-13u+42=0$$
$$(u-6)(u-7)=0 \quad \textbf{Factor}$$
$$u-6=0 \quad \text{or} \quad u-7=0 \quad \textbf{Zero-product rule}$$
$$u=6 \quad \text{or} \quad u=7$$

Then substitute $(z-7)^2$ for u and solve for z.
$$(z-7)^2=6 \quad \text{or} \quad (z-7)^2=7$$
$$z-7=\pm\sqrt{6} \qquad z-7=\pm\sqrt{7}$$
$$z=7\pm\sqrt{6} \qquad z=7\pm\sqrt{7}$$

There are four solutions, $7\pm\sqrt{6}$ and $7\pm\sqrt{7}$.

16. $\dfrac{x^2}{2x+4}=1$
Multiply both sides by the LCD $2x+4$.
$$x^2=2x+4$$
$$x^2-2x-4=0$$
$$x=\frac{-(-2)\pm\sqrt{(-2)^2-4(1)(-4)}}{2(1)}$$
$$=\frac{2\pm\sqrt{4+16}}{2}$$
$$=\frac{2\pm\sqrt{20}}{2}$$
$$=\frac{2\pm2\sqrt{5}}{2}=1\pm\sqrt{5}$$

20. $\dfrac{-2}{v^2-2v}=\dfrac{1}{v}+\dfrac{3v}{v-2}$
$$\frac{-2}{v(v-2)}=\frac{1}{v}+\frac{3v}{v-2} \quad \textbf{LCD = $v(v-2)$}$$
$$\cancel{v(v-2)}\left[\frac{-2}{\cancel{v(v-2)}}\right]$$
$$=\cancel{v}(v-2)\left[\frac{1}{\cancel{v}}\right]+v\cancel{(v-2)}\left[\frac{3v}{\cancel{v-2}}\right]$$
$$-2=v-2+3v^2$$
$$0=3v^2+v$$
$$0=v(3v+1)$$
$$v=0 \quad \text{or} \quad 3v+1=0$$
$$3v=-1$$
$$v=-\frac{1}{3}$$

Since 0 is a meaningless replacement, the only solution is $-\frac{1}{3}$.

25. $\sqrt{x+4}-x+8=0$
$$(\sqrt{x+4})^2=(x-8)^2 \quad \textbf{Isolate radical and square}$$
$$x+4=x^2-16x+64$$
$$0=x^2-17x+60$$
$$0=(x-12)(x-5)$$
$$x-12=0 \quad \text{or} \quad x-5=0$$
$$x=12 \qquad x=5$$

Check:
$$\sqrt{12+4}-12+8 \stackrel{?}{=} 0$$
$$\sqrt{16}-12+8 \stackrel{?}{=} 0$$
$$0=0$$
$$\sqrt{5+4}-5+8 \stackrel{?}{=} 0$$
$$\sqrt{9}-5+8 \stackrel{?}{=} 0$$
$$6\neq 0$$

The only solution is 12.

29. $\sqrt{2z+5}-\sqrt{z+2}=1$
$$(\sqrt{2z+5})^2=(\sqrt{z+2}+1)^2$$
$$2z+5=z+2+2\sqrt{z+2}+1$$
$$(z+2)^2=(2\sqrt{z+2})^2$$
$$z^2+4z+4=4(z+2)$$
$$z^2+4z+4=4z+8$$

$$z^2 = 4$$
$$z = \pm\sqrt{4} = \pm 2$$

Since $+2$ and -2 check, they are both solutions.

31. $(x - 5)^{1/2} = x^{1/2} + 1$

$$(\sqrt{x-5})^2 = (\sqrt{x} + 1)^2$$
$$x - 5 = x + 2\sqrt{x} + 1$$
$$-6 = 2\sqrt{x}$$
$$-3 = \sqrt{x}$$

But since $\sqrt{x}$ represents the principal root of x, that is, the nonnegative root, it cannot equal -3. Thus, we conclude that the equation has no solution.

35. Let x = time for Wanda to do the job,

$x - 3$ = time for Maria to do the job,

4 = time to do the job together,

$\frac{1}{x}$ = amount Wanda does in 1 hour,

$\frac{1}{x-3}$ = amount Maria does in 1 hour,

$\frac{1}{4}$ = amount done together in 1 hour.

Then we must solve:

$$\frac{1}{x} + \frac{1}{x-3} = \frac{1}{4}.$$

$$4x(x-3)\left[\frac{1}{x} + \frac{1}{x-3}\right] = 4x(x-3)\left[\frac{1}{4}\right]$$ **Multiply both sides by the LCD = 4x(x − 3)**

$$4(x-3) + 4x = x(x-3)$$ **Distribute and simplify**

$$4x - 12 + 4x = x^2 - 3x$$
$$0 = x^2 - 11x + 12$$

$$x = \frac{-b \pm \sqrt{b^2 - 4ac}}{2a}$$
$$= \frac{-(-11) \pm \sqrt{(-11)^2 - 4(1)(12)}}{2(1)}$$
$$= \frac{11 \pm \sqrt{121 - 48}}{2}$$
$$= \frac{11 \pm \sqrt{73}}{2}$$
$$\approx \frac{11 \pm 8.5}{2} = \begin{cases} \dfrac{11 + 8.5}{2} = 9.75 \\[2ex] \dfrac{11 - 8.5}{2} = 1.25 \end{cases}$$

But x cannot be 1.25 since $x - 3$ is then negative. Thus, to the nearest tenth, it would take Wanda 9.8 hr and Maria 6.8 hr to do the job alone.

42. Let x = rate of the airplane going,

$x + 25$ = rate of the airplane returning.

Then using $d = rt$ solved for t, $t = \frac{d}{r}$, we have:

$$\frac{2000}{x} = \text{time going,}$$
$$\frac{2000}{x+25} = \text{time returning.}$$

Since the total time of the trip was 9.5 hr, we must solve:

$$\frac{2000}{x} + \frac{2000}{x+25} = 9.5.$$

$$x(x+25)\left[\frac{2000}{x} + \frac{2000}{x+25}\right] = 9.5x(x+25)$$ **Multiply by LCD**

$$2000(x+25) + 2000x = 9.5x(x+25)$$
$$2000x + 50{,}000 + 2000x = 9.5x^2 + 237.5x$$
$$0 = 9.5x^2 - 3762.5x - 50{,}000$$

Clear the decimals by multiplying through by 10.

$$95x^2 - 37{,}625x - 500{,}000 = 0$$

Divide through by 5.

$$19x^2 - 7525x - 100{,}000 = 0$$

Use the quadratic formula.

$$x = \frac{7525 \pm \sqrt{(-7525)^2 - 4(19)(-100{,}000)}}{2(19)}$$
$$= \frac{7525 \pm \sqrt{64{,}225{,}625}}{38}$$
$$\approx \frac{7525 \pm 8014.09}{38}$$

Since the negative solution is meaningless, $x \approx \frac{7525 + 8014.09}{38} \approx 408.92 \approx 409$. Thus, the airplane traveled $x + 25 = 409 + 25 = 434$ mph returning.

48. $\frac{1}{x} + \frac{n}{x+1} = 1$ for x

$$\cancel{x}(x+1)\left[\frac{1}{\cancel{x}}\right] + x\cancel{(x+1)}\left[\frac{n}{\cancel{x+1}}\right] = x(x+1)[1]$$ **Multiply by LCD = x(x + 1)**

$$x + 1 + nx = x^2 + x$$
$$0 = x^2 - nx - 1$$
$$x = \frac{-b \pm \sqrt{b^2 - 4ac}}{2a}$$
$$= \frac{-(-n) \pm \sqrt{(-n)^2 - 4(1)(-1)}}{2(1)}$$
$$= \frac{n \pm \sqrt{n^2 + 4}}{2}$$

8.5

4. $2x^2 + x \geq 15$

$$2x^2 + x - 15 \geq 0$$

First solve for the critical points.

$$2x^2 + x - 15 = 0$$
$$(2x - 5)(x + 3) = 0$$
$$x = \frac{5}{2}, -3$$

The critical points form the intervals $(-\infty, -3)$, $\left(-3, \frac{5}{2}\right)$, and $\left(\frac{5}{2}, \infty\right)$. Use -4, 0, and 3 as test points.

$$-4: 2(-4)^2 + (-4) - 15 = 13 > 0$$
$$0: 2(0)^2 + (0) - 15 = -15 < 0$$
$$3: 2(3)^2 + (3) - 15 = 6 > 0$$

Thus, the solution includes the intervals containing -4 and 3 (where $2x^2 + x - 15 \geq 0$) together with $\frac{5}{2}$ and -3 since the inequality is $\geq$. The solution is

$$x \leq -3 \quad \text{or} \quad x \geq \frac{5}{2},$$

or, in interval notation, $(-\infty, -3]$ or $\left[\frac{5}{2}, \infty\right)$.

9. $3x^2 - 2x \leq 2$
$3x^2 - 2x - 2 = 0$

$$x = \frac{-(-2) \pm \sqrt{(-2)^2 - 4(3)(-2)}}{2(3)} = \frac{2 \pm \sqrt{4 + 24}}{6}$$
$$= \frac{2 \pm \sqrt{28}}{6} = \frac{2 \pm 2\sqrt{7}}{6} = \frac{1 \pm \sqrt{7}}{3}$$

Thus, the critical points are $\frac{1+\sqrt{7}}{3}$ (approximately 1.2) and $\frac{1-\sqrt{7}}{3}$ (approximately -0.5). Using test points -1, 0, and 2 we can see that the solution is all numbers between the critical points. That is, the solution is

$$\frac{1 - \sqrt{7}}{3} \leq x \leq \frac{1 + \sqrt{7}}{3}$$

or, in interval notation, $\left[\frac{1-\sqrt{7}}{3}, \frac{1+\sqrt{7}}{3}\right]$.

13. $x^2 + 9 \leq 0$
First solve for the critical points.

$$x^2 + 9 = 0$$
$$x^2 = -9$$
$$x = \pm 3i$$

Since the solutions are imaginary (not real), there are no critical points. Thus, the solution to the inequality will be all real numbers or no real numbers, depending on what happens at a single test point. Use 0 as the test point to get

$$0 + 9 = 9 \leq 0,$$

which is false. Since 0 is not a solution, no real number is a solution. The inequality has no solution.

17. $$\frac{2x + 1}{x - 1} < 3$$

$$\frac{2x + 1}{x - 1} - 3 < 0$$
$$\frac{2x + 1}{x - 1} - \frac{3(x - 1)}{x - 1} < 0$$
$$\frac{2x + 1 - 3x + 3}{x - 1} < 0$$
$$\frac{-x + 4}{x - 1} < 0$$

The critical points are solutions to:

$$-x + 4 = 0 \quad \text{and} \quad x - 1 = 0$$
$$x = 4 \qquad\qquad x = 1.$$

The critical points separate the number line into $(-\infty, 1)$, $(1, 4)$, and $(4, \infty)$. Using a test point in each, we see that $\frac{-x+4}{x-1} < 0$ when x is in $(-\infty, 1)$ or $(4, \infty)$. The solution is

$$x < 1 \quad \text{or} \quad x > 4$$

or, in interval notation,

$$(-\infty, 1) \quad \text{or} \quad (4, \infty).$$

22. $$\frac{(x - 3)(x + 2)}{x - 1} > 0$$
The critical points are solutions to:

$$x - 3 = 0 \quad \text{or} \quad x + 2 = 0 \quad \text{or} \quad x - 1 = 0$$
$$x = 3 \qquad\quad x = -2 \qquad\quad x = 1$$

The critical points separate the number line into the intervals $(-\infty, -2)$, $(-2, 1)$, $(1, 3)$, and $(3, \infty)$. Choosing a test point in each we discover that $\frac{(x-3)(x+2)}{x-1} > 0$ when x is in $(-2, 1)$ or $(3, \infty)$ which gives the solution in interval notation. Alternatively, the solution could be given as $-2 < x < 1$ or $x > 3$.

9.1

5. $f(x) = x^2 - 4x + 3$
Then $a = 1$, $b = -4$, and $c = 3$.
Since $a = 1 > 0$, the parabola opens up. The x-coordinate of the vertex is

$$-\frac{b}{2a} = -\frac{-4}{2(1)} = \frac{4}{2} = 2.$$

The y-coordinate of the vertex is

$$f\left(-\frac{b}{2a}\right) = f(2) = (2)^2 - 4(2) + 3$$
$$= 4 - 8 + 3$$
$$= -1.$$

The vertex is $(2, -1)$. Since the vertex is below the x-axis and the parabola opens up, there will be x-intercepts found by solving

$$x^2 - 4x + 3 = 0.$$
$$(x - 1)(x - 3) = 0$$
$$x - 1 = 0 \quad \text{and} \quad x - 3 = 0$$
$$x = 1 \qquad\qquad x = 3$$

The x-intercepts are $(1, 0)$ and $(3, 0)$.

12. $g(x) = 3x^2 + 5x + 1$
Then $a = 3$, $b = 5$, and $c = 1$.
Since $a = 3 > 0$, the parabola opens up. The x-coordinate of the vertex is

$$-\frac{b}{2a} = -\frac{5}{2(3)} = -\frac{5}{6}.$$

The y-coordinate of the vertex is

$$f\left(-\frac{b}{2a}\right) = f\left(-\frac{5}{6}\right) = 3\left(-\frac{5}{6}\right)^2 + 5\left(-\frac{5}{6}\right) + 1$$
$$= 3\left(\frac{25}{36}\right) - \frac{25}{6} + 1$$
$$= \frac{25}{12} - \frac{50}{12} + \frac{12}{12}$$
$$= -\frac{13}{12}$$

The vertex is $\left(-\frac{5}{6}, -\frac{13}{12}\right)$. Since the vertex is below the x-axis and the parabola opens up, there will be x-intercepts found by solving

$$3x^2 + 5x + 1 = 0.$$

Since this won't factor, we use the quadratic formula.

$$x = \frac{-b \pm \sqrt{b^2 - 4ac}}{2a} = \frac{-5 \pm \sqrt{(5)^2 - 4(3)(1)}}{2(3)}$$
$$= \frac{-5 \pm \sqrt{25 - 12}}{6}$$
$$= \frac{-5 \pm \sqrt{13}}{6}$$

The intercepts are $\left(\frac{-5 + \sqrt{13}}{6}, 0\right)$ and $\left(\frac{-5 - \sqrt{13}}{6}, 0\right)$.

21. $P = -x^2 + 80x - 1500$ must be maximized.
With $a = -1$ and $b = 80$, $-\frac{b}{2a} = -\frac{80}{2(-1)} = 40$.
Thus, the x-coordinate of the vertex is 40, and the y-coordinate (the value of P when x is 40) is $-(40)^2 + 80(40) - 1500 = 100$. The maximum profit ($P$) is \$100 when $40(x)$ sandwiches are sold.

9.2

8. $f(x) = \sqrt{x^2 - 8x + 12}$
The domain of f is all real numbers x for which $x^2 - 8x + 12 \geq 0$. We need to solve the inequality. First solve the equation $x^2 - 8x + 12 = 0$ to obtain the critical points.

$$x^2 - 8x + 12 = 0$$
$$(x - 2)(x - 6) = 0$$
$$x - 2 = 0 \quad \text{and} \quad x - 6 = 0$$
$$x = 2 \qquad\qquad x = 6$$

The critical points separate the number line into the intervals $(-\infty, 2)$, $(2, 6)$, and $(6, \infty)$. Using test points in each of these we find that $x^2 - 8x + 12 \geq 0$ when x is in $(-\infty, 2]$ or $[6, \infty)$. Thus, the domain of f is all real numbers $x \leq 2$ or $x \geq 6$.

17. Given $f(x) = 4 + 3x$ and $g(x) = x^2 - 3$, then

$$g(-1) = (-1)^2 - 3 = 1 - 3 = -2,$$

and $f[g(-1)] = f[-2] = 4 + 3(-2) = 4 - 6 = -2$.

9.3

16. $f(x) = 2x^3$
Then $y = 2x^3$, so we interchange x and y and solve for y.

$x = 2y^3$ **Interchange x and y**

$\frac{x}{2} = y^3$

$\sqrt[3]{\frac{x}{2}} = y$ **Take cube root of both sides**

Thus, $f^{-1}(x) = \sqrt[3]{\frac{x}{2}}$.

20. $g(x) = x^2 + 1$, $x \geq 0$
Then $y = x^2 + 1$, $x \geq 0$, so interchange x and y and solve for y.

$x = y^2 + 1$

$x - 1 = y^2$

$\pm\sqrt{x - 1} = y$ **Take square root of both sides**

But since $x \geq 0$ in the original function, when x and y are interchanged, $y \geq 0$ now, so we discard the negative root. Thus, $g^{-1}(x) = \sqrt{x - 1}$.

10.1

38. Given $N = (2000)2^t$, when $t = 0$, we have

$N = (2000)2^0.$

$= (2000)(1)$ **$2^0 = 1$**

$= 2000$

Thus there are 2000 bacteria initially (at time $t = 0$). When $t = 7$,

$$N = (2000)2^7$$
$$= (2000)(128) = 256{,}000.$$

In 7 hours there are 256,000 bacteria.

10.3

15. $\log_a xy^2\sqrt{z}$

$= \log_a x + \log_a y^2 + \log_a \sqrt{z}$ **Product rule**

$= \log_a x + \log_a y^2 + \log_a z^{1/2}$ **$\sqrt{z} = z^{1/2}$**

$= \log_a x + 2 \log_a y + \frac{1}{2} \log_a z$ **Power rule**

17. $\frac{1}{2} \log_a y - \log_a x + 2 \log_a z$

$= \log_a y^{1/2} - \log_a x + \log_a z^2$ **Power rule**

$= \log_a \sqrt{y} - \log_a x + \log_a z^2$ **$y^{1/2} = \sqrt{y}$**

$= \log_a \frac{\sqrt{y}}{x} + \log_a z^2$ **Quotient rule**

$= \log_a \frac{z^2\sqrt{y}}{x}$

25. $\log_a 0.2 = \log_a \dfrac{2}{10}$

$= \log_a \dfrac{1}{5}$

$= \log_a 1 - \log_a 5$

$= 0 - 0.6990$

$= -0.6990$

10.4

34. We need to evaluate

$$p = 200 - 0.4e^{0.004x}$$

when $x = 100$.

$$p = 200 - 0.4e^{(0.004)(100)}$$
$$= 200 - 0.4e^{0.4}$$

Using a calculator, we evaluate $e^{0.4}$. The steps are:

0.4 [INV] [ln x] ⟶ 1.4918247

Multiply this by 0.4.

[×] 0.4 ⟶ 0.5967299

Change the sign and add 200.

[±] [+] 200 [=] ⟶ 199.40327

The price is \$199.40, to the nearest cent.

37. We need to evaluate

$$L = 8.8 + 2.2 \ln d$$

when $d = 15$.

$$L = 8.8 + 2.2 \ln 15$$

The steps on a calculator are:

15 [ln x] [×] 2.2 [+] 8.8 [=] ⟶ 14.75771

To the nearest tenth, the limiting magnitude is 14.8.

10.5

5. $\log_2 (x + 1) + \log_2 (x - 1) = 3$

$\log_2 (x + 1)(x - 1) = 3$ **Product rule**

$\log_2 (x^2 - 1) = 3$

$x^2 - 1 = 2^3$ **Change to exponential equation**

$x^2 - 1 = 8$

$x^2 - 9 = 0$

$(x - 3)(x + 3) = 0$

$x - 3 = 0$ or $x + 3 = 0$

$x = 3$ $\quad x = -3$

Check: $x = 3$ $\quad \log_2 (3 + 1) + \log_2 (3 - 1) \stackrel{?}{=} 3$

$\log_2 4 + \log_2 2 \stackrel{?}{=} 3$

$2 + 1 \stackrel{?}{=} 3$

$3 = 3$

$x = -3$ $\quad \log_2 (-3 + 1) + \log_2 (-3 - 1) \stackrel{?}{=} 3$

Since $\log_2 (-2)$ and $\log_2 (-4)$ are not defined, the only solution is 3.

16.

$$3^{x^2}9^x = \frac{1}{3}$$
$$3^{x^2}(3^2)^x = 3^{-1}$$
$$3^{x^2}3^{2x} = 3^{-1}$$
$$3^{x^2+2x} = 3^{-1}$$
$$x^2 + 2x = -1$$
$$x^2 + 2x + 1 = 0$$
$$(x + 1)^2 = 0$$
$$x + 1 = 0$$
$$x = -1$$

18. $2^{2x}3^{x+1} = 5$

Take the common log of both sides.

$\log[2^{2x}3^{x+1}] = \log 5$

$\log 2^{2x} + \log 3^{x+1} = \log 5$ **Product rule**

$2x \log 2 + (x + 1) \log 3 = \log 5$ **Power rule**

$2x \log 2 + x \log 3 + \log 3 = \log 5$

$2x \log 2 + x \log 3 = \log 5 - \log 3$

$x(2 \log 2 + \log 3) = \log 5 - \log 3$

$x = \dfrac{\log 5 - \log 3}{2 \log 2 + \log 3}$

This is the exact answer. We can use a calculator to get an approximate solution, $x \approx 0.206$.

20. If the intensity is *reduced* by 80% of I_0, then we are looking for the depth at which the intensity is 20% of I_0, which is $0.20I_0$. Thus, we need to solve

$0.20I_0 = I_0e^{-0.008x}$.

$0.20 = e^{-0.008x}$ **Divide both sides by I_0**

$\ln 0.20 = \ln e^{-0.008x}$ **Take natural log of both sides**

$\ln 0.20 = -0.008x$ **$\ln e^u = u$**

$\dfrac{\ln 0.20}{-0.008} = x$

Using a calculator we obtain $x \approx 201.17974$. To the nearest tenth, the depth at which the intensity is reduced by 80% is about 201.2 ft.

10.6

9. We need to find y when $x = 50{,}100x_0$ in $y = \log\frac{x}{x_0}$.

$y = \log\dfrac{50{,}100x_0}{x_0}$

$= \log 50{,}100$ **Divide out x_0**

≈ 4.6998377

To the nearest tenth, the magnitude of the earthquake on the Richter scale is about 4.7.

12. Consider $y = 5000e^{0.02t}$. The initial population, when $t = 0$, is

$$y = 5000e^{(0.02)(0)} = 5000e^0 = 5000. \quad \mathbf{e^0 = 1}$$

We want t when the population, y, is double this, which is 10,000.

$$10{,}000 = 5000e^{0.02t}$$
$$2 = e^{0.02t} \quad \text{Divide both sides by 5000}$$
$$\ln 2 = \ln e^{0.02t} \quad \text{Take natural log of both sides}$$
$$\ln 2 = 0.02t \quad \ln e^u = u$$
$$\frac{\ln 2}{0.02} = t$$
$$34.657359 \approx t \quad \text{Use a calculator}$$

Thus, the population will double in about 34.7 years.

13. Consider $y = 250e^{-0.5t}$. The initial amount is when $t =$ 0.

$$y = 250e^{(-0.5)(0)} = 250e^0 = 250.$$

We want t when the amount y is 125 (half the initial amount).

$$125 = 250e^{-0.5t}$$
$$0.5 = e^{-0.5t}$$
$$\ln 0.5 = \ln e^{-0.5t}$$
$$\ln 0.5 = -0.5t$$
$$\frac{\ln 0.5}{-0.5} = t$$
$$1.3862944 \approx t$$

The half-life is about 1.39 years.

11.1

5. $y = x^2 - 5$ This can be written as

$$y = 1 \cdot (x - 0)^2 - 5.$$

The vertex is $(0, -5)$, and since $a = 1 > 0$, the parabola opens up.

9. $x = -2y^2 - 4y - 7$ We complete the square in y to write the equation in standard form.

$$x = -2(y^2 + 2y \qquad) - 7 \quad \text{Factor out } -2$$
$$= -2(y^2 + 2y + 1) - 7 + 2$$

Since 1 is added inside the parentheses to complete the square, and it is multiplied by the factor of -2, we have really added -2 to the expression which is then balanced out by adding $+2$ to the -7. We have

$$x = -2(y + 1)^2 - 5.$$

The vertex is $(-5, -1)$, and since $a = -2 < 0$, the parabola opens left.

11.2

9. $4x^2 + 4y^2 + 40x - 4y + 37 = 0$
We need to complete the square in both x and y.

$$4x^2 + 40x \qquad + 4y^2 - 4y \qquad = -37$$

But to complete the square we must have the coefficient of the squared terms equal to 1. Divide through by 4.

$$x^2 + 10x \qquad + y^2 - y \qquad = -\frac{37}{4}$$

Add $\left[\frac{1}{2}(10)\right]^2 = [5]^2 = 25$ and $\left[\frac{1}{2}(-1)\right]^2 = \left[-\frac{1}{2}\right]^2 = \frac{1}{4}$ to both sides.

$$x^2 + 10x + 25 + y^2 - y + \frac{1}{4} = -\frac{37}{4} + 25 + \frac{1}{4}$$

$$(x + 5)^2 + \left(y - \frac{1}{2}\right)^2 = 16 = 4^2$$

The radius is 4 and the center is $\left(-5, \frac{1}{2}\right)$.

11.4

4. $5x^2 + xy - y^2 = -1$
$y - 2x = 1$

Solve the second equation for y and substitute into the first.

$$y = 2x + 1$$
$$5x^2 + x(2x + 1) - (2x + 1)^2 = -1$$
$$5x^2 + 2x^2 + x - (4x^2 + 4x + 1) = -1$$
$$5x^2 + 2x^2 + x - 4x^2 - 4x - 1 = -1$$
$$3x^2 - 3x - 1 = -1$$
$$3x^2 - 3x = 0$$
$$3x(x - 1) = 0$$

$$x = 0 \quad \text{or} \quad x - 1 = 0$$
$$x = 1$$
$$y = 2(0) + 1 = 1 \quad y = 2(1) + 1 = 3$$

The solutions are (0, 1) and (1, 3).

9. $3xy - y^2 = -13$
$2xy + y^2 = -2$

$$6xy - 2y^2 = -26 \quad \text{Multiply first equation by 2}$$
$$-6xy - 3y^2 = 6 \quad \text{Multiply second equation by } -3$$
$$-5y^2 = -20$$
$$y^2 = 4$$
$$y = \pm 2$$

For $y = 2$, $3x(2) - (2)^2 = -13$

$$6x - 4 = -13$$
$$6x = -9$$
$$x = \frac{-9}{6} = -\frac{3}{2}.$$

For $y = -2$, $3x(-2) - (2)^2 = -13$

$$-6x - 4 = -16$$
$$-6x = -9$$
$$x = \frac{-9}{-6} = \frac{3}{2}.$$

The solutions are $\left(-\frac{3}{2}, 2\right)$ and $\left(\frac{3}{2}, -2\right)$.

13. Let x = one number,
y = the other number.

$x^2 + y^2 = 100$ **The sum of the squares is 100**

$x - y = 2$ **The difference of the numbers is 2**

$x = y + 2$ **Solve the second equation for x**

$$(y + 2)^2 + y^2 = 100$$
$$y^2 + 4y + 4 + y^2 = 100$$
$$2y^2 + 4y + 4 - 100 = 0$$
$$2y^2 + 4y - 96 = 0$$
$$y^2 + 2y - 48 = 0$$
$$(y - 6)(y + 8) = 0$$

$y - 6 = 0$ or $y + 8 = 0$

$y = 6$ $\quad$ $y = -8$

$x = 6 + 2 = 8$ $\quad$ $x = -8 + 2 = -6$

12.1

4. Since $(1)(2) = 2$, $(2)(2) = 4$, $4(2) = 8$, and $(8)(2) = 16$, each element after the first one in the list is obtained by multiplying the preceding element by 2. Thus, the next element should be $(16)(2) = 32$.

14. The problem is *not* that you have 8 apples, take away 5, and have 3 left. The problem states that you *take* 5 apples, and thus you will *have* 5 apples.

12.3

16. Since $\angle EDF = 0°$ and $\angle DFE = 180°$, the sum of their measures is 180°, making them supplementary.

12.4

9. The third angle in the first triangle is $180° - (80° + 40°) = 60°$. Thus no angle in the first triangle is 70°, and the two triangles cannot be similar.

12.5

9. $C = 2\pi r$
$= 2\pi(7.7 \text{ mi})$
$= 15.4\pi \text{ mi}$

12. $C = 2\pi r$
$= 2\pi(0.05 \text{ cm})$
$= 2(3.14)(0.05) \text{ cm}$
$= 0.314 \text{ cm}$

12.6

16. Divide the area into two rectangles. One is 7 yd long and 3 yd wide and the other is 3 yd long and 2 yd wide.

$7 \text{ yd} \times 3 \text{ yd} = 21 \text{ yd}^2$

$3 \text{ yd} \times 2 \text{ yd} = 6 \text{ yd}^2$

$21 \text{ yd}^2 + 6 \text{ yd}^2 = 27 \text{ yd}^2$

$$\frac{\$16.50}{1 \text{ yd}^2} \times \frac{27 \text{ yd}^2}{1} = (\$16.50)(27)$$
$$= \$445.50$$

32. First find the area of the rectangle which is 28 cm long and 14 cm wide.

$$A = (28 \text{ cm})(14 \text{ cm}) = 392 \text{ cm}^2$$

Now find the area of the two triangles and subtract from the area of the rectangle. The base of each triangle is 10 cm and the height is 14 cm − 7 cm = 7 cm.

$$A = \frac{1}{2}(10 \text{ cm})(7 \text{ cm}) = 35 \text{ cm}^2$$
$$2A = 70 \text{ cm}^2$$

Thus, the area is $392 \text{ cm}^2 - 70 \text{ cm}^2 = 322 \text{ cm}^2$.

37. $A = \pi r^2$
$= \pi(7.7 \text{ mi})^2$
$= 59.29\pi \text{ mi}^2$

40. $A = \pi r^2$
$= \pi(0.05 \text{ cm})^2$
$\approx (3.14)(0.0025) \text{ cm}^2$
$= 0.00785 \text{ cm}^2$

41. Calculate the area for each hose and subtract. The radius of the first hose is $\frac{3}{8}$ in and the radius of the second hose is $\frac{1}{4}$ in.

$$A = \pi r^2$$
$$\approx 3.14\left(\frac{3}{8} \text{ in}\right)^2$$
$$\approx 0.442 \text{ in}^2$$
$$A = \pi r^2$$
$$\approx 3.14\left(\frac{1}{4} \text{ in}\right)^2$$
$$\approx 0.196 \text{ in}^2$$

$$0.442 \text{ in}^2 - 0.196 \text{ in}^2 = 0.246 \text{ in}^2$$

12.7

3. Convert each number to decimals:
9.5 in by 3.4 in by 8.75 in

$$V = (9.5 \text{ in})(3.4 \text{ in})(8.75 \text{ in})$$
$$= 282.625 \text{ in}^3$$

10. $A = 2lh + 2wh + 2lw$

$= 2(15.5\text{ cm})(20.2\text{ cm}) + 2(6.4\text{ cm})(20.2\text{ cm}) + 2(15.5\text{ cm})(6.4\text{ cm})$

$= 1083.16\text{ cm}^2$

15. $V = e^3$

$= (22\text{ cm})^3$

$= 10{,}648\text{ cm}^3$

Since water weighs 1 $\frac{\text{gm}}{\text{cm}^3}$ the answer is 10,648 g.

16. Area of bottom $= (22\text{ cm})^2 = 484\text{ cm}^2$

Area of 4 sides $= 4(22\text{ cm})^2 = 4(484)\text{cm}^2$

$= 1936\text{ cm}^2$

Total required $= 484\text{ cm}^2 + 1936\text{ cm}^2$

$= 2420\text{ cm}^2$

19. $V = \pi r^2 h$

$= \pi(12.2\text{ in})^2(30\text{ in})$ **Radius is $\frac{1}{2}$ the diameter**

$\approx (3.14)(12.2)^2(30)\text{in}^2$

$\approx 14{,}020.7\text{ in}^3$

22. $A = 2\pi rh + 2\pi r^2$

$\approx 2(3.14)(12.2\text{ in})(30\text{ in}) + 2(3.14)(12.2\text{ in})^2$

$\approx 3233.2\text{ in}^2$

25. $V = \frac{4}{3}\pi r^3$

$= \frac{4}{3}\pi\left(\frac{3}{4}\text{ in}\right)^3$

$\approx \left(\frac{4}{3}\right)(3.14)\left(\frac{3}{4}\right)^3\text{ in}^3$

$\approx 1.766\text{ in}^3$

29. $A = 4\pi r^2$

$= 4\pi(2.8\text{ yd})^2$

$\approx 4(3.14)(2.8)^2\text{ yd}^2$

$\approx 98.47\text{ yd}^2$

30. Calculate the volume of the can and the volume of the pie pan and compare the two.

$V = \pi r^2 h$

$= \pi(4\text{ cm})^2(12\text{ cm})$

$\approx (3.14)(4)^2(12)\text{ cm}^3$

$= 602.88\text{ cm}^3$

$V = \pi r^2 h$

$= \pi(10\text{ cm})^2(3\text{ cm})$

$\approx (3.14)(10)^2(3)\text{ cm}^3$

$= 942\text{ cm}^3$

$$\frac{942\ \cancel{\text{cm}^3}}{1} \times \frac{1\text{ can}}{602.88\ \cancel{\text{cm}^3}} = 1.6\text{ cans}$$

35. Find the area of the triangle and subtract 3 times the area of one hole.

$A = \frac{1}{2}bh$

$= \frac{1}{2}(9\text{ in})(7.8\text{ in})$

$= 35.1\text{ in}^2$

$A = \pi r^2$

$= \pi(1\text{ in})^2$ **Radius is $\frac{1}{2}$ (2 in) = 1 in**

$\approx 3.14\text{ in}^2$

The area of the three circles is

$3(3.14\text{ in}^2) = 9.42\text{ in}^2$

$35.1\text{ in}^2 - 9.42\text{ in}^2 = 25.68\text{ in}^2$

INDEX